Low Power Semiconductor Devices and Processes for Emerging Applications in Communications, Computing, and Sensing

Devices, Circuits, and Systems

Series Editor Krzysztof Iniewski

Wireless Technologies
Circuits, Systems, and Devices
Krzysztof Iniewski

Circuits at the Nanoscale
Communications, Imaging, and Sensing
Krzysztof Iniewski

Internet Networks
Wired, Wireless, and Optical Technologies
Krzysztof Iniewski

Semiconductor Radiation Detection Systems
Krzysztof Iniewski

Electronics for Radiation Detection
Krzysztof Iniewski

Radiation Effects in Semiconductors
Krzysztof Iniewski

Electrical Solitons
Theory, Design, and Applications
David Ricketts and Donhee Ham

Semiconductors
Integrated Circuit Design for Manufacturability
Artur Balasinski

Integrated Microsystems
Electronics, Photonics, and Biotechnology
Krzysztof Iniewski

Nano-Semiconductors
Devices and Technology
Krzysztof Iniewski

Atomic Nanoscale Technology in the Nuclear Industry
Taeho Woo

Telecommunication Networks
Eugenio Iannone

For more information about this series, please visit: https://www.crcpress.com/Devices-Circuits-and-Systems/book-series/CRCDEVCIRSYS

Low Power Semiconductor Devices and Processes for Emerging Applications in Communications, Computing, and Sensing

Edited by
Sumeet Walia

Managing Editor
Krzysztof Iniewski

CRC Press
Taylor & Francis Group
Boca Raton London New York

CRC Press is an imprint of the
Taylor & Francis Group, an **informa** business

CRC Press
Taylor & Francis Group
6000 Broken Sound Parkway NW, Suite 300
Boca Raton, FL 33487-2742

Printed on acid-free paper

International Standard Book Number-13: 978-1-138-58798-4 (Hardback)

Library of Congress Cataloging-in-Publication Data

Names: Walia, Sumeet, editor.
Title: Low power semiconductor devices and processes for emerging applications in communications, computing, and sensing / [edited by] Sumeet Walia.
Description: Boca Raton : Taylor & Francis, a CRC title, part of the Taylor & Francis imprint, a member of the Taylor & Francis Group, the academic division of T&F Informa, plc, 2018. | Series: Devices, circuits, and systems | Includes bibliographical references and index.
Identifiers: LCCN 2018024678| ISBN 9781138587984 (hardback : alk. paper) | ISBN 9780429503634 (ebook : alk. paper)
Subjects: LCSH: Low voltage integrated circuits.
Classification: LCC TK7874.66 .L6735 2018 | DDC 621.3815--dc23
LC record available at https://lccn.loc.gov/2018024678

Visit the Taylor & Francis Web site at
http://www.taylorandfrancis.com

and the CRC Press Web site at
http://www.crcpress.com

Contents

Preface

Making processing information more energy-efficient would save money, reduce energy use and permit batteries that provide power to mobile devices to run longer or be smaller in size. New approaches to the lower energy requirement in computing, communication and sensing need to be investigated. This book addresses this need in multiple application areas and will serve as a guide in emerging circuit technologies.

Revolutionary device concepts, sensors and associated circuits and architectures that will greatly extend the practical engineering limits of energy-efficient computation are being investigated. Disruptive new device architectures, semiconductor processes and emerging new materials aimed at achieving the highest level of computational energy efficiency for general purpose computing systems need to be developed. This book will provide chapters dedicated to such efforts from process to device.

Series Editor

Krzysztof (Kris) Iniewski is managing R&D at Redlen Technologies Inc., a start-up company in Vancouver, Canada. Redlen's revolutionary production process for advanced semiconductor materials enables a new generation of more accurate, all-digital, radiation-based imaging solutions. Kris is also a president of CMOS Emerging Technologies (www.cmoset.com), an organization of high-tech events covering communications, microsystems, optoelectronics, and sensors. In his career, Dr. Iniewski has held numerous faculty and management positions at the University of Toronto, the University of Alberta, Simon Fraser University, and PMC-Sierra Inc. He has published over 100 research papers in international journals and conferences. He holds 18 international patents granted in the United States, Canada, France, Germany, and Japan. He is a frequent invited speaker and has consulted for multiple organizations internationally. He has written and edited several books for IEEE Press, Wiley, CRC Press, McGraw-Hill, Artech House, and Springer. His personal goal is to contribute to healthy living and sustainability through innovative engineering solutions. In his leisure time, Kris can be found hiking, sailing, skiing, or biking in beautiful British Columbia. He can be reached at kris.iniewski@gmail.com.

Editor

Sumeet Walia is a vice chancellor's fellow at the Royal Melbourne Institute of Technology in Australia. Dr. Walia earned his PhD in the multidisciplinary field of functional materials and devices. His research focuses on low-dimensional nanoelectronics, including micro/nanoscale energy sources, electronic memories, sensors, and transistors. He holds three patents and has been recognized as one of the top 10 innovators under-35 in Asia by *MIT Technology Review*. He has published several high-impact research articles and is a reviewer for a number of international peer-reviewed journals and government grant bodies. He can be reached at waliasumeet@gmail.com and sumeet.walia@rmit.edu.au.

Contributors

Mohammad Karbalaei Akbari
University of Ghent Global Campus
Incheon, South Korea

Mark B.H. Breese
Singapore Synchrotron Light Source (SSLS)
and
Department of Physics
and
NUSNNI-NanoCore
National University of Singapore
Singapore

Daniel R. Brennan
School of Engineering
Newcastle University
Newcastle upon Tyne, UK

Wayne Burleson
Department of Electrical and Computer Engineering
University of Massachusetts Amherst
Amherst, Massachusetts

Taw Kuei Chan
Department of Physics
National University of Singapore
Singapore

Hua-Khee Chan
School of Engineering
Newcastle University
Newcastle upon Tyne, UK

Hong Chen
Institute of Microelectronics
Tsinghua University
Beijing, China

Lawrence T. Clark
School of Electrical, Computer and Energy Engineering
Arizona State University
Tempe, Arizona

Rémi Comyn
Université Côte d'Azur, CNRS, CRHEA
Valbonne, France

Yvon Cordier
Université Côte d'Azur, CNRS, CRHEA
Valbonne, France

Roy Dagher
Université Côte d'Azur, CNRS, CRHEA
Valbonne, France

Bhupendra Nath Dev
Department of Physics and School of Nano Science and Technology
Indian Institute of Technology Kharagpur
Kharagpur, India

C.Y. Fong
Department of Physics
University of California, Davis
Davis, California

Eric Frayssinet
Université Côte d'Azur, CNRS, CRHEA
Valbonne, France

Filippo Giannazzo
Consiglio Nazionale delle Ricerche – Istituto per la Microelettronica e Microsistemi (CNR-IMM)
Catania, Italy

Jonathan P. Goss
School of Engineering
Newcastle University
Newcastle upon Tyne, UK

Giuseppe Greco
Consiglio Nazionale delle Ricerche – Istituto per la Microelettronica e Microsistemi (CNR-IMM)
Catania, Italy

Philipp Gutruf
Center for Bio-Integrated Electronics
Simpson Querrey Institute for BioNanotechnology
and
Department of Materials Science and Engineering
Northwestern University
Evanston, Illinois

Noel Healy
School of Engineering
Newcastle University
Newcastle upon Tyne, UK

Matthew D. Higgins
Warwick Manufacturing Group
University of Warwick
Coventry, UK

Ben R. Horrocks
School of Chemistry
Newcastle University
Newcastle upon Tyne, UK

Alton B. Horsfall
Department of Engineering
University of Durham
Durham, UK

Hanjun Jiang
Institute of Microelectronics
Tsinghua University
Beijing, China

Raghavan Kumar
Intel Lab
Portland, Oregon

Mark S. Leeson
School of Engineering
University of Warwick
Coventry, UK

Shuo Li
Department of Electrical and Computer Engineering
University of Massachusetts Amherst
Amherst, Massachusetts

Yi Lu
Warwick Manufacturing Group
University of Warwick
Coventry, UK

Sinu Mathew
NUSNNI-NanoCore
National University of Singapore
Singapore
and
St. Berchmans College
Changanassery, India

Adrien Michon
Université Côte d'Azur, CNRS, CRHEA
Valbonne, France

Marzaini Rashid
School of Physics
Universiti Sains Malaysia
Penang, Malaysia

Fabrizio Roccaforte
Consiglio Nazionale delle Ricerche – Istituto per la Microelettronica e Microsistemi (CNR-IMM)
Catania, Italy

Shaojie Su
Institute of Microelectronics
Tsinghua University
Beijing, China

John T.L. Thong
Department of Electrical and Computer Engineering
National University of Singapore
Singapore

Vinay Vashishtha
School of Electrical, Computer and Energy Engineering
Arizona State University
Tempe, Arizona

T. Venkatesan
NUSNNI-NanoCore
and
Department of Physics
and
Department of Electrical and Computer Engineering
National University of Singapore
Singapore

Zhihua Wang
Institute of Microelectronics
Tsinghua University
Beijing, China

Nicholas G. Wright
School of Engineering
Newcastle University
Newcastle upon Tyne, UK

Xiaolin Xu
Department of Electrical and Computer Engineering
University of Florida
Gainesville, Florida

Lin H. Yang
Physics Division, Lawrence Livermore National Laboratory
Livermore, California

Y.J. Zeng
School of Physics
Sun Yat-sen University
Guangzhou, China

R.L. Zhang
School of Physics
Nanjing University
Nanjing, China

Peng Zhou
Fudan University
Shanghai, China

Serge Zhuiykov
University of Ghent Global Campus
Incheon, South Korea

1

ASAP7: A finFET-Based Framework for Academic VLSI Design at the 7 nm Node

Vinay Vashishtha and Lawrence T. Clark

CONTENTS

1.1 Introduction

Recent years have seen fin field-effect transistors (finFETs) dominate highly scaled, e.g., sub-20 nm, complementary metal-oxide-semiconductor (CMOS) processes (Wu et al., 2013; Lin et al., 2014) due to their ability to alleviate short channel effects, provide lower leakage, and enable some continued V_{DD} scaling. However, the availability of a realistic finFET-based predictive process design kit (PDK) for academic use that supports investigation into both circuit as well as physical design, encompassing all aspects of digital design, has been lacking. While the finFET-based FreePDK15 was supplemented with a standard cell library, it lacked full physical verification, layout vs. schematic check (LVS) and parasitic extraction (Bhanushali et al., 2015; Martins et al., 2015). Consequently, the only available sub-45 nm educational PDKs are the planar CMOS-based Synopsys 32/28 nm and FreePDK45 (45 nm PDK) (Goldman et al., 2013; Stine et al., 2007). The cell libraries available for those processes are not very realistic since they use very large cell heights, in contrast to recent industry trends. Additionally, the static random access memory (SRAM) rules and cells provided by these PDKs are not realistic. Because finFETs have a three-dimensional (3-D) structure and there have been significant density impacts in their adoption, using planar libraries scaled to sub-22 nm dimensions for research is likely to give poor accuracy.

Commercial libraries and PDKs, especially for advanced nodes, are often difficult to obtain for academic use, and access to the actual physical layouts is even more restricted. Furthermore, the necessary non-disclosure agreements (NDAs) are unmanageable for large university classes and the plethora of design rules (DRs) can distract from the key points. NDAs also make it difficult for the publication of physical design as these may disclose proprietary DRs and structures.

This chapter focuses on the development of a realistic PDK for academic use that overcomes these limitations. The PDK, developed for the N7 node before 7 nm processes were available even in industry, is thus *predictive*. The predictions have been based on publications of the continually improving lithography, as well as our estimates of what would be available at N7. The original assumptions are described in Clark et al. (2016). For the most part, these assumptions have been accurate, except for the expectation that extreme ultraviolet lithography (EUVL) would be widely available, which has turned out to be optimistic. The background and impact on design technology co-optimization (DTCO) for standard cells and SRAM comprise this chapter. The treatment here includes learning from using the cells originally derived in Clark et al. (2016) in realistic designs of SRAM arrays and large digital designs using automated place and route tools.

1.1.1 Chapter Outline

The chapter first outlines the important lithography considerations in Section 1.3. Metrics for overlay, mask errors and other effects that limit are described first. Then, modern liquid immersion optical lithography and its use in multiple patterning (MP) techniques that extend it beyond the standard 80 nm feature limit are discussed. This sets the stage for a discussion of EUV lithography, which can expose features down to about 16 nm in a single exposure (SE), but at a high capital and throughput cost. This section ends with a brief overview of DTCO. DTCO has been required on recent processes to ensure that the very limited possible structures that can be practically fabricated are usable to build real designs. Thus, a key part of a process development is not just to determine transistor and interconnect structures that are lithographically possible, but also to ensure that successful designs can be built with those structures. This discussion is carried out by separating the front end of line (FEOL), middle of line (MOL), and back end of line (BEOL) portions of the process, which fabricate the transistors, contacts and local interconnect, and global interconnect metallization, respectively. The cell library architecture and automated placement and routing (APR) aspects comprise the next section, which with the SRAM results comprise most of the discussion. The penultimate section describes the SRAM DTCO and array development and performance in the ASAP7 predictive PDK. The final section summarizes.

1.2 ASAP7 Electrical Performance

The PDK uses BSIM-CMG SPICE models and the values used are derived from publicly available sources with appropriate assumptions (Paydavosi et al., 2013). A drive current increase from 14 to 7 nm node is assumed to be 15%, which corresponds to the diminished I_{dsat} improvement over time. In accordance with modern devices, the saturation current is assumed to be 4.5x larger than that in the linear region (Clark et al., 2016). A relaxed 54 nm contacted poly pitch (CPP) allows a longer channel length and helps with the assumption of a near ideal subthreshold slope (SS) of 60 mV/decade at room temperature, along with a drain-induced barrier lowering (DIBL) of approximately 30 mV/V. P-type metal-oxide semiconductor (PMOS) strain seems to be easier to obtain according to the 16 and 14 nm foundry data and larger I_{dsat} values for PMOS than those for a n-type metal-oxide semiconductor (NMOS) have been reported (Wu et al., 2013; Lin et al., 2014). Following this trend, we assume a PMOS-to-NMOS drive ratio of 0.9:1. This value provides good slew rates at a fan-out of six (FO6), instead of the traditional four.

Despite the same drawn gate length, the PDK and library timing abstract views support four threshold voltage flavors, viz. super low voltage threshold (SLVT), low voltage threshold (LVT), regular threshold voltage (RVT), and SRAM, to allow investigation into both high-performance and low-power designs. The threshold voltage is assumed to be

TABLE 1.1

NMOS Typical Corner Parameters (Per Fin) at 25°C

Parameter	SRAM	RVT	LVT	SLVT
I_{dsat} (μA)	28.57	37.85	45.19	50.79
I_{eff} (μA)	13.07	18.13	23.56	28.67
I_{off} (nA)	0.001	0.019	0.242	2.444
V_{tsat} (V)	0.25	0.17	0.10	0.04
V_{tlin} (V)	0.27	0.19	0.12	0.06
SS (mV/decade)	62.44	63.03	62.90	63.33
DIBL (mV/V)	19.23	21.31	22.32	22.55

TABLE 1.2

PMOS Typical Corner Parameters (Per Fin) at 25°C

Parameter	SRAM	RVT	LVT	SLVT
I_{dsat} (μA)	26.90	32.88	39.88	45.60
I_{eff} (μA)	11.37	14.08	18.18	22.64
I_{off} (nA)	0.004	0.023	0.230	2.410
V_{tsat} (V)	−0.20	−0.16	−0.10	−0.04
V_{tlin} (V)	−0.22	−0.19	−0.13	−0.07
SS (mV/decade)	64.34	64.48	64.44	64.94
DIBL (mV/V)	24.10	30.36	31.06	31.76

changed through work function engineering. For SRAM devices, the very low leakage uses both a work function change and lightly doped drain (LDD) implant removal. The latter results in an effective channel length (Leff) increase, gate-induced drain leakage (GIDL) reduction, and an overlap capacitance reduction. The drive strength reduces from SLVT to SRAM. The SRAM V_{th} transistors are a convenient option for use in retention latches and designs that prioritize low-standby power. In addition to typical-typical (TT) models, fast-fast (FF) and slow-slow (SS) models are also provided for multi-corner APR optimization. Tables 1.1 and 1.2 show the electrical parameters for single fin NMOS and PMOS, respectively, for the TT corner at 25°C (Clark et al., 2016). The nominal operating voltage is V_{DD}=700 mV.

1.3 Lithography Considerations

Photolithography, hereinafter referred to simply as lithography, in a semiconductor industry context, refers to a process whereby a desired pattern is transferred to a target layer on the wafer through use of light. Interconnect metal, via, source-drain regions, and gate layers in a CMOS process stack are a few examples of the patterns defined, or "printed", using lithography.

A simplified pattern transfer flow is as follows. From among the pattern information that is stored in an electronic database file (GDSII) corresponding to all the layers of a given integrated circuit (IC) design, the enlarged pattern, or its photographic negative, corresponding to a single layer is inscribed onto a photomask or reticle. The shapes on the photomask, hereinafter referred to as mask, define the regions that are either opaque or transparent to light. Light from a suitable source is shone on the mask through an illuminator, which modifies the effective manner of illumination, and passes through the transparent mask regions. Thereafter, light passes through a projection lens, which shrinks the enlarged pattern geometries on the mask to their intended size, and exposes the photoresist that has been coated on the wafer atop the layer to be patterned. The photoresist is developed to either discard or retain its exposed regions corresponding to the pattern. This is followed by an etch that removes portions of the target layer not covered by the photoresist, which is then removed, leaving behind the intended pattern on the layer. Both lines and spaces can be patterned through this approach with some variations in the process steps.

Lithography plays a leading role in the scaling process, which is the industry's primary growth driver, as it determines the extent to which feature geometries can be shrunk in successive technology nodes. Lithography is one of the most expensive and complex procedures in semiconductor manufacturing, with mask manufacturing being the most expensive processing steps within lithography (Ma et al., 2010). Both the complexity and the number of masks used for manufacturing at a node affect the cost, and an increase in either of these can increase the cost to the point of becoming the limiting factor in the overall cost of the product.

As in any other manufacturing process, the various lithography steps also suffer from variability. The lithographic resolution determines the minimum feature dimension, called the critical dimension (CD), for a given layer and is based on the lithography technique employed at a particular technology node. DRs constitute design guidelines to minimize the effects from mask manufacturability issues, the impact of variability and layer misalignment, and ensure printed pattern fidelity to guarantee circuit operation at good yield. Ascertaining these DRs thus requires consideration of the following lithography-related metrics that can cause final printed pattern on a layer to deviate from the intent and/or result in reliability issues.

1.3.1 Lithography Metrics and Other Considerations for Design Rule Determination

1.3.1.1 Critical Dimension Uniformity (CDU)

CDU relates to the consistency in the dimensions of a feature printed in resist. CD variations arise due to a number of factors—wafer temperature and photoresist thickness, to name a few. It is defined by

$$\mathrm{CDU} = \frac{\sqrt{\mathrm{CDU}_E^2 + \mathrm{CDU}_F^2 + \mathrm{CDU}_M^2}}{2}, \tag{1.1}$$

where CDU_E, CDU_F, and CDU_M are the CD variation due to dose, focus, and mask variations (Chiou et al., 2013). The required CDU is typically calculated as 7% of the target CD requirement, but modern scanner systems continue to push the envelope beyond that requirement. The 3σ CDU for 40 nm isolated and dense lines can be as small as 0.58 and 0.55 nm (DeGraff et al., 2016), respectively, for ASML's TWINSCAN NXT:1980Di optical immersion lithography scanner released in 2016. For the ASAP7 PDK, we assumed a CDU of 2 nm for optical immersion lithography, which is in line with Vandeweyer et al. (2010). For EUVL patterned layers, we assumed the CDU to be 1 nm, which is close to the 1.2 nm CDU estimated by Van Setten et al. (2014) and the later CDU specification of 1.1 nm for ASML's TWINSCAN NXE:3400B EUV scanner (ASML, 2017).

1.3.1.2 Overlay

Overlay refers to the positional inaccuracy resulting from the misalignment between two subsequent mask steps and denotes the worst-case spacing between two non-self-aligned mask layers (Servin et al., 2009). Single machine overlay (SMO) refers to the overlay arising from both layers being printed on the same machine (scanner), which results in better alignment accuracy and thus smaller overlay. However, using the same machine for two layers or masks is slower from a processing perspective, and consequently, more expensive. Matched-machine overlay (MMO) refers to that arising from two successive layers being printed on different machines, resulting in a larger value than SMO. As two separate scanners are employed, the overall processing rate in an assembly line setting is faster.

Lin et al. (2015) predicted 3σ SMO and MMO values at N7 to be 1.5 and 2 nm, respectively. ASML's TWINSCAN NXT:1980Di optical immersion lithography scanner released in 2016 has a 3σ SMO and MMO of 1.6 and 2.5 nm, respectively, while its Twinscan NXE:3400B EUV scanner released in 2017 has 3σ SMO and MMO of 1.4 and 2 nm, respectively (ASML, 2017). For the PDK, we assumed a 3σ MMO of 3.5 nm for optical immersion lithography, based on ASML (2015). We assumed a 3σ MMO of 1.7 nm for the EUVL, based on the estimates by Van Setten et al. (2014).

1.3.1.3 Mask Error Enhancement Factor (MEEF) and Edge Placement Error (EPE)

MEEF refers to the ratio of wafer or resist CD error to the mask CD error and is given as

$$\text{MEEF} = \frac{\Delta \text{CD}_{\text{wafer}}}{\Delta \text{CD}_{\text{mask}}}. \tag{1.2}$$

Thus, it denotes the amount by which errors on the mask are magnified when they are transferred to the wafer and it depends on the mask, optics, and the process. Its effects are more pronounced near the resolution limit for a specific patterning technique (Yeh and Loong, 2006). Features such as the metal line-ends or tips are typically more adversely affected in optical immersion lithography (193i) systems. Van Setten et al. (2014) found the MEEF for 193i patterned layers to range from five to seven, but found it to be nearly one for EUVL patterned layers.

The EPE gives the deviation in edge placement of one layer relative to another, while accounting for both CDU and overlay contributions. For two layers, each patterned through SE steps, the EPE is given as

$$\text{EPE} = \sqrt{\left(\frac{3\sigma\text{CDU}_{\text{layer 1}}}{2}\right)^2 + \left(\frac{3\sigma\text{CDU}_{\text{layer 2}}}{2}\right)^2 + \left(3\sigma\text{Overlay}_{\text{layer 1-2}}\right)^2}. \tag{1.3}$$

1.3.1.4 Time-Dependent Dielectric Breakdown (TDDB)

The primary DR limiter for metal layers is TDDB. At very small fabrication dimensions, very high electric fields are generated not just in the gate dielectric, but also in all isolating dielectrics between metals. A key issue in the DTCO process is determining the worst-case spacing between any two metal structures with misalignment, so that the resulting process is reliable against TDDB. TDDB occurs due to the presence of a large (although not as high as in gate dielectrics) electric field between two conductors over a long duration. Its severity is more readily pronounced in conductor layers with large overlay issues, for instance between a via and metal at disparate voltages, or in the MOL layers (Standiford and Bürgel, 2013). Obviously, sharp edges exacerbate the fields and are thus also an important issue.

Although layer self-alignment can alleviate the TDDB to some extent, it does not guarantee complete mitigation and necessitates other measures. One such case is the self-aligned raised source-drain contact to gate separation, that must be increased through the addition of extra spacer thickness and gate cap (Demuynck et al., 2014). This is partially in anticipation of some erosion of the self-aligning spacer material. The final separation between two layers must not, therefore, be based on just overlay and CDU, i.e., EPE, but also on the TDDB requirement, i.e., the expected potential differences between the structures. For the PDK, we assumed a 9 nm spacing requirement for TDDB prevention, a value similar to that assumed by Standiford and Bürgel (2013). Given that operating voltages are well below 1 V V_{DD}, this is conservative, which hopefully covers for any other small errors in the analysis.

1.3.2 Single Exposure Optical Immersion Lithography

The conventional lithography resolution limit, which determines the CD, is given by the Rayleigh equation as follows (Ito and Okazaki, 2000):

$$\mathrm{CD} = k_1 \frac{\lambda}{\mathrm{NA}} \tag{1.4}$$

where λ is the illumination source wavelength and NA is the projection lens numerical aperture. NA is given as

$$\mathrm{NA} = n_1 \sin\theta, \tag{1.5}$$

where θ is the maximum angle of the light diffracted from transparent mask regions, which can be captured by the lens; and n_1 is the refractive index of the material between the projection lens and wafer. The value of processing factor k_1 in Equation 1.4 depends on the illumination method and the resist process.

The term optical lithography has become nearly synonymous with the use of ArF light sources in the industry, employed since the 90 nm technology node (Liebmann et al., 2014a). The use of water to boost the NA leads to the technique being termed as optical immersion lithography. NA for the present 193i toolsets is 1.35 and the k_1 value is 0.28. The present set of values for these terms are a result of enhancements over the years, arising from resist process improvements and resolution enhancement techniques such as optical proximity correction (OPC), off-axis illumination (OAI), and source mask optimization (SMO) among others, each enabling a smaller CD at successive technology nodes. Ultimate limits for NA and k_1 are 1.35 and 0.25, respectively, but operating at these limits is challenging (Lin, 2015). Thus, as it stands, CD or half pitch for the layers patterned using an SE in 193i is about 40 nm. Although, the technical specifications for ASML's TWINSCAN NXT:1980Di optical immersion lithography scanner suggests that it can attain a resolution of about 38 nm (ASML, 2016).

1.3.3 Multi-Patterning Approaches

The use of 193i SE to achieve CD targets for all the patterned layers ended at the 22 nm technology node. This marked a severe restriction to continuing with the scaling trends for the subsequent nodes. Overcoming this limitation requires the use of multi-patterning (MP) techniques.

1.3.3.1 Litho-Etch^x (LE^x)

One of the most straightforward approaches to MP involves using multiple independent lithography and etch steps, where one litho-etch (LE) step refers to patterning shapes on a given layer through SE and etch step. The technique is termed LE^x, where x represents the number of LE steps. It applies to any light source and is not specific to just 193i. LE^2, or LELE, technique is of more immediate use since it is used to pattern the target shapes just below the SE patterning limit. To prepare a design for the LELE process, the design layout (Figure 1.1a)—containing target shapes at a pitch that is smaller than the SE limit—is decomposed into two separate layers, e.g., A and B, with different "colors" (Figure 1.1b). This decomposition or "coloring" must produce shapes assigned to a specific color layer at a pitch that can be patterned through an SE, thereby "splitting" the pitch. Consequently, the two color layers, resulting from the LELE decomposition step, correspond to two masks that are used in consecutive LE steps to pattern all the shapes on a single layer. This approach results in the LELE steps defining the line CD. Yet another approach involves specifying the space CD through LELE steps, instead of the line CD. While the process steps are simpler for the latter, the layout decomposition is more complex compared to the former approach.

The same basic principles used for LELE can be extended to the LE^x process with an x value that is larger than two, where x denotes the number of distinct colors and corresponding masks. As multiple masks are used to pattern a single layer, the cost associated with LE^x is higher than the SE lithography and increases with the number of masks. Unsurprisingly, the greater the number of LE steps to pattern a single layer, the higher the complexity and overlay concerns. Moreover, misalignment between steps must be considered, and jagged edges can result. LE^x is prone to odd-cycle conflicts, whereby decomposition may result in a coloring conflict when more than two shapes geometries exist so as to preclude topology patterning through SE. Such odd-cycle conflicts will be discussed in later sections. Stitching can alleviate these odd-cycle conflicts to some extent by patterning a contiguous shape through different exposures. It requires that the disparate fragments of the same shape have a reasonable overlap to counter EPE. Stitching does not completely

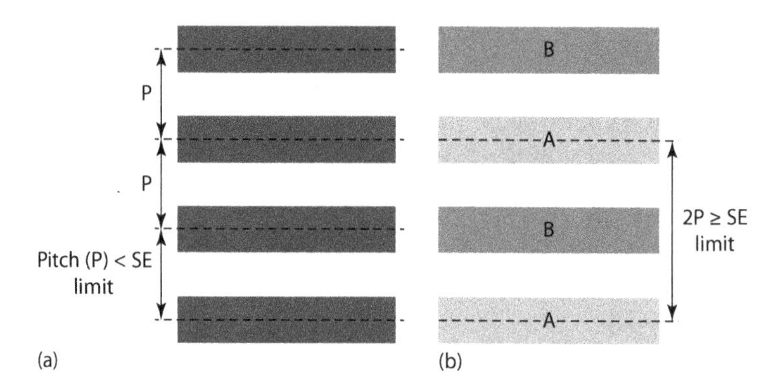

FIGURE 1.1
Litho-etch-litho-etch (LELE) multi-patterning (MP) approach. (a) Target layout shapes with inter-shape pitch below the single exposure (SE) patterning limit. (b) Target layout after decomposition of shapes into separate colors. Same-color shapes are at a pitch above the 193i SE patterning limit.

mitigate odd-cycle conflicts in all topologies. One-dimensional (1-D) patterns are specially challenging from this perspective.

1.3.3.2 Self-Aligned Multiple Patterning

Self-aligned multiple patterning (SAMP) represents another MP approach that seeks to limit the mask-defined line or space CD, thereby reducing the overlay error as compared to LE^x. The technique derives its name from spacers that are deposited along the sidewalls of a mask-defined 1-D or bidirectional (2-D) line, and are thus self-aligned to it. These spacers subsequently define the layer as the actual mask. Self-aligned double patterning (SADP) and self-aligned quadruple patterning (SAQP) are two of the more common forms of SAMP technique and denote whether the pitch is split by a factor of two or four, respectively. In the latter case, a first spacer is used to produce two second spacers, i.e., pitch splitting, that is used to pattern the actual lines.

SAMP can be broadly categorized into "spacer positive tone" and "spacer negative tone" process flows (Ma et al., 2010). In the former, the spacers define the dielectric isolation or space between the lines; therefore, the process is also called "spacer-is-dielectric" (SID). It allows for multiple line and space CDs. In the latter process flow, also called "spacer-is-metal" (SIM), the spacers define the line. However, the latter only allows for two line widths and allows more variability in the intra-layer line spacing, which is a reliability risk for TDDB.

Figure 1.2 shows a generic SADP SID flow. Similar to LE^x, decomposition is also necessary for the SADP process (Figure 1.2b). In decomposition, one of these two masks (e.g., mask A) is selected as a candidate for a derivative photomask, using which the shape is patterned on the resist using an SE (Figure 1.2c). The resist is trimmed in the event the target line CD is under the SE resolution limit, followed by an etch and resist strip for mandrel formation (Figure 1.2d) (Oyama et al., 2015). The mandrel is a sacrificial feature, around which the spacers are deposited as shown in Figure 1.2e. Note that the result includes loops. A second photomask, called the block mask, is then used in conjunction with the spacers to "block" the regions where the feature should not be present. In the case of metal lines, the remaining regions define the trenches in a damascene process and subsequently define the line widths. The SAQP process involves two spacer deposition steps, where the first set of spacers serve as mandrels for the second set of spacers. This is evident in the following fin examples.

Note that the decomposition criterion for SAMP is different from LE^x, since the mandrel is continuous and any discontinuities in it can be marked using the block mask, unlike LE^x, where the lines with such discontinuities must be patterned through a separate exposure, resulting in a potential odd-cycle conflict for certain topologies. It must also be noted that for SAMP, shapes with different colors do not correspond to separate lithography masks, as the deposition of other colored shapes is through spacer deposition and the number of masks employed is lower than the number of decomposition colors.

1.3.3.3 Multiple Patterning Approach Comparison

Before selecting a particular MP technique for layer patterning, consideration must be given to the variability, complexity, and cost associated with it, and the way in which it differs from another MP technique based on these metrics. Being the simplest and most commonly used MP techniques, LELE and SADP are contrasted here to determine their favorability as the preferred MP technique. In terms of cost, the SADP process is more

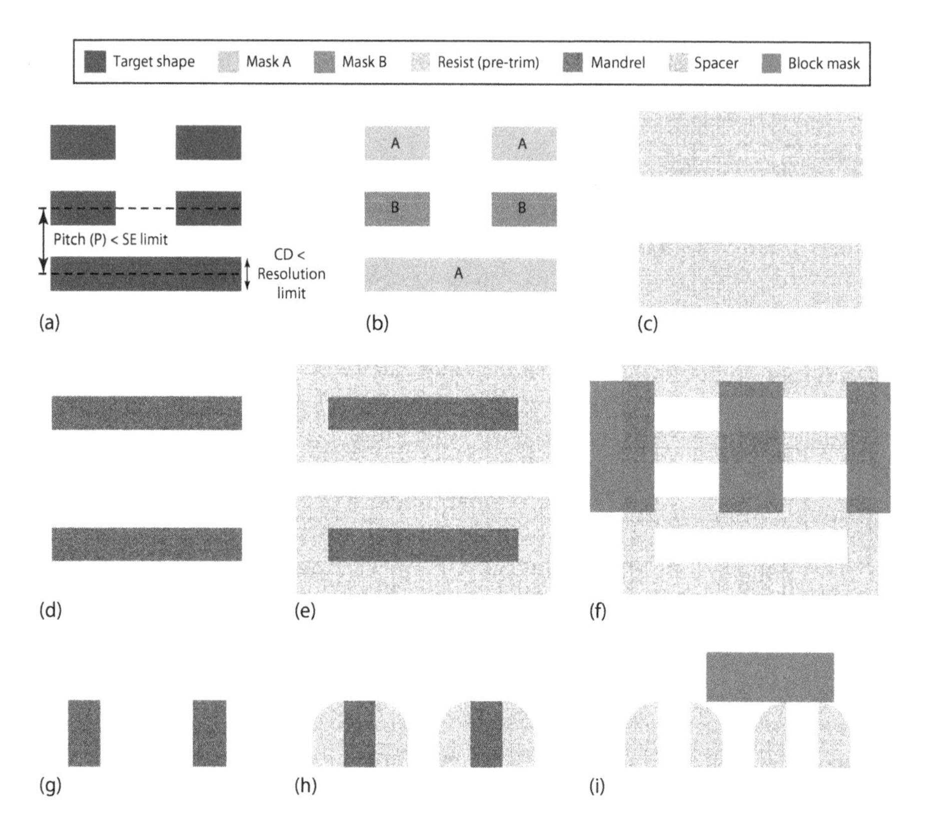

FIGURE 1.2
A "spacer-is-dielectric" (SID) or "spacer positive tone" self-aligned double patterning (SADP) process. (a) Target layout shapes with inter-shape pitch and line CD below the single exposure (SE) patterning limit. (b) Target layout after decomposition of shapes into separate colors. (c) Resist patterning (pre-trim) with CD obtainable through single exposure resolution limit. (d) Mandrel formation (post resist trim and etch). (e) Sidewall spacer deposition. (f) Spacer and block mask–defined trench formation for line patterning. (g) Mandrel cross section. (h) Mandrel and spacer cross section after spacer deposition. (i) Spacer cross section after mandrel strip. Block mask is overlaid on top of the spacer to denote the regions where metal will not be deposited, but is not indicative of the process.

expensive than the LELE due to its sequential etch and deposition steps (Vandeweyer et al., 2010). Liebmann et al. (2015) put the 193i LELE and 193i SADP normalized wafer costs at 2.5× and 3×, respectively, of the 193i SE cost. However, the lower variability and resulting smaller values for similar DRs that determine design density weigh in favor of SADP.

The LELE steps define the line CD and as the two line populations are distinct, their CDU are generally uncorrelated (Arnold 2008). Any overlay between the two masks affects the space CD. When LELE steps define the space CD instead, the CDU between the two space populations is uncorrelated and overlay error affects the line CD. Thus, notwithstanding its use to define either the line or space CD, the CDU is entangled with overlay in the LELE process. The EPE in LELE can be calculated as

$$EPE_{LELE} = \sqrt{\left(\frac{3\sigma CDU_{\text{line 1}}}{2}\right)^2 + \left(\frac{3\sigma CDU_{\text{line 2}}}{2}\right)^2 + \left(3\sigma \text{Overlay}_{\text{line 1-2}}\right)^2} \quad (1.6)$$

for metal lines. On the other hand, in an SADP process, as the line or space CD is mostly defined through a single mask and spacer, the overlay or misalignment does not play a significant role in CD determination unless the target shape edges that determine CD are defined by the block mask. Consequently, block mask edge definition should be avoided. The EPE in SADP, for spacer-defined features, can thus be given as

$$\mathrm{EPE_{SADP}} = \sqrt{\left(3\sigma \mathrm{CDU_{litho}}\right)^2 + \left(3\sigma \mathrm{spacer_{left\_edge}} + 3\sigma \mathrm{spacer_{right\_edge}}\right)^2}. \tag{1.7}$$

The spacer edge–related terms in Equation 1.7 are associated with the spacer 3σ CDU, which can be as small as 1 nm (Ma et al., 2010). The absence of overlay due to the absence of block mask–defined edges, together with the small spacer CDU, gives SADP an advantage over LELE in terms of EPE control for spacer-defined edges, assuming greater than 3 nm LELE overlay and a well-controlled SADP spacer CDU of about 1 nm. However, if LELE overlay can be made small enough (~2.5 nm), then the EPE for the LELE overlay-influenced edges approach that for SADP spacer-defined edges, assuming that the remaining EPE constituent terms remain unchanged (Jung et al., 2007). As mentioned earlier, ASML's TWINSCAN NXT:1980Di 193i scanner has a 3σ MMO of 2.5 nm, but this is accompanied by a change in the CDU value, so that the advantage still lies with SADP. Finally, the spacers do not suffer from much line edge roughness (LER) (Oyama et al., 2012), which is another advantage that SADP offers over LELE. From an electrical perspective, the larger lithography-related variations for LELE patterned layers translate into larger RC variations, which can be nearly twice as much as that for SADP patterned layers employed for critical net routing, primarily due to capacitance variations (Ma et al., 2012). Thus, a choice of LELE creates a potential plethora of metal corners for designers to deal with.

The impact of topology imposed DR constraints must also be considered when selecting an MP technique, so as to limit any density penalty. Double patterning can typically overcome the large tip-to-tip (T2T) or tip-to-side (T2S) spacing requirements inherent to SE lithography by avoiding these features from being assigned to the same color. The SE T2T or T2S spacing values are larger than the minimum SE-defined width to spatially accommodate the hammerheads used for OPC applied to ensure pattern fidelity (Wong et al., 2008). Large spacing also prevents tips from shorting due to bridging by ensuring sufficient contrast, as the regions between tips have low image contrast. However, even with double patterning, certain feature topologies, such as gridded metal routes with discontinuities, can result in T2T or T2S features on the same mask (Ma et al., 2010). This occurs because shapes on adjacent routing tracks must be colored alternatingly, which forces the shapes along the same track to be the same color. For an LELE patterned layer, as illustrated in Figure 1.3a, this color assignment necessitates the use of DR values related to SE T2T spacing, x, that are even larger than the minimum SE-defined width and nullify the advantage of double exposure. By comparison, the T2T and T2S features in SADP are block/cut mask defined (see Figure 1.3b). Although also SE-defined, the block/cut mask width is similar to the SE-defined line width, y, instead of the larger T2T or T2S SE spacing requirement. As identified by Ma et al. (2012), this enables SADP patterned metal routes on the same track to connect to pins on a lower metal interdigitating another lower metal route as shown in Figure 1.4a, while different tracks must be used as in Figure 1.4b to realize the same connections for LELE patterned metal routes. Thus, SADP can lower the density penalty in certain design scenarios.

Furthermore, as shown in Figure 1.5a, LELE can produce odd-cycle conflicts during mask decomposition into two colors. Such a conflict requires increasing the spacing between the features to the SE T2T value (x) in order to resolve the conflict. In contrast, SADP is largely free of odd-cycle conflicts for 1-D topologies with equal metal width along routing tracks,

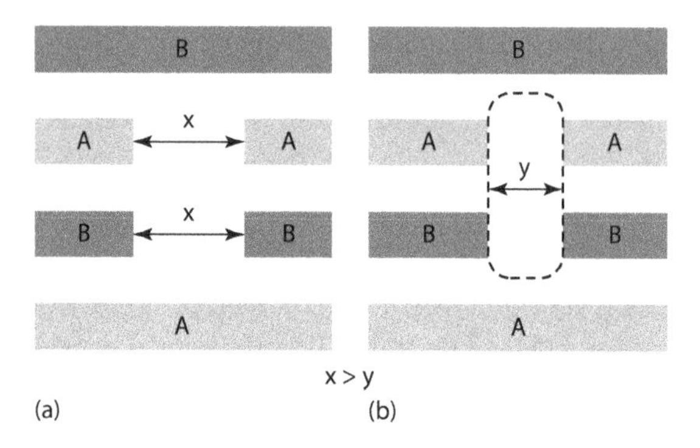

FIGURE 1.3
Comparison of tip-to-tip (T2T) spacing cases for different double patterning approaches. (a) Scenario where T2T spacing, x, is single exposure limited for LELE patterned layers is not too uncommon for designs with 1-D, gridded routing. (b) T2T spacing, y, for SADP patterned layer is defined by the block or cut mask (dotted polygon). Being a line-like feature, the block/cut mask width is similar to the single exposure–defined line width, which is smaller than the spacing required for single exposure–defined tips.

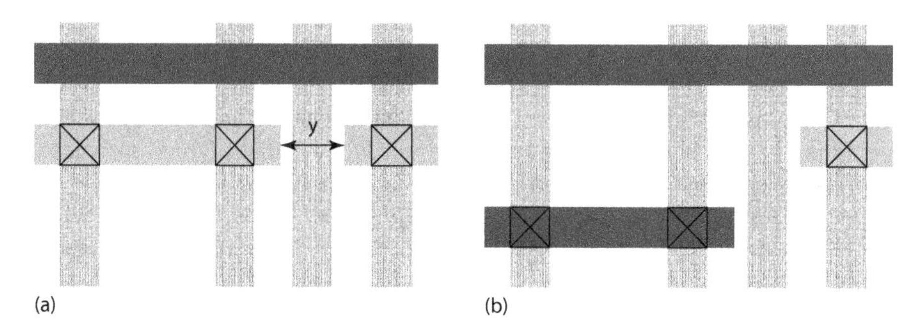

FIGURE 1.4
Pin connection scenarios for double patterned layers. (a) Connections to next neighboring pins can be made through metals on the same track by leveraging the smaller block mask–defined T2T spacing for an SADP patterned metal layer. (b) The same becomes impossible with an LELE patterned metal layer due to larger T2T spacing requirements for same-color shapes and different tracks must be utilized, which results in density penalty.

as evident in Figure 1.5b. However, odd-cycle conflicts may arise during mask decomposition into two colors for SADP, when unequal metal widths or 2-D features are used as shown in Figure 1.5c. Thus, the simpler SADP is adopted for BEOL in the ASAP7 PDK. Some simple DRs, such as limiting line widths to specific values and pitches, make automatic decomposition possible. This is handled by the Calibre design rule checking (DRC) flows automatically, to simplify usage.

1.3.4 Extreme Ultraviolet Lithography (EUVL)

1.3.4.1 EUVL Necessity

Keeping in line with the area scaling trends requires manufacturing capability enhancement through technological advancement, with pitch scaling through photolithographic improvements constituting the major effort. This increases the wafer cost initially by nearly 20%–25% at a new technology node (Mallik et al., 2014). As a particular technology node matures, the subsequent process and yield optimization bring down the wafer cost. These improvements, together with the increased transistor density, eventually result in

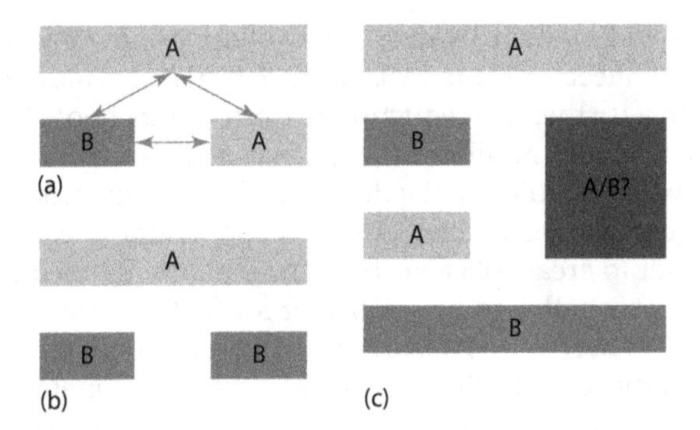

FIGURE 1.5
(a) Even the most common topologies suffer from LELE odd-cycle conflicts. (b) SADP is largely free of such conflicts, given correct coloring and considering 1-D equal width metals. (c) Conflicts with SADP may arise for wide metals used in power or clock routing.

the overall cost reduction per transistor. 193i MP adoption to pattern increasingly small FEOL and BEOL pitches has led to a larger than conventional wafer cost increment. This is due to an increased number of photomasks and process steps required to pattern a given layer with MP—hitherto patterned through a single LE step. MOL layer and finFET introduction have led to further cost escalation as they constitute critical layers that require MP. The transistor cost-reduction trend has slowed down over the previous few nodes, to which MP has contributed to some extent. This contribution will become even larger with an increase in layers patterned through MP techniques with each new node, resulting in a further slow-down in the per transistor cost-reduction trend. Thus, MP can potentially undermine the cost-effectiveness associated with transitioning to a new node.

In addition to cost, MP is also challenging from a variability perspective, which leads to stringent CDU and overlay considerations, as discussed in previous sections. Accommodating these requires guard-bending the projected pitch/spacing targets for relaxed process margins. Patterning a critical BEOL layer through MP is one such case. Patterning a 2-D shape using LE^x necessitates stitching, which complicates overlay requirements and requires that the pitch target be relaxed to ensure sufficient shape overlap. An alternative is using SADP patterning, which is not very amenable to 2-D shapes, and a 1-D patterning approach using SADP becomes the other choice (Ryckaert et al., 2014). This detrimentally affects the cell circuit density, but is easier to manufacture (Vaidyanathan et al., 2014). Mallik et al. (2015) estimate a 16% and 5%–15% lower area penalty for 2-D, instead of 1-D, metal layers for SRAM cell and standard cells, respectively. A pure 1-D approach makes the layout of even relatively simple logic gates more difficult and results in poor input pin accessibility.

EUVL can mitigate some of the MP-related issues. It uses a 13.5nm light wavelength, instead of 193nm used for ArF immersion lithography. This enables patterning the features at a much smaller pitch and resolution, so that a single EUV exposure suffices for patterning at the target pitch only attainable through multiple exposure with 193i, thus greatly simplifying the DRs. Since academic use was a primary goal of the PDK, opting for simpler EUV rules was a key consideration, even at the risk of optimism in EUV availability. However, we are presently working on additions and libraries to support pure 1-D metallization.

1.3.4.2 EUVL Description and Challenges

ASML's EUVL scanners have a 0.33 NA and can operate with a processing factor k_1 of nearly 0.4 (Van Setten et al., 2015; Kerkhof et al., 2017). Using these values, together with a 13.5 nm wavelength, in Equation 1.4 gives a CD of approximately 16 nm. Improvements have led the CD to be reduced to 13 nm (ASML, 2017; Kerkhof et al., 2017). An NA improvement to 0.5 can lead to a further CD reduction to 8 nm (VanSchoot et al., 2015).

EUVL differs from 193i in a number of ways that make it a more challenging patterning approach and have contributed to the delay in EUVL being production-ready. Light at EUV wavelength is generated as follows. A droplet generator releases tin droplets, which are irradiated by a laser to create plasma containing highly ionized tin that emits light at 13.5 nm wavelength that is gathered by a collector for further transmission (Tallents et al., 2010). Light in the EUV spectrum is absorbed in the air, which necessitates manufacturing under vacuum condition. It is also absorbed by nearly all materials, which precludes optical lens use to prevent high energy loss. Instead, reflective optics, i.e., mirrors, are used. These mirrors have a reflectivity of around 70% and an EUV system can contain over 10 such mirrors, resulting in only around 2% of the optical transmission to reach the wafer (Tallents et al., 2010). Inefficiency in the power source reduces the transmitted optical power even further, thus creating a demand for a high-power light source. It is also desirable for the photoresist to have a high sensitivity to EUV light.

Production throughput for a photolithography system, given in wafers per hour (WPH), is closely associated with cost-effectiveness. To a considerable extent, it relies on the optical power transmitted to the wafer, and thus on the EUV light source power. It also depends on the amount of time the system is available for production, i.e., system availability. System downtime adversely affects the cost. For EUVL patterning to be cost-effective, a source power exceeding 250 W is desired for over 100 WPH throughput at a 15 mJ/cm^2 photoresist sensitivity (Mallik et al., 2014). Currently, the source power is around 205 W and droplet generator, hitherto a major factor in EUVL system unavailability, has become a smaller concern in more modern EUVL systems (Kim et al., 2017). These and other improvements have brought EUVL systems close to the HVM production goals by increasing the throughput to 125 WPH (Kerkhof et al., 2017). Collector lifetime is the biggest contributor to the system unavailability at the moment and a number of other issues must be surmounted for further cost-effectiveness (Kim et al., 2017). Overall, EUVL systems continue to improve and are slated to be deployed by some foundries for production at N7 (Ha et al., 2017; Xie et al., 2016).

1.3.4.3 EUVL Advantages

The decision to choose SE EUV over 193i MP approaches comes down to both cost and complexity concerns. The EUV mask cost alone is approximately 1.5× that of a 193i mask (Mallik et al., 2015). Other operational expenses bring up the EUVL cost to nearly 3× of 193i SE (Liebmann et al., 2015). The number of masks used in each technology node has increased almost linearly up until N10, but the continuation of MP use will result in an abrupt departure from this trend at N7 (Dicker et al., 2015). The issue is compounded by an increase in the associated process steps, which further adds to the cost. Liebmann et al. (2015) estimate the normalized LE^2, LE^3, SADP, and SAQP cost to be 2.5×, 3.5×, 3×, and 4.5× that of 193i SE, respectively. Dicker et al. (2015) estimate a 50% patterning cost reduction with EUV as opposed to SAQP. They also estimate a faster time to yield and time to market with EUVL due to cycle time reduction as compared to 193i MP approaches that suffer from a large learning cycle time—as large as 30% compared to 2-D EUV, thus improving EUVL cost-effectiveness.

EUVL SE also reduces the process complexity by virtue of reduced overlay. The small EUV wavelength allows the processing factor k_1 to be relatively large—in the range 0.4–0.5.

This enables EUVL to have a high contrast, given by normalized aerial image slope (NILS), than 193i and allows features to be printed with higher fidelity (Kerkhof et al., 2017; Ha et al., 2017). The high feature fidelity, better corner rounding, and an SE use with EUVL cause fewer line and space CD variations. Consequently, metals and vias patterned through EUVL have more uniform sheet resistance (Ha et al., 2017). They also have lower capacitance as compared to SADP patterned shapes, as EUV SE obviates dummy fills and metals cuts. These improvements contribute to improved scalability and better performance (Kim et al., 2017).

Mallik et al. (2014) estimate the normalized wafer cost to increase by 32% at N10 and by a further 14% at N7 without EUV insertion at these nodes. They also estimate a 27% cost reduction, as compared to the latter case, due to EUVL use at N7 for critical BEOL layer patterning with a 150 WPH throughput as a best-case scenario. Ha et al. (2017) put the number of mask reduction at N7 due to EUVL use at 25%. Dicker et al. estimate over 40% cost per function reduction in moving from N10 to N7, and further to N5 as a consequence of EUV insertion for critical BEOL layers. Thus, EUVL deployment at these nodes will likely help ensure the economic viability of process node transition.

1.3.5 Patterning Cliffs

Patterning cliffs mark the pitch limits for a given lithography technique or MP approach. Table 1.3 summarizes these pitch limits for the metal layers (Sherazi et al., 2016; VanSchoot et al., 2015).

It must be noted that EUV scanners are being continuously refined and their capabilities may vary, resulting in different final patterning cliffs. However, Table 1.3 gives good rule of thumb values.

1.3.6 DTCO

When a process is still in development, designers are faced with "what if?" scenarios, where the process developers naturally wish to limit the process complexity, but excessive limitation may make design overly difficult or lacking the needed density to make a new process node worthwhile. Those who must make decisions regarding cell architecture for future processes face significant challenges, as the target process is not fully defined. Specifically, as bends in diffusions and gates have become increasingly untenable, the

TABLE 1.3

Patterning Cliffs, i.e., Minimum Feature Pitch for a Particular Patterning Technique

Patterning Technique	Minimum Pitch (nm)
193i	80
193i LELE	64
193i LELELE	45
193i SADP	40
EUV SE (2-D, NA = 0.33)	36
EUV SE (1-D, NA = 0.33)	26
EUV SE (2-D, NA = 0.55)	22
193i SAQP	20
EUV SE (1-D, NA = 0.55)	16

MOL layers have been introduced to connect source-drain regions and replace or augment some structures such as poly crossovers. Decisions that were once purely up to the technology developers increasingly affect design possibilities. Consequently, DTCO is used to feed the impact of such process structure support decisions on the actual designs back into the technology decision-making process (Aitken et al., 2014; Chava et al., 2015; Liebmann et al., 2014b, 2015). It is increasingly important as finFET width discretization and MP constrain the possible layouts. Consequently, determining DRs progressed in this predictive PDK development by setting rules based on the equipment capabilities, designing cell layouts to use them, and iterating the rules based on the outcomes. This includes the APR and SRAM array aspects as described later.

1.4 FEOL and MOL Layers

Transistors are assumed to be fabricated using a standard finFET-type process: a high-K metal gate replaces an initial polysilicon gate, allowing different work functions for NMOS and PMOS, as well as different threshold voltages (V_{th}) (Vandeweyer et al., 2010; Seo et al., 2014; Lin et al., 2014; Schuegraf et al., 2013). Fins are assumed to be patterned at a 27nm pitch and have a 7nm drawn (6.5nm actual) thickness. The layer active is drawn so as to be analogous to the diffusion in a conventional process and encloses the fins—over which raised source-drain is grown—by 10nm on either side along the direction perpendicular to the fin run length (Clark et al., 2016). The drawn active layer differs from the actual active layer, which is derived by extending it halfway underneath the gates—perpendicular to the fins. The actual active layer, therefore, corresponds to the fin "keep" mask, with its horizontal extent marking the place where fins are cut and its vertical extent denoting the raised source-drain regions.

Gates are uniformly spaced on a grid with a relatively conservative 54nm CPP. Gates are 20nm wide (21nm actual). Spacer formation follows poly gate deposition (Hody et al., 2015). Cutting gate polysilicon with the gate cut mask, in a manner that keeps the spacers intact with a dielectric deposition following, ensures that fin cuts are buried under gates or the gate cut fill dielectric, so source/drain growth is on full fins. A double diffusion break (DDB) is assumed to be required to keep fin cuts under the gate. Recently announced processes have removed the DDB requirement, improving standard cell density. Adding this into the PDK as an option is under consideration for a future release. A 20nm gate cap layer thickness is assumed. This thickness provides adequate distance to avoid TDDB after self-aligned contact etch sidewall spacer erosion, accounting for gate metal thickness nonuniformity (Demuynck et al., 2014). The dual spacer width is 9nm.

The resulting FEOL and MOL process cross section comprises Figure 1.6. Figure 1.6a shows the (trapezoidal) source/drains grown on the fins. The MOL layers can be used for functions typically reserved for the first interconnect metal (M1) layer (Lin et al., 2014) and serve to lower M1 routing congestion, thereby improving standard cell pin accessibility (Ye et al., 2015). The connection to the MOL local interconnect source-drain (LISD) layer is through the source-drain trench (SDT) contact layer. The minimum SDT vertical width of 17nm is required in a SRAM cell, which necessitates patterning using EUVL. It has a 24nm drawn horizontal width, with the actual width being 25nm. This width is larger than the 15nm gap between the spacers, so as to ensure complete gap coverage and contact with the raised source-drain (RSD), despite the 5nm 3σ EPE for EUV.

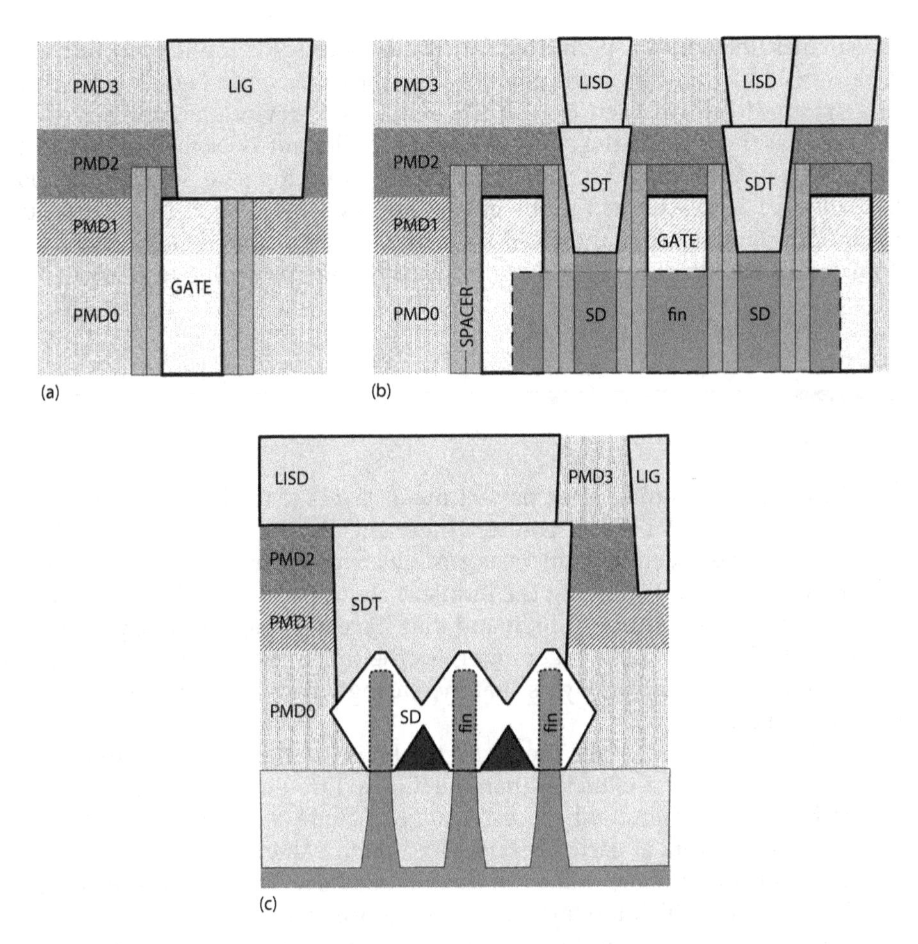

FIGURE 1.6
FEOL and MOL cross sections. (a) LIG connection to the gate. (b) LISD to SDT and SDT to source-drain (SD) connection. LISD location in the stack allows it to cross over gates to be used for routing. (c) Fin and SD cross section. LIG is shown here to illustrate its necessary separation from LISD and SDT. Sub-fin and shallow trench isolation (STI) are evident underneath PMD0.

LISD provides the means of connecting RSD regions to power rails and other equipotential RSDs within a standard cell. It is drawn at the same horizontal width as SDT for lower resistance when connecting to RSD through SDT. LISD may also be used for routing purposes within a standard cell at 18 nm width and 36 nm pitch, as it can pass over gates—further lowering M1 usage. The layer thicknesses are defined by different dielectric layers, so that appropriate etch stops can be used. This implies 2-D LISD routing, which, when combined with width and pitch assumptions, means that EUVL must be used for patterning. LISD connects to M1 through via 0 (V0), which is another MOL layer.

The local interconnect gate (LIG) layer is used for connecting gates to M1 through V0 and for power delivery to standard cells by connection to LISD upon contact. The minimum LIG width of 16 nm is dictated by the LIG power rail spacing from the gate. The width value implies that LIG is also patterned using EUVL. Recent advances in EUVL have demonstrated resolutions lower than 18 nm (Neumann et al., 2015; VanSchoot et al., 2015) even for 2-D patterns, but the PDK restricts this value as the minimum line width for 2-D layers, in accordance with a more conservative 36 nm pitch for 2-D EUVL (Mallik et al., 2015).

However, smaller line width is permitted for LIG as its usage is limited to unidirectional, i.e., 1-D, patterns. LIG connects to the gate through the cap layer in Figure 1.6b, which cuts through a standard cell between the NMOS and PMOS devices. Figure 1.6c shows the same view, but at the fins, so the source/drains are illustrated perpendicular to that in Figure 1.6a. As mentioned, the fin cuts occur under a dummy gate, which comprises 1/2 of a DDB. The SDT is self-aligned to the gate spacer as shown. MOL rules turn out to be limiting for both standard cells and SRAM. The details are described in Sections 1.6.3 and 1.8.1. More details and the transistor electrical behavior are presented in Clark et al. (2016).

1.5 BEOL Layers

The ASAP7 PDK assumes nine interconnect metal layers (M1-M9) for routing purposes and corresponding vias (V1-V8) to connect these metals. Figure 1.7 shows a representative BEOL stack cross section, comprising a lower metal layer (Mx), via (Vx), and an upper metal layer (M$x+1$). Following the industry trend (Lin et al., 2014), all BEOL layers assume copper (Cu) interconnects. Metal and vias have a 2:1 aspect ratio, in line with the ITRS roadmap (ITRS, 2015). Table 1.4 enumerates the thickness of the interconnect layers. Figure 1.7 also shows the barrier layers that increasingly consume the damascene trench, increasing resistance.

In the ASAP7 PDK, metals have the same thickness as the corresponding inter-layer dielectric (ILD), but the vias are thicker than the ILD by 10%—an amount corresponding to the assumed HM thickness. Actual processes at N7 include as many or more than 14 metals, with more metal layers at each thickness and pitch value, culminating in two layers that are much thicker and wider at the top than in our PDK. The layers here are representative of all but the very thick top layers, which are primarily for power distribution. We had initially not foreseen PDK use for large die power analysis, but are considering adding layers to make the PDK more amenable to full die analysis.

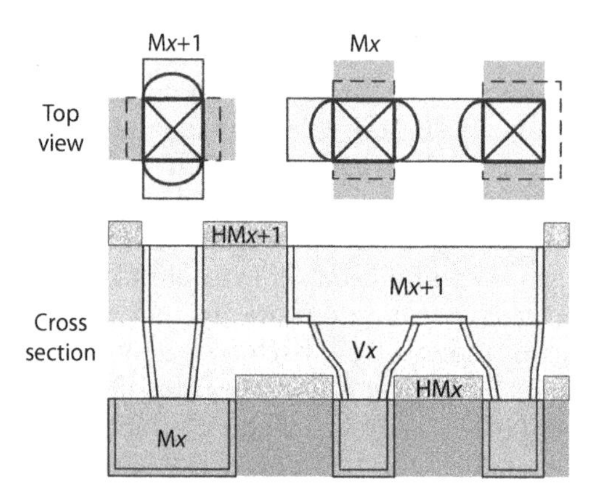

FIGURE 1.7
Representative BEOL cross section. The dotted lines represent the actual self-aligned via (SAV) masks top view. Vias with (left two) and without end-cap (rightmost) are shown. Arcs along M$x+1$ length denote via widening at the non-hard mask edges, evident in the cross section for the two vias at the right.

TABLE 1.4
BEOL Layer Thickness and Metal Pitches

Metal/Via Layer	Thickness (nm)
M1-M3	36
V1-V3	39.6
M4-M5	48
V4-V5	52.8
M6-M7	64
V6-V7	70.4
M8-M9	80
V8	88

1.5.1 SAV and Barrier Layer

In a typical via-first flow, the ILD corresponding to a via sustains damage and erosion—caused by dry etch and cleaning steps, respectively, during both via and metal patterning. The metal width itself is also affected, since via formation goes through the metal damascene trench. This results in via widening that can cause shorts to adjacent, non-equipotential metal lines (Baklanov et al., 2012; Brain et al., 2009). This issue is exacerbated by small metal pitches at lower technology nodes. Consequently, in the ASAP7 DRs, a self-aligned via (SAV) formation flow, as described by Brain et al. (2009), is assumed, whereby vias are patterned after the upper interconnect metal layer is patterned on a hard mask (HM) that is relatively unaffected by the via etch. The HM greatly limits via widening perpendicular to its edges, although some widening occurs along the metal, as evident in Figure 1.7. The resulting via edges are delineated by the upper metal HM, i.e., via self-alignment. Via mask edges tend to extend outward where they are defined by the via rather than a HM. Nonetheless, the via is perfectly aligned perpendicular to the upper metal direction despite via and upper metal overlay errors. The dashed lines in Figure 1.7 show that the actual via mask overlaps so that the HM defines the width even with misalignment. However, for simplicity the vias are drawn conventionally in the PDK. They are sized as part of the DRC flows.

Barrier materials, such as tantalum nitride (TaN), are required at the Cu and ILD interface to prevent Cu diffusion. The thickness of the barrier—composed of more resistive TaN—does not scale commensurately with the interconnect scaling. This, together with diffuse electron scattering at interfaces, causes a greater increase in line resistance than is expected as a consequence of scaling (Im et al., 2005). Additionally, the presence of TaN at the via interface has the undesirable effect of increased resistance (Schuegraf et al., 2013). For this reason, the PDK assumes the use of manganese-based self-forming barriers (SFBs) that assuage the shortcoming of TaN by virtue of their conformity, surface smoothness, smaller thickness in the total interconnect fraction, and high diffusivity in Cu (Schuegraf et al., 2013; Au et al., 2010).

The metal resistivity, as specified for extraction purposes, is calculated based on (Pyzyna et al., 2015)

$$\rho = \rho_0 \frac{3}{8} C(1-p)\left(\frac{1}{h} + \frac{h}{A}\right)\lambda + \rho_0 \left[1 - \frac{3}{2}\alpha + 3\alpha^2 - 3\alpha^3 \ln\left(1 + \frac{1}{\alpha}\right)\right]^{-1} \tag{1.8}$$

where ρ_0 denotes the bulk resistivity; C is a geometry-based constant; p is the electron collision specularity with surfaces; λ is the bulk electron mean free path; h is the line height; A is the cross-section area; α is given as $\lambda R/[G(1 - R)]$, where R is the electron reflection coefficient at the grain boundaries and G is the average grain size. The first and second terms are the resistivity due to surface electron scattering and grain boundary scattering, respectively. The latter dominates for the 7nm node (Pyzyna et al., 2015).

1.5.2 EUV Lithography Assumptions and Design Rules

In addition to some of the FEOL, i.e., fin cut, and MOL layers in the PDK, EUVL is also assumed for patterning M1-M3 and vias corresponding to these metals, i.e., V1 through V3. The choice of an EUVL assumption and the accompanying M1-M3 pitch of 36nm is based on the premise that this pitch may be attained using single EUV exposure (Mallik et al., 2014). Meeting the same target using optical immersion lithography requires the use of MP techniques, such as SAQP with LE or LELE block mask, which also pushes toward all 1-D topologies. While EUVL is costlier than optical immersion lithography when considering an SE, the use of multiple masks in SAQP with block means that the MP approach becomes nearly as expensive as EUVL due to expensive mask tooling and associated processing steps (Liebmann et al., 2015). It is noteworthy that several ITRS target pitch values at N7 are at or over the EUV SE cliff. As a result, beyond N7, MP will be required even for EUV lithography-defined layers. With N7 at the cutoff, we felt EUVL would be appropriate.

As EUVL permits the single mask use, employing it simplifies the design process by circumventing the issues related to MP, such as complicated cell pin optimization issues due to SADP and odd-cycle conflicts as a consequence of LELE use (Xu et al., 2015). As mentioned earlier, EUVL use at a slightly relaxed pitch of 36nm also permits 2-D routing, which has the effect of further simplifying both standard cell and SRAM cell design. Thus, given similar cost and reduced design complexity, the PDK assumed EUVL over 193i MP schemes for a number of layers that complicate standard cell design, viz. SDT, LISD, LIG, V0, and M1. However, subsequent iterations of the PDK will revise this choice, so as to use SAQP for 1-D M1-M3, as a consequence of further delays in EUV readiness for high volume manufacturing. While the aggressive pitch value of 36nm that EUV affords is not required for M2 and M3, we considered these layers to be EUVL patterned due to the ease that this assumption lends when routing to standard cells during APR. Having the vertical M3 match the M2 also allows better routing density vs. a choice of say, the gate pitch for vertical M3. This also allows relative flexibility in metal directions. We foresee vertical M2 as a better choice for 1-D cells.

A 3σ EPE of 5nm for two EUV layers, when determining inter-layer DRs, was calculated assuming a 3σ mixed machine overlay (MMO) of 1.7nm for the EUV scanner, a 3σ error in the placement of 2nm due to process variations and through additional guardbanding (Van Setten et al., 2014). Patterned M1-M3 lines have a 36nm 2-D pitch and a minimum line width of 18nm is enforced by the DRs. T2T spacing for narrow lines is 31nm following Van Setten et al., while wider lines can have a smaller T2T spacing at 27nm (Van Setten et al., 2015; Mulkens et al., 2015). As per the PDK DRs, lines narrower than 24nm are considered thin lines and those wider than this value are considered wide lines. This threshold value was determined based on the minimum LISD width, since LISD routes near the power rails become the limiting cases for T2T spacing. A moderate T2S spacing of 25nm follows the results demonstrated by Van Setten et al. (2014). A corner-to-corner EUV metal spacing of 20nm enables via placement to metals on parallel tracks at the minimum possible via spacing of 26 with a 5nm EUV upper metal end-cap to allow full enclosure.

1.5.3 MP Optical Lithography Assumptions and Design Rules

Metal interconnect layers above the intermediate metal layers, i.e., M3, are assumed to be patterned using 193nm optical immersion lithography and MP. The metal pitches are the same as the thickness values described in Table 1.3 (the 2:1 aspect ratio). The pitch values for M4-M5 and M6-M7 correspond to the targets defined by Liebmann et al. (2015) for 1.5× and 2× metal, respectively. The same metal pitch ratios do not apply to our PDK since the 1× metal pitch is 36nm to ensure 2-D routing instead of the roadmap value of 32nm. Moreover, recent foundry releases seem to be tending to the more conservative 36nm pitch as well. Incrementing the pitch in multiples other than 0.5× does not have any design implication as proven by our DTCO APR experiments. The metal and via aspect ratios are in line with ITRS roadmap projections.

1.5.3.1 Patterning Choice

The M4-M5 pitch target of 48 nm can be attained using either SADP or LE^3, and that of 64nm for M6-M7 using either SADP or LELE. LE^3 was dismissed outright for M4-M5 patterning, since it is costlier than SADP (Liebmann et al., 2015), and even though the lower LELE cost is appealing for M6-M7 patterning, SADP has lower EPE, LER, and hence RC variations as mentioned in Section 1.3.3.3. Therefore, we chose SADP over LELE for patterning the layers M4-M6. Furthermore, Ma et al. (2010) also found that LELE fails the TDDB reliability test at 64nm pitch with 6nm overlay while SADP appears reliable. Although we used a 3.5nm overlay for 193i patterned layers that would allay the TDDB severity, we still assumed SADP patterning for M6-M7 so as to be more cautious, given the TDDB concern in addition to the other aforementioned SADP advantages. We chose SID over SIM as it permits multiple metal widths and spaces, which is beneficial for clock and most importantly, power routing. Patterning 2-D shapes is possible in SADP but presents challenges, as certain topologies, such as the odd-pitch U and Z constructs, may either not be patterned altogether or contain block mask–defined metal edges that result in overlay issues and adversely affect metal CD (Ma et al., 2012). Consequently, we chose to restrict M4 through M7 to 1-D (straight line) routing in the DRs.

The ASAP7 PDK supports DRCs based on actual mask decomposition into two different masks (colors) for the purpose of SADP. The mask decomposition is performed as part of the rule decks using the Mentor Graphics Calibre MP solution. An automated decomposition methodology (Pikus, 2016) is employed and does not require coloring by the designer, which greatly simplifies the design effort. As mentioned, we consider this key in an academic environment. To the best of our knowledge, this is the first educational PDK that offers DRs based on such a decomposition flow for SADP. In contrast, the FreePDK15 also used multi-colored DRs (Bhanushali et al., 2015), but required decomposition by a designer by employing different-colored metals for the same layer.

1.5.3.2 SADP Design Rules and Derivations

DRs must ensure that shapes patterned using the two photomasks, viz. the block and the mandrel, can be resolved. This entails writing DRCs in terms of the derived photomasks and perhaps even showing these masks to the designer. However, fixing DRCs by looking at these masks, and the spacer, is both non-intuitive and confusing. In the flows here, the colors are generated automatically, and the flows can produce the block and mandrel masks created by metal layer decomposition. We formulated restrictive design rules (RDRs) that ensure correct-by-construction metal topologies and guarantee resolvable shapes created using mandrel and block masks referring only to the metals as drawn, so that the rules are color agnostic. Nonetheless, to validate the DRCs, the mandrel, block mask, and spacer shapes are completely derived in our separate validation rule decks (Vashishtha et al., 2017a).

The decomposition criteria for assigning shapes to separate colors for further SADP masks and spacer derivation is as follows. In addition to assigning shapes at a pitch below the 193i pitch limit of 80nm to separate colors, a single CD routing grid–based decomposition criterion is also used. The latter prevents isolated shapes, which do not share a common run length with another shape, from causing coloring conflict with a continuous mandrel through the design extent. Any off-track single CD shapes are marked invalid to enforce APR compatibility. Metals wider than a single CD are checked against the single CD metal grid to prevent them from causing incorrect decomposition. Such cases are essentially odd-cycle conflicts between wide metals and a single CD metal grid.

The ASAP7 PDK supports metals wider than a single CD, but the DRs stipulate that the width be such that the metal spans an odd number of routing tracks. This prevents odd-cycle coloring conflicts between wide metals and single CD metal grids with a preassigned color. Wide metals are particularly useful for power distribution purpose. The PDK also supports rules to ensure correct block mask patterning. The minimum block mask width is considered as 40nm, which corresponds to the minimum 193i resolution. The same T2T spacing value between two SADP patterned metals on the same routing tracks, as well as on the adjacent routing tracks, prevents minimum block mask width violation. A 44nm minimum parallel run length between metals on adjacent tracks is also enforced to ensure sufficient block mask T2T spacing. These rules are described in greater detail in Vashishtha et al. (2017a).

1.6 Cell Library Architecture

A standard cell library has become the usual way to construct general digital circuits, by synthesizing a hardware description language (HDL; typically Verilog) behavioral description and then performing APR of the gates and interconnections. This section describes the cell library. We also explain the cells provided and discuss changes to the DRs based on APR results.

1.6.1 Gear Ratio and Cell Height

As in planar CMOS with gridded gates, which was introduced by Intel at the 45nm node and other foundries at 28nm, the selection of standard cell height in a finFET process is also application specific. Low power application-specific integrated circuits (ASICs) tend to use short cells for density, and high-performance systems, e.g., microprocessors, have historically used taller cells. The cell height constrains the number of usable fins per transistor and thus the drive strength as well as the number of M1 tracks for cell internal routing. Assuming horizontal M2, an adequate number of M2 pitches, with sufficient M2 track access to cell pins, are also required. Thus, the horizontal metal (here M2) to fin pitch ratio, known as the gear ratio, becomes an important factor.

The 3/4 M2 to fin pitch ratio in the ASAP7 predictive PDK used for our libraries allows designing standard cells at 6, 7.5, 9, or 12 M2 tracks and with 8, 10, 12, or 16 fins fitting in the cell vertical height, respectively. Multiple foundries have mentioned "fin depopulation" on finFET processes, i.e., using fewer fins per gate over time. This appears primarily driven by the diminishing need for higher speed, increasingly power-constrained designs, high drive per fin, MP lithography considerations, and the near 1:1 PMOS:NMOS drive ratio, as well as power density considerations. Moreover, for academic use and teaching cell design issues, a large number of tracks make tall cells trivial and their excessive drive capability makes them unsuitable for low power applications. The two fin wide transistors in 6-track library cells have limited M1 routes and pin accessibility, but the best density. 1-D layout, with two metals layers for intra-cell routing, is well suited for 6-track cells, but is not very amenable for classroom use. Therefore, we chose a 7.5-track cell library, which is in line with other publications targeted at N7 (Liebmann et al., 2015). This choice also allows wider M2 power follow-rails at the APR stage, which are preferred for robust power delivery. However, wide M2, while supportable (and indeed required for SAQP M2), requires non-SAV V1 on the power follow-rails. This feature is currently not supported.

Utilizing all of the 10 fins in a 7.5-track library is not possible as sufficient separation is required between the layers SDT and LISD that connect to the transistor source-drain. Also, transistors must have sufficient enclosure by the select layers used for doping. The minimum transistor separation thus dictates that the middle two fins in a standard cell may not be used, as apparent in Figure 1.8, which shows a NAND2 and inverter cell adjacent to each other. Figure 1.8 also illustrates the cell boundaries producing a DDB between the cells. The fin closest to the power rails cannot be used either. Thus, each transistor has three fins and this choice is aided by the nearly equal PMOS-to-NMOS ratio.

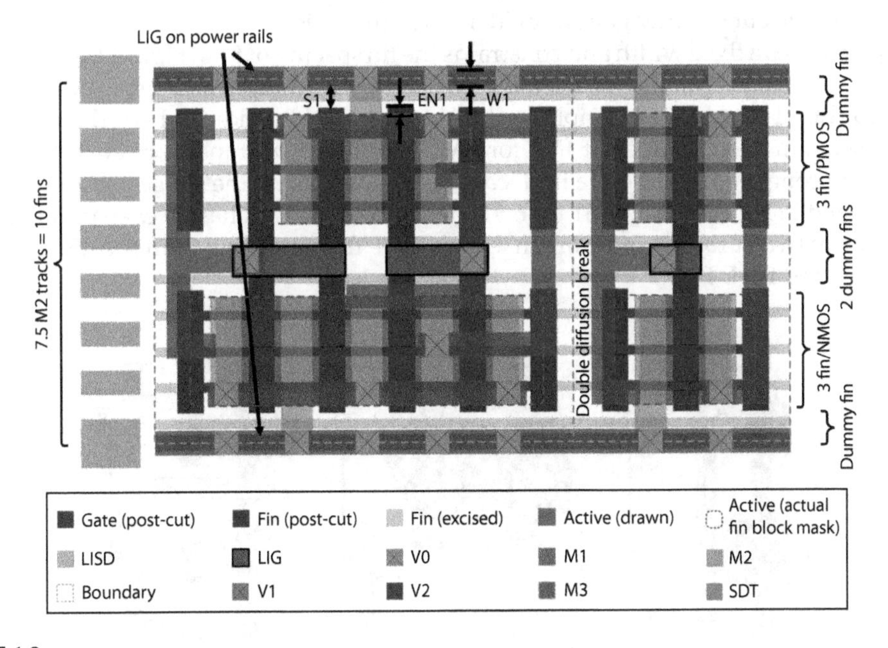

FIGURE 1.8
A 7.5 M2 track standard cell template, with three fins per device type, shows the FEOL, MOL, and M1 in adjacent NAND2 and inverter. A double diffusion break is required between the cells. Fins are tucked underneath the gate when breaking the diffusion. LIG, V0, and M1 are not shown here. S1 = minimum LIG to GATE spacing, EN1 = minimum GATE end-cap, W1 = minimum LIG width.

Notwithstanding the number of fins, the transistors have sufficient drive capability due to the large per-fin drive current value (Clark et al., 2016). ASAP7 standard cells are classified based on their drive strengths, so that those with three-fin transistors are said to have a 1× drive. Those with less drive strength have fraction values, e.g., a one-fin transistor corresponds to a p33× (0.33×) drive strength.

1.6.2 Fin Cut Implications

Again referring to Figure 1.8, the vertical extent of the drawn active layer in the PDK denotes the raised source-drain region. As evident in Figures 1.6, 1.8, and 1.9, the actual active fin "keep" mask extends horizontally halfway underneath the non-channel forming (dummy) gates, where fins are actually cut. Implementing the fin cuts in such a manner necessitates a DDB requirement for the cases where diffusions at disparate voltages must be separated. Therefore, all standard cells end in a single diffusion break (SDB), resulting in a DDB upon abutment with other cells to ensure fin separation. Apart from the cell boundaries, an SDB may exist elsewhere in a cell if equipotential diffusion regions are present on either side of a dummy gate and is equivalent to shorted fins underneath such a gate. Nonetheless, this is a useful structure, particularly in the latch and flip-flop CMOS pass-gates.

As shown in Figure 1.9, block mask rounding effects, when implementing fin cuts, create sharp fin edges that have a high charge density and, consequently, a high electric field (Vashishtha et al., 2017d). This creates a possible TDDB issue between the fin at the cut edge and the dummy gate, potentially causing leakage through the gate oxide. To mitigate this, in the libraries provided with the PDK, the dummy gates are cut at the middle in the standard cells to preclude an electrical path between the PMOS and NMOS transistors through the dummy gate. The gate cut also allows longer LIG for nearby routing without attaching to the cut dummy gate, as evident in Figure 1.9c.

SAQP requires a fixed width and constrains the fin spacing of fins (Figure 1.10). Assuming an EUVL fin cut/keep mask enables the 7.5T library as designed (Figure 1.10a). If MP is assumed, LELE cut/keep is complicated as in Figure 1.10b and c, where the middle fins are excised. An even number of fins for PMOS and NMOS transistors allows the SAQP mandrel to completely define the fin locations, greatly easing the subsequent fin masking as illustrated in Figure 1.10d. A 6T (or 6.5T) or 9T library, which has an even number of fins per device at two or four, is more amenable to SAQP fin patterning than the 7.5T library we primarily used for DTCO.

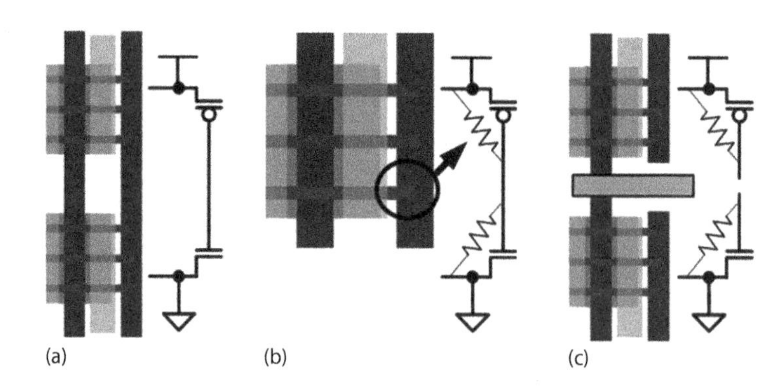

FIGURE 1.9
Dummy gate at DDB (a). Sharp fin edges arise due to mask rounding when cutting fins. This creates a TDDB scenario between the fins and the dummy gate (b). This is avoided where possible by cutting the dummy gates (c).

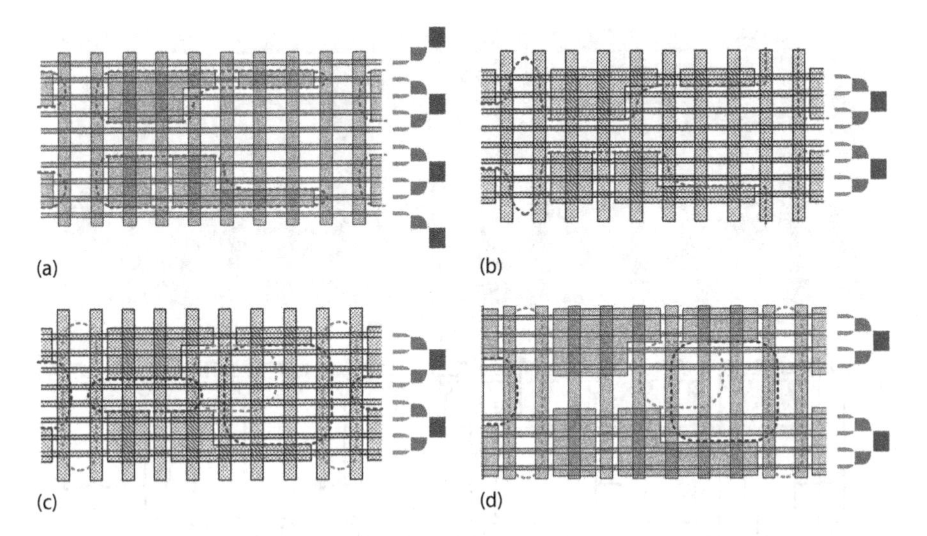

FIGURE 1.10
The original EUV fin keep mask assumptions and SAQP spacer/mandrels (a). Another fin keep option that allows larger patterns (b). LELE fin cut (c) may be problematic, but an even number of fins eases this (d).

1.6.3 Standard Cell MOL Usage

LISD is primarily used to connect diffusion regions to power rails and other equipotential diffusions within the cell through M1, somewhat relieving M1 routing congestion. However, connections to the M1 power rails cannot be completed using LISD alone, since that would require extending LISD past the cell boundary to allow sufficient V0 landing on LISD, or a 2-D layout that would have issues with the T2S spacing. Larger LISD overlap (past the cell boundary) would violate the 27nm LISD T2T spacing by interaction with the abutting cell LISD that is not connected to the power rails. Therefore, an LIG power rail is used within the cells to connect LISD to M1, as LISD and LIG short upon contact, but being different layers, do not present 2-D MEEF issues. The PDK DRs require a 4 nm vertical gate end-cap past active and a 14nm LIG to gate spacing (S1). These rules restrict the LIG power rail to its 16nm minimum width (W1). This width does not provide fully landed V0, but the LIG power rail is fully populated with V0 to ensure robustness and low resistance.

LISD can also serve the secondary role of a routing layer within a cell, but is used sparingly, as local interconnects favor tungsten over copper, and therefore suffer from higher resistivity (Sherazi et al., 2016). However, LISD routing is helpful in complex cells, especially sequential cells, due to M1 routing congestion and allows us to limit M2 track usage. Figure 1.11 shows a D-latch, in which LISD use to connect diffusions across an SDB of a constituent tristate inverter becomes necessary, since M1 cannot be used due to intervening M1 routes or their large T2T and T2S spacing requirements. The SDBs arise due to the inability to accommodate two gate contacts along a single gate track, which also necessitates an M1 crossover. SDBs are used in the cell CMOS pass-gate (crossover) composed of a CLKN-CLKB-CLKN gate combination evident in the schematic at the bottom of Figure 1.11. DDBs are used in the cell, as all diffusions cannot be shared.

1.6.4 Standard Cell Pin and Signal Routing

The cell library architecture emphasizes maximizing pin accessibility, since it has a direct impact on the block density after APR. Where possible, standard cell pins are extended to

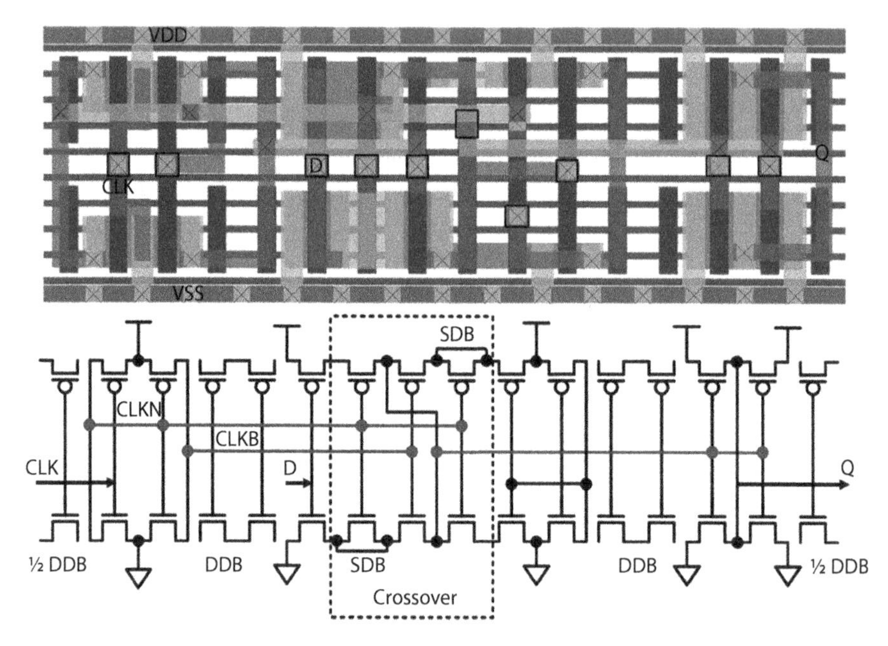

FIGURE 1.11
Cell layout of a transparent high D-latch. Double and single diffusion breaks are shown. The latter require LISD crossovers. Fins shown are prior to the cut.

span at least two and preferably three M2 routing tracks. This ensures that even standard cells with a large number of pins, such as the AOI333 in Figure 1.12, have all of their M1 pins accessible to multiple M2 tracks. The figure demonstrates the use of staggered M2 routes to allow access to the cell pins by M3. Good pin access is forced by our use of the M1 template for cell design (Vashishtha et al., 2017d). The M1 template was employed for rapid cell library development, as the pre-delineated metal constructs take the M1 DRs into consideration, which aids in DRC error-free M1 placement. Most, but not all, of the M1 routing in the standard cells is template based.

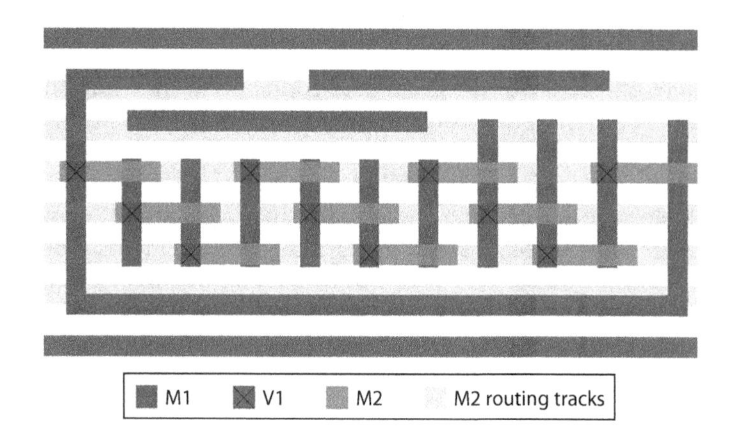

FIGURE 1.12
An AOI333 cell with all of its input M1 pins connected to M2. The staggered M2 allows further connections to M3. Note that this version does not use the bends on M1, following the post-APR DTCO changes. These vias may be slightly unlanded.

Referring to Figure 1.8, the non-power M1 lines closest to the cell boundary are horizontal, which alleviates the T2S DR spacing requirement for such routes as they constitute a side rather than a tip feature. This approach was used for the initial cell version in the library, as it allowed better V1 landing from M2. However, this feature was removed after extensive APR validation, as mentioned later. The 7.5-track tall standard cell results in equidistant M2 routing tracks that are all 18 nm wide, except for the wider spacing of 36 nm around the M2 power follow-rails. The cells accommodate seven horizontal M1 routes that are arranged as two groups of three equidistant tracks between the M1 power rails, with a larger spacing at the center of the cell where vertical M1 is required for pins. Accounting for the extra M1 spacing near the center, instead of the power rails, helps to maximize pin access through M1 pin extension past the M2 tracks adjacent to the power rails. The vertical M1 template locations, corresponding to the pins, match the CPP. Originally, they always ended at a spacing of 25 nm from the nearest horizontal M1 constructs that they do not overlap, so as to honor the M1 T2S spacing. However, L-shaped pins formed using solely vertical M1 constructs on either one of their ends, do not provide the stipulated 5 nm V1 enclosure necessary when connecting to an M2 track. Consequently, the final topology in Figure 1.12, which does not allow full V1 landing to input pins, is adopted for releases after the first, as described in Section 1.6.6.

Connecting to M1 near the power rails is enabled by via merging. While the example presented here is for V0, all layers support via merging. The presence of V0 on the power rails may engender a case similar to that in Figure 1.13a, where the unmerged via mask spacing *"d"* may be smaller than the lithographically allowed minimum metal side-to-side spacing *"y"*. Merging these vias (Figure 1.13b) in such cases permits the patterning of V0 containing M1 tracks at minimum spacing. It is enabled by keeping the vias on grid as evident in Figure 1.13c.

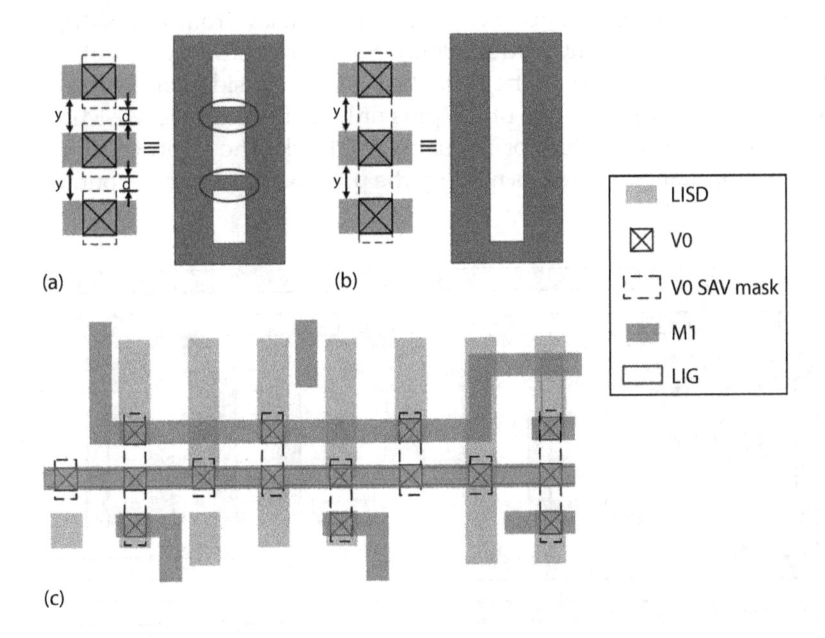

FIGURE 1.13
(a) Sub-lithographic features (ellipses) on an SAV mask. (b) Merging SAV masks precludes such features. (c) Merging enables M1 signal routes to be at the minimum spacing from the power rails, which would otherwise have not been possible due to unmerged SAV mask shapes.

1.6.5 Library Collaterals

A cell library requires liberty timing characterization files (compiled into .db for Synopsys tools) as well as layout exchange format (LEF) of the cell pins and blockages (FRAM for Synopsys). We have focused primarily on Cadence collateral, since we are most familiar with those tools. The technology LEF provides basic DR and via construction information to the APR tool. This is a key enabling file for high quality, i.e., low DRC count, APR results. Additional files are the .cdl (spice) cell netlists. We provide Calibre PEX extracted, and LVS versions. The former allow full gate-level circuit simulation without needing to re-extract the cells. Data sheets of the library cells are also provided. These are generated automatically by the Liberate cell characterization tool. Mentor Graphics Calibre is used for DRC, LVS, and parasitic RC extraction. The parasitic extraction decks allow accurate circuit performance evaluation.

1.6.6 DTCO-Driven DR Changes Based on APR Results

The original M1 standard cell template shown in Figure 1.14a was meant to provide fully landed V1 on M1 in all cases. However, this meant that where an L shape could not be provided, the track over that M1 portion of the pin was lost. APR tools generally complete the routing task, resulting in DRC violations when the layout cannot be completed without producing one. The APR exercises illuminated some significant improvements that could be made, as well as pointing out some substandard cells, which have been revised. A key change to the library templates is illustrated here.

Based on reviewing the APR results described in Section 1.7.3, the V1 landing DR was changed to 2 nm, which does not provide full landing with worst-case misalignment. However, most standard cell pins are inputs and thus drive gates. We decided that gaining the useful pin locations was desirable for multiple reasons. Firstly, a single gate load does not present much capacitance, so the extra RC delay from a misaligned (partially landed) via will not significantly affect delay. Secondly, some gates, notably those with a large number of inputs such as the AOI22 cell, have considerably better pin access, as shown in Figure 1.14b. Initially, two of the pins only had one usable M2-M1 intersection at the middle. These also conflicted for the same M2 track. The improved layout is simpler and has lower capacitance, as well as moving the pins to a regular horizontal grid identical

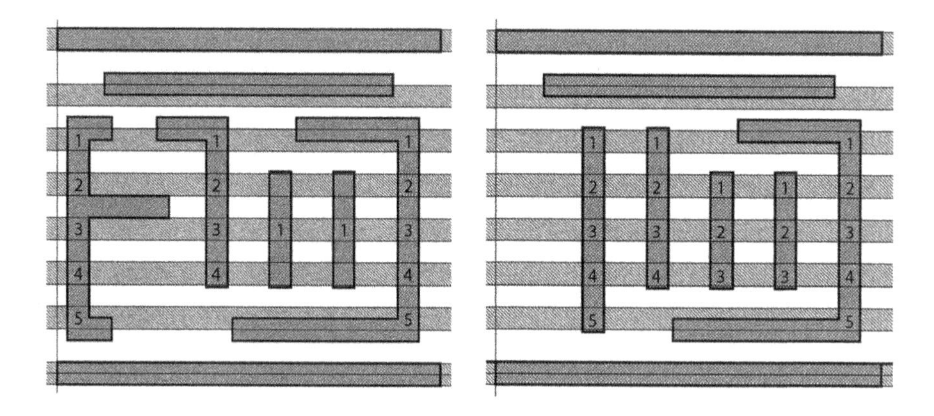

FIGURE 1.14
The original AOI22x1 M1 layout and M2 track overlay (a) and the new layout (b) showing the improved access by relaxing the V1 to M1 overlap rule from 5 nm (full landing) to 2 nm.

to the 54 nm gate grid. The cell output nodes still require 2-D routing to reach the correct diffusions. This maintains the original full 5 nm landing for the output node that drives a much larger load and is subject to self-heating.

1.7 Automated Place and Route with ASAP7

1.7.1 Power Routing and Self-Aligned Via (SAV)

Wide power rails are important in their ability to provide a low resistance power grid. Low resistance alleviates IR drop, particularly the dynamic droop in each clock cycle. The ASAP7 PDK comprehends this by providing wide metal DRs that are compatible with the restrictive metal DRs. While the M2 and M3 are assumed EUV, they can easily be converted to SAQP assumptions and the place and route collateral keeps them on a grid that facilitates this change (Vashishtha et al., 2017b). Consequently, all M2 and M3 routes are, like M4 through M7, on grid with no bends in the provided APR flows. On the SADP M4 through M7, wide metals must be 5×, 9×, 13× of the minimum metal widths, so that the double patterning coloring is maintained. These rules are not enforced by DRCs on M2 and M3 as they are on M4-M7, but the APR flows follow the convention via the techLEF and appropriate power gridding commands in the APR tool.

A portion of the V_{SS} and V_{DD} power grid of the fully routed digital design is shown in Figure 1.15. In Figure 1.15a, the top down view shows the minimum width horizontal M2 power rails over cell top and bottom boundaries. A vertically oriented M3 is attached to each power rail. The V_{SS} on the left and the V_{DD} on the right are connected by a single wide self-aligned V2 to each respective M2 rail. A horizontal M4 is connected to M3 with five SAVs, which can be placed close together for minimum resistance due to the M3 EUV lithography assumption. The five vias fit perfectly in the 9× width M3 power routes. The similar V4 connections from M4 to M5 at the lower right of Figure 1.15a have only three vias, since these are assumed to be patterned by LELE. The technology LEF file provides via definitions at this spacing.

All vias are consistently SAV type, with the same minimum width as other vias on the layer, and are laid out on the routing grid. Ideally, wide M2 could be used for the follow-rails to minimize the resistance. This, however, would require non-SAV V1 on the follow-rails,

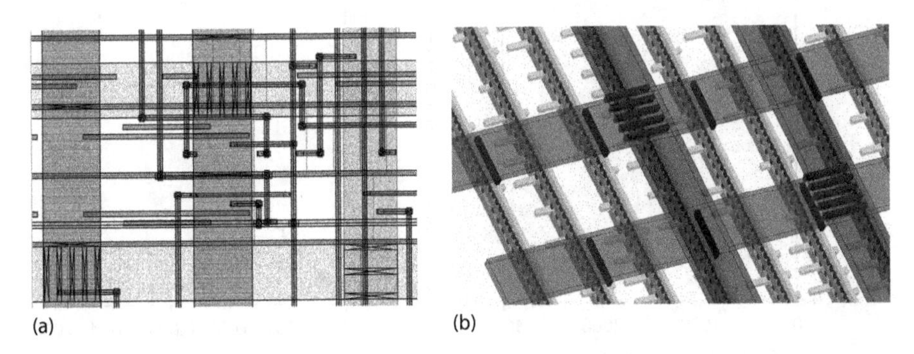

FIGURE 1.15
Layout portion of an APR block (a). The wide vias for power distribution are evident connecting M2, M3, and M4. A 3-D extraction of a section of a fully routed design showing M4 down to the MOL layers (b). The way that LISD connects to LIG beneath the M1 power rails is evident.

since SAVs would be too close to the underlying horizontal M1 tracks. The reader will also note that in an SAQP M3 and M2 scenario, all spaces are identical, so a wider M2 would be required. Figure 1.15b shows a 3-D extracted view of the V_{SS} and V_{DD} power rails. The perspective clearly shows the M2-V1-M1-V0-LIG follow-rail stack, with the LISD protruding into the cell gate areas to provide current paths to the NMOS and PMOS sources from V_{SS} and V_{DD}, respectively. The figure emphasizes the M2 over M1 follow-rails, but the M3 and M4 power distribution grid is evident. There is substantial redundancy in the power scheme—if a break forms in M2, M1, or LIG due to electromigration or a defect, the other layers act as shunts.

1.7.2 Scaled LEF and QRC TechFile

We use Cadence Innovus and Genus for the library APR and synthesis validation, respectively. Standard academic licensing did not allow features below 20 nm in 2016–2017, moving to support 14 nm in 2017. Neither is adequate for the 7 nm PDK. Synopsys academic licensing has the same issue as of this writing. To work around this, APR collateral is scaled 4× for APR, and the output .gds is scaled when streaming into Virtuoso to run Calibre LVS and DRC. Consequently, the technology LEF and macro LEF are scaled up by a factor of four, as is the technology LEF file. Since APR is performed at 4× size but Calibre parasitic extraction is at actual size, a scaled QRC technology file is used in APR. We ran APR on the EDAC design and constrained the metal layers to provide high wire density. The QRC-based SPEF was then compared to the SPEF obtained from Calibre PEX after importing the design into Virtuoso. The QRC technology file was then iteratively "dialed in" to match Calibre. We had expected this to be a straightforward (linear) scaling, but it was not, presumably due to capacitance non-linearity from fringing fields. Figure 1.16 shows that good resistance and capacitance correlation was obtained, with nearly 98% total net capacitance correlation and better than 99% total net resistance correlation.

The technology LEF has fixed via definitions and via generation rule definitions for all layers. As mentioned, they are assumed SAV, allowing upper metal width vias. High-quality (low resistance) multi-cut via generation for wide power stripes and rings, essential to low resistance power delivery, are also provided for one set of metal widths. User modification for different sizes should be straightforward, following the examples in the provided technology LEF file.

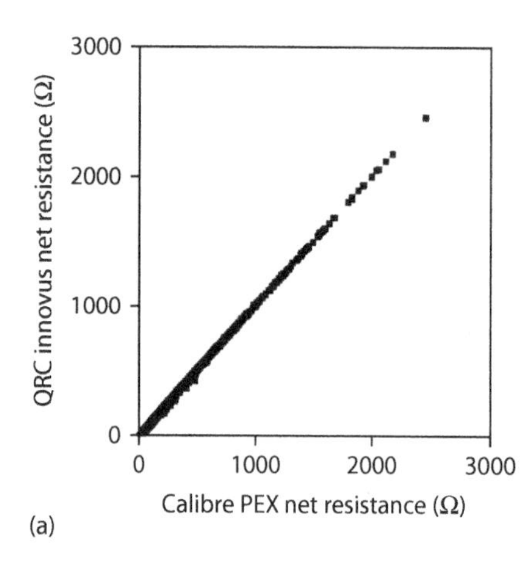

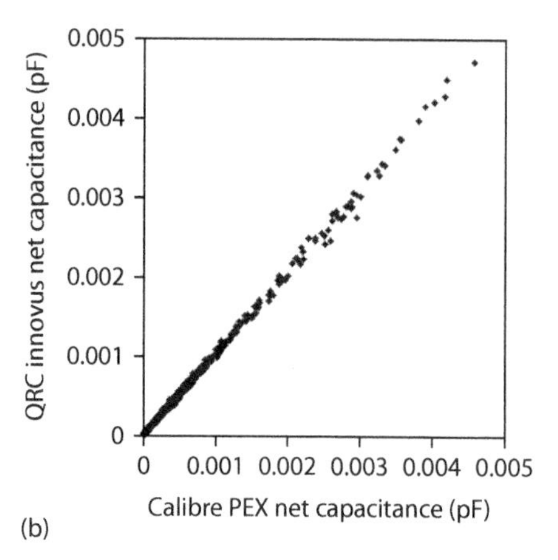

FIGURE 1.16
Correlation of Calibre and scaled QRC techFile resistance (a) and capacitance (b).

1.7.3 Design Experiments and Results

We performed numerous APR experiments in iterative DTCO on the library, to develop the technology LEF to drive routing and via generation in the Innovus APR tool, and to evaluate the library richness and performance. Three basic designs were used. The first is a small L2 cache error detection and correction (EDAC) block, which generates single error correct, double error correct (SEC-DEC) Hamming codes in the input and output pipelines. In the output direction, the syndrome is generated and in the event of incorrect data, the output is corrected by decoding the syndrome. This design has also been used in class laboratory exercises to use the APR (Clark et al., 2017). It has about 2k gates. The second design is a triple modular redundant (TMR) fully pipelined advanced encryption standard (AES), intended as a soft-error mitigation test vehicle (Chellappa et al., 2015; Ramamurthy et al., 2015). This is a large design, with about 350k gates in most iterations. Finally, we used an MIPS M14k, with the Verilog adapted to SRAMs designed on the ASAP7 PDK (Vashishtha et al., 2017c) to test the integration of SRAMs and their collateral (.lib and LEF). This design requires about 50k gates.

Versions of the designs are shown post-APR in Figure 1.17. In general, there are less than about 20 DRCs after importing these designs back into Virtuoso and running Calibre. Most, if not all of the DRCs are found by Innovus. Because they were the first arrays fully designed, 8kB SRAM arrays are used for the M14k instruction cache (left) and data cache

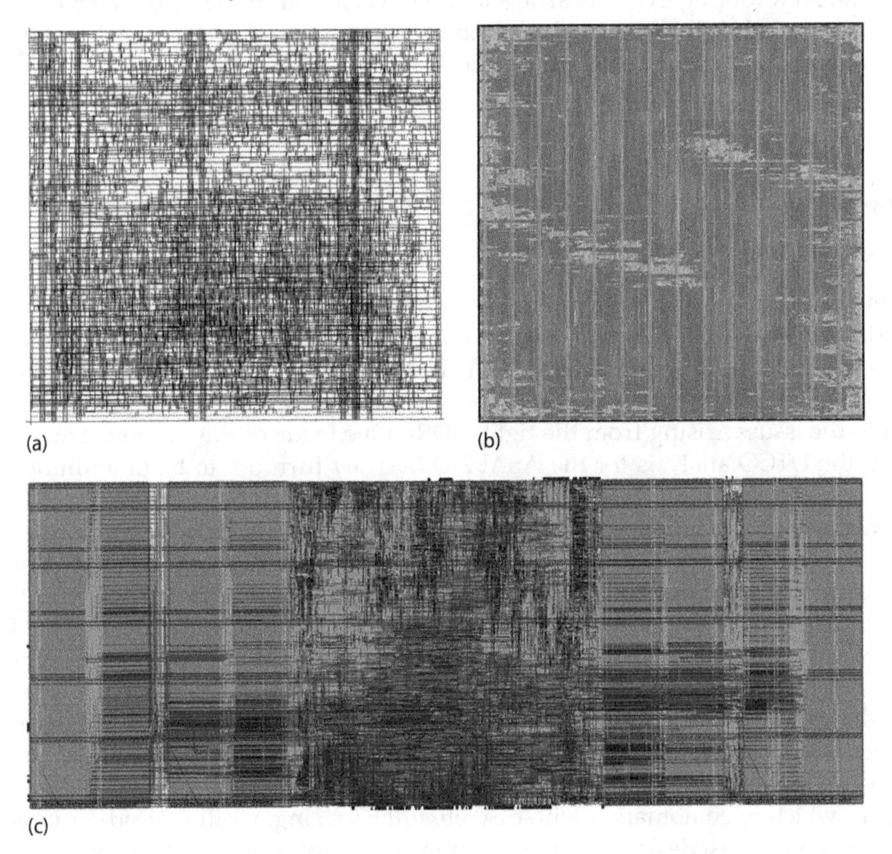

FIGURE 1.17
APR layouts of finished APR blocks. (a) L2 cache EDAC, which is approximately 22 × 20 μm. (b) The TMR AES engine, which is 215 × 215 μm. (c) The MIPS processor, which is 208 × 80 μm in this version.

(right) tag and data arrays. Ideally, these would be smaller, but following field-programmable gate array (FPGA) convention, some address inputs and many storage locations are unused in this example. The EDAC (Figure 1.17a) allows fast debug of APR problems, running in minutes. This design can route in three metal layers (Clark et al., 2017) but the figure includes seven. This design can reach 6 GHz at the TT process corner (25°C) with extensive use of the SLVT cells.

The AES (Figure 1.17b) has 14 pipeline stages with full loop unrolling for both key and data encryption/decryption and requires at least four routing layers, i.e., M2-M5. The design shown uses seven. It has 1596 input and output pins and contains three independent clock domains to support the TMR. TMR storage increases the number of flip-flops to over 15k. This design thus has large clock trees to ensure adequate library support of clock tree synthesis.

The MIPS M14k processor comprises Figure 1.17c and includes large SRAM arrays. As mentioned, the code was adapted from that provided for FPGA implementation by changing the cache arrays from block random access memory (BRAM) to SRAM arrays designed for the ASAP7 PDK. The translation lookaside buffers (TLBs) and register file are synthesized, occupying about one-quarter of the standard cell area. The 8kB SRAM arrays are apparent in the figure. They are too large for the design, which only needs 1kB tag arrays and 2kB data memory arrays for each cache, but confirmed the liberty and LEF files, as well as the APR routing over the arrays to the pins located in the sense/IO circuits at the center. The control logic, between the left and right storage arrays and the decoders, is laid out using Innovus APR, as are the pre-decoders.

1.8 SRAM Design

SRAMs are essential circuit components in modern digital ICs. Due to their ubiquity, foundries provide special array rules that allow smaller geometries than in random logic for SRAM cells to minimize their area. SRAM addressability makes them ideal vehicles for defect analysis to improve yield in early production. Moreover, running early production validates the issues arising from the tighter DRs. One focus of this section is how SRAMs affected the DTCO analysis for the ASAP7 DRs. They turn out to be more limiting, and thus more important than the previously discussed cell library constructs.

Due to the use of the smallest geometries possible, SRAMs are especially prone to random microscopic variations that affect SRAM cell leakage, static noise margin (SNM), read current (speed), and write margin. Consequently, they also provide a place to discuss the PDK use in statistical analysis, as well as our assumptions at N7. Historically, the 6-T SRAM transistor drive ratios required to ensure cell write-ability and read stability have been provided by very careful sizing of the constituent transistors, i.e., the pull-down is largest, providing a favorable ratio with the access transistor to provide read stability, while the pull-up PMOS transistor is smallest, so that the access transistor can overpower it when writing. Improving PMOS vs. NMOS strain has led to near identical, or in some literature greater PMOS drive strengths, which in combination with discrete finFET sizing, requires read- or write-assist techniques for a robust design. The yield limiting cases occur for cells that are far out on the tail of the statistical distribution, due to the large number of SRAM cells used in a modern device. Consequently, statistical analysis is required.

1.8.1 FinFET Implications and Fin Patterning

Besides the constraint on transistor width to discrete fin count, MP techniques, e.g., SADP or SAQP, further complicate the allowed cell geometries. On finFET processes, SRAM cells are divided into classes based on the ratios of the pull-up, pass-gate (access), and pull-down ratios, represented by PU, PG, and PD fin counts. The different-sized cells that we used for DTCO in the PDK comprise Figure 1.18. For instance, the smallest cell is 111 (Figure 1.18a) and has nominally equal drive strength for each device. The cell that most easily meets the read stability and write-ability requirements previously outlined is thus the 123 cell whose layout is illustrated in Figure 1.18d. As in the standard cells, there must be adequate spacing between the NMOS and PMOS devices for well boundaries, as well as active region separation. Thus, at least one fin spacing is lost between adjacent NMOS transistors in separate cells and between the NMOS and PMOS devices. The 112 and 122 cells comprise Figure 18b and c, respectively

The SRAM DR active mask spacing, which is optimistically set at a single fin, is evident in Figure 1.19a and c. The single fin spacing allows the SAQP fin patterning to be uniform across the ASAP7 die. However, the fin patterning can be changed (and often is on foundry processes) by adjusting the mandrels. The EUV assumption drives some of the metal patterning. Referring to Figure 1.19b, note that the M1 is not 1-D, in that the cell V_{DD} connection zigzags through the array on M1. This M1 is redundant with the straight M2 route. This also provides full M1 landing for the via 1 (V1) connection in each cell. The zigzag is forced by the MEEF constraints described earlier (the M1 lines vertical in the figure that connect the V_{SS} and BL to the MOL). Since the MOL provides a great deal of the connections, M1 BL designs are also possible and given the slow actual roll out of EUV, are probably dominant at the 10 nm and possibly the 7 nm nodes. However, they are not compatible with the horizontal M2 direction previously outlined for ASAP7. Consequently, we focus here on M2 BL designs, which maintain the same metal directions that are used in the standard cell areas across the SRAM arrays. The cell gate layers through metallization are illustrated in Figure 1.19d, which emphasizes the high aspect ratios of modern metallization. The fins are omitted from the figure.

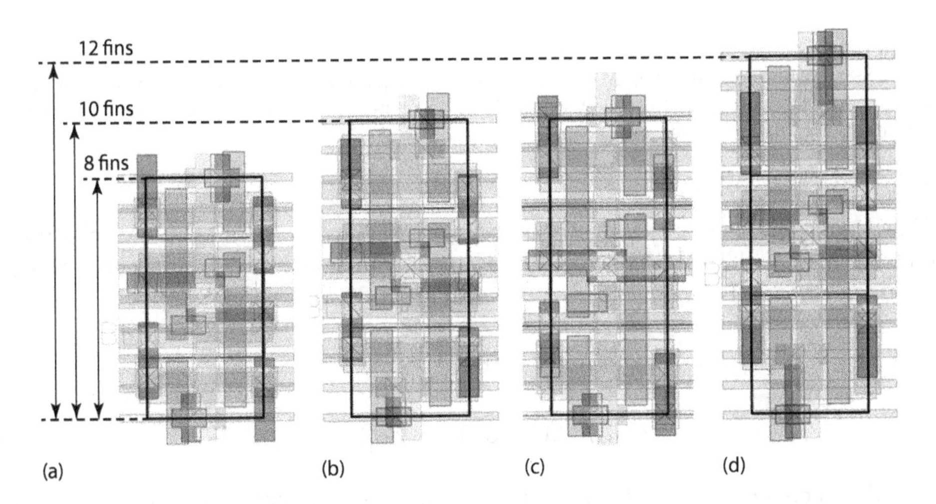

FIGURE 1.18
Layouts for 111 (a), 112 (b), 122 (c), and 123 (d) ASAP7 SRAM cells.

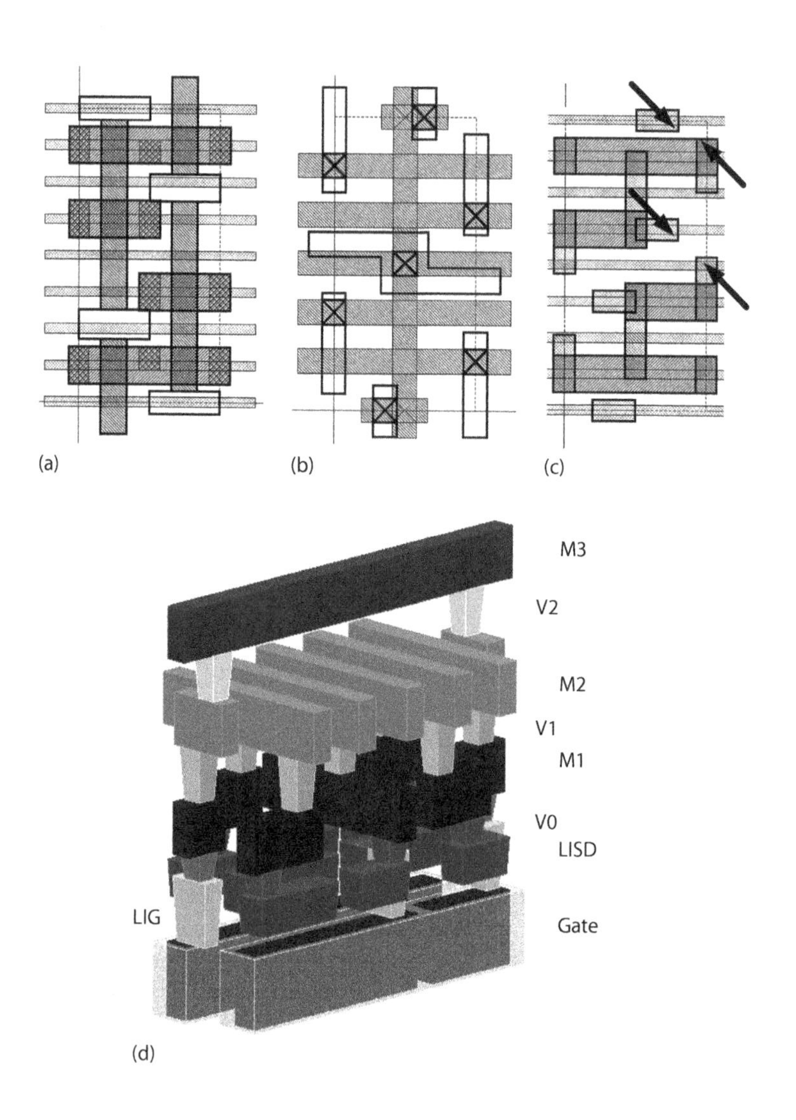

FIGURE 1.19
ASAP7 111 SRAM cell layout showing fin, active, SDT, gate, and gate cut layers (a); M1 and M2 as well as via 1 (b); and active and MOL layers (c). A 3-D cell view with the correct metal aspect ratios (d). The gate spacer is shown as transparent, but fins are not shown.

1.8.2 Statistical Analysis

AVT has long been used as a measure of local mismatch. AVT is a transistor channel area normalized $\sigma\Delta V_T$, where $\sigma\Delta V_T$ is the V_T difference variance as measured between nearby, identical transistors (Pelgrom et al., 1989; Kuhn et al., 2011). Thus,

$$\text{AVT} = \sigma\Delta V_T \times \sqrt{A} \tag{1.9}$$

$$\sigma\Delta V_T = \sigma\left(\left|V_{T1} - V_{T2}\right|\right). \tag{1.10}$$

In non-fully depleted devices, the dopant atoms are proportional to the area but vary statistically as random dopant fluctuations that in aggregate affect the transistor V_{th}, and dominate the mismatch. Other parameters affect the matching, but can often be driven out by improvements in the manufacturing processes and design. SRAM memories use the smallest devices that can be fabricated (in this case a single fin). Consequently, SRAMs are strongly affected and statistical yield analysis is critical to the overall design (Liu et al., 2009). As a result, SRAM yield falls off rapidly as V_{DD} is lowered toward V_{DDmin}, the minimum yielding SRAM V_{DD}. This is due to increasing variability in the drive ratios as the transistor gate overdrive $V_{GS}-V_{th}$ diminishes, due to V_{th} variations. In metal gate and finFET devices, the variability is primarily due to fin roughness and metal gate grain size variations (Liu et al., 2009; Matsukawa et al., 2009). Nonetheless, due to its ease of measurement and historical significance, AVT is still used to characterize mismatch (Kuhn et al., 2011). FinFET devices, while having different sources of variability since they are fully depleted, continue to use the AVT rubric for analysis. In general, AVT tracks the inversion layer thickness, including the effective oxide thickness. It improved by nearly a half with the advent of high-K metal gate processes, which eliminated the poly depletion effect, and again with finFETs, as they are fully depleted (Kuhn et al., 2011). The primary sources of variability in finFETs turn out to be metal gate grain size and distribution, followed by fin roughness and oxide thickness and work function variations. We use AVT = 1.1 mV.μm, which we chose as a compromise between the NMOS and PMOS values published for a finFET 14-nm process (Giles et al., 2015). The literature has reported pretty steady values across key device transistor fabrication technologies, i.e., poly gate, metal gate, and finFET. Of course, the overall variability increases in our 7 nm SRAM transistors vs. the 14 nm devices, since the channel area is less.

For sense amplifier analysis, we use a conventional Monte Carlo (MC) approach, applying different differentials to a fixed variability case to find the input-referred offset, then applying the next fixed variability, to reach the desired statistics. To determine the offset mean and sigma, 1k MC points are used. For SRAM cell analysis, we use a combination of MC and stratified sampling (Clark et al., 2013). In general, unless huge numbers of simulations are used, MC is not adequate for estimations at the tails of the distributions. It is, however, adequate to estimate the standard deviation σ and mean μ of the distributions. We confirm the needed σ/μ using the stratified sampling approach, which applies all of the statistical variability at each circuit (sigma) strata, to all the possible combinations of the transistors within that strata, using a full factorial. Thus, 728 simulations are performed, one for each possible 1, 0, or −1 impact on each combination of devices at each strata, attempting to find failures in that circuit strata (at that circuit sigma). As an example, if the circuit sigma is distributed evenly across six devices, then there is minimal variation in the circuit. However, if all of the circuit-level variability is applied to just the access and pull-down transistors, and in opposite fashion, then the read stability may be jeopardized. In this manner, the technique evaluates the tail of the circuit distribution directly and efficiently. At each strata, the transistor variation applied changes with the number of non-zero variations, following the circuit σ_{DEVICE} as

$$\sigma_{DEVICE} = \frac{\sigma_{CELL}}{\sqrt{\sum a_i^2}}. \tag{1.11}$$

The technique has been successful at predicting the SRAM yield for foundry processes (Clark et al., 2013).

1.8.3 SRAM Cell Design and DTCO Considerations

1.8.3.1 MOL Patterning

Referring to Figure 1.19c, the key DTCO limitation is the corner of LISD and SDT to LIG spacing, as indicated by the arrows at the upper right. EUV single patterning would result in some rounding, which helps by reducing the peak electric field and increasing the spacing slightly. As before, we use conservative assumptions. Another result of maintaining adequate spacing for TDDB with worst-case misalignment of the LISD/SDT and LIG layers is that the SDT does not fully cover the active areas, evident at the top and bottom of the NMOS stacks (Figure 1.19a). This led us to separate the LISD- and LIG-drawn layers. We believe that given the late introduction of EUV, it will most likely be used for MOL layer patterning, as well as vias and cuts. The former is due to projections that MOL layers may require as many as five block masks for SRAM at N7, which makes the expense of EUV more favorable (Sakhare et al., 2015). Foundry presentations have also shown MOL experimental EUV results.

1.8.3.2 1-D Cell Metallization

The 111 cell (Figure 1.19) has 1-D M2 and M3, potentially making it very amenable with a 1-D cell library, and as mentioned, M1 BL designs are compatible with horizontal M1 1-D cell library architectures. The 2-D cell layouts of the 122 cell are shown in Figure 1.20a–d. The similarity to the overall architecture of the 111 cell is apparent. The same MOL limitations exist, but the two fin NMOS devices make the overall source/drain connections to

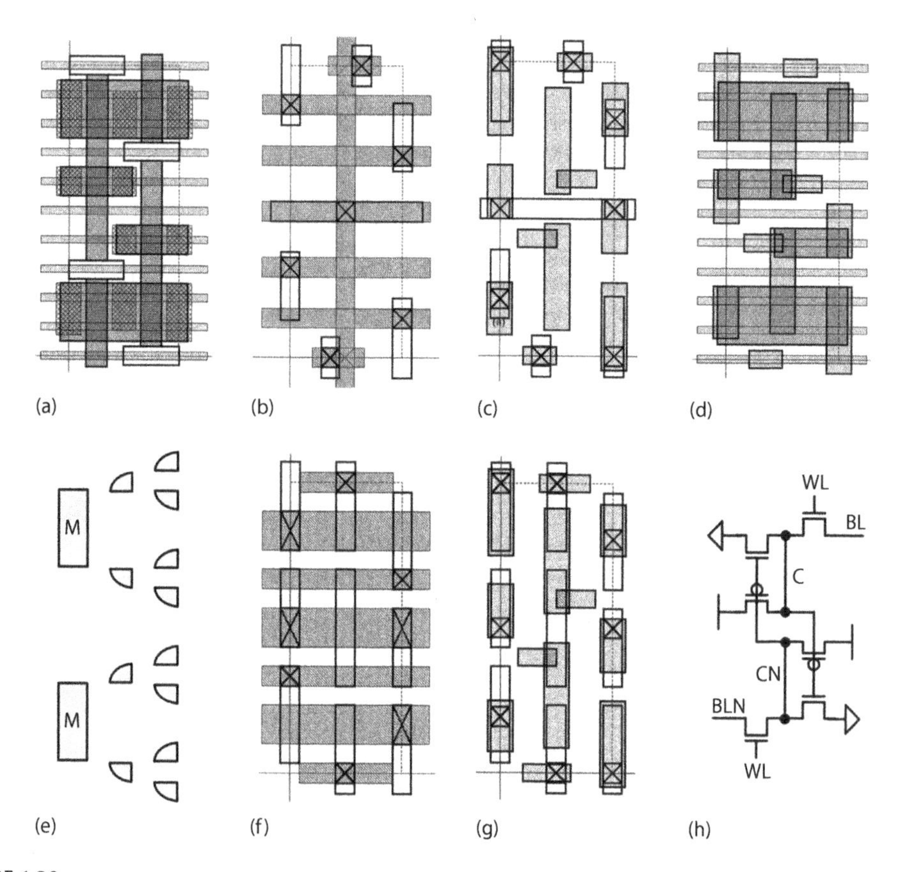

FIGURE 1.20
ASAP7 122 SRAM cell layout showing fin, active, SDT, gate, gate cut, active, and SDT layers (a); M3, V2, M2, V1, and M1 layers (b); M1, V0, and MOL (c); and fin, active, and MOL (d). The SAQP mandrel and spacers produced for SID M2 (e) and the resulting M2, V1, and M1 layers (f). M1 through MOL (g) and the schematic (h).

the NMOS stacks considerably better, as they have more coverage and lower resistance at worst-case layer misalignments. The EUV M2 assumption allows narrow metals with wide spacing. This in turn reduces line-to-line coupling and overall capacitance. Other widths are possible.

Figure 1.20f and g show 1-D metal impact on the cell design at the M2 through MOL layers. Here, M1 and M2 are produced by 1-D stripes using SADP or SAQP, respectively, and then cut to produce separate metal segments. The cuts are assumed the same as SAV, i.e., 16 nm. Cuts are aligned to each other where possible. Wider M2 is required given the SID assumptions. The SAQP mandrel and the first- and second-level spacers to produce this layout are shown in Figure 1.20e. The spacers walk across the cells, but the wide, thin, wide, thin repetitive pattern is easily produced. The lines/cuts 1-D metal approach requires a dummy M1 in the middle of the cell, evident in Figure 1.20f and g. The SAV V1 are wider, but still aligned by the M2 HM, following the convention used for the APR power routes above. Note that these vias are un-landed on M1. Figure 1.20h is provided as an aid to the transistor layout pattern for readers who are not readily familiar with these modern, standard SRAM layouts.

1.8.3.3 Stability and Yield Analysis

Read mode SNM analysis following Seevinck et al. (1978) at the typical process corner is remarkably similar for the four cells (Vashishtha et al., 2017c). The 112 cell has the best SNM due to the 1:2 PG to PD ratio, followed by the 123 cell with its 2:3 ratio. At TT, the 122 cell has the lowest value, but it is very close to the 111 cell. Referring to Table 1.5, the read SNM σ is inversely proportional to the number of fins as expected. The read SNM overall quality can be ranked by the standard μ/σ of each, since it gives an indication of the sigma level at which the margin vanishes. Using this metric, the largest (123) cell is the best, as expected. The 111 cell is the worst, also as expected, entirely due to greater variability. The 112 and 122 cells are nearly even, with the latter making up for a lower baseline SNM with lower variability. The lithography for the 122 cell is easier, and as shown in Table 1.5, it has better write margins, so it becomes the preferred cell. With the possible exception of the 111 cell, all cells have good read margins.

TABLE 1.5

Mean, Variation, and Mean/Sigma of Read SNM, Hold SNM, and Write Margin Simulations for 111, 112, 122, and 123 SRAM Cells at Nominal V_{DD} = 0.5 V (Hold at V_{DD} = 350 mV) and 25°C

	Cell Type				
Margin Type	**Quantity**	**111**	**112**	**122**	**123**
Read SNM	μ	105.2	116	100.4	106.6
	σ	11.2	9.48	8.3	7.5
	μ/σ	9.4	12.2	12.1	14.2
Hold SNM	μ	119.0	126.1	126.1	128.8
	σ	8.2	6.2	6.2	5.4
	μ/σ	14.4	20.3	20.3	23.9
Write Margin	μ	68.1	69.3	103.6	98.4
	σ	20.9	23.8	18.3	15.8
	μ/σ	3.3	2.9	5.7	6.2

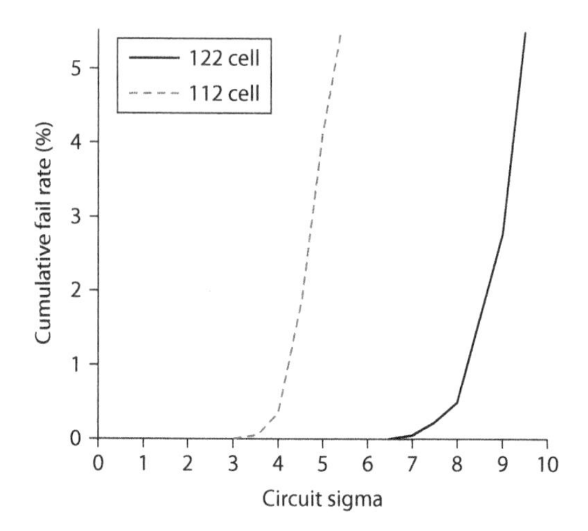

FIGURE 1.21
Comparing the failure rates of the 112 and 122 SRAM cell write margins using the stratified sampling approach.

In the arrays presented here, a CMOS Y-multiplexer allows a single tristate write circuit per column and eliminates the need for separate write enables (Vashishtha et al., 2017c). Write margin is essentially the BL voltage at which the cell writes under the worst-case voltage and variability conditions. For this analysis, we use an approach that includes the write driver and Y-multiplexer variability, since the write signal passes through it. The analysis thus includes not only the write-ability of the cell, but also the ability of the column circuits to drive the BL low. MC-generated V_{th} offsets are applied to the cell, Y-mux, and write driver transistors. The write driver input swings from V_{SS} to V_{DD}, sweeping the driver output voltage low. We tie the gate of the opposing PMOS transistor in the SRAM cell low, so that the circuit ratio varies only with the write driver voltage. The margin is the BL difference between the point at which the far SRAM node (CN) rises past $V_{DD}/2$ and the lowest BL voltage that can be driven. Table 1.5 shows that the 122 cell has good margin. The high sigma margins are confirmed by the stratified sampling approach as shown in Figure 1.21. The first fails occur at the sigma predicted in Table 1.5, after 2.9 μ/σ for the 112 cell, and after 5.7 (with some added margin) for the 122 cell. While five sigma margins have been suggested in academic papers (Qazi et al., 2011) and this value has been used in commercial designs at the column group level, we consider it inadequate at the cell level since a column group can have as many as 8 × 256 cells. The conclusion is that the preferred 122 cell, which has good density and read stability, would benefit from write assist.

The hold margin controls V_{DDmin}, as well as limiting write assist using reduced V_{DD} or raised V_{SS}. Our target V_{DDmin} is 0.5 V, so we use a V_{DDcol} of 350 mV as the hold SNM evaluation point to provide guard band. The results are also shown in Table 1.5. A low SRAM transistor I_{off} allows a good I_{on} to I_{off} ratio at low voltage, providing large hold SNM at 350 mV V_{DD} μ/σ.

1.8.4 Array Organization and Column Design

Further analysis comprehends more than one SRAM cell, so we proceed by describing the column group, which is the unit that contains the write, sense, and column multiplexing circuits, as well as multiple columns of SRAM above and below it. Typical choices are four or eight columns of SRAM per sense amplifier, so we (somewhat arbitrarily) chose four.

The basic circuit is shown in Figure 1.22, including the "DEC" style sense amplifier. This sense circuit is chosen since it is naturally isolated from the BLs during writes and one of the author's prior experience has shown it to have similar mismatch, but simpler timing than the sense amplifier using just cross-coupled inverters. Note that changing the array size merely requires changing the height of the SRAM columns and the number of column groups included in the array. As shown in Figure 1.22, the differential sense amplifier drives a simple set-reset (SR) latch to provide a pseudo-static output from the array. The SR latch has a differential input, so it will be well behaved, and does not need a delayed clock that a conventional D-latch would in this function. There is thus no race condition on the sense precharge.

The sense amplifier input referred offset voltage was determined by SPICE MC simulation. As noted previously, the simulations apply between –100 and 100 mV to V(SA) – V(SAN) in

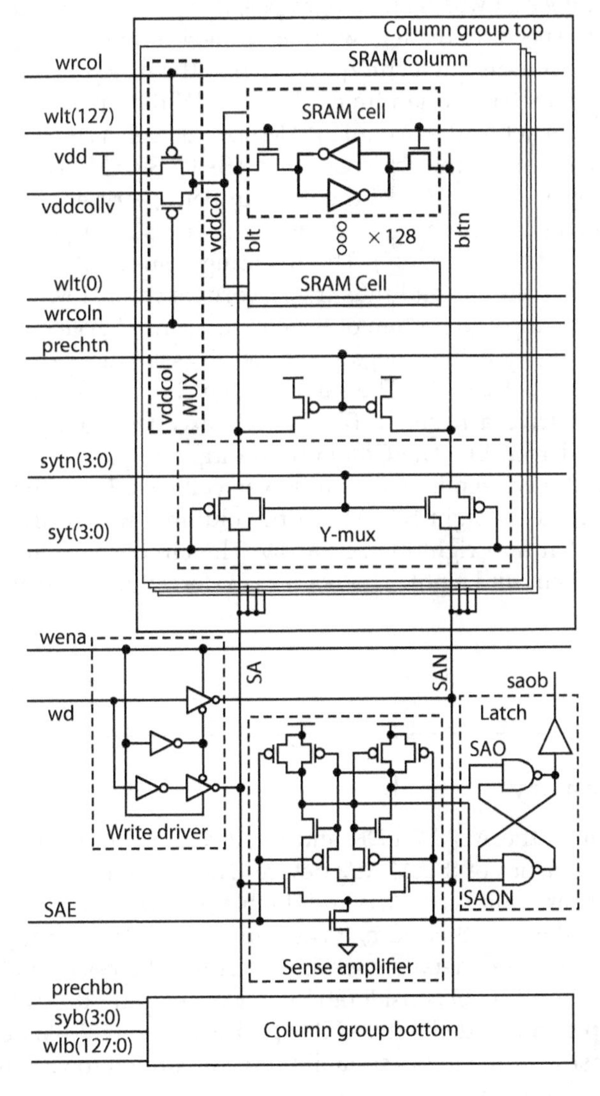

FIGURE 1.22
Column group circuits. Four SRAM columns are attached at the top and bottom of the sense and I/O circuits, sharing a common CMOS pass-gate Y-mux for reads and writes.

1 mV increments for each random (MC chosen) amplifier transistor mismatch selection. The point at which the amplifier output changes direction is the input-referred offset for that mismatch selection. The results produce a mean offset less than 1 mV, which is near the expectation of zero, also indicating no serious systematic offsets due to the layout. The distribution is Gaussian. The input-referred offset standard deviation is 16.5 mV. Using the aforementioned five sigma offset for the sense, 82.6 mV of voltage difference is required at the sense nodes for correct operation. We guard band this up to 100 mV since extra signal is required to provide adequate speed—the sense operation with no residual offset starts the circuit in the metastable state and can produce significant output delays.

1.8.5 Write Assist

The limitations of transistor width and length possible given the MP and other limitations essentially require the need for read- or write-assist techniques in the ASAP7 SRAM memory designs. Thus, circuit design techniques provide the needed statistical yield margins instead of the cell geometries (Chandra et al., 2010). While many assist approaches have been published, the impact on the overall SRAM size, as well as the variability of the assist techniques themselves, drove the choice for our ASAP7 SRAM arrays. We evaluated two approaches. First, lowering the V_{DD} of the column being written. Previous designs have used timed pull-downs to V_{SS}. In the SRAMs here, we use charge sharing V_{DDcol} voltage generation as in Chandra et al. (2010). The resulting column V_{DD} voltage is capacitively matched to the SRAM columns and thus tracks corners well. However, this scheme adds eight poly tracks to the top and bottom of the column sense/write circuits.

This area is recovered by using a negative BL write assist that provides the write driver with a low supply of less than V_{SS}. The same charge sharing circuit is used, with minor polarity changes, to drive a negative BL. We see excellent margin improvement with −150 mV on the low-driven BL. The CMOS BL multiplexer is unchanged and passes the negative voltage well. Leakage increases at lower voltages. For either scheme, sufficient capacitance is provided by eight SRAM width columns, which are integrated into the dummy cells at the left and right of each array. The circuit occupies the same columns in the read/write/IO circuit height. The reader is referred to Vashishtha et al. (2017c) for details.

1.9 Chapter Summary

This chapter described the ASAP7 PDK and its development. It discussed the electrical performance characteristics of the ASAP7 device. The chapter also provided a basic overview of lithography variability concerns and double patterning approaches, and how they compare with each other. Furthermore, it covered the EUV lithographic basics and challenges associated with it. PDK details, such as patterning choices for the various layers, cell library architecture, and DTCO considerations for developing the architecture were also discussed. The chapter concluded with APR experiments and SRAM designs based on the PDK, which demonstrated the PDKs suitability for research into various VLSI circuit and system design–related aspects.

The ASAP7 PDK has been deployed in graduate-level VLSI courses at the Arizona State University since 2015. The PDK is available for free to universities. It has received attention from research groups and faculty at a number of other universities. This indicates a high likelihood of its adoption in the classroom. Thus, we hope that the ASAP7 PDK will fulfill its development intent of enabling sub-10 nm CMOS research in academia at a much larger scale and beyond a few university research groups with access to advanced foundry PDKs. We intend to continue our work for other process nodes and ASAP5 PDK development for the 5 nm technology node is currently underway.

References

Aitken, R., et al., Physical design and FinFETs, *Proc. ISPD*, pp. 65–68, 2014.

Arnold, W., Toward 3 nm overlay and critical dimension uniformity: An integrated error budget for double patterning lithography, *Proc. SPIE*, vol. 6924, pp. 692404-1-9, 2008.

ASML, TWINSCAN NXT:1970Ci, https://www.asml.com/asml/show.do?lang=EN&ctx=46772&dfp_product_id=8036, 2015.

ASML, TWINSCAN NXT:1980Di, https://www.asml.com/products/systems/twinscan-nxt/en/s46772?dfp_product_id=10567, 2016.

ASML, TWINSCAN NXE:3400B, https://www.asml.com/products/systems/twinscan-nxe/twinscan-nxe3400b/en/s46772?dfp_product_id=10850, 2017.

Au, Y., et al., Selective chemical vapor deposition of manganese self-aligned capping layer for Cu interconnections in microelectronics, *J. Electrochem. Soc.*, vol. 157, no. 6, p. D341–345, 2010.

Baklanov, M., P. S. Ho, and E. Zschech, eds, *Advanced Interconnects for ULSI Technology*, Wiley, 2012.

Bhanushali, K., et al., FreePDK15: An open-source predictive process design kit for 15nm FinFET technology, *Proc. ISPD*, pp. 165–170, 2015.

Brain, R., et al., Low-k interconnect stack with a novel self-aligned via patterning process for 32 nm high volume manufacturing, *Proc. IITC*, pp. 249–251, 2009.

Chandra, V., et al., On the efficacy of write-assist techniques in low voltage nanoscale SRAMs, *Proc. DATE*, vol. 1, pp. 345–350, 2010.

Chava, B., et al., Standard cell design in N7: EUV vs. immersion, *Proc. SPIE*, vol. 9427, 94270E-1-9, 2015.

Chellappa, S., et al., Advanced encryption system with dynamic pipeline reconfiguration for minimum energy operation, *Proc. ISQED*, pp. 201–206, 2015.

Chiou, T., et al., Lithographic challenges and their solutions for critical layers in sub-14 nm node logic devices, *Proc. SPIE*, vol. 8683, pp. 86830R-1-15, 2013.

Clark, L. T., et al., SRAM cell optimization for low AVT transistors, *Proc. ISLPED*, pp. 57–63, 2013.

Clark, L. T., et al., ASAP7: A 7-nm finFET predictive process design kit, *Microelectron. J.*, vol. 53, pp. 105–115, 2016.

Clark, L. T., et al., Design flows and collateral for the ASAP7 7 nm FinFET predictive process design kit, *Proc. MSE*, pp. 1–4, 2017.

DeGraff, R., et al., NXT:1980Di immersion scanner for 7 nm and 5 nm production nodes, *Proc. SPIE*, vol. 9780, pp. 978011-1-9, 2016.

Demuynck, S., et al., Contact module at dense gate pitch technology challenges, *Proc. IITC*, pp. 307–310, 2014.

Dicker, G., et al., Getting ready for EUV in HVM, *Proc. SPIE*, vol. 9661, pp. 96610F-1-7, 2015.

Giles, M. D., et al., High sigma measurement of random threshold voltage variation in 14 nm logic FinFET technology, *IEEE Symp. VLSI Tech.*, pp. 150–151, 2015.

Goldman, R., et al., 32/28 nm educational design kit: Capabilities, deployment and future, *Proc. PrimeAsia*, pp. 284–288, 2013.

Ha, D., et al., Highly manufacturable 7 nm FinFET technology featuring EUV lithography for low power and high performance applications, *Proc. VLSIT*, pp. T68–T69, 2017.

Hody, H., et al., Gate double patterning strategies for 10-nm node FinFET devices, *Proc. SPIE*, vol. 9054, pp. 905407-1-7, 2015.
Im, S., et al., Scaling analysis of multilevel interconnect temperatures for high-performance ICs, *IEEE Trans. Electron. Devices*, vol. 52, no. 12, pp. 2710–2719, 2005.
Ito, T., and S. Okazaki, Pushing the limits of lithography, *Nature*, vol. 406, no. 6799, pp. 1027–1031, 2000.
ITRS. http://www.itrs2.net/, 2015.
Jung, W., et al., Patterning with amorphous carbon spacer for expanding the resolution limit of current lithography tool, *Proc. SPIE*, vol. 6520, no. 2007, pp. 1–9, 2007.
Kerkhof, M., et al., Enabling sub-10 nm node lithography: Presenting the NXE:3400B EUV scanner with improved overlay, imaging, and throughput, *Proc. SPIE*, vol. 10143, pp. 101430D-1-14, 2017.
Kim, S.-S., et al., Progress in EUV lithography toward manufacturing, *Proc. SPIE*, vol. 10143, pp. 1014306-1-10, 2017.
Kuhn, J., et al., Process technology variation, *IEEE Trans. Elec. Devices*, vol. 58, no. 8, pp. 2197–2208, 2011.
Liebmann, L., et al., Demonstrating production quality multiple exposure patterning aware routing for the 10 nm node, *Proc. SPIE*, vol. 9053, pp. 905309–1-10, 2014a.
Liebmann, L., et al., Design and technology co-optimization near single-digit nodes, *Proc. ICCAD*, pp. 582–585, 2014b.
Liebmann, L., et al., The daunting complexity of scaling to 7NM without EUV: Pushing DTCO to the extreme, *Proc. SPIE*, vol. 9427, pp. 942701-1–12, 2015.
Lin, C., et al., High performance 14 nm SOI FinFET CMOS technology with 0.0174 μm2 embedded DRAM and 15 levels of Cu metallization, *IEDM*, pp. 74–76, 2014.
Lin, B. J., Optical lithography with and without NGL for single-digit nanometer nodes, *Proc. SPIE*, vol. 9426, pp. 942602-1-10, 2015.
Liu, Y., K. Endo, and O. Shinichi, On the gate-stack origin of threshold voltage variability in scaled FinFETs and multi-FinFETs, *IEEE Symp. VLSI Tech.*, pp. 101–102, 2009.
Ma, Y., et al., Decomposition strategies for self-aligned double patterning, *Proc. SPIE*, vol. 7641, pp. 76410T–1-13, 2010.
Ma, Y., et al., Self-aligned double patterning (SADP) compliant design flow, *Proc. SPIE*, vol. 8327, pp. 832706-1-13, 2012.
Mallik, A., et al., The economic impact of EUV lithography on critical process modules, *Proc. SPIE*, vol. 9048, pp. 90481R-1-12, 2014.
Mallik, A., et al., Maintaining Moore's law: Enabling cost-friendly dimensional scaling, *Proc. SPIE*, vol. 9422, pp. 94221N–94221N-12, 2015.
Martins, M., et al., Open cell library in 15 nm FreePDK technology, *Proc. ISPD*, pp. 171–178, 2015.
Matsukawa, T., et al., Comprehensive analysis of variability sources of FinFET characteristics, *IEEE Symp. VLSI Tech.*, pp. 159–160, 2009.
Mulkens, J., et al., Overlay and edge placement control strategies for the 7 nm node using EUV and ArF lithography, *Proc. SPIE*, vol. 9422, pp. 94221Q-1-13, 2015.
Neumann, J. T., et al., Imaging performance of EUV lithography optics configuration for sub-9 nm resolution, *Proc. SPIE*, vol. 9422, pp. 94221H–1-9, 2015.
Oyama, K., et al., CD error budget analysis for self-aligned multiple patterning, *Proc. SPIE*, vol. 8325, pp. 832517-1-8, 2012.
Oyama, K., et al., Sustainability and applicability of spacer-related patterning towards 7nm node, *Proc. SPIE*, vol. 9425, 942514-1-10, 2015.
Paydavosi, N., et al., BSIM—SPICE models enable FinFET and UTB IC designs, *IEEE Access*, vol. 1, pp. 201–215, 2013.
Pelgrom, M. J. M., et al., Matching properties of MOS transistors, *IEEE J. Solid-state Circ.*, vol. 24, no. 5, pp. 1433–1439, Oct. 1989.
Pikus, F. G., Decomposition technologies for advanced nodes, *Proc. ISQED*, pp. 284–288, 2016.

Pyzyna, A., R. Bruce, M. Lofaro, H. Tsai, C. Witt, L. Gignac, M. Brink, and M. Guillorn, Resistivity of copper interconnects beyond the 7 nm node, *Symp. VLSI Circuits*, vol. 1, no. 1, pp. 120–121, 2015.

Qazi, M., K. Stawiasz, L. Chang and A. P. Chandrakasan, A 512kb 8T SRAM macro operating down to 0.57 V with an AC-coupled sense amplifier and embedded data-retention-voltage sensor in 45 nm SOI CMOS, *IEEE J. Solid-State Circuits*, vol. 46, no. 1, pp. 85–96, 2011.

Ramamurthy, C., et al., High performance low power pulse-clocked TMR circuits for soft-error hardness, *IEEE Trans. Nucl. Sci.*, vol. 6, pp. 3040–3048, 2015.

Ryckaert, J., et al., Design technology co-optimization for N10, *CICC*, pp. 1–8, 2014.

Sakhare, S., et al., Layout optimization and trade-off between 193i and EUV-based patterning for SRAM cells to improve performance and process variability at 7 nm technology node, *Proc. SPIE*, vol. 9427, p. 94270O–1-10, 2015.

Schuegraf, K., et al., Semiconductor logic technology innovation to achieve sub-10 nm manufacturing, *IEEE J. Electron. Dev. Soc.*, vol. 1, no. 3, pp. 66–75, 2013.

Seevinck, E., et al., Static noise margin analysis of MOS SRAM cells, *IEEE J. Solid-State Circuits*, vol. SC-22, no. 5, pp. 748–754, 1978.

Seo, S., et al., A 10 nm platform technology for low power and high performance application featuring FINFET devices with multi workfunction gate stack on bulk and SOI, *Proc. VLSIT*, pp. 1–2, 2014.

Servin, I., et al., Mask contribution on CD and OVL errors budgets for double patterning lithography, *Proc. SPIE*, vol. 7470, pp. 747009-1-13, 2009.

Sherazi, S. M. Y., et al., Architectural strategies in standard-cell design for the 7 nm and beyond technology node, *Proc. SPIE*, vol. 15, no. 1, pp. 13507-1-11, 2016.

Standiford, K., and C. Bürgel, A new mask linearity specification for EUV masks based on time-dependent dielectric breakdown requirements, *Proc. SPIE*, vol. 8880, pp. 88801M-1-7, 2013.

Stine, J. E., et al., FreePDK: An open-source variation aware design kit, *Proc. MSE*, pp. 173–174, 2007.

Tallents, G., E. Wagenaars and G. Pert, Optical lithography: Lithography at EUV wavelengths, *Nat. Photonics*, vol. 4, no. 12, pp. 809–811, 2010.

Vaidyanathan, K., et al., Design implications of extremely restricted patterning, *J. Micro/Nanolith. MEMS MOEMS*, vol. 13, pp. 031309-1-13, 2014.

Vandeweyer, T., et al., Immersion lithography and double patterning in advanced microelectronics, *Proc. SPIE*, vol. 7521, pp. 752102-1-11, 2010.

Van Schoot, J., et al., EUV lithography scanner for sub-8 nm resolution, *Proc. SPIE*, vol. 9422, pp. 94221F-1-12, 2015.

Van Setten, E., et al., Imaging performance and challenges of 10 nm and 7 nm logic nodes with 0.33 NA EUV, *Proc. SPIE*, vol. 9231, pp. 923108-1-14, 2014.

Van Setten, E., et al., Patterning options for N7 logic: Prospects and challenges for EUV, *Proc. SPIE*, vol. 9661, pp. 96610G-1-13, 2015.

Vashishtha, V., et al., Design technology co-optimization of back end of line design rules for a 7 nm predictive process design kit, *Proc. ISQED*, pp. 149–154, 2017a.

Vashishtha, V., et al., Systematic analysis of the timing and power impact of pure lines and cuts routing for multiple patterning, *Proc. SPIE*, vol. 10148, pp. 101480P-1-8, 2017b.

Vashishtha, V., et al., Robust 7-nm SRAM design on a predictive PDK, *Proc. ISCAS*, pp. 360–363, 2017c.

Vashishtha, V., et al., ASAP7 predictive design kit development and cell design technology co-optimization, *Proc. ICCAD*, pp. 992–998, 2017d.

Wong, B. P., et al., *Nano-CMOS Design for Manufacturability: Robust Circuit and Physical Design for Sub-65nm Technology Nodes*, Wiley, 2008.

Wu, S. Y. et al., A 16 nm FinFET CMOS technology for mobile SoC and computing applications, *Proc. IEDM*, pp. 224–227, 2013.

Xie, R., et al., A 7 nm FinFET technology featuring EUV patterning and dual strained high mobility channels, *Proc. IEDM*, vol. 12, pp. 2.7.1–2.7.4, 2016.

Xu, X., et al., Self-aligned double patterning aware pin access and standard cell layout co-optimization, *IEEE Trans. Comput. Des. Integr. Circuits Syst.*, vol. 34, no. 5, pp. 699–712, 2015.

Ye, W., et al., Standard cell layout regularity and pin access optimization considering middle-of-line, *Proc. GLSVLSI*, pp. 289–294, 2015.

Yeh, K., and W. Loong, Simulations of mask error enhancement factor in 193 nm immersion lithography, *Jpn. J. Appl. Phys.*, vol. 45, pp. 2481–2496, 2006.

2

When the Physical Disorder of CMOS Meets Machine Learning

Xiaolin Xu, Shuo Li, Raghavan Kumar, and Wayne Burleson

CONTENTS

While the development of semiconductor technology is advancing into the nanometer regime, one significant characteristic of today's complementary metal-oxide-semiconductor (CMOS) fabrication is the random nature of process variability, i.e., the physical disorder of CMOS transistors. It is observed that with the continuous development of semiconductor technology, the physical disorder has become an important factor in CMOS design during the last decade, and will likely continue in the forthcoming years. The low cost of modern semiconductor design and fabrication techniques benefits the ubiquitous applications, but also poses strict constraints on the area and energy of these systems. In this context, many traditional circuitry design rules should be reconsidered to tolerate the possible negative effects caused by the physical disorder of CMOS devices. As the physical deviation from nominal specifications becomes a big concern for many electronic systems [1], the performance of electronic blocks that have high requirements on the symmetric design and fabrication process is greatly impacted. For example, the resolution of time-to-digital converters (TDC) will be decreased by the physical process variations, which deviates the fabricated delay elements from the designed (nominal) delay length. On the one hand, such process variations introduce uncertainty into the standard CMOS design [2], while on the other hand, it also becomes a promising way to leverage such properties for constructive purposes. One major production of this philosophy is the development of physically unclonable functions (PUFs) in the literature, which extract secret keys from uncontrollable manufacturing variabilities on integrated circuits (ICs).

As electronic designs like PUF and TDC are impacted by the process variations of fabrication, many issues including variability, modeling attacks and noise sensitivity also need to be reconsidered and addressed. Due to the microscopic characteristic of such physical disorder, it is either infeasible or very expensive to measure and mitigate it with tractional electronic techniques. For example, any physical probing into a PUF instance would change the original physical disorder and therefore the measured results are not necessarily correct. In this context, employing the *non-invasive* way to characterize the internal physical disorder becomes a promising solution. This chapter presents some recent work on advancing this physical disorder modeling with the help of machine learning techniques. More specifically, it shows that through modeling the physical disorder, machine learning techniques can benefit the performance (i.e., reliability improvement) of PUFs by filtering out unreliable challenges and responses (challenge–response pairs [CRPs]). As for a fabricated TDC circuitry with a given internal physical disorder, it is demonstrated that a backpropagation-based machine learning framework can be utilized to mitigate the process variations and optimize the resolution.

This chapter is structured as follows. Section 2.1 briefly describes the sources of process variations, i.e., the physical disorder in modern CMOS circuits. An introduction to common PUF terminologies and performance metrics used throughout this chapter is provided in Section 2.2. Section 2.3 provides a brief overview of some of the most popular silicon-based PUF circuits along with their performance metrics. Next, the machine learning modeling method is reviewed as a threat in attacking PUFs in Section 2.4. Our discussion continues on applying machine learning techniques to help with understanding and mitigating the physical disorder of CMOS circuitry in Sections 2.5 and 2.6. More specifically, machine learning techniques can also be employed in a constructive way to improve the reliability of PUFs and the resolution of TDCs. Finally, we provide concluding remarks in Section 2.7.

2.1 Sources of CMOS Process Variations

The sources of process variations in ICs are summarized in this section. The interested reader is also referred to the book chapter by Kim et al. [3] for more details on the sources of CMOS variability. From the perspective of circuits, the sources of variations can be either *desirable* or *undesirable*. The desirable source of variations refers to process manufacturing variations (PMVs) as identified in [3]. Environmental variations and aging are undesirable for some functional circuits like PUFs.

2.1.1 Fabrication Variations

Due to the complex nature of the manufacturing process, the circuit parameters often deviate from their intended value. The various sources of variability include proximity effects, chemical-mechanical polishing (CMP), lithography system imperfections and so on. The PMVs consist of two components as identified in the literature, namely systematic and random variations [4].

2.1.1.1 Systematic Variations

The systematic component of process variations includes variations in the lithography system, the nature of the layout and CMP [4]. By performing a detailed analysis of the layout, the systematic sources of variations can be predicted in advance and accounted for in the design step. If the layout is not available for analysis, the variations can be assigned statistically [4].

2.1.1.2 Random Variations

Random variations refer to non-deterministic sources of variations. Some of the random variations include random dopant fluctuations (RDF), line edge roughness (LER) and oxide thickness variations. Random variations are often modeled using random variables for design and analysis purposes.

2.1.2 Environmental Variations and Aging

Environmental variations are detrimental to PUF circuits. Some of the common environmental sources of variations include power supply noise, temperature fluctuations and external noise. These variations must be minimized to improve the reliability of PUF circuits. Aging is a slow process and it reduces the frequency of operation of circuits by slowing them down. Circuits are also subjected to increased power consumption and functional errors due to aging [5].

2.2 Terminologies and Performance Metrics of PUF

Some terminologies and performance metrics like challenge–response pair (CPR), reliability, uniqueness and unpredictability used to evaluate PUFs are briefly summarized in the following sections.

2.2.1 Challenge–Response Pairs (CRPs)

The inputs to a PUF circuit are known as *challenges* and the outputs are referred to as *responses*. A challenge associated with its corresponding response is known as a *challenge–response pair* (CRP). In an application scenario, the responses of a PUF circuit are collected and stored in a database. This process is generally known as *enrollment*. Under a *verification* or *authentication* process, the PUF circuit is queried with a challenge from the database. The response is then compared against the one stored in the database. If the responses match, the device is authenticated.

2.2.2 Performance Metrics

Three important metrics are used to analyze a PUF circuit, namely uniqueness, reliability and unpredictability.

2.2.2.1 *Uniqueness*

One important application of PUF devices is to generate unique signatures for device authentication. Thus, it is desirable that any two PUF instances can be easily identified from another. Toward this end, a typical measure used to analyze uniqueness is known as *inter-distance* and is given by [3]

$$d_{\text{inter}}(C) = \frac{2}{k(k-1)} \sum_{i=1}^{i=k-1} \sum_{j=i+1}^{j=k} \frac{\text{HD}(R_i, R_j)}{m} \times 100\%. \tag{2.1}$$

In Equation 2.1, $\text{HD}(R_i, R_j)$ stands for the Hamming distance between two responses R_i and R_j of m bits long for challenge C, and k is the number of PUF instances under evaluation. For a group of PUF instances, the desired inter-distance (i.e., uniqueness) of them is 50%. By carefully looking at Equation 2.1, one can correspond the inter-distance $d_{\text{inter}}(C)$ to the mean of the Hamming distance distribution obtained over k chips for challenge C. While designing a PUF circuit, inter-distance is often measured through circuit simulations. A common practice is to perform Monte Carlo simulations over a large population of PUF instances. In simulations, care must be taken to efficiently model various sources of manufacturing variations in CMOS circuits, as they directly translate into uniqueness. During the simulations, manufacturing variations are modeled using a Gaussian distribution. In such cases, the mean and standard deviation of the Gaussian distribution under consideration must correspond to either inter-die or inter-wafer variations' statistics.

2.2.2.2 *Reliability*

As PUFs are built on microscopic process variations, a challenge applied to a PUF operating on an IC will not necessarily produce the same response under different operating conditions. The reliability of a PUF refers to its ability to produce the same responses under varying operating conditions. Reliability can be measured by averaging the number of flipped responses for the same challenge under different operating conditions. A common measure of reliability is *intra-distance*, given by [3,6]

$$d_{\text{intra}}(C) = \frac{1}{s} \sum_{j=1}^{s} \frac{\text{HD}(R_i, R'_{i,j})}{m} \times 100\%. \tag{2.2}$$

In Equation 2.2, R_i is the response of a PUF to challenge C under nominal conditions, s is the number of samples of response R_i obtained at different operating conditions, $R'_{i,j}$ corresponds to the jth sample of response R_i for challenge C and m is the number of bits in the response. Intra-distance is expected to be 0% for ideal PUFs, which corresponds to 100% reliability. The terms intra-distance (d_{intra}) and reliability have been used interchangeably further in this chapter. Given d_{intra}, reliability can always be computed (100 d_{intra}(%)). As an important feature of PUF performance, several contributions exist in the literature to improve the reliability of PUFs from circuit and system perspectives [7–13].

2.2.2.3 *Unpredictability*

For security purposes, the responses from a PUF circuit must be unpredictable. Unpredictability is a measurement that quantifies the randomness of the responses from the same PUF device. This metric can be evaluated using the National Institute of Standards and Technology (NIST) tests [6,14]. Silicon PUFs produce unique responses based on intrinsic process variations, which are very difficult to clone or duplicate by the manufacturer. However, by measuring the responses from a PUF device for a subset of challenges, it is possible to create a model that can mimic the PUF under consideration. Several modeling attacks on PUF circuits have been proposed in the literature [15]. The type of modeling attack depends on the PUF circuit. A successful modeling attack on a PUF implementation may not be effective for other PUF implementations. Modeling attacks can be made more difficult by employing some control logic surrounding the PUF block, which prevents direct read-out of its responses. One such technique is to use a secure one-way hash over PUF responses. However, if PUF responses are noisy and have significant intra-distance, this technique will require some sort of error correction on PUF responses prior to hashing [8].

2.3 CMOS PUFs

This section provides an overview of some PUF instantiations proposed in the literature. Based on the challenge–response behavior, PUFs can be classified into two categories, namely strong PUFs and weak PUFs. Strong PUFs have a complex challenge–response behavior and it is very difficult to physically clone them. They are often characterized by CRPs exponential to the number of challenge bits. So, it is not practical to read out all CRPs within a limited time using measurements. On the other hand, weak PUFs accept only a reduced set of challenges and in the extreme case, they accept just a single challenge. Such a construction requires that the response is never shared with the external world and is only used internally for security operations. We present an overview of some popular constructions of PUFs, namely arbiter PUFs in Sections 2.3.1, static random-access memory (SRAM) PUFs in Section 2.3.2 and one of its variants called a data retention voltage (DRV) PUF in Section 2.3.3.

2.3.1 Arbiter PUF

The concept of arbiter PUFs was introduced in [16,17]. An arbiter PUF architecture is shown in Figure 2.1. An n-bit arbiter PUF is composed of n stages, with each stage employing two 2:1 MUXs, as depicted in Figure 2.1. A challenge vector $\mathbf{C} = \{c_1, c_2,\ldots, c_n\}$ is applied as the

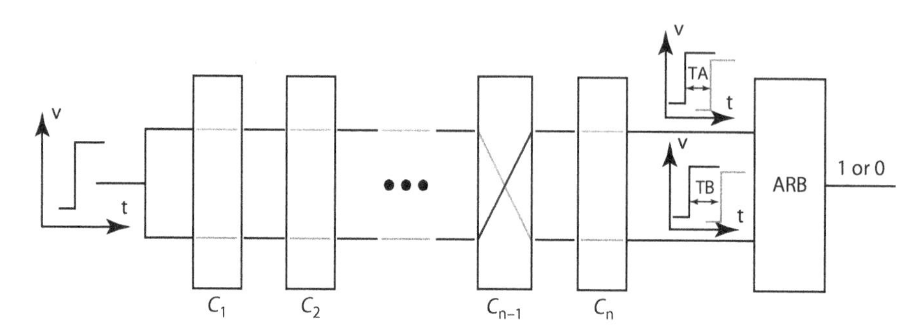

FIGURE 2.1
Schematic of an arbiter PUF. Challenge **C** controls the propagation paths of rising edges that gather the delay mismatch as they propagate toward the final arbiter.

control signals for all stages to configure two paths through the PUF toward the arbiter; at each stage, the paths are configured to be either straight or crossing. Thus, a rising edge applied at the input of the first stage gathers the delay mismatch from the paths through each stage while propagating toward the arbiter. The arbiter is a latch that digitizes the response into "1" or "0" by judging which rising edge is the first to arrive. It is important to note that the number of CRPs is exponential to the number of challenge bits. So, arbiter PUFs fall under the category of strong PUFs.

2.3.2 SRAM PUF

The concept of SRAM PUFs was first introduced in [18–20]. SRAM is a type of semiconductor memory, which is composed of CMOS transistors. SRAM is capable of storing a fixed written value "0" or "1", when the circuit is powered up. A SRAM cell capable of storing a single bit is shown in Figure 2.2. Each SRAM bit cell has two cross-coupled inverters and two access transistors. The inverters drive the nodes q and qb, as shown in Figure 2.2. When the circuit is not powered up, the nodes Q and $\bar{Q}$ are at logic low (00). When the power is applied, the nodes enter a period of metastability and settle down to either one

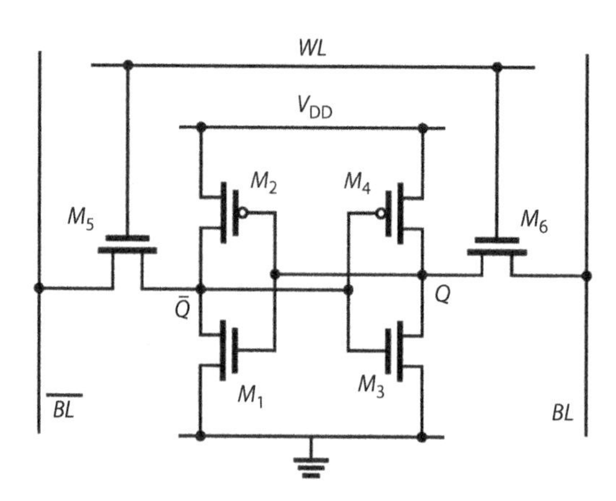

FIGURE 2.2
A SRAM cell using standard CMOS transistors. *BL* is the bit line, *WL* refers to word line, Q and $\bar{Q}$ are the state nodes for storing a single bit value. The transistors that make up inverters are shown in black and the access transistors are shown in gray.

of the states ($Q=0$, $\bar{Q}=1=1$ or $Q=1$, $\bar{Q}=0$). The final settling state is determined by the extent of the device mismatch in driving inverters and thermal noise. For PUF operation, it is desirable to have device mismatch dominant over thermal noise, as thermal noise is random in nature. The power-up states of SRAM cells were used for generating unique fingerprints in [19,20]. Experiments were conducted over 5120 64-bit SRAM cells from commercial chips and d_{inter} was found to be around 43%. Similarly, d_{intra} was found to be around 3.7%. Similar results were obtained in [18]. An extensive large-scale analysis over 96 application-specific integrated circuits (ASICs), with each ASIC having four instances of 8kB SRAM memory, was performed in [21]. The experimental results from [21] show that SRAM PUFs typically have high entropy, and the responses from different PUF instances were found to be highly uncorrelated.

2.3.3 DRV PUF

In the study of SRAM, DRV stands for the minimum supply voltage at which the written state is retained in a SRAM. In [22], DRV is proposed as the basis for a new SRAM-based PUF that is more informative than SRAM PUFs [23,24]. For example, DRV can uniquely identify circuit instances with 28% greater success than SRAM power-up states that are used in PUFs [22]. The physical characteristics responsible for DRV are imparted randomly to each cell during manufacturing, providing DRV with a natural resistance to cloning. It is shown that DRVs are not only random across chips, but also have relatively little spatial correlation within a single chip and can be treated in analysis as independent [25]. The proposed technique has the potential for wide application, as SRAM cells are among the most common building blocks of nearly all digital systems.

However, one major drawback of DRV is that it is highly sensitive to temperature, which makes it unreliable and unsuitable for use in a PUF. In [10], the idea of DRV fingerprinting is further extended to create a PUF based on DRV. Moreover, to overcome the temperature sensitivity of DRV, a DRV-based hashing scheme that is robust against temperature changes is proposed. The robustness of this hashing comes from the use of reliable DRV-ordering instead of less reliable DRV values. The use of DRV-ordering can be viewed as a differential mechanism at the logical level instead of the circuit level as in most PUFs. To help validate the DRV PUF, a machine learning technique is also proposed for simulation-free prediction of DRVs as a function of process variations and temperature. The machine learning model enables the rapid creation of large DRV data sets required for evaluating the DRV PUF approach.

2.4 Machine Learning Modeling Attacks on PUFs

In this section, we review some existing attacks on PUFs; more specifically, we focus on the machine learning–based modeling attacks. Recently, the emergence of unreliable CRP data as a tool for modeling attacks has introduced an important aspect to be addressed. Also, the vulnerabilities posed by strong PUFs toward modeling attacks will pave the way for immense competition between codemakers and codebreakers, with the hope that the process will converge on a PUF design that is highly resilient to known attacks.

2.4.1 Modeling Attacks on Arbiter PUFs

Because of the additive nature of path delay, several successful attempts have been made to attack arbiter PUFs using *modeling attacks* [15,26–28]. The basic idea is to observe a set of CRPs through physical measurements and use them to derive runtime delays using various machine learning algorithms. In [27], it was shown that arbiter PUFs composed of 64 and 128 stages can be attacked successfully using several machine learning techniques, achieving a prediction accuracy of 99.9% by observing around 18,000 and 39,200 CRPs, respectively. These attacks are possible only if the attacker can measure the response, i.e., the output of a PUF circuit is physically available through an input/output (I/O) pin in an IC. If the response of a PUF circuit is used internally for some security operations and is not available to the external world, modeling attacks cannot be carried out unless there is a mechanism to internally probe the output of the PUF circuit. However, probing itself can cause delay variations, thereby affecting the accuracy of the response measurement.

Several nonlinear versions have been proposed in the literature to improve the modeling attack resistance of arbiter PUFs. One version is feed-forward arbiter PUFs, in which some of the challenge bits are generated internally using an arbiter as a result of racing conditions at intermediate stages. However, such a construction has reliability issues because of the presence of more than one arbiter in the construction. This is evident from the test chip data provided in [16]. It was reported that feed-forward PUFs have d_{intra} ≈10% and d_{inter} 38%. Modeling attacks have been attempted on feed-forward arbiter PUFs and the attacks used to model simple arbiter PUFs were found to be ineffective when applied to feed-forward arbiter PUFs. However, by using evolution strategies (ES), feed-forward arbiter PUFs have been shown to be vulnerable to modeling attacks [27]. Non-linearities in arbiter PUFs can also be introduced using several simple arbiter PUFs and using an XOR operation across the responses of simple arbiter PUFs to obtain the final response. This type of construction is referred to as an XOR arbiter PUF. Though XOR arbiter PUFs are tolerant to simple modeling attacks, they are vulnerable to advanced machine learning techniques. For example, a 64-stage XOR arbiter PUF with 6 XORs has been attacked using 200,000 CRPs to achieve a prediction accuracy of 99% [29–32]. All these modeling attacks demonstrate an urgent need to design a modeling attack–resistant arbiter PUF.

2.4.2 Attacks against SRAM PUFs

Since SRAM PUFs based on power-up states only have a single challenge, the response must be kept secret from the external world. Modeling attacks are not relevant for SRAM PUFs. However, other attacks such as side-channel and virus attacks can be employed, but they are not covered in this chapter.

2.5 Constructively Applying Machine Learning on PUFs

Reliability is an important feature of PUFs that reflects their ability to produce the same response for a particular challenge despite the existence of noise. So, it is possible that a PUF operating in different conditions generates a different response to the same challenge vector. The PUF output is therefore a function of not only the challenge and process variations, but also the transient environmental conditions. Therefore, to make a PUF highly reliable, unstable CRPs that are easily flipped by environmental noise and aging should be corrected or even not used. Generally, the reason that PUF responses are unreliable is because supply voltage and temperature variations can overcome the impacts of process variations and flip the responses. Besides the transient noise, device aging is also an important but rarely studied source of unreliability in PUFs. Unlike environmental noise that temporarily flips PUFs' CRPs (PUFs work more reliably when the supply voltage and temperature return to normal), device aging causes a permanent change in the behavior of a PUF. Device aging is usually caused by negative bias temperature instability, hot carrier injection, time-dependent dielectric breakdown and electro-migration [33,34]. In this section, we present some methodologies that use machine learning–related techniques to improve the reliability of PUFs.

2.5.1 Using Machine Learning to Improve the Reliability of Arbiter PUFs

2.5.1.1 Mechanism of Arbiter PUFs

In an n-bit arbiter PUF (Figure 2.1), the propagation delay from the input to the first stage to the top and bottom outputs of the ith stage can be defined as D^i_{top} and D^i_{bottom}, respectively. The delay mismatch between two delay paths is summed up as the timing difference between D^n_{top} and D^n_{bottom} (Figure 2.3). By mapping the original challenge $c_i \in \{0, 1\}$ into $c_i \in \{-1, 1\}$, the path delay can be formulated as

$$\begin{aligned} D^i_{\text{top}} &= \frac{1+c_i}{2}\left(t^i_{\text{top}} + D^{i-1}_{\text{top}}\right) \\ &\quad + \frac{1-c_i}{2}\left(t^i_{u\_\text{across}} + D^{i-1}_{\text{bottom}}\right) \\ D^i_{\text{bottom}} &= \frac{1+c_i}{2}\left(t^i_{\text{bottom}} + D^{i-1}_{\text{bottom}}\right) \\ &\quad + \frac{1-c_i}{2}\left(t^i_{d\_\text{across}} + D^{i-1}_{\text{top}}\right) \end{aligned} \tag{2.3}$$

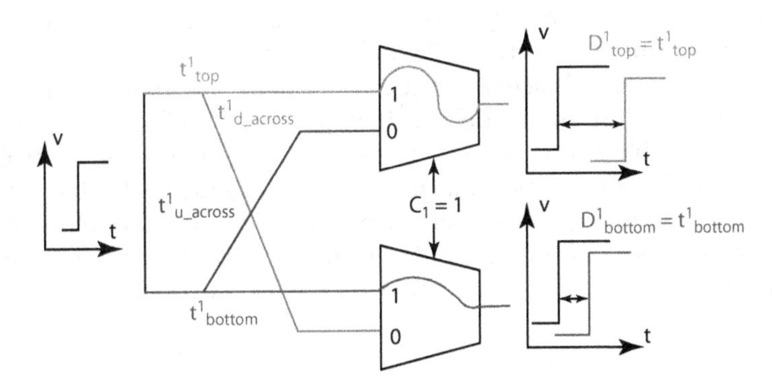

FIGURE 2.3
Propagation paths through the delay cells of an arbiter PUF.

where $D^0_{top} = D^0_{bottom} = 0$ and $t^i_{top}, t^i_{bottom}, t^i_{u\_across}, t^i_{d\_across}$ represent the four possible delays through the ith stage. Denoting the delay difference between the top and bottom arbiter inputs as $T_A - T_B$, following Equation 2.3, the delay difference between the two paths is $T_A - T_B = D^n_{top} - D^n_{bottom}$. The response ($r$) of an arbiter PUF of n-bit length is therefore determined by the sign of $T_A - T_B$ (Equation 2.4):

$$r = \begin{cases} 0, & \text{if } \operatorname{sgn}(T_A - T_B) > 0 \\ 1, & \text{if } \operatorname{sgn}(T_A - T_B) < 0 \end{cases} \tag{2.4}$$

From Equation 2.4, it is clear that a PUF response is flipped when the sign of $T_A - T_B$ changes, either from positive to negative or vice versa.

2.5.1.2 Modeling the $T_A - T_B$ of Arbiter PUFs

Because a PUF operating in different conditions can generate a different response to the same challenge vector, the PUF output is a function of not only the challenge and process variations, but also the transient environmental conditions. Therefore, to make a PUF highly reliable, unstable CRPs that are easily flipped by environmental noise and aging should not be used or corrected. Toward this end, a machine learning–based modeling method is proposed in [35], which helps with generating reliable CRPs without extra circuitry other than a normal PUF. Based on this technique, the unreliability source of a PUF is classified into two aspects: transient noise (e.g., temperature and supply voltage variations) and device aging. A machine learning model can be trained for PUF characterization and utilize the model for identifying and filtering out the unreliable challenge vectors for each PUF, allowing higher reliability to be achieved.

An n-bit arbiter PUF is composed of n stages, with each stage employing two 2:1 MUXs as depicted in Figures 2.1 and 2.3. A challenge vector $\mathbf{C} = \{c_1, c_2, \ldots, c_n\}$ is applied as the control signals for all stages to configure two paths through the PUF toward the arbiter; at each stage, the paths are configured to be either straight or crossing. To explore the impact of noise in more detail, simulations on a set of 64-bit arbiter PUFs show that the flipped responses are from challenges that correspond to $T_A - T_B$ in a small range:

$$\mathrm{DD}_{umin} \le T_A - T_B \le \mathrm{DD}_{umax} \tag{2.5}$$

where DD_{umin} and DD_{umax} represent the minimum/maximum delay difference between T_A and T_B of unreliable challenges. Based on this observation, it can be concluded that if only the challenge vectors satisfying either $T_A - T_B > \mathrm{DD}_{umax}$ or $T_A - T_B < \mathrm{DD}_{umin}$ are applied to the PUF, then the responses of the PUF will be reliable.

Since knowing the $T_A - T_B$ of each challenge vector makes it possible to get reliable CRPs by avoiding challenges with smaller delay difference, these challenges with smaller $T_A - T_B$ are likely to be unreliable and therefore can be discarded. However, as probing inside a PUF to measure $T_A - T_B$ is not practical, since it will introduce extra bias into the circuit operation and impact the original results, the machine learning modeling method is proposed to accomplish this job by building a PUF model. With a PUF model, the $T_A - T_B$ value for each challenge vector can be derived. Two parameters are defined as

$$\begin{aligned} \alpha_i &= \left(t^i_{top} - t^i_{bottom} + t^i_{d\_across} - t^i_{u\_across}\right)/2 \\ \beta_i &= \left(t^i_{top} - t^i_{bottom} - t^i_{d\_across} + t^i_{u\_across}\right)/2 \end{aligned} \tag{2.6}$$

Based on Equation 2.6, for a given challenge vector $\mathbf{C} = \{c_1, c_2, \ldots, c_n\}$, a corresponding response generation can be modeled by accumulating the delay mismatches through delay stages as

$$T_A - T_B = \alpha_1 k_0 + \cdots + (\alpha_n + \beta_{n-1}) k_{n-1} + \beta_n k_n \tag{2.7}$$

where $k_n = 1$ and $k_i = \prod_{j=i+1}^{n} c_j$, reflecting the number of times that the rising edges will change tracks between the ith stage and the arbiter. Thus, knowing the challenge and α_i and β_i, $i \in (1, \ldots, n)$ makes it possible to compute DD for any challenge. By denoting the delay parameters of an arbiter PUF with vector $\mathbf{p}_{\text{model}} = \{\alpha_1, \alpha_2 + \beta_1, \ldots, \alpha_n + \beta_{n-1}, \beta_n\}$, and defining challenge features as $\mathbf{k} = \{k_0, k_1, \ldots, k_n\}$, the model-predicted $T_A - T_B$ of each challenge vector can be denoted as

$$T_A - T_B = \langle \mathbf{p}_{\text{model}}, \mathbf{k} \rangle \tag{2.8}$$

The foregoing equations show that the machine learning modeling technique can be used to model $\mathbf{p}_{\text{model}}$ and predict $T_A - T_B$. To accomplish this, a set of known CRPs can be used to train a support vector machine (SVM) classifier. SVM models are powerful learning tools that can perform binary classification of data. Classification is achieved by a linear or nonlinear separating surface in the input space of the data set. Previously, SVMs have been widely used in attacking arbiter PUFs [16,29,36]. Note that finding the accurate delay difference is not an explicit objective of the SVM model, as the SVM model only seeks a value of $\mathbf{p}_{\text{model}}$ that can accurately predict responses. Fortunately, because the raw PUF responses are determined by the sign of $T_A - T_B$ (Equation 2.4), a model that is good at predicting unknown responses is also accurate in quantifying the values of $T_A - T_B$. Therefore, the flow of using a PUF model to enhance PUF reliability becomes: (1) train a binary classifier for a PUF and (2) use the trained PUF model to quantify the delay difference induced by each challenge. In other words, the model-quantified delay difference can be used to infer whether or not a challenge will generate a reliable PUF response. The proposed method can be evaluated by comparing the model-predicted $T_A - T_B$ with the golden value of $T_A - T_B$ that is extracted from a PUF. If the PUF model works well for response prediction, then it is expected that ρ in Equation 2.9 is close to 1 when the size of the CRP data set used to train the model is large enough.

$$\begin{aligned} &\operatorname{corr}\left((T_A - T_B)_{\text{golden}}, (T_A - T_B)_{\text{model}}\right) \\ &= \operatorname{corr}\left((T_A - T_B)_{\text{golden}}, \langle \mathbf{p}_{\text{model}}, \mathbf{k} \rangle\right) = \rho \end{aligned} \tag{2.9}$$

The correlation coefficient (ρ) between $(T_A - T_B)_{\text{golden}}$ (from simulation) and $(T_A - T_B)_{\text{model}}$, which is calculated as $(\mathbf{p}_{\text{model}}, \mathbf{k})$, is as shown in Figure 2.4a.

As can be seen, although $(T_A - T_B)_{\text{golden}}$ and $(T_A - T_B)_{\text{model}}$ are different in scale, the correlation between them is very high. Note that it is not required to know the exact value of the delay differences to select the reliable challenges, but only the relative magnitudes of the delay differences are needed. From Figure 2.4a and b, it can be inferred that the model-predicted $T_A - T_B$ values from the model are in good agreement with that from the PUF circuit.

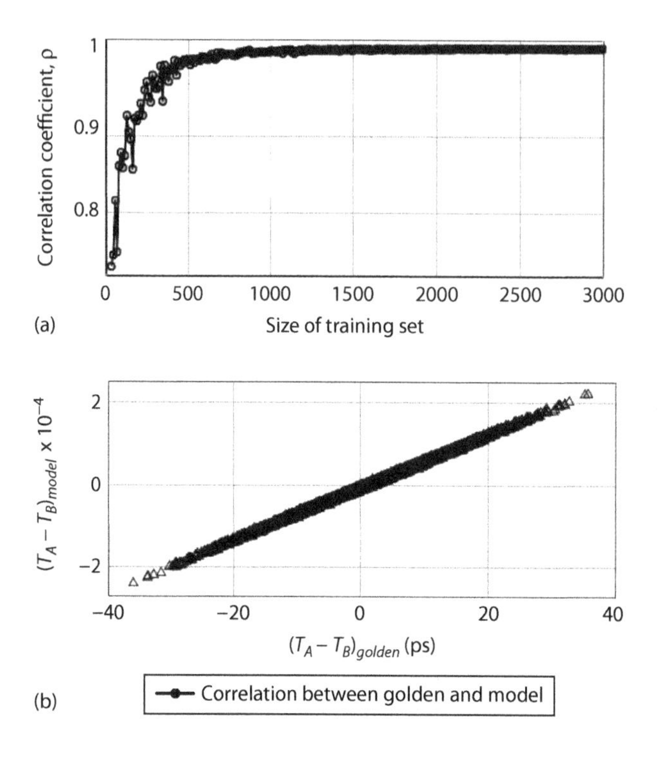

FIGURE 2.4
In Equation 2.4a, the correlation coefficient ρ between golden delay difference and model-predicted delay difference. While the PUF training size is increasing, higher ρ is achieved. In Equation 2.4b, based on the model trained with 3000 CRPs, there is good agreement between $(T_A - T_B)_{\text{golden}}$ and $(T_A - T_B)_{\text{model}}$ for 2000 random challenges. (a) Correlation coefficient increases with training size. (b) Good agreement between golden and model-predicted delay difference.

2.5.1.3 Improving PUF Reliability with PUF Model

As previously concluded, the flipped PUF responses are only those satisfying $DD_{\text{umin}} \leq (T_A - T_B)_{\text{golden}} \leq DD_{\text{umax}}$. However, due to the large size of a PUF CRP space, 2^n for an n-bit arbiter PUF, it is necessary to reconsider how to build the PUF model to predict the delay cutoffs that will rule out approximately fraction discarded ratio (dr) of the overall challenge space. For this purpose, there are still two questions that need to be answered:

Algorithm 1 Use the machine learning model to compute the range of model-predicted delay differences that are likely to be unreliable for a given PUF. Challenges predicted to have delay differences inside this range will not be applied to the PUF, and this will improve the overall PUF reliability. Reliability can be improved by using a larger value of *dr* to discard more challenges.

Input: A discard ratio *dr* and a set of challenges and corresponding responses obtained from a single PUF at nominal supply voltage and temperature.

Output: A range $[DD_{\text{min}}, DD_{\text{max}}]$ of delay differences to consider as unreliable for this PUF.

1: Let **k** be the challenges mapped to challenge features (Equations 2.7 and 2.8)
2: $\mathbf{p}_{\text{model}} \leftarrow$ **SVM(k,** responses) {train the PUF model}
3: $\mu_p = \text{avg}\langle \mathbf{p}_{\text{model}}, \mathbf{k} \rangle$ {mean predicted delay difference}
4: $\sigma_p^2 = \text{var}\langle \mathbf{p}_{\text{model}}, \mathbf{k} \rangle$ {variance of predicted delay difference}
5: the distribution of delay differences across all challenges is modeled to be $N(\mu_p, \sigma_p^2)$
6: $\text{DD}_{\text{min}} = F^{-1}(0.5 - dr/2) = \mu_p + \sigma_p \Phi^{-1}(0.5 - dr/2)$
7: $\text{DD}_{\text{max}} = F^{-1}(0.5 + dr/2) = \mu_p + \sigma_p \Phi^{-1}(0.5 + dr/2)$ {delay difference cutoffs based on PUF model $\mathbf{p}_{\text{model}}$ and selected challenge features **k** (Equation 2.11)}
8: **return** $[\text{DD}_{\text{min}}, \text{DD}_{\text{max}}]$

1. How many CRPs are enough to train an accurate model to characterize the DD_{umax} and DD_{umin} cutoffs for each PUF?
2. For a given PUF model, what is the *dr* of CRPs that must be filtered to achieve an expected reliability level?

The first question can be answer by employing the techniques in Algorithm 1. The physical features of a PUF follow Gaussian distribution; therefore, the $T_A - T_B$ of each applied challenge vector will also follow a Gaussian distribution. A model can be firstly built to model this distribution, as shown in Equation 2.10. By selecting and applying the challenges randomly, the training set of PUF ensures that it covers a more unreliable range.

$$F(\text{DD}, \mu_p, \sigma_p) = \frac{1}{\sigma_p \sqrt{2\pi}} \exp^{-((\text{DD}-\mu p)^2/2\sigma_p^2)} \tag{2.10}$$

Because the unreliable responses are only related to challenges that are generating smaller $T_A - T_B$, even without knowing the exact range of the delay difference for these unreliable CRPS, it can be quantified by applying a quantile function. For example, by denoting the probit function of a standard normal distribution with $\Phi^{-1}(dr)(dr \in (0, 1))$, the range of the delay difference that should be discarded can be expressed in Equation 2.11 (as Steps 6 and 7 in Algorithm 1), where *dr* stands for the ratio of challenges that should be discarded from the CRP database of each PUF. Apparently, there exists a tradeoff between the value of *dr* and the number of usable challenges. For example, a larger *dr* implies that more challenges should be discarded, while a smaller *dr* means less challenges will be filtered out.

$$F^{-1}(dr) = \mu_p + \sigma_p \Phi^{-1}(dr) \tag{2.11}$$

By applying a challenge to the trained PUF model, if its delay difference satisfies $((T_A - T_B)_{\text{model}}\, \varepsilon[\text{DD}_{\text{min}}, \text{DD}_{\text{max}}])$, then it will be marked as reliable and applied on PUFs. By formulating a set of such reliable challenges and comparing the corresponding responses with the golden data set, it is found that as the training size of a PUF model increases, the characterized cutoffs DD_{min} and DD_{max} also become more accurate, as shown in Figure 2.5. As a result, fewer challenges need to be discarded to achieve the same reliability.

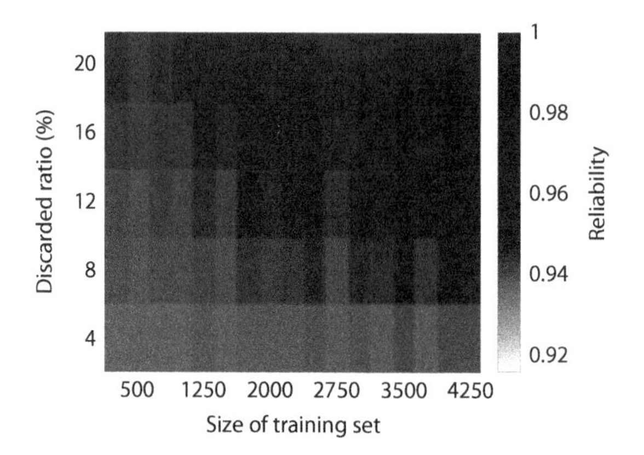

FIGURE 2.5
Validation under aging and environmental noise, across all of the simulated PUF instances. The tradeoff between training size and discarded ratio can be seen in the figure. A larger *dr* is conservative and can compensate for the lower-quality delay predictions of a model trained from a smaller training set.

2.5.2 Using Machine Learning to Model the Data Retention Voltage of SRAMs

DRV is a commonly studied feature of SRAMs in low-power research. A *simulator program* with integrated circuit emphasis (SPICE) simulator is usually employed to characterize the lowest DRV of different CMOS technology nodes. However, the time consumed by a SPICE simulator is relatively high to formulate the DRV of a SRAM cell. This is because multiple supply voltage values should be applied to find the maximum voltage that induces a data retention failure. In [10], it is reported that simulating a single test voltage on a single SRAM cell for 2 ms has a runtime of 0.17 s with an Intel Xeon E5-2690 processor running at 2.90 GHz with 64GB of RAM.

To save the DRV simulation time, an alternative method is to predict the DRV using a device model. According to [37], the DRV of a SRAM cell is determined by the environmental temperature and the process variations of the transistors. Therefore, the DRV value of a SRAM cell can be formulated as a function of physical features like temperature T and transistor width, length and threshold voltage (W, L and V_{th}, respectively). To study this relationship, Qin et al. [37] propose an analytical model as shown in Equation 2.12, where DRV_r is the DRV at room temperature, and DRV_f is defined in Equation 2.13 with ΔT representing the temperature difference from room temperature. Terms a_i, b_i and c in Equation 2.13 are fitting coefficients and their values are determined empirically for each CMOS technology process [37].

$$\text{DRV} = \text{DRV}_r + \text{DRV}_f \tag{2.12}$$

$$\text{DRV}_f = \sum_{i=1}^{6} a_i \times \frac{\Delta(W_i/L_i)}{W_i/L_i} + \sum_{i=1}^{6} b_i \times \Delta(V_{thi}) + c \times \Delta T \tag{2.13}$$

Although this model can be used to estimate the DRV of a SRAM cell, it has two weaknesses that create the need for a more advanced model:

1. To formulate a specified DRV value with Equation 2.13, firstly, the user needs to know the DRV_r for each SRAM cell. However, this physical feature cannot be

expressed as a function of transistor parameters and can only be characterized through hardware measurement or computationally expensive circuit simulation.

2. It is impractical to apply the same coefficients a_i, b_i and c to different SRAM cells. In reality, the DRV of different cells increases according to different coefficients depending on their unique process variations. This distinction is especially important in building the model.

2.5.2.1 Predicting DRV using Artificial Neural Networks

To address the two aforementioned weaknesses, Xu et al. propose to use machine learning for DRV prediction [10]. The proposed technique can predict the DRV value for a SRAM cell by feeding the physical parameters into a machine learning model. The basis of this technique is that for a certain CMOS technology node, the values of process variations only vary over a bounded range; therefore, the DRV values also fall into a bounded range $[DRV_{min}, DRV_{max}]$. Based on the modeling technique, the range of DRVs is firstly divided into K labels with each standing for a smaller range with size DRV (Equation 2.14). By training a machine learning model that maps the physical features into these K DRV classes, the DRV value for unknown SRAM cells can be predicted with the model, as shown in Figure 2.6.

$$\begin{aligned}[DRV_{min}, DRV_{max}] = \{&[DRV_{min}, DRV_{min} + \Delta DRV) \cup \\ &[DRV_{min} + \Delta DRV, DRV_{min} + 2 \times \Delta DRV) \cup \ldots \\ &\ldots \cup [DRV_{max} - \Delta DRV, DRV_{max}]\}\end{aligned} \tag{2.14}$$

An artificial neural network (ANN) is a widely used machine learning method to solve so-called multi-classification problems. More specifically, to model the DRV values of SRAMs, the outputs of an ANN are used for different DRV classes. During the modeling

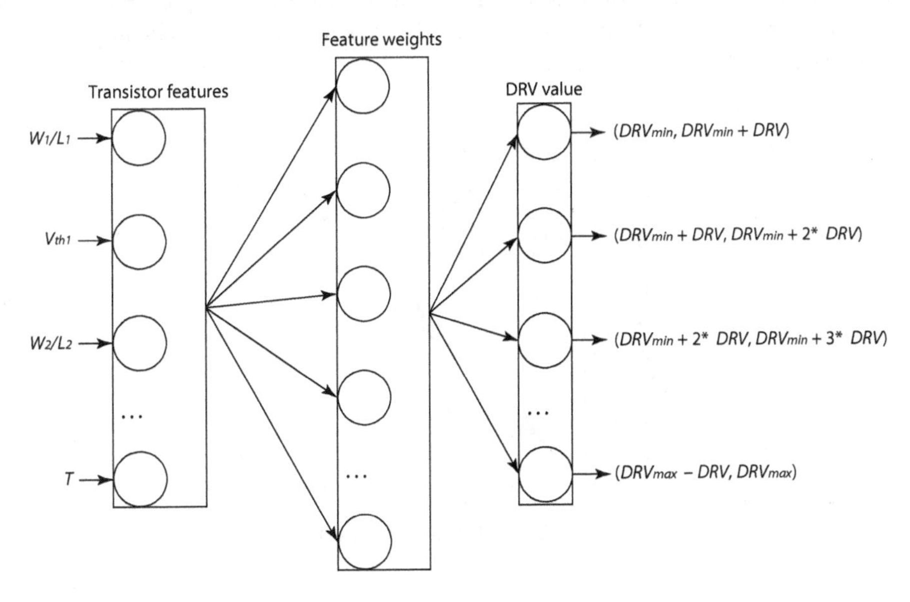

FIGURE 2.6
The layer composition of an artificial neural network for DRV classification and prediction.

training, the SRAM parameters are used to optimize the internal layers and the neuron nodes of an ANN and minimize the general prediction errors. A set of samples (including the DRV values and the corresponding physical characteristics of SRAM cells) are collected by SPICE simulation and used to train an ANN model.

2.5.3 Performance of DRV Model

The ANN model–predicted DRV values are compared with the SPICE simulated values in Figure 2.7, where the R value denotes the correlation between the model outputs and golden values. For the ANN-based DRV model, it is noted that there is a high correlation between prediction and output for all data sets. Before the ANN model proposed in [10], the linear model described in Equation 2.13 was widely used to model the DRV value of SRAM designs, which can be optimized with a linear regression (LR) method that fits a data set with linear coefficients. The most common type of LR is a *least-squares fit*, which can find an optimal line to represent the discrete data points. In an LR model, the same physical parameters of ANN models are defined as input training features $p=\{p_i, p_i \in \{W_1/L_1, W_2/L_2,\ldots, T\}\}$. By denoting the linear coefficients with $\theta=\{\theta_0, \theta_1,\ldots, \theta_n\}$:

$$h_\theta(p)=\theta_0+\theta_1\times p_1+\cdots+\theta_n\times p_n \tag{2.15}$$

where θ stands for the set of coefficients (e.g., a_i and b_i as shown in Equation 2.13). Each training sample is composed of a transistor feature set p and the corresponding golden DRV value, $\mathrm{DRV}_{\mathrm{golden}}$, from SPICE simulation. Based on the least-squares fit rule, the cost function of m training examples can be expressed as

$$J(\theta)=\frac{1}{2m}\sum_{k=1}^{m}\left(h_\theta\left(p^{(k)}\right)-\mathrm{DRV}_{\mathrm{golden}}^{(k)}\right)^2 \tag{2.16}$$

where $p^{(k)}$ corresponds to the training features of a kth training sample, like the transistor sizes and temperature. To obtain the optimal θ, "gradient descent" can be applied simultaneously on each coefficient θ_j, $j\in(1, 2,\ldots, n)$:

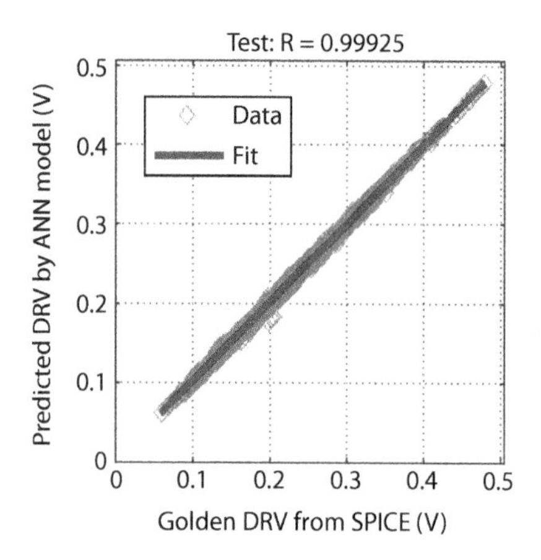

FIGURE 2.7
Training results based on the neural network model, across three data sets. R denotes the correlation between golden DRV data from SPICE simulation and the predicted DRV value from our model.

Repeat{

$$\theta_j := \theta_j - \alpha \frac{\partial J(\theta)}{\partial \theta_j}$$
$$= \theta_j - \alpha \frac{1}{m} \sum_{i=1}^{m} \left(h_\theta \left(p^{(k)} \right) - \mathrm{DRV}^{(k)}_{\mathrm{golden}} \right) p_i^{(k)} \quad (2.17)$$

}

α is the learning rate of the LR model and $p^{(k)}$ is the ith feature of the kth training sample.

The experimental results demonstrate that the neural network model achieves smaller prediction errors than the LR model. According to [10], the mean μ and the standard deviation σ of the prediction error for the neural network model are 0.01 and 0.35 mV, respectively, while those of the LR model are 0.041 and 0.9 mV. Hence, it can be concluded that the neural network model outperforms the linear model in modeling the DRV of SRAMs. This is because varied weights and bias are employed in an ANN model for different feature patterns, whereas the linear model formulates all input features with the same optimized coefficients.

2.6 Using Machine Learning to Mitigate the Impact of Physical Disorder on TDC

This section presents a case study that uses machine learning to help mitigate the negative impact of physical disorder on TDC designs. Instead of discussing the traditional TDC schemes that are based on a delay line, the TDC design scheme presented in this section is the configurable compact algorithmic TDC (CCATDC) proposed in [38,39]. For brevity, this section does not cover the design scheme of this TDC architecture but focuses more on the use of machine learning for mitigating the process variations of this circuitry.

2.6.1 Background of TDC

High-resolution time measurement is a common need in modern scientific and engineering applications, such as time-of-flight measurement in remote sensing [40,41], nuclear science [42], biomedical imaging [43], frequency synthesizer and time jitter measurement for radio frequency transceivers in wireless communication [44]. To advance the post-processing of time signals, a TDC that bridges time measurements and digital electronic devices is proposed. As a measurement system, a TDC quantifies the time interval between two events by digitizing it, which greatly favors the post-processing in a digital way. Various TDC schemes have been proposed, such as the cyclic successive approximation–based TDC, pulse interpolation–based TDC and delta-sigma-based TDCs [45,46].

Most conventional TDC designs are implemented with delay lines, usually composed of two channels for two input signals, as shown in Figure 2.8a. The time difference (TD)

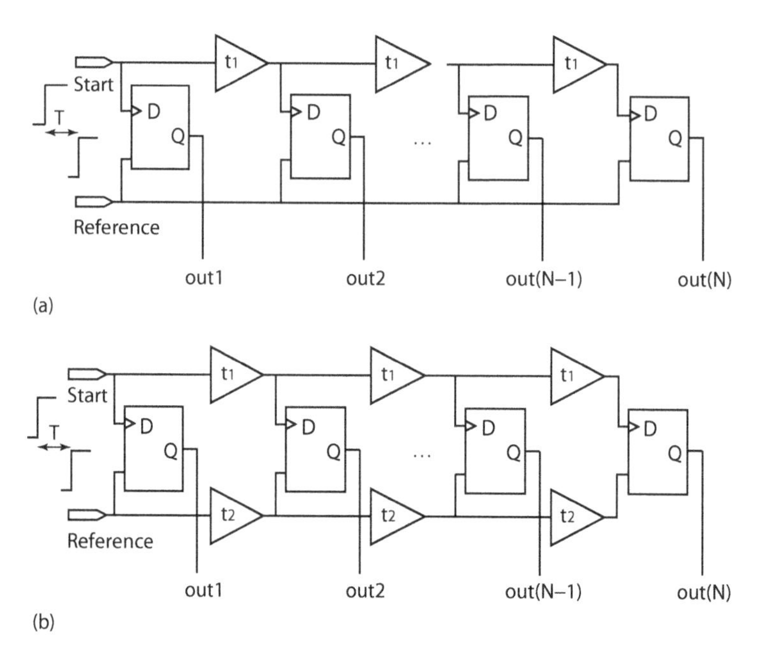

FIGURE 2.8
Schematic of traditional delay line–based TDCs. (a) Delay line–based TDC. (b) Vernier delay line–based TDC.

between the two input signals, "start" and "reference", is digitized into binary codes. To characterize the TD between two signals, the "start" signal is sequentially postponed by the delay elements. Theoretically, each delay element "slows down" the "start" signal by a constant timescale (td1 as in Figure 2.8b); the output of each delay element is then compared with a "reference" signal to generate a binary output. The highest resolution of a delay line–based TDC is the delay of each single element, which is limited by the CMOS fabrication technology node (a more advanced CMOS fabrication technology achieves higher resolution). To improve the resolution and conversion accuracy of such TDCs, several techniques have been proposed, such as a stretching pulse [47], using a tapped delay line [48] and employing a differential delay line [49].

As it becomes more difficult to control the process variations of CMOS fabrication with advanced technology nodes, the physical deviation from the designed value also becomes a big concern that limits the resolution of TDC design. For example, the fabricated delay elements in a TDC circuitry usually deviate from the designed (nominal) delay length, for example, T_{d1} in Figure 2.8a. Such process variations introduce design uncertainty into the time-to-digital conversion and decrease the conversion accuracy [2].

2.6.2 Mitigate the Process Variations by Reconfiguring the Delay Elements

The formulation of the physical disorder of CMOS devices comes from the fabrication stage, i.e., during which the length of the TDC delay element deviates from the designed value. Therefore, one solution to mitigate such impact is designing the delay element with an adjustable length and then regulating the length of the delay chain circuitry to improve its performance. This philosophy is based on the truth that: once an electronic circuit is fabricated, it is infeasible to remove the process variations. Fortunately, the actual in-path delay can be configured by using an adjustable delay line, thereby improving the

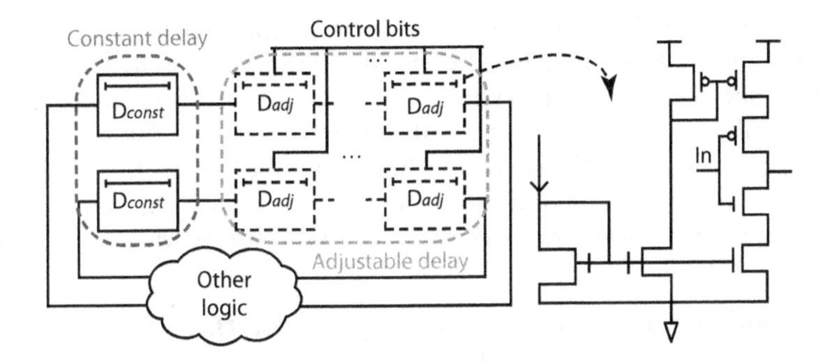

FIGURE 2.9
Schematic of the proposed configurable delay element, composed of two blocks, D_{const} and D_{adj}. There are many possible implementations of the adjustable delay element, and in this figure a current-controlled delay element is shown as an example.

robustness of CCATDC. There are many possible realizations of the adjustable delay element, such as voltage, current and digital controlled.

The schematic of an adjustable delay channel is shown in Figure 2.9, which is composed of two parts: the constant delay line, D_{const}, and the adjustable delay line, D_{adj}. Note that there still exist process variations in the constant delay line that deviate it from the designed delay length. The purpose of adding it is to roughly calibrate the reference time, T_{ref}, for the purposes of course tuning. A high-resolution adjustment is realized with the adjustable delay line, which fulfills the job of fine-tuning. To achieve a higher resolution, two parallel delay chains are utilized in two channels to form a vernier architecture. The control signals of the adjustable delay line can be supplied by an external calibration circuit.

2.6.3 Delay Chain Reconfiguration with Machine Learning

An important purpose of configuring the propagation delay of each element is to minimize the deviation of the process variations of CMOS transistors. However, since all components in the CCATDC design are of nanometer magnitude, it is again impractical to measure such process variations with an external instrument [35]. In [10,35], a machine learning technique was utilized to characterize the microscopic process variations and related physical features. This section continues this thread by employing machine learning techniques to characterize and configure the delay length of CCATDC. More specifically, the backward propagation of errors (backpropagation) algorithm is used. Backpropagation is a widely used technique in training ANNs, and is often used with other optimization methods such as gradient descent. Backpropagation training is usually composed of two phases: propagation and weight updating. Once an input vector is applied to the neural network model, it will be propagated from the input layer to the output layer. The output value of each input vector will be obtained and compared with the desired (golden) value; a loss function will be used to calculate the error for each neuron in the trained model and update the weight parameters correspondingly.

Within the framework, the controlling signals (e.g., the controlling current inputs) of the adjustable delay line (as shown in Figure 2.9) are fed into the backpropagation model as input vectors: $\mathbf{C}=(C_0, C_1,\ldots, C_{n-1})$ in Algorithm 2. To comprehensively configure the delay chain, m different time inputs $\mathbf{T}=(t_0, t_1,\ldots, t_{m-1})$ and corresponding digital conversion outputs $\mathbf{O}=(O_0, O_1,\ldots, O_{m-1})$ are used to train the model. The accuracy of CCATDC is optimized by setting Δ_{err} as the threshold of conversion error. Before the model is trained, weight parameters are initialized as 0. During the model training, the controlling signals are fed into the HSPICE simulation and corresponding conversion outputs are obtained (Line 5); the simulated results are then compared with the golden values (Line 7). The conversion error is then backpropagated to the network (Line 8) and the controlling input vector is optimized to achieve a higher conversion accuracy (Line 9). The configuration procedure will terminate while the conversion error is below the pre-set threshold Δ_{err} (Line 10).

2.6.3.1 Performance of Configurable Compact Algorithmic TDC

The proposed CCATDC architecture is implemented with the predictive technology model (PTM) 45 nm standard cell libraries [50] and its performance is tested by applying random process variations on all transistors of the circuit; transient noise is also added in the simulation. The experimental result demonstrates that CCATDC is robust against process variations and transient noise, and the conversion accuracy is much higher if more conversion bits are utilized.

Algorithm 2 Given a CCATDC chip, configure the delay elements to improve the time digitization accuracy.
Input: a configurable compact algorithmic TDC
Input: n control signals $\mathbf{C} = (C_0, C_1, \ldots, C_{n-1})$
Input: weight vector with k parameters, $\mathbf{\Theta} = (\theta^0, \theta^1, \ldots, \theta^k)$
Input: m time input $\mathbf{T} = (t_0, t_1, \ldots, t_{m-1})$ and corresponding ideal conversion output $\mathbf{O} = (O_0, O_1, \ldots, O_{m-1})$ for the designed T_{ref} and amplification gain
Input: Δ_{err}, the threshold of conversion error

1: Initialize weight parameters $\theta^i \leftarrow 0$
2: Initialize error parameters $\Delta^i \leftarrow \Delta_{err}$
3: **while** $\max(\Delta^i) >= \Delta_{err}$ **do**
4: **for** $i := 0$ **to** $m - 1$ **do**
5: $\hat{O}_i = \text{HSPICE}(t_i, \mathbf{C})$
6: **end for**
7: compute conversion error $\Delta = (\hat{\mathbf{O}} - \mathbf{O})$
8: BackwardPropagateError(Δ)
9: UpdateWeights(Θ, $\mathbf{C}$)
10: **end while**
11: **return C**

2.7 Conclusion

In this chapter, an overview of some of the latest CMOS circuits like PUFs and TDCs in the literature is provided. Notably, the relationship between a physical disorder and these two circuit primitives is different: PUF is built on the physical disorder of CMOS fabrication, while the uncontrollable process variations of CMOS transistors degrade the resolution of TDCs. Apart from constructively leveraging these physical disorders, the modeling of them also provides a new way to study the process and environmental variations on the performance of CMOS circuitry. Based on the case studies presented in this chapter, we can see that machine learning is a good tool to study the microscopic physical disorder of CMOS circuitry and therefore provides a unique perspective to guide the design. The performance metrics used for analyzing these circuits are identified and several machine learning techniques are employed in a constructive way to benefit the performance of these designs.

References

1. S. R. Sarangi, B. Greskamp, R. Teodorescu, J. Nakano, A. Tiwari, and J. Torrellas. Varius: A model of process variation and resulting timing errors for microarchitects. *IEEE Transactions on Semiconductor Manufacturing*, 21(1):3–13, 2008.
2. J.-P. Jansson, V. Koskinen, A. Mantyniemi, and J. Kostamovaara. A multichannel high-precision CMOS time-to-digital converter for laser-scanner- based perception systems. *IEEE Transactions on Instrumentation and Measurement*, 61(9):2581–2590, 2012.
3. I. Kim, A. Maiti, L. Nazhandali, P. Schaumont, V. Vivekraja, and H. Zhang. From statistics to circuits: Foundations for future physical unclonable functions. In A.-R. Sadeghi and D. Naccache, editors, *Towards Hardware-Intrinsic Security*, Information Security and Cryptography, pages 55–78. Springer, 2010.
4. D. Blaauw, K. Chopra, A. Srivastava, L. Scheffer. Statistical timing analysis: From basic principles to state of the art. *IEEE Transactions on Computer-Aided Design of Integrated Circuits and Systems*, 27(4):589–607, 2008.
5. W. Wang, V. Reddy, B. Yang, V. Balakrishnan, S. Krishnan, and Y. Cao. Statistical prediction of circuit aging under process variations. In *IEEE Custom Integrated Circuits Conference*, pages 13–16. IEEE, 2008.
6. D. Forte and A. Srivastava. On improving the uniqueness of silicon-based physically unclonable functions via optical proximity correction. In *IEEE/ACM Design Automation Conference*, pages 96–105. June 2012.
7. V. Vivekraja and L. Nazhandali. Feedback-based supply voltage control for temperature variation tolerant PUFs. In *IEEE International Conference on VLSI Design*, VLSID '11, Washington, DC, USA, pages 214–219. IEEE Computer Society, 2011.
8. M.-D. Yu and S. Devadas. Secure and robust error correction for physical unclonable functions. *IEEE Design Test of Computers*, 27(1):48–65, 2010.
9. C. Bösch, J. Guajardo, A.-R. Sadeghi, J. Shokrollahi, and P. Tuyls. Efficient helper data key extractor on FPGAs. In *Cryptographic Hardware and Embedded Systems*, volume 5154 of *Lecture Notes in Computer Science*, pages 181–197. Springer, 2008.
10. X. Xu, A. Rahmati, D. E. Holcomb, K. Fu, and W. Burleson. Reliable physical unclonable functions using data retention voltage of SRAM cells. *IEEE Transactions on Computer-Aided Design of Integrated Circuits and Systems*, 34(6):903–914, 2015.
11. X. Xu and D. Holcomb. A clockless sequential PUF with autonomous majority voting. In *Great Lakes Symposium on VLSI, 2016 International*, pages 27–32. IEEE, 2016.
12. X. Xu and D. E. Holcomb. Reliable PUF design using failure patterns from time-controlled power gating. In *Defect and Fault Tolerance in VLSI and Nanotechnology Systems (DFT), 2016 IEEE International Symposium on*, pages 135–140. IEEE, 2016.

13. X. Xu, V. Suresh, R. Kumar, and W. Burleson. Post-silicon validation and calibration of hardware security primitives. In *VLSI (ISVLSI), 2014 IEEE Computer Society Annual Symposium on*, pages 29–34. IEEE, 2014.
14. R. Maes and I. Verbauwhede. Physically unclonable functions: A study on the state of the art and future research directions. In A.-R. Sadeghi and D. Naccache, editors, *Towards Hardware-Intrinsic Security*, Information Security and Cryptography, pages 3–37. Springer Berlin Heidelberg, 2010.
15. G. Hospodar, R. Maes, and I. Verbauwhede. Machine learning attacks on 65 nm arbiter PUFs: Accurate modeling poses strict bounds on usability. In *IEEE International Workshop on Information Forensics and Security*, pp. 37–42. IEEE, 2012.
16. D. Lim. Extracting secret keys from integrated circuits. Master's thesis, Massachusetts Institute of Technology, Dept. of Electrical Engineering and Computer Science, 2004.
17. G. Suh and S. Devadas. Physical unclonable functions for device authentication and secret key generation. In *IEEE/ACM Design Automation Conference*, pages 9–14, June 2007.
18. J. Guajardo, S. S. Kumar, G.-J. Schrijen, and P. Tuyls. FPGA intrinsic PUFs and their use for IP protection. In *International Workshop on Cryptographic Hardware and Embedded Systems*, CHES'07, pages 63–80. Springer-Verlag, 2007.
19. D. E. Holcomb, W. Burleson, and K. Fu. Initial SRAM state as a fingerprint and source of true random numbers for RFID tags. In *Proceedings of the Conference on RFID Security*, 2007.
20. D. E. Holcomb, W. P. Burleson, and K. Fu. Power-up SRAM state as an identifying fingerprint and source of true random numbers. *IEEE Transactions on Computers*, 58(9):1198–1210, 2009.
21. S. Katzenbeisser, U. Kocabaş, V. Rožić, A.-R. Sadeghi, I. Verbauwhede, and C. Wachsmann. PUFs: Myth, fact or busted? A security evaluation of physically unclonable functions (PUFs) cast in silicon. In *International Conference on Cryptographic Hardware and Embedded Systems*, CHES'12, pages 283–301. Springer-Verlag, 2012.
22. D. E. Holcomb, A. Rahmati, M. Salajegheh, W. P. Burleson, and K. Fu. DRV-fingerprinting: Using data retention voltage of SRAM cells for chip identification. In *International Workshop on Radio Frequency Identification. Security and Privacy Issues*, pages 165–179. Springer, 2013.
23. J. Guajardo, S. Kumar, G. Schrijen, and P. Tuyls. FPGA intrinsic PUFs and their use for IP protection. *Cryptographic Hardware and Embedded Systems*, 2007.
24. D. E. Holcomb, W. P. Burleson, and K. Fu. Power-up SRAM state as an identifying fingerprint and source of true random numbers. *IEEE Transactions on Computers*, 2009.
25. A. Kumar, H. Qin, P. Ishwar, J. Rabaey, and K. Ramchandran. Fundamental data retention limits in SRAM standby: Experimental results. In *Quality Electronic Design, 2008. ISQED 2008. 9th International Symposium on*, pages 92–97. IEEE, 2008.
26. M. Majzoobi, F. Koushanfar, and M. Potkonjak. Testing techniques for hardware security. In *IEEE International Test Conference*, pages 1–10. Oct. 2008.
27. U. Rührmair, F. Sehnke, J. Sölter, G. Dror, S. Devadas, and J. Schmidhuber. Modeling attacks on physical unclonable functions. In *ACM Conference on Computer and Communications Security*, CCS'10, New York, USA, pages 237–249. ACM, 2010.
28. X. Xu, U. Rührmair, D. E. Holcomb, and W. Burleson. Security evaluation and enhancement of bistable ring PUFs. In *International Workshop on Radio Frequency Identification: Security and Privacy Issues*, pages 3–16. Springer, 2015.
29. U. Rührmair, J. Sölter, F. Sehnke, X. Xu, A. Mahmoud, V. Stoyanova, G. Dror, J. Schmidhuber, W. Burleson, and S. Devadas. PUF modeling attacks on simulated and silicon data. *Information Forensics and Security, IEEE Transactions on*, 2013.
30. X. Xu and W. Burleson. Hybrid side-channel/machine-learning attacks on PUFs: A new threat? In *Proceedings of the Conference on Design, Automation & Test in Europe*, page 349. European Design and Automation Association, 2014.
31. U. Rührmair, X. Xu, J. Sölter, A. Mahmoud, M. Majzoobi, F. Koushanfar, and W. Burleson. Efficient power and timing side channels for physical unclonable functions. In *International Workshop on Cryptographic Hardware and Embedded Systems*, pages 476–492. Springer, 2014.

32. U. Rührmair, X. Xu, J. Sölter, A. Mahmoud, F. Koushanfar, and W. Burleson. Power and timing side channels for PUFs and their efficient exploitation. *IACR Cryptology ePrint Archive*, 2013:851, 2013.
33. S. Khan, S. Hamdioui, H. Kukner, P. Raghavan, and F. Catthoor. Incorporating parameter variations in BTI impact on nano-scale logical gates analysis. In *Defect and Fault Tolerance in VLSI and Nanotechnology Systems (DFT), 2012 IEEE International Symposium on*, pages 158–163. IEEE, 2012.
34. D. Lorenz, G. Georgakos, and U. Schlichtmann. Aging analysis of circuit timing considering NBTI and HCI. In *On-Line Testing Symposium, 2009. IOLTS 2009. 15th IEEE International*, pages 3–8. IEEE, 2009.
35. X. Xu, W. Burleson, and D. E. Holcomb. Using statistical models to improve the reliability of delay-based PUFs. In *VLSI (ISVLSI), 2016 IEEE Computer Society Annual Symposium on*, pages 547–552. IEEE, 2016.
36. S. S. Avvaru, C. Zhou, S. Satapathy, Y. Lao, C. H. Kim, and K. K. Parhi. Estimating delay differences of arbiter PUFs using silicon data. In *2016 Design, Automation & Test in Europe Conference & Exhibition (DATE)*, pages 543–546. IEEE, 2016.
37. H. Qin, Y. Cao, D. Markovic, A. Vladimirescu, and J. Rabaey. SRAM leakage suppression by minimizing standby supply voltage. In *5th International Symposium on Quality Electronic Design*, pages 55–60. IEEE, 2004.
38. S. Li and C. D. Salthouse. Compact algorithmic time-to-digital converter. *Electronics Letters*, 51(3):213–215, 2015.
39. S. Li, X. Xu, and W. Burleson. CCATDC: A configurable compact algorithmic time-to-digital converter. In *VLSI (ISVLSI), 2017 IEEE Computer Society Annual Symposium on*, pages 501–506. IEEE, 2017.
40. D. Marioli, C. Narduzzi, C. Offelli, D. Petri, E. Sardini, and A. Taroni. Digital time-of-flight measurement for ultrasonic sensors. *IEEE Transactions on Instrumentation and Measurement*, pages 93–97. IEEE, 1992.
41. S. Li and C. Salthouse. Digital-to-time converter for fluorescence lifetime imaging. In *Instrumentation and Measurement Technology Conference (I2MTC), 2013 IEEE International*, pages 894–897. IEEE, 2013.
42. N. Bar-Gill, L. M. Pham, A. Jarmola, D. Budker, and R. L. Walsworth. Solid-state electronic spin coherence time approaching one second. *Nature Communications*, 4:1743, 2013.
43. A. S. Yousif and J. W. Haslett. A fine resolution TDC architecture for next generation PET imaging. *IEEE Transactions on Nuclear Science*, pages 1574–1582. IEEE, 2007.
44. J.-P. Jansson, A. Mantyniemi, and J. Kostamovaara. A CMOS time-to-digital converter with better than 10 ps single-shot precision. *IEEE Journal of Solid-State Circuits*, 41(6):1286–1296, 2006.
45. A. Mantyniemi, T. Rahkonen, and J. Kostamovaara. A CMOS time-to-digital converter (TDC) based on a cyclic time domain successive approximation interpolation method. *IEEE Journal of Solid-State Circuits*, 44(11):3067–3078, 2009.
46. W.-Z. Chen and P.-I. Kuo. A $\Delta\Sigma$ TDC with sub-ps resolution for PLL built-in phase noise measurement. In *European Solid-State Circuits Conference, ESSCIRC Conference 2016: 42nd*, pages 347–350. IEEE, 2016.
47. S. Tisa, A. Lotito, A. Giudice, and F. Zappa. Monolithic time-to-digital converter with 20ps resolution. In *Solid-State Circuits Conference, 2003. ESSCIRC'03. Proceedings of the 29th European*, pages 465–468. IEEE, 2003.
48. R. B. Staszewski, S. Vemulapalli, P. Vallur, J. Wallberg, and P. T. Balsara. 1.3 v 20 ps time-to-digital converter for frequency synthesis in 90-nm CMOS. *IEEE Transactions on Circuits and Systems II: Express Briefs*, 53(3):220–224, 2006.
49. P. Dudek, S. Szczepanski, and J. V. Hatfield. A high-resolution CMOS time-to-digital converter utilizing a vernier delay line. *IEEE Journal of Solid-State Circuits*, 35(2):240–247, 2000.
50. W. Zhao and Y. Cao. New generation of predictive technology model for sub-45 nm early design exploration. *IEEE Transactions on Electron Devices*, 53(11):2816–2823, 2006.

3

Design of Alkali-Metal-Based Half-Heusler Alloys Having Maximum Magnetic Moments from First Principles

Lin H. Yang, R.L. Zhang, Y.J. Zeng, and C.Y. Fong

CONTENTS

3.1 Introduction

Half-Heusler alloys have a $C1_b$ structure, which is a variation of the full-Heusler crystalized in the $L2_1$ structure, as schematically shown in Figure 3.1. The Bravais lattice is a face-centered-cubic (fcc) lattice shown as the outer cube in Figure 3.1. The notations of the atomic positions in a unit cell are adopted from Wyckoff (1963). For the $C1_b$ structure, three elements occupy 4a, 4b, and 4c sites, respectively, while 4d sites are vacant. The first $C1_b$ alloy studied was composed of two transition-metal (TM) elements denoted by X (Ni) and Y (Mn) and one pnictogen (Sb) designated as Z (de Groot *et al.*, 1983). Later, alloys with X, a simple metal, were considered (Kieven *et al.*, 2010; Jungwirth *et al.*, 2011; Roy *et al.*, 2012).

To search for other potential half-metallic candidates in half-Heusler alloys, Damewood *et al.* (2015) replaced one of the TM elements, the said X, by an alkali-metal element and the pnictogen (Z) by a Group IV element. The arrangements of the three atoms (X, Y, and Z) at three different sites – 4a, 4b, and 4c (Figure 3.1) – in the $C1_b$ structure define three phases, the alpha (α), beta (β), and gamma (γ) phases (Larson *et al.*, 2000) (Table 3.1).

The unique feature of these half-Heusler alloys is to exhibit half-metallicity (de Groot *et al.*, 1983) in their electronic and magnetic properties. The half-metallicity is characterized by showing metallic properties in one spin channel and insulating behaviors in the opposite spin states. Consequently, the conduction current in these alloys is expected to have a 100% spin polarization according to the Julliére (1975) formula:

$$P = {(N_m - N_i)}\big/{(N_m + N_i)}$$

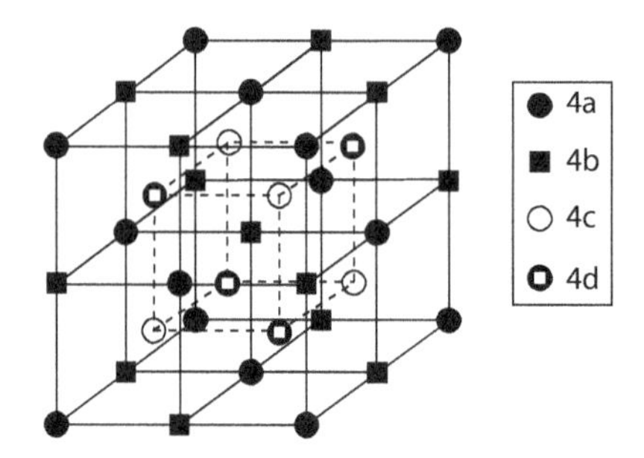

FIGURE 3.1
The $L2_1$ structure. Wyckoff's notations of atomic positions are expressed in rectangular coordinates: 4a=$(0,0,0)a_0$, 4b$=(1/2,1/2,1/2)a_0$, 4c$=(1/4,1/4,1/4)a_0$, and 4d$=(3/4,3/4,3/4)a_0$, where a_0 is the lattice parameter of the large cube. The $C1_b$ structure shares the same picture except 4d sites are vacant.

TABLE 3.1
Arrangements of X, Y, and Z in α, β, and γ Phases

Phase	4a	4b	4c
α	Z	Y	X
β	X	Y	Z
γ	Z	X	Y

Note: The 4d sites are vacant (not listed here).

where P is the spin polarization at the Fermi energy, E_F, so that the spin carrying mobile carriers can be easily transported. In this representation, Julliére defined P in terms of the density of states (DOS) at E_F; N_m represents the DOS of the metallic channel; and N_i is the corresponding DOS of the insulating channel. For half-metals, N_m is finite while N_i is zero because E_F falls in the insulating gap. The spin polarization at E_F, P, is therefore 1. In principle, they can be ideal materials for spintronic devices.

To predict whether a half-Heusler alloy can be a half-metal, two criteria related to its electronic and magnetic properties must be satisfied. The first one concerns the DOS at E_F. The E_F should intersect at least one band in the metallic states and fall in the gap of the insulating spin channel. The second one is the integer value of the magnetic moment/unit cell. Because the valence states in the insulating spin channel are completely occupied, the occupation number should be an integer. Adding to the fact that the total number of electrons is an integer, the remaining number of electrons in the other spin channel should also be integers. The resulting net electrons contribute their spins to the moment. With the electronic spin *g-factor*, g_s to be 2, the predicted value of the magnetic moment/unit cell should strictly be an integer in units of μ_B, where μ_B is the Bohr magneton.

After the seminal publication by de Groot *et al.* (1983), tremendous efforts have been devoted to realizing spintronic devices fabricated by half-Heusler alloys and to searching for new alloys based on theoretical predictions and experimental measurements. However, devices designed to be made of half-Heusler alloys are still illusive due to the complications of growing flawless samples (Otto *et al.*, 1987) and other factors discussed in chapter 3 of the monograph by Fong *et al.* (2013). Although polycrystalline samples of NiMnSb have been reported (Gardelis *et al.*, 2004), a quality single crystal of NiMnSb, on the other hand, has not appeared in the literature. Earlier predictions listed in the monograph (Fong *et al.*, 2013) were not helpful because the stability issue of the half-Heusler alloys was not addressed at the time. Recently, this issue has been studied by Damewood *et al.* (2015) and

Zhang *et al.* (2017). By comparing the acoustic phonon spectra of half-Heusler alloys to the ones in half-metals with a zinc blende structure, they have concluded that half-Heusler alloys are stable even when not at their optimized lattice constants at $T = 0$ K. One should expect, however, that in such samples there will be some stresses that can affect the lifetime of the devices made from these alloys. To reduce the stress, it will be more effective to first carry out a theoretical search of alloys exhibiting half-metallic properties at or near their optimized lattice constants instead of using a trial-error approach.

We can now summarize the criteria for spintronic applications using half-Heusler. They are (a) to have half-metallic properties at lattice constants at or near equilibrium at $T = 0$ K; (b) to have the largest magnetic moment possible for a 3d TM element; and (c) to have the largest moment exhibiting at or near the optimized lattice constant at $T = 0$ K. Among the three conditions, the last one is the most challenging.

In this review, we focus on the following issue: How do we overcome the challenge? This is accomplished by (a) choosing the alkali-metal atoms for the non-TM element in the $C1_b$ structure; and (b) applying the Pauli principle to half-Heusler alloys to have the largest possible magnetic moment (5 μ_B) for 3d TM elements. We restrict our study of half-Heulser alloys to have Cr as the TM element in the β and γ phases because these two phases have lower free energies than the α phase. We will present our findings of 9 out of 30 β- and γ-phase alloys that show favorable half-metallicity and have the largest magnetic moments at their respective equilibrium lattice constants.

In Section 3.2, we present guidelines deduced from physics for how to choose the three elements and how to attain the largest magnetic moment for the alloys. In Section 3.3, a method of calculation will be presented. Results and discussion will be given in Section 3.4. Finally, a summary will be given in Section 3.5.

3.2 Guiding Principles of Designing the Half-Heusler Alloys

Our choices of the three elements in the primitive cell of the half-Heusler alloys are guided by the following physics: (a) the nearest-neighbor configurations of the non-metal atom with the 3d TM element under the tetrahedral environment governing the charge transfer between the two atoms to form bonds; (b) the requirement of the strong electro-negativity of the non-metal atoms on the alkali-metal elements to force a complete charge transfer from the metal atom to the non-metal atom; and (c) the Pauli principle dictates the volume required to align the spins at the TM site.

To address the first guiding principle, we first realize that the arrangement of atoms in the III-V compound semiconductors is closely related to the one in the half-Heusler alloys. A III-V compound, e.g. GaAs, has a zinc blende structure as shown in Figure 3.2. Each atom in the zinc blende structure is surrounded by four neighbors of different species forming the tetrahedral environment. The s- and p-states of the cation under the tetrahedral environment hybridize with the s- and p-states of a possibly different principal quantum number of its neighbors to form the sp^3 covalent bonds. It is clearly seen that the $C1_b$ structure (Figure 3.1) and the zinc blende structure (Figure 3.2) should share some similarity in their bonding features. In Figure 3.3, we show the charge distributions of the β-phase LiMnSi and the zinc blende MnSi; both crystals are half-metals. The sections are defined by the solid lines cutting through the (110) plane in Figure 3.2. The bond charges between Mn and Si are shown and labeled. Their orientations and locations show close similarity

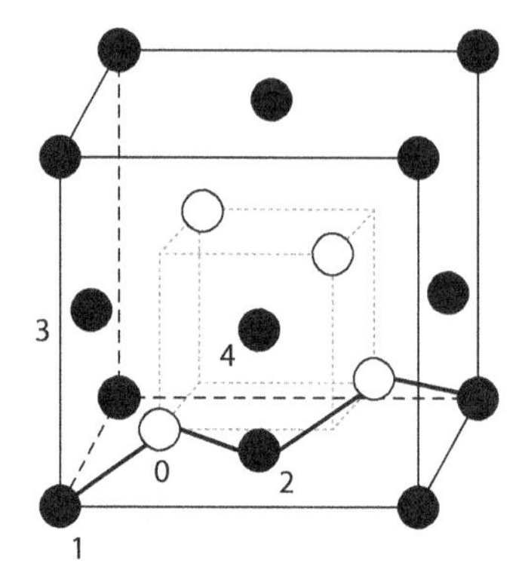

FIGURE 3.2
The zinc blende structure. The filled circles correspond to 4a atoms and the open circles are equivalent to 4c atoms in Figure 3.1. The solid lines indicate where the bonds formed between the atoms. The nearest neighbors of the atom 0 (open circle) are indicated by numbers 1–4.

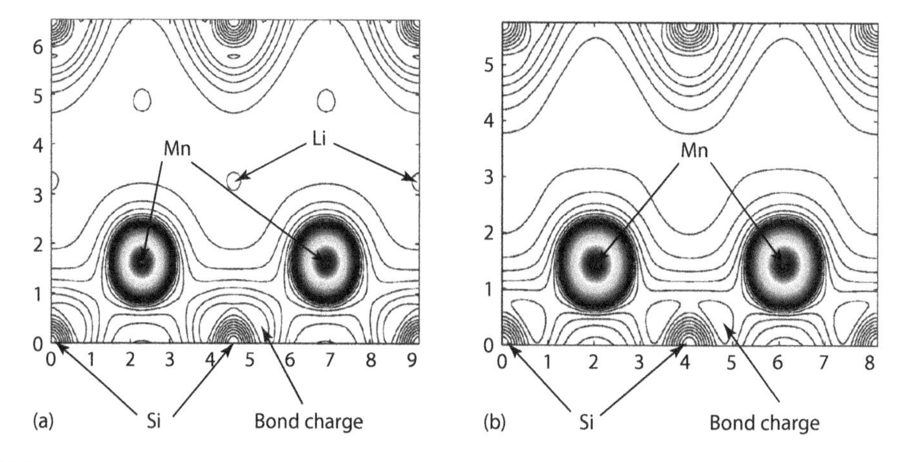

FIGURE 3.3
(a) Charge distribution of β-LiMnSi in a section defined by the solid lines shown in Figure 3.2. (b) Charge distribution of MnSi in the same section. The bond charges for both cases are seen. The shaded regions are the high densities from the Mn d-states.

for the two compounds. In Figure 3.3a, the label "Li" indicates the location of the Li atom in LiMnSi. There is no charge around the Li atom. Its electron is completely transferred to the non-metallic atom.

To further understand the guiding principles, we put forward the notion of electro-negativity. Physically, an element having a strong electro-negativity attracts electrons from its nearest neighbors in a solid to form bonds. Most of these elements are located at the right of the Group IV elements (C, Si, and Ge), such as pnictides, chalcogens, and halides in the periodic table. With respect to the alkali-metal and the TM elements, these species including Group IV elements are called "non-metal elements," which have the strongest electro-negativities among the three species in the compound. We therefore concluded that the important role played by "electro-negativity" in full-Heusler alloys composed of two TM elements – the bonding to hold the crystals together – is by virtue of the nearest-neighbor configuration of one of the TM elements with the non-metal element (Shaughnessy *et al.*, 2013).

In a half-Heusler alloy, the consequences of transferring electrons from the TM element to its nearest-neighbor non-metal element are twofold: (a) to fill the outer electronic shell of the non-metal element; and (b) to determine the magnetic moment of the alloy by the remaining electrons at the TM site. The transferred electrons do not contribute to any magnetic moments of the alloys because their spins are paired due to bond formations. These results suggest that the magnetic moment of a half-Heusler alloy can be simply accounted for by counting the number of electrons being transferred from the TM element and the number of remaining electrons at the site where the TM element is located. This fact has been recognized by Schwarz (1986) who proposed the so-called "ionic model" and has been used by Damewood *et al.* (2015). Based on the physics governed by "electro-negativity," the desired half-Heusler alloys should have the TM atom as the nearest neighbor to the non-metal atom with strong electro-negativity and should have the values of the magnetic moment determined by the ionic model.

In this review, we decided to take Cr as the TM element. Its valence is 6. To have the largest magnetic moment of the alloy, only one electron should be transferred to its neighboring non-metal atom. We then chose an alkali element as the nearest neighbor to the non-metal element to stabilize the crystalline structure. This alkali-metal offers one electron to the most electro-negative element. Thus, we chose Group VI atoms as the non-metal element in the alloys, which takes one electron each from the alkali-metal and TM element to attain the largest magnetic moment/unit cell.

We now turn to the third guiding principle. Depending on the open space or volume available around the TM element, the value of the magnetic moment can vary. The reason is the Pauli principle – electrons with parallel spins need a large volume to avoid each other – they should not occupy the same spatial region. We recognized this physical fact in the study of MnC, a half-metal with the zinc blende structure (Qian *et al.*, 2004). In MnC, its moment/unit cell is 3 μ_B at a larger lattice constant (5.0 Å), while at a smaller lattice constant (4.0 Å) the moment reduces to only 1 μ_B. The larger value is consistent with the prediction of the ionic model but not the one for the smaller lattice constant. The smaller volume causes one of the spins to flip due to the Pauli principle. In order to achieve the maximum magnetic moment for an alloy, we should take the Pauli principle into consideration. However, one should be aware that the free energy of each compound can be influenced by the bonding strength and the exchange interaction as a function of lattice constants. In order to have the optimized lattice constant (at $T=0$ K) and the largest magnetic moments, this has become a challenging optimization problem.

To address the optimization problem, we have applied the guiding principles in α-LiCrS alloy (Zhang *et al.*, 2016). It is the first alloy to demonstrate the power of the guiding principles and to meet the optimization challenge. In this alloy, the S atom has the strongest electro-negativity and thus needs two electrons to fill its $n=3$ electronic shell. The TM atom, Cr, is located at $(1/2, 1/2, 1/2)a_0$, where $a_0=6.0$ Å is the lattice constant of the corresponding fcc cell. With the S atom at $(1/4, 1/4, 1/4)a_0$, the nearest-neighbor configuration causes the Cr atom in principle to lose two of its electrons. That would result in four d-electrons remaining at the Cr site to give the magnetic moment of 4 μ_B. In order to increase the magnetic moment, we then chose Li, the least electro-negative element, as the other neighbor to the S atom, so the two required transferred electrons are equally contributed by Li and Cr atoms – one electron for each element.

Consequently, we found that α-LiCrS has 5 μ_B at its optimized lattice constant at $T=0$ K (Zhang *et al.*, 2017). Unfortunately, compared to the other two phases, this α phase has the highest free energy at $T=0$ K and may not be favorable to grow. Therefore, it is worth the effort to find other half-Heusler alloys in the β and γ phases because they have comparable lower free energies than the α phase at the same lattice constants. These two phases have the potential to be grown in polycrystalline thin-film forms with minimum lattice mismatch at the interface.

3.3 Method of Calculation

To find the desired half-metallic half-Heusler alloys, the fcc primitive cell was used for the two phases (β and γ alloys). There is one TM atom per unit cell. The spin-polarized version of the Vienna ab-initio simulation package (VASP) (Kresse and Furthmüller, 1996) with projector-augmented-wave (PAW) pseudopotentials (Blöchl, 1994) for the elements was used in this review. The generalized gradient approximation (GGA) of Perdew *et al.* (1996) was used to treat the electron–electron exchange–correlation interactions in the PAW pseudopotentials and crystalline calculations. We used plane-wave basis functions with a 1200 eV kinetic energy cutoff, E_{cut}, for all calculations. The Monkhorst and Pack (1976) mesh of (15, 15, 15) was used to calculate the total charge density. The convergences of the total energy and the magnetic moment for each studied alloy are better than 1.0 meV and 1.0 mμ_B, respectively. The values of the E_{cut} and the mesh points should be checked for the convergences of the two quantities. Otherwise, interesting physics will be missed (Zhang *et al.*, 2017).

3.4 Results and Discussion

In Table 3.2, we tabulate first our results according to the order of the β and γ phases of the half-Heusler alloys exhibiting the desired properties – maximum magnetic moment per unit cell, $M = 5\ \mu_B$, at the optimized lattice constant. In addition, the following information is also given in Table 3.2: (a) the range of lattice constants and the corresponding optimized

TABLE 3.2

Range of Lattice Constants (a), Equilibrium Lattice Constant (a_0), Magnetic Moment (M), Free Energy (F), and Energy Gap (E_g) in the Insulating Channel for the Alloys with Integer Magnetic Moments

Alloy	a (Å)	a_0 (Å)	M (μ_B)	F (eV)	E_g (eV)	E_c (eV)	E_v (eV)
β-CsCrS	6.50–8.00	7.35	5	−13.492	1.926	0.107	−1.819
β-CsCrSe	6.40–7.68	7.58	5	−13.062	1.758	0.003	−1.755
γ-KCrS	6.50–7.00	6.65	5	−14.030	3.025	0.251	−2.774
γ-KCrSe	6.40–6.96	6.86	5	−13.692	2.619	0.026	−2.593
γ-CsCrO	6.15–7.70	6.49	5	−13.500	2.153	0.621	−1.532
γ-CsCrS	6.70–7.50	7.06	5	−13.461	2.519	0.582	−1.937
γ-CsCrSe	6.98–7.80	7.27	5	−13.150	2.398	0.593	−1.805
γ-RbCrS	6.40–8.00	6.83	5	−13.710	2.866	0.414	−2.452
γ-RbCrSe	6.60–7.40	7.04	5	−13.390	2.382	0.139	−2.243

Note: The bottom of conduction band energy (E_c) and the top of valence band energy (E_v) relative to Fermi energy (set to be zero) are also listed in the last two columns.

one at $T=0$ K for each alloy that shows the largest integer magnetic moments; (b) the magnetic moment, M; (c) the energy gap, E_g, in the insulating channel; and (d) the energies at the bottom of the conduction bands and the top of the valence bands measured with respect to E_F. We list E_c and E_v relative to E_F to examine whether the spin–orbit interaction can be neglected to maintain their half-metallicity in these alloys.

For the β phase, only two Cs-based half-Heusler alloys, β-CsCrS and β-CsCrSe, are predicted to have maximum magnetic moments at their corresponding equilibrium lattices at $T=0$ K. Their lattice constants are larger than 7.0 Å as a result of the Pauli principle. This means that the Cr–S bond length is more than 3.031 Å, so there is enough space around Cr to align its spin moments. There are, however, seven favorable γ-phase K-, Cs- and Rb-based half-Heusler alloys. Among the seven cases, two of them (γ-CsCrS and γ-CsCrSe) can compete with their counterparts in the β phase. The β-phase CsCrS is lower in free energy by 0.031 eV than the γ phase, while the β-phase CsCrSe is higher in energy than the γ phase by 0.088 eV. We therefore expect that the growth of these two half-Heusler alloys can have mixed phases.

In each phase, the optimized lattice constants, a_0, for the S-composed alloys are smaller than the Se ones. This demonstrates a combination of stronger electro-negativity and smaller ionic radius of the S atom. Except the β-CsCrS and β-CsCrSe as well as the γ-KCrS and γ-KCrSe, we can roughly state that the range of lattice constants for the alloys to have the largest magnetic moment in a unit cell is consistent with the ionic radii of the Group VI elements. Another general statement can be made: the ionic radii of three alkali-elements, K, Cs, and Rb, are larger than Na and Li to form half-Heusler alloys having the most favored half-metallic properties. This is another manifestation of the role played by the Pauli principle. To examine whether the spin–orbit interactions have an effect on the desired half-metallic properties, Table 3.2 lists the values at the top of the valence bands (E_v) and the bottom of the conduction bands (E_c) relative to the Fermi energies, which are set to zero. Based on the conclusion given by Zhang *et al.* (2017): the spin–orbit interaction is negligible when both E_c and E_v are more than 0.1 eV from E_F. We expect that, except β-CsCrSe and γ-KCrSe, the spin–orbit effect on these half-Heusler alloys is negligible (Zhang *et al.*, 2016).

Table 3.3 lists those alloys that have integer magnetic moments in the range not covering their respective equilibrium lattice constants at $T=0$ K. For comparison, their magnetic moments at their equilibrium lattice constants are also listed. They are all in the β phase. We suggest that these alloys can be grown in layered forms.

The remaining 17 alloys predicted not to have half-metallic properties are listed in Table 3.4. In general, these alloys have smaller equilibrium volumes due to the presence of Li and Na, respectively. The unfavorable predictions are understandable based on the Pauli principle.

TABLE 3.3

Equilibrium Lattice Constant (a_0), Magnetic Moment (M), Free Energy (F), and the Range of Lattice Constants (a) Have Integer Magnetic Moments for Half-Heusler Alloys

Alloy	a_0 (Å)	M (μ_B)	F (eV)	a (Å)	M (μ_B)
β-CsCrO	6.50	4.9996	−14.353	5.50–6.40	5
β-KCrS	6.80	4.9982	−14.388	6.50	5
β-RbCrSe	7.30	4.9952	−13.472	6.30–6.60	5
β-RbCrS	7.00	4.9996	−13.936	6.40–6.60	5

TABLE 3.4

Equilibrium Lattice Constant (a_0), Magnetic Moment (M), and Free Energy (F) for Alloys that Do Not Have the Desired Half-Metallic Properties

Alloy	a_0 (Å)	M (μ_B)	F (eV)
β-RbCrO	6.40	4.9716	−14.699
β-LiCrO	5.50	4.8268	−17.666
β-NaCrS	6.30	4.9819	−15.310
β-LiCrSe	6.10	4.7110	−15.691
β-KCrO	6.50	4.9463	−14.931
β-NaCrO	6.40	4.8481	−15.374
β-LiCrS	6.40	4.9817	−16.064
β-KCrSe	7.00	4.9934	−13.885
β-NaCrSe	6.50	4.9763	−14.670
γ-LiCrO	5.50	4.8892	−16.156
γ-RbCrO	6.40	4.9973	−13.633
γ-NaCrS	6.30	4.9817	−14.676
γ-LiCrSe	6.20	4.8972	−15.217
γ-KCrO	6.50	4.9761	−13.763
γ-NaCrO	6.40	4.8642	−14.169
γ-LiCrS	6.40	4.9809	−15.405
γ-NaCrSe	6.50	4.9729	−14.256

3.5 Summary

We have carried out systematic studies in an attempt to search for alkali-metal-based and Cr-related half-Heusler alloys with half-metallic properties desirable for spintronic applications. In order to have their half-metallicity and the maximum magnetic moments happening at their optimized lattice constants, the guiding principles of selecting the elements based on physics were given. A total of 30 cases were studied, 9 of which show the desired properties. Among them, β-CsCrS is best and is worth growing. Two of them, β-CsCrSe and γ-KCrSe, may have a large spin–orbit effect due to E_F and the bottom of the conduction bands is close in energy to mix the up-and-down spin states. The growth of the remaining six alloys may have mixed phases. Four β-phase alloys showing the largest magnetic moment/unit cell at lattice constants away from the respective optimized value should be grown in thin-film forms. The remaining 17 cases may not be worth trying. They mostly involve smaller alkali-metal elements causing smaller lattice constants and do not exhibit half-metallic properties attributed to the role played by the Pauli principle.

Acknowledgments

Work at Lawrence Livermore National Laboratory was performed under the auspices of the U.S. Department of Energy by Lawrence Livermore National Laboratory under Contract DE-AC52-07NA27344. RLZ was supported by grants from the National Natural Science Foundation of China (Grant No. 10904061) and China Scholarship Council. Work at UC Davis was supported in part by the National Science Foundation (Grant No. ECCS-0725902).

References

Blöchl, P. E., 1994. Projector augmented-wave method. *Phys. Rev. B*, Volume 50, p. 17953.

Damewood, L. *et al.*, 2015. Stabilizing and increasing the magnetic moment of half-metals: The role of Li in half-Heusler LiMn Z (Z = N, P, Si). *Phys. Rev. B*, Volume 91, p. 064409.

de Groot, R. A., Mueller, F. M., van Engen, P. G. & Buschow, K. H. J., 1983. New class of materials: Half-metallic ferromagnets. *Phys. Rev. Lett.*, Volume 50, p. 2024.

Fong, C. Y., Pask, J. E. & Yang, L. H., 2013. *Half-Metallic Materials and Their Properties*. London: Imperial College Press.

Gardelis, S. *et al.*, 2004. Synthesis and physical properties of arc melted NiMnSb. *J. Appl. Phys.*, Volume 95, p. 8063.

Julliére, M., 1975. Tunneling between ferromagnetic films. *Phys. Lett. A*, Volume 54, p. 225.

Jungwirth, T. *et al.*, 2011. Demonstration of molecular beam epitaxy and a semiconducting band structure for I-Mn-V compounds. *Phys. Rev. B*, Volume 83, p. 035321.

Kieven, D. *et al.*, 2010. I-II-V half-Heusler compounds for optoelectronics: Ab initio calculations. *Phys. Rev. B*, Volume 81, p. 075208.

Kresse, G. & Furthmüller, J., 1996. Efficient iterative schemes for ab initio total-energy calculations using a plane-wave basis set. *Phys. Rev. B*, Volume 54, p. 11169.

Larson, P., Mahanti, S. D. & Kanatzidis, M. G., 2000. Structural stability of Ni-containing half-Heusler compounds. *Phys. Rev. B*, Volume 62, p. 12754.

Monkhorst, H. J. & Pack, J. D., 1976. Special points for Brillouin-zone integrations. *Phys. Rev. B*, Volume 13, p. 5188.

Otto, M. J. *et al.*, 1987. Electronic structure and magnetic, electrical and optical properties of ferromagnetic Heusler alloys. *J. Mag. and Mag. Mat.*, Volume 70, p. 33.

Perdew, J. P., Burke, K. & Ernzerhof, M., 1996. Generalized gradient approximation made simple. *Phys. Rev. Lett.*, Volume 77, p. 3865.

Qian, M. C., Fong, C. Y. & Yang, L. H., 2004. Coexistence of localized magnetic moment and opposite-spin itinerant electrons in MnC. *Phys. Rev. B*, Volume 70, p. 052404.

Roy, A., Bennett, J. W., Rabe, K. M. & Vanderbilt, D., 2012. Half-Heusler semiconductors as piezoelectrics. *Phys. Rev. Lett.*, Volume 109, p. 037602.

Schwarz, K., 1986. CrO_2 predicted as a half-metallic ferromagnet. *J. Phys. F: Met. Phys.*, Volume 16, p. L211.

Shaughnessy, M. *et al.*, 2013. Structural variants and the modified Slater-Pauling curve for transition-metal-based half-Heusler alloys. *J. Appl. Phys.*, Volume 113, p. 043709.

Wyckoff, R. W. G., 1963. *Crystal Structures*. 2nd ed. San Francisco: John Wiley & Sons.

Zhang, R. L. *et al.*, 2016. A half-metallic half-Heusler alloy having the largest atomic-like magnetic moment at optimized lattice constant. *AIP Advances*, Volume 6, p. 115209.

Zhang, R. L. *et al.*, 2017. Two prospective Li-based half-Heusler alloys for spintronic applications based on structural stability and spin-orbit effect. *J. Appl. Phys.*, Volume 122, p. 013901.

Zhang, R. L., Fong, C. Y. & Yang, L. H., 2017. unpublished.

4

Defect-Induced Magnetism in Fullerenes and MoS_2

Sinu Mathew, Taw Kuei Chan, Bhupendra Nath Dev, John T.L. Thong, Mark B.H. Breese, and T. Venkatesan

CONTENTS

4.1 Magnetism in Allotropes of Carbon

The observation of magnetism in materials made purely of carbon origin [1–5] has created enormous interest both in basic and applied fields of research. The difference in the magnetic susceptibilities of various carbon allotropes and the influence of particle size on susceptibility were first reported by Sir C.V. Raman and P. Krishnamurthy in 1929 [6]. Bulk crystalline graphite is a strong diamagnet, with magnetic susceptibility second only to superconductors. However, graphite containing certain defects can exhibit long-range magnetic ordering [7]. C.V. Raman reported that "*Diamagnetic susceptibility of graphite is markedly a function of particle size, diminishing steadily with increase in sub-division of the substance*" [6]. Later in 1995, Ishi et al. found that microcrystalline graphite shows ferromagnetic ordering [8,9]. The first organic ferromagnet was reported by Korshak et al. in 1987 [10]. Later in 1991, Ovchinnikov and Shamvosky theoretically predicted the possibility of a ferromagnetically ordered phase of mixed sp^2 and sp^3 pure carbon [11]. Interestingly, the calculated saturation magnetic moment in the sp^2–sp^3 mixed phase is higher than that of pure iron (Fe). Makarova et al. reported ferromagnetism in a two-dimensionally polymerized rhombohedral phase of C_{60} [1]. Highly oriented pyrolitic graphite (HOPG) samples were found to show ferromagnetic (or ferrimagnetic) ordering when irradiated with MeV protons [12,13]. Low energy nitrogen ion (a few keV energy)-implanted nano-diamonds have also been reported to be ferromagnetic [14]. According to a theoretical investigation by Vozmediano et al., proton irradiation can produce large local defects that give rise to the appearance of local moments whose interaction can induce ferromagnetism in a large portion of the graphite sample [7].

Ohldag et al. demonstrated that magnetic order in proton-irradiated metal-free carbon films originates from a carbon π-electron system using x-ray dichroism spectromicroscopy [15]. Ferromagnetic ordering has also been observed in doped C_{60} and microcrystalline carbon samples [16–18]. The discovery of ferromagnetism in rhombohedral C_{60} polymers [2,14] has opened up the possibility of a whole new family of magnetic fullerenes and fullerides.

We have found that magnetism is induced in proton-irradiated C_{60} films [19]. The use of ion beams to produce magnetically active areas points toward the possibility of tailoring the size of such magnets and has potential applications in spin electronics [14,20,21].

4.2 Soft Ferromagnetism in C_{60} Thin Films

Magnetization measurements were performed on pristine and 2 MeV proton-irradiated C_{60} films deposited on hydrogen-passivated Si(111) surfaces [19,22,23]. Magnetic measurements were carried out using a superconducting quantum interference device (SQUID) magnetometer.

The results of magnetization vs field (M–H) measurements on a pristine sample and samples irradiated at a fluence of 6×10^{15} H^+/cm^2 are shown in Figure 4.1.

The magnetization vs field (M–H) measurements at 5 K for the irradiated film show a marked increase in magnetization and a tendency toward saturation (Figure 4.1). A weak remnant magnetization of the order of a few tens of μemu is observed (data not shown).

At 300 K, both the as-deposited and the irradiated sample show diamagnetic behavior. Magnetization data in the temperature range of 2 K–300 K, in a 1 T applied field, for the irradiated film shows much stronger temperature dependence compared with that of the pristine film (Figure 4.2).

The magnetization (M) vs field (H) isotherm of irradiated C_{60} obtained at 5 K (Figure 4.1) clearly shows a tendency toward saturation, which is a signature of ferromagnetism. The hysteresis associated with the M–H curve is feeble, as expected for a soft ferromagnetic material. Similar soft-ferromagnetism was indeed observed earlier in an organic fullerene C_{60} material, namely C_{60} $TDAE_{0.86}$, where the ferromagnetic state showed no remanence [17]. An alternative explanation of the M–H curve for the irradiated sample in Figure 4.1 is that it arises from superparamagnetism. In order to demonstrate that the observed M–H curve is not due to superparamagnetism and the observed magnetic behavior in the irradiated C_{60} film is stable over a long period, we present the results of measurements made

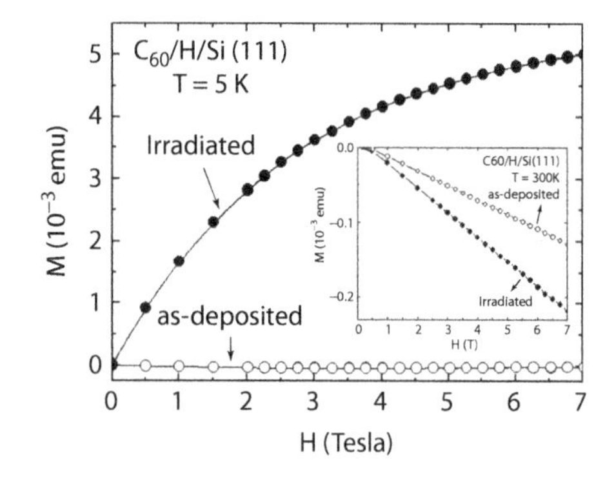

FIGURE 4.1
M vs H at 5 K for an as-deposited and an irradiated C_{60} film after subtracting the substrate [H-Si(111)] contribution (substrate data not shown here). (Reprinted with permission from S. Mathew et al., *Phys. Rev. B* 75, 75426, 2007 copyright (2017) by the American Physical Society.)

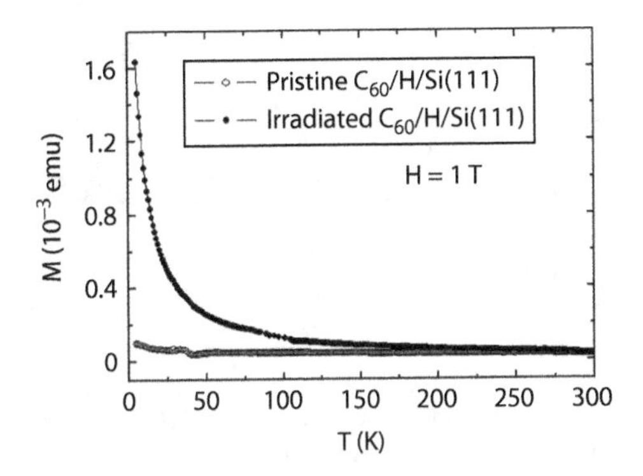

FIGURE 4.2
M vs T for as-deposited and irradiated C_{60} film in an applied field of 1 T (substrate [H-Si(111)] contribution subtracted). (Reprinted with permission from S. Mathew et al., *Phys. Rev. B.* 75, 75426, 2007 copyright (2017) by the American Physical Society.)

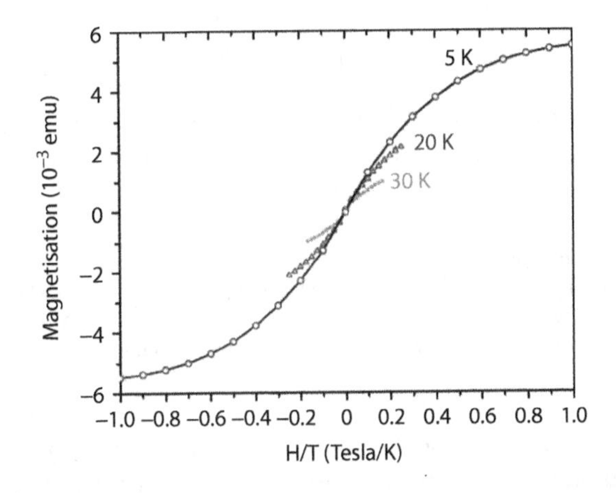

FIGURE 4.3
M vs H/T at 5 K, 20 K and 30 K for the same irradiated C_{60} film as in Figure 4.2 measured again one year after the first measurement (Figure 4.1). (Reprinted with permission from S. Mathew et al., *Phys. Rev. B* 75, 75426, 2007 copyright (2017) by the American Physical Society.)

on the same sample one year after the original measurements (Figure 4.3) had been made. Measurements at three different temperatures (5 K, 20 K and 30 K) have been made to test one criterion for superparamagnetism, that is M–H isotherms should scale as H/T [24]. The magnetization vs H/T curves for different temperatures are shown in Figure 4.3 and from the figure it is clear that they do not superimpose. These isotherms should superimpose for paramagnetic or superparamagnetic materials. From these results, in conjunction with the observed feeble remnant magnetization, we conclude that the irradiated C_{60} films are soft ferromagnets. The results also show that the ferromagnetic behavior in the irradiated film is stable over a long period of time.

The effects of the foregoing irradiation on the nature of the fullerene cage structure and the atomic arrangements can be understood using Raman spectra and high-resolution transmission electron microscope (HRTEM) images. Raman spectroscopy results for a pristine and an irradiated sample at a fluence of 6×10^{15} ions/cm^2 are given in Figure 4.4. An enlarged part of the spectrum in the range of 1400–1550 cm^{-1} is given in Figure 4.4 (b) and (c). In the pristine sample, two dominant peaks are seen at 1458 and 1467 cm^{-1}. In the irradiated sample,

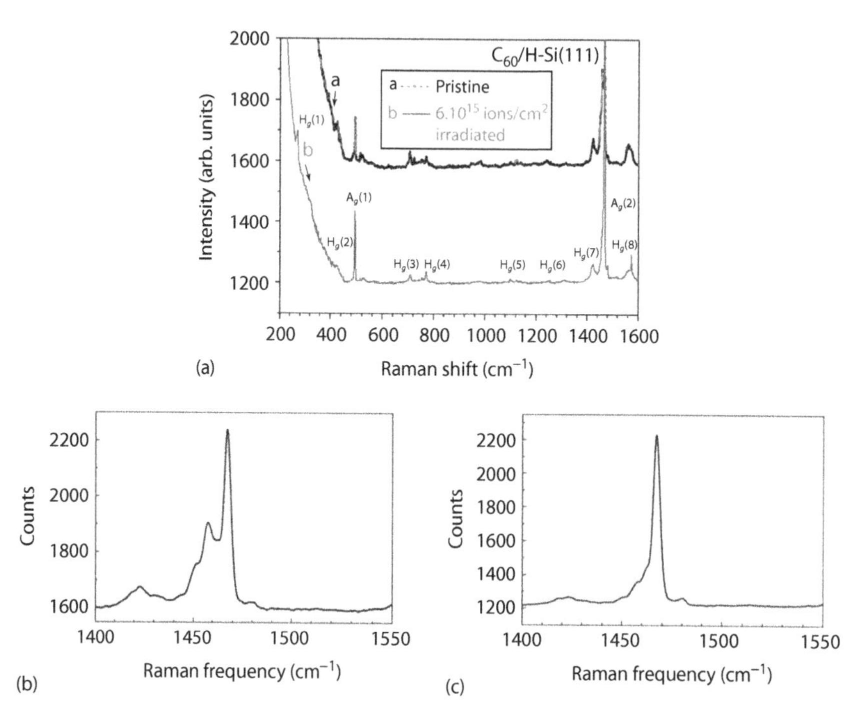

FIGURE 4.4
Raman spectrum of (a) pristine and C_{60}/H-Si(111) film irradiated with 2 MeV H^+ at a fluence of $6\times10^{15}/cm^2$. An enlarged part of $A_g(1)$ mode of pristine and irradiated spectra in (a) is given in (b) and (c). (Reprinted with permission from S. Mathew et al., *Phys. Rev. B.* 75, 75426, 2007 copyright (2017) by the American Physical Society., and S. Mathew et al., 2007, Two mega-electron volt proton-irradiation effects on fullerene films, *Radiat Eff Defect S*, Nov. 2007, Taylor & Francis.)

the first peak height is reduced while the second is enhanced. Sauvajol et al. [25] showed that the appearance of a mode at 1458 cm^{-1} was due to phototransformation (partial polymerization) of pure C_{60}. However, in the irradiated film, the intensity of this mode is reduced along with an enhancement of the 1467 cm^{-1} vibrational mode. The reduction in the intensity of the 1458 cm^{-1} mode in the irradiated sample in Figure 4.4 may be due to the fact that irradiation has caused enough disorder so that the fraction of the phototransformed component is reduced in comparison with the pristine film. Another explanation for the appearance of the mode at 1458 cm^{-1} is given by Duclos et al. [26]. They showed that a C_{60} film grown and kept under vacuum shows only a broad peak at 1458 cm^{-1}. Exposure of the film to O_2 or air gives rise to the peak at 1467 cm^{-1}. Although the authors could not confirm the origin of this peak, they tend to believe that it lies in the structural perturbation of the C_{60} molecule itself. This conclusion was based on, besides other reasons, the fact that a total of 16 modes were observed, while the icosahedral symmetry of C_{60} predicts only 10, signifying a lowering of the symmetry of the C_{60} molecule in the O_2- or air-exposed film. With ion irradiation, a fraction of C atoms is displaced from the C_{60} cage and thus contributes to symmetry lowering. The displacement of C atoms might be responsible for the enhancement of the intensity of the peak at 1467 cm^{-1} in the irradiated film. We believe this is why the intensity of the peak at 1467 cm^{-1} increases in the irradiated film.

The perturbation of the C_{60} molecular cage can be inferred from the HRTEM image in Figure 4.5 (b), although the overall crystalline structure, such as the crystalline planes and the diffraction spots (Figure 4.5), remain that of the fcc C_{60} crystals.

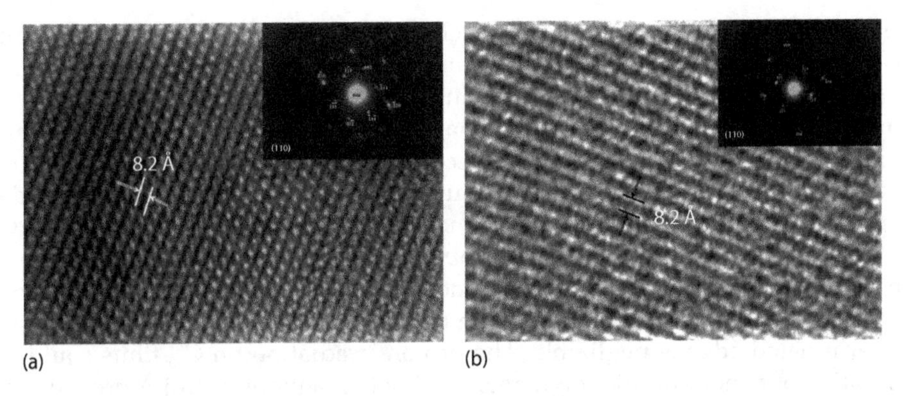

FIGURE 4.5
A HRTEM lattice image from (a) an as-deposited C_{60} film and (b) an irradiated C_{60} film. The corresponding selected area transmission electron diffraction (TED) patterns are given in the insets, respectively. The (111) planar spacing of fcc fullerene is marked in (a) and (b). (Reprinted with permission from S. Mathew et al., *Phys. Rev. B* 75, 75426, 2007 copyright (2017) by the American Physical Society.)

HRTEM lattice images of as-deposited and irradiated samples with corresponding diffraction patterns are shown in Figure 4.5. The transmission electron diffraction (TED) and HRTEM images confirm the crystalline nature of the C_{60} film in the fcc structure. Twin structures are seen in Figure 4.5 (a). The presence of defect structures in the irradiated film is evident from Figure 4.5 (b). In the case of the irradiated film (Figure 4.5 (b)), although the overall crystalline structure, such as the crystalline planes and the diffraction spots, remain that of fcc C_{60} crystals, one can clearly observe the perturbation of the C_{60} molecular cages.

The magnetic moment observed for the irradiated sample in an applied field of 7 T is about 5×10^{-3} emu with a tendency toward saturation. The magnetization curve of the irradiated film in high fields at 5 K compared with that of the as-deposited film and the empty substrate (the latter not shown here) gives clear evidence for the irradiation-induced magnetism in C_{60} films. The total amount of magnetic impurities (Fe, Cr, Ni) was determined by post-irradiation proton-induced x-ray emission (PIXE) experiments and was estimated to be ~50 ppm; the maximum magnetic moment contribution due to all of these impurities in our film will be less than 5×10^{-7} emu. Thus, the contribution to the observed magnetization due to these impurities is negligible.

The range of 2 MeV protons, calculated using the *stopping and range of ions in matter* (SRIM) [27] simulation code for an amorphous carbon target having the density of C_{60}, is found to be ~50 μm. Since our C_{60} film is only ~1.9 μm thick, the protons pass through the film and are buried deep in the Si substrate. The total energy loss of the proton beam in the present 1.9 μm thick C_{60} film is ~45 keV. The energy loss of protons at the top and bottom of the C_{60} film are 24.4 and 24.8 eV/nm, respectively. As the proton energy loss is nearly uniform over the whole thickness of the film, it is reasonable to assume that the irradiation damage is uniformly produced throughout the whole thickness of the film. We have used the total thickness of the film in order to determine the magnetization value in emu/g. The magnetization curve of the irradiated sample in Figure 4.1 has a tendency toward saturation at high fields. The magnetization at 7 T is about 200 emu/g. In the proton-irradiated C_{60} films, although defects are created, the observation of ordered periodic lattice fringes in the irradiated sample and the corresponding TED pattern indicate that irradiation did not cause the disintegration of the C_{60} cage leading to amorphization. Regarding the mechanism for the formation of the magnetic state in all carbon systems, among others, the defect-mediated mechanism appears to be the most general one. The defect-mediated mechanism has been addressed in a number of publications [1,2,7]. The possible origin of magnetism in these irradiated C_{60} films could be (i) irradiation-induced carbon vacancy in the system could lead to a singly occupied dangling sp^2 orbital that can give rise to a magnetic moment; (ii) the presence of nanographitic fragments that have zigzag edges could lead to splitting up of the flat energy bands and lowering the energy of the spin-up band than the spin-down band and hence the appearance of ferromagnetism in the material. Recent theoretical studies have predicted that magnetic ground states are stable in the edges of isolated graphene sheets whether they are hydrogen passivated or not [28]. In the present work, 2 MeV protons (range 50 μm) were used to irradiate a film of thickness 2 μm and hence the possibility of a H-terminated edge is negligible. The proton irradiation on C_{60} films could lead to the formation of nanographitic fragments leading to magnetism. (iii) A broken cage-like structure of polymerized fullerene and consequent broken interfragment C–C bonds could also lead to a magnetic moment. However, the Raman spectrum of irradiated film does not reveal any significant damage to the cage-like structure of C_{60}. The possible presence of magnetic moments in a hexagonal polymeric C_{60} layer has been predicted [1]. In this prediction, the magnetic moments would apparently arise from the formation of radical centers in polymerized C_{60}, with partially broken intermolecular bonds, without damage to the fullerene cages [29].

Although the details may be different for different carbon systems (graphite, polymeric fullerene, nanotubes, etc.), the common feature is the presence of undercoordinated atoms, such as atoms near vacancies [30] and atoms in the edges of graphene-like nanofragments [29–32]. Ohldag et al. have performed x-ray magnetic circular dichroism (XMCD) measurements on 2.25 MeV proton-irradiated graphitized carbon samples of thickness 200 nm and obtained a ferromagnetic signal [15]. They also demonstrated that the intrinsic origin of magnetism is from the π-electrons of carbon [15]. The clear observation of magnetism in graphitized carbon films of thickness 200 nm indicates that the effects are mainly due to the defects produced during the passage of protons through the film. Ion irradiation of any materials generates vacancies. The enhancement in magnetization observed in H^+-irradiated C_{60} samples may be due to defect moments from vacancies and/or the deformation and partial destruction of the fullerene cage. The HRTEM image in Figure 4.5 points to this possibility. Raman spectroscopy and HRTEM measurements on the irradiated sample show the stability of the fullerene crystal structure under the present irradiation condition. Further studies, such as estimating the presence of nanographitic fragments and carbon vacancies in such systems, will provide deeper insight into the origin of magnetism in the ion-irradiated C_{60} films.

According to a density functional study [31] of magnetism in proton-irradiated graphite [13], it is shown that H-vacancy complex plays a dominant role in the observed magnetic signal. For a fluence of 10 μC, the predicted signal is 0.8 μemu, which is in agreement with the experimental signal [3]. The implanted proton fluence in our sample is 77 μC (6×10^{15} ions/cm^2) and all the protons are buried in silicon. So far, we have not come across any report showing magnetic ordering in proton-irradiated silicon. Even if we assume the same kind of magnetism due to H-vacancy complex in proton-irradiated silicon as in proton-irradiated graphite, the expected magnetic signal would be three orders of magnitude smaller than our observed result. Considering this fact, we can safely ignore the contribution of implanted protons in the Si substrate to the observed magnetism, which is predominantly due to atomic displacements caused by energetic protons while passing through the film.

4.3 Magnetism in MoS_2

4.3.1 Introduction

MoS_2 is one of the central members in transition metaldichalcogenide compounds [33]. Its layered structure, held together with van der Waals forces between the layers, along with its remarkable electronic properties, such as charge density wave transitions in transition metal dichalcogenides, make the material interesting from both fundamental and applied research perspectives [33–35]. Furthermore, monolayer MoS_2 is a semiconducting analog of graphene and has been fabricated [36,37].

An interesting weak ferromagnetism phenomenon in nanosheets of MoS_2 had previously been reported by Zhang et al. who attributed the observed magnetic signal to the presence of unsaturated edge atoms [38]. There have been several theoretical efforts in understanding ferromagnetic ordering in MoS_2. Li et al. predicted ferromagnetism in zigzag nanoribbons of MoS_2 using density functional theory [39]. The formation of magnetic moments was also reported in Mo_nS_{2n} clusters [40], nanoparticles [41] and nanoribbons [42–44] of MoS_2 from first-principle studies. The discovery of ferromagnetism in MoS_2 nano-sheets along with various simulation studies and the above reports on magnetism in carbon allotropes raise the possibility of magnetism in ion-irradiated MoS_2 system.

Here, we present the results of magnetization measurements on pristine and 2 MeV proton-irradiated MoS_2 samples [45].We find that magnetism is induced in proton-irradiated MoS_2 samples. The observation of long-range magnetic ordering in ion-irradiated MoS_2 points toward the possibility of selectively fabricating magnetic regions in a diamagnetic matrix, which may enable the design of unique spintronic devices.

4.3.2 Ferrimagnetism in MoS_2

Samples for irradiation were prepared in the following way. MoS_2 flakes, 2 mm in diameter and ~200 μm in thickness, were glued with diamagnetic varnish onto a high purity silicon substrate. Ion irradiations were carried out at room temperature. Magnetic measurements were performed using a SQUID system (MPMS SQUID-VSM) with a sensitivity of 8×10^{-8} emu.

The results of magnetization vs field (M–H) measurements at 300 K and 10 K for the sample irradiated at a fluence of 1×10^{18} ions/cm^2 and the pristine sample are shown in Figure 4.6. The pristine sample is diamagnetic in nature. The appearance of hysteresis along with a clear remanence and coercivity (~700 Oe at 10 K) and its decrease with increasing temperature in the irradiated sample (Figure 4.6 (A) and (B)) clearly indicate that MeV proton irradiation has induced ferro- or ferrimagnetic ordering in the MoS_2. The observed magnetic ordering can be due to the presence of defects such as atomic vacancies, displacements and saturation of a vacancy by the implanted protons. The same sample was subsequently irradiated at cumulative fluences of 2×10^{18} and 5×10^{18} ions/cm^2 to probe the evolution of induced magnetism with ion fluence.

Magnetizations as a function of field isotherms at a fluence of 5×10^{18} ions/cm^2 are shown in Figure 4.7 (a). An enlarged view of the M–H curves near the origin is given in the inset and the decrease of coercivity with increasing temperature is clear from the plot. A plot of coercivity vs ion fluence at various temperatures is given in Figure 4.7 (b). The value of coercivity is found to increase with ion fluence at all temperatures used

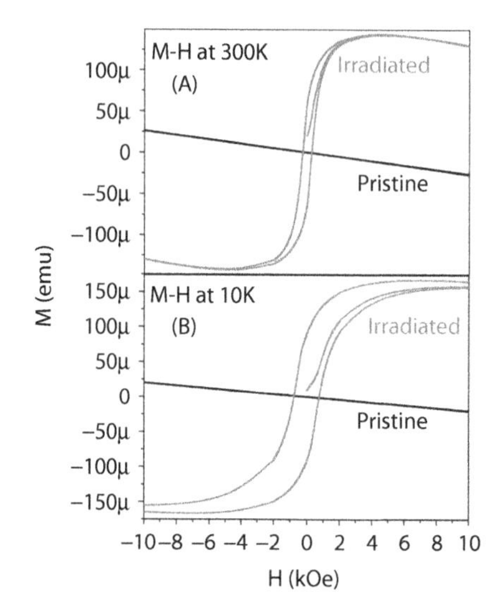

FIGURE 4.6
M vs H curve (A) at 300 K and (B) at 10 K for a pristine and an irradiated MoS_2 after subtracting the substrate Si contribution. (Reproduced from S. Mathew et al., *Appl. Phys. Lett.* 101, 102103, 2012 with the permission of AIP Publishing.)

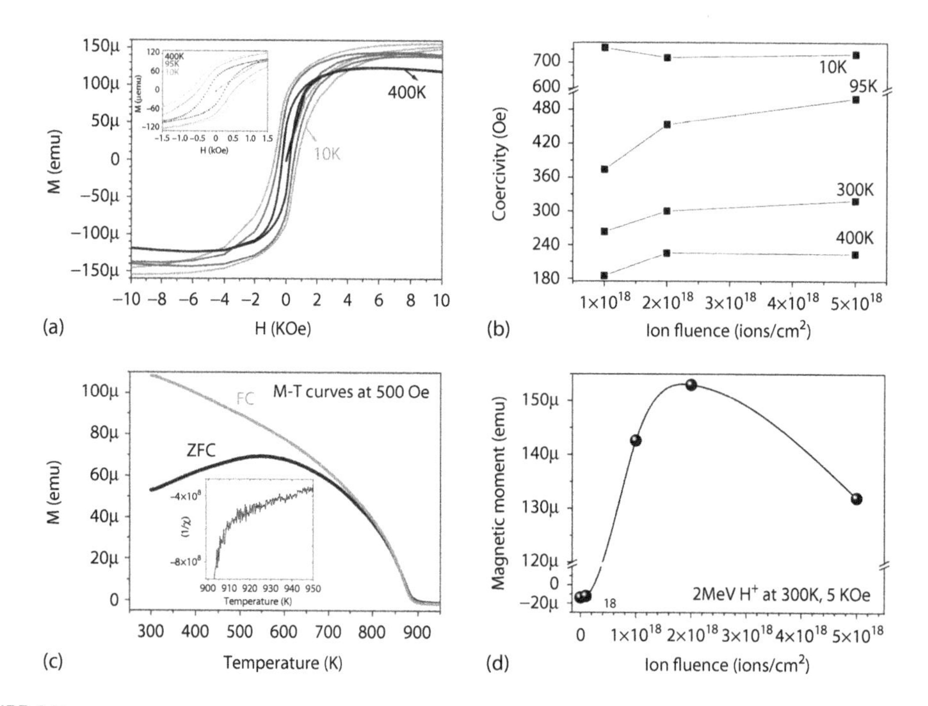

FIGURE 4.7
(a) M vs H curve at various temperatures from 400 K (black), 95 K (blue) and 10 K (red) for a pristine and an irradiated MoS_2 at a fluence of 5×10^{18} ions/cm^2. An enlarged view of the M–H isotherms near the origin is shown in the inset. The variation of coercivity vs ion fluence is shown in (b). (c) Zero field cooled (ZFC) and field cooled (FC) magnetization vs temperature measurements in an applied field of 500 Oe for an irradiated MoS_2 sample at a fluence of 5×10^{18} ions/cm^2. The inverse of the estimated magnetic susceptibility vs temperature plot near T_c (900–950 K) is shown in the inset. (d) The value of magnetization at 300 K with H = 5 kOe as a function of ion fluence from 1×10^{17} ions/cm^2 to 5×10^{18} ions/cm^2. (Reproduced from S. Mathew et al., *Appl. Phys. Lett.* 101, 102103, 2012 with the permission of AIP Publishing.)

in this study, although at 10 K it is almost constant. A large variation of coercivity with temperature indicates the presence of long-range magnetic ordering in the irradiated samples. The zero field cooled (ZFC) and field cooled (FC) magnetizations at an applied field of 500 Oe are given in Figure 4.7 (c). An insight into the nature of magnetic ordering (ferro- or ferrimagnetic) can be gained by analyzing the variation of susceptibility with temperature. The inverse of susceptibility plot is shown as an inset of Figure 4.7 (c). The value of Curie temperature (T_c) is estimated to be ~895 K. The nature of the curve near T_c, a concave curvature with respect to the temperature axis, is characteristic of ferrimagnetic ordering, whereas for a ferromagnetic material this curvature near T_c would be convex [46]. The variation of magnetization at 300 K with an applied field of 5 kOe is plotted in Figure 4.7 (d) for different fluences. The magnetization of the pristine sample and the sample irradiated at a fluence of 1×10^{17} ions/cm^2 is negative, at a fluence of 1×10^{18} ions/cm^2 the sample magnetization becomes positive and at 2×10^{18} ions/cm^2 the magnetism in the sample increases further, and at a fluence of 5×10^{18} ions/cm^2 it has decreased. The dependence of magnetization on the irradiation fluence observed in Figure 4.7 (d) (the bell-shaped curve) indicates that the role of the implanted protons and thus the effect of end-of-range defects for the observed magnetism is minimal, as had been shown in the case of proton-irradiated HOPG [15,47,48]. It was demonstrated that 80% of the measured magnetic signal in the 2 MeV H^+-irradiated HOPG originates from the top 10 nm of the surface [48]. To probe the H^+-irradiated radiation-induced modification in our samples near the surface region, we used x-ray photoelectron spectroscopy (XPS) and Raman spectroscopy.

The modifications in the atomic bonding and core-level electronic structure can be probed using XPS. The XPS spectra of a pristine sample and the sample irradiated at a fluence of 5×10^{18} ions/cm^2 are shown in Figure 4.8(a)–(d). Fitting of the spectra was done by a chi-square iteration program using a convolution of Lorentzian–Gaussian functions with a Shirley background. For fitting the Mo 3d doublet, the peak separation and the relative area ratio for 5/2 and 3/2 spin–orbit components were constrained to be 3.17 eV and 1.5 eV, respectively, while the corresponding constraints for the S 2p 3/2 and 1/2 levels were 1.15 eV and 2 eV, respectively [49,50]. The peaks at 228.5 eV and 231.7 eV observed in the pristine spectrum of Mo are identified as Mo $3d_{5/2}$ and $3d_{3/2}$, while the small shoulder at 226 eV in Figure 4.8 (a) is the sulfur 2s peak [49]. In the irradiated spectrum in Figure 4.8 (b), apart from the pristine Mo peaks, two additional peaks at 229.6 eV and 232.8 eV are visible. The peak observed at 229.6 eV in Figure 4.8 (b) has 18% intensity of the total Mo signal, which could be due to a Mo valence higher than +4. The binding energy positions of 3d levels in Mo (V) have been reported to be 2 eV higher than those of Mo (IV) [51]. We found a peak at 229.6 eV in the irradiated sample that is only 1.0 eV above that of the Mo (IV) level. The pristine spectrum of S consists of S $2p_{3/2}$ and $2p_{1/2}$ peaks at 161.4 and 162.5 eV and another two peaks at 163 eV and 164.2 eV. The peak at 163 eV in the irradiated spectrum of S in Figure 4.8 (d) had increased by 6% in intensity compared to that in the pristine sample.

An indication of the nature of the induced defects and crystalline quality can be gained using Raman spectroscopy, which was a major tool for characterizing ion irradiation–induced defects in graphene and graphite in a recent study [51]. The Raman spectra of the pristine and irradiated samples at a fluence of 5×10^{18} ions/cm^2 are shown in Figure 4.9. The E_{2g}^1 mode at 385 cm^{-1} and A_{1g} mode at 411 cm^{-1} are clearly seen in Figure 4.9 [52]. In the low frequency sides of E_{2g}^1 and A_{1g} phonon modes, the peaks observed in the deconvoluted spectra are the Raman-inactive E_{1u}^2 and B_{1u} phonons; these modes become Raman active due to the resonance effect, as observed by Sekine et al. [53]. These extra phonons are

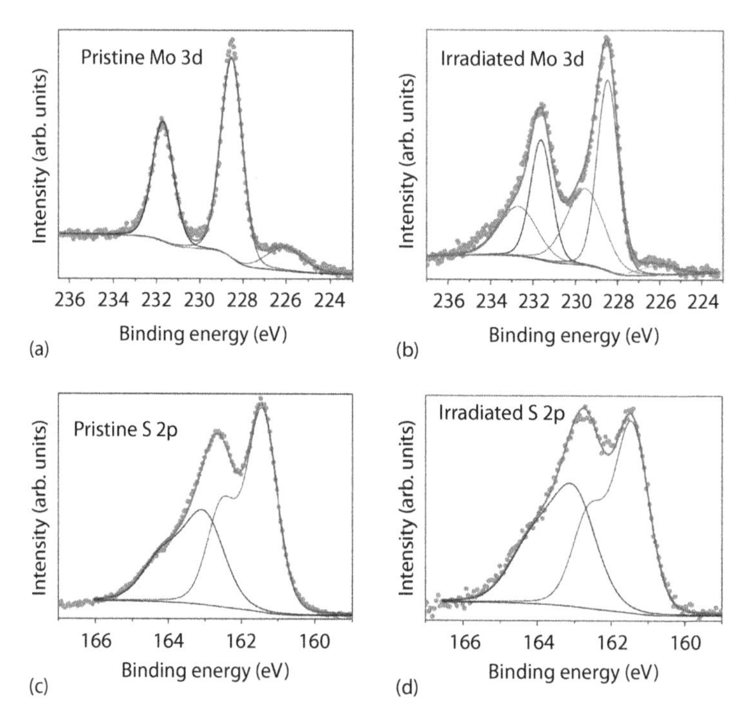

FIGURE 4.8
XPS spectra from pristine and irradiated MoS_2 at a fluence of 5×10^{18} ions/cm^2. The Mo peak is given in (a) and (b) and the S peak in (c) and (d). The fitted spectra along with the constituent peaks and experimental points are also shown. (Reproduced from S. Mathew et al., *Appl. Phys. Lett.* 101, 102103, 2012 with the permission of AIP Publishing.)

Davydov pairs of the E^1_{2g} and A_{1g} modes [54]. The broad peak observed at ~452 cm^{-1} can be the second order of LA(M) phonon [54].

In the irradiated sample, a peak at 483 cm^{-1} is clearly visible. Frey et al. reported a peak at 495 cm^{-1} in chemically synthesized fullerene-like and platelet-like nanoparticles of MoS_2 and assigned this to the second-order mode of the zone-edge phonon at 247 cm^{-1} [54]. Phonon dispersion and the density of the phonon state calculations have shown a peak ~250 cm^{-1} due to a TO branch phonon [56]. The mode at 483 cm^{-1} is close to the above zone-edge phonon observed in nanoparticles of MoS_2 [54]. The appearances of a mode at 483 cm^{-1} along with the broadening of the mode at 452 cm^{-1} indicate the presence of lattice defects due to proton irradiation of the sample. The FWHM of the E^1_{2g} and A_{1g} modes has not increased in the irradiated MoS_2, and this shows that the lattice structure has been preserved in the near-surface region of the irradiated sample. The A_{1g} mode couples strongly to the electronic structure compared to E_{2g} as observed in resonance Raman studies [55]. The ratio of the intensity of A_{1g} to E_{2g} modes (hereafter R) can be a measure of Raman cross section as discussed in high pressure Raman spectroscopy studies [56]. The intensity ratio R is found to enhance by 16% in the irradiated sample compared to the pristine sample as shown in Figure 4.9. This increase in the intensity ratio R in the irradiated sample can be attributed to the induced changes in the electronic band structure, which enhances the interaction of electrons with A_{1g} phonons [56]. The intensities of the E^2_{1u} and B_{1u} phonon modes were also found to be enhanced in the irradiated sample. This enhancement of the Davydov pairs and second-order LA(M) peak along with the appearance of a defect mode indicate a deviation from the perfect symmetry of the system.

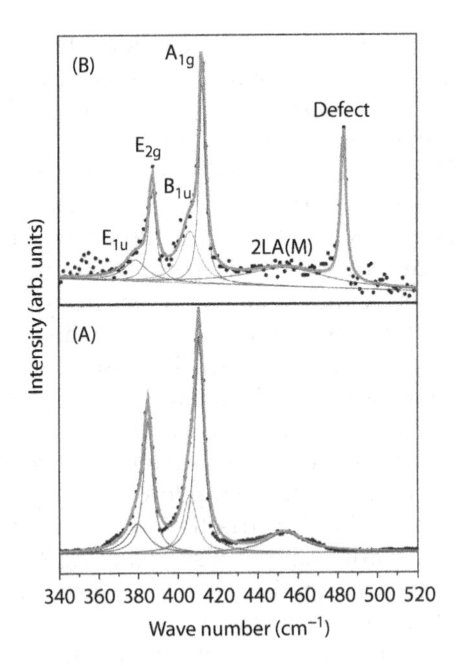

FIGURE 4.9
Raman spectra of (A) pristine MoS_2 and (B) irradiated MoS_2 at a fluence of 5×10^{18} ions/cm^2. The deconvoluted modes are labeled in the spectrum, and the fitted curve with constituent peaks and experimental points are also given. (Reproduced from S. Mathew et al., *Appl. Phys. Lett.* 101, 102103, 2012 with the permission of AIP Publishing.)

The magnetic moment observed for the irradiated sample at an applied field of 2000 Oe is as high as ~150 μemu along with a clear hysteresis and coercive field of 700 Oe at 10 K. The magnetization curve of the irradiated flake compared with that of the pristine sample and the blank substrate (the latter not shown here) gives clear evidence for the irradiation-induced magnetism in MoS_2. The total amount of magnetic impurities present in the sample was determined by post-irradiation PIXE experiments. PIXE results show ~54 ppm of Fe in the sample and if we assume that all of these Fe impurities became ferromagnetic, which is a rather unrealistic assumption, the maximum signal would be only 39 μemu, far short of the observed magnetization that is at least a factor of 4 larger. Thus, the observed magnetism cannot be explained by magnetic impurities. The fact that we observe diamagnetism in the pristine sample and clear magnetic hysteresis loops in the same sample after exposure to MeV protons provides irrefutable evidence for the intrinsic nature of the observed magnetic signal.

Regarding the mechanism for the formation of the magnetic state in MoS_2, among others, the defect-mediated mechanism appears to be the most general one. Possible origins of magnetism in an irradiated MoS_2 system could include the following: (i) irradiation-induced point defects that can give rise to a magnetic moment; and (ii) the presence of edge states as fragments that have zigzag or armchair edges could lead to splitting the flat energy bands and lowering the energy of the spin-up band compared to the spin-down band, leading to ferromagnetism in the material. The Raman spectrum in Figure 4.9 (b) showed a defect mode at 483 cm^{-1} due to the presence of zone-edge phonons in the irradiated sample. The appearance of well-defined E^1_{2g} and A_{1g} modes in Figure 4.9 (b) does not reveal any significant damage to the crystal structure of MoS_2 lattice. The presence of atomic vacancies, predominantly S in our case, can create a loss of symmetry and hence the coordination number of Mo atom would not remain as 6 in the irradiated sample. Tiwari et al. observed sulfur vacancy–induced surface reconstruction of MoS_2 under high-temperature treatment (above 1330 K) [57]. The higher valence of Mo observed in the irradiated sample could be due to a reconstruction of the lattice, apart from the presence of sulfur vacancies [57–59].

An estimate of the defect density created by the proton beam in MoS_2 can be determined from Monte Carlo simulations (SRIM 2008) using full damage cascade. The displacement energy of Mo and S for the creation of a Frenkel pair used for the calculation is 20 eV and 6.9 eV, respectively, as reported by Komsa et al. in a recent study of electron irradiation hardness of transition metal dichalcogenides [60]. According to this calculation, the 2 MeV proton comes to rest at a depth of 31 μm from the surface. The distance between vacancies estimated at the surface and at the end of the range after irradiating with 1×10^{18} ions/cm^2 is 6.3 and 2.7 Å, respectively. These calculated values are overestimates because the annealing of defects and the crystalline nature of the target have not been incorporated in SRIM simulations.

We carried out another experiment involving low-energy proton irradiation where we subjected the MoS_2 sample to 0.5 MeV H^+ irradiation at an ion fluence of 1×10^{18} ions/cm^2 (the fluence at which magnetic ordering was observed using 2 MeV protons) and at a lower fluence of 2×10^{17} ions/cm^2. Ferromagnetism was not observed in the former case, whereas a weak magnetic signal with a clear variation of coercivity with temperature was observed at the lower fluence in the latter case as evident from Figure 4.10. Also, we have observed a weak magnetic signal with a clear hysteresis loop in the former case (the sample irradiated at a fluence of 1×10^{18} ions/cm^2) after annealing (350°C, 1 hour in Ar gas flow). If ion damage were solely responsible for magnetization, then at the end of the range of the 2 MeV ion and of the 0.5 MeV ion there should be very little difference. The electronic energy loss of the ion helps recrystallization while the nuclear energy loss is only responsible for defect creation [61,62]. The ratio between electronic and nuclear energy loss is 42% greater

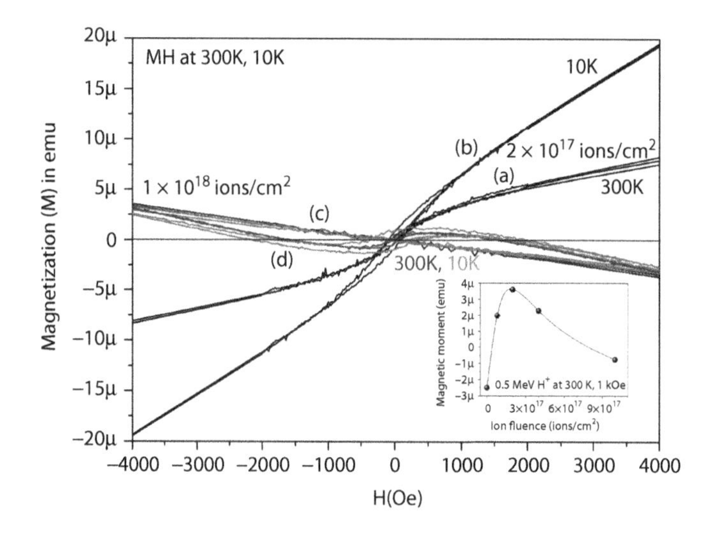

FIGURE 4.10
M vs H curves for an irradiated MoS_2 sample using 0.5 MeV protons (a) at 300 K, (b) at 10 K for a fluence of 2×10^{17} ions/cm^2, (c) for a fluence of 1×10^{18} ions/cm^2 at 300 K (blue curve) and at 10 K (red curve) and (d) the same sample in (c) after annealing in an Ar flow in a tube furnace (350°C, 1 hour). The value of the magnetization at 300 K with H = 1 kOe as a function of ion fluence is shown in the inset. The background contribution from the substrate Si is subtracted in all the curves. (Reproduced from S. Mathew et al., *Appl. Phys. Lett.* 101, 102103, 2012 with the permission of AIP Publishing.)

for a 2 MeV proton compared to a 0.5 MeV proton. This enhanced electronic energy loss component will allow the defective lattice to recover albeit with atomic displacements in the case of 2 MeV ion irradiation [61,62]. In the case of 0.5 MeV ions, the electronic energy loss appears to be insufficient for the required reconstruction of the lattice together with the atomic displacements to induce a strong ferromagnetic signal and hence it is not effective in creating magnetic ordering in MoS_2.

The appearance of magnetism observed in proton-irradiated MoS_2 samples can be due to a combination of defect moments arising from vacancies, interstitials, deformation and partial destruction of the lattice structure, i.e., the formation of edge states and reconstructions of the lattice. To identify the relative contributions of the ion beam–induced defects toward the observed magnetism in MoS_2, first-principle simulations incorporating atomic vacancies, edge states and lattice reconstructions, such as the ones performed in the case of various carbon allotropes, are required.

4.4 Conclusions

We have shown that soft-ferromagnetism can be induced in C_{60} films with 2 MeV proton irradiation, while MoS_2 displays ferrimagnetic behavior with a Curie temperature of 895 K. Magnetism in this irradiated C_{60} film arises due to atomic displacements caused by the energetic protons as they pass through the film. Possible sources of magnetization in the above systems are isolated vacancies, vacancy clusters, formation of edge states and reconstructions of the lattice. The discovery of ion irradiation–induced magnetism in MoS_2 and C_{60} films sheds light on tailoring its properties by engineering the defects using energetic ions and provides a route for future applications of these materials.

Acknowledgments

S. Mathew would like to dedicate this article to the memory of Professor S. N. Behera (IOP Bhubaneswar, India) who had been an inspiration to look for magnetism in unconventional magnetic materials such as in modified carbon allotropes and other layered solids.

I acknowledge the support and care received from my mentors Professors T. Venkatesan and M.B.H. Breese from NUSNNI-NanoCore NUS Singapore, Professor J.T.L. Thong (Dept. ECE, NUS Singapore) and Professor B.N. Dev (IOP Bhubaneswar and IACS Kolkata India and now in IIT Kharagpur). I also thank all my collaborators in NUS Singapore and IOP Bhubaneswar, without their support this work would not have been possible.

References

1. T.L. Makorova, B. Sundqvist, R. Hohne, P. Esquinazi, Y. Kopelevich, P. Scharff, V.A. Davydov, L.S. Kashevarova and A.V. Rakhmania, *Nature* 413 (2001) 716; *Nature* 440 (2006) 707.
2. R.A. Wood, M.H. Lewis, M.R. Lees, S.M. Bennington, M.G. Cain and N. Kita-mura, *J. Phys. Cond. Matter.* 14 (2002) L385.

3. P. Esquinazi and R. Hohne, *J. Magn. Magn. Mater* 290–291 (2005) 20.
4. A.V. Rode, E.G. Gamaly, A.G. Christy, J.G.F. Gerald, S.T. Hyde, R.G. Elliman, B. Luther-Davies, et al., *Phys. Rev. B* 70 (2004) 54407.
5. J. Tuček, K. Holá, A. B. Bourlinos, P. Błoński, A. Bakandritsos, J. Ugolotti, M. Dubecký, et al., *Nat. Commun.* 8 (2017) 14525.
6. C.V. Raman and P. Krishnamurthy, *Nature* 124 (1929) 53.
7. M.A.H. Vozmediano, F. Guinea and M.P. Lopez-Sancho, *J. Phys. Chem. Solids* 67 (2006) 562.
8. C. Ishii, Y. Matsumura and K. Kaneko, *J. Phys. Chem.* 99 (1995) 5743.
9. C. Ishii, N. Shindo and K. Kaneko, *Chem. Phys. Lett.* 242 (1995) 196.
10. Yu.V. Korshak, T.V. Medvedeva, A.A. Ovchnnikov and V.N. Spector, *Nature* 326 (1987) 370.
11. A.A. Ovchinnikov and I.L. Shamovsky, *J. Mol. Struct. (Theochem.)* 251 (1991) 133.
12. P. Esquinazi, D. Spemann, R. Hohne, A. Setzer, K.H. Han and T. Butz, *Phys. Rev. Lett.* 91 (2003) 227201.
13. P. Esquinazi, D. Spemann, K. Schindler, R. Höhne, M. Ziese, A. Setzer, K.-H. Han, S. Petriconi, et al., *Thin Solid Films* 505 (2006) 85.
14. S. Talapatra, P.G. Ganesan, T. Kim, R. Vajtai, M. Huang, M. Shima, G. Ramanath, et al., *Phys. Rev. Lett.* 95 (2005) 97201.
15. H. Ohldag, T. Tyliszczak, R. Höhne, D. Spemann, P. Esquinazi, M. Ungureanu and T. Butz, *Phys. Rev. Lett.* 98 (2007) 187204.
16. B. Narymbetov, A. Omerzu, V.V. Kabanov, M. Tokumoto, H. Kobayashi and D. Mihallovic, *Nature* 407 (2000) 883.
17. P.M. Allemand, K.C. Khemani, A. Koch, F. Wudl, K. Holczer, S. Donovan, G. Gruner, et al., *Science* 253 (1991) 301.
18. T.L. Makarova and F. Palacio (Eds.), *Carbon-Based Magnetism*, Elsevier, Ams-terdam, 2006.
19. S. Mathew, B. Satpati, B. Joseph, B.N. Dev, R. Nirmala, S.K. Malik and R. Kesavamoorthy, *Phys. Rev. B* 75 (2007) 75426.
20. D. Spemannn, P. Esquinazi, R. Höhne, A. Setzer, M. Diaconu, H. Schmid and T. Butz, *Nucl. Instr. Meth. Phys. B* 231 (2005) 433.
21. O.V. Yazev and M.I. Katsnelson, *Phys. Rev. Lett.* 100 (2008) 47209.
22. A.F. Hebard, O. Zhou, Q. Zhong, R.M. Fleming and R.C. Haddon, *Thin Solid Films* 257 (1995) 147.
23. S. Mathew, B. Satpati, B. Joseph and B.N. Dev, *Appl. Surf. Sci.* 249 (2005) 31.
24. A.E. Berkowitz and E. Kneller, *Magnetism and Metallurgy*, Academic Press, New York, 1969 p. 393; W.E. Henry. *Rev. Mod. Phys.* 25 (1953) 163.
25. J.L. Sauvajol, F. Brocard, Z. Hircha and A. Zahab, *Phys. Rev. B* 52 (1995) 14839.
26. S.J. Duclos, R.C. Haddon, S.H. Glarum, A.F. Hebard and K.B. Lyons, *Solid State Comm.* 80 (1991) 481.
27. J.F. Ziegler, J.P. Biersack and U. Littmark, *SRIM 2003—A Version of the TRIM Program: The Stopping and Range of Ions in Matter*, Pergamon Press, New York, 1995.
28. H. Lee, N. Park, Y.W. Son, S. Han and J. Yu, *Chem. Phys. Lett.* 398 (2004) 207.
29. V.V. Belavin, L.G. Bulusheva, A.V. Okotrub and T.L. Makarova, *Phys. Rev. B* 70 (2004) 155402.
30. P.O. Lehtinen, A.S. Foster, Y. Ma, A.V. Krasheninnikov and R.M. Nieminen, *Phys. Rev. Lett.* 93 (2004) 187202.
31. P. Esquinazi, A. Setzer, R. Hohne, C. Semmelhak, Y. Kopelevich, D. Spemann, T. Butz, et al., *Phys. Rev. B* 66 (2002) 24429.
32. K. Nakada, M. Fujita, G. Dresselhaus and M.S. Dresselhaus, *Phys. Rev. B* 54 (1996) 17954.
33. J.A. Wilson and A.D. Yoffe, *Adv. Phys.* 18 (1969) 193.
34. A.H. Castro Neto and K. Novoselov, *Rep. Prog. Phys.* 74 (2011) 082501.
35. H.S.S.R. Matte, A. Gomathi, A.K. Manna, D.J. Late, R. Datta, S.K. Pati and C.N.R. Rao, *Angew. Chem.* 122 (2010) 4153.
36. K.F. Mak, C. Lee, J. Hone, J. Shan and T.F. Heinz, *Phys. Rev. Lett.* 105 (2010) 136805.
37. Y. Zhan, Z. Liu, S. Najmaei, P.M. Ajayan and J. Lou, *Small* 8 (2012) 966.
38. J. Zhang, J.M. Soon, K.P. Loh, J. Yin, J. Ding, M.B. Sullivian and P. Wu, *Nano Lett.* 7 (2007) 2370.
39. Y. Li, Z. Zhou, S. Zhang and Z. Chen, *J. Am. Chem. Soc.* 130 (2008) 16739.

40. P. Murugan, V. Kumar, Y. Kawazoe and N. Ota, *Phys. Rev. A* 71 (2005) 063203.
41. K. Zberecki, *J. Supercond. Novel Magn.* 25 (2012) 2533.
42. A. Vojvodic, B. Hinnemann and J.K. Nørskov, *Phys. Rev. B* 80 (2009) 125416.
43. A.R. Botello-Mendez, F. Lopez-Urias, M. Terrones and H. Terrones, *Nanotechnology* 20 (2009) 325703.
44. R. Shidpour and M. Manteghian, *Nanoscale* 2 (2010) 1429.
45. S. Mathew, K. Gopinadhan, T.K. Chan, D. Zhan, L. Cao, A. Rusdi, M.B.H. Breese, et al., *Appl. Phys. Lett.* 101 (2012) 102103.
46. H.P. Myers, *Introductory Solid State Physics,* Taylor & Francis Group, Oxford, 1997.
47. T.L. Makarova, A.L. Shelankov, I.T. Serenkov, V.I. Sakharov and D.W. Boukhvalov, *Phys. Rev. B* 83 (2011) 085417.
48. H. Ohldag, P. Esquinazi, E. Arenholz, D. Spemann, M. Rothermel and A. Setzer, *New J. Phys.* 12 (2010) 123012.
49. J.R. Lince, T.B. Stewart, M.M. Hills, P.D. Fleischauer, J.A. Yarmoff and A. Taleb-Ibrahimi, *Surf. Sci.* 210 (1989) 387.
50. T.A. Patterson, J.D. Carver, D.E. Leyden and D.M. Hercules, *J. Phys. Chem.* 80 (1976) 1700.
51. S. Mathew, T.K. Chan, D. Zhan, K. Gopinadhan, A.R. Barman, M.B.H. Breese, S. Dhar, et al., *J. Appl. Phys.* 110 (2011) 84309.
52. T.J. Wieting and J.L. Verble, *Phys. Rev. B* 3 (1971) 4286.
53. T. Sekine, K. Uchinokura, T. Nakashizu, E. Matsuura and R. Yoshizaki, *J. Phys. Soc. Jpn.* 53 (1984) 811.
54. G.L. Frey, R. Tenne, M.J. Matthews, M.S. Dresselhaus and G. Dresselhaus, *Phys. Rev. B* 60 (1999) 2883.
55. N. Wakabayashi, H.G. Smith and R.M. Nicklow, *Phys. Rev. B* 12 (1975) 659.
56. T. Livneh and E. Sterer, *Phys. Rev. B* 81 (2010) 195209.
57. R.K. Tiwari, J. Yang, M. Saeys and C. Joachim, *Surf. Sci.* 602 (2008) 2628.
58. R.St.C. Smart, W.M. Skinner and A.R. Gerson, *Surf. Interface Anal.* 28 (1999) 101–105.
59. J.R. Lince, D.J. Carre and P.D. Fleischauer, *Langmuir* 2 (1986) 805.
60. H.P. Komsa, J. Kotakoski, S. Kurasch, O. Lehtinen, U. Kaiser and V. Krasheninnikov, *Phys. Rev. Lett.* 109 (2012) 035503.
61. T. Venkatesan, R. Levi, T.C. Banwell, T. Tomberllo, M. Nicolet, R. Hamm and E. Mexixner, *Mat. Res. Soc. Symp. Proc.* 45 (1985) 189.
62. A. Benyagoub and A. Audren, *J. Appl. Phys.* 106 (2009) 83516.

5

Hot Electron Transistors with Graphene Base for THz Electronics

Filippo Giannazzo, Giuseppe Greco, Fabrizio Roccaforte, Roy Dagher, Adrien Michon, and Yvon Cordier

CONTENTS

5.1 Introduction

The terahertz (THz) frequency region of the electromagnetic spectrum (from 0.1 to 10 THz, corresponding to millimeter and sub-millimeter wavelengths) is the spectral range that separates electronics from photonics. Although access to this spectral range is strategic for application areas like communications, medical diagnostics and security, this has been historically difficult because of the technical challenges associated with generating, detecting, processing and radiating such high-frequency signals. From the electronic devices point of view, controlling and manipulating signals in this portion of the radio frequency (RF) spectrum requires solid-state transistors with a cut-off frequency (f_T) and maximum oscillation frequencies (f_{max}) well above 1 THz.

To date, the high electron mobility transistors (HEMTs) and heterojunction bipolar transistors (HBTs) have been the two main device architectures employed for ultrahigh-frequency applications. In particular, indium phosphide–based HEMTs and HBTs have been demonstrated with f_{max} exceeding 1 THz [1–5]. These record RF figures of merit have been achieved by coupling the advantageous physical properties of the InGaAs/InP material system (i.e. the large heterojunction offset for carrier confinement, the high electron mobilities in quantum well channels for HEMTs and p-doped bases for HBTs and the high achievable doping levels for low Ohmic contact resistivities) with aggressive lateral and vertical scaling of transistors geometries. However, further improvements of HEMTs' performances will ultimately be limited by the saturation velocity of electrons in the channel, whereas the diffusion of the minority carriers (electrons) across the p-type base will limit the performances of HBTs.

In this context, the hot electron transistor (HET) can represent a promising device concept with the potential to overcome the fundamental limitations of HEMTs and HBTs in ultrahigh-frequency applications. The HET is a unipolar and majority carrier vertical device where the base-to-emitter voltage controls the injection and the transport of ballistic hot electrons through an ultrathin transit layer (base) to the collector terminal. Due to its working principles, the HET does not suffer from the intrinsic limitations of HBTs (i.e. the minority carrier diffusion and recombination in the base) and has the potential to reach superior performances in the THz frequency range. However, the practical implementation of this device concept relies primarily on the possibility of achieving ballistic transport in the base, as well as on the efficiency of hot electrons injection from the emitter and finally on the filtering efficiency of the base-collector (B-C) barrier.

The HET device concept was introduced more than 50 years ago by Mead [6]. Since then, several material systems have been used for HET development including metal thin films [6–9], complex oxides [10], superconducting materials [11], III-V semiconductor heterostructures [12] and, more recently, Group III-nitride semiconductors heterostructures [13]. However, the successful demonstration of high-performance HETs has been limited by some technological issues, such as the difficulty to scale the base thickness below the electron mean free path of the carriers. In this context, single-layer, two-dimensional (2D) materials are naturally suitable for applications requiring ultrathin, defect-free films. In particular, monolayer graphene (Gr) with its excellent transport properties (high mobility, from ~10^3 up to ~$10^5 cm^2 V^{-1} s^{-1}$, and micrometer electron mean free path) [14–16] and a dangling-bond-free inert surface is an ideal candidate as a low resistance, scattering-free base material in HETs. Theoretical studies have predicted that with an optimized structure, f_T up to several terahertz [17], I_{on}/I_{off} over 10^5 and high-current gain can be achieved with Gr base HETs (GBHET). The initial experimental demonstrators of GBHETs, reported in 2013, showed successful direct current (DC) operation in terms of current modulation ($I_{on}/I_{off} > 10^5$) but suffered from a low output current density (~μA/cm^2), low current gain, low injection efficiency and high threshold voltage. These limitations were not intrinsic to the use of Gr as a base material and significant improvements have been obtained more recently by the careful choice of the barrier layers' materials and the improvement of the interfaces quality.

In this chapter, the operating principles of an ideal HET will be introduced, illustrating the device's DC characteristics and discussing the impact of the main physical parameters on the DC figures of merit (current transit ratio and current gain) and on the alternating current (AC) figures of merit (f_{max} and f_T). Therefore, an historical perspective on the attempted implementations of this device concept will be provided, starting from the first proposal of a metal base HET [6] to more recent implementations, such as the nitride semiconductors–based HETs and the GBHETs. The theoretical DC and AC performances of GBHETs will be discussed and the state-of-the-art GBHETs will be presented. The last section of the chapter will present open issues and new ideas to improve the performances of GBHETs.

5.2 Hot Electron Transistor (HET): Device Concept and Operating Principles

A general schematic illustration of a HET device and the related energy band diagram under equilibrium conditions are reported in Figure 5.1a and b, respectively. The HET

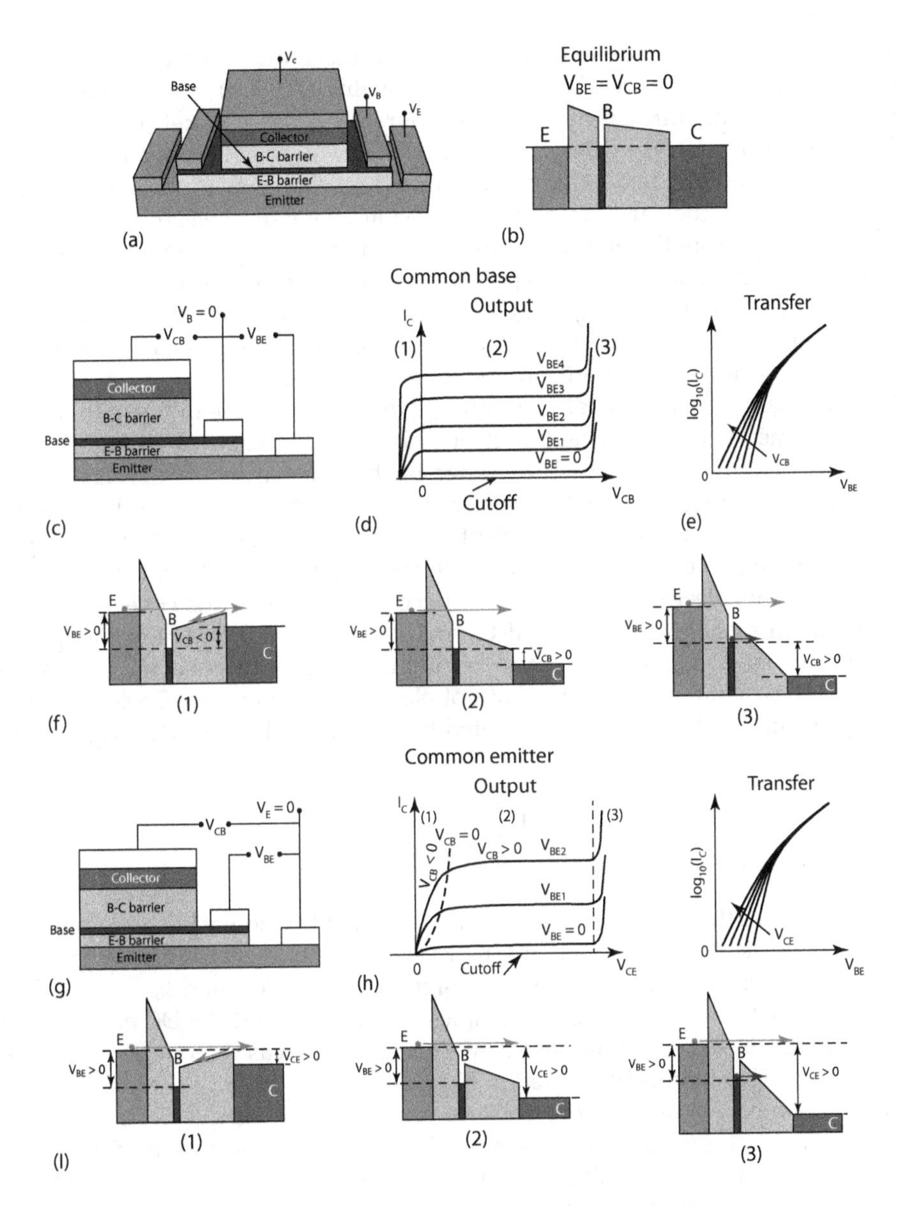

FIGURE 5.1
(a) Schematic illustration of a hot electron transistor HET and (b) energy band diagram of the device under equilibrium conditions ($V_{BE} = V_{CB} = 0\,V$). Cross-sectional schematic of a HET biased in the common-base configuration (c), output characteristics $I_C - V_{CB}$ (d), transfer characteristics $I_C - V_{BE}$ (e) and band diagrams for different V_{CB} values (f). Cross-sectional schematic of a HET biased in the common-base configuration (g), output characteristics $I_C - V_{CB}$ (h), transfer characteristics $I_C - V_{BE}$ (i) and band diagrams for different V_{CB} values (l).

consists of an emitter (E), a base (B) and a collector (C) region, which are separated by two barrier layers, i.e. the emitter-base (E-B) barrier with barrier energy height ϕ_{EB} and thickness d_{EB}, and the B-C barrier with barrier energy height ϕ_{BC} and thickness d_{BC}.

For a sufficiently high forward bias applied between the base and the emitter, electrons are injected into the base. A key aspect of the HET operation is that the injected electrons (hot electrons) have a higher energy compared to the Fermi energy of the electrons thermal population (cold electrons) in the base. Depending on the values of ϕ_{BE} and d_{BE}, as well as on the barrier's material quality, several potential mechanisms can rule the injection of electrons across the E-B barrier, i.e. (i) direct tunneling (DT), (ii) Fowler–Nordheim (FN) tunneling through the thinned triangular barrier, (iii) Poole–Frenkel (PF) traps' assisted emission through the barrier and (iv) thermionic emission (TE) above the barrier. As a matter of fact, FN and TE are the two most favorable mechanisms for hot electron injection into the base. On the contrary, DT and PF emission give rise to electron injection at any energy in the range from the emitter conduction band to the top of the E-B barrier, so that most of the injected carriers will keep part of the cold electron population in the base. As discussed later in this section, high values of the injected hot electrons current density, J_E, are fundamental to achieving high-frequency performances of hot electrons transistors. Hence, ϕ_{BE} and d_{BE} must be properly chosen to maximize J_E.

Ideally, for a base thickness $d_B < \lambda_{mfp}$ (with λ_{mfp} the scattering mean free path of hot electrons), a large fraction of the injected electrons can traverse the base ballistically, i.e. without losing energy, and finally reach the edge of the B-C barrier. This second barrier is aimed to act as an energy filter, which allows the hot electrons to reach the collector and reflects back the electrons with insufficient energy. These reflected electrons eventually become part of the cold electrons population in the base and contribute to the base current (I_B), whereas the hot electrons reaching the collector give rise to the collector current (I_C). Besides transmitting hot electrons, the B-C barrier must be thick and high enough to block the leakage current, I_{BCleak}, of cold electrons from the base to the collector. For a thick B-C barrier, the transmission probability T of hot electrons with energy $E > \phi_{BC}$ is ultimately ruled by quantum mechanics and it is related to the energy difference $E - \phi_{BC}$ as follows:

$$T = \left[\frac{1}{2} + \frac{1}{4}\left(\sqrt{\frac{E}{E-\phi_{BC}}} + \frac{E-\phi_{BC}}{E}\right)\right]^{-1} \tag{5.1}$$

Equation 5.1 implies that, for finite values of ϕ_{BC}, a certain fraction of hot electrons is always reflected back at the B-C barrier. Furthermore, since the hot electrons' energy is comparable to the E-B barrier height, ϕ_{BC} should be significantly lower than ϕ_{EB}.

Figure 5.1c–f further illustrates the principles of operation and the DC electrical characteristics of a HET in the common-base configuration, whereas the operation in the common-emitter configuration is illustrated in Figure 5.1g–l.

In the common-base configuration, the base contact is grounded (see Figure 5.1c), and a potential difference, $V_{BE} = (V_B - V_E) > 0$, is applied between the base and the emitter contacts, in order to allow hot electrons injection in the base. Depending on the values of the potential difference, $V_{CB} = V_C - V_B$, between the collector and base contacts, three current transport regimes can be observed in the common-base output characteristics, $I_C - V_{CB}$, for different values of V_{BE} (Figure 5.1d). For $V_{CB} > 0$ (region 2 in Figure 5.1d), I_C is almost independent of V_{CB}, i.e. all the injected hot electrons are transmitted above the B-C barrier (panel 2 of Figure 5.1f). For $V_{CB} < 0$, the collector edge of the B-C barrier is raised up and part of the hot electrons are reflected back in the base (panel 1 of Figure 5.1f), resulting in a decrease of I_C with increasing negative values of V_{CB}, up to device switch-off (region 1 of Figure 5.1d). For large positive values of V_{CB}, the leakage current (I_{BCleak}) contribution of cold electrons injected by FN tunneling through the B-C barrier becomes large (panel 3 of Figure 5.1f) and this leads to a rapid increase of I_C as a function of V_{CB} (region 3 of Figure 5.1d). The typical shape of the transfer characteristics ($I_D - V_{BE}$) for different values of V_{CB} are also reported in Figure 5.1e.

In the common-emitter configuration (Figure 5.1g), the emitter contact is grounded ($V_E = 0$), and electrons injection from the emitter to the base is achieved due to the potential difference, $V_{BE}=(V_B-V_E)>0$. The common-emitter output characteristics (Figure 5.1h) are obtained by sweeping $V_{CE}=V_C-V_E$ from 0 to positive values, for different values of V_{BE}. Three current transport regimes can be observed in the I_C-V_{CE} characteristics. For a fixed V_{BE}, low V_{CE} values (region 1 of Figure 5.1h) correspond to a $V_{CB}<0$, i.e. to an upward bending of the B-C barrier at the collector edge (panel 1 of Figure 5.1l), resulting in a partial back-reflection of hot electrons from the base. In this transport regime, I_C grows with increasing V_{CE} up to the condition corresponding to $V_{CB}=0$. For larger V_{CE} values (region 2 of Figure 5.1h), I_C is independent of V_{CE}, i.e. all the injected hot electrons are transmitted to the collector (panel 2 of Figure 5.1l). Finally, for very large V_{CE} values (region 3 of Figure 5.1h), the cold electrons leakage current through the B-C triangular barrier (panel 3 of Figure 5.1l) becomes dominant, leading to a rapid increase of I_C. The typical shape of the transfer characteristics (I_D-V_{BE}) for different values of V_{CE} are also reported in Figure 5.1i.

The main figures of merits for the DC operation of a HET are

1. The common-base current transfer ratio $\alpha=I_C/I_E$
2. The common-emitter current gain β defined as $\beta=I_C/I_B$

For good DC performances, $\alpha\approx1$ and β as large as possible are needed. This ensures that most of the electrons injected from the emitter reach the collector and only a very small fraction of them contribute to the base current. We can express α as $\alpha=\alpha_B\alpha_{BC}\alpha_C$, where $\alpha_B=\exp(-d_B/\lambda_{mfp})$ is the base efficiency, α_{BC} is the B-C barrier filtering efficiency and α_C is the collector efficiency, respectively. An ultrathin base with a high electron mean free path is required to achieve $\alpha_B\approx1$.

The high-frequency figures of merit for a transistor are the current gain cut-off frequency, f_T, and the power gain maximum oscillation frequency, f_{max}. Both of these are defined in terms of the parameters of the small-signal (SS) equivalent circuit, which is very similar to that of the HBT.

In particular, f_T is defined as follows:

$$\frac{1}{2\pi f_T}=\tau_d+\frac{C_{tot}}{g_m} \tag{5.2}$$

τ_d is the sum of the delay times associated with electrons transit in the E-B barrier layer, in the base and in the B-C filtering layer, i.e.

$$\tau_d=\tau_{EB}+\tau_B+\tau_{BC} \tag{5.3}$$

In particular, τ_B can be expressed as $\tau_B=d_B/v_B$, where d_B is the base thickness and v_B is the velocity of hot electrons in the base. In the case of an ultrathin base, ensuring ballistic transport, this term is very small and can be neglected. The delay τ_{EB} due to electronic transport in the B-E barrier strongly depends on the current injection mechanisms, i.e. it can be very small for FN tunneling through the barrier and higher for TE over the barrier. The delay τ_{BC} associated with electrons transit in the B-C barrier is typically the largest term in Equation 5.3. It is proportional to d_{BC}/v_s, where d_{BC} is the B-C barrier thickness and v_s is the saturated electron velocity in this region. A thin B-C barrier would be necessary to reduce τ_{BC}. On the other hand, d_{BC} cannot be reduced too much in order to avoid an increase of J_{BCleak} and a low collector breakdown voltage, V_{Cbr}.

The second term in Equation 5.2 is the base charging delay, i.e. the time spent charging the base. It is directly proportional to the total capacitance, $C_{tot}=C_{EB}+C_{BC}$, where $C_{EB}=\varepsilon_0\varepsilon_{EB}/d_{EB}$ and $C_{BC}=\varepsilon_0\varepsilon_{BC}/d_{BC}$ are the capacitances of the E-B barrier layer and the B-C filtering layer connected in parallel, ε_0 is the vacuum permittivity and ε_{EB} and ε_{BC} are the relative permittivities of the E-B and B-C barrier layer materials. Furthermore, it is inversely proportional to the device transconductance $g_m = dJ_C/dV_{BE}$. Clearly, the most effective way to minimize this charging delay is represented by the increase of g_m. In fact, a reduction of the barrier layer capacitances would imply an increase of the thickness d_{BE} and d_{BC}, with a consequent impact on the transit delay times. Under saturation conditions, when all the hot electrons injected from the emitter reach the collector ($J_C \approx J_E$), the transconductance $g_m \approx dJ_E/dV_{BE}$. As the emitter current is injected over a barrier, it exhibits an exponential dependence on V_{BE}, i.e. $J_E \propto \exp(qV_{BE}/kT)$. As a result, $g_m \propto qJ_E/kT$. This means that a high injection current density is one of the main requirements to achieve a high cut-off frequency, f_T.

The maximum oscillation frequency f_{max} can be expressed as

$$f_{max} = \sqrt{\frac{f_T}{2\pi R_B C_{BC}}} \tag{5.4}$$

where R_B is the resistance associated with "lateral" current transport in the base layer from the device active area to the base contact. Hence, R_B is the sum of different contributions, i.e. the "intrinsic" base resistance $R_{B\_int} \propto \rho/d_B$ (with ρ the base resistivity and d_B the base thickness), the resistance of the Ohmic metal contact with the base and the access resistance from this contact to the device active area. All these contributions should be minimized to achieve a low R_B. Of course, the most challenging issue to obtain high f_{max} is to fabricate an ultrathin base (allowing ballistic transport of hot electrons in the vertical direction) while maintaining low enough intrinsic and extrinsic base resistances. However, for most of the bulk materials, reducing the film thickness to the nanometer or sub-nanometer range implies an increase of the resistivity, due to the dominance of surface roughness and/or grain boundaries scattering, as well as to the presence of pinholes and other structural defects in the film. In this context, 2D materials, such as graphene and transition metal dichalcogenides (TMDs), can represent ideal candidates to fabricate the base of HETs, since they maintain excellent conduction properties and structural integrity down to single atomic layer thickness. The state of the art and challenges in the use of graphene as a base material for HETs will be discussed in the final section of this chapter.

5.3 HET Implementations: Historical Perspective

This section presents an historical overview on the attempts made to implement the HET device concept using different materials systems, starting from the metal base transistor originally proposed by Mead [6] in the 1960s, to recent demonstrations of HETs fabricated by bandgap engineering of epitaxial nitride semiconductors [18–20]. Section 5.4 will be devoted to illustrating the very recent efforts to employ graphene or other 2D materials as ultimately thin base films of HETs [21–24].

5.3.1 Metal Base HETs

The first proposal of a three-terminal HET consisted of a metal-oxide-metal-oxide-metal (MOMOM) structure, whose band structure is illustrated in Figure 5.2a. The first

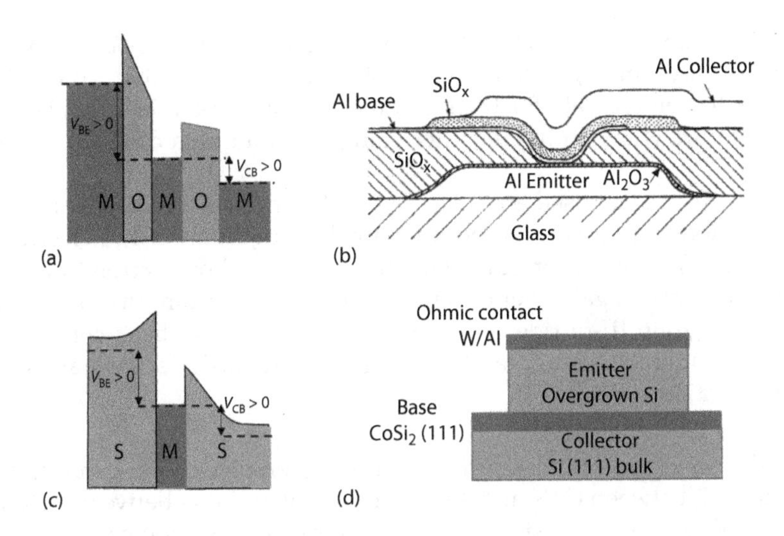

FIGURE 5.2
(a) Energy band diagram of a metal-oxide-metal-oxide-metal (MOMOM) hot electrons transistor and (b) experimental implementation of this device using an $Al/Al_2O_3/Al/SiO_x/Al$ stack. (c) Energy band diagram of a semiconductor-metal-semiconductor (SMS) hot electron transistor and (d) its implementation with a $Si/CoSi_2/Si$ stack. (Panel (b) adapted with permission from Ref. [6] Copyright (1961), American Institute of Physics. Panel (d) adapted with permission from Ref. [8] Copyright (1985), American Institute of Physics.)

prototype was practically implemented using an $Al/Al_2O_3/Al/SiO_x/Al$ stack [6], as shown in the cross-sectional schematic of Figure 5.2b. The emitter terminal was represented by the thick Al stripes on the bottom, the Al_2O_3 E-B barrier (5–10 nm thick) was formed by oxidation of the Al surface and the base was obtained by the evaporation of an Al thin film (~30 nm). Finally, the B-C barrier was formed by the deposition of an SiO_x thin film (≈10 nm), with the collector terminal obtained by evaporating thick Al stripes on top. Electron injection in the base was based on electron tunneling through the thin Al_2O_3 film into a high-energy state in the metal base, and a common-base current transfer ratio $\alpha \approx 0.1$ was reported for this very first HET prototype. This device structure suffered many drawbacks: (i) the difficulty of realizing a pinhole-free metal base thinner than the electron mean free path in the metal; and (ii) the presence of traps in the E-B and B-C barrier layers, resulting in low injection and collection efficiencies. To address this latter issue, the MOMOM device structure was later replaced by the semiconductor-metal-semiconductor (SMS) transistor (see schematic in Figure 5.2c), where the metal base layer was sandwiched between two semiconductors and the injection of hot electrons was primarily due to TE over the E-B Schottky barrier [7]. These devices were practically realized by an $Si/CoSi_2/Si$ stack (see schematic in Figure 5.2d), where the $CoSi_2$ base (with thickness of 5–30 nm) was obtained by a solid-state reaction (silicidation) of Co films deposited on Si, resulting in an epitaxial interface [9]. This epitaxial interface between the emitter and base layers was beneficial to improve the hot electron injection. However, the limitation in the base efficiency still remained, resulting in a poor common-base transfer ratio α.

5.3.2 HETs Based on Semiconductor Heterostructures

Progress in the heteroepitaxial growth of semiconductors also opened the way to the implementation of HETs fully based on semiconductor heterostructures. III-V materials were first employed to fabricate HETs with the E-B and B-C barriers realized by bandgap engineering. As an example, Levi and Chiu [25] employed an $AlSb_{0.92}As_{0.08}$/InAs/GaSb heterostructure to demonstrate for the first time a double heterojunction HET operating at room temperature with common-emitter gain $\beta > 10$ and high collector current density >1200 A/cm^2.

Later on, Moise et al. [26] used an InAlGaAs/AlAs/InGaAs/InAlGaAs/InGaAs heterostructure to demonstrate a HET based on tunneling injection of hot electrons through the AlAls barrier. This device, originally named tunneling hot electron transfer amplifier (THETA), showed current gain over the entire voltage range from turn-on to breakdown.

More recently, Group III-nitride semiconductors (III-N) have been considered for epitaxial HETs fabrication. As compared to III-V, III-N materials offer several advantages for HETs implementation:

1. The wider bandgap modulation range and higher conduction band discontinuities (with ΔE_c between GaN and AlN being 1.8 eV and that between AlN and InN being ~3 eV) enable the design of HETs with higher injection energy and, hence, higher injection velocity.
2. Polarization charges in the III-N systems can be engineered to obtain a high-density and high-mobility 2DEG in the base that can be used to reduce the base resistance R_B in the HET.
3. The III-N systems possess high intervalley separations (~1.3 eV for GaN as compared to ~0.3 eV for GaAs), and hence, a wider range of injection energies is possible in the nitride systems prior to the onset of intervalley scattering events.

In 2000, Shur et al. [13] first simulated a unipolar HET based on a GaN/AlGaN/GaN/AlInGaN heterostructure, where hot electron injection from the emitter (n$^+$-GaN) to the thin (5 nm) base (n$^-$-GaN) occurred by FN tunneling through the AlGaN barrier layer. Those simulations predicted a high emitter current density ($J_E = 10^4 - 10^5$ A/cm^2), high common-base current gain $\beta = I_C/I_B$ up to 60 and low base resistance R_B, thanks to the high mobility (~1500 cm^2/V_s) and high concentration ($n \approx 10^{13}$ cm^{-2}) polarization induced 2DEG.

The first demonstrators of III-N-based HETs have been reported more recently. As an example, Dasgupta et al. [18] demonstrated a HET with an $Al_{0.24}Ga_{0.76}N$ barrier emitter, a GaN base (10 nm in thickness) and an $Al_{0.08}Ga_{0.92}N$ barrier collector structure. The values of the E-B barrier ($\phi_{EB} = 0.27$ eV) and the B-C barrier ($\phi_{BC} = 0.13$ eV) were obtained, respectively, for the used Al mole fractions. TE was the main current injection mechanism from the emitter to the base, resulting in the high drive current capability of this device. A common-base transfer ratio $\alpha \approx 0.97$–0.98 was obtained for this first transistor. More recently, III-N HETs based on tunneling current injection through an ultrathin AlN emitter barrier into a GaN base with sub-10 nm thickness [19,20] have been demonstrated. As an example, Yang et al. [19] fabricated a HET with the heterostructure illustrated in Figure 5.3a and b, where the E-B barrier was represented by 3 nm AlN (capped by 1 nm UID GaN) and the base was formed by 8 nm n$^+$-GaN. A polarization engineered B-C barrier was obtained by stacking several layers, i.e. a 4 nm UID GaN spacer, 6 nm $Al_{0.25}Ga_{0.75}N$, 5 nm $AlGa_{1-x}N$ (with x ranging from 0.2 to 0.25 moving from top to bottom), 42 nm $Al_{0.3}Ga_{0.7}N$ and 28 nm UID GaN. Finally, the sub-collector was represented by 210 nm n$^+$-GaN. The output characteristics of this HET in the common-emitter configuration are reported in Figure 5.3c. The base current density (J_B vs V_{BE}) and the transfer characteristics (J_C vs V_{BE}) are shown in Figure 5.3d, from which a current gain $\beta \approx 1.5$ (at $V_{BE} = 8$ V) was deduced (see Figure 5.3d, right axis).

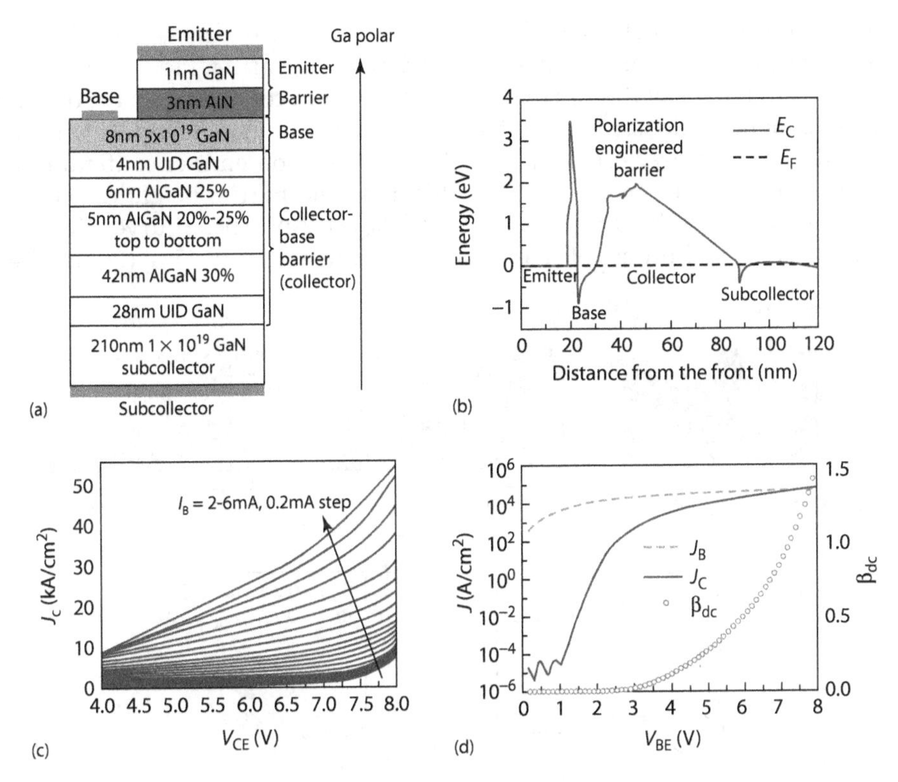

FIGURE 5.3
Schematic cross section (a) and energy band diagram (b) of a HET with a GaN emitter, an AlN (3nm) tunneling barrier, an n^+-GaN (8nm) base, a graded base-collector barrier and an n^+-GaN collector. Common-emitter output characteristics (c) and transfer characteristics (d) of this device. (Figures adapted with permission from Ref. [19] Copyright (2015), American Institute of Physics.)

Despite these very promising performances, these devices suffer from limitation in the base thickness scalability. This is an inherent limitation of the conventional semiconductor growth process, which increases the interface roughness and degrades the transport properties in very thin (<5nm) base layers. Furthermore, the large polarization in III-N materials causes enhanced scattering in alloys and at interfaces. In this context, 2D materials, such as Gr [27] and semiconductor TMDs (MoS_2, $MoSe_2$, WS_2, WSe_2, etc.) [28,29], have the natural advantage of being stable materials down to single-layer thickness, which makes them ideal candidates to overcome the issue of scalability in the base of HETs. During the last few years, several groups have proposed different versions of devices with the base made with 2D materials [30,31]. In the following section, an overview of current strategies followed to implement graphene-base HETs (GBHETs) will be presented.

5.4 HETs with a Graphene Base

5.4.1 Theoretical Properties

A first theoretical estimation of the high-frequency performances of a GBHET was performed by Mehr et al. [32], by solving the 1D Schrödinger equation for the system formed by a metal emitter and collector, a Gr base, a thin insulator working as the E-B tunneling barrier and a thicker insulator working as the B-C filtering barrier. Figure 5.4a reports the schematic band diagrams of the GBHET under equilibrium conditions, in the off-state and in the on-state configurations. Figure 5.4b and c show the simulated output characteristics ($J_C - V_{CE}$) and transfer characteristics ($J_C - V_{BE}$) in the common-emitter configuration for a device with an E-B barrier layer with 3 nm thickness and barrier height ϕ_{EB} = 0.2 eV, and a B-C filtering barrier with 70 nm thickness and ϕ_{BC} = 0.2 eV. These curves illustrate a J_C modulation over several orders of magnitude as a function of the V_{BE} (Figure 5.4c) and

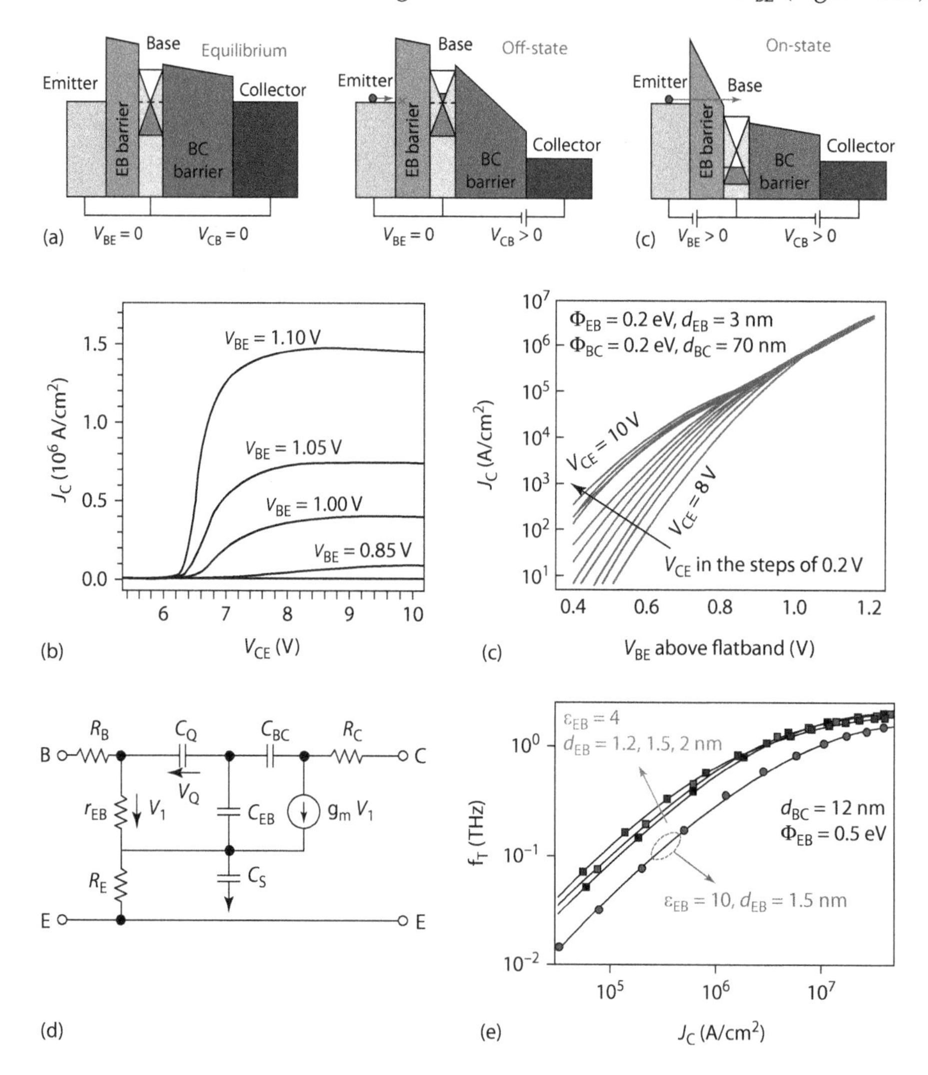

FIGURE 5.4
(a) Schematic band diagrams of a graphene-base hot electron transistor (GBHET) under equilibrium conditions, in the off-state and in the on-state. (b) Simulated common-emitter output characteristic $I_C - V_{CE}$ (for different V_{BE} values) and (c) transfer characteristics $I_C - V_{BE}$ (for different V_{CE} values). (d) Small-signal equivalent circuit of the GBHET and (e) evaluated cut-off frequency as a function of the collector current density J_C for different values of the E-B barrier thickness and permittivity. (Panels (b) and (c) adapted with permission from *Solid State Commun.*, 224, Vaziri, S., et al., Going ballistic: Graphene hot electron transistors, 64–75, Copyright (2015) from Elsevier. Panel (e) adapted with permission from *Microelectron. Eng.*, 109, Driussi, F., et al., Modeling, simulation and design of the vertical graphene-base transistor, 338–341, Copyright (2013) from Elsevier.)

current saturation in the output characteristics (Figure 5.4b). The cut-off frequency, f_T, was evaluated according to Equation 5.2 using the small-signal equivalent circuit illustrated in Figure 5.4d.

The main peculiar aspect of using a Gr base (instead of an ideal metal base) is that the applied bias $V_{CB}>0$ and $V_{BE}>0$ produce not only an increase in the electric field across the B-C and E-B barriers, but they also cause a Gr Fermi level shift with respect to the Dirac point, resulting in a change of the Gr charge density, Q_{gr}. This electrostatic behavior of Gr is expressed by its quantum capacitance $C_Q = |dQ_{gr}/dV_Q|$, where $V_Q=(E_F - E_D)/q$ is the potential drop on Gr.

C_Q is proportional to the density of states $D(E_F - E_D) = 2|E_F - E_D|/(\pi\hbar^2 v_F^2)$, where $\hbar$ is the reduced Planck's constant and v_F is the graphene Fermi velocity [33]. As compared to an ideal metal contact, which is able to accommodate large charge changes in response to small potential variations ($C_Q \longrightarrow \infty$), in the case of Gr C_Q has a finite value due to the finite density of states at the Fermi level. These peculiar properties of the Gr base clearly have an effect on both the DC and AC performances of the GBHET.

As illustrated in the equivalent circuit in Figure 5.4d, C_Q is connected in series to the parallel combination of the E-B capacitance (C_{EB}) and of the B-C capacitance (C_{BC}). As a result, the total capacitance in Equation 5.2 must be expressed as $C_{tot} = C_Q(C_{EB} + C_{BC})/(C_Q + C_{EB} + C_{BC})$. Hence, the finite value of C_Q for Gr results in a reduced C_{tot} (compared to the case of an ideal metal contact) and could be beneficial to the increase of f_T, according to Equation 5.2. On the other hand, since the bias V_{BE} partially falls on Gr, the effective electric field across the E-B barrier is lower with respect to the case of an ideal metal base. This results in a reduced injected current density J_E and, ultimately, in a lower transconductance g_m. Hence, the final effect of the quantum capacitance is a reduction of the estimated maximum f_T value with respect to the case when C_Q is not considered [32]. It should be noted that, notwithstanding this detrimental effect of the finite and bias dependent C_Q of Gr, f_T well above 1 THz have been predicted by these calculations. Figure 5.4e illustrates the behavior of f_T vs the collector current density J_C evaluated considering GBHETs with different values of the E-B barrier thickness and permittivity [34].

It should be noted that these early models did not include scattering effects and charge trapping at interfaces. More refined models for GBHETs including these effects have been subsequently proposed [35–37].

5.4.2 GBHETs Implementation

The experimental implementation of a GBHET firstly requires a selection of insulating or semiconducting materials with proper band alignment with respect to Gr to work as E-B and B-C barriers.

Figure 5.5 shows a summary of the reported literature values for the bandgap and electron affinity of Gr and common semiconductors and insulators [38]. In order to maximize the current injection efficiency from the emitter to the Gr base, the conduction band offset between Gr and the E-B barrier material should be as low as possible. Furthermore, to minimize back-scattering of hot electrons at the B-C filtering barrier, the conduction band offset of this layer with Gr should be lower than the E-B barrier one. Besides these criteria on the theoretical band alignment, high material quality is required for both the E-B and B-C barriers, in order to avoid leakage current through these layers.

The first experimental prototypes of GBHETs were reported by Vaziri et al. [21] and Zeng et al. [22] in 2013. Those demonstrations were fabricated on Si wafers using a complementary metal-oxide-semiconductor (CMOS) compatible technology and were based on metal/insulator/Gr/SiO_2/n^+-Si stacks, where n^+-doped Si substrate worked as the emitter, a thin (5nm) SiO_2 as the E-B barrier, a thicker high-k insulator (Al_2O_3 or HfO_2) as the B-C barrier and the topmost metal layer (e.g. Ti) as the collector.

Figure 5.6 illustrates the band diagrams of this GBHET prototype in different operating conditions, i.e. the flatband (a), the off-state (b) and the on-state (c) conditions. The measured common-emitter output and transfer characteristics of this device are reported in Figure 5.6d and e, respectively. Although the DC characteristics showed several orders of magnitude modulation of J_C as a function of V_{BE}, these first prototypes suffered high threshold voltage and a very poor injected current density (in the order of $\mu A/cm^2$) due

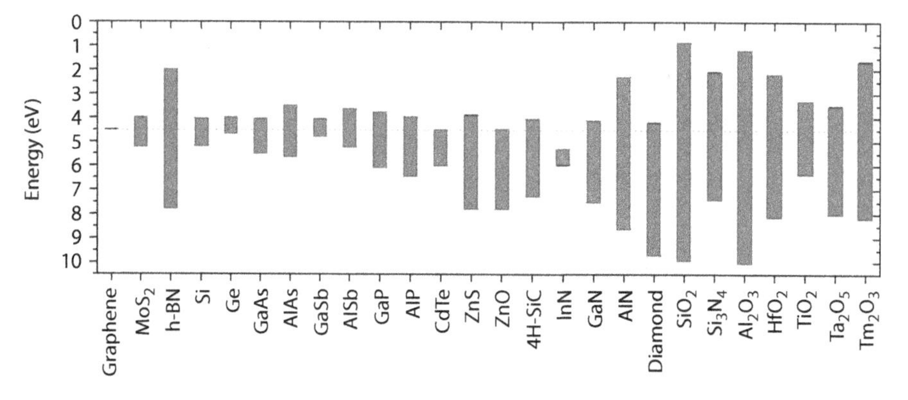

FIGURE 5.5
Band alignment of Gr with common semiconductors or insulators.

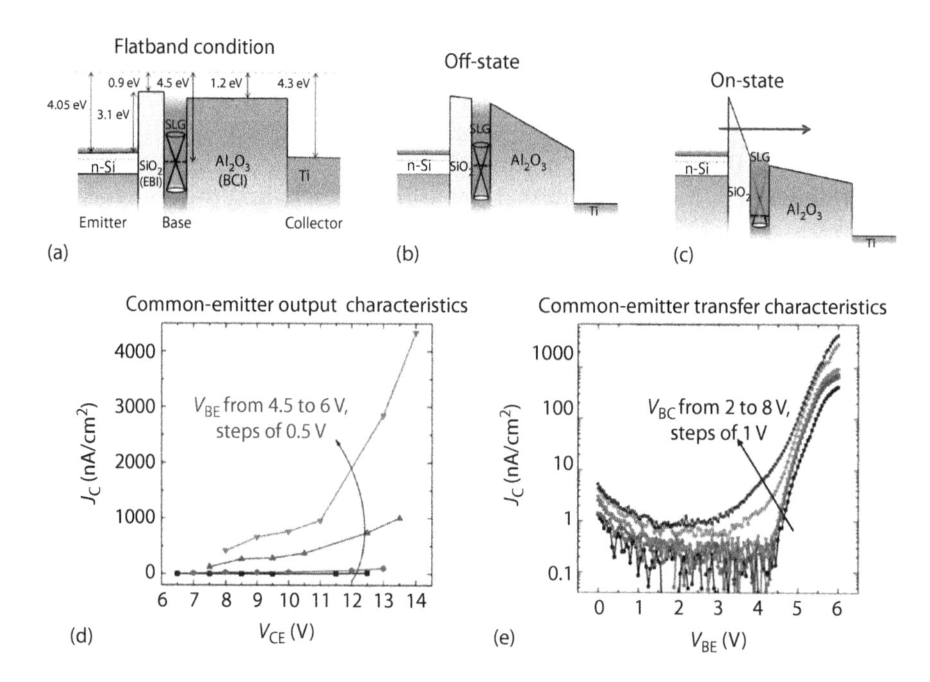

FIGURE 5.6
Band diagram illustrations of the Si/SiO_2/Gr/Al_2O_3/Ti GHET in the flatband (a), off-state (b) and on-state (c) conditions. Common-emitter output characteristics J_C-V_{CE} (d) and transfer characteristics J_C-V_{BE} (e) of the device. (Figures adapted with permission from Ref. [21] Copyright (2013) American Chemical Society.)

to the high Si/SiO_2 barrier, hindering their application at high frequencies. In order to improve the current injection efficiency, other materials have been investigated as E-B barrier layers in replacement of SiO_2 [23].

Figure 5.7 shows the comparison of the injected current density J_{BE} vs V_{BE} characteristics measured on n^+-Si/insulator/Gr junctions with different insulator barrier layers, i.e. a single layer of SiO_2 (5 nm) or HfO_2 (6 nm) (see Figure 5.7a) and a TmSiO (1 nm)/TiO_2 (5 nm) double layer (see Figure 5.7b). As an example, using a 6 nm thick HfO_2 (including a 0.5 nm interfacial SiO_2) deposited by atomic layer deposition (ALD) results in an improved threshold voltage and a higher J_{BE} with respect to the case of an SiO_2 barrier with similar thickness (see Figure 5.7a). Further improvements have been obtained using the TmSiO/TiO_2 (1/5 nm) bilayer, as shown in Figure 5.7b. The role played by the two insulating layers with different electron affinities in the current injection is schematically illustrated in the insert of Figure 5.7b. The thin TmSiO layer (with low electron affinity) in contact with the Si emitter allows high current injection by step tunneling, while the thicker TiO_2 layer (with higher electron affinity) serves to block the leakage current from the Si valence band. A GBHET with this TmSiO/TiO_2 E-B barrier and the B-C barrier made of a 60 nm Si film deposited on Gr has also been fabricated [23]. A collector current density $J_C \approx 4\,A/cm^2$ (more than five orders of magnitude higher than in the first prototypes) was obtained at $V_{BE} = 5\,V$ and for $V_{BC} = 0\,V$ (to avoid the contribution of cold electrons leakage current from the base to the collector). However, the device still suffered from low values of $\alpha \approx 0.28$ and $\beta \approx 0.4$. This can be due to the insufficient quality of the interface between Gr and the deposited B-C barrier, which is a common issue for all insulator or semiconductor thin films deposited on the chemically inert Gr surface.

The above-discussed attempts to implement the GBHET device using Si as the emitter material have been mainly motivated from the perspective of integrating this new technology with the state-of-the-art CMOS fabrication platform. More recently, some research groups in both the United States and Europe have investigated the possibility of demonstrating GBHETs by the integration of Gr with nitride semiconductors. As already discussed in Section 5.3.2, GaN/AlGaN or GaN/AlN heterostructures are excellent systems for use as emitter/emitter-base barriers, due the presence of high-density 2DEG at the interface and to the high structural quality of the barrier layer compared to current oxides.

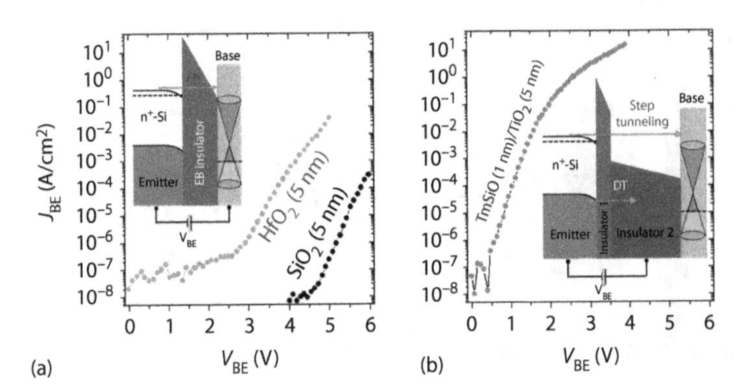

FIGURE 5.7
Injected current density J_{BE} vs V_{BE} measured on n^+-Si/insulator/Gr junctions with a single-layer insulator barrier: SiO_2 (5 nm), HfO_2 (6 nm) (a), and with a bilayer insulator barrier TmSiO (1 nm)/TiO_2 (5 nm) (b). Mechanisms of current injection through the single layer and the bilayer insulator are illustrated in the inserts. (Figures adapted with permission from *Solid State Commun.*, 224, Vaziri, S., et al., Going ballistic: Graphene hot electron transistors, 64–75, Copyright (2015) Elsevier.)

Very efficient current injection by FN tunneling has been shown in the case GaN/AlN/Gr heterojunctions with ultrathin (3 nm) AlN barrier [24], whereas TE has been demonstrated as the main current transport mechanisms in GaN/AlGaN/Gr systems with a thicker (~20 nm) AlGaN barrier layer [39–41].

Figure 5.8a and b illustrates a cross-section schematic and the band diagram of a recently demonstrated GBHET based on a GaN/AlN/Gr/WSe_2/Au stack [24]. A bulk n^+-doped GaN ($N_D \approx 10^{19}$ cm^{-3}) substrate with very low threading dislocation defect density (<10^5 cm^{-2}) was used as the emitter, and a 3 nm AlN tunneling barrier was grown on top of it by plasma-assisted molecular beam epitaxy. The use of a low threading dislocation density substrate ensured a high AlN film quality and minimal leakage current through the dislocations. A 3 nm GaN layer was used as a capping layer between the AlN and the Gr base. In order to circumvent the problems related to the poor interface quality between Gr and conventional insulators or semiconductors deposited on top of it, a layered semiconductor (WSe_2) from the family of TMDs was adopted as a B-C barrier layer. Thin films of WSe_2 were obtained by mechanical exfoliation from the bulk crystal and then transferred onto Gr to form a van der Waals (vdW) heterojunction with a defect-free, sharp interface. The resulting Gr/WSe_2 Schottky junction is characterized by a low barrier height due the small band offset (~0.54 eV) between Gr and WSe_2.

Figure 5.8c shows the common-base output characteristics ($I_C - V_{CB}$) for different values of the emitter injection current, I_E, in the case of a GBHET with a 2.6 nm thick WSe_2 barrier (corresponding to 4 WSe_2 monolayers). Furthermore, Figure 5.8d plots $\alpha = I_C/I_E$ as a function of V_{CB} in the same bias range. Three current transport regimes can be identified

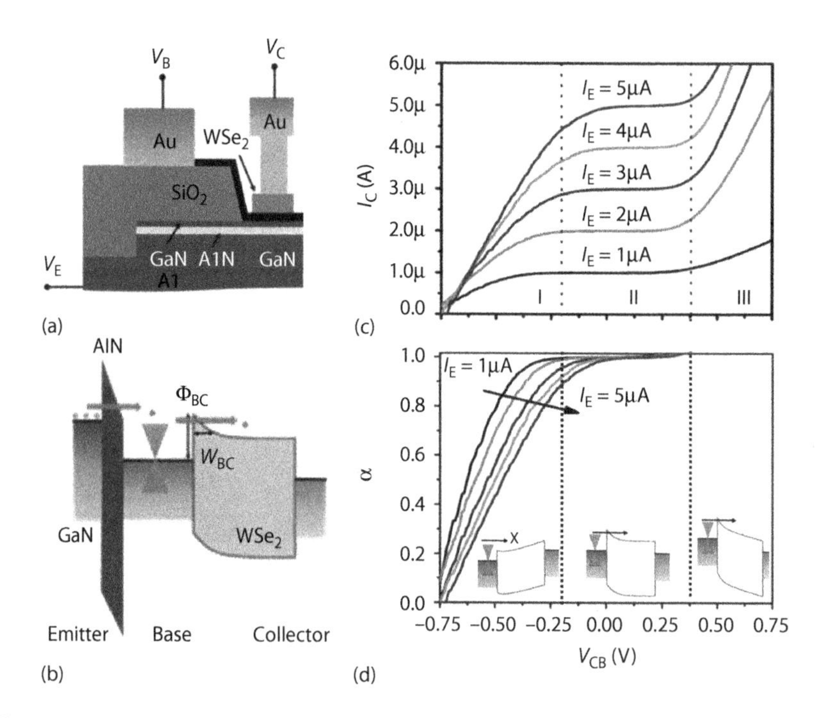

FIGURE 5.8
(a) Cross-section schematic and (b) band diagram of a recently demonstrated GBHET based on a GaN/AlN/Gr/WSe_2/Au stack. (c) Common-base output characteristics ($I_C - V_{CB}$) for different values of the emitter injection current I_E in the case of a GBHET with 2.6 nm thick WSe_2 barrier. (d) Plots of $\alpha = I_C/I_E$ as a function of V_{CB} in the same bias range. (Images adapted with permission from Ref. [24] Copyright (2017) American Chemical Society.)

in the output characteristics. At intermediate V_{CB} bias (region II), I_C is almost independent from the V_{CB} and $\alpha \approx 1$, indicating that almost all the injected hot electrons are able to overcome the Gr/WSe_2 Schottky barrier and reach the collector (as illustrated in the band diagram in the central insert of Figure 5.8c). For $V_{CB} < 0$ (region I), the injected electrons from the emitter are reflected back by the elevated B-C potential barrier (as shown in the band diagram in the left insert of Figure 5.8c), resulting in a reduced I_C and $\alpha < 1$. Finally, at higher positive V_{CB}, the current starts to increase and α becomes superior to 1 due to the increasing contribution of cold electrons leakage current from the base, as illustrated in the band diagram in the right insert of Figure 5.8c. Although this device shows excellent DC characteristics in terms of α, it exhibits a very limited operating V_{BC} window (~0.3 V), as a consequence of the poor blocking capability of the B-C junction with an ultrathin WSe_2 barrier. Increasing the number of layers was demonstrated to improve the blocking capability of the B-C barrier, as the larger interlayer resistance between layers of WSe_2 suppresses the cold electron transport between base and collector. On the other hand, a thicker filtering barrier results in a reduced value of α. As an example, $\alpha = 0.75$ was evaluated for a GaN/AlN/Gr/WSe_2/Au GBHET with 10 nm thick WSe_2 barrier [24].

5.4.3 Open Challenges for the Fabrication of GBHETs

Table 5.1 reports a comparison of the main DC electrical parameters (i.e. the collector current density J_C, the common-base current transfer ratio α and the common-emitter current gain β) for the state-of-the-art GBHETs and nitride semiconductors HETs with sub-10 nm base thickness.

This comparison shows that, in spite of their theoretically predicted superior performances (related to ballistic transport in the atomically thin Gr base), GBHETs still suffer reduced values of J_C, α and β with respect to HETs fabricated by bandgap engineering of III-N semiconductors (even with a thicker GaN base). This degradation of GBHET performances can be due to the not-ideal quality of Gr interfaces with the emitter and collector barriers. Hence, the first challenging task to achieve the optimal behavior of GBHETs is the deposition of a defects-free B-C barrier onto the chemically inert Gr surface. In addition, a suitable barrier height/thickness is required for this layer to minimize at the same time the quantum mechanical reflection of hot electrons and the leakage current of cold electrons from the Gr base. Uniform, i.e. pinhole-free, oxide layers (such as Al_2O_3 and HfO_2) can be deposited on Gr by ALD after proper surface functionalization, with minimal degradation of the Gr electronic/structural properties [42]. Although these oxide films typically exhibit a very low leakage current and a high breakdown field [43], their electron affinity is too low (as shown in Figure 5.5), resulting in a high B-C barrier, ϕ_{BC}. Furthermore, interfaces with ALD-deposited high-k dielectrics typically suffer significant electron trapping.

Semiconductor films are much more suitable as B-C barrier layers for two reasons: (i) the higher electron affinity as compared to oxides (resulting in a lower ϕ_{BC}), and (ii) the possibility of tailoring the barrier width (d_{BC}) by controlling the doping of the semiconductor. As an example, by simple band-alignment consideration, GaN would be the most suitable B-C barrier for a GBHET with a GaN emitter. However, the deposition of GaN thin films with suitable crystalline quality on Gr by MBE or MOCVD currently represents a challenging task [44,45]. The transfer of semiconducting TMDs thin films exfoliated from the parent bulk crystals onto Gr can represent an alternative solution to obtain a high-quality B-C barrier [24]. However, current methods to obtain such vdW heterostructures are not scalable and can be used for fabricating only a few small-size devices, whereas large-scale vdW epitaxy is still in its infancy [46].

TABLE 5.1

Comparison of J_C, α and β for the-State-of-the-Art GBHETs and Nitride-HETs with Sub-10 nm Base Thickness

Emitter/Emitter-Base Barrier	Base (Thickness)	Base-Collector Barrier	JC (A/cm²)	α	β	References
Si/SiO_2	Gr (0.35 nm)	Al_2O_3	$\sim 1\times10^{-5}$	~0.06	~0.06	[21]
Si/SiO_2	Gr (0.35 nm)	Al_2O_3, HfO_2	$\sim 5\times10^{-5}$	~0.44	~0.78	[22]
$Si/TmSiO/TiO_2$	Gr (0.35 nm)	Si	~4	~0.28	~0.4	[23]
GaN/AlN	Gr (0.35 nm)	WSe_2 (10 nm)	~50	~0.75	4–6	[24]
Si/SiO_2	MoS_2 (0.7 nm)	HfO_2	$\sim 1\times10^{-6}$	~0.95	~4	[30]
$GaN/Al_{0.24}Ga_{0.76}N$	GaN (10 nm)	$Al_{0.08}Ga_{0.92}N$	$\sim 5\times10^{3}$	~0.97	—	[18]
GaN/AlN	GaN/InGaN (7 nm)	GaN	$\sim 2.5\times10^{3}$	>0.5	>1	[20]
GaN/AlN	GaN (8 nm)	AlGaN/GaN	$\sim 46\times10^{3}$	~0.93	~14.5	[19]

The second critical task for the future implementation of the GBHET technology is the development of scalable methods for the fabrication of the Gr base. To date, the most used approach to fabricate the Gr base has been the transfer of Gr grown by chemical vapor deposition (CVD) on catalytic metals (such as Cu) to the surface of the E-B barrier layer. Although this is a very versatile and widely used method, it suffers from some drawbacks related to Gr damage and polymer contaminations during the transfer procedure, as well as from possible adhesion problems between Gr and the substrate. Furthermore, it typically introduces undesired metal (Cu, Fe) contaminations [47] originating from the growth substrate and the typically used Cu etchants. An intense research activity is still in progress to optimize Gr transfer procedures to minimize Gr defectivity and contaminations associated with transfer [48–51].

In many respects, the direct growth/deposition of Gr on the E-B barrier would be highly desirable to avoid some of the above-mentioned issues related to Gr transfer. However, to date, high-quality Gr growth has been demonstrated on few semiconducting or semi-insulating materials, such as silicon-carbide [52–56] and, more recently, germanium [57]. Single or a few layers of Gr can be obtained on the Si face (0001) of hexagonal SiC, either by controlled sublimation of Si at high temperatures (typically > 1650°C) in Ar at atmospheric pressure or by direct CVD at lower temperatures (~1450°C) using an external carbon source (such as C_3H_8) with H_2 or H_2/Ar carrier gases [55]. Gr grown on SiC(0001), commonly named epitaxial graphene (EG), generally exhibits a precise epitaxial orientation with respect to the substrate, which originates from the peculiar nature of the interface, i.e. the presence of a carbon buffer layer with mixed sp^2/sp^3 hybridization sharing covalent bonds with the Si face of SiC [58,59]. This buffer layer has a strong impact both on the lateral (i.e. in plane) current transport in EG, causing a reduced carrier mobility, and on the vertical current transport at the EG/SiC interface [60,61]. Hydrogen intercalation at the interface between the buffer layer and the Si face has been demonstrated to be efficient in increasing Gr carrier mobility and tuning the Schottky barrier and, hence, the vertical current transport across the Gr/SiC interface [62]. Recently, a GBHET system based on a metal/AlN/Gr/SiC stack has been presented [63], where SiC worked as the collector, the epitaxial Gr/SiC Schottky barrier worked as the B-C barrier and a thin AlN films deposited on Gr by atomic layer epitaxy worked as the E-B barrier layer [64].

Recently, CVD growth of Gr from carbon precursors on III-N substrates/templates has been investigated. Gr deposition on these non-catalytic surfaces represents a challenging task, as it requires significantly higher temperatures as compared to conventional

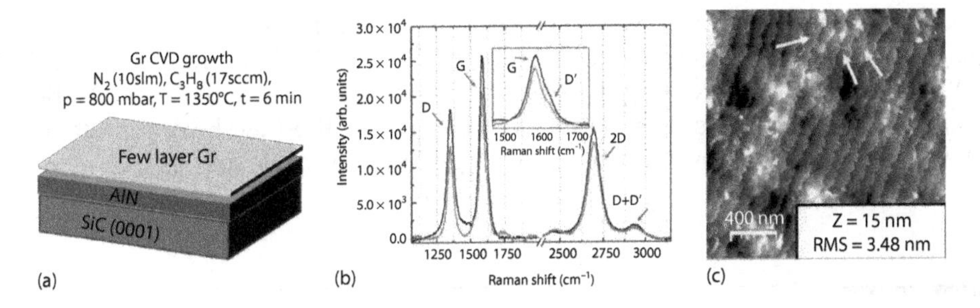

FIGURE 5.9
(a) Schematic illustration of the direct (non-catalytic) CVD growth of Gr at high temperature (T = 1350°C) on the surface of an AlN/SiC template using propane (C_3H_8) as the carbon precursor. (b) Raman spectra collected at two different positions of the AlN surface after Gr deposition. (c) Surface morphology of AlN with deposited Gr. Wrinkles on the Gr surface are highlighted by white arrows.

deposition on metals. The first experimental works addressing this issue showed the possibility of depositing a few layers of Gr both on bulk AlN (Al and N face) and on AlN templates grown on different substrates, such as Si(111) and SiC, at temperatures >1250°C using propane (C_3H_8) as the carbon source, without significantly degrading the morphology of AlN substrates/templates [65,66]. Figure 5.9a schematically illustrates the CVD growth conditions of Gr on the surface of an AlN/SiC template, i.e. the C_3H_8 and N_2 gas fluxes, the pressure p = 800 mbar and the temperature T = 1350°C. Figure 5.9b shows Raman spectra collected at different positions on the AlN surface after Gr deposition, demonstrating the deposition of a few layers of Gr, with Gr domain size in the order of 30 nm. Figure 5.9c illustrates the typical surface morphology of AlN with deposited Gr. Wrinkles, i.e. typical corrugations of the Gr membrane, are highlighted by white arrows. In spite of the very promising results of these experiments, further work will be required to evaluate the feasibility and the effects of CVD Gr growth on AlN/GaN or AlGaN/GaN heterostructures. Moreover, the possibility of integrating these high-temperature processes in the fabrication flow of GBHETs needs to be investigated.

5.5 Summary

In summary, we have discussed the HET device concept for THz applications, starting from the operating principles, to the physical and technological challenges related to its practical implementation. The most recent developments of this technology have been discussed, in particular nitride semiconductors–based HETs and the graphene-base HET, with open issues and new ideas to exploit the expected ultrahigh-frequency performances of this device.

Acknowledgments

The authors want to acknowledge the following colleagues for useful discussions: E. Frayssinet (CNRS-CRHEA, France); M. Leszczynski, P. Kruszewski and P. Prystawko (TopGaN, Poland); S. Ravesi, S. Lo Verso and F. Iucolano (STMicroelectronics, Catania,

Italy); G. Fisichella, E. Schilirò, S. Di Franco, P. Fiorenza, R. Lo Nigro, I. Deretzis, A. La Magna, G. Nicotra, C. Bongiorno and C. Spinella (CNR-IMM, Catania, Italy); R. Yakimova and A. Kakanakova (Linkoping University, Sweden); B. Pecz (HAS-MFA, Budapest, Hungary). This chapter has been supported, in part, by the Flag-ERA project "GraNitE: Graphene heterostructures with Nitrides for high frequency Electronics".

References

1. R. Lai, X. B. Mei, W. R. Deal, W. Yoshida, Y. M. Kim, P. H. Liu, J. Lee, J. Uyeda, V. Radisic, M. Lange, et al., Sub 50 nm InP HEMT device with f_{max} greater than 1 THz, *Proceeding of IEEE Electron Devices Meeting*, Washington, DC, December (2007), pp. 609–611.
2. D. H. Kim, J. A. del Alamo, P. Chen, W. Ha, M. Urteaga, and B. Brar, 50-nm E-mode In0.7Ga0.3As PHEMTs on 100-mm InP substrate with >1 THz, *Proceeding of IEEE Electron Devices Meeting*, San Francisco, CA, December (2010), pp. 30.6.1–30.6.4.
3. M. Urteaga, R. Pierson, P. Rowell, V. Jain, E. Lobisser, and M. J. W. Rodwell, 130 nm InP DHBTs with $f_t > 0.52$ THz and $f_{max} > 1.1$ THz, *Proceeding of 69th Annual Device Research Conference*, Santa Barbara, CA, June (2011), pp. 281–282.
4. V. Jain, J. C. Rode, H.-W. Chiang, A. Baraskar, E. Lobisser, B. J. Thibeault, M. Rodwell, M. Urteaga, D. Loubychev, A. Snyder, et al., 1.0 THz f_{max} InP DHBTs in a refractory emitter and self-aligned base process for reduced base access resistance, *Proceeding of 69th Annual Device Research Conference*, Santa Barbara, CA, June (2011), pp. 271–272.
5. J. C. Rode, H.-W. Chiang, P. Choudhary, V. Jain, B. J. Thibeault, W. J. Mitchell, M. J. Rodwell, M. Urteaga, D. Loubychev, A. Snyder, et al., Indium phosphide heterobipolar transistor technology beyond 1-THz bandwidth, *IEEE Trans. Electron. Devices* 62 (9), 2779–2785 (2015).
6. C. A. Mead, Operation of tunnel-emission devices, *J. Appl. Phys.* 32, 646–652 (1961).
7. M. M. Atalla and R.W. Soshea, Hot-carrier triodes with thin-film metal base, *Solid-State Electron.* 6, 245–250 (1963).
8. J. C. Hensel, A. F. J. Levi, R. T. Tung, and J. M. Gibson, Transistor action in Si/$CoSi_2$/Si heterostructures, *Appl. Phys. Lett.* 47, 151–153 (1985).
9. E. Rosencher, P. A. Badoz, J. C. Pfister, F. Arnaud d'Avitaya, G. Vincent, and S. Delage, Study of ballistic transport in Si-$CoSi_2$-Si metal base transistors, *Appl. Phys. Lett.* 49, 271–273 (1986).
10. T. Yajima, Y. Hikita, and H. Y. Hwang, A heteroepitaxial perovskite metal-base transistor, *Nat. Mater.* 10 (3), 198–201 (2011).
11. M. Tonouchi, H. Sakai, T. Kobayashi, and K. Fujisawa, A novel hot-electron transistor employing superconductor base, *IEEE Trans. Magn.* 23, 1674–1677 (1987).
12. M. Heiblum, D. C. Thomas, C. M. Knoedler, and M. I. Nathan, Tunneling hot-electron transfer amplifier: A hot-electron GaAs device with current gain, *Appl. Phys. Lett.* 47, 1105 (1985).
13. M. S. Shur, A. D. Bykhovski, R. Gaska, M. Asif Khan, and J. W. Yang, AlGaN–GaN–AlInGaN induced base transistor, *Appl. Phys. Lett.* 76, 3298 (2000).
14. A. S. Mayorov, R. V. Gorbachev, S. V. Morozov, L. Britnell, R. Jalil, L. A. Ponomarenko, P. Blake, K. S. Novoselov, K. Watanabe, T. Taniguchi, et al., Micrometer-scale ballistic transport in encapsulated graphene at room temperature, *Nano Lett.* 11, 2396–2399 (2011).
15. S. Sonde, F. Giannazzo, C. Vecchio, R. Yakimova, E. Rimini, and V. Raineri, Role of graphene/substrate interface on the local transport properties of the two-dimensional electron gas, *Appl. Phys. Lett.* 97, 132101 (2010).
16. F. Giannazzo, S. Sonde, R. Lo Nigro, E. Rimini, and V. Raineri, Mapping the density of scattering centers limiting the electron mean free path in graphene, *Nano Lett.* 11, 4612–4618 (2011).
17. B. D. Kong, Z. Jin, and K. W. Kim, Hot-electron transistors for terahertz operation based on two-dimensional crystal heterostructures, *Phys. Rev. Appl.* 2, 054006 (2014).

18. S. Dasgupta, N. A. Raman, J. S. Speck, and U. K. Mishra, Experimental demonstration of III-nitride hot-electron transistor with GaN base, *IEEE Electron. Device Lett.* 32, 1212–1214 (2011).
19. Z. Yang, Y. Zhang, D. N. Nath, J. B. Khurgin, and S. Rajan, Current gain in sub-10nm base GaN tunneling hot electron transistors with AlN emitter barrier, *Appl. Phys. Lett.* 106, 032101 (2015).
20. G. Gupta, E. Ahmadi, K. Hestroffer, E. Acuna, and U. K. Mishra, Common emitter current gain >1 in III-N hot electron transistors with 7-nm GaN/InGaN base, *IEEE Electron. Device Lett.* 36, 439–441 (2015).
21. C. S. Vaziri, G. Lupina, C. Henkel, A. D. Smith, M. Ostling, J. Dabrowski, G. Lippert, W. Mehr, and M. Lemme, A graphene-based hot electron transistor, *Nano Lett.* 13, 1435 (2013).
22. C. Zeng, E. B. Song, M. Wang, S. Lee, C. M. Torres, J. Tang, B. H. Weiller, and K. L. Wang, Vertical graphene-base hot electron transistor, *Nano Lett.* 13, 2370 (2013).
23. S. Vaziri, M. Belete, E. Dentoni Litta, A. D. Smith, G. Lupina, M. C. Lemme and M. Östlinga, Bilayer insulator tunnel barriers for graphene-based vertical hot-electron transistors, *Nanoscale* 7, 13096–13104 (2015).
24. A. Zubair, A. Nourbakhsh, J.-Y. Hong, M. Qi, Y. Song, D. Jena, J. Kong, M. Dresselhaus, and T. Palacios, Hot electron transistor with van der Waals base-collector heterojunction and high-performance GaN emitter, *Nano Lett.* 17, 3089–3096 (2017).
25. A. F. J. Levi and T. H. Chiu, Room-temperature operation of hot-electron transistors, *Appl. Phys. Lett.* 51, 984 (1987).
26. T. S. Moise, Y.-C. Kao, and A. C. Seabaugh, Room-temperature operation of a tunneling hot-electron transfer amplifier, *Appl. Phys. Lett.* 64, 1138–1140 (1994).
27. F. Giannazzo and V. Raineri, Graphene: Synthesis and nanoscale characterization of electronic properties, *Rivista del Nuovo Cimento* 35, 267–304 (2012).
28. F. Giannazzo, G. Fisichella, A. Piazza, S. Agnello, and F. Roccaforte, Nanoscale inhomogeneity of the Schottky barrier and resistivity in MoS_2 multilayers, *Phys. Rev. B* 92, 081307(R) (2015).
29. F. Giannazzo, G. Fisichella G. Greco, S. Di Franco, I. Deretzis, A. La Magna, C. Bongiorno, G. Nicotra, C. Spinella, M. Scopelliti, et al., Ambipolar MoS_2 transistors by nanoscale tailoring of Schottky barrier using oxygen plasma functionalization, *ACS Appl. Mater. Interfaces* 9, 23164–23174 (2017).
30. S. Vaziri, A. D. Smith, M. Östling, G. Lupina, J. Dabrowski, G. Lippert, W. Mehr, F. Driussi, S. Venica, V. Di Lecce, et al., Going ballistic: Graphene hot electron transistors, *Solid State Commun.* 224, 64–75 (2015).
31. C. M. Torres, Y. W. Lan, C. Zeng, J. H. Chen, X. Kou, A. Navabi, J. Tang, M. Montazeri, J. R. Adleman, M. B. Lerner, et al., High-current gain two-dimensional MoS_2-base hot-electron transistors, *Nano Lett.* 15 (12), 7905–7912 (2015).
32. W. Mehr, J. Dabrowski, J. C. Scheytt, G. Lippert, Y.-H. Xie, M. C. Lemme, M. Ostling, and G. Lupina, Vertical graphene base transistor, *IEEE Electron. Device Lett.* 33, 691–693 (2012).
33. F. Giannazzo, S. Sonde, V. Raineri, and E. Rimini, Screening length and quantum capacitance in graphene by scanning probe microscopy, *Nano Lett.* 9, 23 (2009).
34. F. Driussi, P. Palestri, and L. Selmi, Modeling, simulation and design of the vertical graphene base transistor, *Microelectron. Eng.* 109, 338–341 (2013).
35. V. Di Lecce, R. Grassi, A. Gnudi, E. Gnani, S. Reggiani, and G. Baccarani, Graphene base transistors: A simulation study of DC and small-signal operation, *IEEE Trans Electron. Devices* 60, 3584–3591 (2013).
36. V. Di Lecce, R. Grassi, A. Gnudi, E. Gnani, S. Reggiani, and G. Baccarani, Graphene-base heterojunction transistor: An attractive device for terahertz operation, *IEEE Trans. Electron. Devices* 60, 4263–4268 (2013).
37. V. Di Lecce, A. Gnudi, E. Gnani, S. Reggiani, and G. Baccarani, Simulations of graphene base transistors with improved graphene interface model. *IEEE Trans. Electron. Devices* 36, 969–971 (2015).
38. U. K. Mishra and J. Singh, *Semiconductor Device Physics and Design*, Springer, Dordrecht, 2008, ISBN 978-1-4020-6480-7.
39. G. Fisichella, G. Greco, F. Roccaforte, and F. Giannazzo, Current transport in graphene/AlGaN/GaN vertical heterostructures probed at nanoscale, *Nanoscale*, 6, 8671–8680 (2014).

40. F. Giannazzo, G. Fisichella, G. Greco, and F. Roccaforte, Challenges in graphene integration for high-frequency electronics, *AIP Conference Proc.* 1749, 020004 (2016).
41. F. Giannazzo, G. Fisichella, G. Greco, A. La Magna, F. Roccaforte, B. Pecz, R. Yakimova, R. Dagher, A. Michon, and Y. Cordier, Graphene integration with nitride semiconductors for high power and high frequency electronics, *Phys. Status Solidi A* 214, 1600460 (2017).
42. R. H. J. Vervuurt, W. M. M. Kessels, and A. A. Bol, Atomic layer deposition for graphene device integration, *Adv. Mater. Interfaces* 4, 1700232 (2017).
43. G. Fisichella, E. Schilirò, S. Di Franco, P. Fiorenza, R. Lo Nigro, F. Roccaforte, S. Ravesi, and F. Giannazzo, Interface electrical properties of Al_2O_3 thin films on graphene obtained by atomic layer deposition with an in situ seed-like layer, *ACS Appl. Mater. Interfaces* 9, 7761–7771 (2017).
44. T. Araki, S. Uchimura, J. Sakaguchi, Y. Nanishi, T. Fujishima, A. Hsu, K. K. Kim, T. Palacios, A. Pesquera, A. Centeno, et al., Radio-frequency plasma excited molecular beam epitaxy growth of GaN on graphene/Si(100) substrates, *Appl. Phys. Express* 7, 071001 (2014).
45. J. Kim, C. Bayram, H. Park, C.-W. Cheng, C. Dimitrakopoulos, J. A. Ott, K. B. Reuter, S. W. Bedell, and D. K. Sadana, Principle of direct van der Waals epitaxy of single-crystalline films on epitaxial graphene, *Nature Commun.* 5, 4836 (2014).
46. Y. Shi, W. Zhou A.-Y. Lu, W. Fang, Y.-H. Lee, A. L. Hsu, S. M. Kim, K. K. Kim, H. Y. Yang, L.-J. Li, et al., van der Waals epitaxy of MoS_2 layers using graphene as growth templates, *Nano Lett.* 12, 2784–2791 (2012).
47. G. Lupina, J. Kitzmann, I. Costina, M. Lukosius, C. Wenger, A. Wolff, S. Vaziri, M. Östling, I. Pasternak, A. Krajewska, et al., Residual metallic contamination of transferred chemical vapor deposited graphene, *ACS Nano* 9, 4776–4785 (2015).
48. G. Fisichella, S. Di Franco, F. Roccaforte, S. Ravesi, and F. Giannazzo, Microscopic mechanisms of graphene electrolytic delamination from metal substrates, *Appl. Phys. Lett.* 104, 233105 (2014).
49. J.-Y. Hong, Y. C. Shin, A. Zubair, Y. Mao, T. Palacios, M. S. Dresselhaus, S. H. Kim, and J. Kong, A rational strategy for graphene transfer on substrates with rough features, *Adv. Mater.* 28, 2382–2392 (2016).
50. J.-K. Choi, J. Kwak, S.-D. Park, H. D. Yun, S.-Y. Kim, M. Jung, S. Y. Kim, K. Park, S. Kang, S.-D. Kim, et al., Growth of wrinkle-free graphene on texture-controlled platinum films and thermal-assisted transfer of large-scale patterned graphene, *ACS Nano* 9, 679–686 (2015).
51. H. H. Kim, S. K. Lee, S. G. Lee, E. Lee, and K. Cho, Wetting-assisted crack- and wrinkle-free transfer of wafer-scale graphene onto arbitrary substrates over a wide range of surface energies, *Adv. Funct. Mater.* 26, 2070–2077 (2016).
52. C. Berger, Z. Song, X. Li, X. Wu, N. Brown, C. Naud, D. Mayou, T. Li, J. Hass, A. N. Marchenkov, et al., Electronic confinement and coherence in patterned epitaxial graphene, *Science* 312, 1191–1195 (2006).
53. K. V. Emtsev, A. Bostwick, K. Horn, J. Jobst, G. L. Kellogg, L. Ley, J. L. McChesney, T. Ohta, S.A. Reshanov, J. Rohrl, et al. Towards wafer-size graphene layers by atmospheric pressure graphitization of silicon carbide, *Nat. Mater.* 8, 203–207 (2009).
54. C. Virojanadara, M. Syvajarvi, R. Yakimova, L. I. Johansson, A. A. Zakharov, and T. Balasubramanian, Homogeneous large-area graphene layer growth on 6H-SiC(0001), *Phys. Rev. B* 78, 245403 (2008).
55. A. Michon, S. Vézian, E. Roudon, D. Lefebvre, M. Zielinski, T. Chassagne, and M. Portail, Effects of pressure, temperature, and hydrogen during graphene growth on SiC(0001) using propane-hydrogen chemical vapor deposition, *J. Appl. Phys.* 113, 203501 (2013).
56. C. Bouhafs, A. A. Zakharov, I. G. Ivanov, F. Giannazzo, J. Eriksson, V. Stanishev, P. Kühne, T. Iakimov T. Hofmann, M. Schubert, et al., Multi-scale investigation of interface properties, stacking order and decoupling of few layer graphene on C-face 4H-SiC, *Carbon* 116, 722–732 (2017).
57. J.-H. Lee, E. K. Lee, W.-J. Joo, Y. Jang, B.-S. Kim, J. Y. Lim, S.-H. Choi, S. J. Ahn, J. R. Ahn, M.-H. Park, et al., Wafer-scale growth of single-crystal monolayer graphene on reusable hydrogen-terminated germanium, *Science* 344, 286–288 (2014).

58. F. Varchon, R. Feng, J. Hass, X. Li, B. N. Nguyen, C. Naud, P. Mallet, J. Y. Veuillen, C. Berger, E.H. Conrad, et al. Electronic structure of epitaxial graphene layers on SiC: Effect of the substrate, *Phys. Rev. Lett.* 99, 126805 (2007).
59. G. Nicotra, Q. M. Ramasse, I. Deretzis, A. La Magna, C. Spinella, and F. Giannazzo, Delaminated graphene at silicon carbide facets: Atomic scale imaging and spectroscopy, *ACS Nano* 7, 3045–3052 (2013).
60. F. Giannazzo, I. Deretzis, A. La Magna, F. Roccaforte, and R. Yakimova, Electronic transport at monolayer-bilayer junctions in epitaxial graphene on SiC, *Phys. Rev. B* 86, 235422 (2012).
61. S. Sonde, F. Giannazzo, V. Raineri, R. Yakimova, J.-R. Huntzinger, A. Tiberj, J. Camassel, Electrical properties of the graphene/4H-SiC (0001) interface probed by scanning current spectroscopy, *Phys. Rev. B* 80, 241406(R) (2009).
62. F. Speck, J. Jobst, F. Fromm, M. Ostler, D. Waldmann, M. Hundhausen, H. Weber, and T. Seyller, The quasi-freestanding nature of graphene on H-saturated SiC(0001), *Appl. Phys. Lett.* 99, 122106 (2011).
63. A. D. Koehler, N. Nepal, M. J. Tadjer, R. L. Myers-Ward, V. D. Wheeler, T. J. Anderson, M. A. Mastro, J. D. Greenlee, J. K. Hite, K. D. Hobart, et al., Practical challenges of processing III-nitride/graphene/SiC devices, *CS MANTECH Conference*, May 18–21, (2015), Scottsdale, Arizona.
64. N. Nepal, V. D. Wheeler, T. J. Anderson, F. J. Kub, M. A. Mastro, R. L. Myers-Ward, S. B. Qadri, J. A. Freitas, S. C. Hernandez, L. O. Nyakiti, et al., Epitaxial growth of III-nitride/graphene heterostructures for electronic devices, *Appl. Phys. Express* 6, 061003 (2013).
65. A. Michon, A. Tiberj, S. Vezian, E. Roudon, D. Lefebvre, M. Portail, M. Zielinski, T. Chassagne, J. Camassel, and Y. Cordier, Graphene growth on AlN templates on silicon using propane-hydrogen chemical vapor deposition, *Appl. Phys. Lett.* 104, 071912 (2014).
66. R. Dagher, S. Matta, R. Parret, M. Paillet, B. Jouault, L. Nguyen, M. Portail, M. Zielinski, T. Chassagne, S. Tanaka, et al., High temperature annealing and CVD growth of few-layer graphene on bulk AlN and AlN templates, *Phys. Status Solidi A* 214, 1600436 (2017).

6

Tailoring Two-Dimensional Semiconductor Oxides by Atomic Layer Deposition

Mohammad Karbalaei Akbari and Serge Zhuiykov

CONTENTS

6.1 Introduction

The development of the scientific field of two-dimensional (2D) materials was prompted by the discovery of graphene and its extraordinary properties, which opened up new horizons for materials scientists [1–6]. The field evolved with the invention of a number of 2D materials beyond graphene with unexpected and distinctive physical and chemical properties, leading to considerable attention being devoted to understanding the properties of this new class of nanomaterials [7–11]. Ultra-thin 2D semiconductor oxides introduced another class of quasi-2D films whose properties often deviated from their respective microstructural counterparts. The nanoscale dimensions of 2D materials have a much greater level of structural, thermodynamic and kinetic freedom compared with those constraints in microstructural materials, caused by the microscopic extension of three-dimensional structures, making it possible to achieve unprecedented properties in 2D nano-films. The rise of these unexpected 2D semiconductor oxides properties is in close relation with their ultra-thin thickness. However, deviation from regular physical and chemical behavior is not determined by the specific film thickness, but depends on the intrinsic properties of oxide materials, their synthesis methods, structural features and the targeted properties. Accordingly, the addition or reduction of a monolayer can appreciably affect the properties of ultra-thin films and deviations may occur as a function of film thickness [11–15]. The definition of a 2D nanostructure is quite broad since the various synthesis techniques give very different structural and geometrical features to the 2D nanostructures produced. The concept of 2D oxide structures can be assigned to ultra-thin oxide films with a thickness of less than 1 nm (monolayer) up to a few nanometers (several fundamental layers). So far, the attention of researchers has mostly concentrated on the development of ultra-thin oxide films with a thickness of only one atomic layer or a single polyhedron layer of oxide materials, which approaches the behavior of oxide materials at the 2D limit with a high level of approximation [16]. In this context, characterizing and exploiting the properties of 2D oxide materials for practical applications and acquiring knowledge of device fabrication have attracted tremendous attention together with advancing the miniaturization of devices and shrinking the component dimensions [17–19]. Apart from pure scientific exploration of 2D semiconductor oxides, nowadays novel technologies use quasi ultra-thin oxide structures as the main components of novel nano-devices, which can be upgraded and evolved into new configurations. Consequently, promising applications of 2D oxide nanostructures are envisaged. Ultra-thin oxide materials have already found several applications in solid oxide fuel cells [20], catalyst films [21], corrosion protection layers [22], gas sensors [23], spintronic devices [24] and ultraviolet (UV) and visible light sensors [25]. Two-dimensional oxide semiconductors are considered one of the main components of metal-oxide-semiconductor field-effect transistors (MOSFETs) and other novel nanoelectronic devices [26]. Solar energy cells [27], plasmonic devices [28,29], data storage applications [30], biocompatibility and biosensing features [31] and, lately, supercapacitance properties [32,33] and electrochemical sensing [34] are among the recently announced applications of 2D oxide semiconductors.

A large number of ultra-thin layered materials can be synthesized by mechanical and chemical exfoliation, as well as deposition and growth techniques [15,35,36]. Achieving freely suspended, large-area 2D films is extremely challenging since the structural stability and molecular integrity of 2D films can be easily destroyed. Accordingly, one of the main challenges is the versatile and conformal deposition of these quasi-2D oxide nanostructures on convenient substrates including conducting, semiconducting and dielectric sublayers [37–45]. Precise control of thickness, monitoring of conformity and continuity of 2D oxide films over the substrate are among the major technical challenges, especially when the wafer-scale growth of 2D nanostructures is the main scope of discussion [46]. The growth of 2D oxides is achievable by using high-tech deposition techniques mostly exploiting the benefits of physical vapor and chemical vapor deposition (CVD) approaches by exerting ultra-high vacuum conditions during the fabrication process [46–48]. Since thin film growth is inherently a non-equilibrium process, discussion of the stoichiometry of deposited 2D oxide films, the stabilization of meta-stable oxide phases, the structural controllability and phase complexities of oxide–metal interface systems are of vital importance.

Atomic layer deposition (ALD) is a cyclic vapor–phase deposition process that has the advantages of the temporally separated and self-limiting reactions of two or more reactive precursors [49]. The sequential order and restricting nature of the ALD process enable the deposition of ultra-thin films of one atomic layer thickness at a time. Despite the capabilities of the ALD technique, the conformal deposition of 2D semiconductor oxide nanostructures on a desired surface is an extremely challenging process since the complete monolayer film is not deposited even if the surface of the substrate is covered entirely and saturated by precursors. In this case, the reactivity of the substrate surface to precursors and the later sequential ligand exchanges and self-limiting reactions are strongly engaged in the successful deposition of monolayer structures. These are considered as interwoven parameters affecting the successful or unsuccessful outcomes of the ALD process of the 2D nanostructures [50,51]. It turns out to be a more intricate case when the focus of the study shifts from the deposition of simple 2D oxides to ternary or complex oxide structures or heterojunctions.

The aim of this chapter is therefore to provide an overview of the development of 2D nanostructures using the ALD technique. Attention will be concentrated on the deposition of 2D semiconductor oxides (e.g. films with a thickness of a few or several monolayers) on a conducting or semiconducting substrates (e.g. metal or silicon/silicon oxide substrate). The chapter will begin with an explanation of interfacial and structural concepts in deposited 2D oxide films. Then, the principles of ALD will be introduced focusing on the development of 2D nanostructures. In doing so, the ALD development of 2D WO_3, TiO_2 and Al_2O_3 oxides will be discussed as practical cases. The developed recipes can be used as technical examples and adjusted to deal with the other challenges in the development of 2D nanostructured semiconductor oxides. The chapter will end with discussions on the practical applications of ALD-developed 2D oxide semiconductors.

6.2 Interfacial and Structural Concepts in Deposited 2D Oxides

The growth behavior of 2D oxide films is affected by a series of interconnected parameters, including the surface and interfacial energy of the oxide–metal system, the stoichiometry and thickness of 2D oxide films, the lattice mismatch between oxide film and metal substrate and the growth conditions [41,52]. Here, the focus is on interfacial and structural concepts in metal-supported 2D oxide films to provide a deeper understanding for future discussions.

6.2.1 Substrate Effects

The atomic structure is one of the main fundamental characteristics of materials from which many other key properties are determined. From a structural point of view, the crystalline bulk of oxide materials can be considered as the sequential order of alternating crystalline planes or, from another point of view, they can be envisaged as the combination of metal oxygen polyhedral coordination blocks that are connected via shared corners, edges and the plane of polyhedral coordination blocks [14]. The crystalline structures of some simple and complex 2D oxide nanostructures are presented in Figure 6.1 [53]. Taking a typical titanium oxide as an example, the host layer consists of TiO_6 octahedra, which are connected via edge sharing to form a 2D structure. Theoretically, a stoichiometric TiO_2 composition is gained by full occupation of the octahedral site by Ti atoms, while the simple lack of Ti sites or the substitution of Ti atoms by other metal atoms results in a negative charge in the host TiO layer [54]. The complexity of stoichiometry issues will increase as the structure of the 2D oxide nanostructure becomes more complicated.

In 2D structures, the limitations caused by the structural features of bulk materials are removed, alleviated or altered [55]. The epitaxial effects of a substrate on the growth of 2D oxide films alone, the interaction between the oxide ultra-thin films and their substrates experiences both charge distribution and mass transfer [56]. Compared with a bulk oxide,

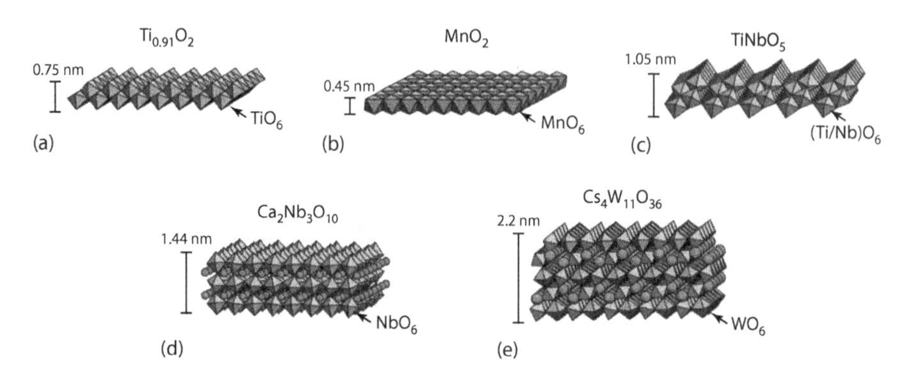

FIGURE 6.1
The structures of selected oxide nanosheets (a) $Ti_{0.91}O_2$, (b) MnO_2, (c) $TiNbO_5$, (d) $Ca_2Nb_3O_{10}$, (e) $Cs_4W_{11}O_{36}$. (Reproduced from Osada, M. and Sasaki, T., *Nanofabrication*, InTech, 2011.)

the chemical bonding, the electronic structure and its levels and interfacial interactions are fundamentally changed in 2D oxide nano-films [57–59], with each individual alteration unpredictably affecting the properties of ultra-thin nanostructures. For practical targets, the deposited monolayer oxides are usually supported by a solid substrate [14,60,61]. The atomically clean surface of inert metals is the most convenient substrate for the deposition of 2D oxide films [14]. The rigid support for a 2D oxide layer can be gained by using noble metals of Group Ib (Cu, Ag, Au) as the substrate. The resulting interface would be abrupt, but not necessarily a chemically inert one, keeping in mind, however, that the interplay and interaction between metal atoms of a substrate and the metal oxide components would be effective in determining the geometrical features of deposited 2D oxide films and overwhelmingly affect the resulting properties of 2D oxide nanostructures [41–43]. The conducting characteristic of metal substrates is beneficial for the charge transfer from an oxide monolayer to a metal substrate for subsequent experimental measurements of the electrical properties of 2D oxide materials [44]. The conductance of metal substrates also gives the opportunity to characterize the structural and chemical features of 2D oxide films by electron-based probe techniques, scanning tunneling microscopy (STM) and electron diffraction techniques [41,45].

Planar mono and few-layer oxides are deposited by various methods on the surface of metal substrates. The stabilization of ultra-thin films can be attained through the interaction between the metal substrate and the deposited oxide films. This new oxide–metal configuration introduces a new class of materials with novel properties that don't exist inherently in nature [60–62]. Depending on the thickness of the 2D oxide film, which is considered a 2D polar structure, the interaction between the oxide and metal films can significantly modify the structural properties, stability and electronic structure of polar oxide films [63]. At the interface of a thick polar oxide film and the metal substrate, the polarity effects are strongly alleviated by the transfer of compensating charges from the oxide film to the metal substrate [64]. For ultra-thin oxide films and especially in the case of uncompensated polarity, the interface of the oxide film and the metal substrate experiences an interfacial charge transfer that no longer originated from the requirements of polarity compensation [63]. In this case, the thin film undergoes a structural disorientation to respond to the electrostatic field created due to the interfacial charge transfer between the ultra-thin oxide film and the metal substrate. The created rumpling causes a separation between the atomic planes of cations and anions of a 2D polar film [65]. The rumpling can be ignored in the case of unsupported oxide films. However, it plays a significant role when the ultra-thin oxide film is grown on a metal substrate [65,66]. The direction and the extent of the charge transfer at the 2D oxide–metal interface are determined by the electronegativity of the metal substrate. When the metal substrate has high electronegativity, the electrons are transferred from the 2D oxide to metal support. Consequently, due to electrostatic forces, the anions of the oxide layer are pushed outward, as presented in Figure 6.2a [67]. For example, in the case of MgO monolayers deposited on transition metals (Ag, Au, Pt, substrates with high electronegativity), dynamic functional theory (DFT) has shown that the electrons flow from the monolayer oxide film into the metal substrate inducing a positive rumpling in which oxygen atoms of 2D films relax outward. Vice versa, in the metal substrate with low electronegativity the electron current will flow from the metal substrate to the 2D metal oxide. The displacement of cations of the oxide layer to the outward face of the 2D films is the feature of these 2D oxide–metal structures, shown in Figure 6.2b [67]. The negative rumpling was predicted by DFT when the 2D MgO film was deposited on the metal surfaces with low electronegativity (Al and Mg) resulting in a closer positioning of the oxygen atoms (anions) to the metal surface [67–69].

The interface of 2D oxide with Au can be considered a practical example of the interaction between 2D oxide films and metal substrates. Once the oxide film has grown, the mass transfer between the oxide film and the metal substrate is usually ignored, since the Au surface is inherently inert. However, the interfacial structure should be considered for the growth of the first monolayer. Thus, the bonding energies between Au and the oxygen (O) or metal (M) component of the 2D oxide film play a significant role in determining the chemical bonding structures at the 2D oxide–Au interface [56,70]. Considering the weak adsorption of oxygen on the Au surface and the facile decomposition of the Au oxide, the preferential bonding between Au and metal atoms of the oxide film is predicted, mostly resulting in the development of Au–M–O interfaces [71–73]. Surprisingly, the work function of the metal substrates can be tuned by the appropriate selection of a 2D oxide–metal interface providing the opportunity to design nano-devices by nano-surface engineering. For instance, while the growth of an MgO thin film decreases the work function of an Au substrate, the deposition of TiO_2 and SiO_2 results in the raised work function of the Au [74–76].

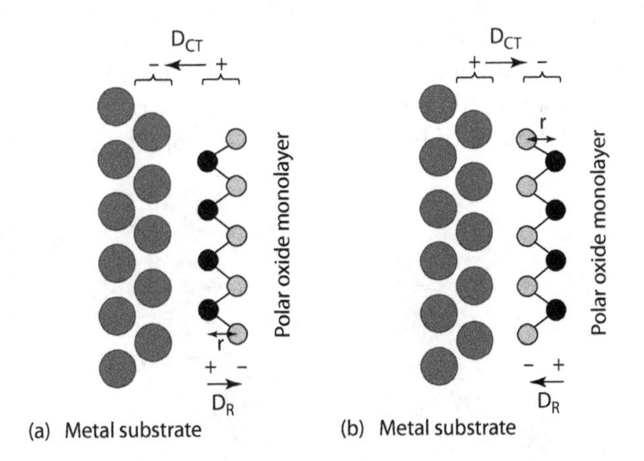

FIGURE 6.2
(a, b) A graphical scheme of the interfacial charge transfer (DCT) and film rumpling (DR) dipole momentum for an oxide monolayer (where the small black and white circles represent cations and anions, respectively) grown on a metal substrate (large gray circle are metal atoms). Regarding the electronegativity of metal substrates, electrons can transfer from or into the 2D oxide film. (Reproduced with permission from Goniakowski, J., et al., *Phys. Rev. B*, 80, 125403, 2009.)

6.2.2 Structural Concepts

The level and quality of the interaction between 2D oxide films and a metal substrate can tangibly affect the structural growth of the developed oxide thin films. A strong interaction between a metal substrate and oxide thin films facilitates the adaptive growth of 2D oxide films from the crystalline structure of a metal substrate [77–79]. A variety of 2D oxide films were grown on Au substrates including TiO_x [71,79], VO_x [80], CoO [81], MoO_3 [82], MgO [74], ZnO [83] and WO_x [84]. The bulk Au has a face-centered-cubic (fcc) crystalline structure, while different crystalline planes of Au can provide different crystalline orientations. As an example, the unconstructed Au(111) consists of hexagonal lattices, while a reconstructed Au(111) plane shows a complex structure [41,52]. It was observed that various deposited oxide films on an Au substrate lifted the herringbone reconstruction, which is caused by the strong interaction between the Au(111) facet and the ultra-thin oxide film. Nevertheless, with the formation of an oxide–Au interface, there is no longer an energetic advantage to adapt from the top layer of Au films. Consequently, the original hexagonal lattice form of the Au(111) facet acts as a template for the growth of ultra-thin oxide films [79]. In the case of TiO_x film grown on the Au(111) substrate, the co-existence of honeycomb and pinwheel structures with different stoichiometries was reported, as displayed in Figure 6.3 [71,79]. When the Ti atoms occupy the three-fold hollow sites of an Au lattice and O atoms position at the bridge sites of Ti atoms, a honeycomb structure with a stoichiometry of Ti_2O_3 is formed, as shown in Figure 6.3a [71,85]. In the other mechanism, the superposing of a metal/O lattice over the Au(111) surface forms moiré patterns resulting in the growth of different appearance pinwheel structures with the stoichiometry of TiO, as demonstrated in Figure 6.3b and c [71,79]. The hexagonal structures are the most commonly observed moiré patterns of ultra-thin 2D oxide films grown on Au(111) substrates. In addition to TiO, the hexagonal moiré patterns have been characterized in FeO [86], CoO [87] and ZnO [83] ultra-thin films developed on the Au(111) substrate.

Generally, the stoichiometry of 2D oxide films is the function of a host number of engaged parameters. Apart from the effects of substrates, the stoichiometry of a film can be altered by changing the growth conditions, oxidation parameters and post treatments. As a rule, the combined effects of various parameters will determine the final stoichiometry of a 2D oxide film. For example, even in the case of a non-metallic substrate, the growth of a distorted hexagonal arrangement of metastable 2D TiO_2 island films was observed on the surface of the single crystal rutile TiO_2, which exhibited completely different symmetry from the rectangular (011) plane of the bulk substrate [88]. The degree of lattice mismatch

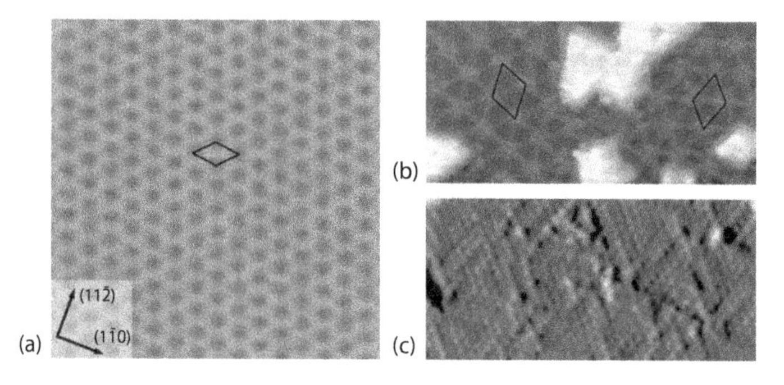

FIGURE 6.3
An STM image of (a) a honeycomb Ti_2O_3 structure, (b) a pinwheel TiO monolayer grown on Au(111) and (c) an atomically resolved STM image of pinwheel structures on Au(111). (Reproduced with permission from Wu, C., et al., *J. Phys. Chem. C*, 115, 8643–8652, 2011.)

is another factor determining the structure of 2D oxide films. The effect of induced stress due to the substrate lattice mismatch can be relaxed and thus becomes weaker as the thickness of the oxide film grows. The TiO_x/Au(111) system can be considered a typical example of the effect of thickness, when the triangular structures start to grow as the second layer on top of the first honeycomb Ti_2O_3 and pinwheel TiO monolayers [71]. While it is difficult to determine the atomic structure of the second monolayer, it can be inferred that the second monolayer is still affected by the lattice structure of the Au substrate. Further discussions on interfacial and structural concepts of a 2D oxide–metal substrate are beyond the scope of this chapter.

6.3 Atomic Layer Deposition of 2D Nanostructures

The utilization of the properties of 2D nanostructures needs the feasibility of fabricating 2D material–based devices. To fulfill this ambition, the atomically thin 2D ultra-thin nanostructures should be tailored to devices. Field-effect transistors (FET) based on 2D materials is a typical example of tailoring 2D nanostructures in practical devices that can be employed in measuring the channel properties of these ultra-thin structures [89]. Advanced fabrication techniques are an essential prerequisite for the fabrication of 2D-based devices. However, existing technologies face serious challenges, especially in the case of conformal deposition of ultra-thin 2D films [46]. The most conventional deposition techniques for the development of an ultra-thin film are sputtering, evaporation and CVD. These techniques are mostly suitable for the fabrication of Si-based devices; however, so far they are not suitable for tailoring 2D materials to device fabrication [46,89]. For example, sputtering uses the harsh environment so achieving flawless 2D structures is impossible. The evaporation and CVD-based techniques cannot achieve high-quality 2D nano-films with precise control of their thickness over a large substrate area. These techniques are developed for thin film deposition in the range of tens to hundreds of nanometers. By employing typical CVD or physical vapor deposition (PVD) techniques it would be rather difficult to achieve continuous films with ultra-thin 2D thickness within a few nanometer range [46,89,90]. At best, the surface of a substrate consists of several island domain structures with thicknesses ranging from a few to several nanometers. These island structures do not theoretically or practically refer to uniformly deposited 2D nanostructures [91]. Furthermore, full coverage of the substrates with a high aspect ratio is not attainable using these methods, which are necessary for practical applications such as various sensors and supercapacitors.

ALD as the ultra-thin film deposition technique is based on controlled and self-limited chemical reactions on the surface of substrates. In practical implementation, it is similar to CVD, except that the precursors are not introduced into the reactor simultaneously, but are typically separated in time by inert gas purges. As distinct from other CVD techniques, in ALD the source vapors are pulsed into the reactor alternately, one at a time, separated by purging, or evacuation periods. Each precursor exposure step saturates the surface with a (sub)monolayer of that precursor [92]. This results in a unique self-limiting film growth mechanism with a number of advantageous features. In fact, the self-saturation mechanism of chemical reactions is the main advantage of ALD, facilitating the atomic-scale control of the thickness of ultra-thin films. As a result, the reaction is limited to the monolayer of a reactant that has been adsorbed on the substrate [93,94]. Through this modification, excellent conformity and uniformity over geometrically complicated substrates and on a

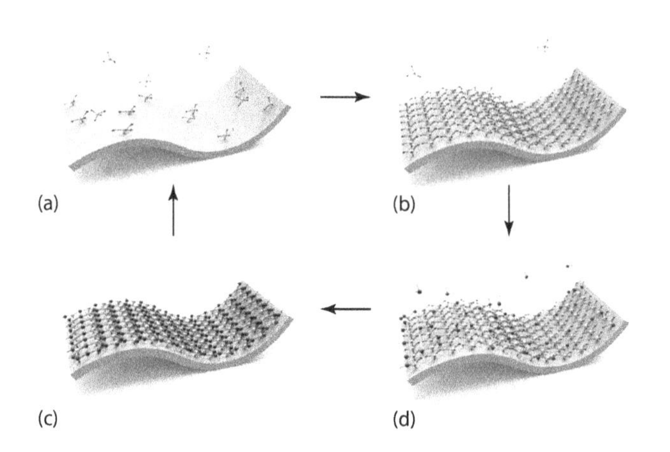

FIGURE 6.4
A schematic of an ALD cycle and its consecutive cycles: (a) exposure of a precursor, (b) removal of unreacted products molecules by purging of inert gas, (c) exposure of counter reactants and (d) purging with inert gas to remove unreacted counter-reactant molecules. (Reproduced with permission from Kim, H. G. and Lee, H. B. R., *Chem. Mater.*, 29, 3809–3826, 2017.)

large spatial resolution can be attained. As a general point of view, one ALD cycle consists of four individual steps, including exposure of precursors, purging of reacted chemicals and reaction products, counter-reactant exposure and finally purging, as demonstrated in Figure 6.4 [89]. Regarding the mechanism of ALD and its sequential exposure of precursors and counter reactants, the chemical reactions are restricted to the substrate surface properties and strongly depend on the surface characteristics. To be clearer, on an individual substrate, the ALD growth behavior shows different characteristics on two areas with different surface properties. The surface sensitivity of ALD is too strong, providing the opportunity for selective deposition, which is routinely called area-selective ALD [95]. The self-limiting nature of ALD also allows excellent composition control and the fabrication of multilayer 2D nanostructures. Compared with the parallel methods of typical thin film deposition including CVD, PVD and sputtering, the ALD process is performed under much milder conditions, which are necessary for the uniform deposition of 2D structures. Another outstanding feature of ALD is its slow growth rate. This ALD feature makes the deposition of high-quality 2D nano-films possible. All the foregoing ALD advantages turn this fabrication method into the most promising technique for wafer-scale deposition of ultra-thin 2D nano-films for practical applications and places ALD in a unique position to address many challenges in the development of 2D semiconductors [46]. Table 6.1 provides a general overview of the existing thin film fabrication techniques and their capabilities ratings [46]. Considering the information presented in Table 6.1, it is evidently understood that the ALD method has significant advantages over other deposition techniques in terms of the deposition process itself and control over various parameters of fabrication. Although several articles report on the ALD of oxide semiconductor films, they mostly deal with the ALD of oxide films whose thicknesses are far beyond 2D limits, so they do not fall under the category of 2D nanostructures. The following sections will concentrate on the ALD of 2D oxide semiconductor films by focusing on the development of 2D semiconductor oxides.

6.3.1 ALD Window

As previously mentioned, the ALD is a surface-sensitive method and its mechanisms are the function of precursors, reactant agents and substrate surface properties; furthermore, it

TABLE 6.1

Properties of Various Modern Techniques for Thin Film Deposition

	Deposition Technique					
Property	**Chemical Vapor Deposition (CVD)**	**Molecular Beam Epitaxy (MBE)**	**Atomic Layer Deposition (ALD)**	**Pulsed Layer Deposition (PLD)**	**Evaporation**	**Sputtering**
Deposition rate	Good	Fair	Poor	Good	Good	Good
Film density	Good	Good	Good	Good	Fair	Good
Lack of pinholes	Good	Good	Good	Fair	Fair	Fair
Thickness uniformity	Good	Fair	Good	Fair	Fair	Good
Sharp dopant profiles	Fair	Good	Good	Varies	Good	Poor
Step coverage	Varies	Poor	Good	Poor	Poor	Poor
Sharp interfaces	Fair	Good	Good	Varies	Good	Poor
Low substrate temp	Varies	Good	Good	Good	Good	Good
Smooth interfaces	Varies	Good	Good	Varies	Good	Varies
No plasma damage	Varies	Good	Good	Fair	Good	Poor

Source: Reprinted with permission from Zhuiykov, S., et al., *Appl. Surf. Sci.*, 392, 231–243, 2017.

depends on ALD fabrication parameters including process design, chamber pressure and temperature, plasma control and other ALD parameters [96,97]. Therefore, the developed ALD recipe for 2D materials is technically regarded as valuable data. The self-saturation phenomena and their mechanisms are the main parameters determining the conformity of ALD film [98–100]. The grow rate of the ALD process is controlled by a process temperature range, in which the ALD reactions occur. This temperature range is called the "ALD window" as shown in Figure 6.5a. To promote or initiate the reactions and to reduce the amount of unreacted precursors on a substrate, an elevated temperature must be selected for the ALD process. An inefficient and low-temperature ALD process may result in the occurrence of the physical adsorption of precursors on the substrate surface, which would consequently lead to precursor condensation on the substrate. However, the ALD window is restricted by excess condensation or thermal activation at the low-temperature range and is generally bounded by precursor degradation or desorption at high temperatures [101,102]. A typical ALD window for a successful ALD process is in the range of 200°C–400°C, as shown in Figure 6.5b [46]. However, it can be changed for specific ALD recipes. The self-controlled growth in the ALD window is the consequence of chemisorption. In some cases, due to lack of thermal energy, the ALD process is assisted by using plasma. Plasma-enhanced atomic layer deposition (PEALD) makes it possible to expand the number of precursors. PEALD is typically employed in the deposition of oxide films to circumvent the need for H_2O as oxygen precursors or to increase the growth rate [104]. However, the ability of conformal coating of 2D structures by a high aspect ratio is probably limited by PEALD because of ozone generation and the inefficient mean free path of plasma ions [103–105].

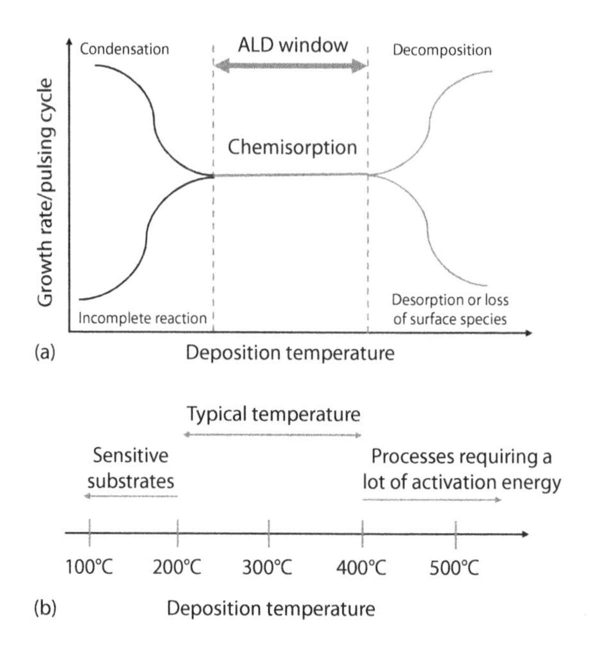

FIGURE 6.5
(a) Dependence of the deposition temperature on the growth rate for ALD and (b) a typical temperature range for the most common ALD precursors. (Reproduced with permission from Zhuiykov, S., et al., *Appl. Surf. Sci.*, 392, 231–243, 2017.)

6.3.2 ALD Precursors

The precursors in an ALD process can be classified into three major types, including inorganic precursors that do not contain any carbon; metal-organic precursors that possess organic ligands, but without metal to carbon bonds; and organometallic precursors that contain organic ligands wherein metal to carbon bands exist [46,96]. The successful design of an ALD process is highly dependent on the proper selection of precursors. As chemical products, precursors should be synthesized by easy routes and need to be conveniently available for practical ALD targets. Self-decomposition, volatility, dissolution and etching effects on substrates, sufficient purity and byproducts are the main parameters that should be considered during synthesis and for the selection of precursors. Despite the significant progress in the synthesis of new chemical compounds, fulfilling the requirement of a suitable precursor for ALD of 2D nanostructures needs a great deal of work. The saturation of the growth rate as a function of the precursor pulse duration is directly attributed to the ALD window temperature. To achieve a high-quality ALD-deposited 2D nanostructure, the self-saturation process should be controlled. Each ALD precursor shows a characteristic sticking coefficient that can be changed on various substrates. By changing the ALD conditions, the sticking coefficient directly affects the precursor purge time. For instance, the growth per cycle (GPC) obtained from an insufficient purge time can be twice as high as when obtained at the optimized ALD recipe [106,107]. The same challenges should be encountered when the second reactant precursor comes into action. For example, in the case of metal oxide, sufficient H_2O pulse time is the essential parameter that should be considered to obtain high-quality 2D oxide nano-films. Evidently, insufficient oxidant agents will change the growth regimes or affect the composition of ALD films [106]. Another important parameter is the reactivity of precursors, especially when considering that most of the strategies use in the design of an ALD recipe are based on thermally driven reactions [108–110]. In this case, the reactivity becomes a more critical issue when thermally stable precursors are used or when low-temperature deposition is required owing to the requirements of an ALD process. Regarding the aforementioned ALD variables, a vital role that has been established in the design of an ALD recipe for 2D semiconductor oxides is the correct selection of ALD precursors [46].

6.3.3 Case Study: ALD of 2D WO_3

Increasing interest in the employment of ultra-thin semiconductor oxides for advanced applications in functional devices has resulted in more attention being paid to the development of an ALD technique. During the last decade, several scientific reports have confirmed the capability of ALD to fabricate thin film semiconductor oxides such as indium oxide [111,112], tantalum oxide [113], ruthenium oxide [114,115], titanium dioxide [116,117], zinc oxide [118,119], iron oxides [120], tin oxide [121], cobalt oxide [122], tungsten oxide [123] and barium titanate [124]. While the ALD of semiconductor oxide films has been well investigated and the process parameters have been extensively studied, our knowledge of the development of ultra-thin 2D oxide nanostructures is deficient. Accordingly, the number of published articles is low, opening up an uncharted area in nanofabrication topics. Take the WO_3 semiconductor oxide as an example. To date, there have only been a few reported ALD processes for the fabrication of thin film WO_3. A recipe was developed by considering the various parameters of ALD to enable the deposition of a monolayer WO_3 on a large surface [46,84]. The bis(ter-butylimido) bis(dimethylamino) tungsten(VI) $(^tBuN)_2W(NMe_2)_2$ as the tungsten precursor and H_2O as the oxygen source were used and the deposition was carried out on an Si/SiO_2 substrate using the Savannah S100 (Ultratech/Cambridge Nanotech) ALD instrument, as graphically demonstrated in Figure 6.6. To confirm the ALD growth of an oxide film, in situ ellipsometry was used during the deposition process. In experiments, the dependence of the growth rate on the precursor pulse times, substrate temperature and cycle numbers was investigated. It must be stressed that the deposition of ultra-thin 2D films with a thickness of less than 1.0 nm is an extremely challenging target. Non-uniformity and a significantly less deposition rate than the desired thickness were observed after the first deposition. However, a noticeable improvement was observed when the temperature was increased to 350°C. Ellipsometry analysis of several depositions also indicated significantly less practical deposition than the intended desired thickness. To use ellipsometry for thickness measurements, a sufficient coating is needed to

FIGURE 6.6
(a) A schematic interpretation of precursors used $(^tBuN)_2(Me_2N)_2W$ and H_2O, (b) an optical image of a wafer-scale developed 2D WO_3 film made and diced on Si/SiO_2 and (c) an image of MCN Savannah S100 ALD apparatus connected to the glove-box used for WO_3 ALD. (Reproduced with permission from Zhuiykov, S., et al., *Appl. Surf. Sci.*, 392, 231–243, 2017 and Zhuiykov, S., et al., *Appl. Mater. Today*, 6, 44–53, 2017.)

build reliable optical constants. As the deposition thickness yielded 10 nm, the comparison between the measured and the literature values of the refractive index showed good agreement. Changing other ALD parameters, especially precursor dosing, and optimizing the device temperature resulted in an improved deposition rate. An example of the ALD recipe for a 2D WO_3 film is presented in Table 6.2 [84].

As shown in Figure 6.7a and b, the growth rate is sensitive to the precursor doses. As expected, the growth rate stabilized to a maximum thickness per cycle when the precursor dosing achieved full monolayer coverage. Growth rate saturation was observed after 50 ms for H_2O and after 2 s for $(^tBuN)_2W(NMe_2)_2$, indicating the self-limiting growth characteristics of the ALD. In addition, it was established that the growth was also highly sensitive to the substrate temperature. The first deposition at an operating temperature of ~300°C confirmed very limited growth of the 2D WO_3 film. Consequently, the optimum growth condition yielded a stable WO_3 growth (see Figure 6.7c and d). The subsequent ellipsometry map analysis of final deposited films confirmed that the obtained thicknesses for 2D WO_3 were ~0.7 nm and ~1.2 nm. It should be considered that the developed recipe is attributed to the ALD of WO_3 on an Si/SiO_2 substrate with its specific growth condition. Due to the natural roughness of an Si/SiO_2 substrate, the development of a more conformal and consistent monolayer was not possible [84]. An investigation of a wafer-scale ALD of WO_3 on an Au substrate showed that the uniformity of the films decreases and their thicknesses gradually deviate from the targeted value, as the films grow thicker. This can be caused by non-uniform spatial distribution of precursors and the temperature discrepancy in the substrate [86].

Figure 6.8a shows a cross-sectional view of a 2D WO_3 film deposited on an Si/SiO_2 using the ALD recipe presented in Table 6.2. A scanning transmission electron microscopy (STEM) image displays the uniform development of the ALD WO_3 film across a large area of an Si/SiO_2 wafer. The film was semi-amorphous without a clearly identified crystalline

TABLE 6.2

Developed ALD Recipe for 2D WO_3 Using $(tBuN)_2W(NMe_2)_2$ and H_2O Precursors

Deposition Parameters	1	2	3	4	5	6	7
Inner heater (°C)	300	350	350	350	350	350	350
Outer heater (°C)	300	300	280	280	280	280	280
W precursor heater (°C)	80	80	80	80	80	80	80
Isolate pump	–	–	*	*	*	*	*
Exposure	–	–	–	–	–	–	–
Initiate pump	–	–	–	*	–	–	–
Pulse H_2O (s)	0.05	0.05	0.06	0.06	0.06	0.06	0.06
Exposure (s)	–	–	5	5	5	5	5
Initiate pump	–	–	*	*	*	*	*
Purge H_2O (s)	4	4	12	12	12	12	12
Isolate pump	–	–	–	*	*	–	*
Pulse W precursor	1	1	1	1	1	2	1
Exposure	–	–	–	6	6	–	6
Initiate pump	–	–	–	*	*	–	*
Purge W precursor (s)	10	6	6	14	14	14	14
Number of cycle	400	300	300	300	250	250	185
Thickness (nm)	8	30.9	30.9	30.9	28.5	28.5	19.1

Source: Reprinted with permission from Zhuiykov, S., et al., *Appl. Mater. Today*, 6, 44–53, 2017.
* Process is carried out; – process is exempted.

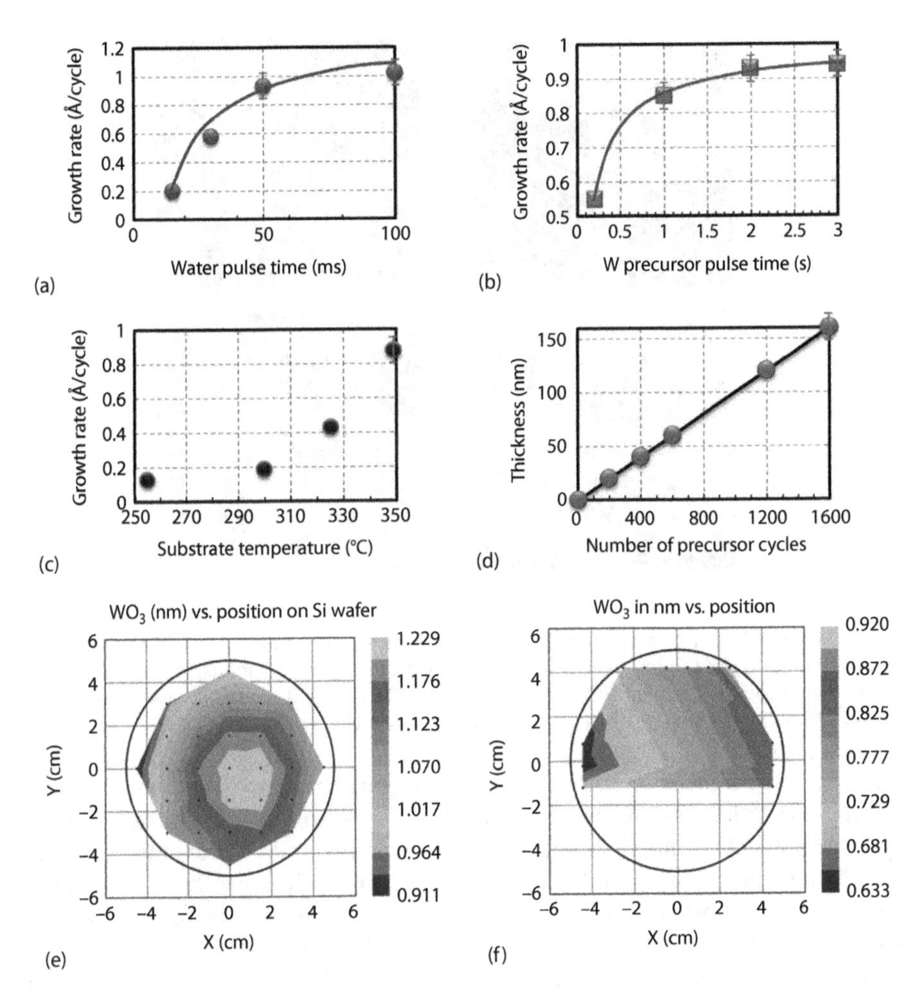

FIGURE 6.7
Experimental data for the growth of a WO_3 film based on the conception of (a) H_2O and (b) $(tBuN)_2W(NMe_2)_2$ precursors. The stabilization of the growth rate to a maximum thickness per cycle after the achievement of a full monolayer coverage. (c) The initial growth rate vs. temperature and (d) the thickness variation of a WO_3 film per ALD cycle number. (e, f) An ellipsometry map for 1.2nm thick and 0.7nm thick WO_3 on Si/SiO_2 after an optimized ALD process. (Reproduced with permission from Zhuiykov, S., et al., *Appl. Mater. Today*, 6, 44–53, 2017.)

structure, as depicted in Figure 6.8b. Atomic force microscopy (AFM) measurements of the 2D WO_3 film and its corresponding scanning electron microscopy (SEM) observations clearly demonstrated the uniformly developed 2D films with a high level of accuracy in thickness, as shown in Figure 6.8c and d. Due to the low conductivity and semi-amorphous nature of the ALD-developed 2D WO_3 film, its surface appeared to be fuzzy and smooth with the visual absence of clearly distinguished grain boundaries. AFM and focused ion beam SEM measurements, which are not presented here, confirmed a high level of correlation of the results obtained by two independent material characterization techniques for the monolayer WO_3.

The surface chemical characterization of ALD 2D WO_3 films by x-ray photoelectron spectroscopy (XPS) gives valuable information about the nature of the bonding and surface chemical composition of 2D films. The study of the XPS peaks corresponding to O 1s-levels of oxygen atoms O^{2-} in the lattice of SiO_2 and WO_3 revealed that, when the samples

FIGURE 6.8
(a) STEM cross-sectional view of a ~3.3nm thick 2D WO_3 film deposited on Si/SiO_2 wafer by ALD and (b) a high magnification image of the interface between Si/SiO_2 and WO_3 film. The inset shows the SAED pattern indicating the crystalline structure of SiO_2, (c) SEM and (d) AFM image of the surface morphology of an ALD-developed WO_3 film. (Reproduced with permission from Zhuiykov, S., et al., *Appl. Mater. Today*, 6, 44–53, 2017.)

consisted of a double layer of the oxygen-octahedron structure, the bottom oxygen was shared with SiO_2. Evidence was provided that metal precursors were adsorbed initially on the surface of SiO_2, thus metal–oxygen bonding was formed on the substrate. XPS spectra depicted that the oxygen-octahedron structure in the 10.0nm thick WO_3 sample was totally free from the substrate effect. The studies also showed that tungsten (W) in both the monolayer and few-layer films was present in the six-valent state (W^{6+}). From the broadening of the core-level spectra of W, it was concluded that the monolayer WO_3 film was not fully oxidized and it was not fully crystalline [125,126]. By increasing the thickness, the crystallinity of thicker 2D films was developed [126]. Further studies of O 1s peaks of XPS analysis confirmed the development of stoichiometric crystalline structures in thicker WO_3 films. The adsorption of OH species on the oxygen vacancies of a monolayer WO_3 and the obtained peaks confirmed the development of a sub-stoichiometric WO_{3-x} compound in the monolayer WO_3 film. Fourier transform infrared spectroscopy (FTIR) further demonstrated the characteristic mode of a WO_3 structure. Modifications in the tungsten oxide framework were suggested by studies of the FTIR spectra of W-O stretching modes, possibly because the ALD-developed surface of a 2D WO_3 was not a fully defect-free surface and this was later confirmed by the absorption of surface-active sites on a 2D film.

6.3.4 Case Study: ALD of 2D TiO_2

The wafer-scale synthesis and conformal growth of a 2D TiO_2 film on an Si/SiO_2 substrate was achieved by ALD using a tetrakis (dimethylamino) titanium (TDMAT) precursor and H_2O as the oxidation agent [127,128]. Regarding the challenging nature of developing 2D oxide nanostructures, the reliability of the fresh TDMAT precursor was initially confirmed. To establish the experimental dependence of the deposition rate on the ALD temperature for the fresh TDMAT precursor, a few blank depositions were done.

Then, the optimum number of cycles, deposition rates and predicted thicknesses were calculated and averaged for each operating temperature. For precise fabrication of extremely thin 2D TiO_2 films, where their thickness must be below ~0.5 nm, the optimum operating temperature of 250°C was selected. To achieve complete coverage of a monolayer TiO_2, two individual ALD super cycles were designed [127]. The first super cycle consisted of 10 consecutive cycles of pulse/purge TDMAT to ensure reliable coverage of the surface by precursors. Afterward, 10 pulse/purge stages of H_2O were performed to ensure the completion of oxidation [127,128].

The reactions that occurred during ALD fabrications of a monolayer TiO_2 are schematically presented in Figure 6.9. From the thermodynamic and kinetic points of view, the development of TiO_2 by TDMAT and H_2O can be separated into two half reactions [128]. The initial growth stage is expected to be influenced by the number of chemisorption sites on the Si/SiO_2 substrate. As graphically demonstrated in Figure 6.9a, the first stage is the chemisorption of TDMAT molecules by active sites of the SiO_2 surface. The next step is the

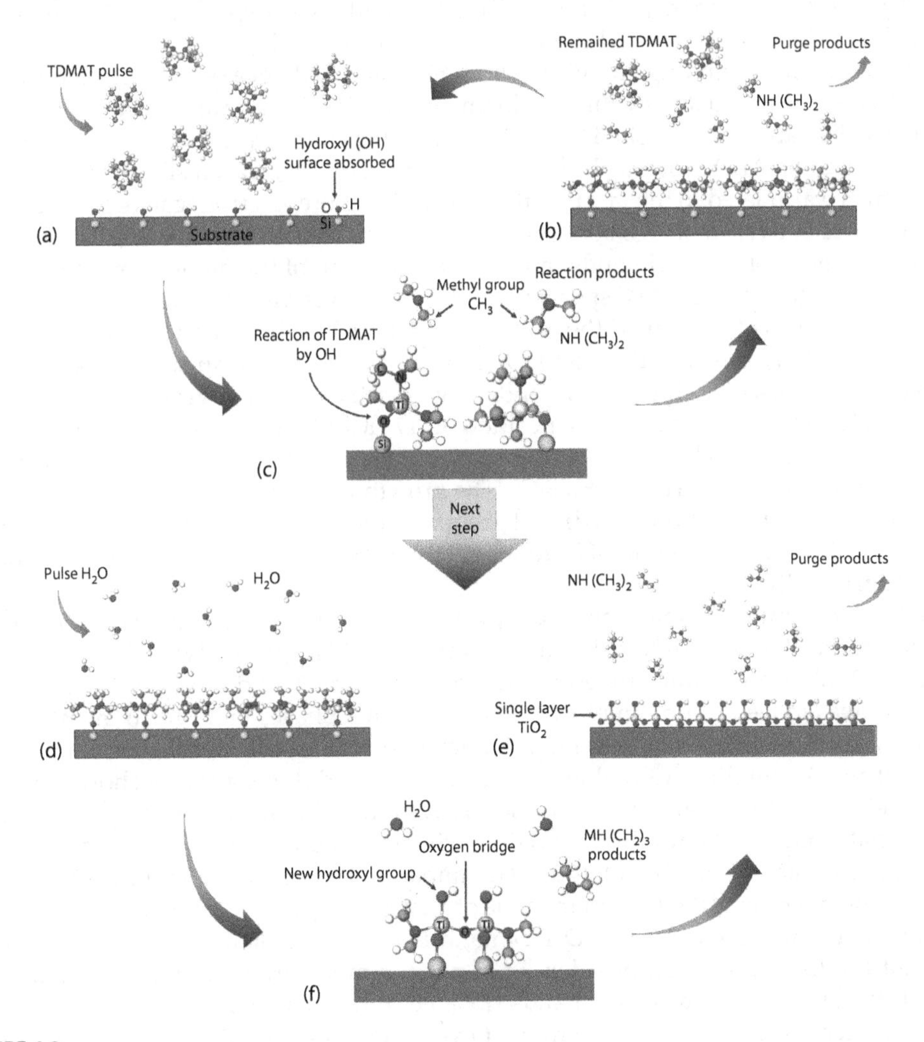

FIGURE 6.9
A graphical scheme of the development of atomic-layered titanium oxide by ALD. (Reproduced with permission from Zhuiykov, S., et al., *Mater. Des.*, 120, 99–108, 2017.)

ligand exchange that usually occurs easily even at very low operating temperatures and it is attributed to the high reactivity of both TDMAT and OH groups [129]. Figure 6.9b shows a scheme of the ligand exchange process. The subsequent step is the purge of the remaining TDMAT and reaction products (NH $(CH_3)_2$) from the reaction chamber [130]. During the next step, by introducing H_2O into the ALD chamber, the oxidation process continues. Figure 9d and e demonstrates the oxidation process and the development of Ti–O–Ti bridge bonding in the deposited film, respectively. On completion of the oxidation process, the 2D TiO_2 film is formed on the surface where all the Ti atoms are connected to each other by oxygen atoms. The interaction between the surface SiO_2 and the deposited TiO_2 was confirmed by observation of the Si-O-Ti spectra in the FTIR measurement of an ALD TiO_2 monolayer film [127]. It confirmed the development of bonding between Ti atoms and Si-O atoms of the SiO_2 substrate. Further studies by XPS measurements showed the characteristic spectra of Si–O–Ti binding energy [127]. Comparative XPS studies confirmed the fact that the bottom oxygen in the atomic-layered TiO_2 film was shared with the SiO_2 substrate [127,131]. The binding energy of Si was also altered after the deposition of the monolayer TiO_2. The position of Si^{4+} shifted to lower energy during ALD of the atomic-layered TiO_2 film. Changes in the binding energies in both Si 2p and Ti 2p peaks after the deposition of 2D TiO_2 could be elucidated by the development of Si–O–Ti bonds at the interface between the native SiO_2 and the deposited 2D TiO_2 film. As titanium was introduced into the Si–O bonds during the ALD process, the binding energy of Si^{4+} decreased slightly from its bulk SiO_2, which was related to the fact that the silicon was more electronegative than titanium. The evidence of oxygen bridge bonding was also demonstrated by characterization of the vibrational mode of Ti–O–Ti bonds in the FTIR spectrum of the monolayer TiO_2 film and also by investigation of the XPS spectrum of the monolayer film. The detection of the FITR characteristics of Ti^{4+} confirmed the existence of Ti(IV)O oxide in a crystalline state [127].

In another study, the growth of 2D TiO_2 films over Si/SiO_2 and Au substrates was investigated by spectroscopic ellipsometry and x-ray reflectivity [132]. A growth rate of 0.75 and 0.87 Å/cycle was measured for the deposition of 2D TiO_2 film on Si/SiO_2 and Au substrates, respectively [132]. In both substrates, the growth rates were linear, excluding a short delay in the first ALD cycles of TiO_2 on Si/SiO_2. The growth incubation in this ALD process was attributed to the preparation of hydroxyl groups on the surface. It seems that it took longer time for TDMAT to react with OH groups on the Si/SiO_2 surface compared with that on the Au surface [133,134].

Studies using micro-Raman spectroscopy revealed the presence of both crystalline anatase and rutile phases of TiO_2 in the ALD-developed 2D TiO_2 films. Typical Raman spectra for 2D TiO_2 films are demonstrated in Figure 6.10a. Raman studies showed that the final structure was indeed a combination of two crystalline phases of TiO_2. To investigate the impact of layer thickness on the Raman characteristic modes of a 2D film, the Raman spectra of 7.0 and 0.7 nm thick TiO_2 films were compared, and the results are shown in Figure 6.10b and c. Compared with a 7.0 nm thick TiO_2 film, a blue shift was observed in the E_g vibrational mode of a 0.7 nm thick TiO_2 film, which was accompanied by a decrease in the peak intensity and width broadening. The impacts of phonon confinement on the blue shift and broadening of the E_g Raman mode of anatase TiO_2 were reported previously [135]. The phonon confinement within 2D TiO_2 nanostructures resulted in incremental phonon momentum distribution and broadening of phonon momentum scattering accompanied by a shift in Raman characteristic bands [136]. From the stoichiometric point of view, the reduced intensity and width broadening of the E_g vibrational mode of anatase can also be attributed to the deviation from the stoichiometric level by reducing the film thickness down to a single fundamental layer [137,138]. XPS studies of 2D TiO_2 films developed on an

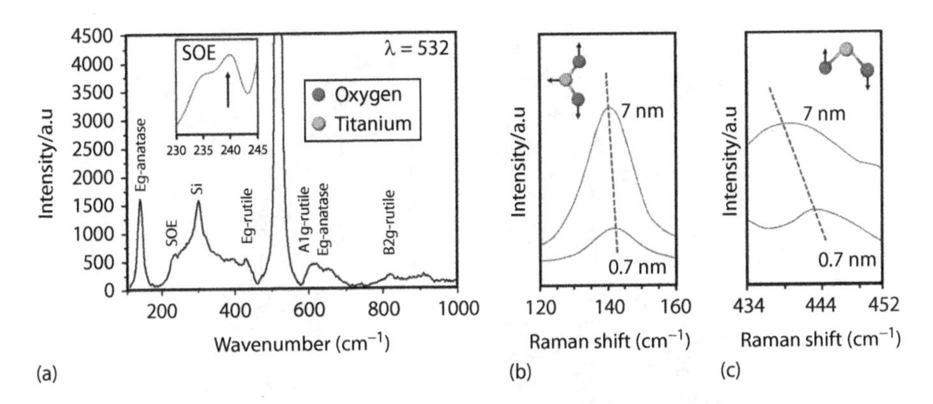

FIGURE 6.10
(a) The Raman spectra of ALD 2D TiO_2 films, (b) and (c) the thickness dependence of Raman peaks. (Reproduced with permission from Karbalaei Akbari, M., et al., *Mater. Res. Bull.*, 95. 380–391, 2017.)

Au substrate provide valuable information for the case of a surface chemical structure of 2D films. The XPS results of 7 and 0.7 nm thick TiO_2 films confirmed that titanium atoms (Ti) exist in a four-valent state (Ti^{4+}). The XPS results indicated that a thicker film (7 nm) was more oxidized than a thinner film (0.7 nm). The development of a stoichiometric oxide crystalline structure in a thicker film (7 nm) was confirmed by the characterization of an O 1s peak of XPS measurements [132]. Furthermore, the presence of weakly adsorbed species on the surface of a 0.7 nm thick 2D TiO_2 film was shown by characterization of OH groups on the surface of a 0.7 nm thick 2D TiO_2 film indicating the development of non-stoichiometric structures with oxygen-deficient sites on the surface of a 0.7 nm thick 2D TiO_2 nanostructure.

6.3.5 Case Study: ALD of 2D Aluminum Oxide

ALD of a monolayer alumina film over Cu and Cu_2O/Cu substrates was investigated using trimethylaluminum (TMA) as the Al precursor and O_2 as the oxidant agent [139]. The interaction of TMA with a Cu surface faces several challenges. Regarding the DFT calculations, the adsorption and dissociation of TMA on the surface of a pure Cu(111) are endothermic, confirming that the interaction of TMA with a pure Cu surface is thermodynamically unfavorable in the absence of hydroxyl groups. XPS measurements and high-resolution electron energy loss spectroscopy did not show any characteristic vibration of Al atoms on a pure Cu surface. On the other hand, first-principle calculations demonstrated that TMA is capable of reacting with a copper oxide surface in the absence of hydroxyl species. The adsorption of Cu by a Cu_2O/Cu surface is limited by the initial amount of oxygen in a Cu_2O lattice structure. This adsorption results in the reduction of some surface Cu^{1+} to metallic copper (Cu^0) and the formation of copper aluminate compounds. The XPS studies revealed that the Al:O atomic percentage ratio at the first deposited layer was approximately 0.46 while the stoichiometric Al_2O_3 yielded an Al:O ratio of 0.66. It confirmed the development of non-stoichiometric 2D alumina layers. Further XPS measurements showed the formation of copper aluminate, most likely $CuAlO_2$ on the basis of the Al:O ratio of 0.5. The STM image of the surface of a TMA-exposed Cu_2O/Cu film demonstrated the growth of 2D islands of $CuAlO_2$ with an average height of approximately 0.19 nm with flat and uniform surfaces, which was close to Cu-O and Al-O bond length, as demonstrated in Figure 6.11a and b. After the second ALD cycle, TMA continued to react with surface Al–O, forming

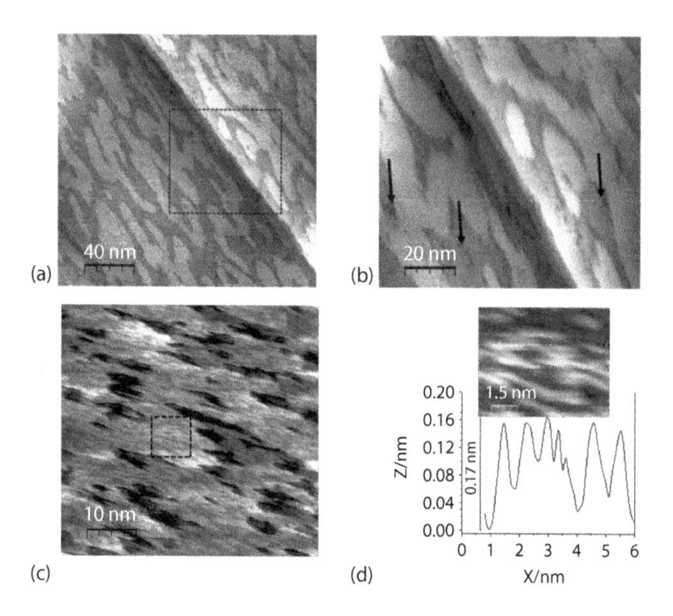

FIGURE 6.11
(a) Low and (b) high magnification STM images of the $Cu_2O/Cu(111)$ surface exposed to first cycle and (c) second cycle of TMA, and (d) zoom-in region of the highlighted section of the image (c) and the line profile along the solid line indicated in the image. (Reproduced from Gharachorlou, A., et al., *ACS Appl. Mater. Interfaces*, 7, 16428–16439, 2015.)

stoichiometric Al_2O_3. The observed morphological changes after consecutive ALD cycles proposed the development of second alumina islands with a thickness of 0.17 nm, as shown by the STM image in Figure 6.11c and d [139]. It was confirmed that TMA readily reacted with oxide surfaces even in the absence of co-adsorbed hydroxyls. For ALD applications on an air-exposed Cu surface, large domains of oxides might still exist that can facilitate the selective ALD of metal surfaces. This is of great importance in thin film applications like microelectronics and catalysis where only a few ALD cycles are desirable.

6.4 The Properties and Applications of ALD 2D Oxide Film

Interest in research relevant to ALD-developed nanostructures is increasing as evident by the steady growth of annual scientific publications [46]. This fact independently acknowledges the great impact of ALD-developed nanomaterials on their various applications. It further confirms the growing interest in the distinguished capabilities of 2D nanostructures that can be obtained in the nano-architecture of the different 2D materials using the ALD technique. Although ALD is currently mainly utilized in the microelectronic industry, its ability for conformal deposition into high aspect ratios offers great potential for a breakthrough in the field of ultra-thin 2D nano-materials [46]. The ALD offers the opportunity to engineer either dielectric or metallic properties of materials by creating nano-laminates or alloy films. This ability of ALD combines the advantageous properties of different materials [140,141]. ALD also has the potential for applications in the field of 2D nano-electrodes of various chemical sensors and biosensors, where high-quality, conformal and ultra-thin films in aggressively shaped nano-electrodes are often used. In fact, the low deposition rate of ALD does not hinder its applications for thin film electrodes [141]. Therefore, in comparison to other ultra-thin film deposition techniques, the ALD technique offers several advantages in the development of 2D nanostructures.

6.4.1 The Catalytic Applications

Metal oxides represent one of the widely employed materials for catalytic applications. Their reducibility is an important property that facilitates the catalysis performance of metal oxides. While the catalytic applications of crystalline thin-oxide films, such as crystalline TiO_2 and MgO, have been the main focus of research activities during the past decades, ultra-thin amorphous oxides have recently confirmed their beneficial role as supports for catalytic applications [142–147]. Despite the modulation of 2D oxide films deposited on metal substrates, the oxide phases can be typically considered in terms of transitional or quasi transitional structures [14,148]. This is due to the incommensurate mismatch between the crystalline structure of the metal supports and 2D oxide films. In actual cases, since the thermodynamic and entropic parameters in the growth process are not precisely controllable, the real structure of ultra-thin oxide films is not fully crystalline [141,148,149]. It especially occurs in the case of 2D oxide films grown using the ALD technique. As a consequence, one of the main drawbacks for the growth of 2D oxides on crystalline metal substrates would be the transition from crystalline-like structures into semi-amorphous or disordered structures. The advantage of the ALD technique is its technological structure, which can be scaled up to meet the requirements of industrial applications. One of the most desired catalyst structures is the embedded catalytic nanostructures in 2D oxide phases [150, 151]. The mentioned catalytic nanostructures have various geometrical shapes, ranging from large nanoparticles to sub-nanometer particles, nano-island structures and nanoclusters. In general, the activity of the catalytic metals can be optimized when they are used in the form of supported small sub-nanoclusters or nanoparticles. To support catalytic nanostructures on the surface, an oxide ultra-thin layer can be used [148,149]. These composite nanostructures, i.e. catalytic nanostructures supported by 2D oxides, offer tremendous advantages. From the theoretical point of view, the charge transfer between catalytic nanostructures and metal supports is facilitated by the tunable conductance of 2D oxide films. From a technological point of view, since the charge transfer is created between a metal substrate and catalytic structures, the deposition of catalytic components is facilitated on 2D structure. Furthermore, the charge states of catalytic particles can be controlled by the precise selection of 2D oxide–metal substrate pairs and the accurate control of 2D oxide film thickness. In doing so, fine-tuning of the catalytic properties of these nanostructures can be gained [149]. Since the trends are inclined toward the application of supported sub-nanometer catalytic structures, the role of 2D oxide films as the support of catalysts has become even more important [152,153]. The direct epoxidation of propylene to propylene oxide by molecular oxygen was confirmed at low temperatures by using sub-nanometer Ag aggregates grown on 2D alumina supports. The 2 nm height silver clusters were grown on three monolayer alumina films developed by ALD [154,155]. Despite the fact that the surface morphology and termination of the ALD-developed oxide films are complex, this 2D oxide structure provides several advantages. The stabilization of catalytic particles against sintering under reaction conditions and the control of structural crystallinity versus amorphicity are just two of them [156]. As an example, high conversion accompanied by high selectivity was gained when 20 nm diameter silver nanoclusters were grown on 7.2 Å thick 2D ALD alumina on a naturally oxidized silicon substrate. The surface of this catalyst structure before and after epoxidation reactions are shown in Figure 6.12a and b, respectively. No noticeable changes were observed in the size of the nanostructured clusters and their distribution before and after the catalytic reactions, while the aspect ratio (height/diameter) of the nanoclusters was changed dramatically [154]. The topographical STM images of the surface, presented in Figure 6.12c, confirmed the amorphous structure of the ALD 2D alumina-support film. X-ray absorption characterization showed that this particular 2D alumina film can be best described as a mixture of tetrahedral and octahedral building units. The same photocatalytic activity was also reported in the case of selective propene epoxidation on immobilized Au clusters by ALD TiO_2 film [153].

Furthermore, the ALD technique provides the opportunity to develop 2D oxide films with a multicomponent structure around the catalyst particles to tune the performance of the system. The ALD of an ultra-thin TiO_2 coat on Pd catalysts grown on an amorphous Al_2O_3 substrate showed the improved resistance of catalyst particles against sintering at high-temperature degradation [155]. A schematic of Pd nanoparticles grown on an Al_2O_3 substrate and a graphical model of the development of a 2D TiO_2 film below and around the same Pd clusters are presented in Figure 6.13a and b. A significant enhancement in the sintering resistance of Pd clusters was observed by grazing-incidence small-angle scattering (GISAXS) (Figure 6.13c). The onset of sintering by Pd clusters supported on Al_2O_3 was around 170°C; an improvement in sintering resistance by 40°C in comparison with the individual Pd clusters. The thin ALD TiO_2 overcoat ($TiO_2/Pd/Al_2O_3$) improved the sintering resistance of the clusters, without any loss in its catalytic activity. This is the central point for the production of new catalytic materials for long-term applications.

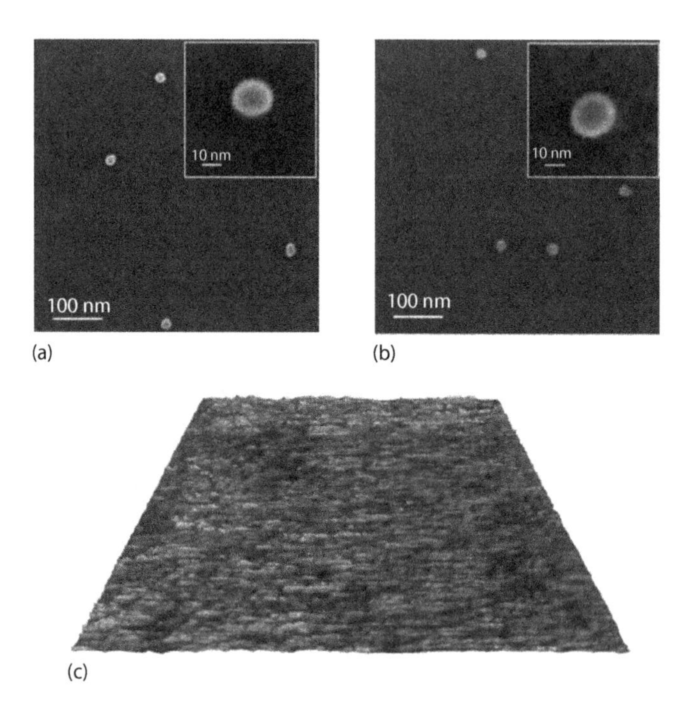

FIGURE 6.12
Representative HRSEM images of silver nanoclusters (a) before the epoxidation reaction and (b) after reaction cycles. (c) Topographic STM image of the amorphous alumina support layer. (Reproduced with permission from Molina, L.M., et al., *Catal. Today*, 160, 116–130, 2011.)

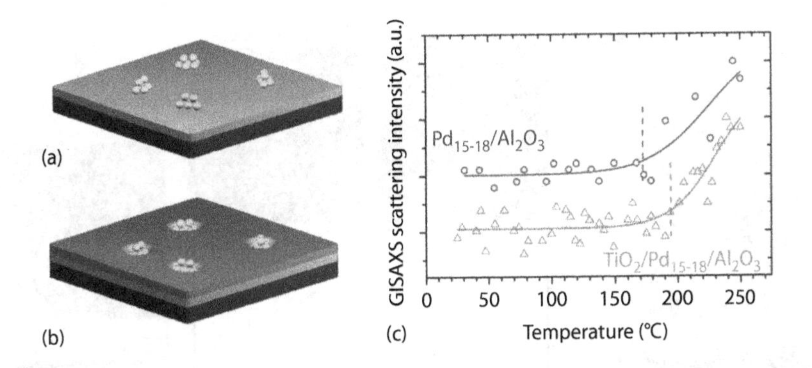

FIGURE 6.13
A graphical image of (a) Pd/Al_2O_3 and (b) titanium oxide over coated cluster sample $TiO_2/Pd/Al_2O_3$ and (c) integrated GISAXS scattering intensities as a function of temperature for Pd clusters without and with a titania protective overcoat. (Reproduced with permission from Lee, S., et al., *J. Phys. Chem. C*, 114, 10342–10348, 2010.)

This structure exhibited further enhanced stability during the reaction, with no indication of particle growth up to about 200°C [155].

The catalytic performance of 2D oxide films is not merely restricted to the embedded photocatalyst nanostructures in 2D oxide films. The efficient photocatalytic degradation of palmitic acid (PA) by 2D TiO_2 films was attained under both UV and visible light illumination [132]. In this study, 2D TiO_2 films with various thicknesses were deposited on Si/SiO_2 and Au substrates. The strong interaction between 2D TiO_2 films and PA molecules was confirmed by an investigation of the FTIR spectrum of 2D TiO_2 films covered by PA molecules. It was shown that a 2D TiO_2 film with a thickness of 3.5 nm can efficiently degrade the PA acid molecules under UV light illumination, while the partial degradation of PA molecules was attained for 0.7 nm thick 2D TiO_2 films. The optical images and FTIR measurements presented in Figure 6.14 show the gradual degradation of PA on the surface of a 2D TiO_2 photocatalyst deposited on Si/SiO_2 and Au substrates. At the beginning of the test, the surface of the photocatalyst film was entirely covered by a thick layer of PA film. By increasing the UV radiation time up to 24 h, integrated island structures with irregular shapes and sizes appeared. Regarding the observations, the photocatalysis process sequentially continued up to the formation of isolated islands and then the separation of PA particles. The surface was finally cleaned from PA molecules after 100 h illumination. The most interesting results were attributed to the efficient degradation of PA molecules under visible light, which were attained by the 2D TiO_2 films deposited on the Au substrate. Heterogeneous photocatalysis can be considered as the main mechanism engaged in the degradation of PA molecules by a 2D TiO_2–Au bilayer. The excitation process, the separation and the transfer of charge carriers under visible light illumination in the Au–TiO_2 bilayer were engaged in the photocatalytic performance [157,158]. Considering the structural and interfacial concepts of 2D oxide–metal heterostructures, the TiO_2–Au interface experiences a charge transfer from Au into TiO_2 semiconductors [149]. This property plays an important role when the plasmonic-assisted visible light electron generation of an Au substrate starts the electron injection into a photocatalyst 2D TiO_2 film [159]. In doing so, the electrons needed for photocatalytic reactions will be provided by the Au sublayer under visible light illumination. The important factors here are the geometrical features and the relation between the thickness of ultra-thin 2D oxide films and the roughness of an Au substrate. The deposition of an ultra-thin 2D oxide on the rough surface of an Au film facilitated plasmonic-assisted electron generation under visible light illumination [160,161].

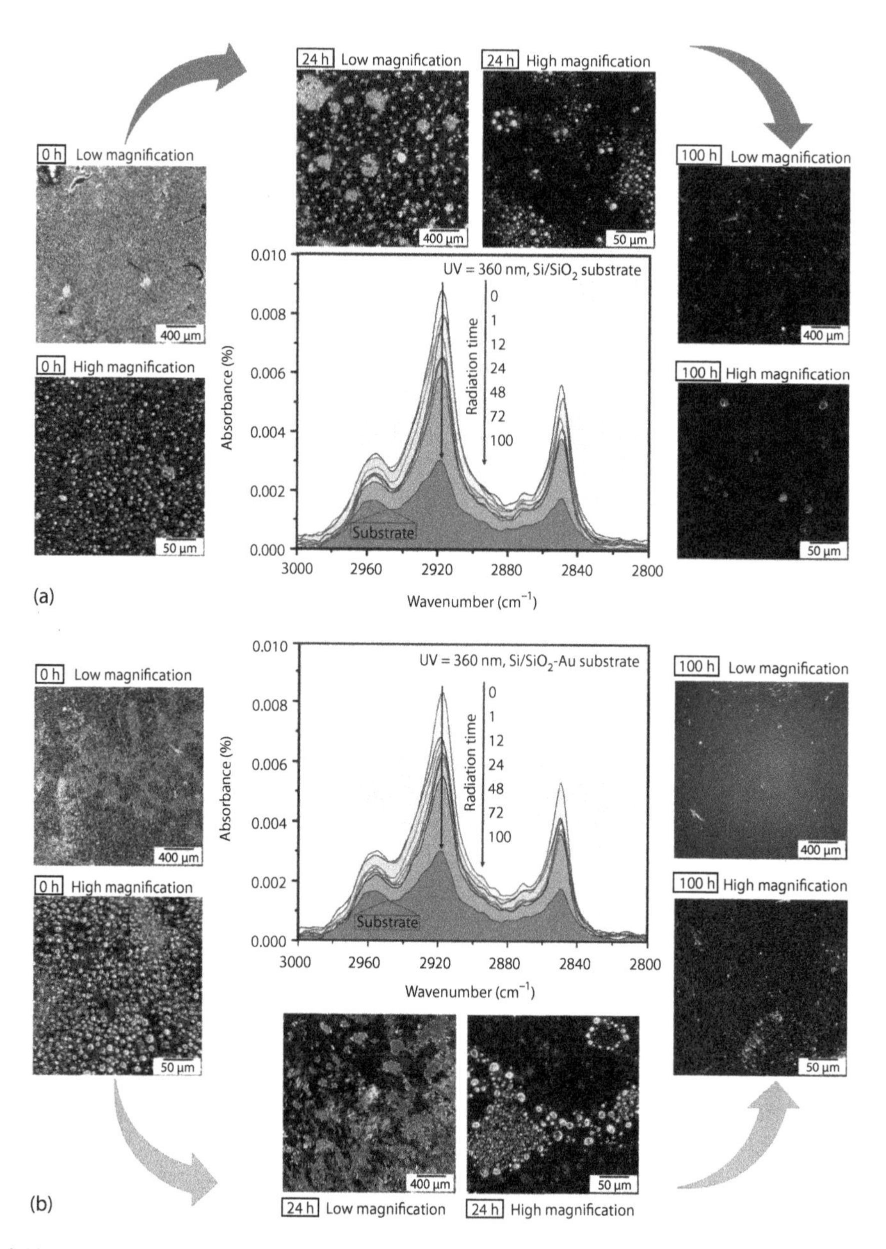

FIGURE 6.14
FT-IR absorbance spectra confirming the degradation of PA coated on TiO_2 film on (a) Si/SiO_2 and (b) Si/SiO_2–Au substrates, under UV illumination, as a function of irradiation time, and its corresponding optical image, representing degradation stages. (Reproduced with permission from Karbalaei Akbari, M., et al., *Mater. Res. Bull.*, 95, 380–391, 2017.)

6.4.2 The Photovoltaic Applications

The idea of employing 2D semiconductor materials in photovoltaic and photodetection applications arises from the fact that 2D films, monolayers or few-layer nanostructures can facilitate the ultra-fast transfer of electrons, and consequently, can pave the way for a new generation of photodetectors with high photoresponding behavior [162,163]. In this regard, although the concept of improving the functional capabilities of 2D nanostructured photodetectors was recently investigated, most of the reported studies only focused on the properties of individual monolayers with micron-sized longitudinal dimensions. A valuable alternative is the ALD technique, which reliably develops 2D nanostructures with precise thickness controllability and with flawless continuity over a large surface area. Furthermore, ALD can be used successfully for device fabrication. Thus, the optimization of an ALD recipe, the geometrical control of 2D oxide films, the stoichiometry and the chemical states of a 2D structure and the electrical contact between the semiconductor 2D oxide film and the metal electrode individually can play a vital role in the photovoltaic and photodetection performance of 2D oxide films.

As a practical example, a UV-A photodetector based on a 0.7 nm thick 2D WO_3 was fabricated by ALD on the surface of 4-inch diameter Si/SiO_2/Au wafers [164]. The Au film was used both as the support for 2D WO_3 and as the electrode for electrical measurements. The fabricated wafer-scale photoelectrode, its graphical scheme and TEM cross-sectional view of a 2D WO_3 film are presented in Figure 6.15a–c, respectively. The UV-A detecting performance of the 2D WO_3 photodetector showed improved photoresponsivity under various bias voltages, as shown in Figure 6.15d. The ultra-fast response time (40 µs, Figure 6.15e and f) of the 2D WO_3-based photodetector was the outstanding feature of this 2D nanostructure.

The measured response time was at least a 400-fold improvement on the previous reports for other WO_3-based UV-photodetectors that have been fabricated [164]. This ultra-fast response was attributed to the 2D nature of the WO_3 film and also to the presence of a metal–semiconductor junction in the fabricated photodetector [164]. The developed 2D photodetectors demonstrated long-term stability and reliable sensitivity to variation in the power density. The results of photoresponsivity measurements proved conclusively that the device has a generally distinguished response to the UV-A. However, the 2D

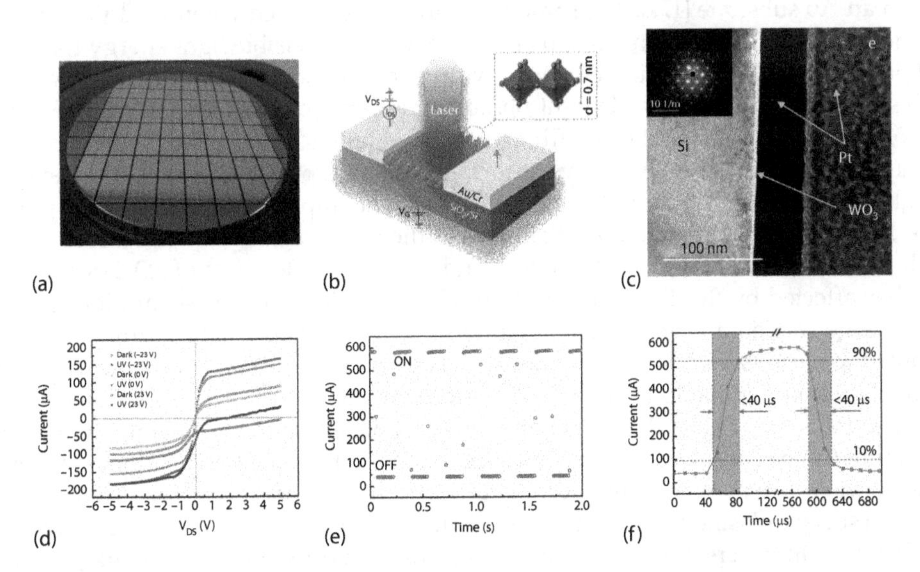

FIGURE 6.15
(a) Optical image of 2D WO_3 thin films deposited on the 4-inch diced Au wafer, (b) three-dimensional schematic view of the 2D WO_3 photodetector, (c) the cross-sectional TEM image of 2D WO_3, (d) the I-V characteristics of device under UV light illumination at different back-gate voltages (increasing from top to down of graph), (e) time-resolved photoresponse of the device under on-off illumination operation, (f) measured photoresponse time of the developed device. (Reproduced with permission from Hai, Z., et al., *Appl. Surf. Sci.*, 405, 169–177, 2017 and Hai, Z., et al., *Sensor Actuat. B Chem.*, 245, 954–962, 2017.)

WO_3-based photodetectors still provided some responses to the incident light with energy levels lower than the bandgap of WO_3. This was attributed to structural defects in the 2D WO_3 film, which can be responsible for photoresponsivity at longer wavelengths of light [165]. A similar ultra-fast photoresponse behavior was also reported when a 0.37 nm thick 2D TiO_2 semiconductor was used as a photodetector [166]. The performance of a 2D TiO_2-based photodetector was investigated under the illumination of various light sources and the measured photocurrent was quick and stable with an ultra-fast response (30 µs) and recovery (63 µs) time. This is a significant improvement on previous results for the response time of other nanostructured TiO_2-based photodetectors [166]. The calculated photoresponsivity (R_λ) of this 2D TiO_2-based photodetector was about 733 times higher than the reported R_λ for a photodetector with a 20 nm thick ALD TiO_2 semiconductor film [166,167]. The fabricated 2D TiO_2 photodetector has demonstrated rewardable long-term stability and a sensitive reaction to the variation of light intensity. A rectifying behavior was observed in the measurement of the photoresponsivity of a 2D TiO_2-based photodetector deposited on an Au substrate. This rectifying response is very similar to the normal metal–semiconductor junction behavior under various back-gate voltages [168,169]. This evidence confirmed that the possible interaction in the 2D oxide–metal interface can change the electrical response of these devices.

Visible light responsivity is a desired property in semiconductor materials, which can facilitate the employment of visible light sources for energy conversion. Due to the large bandgap of oxide semiconductors, the illumination of UV light sources is needed to stimulate the electrons of the valance band of semiconductor oxides [170,171]. Visible light photoresponsivity was observed when 2D semiconductor oxide films were deposited on an Au substrate and then used as a photosensor. A strong photoresponsivity to the $\lambda = 785$ nm visible light laser was measured when 2D WO_3 mono and few-layer films were grown by ALD on an Au substrate [172]. The same behavior was observed when 2D TiO_2 oxide films were deposited by ALD on an Au substrate to harvest the visible light energy for catalytic reactions [133]. The I-V (electrical current to voltage of photodetectors) curves of Figure 6.16a and b show the improved visible light photovoltaic performances of a 2D TiO_2–Au bilayer film compared with that of 2D TiO_2 films developed on an SiO_2 bare substrate. This visible light sensitivity of 2D TiO_2–Au bilayer films also made it possible to recruit the simulated sunlight for photocatalysis of fatty acid molecules. The improved visible light responsivity of the TiO_2–Au bilayer film can be attributed to the effect of TiO_2–Au plasmonic coupling, which is in direct relation to the thickness and geometrical features of 2D oxide films and is further affected by the TiO_2–Au interface. The excitation process, separation and transfer of charge carriers in the Au–TiO_2 bilayer film under visible light illumination can be attributed to various individual mechanisms. The tunneling of high energy carriers or hot electrons into the conduction band of a 2D oxide semiconductor is considered as one of the main mechanisms [173,174]. This mechanism is graphically depicted in Figure 6.16c. In fact, the Au film can be considered as a rough substrate for ultra-thin 2D TiO_2 [160]. This process includes photo-plasmon coupling accompanied by the generation of hot electrons by the decay of a surface plasmon. It continues by the injection or tunneling of hot electrons into 2D TiO_2. It results in the creation of activated electrons with different levels of energy in a TiO_2 conduction band [160,173]. In this mechanism, hot electrons must overcome the metal–semiconductor (MS) Schottky barrier at the interface of the Au–TiO_2 bilayer [160,161]. In another mechanism, the excitation of surface plasmons by visible light enables energy transfer into a 2D TiO_2 film. As graphically shown in Figure 6.16d, in this process the valance-band electrons of 2D TiO_2 are excited resulting in the generation of electrons in TiO_2 and holes in an Au film. In detail, the intense electric field called "hot spot" is effectively introduced into

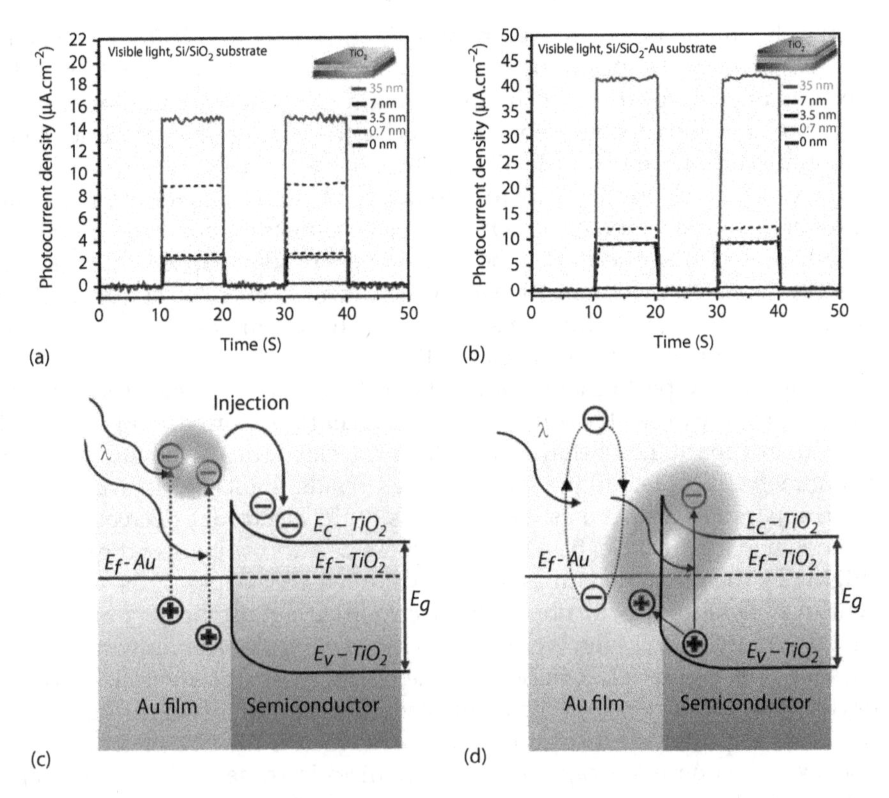

FIGURE 6.16
The I-t plots of photodetectors under visible light illumination. (a) The increasing photocurrent density with increasing the thickness of TiO_2 films deposited on Si/SiO_2 and (b) Au substrates. The mechanisms of photoelectron generation in an Au–TiO_2 bilayer under visible light, (c) the injection of hot electrons into a conduction band of TiO_2, whose potential energy is high enough to overcome the potential barrier between Au–TiO_2 bilayer and (d) the energy transfer by the generation of an electric field in 2D TiO_2 film. (Reproduced with permission from Karbalaei Akbari, M., et al., *Mater. Res. Bull.*, 95, 380–391, 2017.)

the semiconductor layer, exciting TiO_2 electrons from the valence to the conduction band [172,173]. Since a part of a 2D TiO_2 film can be covered by hot spots, the required energy to overcome the Schottky barrier can be provided by the incident of high energy photons of the visible light [153,161]. These mechanisms can adequately explain the improved visible light photocurrent density and photocatalytic performance of a 2D TiO_2–Au bilayer film. The extraction of plasmonic-derived charge carriers from a multilayer stack, comprising a monolayer of Au nanoparticles deposited on the 2D TiO_2–Au bilayer, was independently confirmed by other researchers [174]. This multilayer structure has also demonstrated the broadband and intense absorption of light, which leads to a significant increase in incident photon-to-electron conversion efficiency [174].

6.4.3 Supercapacitance Performance of 2D Oxide Semiconductors

Supercapacitors, as a very promising type of energy storage device, have attracted much attention in recent years owing to their superior properties such as high power density, fast charge/discharge rate, excellent reversibility and long cyclic life [175–177]. While the effects of electrode thickness on capacitor performance are recognized [178], most of them focus on micro and submicron films and only a few scientific reports have focused comprehensively on nanoscale and sub-nanoscale thicknesses [179]. Therefore, an investigation of the effect of the thickness of 2D electrodes at the few-layered level down to the single fundamental layer would be intriguing and influential. Common fabrication methods do not have the capability for such precise thickness and conformity control for wafer-scale deposition of ultra-thin 2D films, which hinders the development of atomically thin 2D supercapacitor electrodes. As a technological advancement, the employment of the ALD technique with special precursors made this investigation possible [180,181]. Despite the short history of ALD application in the development of supercapacitors, the advantages of this precise deposition technique have been proven in several cases. ALD-developed Co_3O_4 [182], NiO [183] and VO_x thin films are some of the examples of using the ALD technique in the development of supercapacitors [184].

The supercapacitance performance of ALD-developed 2D WO_3 electrodes with thickness variations from 6.0nm down to 0.7nm was recently reported [126]. The thickness-dependent super-capacitance behaviors of 2D WO_3 electrodes were studied using the cyclic voltammetry (CV) technique. The obtained results indicated that the capacitance values were apparently changed as the thickness of the electrodes decreased from 6.0 to 0.7nm, as shown in Figure 6.17a. Besides the impressive improvement of the specific capacitance judging by the integrated areas of the CV curves, the shapes of the CV curves also underwent an evident transformation as the electrode turned thinner. For instance, the CV curves of a 6.0nm thick WO_3 film were close to the rectangular shapes without any peak, as clearly shown in Figure 6.17b, indicating the characteristic of an electric double-layer capacitor [185]. The CV curves exhibited almost ideal rectangular shapes along with larger integrated areas as presented in Figure 6.17a–c. This means that the electric double-layer capacitance was the dominant capacitance mechanism in these 2D WO_3 supercapacitors. Even when the thickness of the 2D WO_3 film decreased to 0.7 nm, the general pattern of the CV curves was still shaped like a rectangle, as demonstrated in Figure 6.17d. The appearance of distinct redox peaks demonstrated that the mechanism shifted from the electric double-layer capacitance for a few-layered 2D WO_3 to the pseudo-capacitance for a monolayer WO_3. This phenomenon possibly resulted from the full exposure of the surface atoms during the redox reactions in a monolayer WO_3. The mechanism shift can also be contributed by the sub-stoichiometric structure of a monolayer WO_3, while the few-layered 2D film adopted an adsorption/separation process potentially due to its 2D-layered nature with its rough surface characteristics [186]. The galvanostatic charge-discharge (GCD) measurements confirmed the non-faradic capacitance behavior of the few-layered 2D WO_3, which is consistent with the CV measurements. The pseudo-capacitance behavior of a monolayer WO_3 film was further confirmed by the study of CV curves. The specific capacitances calculations for the 2D WO_3 electrodes suggested a negative correlativity between the specific capacitance of the WO_3 thin film and its thickness. A comparative study in terms of the specific capacitance is summarized in Table 6.3 [126]. Data show that the monolayer WO_3 film (0.7 nm thick) had the highest specific capacitance among most of the WO_3 nanostructured electrodes, demonstrating the superior advantages of the atomically thin 2D WO_3 as supercapacitor electrodes. Cycling stability and electrochemical impedance spectroscopy (EIS) measurements revealed that the monolayer WO_3 film has lower charge transfer resistance compared with that of thicker WO_3 films [126].

It was suggested that the development of heterojunction semiconductors could have distinctive modulation effects on electrical conductivity and ion diffusion behavior [187–189]. The supercapacitance properties of an atomically thin 2D WO_3/TiO_2 heterojunction as an electrochemical capacitor have also been investigated [190]. The ALD-enabled approach to the fabrication of a 2D heterojunction is especially advantageous when the film quality and

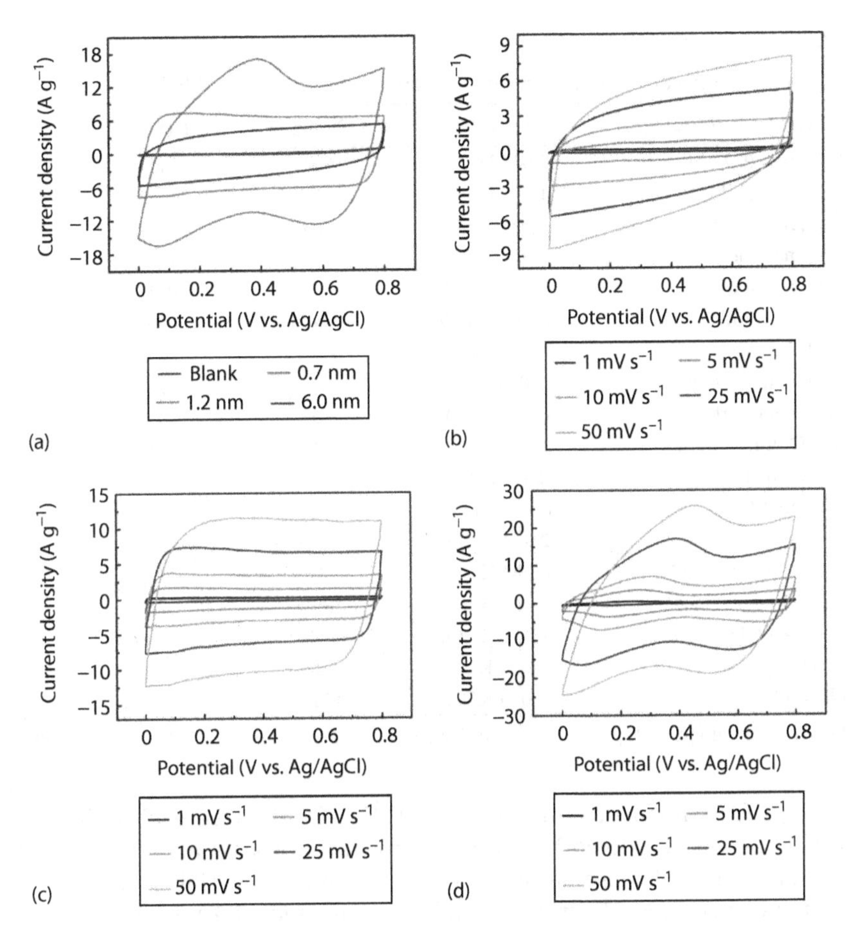

FIGURE 6.17
(a) CV curves of 2D WO_3 samples with different thickness. The CV curves of (b) 6.0 nm, (c) 1.2 nm and (d) 0.7 nm thick WO_3 film at different scan rates. Higher scan rate resulted in more capacitance or larger CV curve. (Reproduced with permission from Hai, Z., et al., *Electrochim. Acta*, 246, 625–633, 2017.)

thickness are critical for the surface/near-surface reactions in high performance supercapacitors [191]. The CV measurements revealed the ideal capacitance behavior of WO_3 2D films while the 2D TiO_2 films showed the pseudo-capacitance behavior [190]. Furthermore, the energy storage mechanism of 2D WO_3/TiO_2 was shown to primarily be a non-faradaic reaction. The GCD studies revealed a longer discharge time for a 2D WO_3/TiO_2 heterojunction compared with that of a 2D WO_3. A longer discharge time indicated the contribution of a 2D TiO_2 film in the improvement of the capacitance of a 2D WO_3/TiO_2 film. The discharge performance of the 2D nanostructures showed that the specific capacitance of a 2D WO_3/TiO_2 heterojunction was much higher than the specific capacitance of individual 2D TiO_2 and 2D WO_3 films. Long-term stability (over thousands of cycles) was another distinguishing characteristic of this 2D supercapacitor electrode. The capacitance retention of a 2D WO_3/TiO_2 heterojunction was slightly lower than that of 2D TiO_2. The long-term cycling stability of 2D WO_3/TiO_2 was improved compared with that of WO_3, which is attributed to the stabilization of the internal interface by the deposition of 2D TiO_2 on 2D WO_3 [126].

The surface functionalization of 2D materials with other nanostructures is one of the proposed strategies to impact the capacitance properties of ultra-thin films [33,192].

TABLE 6.3
Supercapacitor Performance of Various Nanostructured WO_3 Electrodes

Electrode Materials	Electrolyte	Cs (Fg^{-1})	T (a.u.)	Cr (%)
WO_3 film (0.7 nm)	1.0 M H_2SO_4	650.3	2000	65.8
WO_3 film (1.2 nm)	1.0 M H_2SO_4	396.9	2000	75.5
WO_3 film (6.0 nm)	1.0 M H_2SO_4	225.4	2000	91.7
WO_3 nanoflower	1.0 M H_2SO_4	196	5000	85
WO_3 nanoflower	0.5 M H_2SO_4	127	1000	83.7
WO_3 nanorod	1.0 M Na_2SO_4	463	1000	97
WO_3 nanorod	1.0 M H_2SO_4	114	/	/
WO_3 nanoparticle	2.0 M KOH	255	/	/
WO_3 nanoparticle	0.5 M H_2SO_4	54	/	/
h-WO_3 thin film	1.0 M H_2SO_4	694	2000	87
h-WO_3.n.H_2O	0.5 M H_2SO_4	498	/	/
WO_3 thin film	1.0 M Na_2SO_4	530	2000	84
Mesoporous WO_3	2.0 M H_2SO_4	109	/	/
WO_3/WO_3.0.5 H_2O	0.5 M H_2SO_4	290	/	/
GNS/WO_3	1.0 M H_2SO_4	143.6	/	/
rGO/WO_3.H_2O	1.0 M H_2SO_4	244	900	97
Co/WO_3	2.0 M KOH	45	/	/
Ni/WO_3	2.0 M KOH	171.28	/	/

Source: Reprinted with permission from Hai, Z., et al., *Electrochim. Acta*, 246, 625–633, 2017.

In this regard, TiO_2 nanoparticle-functionalized 2D WO_3 films were fabricated by the facile two-step ALD process followed by the post-annealing process at 380°C [33]. The supercapacitance properties of these nanostructured electrode materials were investigated. To fabricate a nanostructured electrode, initially a 6 nm thick 2D WO_3 was deposited on a polycrystalline Au substrate followed by ALD of a 1.5 nm thick TiO_2 film over WO_3. A post-annealing process at air atmosphere resulted in the formation of uniformly distributed TiO_2 nanoparticles on the surface of a 2D WO_3 film. The detailed SEM images of TiO_2 nanoparticle-functionalized 2D WO_3 films are shown in Figure 6.18a and b. XPS and Raman analyses confirmed the development of stoichiometric TiO_2 nanoparticles with an anatase crystalline structure on the surface of a WO_3 film. The TiO_2 functionalized 2D WO_3 electrode demonstrated a distinct improvement in the specific capacitance compared with that of a 2D WO_3 electrode. Studies confirmed that the capacitance in a 2D WO_3 electrode was stored by the accumulation of electrolyte ions between the electrolyte–electrode interfaces, which is known as the electric double-layer capacitance mechanism [193]. The GCD tests also confirmed the dominance of the electric double-layer capacitance mechanism of the galvanostatic charge and the discharge performance of a 2D WO_3 film. After TiO_2 functionalization of a WO_3 2D film, the mechanism shifted to pseudo-capacitance, as demonstrated in Figure 6.18c. The transition of electrochemical mechanisms is ascribed to the following facts. The TiO_2-functionalized WO_3 film contains more active sites. This can result in a higher surface-to-volume ratio and better electron mobility compared with that of pure 2D WO_3 electrodes, facilitating ion adsorption and diffusion on the active material. The improved ion adsorption and diffusion leads to an enhanced electrochemical performance. Furthermore, the improved capacitance performance is also related to the formation of a TiO_2–WO_3 heterojunction with enhanced conductance modulation [194,195]. The characteristics of pseudo-capacitance behavior were also detected by the study of galvanostatic charge and discharge analysis of TiO_2 functionalized 2D WO_3 films. The remarkably 1.5-times enhancement of the specific capacitance and faster charge transfer are the other benefits of the functionalization of 2D WO_3 films with TiO_2 nanoparticles.

6.4.4 Electrochemical Sensors Based on 2D Oxide Semiconductors

The fast and accurate sensing of environmentally hazardous components has been the focus of attention during the last decade [196]. Recently, ALD-developed 2D WO_3 nanostructures were used for the electrochemical detection of hydrazine in aqueous solutions [34]. The main idea behind using 2D nanostructures as sensors is attributed to the employment of the 2D-confined effects of 2D nanostructures to facilitate the efficient sensing of ultra-low concentration chemicals in an aqueous environment [34]. The ALD technique has the capability to develop ultra-thin oxide films with precise thickness controllability. This is important especially when it is known that the electro-catalytic activity of nanostructured materials is significantly dependent on their distribution on the substrate. Traditional methods, such as electrochemical deposition and hydrothermal synthesis, have no reasonable control over the uniform distribution of materials. Furthermore, traditional techniques cannot satisfy the necessity of the long-term stability of sensors, which is due to the gradual detachment and dissolution of the catalyst materials from the substrate [197]. ALD as an alternative technique can fulfill the requirements of a uniformly developed oxide sensor film. The studies of EIS of 2D WO_3 films confirmed that the monolayer WO_3 has the lowest electron transfer resistance compared with that of thicker 2D WO_3 films. The redox peak current of a monolayer WO_3 was higher than that of other thicker 2D WO_3 films. This is due to the largest effective surface area of a monolayer WO_3 resulting in an increase in current and surface charge density. The results of CV tests are shown in Figure 6.19a. The measurements indicate that 2D WO_3 nano-films possess a lower oxidation potential and a higher oxidation peak current for hydrazine electro-oxidation. This can be attributed to faster electron transfer kinetics in 2D WO_3 nano-films owing to a high effective surface area. The efficient electro-catalytic activity of a monolayer WO_3 film without any fouling effects showed the fast electron transfer reactions on the surface of a 2D WO_2 film. From the I-t curves, depicted in Figure 6.19b, fast response and all steady states were achieved within just 2 s for 2D WO_3 films. The monolayer WO_3 film exhibited the largest catalytic response current to changes in the hydrazine concentration, which was considerably higher than that of the other 2D WO_3 nano-films. An evaluation of long-term stability by conducting amperometric experiments indicated that the fabricated monolayer WO_3 films

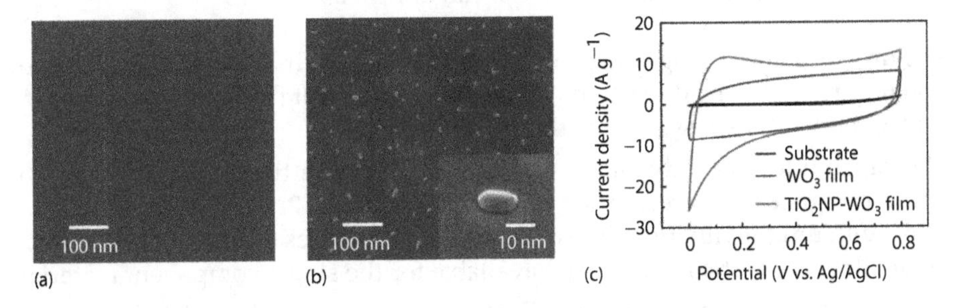

FIGURE 6.18
SEM images of (a) 2D WO_3 film and (b) 2D WO_3 film functionalized by TiO_2 nanoparticles after post-annealing. Inset in (b) is the HRSEM image of TiO_2 nanoparticle and (c) CV curves of the 2D electrodes. (Reproduced with permission from Hai, Z., et al., *Mater. Today Commun.*, 12, 55–62, 2017.)

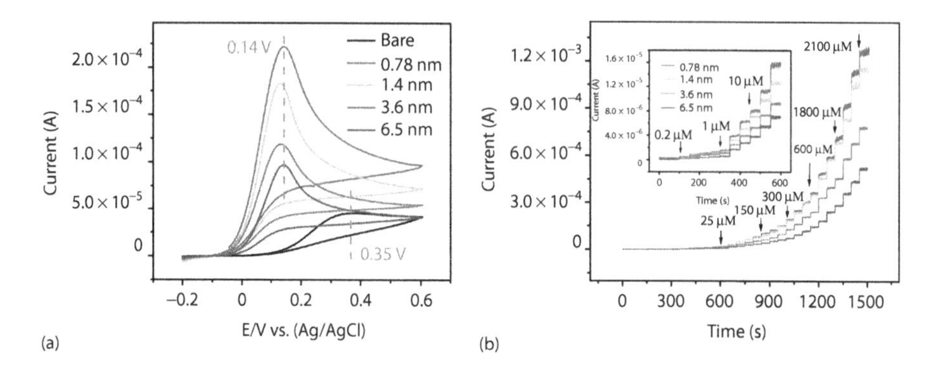

FIGURE 6.19
(a) CV curves of 2D WO_3 films and Au electrodes for hydrazine detection, the lower film thickness demonstrated the higher peak current and the larger catalytic response, (b) the sensitive chronoamperometric response of 2D WO_3 films to the changes of hydrazine concentration. (Reproduced with permission from Wei, Z., et al., *Chem. Electro. Chem.*, 5, 266–272, 2018.)

have excellent long-term stability. As an interesting property, the anti-interfering ability of a monolayer WO_3 was confirmed when the interfering chemicals were added to the aqueous environment containing hydrazine. The amperometric response of monolayer WO_3 nano-films to hydrazine did not change upon the introduction of KCl, $NaNO_3$ and $NaNO_2$, glucose and H_2O_2 into the aqueous solution containing hydrazine [34]. As a practical result, the response of monolayer WO_3 nano-films to hydrazine was not affected by the addition of a 100-fold concentration of inorganic and a 50-fold concentration of biological substances [34]. A sensitivity of 1.23 was recorded for hydrazine sensing achieved by using a monolayer WO_3 in the presence of interfering agents. This value was very close to the value of sensitivity of the monolayer WO_3 (1.24) when there was no interference agent. These observations confirmed the reliability of the 2D WO_3 oxide as electrochemical sensors.

6.5 Summary

The present chapter has highlighted the capabilities of the ALD technique on the development and tailoring of ultra-thin 2D nanostructured semiconductor oxides. Compared with other chemical and PVD techniques, the ALD capabilities for design, characterization and tailoring of these 2D nanostructures have broad appeal as follows:

- The unique self-limiting film growth mechanism originated from the self-saturation nature of chemical reactions in the ALD process. It facilitates the atomic-scale control of the thickness of ultra-thin films.
- The capabilities and advantages of ALD substantiate the employment of a precisely designed ALD recipe for the development of 2D heterostructured oxide films with unique functionalities. These nanostructures offer a combination of 2D oxide films properties that are not available for the single component materials.
- The surface functionalization of 2D oxide films with ALD-assisted complementary processes makes it possible to manipulate the properties of 2D oxide materials. This is considered one of the highly valuable features of ALD for the design and fabrication of the doped and functionalized 2D nanostructures.

Despite the mentioned intrinsic advantage of ALD, the development and monitoring of wafer-scaled 2D oxide films with a high level of uniformity and controlled crystallinity and stoichiometry need the accurate design of an ALD recipe. Herewith, the recent progress in the development of mono and few-layered nanostructured semiconductor oxides was discussed by investigating the practical cases assisted by ALD engineering. The technological ALD parameters were reviewed to provide an overview of the challenges faced during the development of 2D oxide nanostructures. The discussions were then accompanied by complementary material characterization studies to reinforce the scientific aspects of ALD development of 2D nanostructured semiconductor oxides.

Subsequently, the tailoring of ALD-developed 2D semiconductor oxide was introduced by considering the practical applications of these nanostructures. Consequently, the capabilities of ALD for the fabrication of nano-devices based on 2D semiconductor oxides, including photocatalyst nanostructures, photovoltaic and UV and visible light sensors, supercapacitors and electrochemical sensors for hazardous chemicals were presented. In essence, the reliability of the ALD technique for tailoring 2D nanostructured oxide semiconductors for practical applications was confirmed.

References

1. A. K. Geim and K. S. Novoselov, The rise of graphene, *Nat. Mater.* 6 (2007) 183–191.
2. K. S. Novoselov, Z. Jiang, Y. Zhang, S. V. Morozov, H. L. Stormer, U. Zeitler, J. C. Maan, G. S. Boebinger, P. Kim, and A. K. Geim, Room-temperature quantum hall effect in graphene, *Science* 315 (2007) 1379.
3. R. R. Nair, P. Blake, A. N. Grigorenko, K. S. Novoselov, T. J. Booth, T. Stauber, N. M. R. Peres, and A. K. Geim, Fine structure constant defines visual transparency of graphene, *Science* 320 (2008) 1308.
4. F. Schedin, A. K. Geim, S. V. Morozov, E. W. Hill, P. Blake, M. I. Katsnelson, and K. S. Novoselov, Detection of individual gas molecules adsorbed on graphene, *Nat. Mater.* 6 (2007) 652–655.
5. X. Li, W. Cai, J. An, S. Kim, J. Nah, D. Yang, R. Piner, A. Velamakanni, I. Jung, E. Tutuc, et al., Large-area synthesis of high-quality and uniform graphene films on copper foils, *Science* 324 (2009) 1312–1314.
6. S. Bae, H. Kim, Y. Lee, X. Xu, J. S. Park, Y. Zheng, J. Balakrishnan, T. Lei, H. Kim, Y. Song, et al., Roll-to-roll production of 30-inch graphene films for transparent electrodes, *Nat. Nanotechnol.* 5 (2010) 574–578.
7. J. Wrachtrup, 2D materials: Single photons at room temperature, *Nat. Nanotechnol.* 11 (2016) 7–8.
8. M. Chhowalla, D. Jena, and H. Zhang, Two-dimensional semiconductors for transistors, *Nat. Rev. Mater.* 1 (2016) 16052.
9. F. Xia, H. Wang, D. Xiao, M. Dubey, and A. Ramasubramaniam, Two-dimensional material nanophotonics, *Nat. Photonics.* 8 (2014) 899–907.
10. P. L. Cullen, K. M. Cox, M. K. B. Subhan, L. Picco, O. D. Payton, D. J. Buckley, T. S. Miller, S. A. Hodge, N. T. Skipper, V. Tileli, et al., Ionic solutions of two-dimensional materials, *Nat. Chem.* 9 (2017) 244–249.
11. P. Ajayan, P. Kim, and K. Banerjee, Two-dimensional van der Waals materials, *Phys. Today* 69 (2016) 39–44.
12. Y. Liu, N. O. Weiss, X. Duan, H. C. Cheng, Y. Huang, and X. Duan, Van der Waals heterostructures and devices, *Nat. Rev. Mater.* 1 (2016) 16042.
13. D. Jariwala, T. J. Marks, and M. C. Hersam, Mixed-dimensional van der Waals heterostructures, *Nat. Mater.* 16 (2017) 170–181.

14. F. P. Netzer, and S. Surnev, Structure concept in two dimensional oxide materials, In: F. P Netzar and A. Fortunelli (eds.), *Oxide Material at the Two-Dimensional Limit,* (2016) Springer Series in Materials Science, 234, Springer, Cham, Switzerland, pp. 233–250.
15. Z. Lin, A. McCreary, and N. Briggs, 2D materials advances: From large scale synthesis and controlled heterostructures to improved characterization techniques, defects and applications, *2D Mater.* 3 (2016) 042001.
16. S. Surnev, A. Fortunelli, and F. P. Netzer, Structure–property relationship and chemical aspects of oxide-metal hybrid nanostructures, *Chem. Rev.* 113 (2013) 4314–4372.
17. H. Kuhlenbeck, S. Shaikhutdinov, and H. J. Freund, Well-ordered transition metal oxide layers in model catalysis: A series of case studies, *Chem. Rev.* 113 (2013) 3986–4034.
18. J. Shim, H. Park, D. Kang, J. Kim, S. Jo, Y. Park, and J. Park, Electronic and optoelectronic devices based on two-dimensional materials: From fabrication to application, *Adv. Electron. Mater.* 3 (2017) 1600364.
19. F. Wang, Z. Wang, C. Jiang, L. Yin, R. Cheng, X. Zhan, K. Xu, F. Wang, Y. Zhang, and J. He, Progress on electronic and optoelectronic devices of 2D layered semiconducting materials, *Small,* 13 (2017) 1604298.
20. K. Kerman and S. Ramanathan, Performance of solid oxide fuel cells approaching the two-dimensional limit, *J. Appl. Phys.* 115 (2014) 174307.
21. S. Ida and T. Ishihara, Recent progress in two-dimensional oxide photocatalysts for water splitting, *J. Phys. Chem. Lett.* 5 (2014) 2533–2542.
22. G. Pacchioni, Two-dimensional oxides: Multifunctional materials for advanced technologies, *Chem. Eur. J.* 18 (2012) 10144–10158.
23. S. G. Leonardi, Two-dimensional zinc oxide nanostructures for gas sensor applications, *Chemosensors* 5 (2017) 17.
24. E. Kan, M. Li, S. Hu, C. Xiao, H. Xiang, and K. Deng, Two-dimensional hexagonal transition-metal oxide for spintronics, *J. Phys. Chem. Lett.* 4 (2013) 1120–1125.
25. S. P. Ghosh, K. C. Das, N. Tripathy, G. Bose, D. H. Kim, T. I. Lee, J. M. Myoung, and J. P. Kar, Ultraviolet photodetection characteristics of zinc oxide thin films and nanostructures, *IOP Conf. Ser. Mater. Sci. Eng.* 115 (2016) 012035.
26. W. R. Richards and M. Shen, Extraction of two-dimensional metal–oxide–semiconductor field effect transistor structural information from electrical characteristics, *J. Vac. Sci. Technol. B* 18 (2000) 533–539.
27. J. E. T. Elshof, H. Yuan, and P. G. Rodriguez, Two-dimensional metal oxide and metal hydroxide nanosheets: Synthesis, controlled assembly and applications in energy conversion and storage, *Adv. Energy Mater.* 6 (2016) 1600355.
28. M. Alsaif, M. R. Field, T. Daeneke, A. F. Chrimes, W. Zhang, B. J. Carey, K. J. Berean, S. Walia, J. Van Embden, B. Zhang, et al., Exfoliation solvent dependent plasmon resonances in two-dimensional sub-stoichiometric molybdenum oxide nanoflakes, *ACS Appl. Mater. Interfaces* 8 (2016) 3482–3493.
29. Y. Li, Z. Li, C. Chi, H. Shan, L. Zheng, and Z. Fang, Plasmonics of 2D nanomaterials: Properties and applications, *Adv. Sci.* 4 (2017) 1600430.
30. E. Brunet, G. C. Mutinati, S. Steinhaver, and A. Kock, Oxide ultrathin films in sensor applications, In: G. Pacchioni and S. Valeri (eds.), *Oxide Ultrathin Films: Science and Technology,* 1st edn. (2012) Wiley-V-CH, Weinheim, Germany, pp. 239–241.
31. K. Shavanova, Y. Bakakina, I. Burkova, I. Shtepliuk, R. Viter, A. Ubelis, V. Beni, N. Starodub, R. Yakimova, and V. Khranovskyy, Application of 2D non-graphene materials and 2D oxide nanostructures for biosensing technology, *Sensors* 16 (2016) 223.
32. Y. Zhu, C. Cao, S. Tao, W. Chu, Z. Wu, and Y. Li, Ultrathin nickel hydroxide and oxide nanosheets: Synthesis, characterizations and excellent supercapacitor performance, *Sci. Rep.* 4 (2014) 5787.
33. Z. Hai, M. Karbalaei Akbari, Z. Wei, C. Xue, H. Xu, J. Hu, L. Hyde, and S. Zhuiykov, TiO_2 nanoparticles-functionalized two-dimensional WO_3 for high performance supercapacitors developed by facile two-step ALD process, *Mater. Today Commun.* 12 (2017) 55–62.

34. Z. Wei, Z. Hai, M. Karbalaei Akbari, J. Hu, L. Hyde, S. Depuydt, and S. Zhuiykov, Ultra-sensitive, sustainable, and selective electrochemical hydrazine detection by ALD developed two-dimensional WO_3 nano-films, *Chem. Electro. Chem.* 5 (2018) 266–272.
35. R. Ma and T. Sasaki, Two-dimensional oxide and hydroxide nanosheets: Controllable high-quality exfoliation, molecular assembly, and exploration of functionality, *Acc. Chem. Res.* 48 (2015) 136–143.
36. L. Gao, Y. Li, M. Xiao, S. Wang, G. Fu, and L. Wang, Synthesizing new types of ultrathin 2D metal oxide nanosheets via half-successive ion layer adsorption and reaction, *2D Mater.* 4 (2017) 025031.
37. J. Yu, J. Li, W. Zhang and H. Chang, Synthesis of high quality two-dimensional materials via chemical vapor deposition, *Chem. Sci.* 6 (2015) 6705–6716.
38. Z. Yang and J. Hao, Progress in pulsed laser deposited two-dimensional layered materials for device applications, *J. Mater. Chem. C* 4 (2016) 8859–8878.
39. B. Zou, C. Walker, K. Wang, V. Tileli, O. Shaforost, N. M. Harrison, N. Klein, N. M. Alford, and P. K. Petrov, Growth of epitaxial oxide thin films on graphene, *Sci. Rep.* 6 (2016) 31511.
40. T. Tsuchiya, K. Daoudi, I. Yamaguchi, T. Manabe, T. Kumagai, and S. Mizuta, Preparation of tin oxide films on various substrates by excimer laser metal organic deposition, *Appl. Surf. Sci.* 247 (2005) 145–150.
41. C. Wu and M. R. Castel, Ultra-thin oxide film on Au (111) substrate, In: F. P. Netzer and A. Fortunelli (eds.), *Oxide Material at the Two-Dimensional Limit,* (2016) Springer Series in Materials Science 234, Springer, Cham, Switzerland, pp. 149–168.
42. S. Shaikhutdinov and H. J. Freund, Ultrathin oxide films on metal supports: Structure-reactivity relations, *Annu. Rev. Phys. Chem.* 63 (2012) 619–633.
43. S. A. Chambers, Epitaxial growth and properties of thin film oxides, *Surf. Sci. Rep.* 39 (2000) 105–180.
44. N. Nilius, Properties of oxide thin films and their adsorption behavior studied by scanning tunneling microscopy and conductance spectroscopy, *Surf. Sci. Rep.* 64 (2009) 595–659.
45. J. V. Lauritsen and F. Besenbacher, Model catalyst surfaces investigated by scanning tunneling microscopy, In: B. C. Gates and H. Knozinger (eds.), *Advances in Catalysis,* (2006) 50. Elsevier Academic Press Inc, San Diego, CA, pp. 97–147.
46. S. Zhuiykov, T. Kawaguchi, Z. Hai, M. Karbalaei Akbari, and P. M. Heynderickx, Interfacial engineering of two-dimensional nano-structured materials by atomic layer deposition, *Appl. Surf. Sci.* 392 (2017) 231–243.
47. S. Valeri and S. Benedetti, Synthesis and preparation of oxide ultrathin films. In: G. Pacchioni and S. Valeri (eds.), *Oxide Ultrathin Films: Science and Technology,* (2012) Wiley-V-CH, Weinheim, Germany, pp. 1–26.
48. D. C. Grinter and G. Thornton, Characterization tools of ultrathin oxide films. In: G. Pacchioni and S. Valeri (eds.), *Oxide Ultrathin Films: Science and Technology,* (2012) Wiley-V-CH, Weinheim, Germany, pp. 27–46.
49. S. Zhuiykov, Nanostructured two-dimensional materials, In: V. Tewary and Y. Zhang (eds.), *Modelling, Characterization and Production of Nanomaterials,* (2015) Elsevier, Amsterdam, pp. 477–524.
50. D. M. King, X. Liang, and A. W. Weimer, Functionalization of fine particles using atomic and molecular layer deposition, *Powder Technol.* 221 (2012) 13–25.
51. M. Laskela and M. Ritala, Atomic layer deposition chemistry: Recent developments and future challenges, *Angew. Chem. Int. Ed.* 42 (2003) 5548–5554.
52. S. J. Tauster, S. C. Fung, and R. L. Garten, Strong metal-support interactions. Group 8 noble metals supported on titanium dioxide, *J. Am. Chem. Soc.* 100 (1978) 170–175.
53. M. Osada and T. Sasaki, Chemical nanomanipulation of two-dimensional nanosheets and its applications, In: Y. Masuda (ed.), *Nanofabrication,* (2011) InTech.
54. L. Wang and T. Sasaki, Titanium oxide nanosheets: Graphene analogues with versatile functionalities, *Chem. Rev.* 114 (2014) 9455–9486.
55. F. P. Netzer and S. Surnev, Ordered oxide nanostructures on metal surfaces, In: G. Pacchioni and S. Valeri (eds.), *Oxide Ultrathin Films: Science and Technology,* (2012) Wiley-VCH, Weinheim, Germany, pp. 47–73.

56. Q. Fu and T. Wagner, Interaction of nanostructured metal over layers with oxide surfaces, *Surf. Sci. Rep.* 62 (2007) 431–498.
57. F. P. Netzer, F. Allegretti, and S. Surnev, Low-dimensional oxide nanostructures on metals: Hybrid systems with novel properties, *J. Vac. Sci. Technol. B* 28 (2010) 1–16.
58. M. Nagel, I. Biswas, H. Peisert, and T. Chassé, Interface properties and electronic structure of ultrathin manganese oxide films on Ag(001), *Surf. Sci.* 601 (2007) 4484–4487.
59. F. P. Netzer, Small and beautiful: The novel structures and phases of nano-oxides, *Surf. Sci.* 604 (2010) 485–489.
60. A. Dahal and M. Batzill, Growth from behind: Intercalation-growth of two dimensional FeO moiré structure underneath of metal-supported graphene, *Sci. Rep.* 5 (2015) 11378.
61. C. Chang, S. Sankaranarayanan, D. Ruzmetov, M. H. Engelhard, E. Kaxiras, and S. Ramanathan, Compositional tuning of ultrathin surface oxides on metal and alloy substrates using photons: Dynamic simulations and experiments, *Phys. Rev. B* 81 (2010) 085406.
62. G. Pacchioni, Role of structural flexibility on the physical and chemical properties of metal-supported oxide ultrathin films, In: F. P. Netzer and A. Fortunelli (eds.), *Oxide Materials at the Two-Dimensional Limit*, (2016) Springer, Cham, Switzerland, pp. 91–118.
63. C. Noguera and J. Goniakowski, Electrostatics and polarity in 2D oxides, In: F.P. Netzer and A. Fortunelli (eds.), *Oxide Materials at the Two-Dimensional Limit*, (2016) Springer Series in Materials Science 234, Springer, Cham, Switzerland, pp. 201–231.
64. J. Goniakowski and C. Noguera, Microscopic mechanisms of stabilization of polar oxide surfaces: Transition metals on the MgO(111) surface, *Phys. Rev. B.* 66 (2002) 085417.
65. J. Goniakowski and C. Noguera, Polarization and rumpling in oxide monolayers deposited on metallic substrates, *Phys. Rev. B* 79 (2009) 155433.
66. J. Goniakowski, C. Noguera, and L. Giordano, Prediction of uncompensated polarity in ultra-thin films, *Phys. Rev. Lett.* 93 (2004) 215702.
67. J. Goniakowski, C. Noguera, L. Giordano, and G. Pacchioni, Adsorption of metal adatoms on FeO(111) and MgO(111) monolayers: Effects of charge state of adsorbate on rumpling of supported oxide film, *Phys. Rev. B* 80 (2009) 125403.
68. L. Giordano, F. Cinquini, and G. Pacchioni, Tuning the surface metal work function by deposition of ultrathin oxide films: Density functional calculations. *Phys. Rev. B* 73 (2006) 045414.
69. S. J. Tauster, S. C. Fung, R. T. K. Baker, and J. A. Horsley, Strong-interactions in supported-metal catalysts, *Science* 211 (1981) 1121–1125.
70. S. Ma, J. Rodriguez, and J. Hrbek, STM study of the growth of cerium oxide nanoparticles on Au(111), *Surf. Sci.* 602 (2008) 3272–3278.
71. C. Wu, M. S. J. Marshall and M. R. Castell, Surface structures of ultrathin TiO_x films on Au(111), *J. Phys. Chem. C* 115 (2011) 8643–8652.
72. S. Sindhu, M. Heiler, K. M. Schindler, and H. Neddermeyer, A photoemission study of CoO-films on Au(111), *Surf. Sci.* 541 (2003) 197–206.
73. G. Barcaro, E. Cavaliere, L. Artiglia, L. Sementa, L. Gavioli, G. Granozzi, and A. Fortunelli, Building principles and structural motifs in TiO_x ultrathin films on a (111) substrate, *J. Phys. Chem. C* 116 (2012) 13302–13306.
74. Y. Pan, S. Benedetti, N. Nilius, and H. J. Freund, Change of the surface electronic structure of Au(111) by a monolayer MgO(001) film, *Phys. Rev. B* 84 (2011) 075456.
75. E. Cavaliere, I. Kholmanov, L. Gavioli, F. Sedona, S. Agnoli, G. Granozzi, G. Barcaro, and A. Fortunelli, Directed assembly of Au and Fe nanoparticles on a TiOx/Pt(111) ultra-thin template: the role of oxygen affinity, *Phys. Chem. Chem. Phys.* 11 (2009) 11305–11309.
76. J. Goniakowski, L. Giordano, and C. Noguera, Polarity of ultrathin MgO(111) films deposited on a metal substrate, *Phys. Rev. B* 81 (2010) 205404.
77. M. Ritter, W. Ranke, and W. Weiss, Growth and structure of ultrathin FeO films on Pt(111) studied by STM and LEED, *Phys. Rev. B* 57 (1998) 7240–7251.
78. T. U. Nahm, Study of high-temperature oxidation of ultrathin Fe films on Pt(100) by using X-ray photoelectron spectroscopy, *J. Korean Phys. Soc.* 68 (2016) 1215–1220.

79. D. Ragazzon, A. Schaefer, M. H. Farstad, L. E. Walle, P. Palmgren, A. Borg, P. Uvdal, and A. Sandell, Chemical vapor deposition of ordered TiO_x nanostructures on Au(111), *Surf. Sci.* 617 (2013) 211–217.
80. J. Seifert, E. Meyer, H. Winter, and H. Kuhlenbeck, Surface termination of an ultrathin V_2O_3-film on Au(111) studied via ion beam triangulation, *Surf. Sci.* 606 (2012) L41–L44.
81. M. Li and E. I. Altman, Cluster-size dependent phase transition of Co oxides on Au(111), *Surf. Sci.* 619 (2014) L6–L10.
82. S. Y. Quek, M. M. Biener, J. Biener, C. M. Friend, and E. Kaxiras, Tuning electronic properties of novel metal oxide nanocrystals using interface interactions: MoO_3 monolayers on Au(111), *Surf. Sci.* 577 (2005) L71–L77.
83. F. Stavale, L. Pascua, N. Nilius, and H. J. Freund, Morphology and luminescence of ZnO films grown on a Au(111) support, *J. Phys. Chem. C* 117 (2013) 10552–10557.
84. S. Zhuiykov, L. Hyde, Z. Hai, M. Karbalaei Akbari, E. Kats, C. Detavernier, C. Xue, and H. Xu, Atomic layer deposition-enabled single layer of tungsten trioxide across a large area, *Appl. Mater. Today* 6 (2017) 44–53.
85. Z. S. Li, D. V. Potapenko, and R. M. Osgood, Using Moiré patterning to map surface reactivity versus atom registration: Chemisorbed trimethyl acetic acid on TiO/Au(111), *J. Phys. Chem. C* 118 (2014) 29999–30005.
86. X. Deng and C. Matranga, Selective growth of Fe_2O_3 nanoparticles and islands on Au(111), *J. Phys. Chem. C* 113 (2009) 11104–11109.
87. M. Li and E. I. Altman, Shape, morphology, and phase transitions during Co oxide growth on Au(111), *J. Phys. Chem. C* 118 (2014) 12706–12716.
88. J. Tao, T. Luttrell, and M. Batzill, A two-dimensional phase of TiO_2 with a reduced bandgap, *Nat. Chem.* 3 (2011) 296–300.
89. H. G. Kim and H. B. R. Lee, Atomic layer deposition on 2D materials, *Chem. Mater.* 29 (2017) 3809–3826.
90. Y. H. Lee, X. Q. Zhang, and W. Zhang, Synthesis of large area MoS_2 atomic layers with chemical vapor deposition, *Adv. Mater.* 24 (2012) 2320–2325.
91. J. Schoiswohi, S. Surnev, and F. P. Netzar, Reaction on inverse model catalyst surface: Atomic view by STM, *Top. Catal.* 36 (2005) 91–105.
92. C. Detavernier, J. Dendooven, S. P. Sree, K. F. Ludwig, and J. A. Martens, Tailoring nanoporous materials by atomic layer deposition, *Chem. Soc. Rev.* 40 (2011) 5242–5253.
93. R. Warren, F. Sammoura, F. Tounsi, M. Sanghadasa, and L. Lin, Highly active ruthenium oxide coating via ALD and electrochemical activation in supercapacitor applications, *J. Mater. Chem.* A3 (2015) 15568–15575.
94. C. Zhu, P. Yang, D. Chao, X. Wang, X. Zhang, S. Chen, B. Tay, H. Huang, H. Zhang, W. Mai, et al., All metal nitrides solid state asymmetric supercapacitors, *Adv. Mater.* 27 (2015) 4566–4571.
95. N. P. Dasgupta, H. Lee, S. F. Bent, and P. S. Weiss, Recent advances in atomic layer deposition, *Chem. Mater.* 28 (2016) 1943–1947.
96. J. S. Ponraj, G. Attolini, and M. Bosi, Review on atomic layer deposition and application of oxide thin films, *Crit. Rev. Solid State Mater. Sci.* 38 (2013) 203–233.
97. L. Niinisto, J. Paivasaari, J. Niinisto, M. Putkonen, and M. Nieminen, Advanced electronic and optoelectronic materials by atomic layer deposition: An overview with special emphasis on recent progress in processing of high-K dielectrics and other oxide materials, *Phys. Status Solidi A* 201 (2004) 1443–1452.
98. S. M. George, Atomic layer deposition: an overview, *Chem. Rev.* 110 (2010) 111–131.
99. T. Suntola, Atomic layer epitaxy, *Thin Solid Films* 216 (1992) 84–89.
100. I. Jõgi, M. Pärs, J. Aarik, A. Aidla, M. Laan, J. Sundqvist, L. Oberbeck, J. Heitmann, and K. Kukli, Conformity and structure of titanium oxide films grown by atomic layer deposition on silicon substrates, *Thin Solid Films* 516 (2008) 4855–4862.
101. X. Du and S. M. George, Thickness dependence of sensor response for CO gas sensing by tin oxide films grown using atomic layer deposition, *Sens. Actuators B Chem.* 135 (2008) 152–160.

102. M. A. Malik and P. O'Brien, Organometallic and metallo-organic precursors for nanoparticles, In: R. A. Fischer (ed.) *Precursor Chemistry of Advanced Materials: CVD, ALD and Nanoparticles,* (2005) Springer, Berlin, pp. 125–145.
103. S. E. Potts and W. M. M. Kessel, Energy-enhanced atomic layer deposition for more process and precursor versatility, *Coord. Chem. Rev.* 257 (2013) 3254–3270.
104. X. H. Liang, Y. Zhou, J. Li, and A. W. Weimer, Reaction mechanism studies for platinum nanoparticle grown by atomic layer deposition, *J. Nanopart. Res.* 13 (2011) 3781–3788.
105. M. Laskela and M. Ritala, Atomic layer deposition (ALD) from precursors to thin film structures, *Thin Solid Films* 409 (2002) 138–146.
106. A. Philip, S. Thomas, and K. R. Kumar, Calculation of growth per cycle (GPC) of atomic layer deposited aluminum oxide nanolayers and dependence of GPC on surface OH concentration, *Pram. J. Phys.* 82 (2014) 563–569.
107. X. Liang, S. M. George, A. W. Weimer, N. H. Li, J. H. Blackson, and J. D. Harris, Synthesis of a novel porous polymer/ceramic composite material by low-temperature atomic layer deposition, *Chem. Mater.* 19 (2007) 5388–5394.
108. T. Hatanpää, M. Ritala, and M. Leskelä, Precursors as enablers of ALD technology: Contributions from University of Helsinki, *Coord. Chem. Rev.* 257 (2013) 3297–3322.
109. S. W. Lee, B. J. Choi, T. Eom, J. H. Han, S. K. Kim, S. J. Song, W. Lee, and C. S. Hwang, Influences of metal, non-metal precursors, and substrates on atomic layer deposition processes for the growth of selected functional electronic materials, *Coord. Chem. Rev.* 257 (2013) 3154–3176.
110. B. B. Burton, A. R. Lavoie, and S. M. George, Tantalum nitride atomic layer deposition using (tert-butylimido)tris(diethylamido) tantalum and hydrazine, *J. Electrochem. Soc.* 155 (2008) D508–D516.
111. W. J. Maeng, D. Choi, J. Park, and J. S. Park, Indium oxide thin film prepared by low temperature atomic layer deposition using liquid precursors and ozone oxidant, *J. Alloys Comp.* 649 (2015) 216–221.
112. W. J. Maeng, D. Choi, J. Park, and J. S. Park, Atomic layer deposition of highly conductive indium oxide using a liquid precursor and water oxidant, *Ceram. Int.* 41 (2015) 10782–10787.
113. Y. Wan, J. Bullock, and A. Cuevas, Passivation of C-Si surfaces by ALD tantalum oxide capped with PECVD silicon nitride, *Sol. Ener. Mat. Sol. Cells* 142 (2015) 42–46.
114. S. Yeo, J. Park, S. Lee, D. Lee, J. Seo, and S. Kim, Ruthenium and ruthenium dioxide thin films deposited by atomic layer deposition using a novel zero-valent metalorganic precursor (ethylbenzene)(1,3-butadiene)Ru(0), and molecular oxygen, *Microelectron. Eng.* 137 (2015) 16–22.
115. S. H. Kwon, O. K. Kwon, J. H. Kim, S. J. Jeong, S. W. Kim, and S. W. Kang, Improvement of the morphological stability by stacking RuO_2 on Ru thin films with atomic layer deposition, *J. Electrochem. Soc.* 154 (2007) H773–H777.
116. S. Moitzheim, C. S. Miisha, S. Deng, D. J. Cott, C. Detavernier, and P. M. Vereecken, Nanostructured TiO_2/carbon nanosheet hybrid electrode for high-rate thin-film lithium-ion batteries, *Nanotechnology* 25 (2014) 504008.
117. W. Chiappim, G. E. Testoni, R. S. Moraes, R. S. Pessoa, J. C. Sagás, F. D. Origo, L. Vieira, and H. S. Maciel, Structural, morphological, and optical properties of TiO_2 thin films grown by atomic layer deposition on fluorine doped tin oxide conductive glass, *Vacuum* 123 (2016) 91–102.
118. J. L. Tian, H. Y. Zhang, G. G. Wang, X. Z. Wang, R. Sun, L. Jin, and J. C. Han, Influence of film thickness and annealing temperature on the structural and optical properties of ZnO thin films on Si (100) substrates grown by atomic layer deposition, *Superlatt. Microstr.* 83 (2015) 719–729.
119. O. Bethge, M. Nobile, S. Abermann, M. Glaser, and E. Bertagnolli, ALD grown bilayer junction of ZnO:Al and tunnel oxide barrier for SIS solar cell, *Sol. Energy Mater. Sol. Cells* 117 (2013) 178–182.
120. S. C. Riha, J. M. Racowski, M. P. Lanci, J. A. Klug, A. S. Hock, and A. B. F. Martinson, Phase discrimination through oxidant selection in low-temperature atomic layer deposition of crystalline iron oxides, *Langmuir* 29 (2013) 3439–3445.

121. V. Aravindan, K. B. Jinesh, R. R. Prabhakar, V. S. Kale, and S. Madhavi, Atomic layer deposited (ALD) SnO_2 anodes with exceptional cycleability for Li-ion batteries, *Nano Energy* 2 (2013) 720–725.
122. B. Han, J. M. Park, K. H. Choi, W. K. Lim, T. R. Mayangsari, W. Koh, and W. J. Lee, Atomic layer deposition of stoichiometric Co_3O_4 films using bis(1,4-di-iso-propyl-1,4-diazabutadiene) cobalt, *Thin Solid Films* 589 (2015) 718–722.
123. J. Malm, T. Sajavaara, and M. Karpinen, Atomic layer deposition of WO_3 thin films using $W(CO)_6$ and O_3 precursors, *Chem. Vap. Depos.* 18 (2012) 245–248.
124. M. Tsuchiya, S. Sankaranarayanan, and S. Ramanathan, Photon-assisted oxidation and oxide thin film synthesis: A review, *Prog. Mater. Sci.* 54 (2009) 981–1057.
125. R. Sivakumar, R. Gopalakrishnan, M. Jayachandran, and C. Sanjeeviraja, Investigation of x-ray photoelectron spectroscopic (XPS), cyclic voltammetric analyses of WO_3 films and their electrochromic response in FTO/WO_3/electrolyte/FTO cells, *Smart Mater. Struct.* 15 (2006) 877–888.
126. Z. Hai, M. Karbalaei Akbari, Z. Wei, C. Xue, H. Xu, J. Hu, and S. Zhuiykov, Nano-thickness dependence of supercapacitor performance of the ALD-fabricated two-dimensional WO_3, *Electrochim. Acta* 246 (2017) 625–633.
127. S. Zhuiykov, M. Karbalaei Akbari, Z. Hai, C. Xue, H. Xu, and L. Hyde, Wafer-scale fabrication of conformal atomic-layered TiO_2 by atomic layer deposition using tetrakis (dimethylamino) titanium and H_2O precursors, *Mater. Des.* 120 (2017) 99–108.
128. S. Zhuiykov, M. Karbalaei Akbari, Z. Hai, C. Xue, H. Xu, and L. Hyde, Data set for fabrication of conformal two-dimensional TiO_2 by atomic layer deposition using tetrakis(dimethylamino) titanium (TDMAT) and H_2O precursors, *Data Brief* 13 (2017) 401–407.
129. J. Dendooven, S. P. Sree, K. D. Keyser, D. Deduytsche, J. A. Martens, K. F. Ludwig, and C. Detavernier, In situ X-ray fluorescence measurements during atomic layer deposition: Nucleation and growth of TiO_2 on planar substrates and in nanoporous films, *J. Phys. Chem. C* 115 (2011) 6605–6610.
130. M. Bouman and F. Zaera, Reductive eliminations from amido metal complexes: Implications for metal film deposition, *J. Electrochem. Soc.* 158 (2011) D524–D526.
131. R. Methaapanon and S. F. Bent, Comparative study of titanium dioxide atomic layer deposition on silicon dioxide and hydrogen-terminated silicon, *J. Phys. Chem. C* 114 (2010) 10498–10504.
132. M. Karbalaei Akbari, Z. Hai, Z. Wei, J. Hu and S. Zhuiykov, Wafer-scale two-dimensional Au-TiO_2 bilayer films for photocatalytic degradation of palmitic acid under UV and visible light illumination, *Mater. Res. Bull.* 95 (2017) 380–391.
133. T. Dobbelaere, M. Minjauw, T. Ahmad, P. M. Vereeken and C. Detavernier, Plasma enhanced atomic layer deposition of zinc phosphate, *J. Non. Cryst. Solids* 444 (2016) 43–48.
134. K. Kanomata, P. Pansila, B. Ahmmad, S. Kubota, K. Hirahara and F. Hirose, Infrared study on room-temperature atomic layer deposition of TiO_2 using tetrakis(dimethylamino)titanium and remote-plasma-excited water vapor, *Appl. Surf. Sci.* 308 (2014) 328–332.
135. Y. Zhang, W. Wu, and K. Zhang, Raman studies of 2D anatase TiO_2 nanosheets, *Phys. Chem. Chem. Phys.* 18 (2016) 32178–32184.
136. C. Y. Xu, P. X. Zhang, and L. Yan, Blue shift of Raman peak from coated TiO_2 nanoparticles, *J. Raman Spectrosc.* 32 (2001) 862–865.
137. S. J. Park, J. P. Lee, J. S. Jang, H. Rhu, H. Yu, B. Y. You, C. S. Kim, K. J. Kim, Y. J. Cho, S. Baik, et al., In situ control of oxygen vacancies in TiO_2 by atomic layer deposition for resistive switching devices, *Nanotechnology* 24 (2013) 295202.
138. A. L. Bassi, D. Cattaneo, V. Russo, C. E. Bottani, T. Mazza, P. Piseri, P. Milani, F. O. Ernst, K. Wegner, and S. E. Pratsinis, Raman spectroscopy characterization of titania nanoparticles produced by flame pyrolysis: The influence of size and stoichiometry, *J. Appl. Phys.* 98 (2005) 074305.
139. A. Gharachorlou, M. D. Detwiler, X. K. Gu, L. Mayr, B. Klotzer, J. Greeley, R. G. Reifenberger, W. N. Delgass, F. H. Ribeiro, and D. Y. Zemlyanov, Trimethyl aluminum and oxygen atomic layer deposition on hydroxyl-free Cu(111), *ACS Appl. Mater. Interfaces* 7 (2015) 16428–16439.

140. M. Knez, K. Niesch, and L. Niinistoe, Synthesis and surface engineering of complex nanostructures by atomic layer deposition, *Adv. Mater.* 19 (2007) 3425–3438.
141. B. J. O'Neill, D. H. K. Jackson, J. Lee, C. Canlas, P. C. Stair, C. Marshall, J. W. Elam, T. F. Kuech, J. A. Dumesic, and G. W. Huber, Catalyst design with atomic layer deposition, *ACS Catal.* 5 (2015) 1804–1825.
142. A. Beniya, N. Isomura, H. Hirata, and Y. Watanabe, Morphology and chemical states of size-selected Ptn clusters on an aluminium oxide film on NiAl(110), *Phys. Chem. Chem. Phys.* 16 (2014) 26485–26492.
143. M. S. Chen and D. W. Goodman, Interaction of Au with titania: The role of reduced Ti, *Top. Catal.* 44 (2007) 41–47.
144. L. Gragnaniello, T. Ma, G. Barcaro, L. Sementa, F. R. Negreiros, A. Fortunelli, S. Surnev, and F. P. Netzer, Ordered arrays of size-selected oxide nanoparticles, *Phys. Rev. Lett.* 108 (2012) 195507.
145. D. Stacchiola, S. Kaya, J. Weissenrieder, H. Kuhlenbeck, S. Shaikhutdinov, H. J. Freund, M. Sierka, T. K. Todorova, and J. Sauer, Synthesis and structure of ultrathin aluminosilicate films. *Angew. Chem. Int. Ed.* 45 (2006) 7636–7639.
146. H. Sabbah, Amorphous titanium dioxide ultra-thin films for self-cleaning surfaces, *Mater. Express*, 3 (2013) 171–175.
147. M. Sterrer, T. Risse, U. M. Pozzoni, L. Giordano, M. Heyde, H. Rust, G. Pacchioni, and H. Freund, Control of the charge state of metal atoms on thin MgO films, *Phys. Rev. Lett.* 98 (2007) 096107.
148. M. De Santis, A. Buchsbaum, P. Varga, and M. Schmid, Growth of ultrathin cobalt oxide films on Pt(111), *Phys. Rev. B* 84 (2011) 125430.
149. G. Barcaro, L. Sementa, F. R. Negreiros, I. O. Thomas, S. Vajda, and A. Fortunelli, Atomistic and electronic structure methods for nanostructured oxide interfaces, In: F. P. Netzer and A. Fortunelli (eds.), *Oxide Materials at the Two-Dimensional Limit*, (2016) Springer Series in Materials Science 234, Springer, Cham, Switzerland, pp. 39–90.
150. L. Cheng, C. Yin, F. Mehmood, B. Liu, J. Greeley, S. Lee, B. Lee, S. Seifert, R. E. Winans, D. Teschner, et al. Reaction mechanism for direct propylene epoxidation by alumina-supported silver aggregates: The role of the particle/support interface, *ACS Catal.* 4 (2014) 32–39.
151. Y. Lei, F. Mehmood, S. Lee, J. Greeley, B. Lee, S. Seifert, R. E. Winans, J. W. Elam, R. J. Meyer, P. C. Redfern, et al. Increased silver activity for direct propylene epoxidation via sub nanometer size effects, *Science* 328 (2010) 224–228.
152. J. W. Elam and S. M. George, Growth of ZnO/Al_2O_3 alloy films using atomic layer deposition techniques, *Chem. Mater.* 15 (2003) 1020–1028.
153. S. Lee, L. M. Molina, M. J. López, J. A. Alonso, B. Hammer, B. Lee, S. Seifert, R. E. Winans, J. W. Elam, M. J. Pellin, et al., Selective propene epoxidation on immobilized Au6–10 clusters: The effect of hydrogen and water on activity and selectivity. *Angew. Chem. Int. Ed.* 121 (2009) 1495–1499.
154. L. M. Molina, S. Lee, K. Sell, G. Barcaro, A. Fortunelli, B. Lee, S. Seifert, R. E. Winans, J. W. Elam, M. J. Pellin, et al., Size-dependent selectivity and activity of silver nanoclusters in the partial oxidation of propylene to propylene oxide and acrolein: A joint experimental and theoretical study, *Catal. Today* 160 (2011) 116–130.
155. S. Lee, B. Lee, F. Mehmood, S. Seifert, J. A. Libera, J. W. Elam, J. Greeley, P. Zapol, L. A. Curtiss, M. J. Pellin, et al., Oxidative decomposition of methanol on subnanometer palladium clusters: The effect of catalyst size and support composition, *J. Phys. Chem. C* 114 (2010) 10342–10348.
156. M. D. Kane, F. S. Roberts, and S. L. Anderson, Effects of alumina thickness on CO oxidation activity over Pd20/alumina/Re(0001): Correlated effects of alumina electronic properties and Pd20 geometry on activity, *J. Phys. Chem. C* 119 (2015) 1359–1375.
157. Y. Shiraishi, N. Yasumoto, J. Imai, H. Sakamoto, S. Tanaka, S. Ichikawa, B. Ohtanie, and T. Hirai, Quantum tunneling injection of hot electrons in Au/TiO_2 plasmonic photocatalysts, *Nanoscale* 9 (2017) 8349–8361.
158. K. Wu, J. Chen, J. R. McBride, and T. Lian, Efficient hot-electron transfer by a plasmon-induced interfacial charge-transfer transition, *Science* 349 (2015) 632–635.

159. X. Zhang, Y. Chen, R. Liu, and D. Tsai, Plasmonic photocatalysis, *Rep. Prog. Phys.* 76 (2013) 046401.
160. F. Tan, T. Li, N. Wang, S. Lai, W. Yu, and X. Zhang, Rough gold films as broad band absorbers for plasmonic enhancement of TiO_2 photocurrent over 400–800 nm, *Sci. Rep.* 6 (2016) 33049.
161. T. Zhang, D. Su, R. Z. Li, S. J. Wang, F. Shan, J. Xu, and X. Zhang, Plasmonic nanostructures for electronic designs of photovoltaic devices: Plasmonic hot carrier photovoltaic architectures and plasmonic electrode structures, *J. Photon. Energy* 6 (2016) 042504.
162. F. Xia, T. Mueller, Y. Lin, A. Valdes-Garcia, and P. Avouris, Ultra-fast graphene photodetector, *Nat. Nanotechnol.* 4 (2009) 839–843.
163. L. Zheng, L. Zhongzhu, and S. Guozhen, Photodetectors based on two dimensional materials, *J. Semicond.* 37 (2016) 091001.
164. Z. Hai, M. Karbalaei Akbari, C. Xue, H. Xu, L. Hyde, and S. Zhuiykov, Wafer-scaled monolayer WO_3 windows ultra-sensitive, extremely-fast and stable UV-A photodetection, *Appl. Surf. Sci.* 405 (2017) 169–177.
165. W. Tian, C. Zhang, T. Zhai, S. L. Li, X. Wang, J. Liu, X. Jie, D. Liu, M. Liao, Y. Koide, et al., Flexible ultra violet photodetectors with broad photoresponse based on branched ZnS-ZnO heterostructure nanofilms, *Adv. Mater.* 26 (2014) 3088–3093.
166. M. Karbalaei Akbari, Z. Hai, S. Depuydt, E. Kats, J. Hu, and S. Zhuiykov, Highly sensitive, fast-responding, and stable photodetector based on ALD-developed monolayer TiO_2, *IEEE Trans. Nanotechnol.* 16 (2017) 880–887.
167. W. J. Wang, C. X. Shan, H. Zhu, F. Y. Ma, D. Z. Shen, X. W. Fan, and K. L. Choy, Metal–insulator–semiconductor–insulator–metal structured titanium dioxide ultraviolet photodetector, *J. Phys. D: Appl. Phys.* 43 (2010) 045102.
168. F. Jing, D. Zhang, F. Li, J. Zhou, D. Sun, and S. Ruan, High performance ultraviolet detector based on $SrTiO_3/TiO_2$ heterostructure fabricated by two steps in situ hydrothermal method, *J. Alloys Compd.* 650 (2015) 97–101.
169. S. Das, J. Kar, and J. Myoung, Junction properties and application of ZnO single nanowire based Schottky diode, In: A. Hashim (ed.), *Nanowires: Fundamental Research*, (2011) In Tech Publishing, Croatia, pp. 174–178.
170. J. Yang, Y. L. Jiang, L. J. Li, E. Muhire, and M. Z. Gao, High-performance photodetectors and enhanced photocatalysts of two-dimensional TiO_2 nanosheets under UV light excitation, *Nanoscale* 8 (2016) 8170–8177.
171. J. Liu, M. Zhong, J. Li, A. Pan, and X. Zhu, Few-layer WO_3 nanosheets for high-performance UV-photodetectors, *Mater. Lett.* 148 (2015) 184–187.
172. Z. Hai, M. Karbalaei Akbari, C. Xue, H. Xu, S. Depuydt, and S. Zhuiykov, Photodetector with superior functional capabilities based on monolayer WO_3 developed by atomic layer deposition, *Sensor Actuat. B Chem.* 245 (2017) 954–962.
173. M. W. Knight, H. Sobhani, P. Nordlander, and N. J. Halas, Photodetection with active optical antennas, *Science* 332 (2011) 698–702.
174. N. G. Charlene, J. J. Cadusch, S. Dligatch, A. Roberts, T. J. Davis, P. Mulvaney, and D. E. Gómez, Hot carrier extraction with plasmonic broadband absorbers, *ACS Nano*, 10 (2016) 4704–4711.
175. Z. Yu, L. Tetard, L. Zhai, and J. Thomas, Supercapacitor electrode materials: Nanostructures from 0 to 3 dimensions, *Energy Environ. Sci.* 8 (2015) 702–730.
176. G. Wang, L. Zhang, and J. Zhang, A review of electrode materials for electro chemical supercapacitors, *Chem. Soc. Rev.* 41 (2012) 797–828.
177. Y. Liu and X. Peng, Recent advances of supercapacitors based on two-dimensional materials, *Appl. Mater. Today* 7 (2017) 1–12.
178. K. Tsay, L. Zhang, and J. Zhang, Effects of electrode layer composition/thickness and electrolyte concentration on both specific capacitance and energy density of supercapacitor, *Electrochim. Acta* 60 (2012) 428–436.
179. A. Yu, I. Roes, A. Davies, and Z. Chen, Ultrathin transparent, and flexible graphene films for supercapacitor application, *Appl. Phys. Lett.* 96 (2010) 1–4.

180. X. Wang and G. Yushin, Chemical vapor deposition and atomic layer deposition for advanced lithium ion batteries and supercapacitors, *Energy Environ. Sci.* 8 (2015) 1889–1904.
181. R. Warren, F. Sammoura, A. Kozinda, and L. Lin, ALD ruthenium oxide-carbon nanotube electrodes for supercapacitor applications, *Proc. IEEE Int. Conf. Micro Electro Mech. Syst.* 16 (2014) 167–170.
182. C. Guan, X. Qian, X. Wang, Y. Cao, Q. Zhang, A. Li, and J. Wang, Atomic layer deposition of Co_3O_4 on carbon nanotubes/carbon cloth for high-capacitance and ultra stable supercapacitor electrode, *Nanotechnology* 26 (2015) 94001.
183. C. Guan, Y. Wang, Y. Hu, J. Liu, K. H. Ho, W. Zhao, Z. Fan, Z. Shen, H. Zhang, and J. Wang, Conformally deposited NiO on a hierarchical carbon support for high power and durable asymmetric supercapacitors, *J. Mater. Chem. A* 3 (2015) 23283–23288.
184. S. Boukhalfa, K. Evanoff, and G. Yushin, Atomic layer deposition of vanadium oxide on carbon nanotubes for high-power supercapacitor electrodes, *Energy Environ. Sci* 5 (2012) 6872–6879.
185. X. Bai, Q. Liu, J. Liu, H. Zhang, Z. Li, and X. Jing, Hierarchical Co_3O_4@$Ni(OH)_2$ core shell nanosheet arrays for isolated all-solid state supercapacitor electrodes with superior electrochemical performance, *Chem. Eng. J.* 315 (2017) 35–45.
186. K. Kalantar-Zadeh, A. Vijayaraghavan, M. H. Ham, H. Zheng, M. Breedon, and M. S. Strano, Synthesis of atomically thin WO_3 sheets from hydrated tungsten trioxide, *Chem. Mater.* 22 (2010) 5660–5666.
187. N. M. Vuong, D. Kim, and H. Kim, Electrochromic properties of porous WO_3-TiO_2 core shell nanowires, *J. Mater. Chem. C* 1 (2013) 3399–3407.
188. Y. Liu, C. Xie, J. Li, T. Zou, and D. Zeng, New insights into the relationship between photocatalytic activity and photocurrent of TiO_2/WO_3 nanocomposite, *Appl. Catal. A: Gen.* 433–434 (2012) 81–87.
189. Y. R. Smith, B. Sarma, S. K. Mohanty, and M. Misra, Formation of TiO_2-WO_3 nanotubular composite via single-step anodization and its application in photoelectrochemical hydrogen generation, *Electrochem. Commun.* 19 (2012) 131–134.
190. Z. Hai, M. Karbalaei Akbari, C. Xue, H. Xu, E. Solano, C. Detavernier, J. Hu, and S. Zhuiykov, Atomically thin WO_3/TiO_2 heterojunction for supercapacitor electrodes developed by atomic layer deposition, *Comp. Comm.* 5 (2017) 31–35.
191. C. Guan and J. Wang, Recent development of advanced electrode materials by atomic layer deposition for electrochemical energy storage, *Adv. Sci.* 3 (2016) 1500405.
192. B. M. Sanchez and Y. Gogotsi, Synthesis of two-dimensional materials for capacitive energy storage, *Adv. Mater.* 28 (2016) 6104–6135.
193. H. Cesiulis, N. Tsyntsaru, A. Ramanavicius, and G. Ragoisha, The study of thin films by electrochemical impedance spectroscopy, In: I. Tiginyanu, P. Topala, and V. Ursaki (eds.), *Nanostructures and Thin Films for Multifunctional Applications, Nanoscience and Technology* (2016) Springer, pp. 3–42.
194. H. Zhang, S. Wang, Y. Wang, J. Yang, X. Gao, and L. Wang, TiO_2(B) nanoparticle functionalized WO_3 nanorods with enhanced gas sensing properties, *Phys. Chem. Chem. Phys.* 16 (2014) 10830–10836.
195. D. Bekermann, A. Gasparotto, D. Barreca, C. Maccato, E. Comini, C. Sada, G. Sberveglieri, A. Devi, and R. A. Fischer, Co_3O_4/ZnO nanocomposites: From plasma synthesis to gas sensing applications, *Appl. Mater. Interfaces* 4 (2012) 928–934.
196. S. Bagheri, New York/New Jersey Nearshore Waters: A case study in NY/NJ, In: S. Bagheri, (ed.), *Hyperspectral Remote Sensing of Nearshore Water Quality a Case Study in New York/New Jersey*, (2017) Springer, pp. 19–27.
197. A. Salimi and M. Roushani, Non-enzymatic glucose detection free of ascorbic acid interference using nickel powder and nafion sol–gel dispersed renewable carbon ceramic electrode, *Electrochem. Commun.* 7 (2005) 879–887.

7

Tunneling Field-Effect Transistors Based on Two-Dimensional Materials

Peng Zhou

CONTENTS

7.1 Introduction

Although advanced complementary metal-oxide-semiconductors (CMOS) technology has achieved many extraordinary improvements in switching speed, density, functionality and cost, metal-oxide-semiconductor field-effect transistors (MOSFETs) are facing the challenges of scaling down the supply voltage with the increase in power density due to the thermionic emission limitation of carrier injection. Rising leakage currents degrade the I_{ON}/I_{OFF} ratio because the leakage currents will increase exponentially with a reduction in the supply voltage. Only the carriers in the exponential tail of the source Fermi distribution above the channel barrier can migrate to the channel. Considering a large enough bias, the drain current $I_d \propto \int dED(E)v(E)f_s(E)$, where $D(E)$ is the density of states, $v(E)$ is the carrier velocity and $f_s(E)$ is the source Fermi distribution. As $D(E)v(E)$ is a constant for a specific product and $f_s(E) \approx e^{E-E_f^s/kT}$, the SS can be obtained by

$$\mathrm{SS} = \frac{\partial V_g}{\partial(\log_{10} I_d)} = \frac{\partial V_g}{\partial\left(\frac{E}{q}\right)} \frac{kT}{q} \ln 10 \tag{7.1}$$

When considering all the gate voltage has been applied to the channel, we get the ideal case,

$$\mathrm{SS} = \frac{kT}{q} \ln 10,$$

the minimal subthreshold swing of 60 mV/decade at room temperature, which can't be overtaken even if scaled down. Although the power dissipation of a MOSFET, $P = I_{OFF}V_{DD}^3$, the power consumption density increases sharply with the increase in I_{OFF} when scaling down the supply voltage. The leakage power increases by 275-fold in commercial bulk CMOS 45 nm technology when lowering the supply voltage from 0.5 V to 0.25 V [1]. Therefore, it is necessary to resort to other physical mechanisms to solve the power consumption problem, like tunneling field-effect transistors (TFETs).

This review will concentrate on TFETs based on two-dimensional (2D) materials. The history of silicon TFET dates back to 1978 when Quinn et al. proposed the gated p-i-n structure [2], and it was experimentally realized in 1995 by Reddick and Amaratunga [3]. Lateral TFETs on silicon-on-insulator (SOI) observed negative conductance at 90 K [4] and TFETs on SOI with the gate overlapping the depletion region only to minimize capacitance [5]. Later, Group III-V materials [6] and nanotubes were used in TFET. Knoch and Appenzeller proposed nanotube TFETs and an experimentally realized SS below 60 mV/dec at room temperature for the first time in 2005 [7]. Vertical silicon TFETs were fabricated by Hansch et al. [8] using molecular beam epitaxy. TFETs achieved an SS below 60 mV/dec for the first time in a carbon nanotube field-effect transistor (FET) in 2004 [9]. Later, an average SS of ~25 mV/dec was achieved in a GaSb/InAs heterojunction [10] and ~7.8 mV/dec in an Si-based, arch-shaped, gate-all-around TFET [11]. Since its discovery in 2004, graphene has been widely studied and applied to chemical sensors [12], optical devices [13], high-frequency circuits [14] and energy generation and storage [15,16] because of its high mobility of 2.5×10^5 cm^2 V^{-1} s^{-1} [17], excellent thermal conductivity and optical properties, wonderful flexibility, high surface area and so on. Subsequently, many other 2D layered materials have attracted unprecedented attention and multitudinous 2D materials have come to the fore, such as transition metal dichalcogenides (TMDCs), hexagonal boron nitride (h-BN), layered Group-IV and Group-III metal chalcogenides, silicene and germanene. They have been widely investigated in recent decades and have been applied to various devices due to their unique properties different from three-dimensional (3D) materials. Many efforts have been made to achieve TFETs based on 2D materials with low power consumption.

7.2 The Principle of TFET

Taking a p-i-n n-type field-effect transistor (NTFET) as an example to demonstrate the physics of a TFET, a schematic cross section of an NTFET is shown in Figure 7.1a. The source is highly n-doped and the drain is highly p-doped so that the Fermi level of the drain is above its conduction band and that of the source is below its valance band, respectively, as shown in Figure 7.1b. In the OFF state (green lines), the electrons in the valance band can't tunnel into the channel as there is no empty state in the bandgap and the barrier for electrons in the conduction band of the drain is too high to inject into the conduction band of the channel by thermal emission. Therefore, the OFF current is ultralow. Increasing the gate voltage to move the conduction band of the channel below the valance band of the source, the electrons in the energy window ($\Delta\Phi$) of the source (green shading) have a probability of tunneling into the conduction band of the channel as many empty states are available in the conduction band of the channel, switched to the ON state. The electrons above the energy window in the tail of the Fermi distribution cannot tunnel into the channel as there are no empty states in the bandgap. It is effective to achieve an SS below 60 mV/dec by cutting off the tail of the Fermi distribution. In reverse, reducing the gate voltage to move the valance band of the channel above the conduction band of the source, the electrons from the conduction band of the source can tunnel into the valance band of the source, as Figure 7.1b and d shows. Therefore, the p+ doped region is defined as the drain for an NTFET while the n+ doped region is defined as the drain for a p-type field-effect transistor (PTFET) and the surface tunneling junction of an NTFET is at the interface between the p+ region and the channel region (red circle in Figure 7.1a) while that of a PTFET is at the interface between the n+ region and the channel region (red circle in Figure 7.1b). The gate voltage should be higher than a specific voltage to switch on for an NTFET and the gate voltage should be lower than a specific voltage to switch on for a PTFET.

In principle, the TFET is an ambipolar device and both a positive and a negative gate voltage can reduce the tunneling barrier, operating as an NTFET with dominant electron conduction and as a PTFET with dominant hole conduction [18]. The transfer characteristics (channel current versus gate voltage) with the n+ region as the drain and the p+ region as the grounded source are shown in Figure 7.2a. The channel current curves obviously shift with the varying of the drain voltage when the gate voltage is negative while they shift a little when the gate voltage is positive. When $V_g < 0$, $V_g - V_{n+}$ should be lower than V_{TP} to switch on the tunneling junction, where V_{n+} is the potential of the n+ region and V_{TP} is the switch-on voltage of the surface tunneling junction, so the corresponding gate voltage when the channel current is over 10^{-10} A moves right with the increasing drain voltage. When $V_g > 0$, the channel current curves move less prominently as the surface tunneling junction moves to the interface of the channel and the p+ region. Therefore, the varying

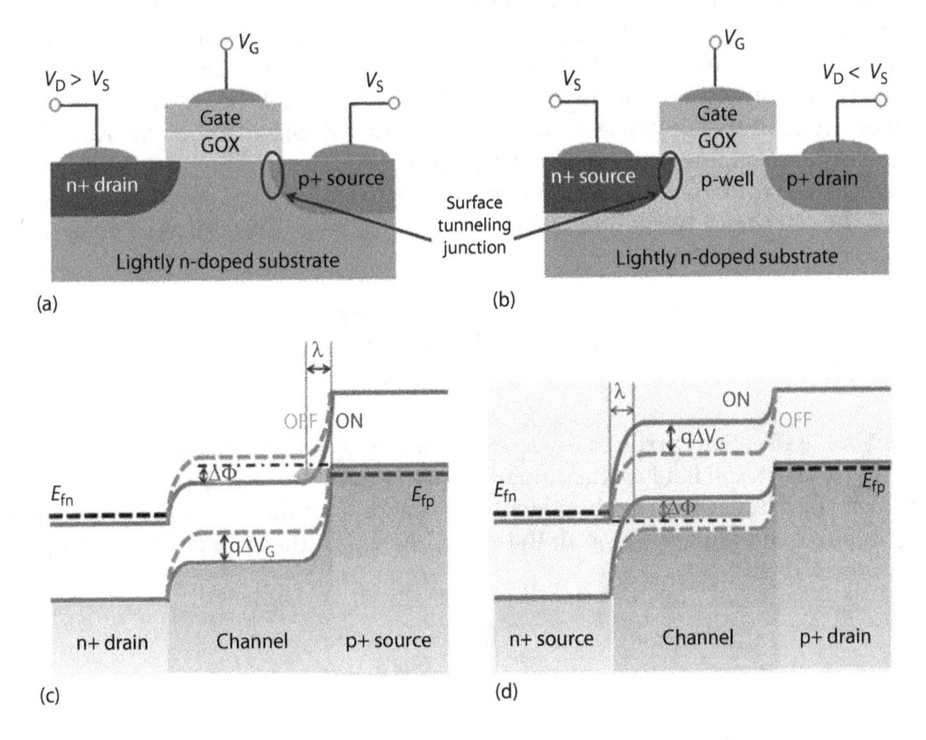

FIGURE 7.1
Principle of complementary p-i-n TFET. Schematic cross section of NTFET (a) and PTFET (b) with applied gate (V_G), source (V_S) and drain (V_D) voltages. Schematic diagrams of the band structures of NTFET (c) and PTFET (d) in ON and OFF states, respectively. The electrons in the energy window (ΔΦ) of the p+ source tunnel into the conduction band of the channel while the electrons in the energy band (ΔΦ) of the channel tunnel into the conduction band of n+ source for PTFET.

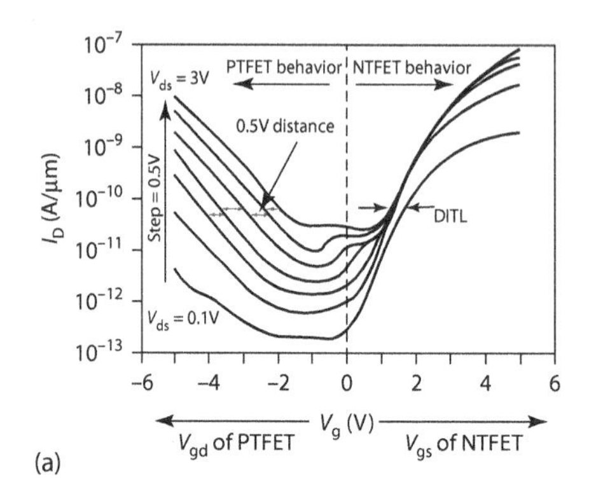

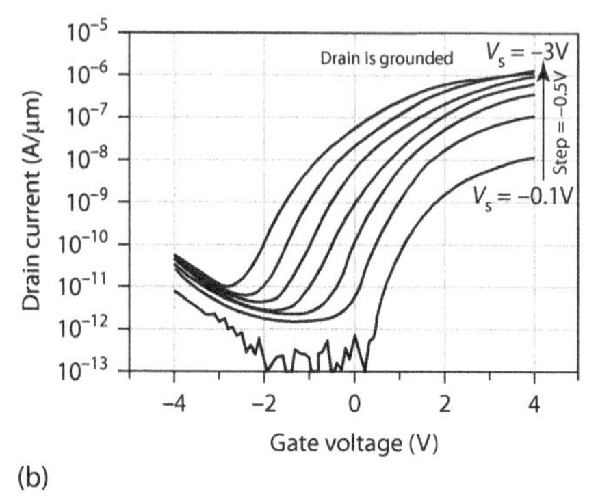

FIGURE 7.2
(a) Transfer characteristics (channel current versus gate voltage) with n+ region as the drain and p+ region as the grounded source. The channel current curves obviously shift with the variety of drain voltage when the gate voltage is negative while they shift a little when the gate voltage is positive due to the changing of the position of the surface tunneling junction. (b) Transfer characteristics with n+ region as the grounded drain and p+ region as the source. The channel current curves obviously shift with the variety of source current when the gate voltage is positive [18].

of the drain voltage cannot influence the switch-on voltage effectively and the V_{TN} (the switch-on voltage of the surface tunneling junction for an NTFET) keeps almost a constant. In converse, if the n+ region is grounded as the drain and the p+ region as the source, the switch-on voltage moves left with the decreasing source voltage, shown in Figure 7.2b. However, the asymmetry of the doping level of the profile and the restriction of the movement of one type of charge carrier with heterostructure (HS) can widen the tunneling barrier at the drain to suppress the ambipolarity and lower the OFF current [18,19].

To describe quantitatively, the tunneling current depends on the transmission probability of the interband tunneling barrier, which can be acquired by Wentzel–Kramers–Brillouin (WKB) approximation,

$$T(E) = \exp\left(-\frac{4\sqrt{2m^* E_g^3}}{3q\hbar\varepsilon}\right),$$

where m^* is the effective carrier mass, E_g is the bandgap, $\hbar$ is Planck's constant divided by 2π and ε is the electrical field in the tunneling junction. Considering an ultrathin body and gate oxides, the screening tunneling length λ (as shown in the schematic diagrams of the band structures of Figure 7.1c and d, the spatial extent of the surface tunneling junction) can be described as

$$\lambda = \sqrt{\frac{\varepsilon_{ch}}{\varepsilon_{ox}} d_{ox} d_{ch}},$$

where d_{ox}, d_{ch}, ε_{ox} and ε_{ch} are the oxide and channel thickness and dielectric constants, respectively [19]. Therefore, the electrical field $\varepsilon = (E_g + \Delta\varnothing)/\lambda$, where E_g is the bandgap of the source and $\Delta\varnothing$ is the energy window as marked in Figure 7.1c and d. The Fermi level of the source is considered several kT lower than its valance band for an NTFET (several kT higher than its conduction band for a PTFET), $f_s(E) \approx 1$ in the energy window [7].

Supposing the states in the energy window of the channel are empty, the tunneling current can be given as

$$I \propto \int dED(E)v(E)T(E)\left[f_s(E) - f_{ch}(E)\right]$$
$$= \int_0^{\Delta\Phi} dED(E)v(E)T(E) \propto \exp\left(-\frac{4\lambda\sqrt{2m^* E_g^{3/2}}}{3q\hbar\left(E_g + \Delta\varnothing\right)}\right)\Delta\varnothing \tag{7.2}$$

It suggests that increasing the tunneling probability to approach unity as much as possible signifies a high ON current. The energy filtering cuts off the high-energy part of the source Fermi distribution, which makes the electrons from this part not tunnel to the channel; thus it is possible to achieve an SS below 60 mV/dec [20]. One of the major advantages of TFETs is their weak dependence on the temperature of the OFF currents, which has to be considered seriously in the operation of highly integrated circuits [19,21,22]. Another advantage of TFETs is the decrease in the SS with the reduction in the gate voltage as the SS is approximately inversely related to the gate voltage if the thickness of the gate dielectrics approaches 1 nm. Thus, TFETs are naturally optimized for low-voltage operation [23].

7.3 Performance Optimization

If TFETs aim to replace CMOS transistors, their ON current should be over 100 mA, an SS far below 60 mV/dec for more than five decades of current beyond several microamperes per micrometer, $V_{DD} < 0.5$ V and $I_{ON}/I_{OFF} > 10^5$. The formula for the tunneling current will be the guide to designing TFETs with competitive properties, although the WKB approximation is more suitable for direct bandgap semiconductors than indirect bandgap semiconductors and the case when quantum effects and phonon-assisted tunneling are dominant.

According to the formula derived from the WKB approximation, a high tunneling probability is expected if m^* and E_g of the source material are as small as possible. Replacing the source material with Ge for silicon n-TFETs and with InAs for silicon p-TFETs can improve λ and result in a steep band edge sharpness. However, the OFF currents will increase greatly with a small energy gap as thermal emission becomes pronounced. So, the drain material should be chosen with a large E_g to form a large energy barrier width at the drain side in the off state to keep the OFF current low. Channel materials with an optimized E_g of (1.1–1.5)qV_{DD} show the best performance in TFETs [24], so WTe_2 with an E_g of 0.75 eV is expected to show best performance when the supply voltage V_{DD} is about 0.5 V [25,26]. Unfortunately, 2h-WeT2 might be not stable in air [27]. The modulation of the channel bans by the gate is represented by λ, and a small λ results in a high transmission probability from the source to the channel, which depends on gate capacitance, doping profiles, device geometry, dimensions, etc. So, a high-permittivity (high-κ) gate dielectric [19] and a low permittivity (low-κ) channel material should be adopted, especially a high-κ gate dielectric, as a small m^* and a proper E_g are more pivotal for the channel materials. Meanwhile, the gate dielectric and channel material should be as thin as possible to reduce λ [7,19]. The high source doping level must be cut off abruptly to achieve as short a tunneling barrier width as possible, which will result in the bandgap narrowing slightly. The types of band alignment also play a crucial role in HSs TFETs [28,29].

The band alignment was adjusted from staggered to broken by varying the Al context in the InAs/Al_xGa_{1-x} HS TFETs and it was found that a staggered band alignment yielded the best I_{ON}/I_{OFF} and an SS below 60 mV/dec [30,31]. However, due to the nonoptimal device structure with an underlap between the gate and the channel and the phenomenological treatment of scattering, Koswatta considered the proposal of Knoch unreasonable and put forward that TFETs with a broken band alignment delivered superior performance [29].

There is a compromise between a small SS and high ON currents in 3D TFETs [32] as λ of the planar device could not be lower than a certain value because of a limited dopant density in the channel and a limited gate oxide. And relatively high Fermi energies are required to ensure a charge density large enough in the electrodes as λ relies on the charge density in the contacts, which have to yield thin tunneling barriers. One-dimensional (1D) TFETs are expected to achieve high ON currents with steep SS because of their improved electrostatics and the one dimensionality of the density of states [32–36]. The quantum capacitance (C_q) will be far smaller than the oxide capacitance (C_{ox}) even in the case of large gate oxide thickness, which will deteriorate the band edge sharpness [7]. Therefore, 1D TFETs have a better gate modulation than 3D TFETs. Both the channel materials and the gate dielectrics using layered 2D materials can be an atom thick, so a high gate modulation can be expected in 2D TFETs.

Another method to improve the gate modulation of TFETs is to use different device structures. Silicon TFETs were achieved with a planar structure for the first time in 1995 [3]. Later, in 1996, SOI surface TFETs with a negative conductance were fabricated to take advantage of the 2D confinement effect [4]. And it was predicted that the gate overlapping the depletion region only can minimize capacitance instead of overlapping the source and drain region and result in a 10 times increase in I_{ON} and a lower SS [37–39]. Besides TFETs with a planar structure, a vertical TFET with the gate oxide on the vertical side walls was invented in silicon to achieve better gate control [8,40], and incorporating pseudomorphic strained $Si_{1-x}Ge_x$ layers between the source and the intrinsic channel results in a significant performance improvement as a direct consequence of tunnel barrier width lowering [41]. Double-gate and gate-all-around structures with high dielectrics can raise ON currents, suppress OFF currents and provide a low SS with a high I_{ON}/I_{OFF} [19,42,43].

7.4 Feasibility of TFETs Based on 2D Materials

Two-dimensional materials are perfect candidates for TFETs mainly due to their rich species, atomic thickness without dangling bonds and high mobility with high ON currents. The discovery of monolayer graphene with unique properties, such as a high mobility and atomic thickness without dangling bonds, gave rise to the prosperity of layered 2D materials in 2004. Layered TMDCs have two hexagonal lattices of MX_2 sandwiches, where M is the transition metal and X is chalcogens, and 2H-MX_2 is honeycombs with a D_{3h} point group symmetry while 1T-MX_2 is center honeycombs with a C_{3v} symmetry [44], as shown in the insets in Figure 7.3. Layered TMDCs can be semiconducting when M=Mo [45,46], W [47], etc., as metallic when M=Nb, Re [48], etc., and even superconducting such as layered CoO_2 [49] and $NbSe_2$ [50]. Group IV-VI chalcogenides exist with a cubic NaCl structure, such as SnTe and PbX (X=S, Se, Te), or a rhombohedral structure, such as SnX and GeX (X=S, Se) with bandgaps for both direct and indirect transitions [51,52]. Layered Group III-VI semiconductors generally has a structure similar to layered GaSe [53,54]. Bulk GaSe has great potential for nonlinear optical application with a direct bandgap of approximately 2.1 eV

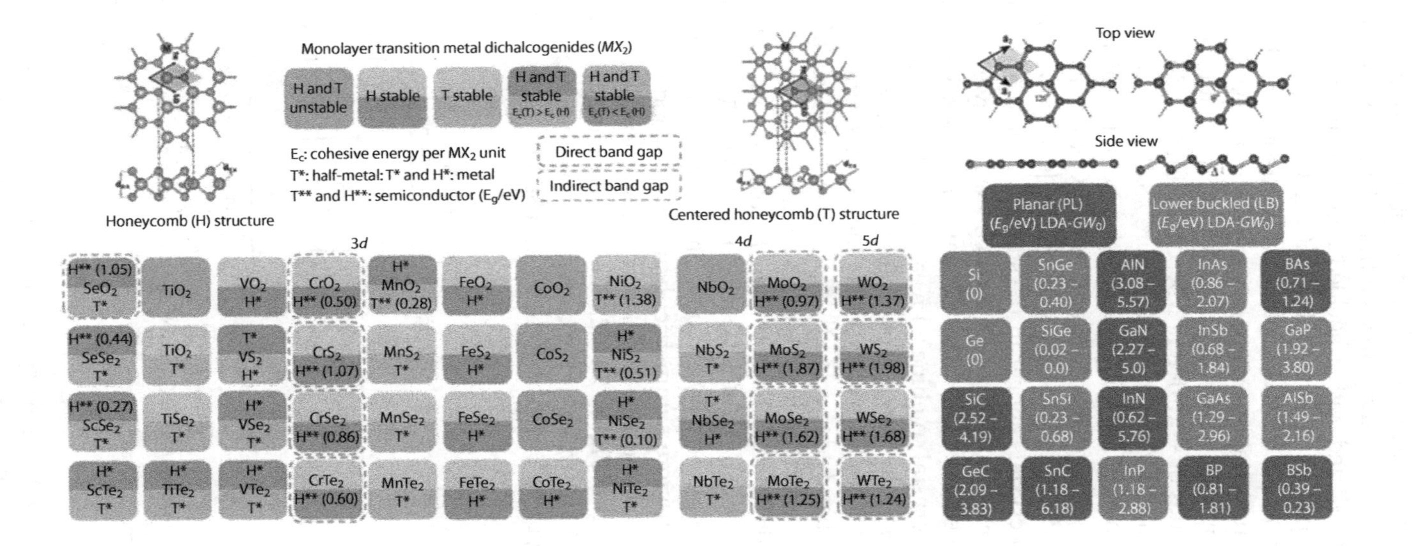

FIGURE 7.3
Summary of the properties of 44 different MX_2 compounds and Group IV elements and Group III-V [60], where M represents transition metal atoms divided into 3d, 4d and 5d groups. +, * and ** represent half-metallic, metallic and semiconducting, respectively, and the numbers in the bracket are the corresponding bandgap of these 2D crystals in semiconducting states. The gray shade means that the 2D crystal is unstable. The lower-lying structure (H or T) is the ground states in each box.

and its nanosheets exhibit a higher responsivity of 2.8 A/W and a higher external quantum efficiency of 1367% at 254 nm than 2D nanosheet devices of MoS_2 and graphene, although its mobility is lower than 1 cm^2/Vs [55–57]. Except graphene, Group IV element graphene-like 2D sheets, like silicene and germanene formed from Si and Ge, have also attracted wide interest; their existence is predicted with a buckled honeycomb structure [58,59]. Layered binary compounds of Group IV elements, such as SiC, GeC and SnC, are predicted to be planar, similar to graphene and BN, while SnSi, SnGe and SiGe are buckled like silicene and germanene [60]. All of these stable compounds were found to be semiconductors with a variety of bandgaps, but graphene is metallic and BN is an insulator. The stability analysis and semiconducting properties of 44 different MX_2 compounds and binary compounds of Group IV elements and Group III-V are shown in Figure 7.3. It is necessary to notice another semiconductor, black phosphorus (BP) with great carrier mobility, outstanding electrostatic modulation, a tunable bandgap of 0.3–2.0 eV and a high ON current [61], which make it widely applied to FET [62,63], battery [64,65], gas sensors [66] and amplitude modulation (AM) demodulator [67]. 2D crystals appear in the stacked layered form in nature and the bandgap of their single layer is usually larger than their stacked layered form [68,69].

Two-dimensional crystal sheets can form a 3D structure by stacking by van der Waals force while the atoms of 3D crystal combine with each other by chemical bonding. Various admixtures of sp^3 bonds result in the electron states at the conduction and valance band edges of 3D semiconductors. The electronic states at the valance band of 3D crystal is more p-like while the conduction band edge is mostly s-like. The directivity of a p-orbital results in the anisotropy of the hole effective mass or the curvature. However, symmetry is desirable for complementary logic devices, like p-type metal-oxide-semiconductor (PMOS) and n-type metal-oxide-semiconductor (NMOS) or PTFET and NTFET, as it could simplify the geometry and layout of complementary circuits. The conduction and valence bands in TMDCs single layer is much more symmetric than 3D semiconductor crystal, although less symmetric than graphene and BN [70,71]. The energy band alignment and relative energy band offset are shown in Figure 7.4 [72]. The relative bandgaps and electron affinity are marked in Figure 7.4 according to the corresponding references [73–75]. A perfect 2D crystal has no dangling bonds on its surface while a 3D crystal necessarily has broken

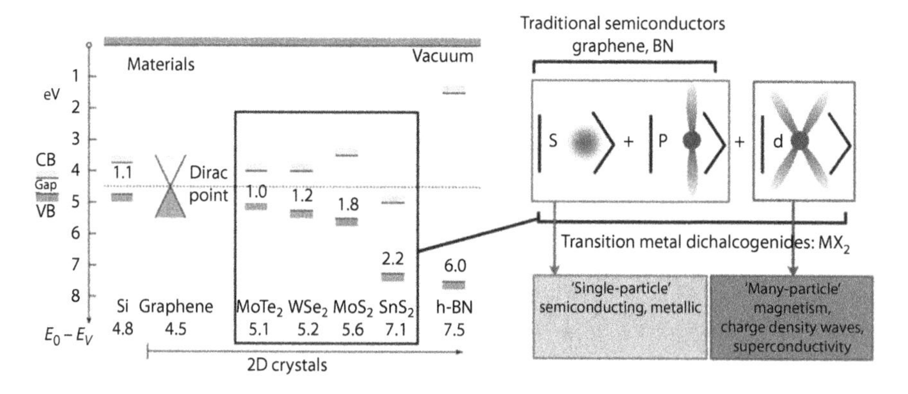

FIGURE 7.4
Energy band alignment of silicon and 2D crystal [72]. The bandgap and electron affinity of various 2D crystals are marked according to the corresponding reports or calculations. The conduction and valance band edge states of TMDCs include s>, |p> and |d> orbitals, while those of traditional semiconductors, graphene and BN only involves the linear combination of |s> and |p> orbitals. Some electronic phenomena of TMDCs require a many-particle effect, such as magnetism, charge density waves and superconductivity.

bonds on its surface, which will be passivated by dielectrics, lattice-matched or -strained HSs. Therefore, the energy gap windows of 2D crystals are not populated by surface states, which are necessary in 3D crystals [74]. It is calculated that the effective masses (m^*) range from $0.34m_0$ to $0.76m_0$ [76], where m_0 is the mass of a free electron, and m^* of 2D BN is ~$0.6m_0$ [77], while graphene is zero-effective mass due to its unique band structure of Dirac cones [78], which have been refined by experimental measurements.

FETs with 2D crystals as the channels, such as graphene, MoS_2 and BP, have shown stable and outstanding electrostatics and transport characteristics. Monolayer MoS_2 transistors with hafnium oxide (HfO_2) as the gate dielectrics have achieved ON current over 10 uA, electron mobility >200 cm^2/Vs and an I_{ON}/I_{OFF} ratio of 10^8 [45]. BP transistors with a thickness of several nanometers have obtained charge-carrier mobility of ~1000 cm^2/Vs with an I_{ON}/I_{OFF} ratio larger than five orders, as shown in Figure 7.5a [79]. Monolayer WSe_2 transistors can work as a p-n junction by electrostatic modulation of the gate voltage [80]. Using high-κ dielectrics as surroundings can damp scattering and improve charge mobility (Figure 7.5b) in transistors with 2D semiconductors as the channel materials because 2D crystals have direct access to the electrons, their spins and atomic vibrations [45,81,82].

To achieve an effective gate electrostatic control over mobile electrons and holes in the channel, the channel thickness has to be reduced when scaling down the source/drain separations in FETs based on 3D crystal semiconductors. However, the surface roughness of a 3D crystal channel will greatly deteriorate when the channels are thinned down. TFETs require a stronger gate control to reduce the screening tunneling length to achieve

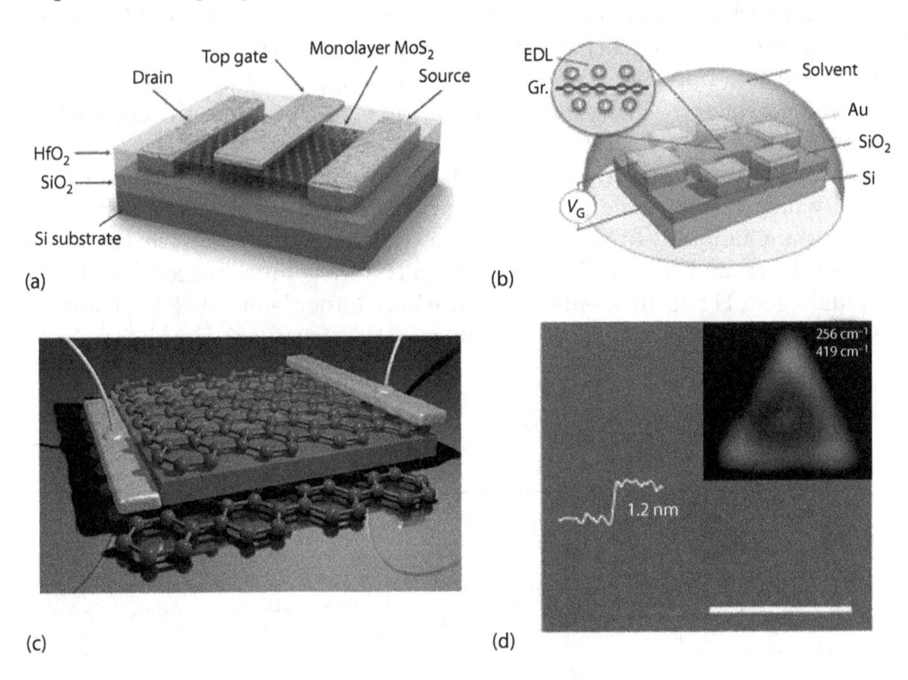

FIGURE 7.5
(a) Three-dimensional schematic view of an MoS_2 transistor with HfO_2 as top-gate dielectrics on a silicon substrate with an SiO_2 layer [45]. (b) Device schematics of a graphene transistor with an ionic electric double layer (EDL) as its dielectric environment, which can greatly improve the charge mobility [81]. (c) Structure of a graphite/BN/graphite device, where the BN layer works as the tunneling layer [85]. (d) AFM image of a triangular domain of a WS_2-WSe_2 lateral HS with a thickness of 1.2 nm. The inset shows the composite image including Raman mapping at 256 cm^{-1} and 419 cm^{-1}. There is no apparent gap or overlap between the WS_2 and WSe_2 signals, meaning that the successive lateral epitaxy of the HS [90].

a high tunneling probability. Therefore, 2D crystals are a wonderful solution to solve these problems as layered 2D can be atomic thick, thinner than 1 nm. And using 2D dielectrics, like h-BN as the gate oxide [83] or tunneling layer (Figure 7.5c) [84,85], can enhance the control of the gate voltage as 2D h-BN can be thinned to a monolayer.

Chemical doping in 2D crystal is still a challenge and not effective as expected, so TFETs of 2D crystal HSs are not experimentally realized as they depend on electrostatic doping to form an energy band offset instead of the differences in chemical composition. So, taking HSs as the channels is a soothing choice for TFETs. There should be few lattice mismatches and defects in the interface as there are no dangling bonds on the surface of layered 2D crystal and the different layers attach to each other by van der Waals force. Various 2D materials can be applied to the permutation and combination of 2D HSs [86–89]. Besides vertical or out-of-plane HSs, in-plane HSs of unpassivated edge growth have been realized by the successive lateral epitaxial growth of a thermal CVD process (Figure 7.5d) [90]. But intrinsic strain will be induced in the lateral epitaxial growth of in-plane HS, which will result in a lattice mismatch in the lateral interface, reduce the coupling strength between p- and d-orbitals and change the band structure [91].

TFETs based on 2D crystals have an electrostatic advantage and simplicity for device structures. The tunneling current flows laterally from the source to the drain in lateral TFETs based on 3D materials (Figure 7.6a). To enhance the gate control over the channel thickness, the channel should be thinned down (Figure 7.6b) as the gate field is vertical, which would result in increasing the bandgap due to quantum confinement and thus decreasing the interband tunneling [72]. The quantum confinement will be suppressed if the channels are replaced by 2D crystal layers (Figure 7.6c) with little consideration of the surface states–related trap states. As tunneling only happens at the interface of the source and the channel in lateral TFETs, the tunneling area is a line when the channel thickness is atomic thin and the tunneling current will be reduced significantly. Therefore, vertical TFETs are taken into consideration. 3D vertical TFETs with side gates are shown in Figure 7.6d, which can be designed with a gate-all-round structure to enhance the gate control over the channels. To further enlarge the tunneling area, the vertical top-gate TFETs can be fabricated (Figure 7.6e). If the HS channels are replaced by 2D-2D stacks, vertical double-gate TFETs are available (Figure 7.6f). Either lateral TFETs or vertical TFETs with 2D crystals as the channels take advantage of the structure's simplicity.

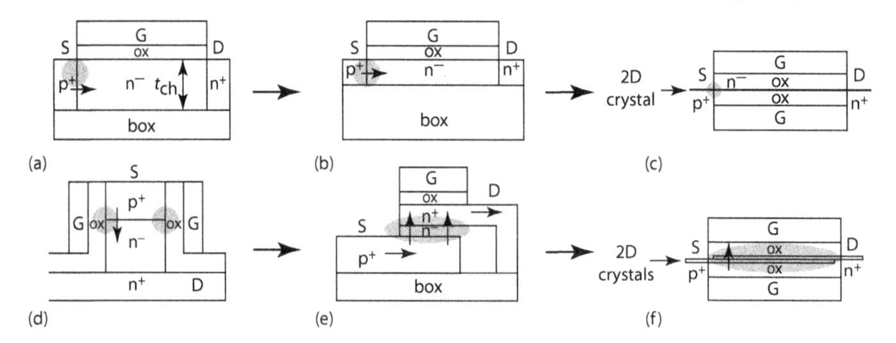

FIGURE 7.6
Device schematics of lateral TFETs based on 3D materials (a), with thinned channel (b) to enhance the gate control over the entire junction thickness. (c) Device schematic of lateral TFETs based on 2D crystal. (d) 3D vertical TFETs with side gates, where the tunneling current flows vertically. (e) 3D vertical TFETs with top gate, where the tunneling current flows in a zigzag way, as indicated by the arrows, and over an area instead of a line, as the shading shows. (f) Vertical double-gate TFETs with p+ and n+ 2D crystal layers. The tunneling junction is marked by the red shading and the arrows indicate the direction of the tunneling current flows [72].

7.5 Current Status of TFETs Based on 2D Materials

The current situation of TFETs based on 2D crystals will be discussed theoretically and experimentally, and can be divided into in-plane tunneling for lateral TFETs and interlayer tunneling for vertical TFETs.

The published simulation works about lateral TFETs based on 2D materials are demonstrated in Figure 7.7. In the case of chemically doped TFET, the source and the drain region is highly doped, especially the source region, as the tunneling junction is formed between the interface of the source region and the electrostatically gated channel. A double-gate structure (Figure 7.8a) is available for TFETs with 2D materials as the channel because of their atomically thin thickness, which makes the gate control over the channel more effective. Currently, most simulation works on TFETs are based on the model structure.

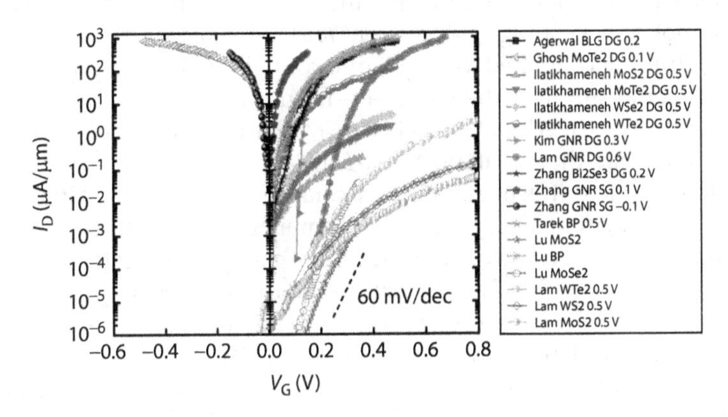

FIGURE 7.7
Simulated transfer characteristics of lateral TFETs based on 2D crystals, such as graphene [100], BP [96,107], $MoTe_2$ [93,98], MoS_2 [93,94,96], WSe_2 [93], WTe_2 [93,94], graphene nanoribbon (GNR) [101,102], Bi_2Se_3 [104], $MoSe_2$ [96] and WS_2 [94].

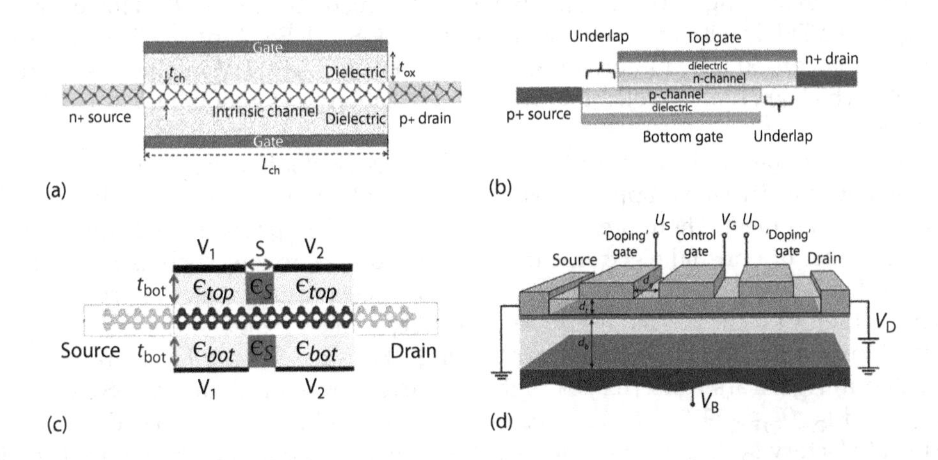

FIGURE 7.8
Two different doping methods for TFETs based on 2D crystals. Schematic cross section of a monolayer 2D crystal-based aligned double gate (a) [98] and a staggered double gate (b) [96] TFET with a chemically doped n+ source, p+ drain and intrinsic channel. (c) Layout of a lateral TFET with electrically doped source and drain region [108]. (d) Electrically doped lateral TFET with control gate [110].

As discussed previously, the source material with a bandgap of about (1.1–1.5)qV_{DD} provides the best performance; therefore, a bandgap ranging from 0.55 to 0.75 eV is the optimal choice for a supply voltage of 0.5 V. The WTe_2 monolayer has great potential for high performance TFET applications because of its extremely small effective mass and a direct bandgap of 0.75 eV [92]. A high ON current over 350 μA/μm and an SS below 60 mV/dec even when the channel current is up to 1 μA/μm in WTe_2 TFET with a source doping level of 10^{20} cm^{-3} and a channel voltage of 0.5 V [93]. The ON current of WTe_2 TFET calculated in another simulation work is of the same order [94], but the SS is larger than that of the former mostly because the bandgap used in the simulation (0.99 eV) is larger than the former one (0.75 eV). Therefore, a small bandgap really counts. By dielectric engineering, lateral WTe_2 TFET can achieve a record ON current of ~1000 μA/μm and an SS below 20 mV/dec, where the gates and the channel material are separated by high-κ dielectrics while the spacing part between the gates is filled with low-κ dielectrics [95]. Simulation work has been done on MoS_2 lateral TFET as MoS_2 has been widely investigated and can be synthetized in large scale. Lateral MoS_2 TFETs cannot achieve a high ON current and an SS below 60 mV/dec with the same doping level of drain and applied voltages due to its large bandgap and its heavy effective mass, which could suppress the source-to-drain tunneling and make it unsuitable for TFETs but ideal for ultra-scaled MOSFETs [93,94,96]. The ON current of a lateral $MoSe_2$ TFET is only a little larger than that of lateral MoS_2 TFETs under the same simulation conditions probably due to its lighter effective mass and smaller bandgap [96]. Although WSe_2 has a larger bandgap (1.56 eV) than $MoTe_2$ (1.08 eV), its smaller reduced effective mass partly compensates for the inferiority resulted from the larger bandgap. Therefore, the ON current of a lateral homogeneous WSe_2 TFET is still larger than that of an $MoTe_2$ TFET [93]. Similarly, a lateral homogeneous WS_2 TFET has a larger ON current than an MoS_2 TFET though a WS_2 has a larger bandgap (1.758 eV) than an MoS_2 (1.66 eV) [94].

Generally, to compete with advanced CMOS technology, $I_{ON} > 1$ mA/μm, an SS far below 60 mV/dec for five orders, $I_{ON}/I_{OFF} > 10^5$ and $V_{DD} < 0.5$ V for TFETs [20]. The performances of lateral homogeneous TFETs based on TMDCs calculated in these works are dissatisfying as their ON currents are not high enough and the SS is below 60 mV/dec only when the channel currents are small. More specifically, the SS should be far below 60 mV/dec for current levels extending to 1–10 μA/μm when switching on the TFET [97]. However, lateral homogeneous TFETs with five monolayer TMDCs (MoS_2, $MoSe_2$, $MoTe_2$, WS_2 or WSe_2) offer an average I_{ON} of 150 μA/μm and an SS of 4 mV/dec at $V_{DD} = 0.1$ V in Ghosh's simulation work [98], which are more excellent than other simulation works about lateral TFETs based on TMDCs.

Due to the unique band structure of graphene whose Brillouin zone boundary is linearly dispersed, its effective mass approaches zero at the point where the valence band and the conduction band meet. Therefore, the carrier mobility of graphene FETs is ultrahigh and a great number of unusual electronic transport properties follow, such as an anomalous quantum Hall effect [99]. TFETs with graphene as their channel material are expected to achieve a high ON current as its ultralight effective mass benefits the tunneling probability significantly. The characteristics of lateral TFETs based on bilayer graphene were calculated in Agarwal's work [100]. A high ON current over 1 mA/μm, an SS as low as 35 mV/dec and $I_{ON}/I_{OFF} > 2910$ with $V_{DD} = 0.2$ V were acquired using contact-induced doping. The unsatisfactory I_{ON}/I_{OFF} is unsurprising due to its extremely small bandgap. Therefore, opening its bandgap controllably is necessary to increase the I_{ON}/I_{OFF} ratio as well as taking advantage of its high mobility and high ON currents. With the width-dependent energy bandgap of graphene, graphene nanoribbon (GNR) widths between 3 and 10 nm correspond to the energy bandgap in the range of 0.46–0.14 eV. The tunability of its bandgap and the light effective mass of its carriers make it a wonderful candidate for TFETs. A lateral TFET based on GNR 5 nm wide was predicted to have a high I_{ON}/I_{OFF} over 10^7 with an ON current of 800 μA/μm and an effective SS of 0.19 mV/dec [101]. Adding a region with a small bandgap between the source and the channel, the ON and OFF currents can be tuned [102], which improves the tunneling efficiency by choosing source material with a smaller bandgap homojunction compared with Ge-Si HS TFETs. It can avoid the lattice mismatch and also the difficulty of forming HSs. However, GNR TFETs suffer from line edge roughness [103], which would enhance the OFF current significantly and reduce the conductance of GNRs. Therefore, it is necessary to optimize the width of GNR to minimize the OFF current with decreasing the ON current as little as possible, while widening the bandgap will suppress the thermal emission but reduce tunneling probability when narrowing down GNR.

Bi_2Se_3 is a topological insulator with a rhombohedral crystal structure, whose energy bandgap ranges from 41 meV to ~0.5 eV for Bi_2Se_3 thin films. A lateral TFET based on two-quintuple-layer Bi_2Se_3 with a thickness of 1.4 nm and a bandgap of 0.252 eV has been simulated [104]. The Bi_2Se_3 TFET with high source/drain doping level and a drain underlap can achieve an I_{ON}/I_{OFF} current ratio of 10^4, an I_{OFF} of 5 nA/μm and an SS of 50 mV/dec with a supply voltage of 0.2 V, which makes it a dynamic power indicator 10 times lower than MOSFET. The topological insulator has a high static dielectric constant ($\varepsilon_r = 100$), not favorable for short-channel FETs because the lateral electric field penetrates into the channel and broadens the source–channel junction [105]. If considering $\varepsilon_r = 20$, the ON current can be enhanced more than twice as the lateral electric field broadens the source–channel junction, reduces the source carrier injection efficiency and degrades the effective SS.

Another unique layered 2D materials is BP, which is especially attractive as it has electronic properties that lie in between graphene and TMDCs. It has a direct bandgap in both bulk (0.3 eV) and monolayer (2 eV), high mobility, light effective mass of carriers ($0.146m_0$), anisotropic effective mass that increases the density of states near the band edges and a low dielectric constant [106]. All these advantages of BP benefit the performance of BP TFETs as discussed previously. Simulation work predicted that multilayered BP TFETs can achieve an ON current as high as graphene TFETs with a small SS of 24.6 mV/dec [96]. Even scaling down the channel length of BP TFETs to 6 nm, the performance is acceptable with a supply voltage of 0.2 V [107]. In conclusion, compared with lateral TFETs based on graphene Bi_2Se_3 and TMDCs, BP is the most promising candidate for lateral TFETs as a BP lateral TFET can achieve a high ON current of ~1 mA/μm with a high I_{ON}/I_{OFF} ratio of ~10^6 and an ultralow SS due to its merits, though lateral TFETs based on other 2D materials are also predicted to be able to achieve an SS below 60 mV/dec.

Most of the simulation work on lateral TFETs on 2D materials are based on the double-gate structure as shown in Figure 7.8a, where the top gate and bottom gate are aligned and the source and drain are formed by chemical doping. A lateral tunneling window along the channel can be suppressed if lateral offset, like underlap double gates, is adopted (Figure 7.8b), which suppresses the parasitic leakage current [96]. It was observed that lateral tunneling was more pronounced in a lateral BP TFET without underlapping compared with its counterpart with underlapping. As chemical doping in 2D materials is currently a great challenge, the n- and p-type potentials can be defined by two gates with opposite polarities at the two sides of the tunneling junction (Figure 7.8c). Electrical doping avoids fluctuations, threshold voltage shifts and bandgap reduction resulting from doping, which reduces the OFF state performance [108]. Given the channel with ultrathin 2D materials, the pacing between the two gates, the thickness of the gate dielectrics, the dielectric constant of the spacing region and the strain greatly affect the performances of electric doping TFETs, especially the first two factors. A thinner dielectric, smaller spacing and a smaller spacer dielectric constant can reduce the SS and enhance the I_{ON} remarkably. Different from chemically doped TFET, the electric field at the tunneling junction is inversely proportional to the total thickness of the top and bottom dielectrics. What really counts in an electrically doped TFET are the physical dielectrics thickness and distance between the gates [109]. As the bottom gate often uses a thick back oxide, the electric field of the tunneling junction will decrease, which can be avoided by using a back oxide with a low dielectric constant compared to the top oxide ($\varepsilon_{bottom} << \varepsilon_{top}$) [108]. If adding another control gate to the channel and the n- or p-type potentials are fixed at a specific level, the electrically doped lateral TFET will be controlled by the control gate similarly to the chemically doped TFET and can be operated as NTFET or PTFET by positive gate or negative voltage, respectively [110].

The strain on the 2D channel of lateral TFETs influences the device performance. For example, applying a biaxial strain of 3% to the WSe_2 channel would increase the ON currents as 3% biaxial strain reduces the effective mass and bandgap by about 10%–20% [108]. At present, there is no experimental realization of lateral TFETs based on layered 2D materials reported mainly due to the challenge of doping in 2D materials and forming multi-gates within a distance of several nanometers. Fortunately, some progress has been made in doping 2D materials [111–116]. At the same time, many measures have been applied to improve the contact between 2D materials and the electrodes, such as annealing, using graphene as electrodes [117], phase engineering [118] and so on. Moreover, dielectric engineers have investigated solutions to enhance the gate control over the channel, like combining high-κ and low-κ oxides by atomic layer deposition and atomic h-BN.

Due to the challenge of chemical doping in 2D materials and the advantage of free dangling bonds on the surface layer of 2D materials, TFETs based on 2D-2D HSs have been investigated at the same time. The structure of vertical double-gate TFETs with p-type and n-type 2D crystal layers is shown in Figure **7.6f**. The simulation work on vertical HS TFETs is demonstrated in Figure 7.9. It is observed that a HS TFET enhances the ON current by an order compared to the homogeneous device, both considered at the same I_{ON}/I_{OFF} ratio in the simulation work, where WTe_2, WS_2 and MoS_2 lateral TFETs and vertical WS_2/

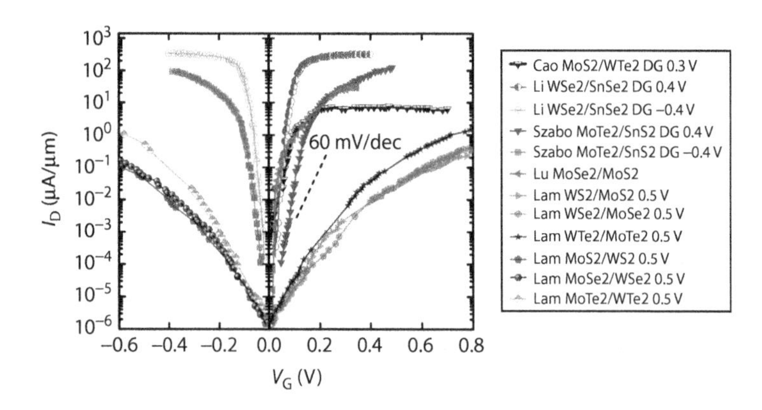

FIGURE 7.9
Simulated transfer characteristics of vertical TFETs based on 2D-2D crystals HSs, such as MoS_2/WTe_2 [119], $WSe_2/SnSe_2$ [121], $MoTe_2/SnS_2$ [120], $MoSe_2/MoS_2$ [96], WS_2/MoS_2, $WSe_2/MoSe_2$ and $WTe_2/MoTe_2$ [94].

MoS_2, $WSe_2/MoSe_2$ and $WTe_2/MoTe_2$ HS TFETs have been simulated, respectively, while $MoTe_2/WTe_2$ HS TFET do not enhance the current significantly compared with a WTe_2 lateral TFET [94]. But the simulation results in this work are not optimistic as the ON current, I_{ON}/I_{OFF} ratio and SS don't meet requirements for current-integrated circuits. The ON current of a vertical $MoS_2/MoSe_2$ HS TFET is also found to be two orders higher than an MoS_2 TFET [96]. However, it is predicted that the vertical TFET based on a monolayer MoS_2 and a WTe_2 heterojunction with a 1 nm thick h-BN interlayer can achieve a low SS of ~40 mV/dec, where the MoS_2 is doped and the h-BN functions as a tunneling barrier [119]. The doping concentration mainly affects the threshold top-gate voltage as mentioned previously [108]. It has been discovered that extending the top gate for a certain length beyond each side of the overlap region of MoS_2 and WTe_2 can improve the SS remarkably because the extending part suppresses the OFF current significantly. For example, the device can attain an ultralow SS of 7 mV/dec with a gate extension length of 20 nm. The ON current of ~10 μA/μm cannot meet the requirement but the TFET is competitive for low power applications. The vertical TFET made of a monolayer $MoTe_2$ and SnS_2 is reported to achieve an ON current >75 μA/μm and an I_{ON}/I_{OFF} ratio of about 10^8 with a supply voltage of 0.4 V [120]. A steep SS of ~14 mV/dec and a high ON current of ~300 μA/μm are estimated theoretically in a vertical TFET based on $WSe_2/SnSe_2$ stacked monolayer HSs [121].

There is a large difference in the work function between VIB- and IVB-TMDCs, providing a solution to form steep HSs with a broken bandgap alignment. It is recommended that VIB-MeX2 (Me=W and Mo; X=Te and Se) as the n-type source and IVB-MeX2 (Me=Zr and Hf; X=S and Se) as the p-type drain can form vertically stacked TFETs due to their unique band edge characters, which results in intervalley scattering during tunneling [92].

Many efforts have been made to experimentally realize vertical TFETs based on 2D-2D HSs. Vertical TFETs based on graphene HSs with h-BN or MoS_2 as the transport barrier exhibit switching ratios of ~50 and ~10,000 at room temperature, respectively [122]. Transistors with a graphene/h-BN/graphene sandwich structure have achieved resonant tunneling and negative differential resistance (NDR), where the tunneling barriers are atomic layers of h-BN [123]. Different from the NDR in the Esaki diode, NDR in these transistors stems from resonant tunneling and appears at both forward and reverse channel bias while NDR in the Esaki diode only appears at the forward bias. Graphene/h-BN/graphene symmetric FETs fabricated by chemical vapor deposition (CVD) at a larger scale are also observed with similar tunneling characteristics, which makes them possible for large-scale applications. Similarly, resonant tunneling is observed in MoS_2/WSe_2/graphene and WSe_2/MoS_2/graphene HSs with NDR characteristics at room temperature, which are synthesized by metal-organic chemical vapor deposition (MOCVD) [124]. Vertical TFETs based on graphene/WS_2/graphene HSs realized an unprecedented I_{ON}/I_{OFF} ratio over 10^6 at room temperature with a very high ON current, where WS_2 functions as an atomically thin barrier between two mechanically exfoliated or CVD-grown graphene layers [125]. More interestingly, NDR can be achieved by a simple three-terminal graphene device due to the competition between electron and hole conduction with the increase in the drain bias [126]. These resonant tunneling devices have potential for applications in high-frequency and logic devices.

A large number of 2D-2D HSs have been investigated, stacked with 2D layers mechanically exfoliated [86,127–130], grown by CVD [131–133] or fabricated by lateral epitaxial growth [90,134,135]. There is no doubt that carriers transfer across the interface very rapidly as the hole transfer from the MoS_2 layer to the WS_2 layer takes place within 50 femtosecond after optical excitation in MoS_2/WS_2 HSs even though there is an air gap of several nanometers in the interface [87] and electrons or holes in the excitons transfer across the interface in subpicosecond time [136]. Some of this work focuses on the application of optoelectronic, photovoltaic or memory devices, while some have developed methods of synthetizing vertical 2D-2D HSs or lateral 2D-2D HSs. Some of these HSs are formed as a p-n diode naturally [127,137] or can be tuned to a p-n junction by the gate voltage [138], which is indispensable for the application of TFETs for ultralow power consumption. As discussed previously, vertical 2D-2D HSs with suitable band alignment can realize tunneling. Dual-gate MoS_2/WSe_2 van der Waals HSs can be gate modulated as an Esaki diode with NDR at 77 K [139]. A type-II alignment between MoS_2 and WSe_2 is determined by micro-beam x-ray photoelectron spectroscopy and scanning tunneling spectroscopy with a valance band offset of 0.83 eV and a conduction band offset of 0.76 eV [140]. However, back-gate MoS_2/WSe_2 van der Waals HSs have also realized NDR even at room temperature [141]. $SnSe_2$/BP van der Waals heterojunctions with broken-gap band alignment behave as an Esaki diode with evident NDR and a peak-to-valley ratio of 1.8 at room temperature and 2.8 at 80 K [142]. Compared with the simulation results of monolayer $WSe_2/SnSe_2$ HS vertical TFETs [121], a high I_{ON}/I_{OFF} ratio of ~10^7 has been realized in 2D-2D TFETs using $WSe_2/SnSe_2$ HSs in the absence of graphene but the ON current and an SS of 100 mV/dec are not as good as expected [143] (Figure 7.10).

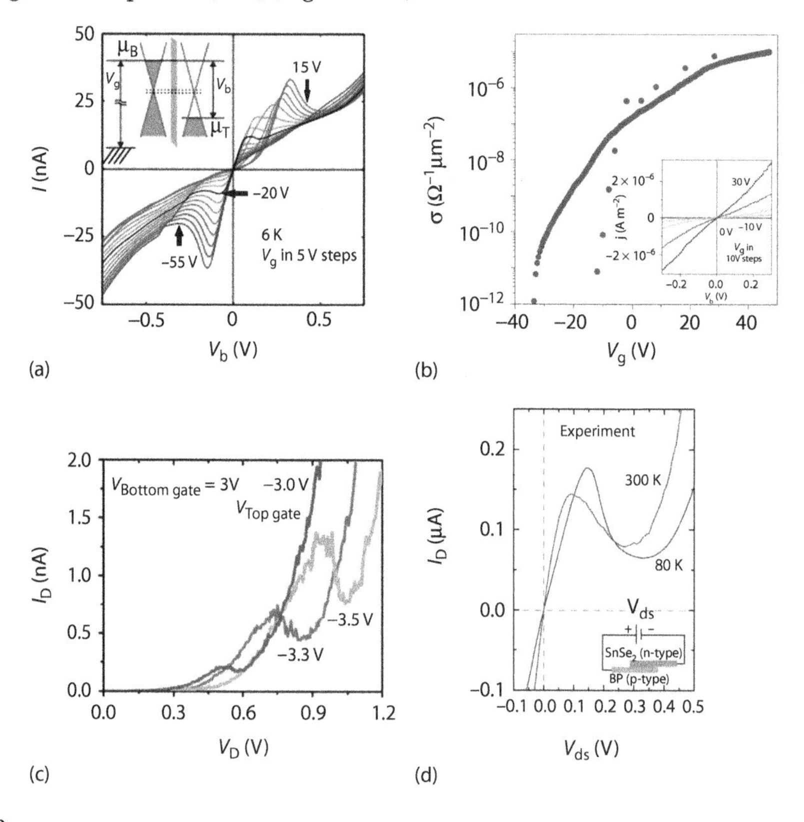

FIGURE 7.10
Tunneling characteristics of vertical TFETs based on different HSs. (a) Current-voltage characteristics of a specific transistor with a graphene/h-BN/graphene sandwich structure at 6 K with gate voltage ranging from 15 V to −55 V in 5 V steps [123]. (b) Conductivity (slope of the current-voltage curve) of vertical TFETs based on graphene/WS_2/graphene HSs as a function of the gate voltage at zero bias (red circles) and 0.02 V bias (blue circles) [125]. (c) Current-voltage characteristics of MoS_2/WSe_2 HS TFET at $V_{gate\text{-}MoS2}=3$ V and $V_{gate\text{-}WSe2}$ varied [142]. (d) Current-voltage characteristics of the BP/$SnSe_2$ Esaki tunnel diode at 80 and 300 K. The peak-to-valley ratio increases as the temperature decreases [143].

Besides 2D-2D HS TFETs, the combination of 2D-3D HSs has also been investigated for solar cell and photodetector applications [144–148], as well as TFETs for ultralow power consumption devices to take advantage of 2D and 3D materials. It is surprising that vertical TFETs based on highly doped p-type germanium (p-Ge) and double-layer MoS_2 HSs have realized promising results with a minimum SS of 3.9 mV/dec and an average SS of 31.1 mV/dec for four decades of drain current from 10^{-13} A to 10^{-9} A at room temperature with a supply voltage of 0.1 V [149]. This is the most significant results of TFETs based on 2D materials. Due to the difficulty of doping in 2D materials and the lack of high p-type 2D materials, it is not easy to form TFETs based on 2D-2D HSs for ultralow power consumption. Therefore, highly p-doped Ge with a relatively low electron affinity and a small bandgap is adopted as the source material and n-type MoS_2 with atomic thickness is used as the channel material modulated by the gate to switch on/off the transistor, which results in a small tunneling barrier width. But the promising gate control ability might derive from its polymer complex gate, resulting in a high gate capacitance [150]. It is seen that the steep SS below 60 mV/dec only appears when the channel current is below 10^{-9} A, which can be improved further to meet the requirement for ultralow power consumption applications. 2D/3D MoS_2/GaN Esaki tunnel diodes are also realized experimentally with a high current density and repeatable NDR at room temperature [151]. The 2D/3D HS TFETs take advantage of both layered 2D materials and 3D materials, which have great potential for high performance device applications.

7.6 Conclusion

TFETs based on 2D materials are currently under intense scrutiny. They have great potential to solve the power consumption problem of MOSFETs with the supply voltage scaled down further. The rich species with different band structures, semimetals, metals and even superconductors, and atomic thickness make them unrivalled candidates for TFETs and the gate dielectrics can be atomically thin 2D materials, like BN, which enhances the gate control remarkably. The simplicity of TFETs based on 2D materials takes advantage of the two-dimensionality of layered 2D materials. WTe_2, GNR and BP are the most suitable materials for lateral homojunction TFETs if the doping technology for 2D materials can be overcome. The combination of 2D materials to form HSs opens a door to a brand new world for electronic device technology, including TFETs. Although the results on TFETs based on 2D HSs acquired are not exciting, the theoretical work inspires us to make more significant progress than the current state of the art.

References

1. D. Kim, Y. Lee, J. Cai, I. Lauer, L. Chang, S. J. Koester, et al., Low power circuit design based on heterojunction tunneling transistors (HETTs), in *IEEE International Symposium on Low Power Electronics and Design*, pp. 219–224, 2009.
2. J. J. Quinn, G. Kawamoto, and B. D. Mccombe, Subband spectroscopy by surface channel tunneling, *Surface Science*, vol. 73, pp. 190–196, 1978.

3. W. M. Reddick and G. Amaratunga, Silicon surface tunnel transistor, *Applied Physics Letters*, vol. 67, pp. 494–496, 1995.
4. Y. Omura, Negative conductance properties in extremely thin silicon-on-insulator (SOI) insulated-gate pn-junction devices (SOI surface tunnel transistors), *Japanese Journal of Applied Physics*, vol. 35, pp. L1401-L1403, 1996.
5. C. Aydin, A. Zaslavsky, S. Luryi, S. Cristoloveanu, D. Mariolle, D. Fraboulet, et al., Lateral interband tunneling transistor in silicon-on-insulator, *Applied Physics Letters*, vol. 84, pp. 1780–1782, 2004.
6. T. Baba, Proposal for surface tunnel transistors, *Japanese Journal of Applied Physics*, vol. 31, pp. L455–L457, 1992.
7. J. Knoch and J. Appenzeller, A novel concept for field-effect transistors: The tunneling carbon nanotube FET, *Device Research Conference Digest, 2005, DRC '05. 63rd*, vol. 1, pp. 153–156, 2005.
8. W. Hansch, C. Fink, J. Schulze, and I. Eisele, A vertical MOS-gated Esaki tunneling transistor in silicon, *Thin Solid Films*, vol. 369, pp. 387–389, 2000.
9. S. O. Koswatta, M. S. Lundstrom, M. P. Anantram, and D. E. Nikonov, Simulation of phonon-assisted band-to-band tunneling in carbon nanotube field-effect transistors, *Applied Physics Letters*, vol. 87, pp. 253107-253107-3, 2005.
10. U. E. Avci and I. A. Young, Heterojunction TFET scaling and resonant-TFET for steep sub-threshold slope at sub-9nm gate-length, in *IEEE International Electron Devices Meeting*, pp. 4.3.1–4.3.4, 2013.
11. J. H. Seo, Y. J. Yoon, S. Lee, J. H. Lee, S. Cho, and I. M. Kang, Design and analysis of Si-based arch-shaped gate-all-around (GAA) tunneling field-effect transistor (TFET), *Current Applied Physics*, vol. 15, pp. 208–212, 2015.
12. Y. Liu, X. Dong, and P. Chen, ChemInform abstract: Biological and chemical sensors based on graphene materials, *ChemInform*, vol. 43, pp. 2283–2307, 2012.
13. M. Liu, X. Yin, E. Ulinavila, B. Geng, T. Zentgraf, L. Ju, et al., A graphene-based broadband optical modulator, *Nature*, vol. 474, pp. 64–67, 2011.
14. R. Cheng and X. Duan, High-frequency self-aligned graphene transistors with transferred gate stacks, *Proceedings of the National Academy of Sciences of the United States of America*, vol. 109, pp. 11588–11592, 2012.
15. K. S. Kim, Y. Zhao, H. Jang, S. Y. Lee, J. M. Kim, K. S. Kim, et al., Large-scale pattern growth of graphene films for stretchable transparent electrodes, *Nature*, vol. 457, pp. 706–710, 2009.
16. Y. Zhu, S. Murali, M. D. Stoller, K. J. Ganesh, W. Cai, P. J. Ferreira, et al., Carbon-based supercapacitors produced by activation of graphene, *Science*, vol. 332, pp. 1537–1541, 2011.
17. A. S. Mayorov, R. V. Gorbachev, S. V. Morozov, L. Britnell, R. Jalil, L. A. Ponomarenko, et al., Micrometer-scale ballistic transport in encapsulated graphene at room temperature, *Nano Letters*, vol. 11, pp. 2396–2399, 2011.
18. P. F. Wang, K. Hilsenbeck, T. Nirschl, M. Oswald, C. Stepper, M. Weis, et al., Complementary tunneling transistor for low power application, *Solid-State Electronics*, vol. 48, pp. 2281–2286, 2004.
19. K. Boucart and A. M. Ionescu, Double-gate tunnel FET with high-κ gate dielectric, *Electron Devices IEEE Transactions*, vol. 54, pp. 1725–1733, 2007.
20. A. M. Ionescu and H. Riel, Tunnel field-effect transistors as energy-efficient electronic switches, *Nature*, vol. 479, pp. 329–337, 2011.
21. Y. Liu, R. P. Dick, L. Shang, and H. Yang, Accurate temperature-dependent integrated circuit leakage power estimation is easy, in *Design, Automation and Test in Europe Conference* and *Exposition*, 2007, pp. 1526–1531.
22. K. K. Bhuwalka, M. Born, M. Schindler, M. Schmidt, T. Sulima, and I. Eisele, P-Channel tunnel field-effect transistors down to sub-50 nm channel lengths, *Japanese Journal of Applied Physics*, vol. 45, pp. 3106–3109, 2006.
23. Q. Zhang, W. Zhao, and A. Seabaugh, Low-subthreshold-swing tunnel transistors, *IEEE Electron Device Letters*, vol. 27, pp. 297–300, 2006.
24. H. Ilatikhameneh, G. Klimeck, and R. Rahman, Can homojunction tunnel FETs scale below 10 nm?, *IEEE Electron Device Letters*, vol. 37, pp. 115–118, 2016.

25. H. Ilatikhameneh, T. A. Ameen, G. Klimeck, and J. Appenzeller, Dielectric engineered tunnel field-effect transistor, *IEEE Electron Device Letters*, vol. 36, pp. 1097–1100, 2015.
26. H. Ilatikhameneh, G. Klimeck, and R. Rahman, 2D tunnel transistors for ultra-low power applications: Promises and challenges, in *Energy Efficient Electronic Systems*, 2015, pp. 1–3.
27. C. H. Lee, E. C. Silva, L. Calderin, M. A. T. Nguyen, M. J. Hollander, B. Bersch, et al., Tungsten ditelluride: A layered semimetal, *Scientific Reports*, vol. 5, p. 10013, 2015.
28. A. S. Verhulst, W. G. Vandenberghe, K. Maex, and S. D. Gendt, Complementary silicon-based heterostructure tunnel-FETs with high tunnel rates, *Electron Device Letters IEEE*, vol. 29, pp. 1398–1401, 2009.
29. S. O. Koswatta, S. J. Koester, and W. Haensch, On the possibility of obtaining MOSFET-like performance and sub-60-mV/dec swing in 1-D broken-gap tunnel transistors, *IEEE Transactions on Electron Devices*, vol. 57, pp. 3222–3230, 2010.
30. J. Knoch and J. Appenzeller, Modeling of high-performance p-type III–V heterojunction tunnel FETs, *IEEE Electron Device Letters*, vol. 31, pp. 305–307, 2010.
31. J. Knoch, Optimizing tunnel FET performance: Impact of device structure, transistor dimensions and choice of material, in *International Symposium on VLSI Technology, Systems, and Applications, 2009. VLSI-TSA*, pp. 45–46, 2009.
32. J. Knoch, S. Mantl, and J. Appenzeller, Impact of the dimensionality on the performance of tunneling FETs: Bulk versus one-dimensional devices, *Solid-State Electronics*, vol. 51, pp. 572–578, 2007.
33. J. Appenzeller, Y. M. Lin, J. Knoch, and P. Avouris, Band-to-band tunneling in carbon nanotube field-effect transistors, *Physical Review Letters*, vol. 93, p. 196805, 2004.
34. Y. Yoon and S. Salahuddin, Barrier-free tunneling in a carbon heterojunction transistor, *Applied Physics Letters*, vol. 97, pp. 033102-033102-3, 2010.
35. R. Gandhi, Z. Chen, N. Singh, and K. Banerjee, Vertical Si-nanowire-type tunneling FETs with low subthreshold swing ( ) at room temperature, *IEEE Electron Device Letters*, vol. 32, pp. 437–439, 2011.
36. R. Gandhi, Z. Chen, N. Singh, and K. Banerjee, CMOS-compatible vertical-silicon-nanowire gate-all-around p-type tunneling FETs with ≤50-mV/decade subthreshold swing, *IEEE Electron Device Letters*, vol. 32, pp. 1504–1506, 2011.
37. C. Aydin, A. Zaslavsky, S. Luryi, and S. Cristoloveanu, Lateral interband tunneling transistor in silicon-on-insulator, *Applied Physics Letters*, vol. 84, pp. 1780–1782, 2004.
38. C. Hu, P. Patel, A. Bowonder, and K. Jeon, Prospect of tunneling green transistor for 0.1V CMOS, in *Electron Devices Meeting*, 2010, pp. 16.1.1–16.1.4.
39. R. Asra, M. Shrivastava, K. V. R. M. Murali, R. K. Pandey, H. Gossner, and V. R. Rao, Tunnel FET for V_{DD} scaling below 0.6V with CMOS comparable performance, *IEEE Transactions on Electron Devices*, vol. 58, pp. 1855–1863. 2011.
40. K. K. Bhuwalka, S. Sedlmaier, A. K. Ludsteck, and C. Tolksdorf, Vertical tunnel field-effect transistor, *IEEE Transactions on Electron Devices*, vol. 51, pp. 279–282, 2004.
41. K. K. Bhuwalka, J. Schulze, and I. Eisele, A simulation approach to optimize the electrical parameters of a vertical tunnel FET, *IEEE Transactions on Electron Devices*, vol. 52, pp. 1541–1547, 2005.
42. A. S. Verhulst, B. Sorée, D. Leonelli, W. G. Vandenberghe, and G. Groeseneken, Modeling the single-gate, double-gate, and gate-all-around tunnel field-effect transistor, *Journal of Applied Physics*, vol. 107, pp. 024518-024518-8, 2010.
43. L. D. Seup, H. S. Yang, K. C. Kang, L. Joung-Eob, L. J. Han, S. Cho, et al., Simulation of gate-all-around tunnel field-effect transistor with an n-doped layer, *IEICE Transactions on Electronics*, vol. 93-C, pp. 540–545, 2010.
44. N. V. Podberezskaya, S. A. Magarill, N. V. Pervukhina, and S. V. Borisov, Crystal chemistry of dichalcogenides MX_2, *Journal of Structural Chemistry*, vol. 42, pp. 654–681, 2001.
45. B. Radisavljevic, A. Radenovic, J. Brivio, V. Giacometti, and A. Kis, Single-layer MoS_2 transistors, *Nature Nanotechnology*, vol. 6, pp. 147–150, 2011.

46. W. S. Yun, S. W. Han, S. C. Hong, I. G. Kim, and J. D. Lee, Thickness and strain effects on electronic structures of transition metal dichalcogenides: 2H-MX2 semiconductors (M = Mo, W; X = S, Se, Te)[J]. *Physical Review B Condensed Matter*, vol. 85, pp. 033305-1-033305-5, 2012.
47. S. Das and J. Appenzeller, WSe_2 field effect transistors with enhanced ambipolar characteristics, *Applied Physics Letters*, vol. 103, pp. 103501-103501-5, 2013.
48. K. K. Tiong, C. H. Ho, and Y. S. Huang, The electrical transport properties of ReS_2 and $ReSe_2$ layered crystals, *Solid State Communications*, vol. 111, pp. 635–640, 1999.
49. K. Takada, H. Sakurai, E. Takayama-Muromachi, F. Izumi, R. A. Dilanian, and T. Sasaki, Superconductivity in two-dimensional CoO_2 layers, *ChemInform*, vol. 422, pp. 53–55, 2003.
50. M. Yoshida, J. Ye, T. Nishizaki, N. Kobayashi, and Y. Iwasa, Electrostatic and electrochemical tuning of superconductivity in two-dimensional NbSe2 crystals, *Applied Physics Letters*, vol. 108, p. 202602, 2016.
51. D. J. Xue, J. Tan, J. S. Hu, W. Hu, Y. G. Guo, and L. J. Wan, Anisotropic photoresponse properties of single micrometer-sized GeSe nanosheet, *Advanced Materials*, vol. 24, pp. 4528–4533, 2012.
52. I. Lefebvre, M. A. Szymanski, J. Olivierfourcade, and J. C. Jumas, Electronic structure of tin monochalcogenides from SnO to SnTe, *Physical Review B*, vol. 58, pp. 1896–1906, 1998.
53. J. Robertson, Electronic structure of GaSe, GaS, InSe and GaTe, *Journal of Physics C Solid State Physics*, vol. 12, p. 4777, 2001.
54. L. Plucinski, R. L. Johnson, B. J. Kowalski, K. Kopalko, B. A. Orlowski, Z. D. Kovalyuk, et al., Electronic band structure of GaSe(0001): Angle-resolved photoemission and ab initio theory, *Physical Review B*, vol. 68, 125304, 2003.
55. P. Hu, Z. Wen, L. Wang, P. Tan, and K. Xiao, Synthesis of few-layer GaSe nanosheets for high performance photodetectors, *ACS Nano*, vol. 6, p. 5988, 2012.
56. K. R. Allakhverdiev, M. Ö. Yetis, S. Özbek, T. K. Baykara, and E. Y. Salaev, Effective nonlinear GaSe crystal. Optical properties and applications, *Laser Physics*, vol. 19, pp. 1092–1104, 2009.
57. M. Schlüter, J. Camassel, S. Kohn, J. P. Voitchovsky, Y. R. Shen, and M. L. Cohen, Optical properties of GaSe and mixed crystals, *Physical Review B*, vol. 13, pp. 3534–3547, 1976.
58. K. Takeda and K. Shiraishi, Theoretical possibility of stage corrugation in Si and Ge analogs of graphite, *Physical Review B Condensed Matter*, vol. 50, pp. 14916–14922, 1994.
59. E. Durgun, S. Tongay, and S. Ciraci, Silicon and III-V compound nanotubes: Structural and electronic properties, *Physical Review B*, vol. 72, p. 075420, 2005.
60. M. Xu, T. Liang, M. Shi, and H. Chen, Graphene-like two-dimensional materials, *Chemical Reviews*, vol. 113, p. 3766, 2013.
61. H. Du, X. Lin, Z. Xu, and D. Chu, Recent developments in black phosphorus transistors, *Journal of Materials Chemistry C*, vol. 3, pp. 8760–8775, 2015.
62. S. Das, M. Demarteau, and A. Roelofs, Ambipolar phosphorene field effect transistor, *ACS Nano*, vol. 8, p. 11730, 2014.
63. F. Xia, H. Wang, and Y. Jia, Rediscovering black phosphorus as an anisotropic layered material for optoelectronics and electronics, *Nature Communications*, vol. 5, p. 4458, 2014.
64. M. Buscema, D. J. Groenendijk, G. A. Steele, V. D. Z. Hs, and A. Castellanos-Gomez, Photovoltaic effect in few-layer black phosphorus PN junctions defined by local electrostatic gating, *Nature Communications*, vol. 5, p. 4651, 2014.
65. T. Hong, B. Chamlagain, W. Lin, H. J. Chuang, M. Pan, Z. Zhou, et al., Polarized photocurrent response in black phosphorus field-effect transistors, *Nanoscale*, vol. 6, p. 8978, 2014.
66. L. Kou, T. Frauenheim, and C. Chen, Phosphorene as a superior gas sensor: Selective adsorption and distinct I-V response, *Journal of Physical Chemistry Letters*, vol. 5, p. 2675, 2014.
67. W. Zhu, M. N. Yogeesh, S. Yang, S. H. Aldave, J. S. Kim, S. Sonde, et al., Flexible black phosphorus ambipolar transistors, circuits and AM demodulator, *Nano Letters*, vol. 15, p. 1883, 2015.
68. K. F. Mak, C. Lee, J. Hone, J. Shan, and T. F. Heinz, Atomically thin MoS_2: A new direct-gap semiconductor, *Physical Review Letters*, vol. 105, p. 136805, 2010.
69. A. Splendiani, L. Sun, Y. Zhang, T. Li, J. Kim, C. Y. Chim, et al., Emerging photoluminescence in monolayer MoS_2, *Nano Letters*, vol. 10, pp. 1271–1275, 2010.

70. K. Wood and J. B. Pendry, Layer method for band structure of layer compounds, *Physical Review Letters*, vol. 31, pp. 1400–1403, 1973.
71. K. Kaasbjerg, K. S. Thygesen, and K. W. Jacobsen, Phonon-limited mobility in n-type single-layer MoS_2 from first principles, *Physical Review B*, vol. 85, p. 115317, 2012.
72. D. Jena, Tunneling transistors based on graphene and 2-D crystals, *Proceedings of the IEEE*, vol. 101, pp. 1585–1602, 2013.
73. R. Yan, Q. Zhang, W. Li, I. Calizo, T. Shen, C. A. Richter, et al., Determination of graphene work function and graphene-insulator-semiconductor band alignment by internal photoemission spectroscopy, *Applied Physics Letters*, vol. 101, 022105, pp. 666–35, 2012.
74. W. Mönch, Valence-band offsets and Schottky barrier heights of layered semiconductors explained by interface-induced gap states, *Applied Physics Letters*, vol. 72, pp. 1899–1901, 1998.
75. R. Schlaf, O. Lang, C. Pettenkofer, and W. Jaegermann, Band lineup of layered semiconductor heterointerfaces prepared by van der Waals epitaxy: Charge transfer correction term for the electron affinity rule, *Journal of Applied Physics*, vol. 85, pp. 2732–2753, 1999.
76. L. Liu, S. B. Kumar, Y. Ouyang, and J. Guo, Performance limits of monolayer transition metal dichalcogenide transistors, *IEEE Transactions on Electron Devices*, vol. 58, pp. 3042–3047, 2011.
77. K. Watanabe, T. Taniguchi, and H. Kanda, Direct-bandgap properties and evidence for ultraviolet lasing of hexagonal boron nitride single crystal, *Nature Materials*, vol. 3, pp. 404–409, 2004.
78. L. A. Ponomarenko, R. V. Gorbachev, G. L. Yu, D. C. Elias, R. Jalil, A. A. Patel, et al., Cloning of Dirac fermions in graphene superlattices, *Nature*, vol. 497, pp. 594–597, 2013.
79. L. Li, Black phosphorus field-effect transistors, *Nature Nanotechnology*, vol. 9, pp. 372–377, 2014.
80. J. Ross, P. Klement, A. Jones, N. Ghimire, J. Yan, D. Mandrus, et al., Electrically tunable excitonic light emitting diodes based on monolayer WSe_2 p-n junctions, *Nature Nanotechnology*, vol. 9, p. 268, 2014.
81. A. K. Newaz, Y. S. Puzyrev, B. Wang, S. T. Pantelides, and K. I. Bolotin, Probing charge scattering mechanisms in suspended graphene by varying its dielectric environment, *Nature Communications*, vol. 3, p. 734, 2012.
82. C. Jang, S. Adam, J. H. Chen, E. D. Williams, S. S. Das, and M. S. Fuhrer, Tuning the effective fine structure constant in graphene: Opposing effects of dielectric screening on short- and long-range potential scattering, *Physical Review Letters*, vol. 101, p. 146805, 2008.
83. G. Lu, T. Wu, Q. Yuan, H. Wang, H. Wang, F. Ding, et al., Synthesis of large single-crystal hexagonal boron nitride grains on Cu-Ni alloy, *Nature Communications*, vol. 6, p. 6160, 2015.
84. L. Britnell, R. V. Gorbachev, R. Jalil, B. D. Belle, F. Schedin, A. Mishchenko, et al., Field-effect tunneling transistor based on vertical graphene heterostructures, *Science (New York, N.Y.)*, vol. 335, pp. 947–950, 2012.
85. L. Britnell, R. V. Gorbachev, R. Jalil, B. D. Belle, F. Schedin, M. I. Katsnelson, et al., Electron tunneling through ultrathin boron nitride crystalline barriers, *Nano Letters*, vol. 12, pp. 1707–1710, 2012.
86. F. Wang, Z. Wang, K. Xu, F. Wang, Q. Wang, Y. Huang, et al., Tunable GaTe-MoS_2 van der Waals p-n junctions with novel optoelectronic performance, *Nano Letters*, vol. 15, 7558–7566, 2015.
87. X. Hong, J. Kim, S. F. Shi, Y. Zhang, C. Jin, Y. Sun, et al., Ultrafast charge transfer in atomically thin MoS_2/WS_2 heterostructures, *Nature Nanotechnology*, vol. 9, pp. 682–686, 2014.
88. N. Flöry, A. Jain, P. Bharadwaj, M. Parzefall, T. Taniguchi, K. Watanabe, et al., A $WSe_2/MoSe_2$ heterostructure photovoltaic device, *Applied Physics Letters*, vol. 107, p. 123106, 2015.
89. Y. Gong, S. Lei, G. Ye, B. Li, Y. He, K. Keyshar, et al., Two-step growth of two-dimensional $WSe_2/MoSe_2$ heterostructures, *Nano Letters*, vol. 15, p. 6135, 2015.
90. X. Duan, C. Wang, J. C. Shaw, R. Cheng, Y. Chen, H. Li, et al., Lateral epitaxial growth of two-dimensional layered semiconductor heterojunctions, *Nature Nanotechnology*, vol. 9, p. 1024, 2014.

91. W. Wei, Y. Dai, and B. Huang, Straintronics in two-dimensional in-plane heterostructures of transition-metal dichalcogenides, *Physical Chemistry Chemical Physics PCCP*, vol. 19, pp. 663–672, 2016.
92. C. Gong, H. Zhang, W. Wang, L. Colombo, R. M. Wallace, and K. Cho, Band alignment of two-dimensional transition metal dichalcogenides: Application in tunnel field effect transistors, *Applied Physics Letters*, vol. 103, p. 053513, 2013.
93. H. Ilatikhameneh, Y. Tan, B. Novakovic, G. Klimeck, R. Rahman, and J. Appenzeller, Tunnel field-effect transistors in 2-D transition metal dichalcogenide materials, *IEEE Exploratory Solid-State Computational Devices and Circuits*, vol. 1, pp. 12–18, 2015.
94. K. T. Lam, X. Cao, and J. Guo, Device performance of heterojunction tunneling field-effect transistors based on transition metal dichalcogenide monolayer, *IEEE Electron Device Letters*, vol. 34, pp. 1331–1333, 2013.
95. H. Ilatikhameneh, T. A. Ameen, G. Klimeck, and J. Appenzeller, Dielectric engineered tunnel field-effect transistor, *IEE Electron Device Letters*, vol. 36, pp. 1097–1100, 2015.
96. S. C. Lu, M. Mohamed, and W. Zhu, Novel vertical hetero- and homo-junction tunnel field-effect transistors based on multi-layer 2D crystals, *2d Materials*, vol. 3, p. 011010, 2016.
97. W. G. Vandenberghe, A. S. Verhulst, B. Sorée, W. Magnus, G. Groeseneken, Q. Smets, et al., Figure of merit for and identification of sub-60 mV/decade devices, *Journal of Applied Physics*, vol. 102, pp. 013510–48, 2013.
98. R. K. Ghosh and S. Mahapatra, Monolayer transition metal dichalcogenide channel-based tunnel transistor, *IEEE Journal of the Electron Devices Society*, vol. 1, pp. 175–180, 2013.
99. T. Ohta, A. Bostwick, T. Seyller, K. Horn, and E. Rotenberg, Controlling the electronic structure of bilayer graphene, *Science*, vol. 313, p. 951, 2006.
100. T. K. Agarwal, A. Nourbakhsh, P. Raghavan, and I. Radu, Bilayer graphene tunneling FET for Sub-0.2 V digital CMOS logic applications, *Electron Device Letters IEEE*, vol. 35, pp. 1308–1310, 2014.
101. Q. Zhang, T. Fang, H. Xing, and A. Seabaugh, Graphene nanoribbon tunnel transistors, *IEEE Electron Device Letters*, vol. 29, pp. 1344–1346, 2009.
102. K. T. Lam, D. Seah, S. K. Chin, and S. B. Kumar, A simulation study of graphene-nanoribbon tunneling FET with heterojunction channel, *IEEE Electron Device Letters*, vol. 31, pp. 555–557, 2010.
103. S. G. Kim, M. Luisier, T. B. Boykin, and G. Klimeck, Computational study of heterojunction graphene nanoribbon tunneling transistors with p-d orbital tight-binding method, *Applied Physics Letters*, vol. 104, pp. 329–337, 2014.
104. Q. Zhang, G. Iannaccone, and G. Fiori, Two-dimensional tunnel transistors based on thin film, *IEEE Electron Device Letters*, vol. 35, pp. 129–131, 2014.
105. J. Chang, L. F. Register, and S. K. Banerjee, Topological insulator Bi_2Se_3 thin films as an alternative channel material in metal-oxide-semiconductor field-effect transistors, *Journal of Applied Physics*, vol. 112, pp. 3045–3067, 2012.
106. Y. Takao and A. Morita, Electronic structure of black phosphorus: Tight binding approach, *Journal of the Physical Society of Japan*, vol. 105, pp. 93–98, 1981.
107. T. A. Ameen, H. Ilatikhameneh, G. Klimeck, and R. Rahman, Few-layer phosphorene: An ideal 2D material for tunnel transistors, *Scientific Reports*, vol. 6, p. 28515, 2015.
108. H. Ilatikhameneh, G. Klimeck, J. Appenzeller, and R. Rahman, Design rules for high performance tunnel transistors from 2D materials, *IEEE Journal of the Electron Devices Society*, 2016.
109. H. Ilatikhameneh, G. Klimeck, J. Appenzeller, and R. Rahman, Scaling theory of electrically doped 2D transistors, *IEEE Electron Device Letters*, vol. 36, pp. 726–728, 2015.
110. G. Alymov, V. Vyurkov, V. Ryzhii, and D. Svintsov, Abrupt current switching in graphene bilayer tunnel transistors enabled by van Hove singularities, *Scientific Reports*, vol. 6, p. 24654, 2016.
111. M. Tosun, L. Chan, M. Amani, T. Roy, G. H. Ahn, P. Taheri, et al., Air-stable n-doping of WSe_2 by anion vacancy formation with mild plasma treatment, *ACS Nano*, vol. 10, pp. 6853–6860, Jul 26 2016.

112. H. Fang, S. Chuang, T. C. Chang, K. Takei, T. Takahashi, and A. Javey, High-performance single layered WSe(2) p-FETs with chemically doped contacts, *Nano Letters*, vol. 12, pp. 3788–3792, Jul 11 2012.
113. H. M. Li, D. Lee, D. Qu, X. Liu, J. Ryu, A. Seabaugh, et al., Ultimate thin vertical p-n junction composed of two-dimensional layered molybdenum disulfide, *Nature Communications*, vol. 6, p. 6564, Mar 24 2015.
114. W. Liu, J. Kang, D. Sarkar, Y. Khatami, D. Jena, and K. Banerjee, Role of metal contacts in designing high-performance monolayer n-type WSe_2 field effect transistors, *Nano Letters*, vol. 13, pp. 1983–1990, 2013.
115. S. Chuang, C. Battaglia, A. Azcatl, S. McDonnell, J. S. Kang, X. Yin, et al., MoS_2P-type transistors and diodes enabled by high work function MoOx contacts, *Nano Letters*, vol. 14, pp. 1337–1342, 2014.
116. D. Mao, X. She, B. Du, D. Yang, W. Zhang, K. Song, et al., Erbium-doped fiber laser passively mode locked with few-layer $WSe_2/MoSe_2$ nanosheets, *Scientific Reports*, vol. 6, p. 23583, 2016.
117. H.-J. Chuang, X. Tan, N. J. Ghimire, M. M. Perera, B. Chamlagain, M. M.-C. Cheng, et al., High mobility WSe_2 p-and n-type field-effect transistors contacted by highly doped graphene for low-resistance contacts, *Nano Letters*, vol. 14, pp. 3594–3601, 2014.
118. R. Kappera, D. Voiry, S. E. Yalcin, B. Branch, G. Gupta, A. D. Mohite, et al., Phase-engineered low-resistance contacts for ultrathin MoS_2 transistors, *Nature Materials*, vol. 13, pp. 1128–1134, 2014.
119. J. Cao, A. Cresti, D. Esseni, and M. Pala, Quantum simulation of a heterojunction vertical tunnel FET based on 2D transition metal dichalcogenides, *Solid-State Electronics*, vol. 116, pp. 1–7, 2016.
120. A. Szabó, S. J. Koester, and M. Luisier, Ab-Initio simulation of van der Waals MoTe 2 –SnS 2 heterotunneling FETs for low-power electronics, *IEEE Electron Device Letters*, vol. 36, pp. 514–516, 2015.
121. M. O. Li, D. Esseni, J. J. Nahas, and D. Jena, Two-dimensional heterojunction interlayer tunneling field effect transistors (thin-TFETs), *Electron Devices Society IEEE Journal of the*, vol. 3, pp. 200–207, 2015.
122. L. Britnell, R. V. Gorbachev, R. Jalil, B. D. Belle, F. Schedin, A. Mishchenko, et al., Field-effect tunneling transistor based on vertical graphene heterostructures, *Science*, vol. 335, pp. 947–950, 2012.
123. L. Britnell, R. V. Gorbachev, A. K. Geim, L. A. Ponomarenko, A. Mishchenko, M. T. Greenaway, et al., Resonant tunnelling and negative differential conductance in graphene transistors, *Nature Communications*, vol. 4, p. 1794, 2013.
124. Y. C. Lin, R. K. Ghosh, R. Addou, N. Lu, S. M. Eichfeld, H. Zhu, et al., Atomically thin resonant tunnel diodes built from synthetic van der Waals heterostructures, *Nature Communications*, vol. 6, pp. 1–10, 2015.
125. T. Georgiou, R. Jalil, B. D. Belle, L. Britnell, R. V. Gorbachev, S. V. Morozov, et al., Vertical field-effect transistor based on graphene-WS_2 heterostructures for flexible and transparent electronics, *Nature Nanotechnology*, vol. 8, pp. 100–103, 2012.
126. Y. Wu, D. B. Farmer, W. Zhu, S. J. Han, C. D. Dimitrakopoulos, A. A. Bol, et al., Three-terminal graphene negative differential resistance devices, *ACS Nano*, vol. 6, pp. 2610–2616, 2012.
127. Y. Deng, Z. Luo, N. J. Conrad, H. Liu, Y. Gong, S. Najmaei, et al., Black phosphorus-monolayer MoS_2 van der Waals heterojunction p-n diode, *ACS Nano*, vol. 8, p. 8292, 2014.
128. P. Rivera, K. L. Seyler, H. Yu, J. R. Schaibley, J. Yan, D. G. Mandrus, et al., Valley-polarized exciton dynamics in a 2D semiconductor heterostructure, *Science*, vol. 351, p. 688, 2016.
129. P. Rivera, Observation of long-lived interlayer excitons in monolayer $MoSe_2$–WSe_2 heterostructures, *Nature Communications*, vol. 6, p. 6242, 2015.
130. S. Bertolazzi, D. Krasnozhon, and A. Kis, Nonvolatile memory cells based on MoS_2/graphene heterostructures, *ACS Nano*, vol. 7, pp. 3246–3252, 2013.
131. K. Wang, B. Huang, M. Tian, F. Ceballos, M. W. Lin, M. Mahjouri-Samani, et al., Interlayer coupling in twisted WSe_2/WS_2 bilayer heterostructures revealed by optical spectroscopy, *ACS Nano*, vol. 10, 2016.

132. B. Li, L. Huang, M. Zhong, Y. Li, Y. Wang, J. Li, et al., Direct vapor phase growth and optoelectronic application of large band offset SnS_2/MoS_2 vertical bilayer heterostructures with high lattice mismatch, *Advanced Electronic Materials*, vol. 2, 2016.
133. Y. Gong, S. Lei, G. Ye, B. Li, Y. He, K. Keyshar, et al., Two-step growth of two-dimensional $WSe_2/MoSe_2$ heterostructures, *Nano Letters*, vol. 15, pp. 6135–6141, 2015.
134. Y. Gong, J. Lin, X. Wang, G. Shi, S. Lei, Z. Lin, et al., Vertical and in-plane heterostructures from WS_2/MoS_2 monolayers, *Nature Materials*, vol. 13, p. 1135, 2014.
135. C. Huang, S. Wu, A. M. Sanchez, J. J. Peters, R. Beanland, J. S. Ross, et al., Lateral heterojunctions within monolayer $MoSe_2$-WSe_2 semiconductors, *Nature Materials*, vol. 13, pp. 1096–1101, 2014.
136. F. Ceballos, M. Z. Bellus, H. Y. Chiu, and H. Zhao, Ultrafast charge separation and indirect exciton formation in a MoS_2–$MoSe_2$ van der Waals heterostructure, *ACS Nano*, vol. 8, pp. 12717–12724, 2014.
137. C. H. Lee, G. H. Lee, V. D. Z. Am, W. Chen, Y. Li, M. Han, et al., Atomically thin p-n junctions with van der Waals heterointerfaces, *Nature Nanotechnology*, vol. 9, pp. 676–681, 2014.
138. P. Chen, J. Xiang, H. Yu, J. Zhang, G. Xie, S. Wu, et al., Gate tunable MoS_2-black phosphorus heterojunction devices, *2d Materials*, vol. 2, p. 034009, 2015.
139. R. Tania, T. Mahmut, C. Xi, F. Hui, L. Der-Hsien, Z. Peida, et al., Dual-gated MoS_2/WSe_2 van der Waals tunnel diodes and transistors, *ACS NANO*, vol. 9, pp. 2017–2019, Jan 2015.
140. M. H. Chiu, C. Zhang, H. W. Shiu, C. P. Chuu, C. H. Chen, C. Y. Chang, et al., Determination of band alignment in the single-layer MoS_2/WSe_2 heterojunction, *Nature Communications*, vol. 6, p. 7666, 2015.
141. A. Nourbakhsh, A. Zubair, M. S. Dresselhaus, and T. Palacios, Transport properties of a MoS_2/WSe_2 heterojunction transistor and its potential for application, *Nano Letters*, vol. 16, 2016.
142. R. Yan, S. Fathipour, Y. Han, B. Song, S. Xiao, M. Li, et al., Esaki diodes in van der Waals heterojunctions with broken-gap energy band alignment, *Nano Letters*, vol. 15, pp. 5791–5798, Sep 9 2015.
143. T. Roy, M. Tosun, M. Hettick, G. H. Ahn, C. Hu, and A. Javey, 2D-2D tunneling field-effect transistors using $WSe_2/SnSe_2$ heterostructures, *Applied Physics Letters*, vol. 108, 083111, pp. 437–5, 2016.
144. M. R. Esmaeili-Rad and S. Salahuddin, High performance molybdenum disulfide amorphous silicon heterojunction photodetector, *Scientific Reports*, vol. 3, p. 2345, 2013.
145. S. Lin, X. Li, P. Wang, Z. Xu, S. Zhang, H. Zhong, et al., Interface designed MoS_2/GaAs heterostructure solar cell with sandwich stacked hexagonal boron nitride, *Scientific Reports*, vol. 5, p. 15103, 2015.
146. M. K. Joo, B. H. Moon, H. Ji, H. H. Gang, H. Kim, G. M. Lee, et al., Electron excess doping and effective Schottky barrier reduction on MoS_2/h-BN heterostructure, *Nano Letters*, vol. 16, 2016.
147. D. Ruzmetov, K. Zhang, G. Stan, B. Kalanyan, G. R. Bhimanapati, S. M. Eichfeld, et al., Vertical 2D/3D semiconductor heterostructures based on epitaxial molybdenum disulfide and gallium nitride, *ACS Nano*, vol. 10, p. 3580, 2016.
148. B. Li, G. Shi, S. Lei, Y. He, W. Gao, Y. Gong, et al., 3D band diagram and photoexcitation of 2D–3D semiconductor heterojunctions, *Nano Letters*, vol. 15, pp. 5919–5925, 2015.
149. D. Sarkar, X. Xie, W. Liu, W. Cao, J. Kang, Y. Gong, et al., A subthermionic tunnel field-effect transistor with an atomically thin channel, *Nature*, vol. 526, pp. 91–95, 2015.
150. M. W. Lin, L. Liu, Q. Lan, X. Tan, K. Dhindsa, P. Zeng, et al., Mobility enhancement and highly efficient gating of monolayer MoS_2 transistors with polymer electrolyte, *Journal of Physics D Applied Physics*, vol. 45, pp. 597–619, 2012.
151. S. Krishnamoorthy, E. W. L. Ii, C. H. Lee, Y. Zhang, W. D. Mcculloch, J. M. Johnson, et al., High current density 2D/3D MoS_2/GaN Esaki tunnel diodes, *Applied Physics Letters*, vol. 109, pp. 147–150, 2016.

8

Surface Functionalization of Silicon Carbide Quantum Dots

Marzaini Rashid, Ben R. Horrocks, Noel Healy, Jonathan P. Goss, Hua-Khee Chan, and AltonB. Horsfall

CONTENTS

Silicon carbide (SiC) nanostructures are appealing as non-toxic, water-stable and oxidation-resistant nanomaterials. Owing to these unique properties, three-dimensionally confined SiC nanostructures, namely SiC quantum dots (QDs), have found applications in the bioimaging of living cells. Photoluminescence (PL) investigations, however, have revealed that across the polytypes: 3C-, 4H- and 6H-SiC, excitation wavelength–dependent PL is observed for larger sizes but deviates for sizes smaller than approximately 3 nm, thus exhibiting a dual feature in the PL spectra. Additionally, the nanostructures of varying polytypes and bandgaps exhibit strikingly similar PL emission centered at approximately 450 nm. At this wavelength, 3C-SiC emission is above the bulk bandgap as expected of quantum size effects, but for 4H-SiC and 6H-SiC the emissions are below bandgap. 4H-SiC is a suitable polytype to study these effects. Density functional theory (DFT) calculations within the ab initio formalism were performed on OH-, F- and H-terminated 4H-SiC-QDs with diameters in the range of 1–2 nm. The chosen surface terminations relate to the HF/ ethanol electrolyte used in the preparation of SiC-QDs and the choice of size coincides with where deviation was observed in experiments. It was found that the absorption onset energies deviate from quantum confinement with -OH and -F terminations, but conform to the prediction when terminated with -H. The weak size-dependent absorption onsets for -OH and -F are due to surface states arising from lone-pair orbitals that are spatially localized to the QD surface where these terminations reside. On the other hand, -H termination shows strong size-dependent absorption onsets due to the delocalization of the electron wave function toward the QD core assisting quantum confinement. It is predicted that the surface-related states dominate up to sizes 2.5 and 2.7 nm for -F and -OH terminations, respectively. As a result, we show that the recombination mechanism involves the interplay between quantum confinement and surface states affecting the resultant energy gap and the resulting PL. Hence, the experimental PL spectra exhibit a dual feature: excitation wavelength independence for small sizes and excitation wavelength dependence for diameters larger than 3 nm, as observed in the experiments reported in the literature.

8.1 The Dual-Feature and Below Bandgap Photoluminescence Spectra in SiC Nanostructures

SiC is a promising material that is both non-toxic and oxidation resistant. 3C-SiC-QDs with an average size of 3.9 nm [1] showed robust PL emission centered at 450 nm with a quantum yield of 17%, comparable to that observed in Si nanoparticles. SiC-QDs were shown to be water stable and oxidation resistant for months without any surface passivation [2]. The properties of SiC such as biocompatibility, chemical inertness, photostability, stability in aqueous environment and resistance to oxidation are highly attractive for a range of applications.

Nonetheless, researchers have observed that the optical properties of SiC-QDs only follow the prediction of quantum confinement for diameters larger than approximately 3 nm, but deviate for smaller sizes. Thus, the expected and required size-tunable optical properties below 3 nm were not achieved. In theory, with the exciton Bohr radii of 2.0, 0.7 and 1.2 nm for 3C-, 6H- and 4H-SiC [2], respectively, strong quantum size effects are expected to occur for diameters smaller than 4 nm in 3C- and 2.4 nm in 4H-SiC-based QDs. The deviation is observed as a dual feature in the PL spectra for the different polytypes of SiC despite the differences in bandgap, along with an unexpected similar emission peak at approximately 450 nm. This 450 nm emission peak is above bandgap for 3C-SiC but below bandgap for 4H-SiC and 6H-SiC, the latter being counterintuitive to quantum confinement. These observations have been attributed to phase transformation into 3C-SiC during ultrasonication or due to surface states, but to date the origin of the dual-feature PL has not been unequivocally determined or explained and is still under investigation. Generally, in order to verify the existence of quantum confinement effects, the red-shifting in PL peak position with increasing excitation wavelengths can be used [2–4]. As the excitation energy is systematically decreased with longer excitation wavelengths, only larger QDs are excited and smaller ones are excluded. Therefore, the PL energy is expected to decrease (red-shift toward longer wavelengths). The majority of studies on the PL spectra of SiC nanoparticles suspended in solvents reported that quantum size effects are observed for larger particles (diameter greater than ~3 nm) but not for smaller diameters [5–7]. In these studies, the PL peak position generally red-shifted with increasing excitation wavelengths for large-sized SiC-QDs, as expected from quantum size effects. In contrast, the PL peak position remained constant (not dependent on excitation wavelength) for small-sized QDs (smaller than 3 nm), which contradicts the predictions from quantum confinement. Mixed results are reported in the literature, but in general the PL emission peak is shown to be insensitive to excitation wavelengths up to around 320 nm. The 'dual-feature' observation is schematically illustrated in Figure 8.1a, where region A depicts the constant PL peak position for excitation wavelengths shorter than 320 nm and region B shows that the PL

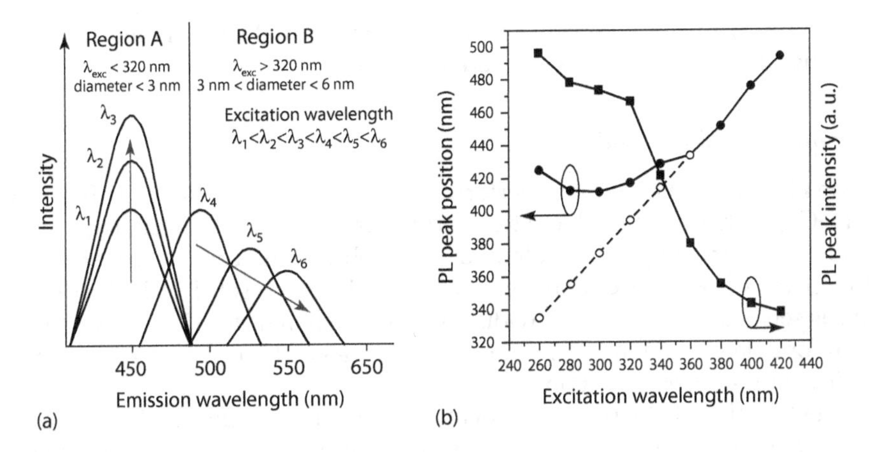

FIGURE 8.1
(a) Schematic illustrating the dual-feature PL spectra of SiC-QDs characterized by the constant PL emission peak in region A and the red-shifting PL peak in region B for excitation wavelengths shorter than or longer than 320 nm. (b) The experimentally observed deviation from quantum confinement (deviate from dotted line) in the PL peak position for excitation wavelengths shorter than 320 nm. (From Zhu et al., *Materials Letters*, 2014, 132, 210–213.)

peak position red-shifts for excitation wavelengths longer than 320 nm. For brevity, the associated regions A and B are referred to as 'constant PL' and 'red-shifting PL' and the combination of both trends is referred to as 'dual feature'.

The dual-feature trend in the PL peak position with respect to the excitation wavelength in the experiment is clearly evident in Figure 8.1b, where deviation from the expected size-dependent PL emission (straight dotted line) occurs at approximately 320 nm [8]. The QD's size distribution in [8] was 2–5.5 nm and has been associated with the excitation of QDs smaller than 3 nm.

8.2 DFT Study on the Optoelectronic Properties of OH-, F- and H-Terminated 4H-SiC Quantum Dots

This section presents the results from DFT calculations within the Ab Initio Modeling Program (AIMPRO) on the optical properties of OH-, F- and H-terminated 4H-SiC-QDs in the diameter range of 10–20 Å. The 4H-SiC polytype serves as a good representative to study both the dual feature and below bandgap PL emissions observed in experiments. The QD diameter range from 10 to 20 Å was chosen in order to investigate the discrepancy in quantum confinement effects between the theoretical prediction and the experimentally observed deviation for SiC-QDs smaller than 30 Å. Practically, biomarkers smaller than 50 Å are required for effective renal excretion and so retaining quantum size effects within this size range is important and warrants further understanding. The deviation in the optical properties observed in 4H-SiC-QDs may be contributed by surface effects as opposed to being primarily due to the bulk (polytype). In order to determine the influence of surface effects, different surface termination groups that are likely to be found in the commonly used HF/ethanol electrolyte are simulated, namely -OH, -F and -H. The effect of and contributions from these surface terminations on the onset energies of absorption cross section (ACS) and joint density of states (JDoS) are analyzed. A model is constructed to explain the experimentally observed dual-feature PL.

8.2.1 Computational Method

Calculations based on the DFT are performed within the AIMPRO code [9,10] for quasi-spherical SiC-QDs having core diameters varying from 10 to 20 Å. The modeling of atoms utilizes norm-conserving pseudo potentials [11]. Kohn–Sham orbitals are expanded using sets of independent real-space, atom-centered Gaussian basis sets [12]. In the case of C, O, F and Si, basis sets with four different widths are used, whereas for H three widths are used. To account for polarization in O, C, Si and F, *d*-Gaussians of one, two, two and four widths are included for the respective atoms.

In constructing the SiC-QDs, a group of atoms within some chosen distance of either an Si or a C atom (Si or C centered) in bulk SiC are selected. Then, atoms at the surface are saturated by either a H atom, an F atom or an OH group.

A plane-wave expansion (Fourier transform) is used to represent the charge density in reciprocal space and the SiC-QDs are modeled using periodic boundary conditions [10], with the cutoff in energy is sufficient to converge the energies to within 10 meV. In this approach, the SiC-QDs are placed in a periodic boundary condition. Following preparation of the SiC core and surface groups, the SiC-QDs are geometrically optimized until all atomic forces are smaller than 0.06 eV/Å.

8.2.2 Results

Firstly, simulations of OH-, F- and H-terminated, 10 Å diameter 4H-SiC-QDs are constructed and subsequently optimized. The corresponding DoS, ACS, JDoS, electron wave function isosurfaces and electron probability densities for this QD size are calculated and compared for the different terminations. Similar analyses for core diameters of up to 20 Å are performed. A model comprising the core and surface highest occupied/lowest unoccupied (HOMO/LUMO) is constructed to explain the dual-feature PL observed in experiments. Finally, a projection of the QD size range influenced by surface states is made based on the calculations.

8.2.2.1 Optical Properties of 10 Å Diameter 4H-SiC-QD Structures

Relaxed geometries of 10 Å diameter QDs terminated with -OH ($Si_{19}C_{20}O_{40}H_{40}$), -F ($Si_{19}C_{20}F_{40}$) and -H ($Si_{19}C_{20}H_{40}$) are illustrated in Figure 8.2. The average Si-C, O-H, Si-X, C-X (X=H, O and F in each case) bond lengths in angstroms (Å) are calculated to be 1.87 Å,

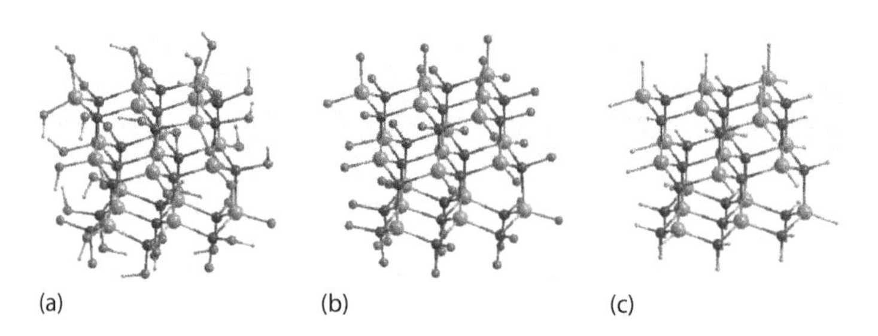

FIGURE 8.2
Geometrically relaxed 10 Å diameter 4H-SiC quantum dots with (a) -OH, (b) -F and (c) -H terminations. Color coding: gray=C, yellow=Si, red=O, green=F and white=H.

1.00 Å, (1.50 Å, 1.65 Å, 1.59 Å), (1.10 Å, 1.46 Å and 1.41 Å) as summarized in Table 8.1. The bond lengths lie within the anticipated range for Si- and C-based single bonds [13,14]. The standard deviation in the Si-C bond lengths across QDs with different surface terminations are within 0.01 Å.

Different surface functionalization (-OH, -F and -H) on the same Si and C core atoms configuration resulted in variation in the HOMO-LUMO energy gap, as shown in Figure 8.3a. The HOMO-LUMO energy gap is clearly distinguishable between different surface terminations, with -H termination showing the largest energy gap (4.50 eV), followed by -F (2.27 eV) and -OH (2.14 eV), respectively.

It is instructive to examine the effects of the surface terminations upon the QDs' ACS, as shown in Figure 8.3b. The absorption onsets correlate with the HOMO-LUMO energy gaps where H termination shows a higher energy optical absorption onset (4.50 eV) compared to -F and -OH in agreement with its larger calculated energy gap. The optical absorption onset of -OH is red-shifted by (2.3 eV) while -F is red-shifted by (1.7 eV) when compared to -H. For optical properties, the JDoS is important as it shows all the possible optical transitions from the valence (HOMO) to the conduction (LUMO) energy levels with energy

TABLE 8.1

10 Å Diameter SiC-QD Bond Lengths

Bond	Si-C	O-H	Si-H	Si-O	Si-F	C-H	C-O	C-F
Average bond length (Å)	1.87	1.00	1.50	1.65	1.59	1.10	1.46	1.41

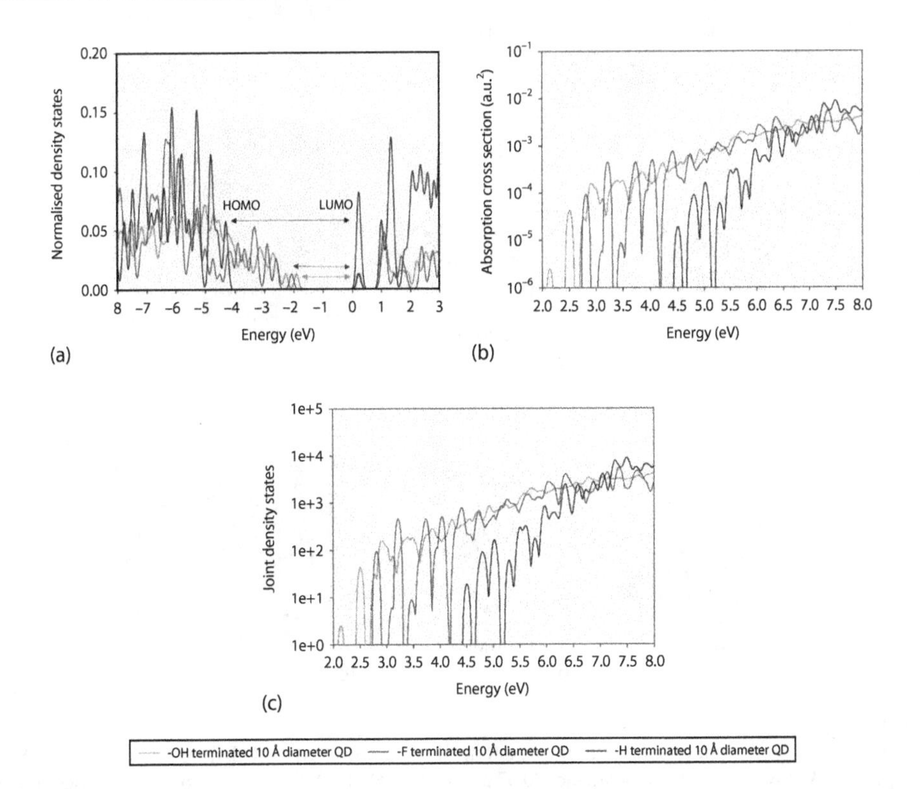

FIGURE 8.3

The respective (a) normalized density of states (DoS), (b) absorption cross section (ACS) and (c) joint density of states (JDoS) for 10 Å diameter -OH, -F and -H functionalized 4H-SiC-QDs.

separation equal to the energy of the absorbed photon. The resemblance in spectral shape and energies of the ACS in comparison to the JDoS indicates that the calculated optical transitions concur with the calculated allowed energies of the electronic DoS.

The difference in the HOMO-LUMO energy gap is further elaborated by examining the spatial distribution of the frontier orbitals (HOMO and LUMO). For -OH termination, Figure 8.4I-a shows HOMO being primarily localized on the surface C and O atoms in the form of C *p*-orbitals and O lone-pairs, respectively. LUMO orbitals are seen primarily on O atoms (Figure 8.4I-b). As shown in Figure 8.4I-c, a two-dimensional slice confirms the electron probability density (within 1 eV of the HOMO) residing mostly on the surface of the QD. The electron probability density examined is the sum of the squared modulus of the wave function:

$$\sum_{n=i} |\psi_i|^2(r)$$

where $(\text{HOMO} - E_i) \leq (1 \text{ eV})$.

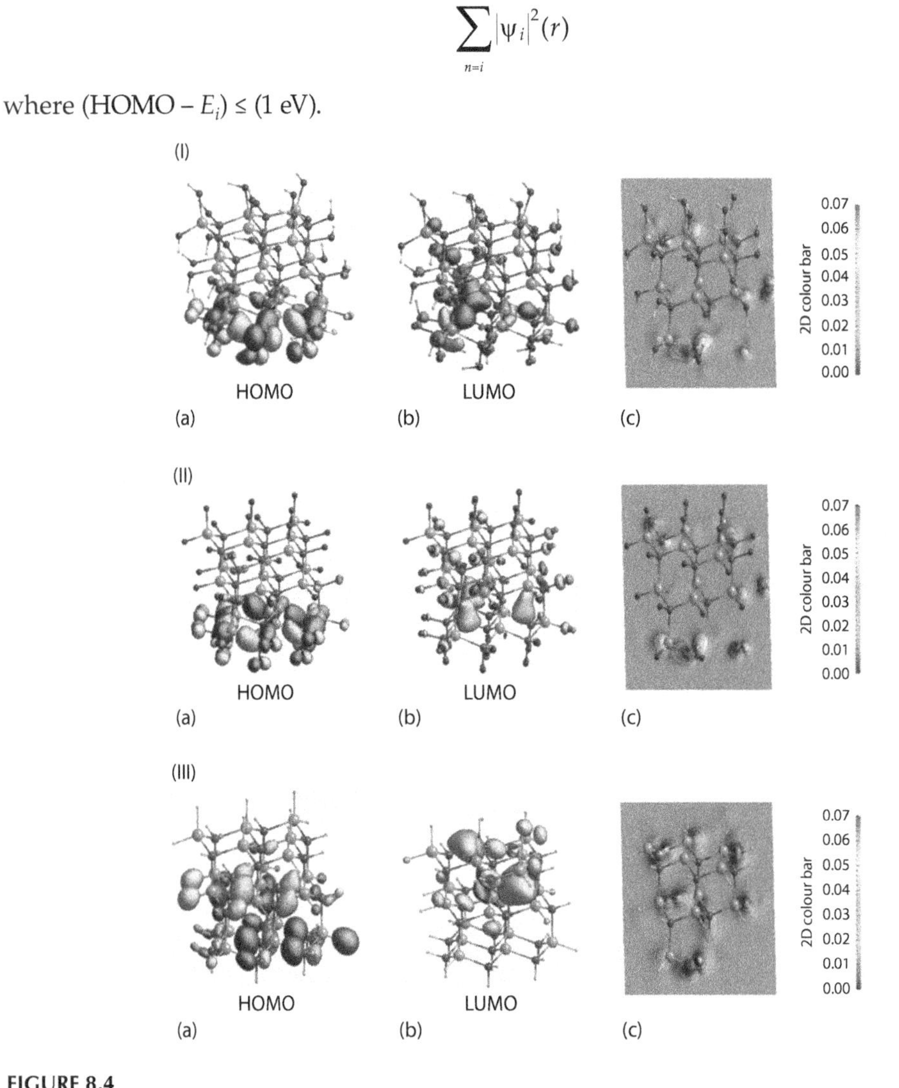

FIGURE 8.4
Wave functions for 10 Å diameter (I) -OH terminated, (II) -F terminated and (III) -H terminated 4H-SiC-QDs with respective 3-D wave function isosurface (0.07 a.u.) for (a) HOMO, (b) LUMO and (c) 2-D slice of electron probability density within 1 eV of the HOMO. The delocalized wave function spatial distribution toward the QD core suggests that the large HOMO-LUMO energy gap for H-termination results from the confining potential of the core of the QD.

For F-termination, lone-pairs on F atoms and p-like wave functions of the nearest neighbors (Figure 8.4II-a and II-c) mainly contribute to the HOMO, while for the LUMO (Figure 8.4II-b), similar to OH-termination, mainly the F atoms contribute. In contrast to -OH and -F terminations, the HOMO orbitals for -H termination (Figure 8.4III-a) are dispersed through the whole cluster, with the electron probability density reaching toward the core of the QD. The LUMO (Figure 8.4III-b) wave function is relatively delocalized to the QD core with minimal contribution from functional H atoms. Relating to the particle in a sphere model, the delocalized wave function spatial distribution toward the QD core suggests that the large HOMO-LUMO energy gap for -H termination results from the confining potential of the core of the QD (quantum confinement) while surface states are dominant and influence the electronic and optical properties of OH- and F-terminated QDs.

While it is shown that the H-terminated 10 Å diameter QD exhibits an optical absorption onset well above that of -F and -OH for a similar cluster size, it would be instructive to investigate the size-dependent quantum confinement effects for a range of QD diameters. In the next sections, the experimentally relevant 10–20 Å diameter range is presented. Full passivation of larger QD surfaces (>10 Å in diameter) with -OH, -F and -H would result in the surface termination species coming closer to each other through which steric repulsion becomes significant. By surface reconstructions with Si and C dimers, these surface species would be better accommodated for larger QD diameters. The effect of surface reconstruction and surface composition on the QDs' electronic and optical properties is presented in the next section.

8.2.2.2 Effect of Surface Composition and Surface Reconstruction

Data relating to optimized SiC-QDs in the diameter range between 10 and 20 Å is presented. The chosen diameter range is within the range where quantum confinement effects are observed in experiments [15], since the exciton Bohr radius in 4H-SiC is 12–18 Å [3,5]. Table 8.2 lists the QD compositions as a function of the QD diameter: N_{Si} (X) and N_C (X) being the number of Si and C atoms for *X*-centered QD (*X* for Si or C at the core center of QD), $N_{surface}$ represents the number of surface sites on unreconstructed clusters and N_{recon} (X) is the number of surface sites for reconstructed clusters with an *X* center (*X* for Si or C). The choice of atom center of either Si or C does not affect the total number of core atoms but the number of Si or C atoms are interchanged as shown in Table 8.2.

TABLE 8.2

SiC-QD Cluster Composition as a Function of QD Diameter in the Range of 10–20 Å

Diameter (Å)	N_{Si} (Si), N_C(C)	N_{Si} (C), N_C(Si)	$N_{surface}$	N_{recon} (Si)	N_{recon} (C)
10	19	20	40	40	40
11	29	29	56	44	44
12	51	41	82	70	70
13	57	62	94	76	76
14	69	74	100	88	88
15	81	83	112	100	100
16	99	96	126	114	114
17	132	118	160	130	136
18	147	151	172	130	142
19	159	175	190	160	148
20	207	199	208	166	172

The number of surface sites on the unreconstructed clusters is independent of whether the QD is Si or C centered. For the reconstructed clusters, the number of surface sites is different for Si- and C-centered cases.

In the case of unreconstructed clusters, only the positions of Si and C atoms are interchanged without changes to bondings with surface sites. In contrast, for the reconstructed clusters, the formation of dimers (Si-Si or C-C) on the surface depends on the choice of C or Si as the QD center. Additionally, Si-Si with a longer bond length may be formed at certain locations where C-C could not, due to its shorter bond length, thus the C atoms are terminated with more surface sites in comparison to Si for this case.

Figure 8.5 shows the 20 Å SiC-QDs with different centers and surface treatments. The average Si-C, O-H, Si-X, C-X (X=H, O and F in each case), Si-Si and C-C bond lengths in

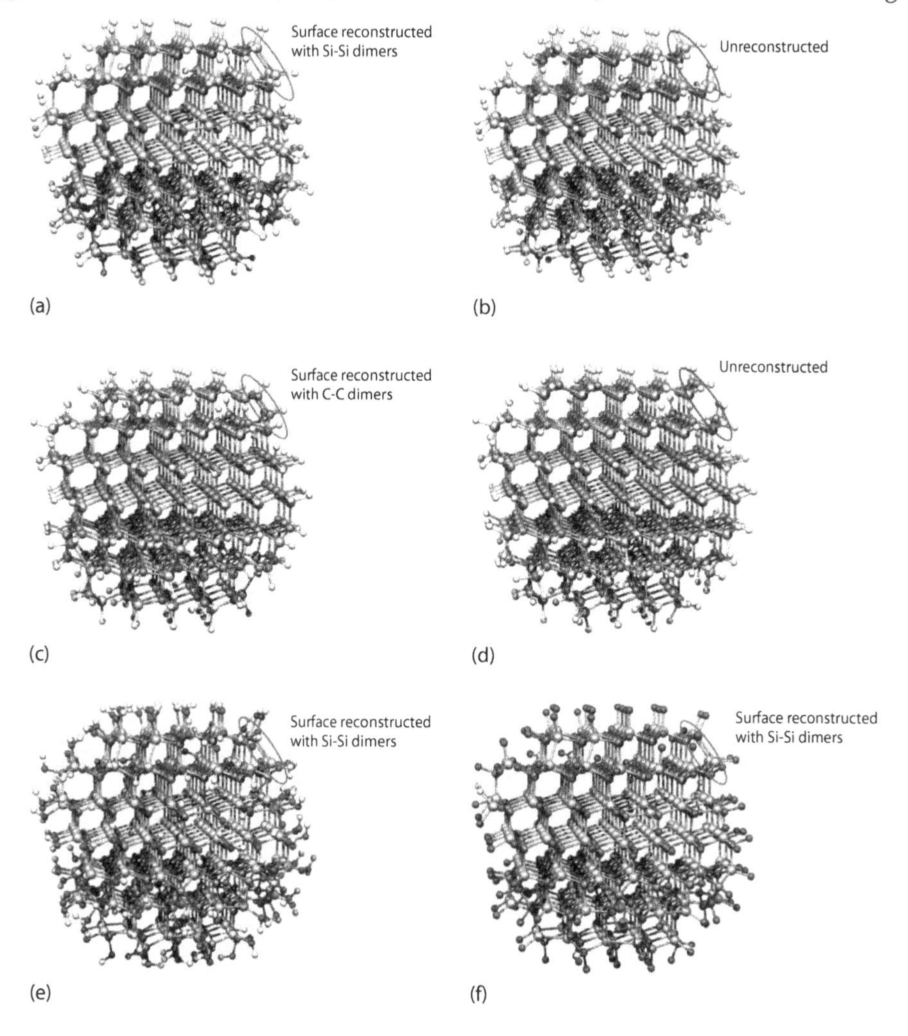

FIGURE 8.5
Geometrically relaxed 20 Å SiC-QDs. (a, b) H-terminated, Si-centered SiC-QDs with and without surface reconstructions, respectively. (c, d) H-terminated, C-centered SiC-QDs with and without surface reconstructions, respectively. (e, f) Si-centered SiC-QDs (with surface reconstructions) terminated with OH- and F-groups, respectively. White, red, yellow, gray and green atoms are H, O, Si, C and F, respectively. The regions indicated by the red and blue ellipses highlight surface sites where there is reconstruction with surface dimers or an unreconstructed surface, respectively.

TABLE 8.3

20 Å Diameter SiC-QD Bond Lengths

Bond	Si-C	O-H	Si-H	Si-O	Si-F	C-H	C-O	C-F	Si-Si	C-C
Average bond length (Å)	1.87	1.00	1.50	1.64	1.59	1.10	1.46	1.42	2.39	1.60

angstroms (Å) are calculated to be 1.87, 1.00, (1.50, 1.64 and 1.59), (1.10, 1.46 and 1.42), 2.39 and 1.60, respectively, which are summarized in Table 8.3. This provides evidence that the optimized structures have reasonable geometries to represent the spherical SiC-QDs observed in experiments. The calculated bond lengths vary only very slightly in comparison to the 10 Å diameter QDs.

To distinguish the impact of surface reconstructions from surface termination, two groups of Si-centered H-terminated QDs within the 10–20 Å size range are compared, one of which includes surface reconstruction.

8.2.2.3 Effect of Surface Termination Groups on Optical Absorption

Figure 8.6a and b shows the ACS for Si-centered, reconstructed, OH- and F-terminated QDs, and can be compared directly with Figure 8.6c. The effect upon the absorption onset for OH- and F-terminations (size independent and flat along ~2.0 eV) in comparison to H-termination (size dependent from 4.5 to ~3.0 eV) is immediately evident. For example, the differences between Figure 8.6a and c (due to different surface termination) outweigh the effects of reconstruction or whether the cluster is centered on the Si or C (Figure 8.6c,d or e,f).

Notably, the quantum confinement effect tends to increase from hydroxyl to fluorine to hydrogen termination, with H-termination showing the clearest size dependence. The absorption onset for H-termination increases from 2.5 to 4.4 eV with reducing size. This is a net increase of ~2 eV in the 10–20 Å diameter range. On the other hand, for F- and OH-terminations, almost no size-dependent change is observed over the same diameter range. As a result, for low core diameters there is a large difference in the onset energy as a function of termination. For example, in the case of 10 Å SiC-QDs, the difference in onset for the ACS for OH- and H-terminations is nearly 2.5 eV.

8.2.2.4 Effect of Surface Termination on Density of States, HOMO/ LUMO Wave Functions and Electron Probability Density

Figure 8.7 shows the DoS for Si-centered, surface-reconstructed, 20 Å diameter SiC-QDs terminated by OH, F and H, with zero energy aligned to the Fermi energy. The significant effect of terminating species is the variation in the HOMO whereby OH, F and H have energies ranging from –0.51 eV to –1.23 eV. In contrast, the LUMO energies are relatively constant, lying in the 1.0–1.1 eV range.

The difference in the energies of the HOMO is related to the degree of wave function localization at the surface. Figure 8.8a shows that the OH-termination HOMO is strongly associated with surface C *p*-orbitals and oxygen lone-pairs. Likewise, HOMOs of F-terminated clusters are associated with F lone-pairs (Figure 8.8c). The LUMOs of OH-termination (Figure 8.8b) and F-termination (Figure 8.8d) consist of mixtures of surface O (F) lone-pairs and core C *p*-orbitals for OH- (F-) terminations, respectively. The lone-pair energies are not significantly affected by the core diameter because they are localized on the surface, leading to the relatively weak size dependence in the ACS (Figure 8.6). It is concluded that

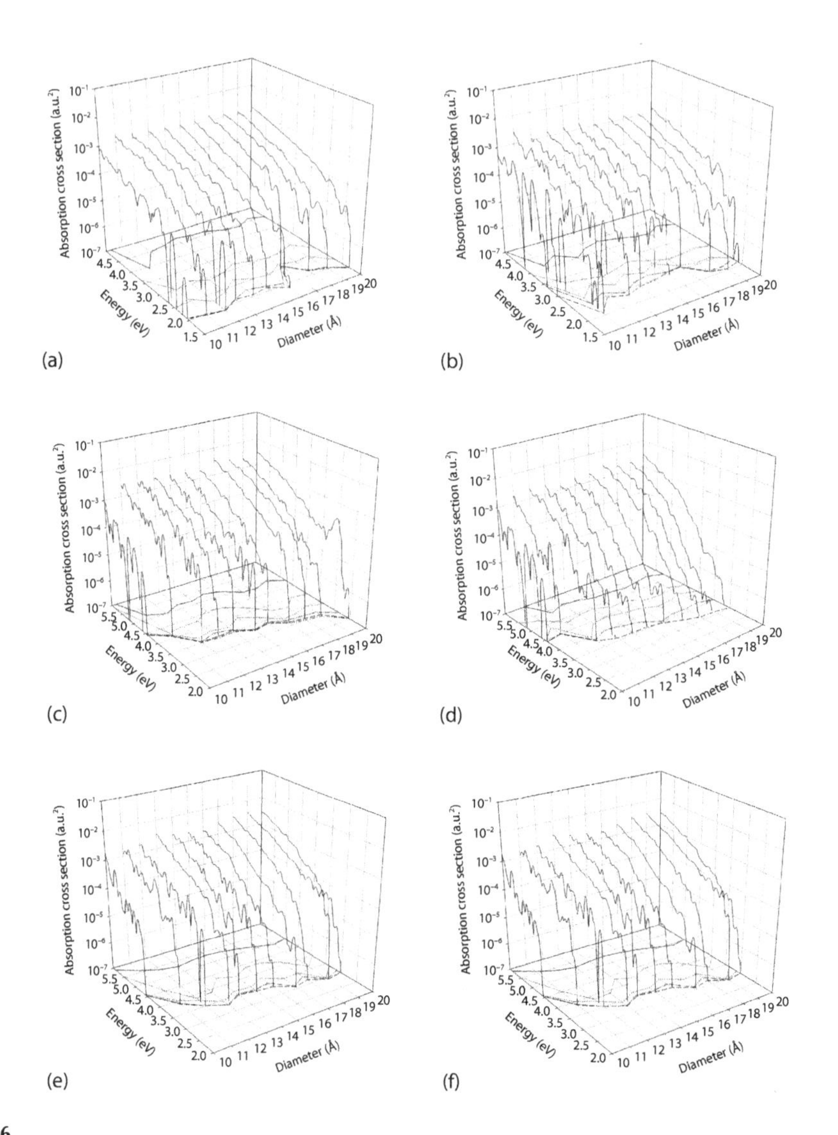

FIGURE 8.6
ACSs for (a) OH-terminated and (b) F-terminated 4H-SiC-QDs. ACSs for Si-centered, H-terminated (c) reconstructed and (d) unreconstructed QD. ACSs for C-centered, H-terminated (e) reconstructed and (f) unreconstructed QDs. The contours indicate 10^{-7} a.u.^{2} to 10^{-3} a.u.^{2}, in factors of 10. The effect of termination species (-OH, -F, -H) outweighs the effects of composition and surface reconstruction.

the seemingly size-independent optical absorption for OH- and F-terminated SiC-QDs is a reflection of the surface states associated with the terminating species.

The influence of surface terminations on the electronic structure is further elucidated by an examination of the energy levels deeper than the frontier electronic levels. Figure 8.9 shows that for -H-terminated Si-centered 20 Å SiC-QDs, the electron probability density of the HOMO (Figure 8.9a) is dominated by electrons within the core of the QD while the LUMO (Figure 8.9b) is surface related arising from the Si dimers. For deeper energy levels, within 0.5 eV of the HOMO (Figure 8.9c), the probability density increases with more electrons populating the QD core while for the LUMO (Figure 8.9d), in addition to the surface, electrons from the core start to contribute. Examining deeper still the energy (within 1 eV) from the frontier electronic levels, it is found that the electron probability density

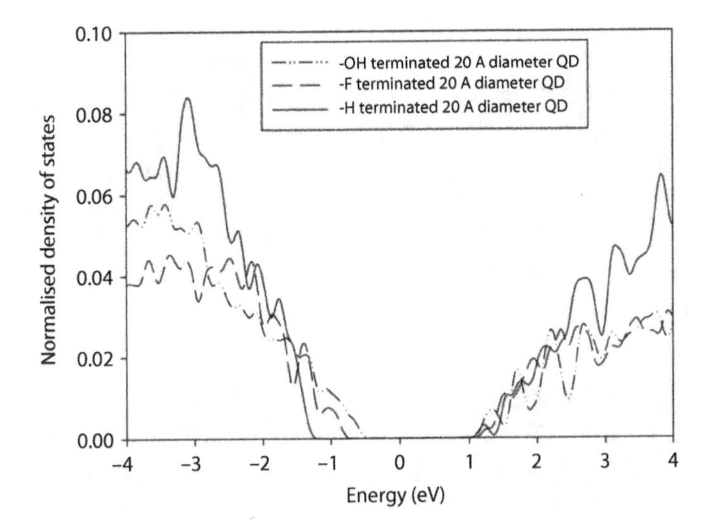

FIGURE 8.7
DoS of 20 Å, Si-centered QDs in the vicinity of the optical gap.

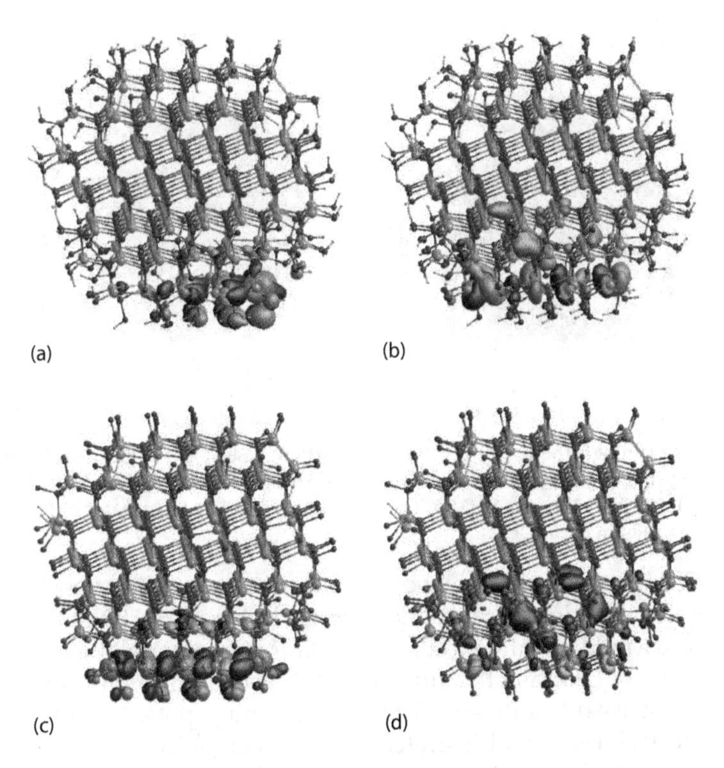

FIGURE 8.8
Wave function iso-surfaces for Si-centered 20 Å SiC-QDs. (a, b) HOMO and LUMO for -OH termination. (c, d) HOMO and LUMO for F-termination. Yellow, gray and white atoms are Si, C and H, respectively. The green and blue iso-surfaces show the molecular orbitals of the positive and negative phase, with an amplitude of 0.04 a.u.$^{-3/2}$. The clusters are shown with the *c*-axis of the underlying 4H-SiC vertical. Lone-pair energies are not significantly affected by the core diameter because they are localized on the surface, leading to the relatively weak size dependence in the ACS for F- and OH-terminations.

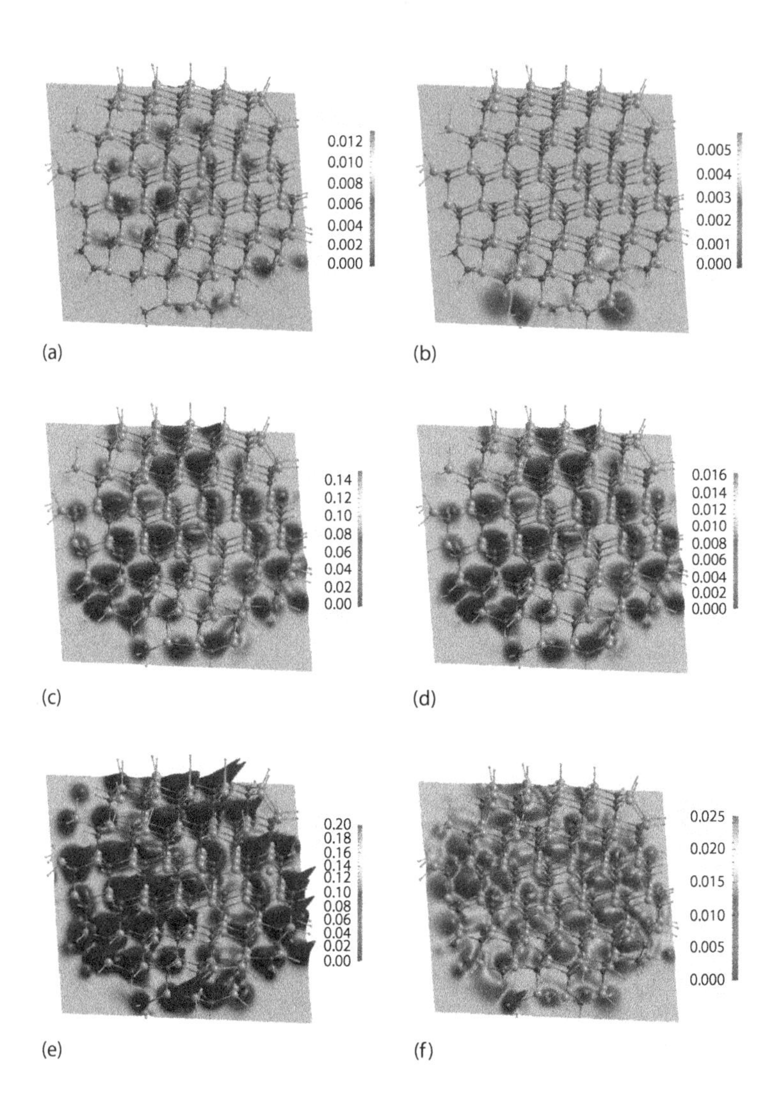

FIGURE 8.9
Two-dimensional slice through the center of an Si-centered 20 Å SiC-QD exhibiting -H electron probability density for (a) HOMO, (b) LUMO, (c) within 0.5 eV of the HOMO, (d) within 0.5 eV of the LUMO, (e) within 1 eV of the HOMO and (f) within 1 eV of the LUMO. The delocalized and localized states of the respective HOMO and LUMO are represented in Figure 8.13a.

is well delocalized throughout the QD core, as can be seen in Figure 8.9e,f. The delocalized electron probability density at higher energy levels (core HOMO and core LUMO) within the QD core would be influenced by the confining potential that supports quantum confinement effects. In contrast, the surface states near the frontal orbital of the LUMO (surface LUMO) are weakly influenced by the confining potential and would obscure quantum size effects. It is clearly illustrated that the energies in the vicinity of the HOMO for H-termination are highly dominated by electrons from the core. On the other hand, for the LUMO, the surface states dominate at lower energies but at higher energies, electrons from orbitals within the core of the QD contribute.

For F-termination, at the frontier orbitals, the HOMO (Figure 8.10a) constitutes electrons that are highly localized to the surface with no contribution from the core, while the LUMO (Figure 8.10b) shows a mix of electrons occupying the surface and core of the QD. Within

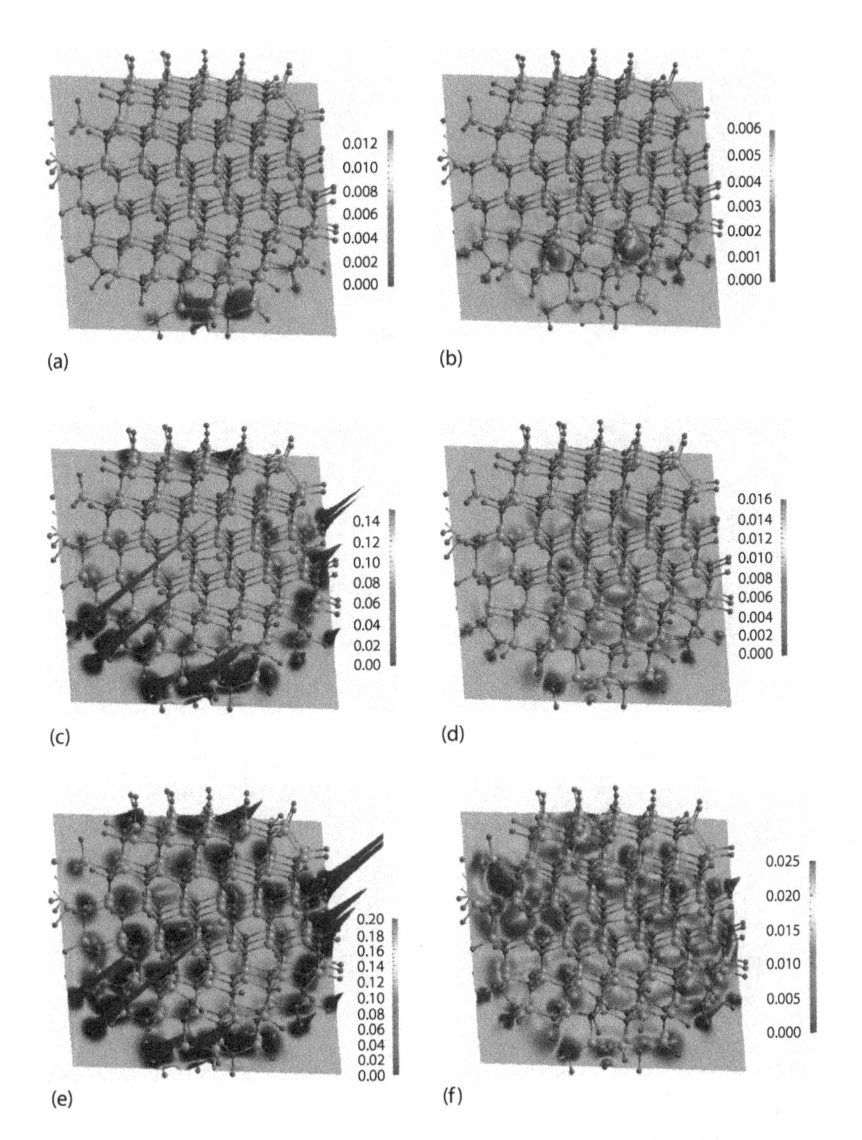

FIGURE 8.10
Two-dimensional slice through the center of an Si-centered 20 Å SiC-QD exhibiting -F electron probability density for (a) HOMO, (b) LUMO, (c) within 0.5 eV of the HOMO, (d) within 0.5 eV of the LUMO, (e) within 1 eV of the HOMO and (f) within 1 eV of the LUMO. The localized states of the respective HOMO and LUMO are represented in Figure 8.13b.

0.5 eV of the HOMO (Figure 8.10c), the electron occupation of surface atoms around the perimeter of the QD is further elaborated and an additional contribution from core-related electrons is observed. The LUMO (Figure 8.10d) in this higher energy interval involves more highly localized electrons at the surface and added contributions from electrons in the core. Higher up in energy to within 1 eV interval from the HOMO and LUMO, more electrons from the QD core participate, such that there is a mix of highly localized surface electrons and core-related electrons in the vicinity of the HOMO (Figure 8.10e), while in the LUMO (Figure 8.10f) the core-related electrons start to dominate.

The trends for -OH termination show that surface localization is greater for -OH, as can be seen for the HOMO (Figure 8.11a), due to surface C *p*-orbitals and oxygen lone-pairs.

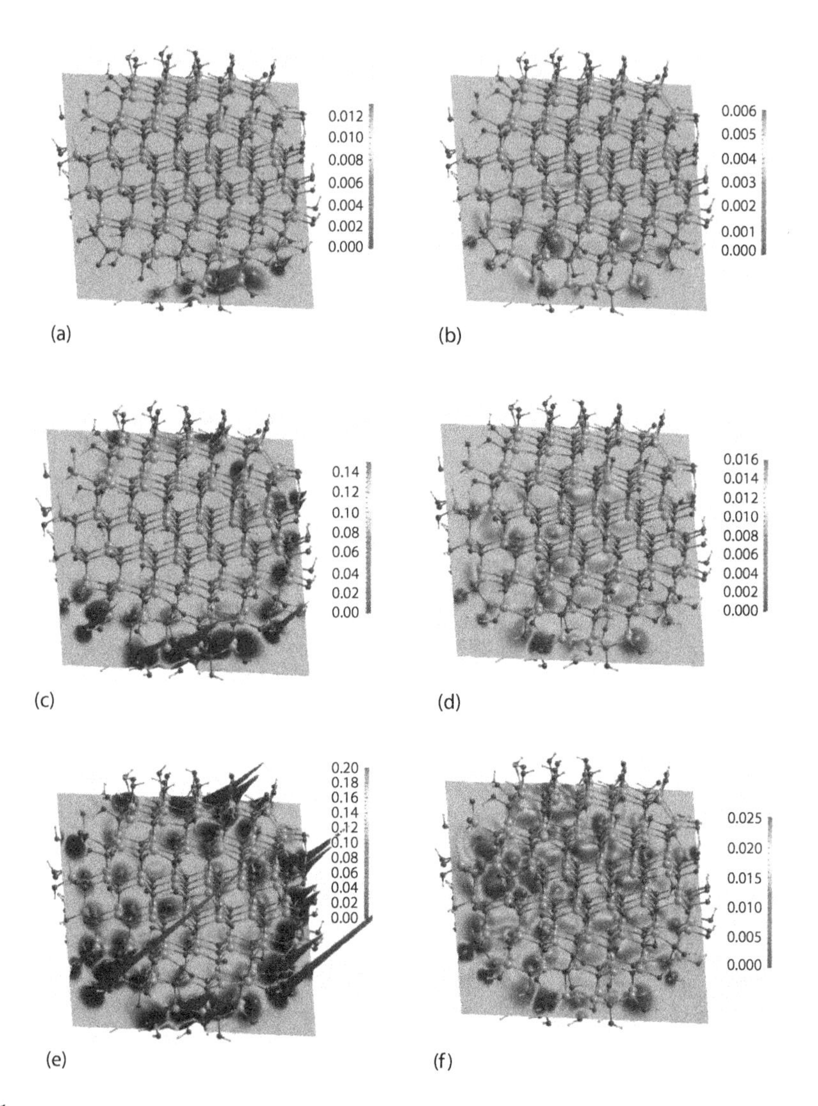

FIGURE 8.11
Two-dimensional slice through the center of an Si-centered 20 Å SiC-QD exhibiting -OH electron probability density for (a) HOMO, (b) LUMO, (c) within 0.5 eV of the HOMO, (d) within 0.5 eV of the LUMO, (e) within 1 eV of the HOMO and (f) within 1 eV of the LUMO. The localized states are represented in Figure 8.13b.

The LUMO (Figure 8.11b) is a mix of surface and core electrons. Toward higher energies of within 0.5 eV of the respective HOMO and LUMO, more electrons occupying the surface of the QD can be observed with still minimal contribution from the QD core in the vicinity of the HOMO (Figure 8.11c), while the electron probability density for energies near the LUMO involves electronic states within the core (Figure 8.11d). It is observed that going higher in energies, the localization of electrons at the surface is balanced out with electrons from within the core, as shown in (Figure 8.11e,f).

The examined electron probability density for -H and -OH terminations spanning the frontier orbital energies toward higher energies indicates that the electrons at the surface and within the QD core are well coupled throughout the investigated energy range for -H. Conversely, for -F and -OH terminations, these are decoupled: the core-related electrons predominantly occupy higher energies whereas surface-related electrons dominate the frontal electronic levels, particularly the HOMO, which causes narrowing of the optical gap. Having inspected the core and the surface electron probability density for 20 Å SiC-QD, further examination of the trends of the JDoS at high energy (core related) versus the JDoS at onset (surface related) for the diameter range 10–20 Å was undertaken. In the next section, the effect of the surface termination groups -H-, -OH and -F upon the optical properties of 4H-SiC-QDs is discussed.

8.2.2.5 Surface-State-Dependent Optical Properties of 4H-SiC-QDs

Figure 8.12a summarizes the energy gap of the QDs as a function of the size and terminating species. H-termination results in a large size dependence, whereas F- and OH-terminations are nearly size independent. For H-termination, the HOMO-LUMO gap increases following the quantum mechanical 'particle in a sphere' picture [16], whereby it is discrete and scales with the QD diameter, d, as d^{-2}. For F- and OH-terminations, additional surface states are present resulting in their energy being governed by the immediate bonding to the SiC surface, and not by the size of the QD. Figure 8.12b allows for a comparison of H-terminated clusters with different compositions and surface reconstructions, which asserts that reconstruction and choice of cluster center have no significant impact upon the energy gap in comparison to the effect of the termination. H-termination consistently exhibits stronger quantum confinement compared to -F and -OH regardless of surface reconstruction. This indicates that the surface termination groups have greater influence than size and surface reconstruction.

To understand the size dependence of SiC-QDs, a simple model, taking into account the different configurations of the surface states, is considered. The states associated with the SiC core exhibiting quantum confinement will result in a narrowing of the optical gap as the core diameter increases. This is represented schematically by the solid lines in Figure 8.13a,b. The second component arises from two surface effects. The first is the introduction of states due to homo-nuclear bonds arising from reconstruction. In the case of H-termination, this introduces an unoccupied state below that of the core, as represented by the higher dashed lines in Figure 8.13a. For a large enough cluster, this surface state is expected to lie outside the core gap, i.e. reconstructions impact the optical properties of relatively small SiC-QDs. However, as previously noted, the impact of the reconstruction is not significant in comparison to surface termination. For F- and OH-terminations (Figure 8.13b), the lone-pair orbitals mean that there are both occupied and empty surface states within the SiC core–related energy gap. Therefore, the predicted optical properties of F- and OH-terminated SiC-QDs can be divided into two regions. In region A, the HOMO and LUMO are associated with the surface states and the energy gap is approximately constant. In region B, the HOMO and LUMO originate from core states, so that quantum confinement effects become significant.

The dual-feature PL spectra of SiC-QDs observed experimentally are represented schematically in Figure 8.13c. In a QD ensemble containing a range of core sizes, short wavelengths excite all QDs [4,5,7], whereas longer wavelengths excite only large QDs. However, 10–60 Å SiC-QDs yield PL peaks that are largely insensitive to excitation wavelength up to around 320 nm. The dual feature of the PL spectra has been discussed in terms of quantum confinement effects only being responsible for the observed size dependence of emission in larger SiC-QDs [4,5,17], but a clear explanation regarding the size-independent features was not established. The observation can be explained if the emission (and excitation) is not purely a core property, but rather has a contribution from the surface states (Figure 8.13a,b). Red-shifts for excitation wavelengths above around 360 nm show

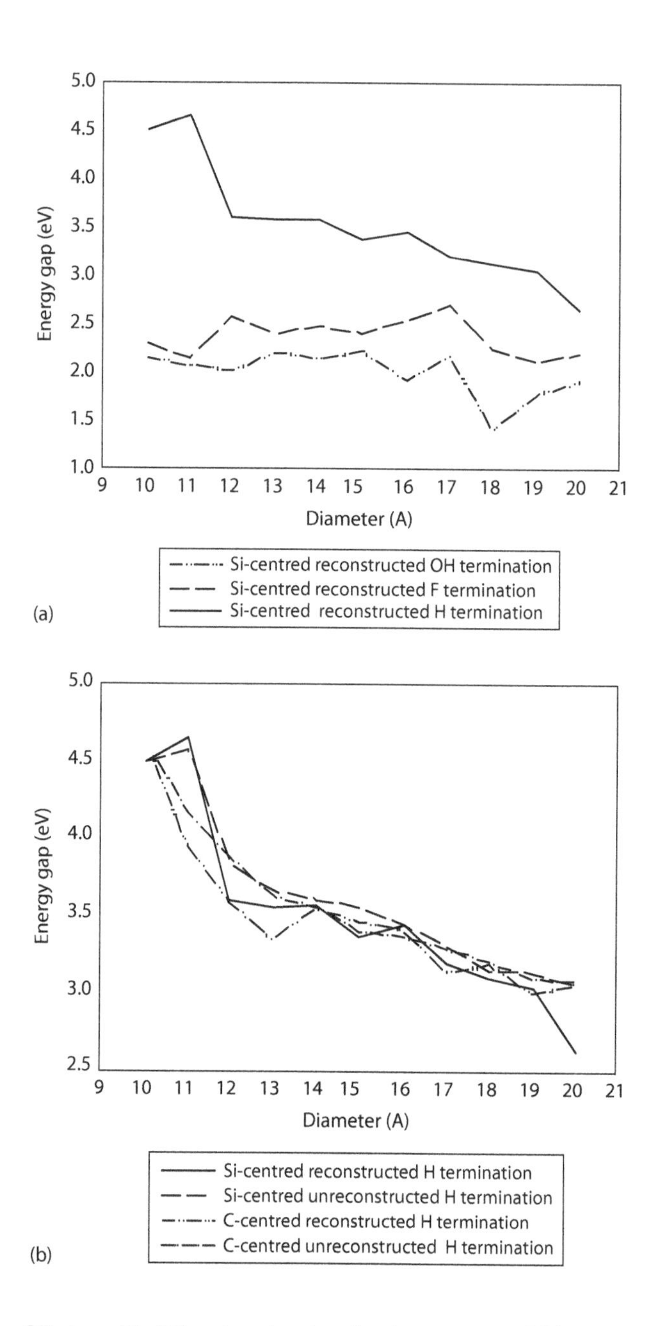

FIGURE 8.12
(a) Energy gap versus QD size with different surface termination groups and (b) energy gap of -H-terminated QD versus QD size by surface reconstruction and surface composition. Surface termination groups (OH, F, H) have a greater influence than size and surface reconstruction.

the onset of quantum confinement effects, as only those SiC-QDs large enough to be in region B in Figure 8.13 are excited.

Experimental findings show that the interplay between surface reconstructions and terminating species is critical below the threshold diameter of <30 Å. In the present context, the threshold size can be defined in a similar fashion, when the QD core–related JDoS start to dominate over those that are surface related.

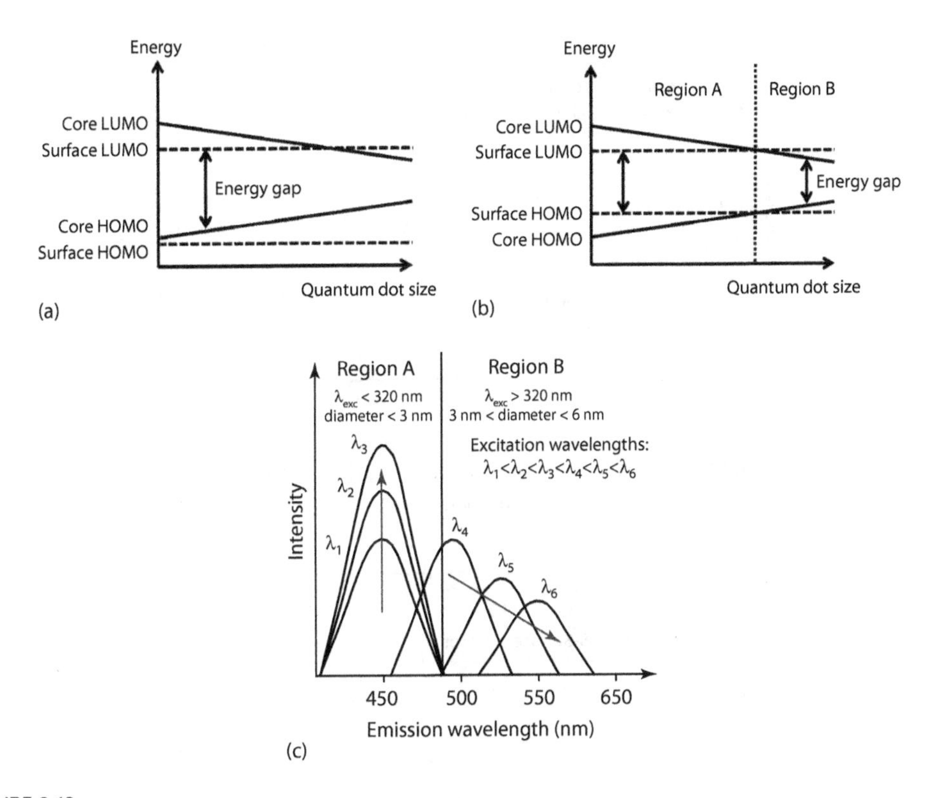

FIGURE 8.13
Illustration of core and surface states related to HOMO and LUMO as a function of size in surface reconstructed 4H-SiC-QD for (a) H-termination and (b) OH- or F-termination with the respective regions A and B. (c) A simplified dual-feature photoluminescence spectrum of SiC-QDs in regions A (when surface states dominate) and B (when core-related states dominate).

To illustrate the relative roles of the surface and core states as simulated in this study, the JDoS contours related to absorption onsets and at 10^3 eV^{-1} representative of the core (Figure 8.14) are plotted. Core atoms–related JDoS show a linear correlation with diameter^{-2} for both OH- and F-terminating groups, consistent with the particle in a sphere model [16,18]. However, the surface state–related JDoS do not increase. Based on the trend, it is predicted that the surface-related states will define the emission up to between 25 and 27 Å diameter for F- and OH-terminations (the intersections of the lines in Figure 8.14) [19]. For other surface groups, the intersection is expected to differ from these values, but the general feature of a dual-feature optical spectrum for systems involving main-group lone-pairs is expected.

8.3 Conclusions

In conclusion, 10–20 Å diameter SiC-QDs with OH-, F- and H-terminations have been investigated. In real experiments, the dual-feature PL was observed for all major polytypes comprising 3C-, 6H- and 4H-SiC. The SiC bulk bandgap increases with polytype in the order: 2.36 eV (525 nm), 3.00 eV (413 nm) and 3.23 eV (383 nm) for 3C, 6H and 4H,

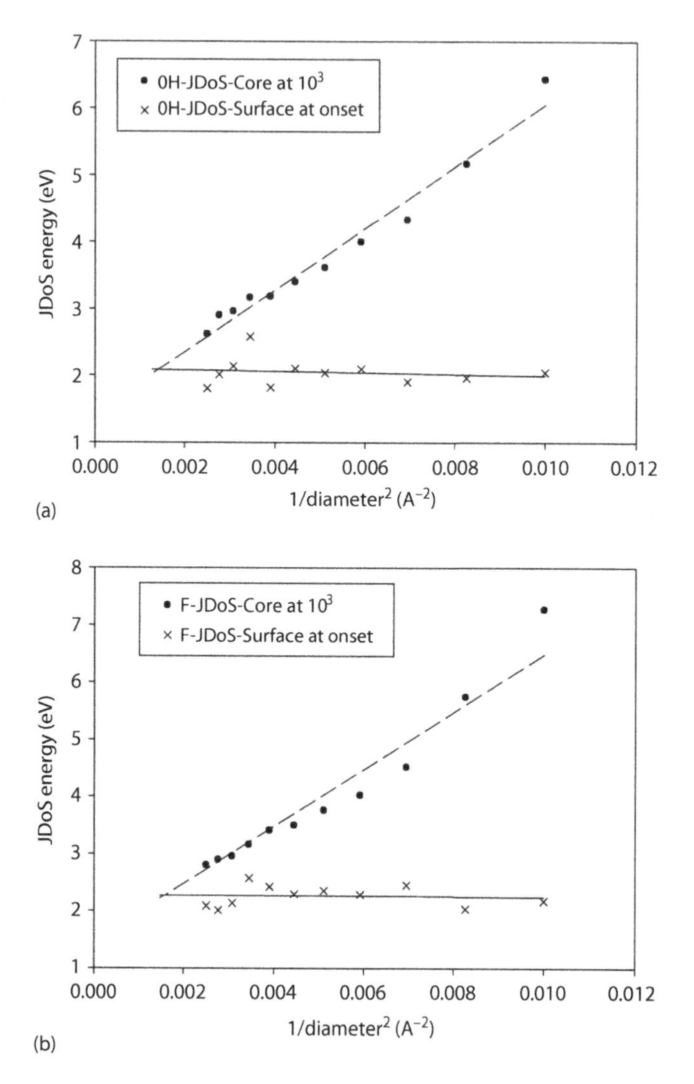

FIGURE 8.14
Change in energy of core-related and surface-related JDoS versus the inverse square of diameter for (a) OH-termination and (b) F-termination. Based on the trend, it is predicted that the surface-related states will define the emission up to between 25 and 27 Å diameter for F- and OH-terminations.

respectively. However, SiC-QDs derived from these polytypes exhibit a strikingly similar constant PL emission wavelength near 450 nm for sizes smaller than 3 nm, which is above bandgap for 3C but below bandgap for 6H and 4H-SiC. The role of polytypes was not observed in the optical properties of SiC-QDs particularly at smaller sizes. Beyond 3 nm, increasing the excitation wavelength results in red-shifting in the PL emission from 450 toward 560 nm, exhibiting PL that is characteristic of 3C-SiC regardless of the starting material. Due to these observations, polytypic phase transformation is frequently given as the explanation.

In this work, by examining the electronic structures of a wide range of compositions and surface treatments, it is concluded that the dual-feature PL spectra observed experimentally is best explained by considering the interplay of core and surface states, where surface states dominate at small scales (smaller than around 25 Å) and core states dominate at larger SiC-QDs.

It is found that the chemical nature of the surface termination groups is a critical factor in controlling the quantum confinement of 4H-SiC-QDs, and hence the optical properties. It is also noted that although surface reconstructions have an impact, they are generally much smaller than that arising from the choice of termination.

Finally, it is noted that SiC-QDs are fabricated using the wet etching methods employing HF, ethanol and water, which will naturally lead to a range of chemical terminations of the surfaces. It is therefore critical to understand the chemical nature of the surface termination, since control of the surface states will be essential in either exploiting or eliminating the size-dependent optical characteristics.

References

1. J. Fan, H. Li, J. Jiang, L. K. Y. So, Y. W. Lam and P. K. Chu, *Small*, 2008, 4, 1058–1062.
2. J. Y. Fan, X. L. Wu, H. X. Li, H. W. Liu, G. G. Siu and P. K. Chu, *Applied Physics Letters*, 2006, 88, 041909.
3. X. L. Wu, J. Y. Fan, T. Qiu, X. Yang, G. G. Siu and P. K. Chu, *Physical Review Letters*, 2005, 94, 026102.
4. X. Guo, D. Dai, B. Fan and J. Fan, *Applied Physics Letters*, 2014, 105, 193110.
5. J. Fan, H. Li, J. Wang and M. Xiao, *Applied Physics Letters*, 2012, 101, 131906.
6. X. L. Wu, S. J. Xiong, J. Zhu, J. Wang, J. C. Shen and P. K. Chu, *Nano Letters*, 2009, 9, 4053–4060.
7. D. Beke, Z. Szekrényes, I. Balogh, Z. Czigány, K. Kamarás and A. Gali, *Nanoscale*, 2015, 7, 10982–10988.
8. J. Zhu, S. Hu, W. W. Xia, T. H. Li, L. Fan and H. T. Chen, *Materials Letters*, 2014, 132, 210–213.
9. P. R. Briddon and R. Jones, *Physica Status Solidi (b)*, 2000, 217, 131–171.
10. M. J. Rayson and P. R. Briddon, *Physical Review B*, 2009, 80, 205104.
11. C. Hartwigsen, S. Goedecker and J. Hutter, *Physical Review B*, 1998, 58, 3641–3662.
12. J. P. Goss, M. J. Shaw and P. R. Briddon, in *Marker-Method Calculations for Electrical Levels Using Gaussian-Orbital Basis Sets*, ed. D. A. Drabold and S. K. Estreicher, 2007, vol. 104, pp. 69–93, Springer, Berlin.
13. D. R. Lide, *Handbook of Chemistry and Physics*, 2006, vol. 87, CRC Press, Boca Raton, FL.
14. M. Vörös, P. Deák, T. Frauenheim and A. Gali, *Applied Physics Letters*, 2010, 96, 051909.
15. X. H. Peng, S. K. Nayak, A. Alizadeh, K. K. Varanasi, N. Bhate, L. B. Rowland and S. K. Kumar, *Journal of Applied Physics*, 2007, 102, 024304.
16. G. Konstantatos and E. H. Sargent, *Colloidal Quantum Dot Optoelectronics and Photovoltaics*, 2013, Cambridge University Press, New York.
17. D. Dai, X. Guo and J. Fan, *Applied Physics Letters*, 2015, 106, 053115.
18. L. E. Brus, *The Journal of Chemical Physics*, 1983, 79, 5566–5571.
19. M. Rashid, A. K. Tiwari, J. P. Goss, M. J. Rayson, P. R. Briddon and A. B. Horsfall, *Physical Chemistry Chemical Physics*, 2016, 18, 21676–21685.

9

Molecular Beam Epitaxy of AlGaN/GaN High Electron Mobility Transistor Heterostructures for High Power and High-Frequency Applications

Yvon Cordier, Rémi Comyn, and Eric Frayssinet

CONTENTS

9.1 Introduction

Since it has been shown that an AlGaN/GaN heterostructure can generate a two-dimensional electron gas (2DEG) [1], GaN-based high electron mobility transistors (HEMTs) have been developed and are now established as the most interesting III-nitride electron devices for high-frequency power amplification as well as power switching. The reason for this is a combination of many factors [2]: the possibility of achieving high sheet carrier concentration in the 2DEG ($\sim 1 \times 10^{13}/cm^2$) with a high saturated velocity ($>1.5 \times 10^7$ cm/s), a quite high electron mobility (up to about 2000 cm^2/V.s at RT) and a breakdown electron field exceeding 3 MV/cm. Moreover, the chemical inertness and the wide energy bandgap of GaN guarantee the thermal stability of the devices.

The most commonly used growth techniques for III-nitride heterostructures are metal-organic vapor phase epitaxy (MOVPE) and molecular beam epitaxy (MBE), each of these techniques having its own advantages and drawbacks. While MOVPE is widely used due to its larger throughput and larger wafer size handling capability, MBE operates at lower temperatures under high vacuum with much less source products consumption and it is equipped with useful in situ inspection tools like reflection high energy electron diffraction (RHEED). MBE production tools for III-nitrides are rare, but many research reactors are used worldwide and are at the origin of demonstrations of optoelectronic and electronic devices. In the following sections, we will first discuss the advantages of using ammonia as the nitrogen source for MBE growth. This will be followed by a description of the growth of AlGaN/GaN HEMT heterostructures on GaN-on-sapphire templates and free-standing GaN, and on foreign substrates like silicon and silicon carbide. The behavior of transistor devices will be described and high-frequency power density results will be presented. For transistor applications on silicon, growth requires even more attention due to possible conductivity through the substrate. In this context, we will see the benefit of reducing the growth temperature of III-nitrides. Finally, we will describe a technological route for the monolithic integration of GaN with silicon electron devices in a complementary metal-oxide-semiconductor (CMOS) first approach.

9.2 Characteristics of Ammonia Source Molecular Beam Epitaxy

The main features of ammonia-MBE (NH_3–MBE) are the following. Ammonia thermally decomposes at the surface of the films at temperatures above 450°C [3]. The typical growth temperature for GaN is 800°C, while thick AlN films necessitate higher temperatures. Usually, AlGaN films can be grown in the 800°C–875°C temperature range, the optimum temperature depending on the Al content and the thickness of the film to be grown. Growth rates in the range of 0.5–1.5 μm/h are quite easy to achieve. However, compared to MOVPE (GaN grown typically at 1000°C), lateral growth is very limited for MBE and whenever a significant roughening develops, it is difficult to recover a flat surface, which is detrimental for electron transport in the channel of lateral devices such as HEMTs. The study described here has been carried out in a Riber Compact 21 MBE reactor equipped with an NH_3 gas injector [4,5] and a plasma source (ADDON RFN50/63) connected to an N_2 gas line [6,7]. The reactor configuration was optimized for uniform films on 2 in. diameter substrates [4]. Nevertheless, even with this configuration, the GaN thickness uniformity deviation is below 3% on 3 in. diameter substrates. As described by Vézian [8], when growth starts, ammonia-MBE-grown GaN exhibits a transition from a spiral growth mode to a mixed growth mode where two-dimensional nucleation is sufficiently active to give rise to kinetic roughening. According to this study, the spiral growth mode occurs at the beginning of the growth, thanks to the step flow growth mode in the presence of screw dislocations. As a result, a coarsening of growth mounds correlated with the decrease in the dislocation density is observed. Thick films (>2 μm) exhibit root mean square (RMS) roughness of the order of 4–5 nm (Figure 9.1a) and an increase in the correlation length saturating around 1 μm.

We discuss here the influence of the nitrogen source flow rate on the growth of GaN. The case of ammonia is quite simple. The optimum growth conditions are N-rich, i.e. they require a large ammonia flow rate (200 sccm in our MBE reactor). At the optimum growth temperature of 800°C and for a fixed Ga flux, the growth rate and the surface morphology slightly change while reducing the ammonia flow rate. Then, a first regime is reached with pits corresponding to the facet development of threading dislocations (TDs) opening at the surface. Moreover, a further reduction of the ammonia flow rate leads to a decrease in the growth rate as well as further development of the surface roughness due to the insufficient amount of available active nitrogen species compared to incoming Ga species [3,9]. The nitrogen plasma growth (PA-MBE) is very different. First, the most frequently reported growth conditions for smooth GaN films are temperatures near 720°C–730°C with fine-tuning of the nitrogen flow rate, radio frequency (RF) cell power and Ga flow rate in order to keep a thin metallic Ga film (2–3 monolayers) floating on the growing surface and a resulting surface roughness of typically 1 nm (Figure 9.1b). Out of this equilibrium, a rough film is obtained or an excess of gallium generates droplets at the surface [6]. In order to mitigate these effects, growth at a higher temperature (780°C–790°C) under a high nitrogen flow rate has been proposed [10], but the sensitivity to threading defects can make the growth more difficult to monitor. A crucial parameter for electron devices such as field-effect transistors is the residual doping level in the channel and the buffer resistivity. As seen in Figure 9.2, the ammonia flow rate has a crucial influence on the residual doping level [9]. Secondary ion mass spectroscopy (SIMS) indicates that silicon and oxygen are the main donors, even in the case of nitrogen plasma growth. Additionally, it seems that the flow rate itself has more influence than the nitrogen species, ammonia or nitrogen molecules. One last point concerns the growth of AlGaN alloys. The desorption of Ga and Al species is negligible when the HEMT AlGaN barrier is grown at 800°C with a large ammonia flow rate, so that composition and growth rate calibrations are facilitated.

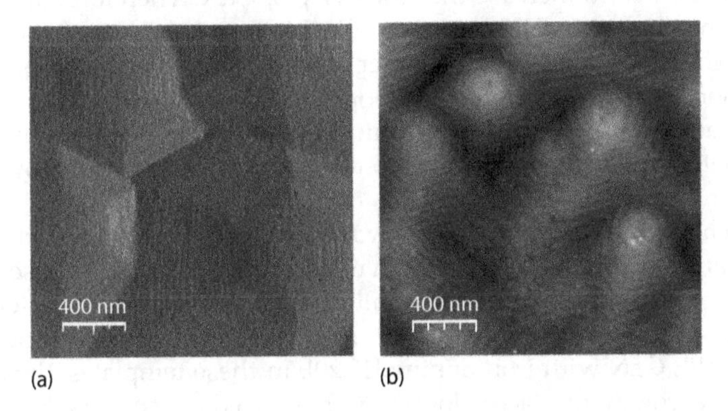

FIGURE 9.1
Tapping mode AFM view of the surface of a GaN layer grown with ammonia (a) and with nitrogen plasma (b). Left picture is a derivative mode image in order to highlight the monolayer height steps present at the surface.

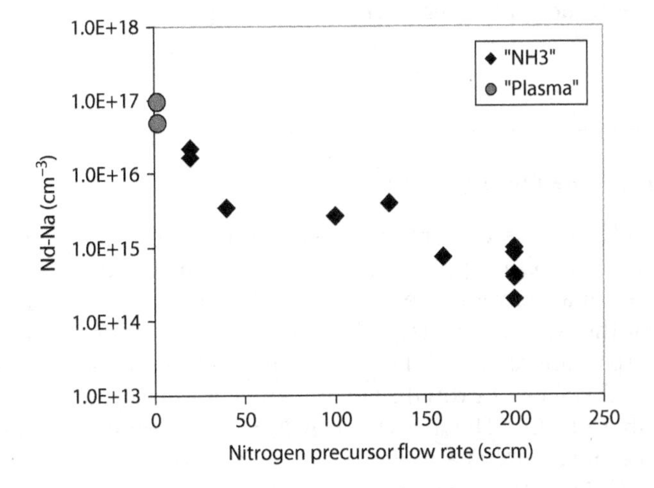

FIGURE 9.2
Donor concentration estimated by C-V as a function of the flow rate of the nitrogen precursor (NH_3 or nitrogen for plasma).

9.3 Homoepitaxy of GaN HEMTs

The availability of high resistivity GaN-on-sapphire templates and free-standing GaN substrates is very helpful for the development of HEMT structures on high crystal quality GaN. Unless significantly thicker GaN is regrown, the regrowth of GaN just replicates the substrate threading dislocation density (TDD) of roughly $1–5\times10^8/cm^2$ in the GaN-on-sapphire templates and $1–5\times10^7/cm^2$ in the free-standing substrates. Recently, bulk GaN substrates grown by the ammonothermal method with an ultralow dislocation density of $10^4 cm^{-2}$ and wafers up to 2 in. in diameter have been shown to be compatible with HEMT structures regrowth by MOVPE [11]. The growers have to face the problems related to the regrowth interface pollution. Even when the substrate/template is highly resistive/semi-insulating, the regrowth interface is contaminated with shallow donors such as silicon. The doping of GaN with elements such as carbon [12,13], beryllium [14], magnesium [15] and iron [16] is efficient to increase the resistivity of GaN. Therefore, the introduction of such elements in a regrown GaN layer is a solution to compensate the source of leakage in transistors. The incorporation rate can depend on the growth technique as well as the growth conditions. For instance, the incorporation of carbon from CBr_4 is highly dependent on temperature and not very efficient in the case of high-temperature MBE [17]. The ionization of methane is preferred for ammonia-MBE [13] since a high growth temperature is desirable.

However, such doping sources are not always available in growth reactors dedicated to HEMTs. Some authors proposed the growth of a thin AlN layer to upraise the conduction band at the regrown interface [18]. Another alternative is to develop GaN templates or substrates ready for epitaxial regrowth with a reduced electrical leakage. In our case, we have developed MOVPE GaN with iron doping [19,20]. In these templates, the amount of iron available at the regrowth interface is low enough to avoid unrecoverable surface roughening but sufficient to compensate the effect of silicon or oxygen contaminants. The transistors fabricated on such epi-ready insulating GaN-on-sapphire templates exhibit drain leakage currents as low as 10 μA/mm at $V_{ds}=10$ V when regrown by ammonia-MBE. As a high temperature favors the diffusion of iron, the leakage can be further reduced by about three orders of magnitude when regrown by MOVPE [20].

9.4 Heteroepitaxy of GaN HEMTs

The growth of GaN HEMTs on foreign substrates requires both insulating and stress-mitigating buffer layers. On sapphire, the difference in thermal expansion coefficients (TEC) induces a residual compressive strain in GaN, which is responsible for a noticeable convex bowing of the wafers. The large lattice parameter mismatch is responsible for a large number of dislocations, which helps in trapping the carriers related to the residual doping at the initial stages of growth. Further thickening of the GaN buffer drastically reduces the number of TDs. The growth of AlN is an alternative to increase the buffer layer resistivity because AlN is a wider bandgap material and the polarization electric field depletes the GaN/AlN regrowth interface from eventual free carriers. In the present study, we chose to grow with NH_3–MBE a HEMT structure on a 1 μm thick MOVPE AlN-on-sapphire template. Contrary to sapphire, GaN grown on SiC or on Si suffers a tensile strain induced by the TEC mismatch. To compensate this effect and the associated risk of layer cracking, GaN is grown on an AlN nucleation layer to benefit from an initial compressive strain related to the 2.5% lattice parameters mismatch strain between both materials. Furthermore, AlN appears to be a useful solution to avoid reactions between gallium and the silicon substrate [21]. But, if the high temperature (800°C) of the NH_3–MBE favors dislocation bending, interactions and elimination, it also promotes the relaxation of this strain [22] so that more complex structures with additional intercalated AlN layers have been grown to obtain 2 μm thick, crack-free HEMT structures on Si(111) [23] and SiC.

9.5 Electrical Properties

Figure 9.3 depicts the main kinds of structures we have grown. Table 9.1 summarizes the best results we obtained on these kinds of structures: the 2DEG carrier density and the low field electron mobility assessed by the Hall effect as well as the buffer residual donor density obtained by exploiting the capacitance-voltage (CV) measurements beyond pinch-off. The TDD assessed by atomic force microscopy (AFM) or x-ray diffraction (XRD) is

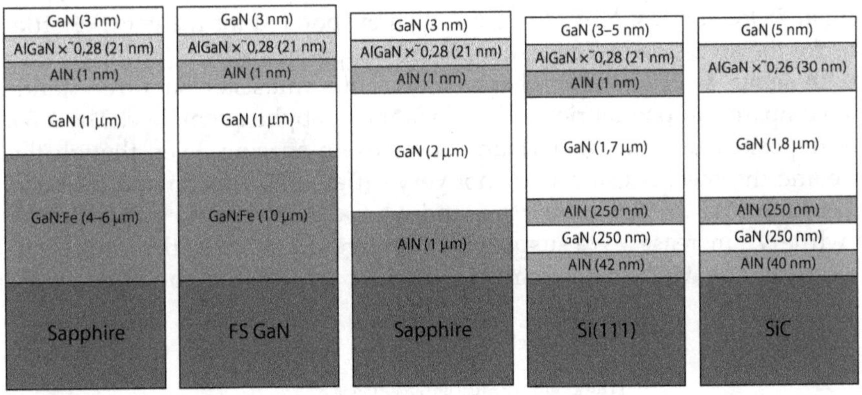

FIGURE 9.3
Schematic cross section of the typical HEMT structures grown by NH_3–MBE.

TABLE 9.1
2DEG Carrier Density and Mobility, Buffer Residual Donor Density and Threading Dislocation Density (TDD) in the Studied HEMTs

	GaN/Al_2O_3	GaN FS	AlN/Al_2O_3	H-SiC	Si(111)
Al content	28%	28%	28%	26%	28%
Ns @ 300 K (cm^{-2})	10^{13}	10^{13}	9×10^{12}	[a]8×10^{12}	9×10^{12}
μ @ 300 K (cm^2/Vs)	2,080	2,140	2,085	[a]1,769	2,000 ([a]1,780)
μ @ <10 K (cm^2/Vs)	30,000 ([a]12,500)	—	—	[a]8,740	12,700 ([a]7,880)
Nd-Na (cm^{-3})	3×10^{13}	3×10^{13}	$<1\times10^{13}$	3×10^{13}	$1–3\times10^{14}$
TDD (cm^{-2})	$0.4–1\times10^9$	$1–2\times10^7$	2.5×10^9	3×10^9	$3–4\times10^9$

[a] No AlN spacer between the AlGaN barrier and the GaN channel.

also reported. At room temperature, electron mobility is mainly limited by optical phonon scattering, interface roughness and alloy scattering. Due to the high carrier density, the screening of TD fields is quite efficient and reduces the influence of the TDD with respect to other scattering mechanisms. Furthermore, it is obvious that the insertion of a thin AlN spacer at the AlGaN/GaN interface enhances room temperature mobility by greater than 200 cm²/V.s. Mobility reaches 2000 cm²/V.s when the TDD is sufficiently low (below 5×10^9 cm^{-2}). A consequence of the good crystal quality is that the carrier density in the 2DEG is mainly determined by the barrier thickness and the Al molar content as well as the GaN cap thickness. At low temperature, the TDD reduction has a noticeable impact on electron mobility enhancement.

9.6 Transistors Evaluation

In order to evaluate the potentialities of these structures, test devices including transmission line model (TLM) and isolation patterns, diodes and transistors have been fabricated by photolithography. The device process starts with mesa definition by a reactive ion etching (RIE) step in a $Cl_2/Ar/CH_4$ mixture. After a short RIE etching, TiAlNiAu stacks are deposited by e-beam evaporation. The ohmic contacts are achieved after rapid thermal annealing at 750°C for 30 s. NiAu films are then evaporated for the gate Schottky contact as well as for the access pads. These devices are not passivated.

Figure 9.4 shows the DC output characteristics of a transistor on a free-standing GaN substrate compared to one fabricated on a GaN-on-sapphire template. The $2.5 \times 150\,\mu m^2$ gates are deposited in a nominal 11 μm source drain spacing. Even though the contact resistance and the sheet resistance are not very different (0.9 Ω.mm and 280 Ω/sq on sapphire, 0.7 Ω.mm and 270 Ω/sq on free-standing GaN), one notices that the I-V curves diverge while V_{gs} increases. We suspect here a large influence of the device self-heating on the electrical behavior. To confirm this, we first analyze the output characteristics of

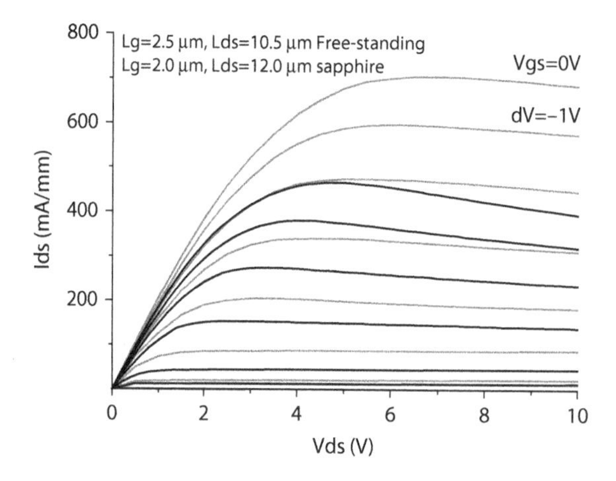

FIGURE 9.4
DC output characteristics of AlGaN/GaN HEMTs regrown on GaN/sapphire template and on a free-standing GaN substrate. Device dimensions are W = 150 μm, L_g = 2.0–2.5 μm and L_{sd}~10.5–12.0 μm.

transistors fabricated with similar dimensions on the different kinds of substrates. We then define a relative DC drain current collapse R as

$$R = (Id_{knee} - Id_{20V})/Id_{knee}$$

with Id_{knee} the maximum drain current and Id_{20V} the drain current obtained at $V_{ds}=20\,V$ for a given gate bias V_{gs}. This is illustrated in the insert of Figure 9.5 on a transistor fabricated on Si(111).

When plotting the relative drain current collapse as a function of the drain current, one clearly sees the increase of the collapse with the drain current, but one also notices the similarity of the collapse for transistors on sapphire with AlN or GaN buffer. Furthermore, the transistors on SiC present the smaller collapse while an intermediate one is obtained on silicon and free-standing GaN. Such a collapse seems not related to the defect density reported in Table 9.1, but to the thermal conductivity of the substrate. However, one type of transistor developed on silicon presents a relative collapse as low as on SiC (dashed line in Figure 9.5). The peculiarity of this transistor is the unusual buffer layer thickness (0.5 µm GaN on a 0.5 µm AlN/GaN stress-mitigating stack).

To go further, pulsed measurements have been performed on transistors fabricated on sapphire and Si(111) [24]. Pulsed current-voltage measurements (600 ns, 100 Hz) performed on devices on sapphire and on silicon substrates confirm the primary importance of thermal effects (Figure 9.6). As expected, the pulsed drain currents obtained after biasing at ($V_{gs}=0\,V$, $V_{ds}=0\,V$) quiescent point are systematically higher than when recorded under DC conditions. The difference increases dramatically on sapphire (Figure 9.6c). However, as shown in Figure 9.6b, it appears that the device on silicon with the thicker GaN buffer layer (here 1.7 µm GaN on a 0.5 µm AlN/GaN stress-mitigating stack) and therefore with a reduced TD density shows an increase of the pulsed drain current collapse with quiescent points at ($V_{gs}=-8\,V$, $V_{ds}=0\,V$) (gate lag) and ($V_{gs}=-8\,V$, $V_{ds}=20\,V$) (drain lag). Note that we

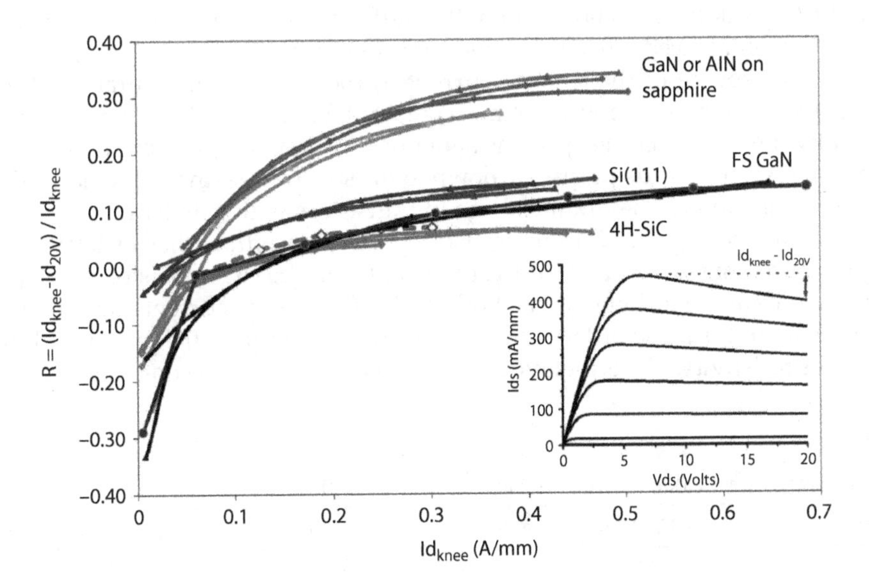

FIGURE 9.5
Relative drain current collapse measured in DC conditions. Device dimensions are $W=150\,\mu m$, $L_g=3–4\,\mu m$ and $L_{sd}=9–12\,\mu m$.

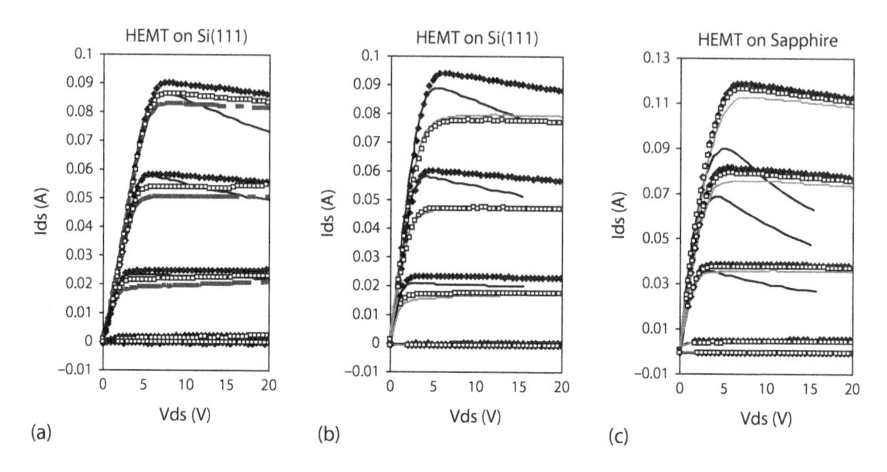

FIGURE 9.6
DC and pulsed measurements performed on HEMTs on silicon with thin buffer (a), thick buffer (b) and on sapphire substrate (c). On silicon, bias conditions are black line (DC, DC), diamonds (0V, 0V), squares (−8V, 0V), gray line (−8V, 20V). On sapphire, bias conditions are black line (DC, DC), diamonds (0V, 0V), squares (−4V, 0V), gray line (−4V, 10V).

previously observed a similar trend for the DC current collapse (Figure 9.5). Thus, it seems that in these bias conditions, the presence of dislocations helps in achieving a more stable device operation. However, GaN transistors generally operate under more severe bias conditions. The development of an efficient surface passivation drastically mitigates the current collapse. For instance, the combination of an N_2O plasma surface treatment with the deposition of SiN/SiO_2 passivation layers leads to small dispersions, as recently demonstrated on 0.25 μm gate transistors developed on similar structures with a resulting state-of-the-art $L_g \times Ft = 15$ GHz.μm product and a power density of 1.5 W/mm at 40 GHz [25].

A critical parameter for transistors and especially power transistors is the buffer resistivity and the breakdown voltage. Often, the buffer resistivity directly impacts the drain leakage current evidenced when the transistor is in the off-state. Sometimes, the gate leakage itself is the source of the leakage current in the off-state configuration. The buffer leakage current measured between ohmic contacts fabricated on HEMT active layers and separated by the mesa etching spacings superior to 10 μm are reported in Table 9.2. With such spacings, electric fields propagate down to the substrate so that all regions of the buffer layer can contribute. The drain leakage in transistors with gate lengths of 2–3 μm and source-to-drain spacings of 12–13 μm is also reported for a drain bias of 100 V and a gate bias V_{gs} of approximately 2 V beyond the threshold voltage. The transistor development W is 150 μm. It is clear from Table 9.2 that both the buffer leakage current and the drain leakage current are of the same order of magnitude, which confirms the primordial influence of the buffer resistivity. Nevertheless, a closer look shows that in some cases, the leakage

TABLE 9.2
Buffer and Transistor Leakage Currents Measured at $V_{ds} = 100$ V

Substrate	GaN/Al_2O_3	AlN/Al_2O_3	SiC	Si(111)	Si(111)
Buffer layer	GaN:Fe	GaN/AlN	GaN/AlN	GaN/AlN	AlGaN/AlN
Buffer leakage at 100 V	<10 μA/mm	<40 μA/mm	<200 μA/mm	20–100 μA/mm	20 μA/mm
Transistor *W* = 150 μm leakage at 100 V	<70 μA	1 mA/mm @ 85 V	<30 μA	4–120 μA	<30 μA

currents of the transistors at pinch-off systematically overpass the buffer leakage currents. This is probably due to the increase in electric field crowding in the vicinity of the gate. The device on Fe-doped GaN-on-sapphire is very satisfying in terms of trade-off between crystal quality and drain leakage, contrary to the device on AlN-on-sapphire whose breakdown voltage is below 100 V, as illustrated in Figure 9.7a.

The transistor leakage on hexagonal SiC better scales with buffer leakage, indicating a more stable behavior with respect to the electric fields and the self-heating. However, the buffer on SiC is not the most resistive, as shown in Figure 9.7a. This can be due to the possible diffusion of silicon from the substrate, as mentioned by Hoke [26] in the case of PA-MBE-grown AlN. The present buffer layer contains a thin AlN layer and a 0.5 μm thick GaN/AlN stack (Figure 9.3) and an increase by two orders of magnitude of the buffer leakage current is obtained in the absence of the AlN interlayer. Thickening of the AlN layers

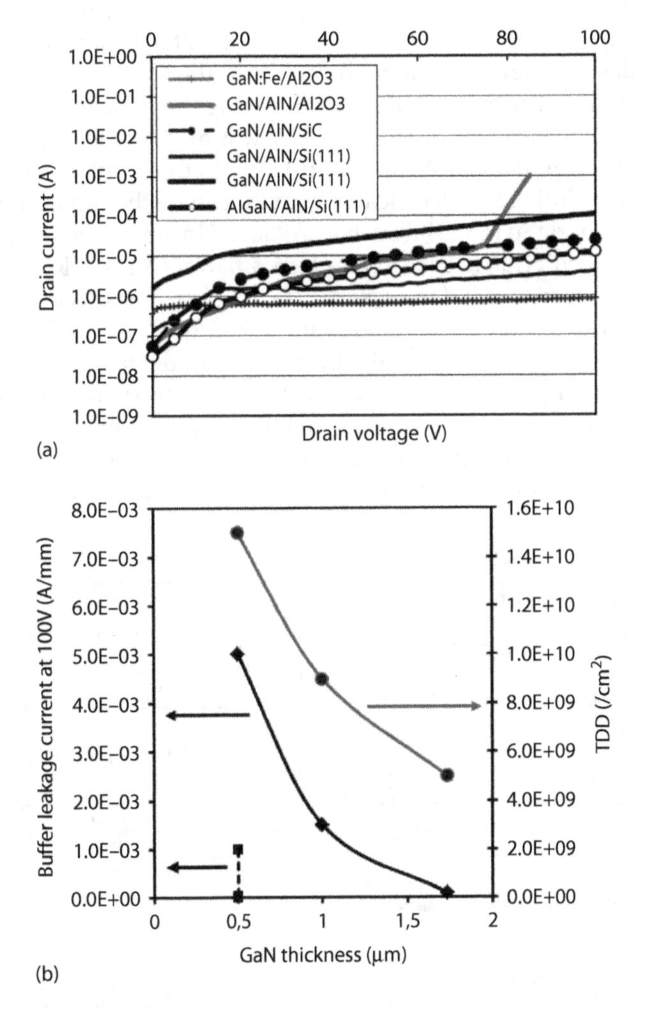

FIGURE 9.7
Transistors drain leakage current under pinch-off conditions (a). Buffer leakage current at 100 V and threading dislocation density in HEMT structures grown on Si(111) with different GaN buffer thicknesses (b). Circles represent TDD, diamonds represent the leakage for buffers on silicon with interlayers and squares represent the leakage for buffers on silicon without interlayers.

as well as reducing the growth temperature of these layers are possible ways to enhance the electrical resistance of buffer layers on SiC.

Our most studied devices are grown on Si(111). Figure 9.7a shows that depending on the growth conditions, the buffer resistivity can vary significantly. Moreover, the GaN thickness in structures like the one described in Figure 9.3 can influence the buffer resistivity in an expected way, increasing the leakage current while reducing the GaN thickness. As this occurs while the dislocation density increases, we suspect that the particular arrangement of dislocations (dislocation loops, bended dislocations, number of screw-type dislocations) in the GaN grown compressively strained on the AlN interlayer is responsible for this. Indeed, for 0.5 μm GaN thickness, the buffers grown without interlayers on Si(111) present much lower leakage currents (Figure 9.7b), but with clearly worse crystal quality (TDD > 2×10^{10}/cm²) and resulting electron mobility of the order of 1300 cm²/V.s. Showing the leakage current recorded at a given bias only is sometimes not enough to describe the behavior of a particular buffer layer. In [27], Leclaire et al. describe the effects of reducing the HEMT buffer layer thickness on the main current-voltage characteristics of transistors and Schottky diodes in order to evaluate the effect on the leakage both through the buffer and through the HEMT barrier layer. It is worth noting that what was deduced previously from the amplitude of the leakage current at 100 V was also true for the breakdown voltage defined at a leakage current of 1 mA/mm. Returning to thick buffer layers with interlayer stacks, a solution to stabilize the electrical resistivity can be the replacement of GaN in the buffer with a larger bandgap material such as AlGaN. However, the growth of such a layer on silicon is more difficult due to the smaller lattice parameter of AlGaN and the resulting stress. Nevertheless, we succeeded in growing up to 1.5 μm crack-free AlGaN with 5%–10% of Al and with a low crystal quality degradation [25,28]. This enabled more stable drain leakage currents below 30 μA at V_{ds} = 100 V in the studied devices.

These devices are not passivated so they are not able to sustain large drain and gate biases. Despite a drain leakage current that can be as low as a few microampere, the breakdown voltage obtained in air is around 200 V (234 V at best) with destruction of the metal contacts. The possible leakage current paths are presented in Figure 9.8. Electrons can be

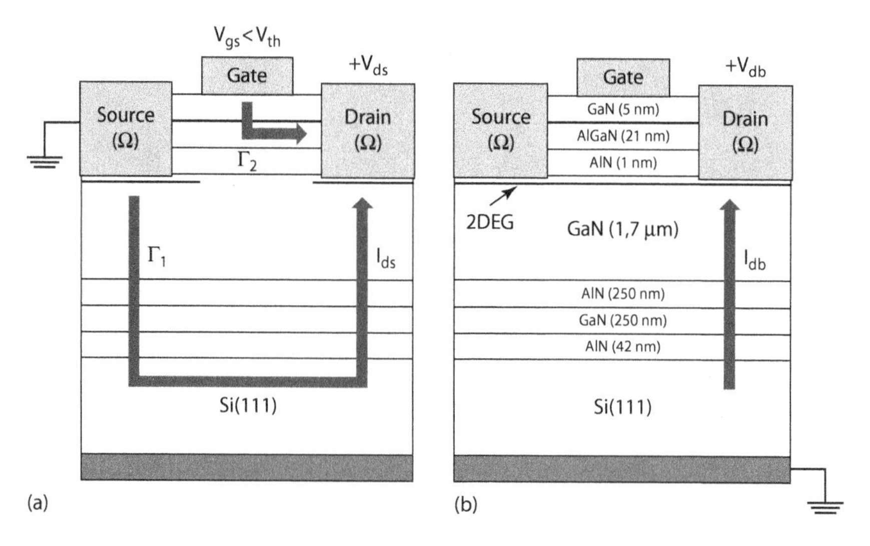

FIGURE 9.8
Leakage paths in the transistor on silicon under pinch-off conditions (a). Vertical leakage path in the device under vertical current configuration (b). (Adapted from Pérez-Tomás, A., et al., *J. Appl. Phys.*, 113, 174501, 2013.)

injected from the source to the substrate through the buffer layer and the nucleation layer, so that a current can flow via the buried part of the buffer or via the substrate (when not insulating) and then reach the positively biased drain contact. When the leakage arising from the gate (path Γ_2) is small enough, drain current leakage at pinch-off can follow path Γ_1, so that measuring the vertical leakage is helpful to identify the detailed mechanisms of the drain leakage.

The behavior of HEMT devices on Si, GaN-on-sapphire template and free-standing GaN substrates has been studied in this vertical configuration by Pérez-Tomás [29]. The drain is positively biased while the substrate is grounded. In this configuration, the leakage through the device on Si is a combination of a Poole–Frenkel (trap assisted) and a resistive conduction with an estimated resistance of 72 kΩ up to a soft breakdown at 420 V. We note here that the total buffer thickness is less than 2.3 µm, leading to an average breakdown field of more than 1.8 MV/cm. Due to the insulating sapphire, more than 350 V is necessary to notice a vertical leakage current with a resistive conduction (85 kΩ) up to 1 kV without any irreversible breakdown phenomena. This similar resistance is explained by the similar growth conditions of the thick GaN grown by MBE either on the silicon substrate or on the GaN-on-sapphire template. The difference in terms of TDD seems insufficient to have a noticeable effect on the vertical conduction. On the other hand, the device on the free-standing GaN substrate follows a resistive behavior only, with a resistance of 7 MΩ up to the destructive breakdown at 840 V. The absence of AlN layers in the buffer and the low TDD are efficiently compensated by the presence of iron in the 10 µm thick GaN layer. However, when the temperature is increased, the vertical leakage current on Si and free-standing GaN rapidly increases with an activation energy of 0.35 eV. In reverse bias conditions (when the drain is negatively biased), the leakage on the silicon substrate is larger with an activation energy of 0.1 eV and a resistance of 20 kΩ, due to the band lineup between the materials. A detailed analysis performed on silicon shows that for a broad range of drain bias, the gate-to-source leakage current is more than one order of magnitude lower than the leakage current between the drain and the source, which itself is the same as the vertical leakage current flowing from the substrate to the drain. This suggests that the preferential leakage path is Γ_1 and then any enhancement of the resistivity of the nucleation layer region will benefit the transistor power DC operation. Before we discuss this point, we come back to recent results we obtained after growing a HEMT with a buffer containing only AlGaN and AlN layers, the only GaN layers being part of the top active region composed of a 10 nm channel and a 0.5 nm cap, as shown in Figure 9.9. The devices are not passivated and $2 \times 150\,\mu m^2$ gate transistors show DC characteristics with a leakage current limited to 1 µA/mm. On the other hand, a monotonous increase of the vertical leakage current is noticed in these devices until the destructive breakdown at bias that can reach 740 V (Figure 9.10), which is particularly high for a structure with a total thickness below 2 µm and results in an average breakdown field of 3.8 MV/cm.

The present results confirm that MBE can provide materials able to sustain high electric fields even when grown on silicon, and show that small Al molar fractions in the buffer layer drastically enhance the breakdown voltage. Compared to GaN, this is probably due to the larger electrical resistivity of AlGaN linked to the increase in bandgap energy, but it may also be linked to the surface quality that presents a lower roughness as evidenced by AFM (generally with RMS below 2 nm vs 4–5 nm for GaN) and which may translate into less propensity to develop defects like hillocks and V-shaped defects, often cited as probable current leakage paths.

As evocated previously, any enhancement in the properties of the AlN nucleation layer on silicon should result in a benefit for the transistors. An interesting example concerns the

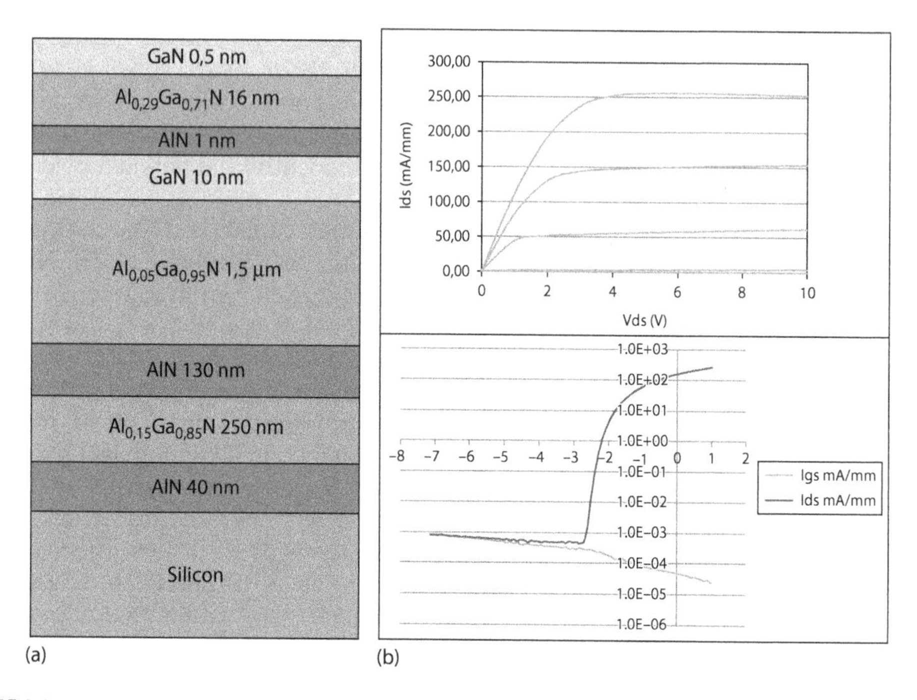

FIGURE 9.9
Schematic cross section of a HEMT structure grown on silicon without any GaN layer in the buffer layer (a); DC output (V_{gs}: 0 V; −4 V by −1 V steps) and transfer characteristics of a transistor (b).

effect of the AlN growth temperature. It is well known that compared to MOCVD, MBE is able to produce compact and smooth AlN nucleation layers at lower growth temperatures. As illustrated by Freedsman et al. [30], it seems more delicate to fabricate reliable, smooth, insulating AlN layers with MOCVD, especially at low temperature. To illustrate the benefits of MBE, the surface morphology of AlN nucleation layers grown on silicon is shown in Figure 9.11. All three samples have been grown well below 1000°C (920°C for

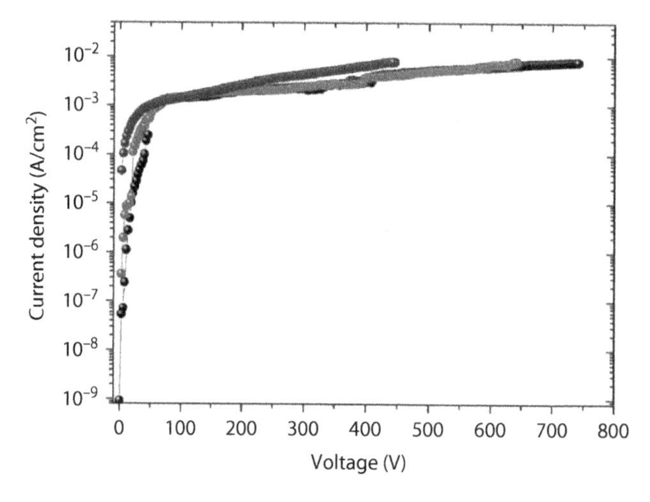

FIGURE 9.10
Leakage current recorded in the vertical configuration for a HEMT structure grown on silicon without any GaN layer in the buffer layer. (Courtesy of Dogmus E., et al., IEMN-CNRS.)

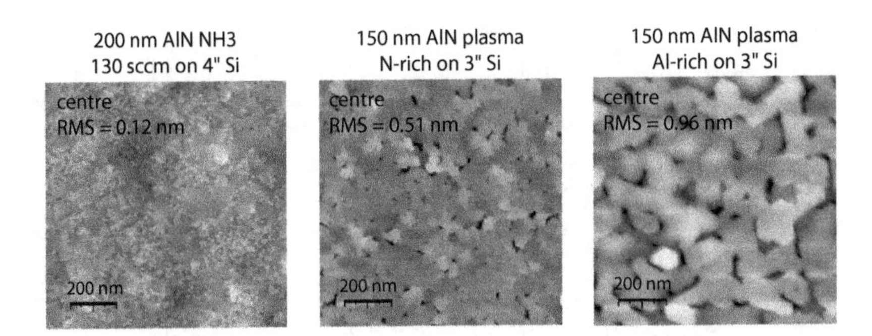

FIGURE 9.11
Tapping mode AFM view of the surface of AlN layers grown with ammonia-MBE and plasma-assisted MBE in the N-rich and Al-rich regimes.

ammonia-MBE and 830°C for N-rich and Al-rich plasma-assisted MBE) and present rather smooth surfaces with RMS roughness below 1 nm without any pits. The thickness is in the 100–200 nm range but similar morphology is obtained for films as thin as 10 nm. However, if temperatures higher than 900°C permit the slight enhancement of the crystal quality in terms of XRD line widths, other phenomena like impurities diffusion/incorporation can mitigate the enthusiasm to grow AlN at higher temperatures; on the contrary, it leads to considering the feasibility of growth at lower temperatures, as Comyn did [31]. The latter demonstrated that within HEMT buffer layer structures grown by ammonia-MBE, it was possible to grow AlN layers at 830°C with limited degradation of the crystal quality. Within a thin buffer HEMT structure (0.5 µm GaN/0.2 µm AlN), Comyn noticed a decrease of the vertical buffer leakage current by two orders of magnitude (Figure 9.12) and an increase of the lateral breakdown voltage while reducing the AlN growth temperature from 920°C to 830°C.

A SIMS analysis of the sample with GaN grown at 800°C and AlN grown at 920°C is shown in Figure 9.13a, while the same structure with AlN grown at 830°C is shown in Figure 9.13b. The only noticeable difference is the silicon profile, which drops more sharply within the AlN nucleation layer grown at low temperature to stabilize at a level about half that measured for AlN grown at 920°C. Despite the huge bandgap of AlN (6.1 eV), silicon

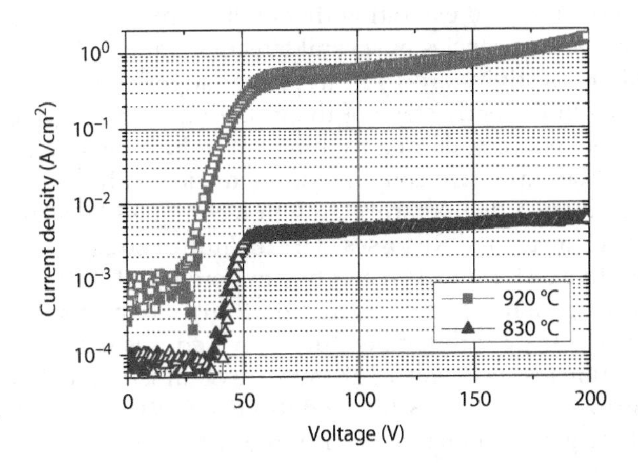

FIGURE 9.12
Leakage current recorded in the vertical configuration for the 500 nm GaN buffer layers with AlN grown at 830°C and 920°C.

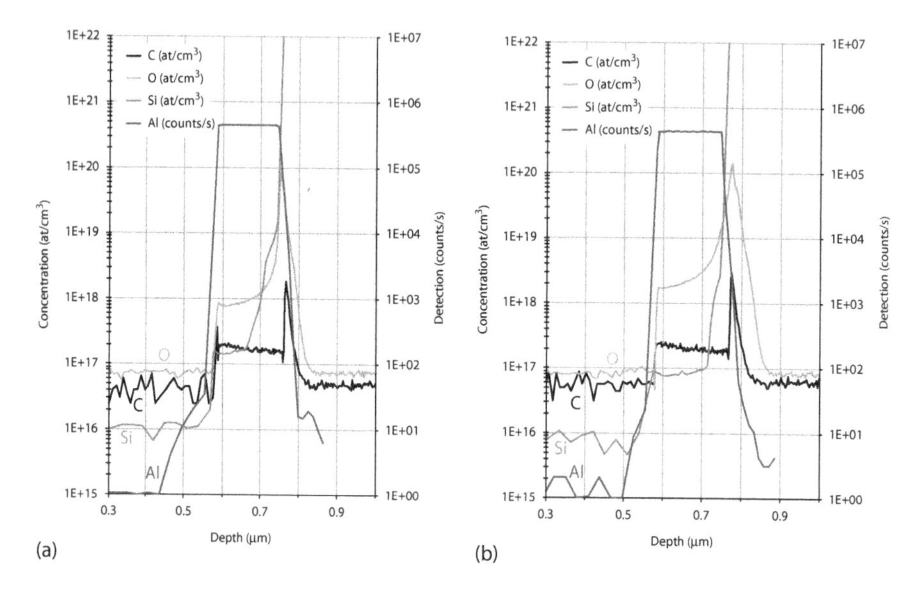

FIGURE 9.13
SIMS profiles recorded for carbon, oxygen, silicon and aluminum elements within buffer layer structures with AlN grown at (a) 920°C and (b) 830°C.

is still an active donor in this material [32], which may explain the difference in electrical resistivity. Another possibility is that AlN grown at a reduced temperature may contain a larger number of punctual defects able to trap any carrier.

9.7 RF Devices

Both the buffer and the substrate resistivity are crucial for RF performances. Even when it is possible to extract the losses in access regions of transistors (de-embedding) [33], performing functional devices and exploiting them within monolithic circuits require minimal losses especially at frequencies of several tens of gigahertz (GHz). For these reasons, substrates with high resistivity are used (typically superior to 3 kΩ.cm). However, the presence of Al atoms in the early stage of the growth and the high temperature usually employed to obtain a good crystal quality film can reduce the overall substrate resistivity. In order to limit the capacitive coupling of the device with the substrate, a thick buffer layer and maybe a lower temperature growth technique like MBE are preferable. For instance, Lecourt et al. measured RF losses of 0.4 dB/mm at 50 GHz on a 2.5 μm thick structure grown on resistive Si(111) ($\rho > 1$ kΩ.cm) by ammonia-MBE [34]. More, it is interesting to note that the propagation losses measured on thin GaN buffer layers grown on AlN are not so sensitive to the growth temperature provided it is below 1000°C. As shown in Figure 9.14, propagation losses below 0.5 dB/mm are obtained up to 70 GHz in 0.5 μm GaN with 0.2 μm AlN grown by MBE at 830°C and 920°C on Si(111) ($\rho > 3$ kΩ.cm). On the other hand, the effect of the growth temperature is more critical when GaN/AlN structures are regrown by MOCVD.

Figure 9.15 shows the state of the art in terms of power densities obtained at 40 GHz for devices with varying gate lengths. These results have been obtained with bias points

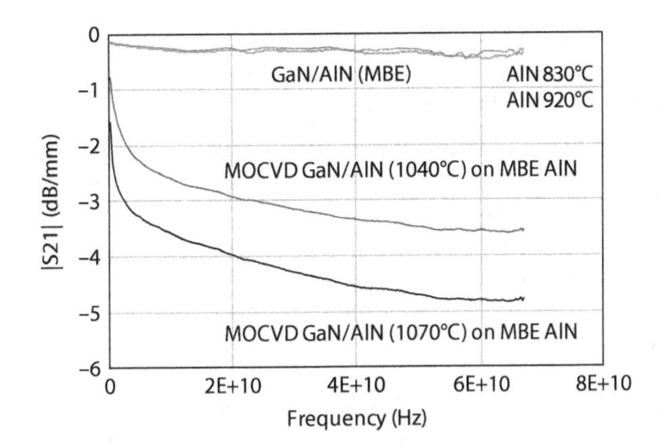

FIGURE 9.14
RF propagation losses in GaN/AlN buffer layers grown on silicon. (Courtesy of Defrance N., et al., IEMN-CNRS.)

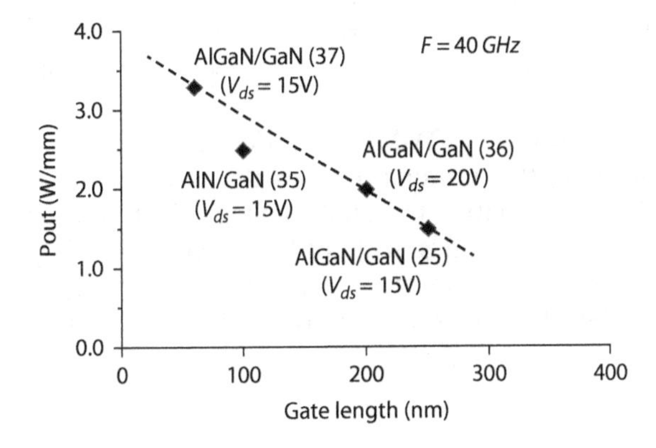

FIGURE 9.15
Gate length dependence of the state-of-the-art output power densities obtained at 40 GHz with Al(Ga)N/GaN HEMTs on silicon substrates.

V_{ds} = 15–20 V and the dependence of the performance is almost linear. The power density results of 1.5 W/mm with 250 nm gates [25] and 3.3 W/mm with 60 nm gates [37] have been obtained with AlGaN/GaN HEMTs grown by ammonia-MBE on silicon. The two intermediate results [35,36] have been obtained with MOVPE-grown structures. As expected, the achievable output power density enhances while lowering the operation frequency (Table 9.3). Results on GaN substrates are still quite rare but at least as good as on silicon. Devices on SiC substrates clearly benefit from a better thermal conductivity, which is crucial for high-frequency power devices. It is often difficult to find direct comparisons between devices made with the same technology on structures grown with different techniques. Palacios did this for HEMT structures grown by PA-MBE and MOVPE on SiC [45]. Ammonia-MBE and PA-MBE produce similar results that are not very far from those obtained using MOVPE. However, the long-term reliability of HEMTs is still an issue, in particular the degradation induced by the electrical stress. An interesting study has been reported by Puzyrev [46] on HEMTs grown either by PA-MBE under Ga-rich and N-rich conditions or by ammonia-MBE or MOVPE (both N-rich conditions). According to this study, hydrogenated Ga-vacancies loosing hydrogen atoms during the stress with hot electrons are responsible for a positive threshold voltage shift and a transconductance

TABLE 9.3

High-Frequency Power Density Results Reported on Silicon, GaN and Silicon Carbide Substrates with Different Growth Techniques

Substrate	Growth Technique	Gate Length (μm)	Frequency (GHz)	Drain Bias (V)	Output Power Density (W/mm)	References
Si	NH_3-MBE	0.25	40	15	1.5	[25]
Si	NH_3-MBE	0.06	40	15	3.3	[37]
Si	NH_3-MBE	0.25	18	35	5.1	[38]
Si	NH_3-MBE	0.3	10	40	7	[39]
GaN	MOVPE	0.15	10	25	4.8	[40]
				50	9.4	
GaN	PA-MBE	0.5	10	25	4.8	[41]
SiC	NH_3-MBE	0.6	10	30	6.3	[42]
				48	11.2	
SiC	PA-MBE	0.25	10	31	6.3	[43]
SiC	NH_3-MBE	0.2	30	40	6.5	[44]
SiC	PA-MBE	0.16	40	30	8.6	[45]
SiC	MOVPE	0.16	40	30	10.5	[45]

degradation in samples grown by PA-MBE (Ga-rich and N-rich). On the other hand, in the case of ammonia-MBE, the author proposes that the dehydrogenation of N-antisite defects is responsible for a negative threshold voltage shift and a reduced transconductance change due to the lower charge state of the defects. The degradation of MOCVD-grown devices is similar to that of ammonia-MBE-grown devices, showing that ammonia favors antisite-related defects themselves passivated by hydrogen.

9.8 Monolithic Integration with Si CMOS Technologies

Compared with MOVPE, another advantage of MBE is the possibility to grow structures at temperatures compatible with the monolithic integration of already processed devices like silicon MOSFETs. This is an alternative to approaches relying on the transfer of GaN epilayers [47] or even devices from the growth substrate to the silicon wafer [48]. In this CMOS first approach, the Si devices can be fabricated in a conventional silicon device process line with no risk associated with contamination of III-V elements. Metal connections for CMOS circuits, e.g. Al, are not present in this step to avoid fusion and are fabricated later. The silicon devices are protected with dielectric masks, and windows are opened to define GaN growth areas.

Questions arise on the choice of the silicon substrate crystal orientation, which has to be compatible with both silicon CMOS device fabrication and good quality III-nitrides. A first possibility is a special silicon-on-insulator (SOI) substrate with an Si(100) top layer for CMOS and a thick Si(111) bulk substrate for IIII-nitride growth [49]. Another possibility is the use of an Si(100) substrate with a perfectly controlled off-cut in order to grow IIII-nitrides with a single in-plane crystal orientation via the formation of only one kind of surface steps (double steps). Even if it is permitted to fabricate high RF performance GaN HEMTs on Si(100) [50], it seems difficult to avoid a high-temperature surface preparation to stabilize the double steps, which can compromise the properties of the silicon devices.

Recently, progress has been made in reducing the thermal budget of preparation for the growth of GaAs on Si(100) [51], but the efficiency of such approaches remains to be demonstrated for GaN on silicon. The growth thermal budget can have a significant impact on the diffusion of dopants in silicon so trade-offs have to be made for monolithic integration. The trade-off we have chosen is the fabrication of both CMOS and GaN HEMTs on Si(111) or on Si(110). The latter allows the growth of III-nitrides with high crystal quality [52] and device performance [50], at least as good as on Si(111), the price to pay being an increased density of interface states at the SiO_2/Si interface of the MOS devices.

The impact of the growth of AlN (at 920°C) and GaN (at 800°C) on the dopants diffusion in silicon is shown in Figure 9.16a. It is clear from the figure that the growth of AlN for the nucleation layer and the stress-mitigating stack is the main contributor to the thermal budget of diffusion. According to simulations, dopants like phosphorus, boron or arsenic should not suffer a huge diffusion if the thermal budget associated with the growth is well below 900°C for several hours, meaning that AlN present in nucleation and stress-mitigating layers has to be grown at a reduced temperature [49,53]. Plasma-assisted MBE usually satisfies such criteria with AlN grown below 830°C and GaN grown at 700°C–730°C. As shown in previous sections, ammonia-MBE can grow AlN with acceptable crystal quality at temperatures as low as 830°C while GaN and AlGaN layers are grown at 800°C.

Another critical aspect of the integration is the mechanical stability of the mask used to perform the local area growth. Indeed, contrary to GaN, the growth of AlN on dielectric masks like SiO_2 or Si_3N_4 is not selective, and consequently the following GaN-based layers stick to these masks. The large tensile thermal mismatch strain between GaN or AlN and silicon rapidly generates cracks and delamination once a critical thickness is reached (Figure 9.16b). To overcome this problem, we have developed a mask covered with a very rough quasi-porous GaN layer on SiO_2 [53]. After the growth of HEMT structures, this sacrificial layer provides easy paths to delamination, preserving the integrity of the underlying SiO_2/Si layers. The delamination is mostly spontaneous on the sample when cooling down and it is totally achieved after cleaning in solvents and buffered hydrofluoric (HF) solution with ultrasonic agitation (Figure 9.16c).

The crystal quality assessed from XRD rocking curves is unchanged compared to planar growth. Furthermore, the chemical stability of the masks used for selective area growth has been confirmed by SIMS measurements showing levels of silicon, oxygen and carbon contaminations that compare well with planar growth. Transistors fabricated on such GaN HEMT structures present low leakage currents and normal output DC characteristics, as

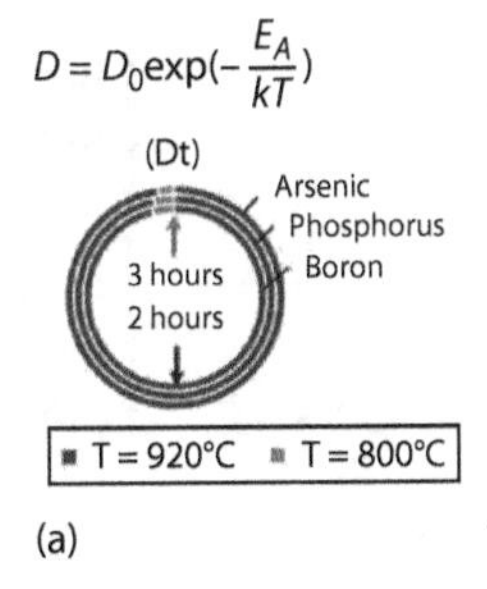

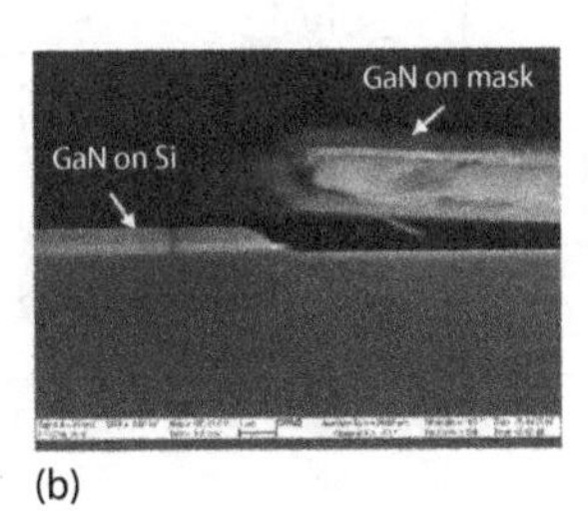

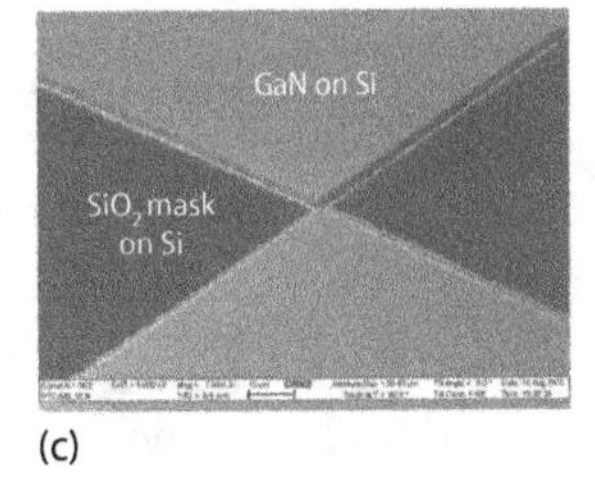

FIGURE 9.16
(a) Growth process impact for thick GaN HEMT on dopant diffusion according to the thermal budgets (D × time); (b) cross-section view of a delaminated part of the dielectric mask on silicon; (c) crack-free local area growth process after polycrystalline GaN removal.

shown in Figure 9.17. This confirms the quality of the local area low-temperature growth process developed for integration.

As mentioned earlier, the necessary trade-off for this monolithic integration involves the degradation of dopants profiles (Figure 9.18) and consequently the threshold voltage and the drain current in silicon CMOS devices. To evaluate the impact of the GaN HEMT growth on these parameters, silicon NMOS have been processed on p-doped Si(110) and Si(100) substrates with a resistivity of 1–10 Ω.cm. A polysilicon gate was defined on a 30 nm thick SiO_2 thermal oxide. The transistor access regions were implanted with phosphorus while the threshold voltage was adjusted with a boron implantation, followed by an activation step in an N_2 ambient for 1 h at 950°C. For the test, the device was covered by a plasma-enhanced chemical vapor deposition (PECVD) SiO_2 film of thickness greater than 200 nm. The wafers were then cut in two pieces; the first halves were annealed in the MBE chamber under NH_3 flow to simulate the different growth process steps while the others halves were used as references. Finally, electrical contacts were fabricated on all samples, consisting of Al pads deposited after opening vias into the SiO_2 layers. Figure 9.18c compares the output characteristics of a 3 μm gate transistor fabricated on a reference Si(100) wafer with a transistor from the same wafer after NH_3 annealing corresponding to the growth of a thick GaN HEMT structure with low-temperature-grown AlN. It is clear that the output current capability is preserved as well as the normally off behavior, but with a change of threshold voltage and source drain resistance.

One proposed mechanism to explain such changes is the oxynitridation-enhanced diffusion (ONED) of dopants due to NH_3. The changes in transistor characteristics could be due to a reduction of the effective gate length (L_{eff}) as described in Figure 9.18b, accompanied by a positive shift of the threshold voltage due to a large increase in the subthreshold slope (SS). However, the latter is significantly reduced by annealing in N_2 at 400°C, which is also necessary in a standard MOS process in order to obtain low SS. Furthermore, the effective gate length variations (Figure 9.19a) determined using the method described in [54] are very large compared to diffusion length simulations. The noticeably larger threshold voltage shift (Figure 9.19b) and SS (Figure 9.19c) obtained on Si(110) compared to Si(100) (Figure 9.19c) leads to the assumption that ONED mainly affects the SiO_2/Si interface properties. In addition, since no significant diffusion length variation occurs when growing thicker GaN buffers (at 800°C), we assume that ONED is activated in the 830°C–850°C temperature range (Figure 19a). According to [55,56], ONED is favored by the presence of

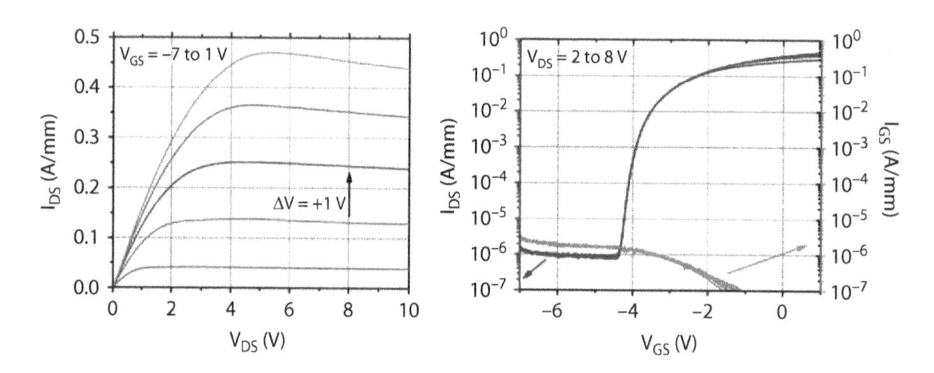

FIGURE 9.17
DC output characteristics and transfer characteristics at V_{ds} = 2–8 V of a transistor fabricated on an AlGaN/GaN HEMT structure after local area growth on silicon. The gate is 3 μm long and centered in a 13 μm source drain spacing.

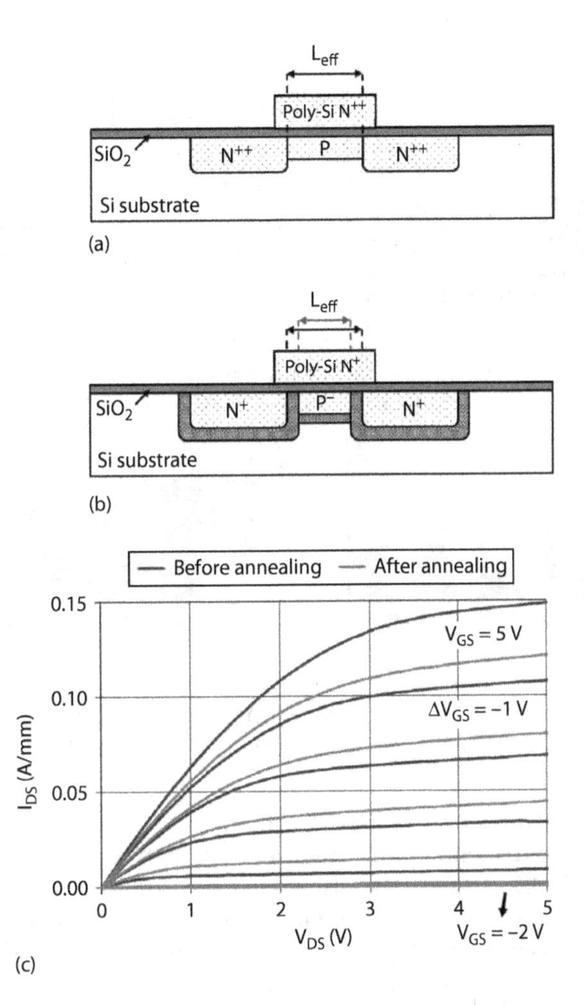

FIGURE 9.18
Cross section of the NMOS before (a) and after (b) annealing; (c) evolution of silicon NMOS output characteristics ($L_g = 3\,\mu m$) after exposure to the process corresponding to the thick structure with low-temperature AlN grown at 850°C.

hydrogen in NH_3; therefore, an alternative mask, e.g. SiN, has to be deposited on SiO_2 in order to effectively inhibit the diffusion of NH_x species, thereby preventing such a drift of MOS device characteristics.

9.9 Conclusion

MBE is able to produce device-quality materials for field-effect transistors such as HEMTs. Compared with ammonia-MBE and MOVPE, the low growth rate of PA-MBE (0.3–0.5 μm/h) has been considered for a long time as a drawback for production. A first attempt to solve this has been to use several nitrogen plasma cells simultaneously allowing to reach more than 0.8 μm/h growth rate [57]. More recently, new developments on plasma cells have enabled to reach more than 2 μm/h growth rate for GaN with unchanged surface

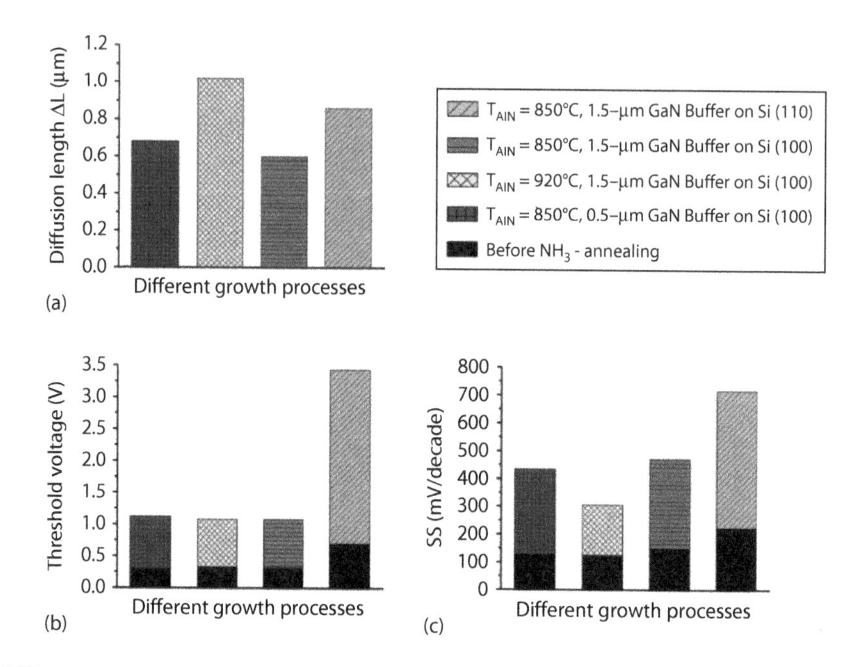

FIGURE 9.19
(a) Extracted effective gate length in NMOS transistors on Si(100) and Si(110) with and without annealing in NH_3 to simulate the HEMT growth processes. (b) Threshold voltage of the same. (c) Sub-threshold slope (SS).

morphology and optical properties [58–60]. However, uniformity has not yet been demonstrated. With a growth temperature intermediate between plasma-assisted MBE and MOVPE, ammonia-MBE presents the advantage of combining a simple N-rich growth regime with an acceptable growth rate and uniformity [4,9]. Moreover, it seems that ammonia-MBE offers the possibility to obtain a better purity material with a lower residual donor concentration, making the incorporation of additional compensation species not always necessary. Compared with MOVPE, one advantage of MBE is the possibility to develop structures at temperatures compatible with the integration of already processed devices, e.g. silicon MOSFETs. Furthermore, the possibility to grow smooth and good crystal quality AlN nucleation layers on silicon at noticeably lower temperatures is an advantage for producing high resistivity AlN/Si interfaces as well as buffer layers with high breakdown voltages and low RF propagation losses. All these points explain why MBE, and in our case why ammonia-source MBE, is so useful to develop electron device structures such as HEMTs.

Acknowledgments

The authors would like to thank all their colleagues in France, Spain, Belgium and Canada who contributed to the results presented here and who will recognize themselves in the references. They are also grateful to RIBER SA and to the agencies that supported a part of these studies. Among these agencies, the French Délégation Générale de l'Armement and the National Research Agency are deeply acknowledged.

References

1. M. Asif Khan, J.N. Kuznia, J.M. Van Hove, N. Pan, and J. Carter, *Appl. Phys. Lett.*, 60, 3027 (1992).
2. B. Gill (Ed.), *III-Nitride Semiconductors and their Modern Devices*, Oxford Science Publications, Oxford University Press, Oxford (2013).
3. M. Mesrine, N. Grandjean, and J. Massies, *Appl. Phys. Lett.*, 72, 350 (1998).
4. Y. Cordier, F. Pruvost, F. Semond, J. Massies, M. Leroux, P. Lorenzini, and C. Chaix, *Phys. Stat. Sol C*, 3, 2325 (2006).
5. Y. Cordier, F. Semond, J. Massies, M. Leroux, P. Lorenzini, and C. Chaix, *J. Crystal Growth* 301–302, 434 (2007).
6. F. Natali, Y. Cordier, C. Chaix, and P. Bouchaib, *J. Crystal Growth*, 311, 2029 (2009).
7. F. Natali, Y. Cordier, J. Massies, S. Vezian, B. Damilano, and M. Leroux, *Phy. Rev. B*, 79, 035328 (2009).
8. S. Vézian, F. Natali, F. Semond, and J. Massies, *Phys. Rev. B*, 69, 125329 (2004).
9. Y. Cordier, F. Natali, M. Chmielowska, M. Leroux, C. Chaix, and P. Bouchaib, *Phys. Status Solidi C*, 9 (3–4), 523 (2012).
10. G. Koblmüller, F. Reurings, F. Tuomisto, and J.S. Speck, *Appl. Phys. Lett.*, 97, 191915 (2010).
11. P. Kruszewski, P. Prystawko, I. Kasalynas, A. Nowakowska-Siwinska, M. Krysko, J. Plesiewicz, J. Smalc-Koziorowska, R. Dwilinski, M. Zajac, R. Kucharski, et al., *Semicond. Sci. Technol.*, 29, 075004 (2014).
12. C. Poblenz, P. Waltereit, S. Rajan, S. Heikman, U.K. Mishra, and J.S. Speck, *J. Vac. Sci. Technol. B*, 22 (3), 1145 (2004).
13. J.B. Webb, H. Tang, S. Rolfe, and J.A. Bardwell, *Appl. Phys. Lett.*, 75, 953 (1999).
14. D.F. Storm, D.S. Katzer, J.A. Mittereder, S.C. Binari, B.V. Shanabrook, X. Xu, D.S. McVey, R.P. Vaudo, and G.R. Brandes, *J. Crystal Growth*, 281, 32 (2005).
15. T.M. Kuan, S.J. Chang, Y.K. Su, J.C. Lin, S.C. Wei, C.K. Wang, C.I. Huang, W.H. Lan, J.A. Bardwell, H. Tang, et al., *J. Crystal Growth*, 272, 300 (2004).
16. A. Corrion, F. Wu, T. Mates, C.S. Gallinat, C. Poblenz, and J.S. Speck, *J. Crystal Growth*, 289, 587 (2006).
17. S.W. Kaun, M.H. Wong, U.K. Mishra, and J. Speck, *Semicond. Sci. Technol.*, 28, 074001 (2013).
18. Y. Cao, T. Zimmermann, H. Xing, and D. Jena, *Appl. Phys. Lett.*, 96, 042102 (2010).
19. Y. Cordier, M. Azize, N. Baron, S. Chenot, O. Tottereau, and J. Massies, *J. Crystal Growth*, 309, 1 (2007).
20. Y. Cordier, M. Azize, N. Baron, Z. Bougrioua, S. Chenot, O. Tottereau, J. Massies, and P. Gibart, *J. Crystal Growth*, 310, 948 (2008).
21. A. Watanabe, T. Takeuchi, K. Hirosawa, H. Amano, K. Hiramatsu, and I. Akasaki, *J. Crystal Growth*, 128, 391 (1993).
22. Y. Cordier, N. Baron, S. Chenot, P. Vennéguès, O. Tottereau, M. Leroux, F. Semond, and J. Massies, *J. Crystal Growth*, 311, 2002 (2009).
23. N. Baron, Y. Cordier, S. Chenot, P. Vennéguès, O. Tottereau, M. Leroux, F. Semond, and J. Massies, *J. Appl. Phys.*, 105, 033701 (2009).
24. Y. Cordier, N. Baron, F. Semond, M. Ramdani, M. Chmielowska, E. Frayssinet, S. Chenot, H. Tang, C. Storey and J.A. Bardwell, *Proceedings of the 35th Workshop on Compound Semiconductor Devices and Integrated Circuits*, Catania (Italy), May 29–June 1, pp. 89–90 (2011).
25. S. Rennesson, F. Lecourt, N. Defrance, M. Chmielowska, S. Chenot, M. Lesecq, V. Hoel, E. Okada, Y. Cordier, and J.-C. De Jaeger, *IEEE Trans. Electron. Dev.*, 60, 3105 (2013).
26. W.E. Hoke, A. Torabi, J.J. Mosca, R.B. Hallock, and T.D. Kennedy, *J. Appl. Phys.*, 98, 084510 (2005).
27. P. Leclaire, S. Chenot, L. Buchaillot, Y. Cordier, D. Theron, and M. Faucher, *Semicond. Sci. Technol.*, 29, 115018 (2014).
28. Y. Cordier, F. Semond, M. Hugues, F. Natali, P. Lorenzini, H. Haas, S. Chenot, M. Laügt, O. Tottereau, and P. Vennegues, *J. Crystal Growth*, 278/1-4, 393 (2005).

29. A. Pérez-Tomás, A. Fontserè, J. Llobet, M. Placidi, S. Rennesson, N. Baron, S. Chenot, J.C. Moreno, and Y. Cordier, *J. Appl. Phys.*, 113, 174501 (2013).
30. J.J. Freedsman, A. Watanabe, Y. Yamaoka, T. Kubo, and T. Egawa, *Status Solidi A*, 213, 424 (2016).
31. R. Comyn, Y. Cordier, V. Aimez, and H. Maher, *Phys. Status Solidi A*, 212, 1145 (2015).
32. S. Contreras, L. Konczewicz, J. Ben Messaoud, H. Peyre, M. Al Khalfioui, S. Matta, M. Leroux, B. Damilano, and J. Brault, *Superlattices Microstruct.*, 98, 253 (2016).
33. E.M. Chumbes, A.T. Schremek, J.A. Smart, Y. Wang, N.C. MacDonald, D. Hogue, J.J. Komiak, S.J. Lichwalla, R.E. Leoni and J.R. Shealy, *IEEE Trans. Electron. Dev.*, 48, 420426 (2001).
34. F. Lecourt, Y. Douvry, N. Defrance, V. Hoel, Y. Cordier, and J.C. De Jaeger, *Proceedings of the 5th European Microwave Integrated Circuits Conference (EuMIC)*, Paris, September 27–28, pp. 33–36 (2010).
35. F. Medjdoub, M. Zegaoui, B. Grimbert, D. Ducatteau, N. Rolland, and P. Rolland, *IEEE Electron. Device Lett.*, 33, 1168 (2012).
36. D. Marti, S. Tirelli, A. Alt, J. Roberts, and C. Bolognesi, *IEEE Electron. Device Lett.*, 33, 1372 (2012).
37. A. Soltani, J.-C. Gerbedoen, Y. Cordier, D. Ducatteau, M. Rousseau, M. Chmielowska, M. Ramdani, and J.-C. De Jaeger, *IEEE Electron. Device Lett.*, 34, 490 (2013).
38. D. Ducatteau, A. Minko, V. Hoel, E. Morvan, E. Delos, B. Grimbert, H. Lahreche, P. Bove, C. Gaquiere, J.C. De Jaeger, et al., *IEEE Electron. Device Lett.*, 27, 7 (2006).
39. D.C. Dumka, C. Lee, H.Q. Tserng, P. Saunier, and R. Kumar, *Electron. Lett.*, 40 (16), 1023 (2004).
40. K.K. Chu, P.C. Chao, M.T. Pizzella, R. Actis, D.E. Meharry, K.B. Nichols, R.P. Vaudo, X. Xu, J.S. Flynn, J. Dion, et al., *IEEE Electron. Device Lett.*, 25 (9), 596 (2004).
41. D.F. Storm, J.A. Roussos, D.S. Katzer, J.A. Mittereder, R. Bass, S.C. Binari, D. Hanser, E.A. Preble, and K. Evans, *Electron. Lett.*, 42 (11), 663 (2006).
42. C. Poblenz, A.L. Corrion, F. Recht, S.S. Chang, R. Chu, L. Shen, J.S. Speck, and U.K. Mishra, *IEEE Electron. Device Lett.*, 28 (11), 945 (2007).
43. N.X. Nguyen, M. Micovic, W.-S. Wong, P. Hashimoto, L.-M. McCray, P. Janke, and C. Nguyen, *Electron. Lett.*, 36 (5), 468 (2000).
44. Y. Pei, C. Poblenz, A.L. Corrion, R. Chu, L. Shen, J.S. Speck, and U.K. Mishra, *Electron. Lett.*, 44 (9), 598 (2008).
45. T. Palacios, A. Chakraborty, S. Rajan, C. Poblenz, S. Keller, S.P. DenBaars, J.S. Speck and U.K. Mishra, *IEEE Electron. Device Lett.*, 26 (11), 781 (2005).
46. Y.S. Puzyrev, T. Roy, M. Beck, B.R. Tuttle, R.D. Schrimpf, D.M. Fleetwood, and S.T. Pantelides, *J. Appl. Physics*, 109, 034501 (2011).
47. J.W. Chung J.-K. Lee, E.L. Piner, and T. Palacios, *IEEE Electron. Device Lett.*, 30, 1015 (2009).
48. Y.-C. Wu, M. Watanabe, and T. LaRocca, *Proceedings of the Radio Frequency Integrated Circuits Symposium (RFIC)*, Phoenix, AZ, May 17–19 (2015).
49. W.E. Hoke, R.V. Chelakara, J.P. Bettencourt, T.E. Kazior, J.R. Laroche, T.D. Kennedy, J.J. Mosca, A. Torabi, A.J. Kerr, H.-S. Lee, et al., *Sci. Technol. B*, 30, 02B101 (2012).
50. A. Soltani, Y. Cordier, J.-C. Gerbedoen, S. Joblot, E. Okada, M. Chmielowska, M.R. Ramdani and J.-C. De Jaeger, *Semicond. Sci. Technol.*, 28, 094003 (2013).
51. R. Alcotte, M. Martin, J. Moeyaert, R. Cipro, S. David, F. Bassani, F. Ducroquet, Y. Bogumilowicz, E. Sanchez, Z. Ye, et al., *APL Mater.*, 4, 046101 (2016).
52. Y. Cordier, J.-C. Moreno, N. Baron, E. Frayssinet, J.-M. Chauveau, M. Nemoz, S. Chenot, B. Damilano, and F. Semond, *J. Crystal Growth*, 312, 2683 (2010).
53. R. Comyn, Y. Cordier, S. Chenot, A. Jaouad, H. Maher, and V. Aimez, *Phys. Status Solidi A*, 213, 917 (2016).
54. K. Terada and H. Muta, *Jpn. J. Appl. Phys.*, 18, 953 (1979).
55. B. Balland and A. Glachant, Silica, silicon nitride and oxynitride thin films: An overview of fabrication techniques, properties and applications. In *Instabilities in Silicon Devices*, G. Barbottin and A. Vapaille (Eds), North-Holland, Amsterdam, pp. 3–144 (1999).
56. P. Fahey, R. Dutton, and M. Moslehi, *Appl. Phys. Lett.*, 43, 683 (1983).

57. R. Aidam, E. Diwo, N. Rollbühler, L. Kirste, and F. Benkhelifa, *J. Appl. Phys.*, 111, 114516 (2012).
58. B.M. McSkimming, F. Wu, T. Huault, C. Chaix, and J.S. Speck, *J. Crystal Growth*, 386, 168 (2014).
59. Y. Kawai, S. Chen, Y. Honda, M. Yamaguchi, H. Amano, H. Kondo, M. Hiramatsu, H. Kano, K. Yamakawa, and S. Den, *Phys. Status Solidi C*, 8, 2089 (2011).
60. Y. Cordier, B. Damilano, P. Aing, C. Chaix, F. Linez, F. Tuomisto, P. Vennéguès, E. Frayssinet, D. Lefebvre, M. Portail, et al., *J. Crystal Growth*, 433, 165 (2016).

10

Silicon Carbide Oscillators for Extreme Environments

Daniel R. Brennan, Hua-Khee Chan, Nicholas G. Wright, and Alton B. Horsfall

CONTENTS

10.1 Introduction

Silicon carbide (SiC)–based electronics have the advantage of being highly radiation resistant and capable of operating at temperatures not possible with conventional, silicon-based technologies [1]. This allows for their utilization as sensor circuits for applications located in hazardous environments, which are often inaccessible or dangerous to personnel. For this reason, it is desirable for the circuit to include some form of wireless communication capability that is realized with SiC-based components. In electronics, there is a standard requirement for repetitive waveforms to perform multiple functions, including the carrier waves required for communications systems. In order to meet this requirement, oscillator circuits are required that, depending on their complexity, can produce any waveform including high-frequency radio waves. These high-frequency radio signals form the basis of wireless communication systems. The challenge in realizing these high-frequency oscillators for deployment in extreme environments is the lack of process maturity in SiC and the shift in performance with temperature. This chapter provides an overview of a range of oscillator circuit topologies that are compatible with analog circuit designs and hence existing SiC technology, before focusing on the realization of a high-temperature Colpitts oscillator, which is used to demonstrate a high-temperature prototype communication circuit. The drift in the central frequency with the temperature of SiC-based oscillators is reported and described in terms of the parametric characteristics of the devices. The circuit is then expanded to enable the first SiC Colpitts oscillator–based voltage-controlled oscillator (VCO), which is presented as a method of enabling the closed loop control of the frequency of an oscillator over a wide temperature range. The chapter concludes with demonstrations of both the amplitude and frequency modulation (FM) of carrier waves from a circuit operating at 300°C, which are subsequently discussed with a view to creating a high-temperature wireless communication system.

10.2 Silicon Carbide Technology Overview

The development of SiC technology has become increasingly rapid in recent years. With significant improvements in wafer growth technology and materials processing, SiC devices have become part of the mainstream in power electronic applications [2,3] and a number of conferences are running sessions dedicated to the deployment of this technology in mainstream applications. Further advantages in terms of small, signal-level devices [4] and sensors [5] have resulted in significant improvements in capabilities; however, these are still only available at research level, where a number of groups are active. The underlying rationale for the continued investment in SiC technology is the excellent material properties, which derive from the high chemical stability of the Si–C bond. SiC is the only technologically relevant semiconductor that has a stable thermal oxide (SiO_2) [6] and can exist in a large number of polytypes – different crystal structures built from the same Si-C sub-unit organized into different stacking sequences. There are over 100 of these polytypes known; however, the vast majority of the research has focused on three: 3C, 6H and 4H. Of these, the 4H polytype is the most common for electronic devices, and commercially available power electronic devices are manufactured almost exclusively from this polytype due to its overall superior material properties. The bandgap of 4H SiC is 3.23 eV at room temperature (compared to 1.12 eV for silicon), which dramatically reduces the intrinsic carrier concentration in comparison to semiconductors such as silicon or gallium arsenide, allowing devices to theoretically operate at temperatures up to 1000°C [7]. SiC also has a high saturation electron velocity, $2 \times 10^7 \text{ cm s}^{-1}$, a thermal conductivity in excess of copper at room temperature and a critical electric field that is almost an order of magnitude higher than that of silicon. These parameters have been exploited in the high-performance power metal-oxide-semiconductor field-effect transistors (MOSFETs) and diodes that are commercially available, but also have the potential to realize high-performance, high-frequency oscillators.

10.3 The Electronic Oscillator

An electronic oscillator is a circuit that produces a repetitive, oscillating signal, converting a direct current (DC) power supply into an alternating current (AC) signal. The shape of the required signal waveform depends on the application for which the oscillator is designed and typical examples include the generation of clock pulses for digital electronics [8], pulse width–modulated switching waveforms for switch mode power supplies [9] and radio frequency (RF) signals for wireless electronic communications [10]. A significant number of oscillator circuits have been realized; however, they can be categorized into two main types: the nonlinear oscillator and the linear or harmonic oscillator.

10.3.1 The Nonlinear Oscillator

The nonlinear oscillator (sometimes referred to as a relaxation oscillator) produces a waveform that is nonsinusoidal, such as a square wave, sawtooth or triangular waveform. The circuit topology comprises an energy storage component in conjunction with a nonlinear switching circuit. The energy storage component is often a capacitor; however, circuits based on the use of inductors have been demonstrated. Alternative implementations of this circuit comprise those that include a negative resistance device that periodically charges and discharges the energy in the storage element, resulting in abrupt changes in the output voltage of the circuit. Square wave relaxation oscillators are used extensively in digital electronics to provide the clock signal for sequential logic circuits such as timers and counters, whereas sawtooth oscillators are used specifically in time base circuits. An example of a single active device relaxation oscillator can be constructed using a unijunction transistor (UJT) device. A UJT comprises one p-n junction, making it similar in structure to a p-n diode; however, it differs in the fact that it has three terminals and so it is also referred to as a double base diode. It is worth noting that while the UJT has a similar construction to a junction field-effect transistor (JFET), the operation in a circuit is significantly different. In a UJT, the current through the device is modulated by the bipolar current injected through the gate under forward bias, rather than by the influence of the depletion region formed in the channel [11] with the gate under reverse bias.

As can be seen from Figure 10.1, the structure of a UJT does show some similarity to that of an N-channel JFET. The device consists of a lightly doped n-type substrate with three p-type terminals, which are labeled as base one (B1), base two (B2) and an emitter (E). B1 and B2 are ohmic contacts formed on the n-epilayer, whereas the E is formed from a heavily doped p-type region. The main physical difference between the UJT and the JFET is that the p-type region completely surrounds the n-type region in the JFET and the surface area of the JFET gate is significantly larger than the area of the emitter junction in the UJT. In a typical UJT, the doping of the emitter region is high, while the dopant concentration in the n-type region is low. This results in the resistance between the base terminals being significantly higher than is typical for a JFET, being in the region of 4–10 kΩ when the emitter is open circuit. In practical devices, the n-type channel has a high resistance and the resistance between the emitter and contact B1 is larger than that between the emitter and B2, resulting from the physical asymmetry in the structure, i.e. the emitter is physically closer to B2 than B1. A UJT does not have the ability to amplify; however, it is capable of controlling a large AC power with a small signal applied to the gate. In operation, the UJT is biased with a positive voltage between the two bases,

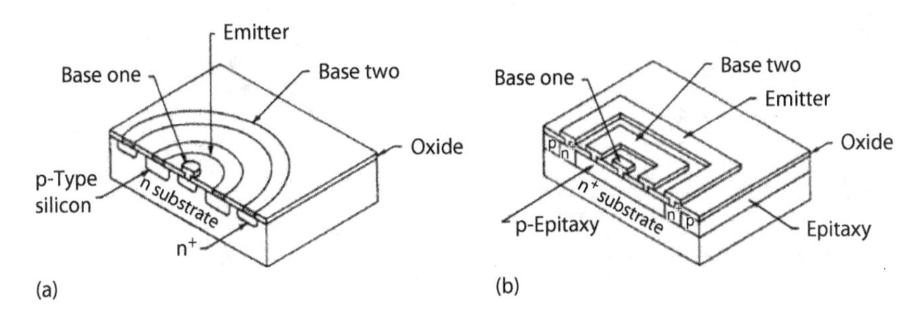

FIGURE 10.1
Example of the physical construction of a UJT. (a) Diffused planar structure and (b) epitaxial planar structure. (From Andronov, A.A., et al., *Theory of Oscillators*, Dover Publications, 2009 [12].)

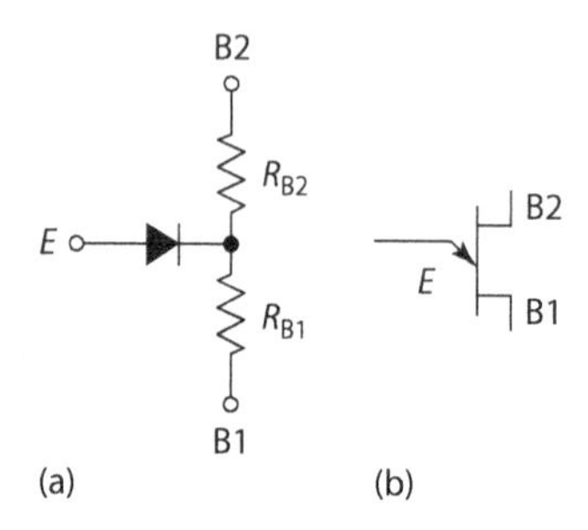

FIGURE 10.2
(a) Schematic representation of a UJT, (b) circuit symbol of a UJT.

whereby B2 is biased more positively than B1, resulting in a potential drop created along the length of the device. This can be seen in the schematic representation of a UJT shown in Figure 10.2a. When the emitter voltage is greater than the voltage applied to B2 plus the built-in potential of the p-n junction (typically 0.7 V for silicon and 2.8 V for SiC [13]), current flows from the emitter into the base region. Because the base region is very lightly doped, the additional charge carriers in the base region, which are injected by the emitter current, result in conductivity modulation. This excess charge reduces the resistance of the base region between the emitter junction and the B2 contact. This reduction in resistance results in the potential difference between the emitter and B1 increasing, which means that the emitter junction is more forward biased, and so even more current is injected. Overall, the effect appears to be similar to a negative resistance at the emitter terminal. This apparent negative resistance is the property that makes the UJT useful as the active component in simple oscillator circuits.

With the emitter open circuit, it is possible to determine the total resistance of the channel, also known as interbase resistance (R_{BB0}), as

$$R_{BB0} = R_{B1} + R_{B2} \tag{10.1}$$

As described previously, practical devices are nonsymmetrical and therefore it is possible to express the ratio between the resistance of the emitter to B_1 to that of the interbase resistance R_{BB0} as the intrinsic standoff ratio, which can be calculated using

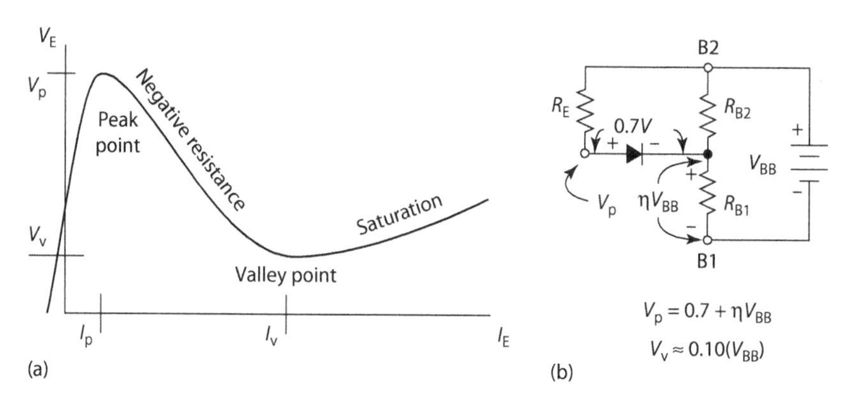

FIGURE 10.3
(a) Emitter characteristic curve, (b) model for V_P.

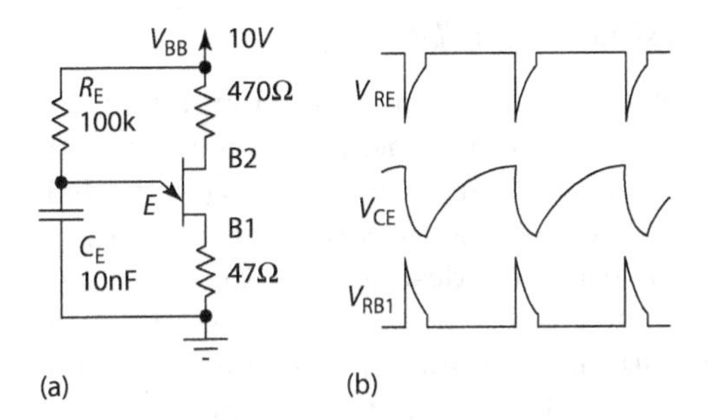

FIGURE 10.4
(a) Nonlinear electronic oscillator circuit using a UJT, (b) waveforms seen at respective points of the circuit.

$$\eta = \frac{R_{B1}}{R_{B1} + R_{B2}} = \frac{R_{B1}}{R_{BBO}} \tag{10.2}$$

Figure 10.3a depicts the emitter current vs voltage characteristic curve for a typical UJT. As the emitter voltage, V_E, increases further, the emitter current, I_E, increases until the peak point, V_P, beyond which the high level of current injection results in a negative resistance region, as can be seen from Figure 10.3.

The emitter voltage reaches a minimum at the valley point, V_V, where the emitter resistance is minimized. By biasing the UJT in the negative resistance region, it is possible to construct a simple electronic oscillator circuit, as shown schematically in Figure 10.4.

The operation of the oscillator circuit shown in Figure 10.4a can be described as follows: the external capacitor, C_E, charges based on the current flowing through the resistor, R_E, until the voltage at the emitter contact reaches the peak point, V_P. The reduction in the UJT resistance between the emitter and base 1 contacts (due to the negative resistance characteristics) discharges the capacitor. This can be seen in the waveform characteristics in Figure 10.4b. Once the potential across the capacitor has discharged below the valley point, the resistance between the emitter and base 1 increases and the capacitor is free to charge again. This results in a repeating oscillation, where the frequency can be controlled by the selection of the external components. Neglecting any parasitics in the circuit, the oscillation frequency of a UJT oscillator can be expressed as

$$f = \frac{1}{\mathrm{RC}\ln\left(\frac{1}{1-\eta}\right)} \tag{10.3}$$

To date, no UJT structures have been realized in SiC technology, which precludes this topology from high-temperature operations. The requirement for negative resistance in the device operation places significant limits on the carrier lifetime in the semiconductor. At present, the typical carrier lifetime in SiC is of the order of 1 μs [14], which is significantly lower than in the state-of-the-art silicon or gallium arsenide wafers, where values closer to 1 ms are commonplace.

10.3.2 The Linear Oscillator

In contrast to the nonlinear oscillator described in the previous section, the linear or harmonic oscillator produces a sinusoidal waveform. Two fundamental types of linear oscillator are possible: the negative resistance oscillator and the feedback oscillator.

10.3.2.1 The Negative Resistance Oscillator

The negative resistance oscillator uses an active component that exhibits negative resistance, such as a magnetron tube or a diode. Previous research has shown the possibility of an impact ionization avalanche transit-time (IMPATT) diode in SiC, which has suitable characteristics for the realization of high-frequency oscillators [15,16].

Similar to the nonlinear oscillator described in the previous section, the operation of these oscillators relies on the concept of negative resistance. Considering the simple tank circuit shown in Figure 10.5a, the initial current pulse is applied at I_{in}. The LCR tank responds with a decaying oscillatory behavior because in every oscillation a fraction of the energy that reciprocates between the capacitor and the inductor is lost as heat in the resistor.

If a device or a circuit that exhibits negative resistance is placed in parallel with R_P, it can be modeled as $-R_P$. If the magnitude of the negative resistance is equal to the parasitic resistance in the tank circuit, R_P, then the circuit continues to oscillate indefinitely, as shown schematically in Figure 10.5b. Such oscillators are used in the production of high frequencies, typically in the microwave region (>300 MHz) and above. The advantage of SiC in such circuits is its ability to operate at higher temperatures than an equivalent silicon device, thus allowing designers to dissipate much higher levels of power in a circuit [18].

Figure 10.6a depicts an example oscillator circuit utilizing an IMPATT diode as the negative resistance element. The circuit topology in the figure contains a resonant circuit, in this example an inductor-capacitor (LC) circuit, but alternatives including those in which a crystal or cavity resonator is connected across an IMPATT diode have identical characteristics.

As shown in Figure 10.5a, a resonant circuit has the possibility to operate as an oscillator, in that it has the ability to store energy in the form of electrical oscillations if excited. However, it will have some form of internal resistance and, as a result, the magnitudes of the oscillations are damped and eventually become zero. The negative resistance of the active device cancels the internal resistance in the resonator, thus creating a resonator where the magnitude of the oscillations shows no evidence of damping. This results in the

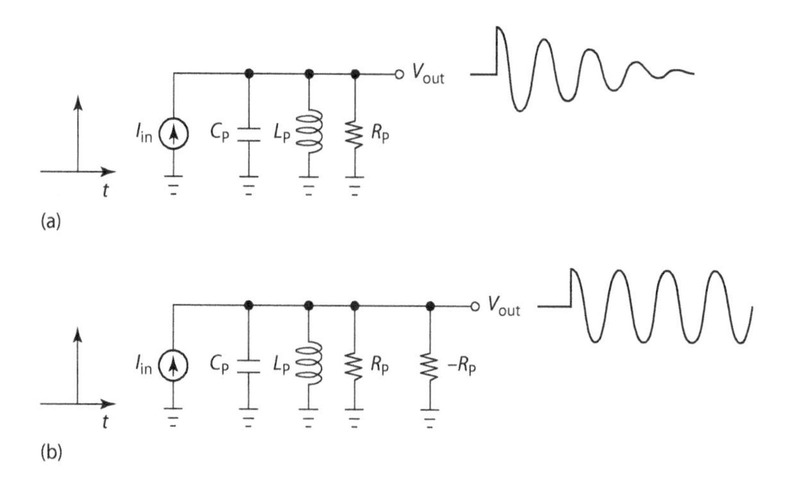

FIGURE 10.5
(a) Decaying impulse response of a tank circuit, (b) the addition of negative resistance to cancel loss in R_P. (From Senhouse, L.S., *IEEE Trans. Elect. Dev.*, 16, 1969, [17].)

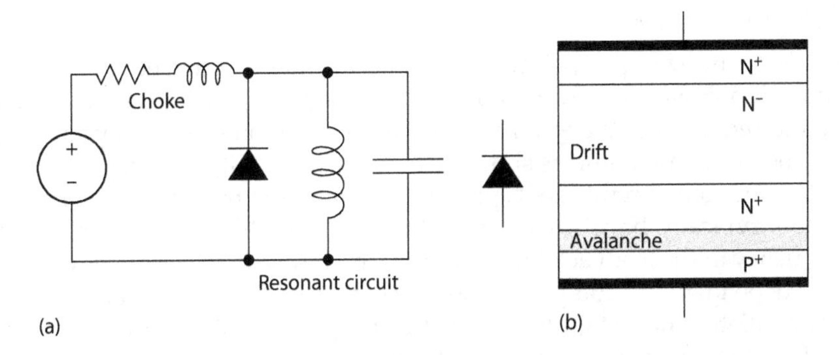

FIGURE 10.6
(a) Circuit of an IMPATT diode–based negative resistance oscillator, (b) structure of an IMPATT diode.

creation of oscillations with a constant magnitude at the resonant frequency, which can be determined using Equation 10.5.

10.3.2.2 The Feedback Oscillator

The feedback oscillator is the most common form of linear oscillator. It requires the use of an active component, typically an electronic amplifier based on a single transistor. The output from the amplifier is connected in a feedback loop to the input through a frequency selective electronic filter, resulting in positive feedback. For the oscillator to operate, the gain of the two stages of the amplifier needs to fulfill the requirement that the product of the gain for the two individual stages must exceed unity, i.e. $BA > 1$. This limitation can be considered in terms of the energy required to ensure a constant magnitude of the output waveform. A schematic representation of the feedback oscillator is shown in Figure 10.7.

When power is first supplied to the amplifier, the electronic noise in the circuit provides the initial signal to initiate the oscillations. The noise travels around the feedback loop, where it is amplified and filtered until it becomes a sine wave at a single specific frequency [20].

Feedback oscillator circuits can be classified according to the type of frequency selective filter they employ in the feedback loop. There are three specific types: crystal oscillators, resistor-capacitor (RC) oscillators and LC oscillators each with their own merits and limitations.

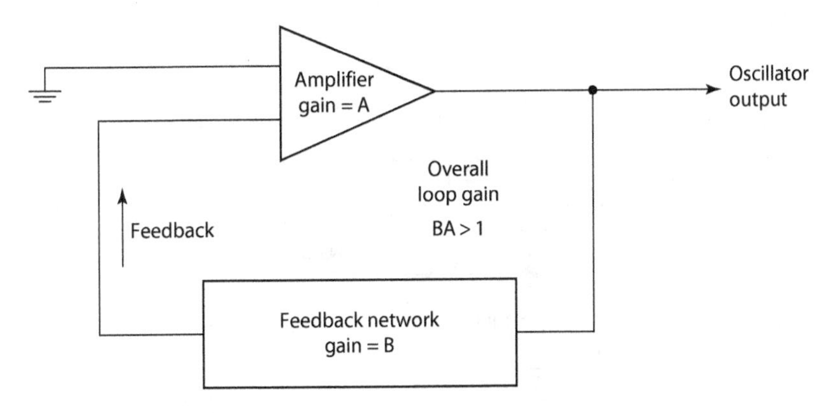

FIGURE 10.7
Schematic representation of a feedback oscillator.

10.3.2.2.1 *Crystal Oscillators*

Crystal oscillators utilize a piezoelectric crystal as the filter in the feedback loop. The crystal mechanically vibrates as a resonator and the frequency of the vibration determines the oscillation frequency of the electronic circuit. A common example circuit for a crystal oscillator is the Pierce oscillator, as shown schematically in Figure 10.8.

Crystals for electronic circuits are typically manufactured from quartz. Quartz mechanical resonators typically have a very high *Q*-factor, resulting in high selectivity to the desired frequency in the feedback loop. In addition, the oscillator frequency shows a weak temperature dependence in comparison to tuned circuits. These two characteristics mean that crystal oscillators have significantly better frequency stability than either LC or RC oscillators. However, quartz is not suitable for the development of high-temperature oscillators because it undergoes a phase transition at 573°C, referred to as the quartz inversion [19]. Above this temperature, the crystals do not operate as electronic filters and the oscillator produces a wide range of frequencies simultaneously. At the current time, no other material has been identified that can demonstrate the mechanical properties of a quartz resonator and operate at temperatures beyond this temperature. While crystal oscillators provide greater frequency stability with variations in temperature, the availability and complexities associated with incorporating a high-temperature crystal component were viewed as a disadvantage in comparison with the simplicity of prototype high-temperature LC oscillator–based circuits. Hence, crystal oscillators have not been experimentally investigated further.

10.3.2.2.2 *RC Oscillators*

RC oscillators utilize a filter network composed of discrete resistors and capacitors to produce lower frequencies (in comparison to the microwave frequencies generated using the negative resistance oscillators described previously), typically in the kilohertz frequency range. A common example circuit topology is the phase shift oscillator, as shown in Figure 10.9.

It can be seen from Figure 10.9 that the JFET is operating as an inverting amplifier, producing an output signal that is phase shifted by 180° in comparison to the signal at the input. In this topology, the RC network shown in the lower part of the image acts as a

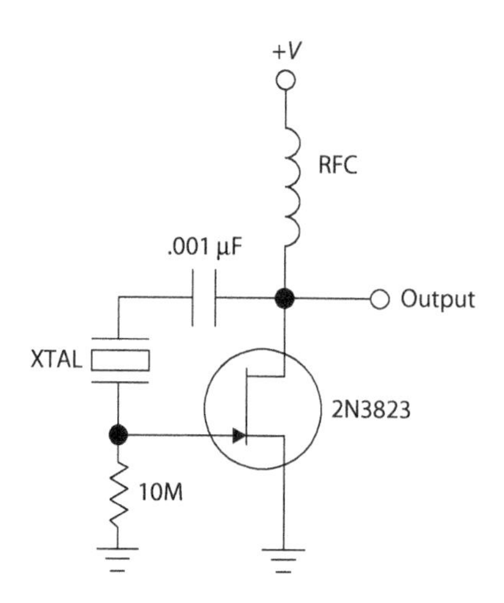

FIGURE 10.8
A Pierce crystal oscillator utilizing a JFET as an active device.

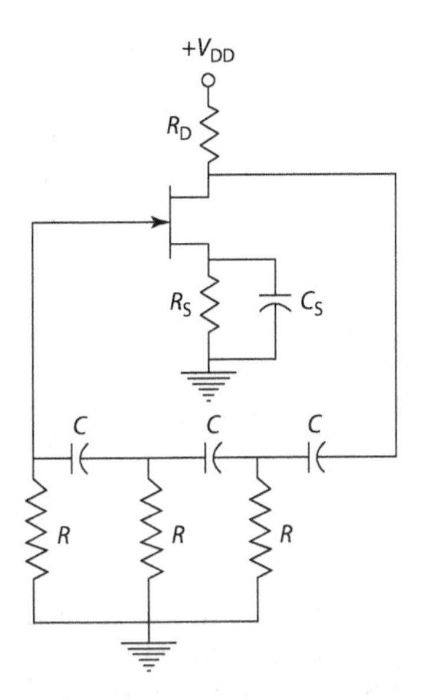

FIGURE 10.9
Phase shift oscillator utilizing a JFET. (From Sanitram, K., *Basic Electronics: Devices, Circuits and Fundamentals*, Dover Publications, 1972 [24].)

phase shift network. This phase shift in this part of the circuit matches that of the transistor, resulting in a fully in-phase signal. The phase shift is generated by the three RC sections in cascade, each of which produces a 60° phase shift. In general, each of the three sections shares common values for the resistance and capacitance of the components. In this case, the frequency of oscillation excluding parasitics can determined from

$$f = \frac{1}{2\pi RC\sqrt{6}} \tag{10.4}$$

Monolithically integrated RC oscillators (where the resistors and capacitors are fabricated on the same semiconductor die as the JFET) are possible. Because of the limited values of capacitance and resistance that can be realized, these circuits typically operate in the megahertz regime; however, the gate-source capacitance and channel resistance of the JFET become comparable to the component values, limiting the maximum oscillation frequency.

10.3.2.2.3 LC Oscillators

In an LC oscillator, the filter is a tuned circuit commonly referred to as a tank circuit, which consists of an inductor and a capacitor connected in parallel. During operation, charge flows back and forth between the plates of the capacitor through the inductor. This transfer of charge occurs at the resonant frequency and so the tank circuit acts as a filter, where the frequency is determined by the capacitance and inductance values. The value is often expressed in terms of the resonant angular frequency, which can be calculated from

$$\omega_0 = \frac{1}{\sqrt{LC}} \tag{10.5}$$

where L is the inductance in Henries and C is the capacitance in Farads.

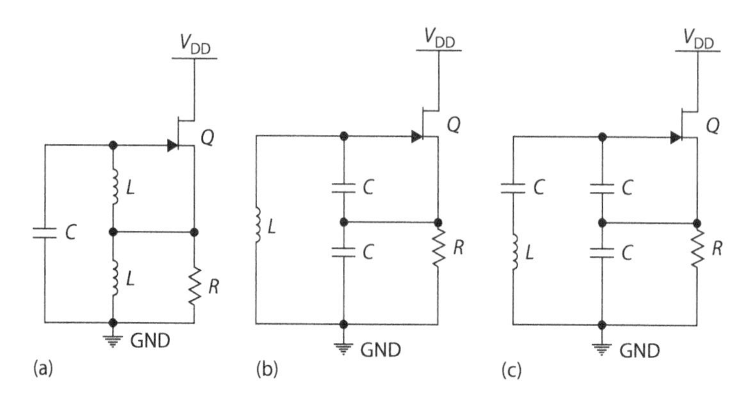

FIGURE 10.10
(a) Hartley, (b) Colpitts and (c) Clapp oscillators.

The internal losses within the tank circuit are compensated by the amplifier, which draws energy from the DC supply used in the circuit, resulting in a constant oscillation magnitude. LC oscillators are most commonly used at radio frequencies, where a tunable frequency source is required. Three commonly used circuit configurations are the Hartley, Colpitts and Clapp oscillators, as shown in Figure 10.10a–c, respectively.

As can be seen from the schematic circuits shown in Figure 10.10, the Colpitts and Clapp oscillators lend themselves more easily to miniaturization than the Hartley oscillator circuit, as they only utilize a single inductor in the design. High-temperature inductors tend to be physically large, as the permittivity and saturation magnetization of magnetic materials reduces with increasing temperature [21] and so the Hartley oscillator will be significantly larger than either the Colpitts or Clapp circuits. In this work, a Colpitts oscillator design was selected due to the ability of the oscillations to self-start utilizing components with lower Q values than the slightly more complicated Clapp design and this offers a more relaxed set of design criteria.

10.4 The Colpitts Oscillator

The Colpitts oscillator is a common form of LC oscillator that utilizes an LC tank circuit and an active device to counteract the damping effect caused by parasitic resistances. The Colpitts oscillator can be realized using a single transistor acting as an amplifier with the addition of a tank circuit. The circuits considered here are based on the use of a depletion mode SiC JFET; however, the analysis can be expanded to include a wide range of alternative transistor families, including MOSFETs and bipolar junction transistors (BJTs) as well as thermionic valves. By feeding the signal back from the output of the amplifier to the input through this LC tank to select a single frequency, it is possible to commission a Colpitts oscillator circuit, as shown by the circuit given in Figure 10.11.

The analysis of operation for a Colpitts oscillator is often performed using a linear systems feedback approach, as described in a number of references including [22]. The main outcome from this approach is a set of expressions for both frequency and the minimum JFET gain in order for the oscillations to maintain a constant magnitude with time. These conditions are described in Equations 10.6 and 10.7, respectively.

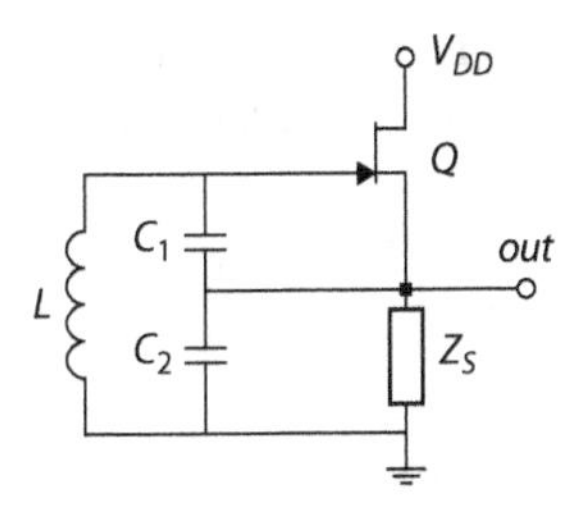

FIGURE 10.11
Circuit schematic of a Colpitts oscillator.

$$f_0 = \frac{1}{2\pi\sqrt{L\dfrac{C_1C_2}{C_1+C_2}}} \tag{10.6}$$

$$g_m r_o = \frac{C_2}{C_1} \tag{10.7}$$

This approach typically results in higher calculated frequencies than those predicted using computer-based simulation tools (for example, simulation program with integrated circuit emphasis [SPICE]) or obtained experimentally due to the exclusion of the gate-source, gate-drain and source-drain capacitances within the JFET. It is also possible to model the characteristics of the Colpitts oscillator using an alternative approach that is based on the concept of negative resistance [9] in one-port oscillators such as the SiC IMPATT diode, replacing the transistor with an ideal entire circuit that has the parasitics explicitly defined as external components, as shown in Figure 10.12.

From Figure 10.12, it can be see that in order to overcome the energy loss from a finite R_P, the active circuit must form a small-signal negative resistance ($R_P<0$), which is required to replenish the energy lost in every oscillation. Hence, the negative resistance can be interpreted as a source of energy. In this example, R_P denotes the equivalent parallel resistance of the tank, and for oscillations to be self-starting it is necessary that $R_P+(-R_P)>0$. As the oscillation amplitude increases, the amplifier will start to saturate, thus decreasing the gain from the feedback loop until it reaches unity. This steady-state condition satisfies the Barkhausen criterion [23] and the oscillations continue. Once the circuit is operating in the steady state, the two resistances, R_P and $-R_P$, must be of equal amplitude. This analytical approach allows for the inclusion of the parasitic capacitances that are inherent in the JFET structure, thus yielding a far more accurate calculation of the oscillation frequency of the circuit. Neglecting the effect of the inductor in the circuit, the AC

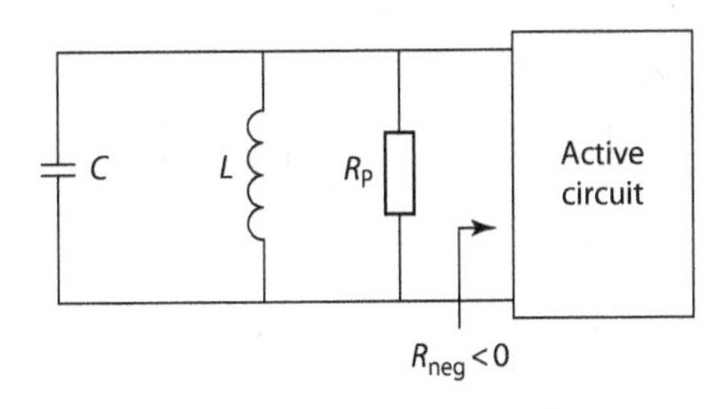

FIGURE 10.12
RLC tank with negative resistance created by the oscillator active network.

equivalent circuit in Figure 10.12 can be used to determine the input impedance, as shown in Figure 10.13. The input impedance of the circuit can be determined from Equation 10.8:

$$|Z_S| \gg \left| \frac{1}{i\omega(C_2 + C_{DS})} \right| \tag{10.8}$$

The parasitic capacitances of the JFET (denoted by Q) have been included in Figure 10.13.

These are the gate-source, C_{GS}, drain-source, C_{DS}, and gate-drain, C_{GD}, capacitances. By assuming the following conditions, it is possible to describe the JFET by means of a small-signal model, which can be utilized within SPICE to accurately describe the behavior of the circuit:

$$r_{DS} \gg \left| \frac{1}{i\omega(C_2 + C_{DS})} \right| \tag{10.9}$$

$$r_{GS} \gg \left| \frac{1}{i\omega(C_1 + C_{GS})} \right| \tag{10.10}$$

where r_{DS} and r_{GS} are the small-signal drain-source and gate-source resistances of the JFET. By replacing the SiC JFET with the simplified small-signal model, it is possible to obtain the equivalent circuit of the oscillator shown in Figure 10.14, which is the small-signal transconductance of the SiC JFET.

From results previously published in the literature and utilizing standard circuit theory, it is possible to determine the limits for the negative resistance required to achieve a steady-state operation for the circuit shown in Figure 10.14:

$$r_{neg} = -\frac{g_m}{\left[\omega^2 (C_1 + C_{GS})(C_2 + C_{DS})\left(1 + \frac{C_{GD}}{C_1 + C_{GS}} + \frac{C_{GD}}{C_2 + C_{DS}}\right)\right]} \tag{10.11}$$

and the equivalent total capacitance in the circuit C_t can be expressed as

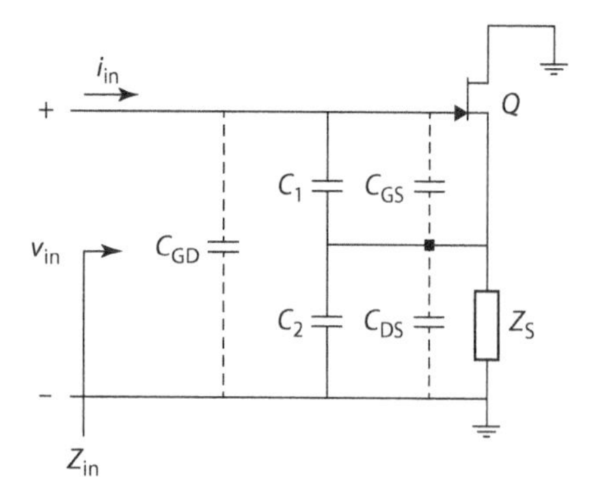

FIGURE 10.13
An AC equivalent circuit of a Colpitts oscillator ignoring the inductor L and including the parasitic capacitances inherent in the JFET transistor structure.

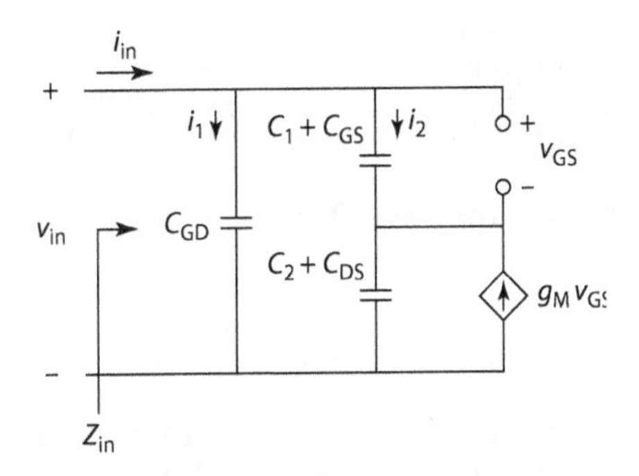

FIGURE 10.14
Small-signal equivalent of the Colpitts oscillator.

$$C_t = \frac{(C_1 + C_{GS})(C_2 + C_{DS})}{C_1 + C_2 + C_{GS} + C_{DS}}\left(1 + \frac{C_{GD}}{C_1 + C_{GS}} + \frac{C_{GD}}{C_2 + C_{DS}}\right) \tag{10.12}$$

It should be noted that combining Equation 10.12 with Equation 10.5 results in the reduction to a well-known expression for the negative resistance [17,24]:

$$r_{neg} = \frac{g_m}{\omega^2 C_1 C_2} \tag{10.13}$$

Utilizing these values and reconnecting the inductor, L, along with a series total loss resistance, r_t, to the input impedance, Z_{in}, it is possible to obtain a series equivalent circuit of the Colpitts oscillator, shown schematically in Figure 10.15.

The total series loss resistance, r_t, can be calculated from the series resistance of both the capacitors and inductors using

$$r_t = r_L + r_{C_1} + r_{C_2} \tag{10.14}$$

Hence, the equivalent parallel resistance of the Colpitts oscillator, R_P, including contributions from the tank circuit can be determined from

$$R_P = \frac{L}{r_t \dfrac{(C_1 + C_{GS})(C_2 + C_{DS})}{C_1 + C_2 + C_{GS} + C_{DS}}\left(1 + \dfrac{C_{GD}}{C_1 + C_{GS}} + \dfrac{C_{GD}}{C_2 + C_{DS}}\right)} \tag{10.15}$$

By substituting Equation 10.12 into Equation 10.5, the oscillation angular frequency for a circuit including the parasitic components can be determined using

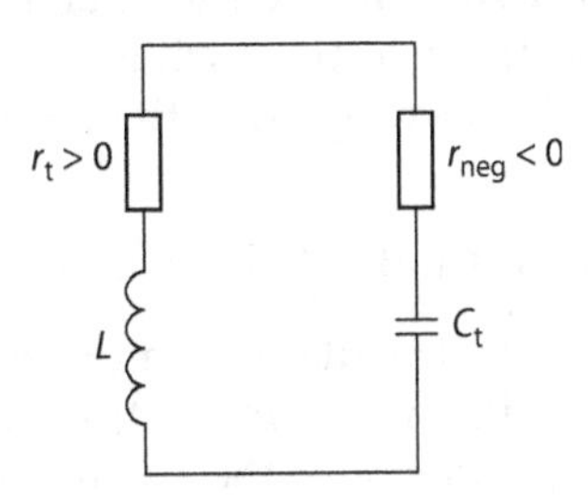

FIGURE 10.15
High-temperature Colpitts oscillator circuit diagram.

$$\omega = \sqrt{\left(L\frac{(C_1 + C_{GS})(C_2 + C_{DS})}{C_1 + C_2 + C_{GS} + C_{DS}}\left(1 + \frac{C_{GD}}{C_1 + C_{GS}} + \frac{C_{GD}}{C_2 + C_{DS}}\right)\right)^{-1}} \tag{10.16}$$

Substituting Equation 10.16 into Equation 10.11 results in

$$r_{\text{neg}} = -\frac{g_m L}{C_1 + C_2 + C_{GS} + C_{DS}} \tag{10.17}$$

Finally, referring back to Figure 10.13, it is possible to determine the negative resistance required for the Colpitts oscillator to self-start, resulting in

$$r_{\text{neg}} = -\frac{(C_1 + C_2 + C_{GS} + C_{DS})^2}{g_m (C_1 + C_{GS})(C_2 + C_{DS})\left(1 + \frac{C_{GD}}{C_1 + C_{GS}} + \frac{C_{GD}}{C_2 + C_{DS}}\right)} \tag{10.18}$$

Equation 10.18 shows that for the circuit to operate, the JFET needs to have a high transconductance, g_m, and low parasitic capacitances, C_{GS} and C_{GD}. This identifies the challenge of designing high-temperature oscillators, to the extent that in addition to the external capacitance values remaining unchanged (the parasitic capacitances within the transistor have a weak temperature dependence), the transconductance of the JFET will limit the upper operating temperature of the circuit. The transconductance of a JFET can be expressed as

$$g_m = \frac{\Delta(I_{DS})}{\Delta(V_{GS})} = \frac{2I_{DSS}}{|V_P|}\left(1 - \frac{V_{GS}}{V_P}\right) \tag{10.19}$$

10.4.1 High-Temperature Colpitts Oscillator

To demonstrate the importance of an accurate model in determining the resonant frequency of an oscillator, a comparison between the different circuit topologies was performed. Experimentally determined parameters for SiC devices operating at high temperatures were used in simulations using both SPICE-based simulations and the theoretical analysis described in the previous section. The characterized SiC components were then packaged into a hybrid module circuit board that was fabricated on an aluminum oxide ceramic substrate. Figure 10.16 shows a schematic representation of the high-temperature Colpitts oscillator, with the room temperature component values shown. The hybrid module was placed inside a Carbolite oven with electronic temperature control during the measurements and allowed to settle for 20 min at each temperature prior to any measurements being taken. The frequency spectrum of the oscillator was measured using an Agilent E4403B spectrum analyzer, which was coupled to the RF output of the circuit under test.

10.4.2 High-Temperature Voltage-Controlled Oscillator

The data in Figure 10.17 compare the calculated, simulated and experimental frequencies of a high-temperature Colpitts oscillator at temperatures up to 573 K. It can be seen from the experimental data that the frequency of the oscillator reduces with increasing temperature. This reduction is due to both the increased capacitance density of the AlN dielectric capacitors, which results from the increasing dielectric constant and the increasing

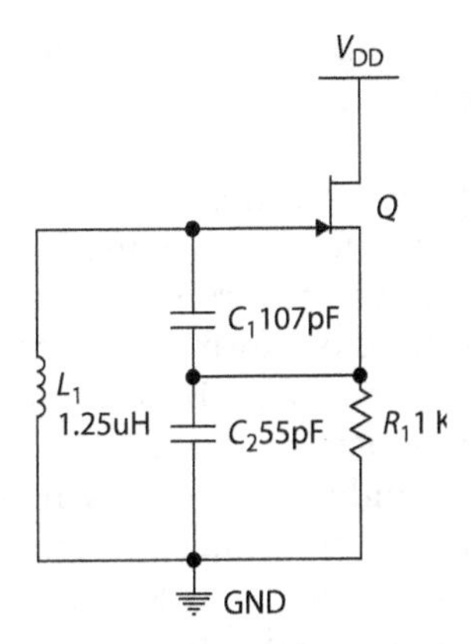

FIGURE 10.16
Comparison of calculated, simulated and experimental frequencies as a function of ambient temperature.

parasitic capacitance of the JFET p-n junction. As can be seen from the data, the predicted oscillator frequency using traditional theory is significantly higher than that obtained experimentally, predominantly because of the parasitic capacitances and inductances in the fabricated circuit. Estimating the parasitic values for the circuit results in modified values, as shown in Figure 10.17; however, the variation of the modified values with temperature does not match the experimental data well. This indicates that the performance of the oscillator circuit is dominated by the shift in the characteristics of the AlN capacitors. The variation in the capacitance density of AlN capacitors has been extracted from the capacitance-voltage characteristics [25] and this can be included as a temperature dependence in the SPICE model of the circuit, resulting in the data set labeled 'Frequency Simulated'

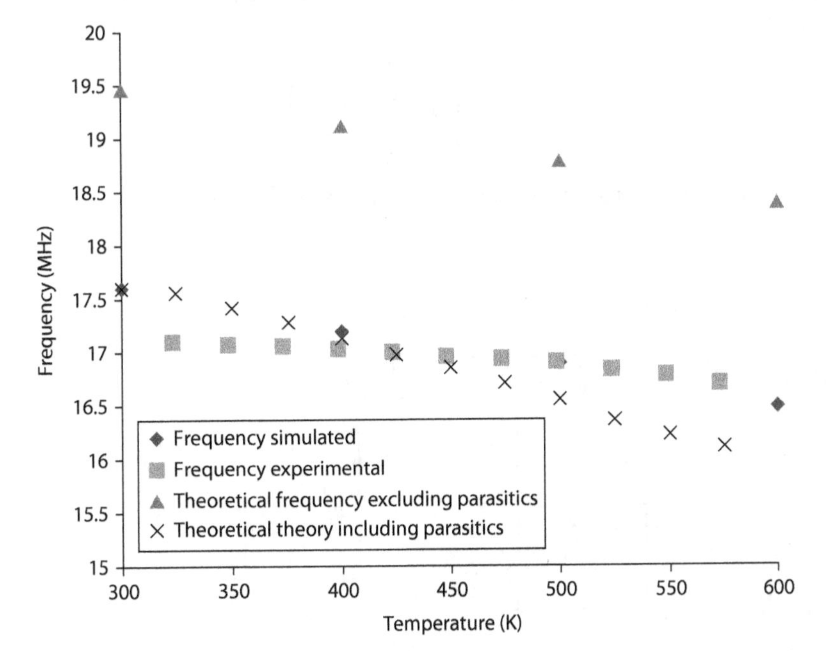

FIGURE 10.17
Comparison of calculated, simulated and experimental frequencies of a silicon carbide Colpitts oscillator as a function of temperature.

in Figure 10.17. This shows a better agreement with the experimental data for temperatures above 400 K; however, the low-temperature frequency is overestimated, similar to the modified theoretical predictions.

A working communication system requires receiver electronics that are tuned to the specific frequency of the carrier waveform generated by the oscillator circuit. While it is possible to create complicated receiver electronics capable of tracking a drifting signal, the challenge of then demodulating the information stored in the carrier signal becomes significantly more difficult. For this reason, it is far more desirable to directly control the frequency of the oscillator in the transmission system. For oscillator circuits based on LC tanks, this is commonly achieved with a device known as a varactor. A varactor is a p-n diode that has been optimized to give a significant capacitance under reverse bias conditions, rather than a conventional p-n diode structure where reverse bias capacitance is minimized to increase the switching speed. Since p-n diodes under reverse bias exhibit a depletion region, the width of which varies with voltage, the capacitance of these devices is inversely proportional to the square root of the applied voltage. Although a small number of reports on SiC varactors can be found in the literature [26,27], it is worth noting that all diodes, including SiC Schottky diodes, exhibit an identical change in capacitance with applied bias. For this reason, it is possible to modify the high-temperature Colpitts oscillator circuit shown in Figure 10.15 to form a high-temperature VCO by the inclusion of a Schottky diode, resulting in the circuit shown in Figure 10.19.

Neglecting the effects of the JFET junction capacitance and any stray capacitance from the circuit on the frequency of the oscillations, the addition of a reverse-biased SiC Schottky diode acting as a varactor can be considered as adding two capacitors, denoted by C_2 and D_1, in series as part of the tank circuit. This results in a capacitance that is voltage dependent in the lower branch of the tank circuit and the oscillator frequency (ignoring the influence of parasitics in the circuit) can now be approximated by

$$f = \frac{1}{2\pi\left(\sqrt{L\frac{C_1 C_{TOT}}{C_1 + C_{TOT}}}\right)} \tag{10.20}$$

where C_{TOT} is the capacitance of the Schottky diode and C_2 in series.

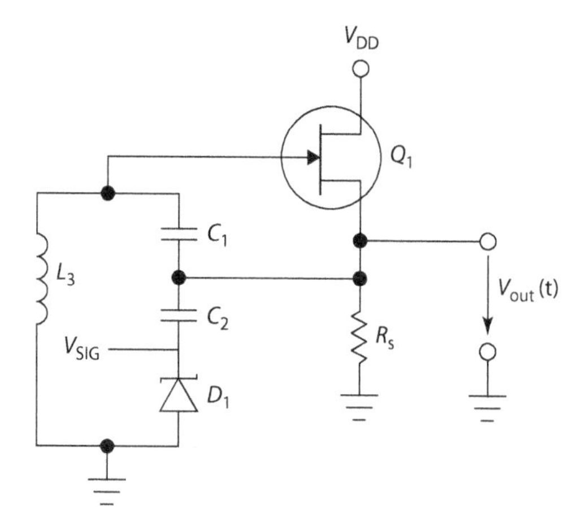

FIGURE 10.18
Circuit schematic of a high-temperature, voltage-controlled oscillator.

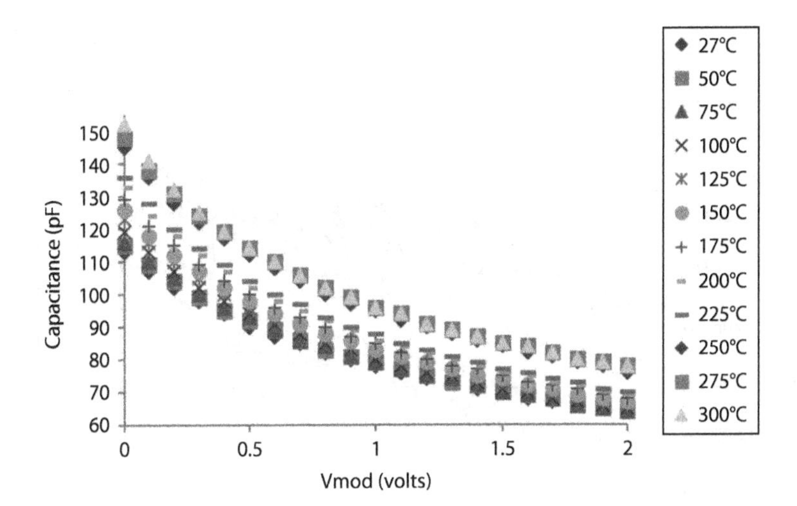

FIGURE 10.19
Capacitance of an SiC Schottky diode with different applied voltage bias as a function of temperature.

The capacitance of the Schottky diode can be decreased by the application of an external bias to the V_{SIG} port shown in Figure 10.18 and can be expressed as

$$C = A\left(\frac{q\varepsilon}{2}\right)^{1/2}\left(\frac{N_D}{\phi_B - V_A}\right)^{1/2} \tag{10.21}$$

where A is the diode area, E is the dielectric constant of the SiC, q is the electronic charge, N_D is the dopant concentration in the n-type region (assuming the abrupt junction approximation), ϕ_B is the barrier height and V_A is the applied bias.

The data in Figure 10.19 show the variation in the capacitance of an SiC Schottky diode as a function of reverse bias for the range of temperatures of interest. The observed decrease in capacitance with increasing temperature is linked to the increasing depletion width formed in the device, which results in the decreasing capacitance of the diode, as is expected from parallel plate capacitor theory [14]. Note that the capacitance of D_1 is in series with the capacitance of C_1 and so it is possible to control C_{TOT} by applying a voltage to D_1. This results in the decrease of C_{TOT} and hence an increase in the resonant frequency of the oscillator, as described by Equation 10.20 and can be seen from the data plotted in Figure 10.20.

As shown by the data in Figure 10.20, it is possible control the frequency of an SiC Colpitts-based VCO across a wide temperature range. The data show the potential to create an electronic feedback loop to stabilize the resonant frequency of the circuit through the application of a bias at the V_{SIG} port of the circuit. This technique can be used to nullify the temperature dependence of the oscillator for temperatures between 273 K and 573 K. This technique can also be used to achieve direct FM of the oscillations and hence open up the possibility of generating modulated signals that can be utilized for data transmission in extreme environments.

10.4.3 Modulation

Modulation is the term used to describe the process of varying one or more properties of a periodic waveform. The high-frequency carrier signal is modulated with a lower-frequency

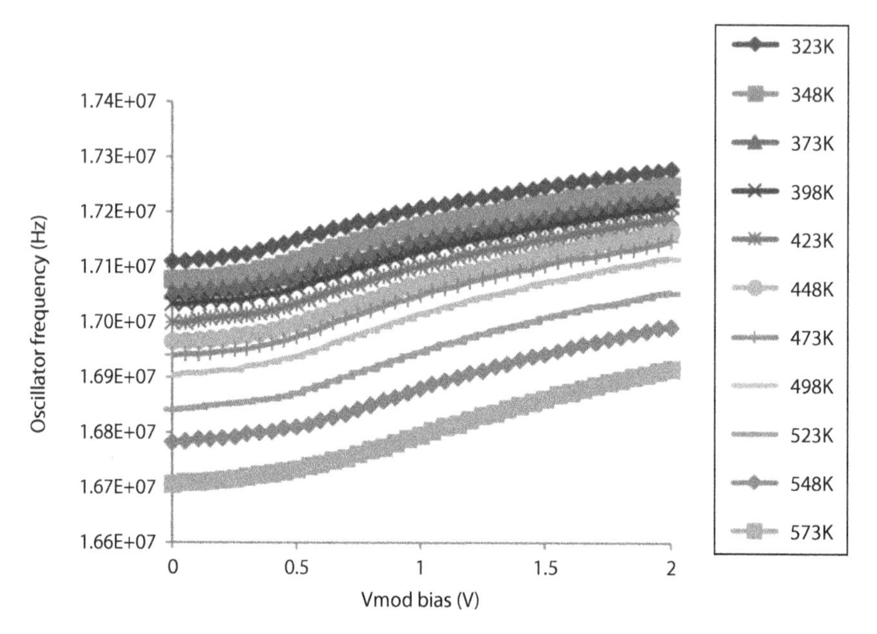

FIGURE 10.20
Frequency range of SiC voltage-controlled oscillator.

signal (often referred to as the secondary signal), which typically contains the information to be transmitted [12]. In principle, three basic methods of modulating the carrier waveform are practical and they can be implemented using either analog or digital methods. Figure 10.21a–c schematically shows examples of amplitude modulation (AM), FM and phase modulation (PM) schemes, respectively.

Considering the AM scheme, shown in Figure 10.21, the carrier waveform of a set frequency is combined with an analog information signal by modulating the amplitude of the carrier signal. This is one of the simplest forms of communication systems; however, data fidelity can be compromised by temporary changes in the efficiency of the communications channel between the transmitter and the receiver, such as changes in weather or people passing by. It is possible to utilize this method in an engineering context for the transmission of low data rate, noncritical digital signals for short distances; however, in order to achieve the performance required for a wireless sensor node, alternative modulation schemes are required. Figure 10.21b is an example of an FM scheme implemented using digital techniques, where the amplitude of the carrier waveform remains constant, while the frequency is varied between two distinct states. Each of these two states is used to represent a binary number. While this modulation system has a greater level of complexity for both the transmitter and the receiver, the significant increase in data fidelity has made this modulation scheme extremely popular for the transmission of data from wireless nodes. An example of a digital PM scheme is depicted in Figure 10.21c. In this scheme, the magnitude and frequency of the signal remains constant, while the phase of the waveform is shifted by 180° to indicate the different binary numbers. This scheme offers the potential of the greatest information density, but with the highest level of complexity – often using microprocessors to control the transmitter and receiver in real time.

Figure 10.22 shows a schematic representation of a simple analog AM communications system that is suitable for implementation in SiC technology. In this example, the analog signal used to modulate the carrier waveform could be produced using a sensor either

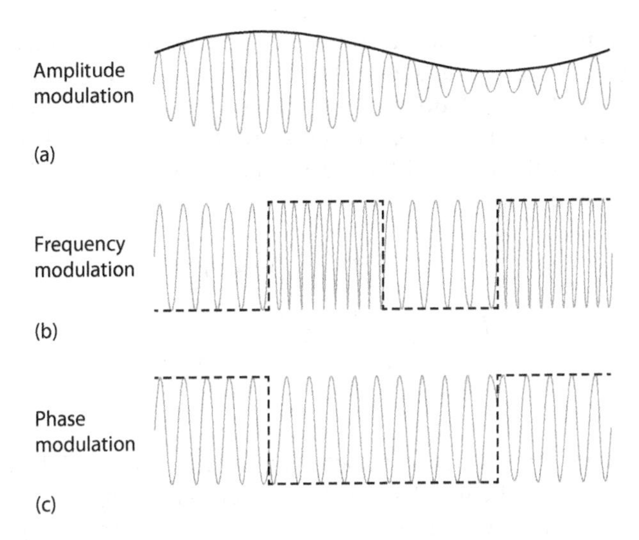

FIGURE 10.21
Example waveforms showing (a) amplitude modulation, (b) frequency modulation and (c) phase modulation.

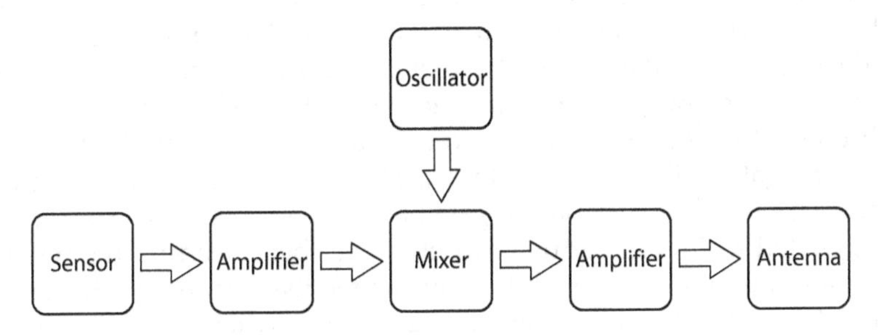

FIGURE 10.22
Block diagram of a simple analog AM communications system.

directly or through previous amplification electronics. The amplitude of the modulation signal is directly proportional to the signal on the sensor and so AM modulation can be achieved by means of a mixer circuit. The mixer ensures that the output amplitude of the oscillator is modulated by the magnitude of the sensor signal. This is the basis of a simple analog communications system where the amplitude of the carrier signal can be directly related to the sensor. The amplifier needs to demonstrate a high linearity at a single frequency to ensure that the data fidelity is maintained and to avoid complex deconvolution at the receiver.

Previously in this chapter, it was shown that by utilizing SPICE models for the SiC JFET and Schottky diode along with the characteristics of the high-temperature passive components, it is possible to design, simulate and commission a high-temperature Colpitts oscillator using SiC devices. Here, this is extended to show that the amplitude of the carrier signal can be directly modulated at 280°C [10]. Using the schematic circuit shown in Figure 10.15 as the Colpitts oscillator, capacitors fabricated from AlN dielectric C_1 and C_2 were selected to have values of 68 and 82 pF, respectively. The choice of AlN as the dielectric was to minimize the leakage current through the capacitors and to enable the integration of the capacitors with the dielectric films commonly used in the fabrication of high-temperature packages. A high-temperature printed circuit board was commissioned from a thick gold film on a ceramic substrate. The approximate frequency of the LC tank was determined from the physical size of the inductor, L. Based on Equations 10.20 and 10.22, the frequency of the oscillations was predicted to be approximately 22 MHz for an inductor value of 1.4 μH. The load resistor in the circuit was set to 1 kΩ, and 12 V DC was supplied to the circuit at the drain of the JFET denoted by Q. The inductance of a planar spiral can be determined from

$$L = \frac{N^2\left(\frac{D_i + N(W+S)}{2}\right)}{30\left(\frac{D_i + N(W+S)}{2}\right) - 11D_i} \tag{10.22}$$

where N is the number of turns on the inductor, D_i is the diameter of the innermost turn, N is the number of turns, W is the width of the tracks and S is the spacing between the turns.

Simulations of the oscillator circuit based on the experimental device parameters extracted indicate that the oscillator is capable of self-starting, as shown by the data in Figure 10.23.

This indicates that the transconductance of the SiC JFETs used in the circuit is sufficient to ensure that the Barkhausen criteria is met and the oscillations continue indefinitely. By modifying the Colpitts oscillator circuit to that shown by the schematic in Figure 10.24, it is possible to utilize a second active component to produce AM of the carrier waveform.

The amplitude of the oscillations is varied by changes in the amplitude of the feedback signal seen at the gate of JFET denoted by Q_1. This is achieved by utilizing JFET Q_2 as a variable resistor. A negative bias on the gate of Q_2 (in respect of the source of the JFET), which is designated as V_{SIG} in Figure 10.24, it is possible to control the resistance of the JFET channel and therefore the magnitude of the feedback signal to the gate of Q_1. This can be clearly observed in the data shown in Figure 10.24, where a 1 MHz sinusoidal signal

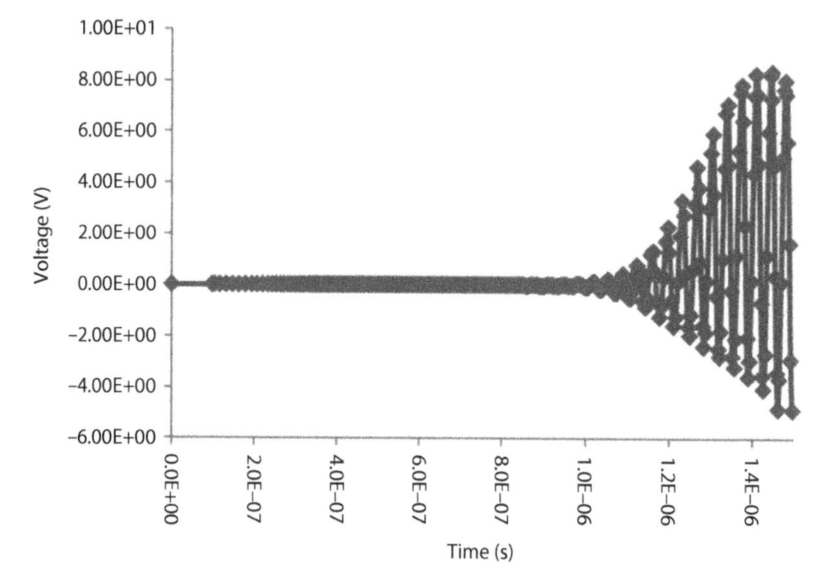

FIGURE 10.23
SPICE simulation of the self-starting of a Colpitts oscillator.

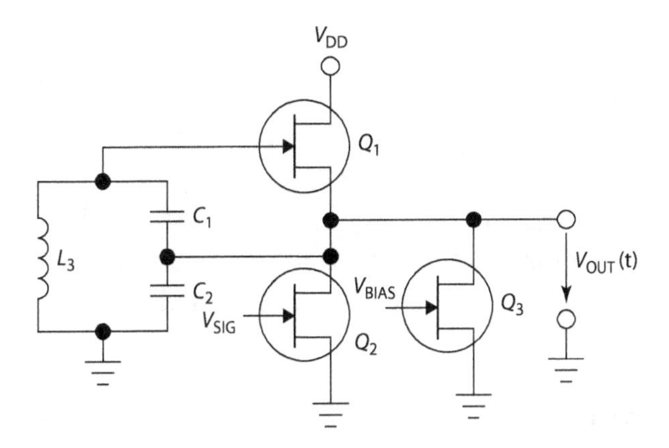

FIGURE 10.24
Circuit schematic used in a high-temperature modulation experiment.

was fed into V_{SIG}. This resulted in changes to the channel resistance of JFET Q_2 and this directly modulated the feedback seen at the capacitance tap. The effect of this change in the capacitance feedback can be observed by increasing the negative bias on V_{SIG}, which results in the amplitude of the carrier signal increasing. This can ultimately be used to create an amplitude-modulated signal at the JFET gate, which is also the signal transmitted through the inductor.

A hybrid module was then assembled using a 1000 μm thick aluminum nitride substrate onto which a seed layer of 10 nm chrome was deposited by physical vapor deposition followed by 250 nm of gold. The gold layer on the substrate was then electroplated to a thickness of approximately 8 μm, to reduce the track resistance. Capacitors fabricated with a 60 nm thick HfO_2 dielectric were selected for their low leakage characteristics (sub μA at 5 V) with values of 68 pF (C_1) and 82 pF (C_2), respectively. The inductor, denoted by L_3, was a gold spiral patterned directly onto the substrate. The frequency characteristics of the inductor were determined at 1 MHz using an Agilent 4284A LCR bridge, which demonstrated an inductance of 1.4 μH and a resistance of 4.8 Ω. The capacitors and JFETs were then attached to the circuit using silver epoxy, which was baked at 150°C for 1 h, prior to the electrical connections being made by gold wire bonding. The frequency spectrum of the oscillator was measured using an Anritsu MS2721B Spectrum Master and the amplitude of the RF signal was measured through an external aerial attached to a Tektronix TDS3045C oscilloscope.

With the gate to Q_3 held at −8 V, the application of a modulation signal to the gate of JFET Q_2 varied the amplitude of the output signal, as can be observed from the data in Figure 10.25. For an applied bias between −4 and −6 V (denoted by region B in Figure 10.26), increasing the magnitude of V_{SIG} increases the resistance of the JFET channel.

As previously described, this increases the magnitude of the output signal, with a behavior that is reasonably linear. For bias levels with a magnitude below 2.5 V, the resistance of the channel reduces the current flowing through the circuit, resulting in a reduction in the amplitude of the oscillations – as can be observed in the region denoted by A in Figure 10.26. The data indicate that it is possible to achieve AM by varying the voltage applied to Q_2 in two regions. However, in region B (the higher bias region) the frequency of the oscillations was observed to vary with the applied bias.

This is a nondesired result from a systems perspective as it will require a high level of sophistication in the receiver circuit and so the performance of the circuit operating in

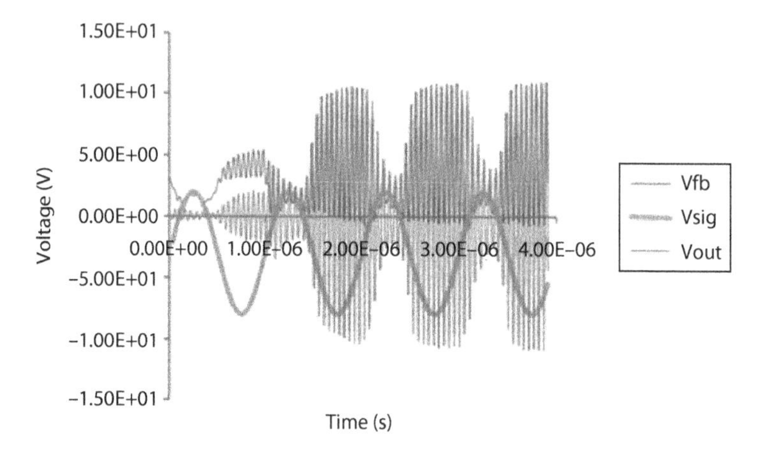

FIGURE 10.25
Simulated output characteristics of the circuit shown in Figure 10.24.

region A was investigated further. As shown by the data in Figure 10.27, the variation of the amplitude with the applied voltage is not linear. This originates in the physics behind the operation of the JFET, where the channel characteristics vary with the square of the applied voltage.

The frequency spectrum of the oscillator while operating at 280°C is shown by the data in Figure 10.28. The gate of Q_2 was held at a constant −3.5 V, resulting in the maximum oscillator amplitude. During the measurements, the amplitude of the peak frequency was 65 dBm above the background noise, with a full half width maximum of approximately 7.2 kHz.

The oscillation frequency was measured as a function of temperature to determine the temperature coefficient of the peak frequency. As can be seen from the data in Figure 10.29,

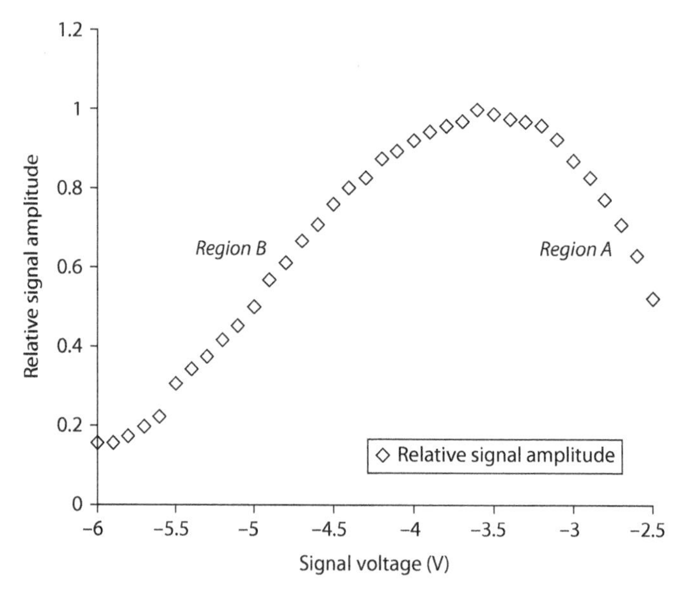

FIGURE 10.26
Relative amplitude of an output signal with varying V bias.

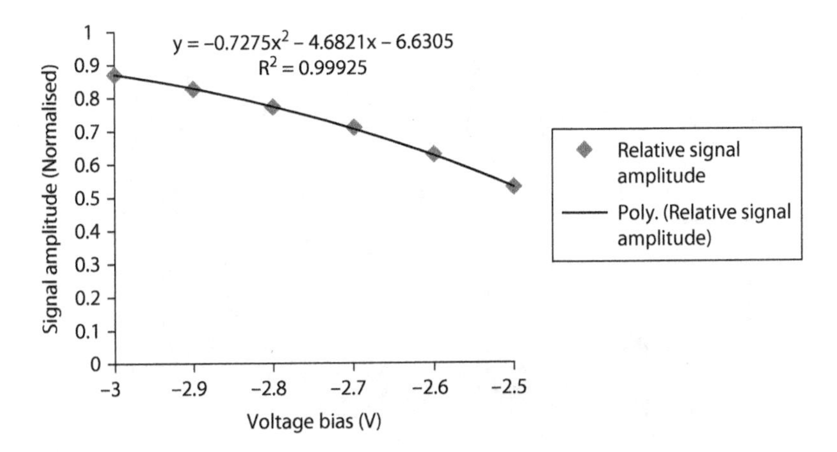

FIGURE 10.27
Signal amplitude vs voltage bias of region A.

the frequency decreases with increasing temperature, linked to the increased capacitance of the HfO_2 metal-insulator-metal (MIM) capacitors [28,29].

The data in Figure 10.30 show that in addition to the frequency shift with temperature, the amplitude of the oscillations also decreases. The data point for 353 K is influenced by a lack of temperature stability in the oven used for the testing, related to the proportional-integral-derivative (PID) parameters being unoptimized for low temperatures. The data for the decrease in signal amplitude with temperature were taken with Q_2 biased at a gate-source potential of −3.5 V to obtain maximum oscillation amplitude and hence maximize the transmitted power. For a high-temperature communications system based on AM to be effective, a feedback system that can maintain the signal amplitude with varying temperatures is essential. With the current maturity level of SiC technology this is not possible, effectively ruling out the possibility of high-fidelity data transmission using an AM communications link.

Extreme environments are not just classified in terms of high ambient temperature, high radiation dose rate and the existence of chemically corrosive species, they also often

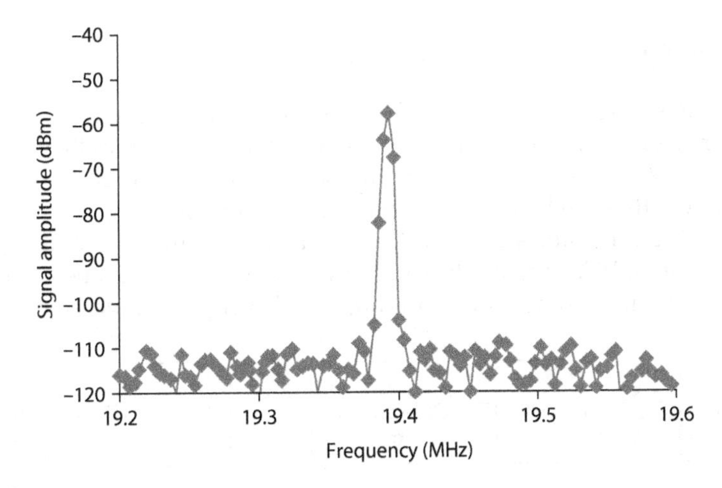

FIGURE 10.28
Frequency spectrum of AM Colpitts oscillator at 280°C.

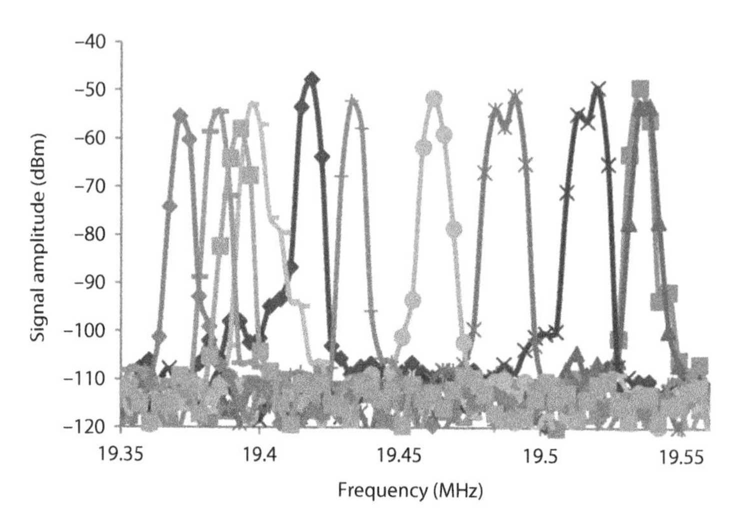

FIGURE 10.29
Change in frequency with increasing temperature.

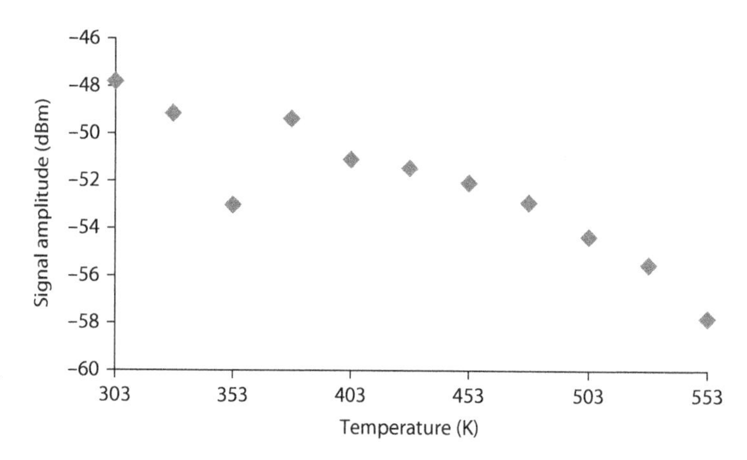

FIGURE 10.30
Maximum signal amplitude as a function of temperature.

include high intensity sources of RF interference or noise. FM communication schemes are inherently more resilient to these sources of noise than AM schemes [30].

Figure 10.31 shows a schematic representation of a simple frequency-modulated communication system that can be implemented in SiC technology.

By replacing the oscillator and mixer blocks in the schematic circuit diagram shown in Figure 10.21 with a VCO circuit, direct FM of the oscillations can be achieved [30]. In a simple sensor system, the sensor output is applied directly to the varactor diode, varying

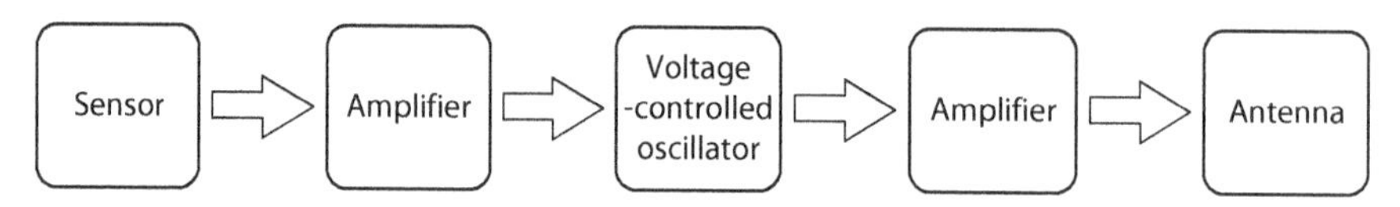

FIGURE 10.31
Block diagram of a simple FM transmitter.

the depletion capacitance and hence the frequency of a VCO. This frequency-modulated signal is then amplified to an antenna, resulting in an analog frequency-modulated communications system.

However, the system is also capable of transmitting data in digital form, as shown by the schematic in Figure 10.20b. The digital transmission of data by means of FM of the carrier wave is called frequency shift keying (FSK), in contrast to the amplitude-modulated transfer of digital data, which is referred to as amplitude shift keying (ASK). FSK can be achieved by choosing and transmitting two distinct frequencies to represent the "1" and "0" binary bits. Similarly, ASK can be achieved by selecting two distinct amplitudes at the same frequency to represent digital information.

The FM circuit shown in Figure 10.17 was demonstrated experimentally and assessed at high temperatures, with a view to estimating the possibility of FSK communications in extreme environments. A hybrid FM circuit, fabricated using the same techniques as the AM circuit previously described, was tested at temperatures up to 300°C. As shown by the data in Figure 10.32, the frequency of the oscillator varies linearly with the increasing magnitude of the reverse bias applied (between −0.75 and −1.5 V) to the Schottky diode (D_1). The shift in oscillator frequency for an applied bias of −0.75 and −1.5 V does not show a significant variation with temperature, which indicates that FM is a suitable technique for data transmission across a wide temperature range. The observed drift in the oscillator frequency with temperature is caused by the shift in the capacitances in the circuit, as described for the AM circuit.

The data in Figure 10.33 show the only reported frequency shift keyed behavior from an all SiC circuit that was operating at 300°C, captured using an Anritsu MS21721B Spectrum Master. The data clearly show two distinct peaks, one located at a frequency of 20 kHz with respect to that of the carrier waveform that represents a digital "1" and the carrier waveform itself, which represents a digital "0". The peaks are formed by the application of a square wave of 1 kHz to the diode in the VCO circuit, shown in Figure 10.19. Hence, it is possible to transmit frequency shift keyed data using an SiC oscillator over a range of temperatures that are beyond the capability of conventional, silicon-based electronic systems.

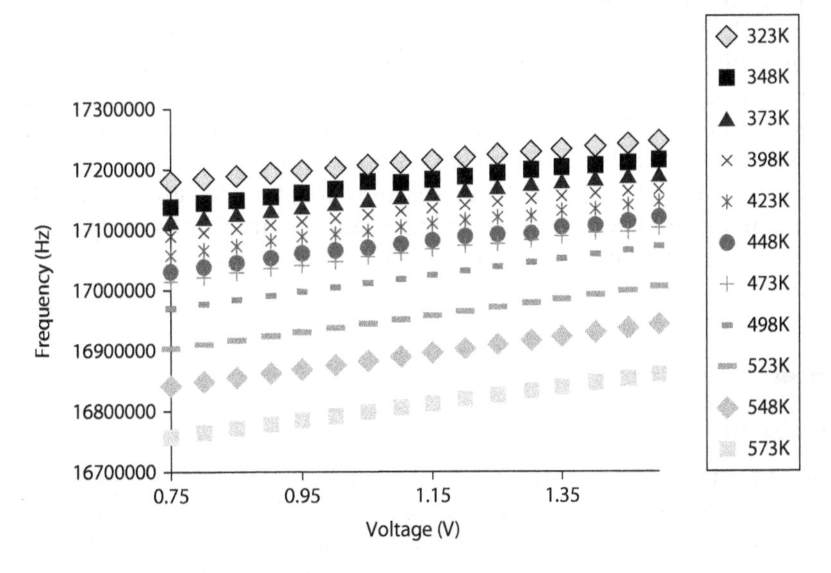

FIGURE 10.32
Linear regions of oscillator frequency change.

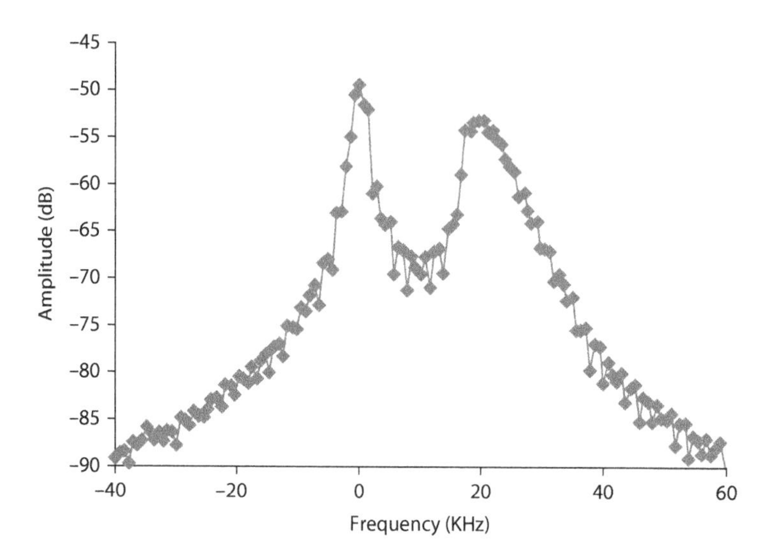

FIGURE 10.33
Spectrum showing frequency shift keyed modulation.

Of the different available modulation schemes, FSK systems are the most power efficient. This is a critical consideration for a wireless sensor node that may be powered by energy harvested from the ambient environment. The disadvantage of this modulation scheme is its low bandwidth efficiency, as it requires significant bandwidth for each sensor to operate in a unique frequency range [31]. However, there is a special case of frequency shift keyed modulation that addresses this issue and can also be viewed as an attractive alternative to the phase shift keyed modulation that commonly prevails in digital communication systems, such as mobile telephones, due to the higher data rates possible. Minimum-shift keying (MSK) has been shown to be a special case of continuous phase FSK, where the frequency deviation equals the bit rate. MSK can also be viewed as a form of offset quadrature-phase shift keying signaling in which the symbol pulse is a half-cycle sinusoid rather than the usual rectangular form. This method has the potential to combine in a single modulation format a number of attractive attributes, including constant envelope, compact spectrum, the error rate performance BPSK and simple demodulation and synchronization circuits. These features make MSK an excellent modulation technique for digital links in which bandwidth conservation and the use of efficient transmitters with nonlinear (amplitude-saturated) devices are important design criteria [31]. However, at this stage of the technology maturity, the requirements for this technique cannot be met by SiC devices and so this must remain an option for future developments.

10.5 Conclusions

This chapter has demonstrated the possibility of fabricating high-temperature electronic oscillators using SiC technology. The low maturity of SiC in comparison to silicon precludes the use of microprocessor-controlled oscillators; however, analog circuits based on discrete components have been shown to offer performance that is suitable for a wide range of applications. Predictions as to the frequency of the oscillations require accurate knowledge of the parasitic capacitances in the circuit, but more crucially the variation with temperature of these parasitics and the discrete components used in the circuit. Utilizing device characteristics that were determined over the temperature range of interest, a high-temperature, Colpitts-based VCO was demonstrated. The circuit was used to show that it is possible to control the frequency drift at high temperature along with AM and FM of two high-temperature oscillators working at temperatures up to 300°C. The physical design of the Colpitts oscillator specifically lends itself to miniaturization, featuring a capacitive feedback path, offering greater frequency stability and physically smaller components than the inductive feedback path found in the Hartley oscillator. The Colpitts oscillator also demonstrates an inherently more powerful self-starting-up ability than the Clapp oscillator, resulting in the allowed utilization of lower components with larger tolerances, which is typical in the low technological maturity SiC devices. LC oscillators provide a simple solution for producing high-frequency sine waves; these circuits contain a tuned LC tank and an active device arranged in an amplifier layout; and they are particularity useful in situations where the energy supply can be intermittent due to their self-starting ability.

References

1. N.G. Wright and A.B. Horsfall, *Journal of Physics D*, vol. 40 (2007) pp. 6345.
2. Littelfuse. Silicon carbide (SiC) diodes. www.littelfuse.com/products/power-semiconductors/silicon-carbide.aspx.
3. Wolfspeed. www.wolfspeed.com.
4. H.K. Chan, N.G. Wood, K.V. Vassilevski, N.G. Wright, A. Peters, and A.B. Horsfall, *Proceedings of IEEE Sensors Conference*, Busan, South Korea, November 1–4 (2015).
5. M.H. Weng, R. Mahapatra, N.G. Wright, and A.B. Horsfall, *IEEE Sensors Journal*, vol. 7 (2007) pp. 1395.
6. J.A. Cooper, M.R. Melloch, R. Singh, A. Agarwal, and J.W. Palmour, *IEEE Transactions on Electron Devices*, vol. 49 (2002) pp. 658.
7. C.M. Zetterling, in *Process Technology for Silicon Carbide Devices*, Inspec Publishing (2002), ISBN 0852969988.
8. M.H. Weng *et al.*, *Semiconductor Science and Technology*, vol. 32 (2017) 054003.
9. O. Mostaghimi, N.G. Wright, and A.B. Horsfall, *Proceedings of ECCE Conference*, Raleigh, North Carolina, September 15–20 (2012) pp. 3956.
10. D.R. Brennan, B. Miao, K.V. Vassilevski, N.G. Wright, and A.B. Horsfall, *Materials Science Forum*, vols. 653–956 (2010) pp. 953.
11. V. Blahm and T.P. Sylvan, *Solid State Design*, vol. 5 (1964) pp. 26.
12. A.A. Andronov, A.A. VItt, and S.E. Khaikin, in *Theory of Oscillators*, Dover Publications (2009), ISBN 0486655083.
13. T. Kimoto, K. Yamada, H. Niva, and J. Suda, *Energies*, vol. 9 (2017) pp. 918.
14. G. Sozzi, M. Puzzanghera, G. Chiorboli, and R. Nipoti, *IEEE Transactions on Electron Devices*, vol. 64 (2017) pp. 2572.
15. J. Jensen, *IRE Trans Circuit Theory*, vol. 4 (1957) pp. 276.
16. I.M. Gottleib, in *Practical Oscillator Handbook*, Newnew (1997) ISBN0750631020.
17. L.S. Senhouse, *IEEE Transactions on Electron Devices*, vol. 16 (1969) pp. 161–165.
18. K.V. Vassilevski, *IJHSES*, vol. 15 (2005) pp. 899.
19. K.V. Vassilevski, *IEEE Electron Device Letters*, vol. 21 (2000) pp. 485–487.
20. B. Razavi, in *Design of Analog CMOS Integrated Circuits*, McGraw-Hill, Boston, MA (2016), ISBN 0072524932.

21. A.H. Morrish, in *The Physical Principles of Magnetism*, Wiley, New York, NY (2001), ISBN 9780780360297.
22. R. Trew, *Proceedings of the IEEE*, vol. 79 (1991) pp. 598–620.
23. D. Leenarts, J. van der Tang, and C. Vacher, in *Circuit Design for RF Transceivers*, Kluwer, Boston, MA (2001), ISBN 9780306479786.
24. K. Sanitram, in *Basic Electronics: Devices, Circuits and Fundamentals*, Dover Publications (1972), ISBN 0486210766.
25. S. Barker, B, Miao, D.R. Brennan, N.G. Wright, and A.B. Horsfall, *Proceedings of IEEE Sensors Conference* (2009), pp. 777.
26. C.M. Anderson, *IEEE Transactions on Electron Devices*, vol. 32 (2011) pp. 788–790.
27. A.P. Knights, A.G. O'Neill, and C.M Johnson, *Proceedings of EDMO Conference* (1999), pp. 301.
28. B. Miao, R. Mahapatra, N.G. Wright, and A.B. Horsfall, *Journal of Applied Physics*, vol. 104 (2008) 054510.
29. B. Miao, R. Mahapatra, R. Jenkins, J. Silvie, N.G. Wright, and A.B. Horsfall, *IEEE Transactions on Nuclear Science*, vol. 56 (2009) pp. 2916.
30. D.R. Brennan, K.V. Vassilevski, N.G. Wright, and A.B. Horsfall, *Materials Science Forum*, vol. 717 (2012) pp. 1269.
31. R. deBuda, *IEEE Transactions on Communications*, vol. 7 (1972) pp. 429–435.

11

The Use of Error Correction Codes within Molecular Communications Systems

Yi Lu, Matthew D. Higgins, and Mark S. Leeson

CONTENTS

Molecular communications (MC) is a promising area with applications in healthcare [1–4], the environment [5,6] and manufacturing [7–9]. As such, it has received an increasing amount of attention in recent years. Considering any kind of communication system, the reliability of the data at the receiver (R_X) is one of the most important factors. Thus, this chapter focuses on the study of error correction codes (ECCs) within diffusion-based MC (DBMC) systems. Introducing ECCs is a fast and efficient way to deal with data reliability and in this chapter it is shown that such codes can be taken forward into the MC system. In addition, due to the energy limitation of nano-machines [10], which are used as the transmitter (T_X) and the R_X in the MC system, here, the energy consumption needed for the introduction of ECCs is also considered.

This chapter begins with a brief introduction to MC systems and the context of the problem. Then, four popular ECCs are introduced into MC systems with an associated analysis of the propagation and communication channel models. Finally, the performance of the coded system with regard to bit error rate (BER) and critical distance is presented.

11.1 Introduction

This section provides an overview of the MC system under consideration. Furthermore, related literature on the use of ECCs in the MC system is also presented.

11.1.1 Architecture of the MC System

To achieve the transmission process between the T_X and the R_X, a carrier signal needs to be generated and released from the T_X and, through propagation, then delivered to the R_X where the original information can be decoded. Considering this transmission process, a conventional communication system should include three main components: the T_X, the communication channel and the R_X. A basic diagram of the communication system is presented in Figure 11.1.

For MC, the molecules themselves are used as the information carrier to be transmitted between the T_Xs and the R_Xs, and these kinds of molecules are called information molecules. The three elements shown in Figure 11.1 can also be defined within the MC system. However, the T_X for the MC system is a bio-nanomachine, such as a modified living cell or a biological nano-robot, which is able to synthesize, store and release information molecules. The original information can be modulated and then released from the T_X.

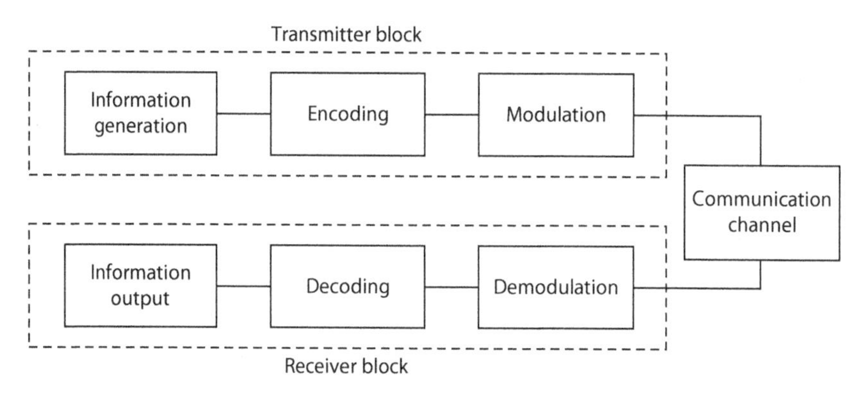

FIGURE 11.1
Block diagram of a basic conventional communication system.

In addition, encoding techniques may be included in the T_X to enhance the reliability of the transmission process.

There are three different types of MC based on different propagation schemes (details are provided in Section 11.1.2). Each propagation scheme can be considered as the information transport between the T_X and the R_X. The R_X in the MC system is another bio-nanomachine. It aims to capture the information molecules from the surrounding area, which are then demodulated and decoded if needed to recover the original information. Here, the R_X is considered an absorbing receiver, where the R_X's surface contains several receptors, the information molecules can be captured by these receptors and these then form chemical bonds to trigger the detection process in the R_X. These information molecules will be destroyed after dissociation.

11.1.2 MC Types

Due to different propagation schemes, the MC types can be divided into three categories [11]: walkway-based MC [12–14], advection-based MC [15] and DBMC [16–18].

In walkway-based MC, the information molecules propagate through active transports. For this type of MC, the pathway between the T_X and the R_X is pre-designed. An example is using molecular motors as the transport mechanism for transmitting information molecules between the T_X and the R_X. Another example of this MC type shows that the information molecules are carried by microtubules, propelled by molecular motors, and are absorbed on a flat surface between the T_X and the R_X. This type of MC is normally used for intra- or inter-cell communications.

For advection-based MC, the molecules propagate through diffusion in a fluidic medium. The use of gap junctions [19–21] and self-propelled microorganisms [22–24] (e.g. bacteria) as the transport mechanisms are two examples of advection-based MC. For the gap junction transport mechanism, the information molecule can be transmitted from the T_X to the R_X through cells that are in contact. The way that the information is inserted into bacteria and the propagation of the bacteria to complete the transmission process is called the bacterial transport mechanism. The bacteria are guided by attractant molecules that are released from the R_X.

In DBMC, the information molecules propagate through their spontaneous diffusion in the medium [11,25,26]. The focus here is on DBMC.

11.1.3 Literature Review of ECCs in the MC System

One of the early works that considered the use of ECCs in the MC system [27] considered the Hamming code as a simple block code for use in MC systems. The results indicated that introducing Hamming codes can improve the performance of the MC system. In [28], Ko et al. proposed a molecular coding distance function that considers the transition probability between codewords. Using this distance function, a suitable code for the MC system can be constructed. Due to the issue of the energy limitation of the nano-machine, the designs of simple codes for the MC system were given in [29]. Minimum energy codes were introduced into MC systems by Bai et al. [30] to reduce the energy consumption by minimizing the average code weights. The work from each study all concluded that the employment of coding techniques in the MC system can improve the BER performance compared to the uncoded system.

In this chapter, four selected ECCs from the block code family and the convolutional code family are introduced with the aim to improve the BER performance and reduce the energy consumption of the MC system. The details for BER and the critical distance of the MC system are considered in Section 11.6.

11.2 Design of the Point-to-Point DBMC System

The basic design of the three-dimensional (3D) point-to-point (PTP) DBMC system is shown in Figure 11.2. This system contains one T_X and one R_X. The center of the T_X (the release point) is at a distance d from the R_X, which has a radius of R.

The T_X can be considered as a functional bio-nanomachine, where the information molecules can be modulated, encoded and released. The propagation process is completed by the free diffusion of molecules. The R_X used in this system is a perfect absorbing receiver, where the information molecules are captured when they reach the capturing area and are removed from the environment. The whole transmission process can be divided into three phases: the emission phase, the diffusion phase and the reception phase.

Before emitting the carrier molecules, the information is modulated and encoded in the T_X, and then the molecules are released into the transmission medium. After emission from the T_X, the information molecules propagate in the environment via diffusion. During this process, they are considered to move randomly and independently. The receptors on the surface of the R_X will detect and absorb information molecules once they arrive in the capturing area. Here, the capturing surface is considered a sphere that has the same radius as the R_X. After that, the information molecules are demodulated and decoded within the R_X where the original information can be recovered.

Three communication scenarios are introduced depending on the different types of T_X and R_X: Scenario 1, a nano-machine communicates with a nano-machine (N2N communication); Scenario 2, a nano-machine communicates with a macro-machine (N2M communication); and Scenario 3, a macro-machine communicates with a nano-machine (M2N communication). The communication links for all the foregoing scenarios are based on the molecular diffusion process.

An intra-body health monitoring system shown in Figure 11.3 is one of the applications that involves all three scenarios. This system contains two sizes of machines, viz. nano-machines and macro-machines. The components of the former, nano-sensors or nano-robots, are all at the scale of a nano-meter. The latter, typically macro-robots, are

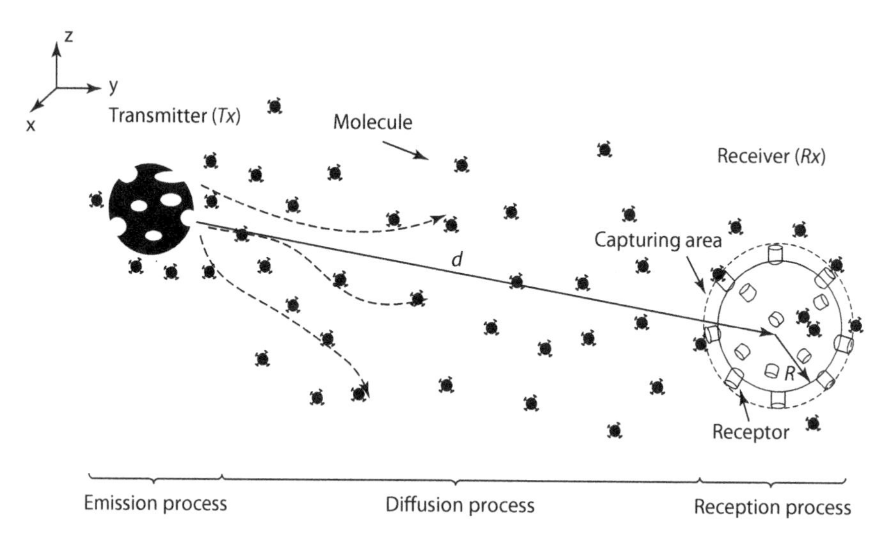

FIGURE 11.2
The 3D PTP DBMC system model.

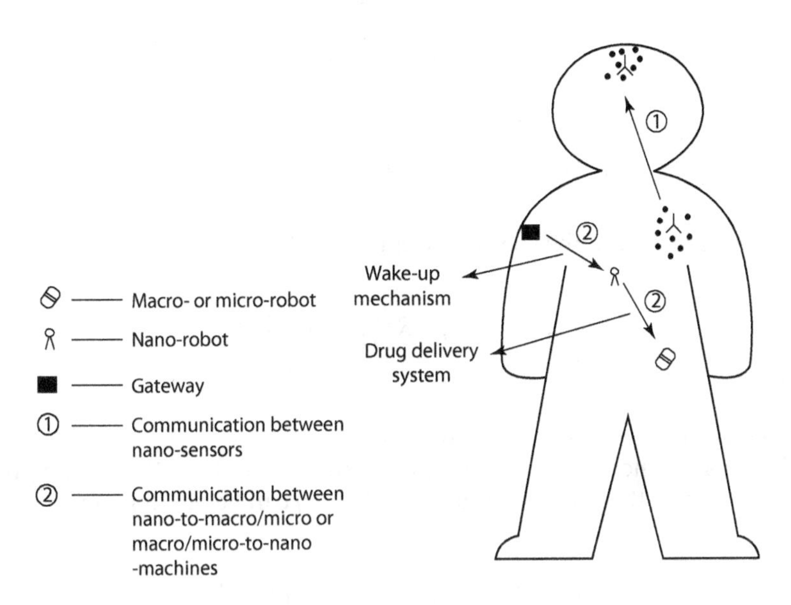

FIGURE 11.3
The communication scenarios of the intra-body nano-network.

manufactured by using a collection of nanoscale components. In drug delivery systems, a set of nano-robots as beacons located around the body can transmit information to guide macro-scale drug delivery robots working around human blood vessels [31,32]. Conversely, some applications, such as macro-machines, which act as gateways need to communicate with nano-robots and transmit information between the outside and inside of the body. This kind of macro-machine is not designed to be mobile and is most likely found on (or just under) the skin. Through emitting the information, nano-robots can be polled to get ready for a specified operation.

Here, the three communication scenarios are considered using the same propagation and communication models, which are introduced in the next two sections.

11.3 The Propagation Model

Modeling molecular propagation is one of the key challenges in predicting MC system performance. To simplify the system analysis, the information molecules are considered to move randomly and independently. The propagation of these information molecules is governed by Brownian motion. As shown in Figure 11.2, when the information molecules reach the capturing area of the R_X, there is a probability that the information molecule escapes the absorption of the R_X; this probability is called the survival probability, which can be denoted as $P_{su}(d, t)$. This probability with time t in the 3D diffusion medium satisfies the backward diffusion equation [27]:

$$\frac{\partial P_{su}(d,t)}{\partial t} = D\nabla^2 P_{su}(d,t), \tag{11.1}$$

where ∇^2 is the Laplace operator, t is the transmission time and D is the diffusion coefficient. The initial condition and the boundary conditions of Equation 11.1 are

$$P_{su}(d,0)=1,\ \forall\ |d|>R, \tag{11.2}$$

$$P_{su}(|d|=R,t)=0 \text{ and } P_{su}(|d|\to\infty,t)=1\ \forall t. \tag{11.3}$$

Exploiting radial symmetry, the solution to Equation 11.1 is

$$P_{su}(d,t)=1-\frac{R}{d}\operatorname{erfc}\left(\frac{(d-R)}{2\sqrt{Dt}}\right), \tag{11.4}$$

where erfc is the complementary error function.

For this MC system model, the capture probability rather than the escape probability is more important. Thus, the capture probability $P_{ca}(d, t)$ can be obtained by

$$P_{ca}(d,t)=1-P_{su}(d,t)=\frac{R}{d}\operatorname{erfc}\left(\frac{(d-R)}{2\sqrt{Dt}}\right). \tag{11.5}$$

Figure 11.4 illustrates that for a certain distance, the capture probability increases with the increase in time, which means that as time increases, more and more molecules will arrive at the receiver; then, by increasing the transmission time, the capture probability will become stable. In addition, for a certain period of time, the capture probability decreases when the distance between the T_X and the R_X increases.

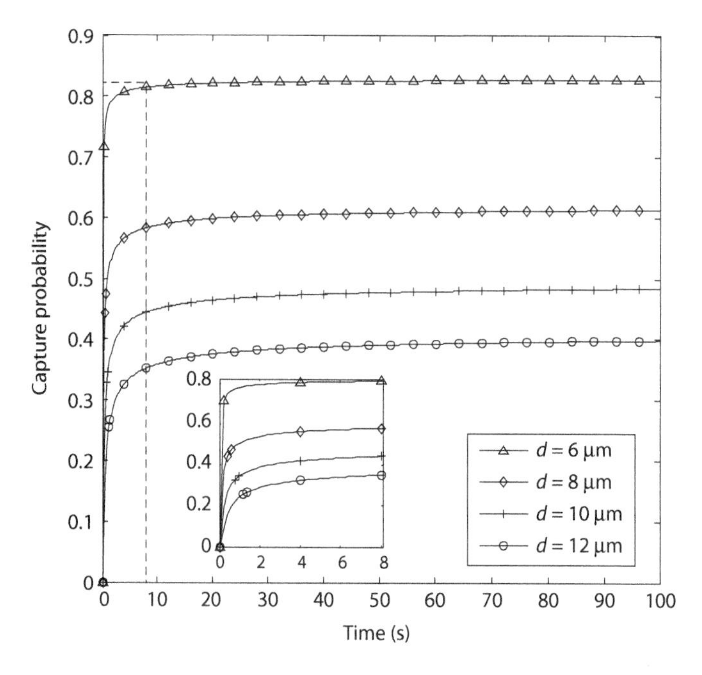

FIGURE 11.4
Capture probabilities vs. time for different distances with $R=5\,\mu m$ and $D=79.4\,\mu m^2\,s^{-1}$.

11.4 The Communication Channel Model

For a PTP DBMC system, the transmitted information is represented by a sequence of symbols that are distributed over sequential and consecutive time slots, which have equal length with one symbol in each slot; the duration of each time slot is denoted as t_s. Here, the on-off keying is used for modulating the information, where "1" represents a specific number of molecules released from T_X, and "0" represents no molecules released. Specifically, if the number of information molecules arriving at the R_X at a certain time slot exceeds a pre-designed threshold τ, the symbol is interpreted as a "1". Otherwise, it will be interpreted as a "0". However, during the transmission process, errors may be caused by intersymbol interference (ISI), which is caused by the remaining molecules from the previously transmitted symbols. The ISI effect is related to the properties of the medium used, the distance of the symbol propagation and the selection of the pre-designed threshold value. Considering a memory limited channel with ISI length I, the current transmitted symbol can be affected by the previous I symbols.

The communication channel used here is a binomial one, where each molecule arrives at the R_X or does not. It has been previously stated that the previous bits can have an influence on the current bit due to ISI. Considering that N_{tx} information molecules are released at the start of the current time slot, the number of molecules received during the current time slot, N_0, follows a binomial distribution given by [33]

$$N_0 \sim \mathcal{B}\left(N_{tx}, P_{ca,0}\right), \tag{11.6}$$

where $P_{ca,0} = P_{ca}(d, t_s)$.

If N_{tx} is large enough, a binomial distribution can be approximated by a normal distribution, thus,

$$N_0 \sim \mathcal{N}\left(N_{tx}P_{ca,0}, N_{tx}P_{ca,0}\left(1 - P_{ca,0}\right)\right). \tag{11.7}$$

The values of t_s for different distance, d, can be selected by the time at which 60% of the information molecules arrive at the R_X [33]. Considering the capture probability shown in Equation 11.5, if t goes to infinity $t \rightarrow \infty$, the analytical result R/d is obtained for the probability that an information molecule is received by R_X. Thus,

$$0.6\frac{R}{d} = \frac{R}{d}\operatorname{erfc}\left(\frac{d-R}{2\sqrt{Dt_s}}\right), \tag{11.8}$$

then t_s can be derived as

$$t_s = \frac{(d-R)^2}{4D\left(\operatorname{erfc}^{-1}(0.6)\right)^2}, \tag{11.9}$$

where erfc^{-1} is the inverse of the complementary error function.

As mentioned previously, the transmitted molecules at the start of the time slot cannot be guaranteed to reach the R_X within the current time slot, and information molecules may still exist in the environment and may arrive in future time slots. Thus, the number of information molecules that are released at the start of the ith time slot before the current one but arrive in the current time slot is denoted as N_i, and described by

$$N_i \sim \mathcal{B}\left(N_{tx}, P_{ca,i} - P_{ca,i-1}\right), \tag{11.10}$$

where $P_{ca,i} = P_{ca}(d, (i+1)\,t_s)$ for $i = \{1, 2,\ldots, I\}$.

The corresponding normal approximation can be obtained as

$$N_i \sim \mathcal{N}\left(N_{tx}\left(P_{ca,i} - P_{ca,i-1}\right), N_{tx}\left(P_{ca,i} - P_{ca,i-1}\right)\left(1 - P_{ca,i} + P_{ca,i-1}\right)\right) = \mathcal{N}\left(\varpi_i, \gamma_i\right), \tag{11.11}$$

where:

$\varpi_i = N_{tx}(P_{ca,i} - P_{ca,i-1})$

$\gamma_i = N_{tx}(P_{ca,i} - P_{ca,i-1})(1 - P_{ca,i} + P_{ca,i-1})$

Thus, the total number of information molecules received in the current time slot, N_T, comprises the number of received information molecules that were sent at the start of the current time slot and the number of received information molecules sent from all I previous time slots [34]:

$$\begin{aligned} N_T &= a_c N_0 + \sum_{i=1}^{I} a_{c-i} N_i \\ &\sim \mathcal{N}\left(a_c N_{tx} P_{ca,0} + \sum_{i=1}^{I} a_{c-i}\varpi_i,\; a_c N_{tx} P_{ca,0}\left(1 - P_{ca,0}\right) + \sum_{i=1}^{I} a_{c-i}\gamma_i \right), \end{aligned} \tag{11.12}$$

where $\{a_{c-i}, i = 0, 1, 2,\ldots, I\}$ represents the binary transmitted information sequence, which includes current and all previous I symbols.

11.5 BER Analysis for the MC System

BER is considered a key metric to assess the performance of systems that transmit information between the T_X and the R_X. Here, the ISI is considered as the main noise source.

11.5.1 BER Analysis

Considering a channel with an ISI length equal to I, error patterns can be obtained by the different permutations of the previous I symbols, so the number of error patterns is 2^I. The errors during the transmission process occur when there is a discrepancy between the transmitter and receiver signals. For a binary transmission, there are two cases, firstly, when a "0" is transmitted, but a "1" is received. Secondly, when a "1" is transmitted, but a "0" is received.

For the first case, the error probability for error pattern j, $P_{e01,j}$, $j = \{1, 2,\ldots, 2^I\}$ shows that the number of received information molecules exceeds τ, which means $N_{T,j} > \tau$, thus,

$$P_{e01,j} = p_{tx}^{\alpha_j}\left(1-p_{tx}\right)^{I-\alpha_j} P\left(N_{T,j}-\tau>0\right)$$
$$= p_{tx}^{\alpha_j}\left(1-p_{tx}\right)^{I-\alpha_j} \Phi\left(\frac{\mu_{01,j}-\tau}{\sigma_{01,j}}\right), \tag{11.13}$$

where:

$$\mu_{01,j} = \sum_{i=1}^{I} a_{c-i,j}\varpi_i, \; \sigma_{01,j} = \sqrt{\sum_{i=1}^{I} a_{c-i,j}\gamma_i}. \tag{11.14}$$

For the second case, the error probability of the normal and Poisson approximations for the error pattern j, $P_{e10,j}$, where $N_{T,j} \le \tau$, thus it can be obtained as

$$P_{e10,j} = p_{tx}^{\alpha_j}\left(1-p_{tx}\right)^{I-\alpha_j} P\left(N_{T,j}-\tau\le 0\right)$$
$$= p_{tx}^{\alpha_j}\left(1-p_{tx}\right)^{I-\alpha_j} \Phi\left(-\frac{\mu_{10,j}-\tau}{\sigma_{10,j}}\right), \tag{11.15}$$

where:

$$\mu_{10,j} = N_{tx}P_{ca,0} + \sum_{i=1}^{I} a_{c-i,j}\varpi_i, \; \sigma_{10,j} = \sqrt{N_{tx}P_{ca,0}\left(1-P_{ca,0}\right)+\sum_{i=1}^{I} a_{c-i,j}\gamma_i}. \tag{11.16}$$

Thus, the average BER of the uncoded MC system, P_e, can be derived as

$$P_e = p_{tx}P_{e10} + \left(1-p_{tx}\right)P_{e01}$$
$$= p_{tx}\sum_{j=1}^{2^I} P_{e10,j} + \left(1-p_{tx}\right)\sum_{j=1}^{2^I} P_{e01,j}, \tag{11.17}$$

where P_{e01} and P_{e10} are represented as

$$P_{e01} = \sum_{j=1}^{2^I} P_{e01,j}, \quad P_{e10} = \sum_{j=1}^{2^I} P_{e10,j}. \tag{11.18}$$

11.5.2 Numerical Results

As mentioned in Section 11.4, the previous transmitted symbols will affect the current transmitted symbol, which causes the ISI. Using an infinite ISI length in the system analysis will substantially increase the complexity of the analysis process. Thus, a limited memory channel is considered here with an ISI length I. In order to obtain an accurate result, the value of I should be evaluated.

Figure 11.5 shows the BER with the number of information molecules per bit for different ISI lengths from 1 to 10 at $d=15$ μm, $R=5$ μm and $D=79.4$ μm²/s. The results show that the longer the ISI length I, the higher the BER. It can also be noted that as the ISI length increases considerably, the effect it has upon the BER becomes less prominent, i.e. the BER

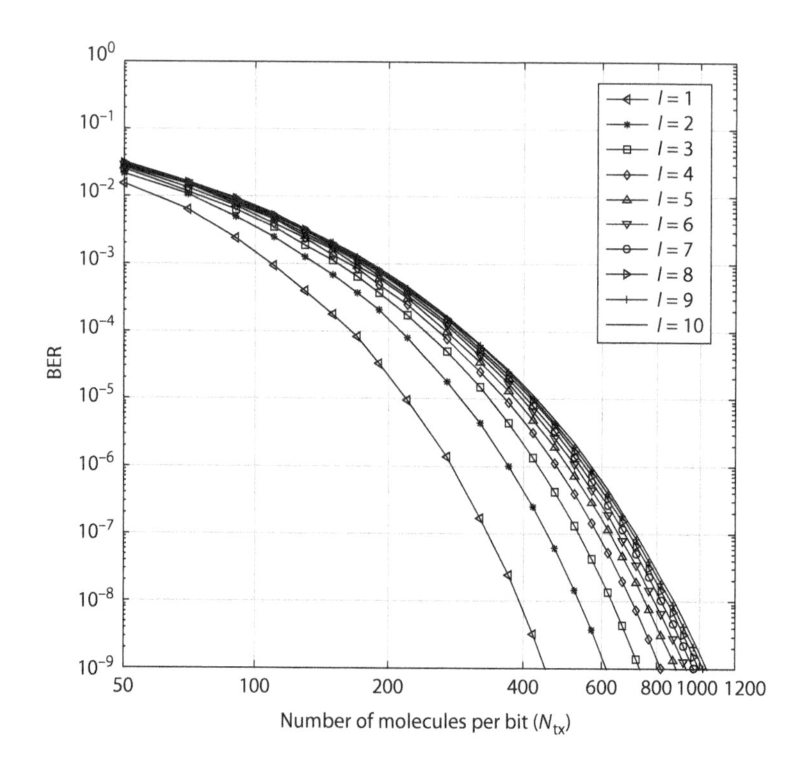

FIGURE 11.5
BER vs. number of molecules per bit for different ISI length $I = \{1, 2, \ldots, 10\}$ at $d = 15\,\mu m$, $p_{tx} = 0.5$.

value begins to converge. Therefore, choosing $I = 10$ for analysis of the channel is enough to produce an accurate result.

The value of the threshold used at the R_X is a pre-designed threshold. It can be obtained by searching the minimum BER for a specific N_{tx} in the range $\tau \in [1, N_{tx}]$.

The BER results shown in all the BER figures from here on are based on a set of parameters that are shown in Table 11.1.

Figure 11.6 demonstrates that the communication system has a better performance if large numbers of molecules are sent at the start of a one-time slot. In addition, for a chosen value of N_{tx}, the BER increases as the distance between the T_X and the R_X increases. Thus, to achieve a better performance of the communication system (lower BER), it is better to have a smaller propagation distance, d, and use a larger number of transmitted molecules per bit, N_{tx}.

TABLE 11.1

Parameter Values

Parameter	Definition	Value
R	Radius of the R_X	$5\,\mu m$
D	Transmission distance	$\{6, 8, 10, 12\}\,\mu m$
D	Diffusion coefficient	$79.4\,\mu m^2 s^{-1}$
N_{tx}	Number of molecules per bit	$50 \sim 10^3$
I	ISI length	10

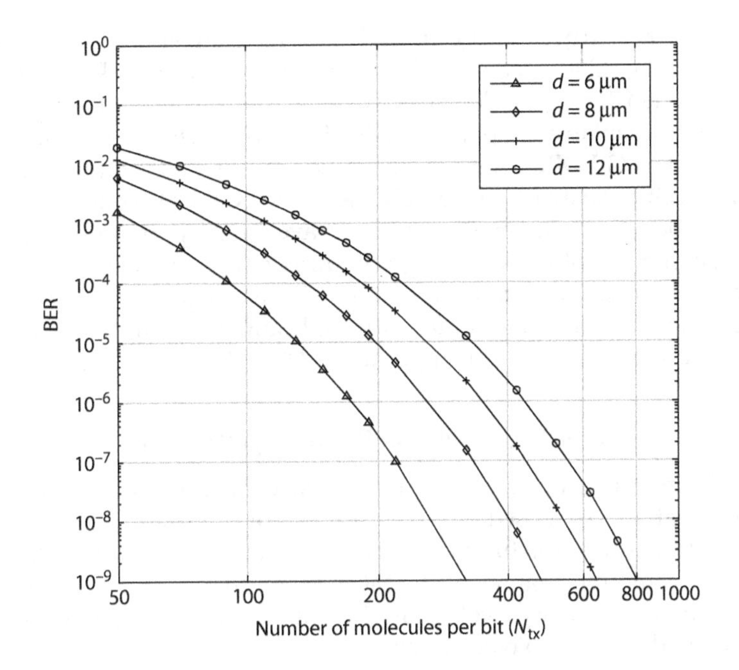

FIGURE 11.6
BER vs. number of molecules per bit.

11.6 ECCs in the PTP DBMC System

11.6.1 Overview

We now introduce four ECCs into the MC system. Three of them are from the block code family: Hamming codes, cyclic Reed–Muller (C-RM) codes and low-density parity-check (LDPC) codes. One of them is from the convolutional codes family, which is self-orthogonal convolutional codes (SOCCs). A brief review of each code is given in the following paragraphs.

Hamming codes, as one of the simplest linear codes devised for error correction, have been widely used in conventional communications systems [35]. Although these simple block codes can only correct a single error and are not powerful codes for the conventional communications system, the encoder and decoder can be implemented much more easily and efficiently in terms of the energy budget for the MC system. Here, the Hamming codes, as the most basic coding technique, are employed in the MC system and are also provided as a comparison.

RM codes [36,37] are a class of linear codes that can be constructed using Galois fields. Here, the RM codes are constructed in cyclic form (C-RM) to show that they are also a subset of Bose, Chaudhuri and Hocquenghem (BCH) codes. By using the shift-register encoder and the majority logic decoder, this kind of C-RM code can be easily encoded and decoded, respectively. From the perspective of energy, the main advantage of such C-RM codes is that the encoder is simpler than the original RM codes, which holds benefits for the above-mentioned applications [38–40].

The random and structured LDPC codes [41,42] are two kinds of LDPC codes. The Euclidean geometry LDPC (EG-LDPC) codes, which are considered for use in the MC system, are one of the structured LDPC codes. Several advantages over random LDPC codes are shown in EG-LDPC codes, such as the existence of several decoding algorithms (cyclic or quasi-cyclic), a simpler decoding scheme and the ability to extend (or shorten) the code in order to adapt to an application [43–45]. A comprehensive account of the implementation of a cyclic EG-LDPC code has been given in [46]. Thus, in this chapter, the focus will be on one specific construction, namely the cyclic EG-LDPC code.

SOCCs are a type of convolutional codes that have the property of being easy to implement, thus satisfying one of the key design requirements, simplicity [47,48]. Furthermore, this kind of convolutional code has been shown to have an equal, or superior, performance to block codes in low-cost and low-complexity applications. Examples can be found detailing their competitiveness in practical applications [49–52].

11.6.2 The Construction of Logic Gates in the Biological Field

In conventional communication systems, the coding techniques can be implemented using electric circuits that are composed of large numbers of transistors; through the interconnection of these transistors, the functions of Boolean logic can be realized. However, considering the T_X, R_X and the channel of the MC system, these electronic circuits cannot be realized due to their complexity level. Thus, the implementation of these logic gates needs to be re-investigated from the biological view.

In [53], the authors considered protein-based signaling networks within biological cells. It is shown that the fundamental motif in all signaling networks is based on the protein phosphorylation and dephosphorylation cycle, which is also called a cascade cycle via kinases and phosphatases, respectively. The cascade cycle can complete this transfer cycle very quickly when operating in ultrasensitive mode, which satisfies the requirement of the logic gates: a fast changing between two states. Thus, it is possible to construct various control and computational analog and digital circuits by combining these cascade cycles.

Figure 11.7 shows how a NAND gate could be formed from a cascade cycle. 'A' and 'B' are the inputs to the kinase step that can cause the cascade cycle to switch. A NAND gate is a universal gate, thus it is possible to build all future logical circuits. In [53], the author presented a basic memory unit, the binary counter and NOT gate by combining cascade cycles.

Furthermore, for the output of electronic circuits, a clear high- or low-level voltage should be clearly seen in order to guarantee the reliability of the transmission process. However, thermal noise and intrinsic distortion of the transistor exist in the circuits, which cause an

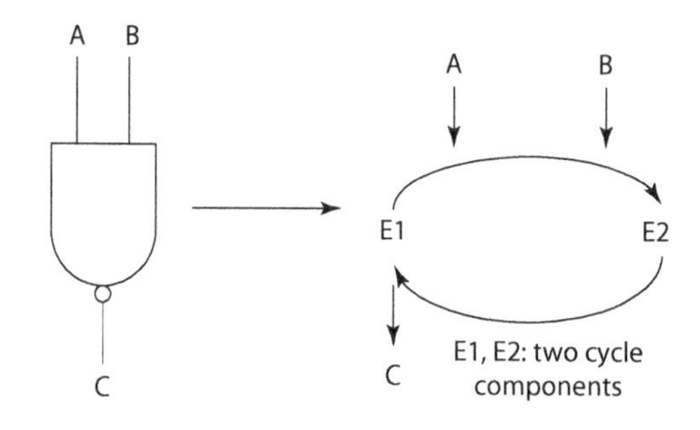

FIGURE 11.7
NAND gate using cascade cycle. (Adapted from Sauro, H.M. and Kholodenko, B.N., *Progress in Biophysics and Molecular Biology*, 86, 5–43, 2004.)

unstable level of output. In the biological field, the circuits formed of cascade cycles also suffer from noise, which is called the cascade cycle intrinsic fluctuation distortion [54]. One of the methods to reduce the effects from this kind of noise is to increase the signal intensity by increasing the number of substrate molecules, N_{sm}, at the input stage. This kind of molecule is the input signal of the cascade cycle. Here, such molecules can be looked upon as the code generation molecules that are required to encode and decode the transmission data. They are different from the information molecules, which are introduced in Section 11.4, as they are internal molecules and are only used for the encoding and decoding processes within the T_X and the R_X, and do not suffer from any effects caused by the diffusion process.

The impact of the number of substrate molecules on the output signal of the cascade cycle has been investigated in [54]. Each substrate molecule is either unmodified or modified at the output stage. The results shown in [54] indicate that for different numbers of substrate molecules used at the input stage, the performance of the output signal is different. When the N_{sm} is small ($N_{sm}=30$), the high and low outputs overlap and blur due to the cascade cycle's intrinsic fluctuation distortion; when N_{sm} increases to 300, a clear transmission output pulse is given. A further increase of N_{sm} will cause a slow output response and information loss. Overall, the selected value of $N_{sm}=300$ is sufficient to reduce the effects that come from the biochemical intrinsic distortion and obtain a clear output signal.

11.6.3 Energy Model

Other than BER, energy consumption is another key metric for a designed system. Introducing coding techniques can reduce the BER; however, in the meantime, an extra cost is also incurred due to the encoding and decoding process within the T_X and the R_X. This cost is proportional to the complexity of the design of the encoder and decoder.

Adenosine triphosphate (ATP) [55] is normally used to measure the energy transfer between cells in living organisms. Here, it is used to calculate the energy requirements of the proposed coding DBMC systems. The energy consumption during the encoding and decoding process can be composed of two parts: the energy cost of the coded circuits construction and the energy cost for generating the substrate molecules.

One cascade cycle can form a NAND gate, and one ATP is cost to activate one cascade cycle [55,56]. The energy cost from one ATP reaction, in Joules, is approximately equal to 8.3×10^{-20} J. Considering such small quantities, here the energy is worked in k_BT. Thus, the energy cost for one ATP reaction is $20k_BT$, where k_B is the Boltzmann constant and it is assumed here that the system is operating at an absolute temperature, $T=300$ K. As mentioned in Section 11.6.2, all further logic circuits can be devised from combinations of NAND gates based on the principles of Boolean algebra [57]. So, the total energy cost for the logic circuits can be computed.

Synthesizing the substrate molecules also results in energy consumption. The energy cost of synthesizing a molecule is approximately $2450k_BT$ [57]. In this case, the total energy cost of synthesizing the substrate molecules is $2450N_{sm}k_BT$.

11.6.4 ECCs in MC Systems

In block coding, the transmitted binary information sequence is segmented into information blocks with a fixed length, k. After the encoding process, each information block is transformed into an n bits codeword [50] by inserting $(n-k)$ parity check bits to improve the reliability of the information. Most known block codes belong to the class of linear codes and the cyclic code is one of the famous linear codes. This class of codes has a strong structural property and is usually used in practice. Compared with the general linear block codes, cyclic codes can be obtained by imposing an additional strong structural element on the code [58]. In this section, Hamming codes, C-RM codes and EG-LDPC codes as the cyclic codes will be described in turn.

11.6.4.1 Hamming Codes

Hamming codes, which are denoted in the form (n_H, k_H), where, $n_H = 2^m - 1$ is the coded length output for the number of parity check bits m, $m \geq 2$, and $k_H = 2^m - 1 - m$ is the number of date bits per block. The minimum distance, d_{minH}, of the Hamming code is 3, which means that only one error can be corrected in each block. The data rate of Hamming codes can be calculated as $r_H = k_H / n_H$.

The Hamming codes constructed as cyclic codes can be encoded by multiplying the information polynomial with the generator polynomial. Here, three Hamming codes are selected for use in MC systems: (7,4), (15,11) and (31,26) Hamming codes. The generator polynomials for these codes are given by $g_{m=3}(x) = x^3 + x + 1$, $g_{m=4}(x) = x^4 + x + 1$ and $g_{m=5}(x) = x^5 + x^2 + 1$, respectively. The Meggitt decoder is used here. Considering the Meggitt theorem, the syndrome polynomial for testing the error patterns of the foregoing Hamming codes are configured as $S_{m=3}(x) = x^2 + 1$, $S_{m=4}(x) = x^3 + 1$ and $S_{m=5}(x) = x^4 + x$, respectively. Figure 11.8 shows a general form of encoders and decoders for Hamming codes.

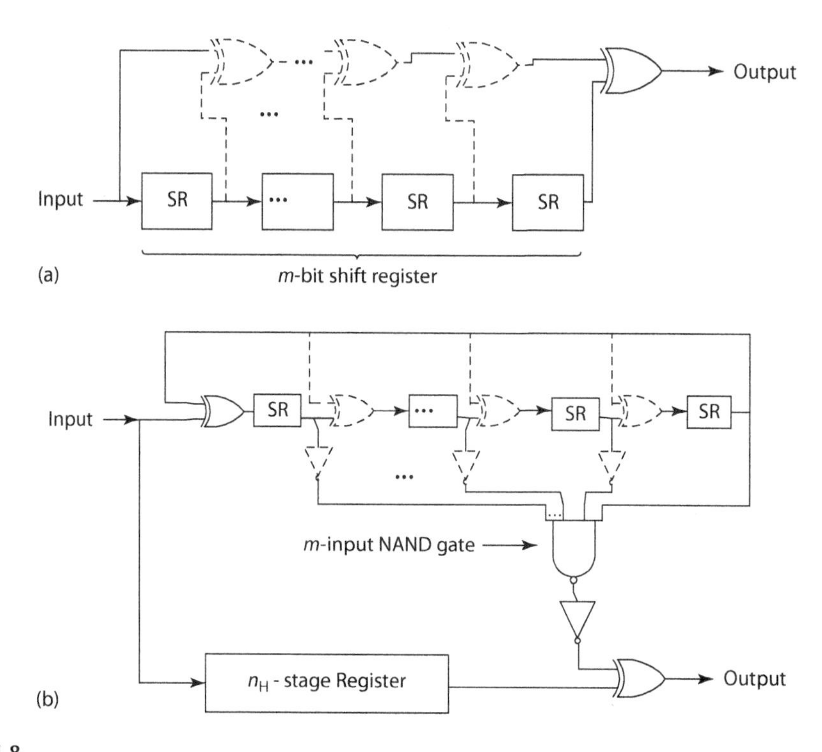

FIGURE 11.8
General nonsystematic encoder (a) and Meggit decoder (b) for Hamming code. (Adapted from Blahut, R.E., *Algebraic Codes for Data Transmission*, Cambridge University Press, 2003.)

11.6.4.2 C-RM Codes

The RM codes considered here are constructed as cyclic codes to achieve multiple error correction capabilities.

For any integer, $l \geq 2$ and $0 \leq z < l-1$, the zth-order C-RM codes can be represented as (z, l) C-RM, with a data rate $r_R = k_R/n_R$, a block length $n_R = 2^l - 1$ and the information length can be calculated as

$$k_R = 1 + \binom{l}{1} + \binom{l}{2} + \cdots + \binom{l}{r} = n_R - \sum_{i=1}^{l-z-1} \binom{l}{i}. \tag{11.19}$$

The minimum distance of a C-RM code is $d_{\mathrm{minR}} = 2^{l-z} - 1$ and the error capability can be calculated as

$$E_{cR} = [d_{\mathrm{minR}} - 1]/2. \tag{11.20}$$

Here, two C-RM codes, (1,4)C-RM and (2,5)C-RM codes are considered for MC systems. The generator and check polynomial for (1,4) and (2,5)C-RM are given by

$$g_{(1,4)\text{C-RM}}(x) = M_1 \cdot M_2 \cdot M_3 = x^{10} + x^8 + x^5 + x^4 + x^2 + x + 1. \tag{11.21}$$

$$h_{(1,4)\text{C-RM}}(x) = (x+1) \cdot (x^4 + x^3 + 1) = x^5 + x^3 + x + 1. \tag{11.22}$$

$$g_{(2,5)\text{C-RM}}(x) = x^{15} + x^{11} + x^{10} + x^9 + x^8 + x^7 + x^5 + x^3 + x^2 + x + 1. \tag{11.23}$$

$$h_{(2,5)\text{C-RM}}(x) = x^{16} + x^{12} + x^{11} + x^{10} + x^9 + x^4 + x + 1. \tag{11.24}$$

A general design of C-RM codes is shown in Figure 11.9, where the encoding process can be achieved using simple shift registers and the decoding process can be realized using a multiple-step majority logic method.

11.6.4.3 EG-LDPC Codes

In this section, a special case called cyclic two-dimensional EG-LDPC codes is considered [43,59]. To simplify the nomenclature, the following LDPC codes are assumed to be cyclic two-dimensional EG-LDPC codes.

LDPC codes can be represented as (n_L, k_L) with a block length $n_L = 2^{2s} - 1$ and an information length $k_L = 2^{2s} - 3^s$. The data rate of LDPC codes can be represented as $r_L = k_L/n_L$, and the error correction capability can be denoted as E_{cL}, where

$$E_{cL} = [d_{\mathrm{minL}} - 1]/2, \tag{11.25}$$

where the minimum distance $d_{\mathrm{minL}} = 2^s + 1$.

The (15,7), (63,37) and (255,175)LDPC codes are considered in this section. The generator polynomial for the above codes is given as

$$g_{(15,7)\text{LDPC}}(x) = x^8 + x^7 + x^6 + x^4 + 1. \tag{11.26}$$

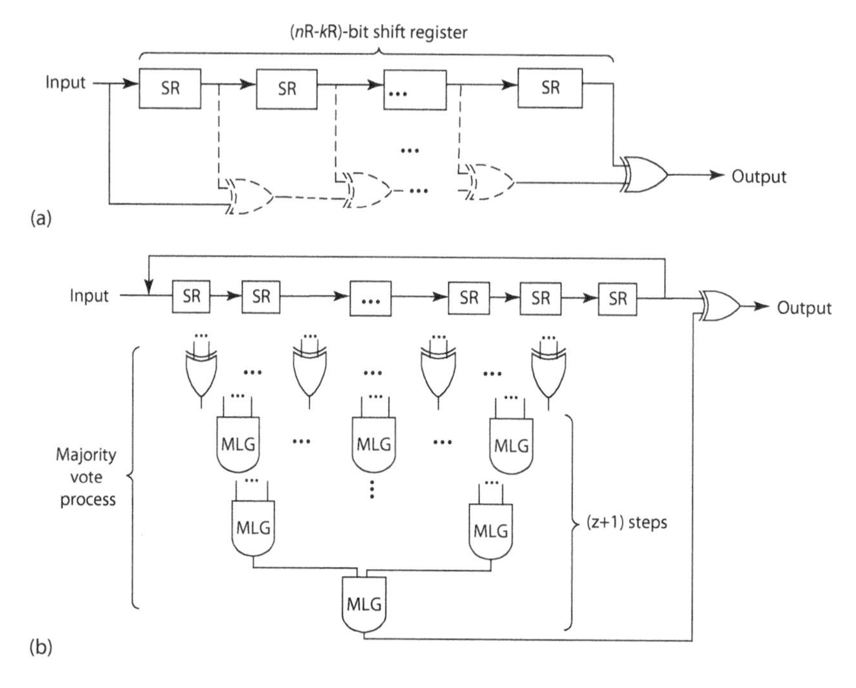

FIGURE 11.9
(a) General encoder and (b) general majority logic decoder for C-RM codes.

$$g_{(63,37)\text{LDPC}}(x) = x^{26} + x^{24} + x^{16} + x^{15} + x^{14} + x^{13} + x^{12} + x^{10} + x + 1. \tag{11.27}$$

$$\begin{aligned} g_{(255,175)\text{LDPC}}(x) = {} & x^{80} + x^{78} + x^{76} + x^{74} + x^{71} + x^{69} + x^{68} + x^{67} \\ & + x^{66} + x^{64} + x^{63} + x^{61} + x^{59} + x^{58} + x^{55} + x^{54} + x^{51} \\ & + x^{49} + x^{47} + x^{45} + x^{42} + x^{40} + x^{39} + x^{38} + x^{37} \\ & + x^{36} + x^{27} + x^{26} + x^{25} + x^{23} + x^{22} + x^{21} + x^{19} \\ & + x^{18} + x^{17} + x^{16} + x^{15} + x^{14} + x^{13} + x^{11} + x^{10} \\ & + x^{9} + x^{7} + x^{6} + x^{3} + 1. \end{aligned} \tag{11.28}$$

Figure 11.10 shows a general encoder and decoder design for LDPC codes. The encoding process can be achieved using simple feedback shift registers and the decoding process can be implemented using a one-step majority logic decoder.

11.6.4.4 SOCCs

One of the convolutional codes considered here is the SOCC. A SOCC is a kind of convolutional code that has the property of being easy to implement, thus satisfying one of the key design requirements of code simplicity [47,48].

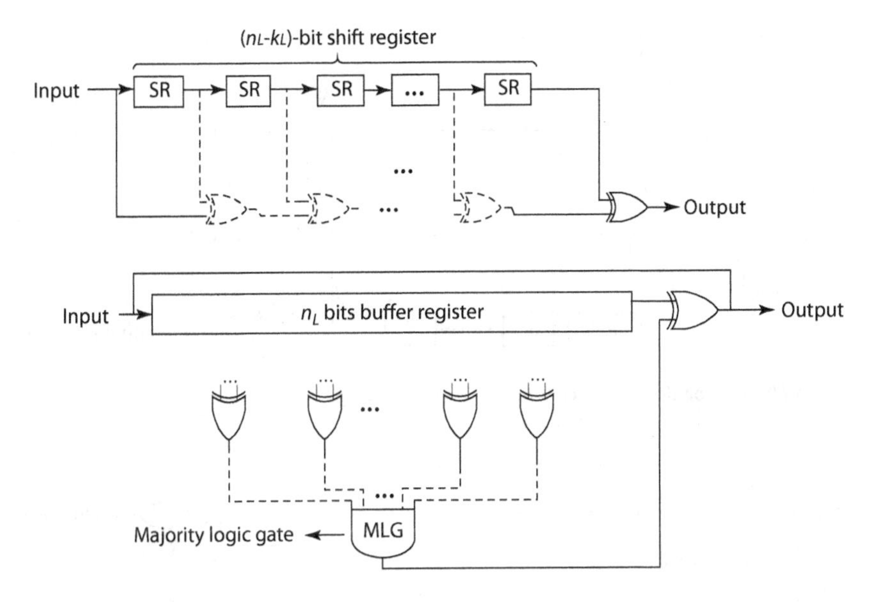

FIGURE 11.10
(a) General encoder design and (b) decoder design for LDPC code.

For an (n_S, k_S, b) SOCC, n_S is the code length, k_S is the information length and b is the number of input memory blocks. The code rate is $r_S = k_S/n_S$ [50]. The error correction capability, E_{cS}, is

$$E_{cS} = \left[J/2 \right], \tag{11.29}$$

where J is the number of check sums that orthogonal on one error.

The effective constraint length, n_E, represents the total number of channel error bits checked by the orthogonal check sum equations, where

$$n_E = \tfrac{1}{2}\left(J^2 + J \right) k_S + 1. \tag{11.30}$$

In this work, only SOCCs with an information length $k_S = n_S - 1$ are considered. Here, four SOCCs are presented: (2,1,6) and (2,1,17)SOCCs, both with $r_S = 1/2$, and (3,2,2) and (3,2,13)SOCCs, both with $r_S = 2/3$.

The generator polynomials of (2,1,6) and (2,1,17)SOCCs are

$$g_{(2,1,6)\text{SOCC},1}{}^{(2)}\left(x \right) = x^6 + x^5 + x^2 + 1, \tag{11.31}$$

$$g_{(2,1,17)\text{SOCC},1}{}^{(2)}\left(x \right) = x^{17} + x^{16} + x^{13} + x^7 + x^2 + 1, \tag{11.32}$$

where $g_{\text{SOCC},i}{}^{(ns)}$ is the generator polynomials with $i = 1, 2, \ldots, k_S$.

For (3,2,2) and (3,2,13)SOCCs, the generator polynomial pairs are

$$g_{(3,2,2)\text{SOCC},1}{}^{(3)}\left(x \right) = x + 1, \tag{11.33}$$

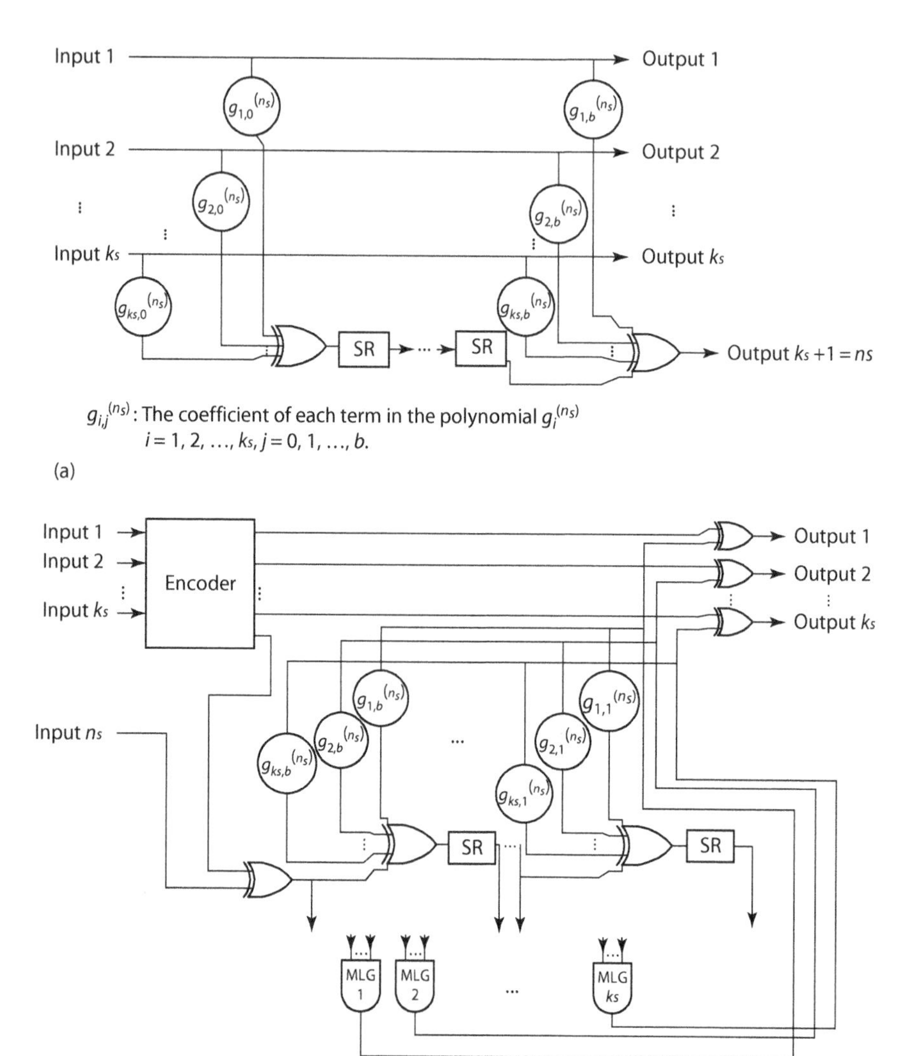

FIGURE 11.11
The general (a) encoder and (b) decoder design for SOCCs.

$$g_{(3,2,2)\mathrm{SOCC},2}{}^{(3)}(x) = x^2 + 1. \tag{11.34}$$

The circuits of the encoder and feedback majority logic decoder for a general SOCC are shown in Figure 11.11.

11.6.5 BER and Coding Gain

Considering a linear block code with a decoder that can correct all errors less or equal to the error correction capacity, the decoded BER for the system with linear block codes can be expressed by the following approximation:

$$P_{\text{e-coded}} \approx \frac{1}{n}\sum_{j=E_c+1}^{n} j\binom{n}{j}\left(P_e^*\right)^j\left(1-P_e^*\right)^{n-j}, \tag{11.35}$$

where n is the block length, E_c is the error correction capacity and P_e^* is the one bit error probability in the uncoded case.

For a fair comparison with the uncoded operation, the number of molecules used for the calculation of P_e^* for the coded system should be evaluated with a reduction in the number of molecules used for an uncoded system, (11.17), by multiplying with the code rate.

On the other hand, for the convolutional code with a feedback majority logic decoder, the theoretical analysis of BER of the convolutional code can be upper bounded by [50]

$$P_{\text{e-coded}} \le \frac{1}{k}\sum_{j=E_c+1}^{n_E} j\binom{n_E}{j}\left(P_e^*\right)^j\left(1-P_e^*\right)^{n_E-j}, \tag{11.36}$$

where k is the information length of the code. $P_{\text{e-coded}}$ can be approximated as [50]

$$P_{\text{e-coded}} \approx \frac{1}{k}\binom{n_E}{E_c+1}\left(P_e^*\right)^{E_c+1}. \tag{11.37}$$

The coding gain is also introduced as a way to measure the BER performance. For MC systems, the coding gain aims to measure the difference between the number of molecules for the uncoded and coded system required to reach the same BER level. It can be directly obtained as

$$G_{\text{coding}} = 10\times\log\left(\frac{N_{\text{uncoded}}}{N_{\text{coded}}}\right), \tag{11.38}$$

where N_{uncoded} and N_{coded} are the number of information molecules for the uncoded and the coded system at a chosen BER level.

11.6.6 Energy Consumption Analysis

The energy consumption for different coding techniques is given in this section with respect to the energy model that was introduced in Section 11.6.3. Based on that model, Table 11.2 shows four basic logic gates and their corresponding ATPs' consumption.

TABLE 11.2

Logic Gates and Corresponding ATPs' Consumptions

Logic Gate	Cost in ATPs
NAND	1
NOT	1
XOR	4
Shift-register unit	4

11.6.6.1 Energy Consumption for Hamming Codes

With reference to Figure 11.8, for $m = \{3, 4, 5\}$ Hamming codes, two XOR gates and m shift-register units are needed for each encoder circuit, which implies that the energy consumption of the encoder circuits is

$$E_{\text{en-H}} = 20\,N_{\text{sm-en}}\left(4m+8\right)+2450\,N_{\text{sm-en}}, \tag{11.39}$$

and three XOR gates, $(m+n_{\text{H}})$ shift-register units, $(m-1)$ NOT gates and one multi-input NAND gate are needed for each decoder circuits, which implies the energy cost of the decoder is

$$\begin{aligned} E_{\text{de-H}} &= 20N_{\text{sm-de}}\left(4\left(m+n_{\text{H}}\right)+\left(m-1+1\right)+12\right)+2450N_{\text{sm-de}} \\ &= 20N_{\text{sm-de}}\left(5m+4n_{\text{H}}+12\right)+2450N_{\text{sm-de}}, \end{aligned} \tag{11.40}$$

where $N_{\text{sm-en}}$ and $N_{\text{sm-de}}$ are the numbers of substrate molecules used for the encoder and decoder, respectively. As referenced in Section 11.6.2, $N_{\text{sm-en}} = N_{\text{sm-de}} = 300$.

11.6.6.2 Energy Consumption for C-RM Codes

Consider the hardware requirement in Figure 11.9. For C-RM codes, the encoder can be achieved using simple feedback shift registers and the decoding process can be completed using a multi-step majority logic method. For a majority logic gate (MLG), an output of one is produced when more than half of its inputs are equal to one, otherwise, the output is zero.

For any J-input MLG, the number of NAND gates can be calculated as

$$N_{\text{NAND-MLGs}} = \begin{cases} \displaystyle\sum_{i=J/2+1}^{J-1}\binom{J}{i}+1, & J \neq 2 \\ 2, & J = 2 \end{cases}. \tag{11.41}$$

For C-RM codes, $(n_{\text{R}}-k_{\text{R}})$ shift registers are used, and the number and the location of the two-input XOR gates in the circuits are dependent upon the generator polynomial of each code. The energy cost of encoding is

$$E_{\text{en-(1,4)RM}} = 20N_{\text{sm-en}}\left(4\left(n_{\text{R}}-k_{\text{R}}\right)+24\right)+2450N_{\text{sm-en}}, \tag{11.42}$$

$$E_{\text{en-(2,5)RM}} = 20N_{\text{sm-en}}\left(4\left(n_{\text{R}}-k_{\text{R}}\right)+40\right)+2450N_{\text{sm-en}}. \tag{11.43}$$

In general, the zth-order C-RM code can be decoded with a $(z+1)$-step majority logic decoder. For these decoding circuits, the total number, N_{ML}, of J-input MLGs used in the circuit can be analyzed as [50]

$$N_{\text{ML}} = 1+\sum_{i=1}^{L-1} J^{i}, \tag{11.44}$$

where $J = d_{\text{minR}}-1$, and $L = z+1$ is the number of steps used in the majority logic decoder.

The multi-input XOR gates used in the majority vote process can be obtained by using the combination of multiple two-input XOR gates, and the number of inputs of the XOR gate is dependent on the check polynomial. In this work, the two-input MLGs are used in (1,3), (2,4) and (3,5) C-RM decoders' design and six-input MLGs are used in the (1,4) and (2,5) C-RM decoders' design. According to (11.41), the two-input MLG and the six-input MLG can be formed by 2 and 22 NAND gates.

In addition, n_R-stage buffer registers and an extra two-input XOR gate are also needed. Here, for C-RM codes, the energy cost of decoding is

$$E_{\text{de-(1,4)RM}} = 20N_{\text{sm-de}}\left(4n_R + 590\right) + 2450N_{\text{sm-de}}, \tag{11.45}$$

$$E_{\text{de-(2,5)RM}} = 20N_{\text{sm-de}}\left(4n_R + 6998\right) + 2450N_{\text{sm-de}}. \tag{11.46}$$

11.6.6.3 Energy Consumption for LDPC Codes

As shown in Figure 11.10, for (15,7), (63,37) and (255,175)LDPC codes, $(n_L - k_L)$ shift registers are used, and the number of two-input XOR gates in the circuits is dependent upon the generator polynomial of each code. The energy cost of encoding is

$$E_{\text{en-(15,7)LDPC}} = 20N_{\text{sm-en}}\left(4\left(n_L - k_L\right) + 16\right) + 2450N_{\text{sm-en}}, \tag{11.47}$$

$$E_{\text{en-(63,37)LDPC}} = 20N_{\text{sm-en}}\left(4\left(n_L - k_L\right) + 40\right) + 2450N_{\text{sm-en}}, \tag{11.48}$$

$$E_{\text{en-(255,175)LDPC}} = 20N_{\text{sm-en}}\left(4\left(n_L - k_L\right) + 180\right) + 2450N_{\text{sm-en}}. \tag{11.49}$$

In addition, for different LDPC codes, the decoding circuits can be modified with ρ-input XOR gates, γ-input MLGs and n_L buffer registers. The multi-input XOR gate can be obtained by using the combination of multiple two-input XOR gates. Here for (15,7), (63,37) and (255,175)LDPC codes, the energy cost of decoding is

$$E_{\text{de-(15,7)LDPC}} = 20N_{\text{sm-de}}\left(4n_L + 57\right) + 2450N_{\text{sm-de}}, \tag{11.50}$$

$$E_{\text{de-(63,37)LDPC}} = 20N_{\text{sm-de}}\left(4n_L + 321\right) + 2450N_{\text{sm-de}}, \tag{11.51}$$

$$E_{\text{de-(255,175)LDPC}} = 20N_{\text{sm-de}}\left(4n_L + 27297\right) + 2450N_{\text{sm-de}}. \tag{11.52}$$

11.6.6.4 Energy Consumption for SOCCs

Referring to the description of SOCCs in Section 11.6.4, for each encoder, the number of shift-register units for the encoder is b and the number of XOR gates is dependent upon on the generator polynomials. The energy cost of the encoding is thus

$$E_{\text{en-(2,1,6)SOCC}} = 20N_{\text{sm-en}}\left(4b + 12\right) + 2450N_{\text{sm-en}}, \tag{11.53}$$

$$E_{\text{en-(2,1,17)SOCC}} = 20N_{\text{sm-en}}\left(4b + 20\right) + 2450N_{\text{sm-en}}, \tag{11.54}$$

$$E_{\text{en-(3,2,2)SOCC}} = 20N_{\text{sm-en}}(4b+12)+2450N_{\text{sm-en}}, \tag{11.55}$$

$$E_{\text{en-(3,2,13)SOCC}} = 20N_{\text{sm-en}}(4b+28)+2450N_{\text{sm-en}}. \tag{11.56}$$

The decoder is composed of two parts, one part is the same as the encoder and the other part contains b register units, k_S MLGs; the MLGs used here are two-input MLGs, where each one can be looked on as an AND gate. The number of XOR gates is dependent on the polynomial generator and the information length. So, the energy cost of the decoding is

$$E_{\text{de-(2,1,6)SOCC}} = 20N_{\text{sm-de}}(8b+37)+2450N_{\text{sm-de}}, \tag{11.57}$$

$$E_{\text{de-(2,1,17)SOCC}} = 20N_{\text{sm-de}}(8b+70)+2450N_{\text{sm-de}}, \tag{11.58}$$

$$E_{\text{de-(3,2,2)SOCC}} = 20N_{\text{sm-de}}(8b+36)+2450N_{\text{sm-de}}, \tag{11.59}$$

$$E_{\text{de-(3,2,13)SOCC}} = 20N_{\text{sm-de}}(8b+74)+2450N_{\text{sm-de}}. \tag{11.60}$$

11.6.7 Critical Distance

In order to analyze when the coding becomes beneficial, the critical distance [60] is utilized as a measure of the real transmission distance at which the coding gain matches the extra energy requirements introduced by the ECCs.

The total energy cost for an uncoded, E_{uncoded}, and a coded, E_{coded}, system can be calculated as

$$E_{\text{uncoded}} = 2450N_{\text{uncoded}}, \tag{11.61}$$

$$E_{\text{coded}} = 2450N_{\text{coded}} + E_{\text{en}} + E_{\text{de}}, \tag{11.62}$$

where N_{uncoded} and N_{coded} are the numbers of molecules used for the uncoded and coded system at a chosen BER level. E_{en} and E_{de} are the energy consumption for the encoding and decoding process.

To reach the same BER level, the energy saving for a coded system compared with an uncoded system can be defined as

$$\begin{aligned}\Delta E &= E_{\text{uncoded}} - E_{\text{coded}} \\ &= 2450(N_{\text{uncoded}} - N_{\text{coded}}) - E_{\text{en}} - E_{\text{de}}.\end{aligned} \tag{11.63}$$

It is clear that when $\Delta E \geq 0$, the use of ECC is beneficial to the MC system. When $\Delta E = 0$, Equation 11.63 reduces to

$$N_{\text{uncoded}} - N_{\text{coded}} = (E_{\text{en}} + E_{\text{de}})/2450. \tag{11.64}$$

Therefore, the relationship between N_{uncoded} and N_{coded} can be obtained by substituting the energy consumption values for different coding schemes that were introduced in Section 11.6.6.

11.7 Numerical Results

The performance of a coded PTP DBMC system is evaluated from two aspects: the BER and the critical distance for N2N, N2M and M2N communication scenarios. The BER results are obtained based on the set of parameters in Table 11.3.

11.7.1 BER for Coded DBMC Systems

Figure 11.12 shows the BER results for the coded PTP DBMC system. Figure 11.12a gives the BER comparison between the system with block codes and the uncoded system, and Figure 11.12b shows the BER comparison between the system with SOCCs and the uncoded system. The coding gains are shown to be 1.35, 1.68, 1.68, 1.68, 2.30, 1.68, 2.43, 2.57, 1.46, 1.68, 1.47, and 1.90 dB for (7,4)Hamming, (15,11)Hamming, (31,26)Hamming, (1,4) C-RM, (2,5)C-RM, (15,7)LDPC, (63,37)LDPC, (255,175)LDPC, (2,1,6)SOCC, (2,1,17)SOCC, (3,2,2)SOCC and (3,2,13)SOCC, respectively, at the BER level of 10^{-3}. And for a BER level of 10^{-9}, the coding gains for the foregoing codes are 1.59, 2.43, 2.78, 2.30, 4.29, 2.76, 5.33, 6.95, 2.89, 4.04, 2.09 and 3.80 dB, respectively. It can be seen that (15,11), (31,26)Hamming codes, (1,4)C-RM code, (15,7)LDPC code and (2,1,17)SOCC have the same coding gain at 10^{-3}; this is because the number of molecules used for achieving that BER level is the same. The results also clearly show that the (255,175)LDPC code gives the largest coding gain compared to other codes.

11.7.2 Critical Distance

For N2N communication, the extra energy requirements introduced by the encoder and decoder need to be taken into account. For N2M communication, the T_Xs are considered much simpler than the R_Xs, so when calculating the energy, only the encoder consumption needs to be taken into account by setting $E_{de} = 0$ in Equation 11.64. Moreover, for M2N communication, the R_X needs to be much simpler than the T_X, so only the decoder consumption needs to be included, so E_{en} is set to zero in Equation 11.64.

As Equation 11.64 shows, the critical distance is affected by two factors. Firstly, the BER performance for different coding schemes at different distances is shown in Figure 11.12. Secondly, the encoder and decoder circuitry for each of the codes is different, with varying levels of complexity such as those in Figures 11.8 through Figure 11.11.

The critical distance for each code can be treated as a baseline level. When the designed parameters fall to the left of this, it is considered worthwhile to apply a particular code in the system designed, otherwise not. This means that the code is worth applying only when the application has a transmission distance equal or larger than the critical distance.

TABLE 11.3

Parameter Setting for BER

Parameter	Definition	Value
R	Radius of R_X	5 μm
D	Transmission distance	6 μm
D	Diffusion coefficient	79.4 μm²s⁻¹
I	ISI length	10
N_{tx}	Number of molecules per bit	10~400

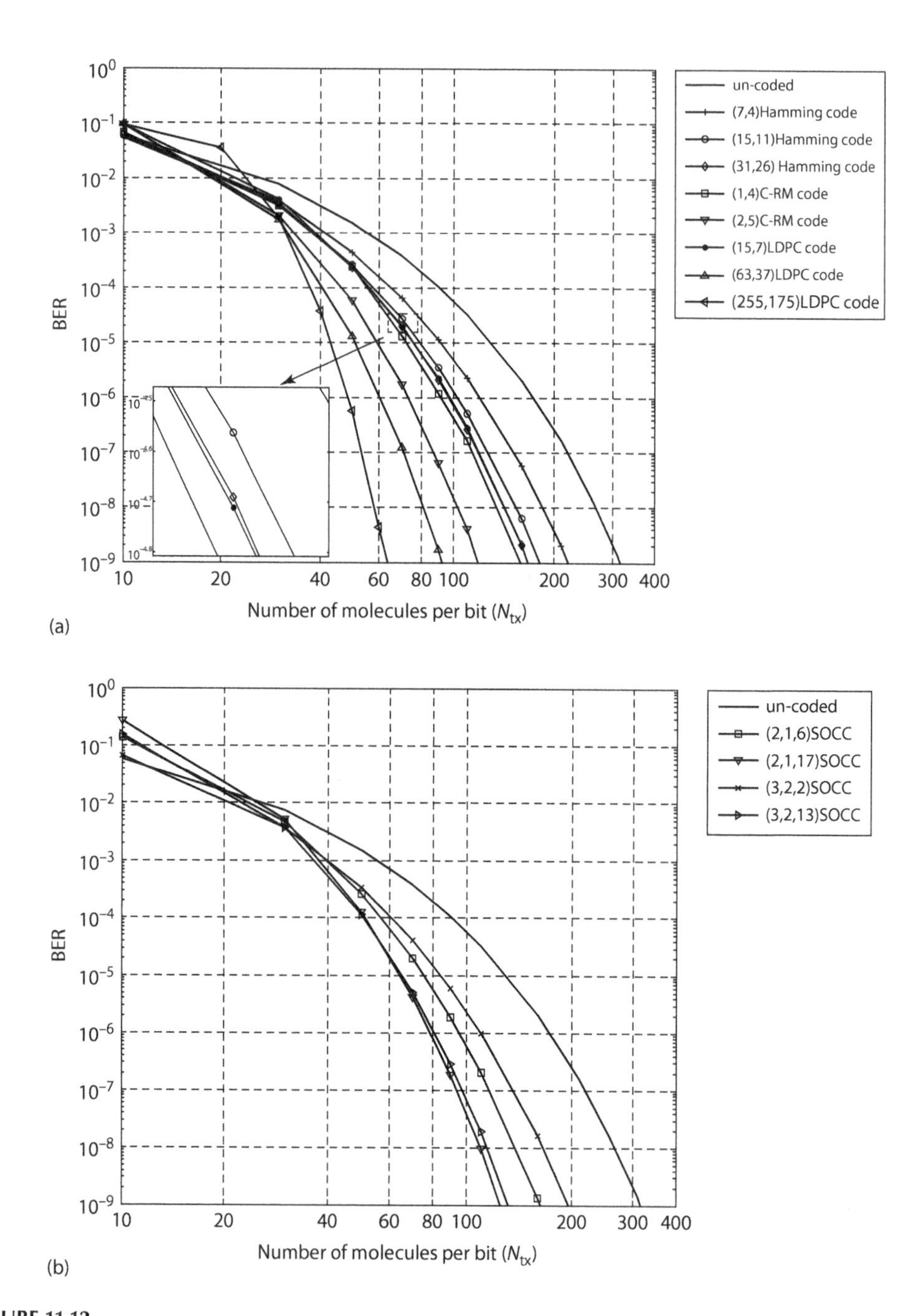

FIGURE 11.12
(a) BER comparison for a coded system with block codes and an uncoded system. (b) BER comparison for a coded system with SOCCs and an uncoded system.

On the other hand, for each communication scenario, there exists a lowest critical distance level for BERs from 10^{-9} to 10^{-3} and the corresponding code is considered as a best-fit ECC.

This best-fit ECC is defined as the code that has a wider application range.

Figures 11.13 through 11.15 show the critical distances of the system with block codes under three different communication scenarios. The results presented in Figure 11.13

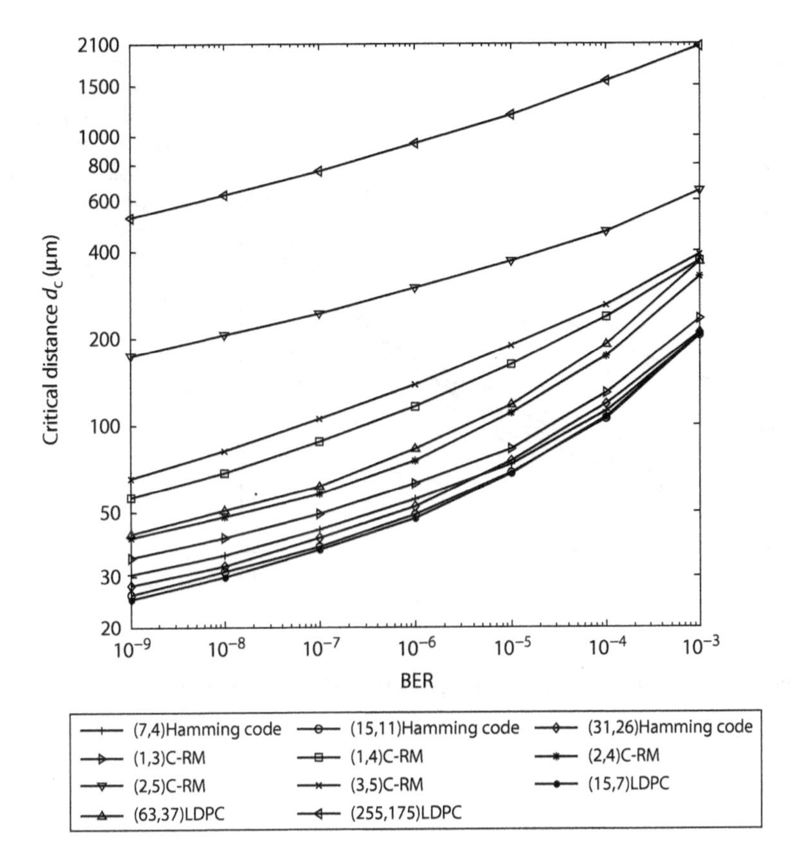

FIGURE 11.13
Critical distance with BER for block codes when considering N2N communication.

consider the N2N communication scenario. In this scenario, the (15,7)LDPC gives the lowest critical distance, and the (255,175)LDPC code from the same code family presents the longest critical distance based on the high complexity of the circuit design. Figures 11.14 and 11.15 illustrate the critical distance for the system when considering the N2M and M2N communication scenarios, respectively. In these cases, the (31,25)Hamming and (15,7) LDPC codes are best with the lowest critical distances, respectively.

Using the concept of the "best-fit ECC" previously introduced, the best-fit block codes for N2N, N2M and M2N communication scenarios are (15,7)LDPC, (31,26)Hamming code and (15,7)LDPC, respectively. These codes are also compared with SOCCs in different communication scenarios, and the results are illustrated in Figures 11.16 through 11.18.

Figure 11.16 considers the critical distance comparison for the N2N communication scenario. In this case, the energy consumption for both the encoder and decoder circuits needs to be considered. The lowest critical distance is given by the use of the (3,2,2)SOCC, which means under this communication scenario that the (3,2,2)SOCC has a wider application range.

Figures 11.17 and 11.18 show the other scenarios: N2M and M2N communications. The lowest critical distance belongs to the (31,26)Hamming code and the (3,2,2)SOCC for the N2M communication scenario, and (3,2,2)SOCC for the M2N communication scenario.

In addition, all the results indicate that the level of critical distance for the N2M communication scenario is lower than the levels of critical distance for the N2N and M2N

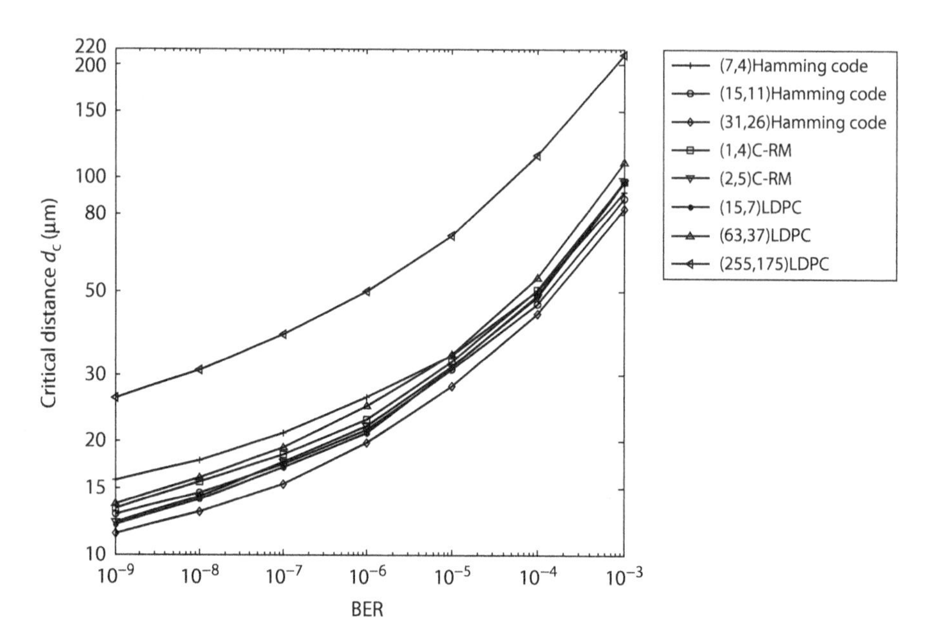

FIGURE 11.14
Critical distance with BER for block codes when considering N2M communication.

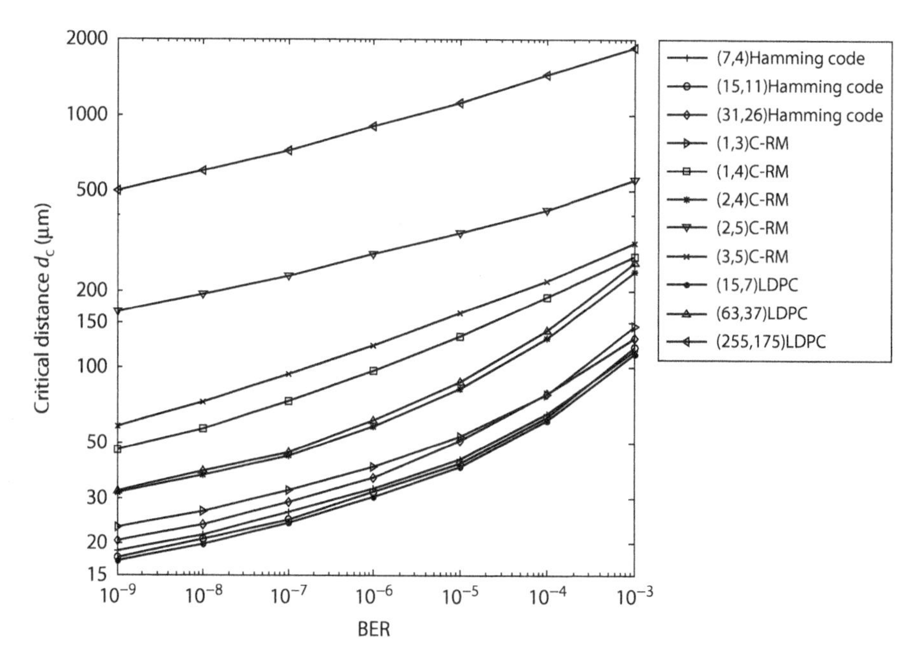

FIGURE 11.15
Critical distance with BER for block codes when considering M2N communication.

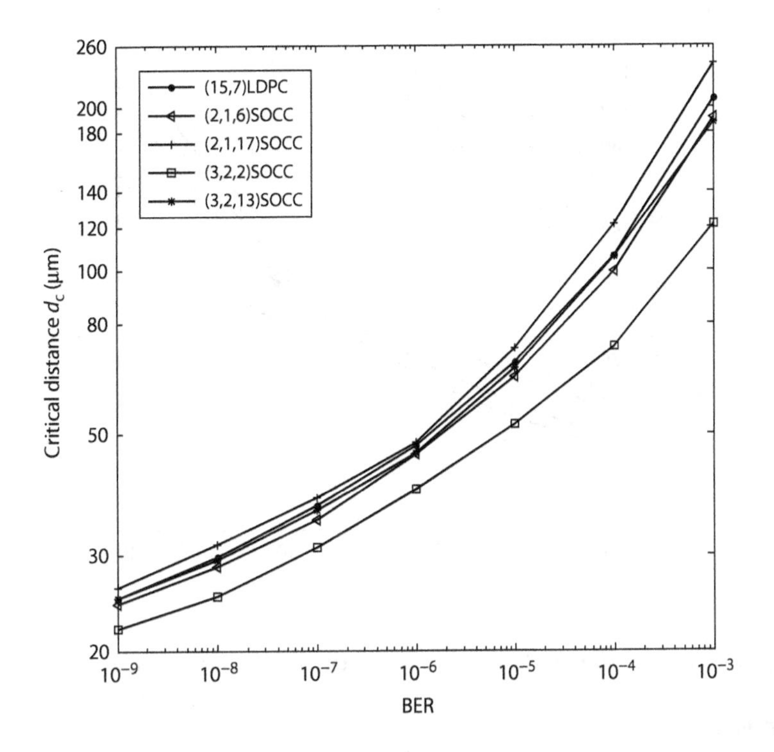

FIGURE 11.16
Critical distance comparisons between (15,7)LDPC and selected SOCCs when considering N2N communication.

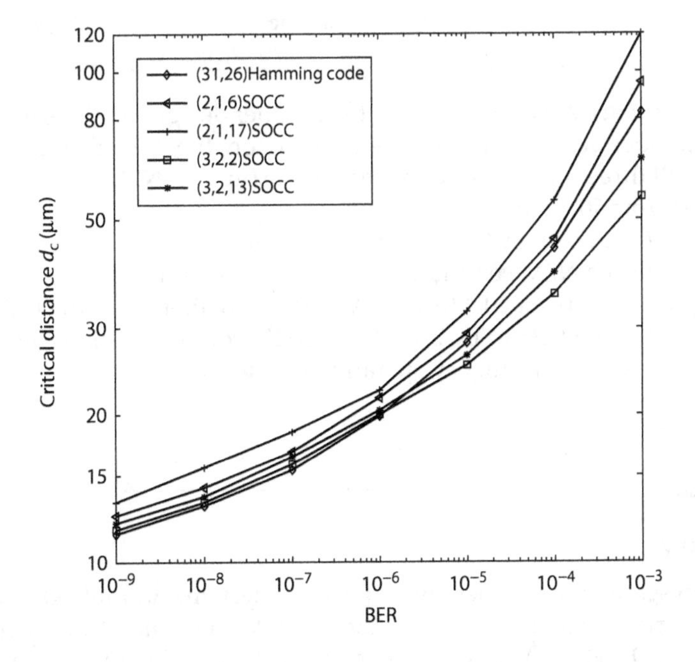

FIGURE 11.17
Critical distance comparisons between (31,26)Hamming code and selected SOCCs when considering N2M communication.

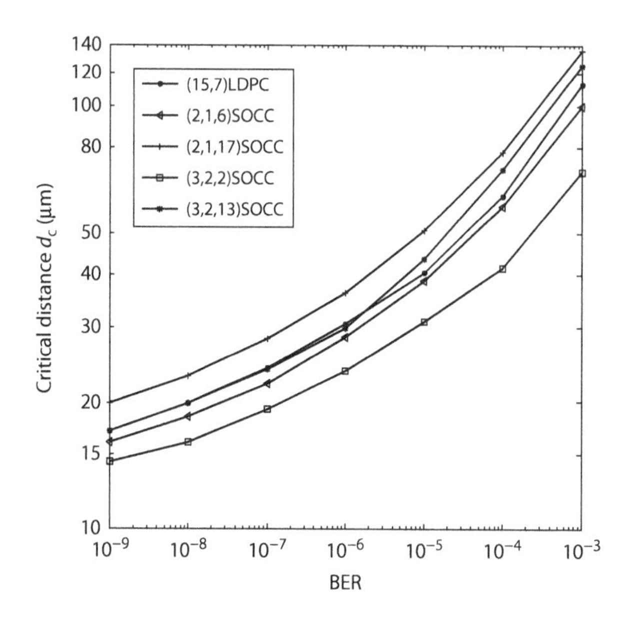

FIGURE 11.18
Critical distance comparisons between (15,7)LDPC and selected SOCCs when considering M2N communication.

TABLE 11.4
Best-Fit ECC for Different DBMC Scenarios

Communication Scenario	Lower BER Operating Region	Higher BER Operating Region
N2N	(3,2,2)SOCC	
N2M	(31,26)Hamming code	(3,2,2)SOCC
M2N	(3,2,2)SOCC	

communication scenarios. This means that the encoder design for all selected ECCs is simpler than the decoder design. Through the comparison, the (3,2,2)SOCC is selected as the best-fit ECC for all three communication scenarios except the N2M communication system with a BER level from 10^{-9} to 10^{-6}.

Table 11.4 gives the best-fit ECCs for different communication scenarios. The lower BER operating region for the N2M communication scenario is from 10^{-9} to 10^{-6}, and the higher BER operating region is from 10^{-6} to 10^{-3}. As shown in Table 11.4, the (3,2,2)SOCC is the superior code as it is the best-fit ECC for N2N, M2N communications and also the N2M communication system with a higher operating BER level.

11.8 Summary

This chapter presented an overview of PTP DBMC systems and introduced coding techniques. After a brief introduction to MC and the relevant state of the art of the ECCs in an MC system, a 3D PTP DBMC system was established. Following this, propagation and communication models were presented. Then, the selected block codes, Hamming, C-RM and LDPC codes, and the selected convolutional code, SOCC, were introduced into the

PTP DBMC system and the performance was compared among them regarding both coding gain and critical distance. The results indicate that the coding techniques do improve the performance of the PTP DBMC systems.

The critical distance for three different communication scenarios was also analyzed by considering the energy consumption caused by the introduction of ECCs. Critical distance was defined as the real transmission distance at which the use of coding techniques becomes beneficial. It was indicated that an increase in the operating BER results in a longer critical distance. For the N2M and M2N communication scenarios, the critical distance decreases compared with the N2N communication scenario. The best-fit ECC for a proposed system can be determined by analyzing these performance metrics.

Overall, this chapter presented the fundamental aspect of the use of coding techniques within the DBMC system. It is believed that this study will help the system designer to understand the basic DBMC system and will be beneficial to the further investigation of new coding techniques within this emerging communication paradigm.

References

1. R. Freitasjr, What is nanomedicine? Nanomedicine: Nanotechnology, *Nanomedicine: Nanotechnology, Biology and Medicine*, vol. 1, pp. 2–9, 2005.
2. Y. Moritani, S. Hiyama, and T. Suda, Molecular communication for health care applications, in *4th Annual IEEE International Conference on Pervasive Computing and Communications Workshops*, Pisa, Italy, 2006, pp. 549–553.
3. T. M. Allen and P. R. Cullis, Drug delivery systems: Entering the mainstream, *Science*, vol. 303, pp. 1818–1822, 2004.
4. B. Atakan, O. B. Akan, and S. Balasubramaniam, Body area nanonetworks with molecular communications in nanomedicine, *IEEE Communications Magazine*, vol. 50, pp. 28–34, 2012.
5. J. Han, J. Fu, and R. B. Schoch, Molecular sieving using nanofilters: Past, present and future, *Lab on a Chip*, vol. 8, pp. 23–33, 2008.
6. T. Nakano, A. W. Eckford, and T. Haraguchi, *Molecular Communication*, Cambridge University Press, 2013.
7. D. Tessier, I. Radu, and M. Filteau, Antimicrobial fabrics coated with nano-sized silver salt crystals, in *Proceedings of the NSTI Nanotechnology Conference*, 2005, pp. 762–764.
8. J. W. Aylott, Optical nanosensors: An enabling technology for intracellular measurements, *Analyst*, vol. 128, pp. 309–312, 2003.
9. M. Kocaoglu, B. Gulbahar, and O. B. Akan, Stochastic resonance in graphene bilayer optical nanoreceivers, *IEEE Transactions on Nanotechnology*, vol. 13, pp. 1107–1117, 2014.
10. J. M. Jornet and I. F. Akyildiz, Joint energy harvesting and communication analysis for perpetual wireless nanosensor networks in the terahertz band, *IEEE Transactions on Nanotechnology*, vol. 11, pp. 570–580, 2012.
11. I. F. Akyildiz, F. Brunetti, and C. Blázquez, Nanonetworks: A new communication paradigm, *Computer Networks*, vol. 52, pp. 2260–2279, 2008.
12. C. Bustamante, Y. R. Chemla, N. R. Forde, and D. Izhaky, Mechanical processes in biochemistry, *Annual Review of Biochemistry*, vol. 73, pp. 705–748, 2004.
13. M. Moore, A. Enomoto, T. Nakano, R. Egashira, T. Suda, A. Kayasuga, et al., A design of a molecular communication system for nanomachines using molecular motors, in *Proceedings of the IEEE Conference on Pervasive Computing and Communications*, Pisa, Italy, 2006, pp. 554–559.
14. V. Serreli, C.-F. Lee, E. R. Kay, and D. A. Leigh, A molecular information ratchet, *Nature*, vol. 445, pp. 523–527, 2007.

15. T. Nakano, T. Suda, M. Moore, R. Egashira, A. Enomoto, K. Arima, Molecular communication for nanomachines using intercellular calcium signaling, in *5th IEEE Conference on Nanotechnology, 2005*, vol. 2, pp. 478–481, 2005.
16. M. Ş. Kuran, H. B. Yilmaz, T. Tugcu, and B. Özerman, Energy model for communication via diffusion in nanonetworks, *Nano Communication Networks*, vol. 1, pp. 86–95, 2010.
17. B. Atakan and O. B. Akan, Deterministic capacity of information flow in molecular nanonetworks, *Nano Communication Networks*, vol. 1, pp. 31–42, 2010.
18. M. Pierobon and I. F. Akyildiz, Diffusion-based noise analysis for molecular communication in nanonetworks, *IEEE Transactions on Signal Processing*, vol. 59, pp. 2532–2547, 2011.
19. T. Nakano, T. Suda, M. Moore, R. Egashira, A. Enomoto, and K. Arima, Molecular communication for nanomachines using intercellular calcium signaling, in *5th IEEE Conference on Nanotechnology*, 2005, pp. 478–481.
20. T. Nakano, T. Suda, T. Koujin, T. Haraguchi, and Y. Hiraoka, Molecular communication through gap junction channels, in *Transactions on Computational Systems Biology X*, Springer, 2008, pp. 81–99.
21. T. Nakano, Y.-H. Hsu, W. C. Tang, T. Suda, D. Lin, T. Koujin, *et al.*, Microplatform for intercellular communication, in *3rd IEEE International Conference on Nano/Micro Engineered and Molecular Systems (NEMS 2008)*, 2008, pp. 476–479.
22. L. C. Cobo and I. F. Akyildiz, Bacteria-based communication in nanonetworks, *Nano Communication Networks*, vol. 1, pp. 244–256, 2010.
23. M. Gregori and I. F. Akyildiz, A new nanonetwork architecture using flagellated bacteria and catalytic nanomotors, *IEEE Journal on Selected Areas in Communications*, vol. 28, pp. 612–619, 2010.
24. M. Gregori, I. Llatser, A. Cabellos-Aparicio, and E. Alarcón, Physical channel characterization for medium-range nanonetworks using flagellated bacteria, *Computer Networks*, vol. 55, pp. 779–791, 2011.
25. M. J. Berridge, The AM and FM of calcium signalling, *Nature*, vol. 386, pp. 759–760, 1997.
26. T. Nakano, M. J. Moore, F. Wei, A. V. Vasilakos, and J. Shuai, Molecular communication and networking: Opportunities and challenges, *IEEE Transactions on NanoBioscience*, vol. 11, pp. 135–148, 2012.
27. M. S. Leeson and M. D. Higgins, Forward error correction for molecular communications, *Nano Communication Networks*, vol. 3, pp. 161–167, 2012.
28. P.-Y. Ko, Y.-C. Lee, P.-C. Yeh, C.-H. Lee, and K.-C. Chen, A new paradigm for channel coding in diffusion-based molecular communications: Molecular coding distance function, in *IEEE Global Communications Conference (GLOBECOM)*, 2012, pp. 3748–3753.
29. P.-J. Shih, C.-H. Lee, P.-C. Yeh and K.-C. Chen, Channel codes for reliability enhancement in molecular communication, *IEEE Journal on Selected Areas in Communications*, vol. 31, pp. 857–867, 2013.
30. C. Bai, M. S. Leeson, and M. D. Higgins, Minimum energy channel codes for molecular communications, *Electronics Letters*, vol. 50, pp. 1669–1671, 2014.
31. A. G. Thombre, J. R. Cardinal, A. R. DeNoto, S. M. Herbig, and K. L. Smith, Asymmetric membrane capsules for osmotic drug delivery: I. Development of a manufacturing process, *Journal of Controlled Release*, vol. 57, pp. 55–64, 1999.
32. S. Nain and N. N. Sharma, Propulsion of an artificial nanoswimmer: A comprehensive review, *Frontiers in Life Science*, pp. 1–16, 2014.
33. M. Ş. Kuran, H. B. Yilmaz, T. Tugcu, and B. Özerman, Energy model for communication via diffusion in nanonetworks, *Nano Communication Networks*, vol. 1, pp. 86–95, 2010.
34. Y. Lu, M. D. Higgins, and M. S. Leeson, Comparison of channel coding schemes for molecular communications systems, *IEEE Transactions on Communications*, vol. 63, pp. 3991–4001, 2015.
35. R. W. Hamming, Error detecting and error correcting codes, *Bell System Technical Journal*, vol. 29, pp. 147–160, 1950.
36. D. E. Muller, Application of Boolean algebra to switching circuit design and to error detection, *Transactions of the IRE Professional Group on Electronic Computers*, vol. EC-3, pp. 6–12, 1954.

37. I. Reed, A class of multiple-error-correcting codes and the decoding scheme, *Transactions of the IRE Professional Group on Information Theory*, vol. 4, pp. 38–49, 1954.
38. S. Boztas and I. E. Shparlinski, Applied algebra, algebraic algorithms and error-correcting codes, in *Proceedings of the 14th International Symposium (AAECC-14)*, Melbourne, Australia, November 26–30, Springer, 2001.
39. W. W. Peterson and E. J. Weldon, *Error-Correcting Codes*, MIT Press, 1972.
40. T. Kasami, L. Shu, and W. Peterson, New generalizations of the Reed–Muller codes—I: Primitive codes, *IEEE Transactions on Information Theory*, vol. 14, pp. 189–199, 1968.
41. R. G. Gallager, Low-density parity-check codes, *IRE Transactions on Information Theory*, vol. 8, pp. 21–28, 1962.
42. D. J. MacKay and R. M. Neal, Near Shannon limit performance of low density parity check codes, *Electronics Letters*, vol. 33, pp. 457–458, 1997.
43. K. Yu, L. Shu, and M. P. C. Fossorier, Low-density parity-check codes based on finite geometries: A rediscovery and new results, *IEEE Transactions on Information Theory*, vol. 47, pp. 2711–2736, 2001.
44. T. K. Moon, *Error Correction Coding: Mathematical Methods and Algorithms*, Wiley, 2005.
45. J. C. Moreira and P. G. Farrell, *Essentials of Error-Control Coding*, Wiley, 2006.
46. P. Reviriego, J. A. Maestro, and M. F. Flanagan, Error detection in majority logic decoding of Euclidean geometry low density parity check (EG-LDPC) codes, *IEEE Transactions on Very Large Scale Integration (VLSI) Systems*, vol. 21, pp. 156–159, 2013.
47. K. Ganesan, P. Grover, and J. Rabaey, The power cost of over-designing codes, in *IEEE Workshop on Signal Processing Systems (SiPS)*, 2011, pp. 128–133.
48. P. Grover and A. Sahai, Green codes: Energy-efficient short-range communication, in *IEEE International Symposium on Information Theory (ISIT)*, 2008, pp. 1178–1182.
49. S. Bougeard, J. F. Helard, and J. Citerne, A new algorithm for decoding concatenated CSOCs: Application to very high bit rate transmissions, in *1999 IEEE International Conference on Personal Wireless Communication*, 1999, pp. 399–403.
50. S. Lin and D. J. Costello, *Error Control Coding: Fundamentals and Applications*, Prentice-Hall, 1983.
51. M. Kavehrad, Implementation of a self-orthogonal convolutional code used in satellite communications, *IEE Journal on Electronic Circuits and Systems*, vol. 3, pp. 134–138, 1979.
52. R. Townsend and E. Weldon, Self-orthogonal quasi-cyclic codes, *IEEE Transactions on Information Theory*, vol. 13, pp. 183–195, 1967.
53. H. M. Sauro and B. N. Kholodenko, Quantitative analysis of signaling networks, *Progress in Biophysics and Molecular Biology*, vol. 86, pp. 5–43, 2004.
54. J. Levine, H. Y. Kueh, and L. Mirny, Intrinsic fluctuations, robustness, and tunability in signaling cycles, *Biophysical Journal*, vol. 92, pp. 4473–4481, 2007.
55. D. L. Nelson, A. Lehninger, M. M. Cox, M. Osgood, and K. Ocorr, *Lehninger Principles of Biochemistry / The Absolute, Ultimate Guide to Lehninger Principles of Biochemistry*, Macmillan Higher Education, 2008.
56. E. Shacter, P. B. Chock, and E. R. Stadtman, Energy consumption in a cyclic phosphorylation/dephosphorylation cascade, *Journal of Biological Chemistry*, vol. 259, pp. 12260–12264, 1984.
57. M. Ş. Kuran, H. B. Yilmaz, T. Tugcu, and B. Özerman, Energy model for communication via diffusion in nanonetworks, *Nano Communication Networks*, vol. 1, pp. 86–95, 2010.
58. R. E. Blahut, *Algebraic Codes for Data Transmission*, Cambridge University Press, 2003.
59. W. Ryan and S. Lin, *Channel Codes: Classical and Modern*, Cambridge University Press, 2009.
60. S. L. Howard, C. Schlegel, and K. Iniewski, Error control coding in low-power wireless sensor networks: When is ECC energy-efficient?, *EURASIP Journal on Wireless Communications and Networking*, vol. 2006, pp. 29–29, 2006.

12

Miniaturized Battery-Free Wireless Bio-Integrated Systems

Philipp Gutruf

CONTENTS

12.1 Introduction: Background and Driving Forces

Contemporary electronic component and devices have recently stagnated in performance increase due to physical limitations in miniaturization and a development al goals solely based on Moore's law are not the only consideration when benchmarking gains of new hardware. A focus on integration has become apparent and the commercial success of new devices relies on form factor and compatibility with biological systems, mostly to improve human–device interactions. This development can be observed by the booming wearables market and health-monitoring devices in various form factors. These devices repurpose and simplify ordinary circuits and technology currently available on the market in wearable housings affixed with flexible bands to limbs. This technology has not changed in the last 20 years and progress is long overdue. Considerable efforts are being made toward the development of a new class of electronics that is intrinsically flexible and stretchable [1]. The research challenge is to overcome the intrinsically rigid nature of contemporary technology. However, the integration of conventional electronic components, passive or active, with biological systems is limited due to the inherent incompatibility in mechanical properties. Most biological systems such as the human skin and the surface of organs are intrinsically soft [2], whereas popular electronic components and their housings are hard. For the creation of these next-generation devices, this fundamental mechanical mismatch has to be overcome. Examples of such soft systems can be found in the literature and cover basic electrical functionality and sensing capabilities [3]. Examples include capacitive epidermal electronics for electrically safe, long-term electrophysiological measurements [4] and human–machine interfaces [5], approaches that leverage the ability of these systems to be laminated directly onto the skin. Other notable applications are the direct lamination of these electronics onto organs, which results in high-quality recordings due to intimate mechanical contact [6,7]. However, for this technology to break through into mainstream electronics for consumers and healthcare, a suitable integration strategy has to be found. Due to the inherent stretchability and soft nature of these circuits, a long-term, chronic wired connection is not possible due to the interfacial stress of the connection and the small force required to break these ultrathin devices, which can be quickly reached by tugging at a wired interface without strain relief. Furthermore, such devices will find their application mainly in intimate monitoring of biological systems either on the skin or on the surface of organs where a wired connection could be a significant infection risk and warrant secondary surgery for device removal. This necessitates a ubiquitous strategy for wireless integration with the currently available infrastructure.

Here, the search for a suitable power supply and communication strategy is critical. The main criteria for the selection of such a power supply and communication strategy are form factor, mechanical compliance and speed.

Table 12.1 compares common wireless standards. We can see that low-power communication protocols are favorable to increase the lifetime of battery-operated devices and for passive device increased access power for peripherals; however, most popular choices such as Bluetooth still require an electrochemical power supply that would not be suitable in form factor for epidermal electronics unless highly complex strain isolation approaches are used to decouple the movement of the target organ from the large stiff energy storage. Thus, the need for energy harvesting that arises for this class of systems. Current technologies that can harvest sufficient power to enable computing and sampling of information are rare. One of these current technology standards that allows for energy harvesting as well as digital communication is near field communication (NFC). Here, magnetic resonant coupling of a primary and secondary antenna allows for considerable transfer of power (up to 20 mW), and data can also be pushed from sender to receiver via amplitude shift keying (ASK) and frequency shift keying (FSK). This makes the technology highly attractive for fully bio-integrated devices. An additional advantage of this technology standard is widely availability due to common integration in smartphones, tablets, pay stations and access control systems.

12.2 NFC: Viable Standard for Highly Integrated Bioelectronics

In direct comparison, NFC is an outstanding technology because it provides digital communication and energy harvesting often in a highly integrated one-chip solution. This avoids large aggregates of passive components and the energy management associated with electrochemical energy storage that increases the device size and limits operational time.

TABLE 12.1

Wireless Connectivity for Bio-Integrated Devices

	NFC	Bluetooth	Wifi
Form factor	1 chip solution + antenna	1 chip solution + antenna	1 chip solution + antenna
Bio-integration	Very good Epidermal form factor	Medium Epidermal form factor with advanced technique	Unknown
Communication distance	Near field (0–1 m)	Far field (<3 m)	Far field (<3 m)
Speed	Low	Medium	High
Energy consumption	+10 mW	–20 mW	–30 mW
Energy harvesting	Yes	No	No

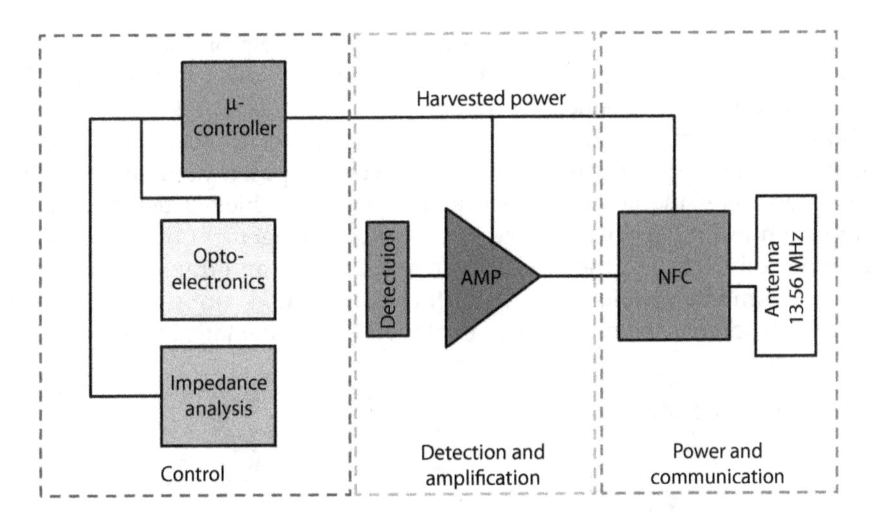

FIGURE 12.1
Electrical working principle scheme of NFC-enabled bio-integrated electronic devices.

In Figure 12.1, a template of a typical NFC-based system for bio-integrated devices is presented. Here, the NFC system-on-a-chip (SOC) is harvesting energy that is passed on to the periphery. A microcontroller is often found in these systems that orchestrates the timing for optoelectronics and impedance analysis for sensing applications such as the fingernail-mounted oximetry presented later in this chapter. Detection systems such as operational amplifiers schemes can also be powered with the harvested energy. The acquired signal is then digitalized by an analog-to-digital converter (ADC) and sent over air to a receiver. The common harvested power available in these systems can exceed 15 mW on a regular basis and provides suitable voltages in the range of 2.5–5 V to power most common optoelectronic, analog and digital electronic components. Receivers can be specialized high-powered systems for long-range applications as well as small, handheld systems commonly present in all households, such as NFC-enabled smartphones. This makes these systems very versatile due to the available infrastructure. The communication range of these systems can vary from a few centimeters up to 3 m for high-powered systems.

12.3 Mechanical Design: Transforming Flexible Electronics to Stretchable Devices

The commercial availability of small form factor–integrated NFC circuits, such as the bare die SOC, allows for the construction of highly stretchable devices that can be intimately bonded to biological systems. Here, the concept is to replace large, traditionally rigid key components of an electrical circuit with intrinsically stretchable elements. For an NFC circuit, this is the antenna, which is spatially the largest part of the device, the highly integrated silicon-based semiconductor that supports data communication, energy harvesting and sensing function can be treated as a rigid island that can be tolerated by soft systems as the strain is accommodated by the surrounding tissue if skin mounted.

An example of such a system is displayed in Figure 12.2. An IC was backpolished resulting in a small form factor 1.5 × 3 mm thin (125 μm) highly integrated system with the capability to yield temperature measurements and analog-to-digital conversion.

The development of a stretchable antenna that is capable of enduring 30% strain is critical to render an epidermal device that conforms intimately to the skin. The device geometry is critical for mechanical compliance; here, a mutual mechanical plane design can be chosen to distribute strain in the sandwich of the polyimide support with the copper core and polyimide encapsulation that also acts as a passivation layer for the copper bridge that completes the antenna. Such a stack can be achieved using a sophisticated microfabrication technique that requires the careful management of adhesion to a sacrificial or carrier substrate and the final release of the device and transfer onto an ultrathin adhesive or low modulus silicone material that bonds the device intimately to the skin. The finalized structure can be seen in an exploded layer schematic in Figure 12.2a.

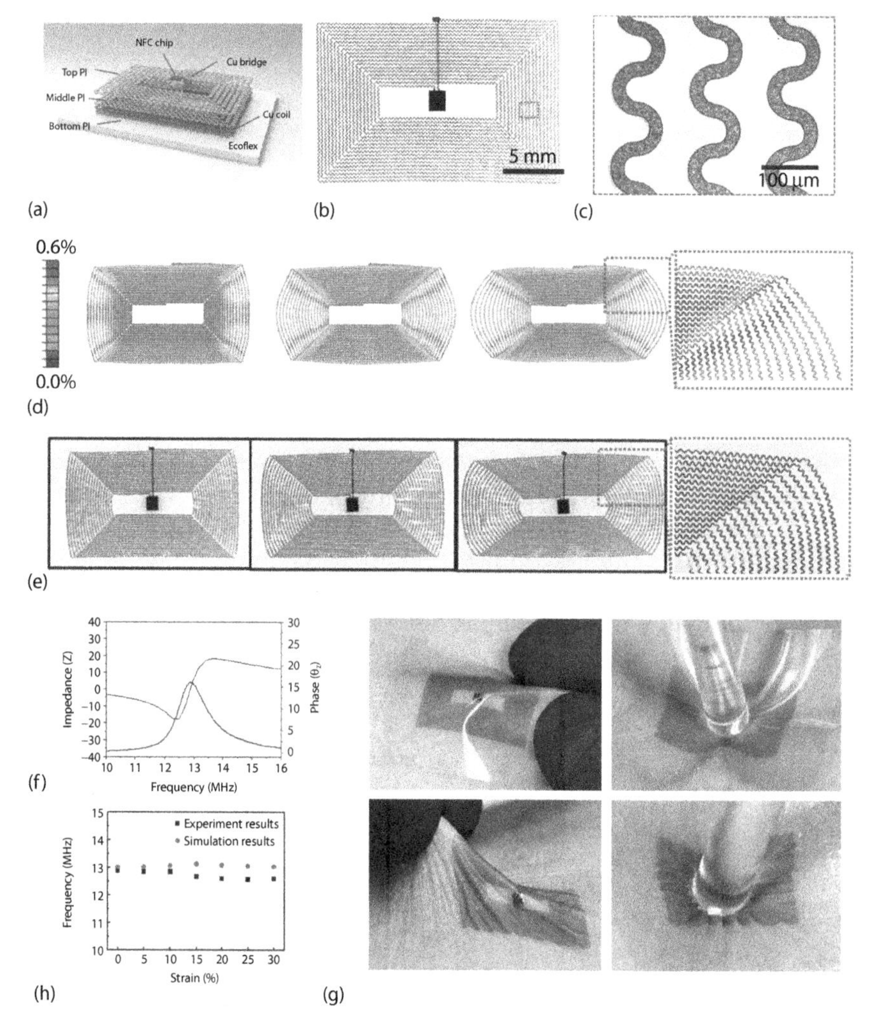

FIGURE 12.2
Stretchable NFC-enabled epidermal electronics. (a) Layered makeup of device, (b) photographic image of the stretchable antenna with closeup and (c) of the serpentine structure facilitating out-of-plane deformation that yields stretchability. (d) FEM simulation of the device with consecutive application of strain up to 30%. (e) Corresponding bench top experiments showing good correlation. (f) Electromagnetic properties of the device. (h) Antenna performance under strain. (g) Application examples with and without flexible backing, demonstrating intimate contact with the skin. (Adapted from Kim J., et al., *Small*, 11, 906–912, 2015 [10].)

The device uses serpentines to facilitate stretchability, a well-documented and explored method to transform flexible substrates into stretchable systems [8]. A photographic image of the device including a magnified view of the serpentines can be seen in Figure 12.2b and c, respectively. The device design can be aided by finite element modeling (FEM), which can be a very useful tool to aid the serpentine geometry to achieve the designed stretchability, which in the case of skin should be around 30% to accommodate the natural motion of the skin. A close match between the computer model and a uniaxial strain testing the application of 10%, 20% and 30% strain states can be observed in Figure 12.2d and e, respectively. Here, we can see that the intrinsic strain occurring in the model does not exceed 0.6%, which is well below the yield strain of copper, the material used for the antenna, demonstrating a successful design and sufficient headroom for on-body applications where strains rarely exceed 30% on the human body. The good correlation between deformations observed in bench top experiments with FEM simulation establishes a valuable tool for future epidermal designs. This method enables a route for computer-aided design that allows for the rapid prototyping of such systems. Mechanical compliance with biological systems is not the only figure of merit for these systems. In order to satisfy the energy requirements of especially optoelectronic systems, the electromagnetic properties of such an antenna are equally important because they define the working distance, the harvesting power and the space occupied by the system. Here, the quality factor (Q factor) is critical. A better Q factor allows for more power to be transferred into the system. At the same time, the bandwidth for the data transmission must also be guaranteed. In the case of the ISO/IEC 15693 standard, it was originally designed for vicinity cards and allows for communication distances of up to 1–1.5 m and is therefore ideal for wearable devices. The base frequency of the resonant magnetic coupling scheme is 13.56 MHz and features an ASK and FSK modulation with a bandwidth of 423.75 kHz. This means that the bandwidth of the antenna has to support this modulation. Therefore, the ideal antenna is perfectly tuned to 13.56 MHz and has a good Q factor supporting a bandwidth of over 423.75 kHz. The successful realization of such electrical properties coupled with an ultrathin layout and system-level stretchability can be seen in Figure 2.2f. Here, the Q factor is good and the bandwidth is around 2 MHz, which allows for good data transmission and high efficiency of power transfer, and was acquired using a passive acquisition method. It can also be seen that the resonant frequency of the system does not shift with strain on the host substrate (Figure 12.2h), a major advantage over systems operating in the far-field regime.

The mechanical compliance of these devices is displayed in Figure 2.2g. Here, the device that is laminated on ultrathin elastomeric substrates is compared to a similar device bonded to a flexible polyethylene terephthalate (PET) substrate, commonly found in radio-frequency identification (RFID)-tagged consumer electronics. It can be seen that the flexible backing is not sufficient for conformal contact. Here, the deformation experienced in the regular use of the device will frustrate the skin interface and cause delamination at the fringes of the flexible backing. The stretchable system, however, offers system-level stretchability and conformal mechanics that work complementary with the skin, showing that there is no mechanical impact of the device on the skin. Therefore, when worn over an extended time period, the device is imperceptible to the subject.

12.4 Miniaturization: Benefits for Bio-Integration

The skin as an interface to access various health status information is very popular because it acts as a window to the underlying vascular structure and as a transmitter of electrical signals giving clinically relevant information about pulse shape, heart rate, heart rate variability and oxygenation. Also, liquids such as sweat and interstitial fluid can be extracted to gather critical information. However, the skin as a mechanical platform is challenging due to its intrinsic movement and its constant renewal of the upper layer, the epidermis. This frustrates interfacial bonding and limits the lifetime of any device attached to it. Therefore, the adhesive plays a critical role in bonding the device to the skin; however, its ultimate lifetime is restricted to a maximum of 1–2 weeks. Adhesives that accommodate device adhesion over an extended period on the skin without causing irritation are currently the subject of investigation in science and industry. To enable a more permanent mounting location that maintains intimate contact to the body, the fingernail or other rigid areas of the body have been identified as a suitable alternative to the skin-mounted devices. Here, a lifetime of up to 4 months is possible, limited solely by the growth rate of the fingernail, which may vary between individual subjects.

Examples of such devices can be seen in Figure 12.3a and b. Here, the NFC electronics are highly miniaturized to minimize the overall footprint of the device to enable application to the fingernail. This is achieved with a multilayer device makeup that increases the inductance of the coil, resulting in a longer communication distance and higher power

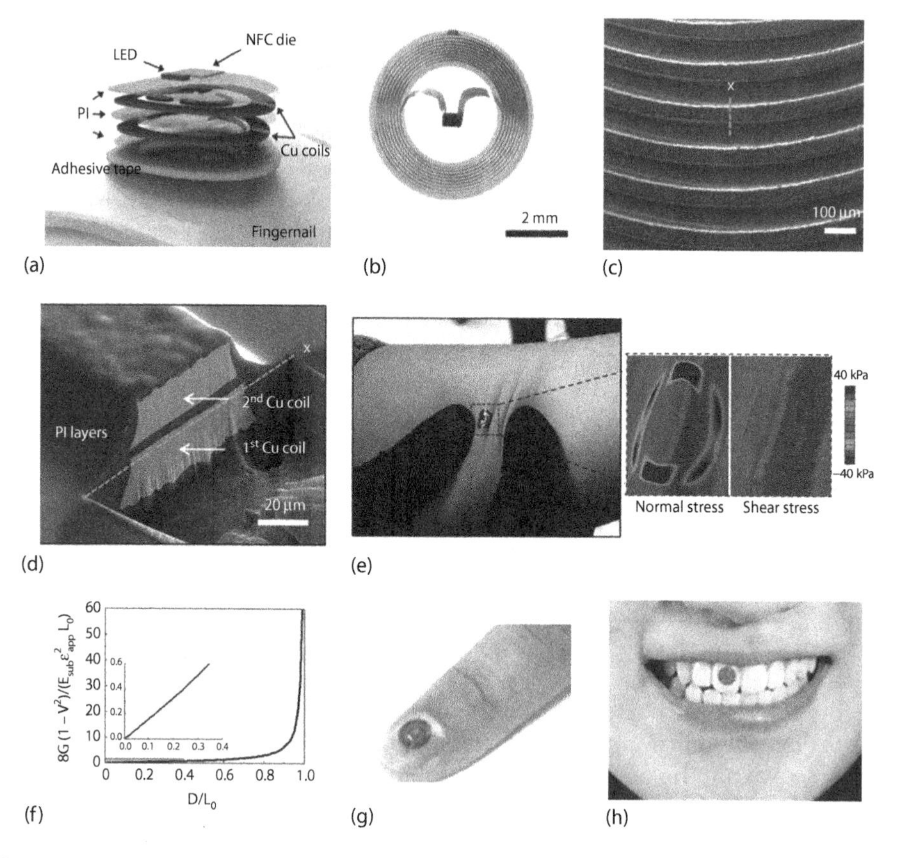

FIGURE 12.3
Highly miniaturized NFC systems. (a) Layered makeup of the device with components and adhesive. (b) Photographic image of the device. (c) SEM picture of the antenna coil with FIB cutaway revealing layered structure (d). (e) Device laminated onto the skin with corresponding normal stress and shear stress FEM simulation. (f) Calculated energy release rates with device size. (g) Application on the fingernail and tooth (h). (Adapted from Kim J., et al., *Advanced Functional Materials*, 25, 4761–4767, 2015 [11].)

reserves for energy harvesting. In this use case, the stretchability of the system is not required; however, flexibility is critical to enable imperceptible devices.

Critically with circuit geometry optimized for such uses, microfabrication strategies allowing for an overall thin device geometry are translatable to commercial-scale manufacturing. The flexibility of the circuitry is ensured with a mutual mechanical plane design, a strategy where in-plane symmetry is used to keep layer stress to a minimum analogous to the stretchable design described earlier in the chapter. Figure 12.3c shows a secondary electron microscope picture with the topographical nature of the coil enabled by a focused ion beam cutaway in Figure 12.3d revealing the layer structure.

A notable benefit of such system miniaturization is the potential use of a flexible device in low strain areas of the skin. The system miniaturization yields favorable mechanics due to the scaling of the underlying physical processes of device delamination and shear stress. Figure 12.3e shows such a device mounted on the forearm, the corresponding FEM simulations showing a low level of shear stress because of the small form factor; the normal stress is also below the sensation threshold, which results in a good user' experience when wearing the device. The small form factor also translates to low energy release rates, which means in practice a device that sticks well to the skin and is not prone to delamination. The scaling of this property can be seen in Figure 12.3f, which demonstrates favorable mechanics at small device diameters.

Critically, the high level of miniaturization allows for mounting locations such as the fingernail (Figure 12.3g) and the tooth (Figure 12.3h), which is a stable mechanical platform and offers rich opportunities for biosensing with potential device lifetimes of multiple years in the case of the tooth.

12.5 Applications: Contactless Vital Information Sensing

As discussed previously, the fingernail is a stable platform for chronic operation, and this presents an opportunity for use in sensing applications. There is a significant need for the continuous recording of vital information such as heart rate, heart rate variability and oxygenation during a chronic illness. Pulse oximetry offers the possibility to capture this vital information. A pulse oximeter operates on the working principle of signal attenuation of two narrowband light sources that interact with hemoglobin in the target region such as the fingers or earlobes, regions chosen due to their easy accessibility for external hardware. Hemoglobin, the main carrier of oxygen in the blood, alters its optical properties based on its oxygenation state. This can be leveraged for sensing by measuring the absorption at wavelengths that are adversely affected by this change. By then measuring the pulsatile component of the signal, the oxygen saturation of the subject can be extracted.

Current sensing solutions are bulky and intrinsically prone to motion artifacts due to heavy sensing strategies such as fingertip clamps that are, due to inertia, inherently unstable because the mass of the wireless hardware and the associated electrochemical energy storage cause relative motion of the clamp relative to the finger. The underlying vascular structure that is nonhomogeneous then results in variation in the optical path length, and a probing volume that shifts the optical signals levels resulting in motion artifacts. Here, ultrathin, flexible, highly miniaturized, lightweight and wireless technologies offer a distinct advantage over contemporary solutions.

Such a solution is presented in Figure 12.4a. Here, we can see the highly integrated layered makeup that consists of the antenna coil, which is, as previously discussed, a dual coil for maximum energy harvesting capability and footprint. The components needed for the pulse oximeter are also carefully chosen to meet the strict requirements for energy consumption, footprint and height requirements.

The electrical working principle is shown in Figure 12.4b, where the NFC scheme introduced at the beginning of the chapter is utilized. In this specific embodiment, a low-power, small outline microcontroller is used to control the red and infrared light-emitting diode (LED) light sources in a time-sequenced manner and a customized analog frontend with a transimpedance amplifier scheme coupled to a broadband semiconductor photodiode is used to capture the back-reflected light emerging from the nail bed. The signal is then digitalized in the bare die NFC chip that also provides power by means of energy harvesting from the ambient electrical field provided by the readout hardware. This makeup allows for a completely battery-free operation. Figure 12.4c shows a photograph of the device,

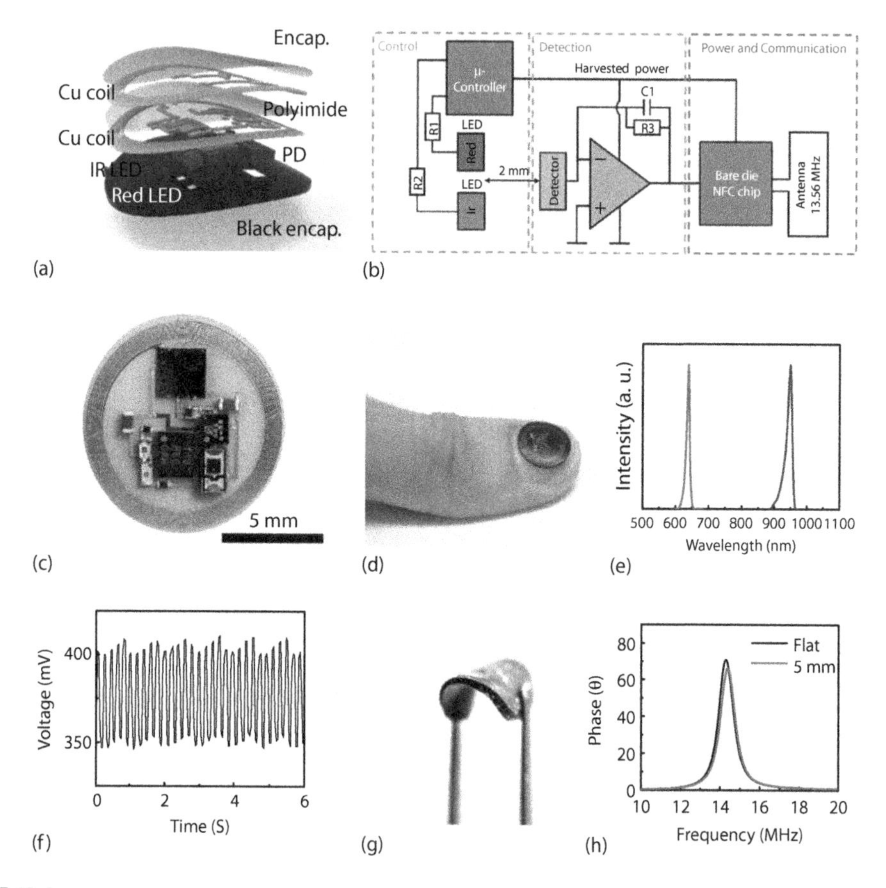

FIGURE 12.4
Highly miniaturized, battery-free, and flexible pulse oximeter system. (a) Layered makeup of the device. (b) Electrical operating scheme. (c) Photographic image of the device before encapsulation. (d) Photographic image of encapsulated device laminated to the fingernail during operation. (e) Spectral analysis of the LED light sources used for this device. (f) Raw signal obtained wirelessly from the device showing the pulsatile component of both excitation wavelengths. (g) Device in bent state demonstrating flexibility. (h) Electromagnetic device performance in flat and bent state. (Adapted from Kim J., et al., *Advanced Functional Materials*, 27, 2017 [12].)

illustrating the compact layout built around the optimized sensor with a light source distance of 2 mm.

To avoid a light short circuit, which manifests itself in a total reflection at the device fingernail or skin interface resulting in no interaction with the target tissue, which is a problem in reflectance-based oximeters, the device is encapsulated in black elastomeric material that also has the added benefit of conformal contact through the low modulus nature of the material. Optically, the light sources are now effectively isolated and suppress direct current light short circuits at the interface.

A characterization of the red and infrared LED can be seen in Figure 12.4e, showing the spectrally narrow bandwidth nature of the light source, which is essential for a good signal-to-noise ratio. Furthermore, a carefully chosen wavelength can yield a higher difference in absorption for the oxygenated and deoxygenated hemoglobin species. The resulting raw signal that is wirelessly acquired is shown in Figure 12.4f, where we can see the distinct pulsatile component of both infrared and red light reflection, which can be used to calculate the pulse oxygenation of the subject.

Figure 12.4g and h shows the bending characteristics of the device. The phase information shows that the device does not suffer from performance loss, judged by evaluating the quality factor of the antenna in its flat and bent state, which shows nearly no decay in performance. This is an important attribute because fingernail curvature can vary greatly from subject to subject, with bending radii of just 5 mm observed in some individuals.

The operational device performance evaluated in Figure 12.5a shows the extracted data from an experiment where the subject was deprived of oxygen intake by holding his or her breath. We can see that the pulsatile amplitude ratio of red and infrared reflectance shifts as the subject's oxygenation decreases. The corresponding extraction of SpO_2 can be seen in Figure 12.5b, with a comparison to a commercially available oximeter shown in Figure 12.5c. We can see that a much higher temporal resolution is achieved with the wireless device. This is possible due to the intrinsically superior signal quality facilitated by the intimate contact of the sensor system with the body yielding better signal-to-noise ratios. The commercial system requires a longer sampling time and averaging to compensate for motion artifacts and higher gain levels due to lower signal amplitudes that result in higher baseline noise.

Another advantage is shown in Figure 12.5d. The SpO_2 extracted from the device mounted on the fingernail is shaken vigorously, and the magnitude of the movement can be seen by simultaneously recorded accelerometer traces. We can observe that the signal is stable throughout the measurement, which is not feasible with commercial systems since they lack mechanical stability due to either high mass or cables associated with the device, which cause motion artifacts.

Application examples are shown in Figure 12.5e and f, where the device is mounted on the fingernail and the data is read out by a computer mouse with NFC functionality. This is an example of a use case that would facilitate continuous readout during daily activity. This capability is critical to address the clinical need of prolonged monitoring of postoperative or chronically ill patients. This information can be relayed via the internet to healthcare providers allowing for much less intrusive care, which enables more flexible surgery recovery scenarios as well as remote diagnosis and treatment. A subtler mounting location is demonstrated in Figure 12.5f, where the device is placed inside the earlobe, allowing for signal extraction when an NFC-enabled smartphone is brought into close proximity, a use case where intermittent oxygenation and heart rate monitoring is sufficient.

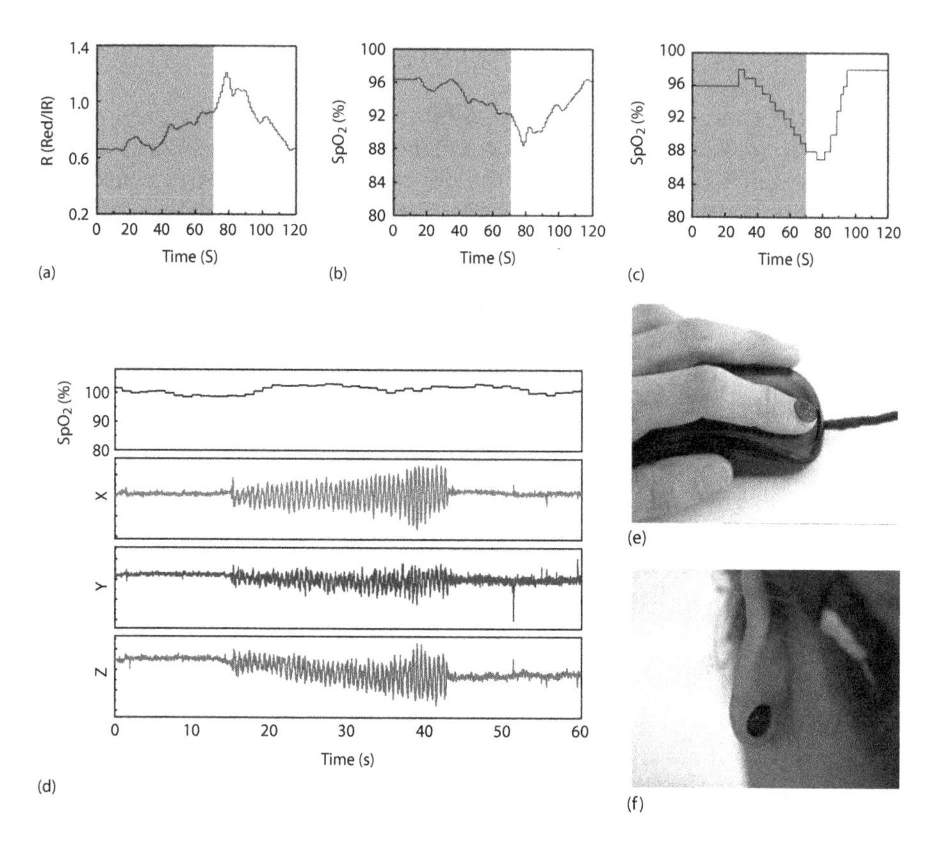

FIGURE 12.5
In situ device characterization. (a) Pulsatile component amplitude ratio over the course of an experiment involving restricted oxygen intake. (b) Corresponding calculation of the pulse oxygen saturation. (c) Control recorded with a commercial pulse oximeter (Apple I health). (d) Oxygenation readout with wireless battery-free device during elevated activity shown by accelerometer traces. (e) Application scenario on an NFC-enabled mouse. (f) Application scenario of the device mounted behind the earlobe. (Adapted from Adapted from Kim J., et al., *Advanced Functional Materials*, 27, 2017.)

12.6 Discussion: Battery-Free Miniaturized Electronics – Applications Beyond the Skin

The utility of the schemes presented in this chapter has a much wider scope than the applications discussed in this chapter. The magnetic resonant power transfer used here is little affected by surrounding moisture, meaning that it is an excellent candidate for the powering and data communication of implants, which will open the door for smart sensor nodes that can assess the health state instantaneously with currently unmet precision in a minimally invasive fashion. The utility of this approach has already been demonstrated in the context of devices for neuroscience [9]. The optogenetic stimulation of targeted neuronal populations has been accomplished. Applications beyond the brain can be envisioned, namely cardiac and peripheral device nodes that evaluate the health status simultaneously and deliver a complete insight into the physiological state of a patient with chronic interfaces that facilitate recordings over multiple years. This would be a leap forward in terms of medical device technology. The data generated from such sensor systems over multiple subjects can build powerful datasets that enable disease diagnostics including early warning systems for high-risk cases with unmatched precision and the removal of human error in data interpretation.

The technology also provides fundamental capabilities in access control and hygiene management where fingernail-mounted devices can give advanced insights into human behavior. Moreover, the technology can be used in industrial applications in hard-to-reach or vacuum-sealed systems where wired connections are prohibitive due to process line content. The advances in this field over recent years present a leap forward in device capability and suggest the acceleration of such sensors in the near future.

References

1. Kim J, Ghaffari R, and Kim D-H: The quest for miniaturized soft bioelectronic devices. *Nature Biomedical Engineering* 2017, 1: s41551-41017-40049.
2. Kim D-H, Lu N, Ma R, Kim Y-S, Kim R-H, Wang S, Wu J, Won SM, Tao H, et al.: Epidermal electronics. *Science* 2011, 333: 838–843.
3. Yeo W-H, Kim Y-S, Lee J, Ameen A, Shi L, Li M, Wang S, Ma R, Jin SH, et al.: Multifunctional epidermal electronics printed directly onto the skin. *Advanced Materials* 2013, 25: 2773–2778.
4. Jeong J-W, Kim MK, Cheng H, Yeo W-H, Huang X, Liu Y, Zhang Y, Huang Y, and Rogers JA: Capacitive epidermal electronics for electrically safe, long-term electrophysiological measurements. *Advanced Healthcare Materials* 2014, 3: 642–648.
5. Jeong JW, Yeo WH, Akhtar A, Norton JJ, Kwack YJ, Li S, Jung SY, Su Y, Lee W, et al.: Materials and optimized designs for human-machine interfaces via epidermal electronics. *Advanced Materials* 2013, 25: 6839–6846.
6. Kim D-H, Ghaffari R, Lu N, Wang S, Lee SP, Keum H, D'Angelo R, Klinker L, Su Y, et al.: Electronic sensor and actuator webs for large-area complex geometry cardiac mapping and therapy. *Proceedings of the National Academy of Sciences* 2012, 109: 19910–19915.
7. Xu L, Gutbrod SR, Bonifas AP, Su Y, Sulkin MS, Lu N, Chung H-J, Jang K-I, Liu Z, et al.: 3D multifunctional integumentary membranes for spatiotemporal cardiac measurements and stimulation across the entire epicardium. *Nature Communications* 2014, 5: 3329.
8. Gutruf P, Walia S, Nur Ali M, Sriram S, and Bhaskaran M: Strain response of stretchable microelectrodes: Controlling sensitivity with serpentine designs and encapsulation. *Applied Physics Letters* 2014, 104: 021908.
9. Shin G, Gomez AM, Al-Hasani R, Jeong YR, Kim J, Xie Z, Banks A, Lee SM, Han SY, et al.: Flexible near-field wireless optoelectronics as subdermal implants for broad applications in optogenetics. *Neuron* 2017, 93: 509–521. e503.
10. Kim J, Banks A, Cheng H, Xie Z, Xu S, Jang KI, Lee JW, Liu Z, Gutruf P, et al.: Epidermal electronics with advanced capabilities in near-field communication. *Small* 2015, 11: 906–912.
11. Kim J, Banks A, Xie Z, Heo SY, Gutruf P, Lee JW, Xu S, Jang KI, Liu F, et al.: Miniaturized flexible electronic systems with wireless power and near-field communication capabilities. *Advanced Functional Materials* 2015, 25: 4761–4767.
12. Kim J, Gutruf P, Chiarelli AM, Heo SY, Cho K, Xie Z, Banks A, Han S, Jang KI, et al.: Miniaturized battery-free wireless systems for wearable pulse oximetry. *Advanced Functional Materials* 2017, 27: 1604373.

13

A Low-Power Vision- and IMU-Based System for the Intraoperative Prosthesis Pose Estimation of Total Hip Replacement Surgeries

Shaojie Su, Hong Chen, Hanjun Jiang, and Zhihua Wang

CONTENTS

13.1 Introduction

13.1.1 Background of Total Hip Replacement

As the average age of the population increases, the number of people suffering from severe pain in their damaged hip joints caused by a variety of human diseases and activities has increased steadily. The last resort to regain full mobility and alleviate the pain is total hip replacement (THR) surgery. According to the Agency for Healthcare Research and Quality (AHRQ), more than 285,000 THRs are performed in the United States annually, and the number is forecasted to double in the next two decades [1]. However, THR surgery still has a failure rate of about 10%, leading to serious complications such as dislocation, prosthetic impingement, intraoperative fracture, infection and leg length discrepancy along with the long-term complications of wearing and loosening [2]. Among them, dislocation and impingement, of which the incidence is reported to range from 0.5% to 11% [3–5], are two common complications.

During THR surgery, the placement of hip prostheses is the key step that will affect the postoperative hip range of motion (ROM), as shown in Figure 13.1. As Widmer pointed out in [6], the malposition of prosthesis placement is a primary cause of dislocation and prosthetic impingement. Researchers consider the hip ROM as an important indicator of joint stability and conclude that THR with a prosthetic hip ROM larger than that for everyday activities can minimize the risk of dislocation and impingement [3,4]. As a result, an intraoperative-assisted system in THR for visualizing the prosthesis placement and estimating the hip ROM for surgeons is strongly needed.

13.1.2 Related Research Work

Many efforts have been made to help the placement of prostheses in the safe zone. Image-guided surgical navigation systems such as computed tomography (CT)-based and CT-free intraoperative navigation systems have brought significant improvements in motion tracking [7]. However, the CT-based navigation system has the disadvantages of high cost, increased operation time and exposure to radiation [8], while the CT-free navigation is even more expensive and requires the affixing of markers to bony anatomical landmarks, leading to more complexity in data collection, calibration and processing, and extra harm to the patients as well [9]. In the 1960s, people began to design instrumented prostheses

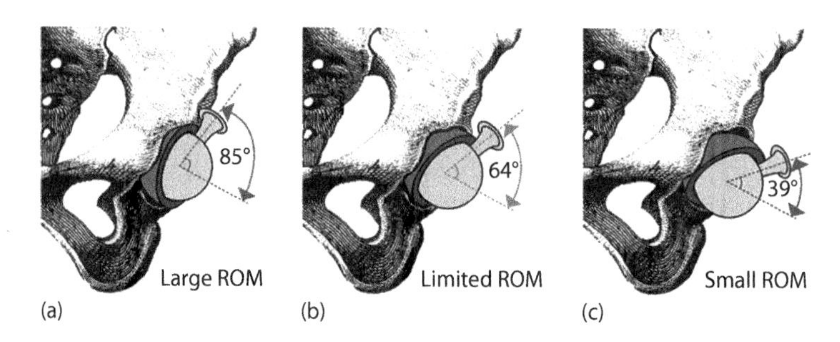

FIGURE 13.1
(a) One can raise the thigh up high if the acetabular cup is implanted in the right orientation, indicating a large hip ROM. (b) Impingement may occur when one raises the thigh up as the abduction of the acetabular cup is a bit larger than the safe zone, indicating a limited hip ROM. (c) Impingement or dislocation occurs during one's everyday activities as the abduction of the acetabular cup is too large, indicating a small hip ROM.

that could collect data in vivo and transmit the data wirelessly outside the body. Graichen et al. [10] have proposed a hip endoprosthesis instrumented with sensors for joint contact forces measurement. Damm et al. [11] have applied six semiconductor strain gauges inside the hollow neck of a new instrumented hip joint prosthesis to measure the deformation of the neck. Bergmann et al. [12] have developed an instrumented hip implant with a thermistor inside to measure the implant temperature. However, these instrumented prostheses are designed for postoperative monitoring and provide no intraoperative help to the surgeons.

Another strong need in THR surgeries is the estimation of hip ROM. One traditional solution is using the standard goniometer during THR surgery [13,14]. However, the use of such a device in current clinical practice is extremely rare and has been taken over by visual estimation [14]. As sensor technology has developed, researchers have tried to deploy miniature inertial measurement units (IMUs) on body segments adjacent to a joint as another solution to calculate the joint angles and ROM [15–18]. The accuracy of IMU drift correction methods are evaluated in [15] by comparing estimates of pelvis, thigh and shank orientation from IMUs and navigation systems during maximal hip flexion and abduction, walking, squatting and standing on one leg. The accuracy of a commercially available IMU system for estimating rotations across the hip, knee and ankle during level walking, stair ascent and stair descent is revealed in [16] with its angle estimation errors ranging from 1.38° to 6.69°. Nevertheless, the relative motion between the soft tissue and the underlying bony anatomy can be a source of error for the wearable IMU solution [18].

13.1.3 Introduction of the Proposed System

In order to help surgeons with the implantation of hip prostheses and the estimation of hip ROM in THR surgeries, a monocular vision- and IMU-based prosthesis pose estimation system has been designed. As shown in Figure 13.2, the system consists of three parts: an acetabular cup with a series of customized patterns printed on its inner surface, a hollow femoral head trial with circuits rigidly mounted inside and a computer with a hip prosthesis pose estimation software [19]. The femoral head trial is the same size as a conventional femoral head prosthesis, and its upper hemisphere is made of transparent materials so that the camera mounted in the center can take images of the reference patterns. During THR

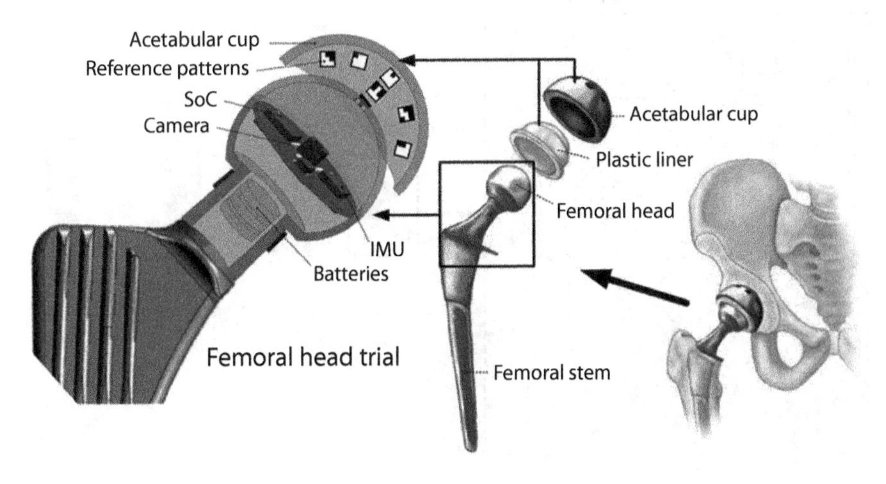

FIGURE 13.2
Structure of the new femoral head trial.

surgery, the sensor data will be obtained in vivo and transmitted wirelessly to a computer where the data will be processed and displayed on a screen. In addition, some low-power technologies are adopted in both the hardware and software implementation to ensure the battery lifetime of the system can fully cover the whole THR surgery. More details about the system will be discussed in the rest of the chapter. Section 13.2 introduces the system architecture. Section 13.3 gives details of the vision- and IMU-based prosthetic pose estimation methods. In Section 13.4, the low-power technologies adopted in the system are described. The implementation and experimental results are presented in Section 13.5. Finally, a conclusion is drawn in Section 13.6.

13.2 System Design

13.2.1 System Architecture

The system architecture is shown in Figure 13.3. The acetabular cup and the femoral head trial are designed for data collection, while the computer is in charge of the data processing and results display. Inside the hollow femoral head trial, a camera, an IMU, a data packer and a customized system-on-a-chip (SoC) are included. The monocular camera has a resolution of 240 × 240 and works with an adaptive frame rate, while the IMU works at a sampling rate of 100 Hz. The data packer synchronizes the data from the two sensors and packs them for transmission. The SoC is composed of three functional blocks: a power management unit, a microcontroller and an radio frequency (RF) transceiver. It manages the sensors and transmits the data packets wirelessly to the computer. The femoral head trial is powered by a battery, which is mounted in the neck of the femoral head trial.

On the computer, the raw sensor data are first extracted from the received data and then used to estimate the pose of the implanted hip prosthesis. The result is displayed to the surgeons in 3D in real time. During this process, an alarm will be given if any potential dislocation or impingement is detected. With the help of the system, the surgeons can

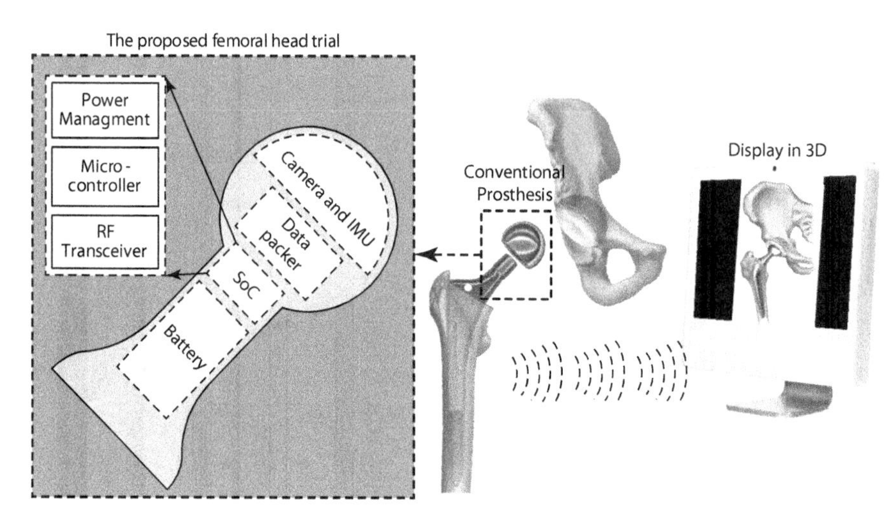

FIGURE 13.3
System architecture.

easily know the poses of both the acetabular cup and the femoral head trial, and make a judgment on the prosthesis placement better than just by their experiences. The femoral head is named as a trial because it will not be left inside the human body. After the prostheses are put within the safe zone, the surgeons will fix the implanted acetabular cup and replace the femoral head trial with a conventional femoral head prosthesis.

13.2.2 Design of the Reference Patterns

Two kinds of reference patterns are designed and printed on the internal surface of the acetabular cup liner [19], as shown in Figure 13.4a. Pattern I is a circular ring that is used as a landmark to adjust the femoral head trial to a preset pose at the beginning of the surgery, while pattern II is designed for relative pose estimation. As shown in Figure 13.4b, pattern II has the following three key design features. Firstly, each pattern II is a square surrounded by a black border. Secondly, there are nine small black or white blocks inside the border, identifying the pattern as black representing "0" and white representing "1", respectively. Thirdly, each pattern II must not be a rotational symmetric figure. A pattern II becomes distinguishable if it has the three features. All pattern II are of the same known size and are printed on the known positions inside the acetabular cup. The four vertexes of each pattern II are used as feature points in the vision-based pose estimation algorithms. Figure 13.3c gives examples of both distinguishable and indistinguishable pattern II.

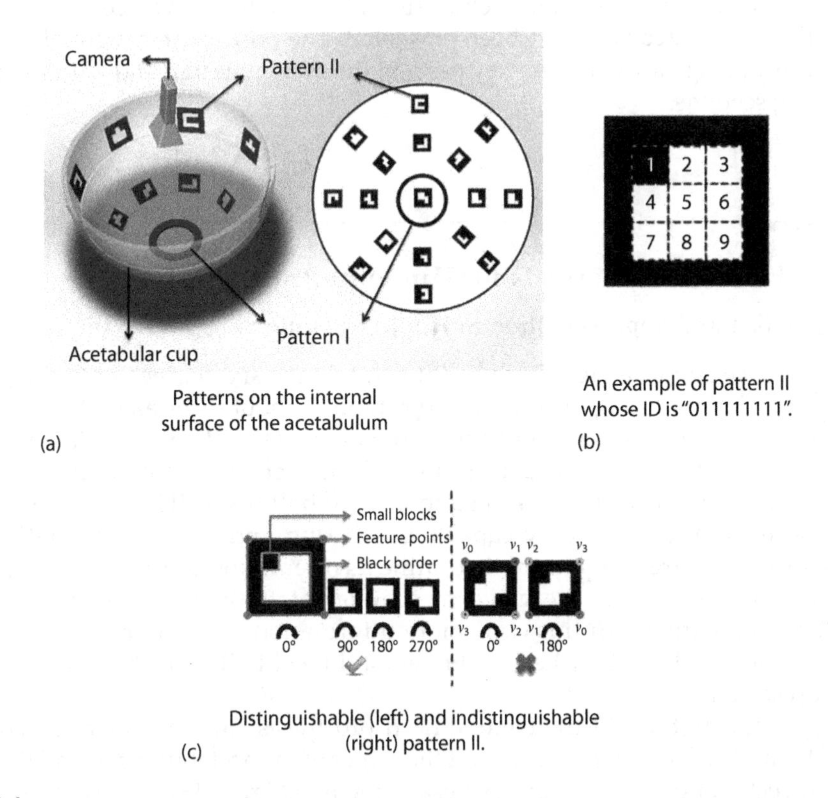

FIGURE 13.4
Reference patterns design on acetabulum cup. (a) Patterns on the internal surface of the acetabulum. (b) An example of pattern II whose ID is "011111111". (c) Distinguishable (left) and indistinguishable (right) pattern II.

13.2.3 Design of the SoC

The functional block diagram of the application-specific SoC with necessary peripheral components can be referred to [20]. As shown in Figure 13.5, the SoC has three main functional blocks. The first one is a power management unit, consisting of three programmable low-dropout (LDO) linear regulators supplying other functional blocks, and one programmable boost charge pump [21] for light-emitting diode (LED) driving. All the regulated voltages are generated by linear regulators except the boosted voltage to drive the LEDs as the flash light, as the switch regulator using inductors is not a good choice due to the number of external components and the consideration of electromagnetic interference (EMI). The second block is a programmable ultra-high frequency (UHF) band transceiver with 3Mbps minimum-shift keying (MSK) transmitting and 64kbps on-off keying (OOK) receiving. The third block is an ultra-low-power digital core, which takes care of the system control, the data processing and the communication protocol. The power management unit and the RF transceiver are controlled by the digital core as well. The digital core architecture is based on the Wishbone bus. Three master modules (including the microcontroller unit [MCU], the inter-integrated circuit [I^2C] controller for test purposes and the watchdog timer) and six slave modules (the medium access control [MAC] controller, the image processor, etc.) are connected to the bus. Since wireless data transmission uses most of the energy, it is necessary to apply low-power technologies to this RF transceiver. The transceiver will be discussed in Section 13.4.1 on low-power technologies.

In this section, the structure and architecture of the system have been introduced and details of the customized SoC have been presented. The pose estimation methods and the low-power technologies are the two key parts in the system design and will be detailed in the following sections.

13.3 Pose Estimation Methods for Hip Joint Prostheses

13.3.1 Modeling and Representation of Hip Joint Motions

In a 3D reference frame, a rigid body is known to have six degrees-of-freedom (DOF), including translations along three axes and rotations about three axes. A hip joint prosthesis is a ball and socket joint, connecting the pelvis with the femur. An artificial hip joint consists of a femoral head prosthesis and an acetabular cup, both of which are supposed to be rigid. Since the round femoral head ball fits well into the fixed acetabular cup, the femoral head is only capable of rotating and only three rotation angles about the three axes are adequate to determine its orientation during surgery. However, dislocation may occur when the surgeons drag or rotate the patient's thigh to test the hip ROM during surgery. In this case, more information other than the three rotation angles is needed to describe whether the femoral head ball is in the center of the acetabular cup socket.

To describe the motions of the femoral head prosthesis, an acetabular cup coordinate system {A} is built, of which the origin is at the center of the acetabular cup and the z-axis is the normal vector of the cup margin plane, as shown in Figure 13.6a. Since the dislocation is taken into consideration, the joint is modeled as a 4-DOF joint, including the rotations about the x-, y- and z-axes, and the translation d along the z-axis. Once d is larger than a

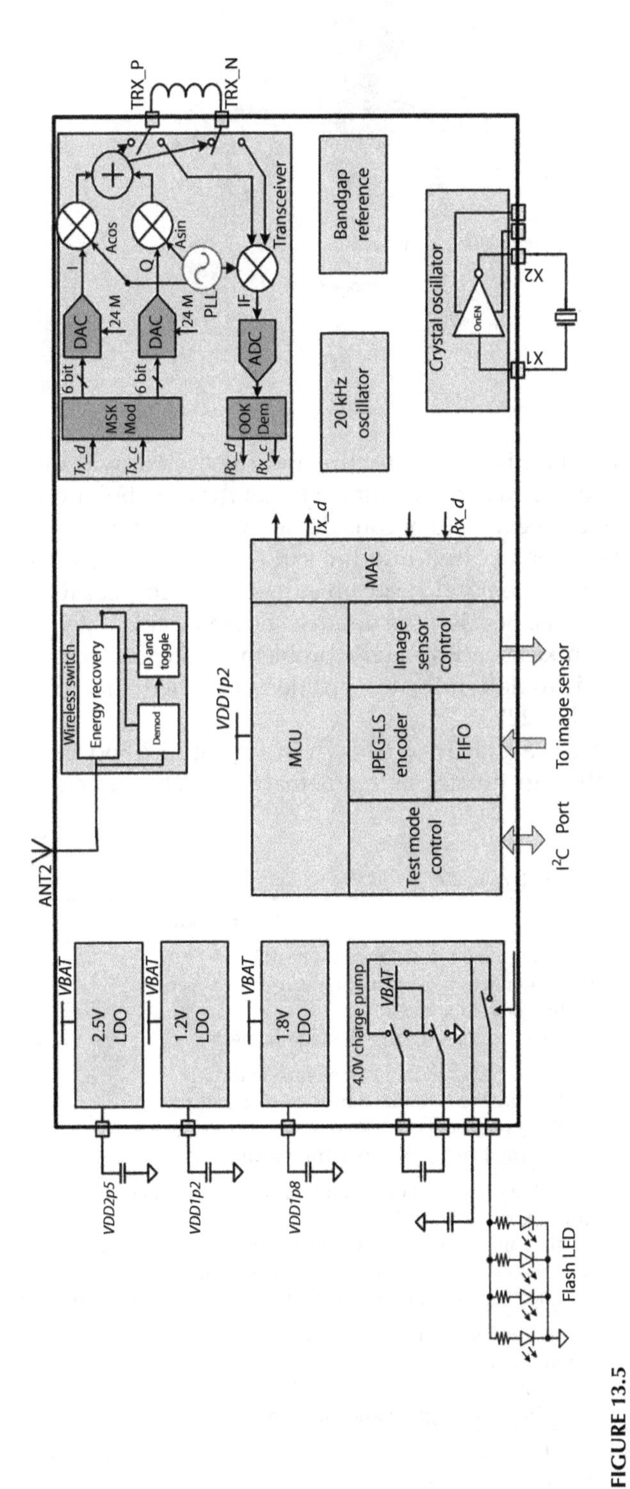

FIGURE 13.5
Block diagram of the SoC (inside the thick line) and the peripheral components.

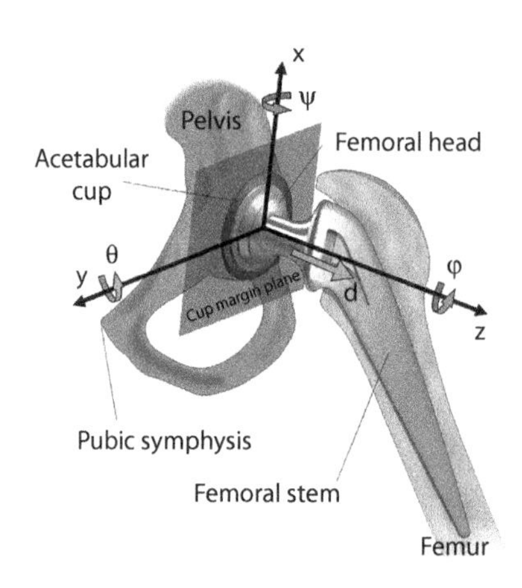

FIGURE 13.6
The 4-DOF model of a hip joint prostheses.

certain value, it means the dislocation occurs and the system will give an alarm to the surgeons until the femoral head trial is put back into the acetabular cup. The translations along the x- and y-axes can be left out of consideration. To help the surgeons judge whether the hip prostheses have been placed into the safe zone, the system should estimate the real-time orientation of the femoral head prosthesis with respect to the acetabular cup and show it to the surgeons. Besides, the depth d should be calculated for the detection of potential dislocations. In kinematics, such a problem is categorized as the forward kinematics problem, which is to determine the position and orientation of the end-effector in terms of the joint variables [22].

Four coordinate systems are defined (shown in Figure 13.7 and listed in Table 13.1) to describe the pose of the hip prosthesis mathematically. The camera inside the femoral

TABLE 13.1

Definitions of Four Coordinate Systems

Name and Notation	Definition
Femoral head coordinate system {F}	Origin: the center of the femoral head ball x_f: orthogonal to z_f, pointing in the medial direction y_f: orthogonal to x_f and z_f z_f: connecting the origin to the femoral neck
Acetabular cup coordinate system {A}	Origin: the center of hemisphere inside the cup x_a: orthogonal to z_a, defined by a specific pattern y_a: orthogonal to x_a and z_a z_a: the normal vector of cup margin plane
Pelvic coordinate system {P}	Origin: the mid-point of left and right anterior superior iliac spine x_p: orthogonal to y_p and z_p y_p: orthogonal to the plane defined with left and right anterior superior iliac spine and the mid-point left and right posterior superior iliac spine z_p: connecting left anterior superior iliac spine to right anterior superior iliac spine
World coordinate system {W}	Origin: a defined point at the objective location x_w: pointing to the east y_w: pointing to the north z_w: the direction of gravitational force

Note: {F} and {A} are used to estimate the relative pose between prostheses, {P} is used by surgeons to judge orientations of prostheses and {W} is used to present absolute poses of prostheses.

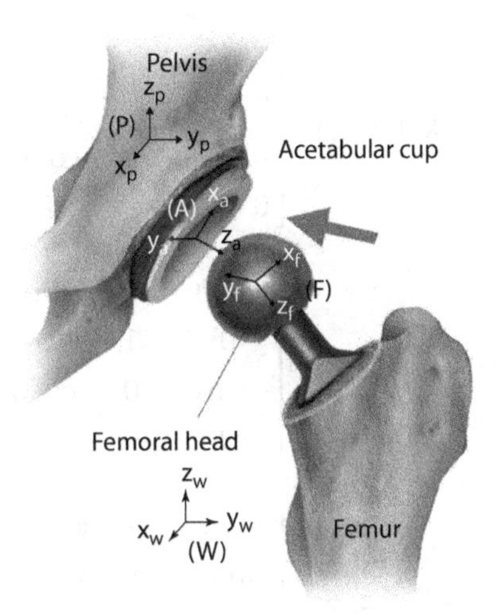

FIGURE 13.7
Definition of the four related coordinate systems.

head estimates the orientation of the femoral head trial with respect to the acetabular cup coordinate system {A} by taking pictures of the inner surface of the acetabular cup, while the IMU estimates the orientation of {F} with respect to {W} (representing the corresponding coordinate system, the same below). Therefore, proper methods are needed to represent the orientation and position of a rigid body with respect to a space-fixed coordinate system.

The femoral head coordinate system {F} is a body-fixed frame attaching to the femoral head trial. Suppose the system always starts with {F} totally overlapping the space-fixed frame {A} (the same below). A point fixed to the femoral head can be represented as $p^T = (x_f, y_f, z_f, 1)$ in both {F} and {A} with homogeneous coordinates applied. {F} has 6-DOF in total, including the 3-DOF translations and the 3-DOF rotations. A translation of the femoral head can be generally represented by three parameters:

$$\text{Trans}(a,b,c) = a\boldsymbol{i}_a + b\boldsymbol{j}_a + c\boldsymbol{k}_a, \tag{13.1}$$

where $\boldsymbol{i}_a$, $\boldsymbol{j}_a$ and $\boldsymbol{k}_a$ are the orthonormal basis of {A}. Since the femoral head is a rigid body, the coordinates of the point in {F} will remain $\boldsymbol{p}$. After the translation of the femoral head trial, the new coordinates of the point in {A}, denoted as $\boldsymbol{p}'$, can be obtained as

$$\boldsymbol{p}' = \begin{bmatrix} x_f + a \\ y_f + b \\ z_f + c \\ 1 \end{bmatrix} = \begin{bmatrix} 1 & 0 & 0 & a \\ 0 & 1 & 0 & b \\ 0 & 0 & 1 & c \\ 0 & 0 & 0 & 1 \end{bmatrix} \boldsymbol{p} = \boldsymbol{H}_T \boldsymbol{p}, \tag{13.2}$$

where $\boldsymbol{H}_T$ is a 4×4 matrix and can be used to represent the translation of the femoral head trial with respect to {A}.

Similar to the translation matrix $\boldsymbol{H}_T$, a rotation can also be represented by a 4×4 matrix, $\boldsymbol{H}_R$. The following three matrices represent three elemental rotations, which are rotations about the x-, y- and z-axes by angles of Ψ, θ and φ, respectively.

$$H_{Rx}(\psi) = Rot(x,\psi) = \begin{bmatrix} 1 & 0 & 0 & 0 \\ 0 & \cos\psi & -\sin\psi & 0 \\ 0 & \sin\psi & \cos\psi & 0 \\ 0 & 0 & 0 & 1 \end{bmatrix}, \tag{13.3}$$

$$H_{Ry}(\theta) = Rot(y,\theta) = \begin{bmatrix} \cos\theta & 0 & \sin\theta & 0 \\ 0 & 1 & 0 & 0 \\ -\sin\theta & 0 & \cos\theta & 0 \\ 0 & 0 & 0 & 1 \end{bmatrix}, \tag{13.4}$$

$$H_{Rz}(\varphi) = Rot(z,\varphi) = \begin{bmatrix} \cos\varphi & -\sin\varphi & 0 & 0 \\ \sin\varphi & \cos\varphi & 0 & 0 \\ 0 & 0 & 1 & 0 \\ 0 & 0 & 0 & 1 \end{bmatrix}. \tag{13.5}$$

Generally, a specific rotation of {F} can be divided into the foregoing three elemental rotations. The three elemental rotations may be extrinsic (rotations about the axes of the space-fixed coordinate system) or intrinsic (rotations about the axes of the body-fixed coordinate system). For example, let $H_R = H_{Rz}(\varphi)H_{Ry}(\theta)H_{Rx}(\Psi)$ be a composition of extrinsic rotations. In this case, H_R represents rotations that first rotate the femoral head Ψ about x_a-axis, then θ about y_a-axis and finally φ about z_a-axis of the space-fixed frame {A}. But if H_R is a composition of intrinsic rotations, then it can also be regarded as rotations of first φ about z_f -axis, then θ about y_f -axis and finally Ψ about x_f -axis of the body-fixed frame {F}. To avoid the confusion of extrinsic and intrinsic rotations, mathematicians have devised several orientation representation methods, among which Euler angles, Tait–Bryan angles and quaternions are the most famous [23].

In 1776, Leonhard Euler introduced the Euler angles to describe the orientation of a rigid body with respect to a space-fixed coordinate system. The Euler angles, typically denoted as (φ, θ, Ψ), are three elemental rotation angles whose first and third rotation axes are the same. The Euler angles following this definition are called proper or classic Euler angles. For instance, the proper Euler angles (φ, θ, Ψ) in Equation 13.6 describe a sequence of rotations that first φ about z-axis, then θ about y-axis and finally Ψ about z-axis if defined by extrinsic rotations.

$$H_R = H_{Rz}(\psi)H_{Ry}(\theta)H_{Rz}(\varphi). \tag{13.6}$$

In the 19th century, Peter Guthrie Tait and George H. Bryan devised a second type of formalism, called Tait–Bryan angles. Usually denoted as (α, β, γ), Tait–Bryan angles represent a composition of three rotations about three different axes, respectively. For instance, defined by intrinsic rotations, the Tait–Bryan angles (α, β, γ) in Equation 13.7 describe a sequence of rotations about z-, y- and x-axes for γ, β and α, respectively.

$$H_R = H_{Rz}(\gamma)H_{Ry}(\beta)H_{Rx}(\alpha). \tag{13.7}$$

The Tait–Bryan angles defined in Equation 13.7 are also called yaw, pitch and roll. They are very useful for aerospace applications. For an aircraft, a pose change can be obtained by three rotations about its principal axes if done in the proper order. A yaw will obtain the bearing angle, a pitch will yield the elevation angle and a roll gives the bank angle.

Both the proper Euler angles and Tait–Bryan angles are easy to understand, but they suffer from a problem called gimbal lock. To avoid this, Irish mathematician William Rowan Hamilton proposed the quaternions in 1843. The quaternion representation obtains a rotation matrix with a single rotation about one axis and does not have a singularity problem [24]. A quaternion $\boldsymbol{q}$ is defined as a four-dimensional vector, consisting of one real number and three imaginary numbers, as shown in Equations 13.8 and 13.9:

$$\boldsymbol{q} = \begin{pmatrix} q_0 & q_1 & q_2 & q_3 \end{pmatrix}^T = q_0 + (q_1\boldsymbol{i} + q_2\boldsymbol{j} + q_3\boldsymbol{k}) = q_0 + \boldsymbol{q}_v, \tag{13.8}$$

$$\boldsymbol{i}^2 = \boldsymbol{j}^2 = \boldsymbol{k}^2 = \boldsymbol{ijk} = -1, \tag{13.9}$$

where q_0, q_1, q_2 and q_3 are real numbers and $\boldsymbol{i}$, $\boldsymbol{j}$ and $\boldsymbol{k}$ are imaginary units.

The quaternion $\boldsymbol{q}$ is called a unit quaternion if it satisfies Equation 13.10:

$$\begin{cases} q_0^2 + (q_1^2 + q_2^2 + q_3^2) = 1 \\ \boldsymbol{q} = q_0 + \boldsymbol{q}_v = \cos(\theta) + \boldsymbol{u}\sin(\theta) \end{cases}, \tag{13.10}$$

where $\boldsymbol{u}$ is a 3D unit vector ($\boldsymbol{u} = u_x\boldsymbol{i} + u_y\boldsymbol{j} + u_z\boldsymbol{k}$). The quaternion components of $\boldsymbol{q}$ are

$$\begin{cases} q_0 = \cos\theta \\ q_1 = u_x \sin\theta \\ q_2 = u_y \sin\theta \\ q_3 = u_2 \sin\theta \end{cases}. \tag{13.11}$$

The point $\boldsymbol{p}$ fixed to the femoral head trial can be regarded as a vector setting out from the origin of {F}, and can also be represented as a quaternion $\boldsymbol{p} = (0, x_f, y_f, z_f)$ in {F}. If the femoral head trial is rotated by an angle of 2θ about the vector $\boldsymbol{u}$, the new coordinates of the point in {A} after the rotation, denoted as $\boldsymbol{p}'$, can be calculated as

$$\boldsymbol{p}' = \boldsymbol{q} \cdot \boldsymbol{p} \cdot \boldsymbol{q}^*, \tag{13.12}$$

where $\boldsymbol{q}$ is a unit quaternion and $\boldsymbol{q}^* = q_0 - q_1\boldsymbol{i} - q_2\boldsymbol{j} - q_3\boldsymbol{k}$ is the conjugate of $\boldsymbol{q}$.

The dot is a specific operation of quaternions called Hamilton product [23], defined as

$$\boldsymbol{a} \cdot \boldsymbol{b} = (a_0 + a_1\boldsymbol{i} + a_2\boldsymbol{j} + a_3\boldsymbol{k})(b_0 + b_1\boldsymbol{i} + b_2\boldsymbol{j} + b_3\boldsymbol{k}) = \begin{pmatrix} a_0b_0 - a_1b_1 - a_2b_2 - a_3b_3 \\ a_0b_1 + a_1b_0 + a_2b_3 - a_3b_2 \\ a_0b_2 - a_1b_3 + a_2b_0 + a_3b_1 \\ a_0b_3 + a_1b_2 - a_2b_1 + a_3b_0 \end{pmatrix}. \tag{13.13}$$

Such a rotation can also be achieved by a rotation matrix H_{Rq}:

$$p' = H_{Rq}p = \begin{bmatrix} 1-2(q_2^2+q_3^2) & 2(q_1q_2-q_3q_0) & 2(q_1q_3+q_2q_0) & 0 \\ 2(q_1q_2+q_3q_0) & 1-2(q_1^2+q_3^2) & 2(q_2q_3-q_1q_0) & 0 \\ 2(q_1q_3-q_2q_0) & 2(q_2q_3+q_1q_0) & 1-2(q_1^2+q_2^2) & 0 \\ 0 & 0 & 0 & 1 \end{bmatrix} p, \tag{13.14}$$

where the relationship between the quaternion and the rotation matrix is shown. As a result, no matter which orientation representation method is applied, a rotation of a rigid body can always be represented by a 4×4 matrix with homogeneous coordinates applied.

13.3.2 Problem Definition

Suppose the system starts with {F} totally overlapping {A}. One of the vertexes of the reference patterns printed on the inner surface of the acetabular cup is expressed as $p_a=(x_a, y_a, z_a, 1)^T$ in {A} and $p_f=(x_f, y_f, z_f, 1)^T$ in {F}; p_f varies with the femoral head trial being moved, while p_a remains unchanged. According to the discussion in the previous section, a 4×4 matrix H can be found to represent the transformation between p_f and p_a, and H can be divided into several translations H_T and rotations H_R as

$$p_f = Hp_a = H_{T1}H_{R1}H_{T2}H_{R2}\ldots H_{Tn}H_{Rn}p_f. \tag{13.15}$$

Because p_f and p_a are actually the same point expressed in different coordinate systems, the matrix H just represents the transformation from the {F} to the {A}, which contains information of relative pose between the femoral head prosthesis and the acetabular cup. Consequently, the prosthesis pose estimation problem of the system is finally defined as the problem of calculating the transformation matrix H in Equation 13.15.

The camera and the IMU in the system provide us with two ways to calculate H. One is to use IMU to directly measure the motion parameters (accelerations and angular velocities) to calculate H. The other is to find the coordinates of the corresponding feature points in the images and solve H with the theory of computer vision. These two solutions will be explained next in detail.

13.3.3 IMU-Based Pose Estimation Method

An IMU is composed of an accelerometer and a gyroscope and is widely used in navigation systems and robots. Mounted inside the femoral head trial, the IMU can be used to estimate the real-time orientation of the femoral head with respect to the world coordinate system {W}. In this part, the accelerometer-based orientation estimation method will be introduced, then the gyroscope-based one and finally the famous data fusion method called extended Kalman Filter (EKF) for the IMU data.

The Tait–Bryan angles (α, β, γ) are selected to explain the principle of the IMU-based orientation estimation method as they are easier to understand. Suppose the system starts with {F} overlapping the world frame {W}. The orientation of the femoral head trial is obtained by the following three intrinsic rotations in sequence of $z_f-y_f-x_f$: first rotates about z_f-axis by an angle γ, called yaw or heading angle ($\gamma\in[-\pi, \pi]$); then about the y_f-axis by an angle β, called pitch or elevation angle $\left(\beta \in \left[-\frac{\pi}{2}, \frac{\pi}{2}\right]\right)$; finally, about x_f-axis by a roll or bank angle α (α∈[−π, π]). The rotation matrix is

$$H = \begin{bmatrix} \cos\gamma & -\sin\gamma & 0 & 0 \\ \sin\gamma & \cos\gamma & 0 & 0 \\ 0 & 0 & 1 & 0 \\ 0 & 0 & 0 & 1 \end{bmatrix} \begin{bmatrix} \cos\beta & 0 & \sin\beta & 0 \\ 0 & 1 & 0 & 0 \\ -\sin\beta & 0 & \cos\beta & 0 \\ 0 & 0 & 0 & 1 \end{bmatrix} \begin{bmatrix} 1 & 0 & 0 & 0 \\ 0 & \cos\alpha & -\sin\alpha & 0 \\ 0 & \sin\alpha & \cos\alpha & 0 \\ 0 & 0 & 0 & 1 \end{bmatrix}$$
$$= \begin{bmatrix} \cos\beta\cos\gamma & \sin\alpha\sin\beta\cos\gamma - \cos\alpha\sin\gamma & \cos\alpha\sin\beta\cos\gamma + \sin\alpha\sin\gamma & 0 \\ \cos\beta\sin\gamma & \sin\alpha\sin\beta\sin\gamma + \cos\alpha\cos\gamma & \cos\alpha\sin\beta\sin\gamma - \sin\alpha\cos\gamma & 0 \\ -\sin\beta & \sin\alpha\cos\beta & \cos\alpha\cos\beta & 0 \\ 0 & 0 & 0 & 1 \end{bmatrix}. \tag{13.16}$$

As the accelerometer is mounted inside the femoral head prosthesis, it measures the projection of the gravitational acceleration in the three axes of {F} if the femoral head trial stays still or moves slowly. Denoted as g_f, the gravitational acceleration vector in {F} is therefore represented as follows:

$$g_f = \begin{bmatrix} g_x \\ g_y \\ g_z \\ 1 \end{bmatrix} = H^{-1} g_w = H^{-1} \begin{bmatrix} 0 \\ 0 \\ g \\ 1 \end{bmatrix} = \begin{bmatrix} -g\sin\beta \\ g\sin\alpha\cos\beta \\ g\cos\alpha\cos\beta \\ 1 \end{bmatrix}. \tag{13.17}$$

By solving the equations in Equation 13.17, the roll α and the pitch β can be obtained in Equation 13.18:

$$\begin{bmatrix} \alpha \\ \beta \end{bmatrix} = \begin{bmatrix} \tan^{-1}\left(\frac{g_y}{g_z}\right) \\ \tan^{-1}\left(\frac{-g_x}{\sqrt{g_y^2 + g_z^2}}\right) \end{bmatrix}. \tag{13.18}$$

However, the yaw γ cannot be determined without the help of other sensors. In addition, if the femoral head trial is in a variable speed motion, the IMU will have a measured acceleration besides the gravitational acceleration, leading to the invalidity of the method.

On the other hand, the gyroscope-based method is good at dealing with the variable motion. The gyroscope measures the angular velocity about the three axes of {F}, denoted as $\boldsymbol{\omega}_f = [\omega_{fx}\ \omega_{fy}\ \omega_{fz}]^T$. The relationship between the Tait–Bryan angles (α, β, γ) and $\boldsymbol{\omega}_f$ is shown in Equation 13.19. The Tait–Bryan angles (α, β, γ) can be obtained by solving the first-order nonlinear differential equations in 13.20.

$$\begin{bmatrix}\omega_{fx}\\ \omega_{fy}\\ \omega_{fz}\end{bmatrix}=\begin{bmatrix}\cos\gamma & -\sin\gamma & 0\\ \sin\gamma & \cos\gamma & 0\\ 0 & 0 & 1\end{bmatrix}\begin{bmatrix}\cos\beta & 0 & \sin\beta\\ 0 & 1 & 0\\ -\sin\beta & 0 & \cos\beta\end{bmatrix}\begin{bmatrix}\dot{\alpha}\\ 0\\ 0\end{bmatrix}$$

$$+\begin{bmatrix}\cos\gamma & -\sin\gamma & 0\\ \sin\gamma & \cos\gamma & 0\\ 0 & 0 & 1\end{bmatrix}\begin{bmatrix}0\\ \dot{\beta}\\ 0\end{bmatrix}+\begin{bmatrix}0\\ 0\\ \dot{\gamma}\end{bmatrix} \tag{13.19}$$

$$=\begin{bmatrix}\cos\beta\cos\gamma & -\sin\gamma & 0\\ \cos\beta\sin\gamma & \cos\gamma & 0\\ \cos\beta & 0 & 1\end{bmatrix}\begin{bmatrix}\dot{\alpha}\\ \dot{\beta}\\ \dot{\gamma}\end{bmatrix}.$$

$$\begin{bmatrix}\dot{\alpha}\\ \dot{\beta}\\ \dot{\gamma}\end{bmatrix}=\begin{bmatrix}\dfrac{\cos\gamma}{\cos\beta} & \dfrac{\sin\gamma}{\cos\beta} & 0\\ -\sin\gamma & \cos\gamma & 0\\ -\cos\gamma & -\sin\gamma & 1\end{bmatrix}\begin{bmatrix}\omega_{fx}\\ \omega_{fy}\\ \omega_{fz}\end{bmatrix} \tag{13.20}$$

Since the gyroscope has drifts, the gyroscope-based method may suffer from a large error if the femoral head trial stays still for a long time. Besides, if the pitch $\beta=\pm(\pi/2)$, the first two terms in the first row in Equation 13.20 will become infinite, leading to a singularity.

Although both the accelerometer and the gyroscope can be used for orientation estimation alone, each has its own limitations. From the foregoing discussion, a complementary property can be found between the accelerometer and the gyroscope. To make use of the complementary property and avoid the problem of singularities, a quaternion-based data fusion algorithm called extended Kalman filter (EKF) [25] is adopted in the system. The block diagram of the IMU data fusion method is shown in Figure 13.8.

In order to use EKF, a state vector $\boldsymbol{X}_k$ and its state transition need to be defined. Let ${}^{F(t)}_{A}\boldsymbol{q}$ be the quaternion that represents the orientation of {F} with respect to {A} at time t. The difference quotient of ${}^{F(t)}_{A}\boldsymbol{q}$ can be computed as

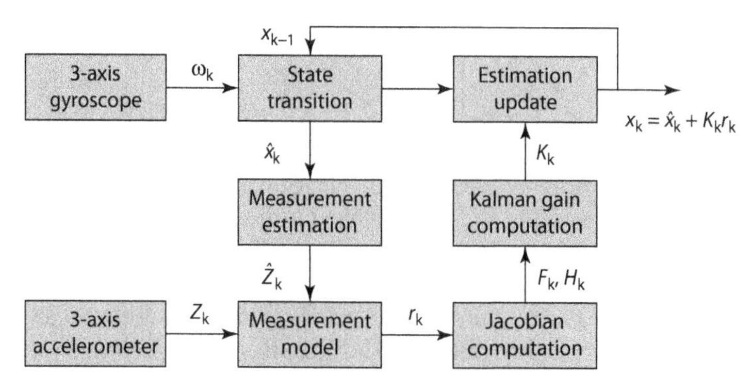

FIGURE 13.8
Block diagram of orientation estimation principles for an IMU.

$$ {}^{F(t)}_{A}\dot{q} = \lim_{\Delta t \to 0} \frac{1}{\Delta t}\left[{}^{F(t+\Delta t)}_{A}q - {}^{F(t)}_{A}q \right] = \lim_{\Delta t \to 0} \frac{1}{\Delta t}\left[{}^{F(t+\Delta t)}_{F(t)}q \cdot {}^{F(t)}_{A}q - {}^{F(0)}_{A}q \cdot {}^{F(t)}_{A}q \right], \tag{13.21} $$

where ${}^{F(t+\Delta t)}_{F(t)}q$ represents the rotation in the period Δt. In the limit of $\Delta t \longrightarrow 0$, the angle of rotation $\delta\theta$ is very small so that the first-order Taylor expansion can be used to approximate the sin and cos functions as

$$ \lim_{\Delta t \to 0} {}^{F(t+\Delta t)}_{F(t)}q = \begin{pmatrix} \cos\frac{\delta\theta}{2} \\ u\sin\frac{\delta\theta}{2} \end{pmatrix} \approx \begin{pmatrix} 1 \\ u\frac{\delta\theta}{2} \end{pmatrix} = \begin{pmatrix} 1 \\ \frac{1}{2}\boldsymbol{\delta\theta} \end{pmatrix}. \tag{13.22} $$

The vector $\boldsymbol{\delta\theta}$ has the direction of the axis of rotation and the magnitude of the angle of rotation. Dividing this vector by Δt, in the limit $\Delta t \longrightarrow 0$, will yield the rotational velocity as

$$ \boldsymbol{\omega} = \begin{bmatrix} \omega_x & \omega_y & \omega_z \end{bmatrix}^T = \lim_{\Delta t \to 0} \frac{\boldsymbol{\delta\theta}}{\Delta t}. \tag{13.23} $$

By combining Equations 13.21 through 13.23, the quaternion derivative ${}^{F(t)}_{A}\dot{q}$ can be derived as

$$ {}^{F(t)}_{A}\dot{q} \approx \left[\begin{pmatrix} 1 \\ \frac{1}{2}\boldsymbol{\delta\theta} \end{pmatrix} - \begin{pmatrix} 1 \\ 0 \end{pmatrix} \right] \cdot {}^{F(t)}_{A}q = \frac{1}{2}\begin{pmatrix} 1 \\ \boldsymbol{\omega} \end{pmatrix} \cdot {}^{F(t)}_{A}q = \frac{1}{2}\boldsymbol{\Omega}(\boldsymbol{\omega})\, {}^{F(t)}_{A}q \tag{13.24} $$

with

$$ \boldsymbol{\Omega}(\boldsymbol{\omega}) = \begin{bmatrix} 0 & -\omega_x & -\omega_y & -\omega_z \\ \omega_x & 0 & \omega_z & -\omega_y \\ \omega_y & -\omega_z & 0 & \omega_x \\ \omega_z & \omega_y & -\omega_x & 0 \end{bmatrix} = \begin{bmatrix} 0 & -\boldsymbol{\omega}^T \\ \boldsymbol{\omega} & -[\boldsymbol{\omega}_\times] \end{bmatrix}. \tag{13.25} $$

The discrete-time model corresponding to Equation 13.24 is

$$ \begin{cases} q_k = \left[I + \frac{T_s}{2}\boldsymbol{\Omega}(\boldsymbol{\omega}_k) \right] q_{k-1},\ k = 1,2,\ldots \\ q_0 = 0, \end{cases} \tag{13.26} $$

where T_s is the sample interval. If the drifts of the gyroscope are taken into consideration, the state vector can be defined as the rotation quaternion of the femoral head trial q_k and bias vector of the gyroscope ${}^{g}b_k$. The state vector X_k is expressed as

$$ \mathbf{X}_{k+1} = \begin{bmatrix} \mathbf{q}_{k+1} \\ {}^{g}\mathbf{b}_{k+1} \end{bmatrix} = \mathbf{f}(\mathbf{x}_k, \boldsymbol{\omega}_{k+1}) + \mathbf{w}_{k+1} = \begin{bmatrix} \mathbf{I} + \frac{T_s}{2}\boldsymbol{\Omega}(\hat{\boldsymbol{\omega}}_{k+1}) & 0 \\ 0 & \mathbf{I} \end{bmatrix} \begin{bmatrix} \mathbf{q}_k \\ {}^{g}\mathbf{b}_k \end{bmatrix} + \mathbf{w}_{k+1}, \tag{13.27} $$

where w_{k+1} is a noise vector and $\hat{\boldsymbol{\omega}}_{k+1} = \boldsymbol{\omega}_{k+1} - {}^{g}b_k$ is the calibrated angular velocity of {F} measured by the gyroscope.

The measurement vector Z_{k+1} is constructed by the output of the accelerometer as

$$Z_{k+1} = a_{k+1} = C_w^f \left[q_{k+1} \right] g + v_{k+1}, \tag{13.28}$$

where g is the gravitational acceleration with respect to {W} and v_{k+1} is a noise vector.
The state vector is estimated iteratively. The specific processing steps are

1. Consider the last filtered state estimate x_k.
2. Calculate the Jacobian matrix F_k for the system dynamics around X_k to linearize the system dynamics, as shown in Equation 13.29:

$$F_k = \frac{\partial f_k}{\partial X_k} \tag{13.29}$$

3. Apply the prediction step of the Kalman filter to the linearized system dynamics obtained in step 2:

$$\hat{x}_{k+1} = \begin{bmatrix} I + \frac{T_s}{2} \boldsymbol{\Omega}\left(\hat{\boldsymbol{\omega}}_{k+1}\right) & 0 \\ 0 & I \end{bmatrix} x_k \tag{13.30}$$

$$P_{k+1|k} = F_k P_k F_k^T + Q_k \tag{13.31}$$

In Equation 13.31, Q_k is the process noise covariance matrix, P_k is the covariance of x_k and $P_{k+1|k}$ is the estimate of the covariance of $\hat{x}_{k+1}$

4. Linearize the observation vector $\hat{z}_{k+1} = h\left(\hat{x}_{k+1}\right)$:

$$H_{k+1} = \frac{\partial h_{k+1}}{\partial \hat{x}_{k+1}} \tag{13.32}$$

5. Apply the filtering cycle of the Kalman filter to the linearized observation dynamics:

$$x_{k+1} = \hat{x}_{k+1} + K_{k+1}\left[z_{k+1} - h\left(\hat{x}_{k+1}\right) \right] \tag{13.33}$$

$$P_{k+1} = \left[I - K_{k+1} H_{k+1} \right] P_{k+1|k} \tag{13.34}$$

In Equations 13.33 and 13.34, K_{k+1} is Kalman gain for iteration $(k+1)$. After a series of iterations, the optimized state vector x_k will converge to a stable value, which is the estimation of the femoral head orientation.

13.3.4 Vision-Based Pose Estimation Method

The vision-based pose estimation method is another method widely used in navigation and automatic control systems. As the camera is rigidly attached to the femoral head trial, the femoral head pose estimation problem can be changed into the camera ego-motion estimation problem.

A camera model, which is used to describe the projection of a point from the 3D coordinate system to the 2D image plane, is the basis of the vision-based pose estimation method. Researchers have proposed several camera models, among which the pinhole model is the most popular. Consider a point $P_w = (x_w, y_w, z_w, 1)^T$ in the world coordinate system {W}. The pinhole camera model in Equation 13.35 shows how P_w is projected to the image:

$$p = \begin{pmatrix} u \\ v \\ 1 \end{pmatrix} = DK_0H_w^cP_w = \begin{pmatrix} \frac{f}{dx} & \frac{f\cot\theta}{dx} & u_0 & 0 \\ 0 & \frac{f}{\sin\theta dy} & v_0 & 0 \\ 0 & 0 & 1 & 0 \end{pmatrix} \begin{pmatrix} R & T \\ 0 & 1 \end{pmatrix} P_w \quad (13.35)$$

In Equation 13.35,

$$D = \begin{pmatrix} \frac{1}{dx} & -\frac{\cot\theta}{dx} & 0 \\ 0 & \frac{1}{\sin\theta dy} & 0 \\ 0 & 0 & 1 \end{pmatrix}, \quad K_0 = \begin{pmatrix} f & 0 & x_0 & 0 \\ 0 & f & y_0 & 0 \\ 0 & 0 & 1 & 0 \end{pmatrix}, \quad and\ H_w^c = \begin{pmatrix} R & T \\ 0 & 1 \end{pmatrix}.$$

As shown in Figure 13.9, the pinhole model is divided into three steps. First, the matrix H_w^c, which contains the camera's rotation matrix R and translation vector T with respect to {W}, transforms the coordinates of P_w from {W} into the camera coordinate system {C}. Then the point is projected from {C} to the 2D image plane by K_0, in which f is the camera focal length and (x_0, y_0) is the optic center on the camera image plane. Finally, the 2D coordinates of the point in the image plane {I} is discretized by D, in which dx and dy represent the size of each pixel in a charge-coupled device/complementary metal-oxide-semiconductor (CCD/CMOS) image sensor with an intersection angle θ. The matrix $K = DK_0$ is

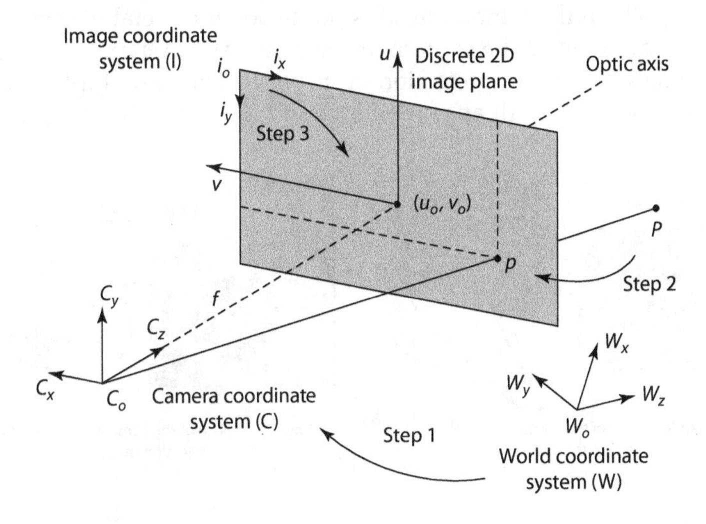

FIGURE 13.9
The general steps of the pinhole model.

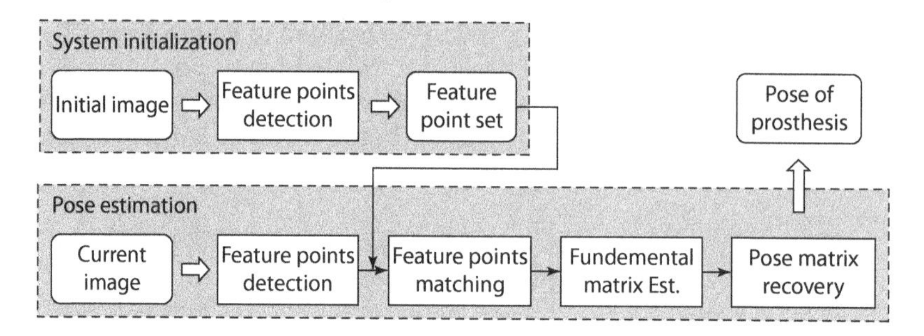

FIGURE 13.10
System initialization and pose estimation pipeline.

called the intrinsic parameter matrix, in which all the parameters are available by the camera calibration, while the matrix H_w^c is known as the extrinsic parameter matrix, containing the ego-motion information of the camera with respect to {W}. If a calibrated camera is given together with a set of 3D points whose coordinates in {W} are known information, the pose of the camera can be acquired by solving Equation 13.35. Researchers have named it a perspective-n-point (PnP) problem.

As introduced in Section 13.2.2, an array of reference patterns is designed on the inner surface of the acetabular cup. Since the pattern IDs and positions of the patterns are prior information, the four vertexes of each pattern II can be used as feature points for camera ego-motion estimation. However, if the patterns are covered by blood during surgery, pattern recognition will become difficult and the outcome of the PnP algorithms may be wrong. To avoid the interference of blood, another method based on epipolar geometry [26] is adopted in the system instead of a PnP algorithm. This method consists of two major steps with its pipeline shown in Figure 13.10.

13.3.4.1 Initialization

Two tasks are performed in initialization: the first is to determine whether the femoral head is inside the acetabular cup; the second is to adjust the femur to the initial pose and start tracking [27]. When the femoral head is put inside the acetabular cup, pattern I may appear in the visual range of the camera. If the camera's optic axis is not perpendicular to the circular ring's plane, the projection of it will become an elliptical ring (shown in Figure 13.11a). A pink circle will appear (shown in Figure 13.11b) when the femoral head

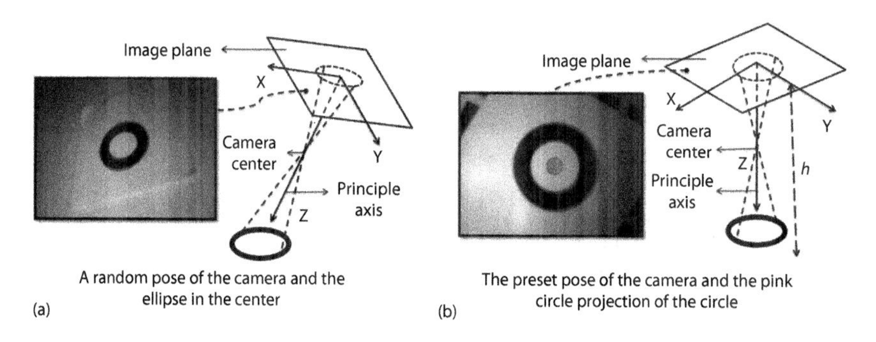

FIGURE 13.11
(a) A random pose of the camera and the ellipse in the center and (b) the preset pose of the camera and the pink circle projection of the circle.

is adjusted to the preset position, where the camera's axis is perpendicular to the circular ring's plane and the distance between the camera image plane and the center of the ring is exactly h (a preset value). The pink circle means that the femoral head is at the preset position inside the acetabular cup and the motion of the femur can be tracked from that initial position.

13.3.4.2 Feature Detection

After initialization, feature detection is done in three steps: contour detection, pattern recognition and vertexes calculation. Each image acquired will be converted from a red-green-blue (RGB) format into a gray scale one and processed by the polygonal approximation algorithm to search all the contours in the image. The quadrilaterals are chosen as candidates from the contours for further pattern detection. To recognize the patterns, samples are taken at the nine points within each possible pattern contour. The pattern is identified by the colors of the nine sample points, with white representing "1" and black representing "0". The 9-bit binary sequence is then converted to a decimal number, which is the ID of the reference pattern. Once a pattern is recognized, the 2D coordinates of its four vertexes will be calculated, which are denoted as (u_0, v_0), (u_1, v_1), (u_2, v_2) and (u_3, v_3). They will be used as feature points. The feature detection process is shown in Figure 13.12.

As the position of each reference pattern is known information, the 3D coordinates of their vertexes in {A} can also be extracted from the database, denoted as (x_0, y_0, z_0), (x_1, y_1, z_1), (x_2, y_2, z_2) and (x_3, y_3, z_3). The correspondences between the 3D feature points and their 2D projections are established according to the ID of each reference pattern:

$$c_i : (u_i, v_i) \leftrightarrow (x_i, y_i, z_i) \quad i = 0,1,2,3.... \tag{13.36}$$

If there are n reference patterns recognized in the image, $4n$ correspondences will be established since each square has four squares. As at least four correspondences are needed to solve the equations in Equation 13.35, one recognized pattern is enough to estimate the camera pose with a PnP algorithm [27]. However, the PnP algorithm strictly relies on the accuracy of the 3D coordinates of the feature points. To get rid of the high dependency on the accuracy of manufacture, a different pose estimation method based on multiple view geometry is applied in this system [28].

13.3.4.3 Feature Matching

Suppose a calibrated camera takes pictures of the same object from two different positions. The camera first takes a picture at Position 1, then shifts to Position 2 by a rotation $\boldsymbol{R}$ and a

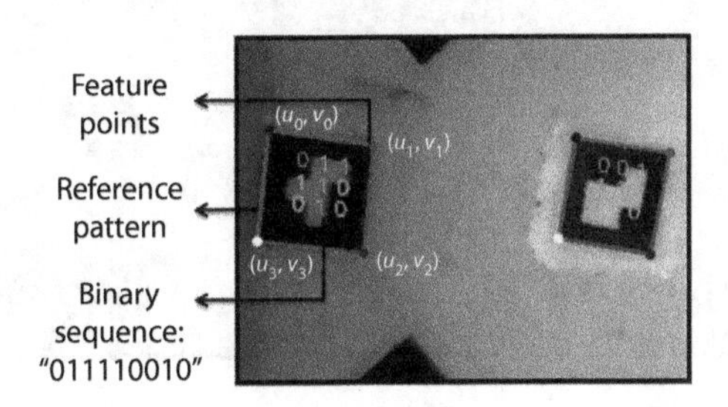

FIGURE 13.12
Find reference patterns and identify them.

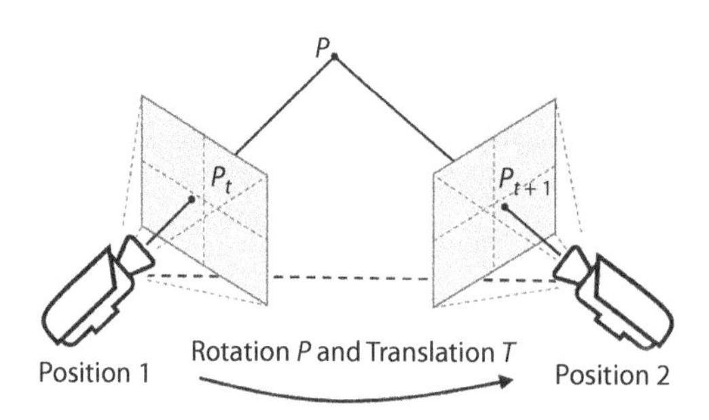

FIGURE 13.13
Principle of two view geometry.

translation T and takes a second picture at Position 2, as shown in Figure 13.13. The point P is fixed in the 3D world coordinate system {W}. Based on Equation 13.35, the projections of P in the two images, denoted as p_t and p_{t+1}, can be represented in Equations 13.37 and 13.38, respectively:

$$p_t = KH_1P, \tag{13.37}$$

$$p_{t+1} = KH_2P, \tag{13.38}$$

where H_1 and H_2 are the camera transformation matrices in Position 1 and 2, respectively, and K is the camera intrinsic parameter matrix.

The relationship between the correspondences in the two images is acquired by combining Equations 13.37 and 13.38:

$$p_{t+1} = KH_2P = K\begin{pmatrix} R & T \\ 0 & 1 \end{pmatrix}H_1P = K\begin{pmatrix} R & T \\ 0 & 1 \end{pmatrix}K^{-1}p_t. \tag{13.39}$$

If more than four correspondences between the two images can be found, the motion of the camera R and T from Positions 1 to 2 can be solved by Equation 13.39. Figure 13.14 shows the feature match between two adjacent image frames.

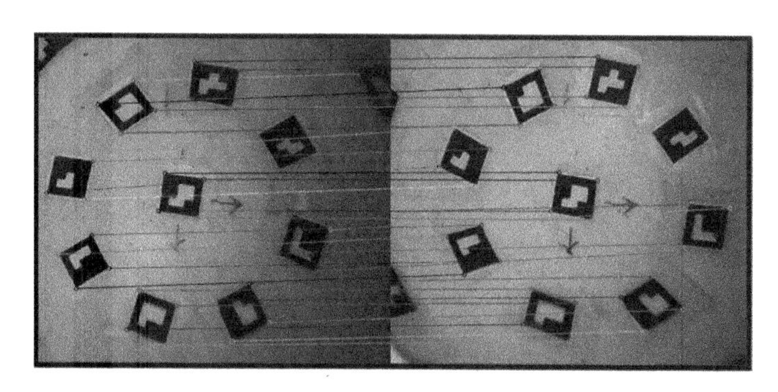

FIGURE 13.14
Matching feature points between two image frames.

13.3.4.4 Estimation of the Fundamental Matrix

After the features are matched, the 3×3 fundamental matrix $\boldsymbol{F}$ is defined as

$$x'^T Fx = 0, \tag{13.40}$$

where $x = (u, v, 1)^T \leftrightarrow x' = (u', v', 1)^T$ are corresponding points. From a set of n point matches, a set of linear equations can be obtained:

$$Af = \begin{bmatrix} u_1'u_1 & u_1'v_1 & u_1' & v_1'u_1 & v_1'v_1 & v_1' & u_1 & v_1 & 1 \\ \vdots & \vdots & \vdots & \vdots & \vdots & \vdots & \vdots & \vdots & \vdots \\ u_n'u_n & u_n'v_n & u_n' & v_n'u_n & v_n'v_n & v_n' & u_n & v_n & 1 \end{bmatrix} f = 0, \tag{13.41}$$

$$f^T = \begin{bmatrix} f_{11} & f_{12} & f_{13} & f_{21} & f_{22} & f_{23} & f_{31} & f_{32} & f_{33} \end{bmatrix}, \tag{13.42}$$

where $\boldsymbol{A}$ is an $n \times 9$ matrix. Given at least eight pairs of correspondences, which means at least two reference patterns should be found in the image, $\boldsymbol{F}$ can be calculated.

13.3.4.5 Recovery of Transformation Matrix

An essential matrix is a 3×3 matrix that captures the geometric relationship (a rotation $\boldsymbol{R}$ and a translation $\boldsymbol{t}$) between two locations of one moving camera [29], and can be calculated from the fundamental matrix $\boldsymbol{F}$ acquired at the previous step and the camera calibration matrices $\boldsymbol{K}$ and $\boldsymbol{K}'$ as

$$E = [t]_\times R = K'^T FK, \tag{13.43}$$

where $[\boldsymbol{t}]_\times$ is a 3×3 antisymmetric matrix defined by vector $\boldsymbol{t}$. The transformation matrix $\boldsymbol{H} = [\mathbf{R} | \boldsymbol{t}]$ can then be acquired by solving Equation 13.43, and the current camera pose can be calculated by

$$P = HP_0, \tag{13.44}$$

where $\boldsymbol{P}$ and $\boldsymbol{P}_0$ are the current and past camera pose matrix.

13.3.5 Data Fusion of the Camera and the IMU

Both the camera and the IMU can be used alone to achieve pose estimation, but each of them has its own limitations. The IMU has large measurement uncertainty at slow motion due to drifts, and can only do pose estimation with respect to the world coordinate system {W} while surgeons are more interested in the relative pose between the femoral head and the acetabular cup. On the other hand, the camera is easily interfered with by blood. Furthermore, if the femoral head trial is moved at high speed, there will be motion blur on the images. Although the motion blur can be relieved by adjusting the camera to a higher frame rate mode, it will also increase the data throughput, leading to a higher power consumption and a higher complexity in the RF transceiver design.

Similar to the relationship between the accelerometer and the gyroscope, the camera and the IMU also share a complementary property in pose estimation. Better results could be obtained by fusing their data, as shown in Figure 13.15. Several algorithms can be used as the optimization filter. In this system, EKF is adopted for the optimization.

As mentioned in Section 13.3.3, a state vector and a measurement vector need to be defined in the EKF and it will iteratively estimate the state vector. The state vector is composed of the rotation quaternion of the femoral head trial q_{k+1}, the bias vector of the gyroscope ${}^g\boldsymbol{b}_{k+1}$ and the accelerometer ${}^a\boldsymbol{b}_{k+1}$. The state vector can be predicted by the IMU data, so $\boldsymbol{X}_{k+1}$ could be expressed as Equation 13.45.

$$\boldsymbol{X}_{k+1} = \begin{bmatrix} q_{k+1} \\ {}^g\boldsymbol{b}_{k+1} \\ {}^a\boldsymbol{b}_{k+1} \end{bmatrix} = \boldsymbol{f}_{k+1}\left(\boldsymbol{X}_k, \boldsymbol{\omega}_{k+1}\right) + w_k \tag{13.45}$$

where w_k is a noise vector and $\boldsymbol{f}_{k+1}$ is the state transition based on the previous state vector x_k and the calibrated angular velocity $\boldsymbol{\omega}_{k+1}$ measured by the gyroscope.

For each image acquired from the camera, a set of features can be detected and the projections of the features are used for pose estimation. If there are N features detected in the current frame, the measurement vector can be defined as

$$\boldsymbol{Z}_k = \begin{bmatrix} z_1^T & \dots & z_N^T \end{bmatrix}^T, \tag{13.46}$$

$$z_i = \begin{bmatrix} u_i \\ v_i \end{bmatrix} + \eta_i = \mathrm{h}_i\left(x, {}^A\boldsymbol{P}_i\right) + \eta_i, \quad i = 1, 2, \dots, N, \tag{13.47}$$

where z_i is the projection of the ith feature point and ${}^A\boldsymbol{P}_i$ is its corresponding 3D coordinates in the acetabular cup frame {A}; η_i is the noise of the feature detection.

The iteration steps are similar to those in Section 13.3.3. The fusion of the camera and the IMU not only provides more accurate pose estimation of the femoral head prosthesis, but it also helps decrease the power consumption, which will be discussed in next section.

13.4 Low-Power Technology

Power consumption is a key factor for electronic devices, especially in medical applications. In this pose estimation system, a 500 mAh Li-ion battery is used to supply the whole system with the constraint of the femoral neck space. In order to ensure that the system

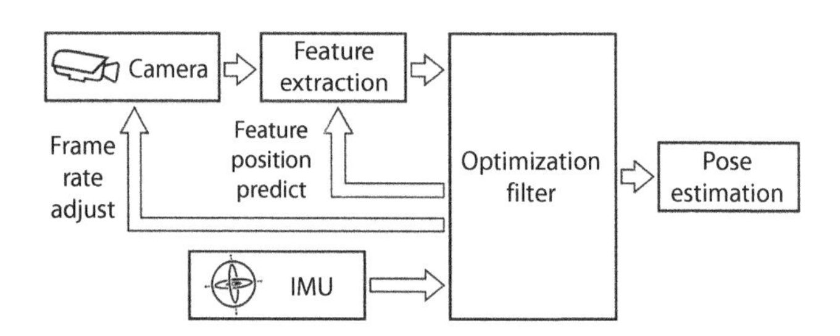

FIGURE 13.15
Fusion of the camera and the IMU data.

works for more than 20 minutes with this battery, low-power technologies are applied in both the hardware and the software design.

13.4.1 Low-Power RF Transceiver

Since most of the energy is consumed by the wireless data transmission, the power consumption of the RF transceiver is the key issue in the SoC design. The overall structure of the transceiver used in this design is shown in Figure 13.16. As the system needs to continuously transmit data to and sparsely receive configuration command from the PC, the system requires a high speed transmitter and a relatively low speed receiver. The MSK modulation has been chosen for the transmitter, since MSK is a constant-envelope modulation that can help to alleviate the design requirement of the data recorder on the other side. A high data rate of 3 Mbps is chosen such that the transmitter will work in burst-mode and the average power consumption of the transmitter will be low [30]. The receiver takes a 64 kbps OOK modulation for circuit simplification. In this system, the effort to lower the circuit power is equivalent to lowering the circuit current. For this reason, the current reusing techniques [31] have been greatly adopted in the transceiver circuit, mainly in the frequency synthesizer and the transmitter.

Figure 13.17 shows the structure of the frequency synthesizer. The voltage-controlled oscillator (VCO) is designed to run at ~800 MHz to generate quadrature-phase local oscillation signals. In this synthesizer, the quadratic frequency divider shares the same direct current (DC) path [32] with the VCO, as shown in Figure 13.3. A classical phase-locked loop (PLL) locks the VCO frequency. The PLL's divider is programmable, giving the SoC the capability of frequency hopping. The VCO has a coarse tuning circuit that helps to calibrate its center frequency automatically. Both the VCO tuning circuit and the phase frequency detector (PFD) take the reference frequency from an on-chip 24 MHz crystal oscillator.

Figure 13.18 shows the structure of the transmitter. The MSK modulator receives the data stream from the MCU, transforms it into the zero-IF baseband waveforms and then sends the 6-bit digital baseband data to the digital-to-analog converters (DACs). There are

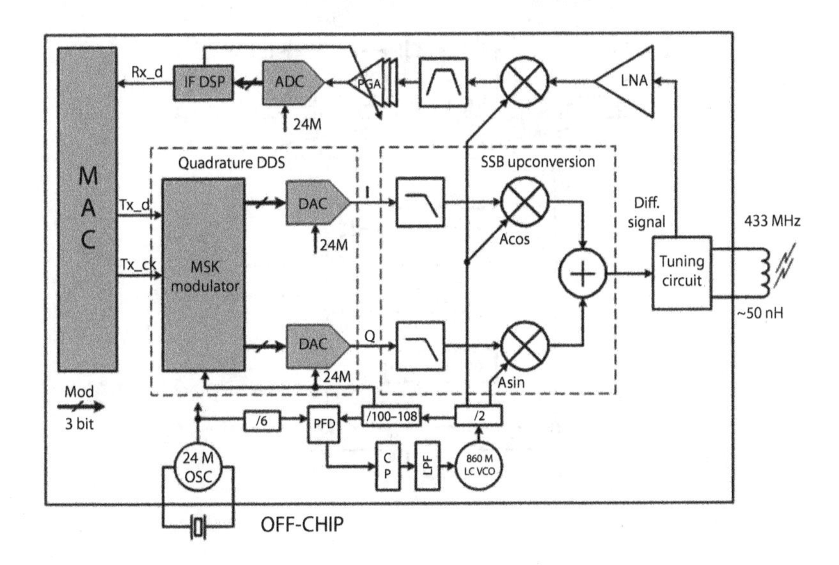

FIGURE 13.16
Transceiver architecture.

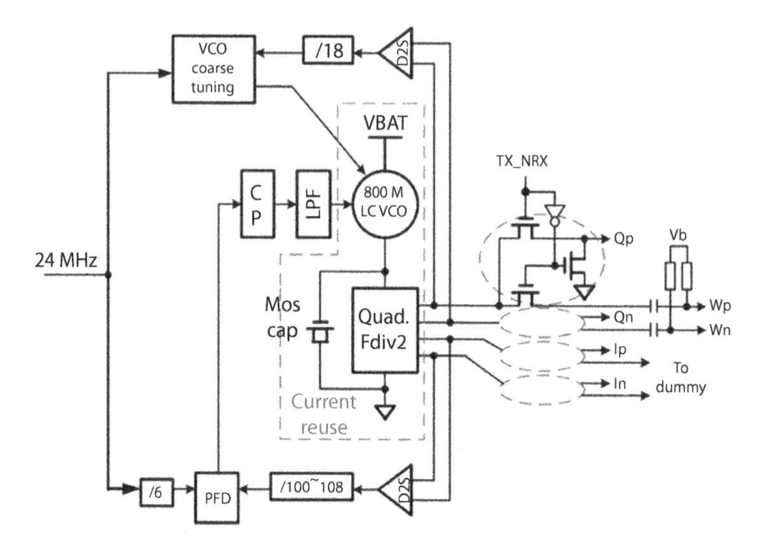

FIGURE 13.17
Quadrature LO oscillator.

two key points in this transmitter structure. First, the DCs of the two DACs are reused by the quadrature mixer (M1 and M2 are used to set the DC path). Secondly, there is no traditional power amplifier in this transmitter, and a coil that serves as the RF energy-emitting component is directly connected to the mixer. This specific structure has been proven effective in the special power-constrained application environment.

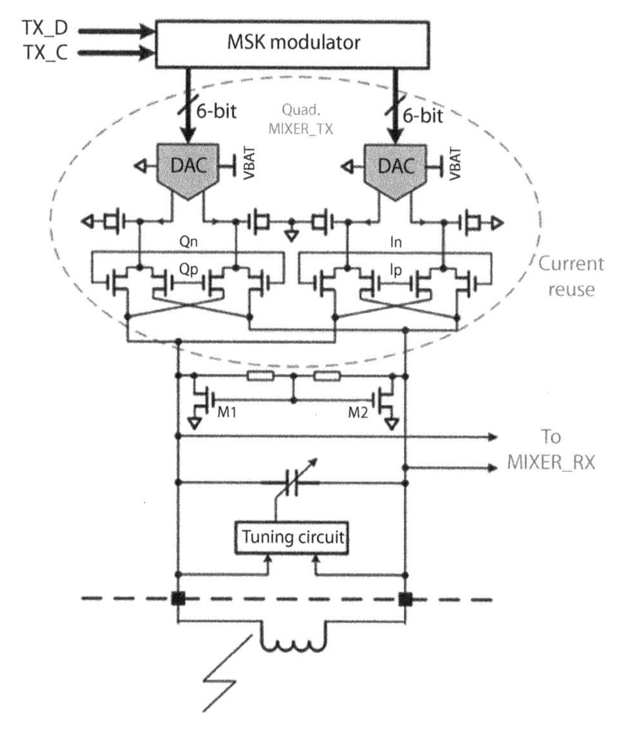

FIGURE 13.18
Transmitter circuit with current reusing.

13.4.2 Adaptive Sensor Control

Motion blur may occur in the images acquired from the camera if the femoral head is rotated at high speed. To avoid this, a simple solution is to set the camera to work at a higher frame rate. However, a higher frame rate means higher power consumption. In this system, an adaptive sensor control algorithm is applied to achieve the tradeoff between the camera frame rate and the power consumption.

When the system is initialized, the camera starts to work at a preset frame rate f_0. Suppose that $\boldsymbol{\omega}_f$ and $\boldsymbol{a}_f$ are the sample of the IMU, representing the angular velocity and the acceleration of the femoral head, respectively. The camera frame rate is determined by the norm of these two vectors as

$$f = \begin{cases} f_0, & \text{if } \boldsymbol{\omega}_f \le \varepsilon_\omega \text{ and } \boldsymbol{\omega}_f \le \varepsilon_\omega \\ \text{floor}\left(\dfrac{\boldsymbol{\omega}_f}{\varepsilon_\omega} f_0\right), & \text{if } \boldsymbol{\omega}_f > \varepsilon_\omega \\ \text{floor}\left(\dfrac{a_f}{\varepsilon_a} f_0\right), & \text{if } \boldsymbol{\omega}_f \le \varepsilon_\omega \text{ and } \boldsymbol{\omega}_f > \varepsilon_\omega \end{cases}, \tag{13.48}$$

where ε_ω and ε_a are two predefined thresholds and floor (x) is the round down function. With this adaptive sensor control solution, the camera will adjust itself to a low frame rate mode if the femoral head is moving at slow motion. This method is a low-power technology applied to the software.

13.5 Implementation and Experimental Results

13.5.1 Implementation Results of the SoC

The SoC has been implemented in the 0.18 μm CMOS technology. The digital core has 30k equivalent gates and 94 kB static random-access memory (SRAM) for image buffering. The die photo of the SoC is shown in Figure 13.19. It occupies a die area of 13.3 mm^2

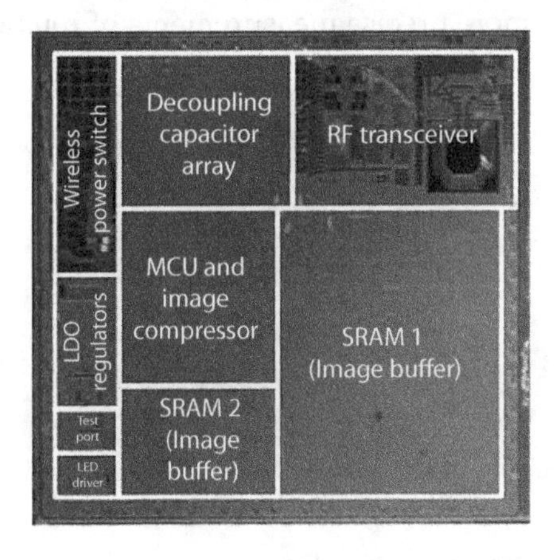

FIGURE 13.19
Die photo of the implemented SoC.

TABLE 13.2
Performance of the SoC

Parameters	Values
Supply voltage	2.5~3.3 V
External components	7
Type of RF link	Bidirectional
TX	
Bit rate	3 Mbps
Power consumption	3.9 mW
RX	
Bit rate	64 kbps
Power consumption	12 mW
MCU power consumption	240 μW
Technology	0.18 μm CMOS
Die area	13.3 mm^2

(3.7 × 3.6 mm). The SoC can work at a battery voltage down to 2.5 V. The power consumption of the SoC has been broken down. The MCU consumes about 200 μA current from a 1.2 V power supply (from the on-chip regulator) when clocked at 24 MHz. The image compressor consumes about 900 μA current from the 1.2 V supply. The MSK transmitter (including all the functional blocks for TX) consumes a total power of 3.9 mW from the 2.5, 1.8 and 1.2 V supplies, while the OOK receiver consumes 12 mW. The performance of this SoC is summarized in Table 13.2.

13.5.2 Experimental Platform for the System

Figure 13.20 shows the experimental platform for the system, which is designed to test the accuracy of pose estimation and is manufactured with ±0.1 mm machining accuracy. The platform consists of a fixed acetabular cup in the center with patterns on its internal surface, a femoral head prosthesis with circuits inside, a rotatable holder and scales printed on the platform. This platform can provide the "femur" with 3-DOFs, including translation along *z*-axis in {F} and rotations around *x*- and *z*-axes in {A}. Vernier scales are designed on the platform to achieve more precise measurements of rotation angles and translation

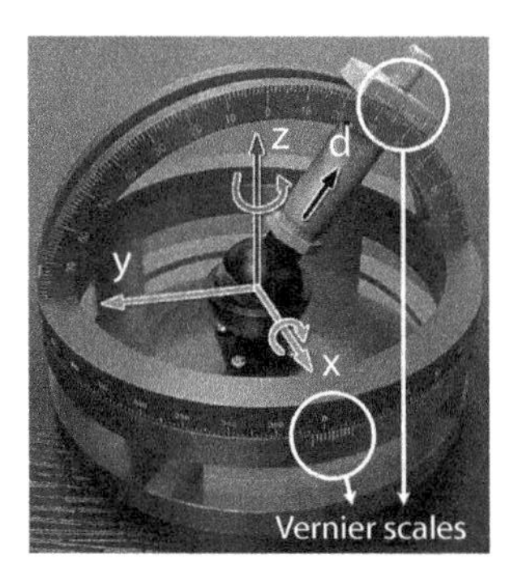

FIGURE 13.20
The experimental platform for system accuracy analysis.

TABLE 13.3
Performance of Methods Using Different Sensors

Method	Relative Error		Power (MW)
	Rotation (%)	Translation (%)	
Monocular vision	7.0	8.0	380
Monocular vision and IMU	4.8	3.2	268

distance. With this platform, the observed values from the scales and the estimation values on the PC are compared and analyzed to verify the pose and depth estimation methods.

Experiments have been conducted to compare the performance of the monocular vision–based pose estimation method and the vision- and IMU-based one. During the experiments, the holder connected with the femoral head trial was first rotated to a preset position to calibrate the system. Next, the holder and the platform were rotated to several specific positions at different speeds. In each position, the system was kept still for a short interval (about 10 seconds), and the values read from the platform scales were recorded. Finally, the estimation errors were calculated by comparing the estimated values of the two methods with the recorded values read from the platform scales. The experimental results are listed in Table 13.3.

The experimental results show that the vision- and IMU-based method has a lower relative error and a lower power consumption than the monocular vision–based method. A sample that records the relative pose estimation results of the two methods is shown in Figure 13.21. As shown in Figure 13.21a, the monocular vision–based method has wavy error curves in rotation estimations, which means that the method suffers from more random errors. The random errors may result from the manufacture accuracy of the reference patterns and the distortion of the macro lens. Benefiting from the data fusion, the vision- and IMU-based method shows better accuracy in estimating rotation about x-axis and translation along z-axis, as shown in Figure 13.21b. As the accelerometer makes no contribution to the estimation of rotation around z-axis, the rotation estimation error of z-axis is larger than that of x-axis. With the adaptive sensor control adopted, the vision- and IMU-based method achieves a 29.5% decrease in power consumption.

13.5.3 Hip Joint Demonstration System

As a preliminary for clinical tests, a demonstration setup for hip ROM estimation has been designed, as shown in Figure 13.22. The demo setup consists of a human hip joint model and a computer. The monocular vision- and IMU-based system is contained inside the hip prosthesis, mounted on the hip joint model [33].

Experiments are conducted in three steps. First, the femur model is placed at a preset position to initialize the system. Second, the six typical femur postures (i.e. extension, flexion, adduction, abduction, internal rotation and external rotation) are simulated with the femur model for hip ROM estimation. Finally, the femur model is rotated to some certain poses, where dislocation is about to happen, to test the alarm signals. The hip poses are displayed in 3D mode on the screen in real time. The system is powered by batteries mounted in the neck of the femoral head prosthesis, and can work for more than 30 minutes, which meets the requirements of THR surgery.

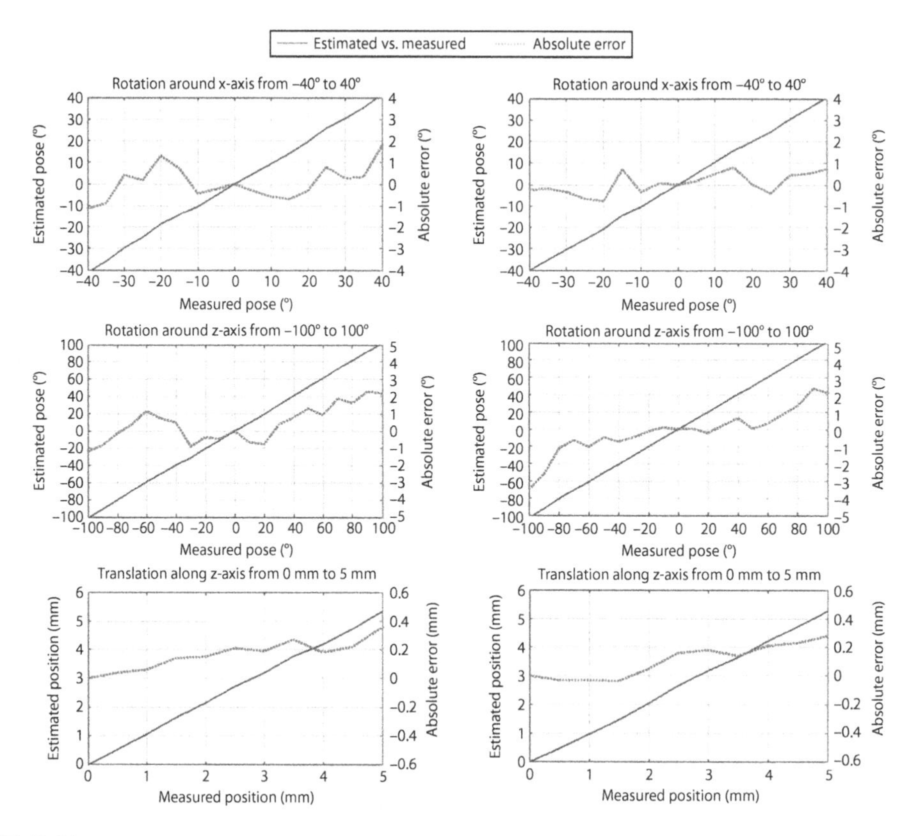

FIGURE 13.21
Comparison between the two proposed methods on estimations of rotations around x- and z-axes, and translations along z-axis. (a) Results of the monocular vision–based pose estimation method. (b) Results of the vision- and IMU-based pose estimation method.

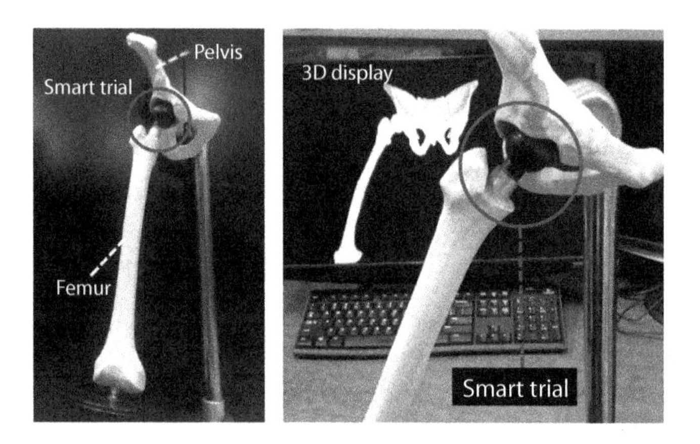

FIGURE 13.22
A demonstration setup of a human hip joint model with the femoral head trial (named smart trial) and a customized acetabular cup on it. The pose of the hip joint is displayed on the computer in real time.

13.6 Conclusion

In this chapter, a monocular vision- and IMU-based system is presented for the intra-operative estimation of the relative pose between a femoral head prosthesis and an acetabular cup. Multiple sensors, a data packer and an SoC with an RF transceiver are mounted inside the femoral head prosthesis. The data are obtained from the sensors and transmitted wirelessly to a computer outside the human body for pose estimation and display. As the sensors are rigidly fixed to the femoral head prosthesis, the prosthesis pose estimation problem is classified as a sensor ego-motion estimation problem and two pose estimation methods are proposed. One is purely based on monocular vision and the other one combines monocular vision with IMU. Experimental results show that the best approach for pose estimation is the method based on the monocular vision and IMU, which has a relative error of 4.8% in the 3-DOF rotation and 3.2% in the 1-DOF translation. In order to reduce the power consumption, the current reusing techniques are applied in the circuit design, and an adaptive sensor control solution is proposed.

References

1. S. Kurtz K. Ong, E. Lau, F. Mowat and M. Halpern, Projections of primary and revision hip and knee arthroplasty in the United States from 2005 to 2030, *The Journal of Bone and Joint Surgery (American)*, vol. 89, no. 4, p. 780, 2007.
2. J. Nutt, K. Papanikolaou and C. Kellett, (ii) Complications of total hip arthroplasty, *Orthopaedics and Trauma*, vol. 27, no. 5, pp. 272–276, 2013.
3. A. Bunn, C. Colwell and D. D'Lima, Bony impingement limits design-related increases in hip range of motion, *Clinical Orthopaedics and Related Research®*, vol. 470, no. 2, pp. 418–427, 2011.
4. M. Ghaffari, R. Nickmanesh, N. Tamannaee and F. Farahmand, The impingement-dislocation risk of total hip replacement: Effects of cup orientation and patient maneuvers, in *Proceeding of 34th Annual International Conference IEEE EMBS*, San Diego, CA, 2012, pp. 6801–6804.
5. O. Kessler, S. Patil, W. Stefan, E. Mayr, C. Colwell and D. D'Lima, Bony impingement affects range of motion after total hip arthroplasty: A subject-specific approach, *Journal of Orthopaedic Research*, vol. 26, no. 10, pp. 1419–1419, 2008.
6. K. H. Widmer, Is there really a 'safe zone' for the placement of total hip components? *Ceramics in Orthopaedics*, vol. 75, no. 5, pp. 249–252, 2006.
7. A. Martin and S. A. Von, CT-based and CT-free navigation in total knee arthroplasty: A prospective comparative study with respects to clinical and radiological results, *Zeitschrift fur Orthopadie und ihre Grenzgebiete*, vol. 143, no. 3, pp. 323–328, 2004.
8. H. Kiefer and A. Othman, OrthoPilot total hip arthroplasty workflow and surgery, *Orthopedics*, vol. 28 no. 28, pp. 1221–1226, 2005.
9. R. McGinnis, S. M. Cain, S. Tao, D. Whiteside, G. C. Goulet, E. C. Gardner, A. Bedi and N. C. Perkins, Accuracy of femur angles estimated by IMUs during clinical procedures used to diagnose femoroacetabular impingement, *IEEE Transactions on Biomedical Engineering*, vol. 62, no. 6, pp. 1503–1513, 2015.
10. F. Graichen, G. Bergmann and A. Rohlmann, Hip endoprosthesis for in vivo measurement of joint force and temperature, *Journal of Biomechanics*, vol. 32, no. 10, pp. 1113–1117, 1999.

11. P. Damm, F. Graichen, A. Rohlmann, A. Bender and G. Bergmann, Total hip joint prosthesis for in vivo measurement of forces and moments, *Medical Engineering and Physics*, vol. 32, no. 1, pp. 95–100, 2010.
12. G. Bergmann, F. Graichen, J. Dymke, A. Rohlmann, G. N. Duda and P. Damm, High-tech hip implant for wireless temperature measurements in vivo, *PLoS One*, vol. 7, no. 8, p. e43489, 2012.
13. R. D. Lea and J. J. Gerhardt, Range-of-motion measurements, *The Journal of Bone and Joint Surgery (American)*, vol. 77, no. 12, pp. 784–798, 1995.
14. I. Holm, B. Bolstad, T. Lütken, A. Ervik, M. Røkkum and H. Steen, Reliability of goniometric measurements and visual estimates of hip ROM in patients with osteoarthrosis, *Physiotherapy Research International*, vol. 5, no. 4, pp. 241–248, 2000.
15. F. Öhberg, R. Lundström and H. Grip, Comparative analysis of different adaptive filters for tracking lower segments of a human body using inertial motion sensors, *Measurement Science and Technology*, vol. 24, no. 8, pp. 50–61, 2013.
16. J. Zhang, A. Novak, B. Brouwer and Q. Li, Concurrent validation of Xsens MVN measurement of lower limb joint angular kinematics, *Physiological Measurement*, vol. 34, no. 8, pp. N63–N69, 2013.
17. J. Favre, R. Aissaoui, B. Jolles, J. de Guise and K. Aminian, Functional calibration procedure for 3D knee joint angle description using inertial sensors, *Journal of Biomechanics*, vol. 42, no. 14, pp. 2330–2335, 2009.
18. T. Seel, J. Raisch and T. Schauer, IMU-based joint angle measurement for gait analysis, *Sensors*, vol. 14, no. 4, pp. 6891–6909, 2014.
19. S. Su, J. Gao, H. Chen and Z. Wang, Design of a computer-aided visual system for total hip replacement surgery, *IEEE International Symposium on Circuits and Systems*, pp. 786–789, 2015.
20. H. Jiang, F. Li, X. Chen, Y. Ning, X. Zhang, B. Zhang, et al., A SoC with 3.9 mW 3Mbps UHF transmitter and 240 μW MCU for capsule endoscope with bidirectional communication, *IEEE International Solid State Circuits Conference*, Beijing, 2010, pp. 1–4.
21. X. Chen, X. Zhang, L. Zhang, X. Li, N. Qi, H. Jiang and Z. Wang, A wireless capsule endoscope system with low-power controlling and processing ASIC, *IEEE Transactions on Biomedical Circuits and Systems*, vol. 3, no. 1, pp. 11–22, 2009.
22. R. Paul, *Robot Manipulators*, Cambridge, MA: MIT Press, 1981.
23. J. Diebel, Representing attitude: Euler angles, unit quaternions, and rotation vectors, *Matrix*, vol. 58, 2006, pp. 14–18.
24. S. H. Won, N. Parnian, F. Golnaraghi and W. Melek, A quaternion-based tilt angle correction method for a hand-held device using an inertial measurement unit, *Industrial Electronics IECON 2008, Conference of IEEE*, Orlando, FL, pp. 2971–2975.
25. J. L. Marins, X. Yun, E. R. Bachmann, R. B. Mcghee and M. J. Zyda, An extended Kalman filter for quaternion-based orientation estimation using MARG sensors, in *2001 IEEE/RSJ International Conference on Intelligent Robots and Systems*, vol. 4, pp. 2003–2011.
26. P. J. Olver and A. Tannenbaum, *Mathematical Methods in Computer Vision*. Springer, New York, 2003.
27. J. Gao, S. Su, H. Chen, H. Jiang, C. Zhang, Z. Wang, et al., Estimation of the relative pose of the femoral and acetabular components in a visual aided system for total hip replacement surgeries, *New Circuits and Systems Conference*, vol. 59, pp. 81–84, 2014.
28. S. Su, Y. Zhou, Z. Wang and H. Chen, Monocular vision- and IMU-based system for prosthesis pose estimation during total hip replacement surgery, *IEEE Transactions on Biomedical Circuits and Systems*, vol. 11, no. 3, pp. 661–670, 2017.
29. Z. Zhang, Essential matrix, in *Computer Vision: A Reference Guide*, K. Ikeuchi, Ed. Springer, New York, 2014, pp. 258–259.
30. Z. Wang, H. Chen, M. Liu and H. Jiang, A wirelessly ultra-low-power system for equilibrium measurements in total hip replacement surgery, *New Circuits and Systems Conference 2012, IEEE*, Montreal, pp. 141–144.

31. K. G. Park, C. Y. Jeong, J. W. Park, J. W. Lee, J. G. Jo and C. Yoo, Current reusing VCO and divide-by-two frequency divider for quadrature LO generation, *IEEE Microwave and Wireless Components Letters*, vol. 18, no. 6, pp. 413–415, 2008.
32. K. Park, C. Jeong, J. Park, J. Lee, J. Jo, and C. Yoo, Current reusing VCO and divide-by-two frequency divider for quadrature LO generation, *IEEE Microwave and Wireless Components Letters*, vol. 18, no. 6, pp.413–415, 2008.
33. S. Su, J. Gao, Z. Weng, H. Chen and Z. Wang, Live demonstration: A smart trial for hip range of motion estimation in total hip replacement surgery, *Biomedical Circuits and Systems Conference 2015 IEEE*, Atlanta, GA, pp. 1–5.

Index

REGIONS OF U. S.
AND CANADA
1
2
3
4
4
5
6
7
14
15

REGIONAL GEOGRAPHY OF THE UNITED STATES AND CANADA

REGIONAL GEOGRAPHY OF THE UNITED STATES AND CANADA

Tom L. McKnight
Professor of Geography
University of California, Los Angeles

PRENTICE HALL
Englewood Cliffs, New Jersey 07632

Library of Congress Cataloging-in-Publication Data

McKnight, Tom L. (Tom Lee), (date)
Regional geography of the United States and Canada / Tom L. McKnight.
p. cm.
Previously published under title: Regional geography of Anglo-America/ C. Langdon White. 6th ed. c1985.
Includes bibliographical references and index.
ISBN 0-13-352956-8
1. United States—Geography. 2. Canada—Geography. I. White, C. Langdon (Charles Langdon), 1897- Regional geography of Anglo-America. II. Title.
E161.3.M35 1992 91-10808
917.3—dc20 CIP

Editor in Chief: Tim Bozik
Acquisition Editor: Ray Henderson
Production Editor: Judi Wisotsky
Copy Editor: James Tully
Interior and Cover design: Lisa A. Domínguez
Cover Photo: Tony Stone/Worldwide/Chicago Ltd.
Prepress Buyer: Paula Massenaro
Manufacturing Buyer: Lori Bulwin
Photograph Coordinator: Lynne Breitfeller

A Simon & Schuster Company
Englewood Cliffs, New Jersey 07632

Printed in the United States of America
10 9 8 7 6 5 4 3 2 1

ISBN 0-13-352956-8

Prentice-Hall International (UK) Limited, *London*
Prentice-Hall of Australia Pty. Limited, *Sydney*
Prentice-Hall Canada Inc., *Toronto*
Prentice-Hall Hispanoamericana, S.A., *Mexico*
Prentice-Hall of India Private Limited, *New Delhi*
Prentice-Hall of Japan, Inc., *Tokyo*
Simon & Schuster Asia Pte. Ltd., *Singapore*
Editora Prentice-Hall do Brasil, Ltda., *Rio de Janeiro*

To my revered mentors
—Edwin J. Foscue and C. Langdon White—
who blazed the trail, established the guideposts, pointed the way, and provided the inspiration. . . .

CONTENTS

5

6

7

8

9

10

11

THE SOUTHEASTERN COAST 231

12

THE HEARTLAND 260

13

THE GREAT PLAINS AND PRAIRIES 300

14

THE ROCKY MOUNTAINS 333

15

THE INTERMONTANE WEST 360

16

THE CALIFORNIA REGION 399

17

THE HAWAIIAN ISLANDS 428

18

THE NORTH PACIFIC COAST 447

19

THE BOREAL FOREST 480

20

THE ARCTIC 510

LIST OF VIGNETTES

PREFACE

Regional geography has recently passed through a period of relative disfavor. Its critics, encouraged by the quantitative revolution in many fields of learning, including geography, have found fault with its alleged lack of precision and methodological rigor. This is not the place to debate such charges, but the author of this book is a strong adherent of the conviction that regional geography meets a basic need in furthering an understanding of the earth's surface and that no surrogate has yet been devised to replace it.

The complexities of human life on the earth are much too vast and intricate to be explained by multivariate analysis and related mathematical and model-building techniques. Such models may be useful, but any real understanding of people and the earth requires that words, phrases, photos, diagrams, graphs, and especially maps be used in meaningful combination. Conceptualization and delimitation of regions are critical exercises in geographic thinking, and the description and analysis of such regions continue as a central theme in the discipline of geography; some would call it the essence of the field.

It is a fundamental belief of the author that a basic goal of geography is *landscape appreciation* in the broad sense of both words, that is, an understanding of everything that one can see, hear, and smell—both actually and vicariously—in humankind's zone of living on the earth. In this book, then, there is heavy emphasis on landscape description and interpretation, including its sequential development.

The flowering of civilization in North America is partly a reflection of the degree to which people have levied tribute against natural resources in particular and the environment in general. From the Atlantic to the Pacific and from the Arctic to the Gulf, people have been destroyers of nature, even as they have been builders of civilizations. Overcrowding of population and overconsumption of material goods have now become so pervasive that reassessment of goals and priorities—by institutions as well as by individuals—is widespread. Reaction to environmental despoliation is strongly developed, and ecological concerns are beginning to override economic considerations in many cases. In keeping with such reaction, environmental and ecologic issues are frequently discussed in this book.

Canada and the United States have a common heritage and have been moving toward similar goals. These factors, along with geographical contiguity and the binational influence of mass media, have produced both commonality of culture and mutual interdependence. Nevertheless, there are clear-cut distinctions between the two countries, and there is a particular concern among many Canadians to define a national character that is separate from both psychological and economic domination of the United States. These national interests and concerns, however, do not mask the fact that the geographical "grain" of the continent often trends north-south rather than east-west; thus, several of the regions delimited in this book cross the international border to encompass parts of both countries.

A renewal of interest in regional geography is now becoming apparent, and nowhere is this more clearly shown than in the increase in publications dealing with aspects of the regional geography of the United States and Canada. There are more journals and more journal articles, and special publications of great significance have appeared. Perhaps most important, the output of state, provincial, regional, and city atlases continues unabated. The chapter-end bibliographies of this book are overflowing with references to useful recent publications; indeed, the bibliographies could easily have been expanded severalfold.

This volume is based significantly on its long-established predecessor, *Regional Geography of Anglo-America,* which flourished through six editions, beginning in 1943, with C. Langdon White and Edwin J. Foscue as senior authors. One of the more popular features of the previous book was the use of boxed vignettes that permitted a more detailed discussion of specific issues and topics. The use of boxes, or vignettes (titled "A Closer Look"), has been significantly expanded in this book, and about half of them are written by guest authors on an invited basis.

ACKNOWLEDGEMENTS

The author has leaned heavily on many geographers and colleagues in associated disciplines. A great many people have contributed thoughts, ideas, information, and critiques of value. Accordingly, the number to whom I am grateful is so large that only a general acknowledgment is possible. To four groups, however, my special thanks are due.

Twenty noted geographers contributed sprightly and insightful vignettes. Their names are listed following the Contents.

Several colleagues went far beyond the norms of collegiality in providing helpful suggestions and comments. Chief among these were:

J. Lewis Robinson of the University of British Columbia;
David W. Lantis of California State University, Chico; and
John Fraser Hart of the University of Minnesota.

Other colleagues commented critically upon specific chapters or issues of major importance. These include:

Gabriel P. Betz, California (Pa.) State College
Joan Clemons, Los Angeles Valley College
Doug Heath, Northampton (Pa.) Community College
Ralph Krueger, University of Waterloo (Ont.)
Richard F. Logan, University of California, Los Angeles
Ron Merckling, a student at the University of Minnesota
Richard L. Nostrand, University of Oklahoma
William H. Wallace, University of New Hampshire
and William C. Wonders, University of Alberta.

The following reviewers provided detailed and helpful critiques of portions of the manuscript:

John L. Anderson, University of Louisville
Saul Aronow, Lamar University
Marshall Bowen, Mary Washington College
James Davis, Western Kentucky University
John Dietz, University of Northern Colorado
Jane Ehemann, Shippensburg University
Val Eichenlaub, Western Michigan University
Philip Gersmehl, University of Minnesota
Charles Gunter, East Tennessee University
John Hudson, Northwestern University
Daniel Jacobson, Michigan State University

Imre Quastler, San Diego State University
Dennis Spetz, University of Louisville

The art work was designed and constructed by the very talented *Patricia Caldwell Lindgren*. Research assistance was cheerfully and competently provided by *June Carter*. The people at Prentice-Hall, especially *Ray Henderson, Dan Joraanstad* and *Judi Wisotsky*, helped to create smooth sailing for this enterprise.

TOM L. MCKNIGHT
Los Angeles

ABOUT THE AUTHOR

Tom McKnight was born and raised in Dallas, Texas. He started out to be a geologist, but discovered geography through the tutelage of Edwin J. Foscue, and soon shifted his interest to the "mother science." His training includes a B.A. degree (geology major, geography minor) from Southern Methodist University, an M.A. degree (geography major, geology minor) from the University of Colorado, and a Ph.D. degree (geography major, meteorology minor) from the University of Wisconsin. He has been lucky enough to live in Australia for extended periods on five different occasions. Most of his professional life has been based at U.C.L.A., but he has also taught temporarily at nine American, three Canadian, and three Australian universities. He served as Chair of the U.C.L.A. Geography Department from 1978 to 1983. His favorite places are Australia, Colorado, Yellowstone Park, and Dallas.

1
THE ANGLO-AMERICAN CONTINENT

The Dominion of Canada and the United States of America are two huge countries that occupy the northern third of the Western Hemisphere. They comprise the bulk of the continent of North America, but share that continent with Mexico and nearly two dozen smaller countries of Central America and the West Indies. The name *Anglo-America* is an appellation that generally has been applied to the United States and Canada as a distinction from *Latin America* to the south (Fig. 1-1). Because of various cultural, social, political, and economic contrasts, this is a meaningful way to subdivide the Western Hemisphere.

The term *Anglo* implies an important attribute of the population of the two countries, as their predominant ancestry is English and most of the economic and political institutions of both countries derive from this heritage. Nevertheless, numerous components of the population within the United States and Canada do not have an English background and the proportion of the non-Anglo populace steadily increases. Some of these minority groups are sufficiently non-Anglicized and occur in sufficient concentrations to provide significant diversity in the human geography of the continent. Most notable is the part of eastern Canada that is occupied predominantly by people of French origin; it encompasses most of the settled portions of the province of Quebec as well as adjacent sections of Ontario and Labrador and much of New Brunswick.

Hawaii, with its prominent Asian and Polynesian elements, is another conspicuous exception to the cultural connotation of the term *Anglo*. Doubtless the reader can think of many other areas in the

FIGURE 1-1 The Americas: Anglo and Latin.

United States and Canada where non-Anglo ethnic groups are notable, such as a broad zone along the Mexican border from the Pacific Ocean to the Gulf of Mexico; the southeastern corner of Florida, where many Cuban refugees have settled; various Indian reservations in western United States and central Canada; concentrations of Inuit and Aleuts in northern and western Alaska and northern Canada; areas of black population concentration in some rural parts of southeastern United States and many United States cities; and various concentrations of other non-Anglo peoples, primarily in urban ghettoes in both Canada and the United States.

Consequently, the term *Anglo-America* increasingly is falling into disfavor. It will be used in this book as a convenient way to refer to a major portion of the earth's surface, and not as a definitive descriptive concept.

CONTINENTAL PARAMETERS

Anglo-America, as defined here, encompasses an area of nearly 7.5 million square miles (19.4 million km^2), which is larger than each of the seven recognized continents except Asia and Africa. It sprawls across 136 degrees of longitude, from 52° west longitude at Cape Spear in Newfoundland to 172° east longitude at Attu Island in the western extremity of the Aleutians. Its latitudinal extent is 64 degrees, from 83° north latitude at Ellesmere Island's Cape Columbia to 19° north latitude on the southern coast of the Big Island of Hawaii.

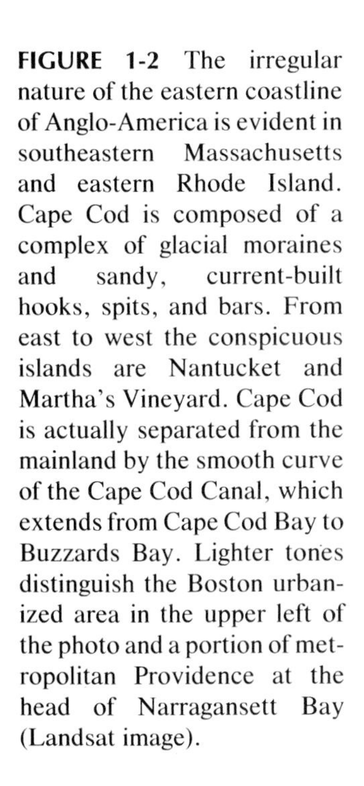

FIGURE 1-2 The irregular nature of the eastern coastline of Anglo-America is evident in southeastern Massachusetts and eastern Rhode Island. Cape Cod is composed of a complex of glacial moraines and sandy, current-built hooks, spits, and bars. From east to west the conspicuous islands are Nantucket and Martha's Vineyard. Cape Cod is actually separated from the mainland by the smooth curve of the Cape Cod Canal, which extends from Cape Cod Bay to Buzzards Bay. Lighter tones distinguish the Boston urbanized area in the upper left of the photo and a portion of metropolitan Providence at the head of Narragansett Bay (Landsat image).

A VIEW FROM SPACE

The continent is roughly wedge shaped, with its broadest expanse toward the north. The great bulk of Anglo-America is thus in the middle latitudes, with a considerable northern extension into the high latitudes and only Hawaii reaching into the tropics.

If we can imagine a view of the entire continent from an orbiting space station on a clear day, certain gross features would appear prominently. Perhaps the most conspicuous configuration is the irregular continental outline; some extensive coastal reaches are relatively smooth, but by and large the margin of the continent is irregular and embayed and there are numerous prominent offshore islands (Fig. 1-2).[1]

The most notable identation in the continental outline is Hudson Bay, which protrudes southward for some 800 miles from Canada's north coast. Despite its vastness, however, Hudson Bay is relatively insignificant in its influence on the geography of Anglo-America. Its surface is frozen for many months and even the open water of summer supports only one sea route of importance and has minimal climatic effects on the surrounding lands.

Much more significant geographically are two extensive oceanic areas whose margins impinge less abruptly on the continent; both the Gulf of Mexico to the southeast and the Gulf of Alaska to the northwest constitute gross irregularities in the continental

[1] The coastline of the high-latitude portions of Anglo-America is much more uneven than that of other sections. For example, Alaska has more coastline mileage than the other 24 coastal states combined.

configuration that have major climatic influences and considerable economic importance. Other coastal embayments that might be conspicuous from the viewpoint of a satellite include the Gulf of St. Lawrence, the Bay of Fundy, and Chesapeake Bay on the east coast; and Puget Sound, Cook Inlet, Bristol Bay, Norton Sound, and Kotzebue Sound on the west.

More than 70,000 islands are another feature that commands attention in the gross outline of the continent. Easily the most prominent island group is the Canadian Arctic Archipelago, an expansive series of large islands to the north of the Canadian mainland that constitutes more than 14 percent of the total area of Canada. The largest islands of the archipelago are Baffin and Ellesmere, and nine of the ten largest islands of Anglo-America are in the group.

Four other sizable islands are clustered around the Gulf of St. Lawrence. Prince Edward Island is a Canadian province, the island of Newfoundland is part of the province of the same name, Cape Breton Island is part of Nova Scotia, and Anticosti Island is part of Quebec (Fig. 1-3).

FIGURE 1-3 Some important place names in Anglo-America.

The islands off the East Coast of the United States are small and sparsely populated with one major exception. The exception is Long Island, whose 1400 square miles (3600 km^2) support a population in excess of 7 million. The other coastal islands of the Atlantic Ocean and Gulf of Mexico are nearly all long, narrow, low-lying sand ridges. Although there are a great many, most are so close to shore and so narrow in width as to be indistinguishable from our theoretical high-altitude viewpoint.

The island pattern off the Pacific Coast is quite uneven. Only a few islands are found off the southern portion of the coast and of these only the Channel Islands of California encompass much acreage. The coast of British Columbia and southern Alaska, on the other hand, is extensively bordered by islands, many of which are large. Most notable is Vancouver Island, which constitutes the extreme southwestern corner of Canada. Other major islands on this coast include the Queen Charlotte Islands of British Columbia, the Alexander Archipelago of southeastern Alaska, Kodiak Island of southern Alaska, and the far-flung Aleutian group.

THE COUNTRIES OF ANGLO-AMERICA

The material in most of this book is presented by regions. As a prelude, it is well to look briefly at the nations as entities, noting a few general facts to serve as a context for regional analysis.

The United States of America

The total area of the United States is 3,615,200 square miles (9,363,400 km^2), a figure exceeded by only three other countries—the Soviet Union, Canada, and China. The 1990 population was about 250,000,000, which also is fourth-ranking among the nations of the world, after China, India, and the Soviet Union.

The United States is a federal republic with a division of power between federal and state governments. Both levels of government have a threefold administration: executive, legislative, and judicial branches. There are 50 states, which vary greatly in size and population. The 48 "old" states, excluding Alaska and Hawaii, are collectively referred to as the *conterminous states*.

The states are subdivided into local governmental units called counties,[2] with the following exceptions: in Louisiana the units are called parishes; in Maryland, Missouri, Nevada, and Virginia there are cities that are independent of any county organization and thus constitute, along with counties, primary subdivisions of these states; and in Alaska the populated parts are subdivided into boroughs. Taken altogether, there is a total of about 3100 counties and county equivalents in the United States.[3]

The national capital is the District of Columbia, which is territory on the northeastern side of the Potomac River that was ceded to the nation by the state of Maryland.[4] In addition to the 50 states and the District of Columbia, the United States governs a number of small islands in the Caribbean Sea, the most important of which is Puerto Rico. Furthermore, the federal government administers or oversees hundreds of tiny islands in the Pacific Ocean, chiefly in Micronesia, under a variety of political relationships.

The Dominion of Canada

Canada is the world's second largest country, with an area of 3,851,800 square miles (9,976,200 km^2). It is not densely populated, however, and its 1990 population of about 26,000,000 ranked only thirty-first among the nations of the world.

The governmental organization of Canada is a confederation with parliamentary democracy that combines the federal form of the United States with the cabinet system of Great Britain. The cabinet system partially unites the executive and legislative branches of government; the prime minister and all, or nearly all, the cabinet are members of the House of Commons. The reigning monarch of Great Britain

[2] Counties vary greatly in size and population. Delaware has the fewest, with 3, whereas Texas has the most, with 254.

[3] The total number of counties is not static, for new ones are added from time to time. In 1981 a new county—Cibola—was created in northwestern New Mexico and in 1983 La Paz County was carved out of Yuma County in southwestern Arizona.

[4] The District of Columbia was created in 1790 when Maryland and Virginia both ceded territory for its establishment. Virginia reannexed its ceded land in 1847, however; so the present District of Columbia consists entirely of land that was originally part of the state of Maryland.

is also the head of the Canadian state and is represented by a governor-general whose duties are formal and rather perfunctory.[5] The prime minister is the active head of the government. Members of the House of Commons are elected by the people of Canada. Members of the Senate, on the other hand, are appointed for life by the cabinet. The House of Commons is the dominant legislative body, with many more powers than the Senate.

The Canadian confederation contains ten provinces and two territories. The easternmost provinces of Newfoundland, New Brunswick, Nova Scotia, and Prince Edward Island are often referred to as the Atlantic Provinces; the latter three are collectively called the Maritime Provinces. Alberta, Saskatchewan, and Manitoba, the three provinces of the western interior, are known as the Prairie Provinces. The other three provinces, Quebec, Ontario, and British Columbia, are not normally considered members of groups. Most of northern Canada is encompassed within the Yukon Territory and the Northwest Territories. The various provinces and territories have different systems for administering local government; each is usually subdivided into counties or districts, which may be further fragmented into minor civil divisions.

The national capital of Ottawa does not occupy a special territory but is within the province of Ontario, adjacent to the border of Quebec. The creation of a federal district (analogous to the District of Columbia) was first officially proposed in 1915, and has been discussed in varying degrees of seriousness ever since, but with no formal action. Nevertheless, a planning district, called the "National Capital Region," has been designated. It comprises some 900 square miles (2300 km^2) and is divided about equally between Ontario and Quebec. Indeed, in some quarters the capital is now referred to as "Ottawa-Hull," since Hull is Ottawa's principal suburb on the Quebec side of the boundary and contains the offices of numerous government agencies.

[5] Despite the "patriation" of the Canadian Constitution in 1982, Canada is still a constitutional monarchy and the roles of the Queen and the Governor-General are unchanged. What is changed is that the Constitution (previously called the British North America Act) has been expanded, principally by the addition of a Charter of Rights and Freedoms, and removed from the legal control of the British Parliament and placed wholly in Canadian hands.

AMICABLE NEIGHBORS

In the early years of their separate political existence, the United States and Canada battled against each other five times, most seriously in the War of 1812. For well over a century, however, neighborliness has prevailed, and their mutual 5525-mile (8840-km) boundary is rightly referred to as "the longest undefended border in the world."

Americans and Canadians are alike in many ways, both as a people and as a society, although the differences are sometimes notable, and often are emphasized by commentators. Life in the two countries is so similar that citizens of neither country experience "culture shock" when visiting the other. Indeed, the movement of people through the 130 border-crossing stations is of enormous magnitude: In the late 1980s Canadians crossed into the United States on about 40 million occasions annually, and the reverse flow amounted to about 33 million annual excursions.

Of at least equal significance as an indicator of an amicable relationship is the amount of commerce between the two countries, which is the most voluminous two-way trade in the world. Each is the other's best customer. In recent years more than $170 billion worth of goods and services were exchanged annually. This total undoubtedly will continue to rise as a result of the U.S.-Canadian Free Trade Agreement, promulgated in 1989, which will eliminate all tariffs between the two countries within a decade.

This is not to say that there are no problems between these neighbors. Significant contentious issues—such as acid rain, Great Lakes pollution, and commercial fishery allotments—continue to strain the relationship.

Despite the similarities of the people and their institutions, it probably is inevitable that tension between these continental neighbors will persist, simply because they are unequal partners; one is 10 times as large (in population and economy) as the other. Canada has always had to face the problem of building a nation in the shadow of a giant. Or, as ex-Prime Minister Pierre Trudeau is purported to have said, "Occupying a continent with the United States is like sleeping with an elephant; no matter how benign the beast is, its slightest wiggle shakes the bed."

This inequality certainly influences attitude. According to a major 1989 public opinion poll, about half the Canadian respondents characterized Americans in negative terms, whereas less than 5 percent of the American respondents felt negatively about Canadians.

Simply stated, it is probably fair to observe that many Canadians view the United States with concern, but most Americans are indifferent to Canada. Canadian novelist Margaret Atwood has noted that what separates the two countries is the world's longest one-way mirror. Canadians gaze south, obsessed with fascination and sometimes fear of the American colossus, but Americans rarely bother to look north. The comment attributed to Chicago mobster Al Capone encapsulates the American viewpoint: "Canada? What street is that on?" Such a statement is not snobbish; it simply represents the inconsistent interest in, and knowledge of, world affairs that is so prevalent in the United States.

SELECTED GENERAL BIBLIOGRAPHY ON CANADA AND THE UNITED STATES

AGNEW, JOHN, *The United States in the World Economy: A Regional Geography*. New York: Cambridge University Press, 1987.

BIRDSALL, STEPHAN, AND JOHN FLORIN, *Regional Landscapes of the United States and Canada* (2nd ed.). New York: John Wiley & Sons, 1981.

BROWN, RALPH H., *Historical Geography of the United States*. New York: Harcourt, Brace and Company, 1948.

CHAPMAN, JOHN D., AND JOHN C. SHERMAN, EDS., *Oxford Regional Economic Atlas: United States and Canada*. London: Oxford University Press, 1975.

CLARK, ANDREW H., "The Look of Canada," *Historical Geography Newsletter,* 6 (Spring 1976), 59–68.

DUNBAR, GARY S., "Illustrations of the American Earth: An Essay in Cultural Geography," *American Studies,* 12 (Autumn 1973), 3–15.

Economic Research Service, U.S. Department of Agriculture, *The Look of Our Land: An Airphoto Atlas of the Rural United States*. Washington, D.C.: U.S. Government Printing Office.
1970. "The Far West"
1970. "North Central"
1971. "East and South"
1971. "The Mountains and Deserts"
1971. "The Plains and Prairies"

ELLIOTT, J. L., *Two Nations, Many Cultures* (2nd ed.). Scarborough, Ont.: Prentice-Hall of Canada, Ltd., 1983.

GERLACH, ARCH C., ED., *The National Atlas of the United States of America*. Washington, D.C.: United States Department of the Interior, Geological Survey, 1970.

GOUGH, B., *Canada*. Englewood Cliffs, NJ: Prentice-Hall, 1975.

HAMELIN, LOUIS-EDMOND, *Canada: A Geographical Perspective*. Toronto: Wiley Publishers of Canada Limited, 1973.

———, *Canadian Nordicity: It's Your North Too*. Montreal: Harvest House, 1979.

HARRIES, KEITH D., *The Geography of Crime and Justice*. New York: McGraw-Hill Book Co., 1974.

———, AND STANLEY D. BRUNN, *The Geography of Laws and Justice: Spatial Perspectives on the Criminal Justice System*. New York: Praeger Publishers, 1978.

HARRIS, R. COLE, AND JOHN WARKENTIN, *Canada Before Confederation: A Study in Historical Geography*. New York: Oxford University Press, 1974.

HART, JOHN FRASER, *The Look of the Land*. Englewood Cliffs, NJ: Prentice-Hall, 1975.

HEADY, EARL, "The Agriculture of the United States," *Scientific American,* 235 (September 1976), 106–27.

JACKSON, RICHARD H., *Land Use in America,* New York: John Wiley & Sons, 1981.

MARSCHNER, FRANCIS J., *Land Use and Its Patterns in the United States*. Washington, D.C.: U.S. Government Printing Office, 1959.

MATTHEWS, GEOFFREY J., AND ROBERT MORROW, JR., *Canada and the World: An Atlas Resource*. Scarborough, Ont.: Prentice-Hall of Canada, 1985.

MCCANN, LARRY D. (ED.), *Heartland and Hinterland: A Geography of Canada* (2nd ed.). Scarborough, Ont.: Prentice-Hall of Canada, Ltd., 1987.

MITCHELL, ROBERT D., AND PAUL A. GROVES, EDS., *North America: The Historical Geography of a Changing Continent*. New York: Rowman & Littlefield, 1987.

PATERSON, J. H., *North America* (8th ed.). New York: Oxford University Press, 1989.

PUTMAN, D. F., AND R. G. PUTMAN, *Canada: A Regional Analysis* (2nd ed.). Toronto: J. M. Dent and Sons, 1979.

ROBINSON, J. LEWIS, *Concepts and Themes in the Regional Geography of Canada.* Vancouver: Talonbooks, 1983.

ROONEY, JOHN F., JR., *A Geography of American Sport.* Reading, MA: Addison-Wesley Publishing Co., 1974.

ROONEY, JOHN F., JR., WILBUR ZELINSKY, AND DEAN R. LOUDER, EDS., *This Remarkable Continent: An Atlas of United States and Canadian Society and Cultures.* College Station: Texas A&M University Press, 1982.

SAUER, CARL O., *Sixteenth Century North America.* Berkeley: University of California Press, 1975.

SHORTRIDGE, BARBARA, *Atlas of American Women.* New York: Macmillan, 1987.

STARKEY OTIS P., J. LEWIS ROBINSON, AND CRANE S. MILLER, *The Anglo-American Realm* (2nd ed.). New York: McGraw-Hill Book Co., 1975.

Statistics Canada, *Canada.* Ottawa: Queen's Printer, annual.

———, *Canada Year Book.* Ottawa: Queen's Printer, annual.

Surveys and Mapping Branch, *The National Atlas of Canada.* Ottawa: Surveys and Mapping Branch, Department of Energy, Mines and Resources, 1973.

United States Bureau of the Census, *Statistical Abstract of the United States.* Washington, D.C.: U.S. Government Printing Office, annual.

WARD, DAVID, ED., *Geographic Perspectives on America's Past.* New York: Oxford University Press, 1979.

WARKENTIN, JOHN, ED., *Canada: A Geographical Interpretation.* Toronto: Methuen Publications, 1968.

WATSON, J. WREFORD, *North America: Its Countries and Regions.* London: Longmans, Green and Co., Ltd., 1968.

———, *Social Geography of the United States.* New York: Longman, 1979.

———, *The United States: Habitation of Hope.* New York: Longman, 1982.

ZELINSKY, WILBUR, *The Cultural Geography of the United States.* Englewood Cliffs, NJ: Prentice-Hall, 1973.

2
THE PHYSICAL ENVIRONMENT

Anglo-America is largely a midlatitude continent, lying entirely north of the Tropic of Cancer, except for Hawaii, and mostly south of the Arctic Circle. It spreads broadly in these latitudes, fronting extensively on all three Northern Hemisphere oceans—Atlantic, Pacific, and Arctic. Physical features vary widely, partly as a result of the great size of the continent; indeed, this diversity of environmental conditions is the keystone to understanding the physical geography of Anglo-America.

THE PATTERN OF LANDFORMS

The basic pattern of physiographic features in Anglo-America is a fourfold division roughly oriented north–south across the continent (Fig. 2-1). In the west is a complex series of mountain ranges and lengthy valleys interspersed with numerous desert basins and plateaus; in the center is an extensive lowland area that widens toward the north; in the east is a broad cordillera of mountains and hills; and along part of the east coast is a coastal plain that swings westward to join the central lowland along the Gulf Coast.

In the conterminous states the western mountain complex consists of two major prongs. In the coastal states a number of steep-sided ranges more or less parallel the coast from Mexico to Canada; they vary in height from only a few hundred feet in parts of the Coast ranges of Oregon and Washington to more than 14,000 feet (4200 m) in the Sierra Nevada of California and the Cascades of Washington.

The Rocky Mountain cordillera consists of a series of southeast-northwest trending ranges that rise abruptly from the central plains and extend with

FIGURE 2-1 Physiographic diagram of Anglo-America (original map by A. K. Lobeck; reprinted by permission of Hammond, Inc.).

only one significant interruption from north-central New Mexico to the Yukon Territory (Fig. 2-2). Between the Rockies on the east and the Sierra Nevada–Cascade ranges on the west is an extensive area of dry lands where plateaus, mesas, desert basins, and short but rugged mountain ranges intermingle.

In western Canada the coastal mountains become more rugged and complex and are separated from the Canadian Rockies by mostly forest-covered, plateaulike uplands. To the northwest the highland orientation changes from south–north to east–west, with the broad lowland of the Yukon–Kuskokwim basins separating the wilderness of the Brooks Range to the north from the massive ranges of southern Alaska. These mountains are the highest and most heavily glaciated of the continent.

The central lowland should not be considered uninterrupted flatland. Much of the terrain is undulating or rolling and there are many extensive areas of low hills, some quite steep-sided. By and large, however, most of the area between the Rockies on the west and the Appalachians on the east is a lowland of gentle relief. Its narrowest extent is toward the south, from which there is notable widening until the longitudinal extent in the north encompasses almost the entire width of the continent. The major exceptions to this pattern are in the rugged and glaciated eastern islands of the Canadian Arctic Archipelago: Baffin, Devon, Ellesmere, and Axel Heiberg.

The mountains of the east are not as high, rough, or obviously glaciated as those of the west and the eastern cordillera is only half as long (Fig. 2-3). Nevertheless, the Appalachian Mountain system extends almost without interruption from Alabama to the Gulf of St. Lawrence, and related ranges carry the trend through most of eastern Que-

FIGURE 2-2 The mountains of western Anglo-America generally are high, steep, rocky, and rugged. This is Hallett Peak above Bear Lake in north-central Colorado (courtesy Union Pacific Railroad).

FIGURE 2-3 The mountains of the eastern part of the continent are not so high and rugged, and most are completely forested. This scene is from central West Virginia (TLM photo).

bec and Labrador. The highest peaks are less than 8000 feet (2400 m) and most crests are less than half that height. The Appalachian system, however, is a broad one and over much of its extent the slopes are steep and heavily wooded. An important outlier of the eastern highlands is found in the tristate area of Arkansas, Missouri, and Oklahoma where the extensive hills of the Ozarks and Ouachitas are found.

The east-coastal lowland is a feature of lesser magnitude than the other three divisions mentioned. It is a classic example of an embayed coastal plain, with many estuaries, bays, and lagoons. From its narrow northern extremity in New England it slowly widens southward and then swings west in Georgia and Florida to link indistinguishably with the southern margin of the central lowland.

Hydrography

Two prominent drainage systems, the Great Lakes–St. Lawrence and the Mississippi–Missouri, dominate the hydrography of Anglo-America (Fig. 2-4). The five Great Lakes (Superior, Huron, Erie, Ontario, and Michigan, the first four shared by the United States and Canada) drain northeastward to the Atlantic via the relatively short St. Lawrence River. Many rivers flow into the Great Lakes, but all are short. The drainage basin of the Great Lakes watershed is remarkably small; it is, for example, only one-sixth the size of the Mississippi–Missouri drainage area (Fig. 2-5).

Most of the central part of the United States and a little of southern Canada are drained by the Mississippi River and its many tributaries. The Mississippi itself flows almost due south from central Minnesota to the Gulf of Mexico below New Orleans. Its principal left-bank tributary is the Ohio River. The Ohio River drains much of the northern Appalachians and the Midwest before being joined by the Tennessee River, which drains much of the southern Appalachians and adjacent areas. The far-reaching Missouri River, emanating from the northern Rockies of Montana and gathering tributaries all across the north-central part of the country, is the major right-bank tributary of the Mississippi.

Between the mouths of the St. Lawrence and Mississippi rivers the well-watered east coast of Anglo-America is drained by a host of rivers, most of moderate length but carrying much water. West of the Mississippi the rivers that enter the Gulf are longer but do not have a large volume of flow.

In the western United States there is great variety in the rivers that reach the Pacific Ocean. The Colorado River is a lengthy desert stream that drains much of the arid Southwest and eventually debouches into the sea in Mexico. The complex Sacramento–San Joaquin system drains much of California. The Columbia, principal river of the Pacific Northwest,

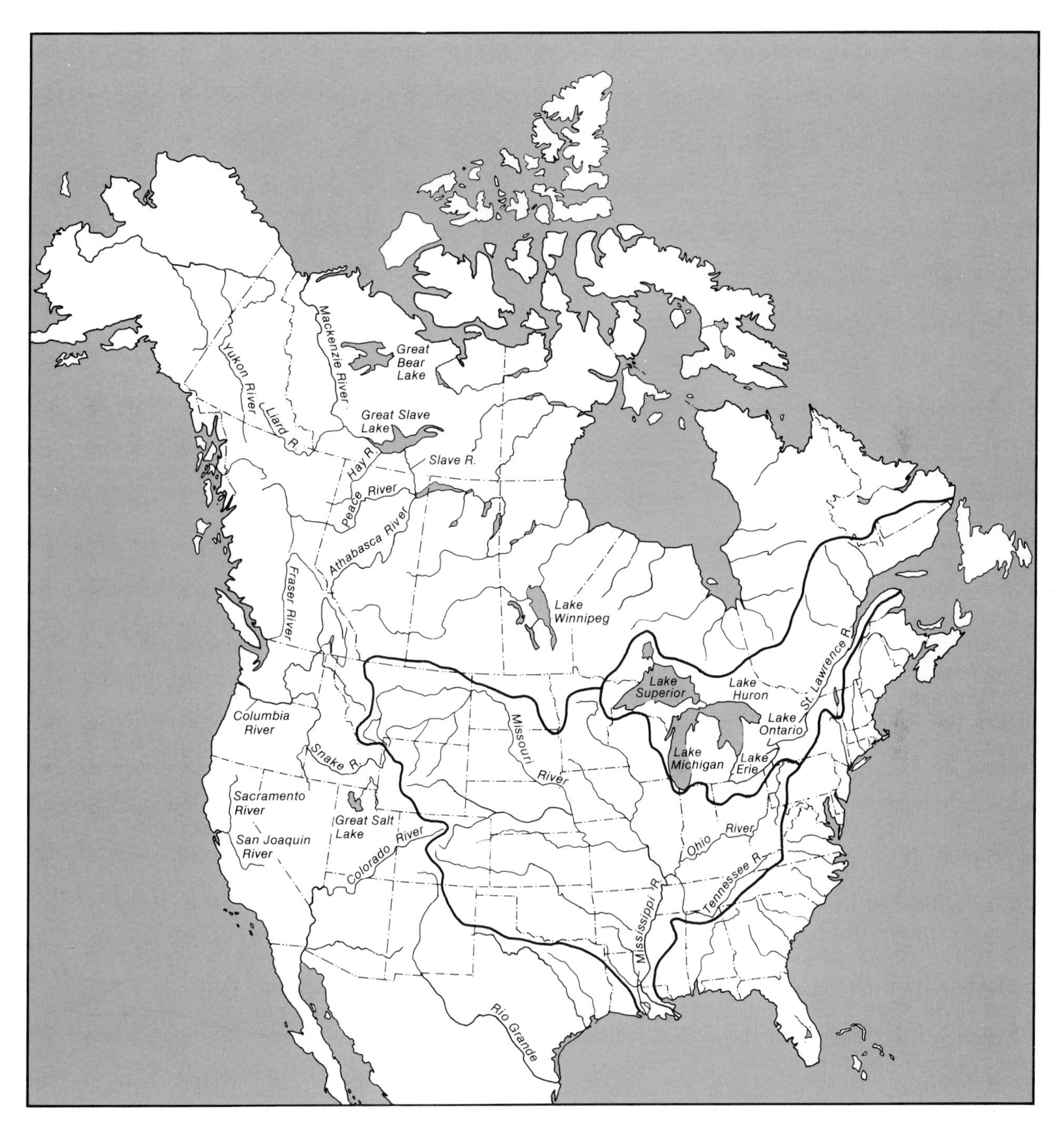

FIGURE 2-4 The major hydrographic features of Anglo-America. The drainage basins of the Mississippi River watershed and of the Great Lakes–St. Lawrence River systems are bounded by a heavy black line.

FIGURE 2-5 The most famous waterfall on the continent is Niagara. The Niagara River bifurcates to flow around Goat Island, with the American Falls on the left of the photo and the more voluminous Canadian (Horseshoe) Falls on the right, immediately beyond the Oneida Observation Tower (courtesy Ontario Ministry of Industry and Tourism).

originates in Canada and traverses some 465 miles (745 km) before crossing into the United States, where it is joined by its main tributary, the Snake, and finally flows for another 300 miles (480 km) as the boundary between Oregon and Washington.

The Pacific drainages of Canada and Alaska encompass many short rivers and a few long ones, but heavy precipitation ensures that the streams have a large volume of flow. The most notable rivers are the Fraser in southern British Columbia and the Yukon in the Yukon Territory and Alaska.

Most of the Arctic drainage of Anglo-America is accomplished by the Mackenzie system or the myriad streams that flow centripetally into Hudson Bay. The Mackenzie system is an unusually complex one; its major water sources are the Liard River and Great Slave Lake, the latter being fed by the extensive watersheds of the Hay, Peace, Athabasca, and Slave rivers.

It should also be mentioned that thousands of square miles in western Anglo-America are not served by external drainage. Particularly in Nevada, Utah, and California, basins of internal drainage abound. Most of these basins contain either shallow or dry lakes in their center (of which Utah's Great Salt Lake is the most conspicuous) and are fed by streams that usually flow only intermittently.

In some parts of Anglo-America lakes are a significant element in the landscape (Fig. 2-6). As a result of more extensive glaciation in the past, Canada has a much larger proportion of its surface area in lakes (as well as marshes and swamps) than does the United States. (There is some truth to Minnesota's claim to be the "land of 10,000 lakes" and Ontario's assertion to be the "land of 100,000 lakes" is equally valid.) Furthermore, many of Canada's lakes are large ones; Great Bear, Great Slave, and Winnipeg, for example, are larger than Lake Ontario, and Canada contains no less than eight lakes that are larger than any wholly United States lake except Lake Michigan.[1] The United States has relatively few natural lakes except in the glaciated sections, primarily New England, New York, and the upper Lakes states, in interior basins of the arid West, and in the flat limestone country of central and southern Florida.

There has been a great proliferation of artificial lakes (reservoirs) in Anglo-America, particularly in the southeastern, south-central, and southwestern states where natural lakes are rare. Most such reser-

[1] There is a total of more than 290,000 square miles (755,000 km^2) of freshwater lakes in Canada, which is an area larger than the state of Texas.

voirs are formed by the simple damming of rivers, producing fingerlike bodies of water that extend for long distances up former stream valleys and are now prominent features on the large-scale maps of almost any area from Carolina to California.

Glaciation

An understanding of the physical geography of Anglo-America requires that some attention be paid to the role of glaciation in creating the contemporary landscape, particularly in Canada. During the most recent "Ice Age" (the Pleistocene epoch) that began 1 million to 2 million years ago and may have ended less than 8000 years ago, extensive continental ice sheets formed in northern and central Canada and made at least four major advances southward (Fig. 2-7). At the height of Pleistocene glaciation the ice sheets extended as far south as Long Island, the Ohio River, the Missouri River, and the middle Columbia River. At the same time, mountain glaciers of considerable size developed throughout the ranges of the West; evidence of Pleistocene glaciation has been discovered as far south as central New Mexico and the San Bernardino Mountains of southern California.

Glacial erosion and deposition completely reshaped the terrain in all areas covered by Pleistocene ice. From the standpoint of geological time the retreat of the ice is so recent that the critical factors of slope, drainage, and surficial material are more directly the result of the action of ice, and meltwater, than of any other landscape-shaping element. This fact is clearly shown by the deranged drainage patterns and the large amount of standing water (lakes, swamps, and marshes) now found in areas once covered by Pleistocene ice sheets.

Except in a few locations, ice is but a minor feature in the contemporary topography of Anglo-America. The only sizable ice sheets, although much smaller than those of Antarctica or Greenland, occur on the four large eastern islands of the Canadian Arctic Archipelago.

Mountain glaciers are also still found in Anglo-America, but they are slight remnants of their past extent. Small living glaciers, only a few acres in size, occur as far south as central Colorado in the Rockies and the Sierra Nevada of California. Mountain glaciers appear with increasing frequency farther north in the western cordilleras, but the only place in the conterminous states where their length is reckoned in miles is in the northern part of the Cascade Range.

Contemporary glaciation is much more extensive in western Canada, and mountain icefields occur in some parts of the Canadian Rockies and British Columbia's Coast Mountains (Fig. 2-8).

FIGURE 2-6 A satellite image of the lower end of the Great Lakes. Shown is the entirety of Lake Ontario and the eastern end of Lake Erie. The Niagara River flows northward to connect the latter with the former, and it is paralleled to the west by the Welland Canal. Leeward clouds are conspicuous around the eastern end of Lake Ontario (Skylab image).

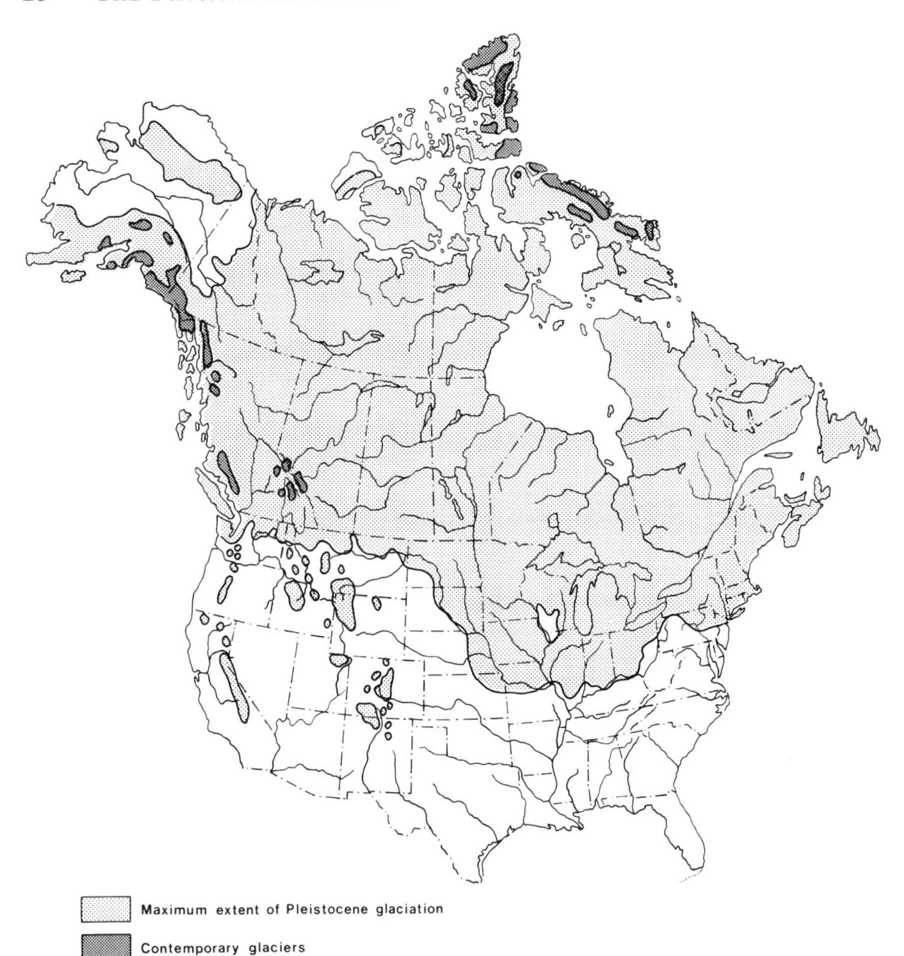

FIGURE 2-7 Recent and contemporary glacial extent in Anglo-America (after *National Atlas of the United States of America,* Washington, D.C.: U.S. Department of the Interior, Geological Survey, 1970, p. 76).

Mountain glaciation reaches its greatest extent in the ranges of southern Alaska and the southwestern corner of the Yukon Territory, where the length of many valley glaciers is measured in tens of miles and the areal extent of one piedmont glacier (the Malaspina) is greater than that of the state of Rhode Island.

Major Landform Regions of Anglo-America

Several geographers and geomorphologists have subdivided the United States and Canada into physiographic, or landform, regions based on various criteria but primarily on the gross distribution of terrain features. Figure 2-9 is a combination and

FIGURE 2-8 Athabaska Glacier is one of the longest in the Canadian Rockies. The distance from the toe to the highest icefall, where the glacier issues from the Columbia Ice Field, is nearly 10 miles [16 km] (TLM photo).

FIGURE 2-9 Major landform regions of Anglo-America are the (1) Southeastern Coastal Plain, (2) Appalachian Uplands, (3) Interior Uplands, (4) Interior Plains, (5) Rocky Mountains, (6) Intermontane Region, (7) Pacific Coast Region, (8) Yukon Basin and Plateaus, (9) Northwestern Highlands, (10) Canadian Shield, (11) Hudson Bay Lowland, and (12) High Arctic Mountains. (N. M. Fenneman, "Physiographic Divisions of the United States," *Annals,* Association of American Geographers, 18 [1928], 261–353; W. W. Atwood, *The Physiographic Provinces of North America* [Boston: Ginn and Co., 1940]; A. K. Lobeck, "Physiographic Diagram of the United States" [Madison: Wisconsin Geographic Press, 1922]; *The National Atlas of the United States of America* [Washington, D.C.: U.S. Department of the Interior, 1970], pp. 61–64; *Atlas of Canada* [Ottawa: Department of Mines and Technical Surveys, 1957], plate 13.)

modification of the work of a number of these scholars.

The *Southeastern Coastal Plain* (1) is one of the flattest portions of North America. Most of the region slopes gently seaward, at the rate of only a few feet or even a few inches per mile, with the gentle slope continuing under water for dozens of miles as a continental shelf. The uninterrupted flatness of the landscape is relieved only sporadically by low (10 to 50 feet [3 to 15 m]) riverine bluffs alongside the broad river valleys or, particularly west of the Mississippi River, by low linear ridges (called cuestas) that parallel the coast.

Most of the region was submerged in relatively recent geologic time and its surface layers consist of loosely consolidated sands, gravels, marls, and clays. The coastal margin is exceedingly irregular as a result of recent submergence and embayment; it is characterized by extensive drowned valleys (estuaries) and many bays, swamps, lagoons, and low-lying sand-bar islands. The portion abutting the Appalachians encompasses the Piedmont, where resistant crystalline bedrock and rising elevation produced a more undulating surface.

The *Appalachian Uplands* (2) extend from Alabama to the Gulf of St. Lawrence and the physiographic trend is continued in the island of Newfoundland. The region includes the "mountainous" part of eastern Anglo-America, although much of the terrain consists of low, forested hills. From Pennsylvania southward most of the surface is underlain with sedimentary rocks. These sediments have been tightly folded in the eastern portion of the region to produce a remarkable sequence of parallel valleys and long, steep-sided ridges, although the easternmost ridge (the Blue Ridge) that contains the highest peaks in the eastern part of the continent consists of ancient crystalline rocks.

To the west of the ridge-and-valley section the sediments are more horizontal; here the so-called Allegheny and Cumberland plateaus are thoroughly dissected by streams and give the appearance of an endless region of low hills. North of Pennsylvania, the underlying rock is mostly crystalline (igneous or metamorphic) and the surface form varies greatly, although hills and low mountains dominate the landscape. The most conspicuous ranges are the Adirondacks in New York, Green Mountains in Vermont, White Mountains in New Hampshire, Notre Dame and Shickshock ranges in Quebec, and Long Range in Newfoundland.

The *Interior Uplands* (3) bear considerable physiographic resemblance to the Appalachians, although they have less altitude and local relief. The Ozark section, mostly in southern Missouri and northern Arkansas, is a dissected plateau that consists of an amorphous pattern of low hills. Separated from the Ozarks by the transverse valley of the Arkansas River is the Ouachita section in western Arkansas and eastern Oklahoma. This is an area of east-west trending, linear ridges and valleys; it is markedly similar to the ridge-and-valley section of the Appalachians.

The *Interior Plains* (4) are a vast area of gentle relief that occupies much of the central portion of the continent. Some portions are remarkably flat, such as the High Plains of west Texas or the prairies of central Illinois; however, most of the region is characterized by undulating terrain or low, even-topped hills. Relatively flat-lying sedimentary rocks underlie the surface in most places. The terrain was conspicuously modified by the action of Pleistocene ice sheets north of the Ohio and Missouri rivers. Numerous lakes, marshes, and ponds occur in the glaciated sections of the region, as well as many long rivers.

The *Rocky Mountains* (5), located just west of the flattish Great Plains portion of the Interior Lowlands, constitute a very abrupt physiographic change when approached from the east. They are characterized by high elevations—more than 50 peaks in Colorado reach above 14,000 feet (4,200 m)—great local relief, rocky ruggedness, and spectacular scenery. Only in the Wyoming Basin area of southern and central Wyoming is there any significant section that is not dominated by mountainous terrain. There was great variety in the mountain-building processes that produced the Rockies. In the southern portions extensive granitic intrusions have been thrust up many thousands of feet whereas farther north sedimentary rocks have been drastically folded, and thrust and block faulting occurred on a large scale. Throughout the region the recent action of mountain glaciers deepened the valleys, steepened the slopes, and sharpened the peaks.

The *Intermontane Region* (6) encompasses a bewildering variety of terrain formed in many different ways. The southern and southwestern portions are basin-and-range country where numerous, discrete, short, rugged mountain ranges are interspersed with flat alluvial-filled valleys. In the "Four Corners" country plateaus, mesas, cliffs, and buttes dominate the landscape, for weakly consolidated horizontal sedimentary rocks were stripped and fret-

FIGURE 2-10 The spectacular starkness of the Intermontane West is demonstrated dramatically in this view of Mitten Butte in Utah's Monument Valley (TLM photo).

ted by arid-land erosion processes (Fig. 2-10). The Columbia and Snake lava plateaus in Idaho, Oregon, and Washington have been deeply incised by major rivers and eroded into rolling hills in some areas. The various plateaus of the Canadian portion of the region (Fraser, Nechako, and Stikine) were severely dissected and in many places appear as hills or mountains.

The *Pacific Coast Region* (7) is largely mountainous, with the trend of the ranges generally paralleling the coast. High and rugged mountains extend the entire length of the region, from the Sierra Nevada of California to the Alaska Range in the north. Several major valleys, particularly the Central Valley of California and the Willamette Valley of Oregon, are sandwiched between massive interior ranges and numerous smaller coastal ranges. The coastline itself is quite regular in the south, where the steep coastal ranges plunge abruptly into the sea; northward the coastline becomes exceedingly irregular, with lengthy bays and fiords interspersed with sinuous peninsulas and numerous islands.

The *Yukon Basin and Plateaus* (8) section occupies most of central Alaska and the southern part of the Yukon Territory, primarily within the drainage basins of the Yukon and Kuskokwin rivers. Hill land predominates in the upstream areas, but much of central Alaska is a broad, flat-floored, poorly drained lowland. The lower courses of the two major rivers have built complex deltas.

The *Northwestern Highlands* (9) section includes the massive barren slopes of the Brooks Range in northern Alaska and a series of rugged mountains in northwestern Canada.

The *Canadian Shield* (10) is an extensive, ancient, stable region floored with some of the world's oldest known crystalline rocks. Except for some relatively rugged hills in eastern Quebec and Labrador, this is a gently rolling landscape typified by many outcrops of bare rock and an extraordinary amount of surface water in summer. There are hundreds of thousands of water bodies, ranging in size from gigantic to minute, connected by tens of thousands of rivers and streams.

The *Hudson Bay Lowland* (11) is a flat coastal plain that slopes imperceptibly toward the sea. It is underlain by recent sedimentary deposits that distinguish it from the surrounding, and underlying, Laurentian Shield.

High Arctic Mountains (12) occupy much of the four large eastern islands of the Canadian Arctic Archipelago. The region is typified by rugged and rocky slopes, large glaciers and ice sheets—several of which are larger than the province of Prince Edward Island—lack of vegetation and soil, and deep permafrost.

THE PATTERN OF CLIMATE

Most of Anglo-America experiences a climate in which seasonal changes are marked, as is characteristic of middle- and high-latitude portions of the world. Summer is generally hot and is the season of maximum precipitation; winter is cold and somewhat drier; and spring and fall are stimulating seasons of transition. There are many exceptions to these generalizations, particularly on the Pacific littoral where winter is the wet season and summer is quite dry.

For the most part the climate of Anglo-America is dominated by weather systems that move across the continent from west to east, in association with the westerly air flow that prevails over the mid-latitudes. Some of these migratory systems are focused around low-pressure cells (extratropical cyclones) that attract unlike air masses, producing frontal disturbances, stormy or unsettled weather, and precipitation that is sometimes abundant (Fig. 2-11). Other systems are dominated by high-pressure cells (anticyclones) that bring clear skies and extreme temperature conditions (usually cold, sometimes hot).

These air circulation patterns (mostly westerly winds) combine with the topographic patterns (north–south trending mountain ranges) and land/sea relationships (oceanic moisture sources to the west, east, and south) to produce plentiful precipitation along much of the west coast, the windward sides of western mountain ranges, and the southeastern part of the continent. Drier conditions prevail in the north of the continent, over much of the central plains area, and particularly in the interior West. This simplified pattern is modified and interrupted by a variety of factors, particularly cold air invasions from the north in winter and onshore flows of warm, moist air from the Gulf of Mexico in summer.

Major Climatic Regions of Anglo-America

There are almost as many classifications of climate as there are climatologists. Although most classifications are fundamentally the same, their minor variations and specialized nomenclature may confuse the reader. No standard classification has been adopted for climates; the map used in this book shows one of the most widely accepted schemes (Fig. 2-12).

The basic pattern that emerges is one of east–west trending climatic zones in the eastern and northern portions of the continent, with north–south

FIGURE 2-11 Winter is a prominent fact of life over most of the continent. This January scene is near Racine in southeastern Wisconsin (TLM photo).

FIGURE 2-12 Major climatic regions of Anglo-America are (1) humid subtropical; (2) humid continental with warm summers; (3) humid continental with cool summers; (4) steppe; (5) desert; (6) mediterranean, or dry summer subtropical; (7) marine west coast; (8) subarctic; (9) tundra; (10) ice cap; and (11) undifferentiated highlands (after Glenn T. Trewartha, Arthur H. Robinson, and Edwin H. Hammond, *Elements of Geography,* 5th ed. [New York: McGraw-Hill Book Company, 1967]).

trending zones in the west. Such a pattern reflects the fundamental significance of latitude, with topographic modification in the west. Generally similar patterns are evident in the succeeding maps of vegetation and soils.

In the southeastern quarter of the United States the climatic type is classified as *humid subtropical* (1). Summer is the most prominent season. Humid tropical air usually pervades most of the region and high temperatures frequently combine with high humidity. Winter is short and relatively mild, although significant cold spells occur several times each year. Precipitation is spread throughout the year, with a tendency toward a maximum in summer, except in Florida where rains associated with hurricanes bring a fall maximum.

In east-central United States there is a large area of *humid continental with warm summer phase* (2) climate. This is a zone of interaction between warm tropical air masses from the south and cold polar air masses from the north, which results in frequent, stimulating weather changes throughout the year. Summers are warm to hot and are the time of precipitation maximums. Winters are cold and there is considerable snowfall.

There is a broad east–west band of *humid continental with cool summer phase* (3) climate along the international border in eastern Anglo-America. It is distinguished from the previous climatic type by its shorter and milder summer and its longer and more rigorous winter.

The *steppe* (4) climate of the Great Plains and Intermontane areas is basically a semiarid climate with marked seasonal temperature extremes. Summers are hot, dry, and windy and are punctuated by occasional abrupt thunderstorms; winters are cold, dry, and windy, with occasional blizzards. Intense and dramatic weather—hail, heat waves, tornadoes, windstorms, and the like—are typical of this region.

The *desert* (5) climate of the southwestern interior is characterized by clear skies, brilliant sunshine, and long periods without rain. Aridity is universal except at higher elevations. Summer is long and scorchingly hot; winter is brief, mild, and delightful.

Central and coastal California is a region of *mediterranean* (6) climate, with its anomalous precipitation regime in which sequential winter frontal storms move in from the Pacific, bringing alternating periods of rain and sunshine. Summer is virtually rainless, for stable high-pressure conditions dominate the atmosphere. Coastal sections have mild temperatures throughout the year whereas inland locations experience hot summers and cool winters.

The *marine west coast* (7) climate of the Pacific littoral is characterized by long, relatively mild, very wet winters and short, pleasant, relatively dry summers. The frequent movement over the region of maritime polar air from the Pacific, and its associated fronts, accounts for the prominence of winter conditions. Exposed slopes at higher elevations receive some of the greatest precipitation totals (both rain and snow) in the world.

The climate of most of central Canada and Alaska is classified as *subarctic* (8). Winter, the dominant season, is long, dark, relatively dry, and bitterly cold. Summer occupies only a brief period, but the succession of long hours of daylight produces several weeks of warm-to-hot weather. Summer rainfall is scanty, but evaporation rates are low and so moisture effectiveness is high. Except in Quebec and Labrador, winter snowfall is also scanty, but there is little melting from October to May.

The *tundra* (9) climate of arctic Canada and Alaska is virtually a cold desert. There is little precipitation at any season; however, evaporation rates are also quite low. Winter is very long, although not as cold as in the subarctic areas. Summer is short and cool.

The *icecap* (10) climate of the High Arctic is rigorous in the extreme. Low temperatures and strong winds make these areas unendurable for humans and animals, except briefly.

The major mountain areas of Anglo-America are classified as having a *highland* (11) climate. Generalizations concerning weather and climate in such areas can be made only with reference to particular elevations or exposures. On the average, temperature and pressure decrease with altitude whereas precipitation and windiness increase. Thus vertical zonation is the key to understanding highland climates.

THE PATTERN OF NATURAL VEGETATION

Unlike landforms and climate, the natural vegetation of Anglo-America underwent major changes caused by humankind. The native flora has been so thoroughly removed, rearranged, and replaced that a discussion of natural vegetation in many areas is largely a theoretical or historical exercise. Still, from a broad standpoint it is possible to reconstruct the major vegetation zones and make some meaningful generalizations about them (Fig. 2-13). The reader should keep in mind that in many parts of the continent introduced plants—particularly crops, pasture grasses, weeds, and ornamentals—are much more conspicuous than native flora.

Anglo-America's natural vegetation associations can be divided into three broad classes: forests, grasslands, and shrublands.

Forests

The forests occur in six zones:

1. Northern coniferous forests, or taiga
2. Eastern hardwood zone[2]

[2] The terms *hardwoods* and *softwoods* are the most generally accepted popular names for the two classes of trees, the *Angiosperms* and the *Gymnosperms*. Most Angiosperms, such as oak, hickory, sugar maple, and black locust, are notably hard woods and many Gymnosperms, such as pines and spruces, are rather soft woods. But there are a number of outstanding exceptions. Basswood, poplar, aspen, and cottonwood, all classified as hardwoods, are actually among the softest of woods. Longleaf pine, on the other hand, is about as hard as the average hardwood, although classified as a softwood. The most accurate popular descriptions for the two groups are *trees with broad leaves* for the Angiosperms and *trees with needles and scalelike leaves* for the Gymnosperms (from Forest Products Laboratory, *Technical Note 187*, Madison, Wisconsin).

FIGURE 2-13 Major natural vegetation regions of Anglo-America are (1) broadleaf deciduous forest, (2) mixed broadleaf deciduous and needleleaf evergreen forest, (3) needleleaf evergreen forest, (4) grassland, (5) mixed grassland and mesquite, (6) broadleaf evergreen shrubland, (7) mediterranean shrubland, (8) tundra, and (9) little or no vegetation (after *The National Atlas of the United States of America,* pp. 90–92; and *Atlas of Canada,* plate 38).

3. Eastern mixed forests
4. Southern pineries
5. Rocky Mountain forests
6. Pacific Coast forests

The taiga, called *boreal forest* in Canada, is composed mostly of coniferous trees growing under conditions of long, bitterly cold winters. It covers about three-fourths of Canada's productive forest land, is widespread in central and southern Alaska, and dips slightly into the states of Minnesota and Wisconsin. For the most part, the trees grow in relatively pure stands, with white and black spruce being the most widespread. Also common are tamarack, balsam, and alpine fir, and jack and lodgepole pine. Sometimes nonconifers—particularly aspen, white birch, and balsam poplar—cover extensive areas. The short growing season and relatively poor drainage of the region inhibit rapid growth so that even old trees are not very tall.

Three belts of forested land extend southward from the taiga. One prong follows the Appalachian Mountains and adjacent lowlands, another the Rocky Mountains, and a third the ranges of the Pacific coastal states. The easternmost prong grades southward from the relatively pure softwood stands of the taiga through a zone of mixed broadleaf and needleleaf trees to an expansive region where hardwoods constituted the original vegetation. South of the hardwood region is another area of mixed forest that gives way to extensive forests dominated by southern yellow pine in the Southeast.

The Rocky Mountain prong is mostly mountain forest, with various species occupying altitudinal zones vertically arranged on the mountainsides. In the southern Rockies trees grow only in the uplands, but in the northern Rockies many of the

FIGURE 2-14 The evergreen forests of the Pacific Northwest are dominated by big trees, as shown here in western Washington (courtesy Weyerhaeuser Company).

lowlands between the ranges are also forested. Most trees are conifers, with spruce and fir occupying the wetter areas; pine and juniper grow in drier localities. Aspen, the first species to occupy an area after a fire, is one of the few deciduous trees and is quite widespread in some places.

The Pacific Coast prong also consists primarily of softwoods. Although the forested zone follows the mountain trend, most of the lowlands north of the Sacramento Valley (except the Willamette Valley) were also originally forested. This is a region of huge trees, generally the largest to be found on earth (Fig. 2-14). Fir, spruce, and hemlock dominate in the north whereas redwoods become conspicuous in northern California. The drier slopes are almost invariably pine covered.

Grasslands

Grasslands are usually found in areas where the rainfall is insufficient to support trees and most have never had any other type of vegetation. Where grass is tall, it is often called *prairie;* where short, *steppe*. Toward the drier margin the short-grass association grades into bunchgrass.

In Anglo-America the eastern portion of the Great Plains was originally clothed in prairie grasses, with a significant eastward extension in the so-called Prairie Wedge of Illinois. The western part of the Plains had a steppe association and there were other significant areas of grassland in the Central Valley of California and in the northern part of the Intermontane Region.

The tundra of the far north can be viewed as a sort of pseudograssland. It is characterized by a low-growing mixture of sedges, grasses, mosses, and lichens, occurring in an amazing variety of species.

Shrublands

Shrublands vary considerably in their characteristics but usually develop under an arid or semiarid climate. The typical plant association is a combination of bushes or stunted trees and sparse grasses. In southern and central Texas the shrubby mesquite tree is the dominant plant. In the area from Wyoming to Nevada sagebrush provides the most common ground cover. In the Southwest cacti and other succulents are characteristic (Fig. 2-15), although some large expanses are dominated by a sparse covering of spindly creosote bush and burroweed. The chaparral-manzanita association of southern and central California is interspersed with some broad reaches of grassland and open oak woodland.

THE PATTERN OF SOIL

The most complex feature of the environment is the soil. This complexity is largely attributable to complicated chemical factors and reactions in soil development that are not only imperfectly understood but also leave little or no mark on the landscape and so are difficult to ascertain without detailed microstudies. Also, minor differences in such related features as slope or drainage can be much more important than broader environmental parameters in determining significant soil characteristics; thus

FIGURE 2-15 Desert vegetation in southern Arizona. Spindly shrubs are characteristic, but the floristic landscape is often dominated by such conspicuous cacti as the organ pipe (left) and giant saguaro (right) (TLM photo).

soil variations over short distances are often much more significant than broader regional variations.

The identification and distribution pattern of meaningful soil categories is consequently difficult to determine except on a very large scale. Most maps of soil categories are hopelessly complex for purposes of macrostudy. Moreover, the past few years have been a period of fundamental change in the way that soils are classified and categorized by pedologists and other scholars interested in soil distribution. Thus an entirely new classification scheme was adopted and many changes in nomenclature and terminology were accepted.

With these caveats in mind, we can make only broad generalizations about the distribution of soil categories in Anglo-America. Figure 2-16 is based on the United States Comprehensive Soil Classification System—now officially called *Soil Taxonomy*—that was developed slowly and meticulously during the 1950s and 1960s by the Soil Survey Staff of the U.S. Department of Agriculture.[3] The principal *orders* (the major categories in the hierarchical classification) are briefly described here.

Alfisols (A) are soils with mature profile development that occur in widely diverse climatic and vegetation environments. They have gray-to-brown

[3] Soil Survey Staff, *Soil Classification: A Comprehensive System—7th Approximation* (Washington, D.C.: U.S. Department of Agriculture, Soil Conservation Service, 1960); Soil Survey Staff, *Supplement to Soil Classification System—7th Approximation* (Washington D.C.: U.S. Department of Agriculture, Soil Conservation Service, 1967); Soil Survey Staff, *Soil Taxonomy: A Basic System of Soil Classification for Making and Interpreting Soil Surveys* (Washington, D.C.: U.S. Department of Agriculture, Soil Conservation Service, 1976); Donald Steila, *The Geography of Soils* (Englewood Cliffs, NJ: Prentice-Hall, 1976); J. S. Clayton et al., *Soils of Canada,* 2 vols. (Ottawa: Canada Department of Agriculture, 1977).

FIGURE 2-16 Major soils regions of Anglo-America: (A) Alfisols, (D) Aridisols, (E) Entisols, (H) Histosols, (I) Inceptisols, (M) Mollisols, (S) Spodosols, (U) Ultisols, (X) complex soil regions, and (Z) areas with little or no soils (after maps of Soil Geography Unit, Soil Conservation Service, U.S. Department of Agriculture).

surface horizons and a clay accumulation in subsurface horizons. They are most widespread in the Great Lakes area, the Midwest in general, and the northern part of the Prairie provinces.

Aridisols (D) are mineral soils that are low in organic matter and dry in all horizons most of the time. Associated primarily with arid climatic regimes, they are most extensively found in the western interior of the United States, particularly the Southwest.

Entisols (E) are primarily of immature development, with a low degree of horizonation. Characteristically they are either quite wet, quite dry, or quite rocky. They occur most widely in northern Quebec and Labrador but are also found in some of the High Arctic islands, in scattered localities in the West, and in southern Florida.

Histosols (H) represent the only order composed primarily of organic rather than mineral matter. They are often referred to by such terms as *bog, peat,* or *muck*. They can be found in any climate, provided that water is available. In Anglo-America they occur most extensively in subarctic Canada, particularly south of Hudson Bay and in the Great Bear Lake area.

Inceptisols (I) also occur in widely differing environments. They are moist soils with generally clear-cut horizonation. Leaching is prominent in their formation. They lack illuvial horizons and are primarily eluvial in character. They are widespread in tundra areas of Canada and Alaska, where they are associated with permafrost, and are also notable in the Appalachians, the Lower Mississippi Valley, and the Pacific Northwest.

Mollisols (M) are mineral soils with a thick, dark surface layer that is rich in organic matter and bases. Their agricultural potential is generally high. They are chiefly found in subhumid or semiarid areas and are the principal soils of the western Corn Belt and the Great Plains, as well as in interior portions of the Pacific Northwest.

Spodosols (S) have a conspicuous subsurface horizon of humus accumulation, often with iron and aluminum. They are commonly moist or wet and heavily leached. They are often associated with the soil-forming process called podzolization and have limited agricultural potential. Their most extensive occurrence in Anglo-America is in southeastern Canada and New England, but they are also found in such divergent locations as northern Florida and the Great Slave Lake area.

Ultisols (U) are thoroughly weathered and extensively leached and so have experienced considerable mineral alteration. Typically they are reddish in color owing to the considerable amounts of iron and aluminum in the surface layers. Their principal locations are in the southeastern quarter of the United States.

Complex (X) soil associations are delineated on the map in many mountain areas and in much of California. Their variety is too complicated to allow generalization at this scale.

Areas with *little or no soil* (Z) are recognized in rugged mountain areas or where permanent icefields exist.

This soil classification differs from most previous ones in that it is generic (based systematically on observable soil characteristics) rather than genetic (based on soil-forming conditions and processes). The resultant distribution pattern is less easy to comprehend because soils with similar characteristics sometimes are found in widely differing environments; for example, Inceptisols dominate in both southern Louisiana and the Northwest Territories. Nevertheless, genesis is not completely ignored, for soil properties are directly related to soil development. Thus the pattern of Figure 2-15 reflects some environmental relationships. The zonation of soils in the eastern and northern parts of Anglo-America is generally in east–west bands whereas that in the western part of the continent is banded north–south, emphasizing the roles of topography, climate, and vegetation in soil formation.

THE PATTERN OF WILDLIFE

Unlike the environmental elements previously discussed, the faunal complement of Anglo-America is generally insignificant in the total geographic scene. The spread of civilization generally has been hostile to native wildlife, resulting in contracting habitats and decreasing numbers. A few species—such as opossum, coyote, armadillo, and raccoon—have withstood the human onslaught and actually expanded their ranges, but such examples are limited. The larger carnivores (e.g., grizzly bear, wolf, mountain lion) have particularly suffered, and continue to dwindle despite significant mitigation endeavors in some areas. Interestingly enough, most ungulate species (such as deer, elk, moose, pronghorn, and bison) have prospered under enlightened wildlife management efforts and the decline of native predators; thus, they now occur in much greater numbers than they did a century ago.

People also have influenced wildlife patterns in Anglo-America in other ways. They introduced new species, such as nutria, ring-necked pheasants, feral pigs, and feral horses, and artificially rearranged the distribution of native species—for example, ex-

FIGURE 2-17 Wildlife is usually inconspicuous in the landscape, but not in the case of this Rocky Mountain bighorn sheep in Alberta's Jasper National Park (TLM photo).

panding the range of mountain goats by deliberate translocation.

Wildlife in most parts of Anglo-America is an inconsequential element in the landscape and attracts little attention from students of geography (Fig. 2-17). But in some areas, usually sparsely populated and with limited economic potential, wildlife assumes a more important role. Where such conditions pertain, the discussion takes place in the appropriate regional chapter of this book.

ECOSYSTEMS

In summing up our discussion of environmental patterns, it is perhaps most useful to turn to the concept of ecosystems. As a term, *ecosystem* is a contraction of the phrase "ecological system." The ecosystem concept is functional, encompassing not only all the organisms (plants and animals) in a given area but also the totality of interactions among the organisms and between the organisms and the nonliving

portion of the environment (soil, rocks, water, sunlight, atmosphere, etc.), which can be thought of as nutrients and energy. Thus an ecosystem is a biological community expressed in functional terms.

It is both a virtue and a complication that the ecosystem concept can be used at almost any scale. In other words, we can speak of a world ecosystem, or the ecosystem of a drop of water, or any level of generalization between these extremes. For our purposes, it is appropriate to consider broad-scale ecosystems that encompass large parts of the continent.

One of the most carefully constructed regionalizations of ecosystems was developed by representatives of the U.S. Forest Service and the U.S. Fish and Wildlife Service in the early 1980s. Called "Ecoregions of North America," it represents an effort to incorporate all major environmental aspects (topography, climate, soils, flora, and fauna) into a single hierarchy of ecosystem regions. Figure 2-18 depicts the two highest levels of this hierarchy.

The *Polar Domain* encompasses the high-latitude portions of the continent where long, cold winters provide the dominant environmental factor. This domain includes four divisions: *Tundra, Tundra Highlands, Subarctic,* and *Subarctic Highlands*.

The *Humid Temperate Domain* includes the eastern half of the conterminous states, the southern fringe of Canada, and the entire Pacific Coast of the continent. It is subdivided into nine divisions.

The *Dry Domain* consists of the interior West of the United States and a small adjacent part of Canada. Its subdivisions are *Semiarid Steppe, Semiarid Steppe Highlands,* and *Arid Desert*.

FIGURE 2-18 Major Ecoregions of Anglo-America.

1	Polar Domain	2F	Humid Maritime Highlands Division
1A	Tundra Division	2G	Subhumid Prairie Division
1B	Tundra Highlands Division	2H	Mediterranean Division
1C	Subarctic Division	2I	Mediterranean Highlands Division
1D	Subarctic Highlands Division	3	Dry Domain
2	Humid Temperate Domain	3A	Semiarid Steppe Division
2A	Humid Warm-Summer Continental Division	3B	Semiarid Steppe Highlands Division
2B	Humid Warm-Summer Continental Highlands Division	3C	Arid Desert Division
2C	Humid Hot-Summer Continental Division	4	Humid Tropical Domain
2D	Humid Subtropical Division	4A	Tropical Savanna Division
2E	Humid Maritime Division	4B	Tropical Rainforest Highlands Division

(After R. G. Bailey, *Ecoregions of the United States* (map), Ogden, Utah: U.S. Department of Agriculture, Forest Service, Intermountain Region, 1976; R. G. Bailey, *Description of the Ecoregions of the United States,* Ogden, Utah: U.S. Department of Agriculture, Forest Service, Intermountain Region, 1978; R. G. Bailey and Charles T. Cushwa, *Ecoregions of North America* (map), Reston, Virginia: U.S. Fish and Wildlife Service, 1981).

FIGURE 2-19 Ecoregions of the Conterminous United States

1 Coast Range
2 Puget Lowland
3 Willamette Valley
4 Cascades
5 Sierra Nevada
6 Southern and Central California Plains and Hills
7 Central California Valley
8 Southern California Mountains
9 Eastern Cascades Slopes and Foothills
10 Columbia Basin
11 Blue Mountains
12 Snake River Basin/High Desert
13 Northern Basin and Range
14 Southern Basin and Range
15 Northern Rockies
16 Montana Valley and Foothill Prairies
17 Middle Rockies
18 Wyoming Basin
19 Wasatch and Uinta Mountains
20 Colorado Plateaus
21 Southern Rockies
22 Arizona/New Mexico Plateau
23 Arizona/New Mexico Mountains
24 Southern Deserts
25 Western High Plains
26 Southwestern Tablelands
27 Central Great Plains
28 Flint Hills
29 Central Oklahoma/Texas Plains
30 Central Texas Plateau
31 Southern Texas Plains
32 Texas Blackland Prairies
33 East Central Texas Plains
34 Western Gulf Coastal Plain
35 South Central Plains
36 Ouachita Mountains
37 Arkansas Valley
38 Boston Mountains
39 Ozark Highlands
40 Central Irregular Plains
41 Northern Montana Glaciated Plains
42 Northwestern Glaciated Plains

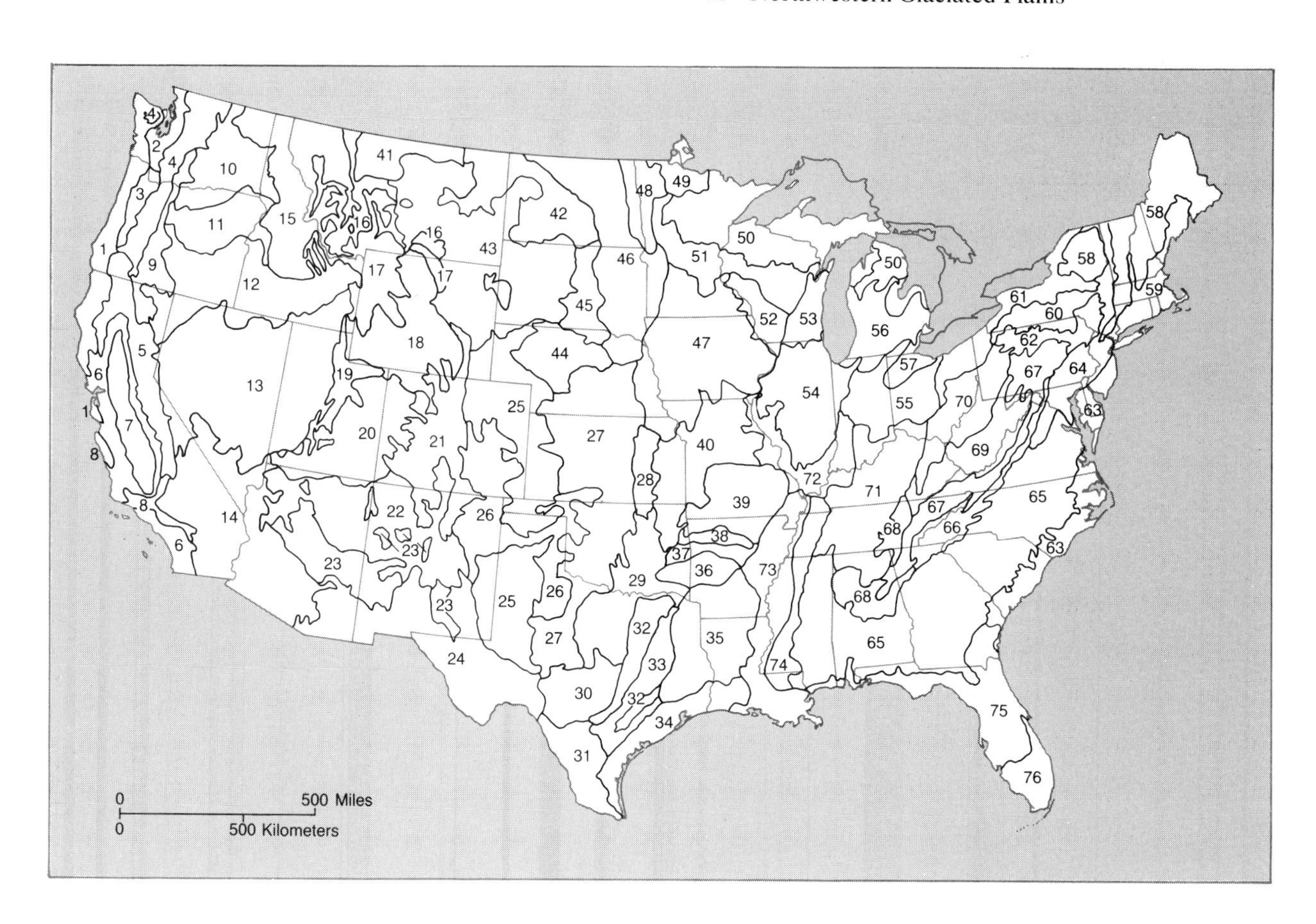

43 Northwestern Great Plains
44 Nebraska Sand Hills
45 Northeastern Great Plains
46 Northern Glaciated Plains
47 Western Corn Belt Plains
48 Red River Valley
49 Northern Minnesota Wetlands
50 Northern Lakes and Forests
51 North Central Hardwood Forests
52 Driftless Area
53 Southeastern Wisconsin Till Plains
54 Central Corn Belt Plains
55 Eastern Corn Belt Plains
56 Southern Michigan/Northern Indiana Till Plains
57 Huron/Erie Lake Plain
58 Northeastern Highlands
59 Northeastern Coastal Zone
60 Northern Appalachian Plateau and Uplands
61 Erie/Ontario Lake Plain
62 North Central Appalachians
63 Middle Atlantic Coastal Plain
64 Northern Piedmont
65 Southeastern Plains
66 Blue Ridge Mountains
67 Central Appalachian Ridges and Valleys
68 Southwestern Appalachians
69 Central Appalachians
70 Western Allegheny Plateau
71 Interior Plateau
72 Interior River Lowland
73 Mississippi Alluvial Plain
74 Mississippi Valley Loess Plains
75 Southern Coastal Plain
76 Southern Florida Coastal Plain

(After James M. Omernik, "Ecoregions of the Conterminous United States," *Annals*, Association of American Geographers, 77 (1987), map supplement. By permission of the Association of American Geographers.)

The *Humid Tropical Domain* includes only the southern tip of Florida (the *Tropical Savanna Division*) and the Hawaiian Islands (*Tropical Rainforest Highlands Division*).

A more recent effort by a geographer of the U.S. Environmental Protection Agency delineates ecoregions for the conterminous states on a somewhat different basis (including land use as well as environmental parameters) and for a different purpose (to assist managers of aquatic and terrestrial resources in recognizing more efficient management options). This system is not hierarchical, and it identifies 76 ecoregions in the 48 conterminous states (Fig. 2-19).

SELECTED BIBLIOGRAPHY

ATWOOD, WALLACE W., *The Physiographic Provinces of North America*. Boston: Ginn & Company, 1940.

BIRD, J. BRIAN, *The Natural Landscapes of Canada: A Study in Regional Earth Science*. Toronto: Wiley Publishers of Canada, Ltd., 1972.

BRYSON, REID A., AND F. KENNETH HARE, EDS., *World Survey of Climatology: Vol. 2, Climates of North America*. New York: American Elsevier Publishing Company, 1974.

CLAYTON, J. S., ET AL., *Soils of Canada*. Ottawa: Research Branch, Canada Department of Agriculture, 1977.

FALCONER, A., ET AL., *Physical Geography: The Canadian Context*. Toronto: McGraw-Hill Ryerson, Ltd., 1974.

FENNEMAN, NEVIN M., *Physiography of Eastern United States*. New York: McGraw-Hill Book Co., 1938.

———, *Physiography of Western United States*. New York: McGraw-Hill Book Co., 1931.

GERSMEHL, PHILIP J., "Soil Taxonomy and Mapping," *Annals*, Association of American Geographers, 67 (1977), 419–428.

HARE, F. KENNETH, AND M. K. THOMAS, *Climate Canada*. (2nd ed.). Toronto: Wiley Canada, Ltd., 1980.

HUNT, CHARLES B., *Natural Regions of the United States and Canada*. San Francisco: W. H. Freeman & Company, 1973.

IVES, J. D., "Glaciers," *Canadian Geographical Journal*, 74 (April 1967), 110–117.

KUCHLER, A. W., *Potential Natural Vegetation of the Conterminous United States*. New York: American Geographical Society, Special Publications 36, 1964.

MCKNIGHT, TOM, *Feral Livestock in Anglo-America*. Berkeley: University of California Press, 1964.

MACLENNAN, HUGH, *Rivers of Canada*. Toronto: Macmillan of Canada, 1974.

MARKHAM, CHARLES G., "Seasonality of Precipitation in the United States," *Annals*, Association of American Geographers, 60 (1970), 593–597.

NELSON, J. G., AND M. J. CHAMBERS, EDS., *Vegetation, Soils, and Wildlife*. Toronto: Methuen Publications, 1969.

NELSON, J. G., M. J. CHAMBERS, AND R. E. CHAMBERS, EDS., *Weather and Climate*. Toronto: Methuen Publications, 1970.

PIRKLE, E. C., AND W. H. YOHO, *Natural Regions of the United States* (4th ed.). Dubuque, IA: Kendall/Hunt Publishing Company, 1985.

RITCHIE, J. C., *Postglacial Vegetation of Canada*. New York: Cambridge University Press, 1988.

STORRIE, M. C., AND C. I. JACKSON, "Canadian Environments," *Geographical Review*, 62 (1972), 309–332.

THOMAS, M. K., "Canada's Climates Are Changing More Rapidly," *Canadian Geographical Journal*, 88 (May 1974), 32–39.

THORNBURY, WILLIAM D., *Regional Geomorphology of the United States*. New York: John Wiley & Sons, 1965.

TULLER, S. E., "What Are 'Standard' Seasons in Canada?" *Canadian Geographical Journal*, 90 (February 1975), 36–43.

VALE, THOMAS R., *Plants and People: Vegetation Change in North America*. Washington, D.C.: Association of American Geographers, 1982.

VISHER, S. S., *Climatic Atlas of the United States*. Cambridge: Harvard University Press, 1954.

ZWINGER, A. H., AND B. E. WILLARD, *Land Above the Trees: A Guide to American Alpine Tundra*. New York: Harper & Row, Publishers, 1972.

3
POPULATION

The areally vast and physically varied subcontinent of Anglo-America was sparsely populated until relatively recent times. Its aboriginal peoples were numerically few, geographically scattered, and culturally diverse. The penetration and settlement of the continent by Europeans signaled an almost total change in its human geography; the Native Americans were decimated and displaced and in a few short decades almost every vestige of their lifestyle was erased. The contemporary human geography of the United States and Canada, then, has been largely shaped almost completely by non-native people. Although the present population is mostly European in origin, it also contains other important elements. The saga of the blending of these elements is a chronicle of great complexity that is treated only briefly in this book. (Fig. 3-1).

MELTING POT OR POTPOURRI?

At the time of establishment of the first European settlements it is probable that the total population of the area now occupied by the United States and Canada did not exceed 5 million.[1] The population today is more than 55 times that figure.

It has often been stated that the United States and Canada are prime examples of a melting pot wherein people of diverse backgrounds are shaped in a common mold, thus becoming new citizens of new countries. There is some factual basis to this image, for tens of millions of immigrants settled in

[1] Estimates from various reputable scholars range from a low of 1 million to a high of 18 million. Hard evidence to support any of these estimates is difficult to come by.

the Anglo-American countries and, with the passage of time, many of their ethnic distinctions were blurred into obscurity. But the melting pot concept is only partially apt; in reality, the population of the two countries consists of an imperfect amalgam of diverse groups. In other words, the melting pot contains a lumpy stew.

By and large, the United States and Canada are countries strongly influenced by northwestern Europe. The historic ties to Great Britain are evident in cultural traits of language, religion, political system, technological achievement, and many other things. Yet, these ties were greatly modified by the importance of the French in Quebec and parts of the American South, the Spanish in southwestern United States, and the African in southeastern United States.

Each group of immigrants has been changed by the North American cultures, and, through time, has altered the North American landscape. New groups arrive and the exchange of cultural traits takes place. This sets into motion the acculturation process and a slightly different North America emerges. The imprint of Latin American and Asian immigrants is presently developing throughout the North America.

However long, complete, or painful the process of assimilation may be, the melting pot concept is probably less apt today than ever before. The disparities among Korean merchants, Hmong

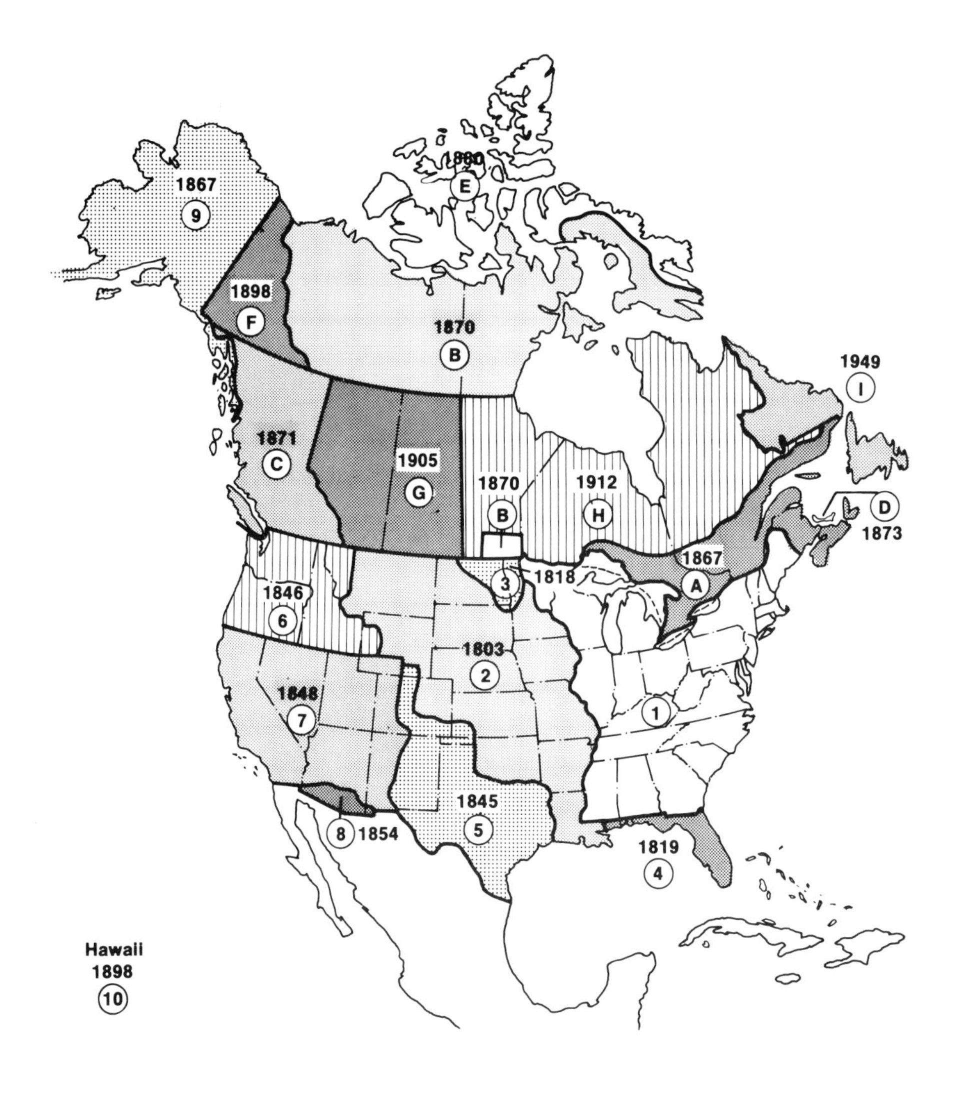

tribesmen, Ukrainian Jews, Nicaraguan refugees, French socialites, Filipino peasants, and Haitian boat people probably are greater than any country has ever confronted. Yet the pluralistic society of the United States and Canada seems to be more tolerant of diversity than once was the case. No longer is there a concerted effort to homogenize the population; that precept has been overtaken by reality.

THE PEOPLING OF ANGLO-AMERICA

The current population mix of the United States and Canada is continually being modified by the influx of immigrants and the egress of emigrants as well as by the rate of natural increase. It is a blend of varied origins and its diverse patterns through the years continue in dynamic flux today. There were five major original source regions for the peopling of these two countries: North America itself, Europe, Africa, Asia, and Latin America.

The Native Americans

The indigenous population of the continent at the time of European contact consisted of a great variety of tribes, thinly scattered. (Fig. 3-2). The largest concentrations were in the area presently called California, in the Southwest, adjacent to the Gulf Coast, and along the Atlantic coastal plain.

The contrast with the relatively large population of *Indians* in Latin America is very marked and apparently has long been a feature of the Americas. There is nothing in North American archaeological evidence to suggest any developed Indian civilizations comparable with those of Mexico, parts of Central America, or the central Andes. The thin population density probably reflected a reliance on simple hunting and fishing or primitive agricultural economies.

It is generally accepted that the "first Americans" were relatively recent (within the last 35,000 years or so) arrivals from Asia, having entered the Western Hemisphere via Alaska and diffusing widely throughout the New World. There were many physical and cultural variations among the Indians; their diversity was at least as great as that of the Europeans who later were to overwhelm them. Their only common physical attributes were black hair, brown eyes, and some shade of brown skin. There was much variety in both material and nonmaterial aspects of Indian culture and hundreds of mutually unintelligible dialects (divided by scholars into six major linguistic groups) were spoken.

FIGURE 3-1 Territorial evolution of the United States and Canada (simplified).

United States
1 Territories and claims of the original thirteen states
2 Louisiana purchase
3 Red River cession
4 Florida purchase
5 Texas annexation
6 Oregon compromise
7 Mexican cession
8 Gadsden purchase
9 Alaska purchase
10 Hawaii annexation

Canada
A Formation of the Dominion of Canada
B Acquisition of Northwest Territories and creation of Manitoba
C Creation of British Columbia
D Unification of Prince Edward Island
E Addition of the Arctic islands
F Formation of Yukon Territory
G Creation of Alberta and Saskatchewan
H Expansion of Manitoba, Ontario, and Quebec
I Annexation of Newfoundland

FIGURE 3-2 Generalized distribution of major tribal groupings in aboriginal Anglo-America.

Inuit[2] and *Aleuts* are thought to be Asian immigrants of much more recent vintage, having crossed from Siberia as little as 1000 to 3000 years ago. The Inuit spread all across the Arctic from the Bering Sea to Greenland, hunting and fishing for a living. The Aleuts, apparently branching off from the mainstream of Inuit life, developed a distinctive culture of their own, based on a fishing economy in the Aleutian Islands.

Initial relations between Europeans and Native Americans were not always unpleasant. Many tribes developed profitable trading patterns with the Europeans. Indeed, such tribes as the Iroquois in

[2] The term *Eskimo* was generally accepted until recently as the name for the indigenous people of the North American tundra. In the last few years many Eskimo people objected to this label as having been applied to them by outsiders. They have not reached total agreement on terminology, but the term "Inuit" is now the most widely approved appellation for the tundra dwellers of North America. It is universally used in Canada, and is slowly coming into favor in Alaska. In this book "Inuit" will be used in preference to "Eskimo." The matter of native names will be explored more fully in Chapter 20.

the East, Crees south of Hudson Bay, Comanches in the Great Plains, and Apaches in the Southwest acquired metal weapons and horses and established short-lived empires at the expense of less fortunate neighboring tribes.

Within a few years or a few decades, however, the insatiable appetite of the land-hungry Europeans led to inevitable conflict. Although some tribes were fierce, brave, and warlike, white settlers subdued and relocated them in relatively short order. The Native Americans were cruelly decimated by warfare, but introduced diseases were even more potent destroyers. For example, a smallpox epidemic wiped out more than 90 percent of the Mandans in the mid-1800s at the same time that cholera was eliminating 50 percent of the Kiowas and Comanches.[3] An "Indian Territory" was established between Texas and Kansas and nearly 100,000 Indians from east of the Mississippi were crowded onto the overutilized hunting grounds of the plains Indians.[4] Even this area was sporadically whittled down to size until it was finally thrown open to white settlement and became the state of Oklahoma.

Although there were occasional later outbursts by renegade Apaches in the Southwest, the last significant Indian conflict ended in the Wounded Knee massacre of 300 Sioux in 1890. By the turn of the century only several tens of thousands of Native Americans survived north of Mexico.

Presently, most Indians either live on reservations or have abandoned tribal life entirely. Native Americans have a rapid population growth rate but represent less than 0.8 percent of the total population of the two countries. Today some half a million Indians live on 280 reservations (including pueblos in New Mexico, colonies in Nevada, and rancherias in California) in the United States; the largest concentrations are in Arizona, New Mexico, and Oklahoma. Another 1 million or so U.S. Indians are living outside reservations; most are in urban areas and approximately one-third of the total is concentrated in Los Angeles.

In Canada there were about half a million Indians in 1990, of whom some 55 percent were status Indians.[5] Two-thirds of them live on nearly 2300 reserves that are widely scattered across the country. The province of Ontario contains about 100,000 Indians; British Columbia ranks second with some 75,000.

In addition, some 65,000 Inuit and Aleuts live in Alaska and approximately 28,000 Inuit reside in Canada.

The Pattern of European Immigration

North America has been by far the principal destination of European immigrants. It is estimated that between 1600 and 1990 nearly 85 million people left Europe to settle elsewhere, and more than four-fifths of them went to the United States and Canada. Although the social process of immigration has varied little during Anglo-American history, the pattern of immigration can be clarified if chronological divisions are established. These divisions are strongly affected by the impact of major immigration legislation that, through time, favored immigrants from particular parts of Europe, and for long periods inhibited immigration from Asia and Latin America.

Before 1815 This is the period of primary European colonization of North America, when nearly all migrants—with the important exception of early French settlers in Acadia (Nova Scotia) and Lower Canada (Quebec)—were of Anglo-Saxon ethnic stock and followers of reformed churches subsequent to the Reformation. They varied from High Church Anglican aristocracy to Puritan and Lutheran peasant dissenters. The majority were English, but there were important groups of Germans, Dutch, and folk from Ulster, Scotland, and Wales.

In total numbers they probably did not greatly exceed 1 million, but their descendants in all parts of the United States and Canada now form an important segment of the population. More significant

[3] William T. Hagan, *American Indians* (Chicago: University of Chicago Press, 1969), p. 94.

[4] The continual forced rearrangement of tribes caused great misery and frustration, which is indelibly echoed in the pointed question of the Sioux chief Spotted Tail: "Why does not the Great Father put his red children on wheels, so he can move them as he will?"

[5] Status Indians are those registered with the federal government as Indians according to the terms of the Indian Act. Nonstatus Indians are native people self-identified as Indians but not registered for the purposes of the Indian Act.

than their actual numbers is the effect that they had in establishing the foundations of the social and economic patterns of the continent for generations. Different regional frameworks were established in the Maritimes, Lower Canada, Upper Canada, New England, the Middle Colonies, and the Southern Colonies, many aspects of which are still evident today.

After the American Revolution immigration from England declined drastically and the most numerous of the newcomers, in most years, were Scotch-Irish. During the 1790s European immigration to Anglo-America often exceeded 10,000 per year, but the flow slowed appreciably during the Napoleonic Wars and almost ceased during the War of 1812.

1815 to 1860s This period represented the first of the great waves of immigration from Europe, a movement unprecedented in world history. Some 6 million immigrants were involved; the rate accelerated from about 200,000 in the 1820s to nearly 3 million in the decade of the 1850s. Compared to the resident population at the time, this influx was enormous.

These were people drawn by the economic opportunities available in a virgin land or driven from their homes by religious or political persecution or by revolution. Every country in Europe was represented, but the great majority were from the North Sea countries. More than half the migrants in this period came from the British Isles, especially Ireland. Germany was second only to Ireland as a source of immigrants; smaller numbers came from France, Switzerland, Norway, Sweden, and the Netherlands. Thus the immigrant mix of this period was less exclusively British and there also was a larger proportion of Catholics (principally Irish and German) than previously.

1860s to 1890s The second great wave of European immigration occupied the three decades following the American Civil War. In composition and character it was much like the first wave, but it was nearly twice as great in magnitude. Most migrants were still from northwestern Europe—Germany (250,000 German immigrants coming to the United States in 1882 constituted the largest number from any single country in any single year prior to the twentieth century) and the British Isles—and from Scandinavia, Switzerland, and Holland.

A significant feature of this period is that it was the only lengthy segment in Canadian history in which net out-migration occurred. Some 800,000 Canadian-born people moved south of the international border during this era.[6] A postwar wave of prosperity in the northern states coincided with an economic depression in Canada and drew a great many people, particularly French-Canadians, to U.S. cities. In addition, farmlands in the Midwest were available for settlement, a further inducement to Canadian immigration. In the later years of this period the Canadian prairies were opened for settlement; but even the flood of Swedes, Ukrainians, Mennonites, Finns, Hungarians, and other Europeans to the Canadian west did not compensate for the southward flow.

1890s to World War I Numerically this was the most significant period, for in these two decades an average of nearly 1 million immigrants per year came from Europe to Anglo-America. Perhaps of equal significance was the change in origin of the flow. From a predominantly northwest European source in the 1880s (87 percent in 1882, for example), the tide shifted to a predominantly southern and eastern European source after the turn of the century (81 percent in 1907).[7] Italy, Austria-Hungary (which included most of eastern Europe), and Russia were the homelands of the bulk of the immigrants,[8] but considerable numbers also came from such countries as Greece and Portugal. The change in Canadian immigration sources was almost as dramatic. The Canadian government had instituted a policy to induce and broaden the base of immigration in 1896 and during the succeeding decade immigrants to Canada from continental Europe were twice the number that came from the British Isles.

[6] T. R. Weir, "Population Changes in Canada, 1867–1967," *The Canadian Geographer,* 11 (1967), 201.

[7] Maldwyn Allen Jones, *American Immigration* (Chicago: University of Chicago Press, 1969), p. 179.

[8] Peak-year immigration to the United States was 286,000 from Italy in 1907; 340,000 from Austria-Hungary in 1907; and 291,000 from Russia in 1913.

This was the period of the melting pot in Anglo-America, but the assimilation of vast numbers of people of different cultures and languages was a slow and difficult process. By the time of World War I there was considerable agitation in the United States, and some in Canada, to slow the pace of immigration.

World War I to 1960s Immigration to Anglo-America decreased markedly during World War I and then began a rapid climb in 1919. Concern mounted that "the greatest social experiment in human history" had become the "melting-pot mistake." Immigrants had contributed to material progress by opening new lands and providing a cheap labor pool for factories, mines, forestry, and construction, but it was becoming clear that the rate of influx was getting out of hand. There was further concern, particularly in the United States, about the "mix" of the immigrants, with increasing demand for restricting immigration from the Latin and Slavic portions of Europe and from Asia.

As a result, legislation was enacted in the United States in the early 1920s to restrict the immigrant flow. Canada did not promulgate significant restrictions for another decade. The basis of the restrictions was the ethnic composition of the United States population prior to the war; thus the countries of northwestern Europe were given relatively large annual quotas (65,000 for the United Kingdom and 25,000 for Germany, for example) whereas other countries received much smaller allotments (6000 for Italy, 850 for Yugoslavia, 0 for Japan). Africa was ignored in the quota system.

The total volume of immigration to the United States declined rapidly (down from 800,000 in 1921 to less than 150,000 by the end of the decade) whereas Canadian immigration maintained a relatively steady pace (between 100,000 and 150,000 per year during most of the 1920s). The Great Depression of the 1930s and war in the early 1940s reduced immigration to insignificant totals for a decade and a half.

Although the quota system was still operating in the United States, after World War II there was a significant upturn in immigration to the United States and Canada. The heaviest influx to both countries was from Great Britain, Italy, and Germany. Bulking large in the postwar migrant flow was the admittance of displaced persons and refugees (without regard to quotas in the United States) in considerable numbers, which significantly increased the influx from eastern Europe until the Iron Curtain was closed.

Late 1960s to Present Drastically altered immigration laws in both countries (since 1965 in the United States and 1967 in Canada, with further significant changes in both countries in the late 1970s) ushered in a new period in the immigration history of North America. In both cases, immigration restrictions were greatly liberalized and "universalized." In the United States system immigration is on a modified preferential basis, up to a maximum of 20,000 per country and 270,000 for the world in a single year. Additional provisions (particularly for refugees and close relatives of U.S. citizens) allow this 270,000 total to be exceeded by well over 100 percent in a given year. For example, 643,000 immigrants were admitted in 1988.

Canada's system is more complicated and flexible, making no reference to national origin. Instead, it is based on annual evaluations of the nation's prevailing economic conditions and demographic "needs."

These revised immigration policies resulted in major changes in the flow of migrants to both Canada and the United States. In the first place, total immigration increased significantly in the latter and moved upward erratically in the former. During the 1980s legal immigration to the United States averaged 590,000 people annually, which is one-third higher than the average for the 1970s, and is substantially more than any time since the early 1920s (Fig. 3-3). In addition, refugees and asylees[9] were admitted at a rate of more than 90,000 each year, and illegal immigrants were estimated to total at least 500,000 annually. Thus for every recent year the total number of people taking up residence in the United States has approached the record high years

[9] Foreigners seeking to escape from oppression in their homelands are admitted to the United States under special legislation, especially the Refugee Act of 1980, that is separate from normal immigration quotas. This procedure has particularly benefitted people from Cuba, Southeast Asia, Haiti, and Hungary.

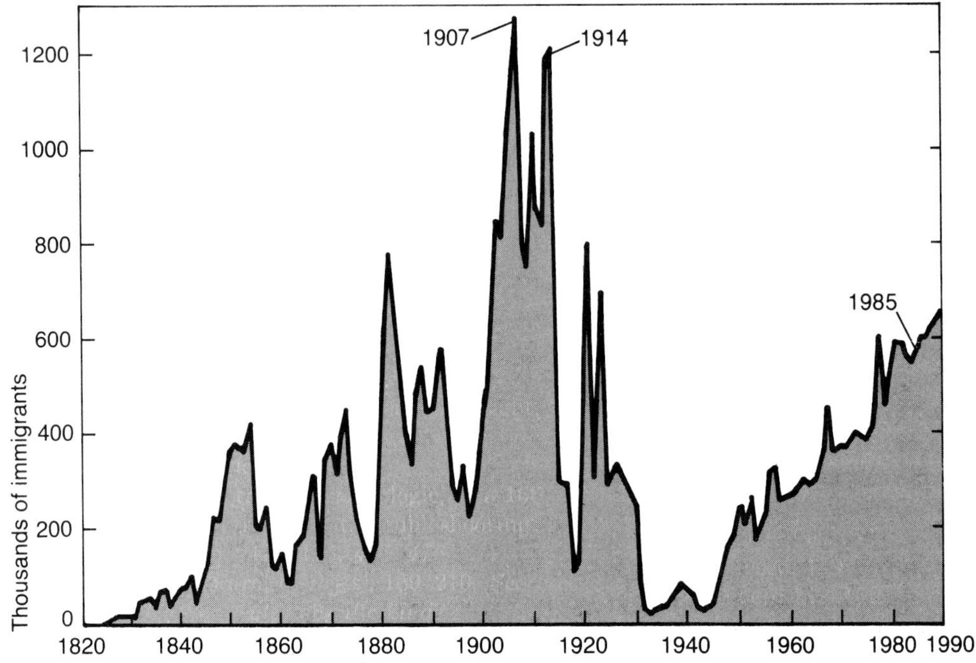

FIGURE 3-3 The historical sequence of legal immigration to the United States, 1820–1990 (based on data from U.S. Department of Justice, Immigration and Naturalization Service).

of history (1907 and 1914) when more than 1.2 million immigrants were admitted. Net immigration now accounts for nearly one-third of total annual population growth in the United States; in Canada the proportion is even higher (Table 3-1).

Another notable result of the changed immigration policies has been a marked decrease in the European component of total immigration (Fig. 3-4). Whereas 80 percent of the immigrants to Canada in the early 1960s were from Europe, they now comprise less than 40 percent of the total. The decrease was even more striking in the United States—from 50 percent European in the early 1960s to 10 percent European in the late 1980s. Although Britain is still the single leading source of immigrants to Canada, the great bulk of European immigrants to North America now emanates from Mediterranean countries, particularly Portugal, Italy, Greece, and Yugoslavia.

Legislation enacted by the U.S. Congress in 1991 produced a major restructuring of national immigration policies, permitting at least 45 percent

TABLE 3-1

Net immigration proportion of total U.S. population growth

Period	Net Immigration Component of Total Population Growth (Percent)
1901–10	39.6
1911–20	17.7
1921–30	15.0
1930–34	−0.1
1935–39	3.2
1940–44	7.4
1945–49	10.2
1950–54	10.6
1955–59	10.7
1960–64	12.5
1965–69	19.7
1970–74	16.2
1975–79	19.5
1980–85	28.4

[After Leon F. Bouvier and Robert W. Gardner, "Immigration to the U.S.: The Unfinished Story," *Population Bulletin*, 41 (November 1986), 20.]

more foreigners to enter the country in each of the next three years and about 35 percent more in every year thereafter. The previous immigration law, which had been in effect for a quarter of a century, emphasized family reunification to such an extent that about 85 percent of visas had gone to Asians and Latin Americans. The new policy does not diminish the absolute number of "family" visas. However, by raising overall numbers substantially, it also reserves a general quota for people from "low-immigration" regions, which is expected to provide a significant boost to immigration from Europe.

Involuntary Immigration: The African Source

Soon after the first white settlers occupied the land of coastal Virginia the problem of labor for clearing the forests, cultivating the soil, and harvesting the crops arose. Because land was free, few settlers would consider working for others when they could have their own land. The local Indian population was too small and resisted subjugation easily since they were far more at home in their environment than the white colonizers. At first, indentured workers—Englishmen who temporarily sold their services for the price of ship passage to the New World—met the labor requirements. But they did not prove satisfactory because they were not numerous enough and because it was difficult to keep them as workers once they reached the frontier.

To help solve the labor problem, Negro slaves were imported from West Africa, initially via already established slaving areas in the West Indies. The first African slaves in the English colonies landed at Jamestown in 1619. By the 1630s slaves were being brought in each year. They were not too popular at first and by the end of the century there were less than 10,000 in the tobacco colonies. But about then the British government began to restrict the sending of convicts (another source of labor) to

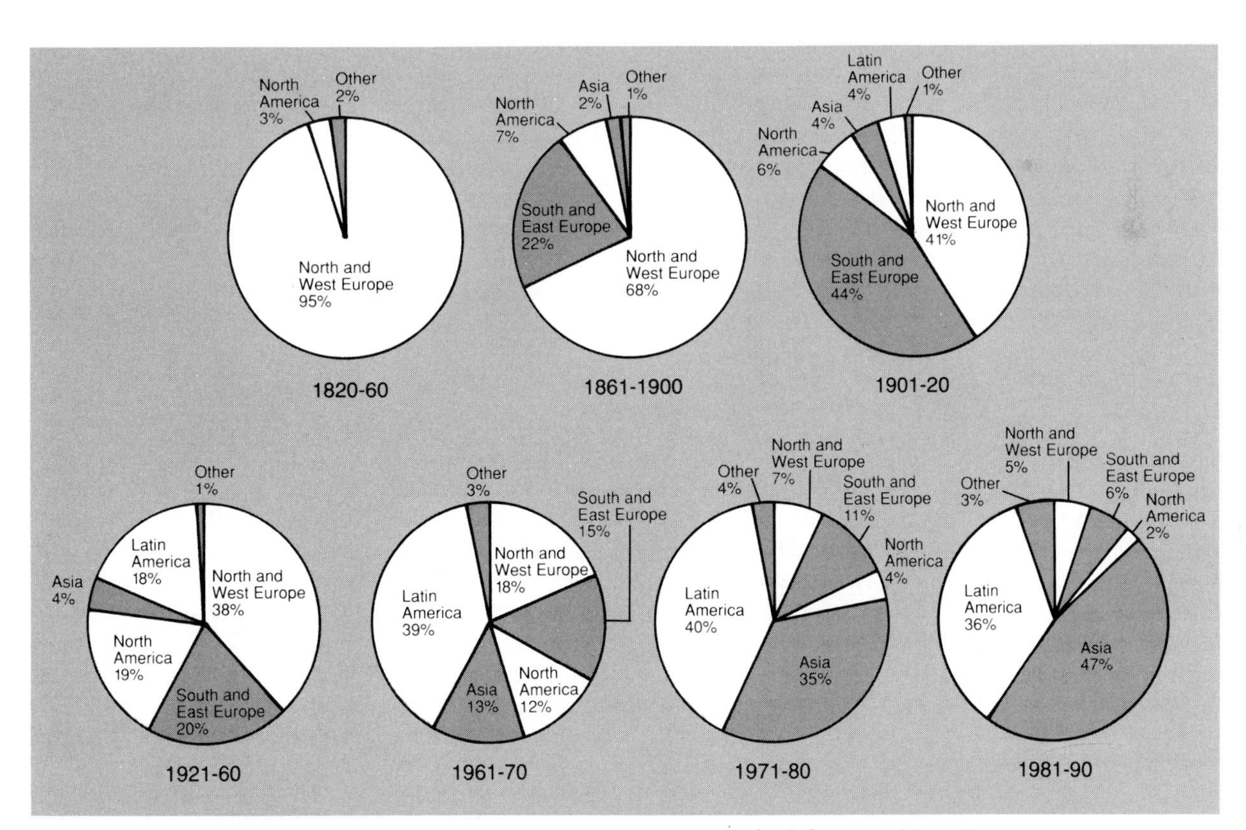

FIGURE 3-4 Sources of immigrants to the United States, 1820–1920 (based on data from U.S. Department of Justice, Immigration and Naturalization Service).

America and cotton became an important crop as a result of Eli Whitney's cotton gin. Thus slaveholding began in earnest and direct slave trade with the Guinea coast of West Africa developed.

On the eve of the American Revolution there were about half a million slaves in British North America, nine-tenths in the southern colonies. Despite their status, the slaves exerted a significant influence on southern life, from language and social customs to agricultural techniques. By 1808, when further importation of slaves was prohibited, nearly 20,000 were being brought into the United States every year.

Slavery was abolished at the time of the Civil War in the 1860s and few African immigrants have entered North America since then, until very recently. The descendants of those involuntary immigrants of the seventeenth to nineteenth centuries numbered nearly 9 million by 1900, however, and more than 30 million by 1990.

Throughout most of these centuries the black population was concentrated in the southeastern part of the United States. In this century, however, a major movement northward and westward occurred. As recently as 1950, blacks lived in approximately equal numbers in three situations: one-third in the rural South, one-third in the urban South, and one-third in cities of the North and West coasts. But the continued rural-to-urban movement has significantly decreased the proportion of nonurban blacks. Most are now city dwellers, with the largest concentrations in New York, Chicago, Los Angeles, Detroit, Philadelphia, and Washington. Ever since the mid-1970s there has been a major and continuing migration of blacks from the cities of the Northeast to the cities of the South.

Only a small number of blacks reside in Canada, largely in Toronto, Montreal, Halifax, and Windsor. Recent immigration from several black Caribbean nations—particularly Jamaica, but also Trinidad, Haiti, and Guyana—has made a large proportional increase, however.

The Irregular Sequence of Asian Immigration

Until the 1960s the immigration of people from Asia to North America was quite limited in number and extremely sporadic in occurrence. Occasionally Asian workers were imported or came of their own volition in large numbers, but throughout most of history their entry was either severely curtailed or totally prohibited.

The earliest Asian immigration was also the largest; some 300,000 Chinese came to California between 1850 and 1882, when the passage of the Chinese Exclusion Act halted the flow. During this same period several thousand Chinese entered British Columbia; originally they relocated from California and later directly from China. The great majority of these immigrants were males; after working in railway construction or gold mining for a while, many returned to China. Most remained on this continent, however, settling in "Chinatowns" in San Francisco, Los Angeles, Seattle, Vancouver, New York, and Chicago.

The first Japanese immigrants came to Hawaii as contract laborers to work on sugar-cane plantations in the 1880s. After Hawaii became a territory of the United States in 1898, the Japanese were free to come directly to the mainland and several tens of thousands did so during the early years of this century. A few thousand Japanese also immigrated to British Columbia.

The quota system stopped Japanese immigration to the United States after World War I, and Canada effectively stopped Chinese immigration by legislation in 1923.[10]

The new immigration policies of the 1960s once again opened the door to Asian settlers and resulted in a veritable flood of Asian immigrants. In the first few years under the new laws more Chinese (from Taiwan and Hong Kong) and Filipino immigrants entered the United States than citizens of any other Eastern Hemisphere countries; there has also been a great upsurge of immigrants from Southeast Asia, Korea, and India (Fig. 3-5). Overall, Asians constituted nearly half of total immigrants to the United States in the 1980s compared with only one-

[10] Some 4300 Chinese had immigrated to Canada in 1919, a larger total than from any other countries except the United Kingdom and the United States. The Chinese Immigration Act of 1923 resulted in such restriction that an average of only one Chinese immigrant per year entered Canada during the next two decades. See W. H. Agnew, "The Canadian Mosaic," in *Canada One Hundred: 1867–1967*, ed. Dominion Bureau of Statistics (Ottawa: Queen's Printer, 1967), p. 89.

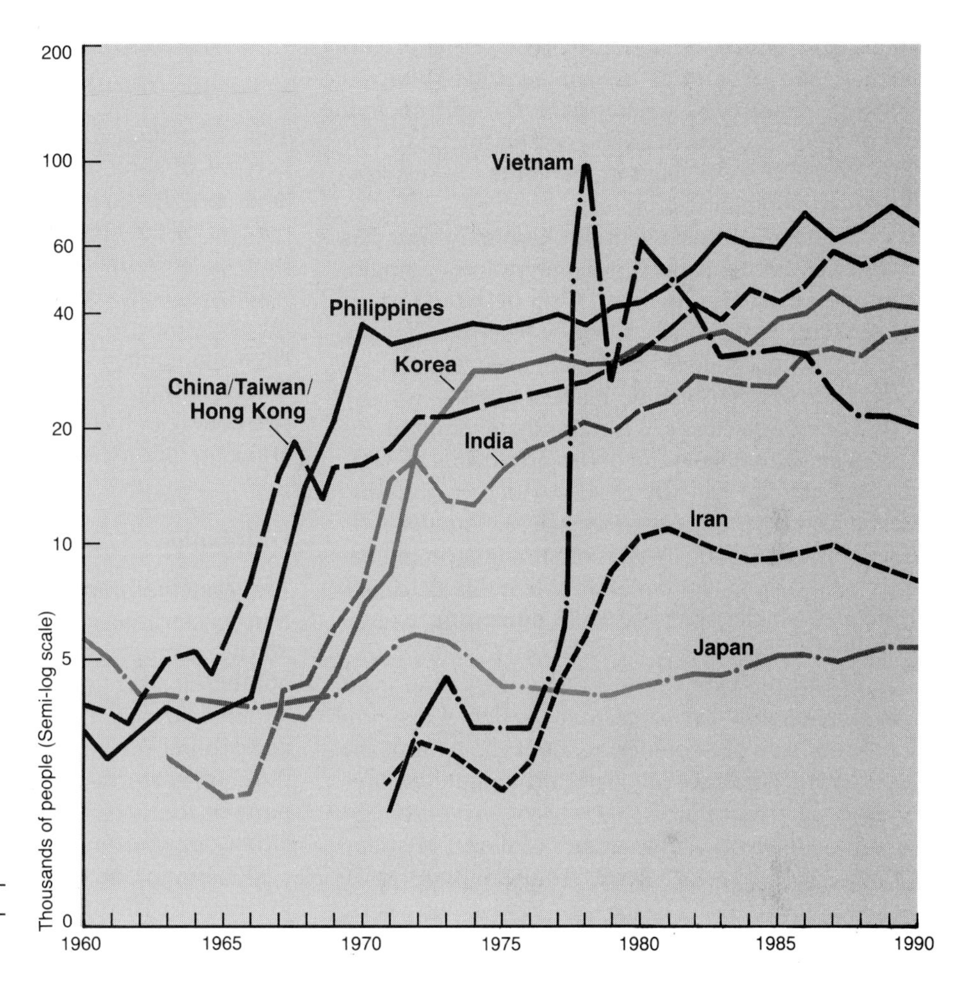

FIGURE 3-5 Principal components of recent Asian immigration to the United States.

fifteenth in the mid-1960s. The Asian proportion in Canada is similar, with notable flows from Vietnam, Hong Kong, India, and the Philippines.

The great majority of all Asian immigrants have settled on the West Coast, particularly in California. Most are urban dwellers, joining the swelling Chinese, Japanese, Filipino, and Korean minorities of Los Angeles, San Francisco, Seattle, and Vancouver, with some spillover into Arizona and Alberta and conspicuous nodes in New York, Chicago, and Toronto.

It should be noted that Hawaii has long had a significant Asian population. The majority of the contemporary populace of that state is of Asian extraction, particularly Japanese and Chinese but with large numbers of Filipinos and Koreans.

Latin American Immigrants

The ancestors of Mexican-Americans were the first Europeans to settle what is now the southwestern United States. A majority of their descendants still reside in the borderlands from Texas to California. Between 1598 and 1821, parts of the states of California, Arizona, New Mexico, Colorado, and Texas were settled by Spanish-speaking peoples from New Spain, which is present-day Mexico. After 1821, portions of this region became part of the Republic of Mexico, and each of the settled areas developed its own regional identity. The transformation of these areas from Spanish outposts to Mexican provinces to Mexican-American subregions represents the planting, germination, and rooting of Mexican-

American cultures in the United States. The origins of many of the features associated with Mexican-American culture, from place names to architecture and numerous social customs, can be traced to this early period of settlement in the United States borderlands.

Throughout the twentieth century there has been a fluctuating pattern of immigration to Anglo-America from Latin America. With the exception of the massive influx from Cuba, it was primarily a move toward economic betterment and was often on a short-term rather than permanent basis.

Latin Americans were totally exempted (as were Canadians) from the quota provisions of U.S. immigration laws in the 1920s, which meant that they essentially enjoyed unrestricted immigration most of the time. Under current immigration regulations there is a maximum allowable influx of 20,000 from any single country, with numerous exceptions.

By far the largest, most continuous, and most conspicuous flow has been from Mexico. Legal immigration from Mexico began about the turn of the century and several million legal immigrants have entered since that time.

In addition, a great number of illegal Mexican aliens ("guesstimates" range from 2 million to 10 million) came into the United States in recent years. Most Mexican immigrants have settled in the southwestern border states, Texas to California, plus Colorado, although there is a significant concentration in Chicago.

Immigration from the West Indies is also of long-standing duration. Puerto Rico has been the principal source because of its political affiliation with the United States. In most recent years Jamaica has furnished more immigrants to the United States than any other country except the Philippines, Italy, Greece, and Hong Kong–Taiwan. More than three-quarters of a million Cubans entered the United States in the last two decades (over 150,000 in 1980 alone), mostly with refugee status. Several thousand Haitian "refugees" also came to the United States in the early 1980s. Despite strong official efforts at dispersal, some 75 percent of these Cuban and Haitian immigrants settled in southern Florida, chiefly in the Miami area.

THE CONTEMPORARY POPULATION OF ANGLO-AMERICA

The 275 million Anglo-Americans represent about 6 percent of total world population. They occupy a land area of some 8.3 million square miles, or 14.5 percent of the land area of the planet. This population is very unequally divided between the two countries, with about ten people in the United States for every one person in Canada. The 10 to 1 ratio has been maintained throughout the past century and indicates that the rate of population growth in the two countries has been approximately equal for several generations.

Distribution

The principal population concentrations are in the northeastern quarter of the United States (Fig. 3-6) and adjacent parts of Ontario and Quebec. The most notable clusters are in the Megalopolis Region of the Atlantic seaboard, around the shores of Lake Erie, and around the southern end of Lake Michigan. In the southeastern quarter of the United States and in parts of southern Ontario there is a moderate population density, fairly evenly distributed except for agglomeration around urban nodes. In the Atlantic Provinces an irregular pattern of moderate density alternates with patches of little population. In the central plains and prairies there is a generally decreasing density from east to west, with obvious concentrations along the major river valleys and at the eastern edge of the Rocky Mountains. The Intermontane West is sparsely populated, with a few minor concentrations around cities.

The Pacific Coast has moderate-to-heavy population density in the valleys, with conspicuous agglomerations around the six principal urban areas. The vast expanses of central and northern Canada and most of Alaska are almost unpopulated; indeed, 72 percent of Canada's population lives within 150 miles (240 km) of the international border.

As it has for some years, the greatest *absolute* population growth continues to be in the states of California, Texas, and Florida and in the provinces of Ontario, Alberta, and British Columbia. The fastest *rates* of increase are in Alaska, the southwestern

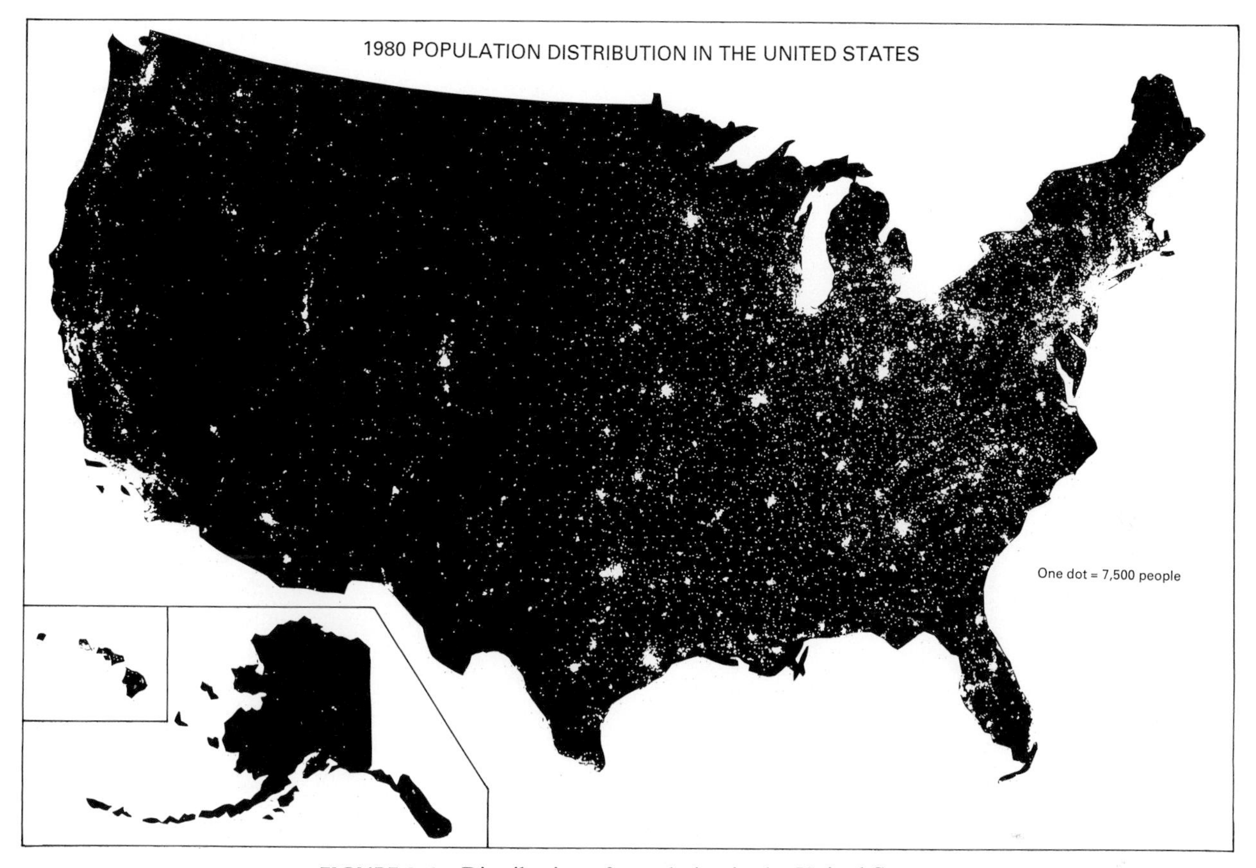

FIGURE 3-6 Distribution of population in the United States.

states (especially Nevada and Arizona), Florida, and Texas, and the provinces of Alberta, Ontario, and British Columbia.

The population shift from Snowbelt to Sunbelt states is absolutely clear-cut. For several years every state from Virginia to California grew more rapidly than the national average whereas no state in the north-central or northeastern part of the country (except New Hampshire) achieved that distinction.

During the 1980s many rural areas lost population. More than half of the rural counties across the nation experienced population decreases; this contrasts sharply with the decade of the 1970s, when less than 20 percent of American rural counties declined in population. Depopulation was most severe in the agricultural states of the Midwest and Great Plains; in Iowa, for example, all but six counties lost population during the decade. Energy-producing states also experienced either decline or very slow growth, as did some northeastern states that endured a continuing dwindling of manufacturing. Nine states (Illinois, Iowa, Louisiana, North Dakota, Michigan, Ohio, Pennsylvania, West Virginia, and Wyoming) decreased in population during the decade.

Ethnic Components

Most of the people in both countries are of northern European background, but there are many significant minorities, and the rapid development of a pluralistic society in both Canada and the United States is striking. The most prominent minority in Canada is the French, who live mostly in Quebec and in adjacent portions of New Brunswick and Ontario (Fig. 3-7). In recognition of the long-standing histori-

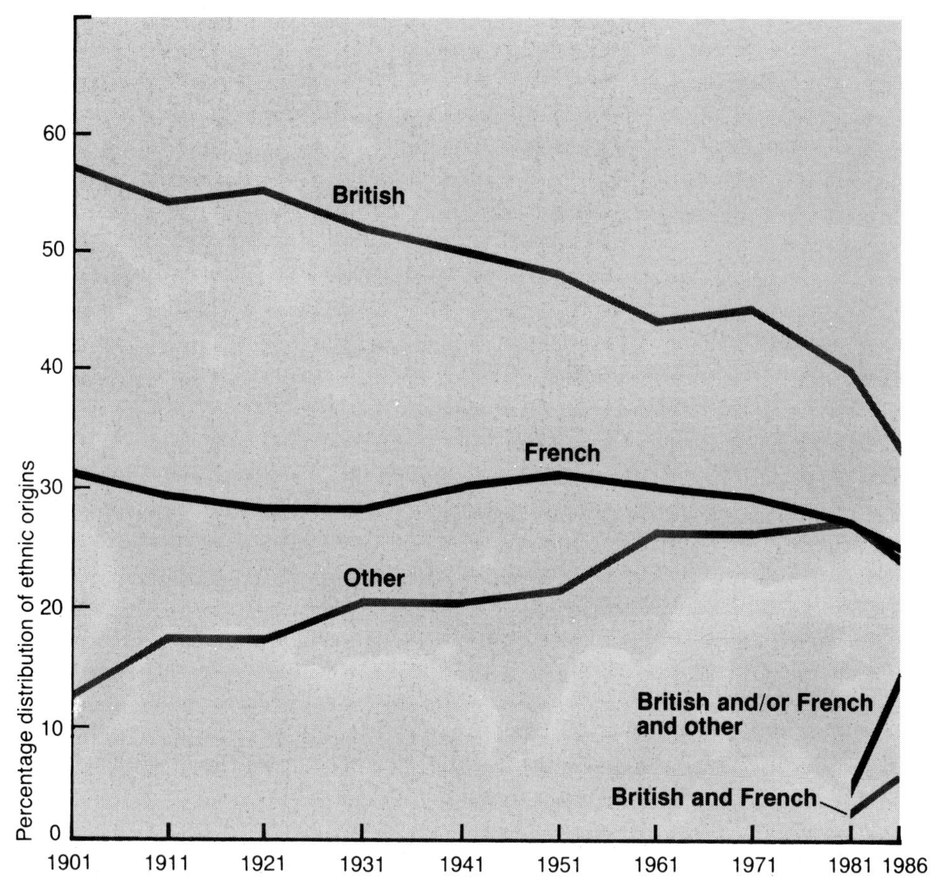

FIGURE 3-7 Changing Canadian ethnicity. Multiple responses were permitted beginning in 1981. (After Pamela Margaret White, *Ethnic Diversity in Canada,* 1986 Census of Canada; and recent estimates by Statistics Canada.)

cal significance of the Franco-Canadians, Canada is officially a bicultural nation. The relative position of French-Canadians, however, is on the wane. For the first century of Canada's existence as a nation, French-Canadians constituted a steady 30 percent of the national population total, but in the last few years this proportion has decreased noticeably. Indeed, both British and French ethnicity is on a declining trend. By the early 1990s only one-third of all Canadians were of distinctly British origin, less than one-fourth were of distinctly French origin, and more than one-fourth were of *neither* British nor French origin (the remaining fraction of the populace was of "multiple" origin). Canadian ethnic diversity varies considerably by region: Newfoundland is the most dominantly British (89 percent), with Prince Edward Island and Nova Scotia not far behind; Quebec is most dominantly French (78 percent); the Northwest Territories, with its sizable population of native Canadians, has the highest proportion of people with non-British, non-French origins (64 percent).

Although people of German, Italian, Dutch, Scandinavian, Polish, Russian, and other European nationalities are significant components of the population in both countries, for the most part they are relatively inconspicuous as minorities. They are sometimes prominently associated with particular areas, such as Scandinavians in Minnesota and Wisconsin or Ukrainians in the Canadian prairies, but a detailed consideration of their distribution is beyond the scope of this general treatment. Some hints can be derived from Table 3-2, however, which lists the specific states with the largest populations of the principal ancestral groups. New York, the traditional "melting pot" state and port of entry for European immigrants, is seen to be especially prominent with regard to people of eastern European, southern European, and West Indian extraction. California, with its enormous total population, mirrors the ethnic diversity of the nation, with particular concentrations of people of Asian and Hispanic origin.

The group defined by the U.S. Census Bureau as "Spanish origin" is a European ancestry minority that is sufficiently prominent in terms of population concentration and distinctive cultural attributes to merit particular attention. The number of people involved is unknown, for the Census Bureau readily admits that this group was underenumerated in official counts. As of 1990, a conservative population estimate is about 20 million citizens and legal aliens of Spanish origin, with several million others who lead a surreptitious existence as illegal aliens. Of this total, about two-thirds of the legal residents and well over half of the illegal ones are of Mexican origin. About nine out of every ten Hispanics of Mexican origin live in the five southwestern states of California, Arizona, New Mexico, Colorado, and Texas.

The second largest component of the Hispanic minority consists of emigrants from Puerto Rico, of whom there are some 2 million in the United States. Puerto Ricans primarily inhabit the metropolises of the Northeast, with the greatest concentration by far

TABLE 3-2

States with largest population of specific ancestry groups, as self-identified in 1980 census (in percentages)

California		***New York***		***Pennsylvania***	
English	10%	Afro-American	7%	Welsh	13%
German	9	Italian	23	Slovak	34
Irish	9	Polish	14	Ukrainian	20
French	10	Russian	24	Croatian	21
Scottish	16	Hungarian	14	Serbian	20
Mexican	44	Puerto Rican	49		
American Indian	11	Greek	17	***Michigan***	
Dutch	10	Austrian	21		
Swedish	13	Rumanian	23	Finnish	18%
Spanish/Hispanic	20	Indian	18	Belgian	17
Danish	16	Jamaican	54		
Portuguese	31	Dominican	78	***Massachusetts***	
Swiss	13	Colombian	34		
Chinese	38	Syrian	16	Canadian	17%
Filipino	44				
Korean	27	***Illinois***		***Minnesota***	
Lebanese	11				
Vietnamese	32	Czech	11%	Norwegian	21%
Armenian	38	Lithuanian	15		
Iranian	35			***Florida***	
		Ohio			
Hawaii				Cuban	57%
		Slovene	45%		
Hawaiian	67%				

Notes: All specific ancestry groups proclaimed by at least 100,000 persons are listed. Under each state the ranking is in order of number of people who proclaimed each ancestry group. Percentage figure refers to the proportion of total U.S. population of that ancestry group living in the listed state.

FIGURE 3-8 Ethnic neighborhoods are notable in many North American cities. This is the well-known sidewalk market along 125th Street in New York City's Spanish Harlem, where the bulk of the population is of Puerto Rican origin (TLM photo).

in New York City (Fig. 3-8). There are perhaps 1 million people of Cuban extraction in the United States, chiefly in southern Florida.

Although most Hispanic-Americans are fluent in English, the majority consider Spanish to be their mother tongue. Most are at least nominally Roman Catholic and they tend to be disadvantaged socially, economically, and politically, although they have developed considerable political leverage in south Texas, New Mexico, southern Arizona, and southern Florida.

Blacks are only a tiny fraction of the Canadian population, but they make up more than 12 percent of the populace of the United States. As in the past, blacks are more numerous in southern states than elsewhere; more than half the nation's blacks live in the South. Morever, the proportion is now increasing. After the decades of massive out-migration to the North and West, a reverse trend has developed since the mid-1970s. More blacks are now moving to the South from the West, and particularly from the North, than are leaving the South.

The black population has become highly urbanized. The long-continuing rural-to-urban flow has slowed, but only because there are not many blacks remaining in rural areas. Washington, Detroit, Baltimore, New Orleans, Atlanta, Newark, Birmingham, and Richmond now have a black majority population within their political limits and in a dozen other large cities the black proportion is above 40 percent. In absolute terms, New York City contains far more blacks than any other city: 1.9 million, or 25 percent of its total population. In most instances, the black population is concentrated in central areas and is a very minor element in the suburbs. More than 25 percent of the total population of *all* central cities in the United States is black whereas less than 5 percent of the suburban population is black.

Other ethnic groups are prominent in the contemporary population only in the major West Coast cities and a few western farming areas and in the largest eastern cities. The total population of Asian origin in Anglo-America is estimated at about 4 million, with Japanese, Chinese, and Filipinos dominating in roughly equivalent total numbers. Hawaii has long been the major North American domicile of people of Asian extraction, but the great flood of immigrants since the late 1960s has given California the principal concentrations of all significant Asian minorities except Japanese, who are still slightly more numerous in Hawaii.

The changing ethnicity of the populace is highlighted by data from the 1980s (Fig. 3-9). During that decade in the United States, the population of Asian extraction grew by about 87 percent, with about two-thirds of the increase due to immigration. During that same period the Hispanic population (not counting undocumented aliens) expanded by 42 percent, about half from immigration and half from natural increase. The Native American population grew by about 22 percent, the black populace by 16 percent, and the non-Hispanic white population by 7.5 percent. Even so, about three-quarters of the nation's population consists of whites of non-Hispanic origin.

Related Cultural Characteristics

The cultural geography of Anglo-America is diverse, complex, and imperfectly understood. No attempt will be made to explore it in systematic fashion here, although the accompanying map of culture areas (Fig. 3-10) is offered as a summary.

It is, however, appropriate to discuss a few details about certain nonmaterial culture elements that are closely associated with population: language, religion, and politics.

In terms of language, the United States is one of the least complex areas in the world. Well over 90 percent of the population is fluent in English and no sectional dialect is different enough to cause any problems in intelligibility. However, pluralism of the population has resulted in bilingualism with significantly large minority populations that have a high growth rate. These conditions are effecting changes in various urban areas. In areas near the Mexican border a large proportion of the people uses Spanish as either a primary or secondary language. In southern Louisiana, French is important, normally in a creolized form, but only in remote rural areas is English not dominant.

On some of the larger Indian reservations, particularly in Arizona, Indian languages are in everyday use. The larger metropolitan areas have various ethnic enclaves where some non-English language dominates. This is most conspicuous in Spanish Harlem and other parts of New York City where Puerto Ricans and Dominicans have settled, in several big-city Chinatowns, in areas of Italian and Pol-

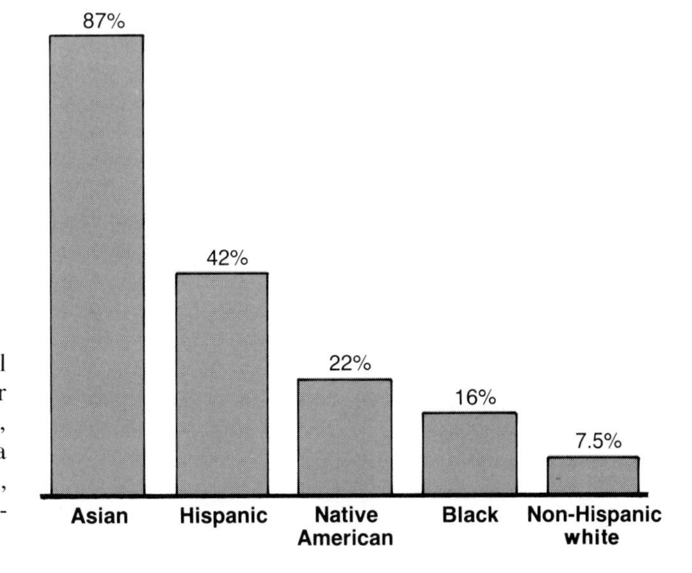

FIGURE 3-9 Proportional population increase by major ethnic groups, United States, 1980–1990 (based on data from Department of Justice, Immigration and Naturalization Service).

ish settlements, in the Cuban settlements of Florida, and in Southeast Asian communities in large western cities.

The linguistic pattern in Canada is much more heterogeneous; English and French are dominant, although many people prefer another language. Census statistics show that English is the mother tongue for 62 percent of the population but is the principal language for 66 percent. French is the mother tongue for 25 percent (but the principal language for only 23 percent), whereas the mother tongue of 13 percent of the population is neither English nor French (Ital-

A CLOSER LOOK The Hispano Homeland

In the United States, people who speak Spanish are culturally diverse, just as people who speak English are not all the same. In the Spanish and Mexican periods, for example, there were three subcultures in the present-day Southwest. Two of them, Tejanos and Californios, in the American period were absorbed into the Mexican immigrant subculture, whose large region today stretches from Texas to California. On the other hand, Spanish-Americans or Hispanos, the oldest and largest of the three colonial subcultures, were not engulfed by Mexican immigrants. Their New Mexico-centered homeland instead is the product of nearly four centuries of interplay with Pueblo Indians, Plains Indians, and Anglos. This is what happened:

In 1598, Spaniards encroached on the Pueblo Indians of the upper Rio Grande basin and created a missionary frontier whose outer boundary by 1680 was marked by Pueblo villages containing Spanish friars (Fig. 3-a). Santa Fe, the only authorized Spanish community, was an enclave in this Pueblo realm. After the Pueblo Revolt of 1680 Hispanos reoccupied the area, founded new settlements, and by 1790 had transformed the old Pueblo realm to their own colonist province, whose expansion had all the while been effectively contained by Plains Indians. The Pueblo villages were now enclaves in an Hispano realm. After 1790, relatively peaceful times allowed Hispano sheepmen to expand in every direction, pushing the region's outer limits to their greatest areal extent by about 1900. Anglos blunted this expansion. Except in Colorado, areal gains and losses around the periphery of the region if the twentieth century have been minor, as shown by the homeland's outer boundary in 1980.

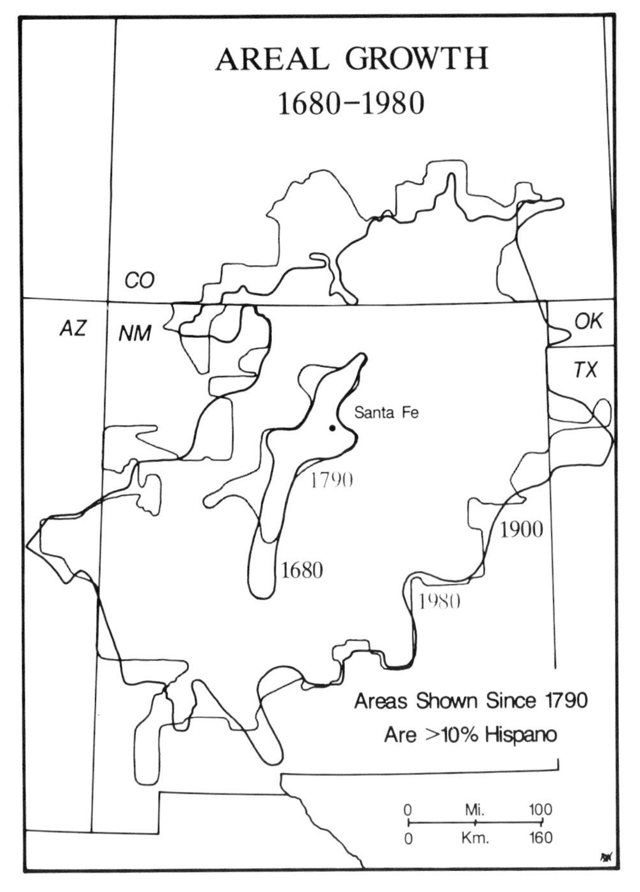

FIGURE 3-a

Anglos played the lead role in the interplay that brought about the relative decline in Hispano dominance. Anglo traders and trappers arrived after 1821, then came soldiers in ian, German, Chinese, and Ukrainian are the leaders). Canada is officially a bilingual country, and this cultural dualism has posed a major stumbling block in any attempt at recognizing a national identity.

Another dimension of Canadian languages was shown in a study that identified 53 distinct indigenous (Indian and Inuit) languages that are still spoken.[11] The total number of speakers of indigenous languages, however, was estimated at 154,000,

[11] *Indigenous Languages of Canada*, Commissioner of Official Languages, Ottawa, 1983.

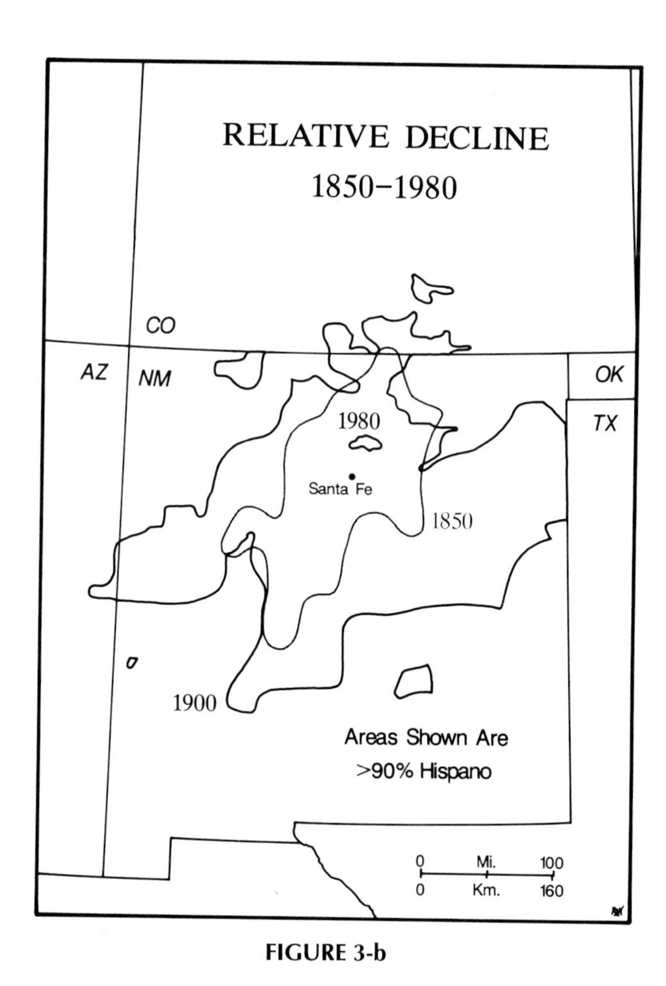

FIGURE 3-b

1846, and by 1850 modest numbers of each were found in the larger Hispanos communities, yet in the region as a whole Hispanos constituted more than 90 percent of the population (Fig. 3-b). After mid-century the number and variety of Anglo intruders increased, especially in the periphery of the Homeland where, by 1900, Anglos were a majority. Yet the interior half of the region, as shown in Fig. 3-b, was still more than 90 percent Hispano in 1900. After 1900, Hispanos were increasingly pulled by Anglo economic opportunity from the interior to the periphery of their region, which to a degree evened out their numbers and reduced their percentages in the interior. And after the Great Depression, out of desperation, Hispanos left their villages and relocated in urban centers, which further lowered their percentages in the countryside. Anglo intrusion meanwhile continued to be massive, and by 1980 the exceptionally high Hispano percentages of 1900 were everywhere eroded except in the one census county division shown.

Many Hispanos left their homeland for states like California after about 1940. In 1980, however, some 400,000 were still living in the region, a number that represented fully 20 percent of the total population. Like most Mexican- , Cuban- , and Puerto Rican-origin Hispanic-Americans, Hispanos are bilingual and bicultural. They are separated from the rest by peculiar cultural attributes including an archaic Spanish language and indigenous folklore forms, and by their strong Spanish ethnic identity. The legacy of all this, then, is a unique people—our only surviving Spanish colonial subculture—whose homeland is both a distinctive part of America and an enclave embedded in the Hispanic Southwest.

Professor Richard L. Nostrand
University of Oklahoma
Norman

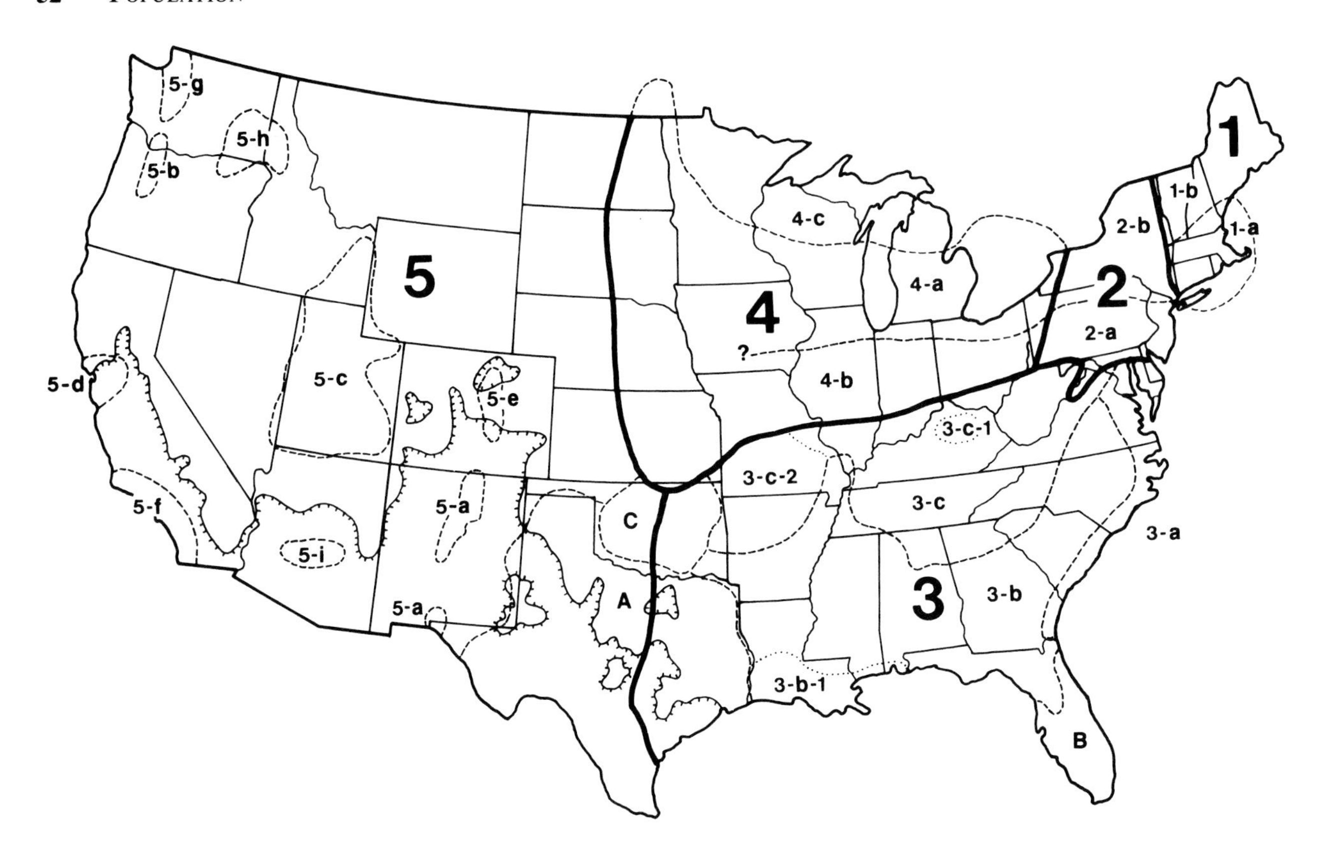

Northern boundary of significant Hispanic-American settlement (after Nostrand)

FIGURE 3-10 Zelinsky's view of the principal culture areas of the United States.

1 New England
- 1–a Nuclear New England
- 1–b Northern New England

2 The Midland
- 2–a Pennsylvanian Region
- 2–b New York Region, or New England Extended

3 The South
- 3–a Early British Colonial South
- 3–b Lowland, or Deep South
  - 3–b–1 French Louisiana
- 3–c Upland South
  - 3–c–1 The Bluegrass
  - 3–c–2 The Ozarks

4 The Middle West
- 4–a Upper Middle West
- 4–b Lower Middle West
- 4–c Cutover Area

5 The West
- 5–a Upper Rio Grande Valley
- 5–b Williamette Valley
- 5–c Mormon Region
- 5–d Central California
- 5–e Colorado Piedmont
- 5–f Southern California
- 5–g Puget Sound
- 5–h Inland Empire
- 5–i Central Arizona

Regions of uncertain status of affiliation are
- A Texas,
- B Peninsular Florida
- C Oklahoma

(After Wilbur Zelinsky, *The Cultural Geography of the United States,* © 1973, pp. 118–119. Reprinted by permission of Prentice-Hall, Inc.)

which is about the same as the number of Canadians who claim Dutch as a mother tongue. Moreover, only three indigenous languages—Cree, Ojibwa, and Inuktitut (the principal Inuit tongue)—are considered to have enough speakers so that the languages are not in danger of dying out.

Religious affiliation is more varied (Fig. 3-11). In the United States approximately 67 percent of the population professes some branch of Protestantism, with Baptists and Methodists as the largest denominations. The southeastern states, the Midwest, and much of the West are dominantly Protestant. Roman Catholics constitute about 25 percent of the total population, with particular concentrations in the Southwest, southern Louisiana, parts of New England, and many larger cities of the Northeast. Only about 3 percent of the population is Jewish and it is distinctly concentrated in the large cities; half the nation's Jews live in New York City, with other major concentrations in Los Angeles, Philadelphia, Chicago, and Boston.

Some 48 percent of the Canadian people profess Roman Catholicism; Catholic strongholds are in Quebec and the Atlantic Provinces. Most of the remainder of the population is Protestant, of which the two dominant denominations are United Church of Canada and Anglican.

Political affiliations have changed markedly in the past three decades in both countries. Two parties, Democratic and Republican, dominate the political scene in the United States at federal, state, and local levels. Democratic strongholds have traditionally been in the southeastern states, in the big cities of the Northeast, and among certain minority

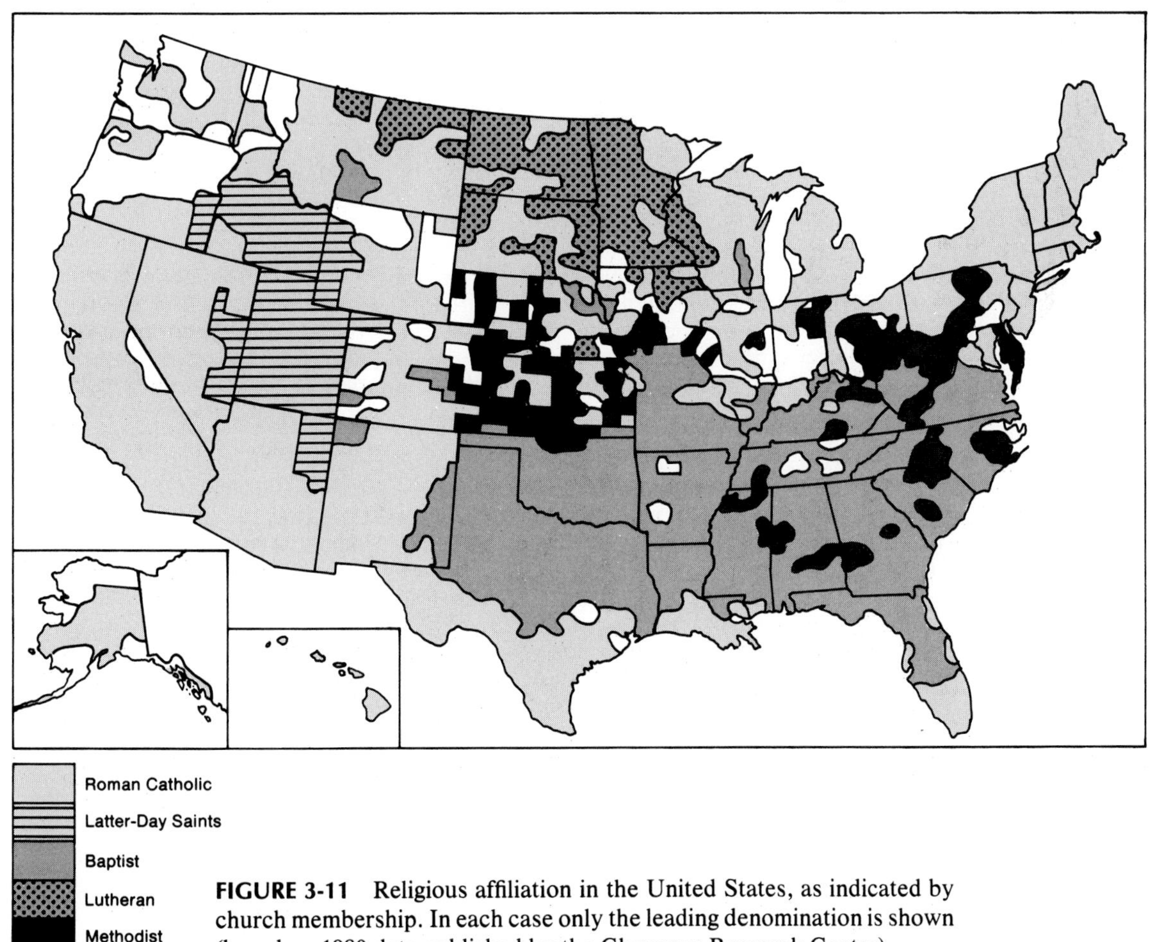

FIGURE 3-11 Religious affiliation in the United States, as indicated by church membership. In each case only the leading denomination is shown (based on 1980 data published by the Glenmary Research Center).

A CLOSER LOOK Season of Marriage in North America

If marriage still remains a well-nigh universal institution in contemporary America, it is a topic to which few geographers have devoted serious thought. (And even fewer have studied divorce.) One reason may be the poor quality and limited quantity of information for small areas, or even states, as to who marries whom when and where. But when we do plot the existing material on maps and graphs for just the single question of marriage dates—the choice of month or day of the week for the ceremony—some unexpectedly fascinating results show up. As happens with other private decisions where we supposedly enjoy total freedom of choice, as in the selection of names for children, candidates for political office, specific types or styles of food, drink, houses, clothing, and hairdos, certain grandly enveloping, but subtle, forces of which we are seldom aware seem to nudge us along certain directions.

It is obvious enough that certain months (and days of the week) are much more popular than others for scheduling weddings. After analyzing all the available 2262 "state-years" (figures on weddings per month for specific years in a given state) for the 61 U.S. and Canadian states and provinces during the period 1844 to 1974, a sample of roughly 30 percent of the possible total, a sequence of six distinct annual cycles materializes. Type 1, the earliest, has a late fall maximum (i.e., the harvest season, in which October, November, and December experience by far the greatest number of marriages). Evidently it was totally dominant throughout North America from colonial times until the 1890s; but by the 1960s this pattern had become extinct (except perhaps among such traditional rural groups as the Hutterites and Amish). Monday through Thursday was the preferred period within the week, but Thanksgiving Day, Christmas, New Year's Day, Easter, and July 4 were heavily favored whatever the day of the week. This sort of wedding calendar made good sense when American society was still predominantly agrarian, churchgoing, and uninhibitedly patriotic.

The other five types are really all variants of a single master pattern: one in which weddings are most frequent during the warmer months, from June through September, with an overwhelming concentration of events during weekends, from Friday through Sunday. Indeed, in some states in some years there may be more ceremonies on a single June Saturday than for the entire month of February. National and church holidays are now avoided, but Valentine's Day is the most popular February date. The most prevalent of these summer-maximum types from the 1890s to the present has been one where June is the leading month. Why June? There is no plausible explanation. Etiquette books and shops specializing in bridal merchandise have never promoted any particular month. Although other advanced countries also report a heaping of marriages during the summer season, only the Dutch and Swedes show a special predilection for June. In addition to the pattern featuring a single June maximum, we find two others in which there is marked secondary peaking in September or August. In the final pair of types, June yields priority, first to September during the period 1910 to 1939 and, then, more recently, to August.

What makes all this geographically exciting is the way these various patterns have moved back and forth across the map with the passage of time. The summer-maximum pattern first emerged in the 1890s precisely where one would have predicted on theoretical grounds: in southern New England, the states that were then the most urbanized, industrialized, and socially modernized. Then, following those general contours of the historical geography of modernization within the continent we can document in a variety of ways, the summer pattern moved outward into the central states and provinces and beyond, and ultimately southward, until finally even the states of the deepest South succumbed.

The accompanying map (Fig. 3-c) documents the situation, in midstream, so to speak, at a time (1920) when the diffusion of the summer maximum pattern had already run its course throughout the northeastern states and the eastern provinces of Canada, along with much of the central and western portions of the continent. We see the new marriage calendar poised on the threshold of the American South ready to complete its sweep of North America.

One must assume some cause-and-effect relationship between general socioeconomic developmental level and the onset of a modern wedding calendar. But it has not been a simple, automatic case, and certainly the effect was not immediate. In the case of Massachusetts, Rhode Island, and Connecticut, the reversal of preferred seasons took place some decades *after* these states had convincingly crossed the threshold into advanced socioeconomic status. Elsewhere as well one can observe a "cultural lag" of some years between the

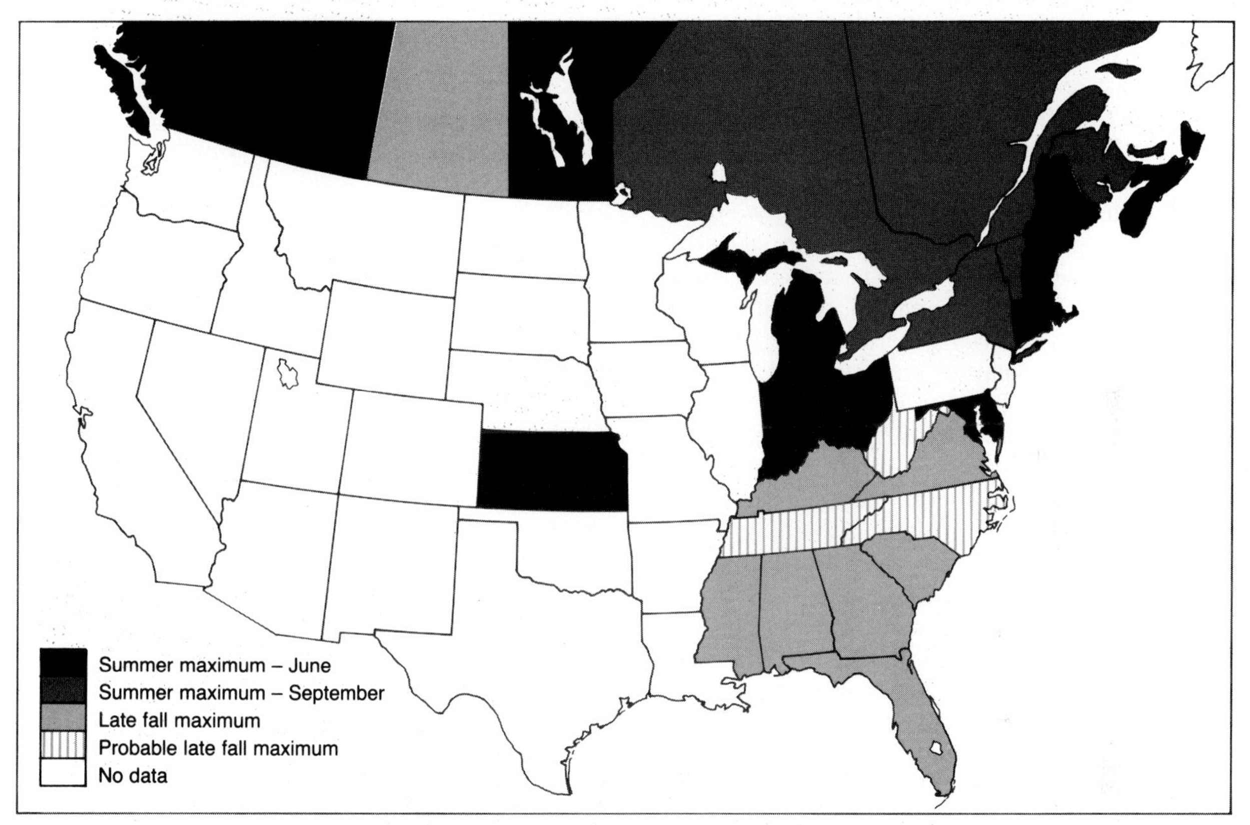

FIGURE 3-c Season of marriage in transition: 1920.

general attainment of modernity and a quiet revolution in wedding practices. We can dismiss the possibility of any standard diffusion of innovations over space by means of personal contact or observation, a mechanism so beloved among geographers. Rather, the explanation would seem to be an unconscious response, albeit delayed, to deep structural change in the organization of society.

The map teases us with one mystery that remains totally baffling: the September phenomenon, something of which the participants were apparently quite unaware. For roughly 50 years, from about 1910 to the 1960s, September scored either first or second in many states and provinces outside the American Southeast. There is no rational explanation as to why September became the leading month in Quebec, Ontario, the Maritimes, and the northeastern United States, but nowhere else, especially between 1925 and 1939, and then completely lost its dominance. Perhaps we shall never know.

Professor Wilbur Zelinsky
Pennsylvania State University
University Park

groups, particularly blacks and Jews. Republican strength has been concentrated in the Midwest, in the interior states of the West, and in rural New England. Significant regional trends in the last few years have been the resurgence of Republicans in the Southeast and of Democrats in New England and parts of the Midwest.

There is intense provincialism in Canadian politics, and dominant parties at the provincial level are often much less important nationally. The two principal national parties are Liberal and Progressive Conservative, but often neither is able to gain a majority in the House of Commons, resulting in either a precarious minority government or a shaky coalition. Two other parties are notable at the federal level: the New Democratic Party has strong social democratic leanings whereas the Social Credit Party is conservative and populist.

With limited exceptions, the major national parties in Canada are also the principal provincial ones, but the provincial parties often espouse different policies from their federal counterparts because of regional interests. The New Democratic and Social Credit parties have been particularly strong in the west (British Columbia and Alberta) and Quebec has been governed since 1976 by the *Parti Québécois,* the political embodiment of the French Canadian separatist movement.

TRENDS AND QUESTIONS

The population of Anglo-America will undoubtedly continue to grow, but the predicted rate of growth is a matter for considerable debate. Projections based simply on the changing age structure of the population would indicate an upsurge in the rate of increase, for the number of young adults in the population is growing rapidly and these are the prime childbearing years. But fertility has been declining at a record rate, presumably as a result of changing attitudes by young adults, and this is a huge imponderable for prognostication. Some authorities have gone so far as to predict the possibility of achieving zero population growth within this century, which is considered by many as a most worthy goal. It is logical to expect a continued high rate of net immigration to both countries; but natural increase is the principal source of population growth and it is difficult to predict natural increase rates.

The striking fact—and it dominates all other continental demographic trends—is the continuing long-term decline in the rate of population increase for both countries. During the decade of the 1950s the U.S. population grew by 19 percent (30 percent in Canada); in the 1960s the U.S. growth rate was 13 percent (18 percent in Canada); during the 1970s the rate declined to 8 percent in the United States (13 percent in Canada). In the decade of the 1980s the U.S. rate increased slightly, to 10 percent, whereas Canada's rate dipped to 8 percent, the lowest for any decade since Confederation. Thus, as we enter the 1990s, the U.S. population is increasing at the slow rate of about 1 percent annually, and Canada's growth rate is only about 0.8 percent annually.

The regional pattern of population change is more clear-cut. The westward movement of people in both countries has been pronounced, if irregular, for years, and it gives every evidence of continuing, although perhaps differing in detail. Although their growth rates slowed in the 1980s, Alberta and British Columbia continue to be (along with Ontario) the Canadian leaders. California persists as the state with the largest absolute population gains on an annual basis, and its rate of growth is exceeded by only five other states. Most other western states, from Texas to Alaska, are growing at a rate that is more than twice the national average.

Growth in the so-called Sunbelt states (the southern tier from Virginia to California and Hawaii) has been notable and relatively continuous in recent years, reflecting the desire of an affluent and footloose population to settle in warmer areas. Related but less pronounced movements to scenic, high-amenity western states, such as Colorado, Alaska, and Washington, are also conspicuous. However, an economic slowdown in several western states—particularly Wyoming, Montana, and Oregon—has greatly retarded their population growth in the last few years.

One of the most striking, and unexpected, demographic trends in the United States and to a lesser extent in Canada was a net in-migration to nonmetropolitan areas during the 1970s. For well over a century there had been a pronounced rural-to-urban movement of people in both countries. Especially since the farm population reached its peak during World War I, most rural counties have experienced a net emigration of people. In hundreds of counties this outmovement exceeded the natural increase (excess of births over deaths), resulting in actual population declines. During the 1970s, however, a trend referred to as "nonmetropolitan population turnaround" occurred, wherein small towns and rural areas experienced notably faster rates of population growth than did metropolitan areas.

Such "counterurbanization" had never before taken place and persisted for the entire decade; a nonmetropolitan growth rate of 14.4 percent easily surpassed a metropolitan growth rate of 10.5 percent. Explanation of the turnaround seems to involve a variety of factors, including more jobs in mining, decentralization of manufacturing and service employment to nonmetropolitan areas, slackening in the exodus from agriculture, the growing function of rural areas as recreational and/or retirement communities, increasing attractiveness of amenity-rich locations, desire for a simpler life away from the traumas of cities, and spillover of residential sprawl around urban areas. It seems clear that different factors pertain to different areas. Whatever the causes, the results were remarkable, if short-lived. Areas previously noted as sources of population flight, such as southern Appalachia, the Arkansas Ozarks, California's Sierra Nevada, and north-country Michigan, became destinations for settlement.

In the 1980s, however, the turnaround was reversed. The urban-to-rural movement dwindled. There was a continuing farm crisis, a recession, a restored reputation for cities, and a waning popularity among urbanites for an alternative lifestyle. During the decade of the 1980s the majority of nonmetropolitan counties actually lost population. Both the United States and Canada became increasingly urbanized.[12] Even so, there has not been a reversion to the pell-mell urban growth that typified the pre-1970 years. This underscores the need to be cautious about generalizations concerning demographic trends; even the most clear-cut geographical patterns may be subject to radical change.

[12] One state—New Jersey—is now officially totally urban. Every county in the state is classed as a metropolitan area because each has a high-density population cluster of at least 50,000 people.

SELECTED BIBLIOGRAPHY

ALLEN, JAMES P., "Recent Immigration from the Philippines and Filipino Communities in the United States," *Geographical Review*, 67 (1977), 195–208.

ALLEN, JAMES P., AND EUGENE J. TURNER, *We the People: An Atlas of America's Ethnic Diversity*. New York: Macmillan, 1987.

Anonymous, "Counting the Uncountable: Estimates of Undocumented Aliens in the United States," *Population and Development Review*, 3 (December 1977), 473–481.

BALLAS, DONALD J., "Geography and the American Indian," *Journal of Geography*, 65 (1966), 156–168.

BARRY, J. L., AND D. C. DAHMANN, "Population Redistribution in the United States in the 1970s," *Population and Development Review*, 3 (December 1977), 443–471.

BEALE, CALVIN L., *The Revival of Population Growth in Nonmetropolitan America*. Washington, D.C.: U.S. Department of Agriculture, Economic Research Service Report 605, 1975.

BEALE, CALVIN L., AND GLENN V. FUGUITT, *Metropolitan and Nonmetropolitan Growth Differentials in the United States since 1980*. Madison: Center for Demography and Ecology, University of Wisconsin–Madison, ca., 1985.

BEAUMONT, R., AND K. MCQUILLAN, *Growth and Dualism: The Demographic Development of Canadian Society*. Toronto: Gage, 1982.

BIGGAR, JEANNE C., "The Sunning of America: Migration to the Sunbelt," *Population Bulletin*, 34 (March 1979), 126–137.

BIKALES, GERDA, "Immigration Policy: The New Environmental Battlefield," *National Parks and Conservation Magazine*, 51 (December 1977), 13–16.

BORCHERT, DAVID J., AND JAMES D. FITZSIMMONS, *Recent Population Change in the United States, A Series of Maps*. Minneapolis: University of Minnesota Press, 1978.

BOSWELL, THOMAS O., AND TIMOTHY C. JONES, "A Regionalization of Mexican Americans in the United States," *Geographical Review*, 70 (January 1980), 88–98.

BROWNING, C. E., "The Shifting Winds of Population Change in the United States," *Geographical Review*, 66 (1976), 94–95.

DAVIS, GEORGE A., AND O. FRED DONALDSON, *Blacks in the United States: A Geographical Perspective*. Boston: Houghton Mifflin Company, 1975.

DENEVAN, WILLIAM M., *The Native Population of the Americas in 1492*. Madison: University of Wisconsin Press, 1976.

DRIVER, HAROLD E., *Indians of North America*. Chicago: University of Chicago Press, 1961.

FUGUITT, GLENN V., AND CALVIN L. BEALE, *Changing Patterns of Nonmetropolitan Distribution*. Madison: Center for Demography and Ecology, University of Wisconsin-Madison, ca., 1986.

GRAFF, THOMAS O., AND ROBERT F. WISEMAN, "Changing Concentrations of Older Americans," *Geographical Review*, 68 (October 1978), 379–393.

HAGAN, WILLIAM T., *American Indians*. Chicago: University of Chicago Press, 1961.

HANSEN, MARCUS LEE, *The Immigrant in American History*. New York: Harper & Row, Publishers, 1964.

HARRIS, R. COLE, ED., *Historical Atlas of Canada, Vol. 1: From the Beginning to 1800*. Toronto: University of Toronto Press, 1987.

JOHNSON, D. W., ET AL., *Churches and Church Membership in the United States*. Washington, D.C.: Glenmary Research Center, 1974.

JONES, MALDWYN ALLEN, *American Immigration*. Chicago: University of Chicago Press, 1960.

JONES, RICHARD C., "Undocumented Migration from Mexico: Some Geographical Questions," *Annals of the Association of American Geographers*, 72 (March 1982), 77–87.

JOSEPH, ALUN E., PHILIP D. KEDDIE, AND BARRY SMIT, "Unravelling the Population Turnaround in Rural Canada," *Canadian Geographer*, 32 (Spring 1988), 17–30.

KOSINSKI, LESZEK A., "How Population Movement Reshapes the Nation," *Canadian Geographical Journal*, 92 (May–June 1976), 34–39.

LEWIS, PEIRCE F., "Common Houses, Cultural Spoor," *Landscape*, 19 (January 1975), 1–22.

LIEBERSON, STANLEY, AND MARY C. WATERS, *From Many Strands: Ethnic and Racial Groups in Contemporary America*. New York: The Russell Sage Foundation, 1988.

LONG, JOHN F., *Population Deconcentration in the United States*, Bureau of the Census, Special Demographic Analyses: CDS-81-5. Washington, D.C.: Government Printing Office, 1981.

LONG, LARRY, *Migration and Residential Mobility in the United States*. New York: The Russell Sage Foundation, 1988.

MARSDEN, L. R., "Is Canada Becoming Overpopulated?" *Canadian Geographical Journal*, 89 (November 1974), 40–47.

MCHUGH, KEVIN H., "Hispanic Migration and Population Distribution in the United States," *Professional Geographer*, 41 (November 1989), 429–439.

MCKEE, JESSE O., ED., *Ethnicity in Contemporary America: A Geographical Appraisal*. Dubuque, IA: Kendall-Hunt Publishing Co., 1985.

NASH, GARY B., *Red, White, and Black: The Peoples of Early America*, (2nd ed.). Englewood Cliffs, NJ: Prentice-Hall, 1982.

PALMER, H., ED., *Immigration and the Rise of Multiculturalism*. Toronto: Copp Clark Publishing Company, 1975.

PRICE, JOHN A., *Indians of Canada: Cultural Dynamics*. Englewood Cliffs, NJ: Prentice-Hall, 1979.

RAITZ, KARL B., "Ethnic Maps of North America," *Geographical Review*, 68 (July 1978), 335–350.

ROSEMAN, CURTIS C., *Changing Migration Patterns within the United States*. Washington, D.C.: Association of American Geographers, Resource Paper 77-2, 1977.

ROSS, THOMAS E., AND TYREL G. MOORE, EDS., *A Cultural Geography of North American Indians*. Boulder, CO: Westview Press, 1987.

SAUER, CARL O., "European Backgrounds," *Historical Geography Newsletter*, 6 (Spring 1976), 35–58.

SHORTRIDGE, JAMES R., "Patterns of Religion in the United States," *Geographical Review*, 66 (1976), 420–434.

SUTTON, IMRE, *Indian Land Tenure: Bibliographical Essays and a Guide to the Literature*. New York: Clearwater Publishing Co., 1975.

TAYLOR, J. GARTH, "Trying to Preserve Our Aboriginal Cultures," *Canadian Geographic*, 100 (December 1980–January 1981), 52–58.

WAREING, J., "The Changing Pattern of Immigration into the United States," *Geography*, 63 (July 1978), 220–224.

WINKS, ROBIN W., *The Blacks in Canada*. New Haven, CT: Yale University Press, 1971.

ZELINSKY, WILBUR, "Changes in the Geographic Patterns of Rural Population in the United States 1790–1960," *Geographical Review*, 52 (1962), 492–524.

———, "Selfward Bound? Personal Preference Patterns and the Changing Map of American Society," *Economic Geography*, 50 (1974), 144–179.

4 THE ANGLO-AMERICAN CITY

The keynote of Anglo-American geography for many decades involved the proliferation and spread of urbanism. An almost invariable growth of individual cities occurred, along with a continuing increase in the number of cities. During the 1970s a remarkable countertrend set in: many cities experienced a declining growth rate, some cities actually diminished in population, and there was an overall faster rate of population expansion in nonmetropolitan than metropolitan areas. As seen in Chapter 3, that aberrant trend has now disappeared, and urban growth has been resumed, if less exuberantly. An urban way of life is an incontrovertible fact for most people in contemporary Anglo-America. By the early 1990s there were 50 metropolitan areas in the United States and Canada with populations exceeding 1 million and another 100 with populations in excess of 250,000.

Despite the dominance of cities in the lifestyle of Anglo-Americans (Fig. 4-1), this book does not emphasize urbanism in its regional treatment for several reasons. The most important is that most Anglo-American cities are quite similar. They have developed at approximately the same time and in roughly the same fashion in countries characterized by a relatively high standard of living and in which people, ideas, and money are shifted easily from one region to another (mobility of population, pervasiveness of mass media, and fluidity of capital).

There are some exceptions to this generalization. Several cities have a character of their own; nobody would accuse New Orleans, Quebec City, or San Francisco of being undistinctive. Nevertheless, the vast majority of Anglo-American cities within the general size-category have a sameness in appearance, morphology, and function that is almost

FIGURE 4-1 The 50 largest metropolitan areas in Anglo-America.

bewildering to a visitor from another continent where cities have grown up more individualistically over a much longer period of development.

Even though the two nations of Anglo-America have been politically separate for more than two centuries, cities on both sides of the international border tend to be remarkably alike. In both countries it is the urban areas that absorb most of the continuing high rate of population and economic growth; in both countries the automobile has assumed a dominant role in shaping city form and patterning urban life; and in both countries the similarity in the standard of living and personal motivation and aspiration is reflected in urban function and morphology. Canadian cities are spread in a long east–west band that has a very narrow north–south breadth and each tends to be more like American cities of similar longitude than like Canadian cities significantly farther east or west; thus Calgary is more like Denver than like Toronto, and Winnipeg resembles Minneapolis more than it does Vancouver.

There are, however, some differences between cities on opposite sides of the border. These differences reflect variation in political and settlement history, ethnic mix, and urban institutional patterns of the two countries, among other things. For example, Canadian cities are denser and more compact, have less variation between their central-city and suburban populations, make greater use of public transit, and have much less local government fragmentation. These are not major structural differences, but they are significant in many urban areas.

This chapter considers Anglo-American cities in general, commenting on the major characteristics of their urban geography. Each regional chapter that follows has a section devoted to the leading cities of the region or to atypical aspects of urbanism in the region or to a particular urban theme that is pertinent to the region.

HISTORICAL DEVELOPMENT OF ANGLO-AMERICAN CITIES

Prior to 1800

In early colonial days small towns sprang up along the Atlantic seaboard, mostly in what are now New England and the Middle Atlantic States in the United States and in the Maritime Provinces and Lower St. Lawrence Valley in Canada (Fig. 4-2). This development was natural because these areas were nearest England and France, particularly the former, whence came immigrants, capital goods, and many consumer goods. Accordingly, merchandising establishments were more advantageously located in port cities from which goods could be more readily distributed to interior settlements. Here, too, were the favored locations for assembling raw materials for export and for performing what little processing was necessary for shipment abroad. A number of small ports existed, but Baltimore, Boston, Philadelphia, New York, and Montreal soon began to exert dominance (although for several decades Quebec City grew almost as rapidly as Montreal).

Urban growth was less impressive in the colonial South, where life centered around the plantation rather than the town. The local isolation and economic self-sufficiency of the plantation were inimical to the development of towns. Thus, nearly all southern settlements were located on navigable streams and each planter owned a wharf accessible to the small shipping of that day. Both Charleston and Savannah were founded early and developed various urban functions, but after a short time neither rivaled the North Atlantic cities in urban development.

At the time of the first census of the United States in 1790, not a city in Anglo-America had as many as 50,000 inhabitants and only Baltimore, Boston, Charleston, New York, Philadelphia, and Montreal exceeded 10,000. No city yet showed any indication of urban dominance; each Atlantic port

FIGURE 4-2 The largest cities in Anglo-America in 1830–31. The minimum size city shown has a population of 4000. (After Maurice Yeates and Barry Garner, *The North American City,* 3rd ed. New York: Harper & Row, Publishers, 1980, p. 51. By permission of Harper & Row.)

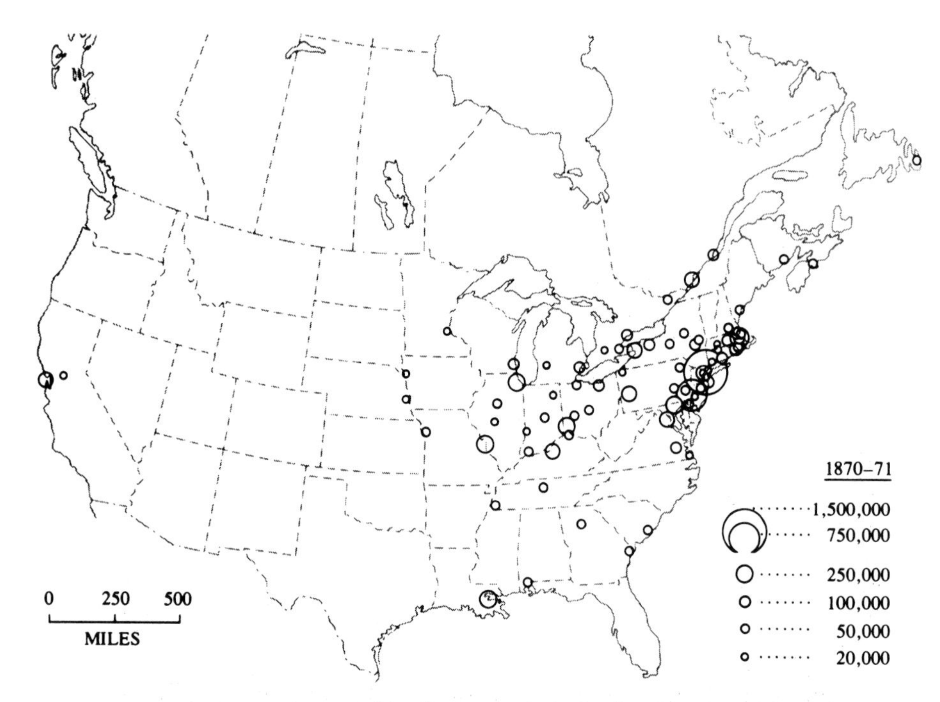

FIGURE 4-3 The largest cities in Anglo-America in 1870–71. The minimum size city shown has a population of 20,000. (After Maurice Yeates and Barry Garner, *The North American City,* 3rd ed. New York: Harper & Row, Publishers, 1980, p. 52. By permission of Harper & Row.)

served a small hinterland and was primarily oriented toward the sea and Europe.

1800–1870

Penetration of the interior and the use of inland waterways (Ohio River, lower Great Lakes, Erie Canal) began to produce a few inland urban centers: Richmond, Lancaster, Pittsburgh, Albany, and a bit later, St. Louis, Cincinnati, Louisville, Buffalo, and Rochester. But the Atlantic ports retained their regional primacy and the Louisiana Purchase added another primary port, New Orleans.

Montreal had grown to dominate a large share of the trade of the interior of the continent and challenged New York, Philadelphia, and New Orleans (Fig. 4-3). The partitioning of British North America into Upper Canada (Ontario) and Lower Canada (Quebec) and the choice of York (later to be named Toronto) as capital of Upper Canada added a significant dimension to the urban scene in the north. For Toronto, the "law of initial advantage operated fully, and by 1830 all rivals to regional control had been subdued."[1]

Still, rural dominance was clear-cut until the 1840s when railways began to develop, the factory system became established, and the industrial function of cities began to grow. The mechanization of spinning and weaving had set the pace in the previous two decades, but other types of manufacturing were oriented mostly toward households and workshop, often in rural locations, until the 1840s.

The large port cities grew especially in size with the building of canals, roads, and railways to the interior. A series of regional rail networks developed, with the larger networks converging at important inland waterway connections.[2] Each important coastal port began to organize its own railway and

[1] Donald P. Kerr, "Metropolitan Dominance in Canada," in *Canada, A Geographical Interpretation,* ed. John Warkentin (Toronto: Methuen Publications, 1968), p. 540.

[2] John R. Borchert, "American Metropolitan Evolution," *Geographical Review,* 57 (July 1967), 315.

push it inland. Of the major American ports, Boston alone was forced to content itself with connecting lines; its Boston and Albany Railway never got beyond the Hudson River. There was little railway development at this time in Maritime Canada. The Grand Trunk Railway, built after 1850 from Chicago through Toronto and Montreal, reached the Atlantic in the United States at Portland, Maine. Consequently, Canada's Atlantic port cities grew very slowly.

The midcentury period was a transition time for city development. Before then, industrialization was quite subordinate except in the five great Atlantic port cities that dominated trade relationships between the Anglo-American agricultural economy and Europe. The mercantile syndrome continued to pervade most other cities for some years, but industrial development and urban growth became much more closely linked. Labor supply was enhanced by the attraction of farm youths to cities and by accelerating immigration; after 1840 foreign immigrants began to concentrate on the edge of the expanding central business district in many urban areas, presaging the large-scale development of ethnic ghettos in decades to come.[3]

New York City rose to undisputed continental primacy at this time, partly because the opening of the Erie Canal cemented its western trade advantages but also because it was able to control much of the external trade of the South. "Indeed, it was largely because the merchants of New York and their itinerant factors controlled the cotton trade that urbanization in the South was extremely slow."[4] Thus Charleston and New Orleans were unable to wrest control of the cotton trade from New York. From roughly equal size with Philadelphia and Boston at the turn of the century, New York's population exceeded half a million by 1850; this population was more than twice that of any other city on the continent.

Inland regions of urbanization were limited mostly to the Ohio Valley (Cincinnati, Pittsburgh, Louisville) and upstate New York (Albany, Troy, Rochester, Buffalo), although Toronto almost kept pace with Montreal's *rate* of growth with nearly a 50 percent increase during the 1850s.

The growth rate of Canadian cities declined sharply during the 1860s, apparently owing to large-scale emigration to the United States. Urban populations were booming in the Midwest. Water transport helped Cincinnati and St. Louis to grow rapidly, but by 1870 they were being challenged by the swiftly growing Great Lakes cities of Detroit, Milwaukee, Cleveland, and especially Chicago. There was also continued fast growth in New England and the Middle Atlantic states, especially near New York. Brooklyn was the third largest city in the nation and Newark and Jersey City were both sizable. In the South only New Orleans had continued major growth and numbered nearly 200,000 people whereas the only western city that had developed to more than 25,000 was San Francisco, with a population of 150,000.

Thus the development of a more-than-regional transportation system, combining regional rail networks with inland waterways, revolutionized urban development in this period. Industrial growth was an important stimulant in the larger cities, but major industrial development came later.

1870–1920

This was an era of maturing for Anglo-American cities, the previous period being a formative one. National transportation systems were completed. National accessibility was extended to the South, the Southwest, and the Far West (Fig. 4-4). The remaining agricultural lands of the West, from Texas to British Columbia, were opened. A variety of major mineral deposits was developed. But the principal stimulus to urban growth was industrial development. The economy of the two countries changed from a commercial-mercantilistic base to an industrial-capitalistic one.

The coastal cities were increasingly important, but much of the growth was concentrated in the industrial belt of northeastern United States and southern Ontario and Quebec. The geographical division of labor, a basis for present-day regionalism, was beginning to be apparent by the end of the Civil

[3] David Ward, "The Emergence of Central Immigrant Ghettoes in American Cities: 1840–1920," *Annals,* Association of American Geographers, 58 (June 1968), 343.

[4] David Ward, *Cities and Immigrants: A Geography of Change in Nineteenth Century America* (New York: Oxford University Press, 1971), p. 29.

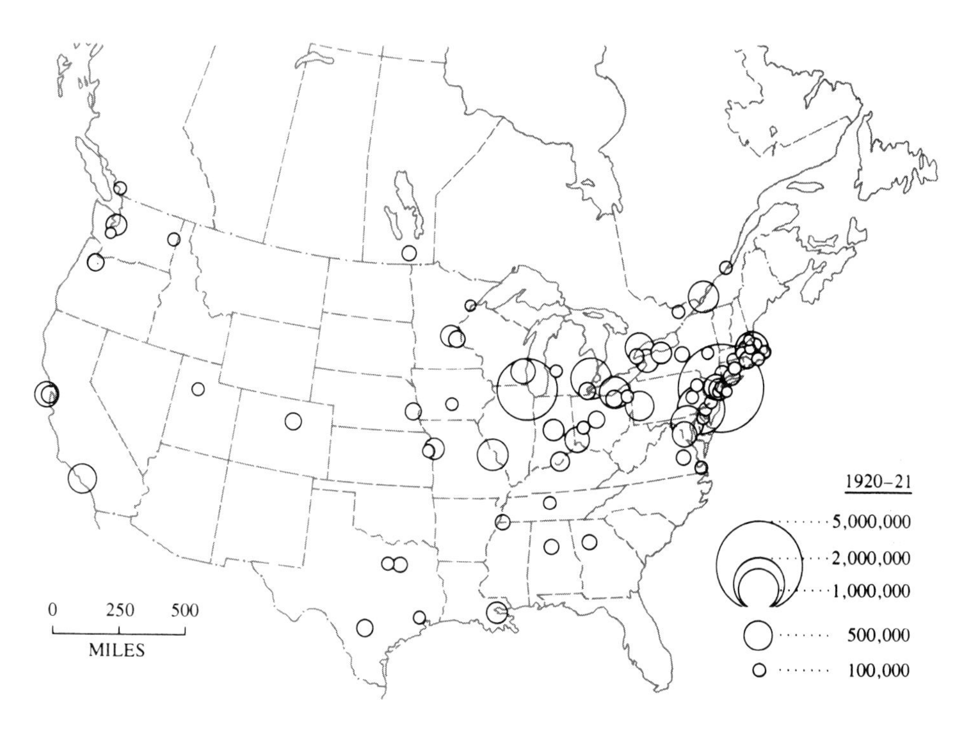

FIGURE 4-4 The largest cities in Anglo-America in 1920–21. The minimum size city shown has a population of 100,000 (After Maurice Yeates and Barry Garner, *The North American City*, 3rd ed. New York: Harper & Row, Publishers, 1980, p. 54. By permission of Harper & Row.)

War period. There followed a quarter century of accelerated westward movement, rapid population increase, heavy immigration, and burgeoning urban growth. Cities found their functions multiplying, but the growth of manufacturing was usually at the core.

The big cities became bigger, but there were also developments among smaller centers. Toronto, only half the size of Montreal at the beginning of the period, began to capture trade territory to the north and west and by the turn of the century had approached parity in size; the growth rates of the two cities have been quite similar ever since. Winnipeg began to grow in the Prairies, and Los Angeles experienced the early stages of its spectacular population increase. There were boom times in Florida urban areas, in Appalachian coal towns, and in cities of the Carolina Piedmont.

This was a tumultuous period in Anglo-American urban development, but at its conclusion our basic urban system was firmly established.

1920–1970

By 1920 half the population of Anglo-America was urban and the proportion continued to increase. There was substantial and continued growth in the older urban areas of the United States and Canada, but the most flamboyant developments were in new cities far removed from the traditional urban centers. The most spectacular growth occurred on the West Coast, from San Diego to Vancouver, and similar trends took form in Florida, Texas, the desert Southwest, and on the inland sides of the Rockies (Colorado and Alberta).

By the beginning of the 1960s approximately 25 percent of the world's population lived in cities of 20,000 or more inhabitants; the comparable figure for Anglo-America was 55 percent, the highest proportion of any populous part of the globe.[5] Thus the

[5] Derived from data in Leroy O. Stone, *Urban Development in Canada* (Ottawa: Dominion Bureau of Statistics, 1967), p. 16.

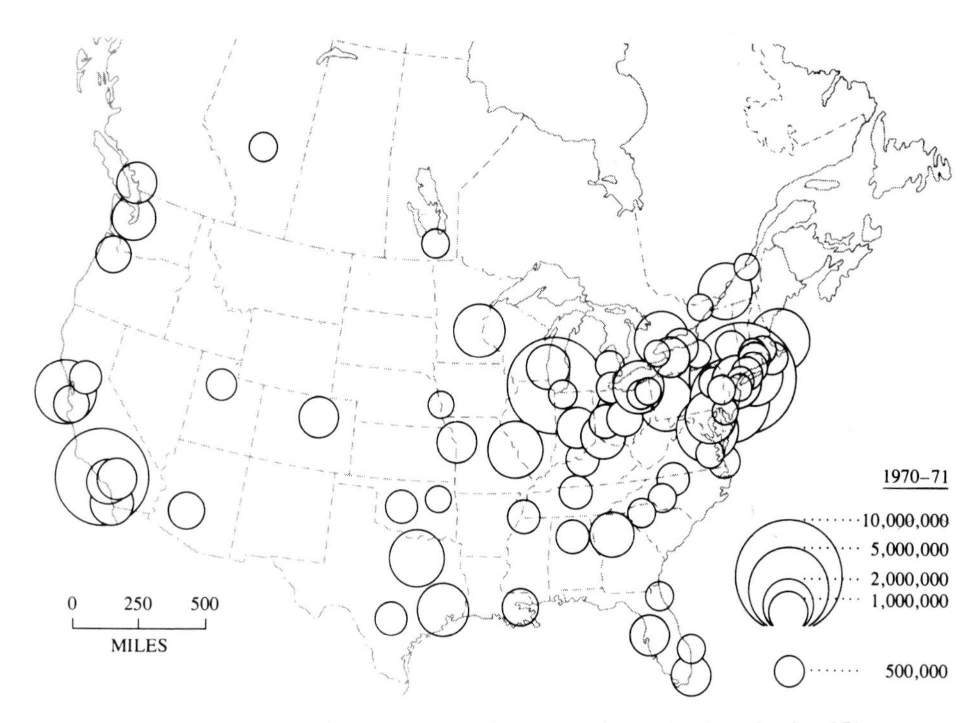

FIGURE 4-5 The largest metropolitan areas in Anglo-America in 1970–71. The minimum size metropolitan area shown has a population of 500,000. (After Maurice Yeates and Barry Garner, *The North American City*, 3rd ed. New York: Harper & Row, Publishers, 1980, p. 57. By permission of Harper & Row.)

concentration of urban population on this continent is not only of recent vintage and sizable magnitude but also has occurred at a remarkably rapid rate (Fig. 4-5).

This was a period in which the automobile shaped newer cities and reshaped older ones and the remarkable development of air travel reinforced many interrelationships of the existing urban system besides adding some new dimensions. Typical characteristics included continued agglomeration, urban sprawl, freeway construction, central-city decay, urban renewal, massive air pollution, suburban high rises, concerted desegregation efforts, planned industrial districts, and extensive neighborhood and regional shopping centers.

1970 to Present

Significant changes in urban population trends appear to have ushered in a new period of development in recent years. It is a period of counterurbanization, one that has been termed a "slow growth epoch," and seems to be most prominently related to a sluggish rate of population growth and a diminishing increase in energy supplies.[6]

Recognizable trends include the following:

1. Larger metropolitan areas are growing at a rate somewhat slower than the general population.
2. Some larger metropolitan areas, especially in the Northeast, are actually decreasing in population for the first time.
3. Nearly all central cities are decreasing in population, which is generally accounted for by white outmigration.

[6] Phillip D. Phillips and Stanley D. Brunn, "Slow Growth: A New Epoch of American Metropolitan Evolution," *Geographical Review*, 68 (July 1978), 274.

4. Rapid population growth has taken place in many smaller cities, particularly in the South and West.[7]

Characteristics of this most recent period are discussed at greater length later in this chapter.

URBAN MORPHOLOGY: CHANGING PATTERNS

Viewed from the air, a typical Anglo-American city appears as a sprawling mass of structures of varying size, shape, and construction, crisscrossed by a checkerboard street pattern that here and there assumes irregularities. The general impression is one of stereotyped monotony. The pattern of form and structure is so repetitive that one can anticipate a characteristic location of specialized districts and associations of activities within them.[8] The stylized arrangement and predictable interrelations make it possible to formulate broad generalizations about Anglo-American urban anatomy that are particularly valid if confined to cities of similar size, function, and regional setting.

The Pattern of Land Use

When considered in detail, the pattern of land use varies with every city. There are, however, such basic similarities that general patterns can be described and, to some extent, explained. The resulting generalizations are broadly valid for most cities whether they are older and slower-growing cities with rigid zoning restrictions, such as Buffalo, or newer, burgeoning cities with only limited land-use zoning regulations, such as Tucson.

Commercial Land Use Most cities are primarily commercial centers. Their attraction as a place for people to live is largely predicated on the concentration of commercial or business activities. For Anglo-American cities as a group, more than half of all jobs are in commercial fields: wholesale trade, retail trade, finance, insurance, real estate, and various kinds of services. Although the proportion of a city's land area devoted to these activities is small—generally less than 4 percent of the total—the structures in which the activities take place are often conspicuous and involve the tallest and most obtrusive buildings in the urban area.

The *central business district* (CBD) is the commercial heart of the city. It normally occupies an area near but slightly removed from the original town site (Fig. 4-6). It usually has a geographically central position in relation to the urban area as a whole; in some cases, however, it may be situated well off center, particularly where prominent physical features, such as coastline, mountain front, or river, are involved. The CBD is normally characterized by the greatest intensity of urban activity: highest daytime population density, most crowded sidewalks, most used surface streets, focus of mass transit routes, principal concentration of taxis, and greatest concentration of high-rise buildings. It also contains the most valuable land in the city and is the principal location of office space, large department stores, restaurants, theaters, hotels, government offices, financial institutions, corporation headquarters, and auto parking facilities.

By the 1960s most Anglo-American CBDs were experiencing a decline in mass retailing and personal services due to the almost overwhelming economic challenge of sparkling new outlying shopping centers. In some CBDs this challenge was met by concerted efforts at downtown retail revival. Downtown merchants began to develop extensive, often climate-controlled shopping malls within the CBD. Significant initial efforts included San Francisco's Ghirardelli Square and Chicago's Brickyard. Frequently another dimension was added by locating many of the facilities underground—multiple underground levels beneath large buildings or street-level plazas, with direct connections both to major buildings on the surface and to subsurface rapid transit facilities, if any exist. They provide an almost fully self-contained environment for urbanites that is a long overdue and eminently logical adjustment to winter in northern cities. Montreal's pioneering example in Place Ville Marie (opened in 1962) has stimulated similar developments in many other cities, even in such mild-winter locations as Los Angeles.

[7] Maurice Yeates and Barry Garner, *The North American City*, 3rd ed. (New York: Harper & Row, Publishers, 1980), p. 64.

[8] Howard J. Nelson, "The Form and Structure of Cities: Urban Growth Patterns," *Journal of Geography*, 68 (April 1969), 199.

FIGURE 4-6 The central business district shows up conspicuously in almost every city, and particularly in the larger ones. This view is toward central Philadelphia (courtesy Philadelphia Convention and Visitors Bureau).

Montreal continues to set the pace for underground urban activities (Fig. 4-7). By the early 1990s there existed a 7-mile (12-km) network of tunnels and shopping malls that connected 1000 stores and 100 restaurants and bars with theaters, hotels, office buildings, residential high rises, and rail, bus, and subway stations. A somewhat different style of subsurface development has been produced in Seattle and Atlanta where antiquated sections of the central city had literally been buried by subsequent construction. Portions of these historic districts have been unearthed and refurbished as underground touring, entertainment, and shopping centers, primarily aimed at tourist business.

Vertical expansion in the other direction has been pioneered by development of "skyways" in Minnesota's Twin Cities (Fig. 4-8). These are enclosed walkways that connect downtown buildings at the second-floor level and comprise the largest network of elevated indoor sidewalks and concourses in the world. In the quarter century since their inception, the skyway networks have expanded to encompass nearly 50 blocks of downtown Minneapolis and more than 30 blocks of downtown St. Paul (as of 1990). Their success, which is not unvarnished,[9] has led to similar programs on a more modest scale in several other cities (e.g., Edmonton, Calgary, Winnipeg, Cincinnati, Houston).

Despite the varied fortunes of the retail-service function, most CBDs persist prominently as the core of the city. Although some offices were lured to the suburbs, the 1970s were a boom period for downtown office-building construction; ever-larger and more complex skyscrapers sprouted in the downtown skyline. Some buildings experienced considerable difficulty in finding sufficient tenants, but the

[9] The principal objections to skyways are: (1) They draw people and businesses from the ground level, leaving that level impoverished and stagnant; (2) they have a negative effect on urban architecture and reduce the stimulating vitality of a varied environment; and (3) the hours of operation are uneven in Minneapolis (not so in St. Paul, where the city owns the skyways, and maintains uniform hours), so that some segments may be closed while others are open.

FIGURE 4-7 Montreal's Metro (rapid-transit system) serves a vast expanse of underground shops and restaurants beneath the center of the city (courtesy Mark Antman, The Image Works).

FIGURE 4-8 The skyways of Minneapolis insulate pedestrians from both winter weather and automobile traffic (courtesy Stock, Boston).

construction trend carried through the 1980s, albeit on a somewhat reduced scale.

A significant element of vigor for CBDs in the United States has been provided by foreign capital. As of the late 1980s foreign investors owned about 23 percent of the total office space in the 10 largest American cities (see Table 4-1). The greatest total investment—nearly $8 billion—was in New York City, but the highest proportion—46 percent of total downtown property—was in Los Angeles. Generally speaking, international investment has been highly selective, concentrating on relatively new, well-located, fully leased office towers. The principal source of foreign capital has been Canada and Japan.

TABLE 4-1

CBD real estate owned by foreigners

Rank	*City*	*Value in Millions*	*Percent of Downtown Property Owned by Foreigners*
1	New York	$7650	21
2	Los Angeles	$2600	46
3	Washington, D.C.	$1560	12
4	Houston	$1370	39
5	San Francisco	$1210	17
6	Chicago	$1200	n.a.
7	Boston	$1020	n.a.
8	Dallas	$915	17
9	Denver	$850	19
10	Minneapolis	$800	32

Source: 1988 data from Coldwell-Banker Commercial Real Estate.

Although there is a centrifugal dispersal trend for most other CBD activities, many aspects of urban life are still prominent in or near the CBD: manufacturing is often represented by printing-publishing plants and a concentration of loft-type garment factories; theaters, hotels, nightclubs, and bars highlight the recreational function; and a surprising number of people is likely to reside near, if not actually within, the area.

Marginal to the CBD is the so-called transition zone. This is a discontinuous area of irregular shape and unpredictable size that has an almost continually changing land-use pattern.[10] Its commercial prominence is often more oriented to wholesaling than to retailing and industrial activities, in the broad sense, are usually notable. Still, much of the land in a typical transition zone is occupied by residences; this is a characteristic location for slum and ghetto development. In general, the transition zone is seedy and dilapidated, although some sections may be uncharacteristically bright and even prosperous owing to public or private urban renewal and slum clearance projects. Grandiose high-rise office buildings or apartment houses sometimes tower above the general obsolescence.

Another significant proportion of a city's commercial land use is found along *string streets,* which are usually major thoroughfares of considerable length, lined on both sides by varied businesses. The development may be patchy and discontinuous, but in larger cities the extent of string-street commercial zones may be measured in consecutive miles. Characteristically the businesses along a string street are small and diverse; however, there may be concentrations of specific types of enterprises, the best-known of which is "automobile row" where new and used car lots are clustered. A growing trend is the construction of high-rise office buildings along string streets, away from the CBD.

A modern expression of these commercial string developments is associated with outlying freeway locations. "The suburban freeway corridor now houses a complete mix of the business establishments regularly frequented by the geographically mobile middle- and upper-class residents of the modern metropolis."[11] These freeway ribbons permit relative ease of redevelopment due to their linear form and low density; thus incremental redevelopment is easier than in CBDs. They, of course, depend wholly on the automobile for customer access.

The most remarkable change in commercial land use in Anglo-American cities since World War II is the emergence of planned *suburban shopping*

[10] There is considerable literature on the transition zone. One of the more definitive statements of the concept is found in Donald W. Griffin and Richard E. Preston, "A Restatement of the 'Transition Zone' Concept," *Annals,* Association of American Geographers, 56 (June 1966), 339–350.

[11] Thomas J. Baerwald, "The Emergence of a New 'Downtown,'" *Geographical Review,* 68 (July 1978), 308.

centers. Continually increasing amounts of a city's retail and service business are being carried out in these outlying centers, which are geared to the automobile era, with much more acreage devoted to parking spaces than to shopping areas. Apparently the first complete shopping center opened in Kansas City's Country Club district in the early 1920s. In the early decades such centers emerged gradually; later, however, the planned shopping plaza blossomed full-grown at birth.

Planned shopping centers in the United States grew from about 100 in 1950 to a total of more than 20,000 in the early 1990s, although the pace of new construction has slowed considerably in recent years. Indeed, in the 1980s there was more expansion of existing shopping centers than the building of new ones.

It is estimated that only about 10 percent of all Anglo-American shopping centers are enclosed and climate controlled, but the great majority of the larger centers do have this characteristic. They are enclosed under a single massive roof that towers high enough to encompass three or four levels of walkway- and escalator-connected shops of varying sizes that are clustered around one or more spacious atria containing fountains, waterfalls, resting areas, and other attractions for the weary shopper.

At the time of this writing, the largest enclosed shopping center in the world is the West Edmonton Mall in Edmonton, Alberta, which is considered to be the world's first and (thus far) only *mega-mall*, with 5.2 million square feet (470,000 m^2) literally under one roof (Fig. 4-9). It contains more than 600 shops, 19 theaters, an ice rink, a 7-acre (2.8 ha) water park, a 3000-bird aviary, an enormous amusement park, a church, a hotel, jobs for 18,000 people, and parking space for 12,000 vehicles. The mall contains nearly one-fourth of Edmonton's retail space and accounts for more than 1 percent of all retail sales in Canada.

The largest shopping mall in the United States is Del Amo Fashion Plaza in Torrance, California (a suburb of Los Angeles), with 2.65 million square feet (240,000 m^2). However, a 4.2 million square foot (380,000 m^2) mall, complete with a Camp Snoopy theme park, is now under construction at Bloomfield, Minnesota (a suburb of Minneapolis), and the developers eventually plan to add up to another 5.3 million square feet (480,000 m^2).

The latest development surge in shopping cen-

FIGURE 4-9 A view down one of the many galleries of the gigantic West Edmonton Mall in Edmonton, Alberta (courtesy V. Wilkinson, Valan Photos).

FIGURE 4-10 Mini-malls have proliferated in most American cities, particularly in California. This scene is in west Los Angeles (TLM photo).

ters is *mini-malls,* which are small, **L**-shaped corner buildings of 5 to 15 retail shops fronted by a small parking lot (Fig. 4-10). Mini-malls first appeared in the late 1970s when the oil crisis persuaded many corner service station owners to sell their properties to the major oil companies, who in turn sold the parcels relatively cheaply and in bulk to developers. Thousands of mini-malls have sprung up across the country, with the greatest concentration in southern California. Criticized for creating parking and traffic problems and for egregious architecture, the mini-mall phenomenon may have run its course, but it is now a well-established part of the commercial scene.

Residential Land Use By far the most extensive use of land in Anglo-American cities is for residences, which occupy 30 to 40 percent of an average city's area.[12] Single-family dwellings (normally separate but sometimes attached to one another as row houses) are not always numerically in the majority, but they occupy more than three-fourths of the area devoted to housing.

Residences are scarce within the CBD, although some housing units are often found on the upper floors of buildings. The greatest residential density in any city normally occurs in and near the transition zone, where grand old homes of the past (frequently converted to rooming houses) are mixed with vast expanses of low-quality residences and occasional redevelopment pockets of high-rise apartments. Moreover, there has been a notable recent trend, particularly in the older cities of eastern United States and Canada, toward *gentrification.* This involves the purchase and restoring and/or refurbishing of old homes, usually by upwardly mobile young adults. The result is a shifting of population back toward, and a reinvigoration of life in, the central city.

The transition zone functions more traditionally as a tenement section that may include a large proportion of the city's slums and ghettos. The inhabitants of such areas are normally blue-collar workers with relatively low incomes and a large proportion is likely to consist of ethnic minorities, particularly blacks.

The *black ghetto*[13] has almost within the space of a single generation become one of the two most rapidly expanding spatial configurations in large cities of the United States (the suburb is the other). Ghettos have been a prominent part of the Anglo-American metropolitan scene for many decades, but the rapid expansion, consolidation, and conspicuous social isolation of the black ghetto in recent years produced what amounts to a new urban sub-

[12] Jerome D. Fellmann, "Land Use and Density Patterns of the Metropolitan Area," *Journal of Geography,* 68 (May 1969), 265.

[13] Harold M. Rose, "The Origin and Pattern of Development of Urban Black Social Areas," *Journal of Geography,* 68 (September 1969), 328.

culture. In most cities of the United States and in the few Canadian cities where blacks reside in any numbers there is very strong de facto segregation between areas of white and black households; this holds true in central cities as well as in suburbs, in the North or the South, and in large cities or small. But it is in central-city locations that blacks find easiest access to housing and it is here that ghetto formation is pronounced and growing.

These ethnic enclaves usually occupy the least desirable parts of the city and strongly tend to be blighted zones except in newer cities where there is a prevalence of relatively new single-family residences located in the path of ghetto expansion (as in Denver, Phoenix, and some California cities). Black ghettos tend to be poverty areas, although there is often a concentration of blacks with an income above poverty level. The ghetto is normally but not always a forced development; nevertheless, it performs the important function of providing a sense of community to its residents, who are often at a critical stage in their lives. Ghettos obviously offer more disadvantages than advantages, but with white abandonment of the central city as a place of residence, territorial dominance has clearly been relinquished to ghetto residents.

Other types of ghettos are also found in and around the transition zone of Anglo-American cities, but rarely are they as large or conspicuous as the black ghettos. Most notable are Hispanic ghettos, usually referred to as *barrios*. They are most prominent in many cities of the Southwest, in New York and other cities of the Northeast where Puerto Ricans and Dominicans are concentrated, and in Miami and other cities of southern Florida that contain large Cuban minorities.

With the general outward shift of population distribution in cities, there tends to be a similarly centrifugal displacement of residential zones based on economic factors. The vast expanse of middle-class housing, normally situated beyond the transition zone, has an inner boundary that moves outward with pressures from the central city. This usually creates a "gray" area that serves as buffer between middle- and lower-class housing, attracts an upwardly mobile segment of the central-city population, and is often a determinant of the direction of ghetto expansion.

There is also a general centrifugal gradation in population density throughout the residential areas from the transition zone to the outer suburbs. Multi-family housing is more common near the city center and lowest population densities are in the outer areas where relative remoteness reduces the price of land and subdivision ordinances require greater spacing between houses. There are, however, numerous variations from this pattern. The unremitting monotony of detached single-family dwellings has been significantly leavened by garden apartment complexes, cluster housing, and contemporary townhouse variations of the old row house form. Such variety produces not only higher population densities but also diversity of residents. The old stereotypes of "suburban sameness" are less valid today than they were a couple of decades ago, when an entire school of social criticism was nurtured on attacking the aesthetics of suburbs.[14]

In the suburban fringe there are initially independently planned street systems in separate communities or subdivisions, often with curvilinear pattern and considerable sprawl. These discrete areas are subsequently joined and the open spaces among them are filled in with further urban (usually residential) development, producing irregular but not necessarily displeasing patterns of urban sprawl. The disparate densities between close-in and more remote residential areas tend to lessen with time as the settlement pattern ages and intensifies.

The *suburb* has a special place in the contemporary folk history of Anglo-America; it is The Place that connotes status, security, comfort, and convenience—the calm of country living with the amenities of the city within easy reach. The surge to the suburbs is nothing new; since at least the 1920s there has been a well-established tradition for the maturing generation, provided that it had the income and the means of internal transportation, to move to the edge of the city and establish yet another peripheral band of housing.[15] The post–World War II expansion of the suburbs, however, is on a heretofore

[14] Dennis J. Dingemans, "The Urbanization of Suburbia: The Renaissance of the Row House," *Landscape*, 20 (October 1975), 31.

[15] James E. Vance, Jr., "Cities in the Shaping of the American Nation," *Journal of Geography*, 75 (January 1976), 50.

FIGURE 4-11 Some suburban housing developments seem to sprawl endlessly, as here in Ft. Lauderdale, Florida (TLM photo).

undreamed-of scale. The more than 125 million Anglo-American suburbanites of the early 1990s amount to nearly half the total population (Fig. 4-11).

Suburbanites are generally but by no means universally well-off financially and are somewhat insulated from the decay and social trauma of the central city. To move to the suburbs is not to escape the problems of the city, because congestion, soaring taxes, crime, drugs, and pollution tend to follow, although generally to a lesser extent.

Although historically considered bedroom zones for a population that commuted to the central cities to work, suburbs increasingly became more complex in structure and function as their size and extent burgeoned. Jobs followed people to the suburbs and numerous nodes of high-density commercial and industrial development—high-rise office buildings, sprawling industrial parks, and immense shopping centers—scattered throughout suburbia are now typical. The usual movement of people, moreover, is from one suburb to another rather than commuting to the CBD, and a certain amount of reverse commuting has become established as central-city blue-collar workers increasingly must go to the suburbs to find work.

Industrial Land Use Although land devoted to industrial usage occupies only an average of 6 to 7 percent of the city's area,[16] in most cities the significance of factories as employment centers is so great that industrial activity is critical to the local economy. Industrial areas may be widely scattered, but generalizations can be made about their location pattern. Most cities have one or more long-established and well-defined factory areas near the CBD, often containing several large firms as well as many smaller ones. These districts were usually established during the era of railway dominance and are characterized by the presence of rail lines and flat land.

If there is a functional waterfront (ocean, canal, or navigable river or lake) in the city, another old industrial area is likely to be located there. Heavy industry may be congregated in such an area, typically primary metals plants, oil refineries, and chemical plants. Often these areas were originally swampy or marshy and were then reclaimed by drainage, landfilling, or both.

Planned industrial districts are the product of more recent years. They are variously located, but usually the site was chosen with care so that ample space is available and access to a major transport route is ensured. Many planned industrial districts were sited along railway lines, but later the critical site factor became a prominent road or highway, for motor trucks are used more often than railways in transporting goods to and from factories. Perimeter or belt-line highways are particularly attractive to the builders of planned industrial parks.

Factories may be found in many other kinds of locations in Anglo-American cities. The principal industrial areas, however, tend to fall into one of the

[16] Fellmann, "Land Use and Density Patterns of the Metropolitan Area," p. 265.

preceding categories. As a useful generalization, it can be stated that most cities have experienced a decentralization of industrial land use; the suburban share continues to grow while the central-city share continues to decline.

Transportation Land Use A surprisingly large amount of land in most cities is devoted to transportation of one sort or another, including the storage of vehicles. This is the second greatest consumer of city space, exceeded only by residential land use.

A CLOSER LOOK The Large Apartment Building in the American City

The apartment building as an architectural form came to America relatively late. Although it was pioneered in northern Italy in the Late Renaissance and had become a fully accepted and even preferred type of shelter in many European cities by the nineteenth century, it was still a very controversial form of housing when it was introduced into New York City during the 1870s. While American cities had long featured a variety of multiunit housing options for those who could not afford an individual home, most of these were rather makeshift. Indeed, most "tenements" in America were simply subdivided houses or three- or four-story multipurpose structures designed to accommodate shops, storage, or "rooms." The apartment building was something else altogether. The apartment building, as introduced from France, featured luxury living in a private suite of rooms along with the very latest in centralized services.

While the large apartment building was accepted in New York City before World War I, it was not widely accepted elsewhere in America until much later. Many complained of the "indecent propinquities" of apartment life, and the single-family house was promoted as the appropriate place for the American family to reside. Burgeoning populations coupled with a reluctance to build apartment buildings meant that many American central cities were overcrowded by the 1920s, and the lack of construction during the Great Depression and World War II only made matters worse. Boarders found "lodgings" where they could. To a very real degree, the massive population losses experienced by many American "inner cities" over the past few decades have been the result of less overcrowding in the housing stock. The relationship between houses and households may thus be getting back to normal after an extended abnormal period.

As late as 1960, the New York Metropolitan Area had as many units in large apartment buildings (using the U.S. Census figures for buildings with at least 10 housing units) as the next 23 largest metropolitan areas combined (Fig. 4-a). By 1980, the ratio was only five to one as apartment construction in metropolitan areas such as Los Angeles and Houston boomed in comparison with New York. It has only been since about 1960 that the apartment building, and more recently the condominium, has become an accepted feature in the typical American city. The high visibility of such "dazzling" projects as the Marina Towers in Chicago and the general revitalization of many American downtowns have helped to make an "urban lifestyle" more acceptable beyond the confines of New York City (Fig. 4-b).

Because the acceptance of the apartment building has coincided with the rise of the Sunbelt, it stands to reason that large multiunit complexes are now more associated with the younger cities of the South and the

FIGURE 4-a High-rise multiunit apartment buildings on Manhattan Island (photo by Larry Ford).

Indeed, in the "newer" Anglo-American cities it is not unusual for streets and parking areas to occupy half the total area of the CBD. Relatively small proportions of the city area are required by other modes of transport, although airports can be very expansive and new container ship terminals require extensive dockside storage space.

Other Types of Land Use Many other types of activities are carried on in a city, but their locational

West than the older cities of the Northeast. Many of our stereotypical notions of crowded, compact eastern cities and sprawling, low-density western cities may need to be revised. The look of tomorrow may be apartment towers on the beach.

The belated acceptance and differential location of the large, multiunit structure may require some rethinking about how American cities are put together. For example, since 1962, more apartment units have been constructed in the suburbs than in central cities, and by 1980 many American metropolitan areas had more large complexes outside the central city than in it. As average household size becomes smaller and overcrowding abates in city houses, the classic urban density gradient could be reversed.

Percentage of total units in complexes with 10 or more units

Metro Area	*1960*	*1980*
New York	43	50
Houston	4	27
Los Angeles	11	24
San Diego	6	20
Atlanta	4	17
Philadelphia	4	13
Detroit	8	11

FIGURE 4-b

Perhaps in the future, American cities will evolve toward a more European form with smaller, quaint houses in the center and large apartment complexes out by the malls and industrial parks. Perhaps other American cities will become more like those of Latin America with large, luxury apartment buildings in highly desirable and accessible locations and small tract houses in remote peripheral zones. By monitoring the locational changes in particular building types such as the large apartment complex, we can understand more clearly the ongoing changes in general urban form.

Professor Larry Ford
San Diego State University

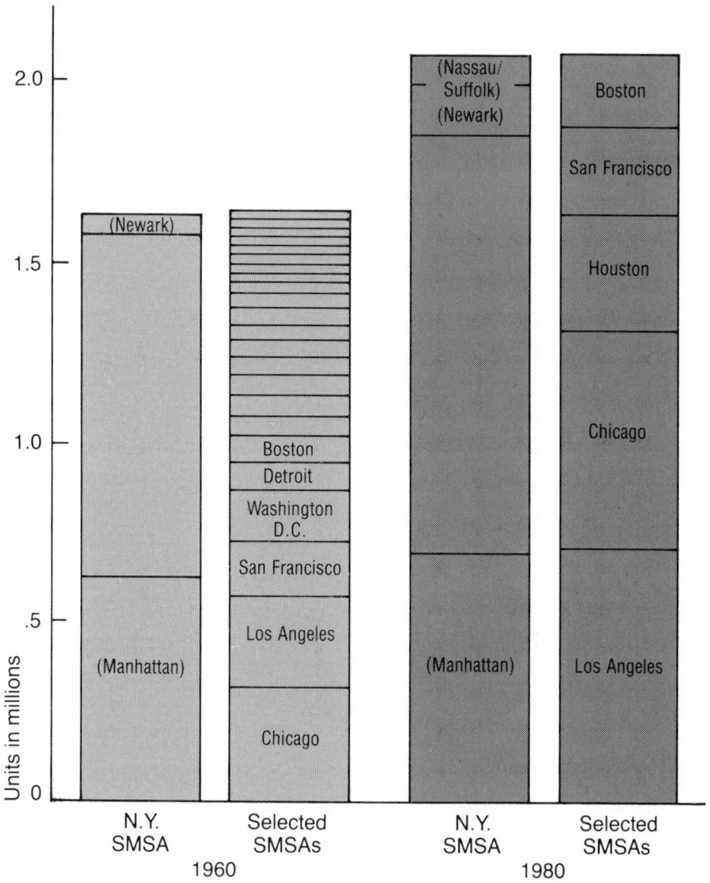

patterns are less predictable. Parks and other types of green spaces are found in all cities and in some cases occupy a large share of the total city area. Institutions of various kinds tend to be widespread but scattered—for example, schools, cemeteries, museums. Government office complexes are relatively insignificant in most cities but may be particularly prominent in a national (Washington and Ottawa) or state (such as Albany or Sacramento) capital or in a city that is a significant regional headquarters for federal activity (such as Denver or San Francisco). Vacant land is another category that occupies varying amounts of space; even the most crowded cities have a certain amount of land that is not being used at present for any purpose.

The Pattern of Transportation

There are two different but overlapping facets to transportation in cities: internal movement within the city and external movement to or from the city. For either facet, the dominant fact of life is the pervasiveness of the automobile. Of all the money spent in the United States for freight transportation, 75 percent is for motor trucking, and more than 90 percent of the total outlay for passenger movement goes to automobiles and buses. Although facilitating the movement of goods and people, the massive increase in rubber-tired transport threatens to overwhelm the system of streets and highways. The wastes of congestion become progressively worse despite every effort to facilitate traffic flow. In the New York area, for example, 33 percent of the trucks do not move at all on a given day and 20 percent of the remainder move empty.

Internal Transport Movement within Anglo-American cities depends primarily on the traffic flow of streets and highways. A large and increasing share of total city area must be devoted to routeways and storage lots for cars and trucks. Essentially every Anglo-American city has a rectangular grid as the basic pattern for its street network—at least in the older portions of the urban area. The pattern has nearly always been modified by subsequent departures from the original layout; in many cases, there is a series of separate grids, adjusted to topography or to surveying changes, that are joined in variable fashions. The grid scheme, with its right-angled intersections, dominates the layout of Anglo-American CBDs. The streets were usually established in the preautomobile era and are normally too narrow to facilitate traffic flow; thus they engender massive downtown congestion and tax the ingenuity of traffic specialists to devise techniques to unclog the streets.

Away from the city center there is usually a greater diversity in the street pattern, particularly in newer subdivisions. Even so, it is rare for regularity to be maintained over a very large area because varied sequences of development and annexation lead to heterogeneity in planning.

Superimposed on the pattern of surface streets in all large Anglo-American cities as well as many small ones is a network (or the beginning of a network) of freeways or expressways (Fig. 4-12). These high-speed, multilaned, controlled-access traffic ways are laid in direct lines across the metropolis from one complicated interchange to another, making functional connection with the surface street system at sporadic intervals by means of access ramps.

Modern freeways carry a large share of travel in most cities, permitting rapid movement (except at rush hours) at about one-third the accident risk of surface streets. Characteristically the freeway network of a city radiates outward from the CBD. Even though functional connection to a freeway is restricted, its route is often an axis along which urban development takes place at higher densities than in the intervening wedges, with the most intensive development likely to occur near access ramps and freeway interchanges.

Almost every large city has one or more beltline routes, often but not always a freeway, that roughly circle the city at a radius of several miles from the CBD. Probably the most famous example of this phenomenon is Boston's Circumferential Highway (Route 128), but the pattern is now common, from Miami's Palmetto Expressway to Seattle's Renton Freeway. Such perimeter thoroughfares often serve as magnets that attract factories and other businesses to locate alongside them. This very attractiveness inhibits the proper functioning of the beltway concept. These routes are intended to move traffic around urban centers, but they fre-

FIGURE 4-12 Freeways usually cut directly across all other forms of land use without regard to the previous transportation route pattern. This is a typical freeway scar across north Dallas (TLM photo).

quently become so overloaded with local traffic that they generate a need for new circumferential routes still farther out to carry the through traffic.

One approach to dealing with urban traffic congestion is a revival of interest in constructing toll roads. In previous eras most toll roads were long-haul intercity highways (such as the Pennsylvania Turnpike and the Ohio Turnpike), but contemporary development focuses on urban outskirts to enable suburban dwellers to reach their jobs and other destinations. Such commuter routes often do not connect with the CBD, as that is not necessarily the most needed connection. Indeed, the 1980 Census discovered that 25 million Americans commuted from one suburb to another to reach their jobs, but only half that many commuted from suburbs to central cities, and the gap has continued to widen since then. New commuter-related toll roads have recently been built or extended in a dozen cities, particularly in Florida and Texas, and are being actively planned in a number of other locales, most notably in California's Orange County (just southeast of Los Angeles).

Despite freeways and toll roads, however, congestion continues to choke the cities of the continent. It is often suggested that the only hope for urban survival is *mass transit* with emphasis on *rapid transit*. Unfortunately, the panacea effect seems to be overestimated. Mass-transit patronage in the United States has absolutely declined more than one-third over the last half century, in a period when the urban population virtually tripled. Patronage of rapid transit, on the other hand, declined only slightly over the same period, although urban governments either subsidize or own (and operate at a loss) all systems that are in operation.

Public transit continues to lose riders in most large U.S. cities, especially the Northeast. Larger Canadian cities, however, have an increasing ridership. New light rail systems are operating in Edmonton, Calgary, and Vancouver and the subway systems of Montreal and Toronto are being expanded (Fig. 4-13).

Enthusiasm for rapid transit in the United States is variable (for example, eagerly accepted by

FIGURE 4-13 New rapid transit systems have been established in several North American cities in recent years. This light rail transit car is in Toronto (TLM photo).

voters in Pittsburgh, thoroughly rejected by citizens in Houston), but the last two decades have seen a flurry of construction. San Francisco's BART system, opened in 1972, was the first of the modern, high-tech rail systems. Since then more than a dozen other major cities have opened new rapid transit systems (most conspicuously in Washington, D.C., and Atlanta) or have lines under construction (even in Los Angeles!). Some are doing better than others, but all require large government subsidy.

Anglo-American urbanites are generally still wedded to their automobiles and their willingness to shift to transit facilities is limited. Moreover, the relatively low population densities in most cities means low volumes of traffic, which works against the feasibility of substantial investments in rapid transit facilities.

There is no way to avoid the fact that rapid transit systems are relatively idle most of the time and in demand only at rush hours. Furthermore, who is served by rapid transit systems? New lines in Chicago and Toronto were found to draw 90 percent of their passengers from bus lines and only 10 percent from automobiles. Most systems accommodate commuters rather than the poor, the aged, and the handicapped, who, some would argue, need public transit most of all.

If not mass transit, and not rapid transit, then what will prevent the Anglo-American city from grinding to a halt someday under the sheer bulk of its street and freeway traffic?

External Transport The movement of people and goods into and out of cities is accomplished in a great variety of ways, although auto and truck transportation dominate. Cities are hubs in the cross-country highway networks of the United States and Canada, with routes converging to join the internal street system of the hubs. Despite the construction of bypasses and beltways, there is much mixing of a city's internal and external roadway traffic, with each contributing to congestion for the other. Even the building of the unprecedented ($70 billion in construction over a 21-year period) Interstate Highway System in the United States has done little to improve traffic within cities; it has immensely facilitated cross-country travel but clearly failed to alleviate urban congestion, which was one of its principal objectives.

In general, railways were important factors in the founding and growth of Anglo-American cities, but in most cases they are relatively less significant today. Nevertheless, railroad facilities are still quite conspicuous in most cities. Rail lines converge on cities in the same fashion and sometimes in the same pattern as highways. For large cities, there are also railway belt lines to facilitate the shifting of rail cars from one line to another. Major passenger and freight terminals are usually located in the transition zone near the CBD, but in some cases the latter have been shifted to more distant sites. Most cities have a single passenger terminal ("Union Station"). An important specialized feature of railway transport is the classification yard where freight trains are assembled and disassembled; yards of this kind are now usually located on the very outskirts of the urban area.

The rapid expansion of air travel in Anglo-America means that most cities are now engaged in the construction of new or the expansion of old airports. There is little in the way of a predictable location pattern except that an airport is usually located several miles (in some cases, dozens of miles) from the CBD, where an extensive area of flat and relatively cheap land is available. Normally a major freeway or other roadway thoroughfare (and sometimes a rapid transit line) is designed to give the airport a direct connection with the CBD. The long-predicted development of intracity helicopter travel in the larger metropolitan areas is as yet relatively insignificant.

Water transportation may be important to the economy of many ocean, river, and Great Lakes ports, but the amount of space used for port facilities is usually small compared to the total area of the city. Piers, docks, and warehouses are normally the most conspicuous permanent features along a port's waterfront, although the recent rapid change to containerization of cargo has led to enormous aggregations of container vans in open spaces adjacent to the docks.

Pipeline transportation is highly specialized and relatively inconspicuous in most cities. Internal networks distribute water and gas and collect sewerage, but these are ubiquitous and largely underground features. External pipeline systems are normally associated with liquid or gaseous fuels, bringing petroleum or gas into the city for either refining or distribution to consumers. In any case, most pipelines are buried and the prominent landscape features associated with this activity are huge storage tanks at the terminals.

Vertical Structure

The building of skyscrapers and other high-rise[17] buildings is not peculiar to Anglo-America, but the concept achieved its first real prominence in New York City and the vertical dimension of the Anglo-American skyline has continued to be significant in any consideration of city form. In the past, the vertical structure of cities was predictable. Within the CBD would be an irregular concentration of tall buildings, with a rapid decrease in height centrifugally in all directions to the very low profile that characterized the vast majority of the urbanized area. The only significant exception to that generalized scheme was New York, with its prominent dual concentration of skyscrapers: the major one in midtown Manhattan and the secondary, but still very impressive, one in lower Manhattan.

Later the pattern of high-rise building became more diffuse. The CBD still has the conspicuous skyline, but tall buildings are being built ever more widely throughout the urban area. Secondary aggregations of high-rise structures are often associated with major suburban shopping centers, planned industrial districts, and even airports. Tall buildings are also being built increasingly along principal string-street thoroughfares, usually in a very sporadic pattern. Los Angeles's famous Wilshire Boulevard, for example, is now marked along its entire 15-mile (24-km) length from the CBD to the Pacific Ocean by an irregular string of high-rise office and residential buildings.

High-rise buildings became practical after the first primitive elevator was invented in 1852, but the skyscraper is a creature—and a symbol—of the twentieth century. On this continent the greatest concentration of tall buildings has always been in the older, more crowded cities of the Northeast; the newer, less intensively developed cities of the West tended to sprawl outward rather than upward. But since World War II, this pattern has changed; prominent skylines now sprout from such plainsland cities as Dallas, Denver, and Winnipeg and the trend extends to almost all large cities of the continent.

New York City still contains the greatest concentration of skyscrapers in the world; two-thirds of the world's tallest buildings and more than half of the nation's buildings over 500 feet (150 m) in height are located on Manhattan Island. The first of the world's "superskyscrapers" was erected in Lower Manhattan—the twin-tower, 110-story, 1350-foot (405-m) high World Trade Center; its 10 million square feet (0.9 million m^2) of office space are served by nearly 200 elevators. The world's tallest office facility, however, is the 1450-foot (435-m) Sears Tower in Chicago.[18] Indeed, Chicago is the only other Anglo-American city with a skyline that includes a number of very tall buildings, but its total is only about 20 percent as large as New York's. The construction of skyscrapers continues apace and the present roster of cities with tallest buildings (Boston, Chicago, Dallas, Houston, New York, Pittsburgh, San Francisco, and Toronto) will undoubtedly change from year to year (Fig. 4-14).

The skyscraper is a visible symbol of high land values[19] and of congestion. Although the skyscraper permits many more people to live and work in a restricted area, it also adds to traffic confusion. The streets of most cities were designed for smaller populations, lower buildings, and more limited movement, so they cannot carry the present traffic load without friction and delays. Traffic slows to a snail's pace in the very places where speed and promptness are most desired. The diffusion of high-rise building away from the CBD is a partial response to this problem.

[17] Although legal definitions vary, any structure over 75 feet (22.5 m) generally is considered to be "high rise" for purposes of fire and other codes.

[18] The tallest human-built structure on the continent is the CN Tower in Toronto, the topmost point of which is 1815 feet (545 m) above its base.

[19] A prestigious new skyscraper can command a rent in excess of \$1000 per square foot (0.09 m^2)

FIGURE 4-14 The tallest structure in any North American city is the CN Tower in Toronto (TLM photo).

General Appearance

Most Anglo-American cities are visually similar, a generalization that is a logical outgrowth of the morphological similarity that is chronicled on preceding pages. Within the CBD, tall buildings dominate the scene. Elsewhere in the urban area, even in the transition zone, the most conspicuous visual element consists of trees, generally rising above low-level residential and commercial rooftops. The visual dominance of trees is interrupted wherever there are extensive special-use areas, such as airports or planned industrial districts, but, in general, their pervasiveness can be seen in cities throughout Anglo-America.

Building ordinances and zoning restrictions, which tend to be similar from city to city, are another reason for the visual similarity of cities in the United States and Canada. Many examples could be cited, but perhaps the most prominent is the requirement that residences be set back from the street; thus in most parts of most cities front yards are required even though their functional role is largely a thing of the past.

A more detailed look at cities shows their many differences in appearance. Every city has a certain visual uniqueness on the basis of street pattern, architecture, air pollution, degree of dirtiness, and a host of other elements. But often such distinctiveness is a function of site (slope land versus flat land or coastal versus inland, for example), regional location (as the widespread adoption of "Spanish" architecture in the Southwest), or relative age.

The federal program of *urban renewal* in the United States, along with its other consequences, had a marked effect on the appearance of many cities. There was no similar national program in Canada, although provincial and municipal authorities inaugurated various urban renewal efforts on a much smaller scale. The idea behind urban renewal is simple enough: communities acquire large parcels of slum property (using the power of eminent domain where necessary) and sell or lease them for massive public or private redevelopment, using mostly federal funds for capital requirements. Slum clearance was often involved, but some projects provided for the conservation and rehabilitation of areas that did not require demolition.

An overriding consideration in urban renewal philosophy was the acceptance of the need for federal aid to revitalize the economic base and taxable resources of cities. Several thousand urban renewal projects were carried out under the program, with erratic results. The principal objections are that more low-income housing is removed than replaced and that costs are much greater than benefits. In any event, the replacement of slums by modern high-rise buildings and green spaces has changed the face of many American cities, from Boston's Bunker Hill to Los Angeles's West End. The greatest emphasis on urban renewal has been in the Northeast, particularly in the cities of Pennsylvania.

Extensive urban renewal activity was phased out in both countries in the 1960s. Government-

FIGURE 4-15 Old waterfront areas have been renovated and redeveloped in many cities. This is part of the Inner Harbor in Baltimore (TLM photo).

sponsored successor programs—called Community Development Block Grants in the United States and Neighborhood Improvement Programs in Canada—are on a much smaller scale.

A problem that most waterside cities have faced with continuing despair is the decay of their traditional waterfront area as the water transport function diminished or changed. Waterfronts often became seedy, disreputable, inhospitable districts that were largely avoided by the citizenry and ignored by the authorities. Despite the shining example of riverfront development (Paseo del Rio) begun in San Antonio in the 1930s, there were few concerted efforts at waterfront rehabilitation until the 1970s. In the last few years, however, many Anglo-American waterfronts, such as Chicago's Navy Pier, Baltimore's Inner Harbor, Boston's Lewis Wharf, Halifax's Historic Properties, Jersey City's Liberty Park, and Toronto's Harbourfront, have been transformed into shopping-recreational-residential complexes of considerable charm and attractiveness (Fig. 4-15).

URBAN FUNCTIONS: GROWING DIVERSITY

Besides being a morphological form that occupies space, a city is a functional entity that performs services for both its population and the population of its hinterland. In doing so, it may also provide services for people in more distant regions or even in foreign lands. Every Anglo-American city is multi-functional in nature, being involved in several different kinds of economic activities.

The Commercial Function

Most cities came into being as trading centers and the commercial aspect of the urban economy is almost invariably a major one regardless of city size or location. This function encompasses both retail and wholesale trade. In almost every major Anglo-American metropolitan area between 22 and 27 percent of the work force is employed in trade. There is a close correlation between a city's population and its component of retail trade, for sales from most stores are usually made to local residents. Wholesale trade, on the other hand, may vary more widely from a predictable norm. Wholesale trade is a minor element in many small cities whereas others, particularly larger metropolises, may be well situated to supply not only the retailers within the city but also those in smaller cities within their hinterland. Cities with gateway positions (either transportational or entrepreneurial gateways), such as Atlanta, Dallas, Memphis, Kansas City, Denver, Minneapolis–St. Paul, and Winnipeg, have an unusually large wholesale trade component in their economy.

The Industrial Function

A certain amount of manufacturing is carried on in any city worthy of the name, although it may largely consist of such prosaic factories as printing shops, ice-cream plants, and bakeries (Fig. 4-16). Larger

FIGURE 4-16 Many contemporary industrial areas are designed in attractive parklike settings. This scene is in Phoenix (TLM photo).

cities generally have a much broader range of manufactures. Some urban areas, of course, may specialize in one of several kinds of manufacturing; then this function becomes unusually important in the city's economy. Manufacturing employs from 15 to 30 percent of the work force in most major metropolitan areas, although there are significant variations above and below this range. Since about 1970, proportional employment in manufacturing (i.e., the percentage of the work force that draws its paychecks from factories) has been on a continual decline in virtually every major urban place in the United States and Canada, as both countries move toward a more service-oriented economy. No significant city has as much as 40 percent of its labor force employed in factories and in no million-population city does this figure exceed 30 percent. Many smaller cities have at least one-third of their total employment in manufacturing, especially in southern New England, the Carolina Piedmont, southern Michigan, and eastern Pennsylvania. Washington, D.C., and Honolulu are at the other end of the scale, with less than 5 percent of their employees working in factories.

The Service Function

The second fastest-growing component of the urban economy is the service sector, which includes a variety of professional, personal, and financial services, such as those provided by attorneys and real estate dealers. Most services do not produce tangible goods; thus, the service output is usually incapable of being stored and is consumed at the time it is produced. Except for the rapidly growing, high-tech oriented, computer-based professional services, service activities tend to be the least mechanized and least automated part of the economy; therefore a rapid increase in service output means a significant growth in the number of available jobs. Total employment in the service component of the economy now exceeds that of the trade component in both the United States and Canada.

The Administration Function

The rapid increase in government employment in both the United States and Canada has overshadowed all other growth sectors of the economy. Government employment expanded by approximately 50 percent in the 1950s, and roughly another 50 percent between 1960 and 1970, although the growth rate slowed somewhat in the 1970s and 1980s. The proportion of the labor force employed in government activities varies widely from city to city. It is expectably high in the national capitals of Washington and Ottawa and in such small-city state and provincial capitals as Lansing, Austin, Charlotte-

A CLOSER LOOK The Not-So-New Laws of Corporate Geography

If you believe the newspapers, the geography of corporate America is coming unglued. The reports seem unanimous: Corporate headquarters are departing the central city for locations in the green countryside, or fleeing the great northeastern cities altogether in favor of more agreeable climes in the air-conditioned Sunbelt. The very architecture of new headquarters underscores the magnitude of change. By now, the model is familiar to us all: a low-slung complex of lushly landscaped postmodern buildings, located in the countryside somewhere between the Carolina Piedmont and San Francisco Bay.

These new corporate complexes herald the arrival of a postindustrial economy: high-tech—service-oriented—clean—laid-back—seemingly free of traditional geographic rules. Theoretically, corporate headquarters can now be located anywhere—or, to paraphrase a recent report in *Business Week*, anywhere with access to a fax machine and a good airport.

In fact, that is not happening. Two geographic rules continue to govern the location of company offices, just as they have for a long time in America. One is *the rule of urban magnetism*. The other is *the rule of intra-urban clustering*.

The rule of urban magnetism. Recently, amid considerable hulabaloo, several major corporate headquarters have been built in geographic settings that are so bosky that a casual observer might conclude that American corporations are going rural. That image, however, is deceptive. Notwithstanding all the corporate verdure, American company headquarters are overwhelmingly located in urban areas, and the biggest corporations are mostly located in *big* urban areas. Indeed, a recent study has shown that when small local companies get successful and grow big, they do what ambitious adolescents have always done: They leave home and move to the nearest big city.

This idea runs counter to much of today's popular mythology and a fair number of newspaper stories. Several years ago, for example, it was widely reported that between 1968 and 1984, Manhattan lost over half of its Fortune 500 companies. But the report also noted that most of the departed firms did not move to Texas or Arizona, but instead to the city's exurbs. In short, they did not leave New York; they simply shifted locations *within* the metropolitan area. Indeed, of America's top 100 corporations listed in *Dun's Business Rankings* for 1988, one-third are still located in greater New York, and 17 are still quartered within a few blocks of each other on Manhattan Island.

The urban pull, furthermore, is not limited to New York—although New York remains the undisputed primate city of corporate America, especially in banking. The overwhelming majority (84 percent) of America's largest corporations have headquarters in metropolitan areas with populations over a million, including a fair number in cities that are popularly dismissed as Rustbelt basket cases: St. Louis, Pittsburgh, Detroit, and Cincinnati.

Even the spacious Sunbelt feels the pull of urban magnetism. In Alabama, for example, 40 of the state's 50 largest corporations maintain headquarters in or near Birmingham, Montgomery, or Mobile; none are located in the rural outback. Of Arizona's top 50 companies, 46 are situated in metropolitan Phoenix or Tucson. And so it goes, all across the country.

The rule of intra-urban clustering. Not only are American corporate headquarters overwhelmingly concentrated in urban areas, they are also highly clustered in select locations *within* those urban areas. Clustering is still most common in traditional downtown locations, sometimes because of historical inertia, sometimes to achieve corporate visibility. Very commonly, corporate headquarters are heavily concentrated in one sector of the city, as in northwest Dallas, for example, and that sector almost always includes the metropolitan areas' wealthiest neighborhoods. (America's corporate officers resemble other people in their dislike of commuting, but, unlike their proletarian fellow-citizens, have the power to enforce their wishes.)

Almost without exception, corporate offices are found clustered near places of maximum accessibility within urban networks. Sometimes they are lined up along interstate highway corridors that lead from the central city to well-heeled exurbs, such as Montgomery County, Maryland's I-270 corridor northwest of Washington. Urban beltway interchanges are notorious breeders of corporate clusters. Boston's Route 128 and the Washington Beltway are classic examples.

Theoretically, such things should not be happening. Given our newfound ability to transmit information and money rapidly and cheaply over long distances, postindustrial offices do not need to locate close to one another. Fax machines are everywhere, and Bangor, Maine, has a superlative airport.

Maybe so, but don't hold your breath, waiting for the stampede to Bangor. Corporate headquarters are not located by the iron hand of the marketplace, but rather by very human decisions. The fact remains that corporate decision-makers are social animals, slow to abandon old habits or old property, and slower still to move very far away from others of their species. As long as that is true, corporate power will remain concentrated in a small number of select locations. Meantime, it is hard to imagine that corporate headquarters will move very far away from big cities, or away from one other.*

Professor Peirce Lewis
Pennsylvania State University
University Park

* A version of this essay appears in the March, 1990 issue of *Landscape Architecture*.

town, and Fredericton. But there are many categories of government employment and a significant proportion of the jobs in this sector are in municipal and county governments; thus most Anglo-American cities have a prominent component of employment in public administration.

Other Urban Functions

Two other urban functions are common: transportation-communication and contract construction. Employment in both has been increasing through the years but at a slower rate than most other sectors of urban economy; consequently, the relative position of the two slowly continues to decline. In most cities each activity employs about 5 percent of the labor force. A few other specialized categories of employment are found in some cities, although only "mining" (which generally means office employees of mining companies) shows up enough to be prominent in individual cases.

URBAN POPULATION: VARIETY IN ABUNDANCE

Whatever else it is, a city is primarily a group of people who have chosen to live in an urban environment because of economic opportunities, amenities that are close at hand, or inertia. The pell-mell rate of urbanization that characterized the 1950s and 1960s slowed down subsequently, partly because of a trend toward nonmetropolitan living but partly because the vast majority of the populace already lived in cities.

We already noted that the population of central cities in metropolitan areas, particularly the largest metropolitan areas, stagnated or declined in recent years, with all or most growth being confined to the expanding suburbs. Indeed, it was during the 1950s that many large cities had their last great growth, having actually lost population within their political limits since that time. One notable result is that contemporary urbanites have more living space than they did a decade or two ago: The population per square mile in Anglo-American urbanized areas decreased by more than 50 percent in the last half century.

The population structure of cities has many aspects. One of the most important is ethnicity.

Anglo-American cities have a remarkably varied array of ethnic mixes. The variety generally follows a regional pattern, but some cities show significant variations from the regional trend. It should be noted that it does not require a very large number of people of a particular ethnic group to put a significant imprint on a city or on that city's image.

The situation in Canadian cities is quite different from that in U.S. cities. Every Canadian city of any size is at least 40 percent British or 40 percent French in terms of the lineage of the population. This bicultural bifurcation identifies the two main groups in any ethnic classification of Canadian urban areas. Figure 4-17 illustrates the magnitude of the bifurcation. The average Canadian urbanite is an Anglo-Saxon Protestant, normally of British descent but sometimes of German or Dutch extraction. In the province of Quebec, on the other hand, urban dwellers are overwhelmingly French; only in Montreal is the population less than 80 percent French.

There is a varied although largely European ethnic mix among the non-British, non-French minority in Canadian cities; however, there is no large, underprivileged, vocal ethnic group around which racial antagonisms cluster. Canadian urban minority groups tend to be upwardly mobile in both social and economic status. Ethnic ghettos occur in Canadian cities, but they are not—except for some French-Canadian areas in predominantly Anglo-Canadian cities—particularly conspicuous in most cases. The most clustered and easily identified small ethnic minorities are probably the Chinese (particularly in Vancouver and Toronto), Hungarians, Greeks, Portuguese, and West Indians (Fig. 4-18).

In cities south of the international border black Americans are often a prominent minority that is all the more conspicuous because of the racially segregated housing pattern that usually prevails and the generally low economic level of the group as a whole. We have noted that blacks live predominantly in ghettos throughout the nation. The central-city ghettos continue a rapid outward expansion, but only very recently is the black populace beginning to deconcentrate from the central cities, due to an increased rate of black suburbanization and a declining central-city growth rate. Blacks are particularly concentrated in the large cities; one-third of all black Americans reside in ten cities and more than one-

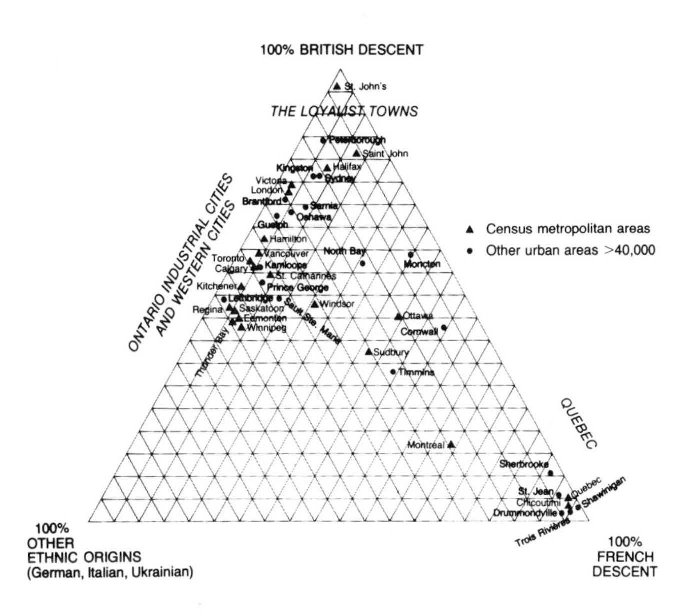

FIGURE 4-17 The ethnic pattern of Canadian cities (1971). This chart indicates the proportion of the city's population composed of each of three different ethnic origins. The closer a city is to a point of the triangle, the more homogeneous is its origin. Note the extreme homogeneity of most Quebec cities, and the spectrum of British and other origins in the remainder of the country. (From J. and R. Simmons, *Urban Canada,* 2nd ed. [Toronto: Copp Clark Publishing Company, 1974], p. 40. By permission of Statistics Canada.)

fourth are concentrated in only six metropolitan areas.

Hispanic-Americans are another prominent urban minority, although their numbers are fewer and their distribution is less widespread than that of blacks. In cities of the Southwest, from California's Central Valley to the Piedmont of north-central Colorado to the east Texas metropolises of Dallas and Houston, Hispanics are frequently found in large numbers. There are other concentrations of this minority in Chicago (largely of Mexican origin), the major cities of Megalopolis (primarily Puerto Ricans), and various urban areas in Florida (mostly Cubans). Although less ghettoized than urban blacks, Hispanics are often clustered in distribution and their generally low economic status is likely to be reflected in slum housing.

FIGURE 4-18 A sidewalk market scene in North America's largest Chinatown, on New York's Lower East Side (TLM photo).

URBAN ILLS: MASSIVE MALADJUSTMENTS

The big city everywhere is the object of criticism. Critics insist that all cities are ailing and are not good places in which to live. They point to smog, crowding, strain on family life, snarled traffic, segregation of minorities, crime, drugs, youth gangs, violence, impersonality, and a host of other urban evils. The long-continued growth of our cities begets massive growing pains.

And urban sprawl continues apace. For example, a two-decade (1966–1986) study of the 70 largest cities in Canada showed that 744,500 acres (301,400 ha) of rural land—an area three times the size of urbanized Toronto—were converted to urban uses.

A CLOSER LOOK Cities and Towns with the Greatest Ethnic Diversity

Most Americans are interested in how the local areas they know are like or unlike other areas, and the rating of cities as to their quality of life has been popular topic for conversation and news reporting. One characteristic of places is their ethnic composition. Because different ethnic groups have long been concentrated in certain regions and cities and because migrations have elaborated older settlement patterns, ethnic variety has become especially intensified in certain cities and towns. Let's ask the question: What places in the United States have the most ethnic diversity? My colleague, Professor Eugene Turner, and I have studied this matter, and our statistical analyses permit some fairly precise answers.*

In modern America there appear to be five basic ethnic groupings: white, black, American Indian, Asian, and Hispanic. Although elsewhere we have examined patterns of diversity when ethnicity is defined in terms of more detailed groupings like Asian Indian and Mexican, here we focus on only the five categories. Diversity can be measured by a statistic called the *entropy index,* which calculates the relative evenness of ethnic group sizes and gives the highest possible value when all groups are present in equal numbers. (However, the index tells us nothing about the social status of the groups or how much the different groups mix socially or occupy the same neighborhoods.) Computer tapes from the U.S. Census of 1980 contain detailed information of the ethnic identities of local populations. Because cities and towns are the most basic local units with which people easily identify, this research analyzes the diversity in all the 2903 urban places with over 10,000 inhabitants. The accompanying map shows those 60 places that ranked highest (Fig. 4-c).

* For a more detailed presentation of this research see James P. Allen and Eugene Turner, "The Most Ethnically Diverse Urban Places in the United States," *Urban Geography,* 10 (1989), 523–539.

Perhaps most Americans would say that ethnic diversity is greatest in the largest cities, and indeed, some of the most diverse urban places are those largest cities (shown in bold type on the map). Los Angeles and San Francisco rank 8th and 10th in diversity among all places, and New York City ranks 27th, although two cities of over a million people each—Philadelphia and Detroit—are much less diverse than one might expect from their size.

If we next look at the locations of the highly diverse places, the map shows that over half are in California, with many of the rest in Texas, Florida, and New Jersey. These states (together with New York City) have been the leading destinations for immigrants in recent years, with California alone receiving about a quarter of all immigrants to this country. Because Asians and Latin Americans make up such a large portion of our recent immigrants, places that become homes to immigrants obviously become more diverse.

Our research demonstrated that highly diverse urban places are often not large and typically represent several different types of places: older industrial cities, cities in metropolitan areas (including newer suburban cities), farm labor towns, and military installations.

The older industrial cities include Hartford, Chicago and East Chicago; all five New Jersey cities; and Lynwood, Richmond, San Pablo, and Pittsburg in California. These places were once prominent manufacturing centers, and those cities still provide homes for many workers and former workers, both whites and blacks, as well as Puerto Ricans and Mexicans depending on the region of the country. For example, Chicago and East Chicago first received large numbers of black workers from the South and Mexican workers from Texas when jobs in steel mills, oil refineries, etc., opened as the United States geared up for World War I and the supply of European immigrant laborers was cut off by German submarine attacks in the Atlantic Ocean. Hartford's ethnic diversity is particularly due to the many Puerto Ricans and Jamaicans who began to settle there in the 1940s; and more recently Cubans, Filipinos, and Asian Indians have moved into New Jersey cities.

Most of the highly diverse California places are independent cities in either the San Francisco or Los Angeles areas, just as the most diverse places in Florida are part of the greater Miami area. In California these places have received many Asians in recent decades, as well as whites, blacks and people of Mexican origin, while in Florida there have been fewer Asians and Mexicans but more blacks, Cubans, and Puerto Ricans. For example, Daly City, California, was one of the first suburban destinations for Chinese from San Francisco's Chinatown, and the Japanese in Gardena originally worked on farms and operated nurseries in what was formerly an agricultural area but is now home to many workers with jobs in nearby industries. Cerritos is a new independent suburb in the Los Angeles area, unusual in having by far the highest median income of all the 60 highly diverse places. In Florida the populations of Carol City and Pinewood have approximately equal numbers of blacks and whites in addition to many Puerto Ricans and Cubans.

Two places—Immokalee in Florida and Delano in California—are diverse primarily because so many inhabitants counted in the Census were agricultural laborers, and in both Oxnard and Stockton former or current farm workers accentuate the ethnic diversity. Farm workers represent a racially varied group, with people of Mexican origin predominating in California and often in Florida but with blacks typically numerous among East Coast migratory workers and Filipinos often retired from or occasionally still part of California crews.

The last group of highly diverse places are military installations and areas associated with them. Army

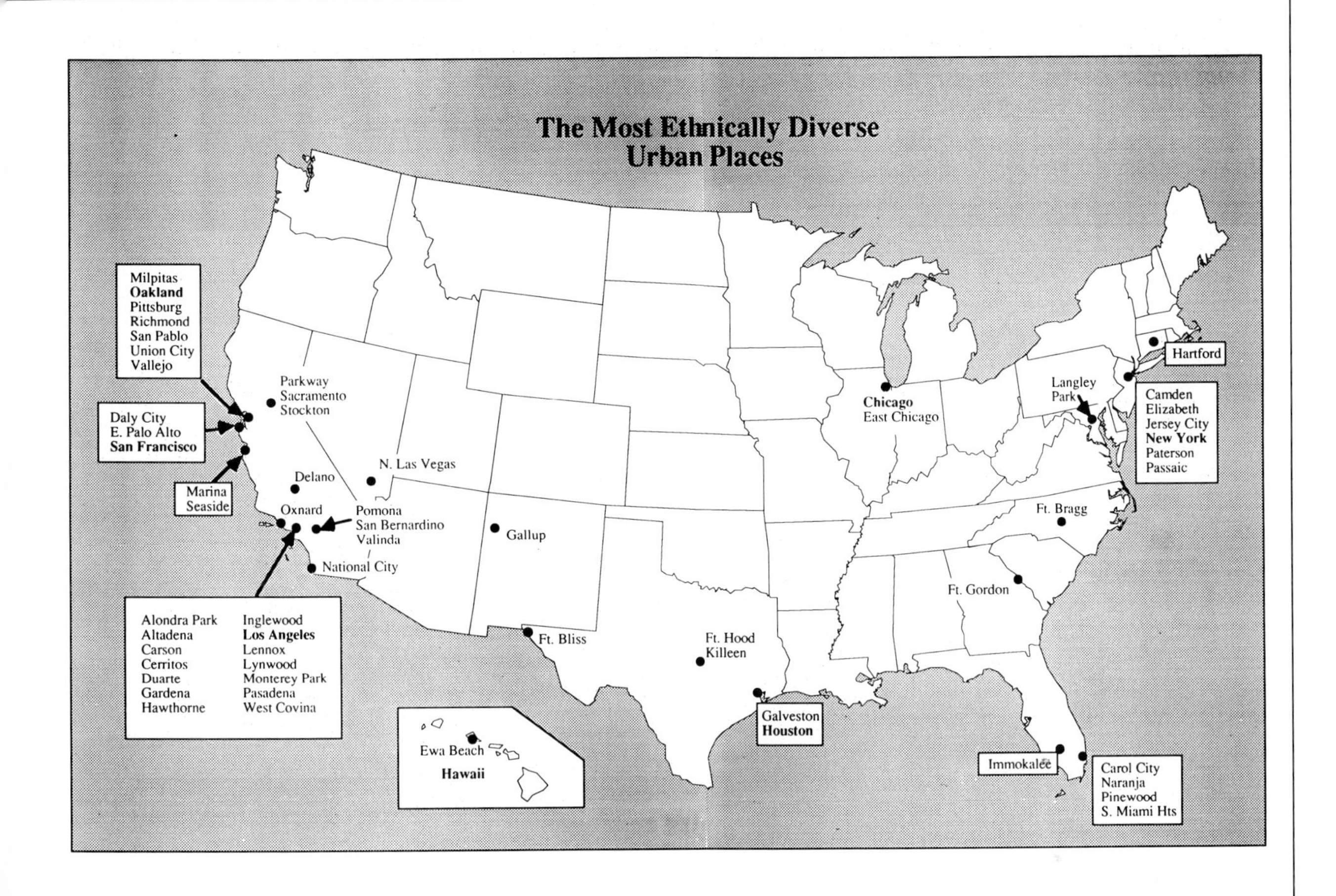

FIGURE 4-c

posts are often especially diverse because blacks usually represent over 30 percent of the population, soldiers of Mexican or Puerto Rican origin make up another minority, and some soldiers have married Asian women while overseas. Navy bases are also sometimes highly diverse, typically with attached Filipino communities because the U.S. Navy has long provided a special avenue of advancement and immigration for young men who first sign up in the Philippines as stewards.

Army posts are evident on the map as forts, such as Fort Gordon and Fort Bragg, but other places are not so readily identified as being connected with the military. In California, many people stationed at Fort Ord live in adjacent Marina and Seaside. Also, National City is integrally tied to the large San Diego navy base, Oxnard has two navy installations, and Vallejo is associated with a major naval shipyard.

Altogether, this research has shown that several different types of places, including many small cities and towns, often exhibit a high degree of ethnic diversity.** However, there are no data to indicate whether interethnic relations in any place are relatively harmonious or whether there are substantial tensions between groups because of cultural or social class differences or economic competition between groups for jobs and homes. Perhaps you could suggest some hypotheses concerning what local situations might tend to produce interethnic harmony as opposed to tensions and suspicions in a town or city. Do you think that a list of most ethnically diverse places will have changed much during the 1980s, or do immigrants and minorities tend to settle in places that are already diverse?

** At the other end of the scale, the least diverse places in the United States tend to be small in size, nearly all white, and often nonmetropolitan. The 10 least diverse urban places are Berlin, N.H.; Saco, Me.; Swampscott, Mass., Nanticoke, Shamokin, Warren, and Whitehall, Penn.; Morton, Ill.; Monroe, Wisc.; and Sun City, Ariz.

Professor James P. Allen
California State University, Northridge

Similar conditions undoubtedly prevail in the United States.

As urban areas expand, provision of such necessities as water, sewerage, paved streets, utilities, refuse collection, police and fire protection, schools, and parks is a continuing headache, particularly when more than one municipal governing body is involved. Also, as the flight to the suburbs continues, it is generally accompanied by a degeneration of much of the core of the city; the results are intensified slums, loss of merchandising revenue, and a decline in the tax base.

As the metropolitan area expands, local transportation becomes more complicated. As many more cars drive many more miles on only a few more streets, relatively speaking, traffic congestion becomes intense, the journey to work lengthens, and parking facilities become inadequate. The big city must maintain constant vigilance to keep from choking.

The maintenance of enough good domestic water also challenges the exploding metropolis. In subhumid regions cities must sometimes reach out dozens or hundreds of miles to pipe in adequate water; even such humid-land cities as New York and Boston must extend lines farther and farther to tap satisfactory watersheds.

Where humans congregate, the delicate problem of pollution is accelerated. Rare indeed is the stream in any urban area that is not heavily infiltrated by inadequately treated liquid waste from home and factory. The shocking condition of American waterways has caused some civic groups to wage stringent cleanup campaigns, with emphasis on adequate sewerage treatment. The result has been heartening improvements in such infamous rivers as the Ohio and Philadelphia's Schuylkill—improvements showing that this problem can be solved in other areas. Today the menace of atmospheric pollution is recognized as a major problem (Fig. 4-19). The highly (and justifiably) publicized smog of Los Angeles is the most striking instance, but "smust" in Phoenix, "smaze" in Denver, and smoke in Montreal are further examples of an undiminishing phenomenon in most large cities. Industrial vapors and burning refuse contribute, but auto-

FIGURE 4-19 Very few large North American cities have escaped the problem of smog. This is Denver on a quiet winter day (courtesy NOAA).

mobile exhaust fumes are generally believed to be the major cause. The air pollution problem will undoubtedly get worse before it gets better.

Wherever people live close together, social friction escalates and the city is the seat of continually burgeoning social problems. Crime statistics increase, with the annual rate climbing above one crime for every 30 urbanites and one violent crime for every 250 urbanites. Similar, or worse, trends can be found for gang membership, substance abuse, and alcoholism.

Coping with the ills of the city and planning to avoid unending escalation of these problems are immensely complicated by the fragmentation of administrative responsibility that is so widespread in the United States, although less so in Canada. Broad planning is inhibited and effective implementation of general solutions is prevented by the multiplicity of municipalities, townships, counties, school districts, and other special districts (everything from cemetery districts to mosquito abatement units). There are more than 85,000 units of government in the United States, most of them empowered to levy taxes. The average 1-million-population metropolitan area in the United States encompasses nearly 300 separate governmental jurisdictions. A few attempts at formation of a metropolitan government for large cities and their suburbs have been mooted, but as yet nothing very effective has resulted.

The situation is less chaotic in Canada, primarily because provincial governments have considerable power to regulate municipal institutions. Canadian cities are not surrounded by as many small independent municipalities as their counterparts in the United States and urban problems can be approached on a broader front. The most far-reaching attempt at metropolitan government was established in Ontario in 1953, when Metro Toronto was created. It has been quite successful in providing integrated services to the metropolitan area for which high capital expenditures were necessary; other aspects of its operation have been less satisfactory. Metro Toronto is at least an innovative guidepost for the future; subsequent experiments along the same line, as in Winnipeg Unicity, will be watched with interest by urban administrators throughout Anglo-America.

URBAN DELIGHTS: THE PROOF OF THE PUDDING

If cities are such bad places, why do so many people live there? Anglo-Americans still flock to urban areas and most metropolitan areas continue to increase in size. Clearly the number of urban critics is significantly less than the number of people who choose to live in urban areas. Disadvantages may be legion, but the attractions are also numerous. Most notable, probably, is the ready availability of material and nonmaterial satisfactions. A vast quantity of goods and services can be purchased, a variety of entertainment may be sampled, and a plethora of mass media provides almost unlimited information and mental stimulation. For many urbanites, however, it appears that the opportunity for social interaction may be even more important than economic advantage or amenity attraction: In cities one can expect to find like-minded people with whom to interact (in a process that Vance called "congregation"). For most people, this combination of positive attributes apparently far outweighs the detrimental aspects of urban life.

URBAN DICHOTOMY: CENTRAL CITIES VERSUS SUBURBS

For several decades Anglo-American urban areas were dichotomized into a dualistic division between the cores and the suburbs. It is only in the last few years, however, that the suburban peripheries experienced such spectacular growth, not only in residential population but also in many other functions that were traditionally restricted to the central areas. As a result, Anglo-American metropolitan areas "have undergone a remarkably swift spatial reorganization from tightly focused single-cores to decentralized multinodal" systems.[20] At first there was merely a cautious outward drift of store clusters in the wake of new residential subdivisions, but the almost instantaneous success of these retailing pioneers encouraged bolder innovations, which were epitomized by the development of immense regional

[20] Peter O. Muller, "Toward a Geography of the Suburbs," *Proceedings*, Association of American Geographers, 6 (1974), 36.

shopping complexes that attracted other kinds of employers to their vicinity. The transportational convenience of suburban freeway and beltway locations encouraged successive waves of retailing, wholesaling, manufacturing, and service-oriented activities to abandon the CBD and shift to the suburbs.

Thus suburbs no longer function merely as bedroom communities, but as "integrated complexes of tall office structures, industrial parks, regional shopping malls, and residential subdivisions."[21] The process involves a reclustering of office, manufacturing, retail, and service activities into more or less distinct suburban nucleations.

Central cities thus provide a continually decreasing variety and magnitude of functions. "The leading residual activity is office functions, a set of interdependent activities requiring face-to-face contact and the external economies of the CBD's specialized services."[22] Increasingly, however, office activities are also moving to the suburbs. Not that the CBDs are dying. On the contrary, a resurgence of downtown construction activity in recent years has increased the availability of white-collar jobs in central cities. In a sense this represents the "suburbanization" of the CBD—a "diffusion toward the downtown of low-perceived-density, auto-convenient landscape elements and forms initially tested and found successful and tasteful in suburbia."[23] A significant proportion of this downtown investment is seen as "defensive," a civic responsibility to the wider urban community to maintain the economic viability of the core. "These new investments represent what could well be the last major defense of the central city against the persistent and ever-growing challenge of the suburbs."[24]

The crux of the dichotomy revolves around money and jobs. A very large share of suburban growth is beyond the political limits of the central cities, which severely undercuts their tax base and payrolls. Thus there are relatively prosperous suburban communities and destitute central cities. Suburbanites largely ignore the central cities and inner-city blue-collar workers are often unable to find replacement jobs downtown and must face either unemployment or the prospect of long-distance reverse commuting to outlying industrial concentrations. The heightened unemployment problem is only one aspect of the increasing social unheavals resulting from inner city economic stagnation, or what George Sternlieb called the "defunctioning" of the central city.

Two-thirds of our metropolitan population now resides in suburbs. Urban analyst Peter Muller concluded that the economic decline of central cities "has become irreversible" and that "suburban dominance within the metropolis will persist and intensify."[25]

Clearly our cities will never again be the same. An understanding of the varied geography of the suburbs is critical to any comprehension of the future geography of Anglo-America.

URBAN DECLINE? THE POPULATION TURNAROUND

Suburban growth and central-city decline may not be the most significant urban trend of recent years. The 1970s experienced a reversal of the metropolitan growth syndrome, something that had never happened before. For the first time nonmetropolitan population grew at a faster rate than metropolitan population: 15 percent versus 10 percent for the period between 1970 and 1980. For the first time there was a relative decline in metropolitan dominance: 75 percent of our populace resided in metropolitan areas in 1980 compared to 76 percent in 1970.

This counterurbanization trend was much less noticeable in Canada for various reasons.[26] From the 1971 to the 1981 census the metropolitan area population of Canada increased from 55 to 56 percent.

[21] Joseph S. Wood, "Suburbanization of Center City," *Geographical Review*, 78 (July 1988), 325.

[22] Muller, "Toward a Geography of the Suburbs," 38.

[23] Wood, "Suburbanization of Center City," 325.

[24] Gerald Manners, "The Office in Metropolis: An Opportunity for Shaping Metropolitan America," *Economic Geography*, 50 (April 1974), 95.

[25] Peter Muller, "Suburbia, Geography, and the Prospect of a Nation without Important Cities," *Geographical Survey*, 7 (January 1978), 16.

[26] For an exposition of this situation, see Yeates and Garner, *The North American City*, p. 235.

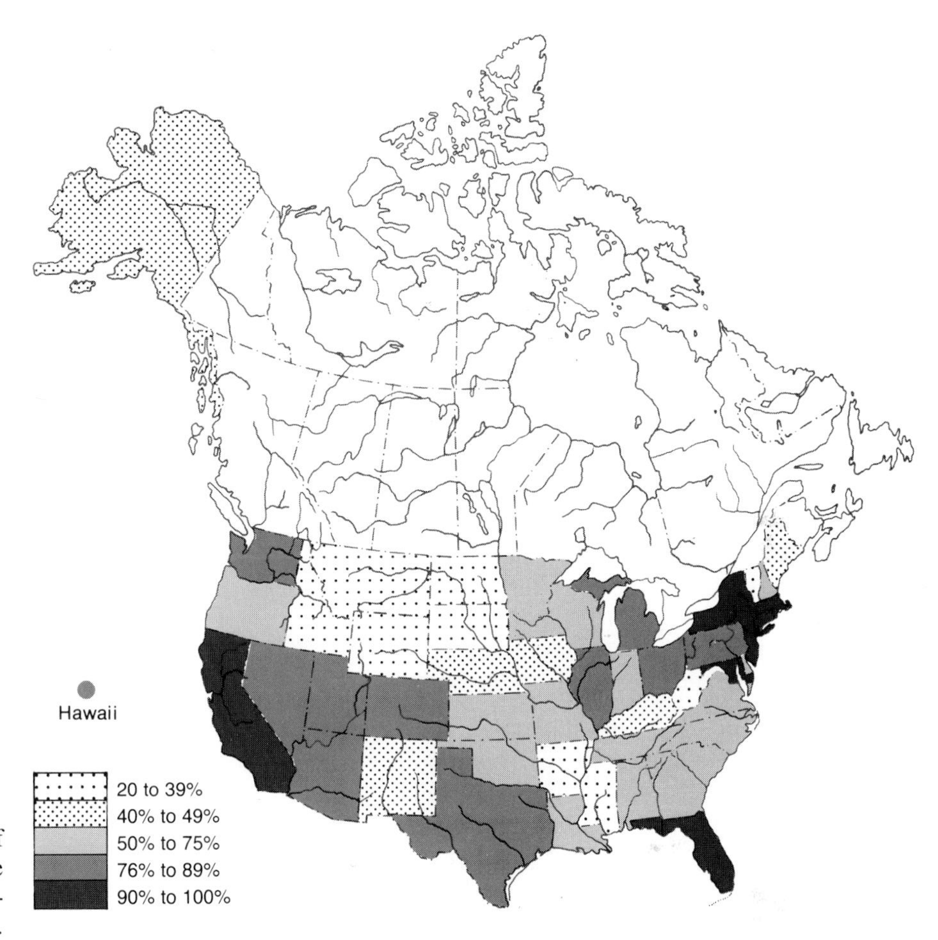

FIGURE 4-20 Proportion of state population living in large (more than 1 million population) metropolitan areas, 1990.

The population turnaround varied considerably in different situations. It was felt most strongly in northeastern United States and eastern Canada; in Canada metropolitan areas grew slowly during the 1970s whereas almost all metropolitan areas in northeastern United States actually declined in population.

After a decade, the nonmetropolitan population turnaround turned around again. Beginning about 1980, the rate of metropolitan population growth once again exceeded that of nonmetropolitan population, and metro growth has surpassed nonmetro growth at an accelerating rate ever since, although not as predominantly as in the pre-1970 period. Urban growth is obviously changing, but most people are still urbanites (Fig. 4-20).

URBAN TOMORROW: THE OUTREACH OF CITY LIFE

What will tomorrow's city be like? Many radical designs and grandiose predictions have been made, but only one thing is clear: The urban place of the future will be a combination of elaborate planning and unstructured eventuality.

The surest bet for the short run is the continuance of urban sprawl. The areal expansion of individual cities will result in the increasing coalescence of adjacent metropolitan areas and the creation of more supercities around the western end of Lake Ontario, around the margin of Lake Erie, around the southern end of Lake Michigan, and along the coast of southern California. The metropolis expands and so do the suburbs and ghettos.

Innovations in the use of urban space will surely proliferate, although surrealistic cities of the future may still be a few decades away. Many cities have experimented with pedestrian malls and walkways, a feature that is likely to continue. The advantages of separating pedestrian and vehicular traffic are clear and in intensive-use areas such separation is worth almost any cost. Increasingly, buildings will also be constructed above transportation routes. The concept of air rights above railway tracks, highways, and transport terminals is now well accepted and will be resorted to more and more.[27] Skyscrapers will continue to grow; buildings of 250 to 300 stories are freely predicted. Subterranean expansion is also to be expected—shopping areas and underground delivery systems for people, goods, and garbage.

New forms of city structure continue to develop. Here are some (futuristic ???) examples.

1. "Outer city" development, also called "satellite downtowns." This represents a centrifugal force in urban development in which a sprawling urban zone develops peripheral to an established metropolis. Prime examples are Orange County near Los Angeles, "Silicon Valley" south of San Francisco, Nassau–Suffolk on outer Long Island, the area between Dallas and Forth Worth, and the "gold coast" north of Miami. These are not commuter locations; they represent a clear displacement of economic and political power away from the adjacent metropolis and become job magnets in their own right.
2. "Community associations." These are clustered congregations of homeowners that function as small democratic units that assess fees and levy fines, contract with maintenance and security firms, and perform many duties normally associated with small-town governments. It is estimated that some 15 million Americans already belong to such associations, ranging from huge planned communities (e.g., Sun City, Arizona) to small clusters of condominiums (Fig. 4-21). There are now more than 50,000 such associations in the country (almost all spawned since 1970). The development of planned communities is guided chiefly by economic and security factors; they are usually access controlled. The greatest proliferation thus far has been in Florida, where perhaps a fourth of the population lives in condominiums or planned communities.
3. "Movable towns." Much of our population is footloose and shifts from place to place, often following the seasons. This rootless population frequently resides in mobile homes or campers and may live in several different "towns" in a single year. These movable towns are particularly notable along the Colorado River in western Arizona, in the Lower Rio Grande Valley of Texas, and in parts of Florida. Residents are often senior citizens.

Regardless of futuristic trends, all indications are that the relative importance of the major metropolis is going to decline as single-city foci of the past become metamorphosed into polycentric metropolitan forms in which multifarious outlying activity concentrations rival, or outstrip, the traditional CBD. The term "galactic city" is increasingly used to refer to such a loose, separated network of urban activity clusters. It also seems clear that the city of the future will be a place where, increasingly, communication is substituted for the physical movement of people. This would portend further deconcentration because the activity centers presumably could be located wherever there was access to global computer and satellite networks.[28]

If the morphology of Anglo-American urban areas is changing, so is the lifestyle of its citizens. "A new kind of large-scale urban society is emerg-

[27] Prominent examples of the use of air rights include one of New York's largest office buildings, the Pan-American Building, over a railway station; Chicago's Merchandise Mart, the world's most spacious office building, over railway lines; and part of the grounds of the United Nations complex over a major Manhattan parkway.

[28] Peter O. Muller, "Transportation and Urban Growth: The Shaping of the American Metropolis," *Focus*, 36 (Summer 1986), 17.

FIGURE 4-21 One of the most famous of the entirely planned retirement communities is Del Webb's Sun City near Phoenix (courtesy Del E. Webb Development Company).

ing that is increasingly independent of the city.[29] The distinction between urban and rural is blurred. Cities can no longer be understood as configurations of population density, for the functional linkages among their far-flung parts and the interactions with nonmetropolitan peripheries are so complex that urban influence extends well beyond the traditional boundaries of metropolitan areas (Fig. 4-22). Urban researchers have increasingly turned to wider reaches. Such concepts as *urban field* and *daily urban system* have been developed to incorporate these vaster functional zones that include not only the central cities and the suburbs but also the extensive, urban-related commuting fields in the nearby hinterlands.[30] In such conceptualizations, well over 90 percent of all Anglo-Americans are urbanites, for better or for worse.

[29] Melvin M. Webber, "The Post-City Age," in *Suburbanization Dynamics and the Future of the City*, edited by James W. Hughes, p. 246. New Brunswick, NJ: Center for Urban Policy Research, Rutgers University, 1974.

[30] The term *urban field* was introduced in John Friedmann and John Miller, "The Urban Field," *Journal of the American Institute of Planners*, 31 (November 1965), 310–321; the term *daily urban system* was coined by C. A. Doxiadis, but the concept has been most thoroughly developed in Anglo-America by Brian Berry (see, for example, *Growth Centers in the American Urban System*, 2 vols. [Cambridge, MA: Ballinger Publishing Co., 1973]).

FIGURE 4-22 The dichotomy of urban Anglo-America is symbolized by the massive edifices of Midtown Manhattan (occupying most of the photo) and the distant sprawl of conspicuous buildings on the New Jersey side of the Hudson River (courtesy New York State Department of Commerce).

SELECTED BIBLIOGRAPHY

ABLER, RONALD, JOHN S. ADAMS, AND KI-SUK LEE, *A Comparative Atlas of America's Great Cities: Twenty Metropolitan Regions*. Minneapolis: University of Minnesota Press, 1976.

ADAMS, JOHN S., ED., *Urban Policy-Making and Metropolitan Dynamics: A Comparative Geographical Analysis*. Cambridge, MA: Ballinger Publishing Company, 1976.

ADAMS, RUSSELL B., "Metropolitan Area and Central City Population, 1960–1970–1980," *Annales de Geographie*, 81 (1972), 171–205.

ARTIBISE, ALAN F. J., *Town and City: Aspects of Western Canadian Urban Development*. Regina, Saskatchewan: University of Regina, Canadian Plains Research Center, 1981.

BAERWALD, T. J., "Emergence of a New 'Downtown'," *Geographical Review*, 68 (July 1978), 308–318.

BERRY, BRIAN J. L., AND QUENTIN GILLARD, *The Changing Shape of Metropolitan America: Commuting Patterns, Urban Fields, and Decentralization Processes*. Cambridge, MA: Ballinger Publishing Company, 1977.

BORCHERT, JOHN R., "America's Changing Metropolitan Regions," *Annals*, Association of American Geographers, 62 (1972), 352–373.

BRADBURY, KATHARINE L., ANTHONY DOWNS, AND KENNETH A. SMALL, *Urban Decline and the Future of American Cities*. Washington, D.C.: The Brookings Institution, 1982.

BRUNN, STANLEY D., AND JAMES O. WHEELER, EDS., *The American Metropolitan System: Present and Future*. Washington, D.C.: V. H. Winston & Sons, 1980.

BUNTING, TRUDI E., AND PIERRE FILION, *The Changing Canadian Inner City*. Waterloo, Ont.: University of Waterloo, 1988.

CERVERO, ROBERT, *Suburban Gridlock*. New Brunswick, NJ: Rutgers University, Center for Urban Policy Research, 1986.

CONZEN, MICHAEL P., "American Cities in Profound Transition: The New City Geography of the 1980s," *Journal of Geography*, 82 (May–June, 1983), 94–101.

———, "The Maturing Urban System in the United States, 1840–1910," *Annals*, Association of American Geographers, 67 (1977), 88–108.

DINGEMANS, DENNIS J., "The Urbanization of Suburbia: The Renaissance of the Row House," *Landscape*, 20 (October 1975), 20–31.

DUNN, EDGAR S., JR., *The Development of the United States Urban System*. Baltimore: Johns Hopkins University Press, 2 vols., 1980 and 1983.

FORD, LARRY R., "The Urban Skyline as a City Classification System," *Journal of Geography*, 75 (1976), 154–164.

GRIFFIN, DONALD W., AND RICHARD E. PRESTON, "A Restatement of the 'Transition Zone' Concept," *Annals*, Association of American Geographers, 56 (1966), 339–350.

HARRIS, CHAUNCY D., "A Functional Classification of Cities in the United States," *Geographical Review*, 33 (1943), 89–99.

HART, JOHN FRASER, "The Bypass Strip as an Ideal Landscape," *Geographical Review*, 72 (April 1982), 218–223.

HARTSHORN, TRUMAN A., "Inner City Residential Structure and Decline," *Annals*, Association of American Geographers, 61 (1971), 72–96.

———, *Interpreting the City: An Urban Geography*. New York: John Wiley & Sons, 1980.

JACKSON, JOHN N., *The Canadian City: Space, Form, Quality*. Toronto: McGraw-Hill Ryerson, 1973.

JOHNSON, R. J., *The American Urban System: A Geographical Perspective*. New York: St. Martin's Press, 1982.

JONES, EMRYS, "American Big Cities," *Geographical Journal*, 145 (March 1979), 106–108.

LAI, DAVID CHUENYAN, *Chinatowns: Towns Within Cities in Canada*. Vancouver: University of British Columbia Press, 1988.

LANG, MICHAEL H., *Gentrification Amid Urban Decline: Strategies for America's Older Cities*. Cambridge, MA: Ballinger Publishing Company, 1982.

LAW, CHRISTOPHER M. WITH E. K. GRIMES, C. J. GRUNDY, M. L. SENIOR, AND J. R. TUPPER, *The Uncertain Future of the Urban Core*. New York: Routledge, Chapman and Hall, 1988.

MANNERS, GERALD, "The Office in Metropolis: An Opportunity for Shaping Metropolitan America," *Economic Geography*, 50 (1974), 93–110.

MARSHALL, JOHN U., "City Size, Economic Diversity, and Functional Type: The Canadian Case," *Economic Geography*, 51 (1975), 37–49.

———, "Industrial Diversification in the Canadian Urban System," *Canadian Geographer*, 25 (Winter 1981), 316–332.

MAYER, HAROLD M., "Cities: Transportation and Internal Circulation," *Journal of Geography*, 68 (1969), 390–408.

MERCER, JOHN, "On Continentalism, Distinctiveness, and Comparative Urban Geography: Canadian and American Cities," *Canadian Geographer* 23 (Summer 1979), 119–139.

MONKKONEN, ERIC H., *America Becomes Urban: The Development of U.S. Cities 1780–1980*. Berkeley and Los Angeles: University of California Press, 1988.

MULLER, PETER O., *Contemporary Suburban America*. Englewood Cliffs, NJ: Prentice-Hall, 1981.

———, *The Outer City: Geographical Consequences of the Urbanization of the Suburbs*. Washington, D.C.: Association of American Geographers, Resource Papers for College Geography, 72–75, 1976.

———, "Suburbia, Geography, and the Prospect of a Nation without Important Cities," *Geographical Survey*, 7 (1978), 13–19.

———, "Transportation and Urban Growth: The Shaping of the American Metropolis," *Focus*, 36 (Summer 1986), 8–17.

MURPHY, RAYMOND E., *The American City: An Urban Geography* (2nd ed.). New York: McGraw-Hill Book Co., 1974.

NELSON, HOWARD J., "The Form and Structure of Cities: Urban Growth Patterns," *Journal of Geography*, 68 (1969), 198–207.

———, "A Service Classification of American Cities," *Economic Geography*, 31 (1955), 189–210.

PALM, RISA, *The Geography of American Cities*. New York: Oxford University Press, 1981.

ROSE, HAROLD M., "The Development of an Urban Subsystem: The Case of the Negro Ghetto," *Annals*, Association of American Geographers, 60 (1970), 1–17.

———, ed., *Geography of the Ghetto: Problems, Perception, and Alternatives*. Dekalb: Northern Illinois University Press, 1972.

SCOTT, ALLEN J., "Production System Dynamics and Metropolitan Development," *Annals of the Association of American Geographers,* 72 (June 1982), 185–200.

SEMPLE, KEITH R., AND W. RANDY SMITH, "Metropolitan Dominance and Foreign Ownership in the Canadian Urban System," *Canadian Geographer,* 25 (Spring 1981), 4–26.

SIMMONS, JAMES, AND ROBERT SIMMONS, *Urban Canada* (2nd ed.). Toronto: Copp Clark Publishing Company, 1976.

SMITH, NEIL, AND PETER WILLIAMS, EDS., *Gentrification of the City*. Boston: Allen & Unwin, 1986.

VANCE, JAMES E., JR., "The American City: Workshop for a National Culture," in *Contemporary Metropolitan America, Vol. I, Cities of the Nation's Historic Metropolitan Core,* ed. John S. Adams, pp. 1–49. Cambridge, MA: Ballinger Publishing Company, 1976.

———, "Cities in the Shaping of the American Nation," *Journal of Geography,* 75 (1976), 41–52.

WARD, DAVID, *Cities and Immigrants: A Geography of Change in Nineteenth Century America*. New York: Oxford University Press, 1971.

———, "The Emergence of Central Immigrant Ghettoes in American Cities: 1840–1920," *Annals,* Association of American Geographers, 58 (1968), 343–359.

WOOD, JOSEPH S., "Suburbanization of Center City," *Geographical Review,* 78 (July 1988), 325–329.

YEATES, MAURICE, *North American Urban Patterns*. New York: John Wiley & Sons, 1980.

YEATES, MAURICE, AND BARRY GARNER, *The North American City* (3rd ed.). New York: Harper & Row, Publishers, 1980.

5

REGIONS OF THE UNITED STATES AND CANADA

It is convenient and, in many ways, useful to refer to Canada and the United States as discrete units that are sufficiently similar to support generalizations that apply to both nations. But such unitary consideration may, in fact, be misleading or inaccurate. Generalizations about either nation usually force a vast number of unlike areas into the same category. It is more meaningful to think of these countries as composed of a number of parts that are more or less dissimilar. Geographers call these parts *regions*. Although the concept of regions leaves something to be desired, it continues "to be one of the most logical and satisfactory ways of organizing geographical information."[1]

[1] Peter Haggett, Andrew D. Cliff, and Allan Frey, *Locational Analysis in Human Geography*, 2nd ed. (New York: John Wiley & Sons, 1977), p. 451.

It is well known that people vary—in speech, customs, habits, mores, and other ways—from region to region; however, there is no simple explanation for such variations (Fig. 5-1). In some cases, it may be attributable to aspects of the environment, whereby an element or complex of the physical realm exerts a pervasive influence on the population. Such effects appear to be most significant in areas where human activity is hindered by environmental extremes, such as deserts, mountains, or swamplands. In other cases, the regional variation of people is more closely related to culture than to nature. Whereas land is relatively changeless, humankind is the active agent; movements of people, changing stages of occupance, or different assessments of the resource base may occasion significant variations in the geography of regions (Fig. 5-2).

FIGURE 5-1 A pristine environment is rare indeed in Anglo-America. This scene, from Wind Cave National Park in South Dakota, is a reminder of the way things were before the blessings of civilization were introduced (TLM photo).

The United States and Canada are nations that consist of many political units, but they are also large portions of the earth's surface and include many varying regions. Each region differs from the others in one or more significant aspects of environment, culture, or economy; indeed, the magnitude of regional differences is often vast and complex. This book treats the continental expanse by regions, with broad sweeps of the brush. It is hoped that such treatment will provide not only a sounder comprehension of the geography of the two countries but also a fuller appreciation of the notion of regionalism, a vital concept to those who seek a better understanding of the world.

Even in such affluent and economically integrated nations as Canada and the United States regional considerations have long been recognized in business and government. Federal Reserve banking districts were formalized as early as 1913, and there are now more than 50 U.S. government bureaus in the executive branch alone that are organized with regional divisions. Businesses have similarly created regional offices, mail-order houses, and magazines to cope with the varied problems and opportu-

FIGURE 5-2 Where people congregate, the landscape is totally changed. This urban sprawl is in Los Angeles (TLM photo).

nities faced in operations covering such extensive areas as the United States and Canada. By working with large units, regional geography can make a more meaningful contribution to understanding national life. The often unsatisfactory artificiality of political boundaries becomes particularly apparent when attention is focused on regions.

Geography, which links the data of the social sciences with those of the natural sciences, is the logical discipline for dealing with regions. It sees in the region not only the physical, biological, social, political, and economic factors, but it also synthesizes them. In short, it considers the region in its totality—not merely the elements that are there but also the processes and relationships that have operated, are operating, and presumably will operate in the future. As J. Russell Smith once said, "The geographer is like the builder of a house who takes brick, stone, sand, cement, nails, wire, lath, boards, shingles, and glass—the products of many industries—and builds them into a symmetrical structure which is not any one of the many things that have entered into it, but is instead a house for the occupancy of man."[2]

THE GEOGRAPHICAL REGION

Geographers generally recognize two kinds of regions: *uniform* regions and *nodal* regions. The former possess significant aspects of homogeneity throughout their extent whereas the latter are very diverse internally and are homogeneous only with respect to their internal structure or organization. A nodal region always includes a focus, or foci, and a surrounding area tied to the focus by lines of circulation. A city and its hinterland illustrate, in crude fashion, the concept of a nodal region.

The prevailing view of this book is that at the macro-scale the geography of the continent can best be understood by dealing with uniform regions; consequently, the ensuing discussion and the regional divisions are based solely on the concept of uniform regions. A recognizable uniform region should have some characteristics that provide a measure of distinctiveness. As expressed by Dickinson, "Regional Geography is normally regarded as the treatment of the variety of spatially distributed phenomena in a particular area . . . [and] the problem for the regional geographer is to discover integrating processes that give some measure of identity and uniqueness to an area."[3]

A uniform region normally contains a core area of individuality in which regional characteristics are most clearly exemplified. The core possesses two distinct qualities that may be blurred in the periphery:

1. It differs noticeably from neighboring core areas.
2. It exists as a recognizable and coherent segment of space defined by the criteria whereby it is selected.

Beyond the core lies a marginal area. Regional boundaries are usually not lines but rather transitional zones that assume the character of adjoining regions or cores. The width may vary from a few feet to many miles; thus the field geographer making a reconnaissance survey seldom knows at what point one region is left and another entered. At some point, of course, the geographer passes from one to the other, but the human eye usually cannot perceive it at the moment of change. The distinguishing features of one region melt gradually into those of the neighboring region except perhaps along a mountain front or the shore of a large body of water.

THE PROBLEM OF REGIONAL BOUNDARIES

Consider, for example, the problem of delimiting the boundaries of one of the most universally recognized regions of the continent, the central flatland known as the Great Plains in the United States and the Prairies in Canada. The eastern boundary of this region is the least exact; in this book it is considered to be an irregular north-south zone extending from Texas to Manitoba. Several criteria are used in positioning the boundary line shown in Figure 5-3, but most significant is the change from a predominantly Corn Belt-type agriculture on the east to a predomi-

[2] J. Russell Smith, *School Geography and the Regional Idea* (Philadelphia: John C. Winston Co., 1934), p. 2.

[3] R. E. Dickinson, *Regional Ecology: The Study of Man's Environment* (New York: John Wiley & Sons, 1970), p. 36

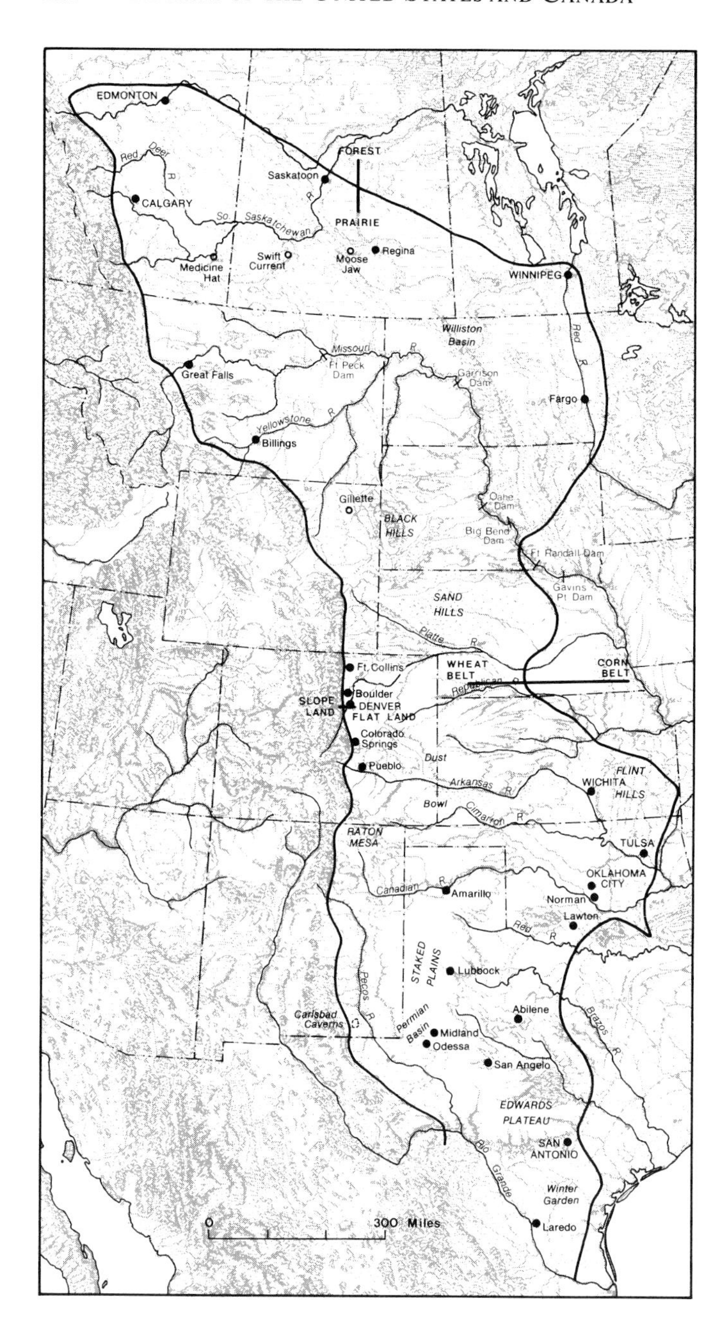

FIGURE 5-3 The Great Plains and Prairies Region, showing the boundary transects discussed in this chapter.

nantly Wheat Belt-type agriculture on the west. To locate this boundary at any particular point is an exercise in frustration. Southeasternmost Nebraska is clearly Corn Belt and southwesternmost Nebraska is clearly Wheat Belt, but where is the boundary?

In an east-west transect of this area one passes slowly from a corn-and-soybeans-dominated crop pattern to a wheat-and-grain sorghum-dominated crop pattern before reaching the midpoint of the transect. But then corn appears significantly again and there is an erratic alternation of the two crop combinations for many tens of miles until the wheat-and-grain sorghum pattern clearly prevails in southwesternmost Nebraska; thus the eastern boundary of the Great Plains—as exemplified by this single transect—that is portrayed as a black line in Figure 5-3 is actually a broad transition zone that approaches 300 miles in width in some places.

The northern boundary of the region is somewhat easier to delimit but is still far from precise. Once again, several criteria are used, but the most prominent criterion is the change from grass and grain in the south to forest in the north. In theory it is reasonable to expect an abrupt and conspicuous demarcation between such different vegetation associations, but actually the change is again transitional. A south-north transect through central Saskatchewan would find a significant interfingering of grass and grain with forest, the whole pattern being complicated by enclaves of forest well to the south and patches of grassland scattered deep in the forest. Again, the map shows a precise line as boundary and, again, the boundary (as generalized from this transect) is erratically transitional, embracing a zone that is in some places almost 100 miles in breadth.

The western boundary of the region is much more clear-cut, but even it is imprecise. The principal criterion for pinpointing the boundary is the change from flat land on the east to sloping land on the west; this change is particularly pronounced where the Great Plains meet the front ranges of the Rocky Mountains. An east-west transect in northern Colorado finds essentially flat land around Denver becoming virtually all slopeland west of Golden, only 25 miles away. But the intervening area fits well in neither region, for the pervasive flatness of the plains is interrupted by the gentle slopes of the Colorado Piedmont, the somewhat steeper slopes of scattered foothills, and the varied terrain of mesas and hogbacks before it is replaced by the more precipitous slopes of the mountain ranges. Thus even one of the most abrupt regional boundaries that can be found has a transition zone several miles in width.

THE PROBLEM OF REGIONAL STATISTICS

Although many geographers agree that regional geography is a fundamental part of their subject—that it is even the heart of the discipline—students of regional geography usually face an enormous handicap for precision study because of the unavailability of quantitative data. Statistics produced by the various data-collecting agencies (government and otherwise) normally are applicable only to political units (states, provinces, counties, townships, and other civil divisions) and so are difficult to apply to the more functional and less precise geographical regions. As a partial response to this problem, more refined statistical units, such as enumeration districts, urbanized areas, metropolitan statistical areas, consolidated metropolitan statistical areas, state economic areas, industrial areas, and labor market areas, have been designated; statistics applicable to such areas are often more meaningful for the regional geographer to use. The difficulty of fitting statistics to regions still persists, but improvements continue to be made.

DETERMINATION OF REGIONS

Dividing a continent into regions is a matter of "scientific generalizing." It serves the dual purpose of facilitating the assimilation of large masses of geographic information by students and organizing data into potentially more meaningful patterns for researchers. One recognizes a region by noting the intimate association existing between the people of an area and their occupance and livelihood pattern within that area. Similarity of interest and activity sometimes indicates a similarity of environment but may be totally unrelated to environmental implications.

Geographers deal with both natural and cultural landscapes. Everyone who has traveled, even slightly, has noted that the natural landscape changes from one part of the country to another. When two unlike areas are adjacent, the geographer may separate them on a map by a line and thereby recognize them as separate natural regions. Similarly, the cultural (human-created or human-modified) landscape may be studied and even delineated into separate cultural regions.

Whenever people come into any area, they promptly modify its natural landscape, "not in a haphazard way but according to the culture system

FIGURE 5-4 Land Resource Regions of the United States. (After James A. Lewis, *Landownership in the United States, 1978,* Agriculture Information Bulletin No. 435. Washington: U.S. Department of Agriculture, 1980, p. 43.)

which [they bring with them]. . . . Culture is the agent, the natural area is the medium, the cultural landscape the result."[4] They cut down the forest, plow under the native grass, raise domesticated animals, erect houses and buildings, build fences, construct roads and railroads, put up telephone and telegraph lines, dig canals, build bridges, and tunnel under mountains. All this constitutes the cultural landscape of a region.

Regions can be defined, recognized, and delineated on the basis of single or multiple characteristics, which can be either simple or complex in concept and determination. In preceding chapters, especially Chapter 2, we noted a variety of regional systems. Four other systems are displayed here as samples of the extraordinary diversity that is possible when dealing with the regional concept.

1. Figure 5-4 shows land resource regions as recognized by the U.S. Soil Conservation Service. The regions are based on agricultural-pastoral-forestry use patterns.
2. Figure 5-5 portrays water resource regions of the United States as recognized by the U.S. Department of Agriculture. These are essentially watershed regions.
3. Figure 5-6 shows "popular regions" as determined by "names of metropolitan enterprises," based on research by cultural geographer Wilbur Zelinsky.
4. Figure 5-7 demonstrates the regional pattern of Anglo-America as devised by popular journalist Joel Garreau, on the basis of personal research.

[4] Isaiah Bowman, *Geography in Relation to the Social Sciences* (New York: Charles Scribner's Sons, 1934), p. 150.

FIGURE 5-5 Water Resource Regions of the United States. (After James A. Lewis, *Landownership in the United States, 1978,* Agriculture Information Bulletin No. 435. Washington: U.S. Department of Agriculture, 1980, p. 43.)

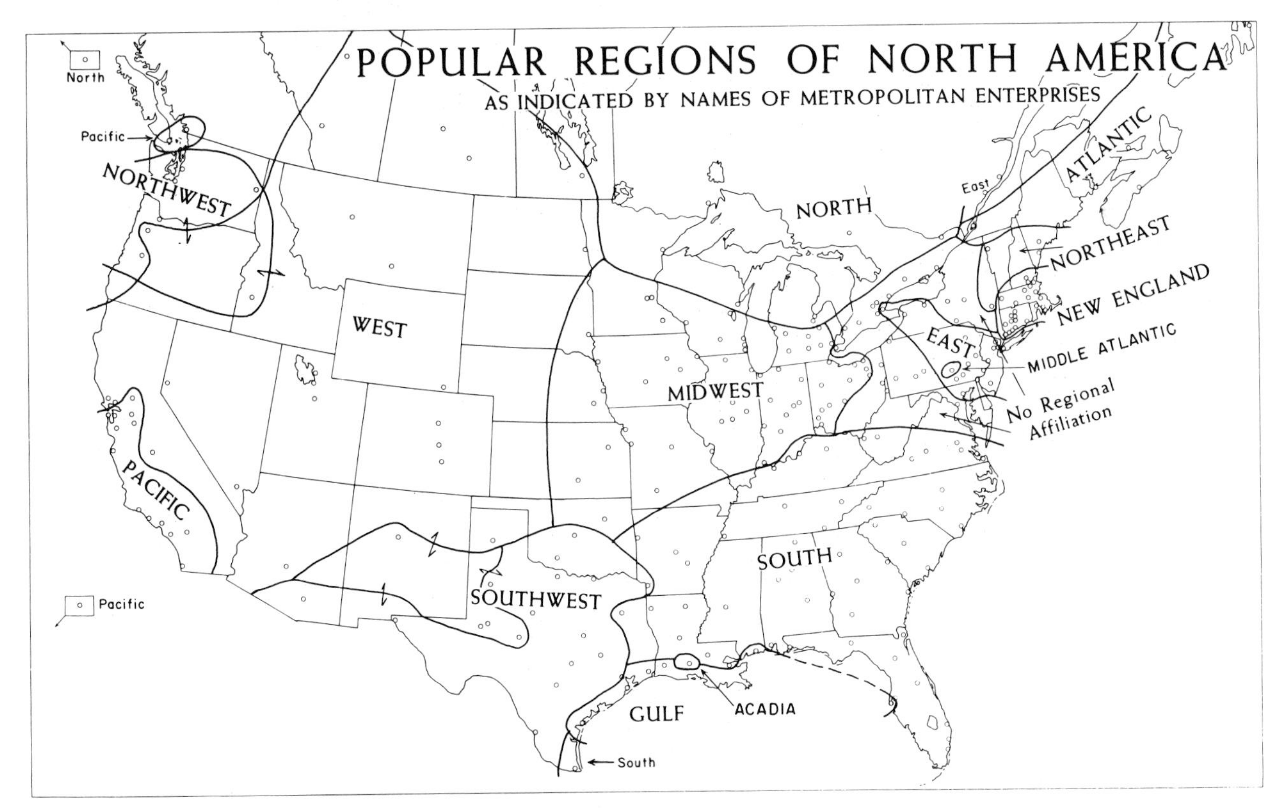

FIGURE 5-6 Vernacular or ''popular'' regions of Anglo-America, based on the frequency of regional and locational terms in the names of enterprises listed in telephone directories of the metropolitan areas of the United States and Canada. (After Wilbur Zelinsky, ''North America's Vernacular Regions,'' *Annals,* Association of American Geographers, vol. 70, March 1980, p. 14. By permission of the Association of American Geographers.)

CHANGING REGIONS

In several disciplines closely related to geography the regional systems are fixed by nature; climatic, pedologic, physiographic, and vegetation regions are all based on relatively static natural boundaries. But geographical regions are not fixed; instead of hard-and-fast boundary lines, they have ever-changing ones. When people push wheat culture farther north in Canada or farther west in Kansas, they are responsible for changes in geographical regions because raising wheat may be of such significance in the regional totality that a shifting of cultivation limits requires a similar shifting of regional boundaries.

As Gentilli stated, "regions are . . . in continuous evolution. . . ."[5] An inherent characteristic of a geographical region is that it is dynamic. It

[5] J. Gentilli, "Regions and Landscapes: Nature and Size, Function and Change," in *Western Landscapes* ed., J. Gentilli (Perth: University of Western Australia Press, 1979), p. 13.

A CLOSER LOOK Regions Are Devised, Not Discovered

It is important to keep in mind that regions are not naturally occurring phenomena awaiting discovery. The landscape does not consist of a God-given mosaic of regions, awaiting recognition by an adroit researcher. Rather, regions are concepts devised in the human mind for some purpose, and they are useful only insofar as they serve that purpose.

The regional system used in this text is designed for pedagogic use. Its purpose is to facilitate learning by the artful generalization of regional characteristics. It is but one of many systems that might be invented.

Anyone can play the regionalization game; imagination is all that is needed. However, the success of any system of regionalization will depend in large measure upon the care with which it is conceived and crafted.

Presented here is a sample of another kind of regional system, designed by a newspaper columnist to entertain readers. It is indicative of the principle that the regional concept has many uses.

A Plan for California in the Nineties: Break It Up

Robert A. Jones

Welcome to the '90s. If you have been reading the predictions for California over the next decade, perhaps you share my fear that we are fast approaching the end of civilization as we know it. Vicious water wars are scheduled to break out between the north and south, the last redwoods will be chopped down, and 19,000 more people will be arriving each week to enjoy it all. Very gloomy.

So I have a modest proposal. Let's face the fact that California has grown far too big, that it makes about as much sense for California to be a single state as it does for the Soviet Union to be a single country. Let's deal with the reality that the cotton farmers of Visalia don't give a fig for the TV execs in Burbank, and vice versa. Let's break up California.

I'm not just talking about the old strategy of drawing a line between the north and the south. Things have gotten much trickier than that. We need a Plan for the Nineties, and here it is.

As you can see from the accompanying map, the Plan provides for three separate states in the new Californias, plus "Oregon." We will get to "Oregon" later. Right now let's take the states one at a time:

• In the south, we must recognize that Los Angeles has become a separate world in California, a city-state only dimly aware of the nether regions to the north. A recent survey showed that Los Angeles makes approximately 500 times more telephone calls to New York than it does to Fresno. The fact that this survey surprised no one is evidence of L.A.'s estrangement from its geographic neighbors.

Creating a state of Los Angeles would liberate the region from the nattering influence of the environmentalists in Northern California. We in the south could get on with making Los Angeles the richest and ugliest city on Earth. We will require lots of desert to convert into subdivisions, and that has been provided. Ditto with coastline. We get the unspoiled stretch from Santa Barbara to San Simeon so we can make it look like Redondo Beach.

If anyone gets nostalgic about open space, the Plan offers a rental program from the state of Lettuce.

Since Los Angeles itself would be too crowded to accommodate the state capitol, we might want to declare San Bernardino a tear-down site and build a new one from scratch, Brasilia-style. As for the name, we should probably recognize the new realities of our time and call it "Sony." They might even chip in on construction costs.

• In the Central Valley of California, the state of Lettuce would provide changes with time as people learn to assess and use their natural environment in different ways, with changing economic, political, or social conditions or with the advance of technology. It may contract, expand, fragment, or drastically change its character through the years.

Regional analysis can be carried out at different levels of generalization, and conclusions derived at one scale may be invalid at another. Various geographers have pointed out that a change in scale often requires a restatement of the problem and "there is no basis for presuming that associations existing at one scale will also exist at another."[6] One implication of this situation is that a hierarchy of regions can be designed. The creation of such a hierarchy is a useful exercise of geographical scholarship. It is be-

[6] H. H. McCarty, J. C. Hook, and D. S. Knos, *The Measurement of Association in Industrial Geography* (Iowa City: University of Iowa, Department of Geography, 1956), p. 16.

a sense of place to our heartland. The valley has always shared more with Nebraska than coastal California. This way the farmers could listen to Tammy Wynette and eat chicken-fried steak in peace. Lettuce would be all theirs.

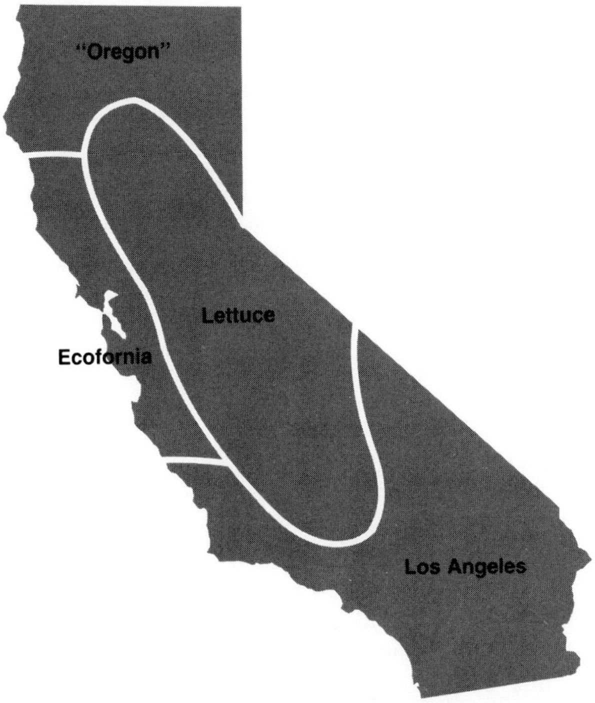

To provide some needed revenge for all the cultural slights suffered over the years, Lettuce would also get the Sierra Nevada. When the coastal folks got sick of their cities, the people of Lettuce could rent them chunks of the Sierra for breath-taking user fees.

• Around the Bay Area we would create what you might call a boutique state. Ecofornia would be small because not much would happen there. San Francisco could convert entirely to tourism and stop worrying about its declining position in California. Within this tiny empire, San Francisco would be forever the center of things.

For territory, Ecofornia would acquire Big Sur to the south and Napa/Sonoma to the north. This would provide some degree of employment diversity. Anyone who got tired of mixing Irish coffees for Pennsylvania optometrists in San Francisco could go to the country and mix Irish coffees at a bed-and-breakfast.

As for the far north, the harsh truth is this region does not belong to California and never has. Does anyone know what goes on in Alturas? That's what I thought, so let's give the far north to Oregon. In return all we ask is that they leave a few redwoods standing, just for old time's sake.

A brilliant plan, I hear you saying, just brilliant, but reality-wise a little unlikely. I understand these doubts. Just keep in mind that they were saying the same thing six months ago in Prague.

FIGURE 5-a A suggested regionalization of California.

(Reprinted from the *Los Angeles Times*, January 3, 1990. By permission.)

FIGURE 5-7 The regions of Anglo-America, as recognized by a popular writer. (After Joel Garreau, *The Nine Nations of North America*. Boston: Houghton Mifflin, 1981.)

yond the intent of this book to do so; nevertheless, understanding some regions appears to be enhanced by the designation of subregions, and subregions are delimited in some chapters.

REGIONS OF THE UNITED STATES AND CANADA

In this book the continent is divided into 15 regions. The criteria for making the regional divisions are both multiple and varied. In some cases, physical considerations were dominant and in others cultural factors played a more important role. But in all instances, the broad regions as delineated reflect as accurately as possible the basic features of homogeneity inherent in various parts of Anglo-America at present. In general, the principal criteria used in determining regional boundaries are the socioeconomic conditions currently characteristic, which have often been decisively influenced by the physical environment and historical development. The regions, as numbered in Figure 5-8, are as follows:

1. the Atlantic Northeast
2. French Canada
3. Megalopolis
4. the Appalachians and the Ozarks
5. the Inland South
6. the Southeastern Coast
7. the Heartland
8. the Great Plains and Prairies
9. the Rocky Mountains
10. the Intermontane Region
11. the California Region
12. the Hawaiian Islands
13. the North Pacific Coast
14. the Boreal Forest
15. the Arctic

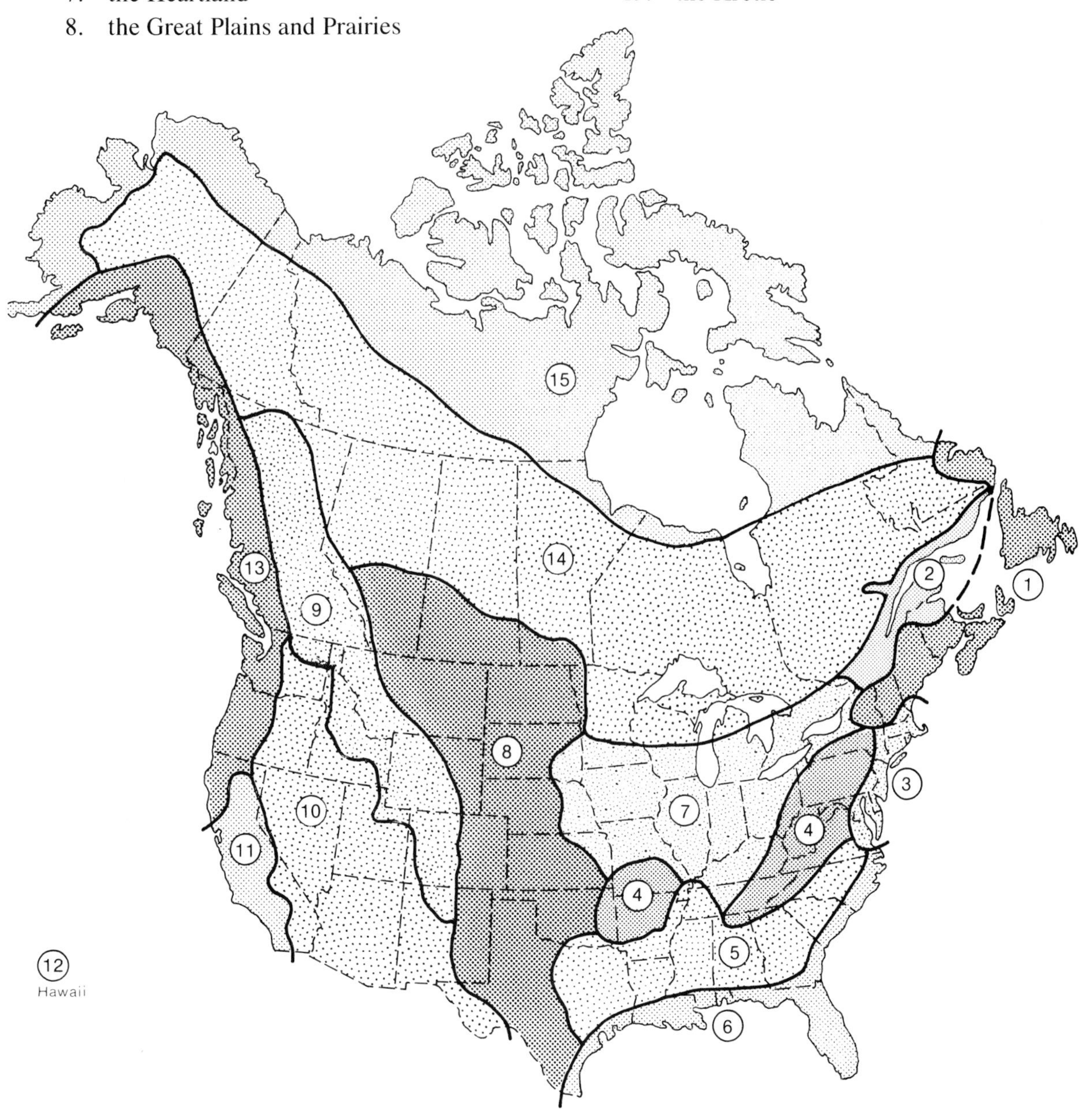

FIGURE 5-8 The geographic regions of the United States and Canada.

SELECTED BIBLIOGRAPHY

BERRY, BRIAN J. L., "Approaches to Regional Analysis: A Synthesis," *Annals,* Association of American Geographers, 54 (1964), 2–11.

BRADSHAW, MICHAEL, *Regions and Regionalism in the United States.* Jackson: University Press of Mississippi, 1988.

BROWNING, CLYDE E., AND WIL GESLER, "The Sun Belt—Snow Belt: A Case of Sloppy Regionalizing," *The Professional Geographer,* 31 (February 1979), 66–74.

GARREAU, JOEL, *The Nine Nations of North America.* Boston: Houghton Mifflin, 1981.

GASTIL, RAYMOND D., *Cultural Regions of the United States.* Seattle: University of Washington Press, 1975.

GEULKE, LEONARD, "Regional Geography," *Professional Geographer,* 29 (1977), 1–7.

HARRIS, R. COLE, "The Historical Geography of North American Regions," *American Behavioral Scientist,* 22 (1978), 115–130.

HART, JOHN FRASER, "The Highest Form of the Geographer's Art," *Annals,* Association of American Geographers, 72 (1982), 1–29.

JAMES, PRESTON, "Toward a Further Understanding of the Regional Concept," *Annals,* Association of American Geographers, 42 (1952), 195–222.

JENSEN, MERRILL, ED., *Regionalism in America.* Madison: University of Wisconsin Press, 1965.

KOHN, CLYDE F., "Regions and Regionalizing," *Journal of Geography,* 69 (1970), 134–140.

KRUEGER, RALPH R., "Unity out of Diversity: The Ruminations of a Traditional Geographer," *Canadian Geographer,* 24 (1980), 335–348.

MCDONALD, JAMES R., *A Geography of Regions.* Dubuque, IA: William C. Brown Co., Publishers, 1972.

MINSHULL, ROGER, *Regional Geography, Theory and Practice.* London: Hutchinson Libraries, 1967.

NICHOLSON, N. L., AND Z. W. SAMETZ, "Regions of Canada and the Regional Concept," in *Regional and Resource Planning in Canada,* ed. Ralph R. Krueger et al., pp. 6–23. Toronto: Holt, Rinehart & Winston of Canada, Ltd., 1963.

RAY, D. MICHAEL, *Dimensions of Canadian Regionalism.* Ottawa: Department of Energy, Mines and Resources, Policy Research and Coordination Branch, 1971.

SCHWARTZ, M. A., *Politics and Territory: The Sociology of Regional Persistence in Canada.* Montreal: McGill-Queens University Press, 1974.

STEEL, R. W., "Regional Geography in Practice," *Geography,* 67 (1982), 2–8.

ULLMAN, STEPHEN H., "Regional Political Cultures in Canada: Part I," *American Review of Canadian Studies,* 7 (Autumn 1977), 1–22.

———, "Regional Political Cultures in Canada: Part II," *American Review of Canadian Studies,* 8 (Autumn 1978), 70–101.

WADE, MASON, ED., *Regionalism in the Canadian Community.* Toronto: University of Toronto Press, 1969.

WALTER, BOB J., AND FRANK E. BERNARD, "Ash Pile or Rising Phoenix? A Review of the Status of Regional Geography," *Journal of Geography,* 77 (1978), 192–197.

WONDERS, WILLIAM C., "Regions and Regionalisms in Canada," *Zeitschrift der Gesellschaft für Kanada-Studien,* 1 (1982), 7–43.

ZELINSKY, WILBUR, "North America's Vernacular Regions," *Annals,* Association of American Geographers, 70 (1980), 1–16.

6

THE ATLANTIC NORTHEAST

A sharp contrast exists between the predominant rural character of the Atlantic Northeast and the highly urbanized region to its south. The Atlantic Northeast is a region in which the forest and the sea have been pervasive influences on the lifestyle of its inhabitants. There is a sense of history and tradition, reflecting the relatively limited resources of the land and the relative richness of the adjacent ocean.

Included within the region are the less populated, rougher parts of northern New England and New York State, all the Maritime Provinces except that portion of New Brunswick that is predominantly French Canadian in population and culture, and the island of Newfoundland with a portion of the adjacent Labrador coast that is oriented toward "Newfoundland-type" commercial fishing (Fig. 6-1). Interior Labrador is not included because of its sparse population and similarity to the Boreal Forest Region. Northern New Brunswick and adjacent Quebec are excluded because of the significant cultural differences in French Canada.

The most difficult decision to be made in delimiting the Atlantic Northeast as a region is the separation of northern and southern New England. Throughout most of its history the people of New England showed a regional consciousness in greater degree, perhaps, than in any other part of North America. Their traditions, institutions, and ways of living and thinking exhibited considerable uniformity; however, in recent years there has been an increasing divergence between the way of life of the urbanites in southern New England and the people of the small towns and rural areas that characterize northern New England. Furthermore, the economic and psychological orientation of much of southern New England is increasingly dominated by New York. It seems clear that urbanized southern New England is more logically a portion of the megalopol-

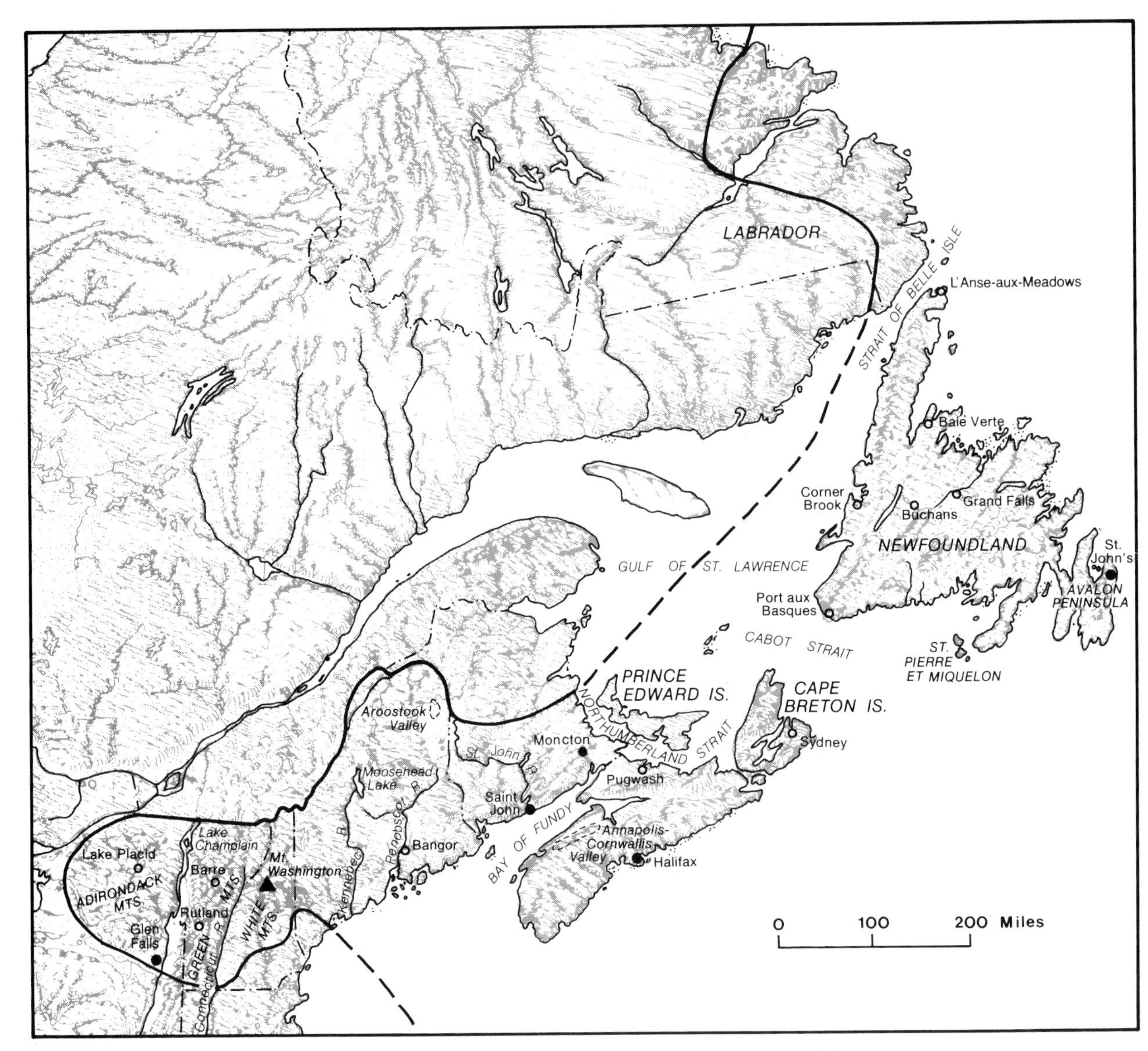

FIGURE 6-1 The Atlantic Northeast (base map copyright A. K. Lobeck; reprinted by permission of Hammond, Inc.).

itan region to the south than it is of the rurally oriented region to the north. Accordingly, the southern boundary of the Atlantic Northeast is considered to lie just north of the urbanized areas of Portland in Maine, the Merrimack Valley in New Hampshire, the small cities of western Massachusetts, and the Mohawk Valley of New York State.

A REGION OF SCENIC CHARM AND ECONOMIC DISADVANTAGE

The Atlantic Northeast is a region that abounds in scenic delights: the verdant slopes of the Adirondacks, the rounded skyline of the Green Mountains, quaint covered bridges and village greens, sparkling

FIGURE 6-2 Much of the Atlantic Northeast is still covered with forest, as in this scene near Truro in Nova Scotia (TLM photo).

blue lakes, Maine's incredibly rockbound coast, the tide-carved pedestal rocks of the Bay of Fundy, the classical symmetry of Prince Edward Island's farm landscape, the ordered orchards of the Annapolis-Cornwallis Valley, the splendid harbors of Halifax and St. John's, the Old World charm of Newfoundland outports. Yet this region of delightful views has suffered longer and more continually from economic handicaps than any other on the continent (Fig. 6-2).

Much of the problem is environmental. Soils are generally poor, the growing season is short and cool, the winter is long and bleak, mineral resources are scarce, and second-growth timber grows slowly. Timber, fish, and running water were the chief physical advantages, but all three have handicaps. The readily accessible forests were long since cut and second-growth replacement takes many decades. Moreover, various parasites took an inordinate toll of the timber resources in recent years. Fishing was the mainstay of the economy for a great many years and has continued in declining importance, but the sea's contribution to the regional economy has been erratic and unstable. The rushing streams of the region were early sources of hydraulic and later hydroelectric power; but their productive capacity is limited, and today the Atlantic Northeast is a power-deficient region.

The foreland location of the region gives it a closest-to-Europe position that provided early economic advantage; indeed, the port of St. John's was referred to as being located halfway across the Atlantic. The foreland, however, can be easily bypassed in contemporary times. The more westerly location of the commercial-industrial heartland of the continent has drawn most trade either into the Boston–New York–Philadelphia axis or up the St. Lawrence River. The foreland position regained some of its former importance in war time and there is an annual regeneration of activity for the Maritime ports when the St. Lawrence is closed by ice; but the long-run commercial and strategic value of the foreland location is in marked decline.

This region experienced slow economic growth even in periods of national affluence. Its resources are too small and too scattered to tempt external investors. The region has been described as one of "effort" rather than of "increment."[1] Even government infusion of capital tends to be easily dispersed without causing much growth.

The population of the region is scattered and there are only a few nodes of urban-industrial concentration. Essentially it is a region of emigration. The rates of population increase for the portions of four states and four provinces included in the region are consistently among the lowest recorded on the continent. There has been a continual trend of Maritimers moving to Ontario and northern New Englanders moving south. And no large influx of immigrants occurred to compensate. Northern New England absorbed a steady stream of French Cana-

[1] David Erskine, "The Atlantic Region," in *Canada: A Geographical Interpretation*, ed. John Warkentin (Toronto: Methuen Publications, 1968), p. 233.

dians into its forest industries, but generally the region has not shared in the continent's high rate of net immigration or net population growth.

Northern New England is economically depressed and the four Atlantic Provinces have a long record as Canada's economic problem region. The extremes are all there, including lower income per person, lower goods output per capita, lower average investment of new capital, and higher unemployment. The Atlantic Provinces overall average nearly 30 percent below the per capita income for all Canada, a disparity that has persisted for decades.

Many study committees were established in both countries to prescribe remedies for the ills of the region. Various assaults on economic deprivation were mounted and many palliatives tried. Some local benefits resulted from these efforts, but the region as a whole continues to suffer.

Two relatively recent developments improved the economic position in the Canadian portion of the region and engendered considerable hope where little existed before. The fishing industry was stimulated by the elimination of much foreign competition and the discovery of enormous offshore oil and gas resources provided a considerable fillip to overall economic development.

THE PHYSICAL SETTING

The Atlantic Northeast is a region of slopeland, forest, and coast; of bare rock, cold waters, and leaden skies; and of thin soils, swift streams, and implacable tides. It is a land where beavers still build dams, moose feed on lake bottoms, and Atlantic salmon come upstream to spawn.

Surface Features

The coastal area consists primarily of low rounded hills and valleys. Most of the region is traversed by fast-flowing streams and much of it is dotted with small lakes. The coastal area has been slightly submerged; consequently, ocean waters have invaded the lower valleys, creating bays or estuaries. Often branch bays extend up the side valleys. The coast is characterized by innumerable good, if small, harbors. Superficially the coasts of Maine and Nova Scotia appear to be fiorded, but they are probably drowned normal river valleys that were slightly modified by ice action. The restricted and indented nature of much of the coastline gave rise to some of the greatest tidal fluctuations to be found anywhere (Fig. 6-3). Many rugged and rocky headlands extend to the water's edge, especially in Maine and Newfoundland. Beaches are relatively small and scarce.

All mountains of the upland area are geologically old and worn down by erosion (Fig. 6-4). The Green Mountains of Vermont have rounded summits, the result of the great ice sheets that covered ridges and valleys alike. The highest peaks are less than 4500 feet (1350 m) above sea level. The White Mountains are higher and bolder and were not completely overridden by the ice, as is shown by the occurrence of several cirques (locally called "ravines"); their highest points are in the Presidential Range, where Mount Washington reaches an elevation of nearly 6300 feet (1890 m). Northeastward the mountains become less conspicuous, although their summits remain at an approximate elevation of 5000 feet (1500 m).

The mountains of eastern Canada are lower and more rounded, having been subdued through long periods of erosion. Their general elevation is slightly more than 2000 feet (600 m) above sea level. The Adirondacks, although an older upland mass, underwent changes at the time of the Appalachian mountain-building movement that caused a doming of the upper surface; furthermore, they were eroded profoundly during glacial times. Although not as high as the White Mountains, the Adirondacks cover more area.

The entire upland is composed largely of igneous and metamorphic rocks—granites, schists, gneisses, marbles, and slates—so valuable that this has become the leading source of building stones on the continent.

The stream courses of the upland area were altered by glaciation. Many water bodies, such as Lake Placid, Lake Winnipesaukee, and Moosehead Lake, characterize the region. They were of inestimable value in the development of the tourist industry.

The Aroostook Valley, occupying the upper part of St. John River drainage, is the result of stream erosion in softer rocks. The Lake Champlain

FIGURE 6-3 The phenomenal tidal fluctuations of the Bay of Fundy serve as significant erosive agents on softer rocks. The 20-foot-high evergreen trees growing on this tide-carved pedestal rock along the New Brunswick coast give an indication of scale (TLM photo).

Lowland and the Connecticut Valley were severely eroded by tongues of ice that moved southward between the Green Mountains and the Adirondacks and between the Green Mountains and the White Mountains. Many valleys and lowlands are underlain by sedimentary rocks.

Newfoundland consists of a combination of moorland and forest, with an abundance of rocks, ponds, and shrubby barrens. It has been described as "a queer dishevelled region where the Almighty appears to have assembled all the materials essential to a large-scale act of creation and to have quit with the job barely begun. Ponds are dropped indiscriminately in valleys and on hilltops, rocks strewn everywhere with purposeless prodigality".[2]

Much of the island is a rolling plateau 500 to 1000 feet (150 to 300 m) above sea level, with elevations above 3000 feet (900 m) in the Long Ranges. The coastline is severely indented; the juxtaposition of bay and peninsula is the most conspicuous feature

[2] Edward McCourt, *The Road Across Canada* (Toronto: Macmillan of Canada, 1965), p. 16.

FIGURE 6-4 The physical landscape of this region is dominated by bare rock, green forest, and blue water. This view is of Mt. Penobscot, near the coast of central Maine (TLM photo).

of the littoral landscape. The coast of southern Labrador is the rugged, elevated, fiorded edge of the Canadian Shield and is strewn with small offshore islands.

Climate

Because of the maritime influence, the coastal areas, particularly in Maine and Nova Scotia, generally have a milder and more equable climate than might be expected in these latitudes. Nova Scotia's mean January temperature is about the same as that of central New York despite being 2 to 5 degrees of latitude farther north. Winters, although long and cold, are not severe for the latitude. Temperatures may fall below zero, however, and snow covers the ground throughout most of the winter. Spring surrenders reluctantly to summer because of the presence of ice in the Gulf of St. Lawrence and because of the cold Labrador Current. Summers are cool, temperatures of 90°F (32°C) being extremely rare. The growing season varies from 100 to 160 days in Nova Scotia.

The precipitation of 40 to 55 inches (1015 to 1400 mm) is well distributed throughout the year. Southeast winds from the warm Gulf Stream blowing across the cold waters between the Gulf of Maine and Newfoundland create the summer fogs that characterize the coasts of New England, New Brunswick, and Nova Scotia. Newfoundland itself is famous for its foggy coasts (Fig. 6-5).

The upland area lies within the humid continental climatic regime, with the Atlantic Ocean exerting little influence. The growing season is short, averaging less than 120 days. Summers are cool and winters extremely cold, temperatures dropping at times to 30° below zero Fahrenheit (−34°C). The abundant precipitation is evenly distributed throughout the year.

In Newfoundland and south coastal Labrador the climate is largely the result of a clash between continental and oceanic influences, with the former dominating. The winters are much colder than those in British Columbia or Britain in the same latitude. Altitude is a factor, as shown by the replacement of forest by tundra at elevations exceeding about 1000 feet (300 m).

No point in Newfoundland is more than 90 miles (144 km) from the sea, but the ocean's relative coldness and the prevailing westerly winds, which bring continental influences, do not permit much amelioration of the temperatures. In the Gulf of St. Lawrence all harbors freeze over in the winter, the Strait of Belle Isle sometimes being completely

FIGURE 6-5 Fog is commonplace in the region, particularly in Newfoundland. This is Logy Bay on the Avalon Peninsula (TLM photo).

blocked by ice. The bays on the Labrador coast and large areas of the adjacent sea freeze solid by October or November. Summers everywhere are cool because the Labrador Current, laden with ice floes and icebergs, moves southward along the east and south coasts.

Natural Vegetation

Most of the land included within this region was once covered with trees. The principal treeless localities were small areas of dunes, marshes, meadows, bogs, and exposed mountain summits except in Newfoundland and coastal Labrador, where temperature, wind, and moisture conditions of the more extensive reaches of low scrubby barrens and modified tundra were inhibitory to tree growth.

The New England section was originally covered with a relatively dense forest of mixed deciduous and coniferous species. Even today more than 80 percent of the land is forested by a mixture of northern hardwoods, with white pine, spruce, and fir. Originally white pine was the outstanding tree; attaining a height of 240 feet and a diameter of 6 feet at the butt, it dwarfed even the tall spruce. It was sometimes called the "masting pine" because the larger trees were marked with the Royal Arrow and reserved for masts for the Royal Navy. Maine is still referred to as the "Pine Tree State." Most of the existing forest is second-growth; the original timber was cut for lumber or fuel or cleared for agriculture many decades ago. Only in the more remote parts of the uplands, primarily in Maine, are there still virgin stands of trees.

In the Maritime Provinces a mixed forest cover is also still widespread. Although it has been thoroughly removed from Prince Edward Island, most of the land is still forested in Nova Scotia, and forest clearing was even less common in New Brunswick. Some large areas of relatively pure hardwood or softwood species may be found, but mixed growth is much more common. Spruce, hemlock, fir, pine, maple, and birch are the typical trees.

The forests of Newfoundland are much more predominantly coniferous, with balsam fir and black spruce dominating. Birch, poplar, and aspen are the principal hardwood species. Nowhere are the trees large; as a result, little lumbering is carried on, although pulping is important. The coast of Labrador is almost completely lacking in forest; but the sheltered stream basins support some tree growth.

Soils

Owing to differences in parent rock, slope, drainage, and previous extent of glaciation, the soils of the Atlantic Northeast are varied; nevertheless, the dominant soils throughout the region are Spodosols. The only other soil order represented importantly in the region is Alfisol. These are soils that developed under cool, moist conditions and are thus leached and acidic in nature. Usually there is a layer of organic accumulation near the surface and, more characteristically, considerable accumulation of compounds of iron and aluminum. A layer of clay accumulation is often found in the Alfisols of the valleys.

Agricultural productivity of these soils is undistinguished. Those that are derived from shales, especially in New Brunswick and northern Nova Scotia, are heavy and poorly drained. The sedimentary floored lowlands give rise to soils that are more fertile for farming; the shales of Vermont and the sandstones of the Annapolis Valley and Prince Edward Island yield more productive agricultural soils than are found elsewhere. Rockiness, poor drainage, and peat formation are major and widespread handicaps to crop and pasture development. Alluvial soils, occurring on narrow-valley floodplains, are extremely important to agriculture even though their total acreage is small.

Fauna

There is nothing out of the ordinary about the wildlife in most of the region; it has a "north woods" environment that contains a predictable faunal complement. One anomalous situation does prevail in Newfoundland. This large island was singularly lacking in a number of the common mainland terrestrial species. There were, for example, almost no rodents, especially the smaller varieties. Three very characteristic denizens of the north woods, porcupine, mink, and moose, were also missing. To remedy these deficiencies, exotic animals were introduced at various times during the past century. The most spectacular success resulted from the in-

A CLOSER LOOK Acid Rain

One of the most vexing and perplexing environmental problems is the rapidly increasing intensity, magnitude, and extent of *acid rain.* This term refers to a phenomenon that involves the deposition of either wet or dry acidic materials from the atmosphere on the earth's surface. Although most conspicuously associated with rainfall, the pollutants may fall to earth with snow, sleet, hail, or fog, or in the dry form of gases or particulate matter.

Sulfuric and nitric acids are the principal culprits recognized thus far. Although there is no universal agreement on the exact origin and processes involved, evidence indicates that the principal human-induced sources consist of sulfur dioxide emissions from smokestacks and nitrogen oxide exhaust from motor vehicles. These and other emissions of sulfur and nitrogen compounds are expelled into the air, where they may be wafted hundreds or even thousands of miles by winds. During this time they may mix with atmospheric moisture to form sulfuric and nitric acids that sooner or later are precipitated (Fig. 6-a).

Acidity is measured on a pH scale based on the relative concentration of active hydrogen ions. The scale ranges from 0 to 14; the lower end representing extreme acidity (battery acid has a value of 1) and the upper end extreme alkalinity (lye has a value of 14). It is a logarithmic scale, which means that a difference of one whole number on the scale reflects a tenfold increase or decrease in absolute values. Rainfall in clean, dust-free air has a pH of about 5.6; thus it is slightly acidic because of the reaction of water with carbon dioxide to form a weak carbonic acid. Increasingly, however, precipitation with pH less than 4.5 (the level below which most fish perish) is being recorded and an acid fog with a record low of 1.7 (8000 times more acidic than normal rainfall) was experienced in southern California in 1982.

Many parts of the earth's surface have naturally alkaline conditions in soil or bedrock that buffer or neutralize acid precipitation. In most of the Atlantic Northeast, however, the buffering capability is limited because of shallow soils and crystalline bedrock. Consequently, the deleterious effects of acid rain were first detected in this region. Later the problem spread much more widely.

The most conspicuous damage is done to aquatic ecosystems (more than 200 lakes in the Adirondack Mountains alone are virtually "biologically dead"), but evidence continues to mount of ill effects on forests, buildings and other structures, crops, and human health (Fig. 6-b).

The problem is exceedingly complex and the ramifications multifarious. It is known, for example, that the Midwest is a major source of toxic emissions that cause damages in the

FIGURE 6-a Acid rain is not readily recognized by the average person (courtesy *Alberta Environment Views*, March/April 1982, p. 3).

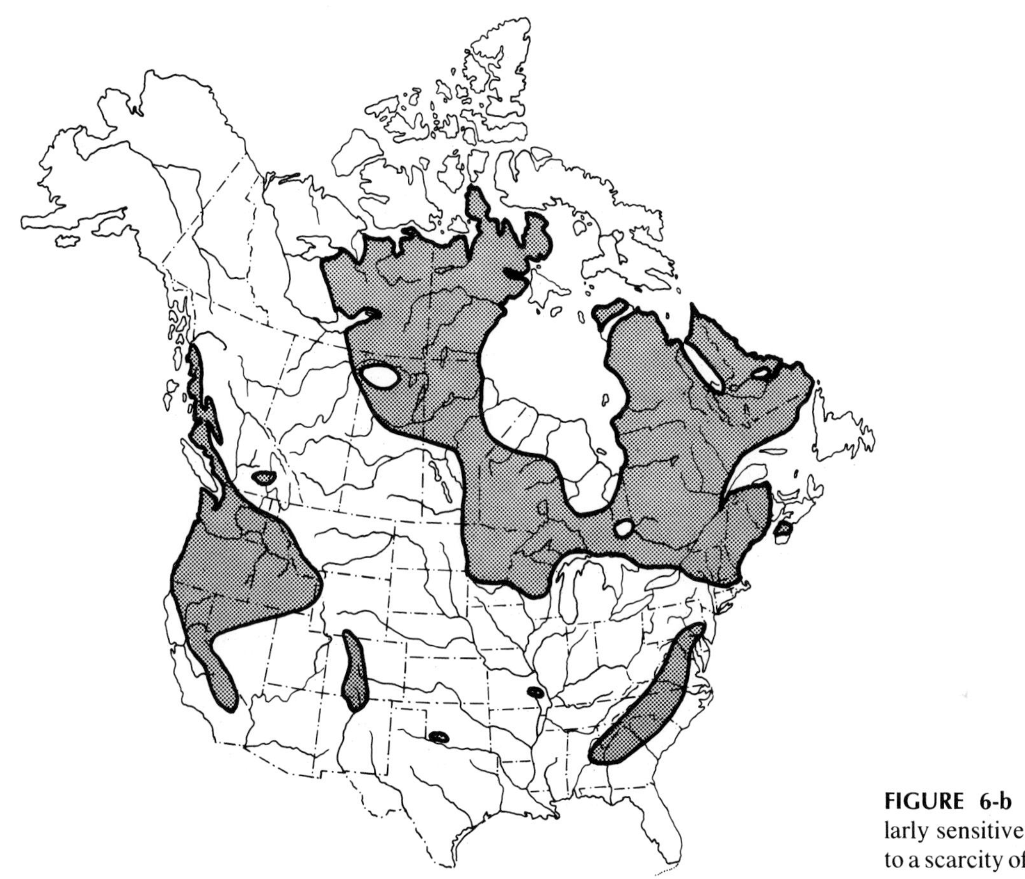

FIGURE 6-b Areas particularly sensitive to acid rain due to a scarcity of natural buffers.

Northeast. But it is understandably difficult to persuade people in Illinois to finance expensive clean-up costs that primarily benefit people in Vermont. Moreover, much of the acid rain problem in eastern Canada originates in the United States; so the relationships are not only interregional but also international. Actually, there is a two-way exchange of acid deposition, but it is highly unbalanced. Emissions from the United States are responsible, on average, for half of the acid deposition that falls on eastern Canada whereas Canadian emissions cause 15 percent of the acid rain that falls on the northeastern United States. In the late 1980s the Canadian government started a comprehensive (and expensive) acid rain control program. No similar initiative has been launched in the United States.

Acid rain is now a matter of concern throughout North America, but it continues to attract most attention in the Northeast. More than half the towns in New Hampshire have their own acid rain monitoring systems. Most television stations in Maine include acid rain measurements in their regular weather reports. Rainfall pH in the Adirondacks now *averages* about 4.2. Government and citizen groups are becoming increasingly mobilized. It has become abundantly clear that the costs of reducing the acid rain problem will be enormous but that the costs of not doing so will be exponentially greater.

TLM

troduction of six moose about the turn of the century; the moose is now more numerous and widespread than the native caribou and has yielded more than 100,000 legal kills to hunters. Chipmunks and mink were also successfully introduced and shrews were brought to Newfoundland to control a larch-destroying sawfly.

SETTLEMENT AND EARLY DEVELOPMENT

The Atlantic Northeast was not settled by immigrants from a single country. The earliest settlers in New England were British, those in the Maritime Provinces were French, and both British and French originally settled in Newfoundland.

The New England Segment

The first important settlement in what is now New England was founded at Plymouth in 1620. All the early colonies, including those before and immediately after the landing of the Pilgrims, were planted on the seaboard. The coast was, then, the first American frontier. Its settlements were bounded by untamed hills on the west and by the stormy Atlantic on the east. Beckoned by the soil and the sea, its shore-dwelling pioneers obeyed both and their adjustments to the two environments laid the foundations for the land and sea life of the nation.

As population became more dense in maritime New England and Canada, the more venturesome settlers trekked farther into the wilderness. As long as the French controlled the St. Lawrence Lowlands, the Indians of the upland remained entrenched in this so-called neutral ground, thus restricting white settlements to the seaboard. In the Adirondacks hostile Iroquois kept the English confined to the Hudson and Mohawk valleys until the close of the Revolutionary War. Feeling that at last the power of the Indian had been broken after the conquest of Canada by the British, pioneers from the older parts of New England penetrated the upland. By the 1760s most lower valleys in New Hampshire and Vermont were occupied. The clearing of the forest for farms led to an early development of logging and lumbering, which could be carried on in winter when farm work was not feasible. The logs, dragged on the snow to frozen streams, were floated to mills in the spring when the ice melted; thus a supplemental source of income was provided for the pioneers, and it continued to be important until the latter part of the nineteenth century.

Maritime Canada

The first permanent settlement in North America north of Florida was at Port Royal on the Bay of Fundy in 1605. Here the French found salt marshes that needed no clearing. This environment was attractive to people from the mouth of the Loire, whose forebears for generations had reclaimed and diked somewhat similar land. These French called their new home *Acadie* (Acadia). The Acadians converted the river marshes into productive farmland that characterized the cultivated area almost exclusively for a century.

The French population grew rapidly after the Treaty of Utrecht in 1713. Louisburg was fortified to guard the mouth of the St. Lawrence and the fishing fleet. The Acadians remained until the outbreak of hostilities between the British and French preceding the Seven Years' War in the 1750s. Then more than 6000 were rounded up and banished; they were scattered from Massachusetts to South Carolina. Some fled into the forests of what are now New Brunswick, Prince Edward Island, and Quebec; some made their way to Quebec City, the Ohio Valley, and even to Louisiana; and others joined the French in St. Pierre, Miquelon, and the West Indies. Many starved. The reason most commonly given for their expulsion was fear on the part of England that such a heavy concentration of French in this part of the continent was a menace to English safety.

This dispersal continued for eight years, ending in 1763; then individuals and groups began to trickle back. Although denied their old properties, they found abodes here and there, mostly on the gulf coast of New Brunswick, the western part of Prince Edward Island, and at scattered localities in Nova Scotia. Frenchmen never returned to Newfoundland in any meaningful numbers.

The first significant arrival of British settlers in the Atlantic Provinces was in 1610 in Newfoundland. Despite a hard and lonely life, oriented almost exclusively toward export of salted codfish, increas-

A CLOSER LOOK L'Anse-aux-Meadows—North America's Earliest European Settlement

Long before Columbus made his historic voyage to the New World in 1492, Norse Vikings had "discovered" North America by voyages from their colonies in Iceland and Greenland. Although there are many hints and inferences of a Viking presence as far south as Virginia and as far west as Minnesota, the only authenticated site of a Norse settlement in North America is near the most northeasterly point of the island of Newfoundland, at a place called L'Anse-aux-Meadows ("the cove of the jellyfish") on Epaves Bay.

Discovered in 1960, the locale is set aside as a Canadian National Historic Park and has been declared a World Heritage Site. The site contained the foundations of seven sod homes, a smithy, four small boat houses, and associated work sheds and cooking pits (Fig. 6-c). The houses and artifacts are clearly of Norse origin, and carbon dating established that they were from about the year A.D. 1000, the time of Viking exploratory voyages.

FIGURE 6-c A restored Viking hut at L'Anse-aux-Meadows, Newfoundland (TLM photo).

The settlement apparently was not of long duration, probably serving as an over-wintering station and used as a base for exploring and seeking resources. It is believed that L'Anse-aux-Meadows was occupied on from two to ten occasions between the years 1000 and 1011.

Until some other settlement site is discovered, L'Anse-aux-Meadows is the only viable candidate for identification as the fabled Vinland of Leif Ericsson's Sagas.

TLM

ing numbers of English and Irish settlers occupied the numerous bays of southeastern Newfoundland through the seventeenth century. By 1750 British settlers and New Englanders were beginning to come in large numbers to Nova Scotia and New Brunswick. Shortly afterward 2000 Germans founded Lunenburg. Large migrations of Scottish people, principally from the Highlands, came after 1800, dominating the population of Nova Scotia and, to a lesser extent, Prince Edward Island.

THE PRESENT INHABITANTS

There is a remarkable homogeneity to the contemporary population of the region. The vast majority of the people are of white Anglo-Saxon ancestory. English is the only language of importance except along the northern margin of the region and in scattered pockets of French settlement; the religious affiliation over large areas is dominantly Protestant except in locales of French or Irish Catholic settlement; and most of the people who live in the region were born in the region, often the offspring of several generations in the region.

The taciturn, traditionally conservative Yankee stereotype is still dominant in northern New England. Yankees descended from antecedents who during a "period of poverty and struggle . . . beat down the forest, won the fields, sailed the seas, and went forth to populate Western commonwealths."[3]

[3] J. Russell Smith and M. Ogden Phillips, *North America* (New York: Harcourt, Brace & Co., 1940), p. 113.

Similar British stock, often referred to as Loyalists, originally peopled much of the Atlantic Provinces of Canada and remains dominant today.

The population of the Atlantic Provinces is somewhat less uniform than that of northern New England because large blocks of immigrants settled in groups and generally tended to occupy the same areas for generations. The origins are usually English, Scottish, or Irish. For example, in eastern Newfoundland in general and the Avalon Peninsula in particular the population is about 95 percent Irish Catholic except for the city of St. John's. Similarly, most of Cape Breton Island is a stronghold of Highland Scots and people of Scottish ancestry predominate throughout northeastern Nova Scotia, eastern New Brunswick, and Prince Edward Island.

The French component of the population, so important in the early days of Acadian settlement, is today largely marginal. A large proportion of the New Brunswick population is of French origin, but their dominance is in the northern part of the province, which is considered to lie outside this region. With increasing distance from Quebec, the French element in the New Brunswick population rapidly decreases. Along the northern margin of New England, from Maine's Aroostook Valley to the northeastern New York counties of Clinton and Franklin, there are a significant number of French Canadians who migrated across the international border to work in forests and mills. Otherwise, clusters of French Canadians are found at several places in Nova Scotia and Prince Edward Island. Wherever the French occur in the region, the traditional manifestations of their culture, typified by the French language and the Roman Catholic religion, are prevalent.

As a general rule, cities of the Atlantic Northeast have a more variegated populace than small towns and rural areas. The larger the urban place, the more cosmopolitan its population is likely to be; thus all cities of the region have various minority ethnic groups, although their numbers are few.

The people of this region are mostly longtime residents. Emigration takes place to Megalopolis, Toronto, and the Prairie Provinces, but there is little immigration to balance it. This fact can be illustrated by comparing the birthplaces of the people of Atlantic cities with those of other Canadian cities. Of the larger Canadian metropolitan areas, only one (Quebec City) has a higher proportion of its population born in the same province than do the three Atlantic cities of St. John's, Saint John, and Halifax.

Much of a region's character is asserted by the attitudes, traditions, and values of its people and by the imprint of these intangibles on the regional landscape. The Atlantic Northeast is singularly rich in this regard. Although the elements are difficult to define and the cause-and-effect relationships are obscure, some of its manifestations are conspicuous. As Edward McCourt has observed about "Newfies," the people of this region "have evolved their own mores, created their own culture, made and sung their own songs, and added to the language according to their need."[4]

One aspect of the cultural character of the region relates to the form and charm of the small units of settlement, the hamlet and village. The name of the village is likely to sound as pleasant as it appears. Thus travelers in central New Hampshire can stop at the hamlet of Sandwich; if unsatisfied there, they can drive on to the next village, Center Sandwich; or the next, North Sandwich. Vivid place names persist throughout the region. Consider a selection from Newfoundland: Harbour Grace, Heart's Delight, Witless Bay, Maiden Arm, Uncle Dickies Burr, Come By Chance, Mount Misery, Hit or Miss Point, Right-in-the-Road Island, Holy Water Pond, Damnable Bay, Lushes Bight, Cuckhold's Head, Bleak Joke Cove, Famish Gut, Horse Chops, Sitdown Pond, and Hug My Dug Island.

The village form has a certain uniformity: the fishing hamlet clustered on a tiny beach about the head of a bay and the farming town rectangularly arrayed around the village green. Actually, village greens are present in almost every town in northern New England but very uncommon in Canada's Atlantic Provinces. White, tall-spired churches with their adjacent cemeteries are, however, the dominant edifices in villages throughout the region.

The rural landscape also has characteristic regional forms. The covered bridge, for example, is a famous feature in northern New England (Fig. 6-6) and is often carefully cultivated as a tourist attraction. Covered bridges are common throughout the

[4] McCourt, *The Road Across Canada*, p. 30.

FIGURE 6-6 A covered bridge over a tidal stream in southern New Brunswick (TLM photo).

Maritimes and are more prevalent in New Brunswick than in any part of New England. House-and-barn arrangements are also interesting. The attached house and barn is notable in New Hampshire and Maine. In Vermont, on the other hand, the house and barn are often located adjacent to the road but on opposite sides, presumably to take advantage of the snow-clearing efforts of the road crews.

AGRICULTURE

Although Newfoundlanders were fishermen from the start, farming usually was the first occupation of settlers elsewhere in the region. In most places it was small-scale and difficult, but was long a dominant activity, reaching its peak in about the 1820s and diminishing greatly in relative importance since then (Fig. 6-7).

New England

Dairying is the principal farm activity in northern New England; and with excellent transportation facilities, the emphasis is on whole-milk production. In addition to hay, corn is widely grown as a feed crop for the dairy herds. It is mostly cut green for silage. The better soils of Vermont are reflected in corn production statistics; Maine and New Hampshire combined yield only two-thirds as much silage corn as Vermont.

The most dynamic aspect of northern New England agriculture in recent years has been the poultry industry. Although there is some emphasis on egg production, the principal focus is the raising of broilers (young, tender-meated chickens weighing about 3 pounds). Income from poultry exceeds that from dairying in both New Hampshire and Maine (Fig. 6-8).

Aside from corn, field crops are relatively minor on the farm scene except in a few specialty areas. Berries and apples are moderately widespread, but potatoes constitute the principal cash crop, especially in Maine, which is the leading producer in the eastern half of the nation. Aroostook County, in the northeastern corner of Maine, is one of the leading potato specialty areas of the nation.

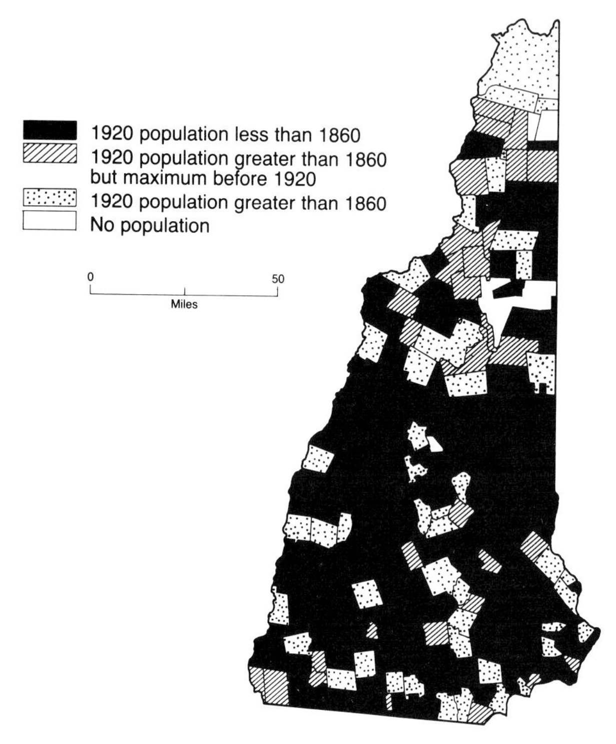

FIGURE 6-7 The decline of agriculture in the 19th century in New England was accompanied by population decrease. This map, showing population change in New Hampshire from 1860 to 1920, is representative of all of northern New England. (After William H. Wallace, "A Hard Land for a Tough People: A Historical Geography of New Hampshire," *New Hampshire Profiles*, April 1975, p. 28.)

The Aroostook emphasis is on seed potatoes in large-scale enterprises with highly mechanized production.

The Atlantic Provinces

Cultivated land in Nova Scotia lies mostly in the western lowland, the granitic interior and the Atlantic Coast tending to discourage settlement; 75 percent of the land is still in forest. The diked lands of Old Acadia, made famous by Longfellow in *Evangeline*, are still fertile, easily cultivated, and productive. Farm crops are primarily those that mature in a short growing season: forage crops, potatoes, vegetables, and fruits.

Nova Scotia's only outstanding area of commercial farming is the Annapolis–Cornwallis Valley in the southwest, which produces a large proportion of Canada's leading fruit crop, the apple. Despite favorable production factors, the growers of the valley have faced an abnormal number of marketing problems, due in part to an individualized approach to selling rather than cooperative or centralized marketing. The output is mostly sold as processed apple products rather than being marketed fresh.

In New Brunswick the area in farms is about equal to that in Nova Scotia, but the area in field crops is nearly twice as great. That the province is not outstanding agriculturally seems proved by the fact that forest still covers nine-tenths of the land. An exception is the Saint John River Valley, which

FIGURE 6-8 Enormous broiler "hotels" are not unusual in northern New England. This is near Waterville, Maine (TLM photo).

is farmed almost as intensively as the Connecticut Valley, although the crop possibilities are more limited. The upper part of the valley (adjacent to the Aroostook area of Maine) is a major potato-growing area.

About two-thirds of the total area of Prince Edward Island is in farms, a much higher ratio than in any other political unit in Anglo-America. The countryside is a delight to the eye, with alternation of field and woodlot, unbelievably green cultivated land, neat farmsteads, and a frequent view of water (Fig. 6-9). The only significant commercial crop is potatoes; Prince Edward Island is one of the two leading potato-growing areas in Canada (the upper St. John Valley of New Brunswick is the other). Fur farming provides an interesting story in Prince Edward Island. Both fox and mink farming got their start in North America on the island and for many decades the former was the mainstay of the island's agricultural economy. In the early years of the twentieth century, however, fur farming shifted chiefly to Ontario and Quebec; by the 1950s almost no active fur farms were left on the island. Within the last decade or so, however, fox farming has returned as a modest enterprise.

FIGURE 6-9 The rural landscape of Prince Edward Island consists mostly of green meadows and pastures, interspersed with small clumps of trees. This scene is near Kensington (TLM photo).

FOREST INDUSTRIES

The Atlantic Northeast, the continent's pioneer logging region, possessed an almost incomparable forest of tall straight conifers and valuable hardwoods. Perhaps nine-tenths of the region was covered in forests. For 200 years or more after the landing of the Pilgrims in 1620, the settlers uninterruptedly continued the removal of trees.

The heyday of forest industries in New England is long since past. There is still considerable woodland, but much of the good timber is inaccessible and much of the accessible timber is of poor quality. Even so, a moderate quantity of sawtimber is cut each year, although no state in the region ranks among the 20 leaders in annual lumber production. Pulpwood production is more significant; in most years only a half dozen states yield more than Maine.

Most forest land in Maine is owned by large corporations. In the northern two-thirds of the state some 75 percent of the land is in only 20 ownerships. The operations of the logging companies enhanced forest recreation by permitting increased accessibility over logging roads, improved wildlife habitats by opening up the dense forest stands, and provided artificial lakes. But the damming of Maine's few remaining wild rivers is a source of great controversy among conservationists, and debates over the value of wildland development policies will probably continue.

After a long period of decline, lumber production is now increasing in northern Maine, although pulpwood is still the principal product. Much of the logging operation in recent years has shifted to virtual clearcutting (Fig. 6-10), largely as a response to the increasing depredations of the spruce budworm, which reduces a mature fir (its preferred host) to a dead stick in five or six years.[5] Throughout the Atlantic Northeast budworm infestations are at abnormally high and growing levels.

Logging is also significant in the Canadian portion of the region, especially in New Brunswick, which produces nearly half of all lumber and pulpwood in the Atlantic Provinces. Moreover, its sawmills and pulp mills are generally larger than

[5] Bret Wallach, "Logging in Maine's Empty Quarter," *Annals of the Association of American Geographers*, 70 (December 1980), 551.

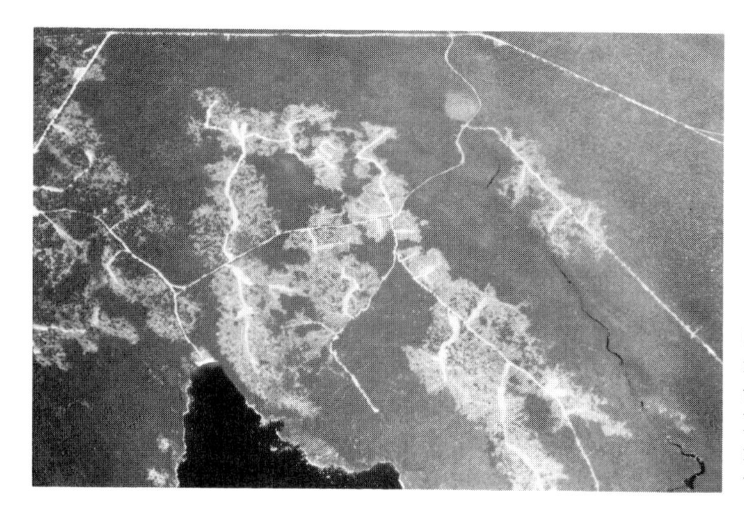

FIGURE 6-10 Clearcutting is now a widely used logging technique in northeastern North America. This area is near Fredericton, New Brunswick (TLM photo).

those in the other provinces. Nova Scotia's forest land, unlike that of New Brunswick and Newfoundland, is largely in private ownership; there is a roughly even balance between output of lumber and pulpwood, with processing primarily in small mills. In Newfoundland most of the productive forest land is owned or leased by three large pulp and paper companies; commercial output is heavily oriented toward pulpwood (Fig. 6-11), although there is also a great deal of subsistence logging for firewood. The limited forest land of Prince Edward Island is found chiefly in small, individually owned woodlots. For the Atlantic Provinces as a whole, the cutting of pulpwood is nearly twice as great as timber harvesting for all other purposes combined.

FIGURE 6-11 A wood-chipping machine in action near Corner Brook, Newfoundland (TLM photo).

FISHING

Fish was the first export from the New World. From Newfoundland to Cape Cod lie offshore banks that are one of the richest fishing grounds in the world (Fig. 6-12). These banks were frequented by Scandinavian, Portuguese, Dutch, English, and French fishermen before the period of colonization in America. As early as 1504 Breton and Norman fishermen were catching cod in the western North Atlantic; and by 1577 France had 150 vessels, Spain 100, Portugal 50, and England 15 fishing for cod on the banks.[6]

Early fisheries concentrated largely on cod, which was salted, pickled, and especially dried for export to European, largely Mediterranean, and tropical markets. Maximum production of dried salted cod was reached in the 1880s, but changing

[6] R. H. Fiedler, "Fisheries of North America," *Geographical Review*, 30 (April 1940), 201.

FIGURE 6-12 Piles of lobster pots awaiting the beginning of the season on a coastal pier in Maine (TLM photo).

fishing conditions and market requirements caused a gradual decline after that. During the present century freezing replaced drying, salting, and pickling as the major technique of preserving fish enroute to market.

The Catch

Although the regional catch includes a great variety of species, cod and lobster are by far the most important, each comprising about 20 percent of the total. Lobster is the most valuable fishery product in every political unit in the Atlantic Northeast except Newfoundland, with cod ranking second except in Newfoundland, where it is first. Cod and other demersal (bottom-feeding) species—such as haddock, pollack, and turbot—mostly are taken by trawling, although a considerable quantity is caught in shallow coastal waters by hook-and-line and by the use of traps. Pelagic (surface-feeding) species, such as herring and mackerel, are caught with a variety of techniques, but particularly with seines.

Lobsters and crabs traditionally have been caught in baited traps (called pots) as they crawl around in shallow water looking for food. Traps are still much in use, but an increasing share of the total catch is taken by deep trawling. Open season on these crustaceans is very short, typically only two or three months per year. Today, however, there is considerable raising of lobsters in "pounds," where they are held captive in the sea. The latest development is the "dryland pound," wherein lobsters are retained indoors in drawers, in near-freezing seawater, for up to a year; this allows fresh lobsters to be marketed year-round.

The Special Case of Newfoundland

Newfoundland, from its earliest days, was a staple product colony, the staple being codfish. For three centuries the bulk of the populace consisted of fishermen. They sold their catch to local merchants, who passed it on to exporters in St. John's, who depended on distant and unstable markets. The result was a precarious economy with continuing poverty for most fishermen and an uncertain opulence for the St. John's exporters. Catching cod in inshore waters remained the unchallenged base for Newfoundland's limited economy until the end of the nineteenth century.

In this century Newfoundland's seaward outlook weakened and the contribution of fishing to the provincial income declined steadily; it is now substantially below that from mining and forestry. But there are still more than 12,000 inshore fishermen in Newfoundland, living primarily in small villages (called outports) at the heads of innumerable bays scattered around the island (Fig. 6-13). Most of their fishing is done within the bays or slightly beyond them; they set cod traps in the bay mouths, use a variety of nets, and use hook-and-line farther out. Before being marketed in Europe, most of the cod was cleaned and then dried on racks (called flakes) in the uncertain Newfoundland sun. Nowadays a small

FIGURE 6-13 A representative Newfoundland outport, The Battery (TLM photo).

percentage of the catch is still handled in this fashion, but the great bulk is filleted, frozen, and sold in the United States.

Value of the Fisheries

Despite continuous exploitation for more than four centuries, western North Atlantic fisheries continue to provide a large and important supply of food—more than 2 billion pounds a year. Two-thirds of the total annual Canadian catch and about one-tenth of the total U.S. catch are from this area. Nova Scotia and Newfoundland vie with British Columbia as the leader in total value of landings among Canadian provinces, with New Brunswick ranking fourth.

In Newfoundland more than 15 percent of the value added in all "commodity-producing industries" (essentially all primary and secondary production) is from the catching and processing of fish. In Nova Scotia and Prince Edward Island the proportion exceeds 10 percent; in no other province is the figure as high as 5 percent. To Canada, the world's leading exporter of fish products, the northwestern Atlantic fisheries continue to be very significant. They are less important to the United States; Massachusetts (with fishermen operating mostly in this area) ranks third among the states in total catch and Maine ranks seventh.

The commercial fisheries of the Atlantic Northeast have a long history of erratic instability, which shows no sign of changing. After three decades of continuing decline, a revival was triggered in the late 1970s, particularly in Canada, when the governments of the United States and Canada declared a 200-mile offshore jurisdiction, which significantly reduced foreign competition in this rich fishing area, especially over most of the Grand Banks, which are now under Canadian control.[7]

By the beginning of the 1990s, however, a serious decline in numbers, particularly of the once abundant cod, caused another major retrenchment in the industry. Apparently, Canada's Atlantic fishing industry had become too efficient. Sophisticated fishing vessels were catching many more fish than the fishery could handle. Most of the problem was generated by offshore draggers that tow a huge bag of net along the bottom, scooping up everything in its path. A dragger collects much more than the target fish (the "bycatch" mostly is killed in the operation, and is simply discarded); it also does damage to the sea bottom environment in the process.

In Canada the allowable quota of cod (set by the federal government) was reduced by 28 percent in only two years, resulting in layoffs of fishermen and closure of processing plants. Particularly hard hit was Newfoundland, where nearly one-fourth of the labor force is employed by the fishing industry. In the early 1990s Newfoundland had an unemployment rate approaching 20 percent, nearly three times the national average.

[7] In 1984 a long and acrimonious controversy between the United States and Canada over the "ownership" of Georges Bank (second most valuable of the fishing banks) in the Gulf of Maine was resolved by the International Court of Justice. The compromise solution gave the United States control of about 75 percent of the prolific waters, with the other 25 percent going to Canada.

MINING

There is a long history of mining in the region. The mineral industries have been cursed with irregular productivity, however, and prosperity has been limited. Still, mining is an important contributor to regional income, some notable expansion of output has occurred in recent years, and the immediate prospects of offshore petroleum production are impressive.

Coal has been mined in Nova Scotia for more than a century, chiefly on the north shore of Cape Breton Island. After three decades of irregular decline, production was rejuvenated in the 1980s, partly for the Cape Breton steel industry, but especially to replace oil in electricity generation.

Another traditional mining activity of the region is the production of *iron ore* on the flanks of the Adirondacks, where initial output began about 1800. Production has been erratic for the last century, but considerable reserves of magnetite are still present.

There are a half dozen notable mining localities on the island of Newfoundland. Most significant is Buchans, in the western interior, where sulfide ores holding copper, lead, zinc, silver, and gold are extracted. At Baie Verte on the north coast are large mines yielding asbestos and copper.

Most of Canada's gypsum and barite, plus a considerable amount of its rock salt, is produced in Nova Scotia. New mines have been opened in New Brunswick for base metals, gold and potash.

The *quarrying* of building stone has for many decades been the only significant "mining" activity in northern New England. The hard-rock complex underlying most of this subregion has long been an important supplier of granite, marble, and slate, particularly from several locations in Vermont. Demand for these high-quality stones decreased in recent years, however, and the prominence of quarrying declined.

The region's most exciting, and potentially most profitable, mineral activity is in the deep and treacherous waters far offshore from the Atlantic Provinces where major reserves of *petroleum* and *gas* were discovered. After years of desultory exploration a major oil field, called Hibernia, was found in 1979 about 170 miles (270 km) southeast of St. John's. More than 100 exploratory wells were drilled and it is estimated that eventual production might be equal to half of Canada's presently known reserves. Staggering engineering problems—deep producing horizons, shifting currents, frequent storms, much fogginess, the danger of collision with floating icebergs—had to be overcome before commercial production began in the late 1980s.

Moreover, there is disagreement between the governments of Canada and Newfoundland as to which is the legal owner of the oil. It is fairly clear under both international and national jurisprudence that the federal government has jurisdiction over far offshore resources, but Newfoundland argues that it joined Canada under special circumstances that allowed it to retain control. The Newfoundland government also bases its case on moral grounds: Newfoundland has always been the poorest and remotest of the provinces and it "must have" the oil revenue in order to survive. After years of discussion, the federal and provincial governments signed a joint development and management agreement in 1985. Meanwhile, St. John's is enjoying a newfound prosperity as a growing commercial and energy-servicing focal point.

The offshore excitement also extends to Nova Scotia. Deep-water exploration revealed evidence of oil over a wide area south and east of this province and in 1983 promising natural gas fields were found near Sable Island, about 200 miles (320 km) east of Halifax (Fig. 6-14).

RECREATION

Tourist-oriented recreation has been a mainstay of northern New England's economy since World War II. The relatively cool summers are attractive to most tourists and the snowy winters are popular with skiers. Moreover, the area is immediately adjacent to the huge population of Megalopolis and there are well-developed highways to enable urbanites to reach secluded rural retreats and wildlands in a relatively short time.

The last three decades have seen a great upsurge in interest in owning second homes in northern New England. Rural land values, which had remained remarkably depressed considering their nearness to large population clusters, have now sky-

FIGURE 6-14 An offshore oil drilling rig at work in a somber sea well to the east of the Nova Scotian coast (courtesy N.F.B. Phototheque/George Hunter photo).

rocketed. Condominium clusters and other kinds of resort living have mushroomed on mountainsides and lakeshores.

Tourism is less significant and more seasonal in the Atlantic Provinces. Most visitors during the relatively short summer tourist season come from nearby localities in the Maritimes or from neighboring Quebec. There are six national parks in the Canadian section of this region, but the visitors seem to be particularly attracted to the beach resorts of Prince Edward Island. The two ferry systems connecting the island to the continent are drastically overcrowded during the summer.

URBANISM AND URBAN ACTIVITIES

We have seen that the Atlantic Northeast is largely a region of rural charm, rural activities, and economic impoverishment. The people of the region, as with almost all regions on the continent, are now mostly urbanites, however, and the endeavors that provide most of the jobs and contribute most to the economy of the region are urban-oriented endeavors. The primary activities of farming, mining, and fishing are quite important and are conspicuous in the landscape; but the secondary and tertiary activities of manufacturing, trade, services, construction, and governmental functions actually dominate the economy and their locus is mostly in urban areas.

There are no large cities in the region. There are, however, many long-established urban places and it is here that economic growth is concentrated and population growth is taking place. As of the early 1990s, nearly two-thirds of the inhabitants of the Atlantic Northeast were urban dwellers.

Much of this pattern of urban growth and rural stagnation is also common to other North American

A CLOSER LOOK A Bridge to Prince Edward Island

Although situated virtually in the middle of the Atlantic Provinces, Prince Edward Island has a remote location because it is separated from the mainland by the broad waters of Northumberland Strait. Its surface transportation links with the rest of the continent consist primarily of a 9-mile (15 km) government-run ferry service to New Brunswick and a 16-mile (25 km) privately owned ferry service to Nova Scotia. These ferries transport some 2 million passengers and 900,000 vehicles annually, but stoppages are not unusual (particularly due to winter ice) and lengthy delays are commonplace (most notably during the summer tourist season).

Ever since the time Prince Edward Island became a Canadian province in the 1870s there have been dreams and plans to provide a "fixed link" with the mainland. Tunnels, causeways, and bridges all have been considered, and construction of a combined causeway-bridge actually was begun in the 1960s, but the federal government soon cancelled the project because of high costs.

Debates about both the desirability and feasibility of a fixed link have raged for years. Proponents focus on convenience and lower transportation costs; opponents worry about environmental degradation and the loss of an unhurried pastoral way of life. In 1988 a referendum was held on the island, resulting in a 60–40 vote endorsing construction of a fixed link.

Federal feasibility studies strongly favor the building of a toll bridge roughly parallel to the shorter ferry route, and detailed planning is now under way (Fig. 6-d). At the present writing, an Environmental Assessment and Review Panel has been formed to review the project, which would be the world's longest bridge over winter ice-covered waters. If the review is favorable, construction on the billion-dollar project might begin as early as 1992.

TLM

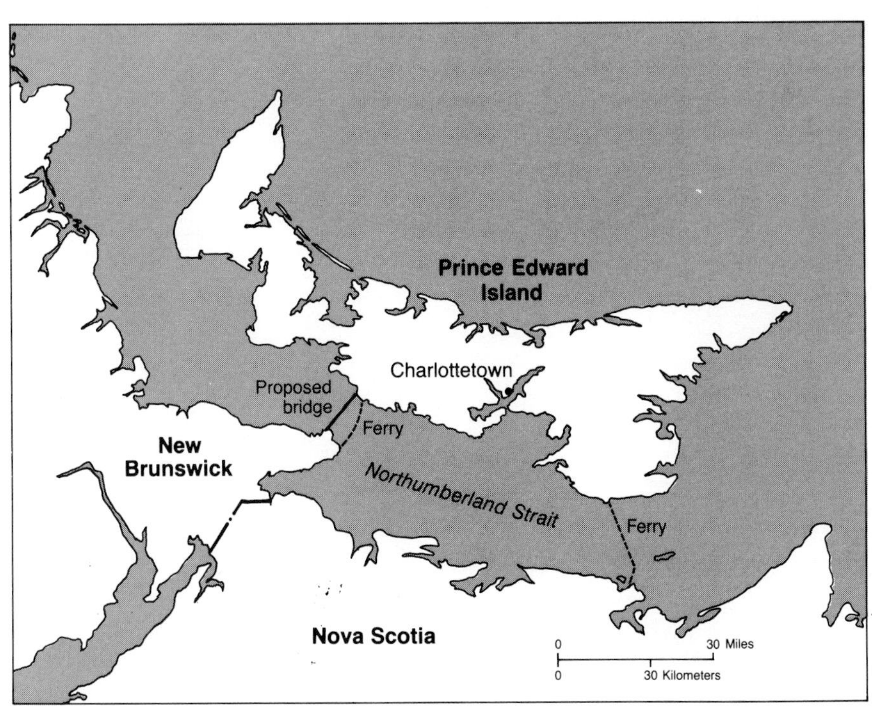

FIGURE 6-d The existing ferries and potential bridge to Prince Edward Island.

regions. Both economic opportunities and the material amenities of life tend to be prevalent in cities; consequently, immigrants from other regions or other continents usually settle in the cities. And as rural opportunities diminish, particularly in agriculture and fishing, the rural-to-urban population drift, especially among young adults, continues.

The industrial functions normally associated with cities are limited and specialized in the region. Among the cities of this region, Halifax alone is large enough to encompass a relatively full range of manufacturing activities, but it is by no means a major industrial center. Notable specialized industrial centers include Sydney–North Sydney in Nova Scotia (iron and steel), St. John's in Newfoundland (oil field equipment), and Bangor in Maine (lumber and pulp).

Many of the more prominent manufacturing facilities in the Atlantic Northeast are located in small towns or even in rural localities several miles from any town. This is particularly true of wood-processing plants, both those producing sawn lumber and those whose output is pulp and/or paper. Some major mills are located in medium-sized "cities," such as Corner Brook, Newfoundland; but more often the mill site is associated with much smaller urban places, such as Berlin, New Hampshire, or Millinocket, Maine. Fish-processing facilities, too, are often found in small coastal settlements rather than in larger cities.

The major urban centers of the region are ports and most of the larger ones experienced a considerable increase in traffic in recent years. *Halifax* is clearly the most important port in the region, besides being the largest city by far. It normally accommodates more overseas shipping than any other eastern Canadian city except Montreal. It has developed a major container terminal, as well as housing both Canada's largest naval base and largest coast guard base. Winter is its most active season, for Canada's St. Lawrence and Great Lake ports usually are closed by ice for several months. Halifax is not really a primate city for the region, but its magnificent harbor and reasonably diversified economic base ensure its relative prosperity (Fig. 6-15).

St. John's also has a splendid protected harbor, although its mouth (the Narrows) is occasionally choked with ice. Its extreme foreland location

FIGURE 6-15 New developments and high-rise buildings cluster around the harbor in Halifax, Nova Scotia (courtesy John Elk III, Stock, Boston).

TABLE 6-1

Largest urban places of the Atlantic Northeast

Name	*Population of Principal City*	*Population of Metropolitan Area*
Bangor, Me.	30,750	84,800
Burlington, Vt.	37,130	129,100
Charlottetown, P.E.I.	15,600	—
Corner Brook, Nfld.	25,100	—
Dartmouth, N.S.*	66,100	—
Fredericton, N.B.	44,700	—
Glens Falls, N.Y.	—	116,100
Halifax, N.S.	113,800	302,100
Moncton, N.B.	55,800	—
Saint John, N.B.	78,400	122,000
St. John's, Nfld.	100,800	162,000
Sydney, N.S.	30,100	—

* A suburb of a larger city.

provides little opportunity for service to anything more than a provincial hinterland and even this port-of-entry and distribution function is challenged by smaller Newfoundland ports. But the harbor facilities have been modernized and St. John's is magnificently situated to serve as a base for offshore fishery fleets, offshore oil activities, and as the only effective storm port in a region of stormy seas.[8]

Saint John is somewhat smaller than St. John's, but its situation is quite different. It serves as one Atlantic terminus (Halifax and Portland are the others) of Canada's transcontinental railway system and thus has a busy season of general cargo activity in winter; however, its somewhat more remote location than Halifax from the Atlantic side and its less suitable harbor give it a secondary rank among Atlantic ports.

These and other cities of moderate size in the Atlantic Northeast are listed in Table 6-1. Each regional chapter in this text includes a table that gives the population for the metropolitan area and the political city, as estimated for 1990 by the United States Bureau of the Census and by Statistics Canada. The cities in each table are arranged in alphabetical order; only the more important ones, based on population and lying within the region under consideration, are included.

On the regional maps at the beginning of each chapter, the following urban places are shown:

1. Those with a 1990 population exceeding 250,000 are indicated by a large dot and their names are shown in all capital letters (as BOSTON).
2. Those with a population between 50,000 and 250,000 are shown by a large dot and their names are in capital and lowercase letters (as Halifax).
3. Those with a population of less than 50,000 that are mentioned prominently in the text are indicated by an open circle and their names are in capital and lowercase letters (as Bangor). To ensure legibility, cities and towns that are distinctly suburbs of larger urban places are not shown.

THE OUTLOOK

In spite of a long history and a distinguished heritage, the Atlantic Northeast has not been a favored region from an economic standpoint. Its natural resources are limited and its position in the northeast and southeast corners of the two countries involved cut it off from the mainstream of activity just enough to inhibit its commercial vitality. In few other significant parts of the continent are there so many part-time workers—part-time farmers, part-time fishermen, part-time loggers. The rest of their time is spent in another wage-earning capacity, in doing odd jobs, or in subsisting on welfare payments (the latter is often a significant portion of total income). There is no reason to expect a change in this broad pattern.

The region has a disproportionately large number of people engaged in marginal activities in the fields of farming, fishing, and logging, even though recent decades showed a steady decline in employment in all three. A dearth of arable land, a restricted market, and a limit on crop options all circumscribe agricultural growth.

[8] When storms, particularly hurricanes, come to the banks area or other parts of the western North Atlantic, less durable craft, particularly fishing vessels, often make for a protected port. St. John's is conspicuously the most suitable anchorage in the area and may become crowded with ships seeking refuge.

There is a general trend toward larger-scale enterprises in farming, forestry, and fishing. The prominent growth in mechanized offshore fishing operations is salutary to the regional economy but does not answer the core problem of local dependence on inshore fisheries. Many factors lie behind the rebirth of the fishing industry in the Atlantic Provinces (the upturn is much more limited on the U.S. side of the border); most important is the restriction on foreign competition, but diversification of catch (for example, herring and snow crab), improvement of equipment, and aggressive marketing have played important roles. Even so, the recent drastic decline in the cod fishery underlines the economic instability that is inherent in this industry.

Mining is prospering erratically, and great hopes are focused on offshore oil and gas activity. Even as the Grand Banks and Sable Island discoveries are being developed, many other favorable geologic structures east of Newfoundland, Nova Scotia, and Maine await exploration—a difficult undertaking, but one that undoubtedly will be accomplished.

Despite favorable trends in mining and some aspects of the fishing industry, the fact is that the Atlantic Provinces are still the poorest part of Canada, with a per capita income 40 percent below that of Ontario. An inordinate share of the budgets of these four provinces is still provided by federal assistance ("equalization payments"). To anticipate a significant change in the long-standing reality of regional (relative) poverty would be illogical.

Northern New England has undergone repeated economic readjustments. Pulp and paper production should continue to be important and hydroelectric developments will probably continue; however, both activities are coming under increasingly severe scrutiny by advocates of wildland preservation. Agricultural specialization should continue but not on a large scale. Poultry raising, dairying's only rival to agricultural dominance, may not grow, but it does provide an element of stability and diversification in the agrarian scene.

Urban activities throughout the region, especially manufacturing, can be expected to grow steadily—if slowly. The rate of expansion will remain well below the average growth rate for both nations and most industrial growth will be in production of items for the local or regional market. But despite the prospect of increased urbanization, the Atlantic Northeast is never likely to be famous for its urban-industrial advantages. Rather, rural amenities are emphasized as the regional attraction. It has been said that "Vermont is the only place within a day's drive of New York that is fit to live in."[9] Or, stated differently, David Erskine has noted that the region's "compensation must be in a life that is freer, or lived in more attractive surroundings, or with closer family ties."[10]

The lure of the Atlantic Northeast is being increasingly appreciated by nearby urbanites. The tourist, the second-home owner, the developer, and the speculator have landed in force. Short- and long-term visitors arrive in ever greater numbers during the summer, and winter recreation also continues to expand. The wider dissemination of such activities over the region augurs well for a more balanced economic gain.

It is probable that for the next few years, major regional concerns will revolve around three issues—international demand for fishery products, development of offshore petroleum and gas, and depredations of the spruce budworm.

[9] Joe McCarthy and the Editors of Time-Life Books, *New England* (New York: Time Incorporated, 1967), p. 163.

[10] Erskine, "The Atlantic Region," p. 280.

SELECTED BIBLIOGRAPHY

BACKUS, RICHARD H., AND DONALD W. BOURNE, EDS., *Georges Bank*. Cambridge: MIT Press, 1987.

BAIRD, DAVID M., "Down the Saint John," *Canadian Geographic*, 109 (October–November 1989), 68–75.

BELLIVEAU, J. E., "The Acadian French and Their Language," *Canadian Geographical Journal*, 95 (October–November 1977), 46–55.

BLACK, W. A., "The Great Northumberland Ice Barrier," *Canadian Geographical Journal*, 87 (October 1973), 30–39.

CALHOUN, SUE, "Snow Crab Bonanza Faltering in Acadia," *Canadian Geographic*, 108 (December 1988–January 1989), 50–59.

CAMERON, SILVER DONALD, "Almighty Cod! It Reigns

Supreme Over Our Atlantic Fishery," *Canadian Geographic,* 108 (June–July 1988), 32–41.

———, "Lure and Lore of the Lobster," *Canadian Geographic,* 109 (June–July 1989), 66–72.

———, Net Losses: The Sorry State of Our Atlantic Fishery," *Canadian Geographic,* 110 (April–May 1990), 28–37.

———, "Port of Halifax Living Up to Its 'Greatest Harbour' Name," *Canadian Geographic,* 108 (December 1988–January 1989), 12–23.

CLARK, ANDREW H., *Acadia: The Geography of Early Nova Scotia*. Madison: University of Wisconsin Press, 1968.

———, *Three Centuries and an Island*. Toronto: University of Toronto Press, 1954.

CLARK, R. H., "Energy from Fundy Tides," *Canadian Geographical Journal,* 85 (November 1972), 150–163.

CRABB, PETER, "Is Newfoundland One of Canada's Maritime Provinces?" *The Professional Geographer,* 33 (November 1981), 489–490.

DAY, DOUGLAS, ED., *Geographical Perspectives on the Maritime Provinces*. Halifax: St. Mary's University, 1988.

GREEN, DAVID F., "Chignecto Marshes: Bird and Hay Country," *Canadian Geographic,* 101 (June–July 1981), 44–49.

HALLIDAY, H. A., "The Lonely Magdalen Islands," *Canadian Geographical Journal,* 86 (January 1973), 2–13.

HOGGART, K., "Resettlement in Newfoundland," *Geography,* 64 (July 1979), 215–218.

IRLAND, LLOYD C., *Wildlands and Woodlots: The Story of New England's Forests*. Hanover, NH: University Press of New England, 1982.

LEBLANC, ROBERT G., "The Acadian Migrations," *Canadian Geographical Journal,* 81 (July 1970), 10–19.

LOTZ, J., "200 Mile Limit Revives Atlantic Fisheries," *Canadian Geographic,* 97 (October–November 1978), 40–45.

MACPHERSON, ALAN G., ED., *The Atlantic Provinces*. Toronto: University of Toronto Press, 1972.

———, AND JOYCE MACPHERSON, *The Natural Environment of Newfoundland, Past and Present*. St. John's: Memorial University of Newfoundland, Department of Geography, 1981.

MANNION, JOHN, ED., *The Peopling of Newfoundland: Essays in Historical Geography*. St. John's: Memorial University of Newfoundland, Institute of Social and Economic Research, 1977.

MCCALLA, ROBERT, *The Maritime Provinces Atlas*. Halifax: Maritext Limited, 1989.

MCCALLA, ROBERT J., "Separation and Specialization of Land Uses in Cityport Waterfronts: The Cases of Saint John and Halifax," *Canadian Geographer,* 27 (Spring 1983), 48–61.

MCGHEE, ROBERT, "The Vikings Got Here First, but Why Did They Leave?," *Canadian Geographic,* 108 (August–September 1988), 12–21.

MCLAREN, I. A., "Sable Island: Our Heritage and Responsibility," *Canadian Geographical Journal,* 85 (September 1972), 108–114.

MCMANIS, DOUGLAS R., *Colonial New England: A Historical Geography*. New York: Oxford University Press, 1975.

MEEKS, HAROLD A., *The Geographic Regions of Vermont: A Study in Maps*. Hanover, NH: Dartmouth University, Geography Publications at Dartmouth 10, 1975.

REES, RONALD, "Changing Saint John: The Old and the New," *Canadian Geographical Journal,* 90 (May 1975), 12–17.

REID, JOHN G., "The Beginnings of the Maritimes: A Reappraisal," *The American Review of Canadian Studies,* 9 (Spring 1979), 38–51.

ROBINSON, J. LEWIS, "Changing Settlement Patterns in Newfoundland," *Geographical Review,* 65 (1975), 267–268.

SCHMANDT, JURGEN, AND HILLIARD RODERICK, EDS., *Acid Rain and Friendly Neighbors: The Policy Dispute between Canada and the United States*. Durham, NC: Duke University Press, 1985.

SURETTE, RALPH, "The Shelf: A Hotbed of Geology and Marine Life Lies beneath Our East Coast Waters," *Canadian Geographic,* 109 (June–July 1989), 40–51.

WALLACE, WILLIAM H., "A Hard Land for a Tough People: A Historical Geography of New Hampshire," *New Hampshire Profiles* (April 1975), 21–32.

WALLACH, BRET, "Logging in Maine's Empty Quarter," *Annals of the Association of American Geographers,* 70 (December 1980), 542–552.

WELSTED, JOHN, "Post-Glacial Emergence of the Fundy Coast: An Analysis of the Evidence," *Canadian Geographer,* 20 (1976), 367–383.

WYNN, GRAEME, *Timber Colony: A Historical Geography of Early Nineteenth Century New Brunswick*. Toronto: University of Toronto Press, 1981.

7

FRENCH CANADA

The most culturally distinctive region of the continent is French Canada. For more than three and a half centuries it has been occupied by people whose primary cultural attributes are different from the settlers of the other regions of the continent.It has been a Franco-culture island in an Anglo-culture sea and this unique cultural expression has been maintained without significant external reinforcement. In the last two centuries there has been very little immigration, or even much tangible support, from France; yet the settled southern portions of Quebec continue to be dominated by a solidly French culture and the areal extent of this influence has continued a slow expansion into adjacent parts of New Brunswick and Ontario, even into New England.

The predominant expression of French-Canadian culture is, of course, the use of the French language. More than 80 percent of the people of the province of Quebec consider French their mother tongue; and if the cosmopolitan city of Montreal is excluded from the statistics, more than 90 percent of the people are Francophones (linguistically French), with many of the remainder (the Anglophones) being bilingual. Indeed, in most of the St. Lawrence Valley and estuary downstream from Montreal the proportion of French speakers exceeds 98 percent of the total population. Hand in hand with the French language in the region goes the Roman Catholic religion. Catholicism in Canada is not restricted to Francophones, but more than 85 percent of the inhabitants of Quebec profess to be followers of the Roman Catholic faith. It is in the French-Canadian region, then, that the Catholic Church has its firm Canadian base, largely accounting for the fact that some 45 percent of the Canadian population professes to be Roman Catholic.

These two important social attributes—language and religion—are not the only nonmaterial elements of the distinctive French-Canadian culture, but they are clearly the most conspicuous.

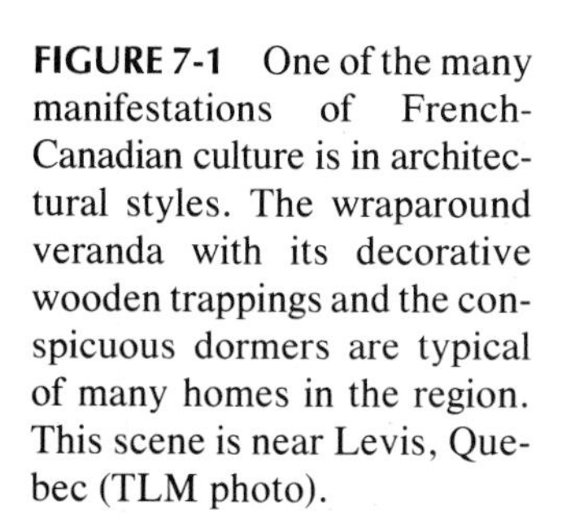

FIGURE 7-1 One of the many manifestations of French-Canadian culture is in architectural styles. The wraparound veranda with its decorative wooden trappings and the conspicuous dormers are typical of many homes in the region. This scene is near Levis, Quebec (TLM photo).

Other intangible culture traits, such as cohesiveness of family life or dietary preferences, are less easily identified and quantified but may be equally significant in some areas.

Although recognizing the importance of nonmaterial culture elements, a geographer is continually seeking to identify and interpret expressions of culture in the landscape. In French Canada this search is amply rewarded. Physical manifestations of the dominant culture can be recognized in farmscapes and townscapes, fences and signs, field patterns and architectural styles, general appearance, place names, and many other facets of the landscape (Fig. 7–1).

Consider, for example, the matter of place names in French Canada. The great majority of all places—towns, rivers, streets, mountains—are named for saints. Any map of the region shows this dominance: St. Laurent, St. Maurice, Ste. Anne de Bellevue, Ste. Foy, St. Hyacinthe, St. Jérôme, St. Félix-de-Valois. The principal areas where such names are not dominant are those in which a later spread of French settlement was superimposed on an already established Anglo framework, as in the Eastern Townships of Quebec or parts of New Brunswick.

The prominence of the church in the French-Canadian way of life is manifested in the landscape by many features other than place names. Roman Catholic religious institutions are numerous and conspicuous. Churches, seminaries, monasteries, convents, shrines, retreat houses, cemeteries, and other edifices are prominent throughout the settled parts of the region. They are usually large in size compared to other structures in the area and often solid and massive in style. Their prominence is somewhat subdued in cities, where secular buildings may also be large and massive. But in smaller towns and villages church-related structures commonly dominate the scene; indeed, most towns in the region cluster about a large, stone Roman Catholic church.

The rural landscape, too, has its characteristic features. Most notable is the pattern of land ownership and field alignment. Most farms are long, narrow rectangles and fields within the farm repeat the pattern on a smaller scale. The background to this unique pattern is discussed subsequently.

Centrally located within the fields are often long heaps of stones that were gathered by the farmer after years of winter frost-heave, and accumulated in piles. Often a cedar pole fence surrounds the field, although wire fences have been increasingly adopted in recent years, particularly in the upstream portions of the region. The farmstead, too, often has predictable characteristics. The buildings are likely to be constructed of unpainted wood, somewhat unkempt, gray, and bleak in appearance. Certain architectural styles are common: the farmhouse often has a lengthy veranda or porch, a certain amount of "gingerbread" on the exterior, and several high dormers; the barn is likely to be of the inclined-ramp variety with livestock housed on the lower floor and a wooden ramp leading to the second story where machinery, tools, and feed are kept.

A CULTURALLY ORIENTED REGION

The designation of French Canada as a major continental region is primarily in recognition of its cultural uniqueness and the significance of this cultural imprint on its geography. Other factors, environmental and economic, contribute to regional unity, but it is the manifestations of French-Canadian culture that are the distinctive shapers of the total geography of this region.

No other region, as delineated in this book, is recognized principally on the basis of its cultural components. Some other sections of the continent have important elements of non-Anglo culture, but in every case the designation of a culturally oriented region is felt to be unwarranted either because the culture does not sufficiently permeate the geography of the region or because the area and population involved are too small to justify separation as a region.

A case might be put, for example, that the southwestern borderlands of the United States—those parts of Texas, New Mexico, Arizona, and California that are close to Mexico—contain millions of people of Mexican extraction who have an Hispanic culture that is analogous in many ways to the French-Canadian culture region in eastern Canada.[1] Although recognizing the validity of this assertion, the author feels that the form and function of life in the borderlands are clearly dominated by an Anglo pattern that is insufficiently different to justify regional recognition. Similar reasoning holds true for the Hawaiian Islands, with their significant Oriental and Polynesian culture complexes, although Hawaii is delineated as a separate region for other reasons.

True cultural distinction is also shown in certain parts of Anglo-America where the bulk of the population is aboriginal in origin. The Navaho–Hopi–Ute Indian complex of the Four Corners country, for example, has recognizable cultural uniqueness. Also, in large stretches of subarctic and arctic Canada and Alaska varying degrees of distinctive Indian, Inuit, and Aleut culture predominate. But in each instance regional recognition does not seem warranted on a largely cultural basis because of the small size of the population involved or its highly fragmented settlement pattern over broad areas where environmental factors seem more important as regional delineators.

FRENCH CANADA AS A REGION AND A CONCEPT

The region of French Canada, as delineated here, does not include all the province of Quebec and is not limited by the borders of Quebec (Fig. 7–2). It is considered to encompass that portion of eastern Canada that is dominated by French-Canadian culture except where significant areas of non-French-Canadian culture or nonsettlement intervene. Thus the region includes most of the southern settled parts of Quebec: from the lower Ottawa River Valley in the west; down the St. Lawrence Valley to include the Gaspé Peninsula, Anticosti Island, and the north shore of the Gulf of St. Lawrence in the east; as well as the relatively densely settled portion of the Shield that encompasses the Lake St. John lowland and the Saguenay River Valley.

It is also considered to include the French-dominated portions of the province of New Brunswick, largely the area north of 47° north latitude. In the seven counties of northern New Brunswick about three-fifths of the population speak French as a mother tongue and in three of the counties more than 80 percent of the people are Francophone.

With a few minor exceptions in northeasternmost New York State, northernmost Vermont, and the Aroostook Valley of Maine, the area of French-Canadian dominance ends abruptly at the international boundary; so the U.S. border can be considered the southern margin of the region.

As thus delimited, the region of French Canada includes more than 90 percent of the population of Quebec and about 35 percent of the population of

[1] See, for example, Richard L. Nostrand, "The Hispanic-American Borderland: Delimitation of an American Culture Region," *Annals*, Association of American Geographers, 60 (1970), 638–661.

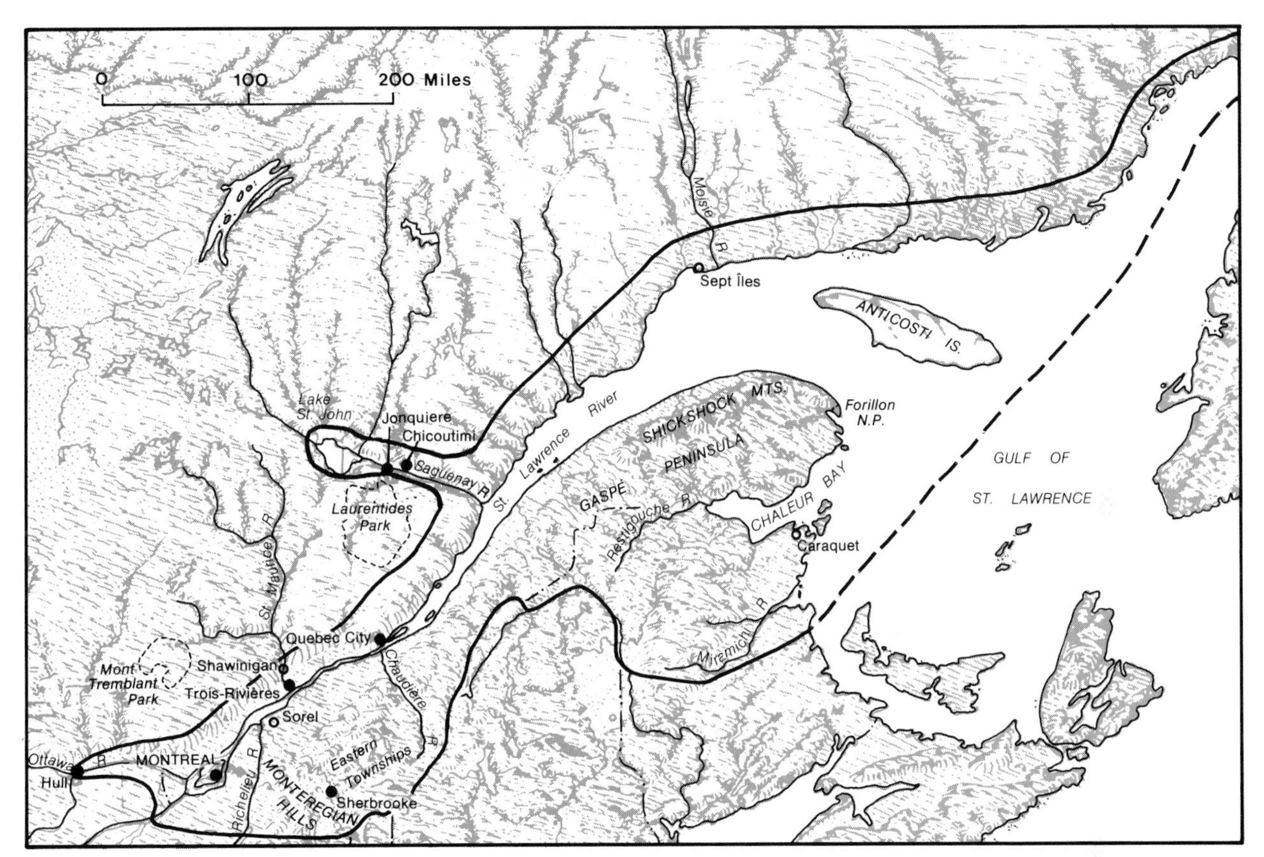

FIGURE 7-2 The French Canada Region (base map copyright A. K. Lobeck; reprinted by permission of Hammond, Inc.).

New Brunswick. This amounts to about one-fourth of the population of Canada or about 6.5 million people as of the early 1990s.

It is important to note that a few hundred thousand French Canadians live outside the region of French Canada just described. They are found throughout the settled parts of Canada, although usually in small numbers. The principal concentrations of French Canadians who live outside the French Canada Region are in the Abitibi–Timiskaming area of west-central Quebec,[2] in the iron mining towns of the Labrador–Quebec border country, in scattered mining communities of east-central Ontario, in southeastern Manitoba (especially the Winnipeg suburb of St. Boniface), and in various parts of the Maritime Provinces mentioned in the previous chapter.

The French Canadians living in these other parts of Canada maintain most of the same cultural attributes as those who live within the French Canada Region except that the French language becomes decreasingly important with increasing distance from Quebec.[3]

[2] It is tempting to expand the boundaries of the French Canada Region so as to encompass the Abitibi–Timiskaming area because of its lengthy historic relationship with the French Canadians of the St. Lawrence lowland and because more than 150,000 French Canadians live there today. However, there are nearly 200 miles of sparsely settled forest land between these two areas at their closest points; this seems to be too great a gap for logical expansion to include a relatively small outlying area.

[3] Ludger Beauregard, "Le Canada Francais par la Carte," *Revue de Géographie de Montréal*, 22 (1968), 35.

THE REGION AND ITS PARTS

The location of the region is a great paradox; it is isolated from surrounding regions by natural barriers and yet has served throughout history as the principal thoroughfare connecting Canada with the Old World. To the north of French Canada is the rocky forested fastness of the Canadian Shield, largely unpopulated. To the south the Appalachian highlands serve as a very effective barrier, with the single important exception of the Lake Champlain lowland, to connection with the populous parts of the United States. To the east is the vastness of the Atlantic Ocean, ice-locked for part of the year. Only to the west is there a relatively easy route to connect "Lower Canada" (as Quebec was formerly called) to "Upper Canada" and even here the zone of settlement is a narrow one, pinched between the southward extension of the Shield and the Adirondack Mountains.

Despite the difficulties imposed by whitewater rapids on the rivers, an estuary and gulf with seasonal ice problems, and rocky impediments to land transport, the St. Lawrence corridor has been the major routeway providing the Canadian heartland with access to Europe and the rest of the world. The development of the St. Lawrence Seaway system over the past several decades has reinforced the importance of the corridor.

Even a region as small as French Canada has various parts, and the parts demonstrate differing characteristics. The St. Lawrence Valley is the central part and core of the region; it is a broad lowland in the southwest that narrows progressively downstream until the estuary of the river occupies almost all the flat land below the Ile d'Orléans. Southeast of the valley and extending to the international border lie the gently rolling and hilly lands of the Eastern Townships where early English settlers have been largely replaced by more recent French-Canadian arrivals. South of the Gulf of St. Lawrence is the rocky, tree-covered peninsula of Gaspé, occupied by marginal farmers and hardy fishermen. The portion of the region in New Brunswick encompasses a peripheral circle of fishing, farming, and forestry around an interior that is almost totally unpopulated. The pattern of life in each section has much in common, but there are important differences, the most striking being between the upstream and downstream portions of the region.

THE ENVIRONMENT

In many ways, the environment of French Canada is similar to that of the Atlantic Northeast: rocky uplands, extensive forests, bleak winters, rushing streams, and limited mineral resources. There are also important differences: the most significant is the amount of flat land and relatively productive soil that provided the region's economy with a widespread agricultural base.

Topography

The St. Lawrence corridor consists of a long stretch of flat, valley-bottom land that varies greatly in width. It is broadest in the Montreal plain and on the right bank of the river between Montreal and Quebec City. Former beach terraces and strand lines, indicating relatively recent emergence of the lowland from beneath the sea, are commonplace. The surface materials of the lowland are mostly sands and other recent deposits of marine, fluvial, or glacial origin, which cover the bedrock foundation quite deeply in places. The major irregularities in the plain are the Monteregian Hills, scattered remnants of old volcanic stocks that rise several hundred feet above their surroundings; most famous is the westernmost, Mount Royal, in the heart of the city of Montreal.

On the left-bank side of the St. Lawrence River the flat land soon gives way to the sloping edge of the Canadian Shield, which rises rockily and in many places abruptly in steep hills or complex escarpments. The only significant extensions of the lowland on this northwest side of the St. Lawrence are where major tributaries, such as the Ottawa, the St. Maurice, and the Saguenay, have breached the shield escarpment.

Southeast of the St. Lawrence there is much flat land that extends well into the Eastern Townships. But slopelands are more characteristic, with stretches of rolling hills becoming higher and more complicated mountain ranges near the U.S. border. The mountains become higher toward the northeast, forming a sort of rolling plateau that in places is deeply dissected by rivers. Elevations of more than 4000 feet (1200 m) are reached in the rocky fastness of the Shickshock Mountains of the Gaspé Peninsula. Northern New Brunswick is a mixture of rolling lowlands and rocky hills, with major valleys carved by the Restigouche, Nipisguit, and Miramichi rivers.

Climate

Winter is the memorable season in French Canada, both because of its coldness and its length. Qualitative descriptions may vary, but there is no dispute that there is a long period of low temperature in the region. Even the areas of mildest climate have a winter period of 5½ or 6 months, with early frosts beginning in October and streams not running freely in the new year until April. January temperatures average below 20°F (−7°C) throughout the region, with Montreal recording a mean of 14°F (−10°C) and Quebec City 10°F (−12°C). Snowfall is heavy (100 inches [2540 mm] annually is not unusual even at sea-level localities) and weather changes are frequent, although warm periods are brief (Fig. 7–3). Storm tracks converge in the region in winter, and migratory cyclones and anticyclones pass with frequency.

But winter's icy grip is not an unmixed blessing. The deep snow cover allows both humans and logs to move more easily in the forest and facilitates the accumulation of logs on frozen waterways for floating downstream in the spring thaw. The "dead" navigation season on the St. Lawrence, occasioned by the winter freeze, has sometimes lasted from December to April. Today the energetic use of icebreakers permits the port of Montreal to function for all but a few days of the year; Quebec City's port is closed only occasionally, for high tides keep the ice broken and shifting in the estuary.

Summer weather is quite varied in the region. The upstream portion of the St. Lawrence Valley experiences much warmth and humidity and has a growing season of approximately five months. Higher elevations and the most easterly coastal areas (Gaspé, both sides of the estuary, and Anticosti Island) have mild-to-cool summers, with a frost-free period of less than 100 days. Summer is a time of generally abundant rainfall, which is a distinct agricultural advantage to the warmer upstream areas. Annual precipitation totals about 40 inches (1016 mm) at both Montreal and Quebec City.

Soils

Soils are quite variable, but most are heavily leached, acidic, deficient in nutrients, and poorly drained. Those that developed on glacial deposits are very stony, and sands and clays are prominent where marine deposits are the parent material.

FIGURE 7-3 Winter is a prominent fact of life in French Canada. Here the frozen St. Charles River in Quebec City serves as a temporary walkway and skating rink (courtesy Government of Quebec, Tourist Branch).

Spodosols dominate in upland areas and in downstream parts of the region. They are poor soils for agriculture—totally leached in the upper horizons, quite acidic, and ash-gray in color. Alfisols are the chief soils of the upstream areas, especially on the terraces, and although somewhat leached and acidic in nature, they respond well to fertilizer and are important agricultural soils. Small areas of alluvial soils in the valley bottoms are the most productive for farming in the region.

Natural Vegetation

Originally almost the entire region was forested except for poorly drained, marshy areas in the lowlands and some higher, rocky upland slopes. Softwoods dominated, with magnificent stands of white pine and fir, particularly in downstream areas and northern New Brunswick. The upstream sections and much of the Eastern Townships were covered with a mixed forest in which maple, elm, beech, and birch were prominent. Most uplands are still forested, although logging has removed all the accessible timber at least once. The lowlands have long since been cleared for farming; yet extensive woodlots have been maintained in the agricultural areas throughout the region.

SETTLING THE REGION

At the time of European contact most of the St. Lawrence Valley and associated lowlands were occupied by Iroquoian tribes. Gaspé and northern New Brunswick were inhabited by the unobtrusive Micmacs and Malecites, who fished along the coast in summer and hunted in the forest in winter.

The Huron tribe occupied a territory in southern Ontario and was a very active trading tribe. Furs were the basis for their trading relationship with the French, with whom the Hurons acquired a trade monopoly from the earliest days of settlement. In 1649–1650 a series of Iroquois attacks, combined with appalling winter starvation, virtually eliminated the Huron people and wiped out their tribal identity.

During the subsequent decade the Iroquois annihilated or adopted most other small tribes in this region except for the Ottawas. The Iroquois signed a treaty with the French at Montreal in 1688 and for some decades continued to serve as a buffer between the settlements in Quebec and those in the American colonies to the southeast.

European settlement in the region dates from 1608 with the founding of Quebec City by Champlain.[4] "New France" was to be an agricultural colony, but throughout the early decades it was fur trading that attracted the most interest and enthusiasm and pulled French explorers ever deeper into the interior of the continent. Agricultural settlement moved slowly up the St. Lawrence and some of its major tributaries; Trois-Rivières was founded as a trading center in 1634 and the first settlers disembarked on Montreal Island in 1642.

The initial settlement and land ownership pattern in rural Quebec was quite different from that in most of the rest of the continent. Today its imprint on the landscape is both unique and notable (see vignette).

In 1763 France gave up its claim to New France and the 65,000 *Canadiens* came under British rule. A land survey was soon carried out and rectangular townships were laid out between the seigneuries of the St. Lawrence and the international border (the Eastern Townships), with the area being thrown open to Anglo-Saxon colonists. Other townships were surveyed sporadically around the margin of the seigneurial territory until it was completely surrounded by lands earmarked for British settlement. Many of the early settlers were "Loyalists" immigrating from the new American republic to the south; English, Scottish, and Irish immigrants also came. Farms were typically square or rectangular and the pattern was totally unlike the long-lot system, although farmsteads sometimes appeared to be aligned because they were built along access roads.

For the better part of a century after France's retirement from Canada, the French populace remained within the seigneurial domain; as the population expanded rapidly, it filled in the as yet unsettled parts of the seigneuries. But by the middle of the

[4] The city's proper name is simply Quebec. But in order to spare confusion, in this book Quebec refers to the province and Quebec City refers to the city.

FIGURE 7-4 Linearity rampant! The farmsteads are lined up along the roads, and the narrow fields extend perpendicularly in either direction. This typical long-lot pattern is southeast of Montreal (TLM photo).

nineteenth century French Canadians were more than half a million strong and overflowed into Anglo-Saxon townships.

In the two decades prior to Canadian confederation (1867) the French population of the Eastern Townships increased by 120 percent whereas the British population of the same area increased by 6 percent. The total numbers of the two groups were approximately equal by the date of confederation, but the continuing rapid increase of the French population soon greatly exceeded that of the British. On the other side of the St. Lawrence, on the edge of the Laurentians, the French-Canadian settlement also rapidly expanded, filling in the interstices among the predominantly Irish settlers in those townships and in townships in northeastern Ontario. Soon there was a French-Canadian majority on the edge of the Shield and a ''spillover'' of farmers and loggers occupied the Saguenay Valley and Lake St. John lowland in considerable numbers (Fig. 7–4).

At the time of confederation the population of the French Canada Region was overwhelmingly rural. Montreal and Quebec City were the only urban centers of note, with populations of about 100,000 and 60,000, respectively. By the time of confederation the occupance of French Canada was virtually complete.

Few new areas of rural settlement were established since then except in the area northwest of Montreal and north of Ottawa—particularly in Terrebonne and Labelle counties—which reached its maximum extent in the 1930s and 1940s. In some other sections the rural population density increased in this century, but in most it actually declined because of farm abandonment. The major change in population distribution has been the growth of cities. French Canada today, like the rest of the country, is overwhelmingly urban.

THE BILINGUAL ROAD TO SEPARATISM

Virtually every city in the region, with the notable exception of Montreal, is overwhelmingly French. The cultural diversity of Montreal provides a dynamic focus for what is certainly the most significant social and political problem facing Canada today: the accommodation of a large and vibrant French-Canadian minority within a predominantly Anglo-Canadian nation. Probably the most important political fact in Canada is that a quarter of its citizens do not speak the majority language as a mother tongue, if at all. The special position of the French-Canadian minority has been legally recognized since the estab-

A CLOSER LOOK A Long-Lot Landscape

The most striking feature of the landscape in rural French Canada is the almost limitless array of long, narrow rectangles that subdivide the agricultural land. With the notable exception of the Eastern Townships, property lines and field boundaries replicate the pattern with faithful precision, disdainful of topographic variation, throughout most of the region. Moreover, farmsteads are almost invariably positioned at the same ends of adjacent rectangles so that neighboring farmhouses exist in close proximity to one another with remarkable linearity of location. This distinctive contemporary landscape morphology is a heritage of the earliest days of French settlement in Lower Canada and survived with tenacious persistence and little change for three centuries.

The first element in the pattern was the establishment of a seigneurial system. In the seventeenth century the kings of France awarded land grants (called *seigneuries*) with feudal privileges to individual entrepreneurs (*seigneurs*). The seigneurs, in turn, were expected to subgrant parcels of land to peasant farmers (*habitants*). The seigneuries varied greatly in size, but each fronted on a river and extended inland for a mile or two in some cases and up to almost 100 miles (160 km) in others. A total of about 240 seigneuries was created, involving about 8 million acres (3.2 million ha) of land, mostly along the St. Lawrence River.

The typical land grant (called a *roture*) within a seigneury was a long, narrow rectangle, fronting for 150 to 200 yards (135 to 180 m) along the river and extending inland for a mile or more.[a] This gave each farm direct access to the river, which was the only transportation route in the early years. When all riverside rotures had been granted, a road would be built along their inland margin, paralleling the river, and a second rank (or *rang*) of rotures would be developed. In some cases, up to a dozen rangs were successively arrayed back from the river, separated from one another by parallel concession roads that ran without break for dozens, or occasionally even hundreds of miles (Fig. 7–a).

The habitants invariably built their farmsteads at the end of their rotures

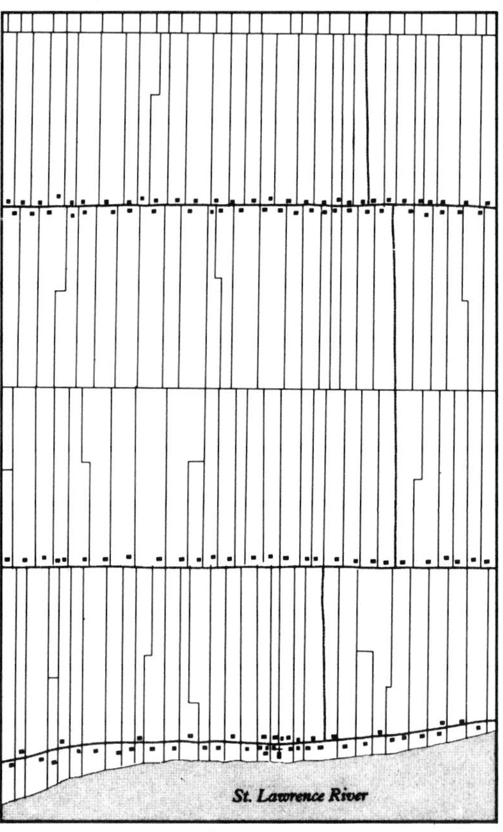

FIGURE 7-a A hypothetical model of rang settlement patterns in Quebec. (From *Canada Before Confederation: A Study in Historical Geography* by R. Cole Harris and John Warkentin, p. 74. Copyright © 1974 by Oxford University Press, Inc. Reprinted by permission.)

[a] R. Cole Harris, "Some Remarks on the Seigneurial Geography of Early Canada," in *Canada's Changing Geography*, ed. R. Louis Gentilcore (Scarborough, Ont.: Prentice-Hall of Canada, 1967), p. 31.

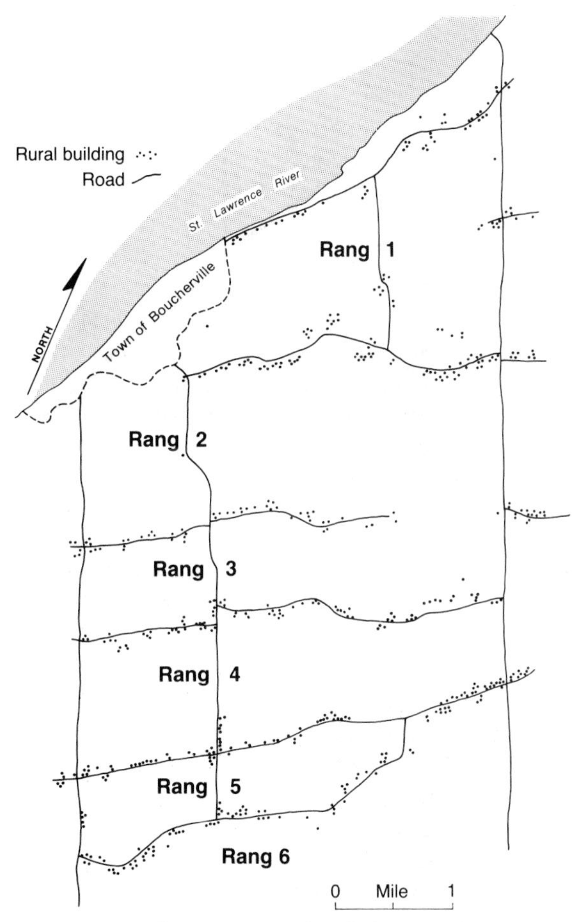

FIGURE 7-b An example of the rang-long lot pattern of rural settlement in a portion of the St. Lawrence Valley. The land ownership pattern of such an area can be visualized as long, narrow properties running at right angles to the roads (map data from Army Survey Establishment 1:50,000 series, Beloeil 31H/11W sheet, 1965).

adjacent to the river or road; thus an almost continuous string of individual settlements grew up along transportation routes.[b] The most common settlement pattern was the *rang double,* in which houses were built on either side of the concession road, thus serving rotures that extended in opposite directions from the road. By the early 1800s many seigneuries contained six or seven rangs double along roads roughly parallel to the river and linked by occasional crossroads with no settlement along them.[c] The custom of equal inheritance rights resulted in increasing fractionization of the rotures along the lines of the original subdivision, with each succeeding fraction becoming narrower so that each farmer still had access to river or road (Fig. 7–b). The original "long-lot" farms were designed to be about ten times as long as they were wide, but repeated linear subdivision sometimes created units that were virtually too narrow to farm economically.[d]

The French population grew slowly at the outset, reaching about 3000 by 1660, but the seigneurial domain continued to expand along the St. Lawrence, reaching downstream to the Gaspé Peninsula and upstream to the border of Upper Canada; it also extended up a number of tributaries, particularly the Chaudière, the Richelieu, and the Ottawa.[e]

TLM

[b] Deffontaines noted that at the end of the French rule a traveler could have seen almost every house in Canada by making a canoe trip along the St. Lawrence and Richelieu rivers. P. Deffontaines, "Le Rang: Type de Peuplement Rural du Canada Francais," *Proceedings,* 17th International Congress of the International Geographical Union (1952), p. 723.

[c] R. Cole Harris and John Warkentin, *Canada Before Confederation* (New York: Oxford University Press, 1974), p. 73.

[d] Peter Brooke Clibbon, "Evolution and Present Pattern of the Ecumene of Southern Quebec," in *Quebec,* ed. Fernand Grenier (Toronto: University of Toronto Press, 1972), p. 17.

[e] This long-lot arrangement of properties and fields was developed in other areas of French settlement on the continent, notably in sections of the Maritimes, in Manitoba, in Louisiana, and along the Detroit River, but nowhere did it reach anything like the magnitude or the permanence of its extent in Quebec.

lishment of the Canadian nation; the British North America Act of 1867, which established the federation, included various irreducible obligations to the province of Quebec. Quebec was guaranteed its civil law, its religious liberty, jurisdiction over its educational system, and the equality of its language in both Parliament and federal courts of the nation as well as in the Legislature and courts of the province. It is unlikely, however, that the framers of this remarkably tolerant legislation anticipated the cultural tenacity of the French Canadians.

The cultural dichotomy persists, with foreboding overtones. Much lip service has been given to the principle of Canada as a bilingual and bicultural nation. In the real world of government and business, however, Canada functioned as an English-oriented country with Quebec as a French-oriented enclave. French-Canadian objections to the status quo are increasingly strident since World War II, the summary complaints being that English is the language of business so that Francophones must use English in order to advance and that the Franco community is being assimilated by the Anglo community.

Quebec's *revolution tranquille* (quiet revolution) of the 1960s brought French- and English-speaking Canadians into direct large-scale competition for jobs and power in modern business and government. This effort was slow in attracting national attention, but the abrupt actions of a fanatic extremist group (the FLQ or *Front de Libération du Québec*) shocked the nation with bombing, kidnapping, and murder. Strong government and private efforts were then instituted to assuage the situation. The "special relationship" of Quebec to the confederation was heartily affirmed, bilingualization of the federal civil service was accelerated, and the prime minister (a French Canadian) actively supported not only bilingualism but a government policy of multiculturalism under a bilingual framework as well.

In the 1970s, however, political polarization based on linguistic polarization became increasingly pronounced (Fig. 7–5). A strict provincial language law that curtailed the use of English in government, business, and education was enacted, making Quebec, in effect, a unilingual province.[5] The *Parti Québécois* was elected to power, with a prominent plank in its platform being the eventual political secession of Quebec from Canada. Indeed, in 1980 a referendum was held in Quebec on whether the provincial government should "negotiate a proposed agreement" of "sovereignty-association" between Quebec and Canada. The federal government urged a negative vote whereas the provincial government obviously favored a positive vote. The electorate of Quebec rejected the proposal by a 60–40 majority.

[5] As a result of this legislation. New Brunswick is now the only province that is legally bilingual on the basis of provincial law, although federal law declares that the nation is legally bilingual.

FIGURE 7-5 This bilingual stop sign on the outskirts of Quebec City has been defaced by painting out the English word; it is a mild, visible expression of the deep passions that underlie French Canada's bilingual/bicultural problem (TLM photo).

For a decade the issue of separatism for Quebec was less conspicuous, but in the early 1990s it surfaced again as a volatile issue. After exhaustive negotiations, a package of constitutional amendments, termed the Meech Lake Accord, was proposed in 1987, but required ratification by all 10 provinces by 1990 in order to become law. Various issues were involved, most of which promised increased powers to the provinces at the expense of the national government. Several special provisions would apply specifically to Quebec, recognizing it as a "distinct society." In the final vote, two provinces refused to ratify the Meech Lake Accord, the measure was defeated, and most Quebecers were antagonized.

Contemporary polls show that about half the population of Quebec is seriously interested in achieving some new political arrangement with the rest of Canada, and a sizable minority wants total secession. It seems clear that a slow unraveling is occurring in the country, but the final political solution is very murky.

THE PRIMARY ECONOMY

Although most of the population is urban and the economy is now basically urban related, the image of French Canada has always been a rural one. In many parts of the region rural activities are prominent and several primary industries contribute significantly to both employment and income.

Agriculture

There are still some 400,000 farmers in French Canada; most carry on a mixed farming enterprise with emphasis on diary production, although the average number of cows per farm is small. Throughout the region more than half the cultivated acreage is in hay, with oats the other principal crop.

The average farm is 150 acres (60 ha) in size and shaped in the traditional long-lot pattern everywhere except in parts of the Eastern Townships and northern New Brunswick. The wooden farmhouse, with its kitchen garden, faces the road. Cordwood is generally stacked beside the house and in the rear are other farm buildings—garage, chicken house, woodshed, piggery, two-story inclined-ramp barn, a silo or two. The narrow rectangular fields usually contain a long linear "centerpiece" of stones that were gathered and stacked by the farmer over many years. The land is divided into several parts, with hay and pasture crops alternating in rotation with row crops (corn for silage, potatoes, other root crops such as rutabagas) or grain (mostly oats). The significant farm animals are dairy cattle, principally Holsteins or Ayrshires; in addition, there are usually poultry and hogs (Fig. 7–6). At the far end of the

FIGURE 7-6 The dominant farm activity in French Canada is dairying. This Holstein herd is in the valley of the Ottawa River between Montreal and Ottawa (TLM photo).

FIGURE 7-7 Many of the region's rivers still are heavily used for transportation and storage of logs. This is the St. Maurice River upstream from Trois-Rivières (TLM photo).

farm is often the woodlot and the sugar bush (sugar maple trees).

The most prosperous farm area in the region is in the upstream part of the St. Lawrence Valley and the southwestern corner of the Eastern Townships. Farms in this section are 25 percent smaller than the provincial average, but a much higher proportion of land is under cultivation and a much more intensive farm operation exists. Machinery is more numerous, fertilizer more frequently applied, and modern techniques more abundant. Most commercial orchards and all the sugar beets of the region are grown here. Other specialty products of significance include vegetables, poultry, tobacco, and honey. Still, most farms are diary farms and the output of milk, butter, and cheese is enormous.

Forest Industries

Quebec has the largest volume of timber resources in Canada; however, most of it is located on the Shield, outside the French Canada Region. There has always been, and continues to be, production from the relatively small woodlots of the St. Lawrence Valley for lumber, pulpwood, posts, and firewood. This production is mostly small scale and only locally important. In the Eastern Townships, Appalachian uplands, Gaspé Peninsula, and northern New Brunswick there has been important commercial output of lumber and pulpwood since the middle of the last century. But large-scale logging enterprises are chiefly located on the Shield and the great bulk of Quebec's production comes from there.

Forest exploitation in the past was restricted almost exclusively to winter. Trees were cut and hauled by various means to the rivers, where they were dumped on the ice. At the time of the spring thaw vast flotillas of rough logs came churning down to the mill sites, where they were caught and processed. This seasonality made part-time logging an important source of income for many residents of the region, especially small farmers whose winter labors were few. But in the last two decades forest exploitation became virtually a year-round operation and opportunities for part-time employment are quite limited.

The rivers still serve as log thoroughfares, with bumpers and booms to guide the logs to the proper catchment areas in quiet water (Fig. 7–7). A large proportion of the cut nowadays is, however, transported by truck or train; thus the gathering of logs can go on every month. Still, the important milling sites are along the streams, especially in the St. Maurice and Saguenay valleys and where major tributaries, such as the Ottawa and the St. Maurice, join the St. Lawrence; indeed, every river junction on the left bank of the St. Lawrence has a sawmill, pulp mill, or both nearby. Other types of wood-processing plants, such as those making doors, plywood, and pressed wood products, tend to be located in the major industrial centers.

Fishing

On the whole, fishing makes a trivial contribution to the regional economy. The principal species caught by French-Canadian fishermen are cod, redfish, and herring. Less than 10 percent of the total value of Canadian Atlantic fisheries is landed at ports in French Canada and less than 1 percent of the regional work force is employed in catching and processing fish. Nevertheless, around the Gaspé Peninsula and in northern New Brunswick fishing is a significant enterprise, and in many places fishing villages and fishermen's houses literally line the bay shores. The village of Caraquet, New Brunswick, for example, is said to be the longest town in Canada. It consists of an almost-continuous line of homes along the south shore of Chaleur Bay, some 20 miles (32 km) long and less than one block wide.

Mining

Mining is another primary activity with significance only in a limited part of the region. The world's

A CLOSER LOOK Anticosti—the Largest Island Nobody Knows

Anticosti Island, located in the principal oceanic shipping lane between Canada and Europe, and only 325 miles (520 km) from Quebec City, is the fourth largest non-Arctic island in North America. Yet its population is only about 350, and very few people ever visit it. With an area of 3045 square miles (7890 km^2), it is nearly half again as large as Prince Edward Island, which is 150 miles (240 km) due south of Anticosti (Fig. 7–c).

The island is a rocky, uplifted horst block, rolling to rugged in topography, with a coastline dominated by steep limestone cliffs and broad reefs. A large bog occupies the eastern end, but most of the island is covered with a fairly dense forest of spruce and balsam.

Anticosti was discovered by Jacques Cartier in 1534, and its first owner was Louis Jolliet, the noted Mississippi River explorer, who received it as a royal gift in 1680. It became a *seigneury,* and for more than two centuries was owned by absentee landlords, with only a handful of permanent residents. It was purchased in 1895 by a wealthy Frenchman to serve as a hunting reserve, and for three decades was the largest private game preserve in North America.

A pulp and paper company acquired Anticosti in 1926, and soon there was a pulpwood forestry boom that supported 3000 residents. Within a few years the logging operations declined, and the Quebec government purchased the island in 1974. It is now a provincial park, its principal attractions being wilderness, a herd of about 100,000 whitetail deer, and numerous fishable streams.

Visitors must obtain a permit to set foot on the island, and another permit if they intend to travel around. There is one small village, called Port Menier, on Anticosti; it is linked to the mainland by daily air service and weekly boat service.

TLM

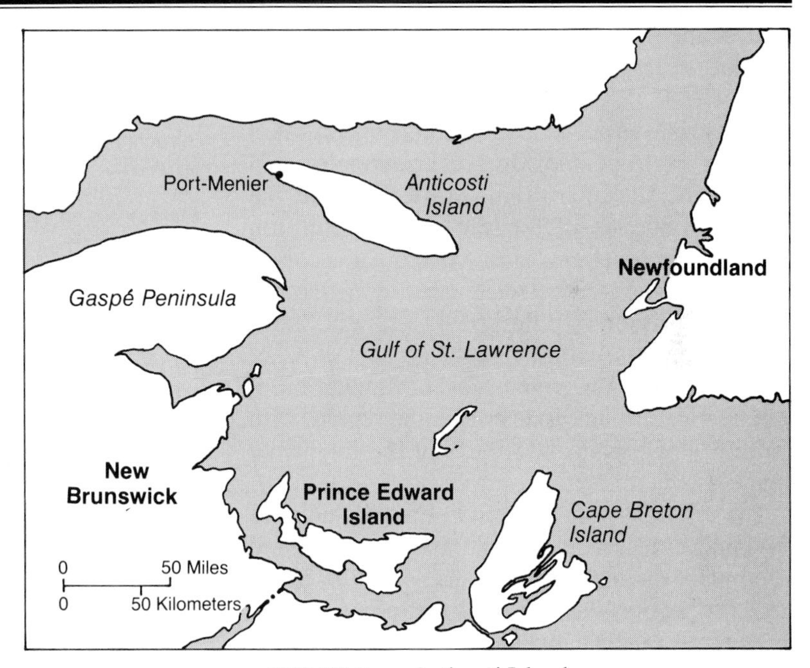

FIGURE 7-c Anticosti Island.

largest commercial deposits of asbestos, with an annual output amounting to more than one-fourth of the world total, are located in the "Serpentine Belt," which extends in an arc northeastward from the Vermont border through the Eastern Townships. Economic hardship has come to the Serpentine Belt in the last few years, as concern for health hazards in the use of asbestos has significantly diminished demand for this mineral.

URBAN-INDUSTRIAL FRENCH CANADA

As in most of the continent, the true dynamism of this region's geography is found in its urban-industrial development. The rapid and significant growth of urbanism, based on solid industrial development, made the French Canadian as much an urbanite as the New Englander or the southern Californian.

Industrial output in the region, concentrated primarily in metropolitan Montreal, amounts to 25 percent of the Canadian total. More than half the Canadian production of tobacco products, cotton textiles, leather footwear, aircraft, and ships comes from French Canada. The single most important industry in the region is pulp and paper, with more than 50 large plants producing 40 percent of Canada's total output. Moreover, most of Canada's aluminum manufacturing capacity is in the region, with the largest single plant at Arvida in the Saguenay Valley.

The urban system of French Canada includes many small cities, a few medium-sized ones, and two dominant metropolises (see Table 7–1 for a listing of the region's largest urban places). The medium-sized centers include the Trois Rivieres–Cap-de-la-Madeleine conurbation of more than 100,000 people at the confluence of the St. Maurice and St. Lawrence rivers; the notable industrial (especially aluminum refining and forest products) and commercial complex of Chicoutimi–Arvida–Jonquiere in the upper Saguenay Valley; Sherbrooke, the subregional center of the Eastern Townships; the prominent industrial city of Shawinigan in the St. Maurice Valley; and Hull, essentially an industrial suburb of Ottawa but located on the Quebec side of the Ottawa River.

TABLE 7-1

Largest urban places of French Canada

Name	Population of Principal City	Population of Metropolitan Area
Beauport, Que.*	63,100	
Brossard, Que.*	57,600	
Charlesbourg, Que.*	69,100	
Chicoutimi, Que.	61,300	157,700
Gatineau, Que.*	81,700	
Granby, Que.	38,500	
Hull, Que.	59,300	
Jonquiere, Que.	58,500	
LaSalle, Que.*	75,700	
Laval, Que.*	285,200	
Longueuil, Que.*	125,700	
Montreal, Que.	1,022,300	2,995,600
Montreal-Nord, Que.*	90,500	
Quebec, Que.	164,600	612,000
Sherbrooke, Que.	74,500	132,500
St-Hubert, Que.*	67,100	
St-Laurent, Que.*	67,200	
St-Leonard, Que.*	76,100	
Ste-Foy, Que.*	69,700	
Trois-Rivières, Que.	50,200	130,100
Verdun, Que.*	60,200	

* A suburb of a larger city.

But it is the bipolar axis of Montreal and Quebec City that dominates the region. Quebec City is the hearth of French-Canadian culture and serves as the political, religious, and symbolic center of French Canada. Montreal is French Canada's contribution to the world, a vibrant and exciting commercial, industrial, and financial node that shares with Toronto the primacy of all Canada.

Montreal

The city of Montreal spreads over most of the island of the same name that is located adjacent to the first major rapids on the St. Lawrence River. The city also overflows onto the nearby Île Jésus and eastward across the river into Chambly County.Its site is dominated by the hill of Mount Royal, which rises directly behind the central business district (Fig. 7–8).

Jacques Cartier discovered an Indian fort and settlement on Montreal Island. Soon a French town

FIGURE 7-8 Looking southeastward across the central business district of Montreal, from Mount Royal toward the St. Lawrence River (TLM photo).

was sited there, because the strategic and trading potentials were superb. Montreal was a fur and lumber trading center for the French, but its principal business and industrial growth occurred under the British. It was long the largest city in Canada, besides being by far the nation's leading port (Fig. 7–9).

With its sprawling suburbs, numerous industrial districts, massive skyscrapers, and heavy traffic, Montreal has much the look of any Anglo-American metropolis. But it is about two-thirds Francophone and is called the second largest French-speaking city in the world. The remainder of its population mix is quite varied, and the life of the city is cosmopolitan. Old World charm shows in the streets and squares of the older sections; the excitement of modern architecture dominates Place Ville Marie and its downtown surroundings; the attractive and efficient Metro subway system gives a fresh dimension to internal transport; and Mirabel airport is one of the largest and most modern in the world.

Montreal does, of course, have problems; they are profuse. Because of various economic and political factors, the most conspicuous being the imposition of French culture on local business by the provincial government, the population and economic growth of Montreal stagnated significantly over the past decade. Toronto has now surpassed Montreal as both the leading population and financial center of the nation, and today the Montreal Stock Exchange

FIGURE 7-9 During most winters the harbor of Montreal is closed by ice for a few weeks. Tugs and ice-breakers help to extend the navigation season (TLM photo).

ranks third behind those of Toronto and Vancouver. Some 200 significant business firms have shifted their operations from Montreal to another province, usually Ontario.

Both the city of Montreal and the province of Quebec are struggling with massive budget deficits, a significant part of which is a holdover from the huge debts incurred in constructing facilities for the 1976 Olympics.

Yet imaginative, and usually expensive, efforts continue to be made for the purposes of improving the livability and ambience of the city and shoring up its economy. One of the most successful has been a downtown rejuvenation that may represent the continent's most successful approach to urban renewal. A covered-city (largely underground) development now links some 80 acres (32 ha) in a multilevel complex that connects 1000 stores and 100 restaurants and bars with theaters, hotels, office buildings, residential high rises, and rail, bus, and Metro stations.

The port of Montreal has more than maintained its position as Canada's busiest, primarily by aggressive development of container facilities. Montreal handles about half of all Canada's container traffic, and it is the third-ranking (after New York and Baltimore) east coast container port of the continent.

Quebec City

Quebec city is much smaller than Montreal, and its importance to Canada is therefore much less, although its significance to French Canada can hardly be overstated. Its imposing site crowns 300-foot (90-m) cliffs that rise abruptly from river level just upstream from the Île d'Orléans, where the St. Lawrence River opens out into its estuary. The historic fort of the Citadel still stands above the cliff ramparts, connected by Canada's most famous boardwalk with a castlelike hotel, the Château Frontenac, whose imposing turrets command a breathtaking view downriver (Fig. 7–10).

The old walled portion of the city (called Upper Town) with its narrow, cobbled streets adjoins the Citadel. Below the cliffs lies Lower Town, also with a large older section of narrow streets and European charm. More modern residential and commercial areas sprawl to the north and west. Indus-

FIGURE 7-10 The spired turrets of Chateau Frontenac (on the right) and the sprawling stone walls of the Citadel (on the left) dominate the cliffs of Quebec City. Modern office towers rise in the distance, beyond the walled premises of Upper Town (courtesy Government of Quebec, Tourist Branch).

trial districts are chiefly close to the harbor, which is increasing its shipping through expanded container facilities and efforts to keep the port open all winter. Although the importance of manufacturing to the economy is growing, Quebec City's major functions are administrative, commercial, and ecclesiastical. Tourism is another important activity in this city, which is probably the most picturesque of the continent.

TOURISM

The unique cultural attributes of French Canada make it a most attractive goal for tourists from non-French parts of the continent, and the large population centers of southern Ontario and northeastern United States are near enough to make accessibility no problem. The landscape, architecture, institutions, and cuisine are the principal attractions; tourist interest centers on things to see rather than things to do except perhaps in Montreal.

There are a number of areas to attract the nature lover and outdoor enthusiast in and near the region. The provincial parks of Mont Tremblant and Laurentides, on the nearby Shield margin, draw many visitors, especially from Montreal and Quebec City; many come for winter sports. New national parks have been established at Forillon in the Gaspé and La Maurice north of Trois Rivières.

Beaches are few and inadequate. But along the St. Lawrence estuary there is a surprising amount of tourist development, often consisting of small cottages situated above mud flats, and summer visitors from nearby inland areas are accommodated in large numbers. Still, the principal attractions for tourists in the region are cultural. Essentially all visitors go to Montreal and most also visit Quebec City; if there is time left over, they may visit some areas of rural charm.

THE OUTLOOK

French Canada has long been an economically disadvantaged region. Although not as serious as in the Atlantic Provinces, a high rate of unemployment and an even greater amount of seasonal unemployment persisted. These factors are most notable in the eastern part of the region—the shores of the estuary, the Gaspé Peninsula, and northern New Brunswick. In the last few years the regional economic disparity decreased somewhat, although much of the improvement was focused in the Montreal area. This trend is likely to continue, although slowly and irregularly.

Quebec's manufacturing expansion since World War II was dramatic. The region, overall, changed from agrarian to industrial in three decades, and the general advantages of market, labor, power, and transportation are still there. Most of the industrial prosperity and growth will continue to occur in the already established centers, especially Montreal, but also Quebec City, Trois Rivières, Hull, and Shawinigan.

The tertiary sector of the economy—trade, finance, services—has grown rapidly and accounts for much of the improvement in average per capita income in the region. Such growth should continue, although it will be focused principally in Montreal; indeed, Montreal will continue as the dominant growth center in almost all phases of the regional economy despite the problems previously discussed.

The primacy of Montreal is such that the provincial economy has been summed up as "Montreal and the Quebec desert." Certainly the upstream part of French Canada participates much more fully in the advantages of urbanization and industrialization than does the rest of the region. This economic disequilibrium will undoubtedly persist and perhaps become even more unbalanced as time passes.

SELECTED BIBLIOGRAPHY

BALDWIN, BARBARA, "Forillon—The Anatomy of a National Park," *Canadian Geographical Journal,* 82 (1971), 148–157.

BRADBURY, JOHN H., "State Corporations and Resource Based Development in Quebec, Canada, 1960–1980," *Economic Geography,* 58 (January 1982), 45–61.

CERMAKIAN, JEAN, "The Geographic Basis for the Viability of an Independent State of Quebec," *Canadian Geographer,* 18 (1974), 288–294.

FRASER, WINSTON, "Anticosti: Big, Unknown, and Now a Park," *Canadian Geographic,* 103 (April–May 1983), 30–39.

GALT, GEORGE, "Westmount Holds On," *Canadian Geographic,* 103 (December 1983–January 1984), 8–19.

GILL, ROBERT M., "Bilingualism in New Brunswick and the Future of L'Acadia," *The American Review of Canadian Studies,* 5 (Autumn 1980), 56.

GRENIER, FERNAND, ed., *Quebec.* Toronto: University of Toronto Press, 1972.

HAMELIN, LOUIS-EDMOND, "French Soul in a British Form," *The Geographical Magazine,* 44 (1972), 744–752.

HAMILTON, JANICE, "Montreal's Underground City," *Canadian Geographic,* 107 (December 1987–January 1988), 50–57.

HARRIS, RICHARD C., *The Seigneurial System in Early Canada.* Madison: University of Wisconsin Press, 1966.

LANG, K. M., "The Gaspé: A Naturalist's Wonderland," *Canadian Geographic,* 98 (February–March 1979), 36–39.

LE BOURDAIS, CELINE, AND MICHEL BEAUDRY, "The Changing Residential Structure of Montreal, 1971–81," *Canadian Geographer,* 32 (Summer 1988), 98–113.

MACKAY, DONALD, *Anticosti, the Untamed Island.* Toronto: McGraw-Hill Book Co., 1979.

SQUIRE, W. A., "New Brunswick's Hills and Mountains," *Canadian Geographical Journal,* 77 (1968), 52–57.

8
MEGALOPOLIS

The essence of the Megalopolis Region is the concentration of a relatively large population in a relatively small area. This is one of the smallest major regions in Anglo-America, encompassing only about 50,000 square miles (130,000 km^2), or less than 1 percent of the continent. Its population of some 44 million people, however, is the second largest population of any region, amounting to 16 percent of the Anglo-American total.

It is a coastal region, and although not now primarily oriented toward the sea, it serves as the major western terminus of the world's busiest oceanic route that extends across the North Atlantic to Western Europe. Its role as eastern terminus of transcontinental land transportation routes and as two-way terminal (international to the east and transcontinental to the west) for airline routes is equally significant.

The region was a major early destination of European settlers. Thus most cities have a long history compared to others on the continent and the region is rich in historical tradition. Yet it is pulsing with change.

It is the premier region of economic and social superlatives to be found in North America. It represents the greatest accumulation of wealth and the greatest concentration of poverty; it has the greatest variety of urban amenities and the greatest number of urban problems; it has the highest population densities and the most varied population mix; and it is clearly the leading business and governmental center of the nation. Its economic and social maladjustments are legion and yet its attempts at alleviating them are imaginative and far-reaching.

Most of all, however, it is an urban region. Its geography is that of cities and supercities. It is one of the most highly urbanized parts of the world and its lifestyle is geared to the dynamic bustle of urban processes, problems, and opportunities.

EXTENT OF THE REGION

Megalopolis, the world's greatest conurbation, developed along a northeast-southwest axis approximately paralleling the Middle Atlantic and southern New England coast of the United States (Fig. 8–1). Its core is the almost completely urbanized area extending from metropolitan New York across New Jersey to metropolitan Philadelphia. It extends northeastward from New York City through a number of smaller cities to metropolitan Boston and southwestward from Philadelphia to Baltimore and Washington.

This metropolitan complex from Boston to Washington has been recognized as a major urban region for a number of years. The concept of a unified Atlantic seaboard metropolitan region was notably publicized by Jean Gottmann's epic study of 1961, in which he adopted the ancient Greek name *Megalopolis* to apply to the region.[1] Since the time of Gottmann's work, continued urban expansion at either end has made it logical to extend the area under consideration northward in New York and New England and southward in Virginia.

The Atlantic coastline marks the eastern margin of the region, which means that fairly extensive rural areas, particularly in southeastern Massachu-

[1] Jean Gottmann, *Megalopolis, The Urbanized Northeastern Seaboard of the United States* (New York: Twentieth Century Fund, 1961).

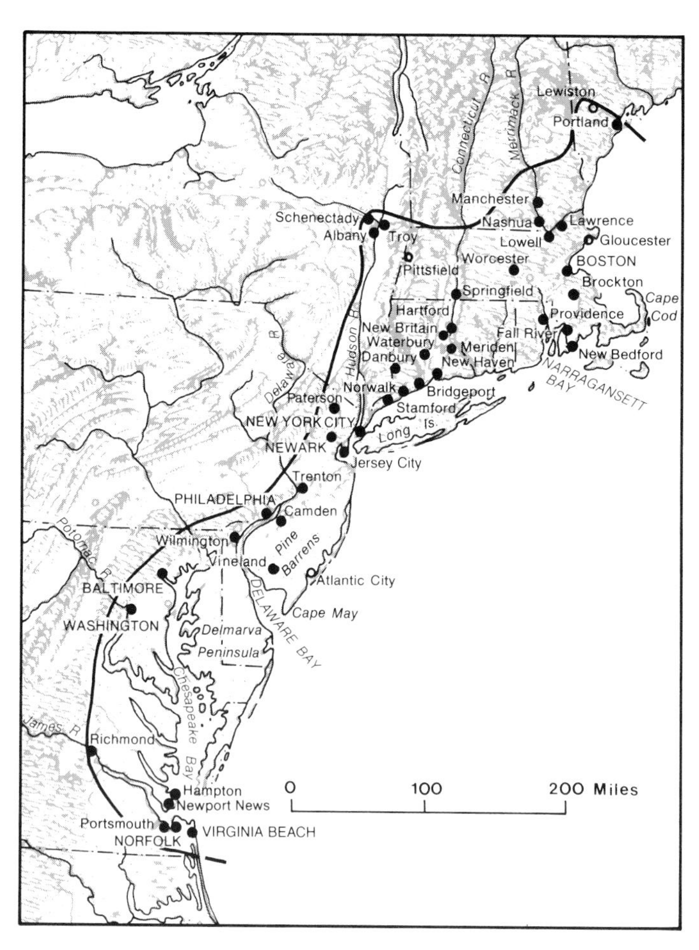

FIGURE 8-1 The Megalopolis Region (base map copyright A.K. Lobeck; reprinted by permission of Hammond, Inc.).

setts, southern New Jersey, the Delmarva Peninsula, and eastern Virginia, are included within the confines of Megalopolis. These lands, however, are used primarily for urban-serving agriculture or recreation and can thus logically be accepted as part of the region.

The western boundary of Megalopolis is considered to be where urban population densities and land-use patterns fade and are replaced by rural densities and patterns to the west of the principal urban nodes of the region. This zone is fairly easy to demarcate in the south, where the Hampton Roads cities, Richmond, Washington, and Baltimore, are involved. West and northwest of Philadelphia the distinction is less clear, as is the case north of New York City and north of Boston. In the first-mentioned district the megalopolitan boundary is drawn to exclude most of eastern Pennsylvania beyond the immediate metropolitan area of Philadelphia, the reasoning being that the Pennsylvania Dutch farming counties of Lancaster and York and the small industrial cities of Reading and Allentown–Bethlehem are relatively independent of Megalopolis and more logically associated with the Appalachian Region to the west. North of New York City the boundary is drawn to include that portion of the Hudson Valley as far as the Albany–Schenectady–Troy metropolitan area. North of Boston the extent of Megalopolis is considered to include the urbanized lower portion of the Merrimack Valley as far as Concord and the urbanized coastal zone as far as Portland and Lewiston.

As thus delimited, the Megalopolis Region encompasses a number of significant urban nodes and complexes, most of which have important interconnections while remaining relatively discrete urban units.

1. In southern New Hampshire and northern Massachusetts are the old but still highly industrialized mill towns of the Merrimack Valley.
2. Metropolitan Boston, with its many suburbs and related towns, spreads over much of eastern Massachusetts.
3. Metropolitan Providence includes most of Rhode Island and small cities of related economic structure in adjacent Massachusetts.
4. The Lower Connecticut Valley area includes a number of closely spaced, medium-sized cities, especially Springfield, Hartford–New Britain, Waterbury, and New Haven.
5. The urbanized node of Albany–Schenectady–Troy.
6. The New York City metropolitan area sprawls over parts of three states and encompasses such adjacent major cities as Newark, Jersey City, Paterson, Elizabeth, and Stamford.
7. Metropolitan Philadelphia includes an extensive area along the lower Delaware River Valley, including Trenton and Camden in New Jersey and Wilmington in Delaware.
8. Metropolitan Baltimore.
9. Metropolitan Washington, which spreads widely into Maryland and Virginia.
10. Metropolitan Richmond.
11. The Hampton Roads urban complex includes the urbanized areas on either side of the mouth of the James River estuary.

CHARACTER OF THE REGION

Nowhere else on the continent is there such a concentration of the physical works of humans to dominate the land and the horizon for square mile after square mile (Fig. 8–2); yet there is a surprisingly rural aspect to much of the region. Despite the prominence of skyscrapers, controlled-access highways, massive bridges, extensive airports, and noisy factories, the green quietness of the countryside is also widespread.

There are many places in Rhode Island, a tiny state but one with the second greatest population density, where a person can stand on a viewpoint and look in all directions and see no evidence of people or their works; there is nothing but trees and clouds and sky (Fig. 8–3). In an extensive area in southern New Jersey, that most urbanized of states, birds nest, streams run pure, and forests are virgin. The deer population of the region is now greater than it was a century ago. Many areas actually have more woodland than they did 10 or 20 years ago because of farm abandonment. For Megalopolis as a whole it is estimated that only about 20 percent of the land is in urban use.

From the viewpoint of an orbiting satellite it would probably be the interdigitation of urban and rural land uses that would be the most prominent aspect of the geographical scene. But in no way is the region entirely urbanized; neither does it form any sort of a single urban unit. Instead, there are major nodes of urbanization that are growing toward one another, usually along the radial spokes of principal highways, with much rural land between the spokes. In some cases, particularly around New York, Philadelphia, and Boston, nodes coalesce; but in most parts of the region the nodes remain discrete and a great deal of nonurban land is interlocked.

Another important facet of the regional character is its dependence on interchange with other regions and countries. Although the full range of urban functions occurs in multiplicity in Megalopolis, the region generates a monstrous demand for primary goods (foodstuffs and industrial raw materi-

A CLOSER LOOK *The Fragility of Regional Delimitation—The Case of Eastern Pennsylvania*

Regions do not "exist"; they are constructs of the human mind. They are not entities awaiting discovery; they are generalized abstractions designed for some purpose. In this book the purpose is pedagogic, to aid in comprehending the total geography of the continent by understanding major portions of the whole.

Within this text the regions are pragmatically defined and their boundaries are portrayed by bold black lines on maps. Such explicitness is in some ways misleading because it implies a precision that is contrary to fact. Moreover, regional delimitation varies with time as geographical data and relationships change.

As a concrete example of these two important principles—imprecision of delineation and variation over time—let us consider the west-central boundary of the Megalopolis Region in eastern Pennsylvania. As recognized in the text and portrayed on the regional map in this chapter, essentially all of Pennsylvania except metropolitan Philadelphia is excluded from the Megalopolis Region, for stated reasons.

An esteemed colleague who lives in the Lehigh Valley of eastern Pennsylvania has entered a dissenting

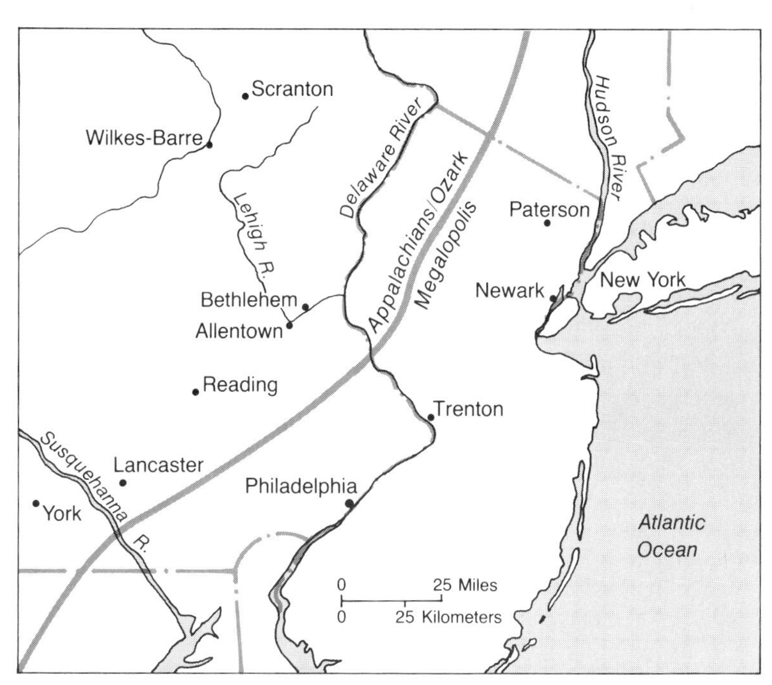

FIGURE 8-a The location of the Lehigh Valley and other parts of eastern Pennsylvania with regard to the Megalopolis/Appalachians boundary as designated in this book.

als), most of which it cannot produce. There is an almost equally significant need for the movement of regional products to extraregional markets; thus major transportation terminals, which are numerous, are loci of much activity and of extensive storage and processing facilities.

Most significant of all, in assessing the geographic character of Megalopolis, is probably the intensity of living that prevails there. This factor is shown in its crudest form by the high population density, both as a general average for the whole region and more specifically for the urbanized nodes. The average population density for the region is nearly 900 persons per square mile. New Jersey, in the heart of the region, is the first state to be classified as "totally urban" by the Bureau of the Census.[2] Its average density is nearly 1000 per square mile. Population density reaches remarkable levels in some cities, the highest being on Manhattan Island with a nighttime density of more than 80,000 and a daytime density of almost 200,000 per square mile.

[2] This means that every county in the state has reached the Census Bureau's definition of "urban"; that is, it includes a central city and a surrounding closely settled urban fringe that together have a population of 50,000 or more.

opinion and buttressed it with strong supporting rationale. He contends that much of eastern Pennsylvania should be included within the confines of the Megalopolis Region. His remarks follow:

As a resident of the Lehigh Valley . . . I have observed accelerating change that leads me to doubt that the area is still sufficiently "independent of Megalopolis" to warrant its exclusion from Megalopolis and therefore its inclusion in Appalachia . . . (Fig. 8-a).

When I first asked [my students] in the late 1970s to evaluate the Pennsylvania segment of your Megalopolitan/Appalachian boundary, I encountered the full range of opinion from strong agreement to strong disagreement. . . . Over the past decade these class discussions have wandered far afield from your stated criterion of relative economic independence, as we have often focused on attitudes, values, and behaviors varying from the relatively trivial (e.g., driving habits) to the more profound (e.g., traditionalism in its many manifestations). Last semester, to my shock, they expressed unanimous disagreement with the passage [that the area is relatively independent of Megalopolis and more logically associated with the Appalachian Region to the west]. . . .

What follows is a brief summary of the growing economic linkage with Megalopolis, which probably is the underlying cause of this rapid change in regional identity perceived by my students. . . . During 1984, 1985, and 1986, fourteen companies, employing approximately 3000 people, based in New York or Philadelphia opened offices in the Lehigh Valley. . . . The cause of this proliferation of back office operations is the substantial cost reductions that can be achieved while remaining within easy striking distance of Philadelphia (about 60 miles) and Manhattan (80 to 100 miles). . . . The effect has been to accelerate the growing economic linkage to Megalopolis. . . . Real estate prices and housing starts have risen dramatically during the last half decade as young New Jerseyans unable to buy a first house in the New York metropolitan area have bought much cheaper houses in Pennsylvania and accepted very long commutes as a necessary consequence. . . . Jobs are also migrating . . . ; if most of these jobs do not come all the way to Bethlehem, they move at least a substantial distance outward from the Megalopolitan urban cores . . . through the now suburbanizing area that used to be the rural buffer between Megalopolis and the Lehigh Valley.

Of equal importance is the fact that this growth has coincided with the rapidly accelerating decay of the largest and oldest components of the "relatively independent" industrial base of the Lehigh Valley economy (e.g., the near bankruptcy of Bethlehem Steel in the early 1980s with 8200 jobs lost and closure of Mack Truck's largest plant in 1987 with 6000 jobs lost). . . . The number of Lehigh Valley workers employed in smokestack industries plunged from a high of 110,000 in 1979 to 77,800 in 1987.

Professor Doug Heath
Northampton Community College
Bethlehem, Pennsylvania

Professor Heath's cogent arguments highlight the fragility of regional delimitation.

TLM

FIGURE 8-2 Megalopolis is a hive of urban activity, as indicated by this north-facing view of Manhattan Island (TLM photo).

Many other elements are involved in "intensity of living." Where people live close together, we also find a clustering of structures, activities, and movements. Such agglomeration is advantageous in providing concentrated opportunities for variable want satisfactions within a limited space. Yet it also engenders the handicaps associated with crowding: waste of space and time, frustration and psychologi-

FIGURE 8-3 Even in Megalopolis there are spacious areas of natural landscape, with little evidence of a human imprint. This forested scene is in southeastern Massachusetts (TLM photo).

cal trauma, pollution and health problems, stifled transportation, and so on.

THE URBAN SCENE

The urban complex of Megalopolis is almost overwhelming in its magnitude and diversity. Within the region are over 1000 places classed as urban by the Census Bureau, more than 4000 separate governmental units, and a population of about 44 million, of which nearly 90 percent is urban (see Table 8–1 for a listing of the region's largest urban places). For convenience, the region is subdivided into 11 urban groupings, to be considered successively from north to south (Fig. 8–4).

The Merrimack Valley

With a total population of about 750,000, the Merrimack Valley of New Hampshire and Massachusetts includes five small cities: Manchester, Nashua,

TABLE 8-1

Largest urban places of the Megalopolis Region

Name	Population of Principal City	Population of Metropolitan Area
Albany, N.Y.	94,540	850,800
Alexandria, Va.*	108,400	
Atlantic City, N.J.	35,060	309,200
Baltimore, Md.	751,400	2,342,500
Bayonne, N.J.*	60,950	
Bergen, N.J.		1,292,300
Boston, Mass.	577,830	2,845,000
Bridgeport, Conn.	139,770	444,000
Bristol, Conn.	60,660	79,000
Brockton, Mass.	92,410	186,800
Cambridge, Mass.*	90,290	
Camden, N.J.	82,180	
Chesapeake, Va.*	147,800	
Chicopee, Mass.*	57,650	
Clifton, N.J.*	76,090	
Cranston, R.I.*	75,800	
Danbury, Conn.	64,420	190,800
East Orange, N.J.*	77,240	
Elizabeth, N.J.	105,150	
Fall River, Mass.	88,920	153,800
Fitchburg, Mass.	38,900	98,100
Hampton, Va.	130,800	
Hartford, Conn.	131,300	755,400
Jersey City, N.J.	217,630	542,200
Lawrence, Mass.	61,500	380,600
Lewiston, Maine	38,510	87,000
Lowell, Mass.	94,070	261,600
Lynn, Mass.*	77,890	
Manchester, N.H.	98,320	149,900
Medford, Mass.*	56,580	
Meriden, Conn.	58,660	
Mount Vernon, N.Y.*	68,840	
Nashua, N.H.	80,440	177,300
Nassau–Suffolk, N.Y.		2,639,000
Newark, N.J.	313,800	1,886,200
New Bedford, Mass.	94,330	167,900
New Britain, Conn.	71,780	147,500
New Haven, Conn.	123,840	523,700
New London, Conn.	27,870	259,300
Newport News, Va.	160,100	
New Rochelle, N.Y.*	68,540	
Newton, Mass.*	82,230	
New York, N.Y.	7,352,700	8,567,000
Norfolk, Va.	286,500	1,380,200
Norwalk, Conn.	76,130	125,700
Paterson, N.J.	138,620	
Pawtucket, R.I.	73,680	324,800
Philadelphia, Pa.	1,647,000	4,920,400
Pittsfield, Mass.	48,120	79,300
Portland, Maine	61,280	212,200
Portsmouth, N.H.	25,730	220,400
Portsmouth, Va.	107,500	
Poughkeepsie, N.Y.	29,940	262,200
Providence, R.I.	156,190	646,800
Quincy, Mass.*	82,640	
Richmond, Va.	213,300	844,300
Salem, Mass.		258,700
Schenectady, N.Y.	66,630	
Somerville, Mass.*	70,070	
Springfield, Mass.	150,320	522,500
Stamford, Conn.	100,260	191,800
Trenton, N.J.	90,790	331,000
Union City, N.J.*	53,630	
Vineland, N.J.	54,750	138,400
Virginia Beach, Va.	365,300	
Waltham, Mass.*	56,440	
Warwick, R.I.	86,740	
Washington, D.C.	617,000	3,734,200
Waterbury, Conn.	104,500	215,800
Wilmington, Del.	70,210	573,500
Worcester, Mass.	156,190	415,700
Yonkers, N.Y.*	183,000	

* A suburb of a larger city.

FIGURE 8-4 The urban complexes of Megalopolis: (1) the Merrimack Valley, (2) Metropolitan Boston, (3) the Narragansett Basin, (4) the Lower Connecticut Valley, (5) Albany-Schenectady-Troy, (6) Metropolitan New York City, (7) Metropolitan Philadelphia, (8) Metropolitan Baltimore, (9) Metropolitan Washington, (10) Metropolitan Richmond, and (11) Hampton Roads.

Lowell, Lawrence, and Haverhill. These former mill towns, highly industrialized and specialized toward textiles and shoes, mirrored the economic pattern and problems of southern New England for many decades (Fig. 8–5). As New England's advantages for making these products declined, areas that could not find satisfactory replacements were subjected to increasing unemployment and economic distress.

Since the prosperous years of the 1920s, the textile industry, which employed well over half of all industrial workers in the valley, declined to a fraction of its former magnitude. The manufacture of leather goods, primarily shoes, decreased much less precipitously and is still the leading type of manufacturing in the valley. Electrical machinery production was the principal growth industry and there has been increasing diversification. The valley is a low-wage area with generally depressed incomes, but during the last several years rapid economic and population growth occurred, particularly in the New Hampshire portion.

Metropolitan Boston

The dominant city of New England is located at the western end of Massachusetts Bay, a broad-mouthed indentation that is generally well protected by the extended arms of Cape Ann and Cape Cod. The site of the original city was a narrow-necked peninsula of low glacial hills and poorly drained swamps. The Charles and Mystic rivers flow into the bay, and much of the older part of the city now occupies land that had to be drained before construction could take place.

Shortly after the founding of Boston in 1630, other towns, such as Cambridge and Quincy, were established on better-drained land nearby. Boston with its numerous protected wharves was, however,

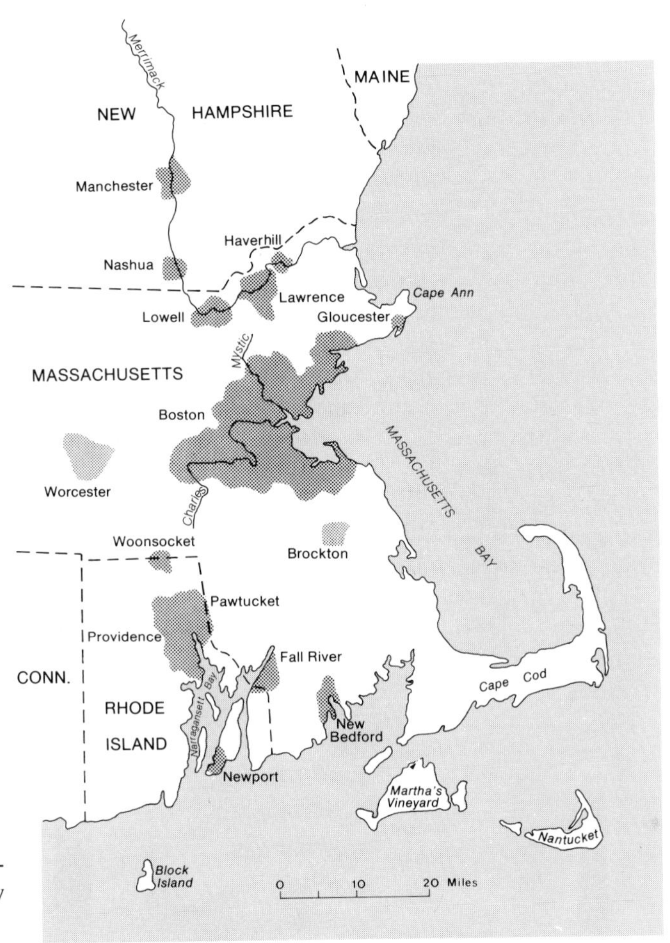

FIGURE 8-5 The metropolitan areas of southeastern New England.

the dominant settlement from the outset. By the end of the colonial period it was exceeded in size nationally only by Philadelphia and New York and it maintained this third-ranking position for more than a century.

Boston's central business district was extended after further drainage projects were completed but remains concentrated chiefly near the original nucleus. Urban sprawl widely expanded the city to the north, west, and south of the original area and many ancillary settlements were absorbed into the metropolis.

Although early significant as an industrial center, the economy of Boston became increasingly oriented toward tertiary and quaternary activities—trade, services, and government employment—which provides about three-fourths of the jobs in the metropolitan area. The port of Boston is a busy one, but the lack of an extensive hinterland resulted in a relative decline in its importance. It is a major importing port, but in total trade it is outranked by four other ports in the Megalopolis Region alone.

The industrial component of Boston's economy is a prominent one and, as with most cities, tells much about the distinctiveness of the city's function. Although textiles and leather goods were the traditional mainstays of manufacturing in New England, the former has not been important in Boston for several decades and the latter has been on an erratic but declining trend for several years, even in the historic shoemaking suburbs of Brockton and Beverley.

Boston has long been a major industrial center, and, despite problems, continues to rank among the national leaders in this regard. Virtually all cities of the northeastern United States have experienced significant relative declines in their industrial importance in the last quarter century, but Boston less than most. As traditional industries declined, eastern Massachusetts experienced a significant industrial resurgence in "high-technology" manufacturing, which involves electronics, drugs, aerospace, instrumentation, weaponry, and especially computers. By the mid-1980s more than one-third of all manufacturing in Massachusetts consisted of the high-technology category, a higher proportion than that of any other state. Many new firms represent spinoffs from older companies that are attracted to the Boston area by the availability of skilled labor, managerial competence, inventive genius, venture capital, and the advantage of technological interrelationships (agglomeration economies) with other local firms. High-technology industries are now as firmly established in Boston as in California, a development that contributed (along with rapid growth in the service sector) to the lowest unemployment rate in the nation (2 percent in 1987). Since then, however, Boston's economy has lost much of its vigor. Many manufacturing jobs have disappeared, the unemployment rate is climbing, and the state is mired in deficit financing. Still, there is a solid base of high-tech industry and considerable optimism for another economic turnaround.

As one of the oldest major cities on the continent and one that was quite prominent in colonial times, Boston maintains an air of historic charm in many of its older sections. Narrow twisting streets, colonial architecture, and tiny parks are widespread. But modernity is increasingly evident even in the central city, as evidenced by skyscrapers, freeways, an airport on reclaimed land in the bay, and even a multilevel parking garage under the famed Boston Common Fig. 8–6). Boston also has one of the first beltways completed around a major American city; the Circumferential Highway (Route 128), at a radius of about 12 miles (19 km) from the

FIGURE 8-6 Boston is the major metropolis of southern New England (courtesy of Greater Boston Convention and Visitors Bureau).

city center, has become a major attraction for light industry and suburban office buildings (it is sometimes referred to as the "electronics parkway"). The pattern is now being repeated on Interstate 495, an outer beltway some 25 miles (40 km) from the central business district, with many planned industrial parks at intersections of the freeway and express arterials from downtown.

Boston has long been known as an elitist city and cultural center with many attractions for living; yet numerous problems exist. Lack of money to meet governmental expenses is at the root of many of them. Most of the city budget comes from property taxes and yet almost half of all municipal property, such as government buildings, churches, and educational institutions,[3] is untaxed. The city population in the past was dominated by Europeans, especially Irish and Italians. Recent years have seen a rapid influx of other ethnic groups, particularly blacks and Puerto Ricans, and intergroup hostility has become a major social problem. Black and Puerto Rican residential areas are more highly segregated than in perhaps any other major Anglo-American city and the areas of lowest per capita income coincide precisely with these districts.

Even so, metropolitan Boston continues to grow, and the average family income is one of the highest of any large city. Imaginative attempts are underway to assuage two urban problems for which Boston has long been famous: reorganization and extension of mass transit facilities to prevent traffic strangulation of an auto-oriented city that is plagued by an infamous maze of narrow, twisting streets and a major start toward metropolitan government. Boston remains the primate city of northern and eastern New England and continues as the northern bastion of Megalopolis.

Narragansett Basin

New England's second largest city and second-ranking industrial center is Providence, situated where the Blackstone River flows into the head of Narragansett Bay. The Providence area, with its industrial satellites of Pawtucket and Woonsocket to the north and Fall River and New Bedford to the southeast, represents in microcosm the industrial history of southeastern New England, demonstrating graphically the economic displacement that occurs in a heavily industrialized area when its leading industry falters.

From colonial days southern New England was the center of the American cotton textile industry. Initial advantages included

1. Excellent water power facilities.
2. A seaboard location for importing cotton.
3. Damp air, which was essential to prevent twisting and snarling during spinning and to reduce fiber breakage in weaving.
4. Skilled labor.
5. Clean, soft water.
6. Location in a major market area.

Since the 1920s, most of the cotton textile industry has relocated in other areas, particularly in the southern Appalachian Piedmont where labor, taxes, power, and raw materials were all less expensive. The greatest concentrations of cotton textile factories in New England were in the Narragansett Basin and the Merrimack Valley; their attrition has been a damaging blow to the local economy. The last cotton-weaving mill in Rhode Island ceased operation in 1968 and both cotton spinning and weaving are almost gone from adjacent parts of Massachusetts.

The woolen textile industry of New England prospered longer. In the early days there was local wool in addition to water power, skilled labor, and an appreciable nearby market. Moreover, Boston has always been the major wool-importing port of the nation. Rapid decline engulfed New England's woolen textile industry in the late 1940s and the downward trend has continued at a slower pace. The Narragansett Basin has been particularly hard hit. Fall River, once the single leading producer of cotton textiles in the country, now has fewer than 3000 textile workers and nearly all are in the woolen industry. The textile industry once employed more than half of all manufactural workers in Providence; it now provides jobs for barely one-tenth of the total.

It should be noted that even with a declining rate of industrial employment, the Narragansett Ba-

[3] Metropolitan Boston has more students per capita than any other American conurbation.

sin cities are still heavily dependent on manufacturing. Indeed, Rhode Island is exceeded only by North Carolina and South Carolina in the proportion of labor force employed in factories. Providence itself, with a preeminent rank as producer of jewelry and silverware, is still among the two dozen leading industrial employment centers in the nation.

Connecticut Valley

The broad lowland drained by the lower Connecticut River contains a number of prosperous, medium-sized cities, with a total population of about 2.5 million (Fig. 8–7). The Springfield–Chicopee–Holyoke complex in Massachusetts and Hartford in Connecticut are located on the river; the other cities of New Britain–Bristol, Waterbury–Naugatuck, Meriden, New Haven, and Bridgeport are a bit farther west.

This is a long-standing industrial district of importance. Factories specialize in diversified light products requiring considerable mechanical skill: machinery, tools, hardware, firearms, brass, plastics, electrical goods, electronics equipment, precision instruments, watches, and clocks. These are mostly products of high value and small bulk, which require little raw material and can easily stand transport charges to distant markets.

This southwestern part of New England has adjusted fairly well to changes in the economy. The flight of textiles has not resulted in nearly so many abandoned factories; instead, other types of manufacturing were attracted by the cadre of trained workers and the available buildings. The growth in electrical machinery production has been steady, partly because of the research and product development facilities of the district. Aircraft engine manufacture is significant in Connecticut even though airframe assembly is accomplished in other parts of the country. Various kinds of machinery and hardware and other light metal goods are produced in quantity in several cities.

Although there is heavy dependence on manufacturing (Connecticut ranks fourth among the states in percentage of its labor force employed in factories), the economy of the subregion is broadly diversified and includes a notable concentration of the nation's insurance industry in Hartford. The

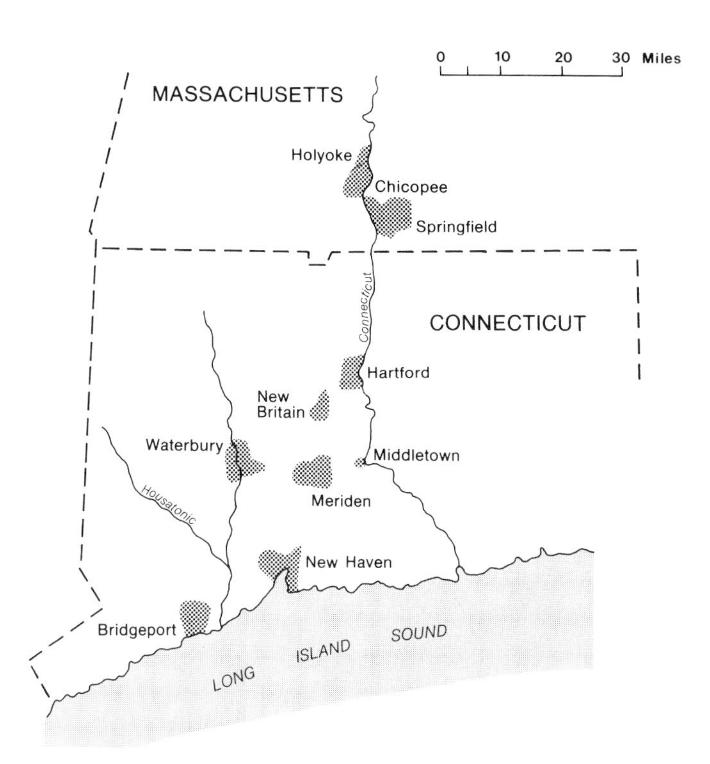

FIGURE 8-7 Metropolitan areas of the lower Connecticut Valley.

commercial orientation of the valley is largely westward toward New York, which apparently is an important catalyst.

This is a district with no major cores. Nineteenth-century industrial cities have persisted but have not dominated. Instead, many nodes of specialized activities emerged. The result is a noncentric population pattern, with much daily movement among a wide range of foci.

Albany–Schenectady–Troy Metropolitan Area

Situated at the commercially strategic confluence of the Mohawk and Hudson rivers, the old Dutch settlement of Albany has become a metropolitan area of 850,000 people clustered about multiple nuclei in parts of four counties (Fig. 8–8). As the upstream end of the Hudson River axis and the eastern end of the low-level Mohawk corridor to the Great Lakes, the Albany area had major crossroads significance from its earliest days. By late in the 1600s Albany was the leading fur-trading center for the English colonies and a century later it was declared the state capital. The completion of the Erie Canal in 1825 was a major stimulus to economic prosperity and population growth. Despite many changes in transportation orientation, the urban node still functions as a transshipment point between river or ocean traffic on one hand and canal or overland traffic on the other.

The area has been an important industrial center for many decades. Its initial raw material advantages (sand, limestone, and nearby iron ore) provided an early start for manufacturing and the industrial component of the economy remained strong through the years. Troy has been noted as a center for making men's shirts; Albany was a wood-manufacturing center; and Schenectady was most famous for General Electric and the American Locomotive Company.

The area's industrial importance later declined, with a commensurate expansion in trade, services, and, especially, government employment.

Metropolitan New York

The metropolitan area of New York City, which occupies two dozen counties in parts of three states, is the core of Megalopolis. It is the principal city of the United States and, in many ways, the most important if not dominant city in the world. The massing of people and activities around the mouth of the Hudson estuary represents Megalopolis at its most intensive.

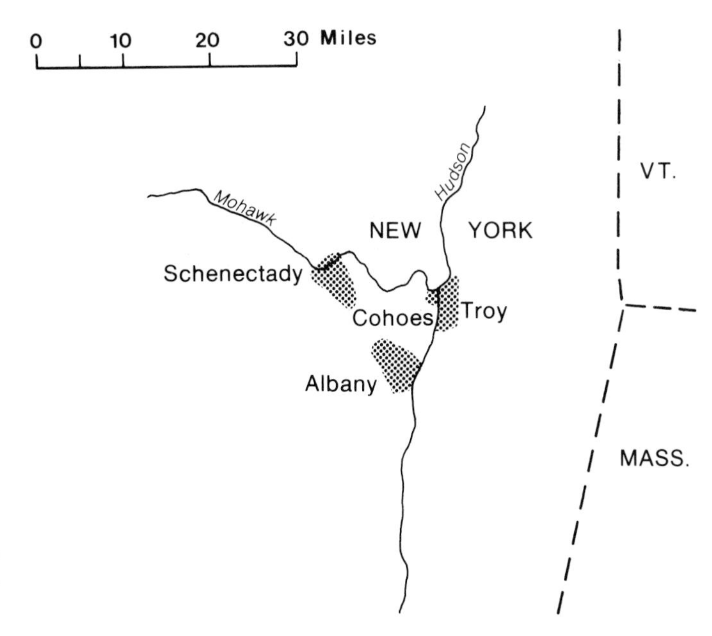

FIGURE 8-8 The Albany-Schenectady-Troy metropolitan area.

The economic primacy of New York cannot be satisfactorily explained in any simple fashion. It is the result of a continuum of complex interactions that are worldwide in scope and spread over several centuries in time. In elementary terms, it can be viewed as the result of an economic struggle among several competing ports to dominate the juncture of two major trade routes: the North Atlantic shipping lanes to Europe and the continental connections to the Anglo-American interior. New York's location did not initially appear to be more advantageous than that of its major competitors, notably Baltimore, Boston, Halifax, Montreal, and especially Philadelphia. The others, however, suffered from various handicaps. Baltimore, Philadelphia, and Montreal all had to be reached by tortuous navigation up narrow estuaries and Montreal was iced in for several months each winter. Both Boston and Halifax had less extensive harbors than New York, and Boston, in particular was plagued by shallow water in its harbor.

New York's major advantage was access to the interior. The availability of an easy water-level route to the Midwest via the Hudson River and the Mohawk Valley gave New York a preeminence over its nineteenth-century competitors that was never relinquished.

New York has a magnificent site for water-borne traffic. It is located at the mouth of the navigable Hudson River on a well-protected harbor that is continually scoured by the river, which keeps the channel deep. The channel leading to the Atlantic is direct and broad. The tidal range is so small that ships can come and go at almost any time. The harbor is well protected from storms and is never blocked by ice. The principal handicaps involve rapid and tricky currents and the occasional presence of dense fog.

A look at any large-scale map shows the exceeding fragmentation of the city's site (Fig. 8–9). This fragmentation was a boon to water-borne commerce but constitutes a major problem to internal communication. Manhattan Island is a narrow finger of land (13.5 miles [21.6 km] long by 2.3 miles [3.7 km] wide at its broadest point) situated between the wide Hudson River on the west, the narrow Harlem River on the north, and the East River (really a tidal estuary connecting New York Harbor

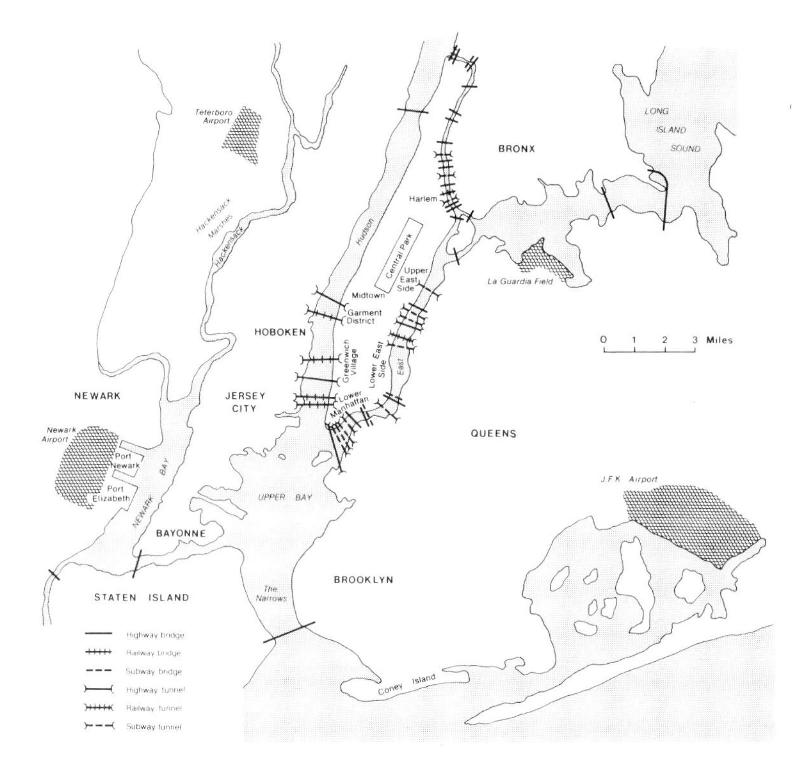

FIGURE 8-9 Manhattan Island and its connections with other parts of the metropolitan area.

FIGURE 8-10 The island of Manhattan extends southward from the distant Bronx mainland, sandwiched between the Hudson River on the left and the East River on the right. The twin towers of the World Trade Center dominate the Lower Manhattan scene at the southern tip of the island (courtesy Port Authority of New York and New Jersey).

and Long Island Sound) on the east (Fig. 8–10). Long Island extends eastward for 100 miles (160 km) between Long Island Sound on the north and the Atlantic Ocean on the south. Staten Island, nearly three times the size of Manhattan, huddles close to the New Jersey shore to the southwest of New York Harbor. Rivers, estuaries, and marshes add to the fragmentation along the Jersey shore.

Function Owing to the immense size of its population and its location at a focus of land, water, and air transport routes, New York City is preeminent in most fields of urban activity in the United States. By almost any measure—retail sales, wholesale sales, foreign commerce handled—of commercial activity, New York is the unchallenged national leader. Indeed, the gross regional product of metropolitan New York is greater than that of such countries as Canada or Brazil.

Although with diminished dominance, New York is still by far the leading U.S. port. The bi-state Port Authority of New York and New Jersey operates this largest port in the nation, as well as the largest air terminal system in the world, the most heavily used tunnels and bridges in the world, and the world's largest office complex (the World Trade Center in lower Manhattan).

Although commerce contributed more to the city's primacy than industry, New York is—almost incidentally—the nation's and the world's largest manufacturing center. The metropolitan area handles more than 7 percent of the country's factory output, which, for example, exceeds the combined total for the six New England states. Despite its magnitude, the industrial structure is not highly diversified. One type of manufacturing, the making of garments, predominates. More garments are made in metropolitan New York than in the next seven largest garment-manufacturing centers combined. Most clothing factories are concentrated in an area of 200 acres (80 ha) in central Manhattan, where they occupy the upper floors of moderate-height buildings and are so inconspicuous as to pass unnoticed by the casual visitor.

New York City continues as the leading financial center of the world. The great money market is focused in the Wall Street district of Lower Manhattan, where a proliferation of financial specialists can provide external economies to businesses at all scales of operation. As a result of the great concentration of financial institutions, the velocity of demand deposits (relative frequency of use of a deposited dollar) is greater than elsewhere in the nation. Furthermore, the two securities exchanges (stock markets) in this district handle nearly nine-tenths of the organized stock and bond transactions of the country.

As a headquarters and managerial center, New York is also unsurpassed. It houses far more corporate headquarters than any other city despite recent decentralization tendencies that have seen literally thousands of offices, and tens of thousands of white-collar jobs, shifted to the suburbs and beyond. Still, the wide variety of business services and skills available in New York provides a spectrum of expertise that is not remotely approached by any other city. Indeed, the conglomeration of office space is easily the world's largest. Within 50 miles (80 km) of Midtown Manhattan is found 40 percent of the total U.S. office space; Manhattan alone has four times that of Chicago.

New York's entertainment, cultural, and tour-

A CLOSER LOOK Global City—New York in the 1990s

New York City has long enjoyed a cosmopolitan, worldly image in the United States as a center of entertainment, the media, and finance. In the 1970s, like many northeastern cities, it was plagued by a wave of deindustrialization and the exodus of its middle class, and its image was severely tarnished. In the 1980s, however, New York bounced back with a vengeance to become one of the healthiest regions in the United States. Its unemployment rates are low, wages are high, and its real estate markets are booming. New York's resurgence is primarily due to the explosion of financial services available there since the mid-1970s. Wall Street has long symbolized the world of commerce, but the 1980s enlarged New York's banking and securities sectors and heavily increased the city's reliance upon them.

The growth of finance in New York is mostly due to the changing international environment. First, the influx of petrodollars caused by the rapid rise of oil prices in the 1970s disproportionately favored New York banks, which promptly lent large quantities to Latin American nations; thus, the debt crisis of the 1980s, which arose largely from the "recycling" of petrodollars, made New York highly susceptible to the threat of default by Third World nations. Second, the deregulation of the securities markets (e.g., allowing foreign firms to trade stocks) and new sources of investment capital (e.g., pension funds) contributed to the great bull market on the New York Stock Exchange. The switch from fixed to floating exchange rates created a lucrative trade in international currencies. Third, New York has become the largest center of foreign banking in the United States, particularly for Japanese financial institutions with enormous assets that dwarf those of American companies. A wave of corporate takeovers, mergers, and leveraged buyouts has created a boom in investment banking. Finally, new telecommunications systems allow banks to switch vast quantities of funds instantaneously around the globe. New York's capital markets have become firmly linked to other "global cities" such as Tokyo and London, and the city is very susceptible to global business cycles.

Associated with the growth of financial firms has been an expansion of business services, including legal firms, advertising, accounting, public relations, engineering, architecture, and consulting. Many of these companies are also tied to the global marketplace. New York law firms, for example, may assist in corporate takeovers in Western Europe or advise clients investing in the United States. British advertising firms have become a major presence in the city. Additionally, foreign tourism to New York has risen dramatically, stimulating demand for hotels and restaurants. These trends have led many observers to wonder whether New York is not more connected to the international economy than to the domestic U.S. economy.

The rapid growth of New York's financial and business services creist functions are outstanding. The city is national leader in theaters, museums, libraries, art galleries, mass media headquarters, higher education institutions, hotel rooms, and most other significant urban amenities.

Morphology The political city of New York is composed of five boroughs: Manhattan, Bronx, Queens, Kings (Brooklyn), and Richmond (Staten Island). The metropolitan area, however, sprawls widely on the mainland and on Long Island (Fig. 8–11).

Manhattan Island is the nerve center of the city. The island consists of an ancient, stable rock mass that is rigid enough to support the tall buildings for which the city is famous. In the middle of Manhattan is Central Park, one of the largest and best-known green spaces in the world. Just to the south is the Midtown district, which contains the world's greatest collection of skyscrapers (Fig. 8–12). Slightly southwest of this area and across Times Square is the garment district, an area of moderate-sized buildings, clogged streets, and jampacked sidewalks. Farther south is the old "Bohemian" district of Greenwich Village, now fashionable again, and the more recent, art-oriented neighborhood of Soho. At the southern end of the island is Lower Manhattan, dominated by the Wall Street financial district and another cluster of skyscrapers. The southeastern section is called the Lower East Side, an area with a lengthy history of crowded tenements, low incomes, and ghettoized ethnic groups (Chinatown, Little Italy, the Bowery, the East Village, and so on). To the east of Central Park is the

ated major changes in the city's labor markets. Many of the occupations in banking and securities paid handsomely (in 1988, the average annual salary in securities exceeded $77,000). The computerization of the office raised entry-level clerical skills, leading to growing shortages of secretaries. Suburban areas in particular have felt the effects of the "baby bust" through very low unemployment rates.

New York's office and housing markets have also changed dramatically since the 1970s. A wave of construction has swept the city, creating innumerable new office towers, hotels, convention centers, and trendy stores. Foreign firms, particularly from Canada and Japan, have purchased many new offices and hotels. With high disposable incomes, employees in financial and business services have gentrified many areas. A forest of luxury condominiums has turned large parts of Manhattan into a "yuppie dormitory." Western Brooklyn has witnessed a "brownstone revival" movement. Even parts of northern New Jersey have become very expensive. As the demand for housing has soared, so have prices, forcing many young couples to flee the city in search of less expensive housing elsewhere in the country.

It is important to note that New York's renaissance has not benefited all social groups equally. New York is a city of extremes, and it shows signs of a growing schism between rich and poor. Large parts of the city are very poor, and one quarter of the residents are on welfare. An estimated 100,000 New Yorkers are homeless, leading to the unwelcome label "New Calcutta." Many minorities in particular have not enjoyed the fruits of the region's recovery. New York's poor public education system has done little to prepare its largely black underclass for work in the service sector or to provide them with the skills necessary to find employment.

Finally, New York in recent years has seen tremendous numbers of immigrants arrive from other nations. Southern Brooklyn boasts the largest Russian population outside the Soviet Union. Many Koreans have set up small fruit and vegetable markets. An exodus from Hong Kong has created a new "Chinatown" in Queens. The large numbers of Puerto Ricans and Cubans have been joined by many Dominicans and Colombians. New York is now also home to more than 700,000 black (non-Hispanic) Caribbeans, who arrived from Haiti, Jamaica, Trinidad, and other islands to make New York the largest West Indian city in the world. These newcomers, like those before them, add another tile to the city's diverse ethnic mosaic and enhance the vitality of a rapidly changing metropolis.

Professor Barney Warf
Kent State University
Kent, Ohio

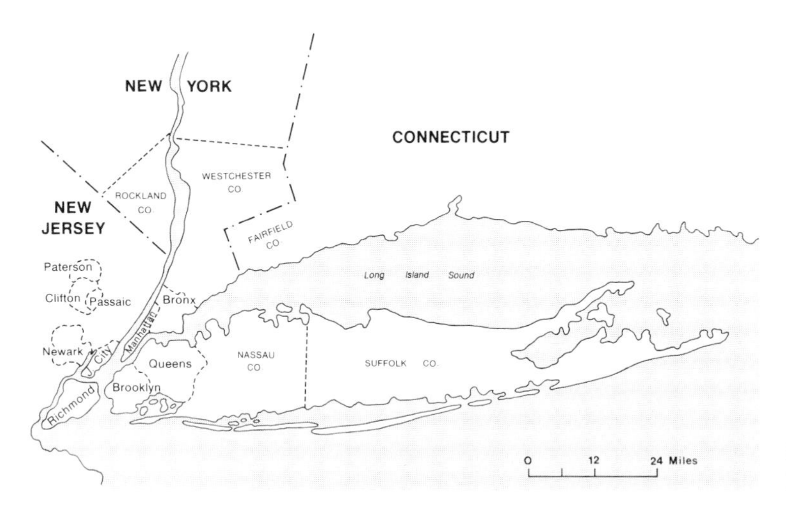

FIGURE 8-11 Metropolitan New York City.

Upper East Side, an area known for its affluence. Northeast of Central Park is the expanding Puerto Rican ghetto of Spanish Harlem (Fig. 8–13) and north of the park is the long-established black Harlem district (Fig. 8–14).

Manhattan is a veritable hodgepodge of functional areas. It includes the greatest concentration of manufacturing facilities (the garment district), commercial buildings (Midtown), and financial institutions (Lower Manhattan) on the continent; yet its residential function is a major one. Detached, single-family dwellings are unknown and the 1.5 million inhabitants of the island live in multistoried apartments. Commuters swell this population total considerably during daylight hours; it is estimated that there are more than 2 million jobs between Central Park and the Battery (southern tip of the island) and that 1.5 million of those are filled by commuters.

The *Bronx* is the only borough of the city that is on the continental mainland. A moderately hilly dis-

FIGURE 8-12 The typical bustle of a Midtown Manhattan street corner (TLM photo).

FIGURE 8-13 One of the "traditional" sidewalk markets on Lexington Avenue in New York City's Spanish Harlem (TLM photo).

trict, it is primarily residential in function. Close settlement in apartments characterizes the southern portion, but farther north there is a sparser density and some single-family homes. The population of the Bronx is approximately equal to that of Manhattan.

The borough of *Queens* occupies more than 120 square miles (310 km^2) at the western end of Long Island, extending from Long Island Sound across the island to Jamaica Bay. It is primarily a residential borough but also has large commercial and industrial districts. Individual homes are more commonplace than in the other boroughs, but multiple residences are in the majority. Kennedy International and La Guardia, two of the four principal metropolitan airports, are located in Queens. The borough's population is about 2.1 million.

There are more people in *Kings* (*Brooklyn*)—about 2.3 million—than in any other borough. It occupies the southwestern tip of Long Island and is closely connected to lower Manhattan by bridges (Fig. 8–15) and tunnels. The business center of Brooklyn is near the eastern end of the Brooklyn Bridge and would be an imposing central business district if it were not under the shadow of Manhattan skyscrapers. Brooklyn contains considerable industrial land, but it is primarily a residential area, largely developed for multifamily apartments.

FIGURE 8-14 A street scene in one of the tenement districts of Harlem (TLM photo).

FIGURE 8-15 Many bridges span the waterways of New York City. This view from Lower Manhattan toward Brooklyn shows the Brooklyn (center) and Manhattan (left) bridges (TLM photo).

The borough of *Richmond* occupies Staten Island, which until the early 1960s had no direct connection with the rest of the city except by ferry. As a result, much of Staten Island is still suburban or semirural. The completion of the Verrazano Narrows Bridge (the world's longest suspension bridge) to Brooklyn changed all that and the population of Richmond has now passed 400,000 and is growing steadily.

The urban spillover into *outer Long Island* has reached beyond Queens and spread throughout most of Nassau County (population 1.4 million) and deep into Suffolk County (1.3 million). The eastern end of Long Island is still rural, but urban sprawl reaches ever farther eastward along the main transport routes.

North and east of the Bronx, the metropolitan area extends into *mainland New York*. Westchester County, with its numerous elite residential areas, has a population of nearly 1 million. Rockland County (300,000) and Fairfield County (in Connecticut) are on the fringes of growth.

On the *Jersey side* of the Hudson, the urban agglomeration extends north, south, and west so that essentially all of northeastern New Jersey is encompassed. Urbanization is not complete in the area, but it is both expanding and intensifying. The Jersey City peninsula has a population of some 600,000; the Newark area, which occupies three counties, has more than 2 million; and the Paterson–Clifton–Passaic area claims over 1.5 million.

New Jersey is a long-maligned state[4] that is now experiencing a prospering economy based on high-tech financial services, defense contracts, booming office construction, and mushrooming suburbs. The state's unemployment rate is well below the national average. Much of the prosperity is within the New York metropolitan area, where even the long-blighted Jersey shore of the Hudson River is now referred to as the "Gold Coast," with rotting piers and aging warehouses giving way to luxury condominiums and high-tech offices.

Of great significance to the entire metropolitan area are the reclaimed marshlands along the western shore of the Hudson estuary and New York Harbor, which are used for heavy industry and nuisance industries (such as oil refineries and chemical plants), railway marshaling yards, new sports facilities, and the predominant share of New York's present port facilities. The containerized cargo-handling facilities of *Port Elizabeth* and *Port Newark* are among the world's finest and most extensive, and most of New York's general cargo traffic funnels through these facilities. In addition, there are a scrap steel termi-

[4] Comedians have noted that the state gem is concrete, the state bird is the mosquito, the state tree is dead, and the state motto is "New Jersey—Landfill of Opportunity."

nal, a pelletized lumber port, an automated (pipeline) wine terminal, and an expansive area for imported automobiles.

Metropolitan Philadelphia

Metropolitan Philadelphia, the fourth largest urban complex in Anglo-America, is not as densely populated or sprawling as New York. Its principal axis lies parallel to the lower course of the Delaware River, extending from Wilmington, Delaware, in the southwest to Trenton, New Jersey, in the northeast, a distance of some 50 miles (80 km) (Fig. 8–16). Its southeast-northwest dimension is less than half as great, reaching from Camden, New Jersey, up the Schuylkill Valley to Norristown, Pennsylvania.

William Penn selected and planned the initial settlement in 1682 on well-drained land a short distance north of the marshy confluence of the Schuylkill and Delaware rivers. Penn's rectangular street pattern with wide lanes and numerous parks was well designed for a relatively small city. As Philadelphia grew, however, it absorbed other settlements so that the street layout is heterogeneous and irregular away from the present central business district. Hills, marshes, and valleys further complicated the pattern in outlying areas. Converging highways and railways produced a radial design for the major thoroughfares.

Much has been written about Philadelphia's misfortune in not having an easy access route to the interior of the continent. It is true that the ridge-and-valley section of the Appalachians, lying athwart the direction of westward penetration, inhibited Philadelphia's competitive position vis-à-vis New York. Nevertheless, Philadelphia was the largest city in Anglo-America for most of the eighteenth century and the Pennsylvania Railroad, completed in the mid-1800s, gave Philadelphia a much more direct (if more costly) route to the Midwest than New York's Hudson-Mohawk corridor provided. For many years the "Pennsy" was one of America's busiest and most prosperous railways.

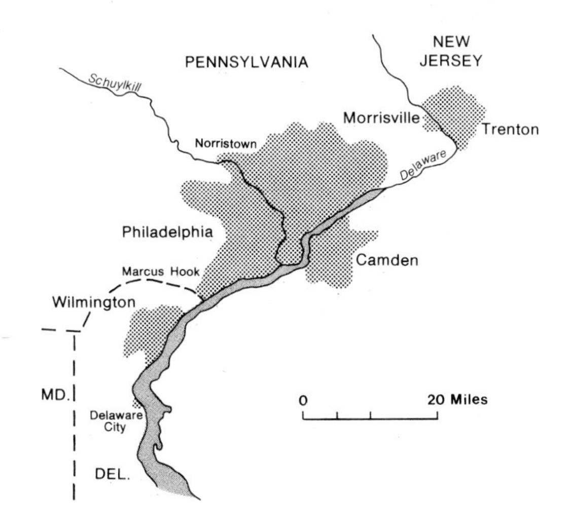

FIGURE 8-16 Metropolitan Philadelphia.

The harbor activities of the lower Delaware are administered by a tristate authority that coordinates the lengthy port facilities and maintains a dredged channel as far upstream as Morrisville. Both banks of the river are lined with piers and wharves downstream from Philadelphia for several miles, continuing on the right bank below Wilmington to New Castle and Delaware City. Oil refineries, chemical plants, shipyards, and other heavy industrial facilities are also numerous in riverside sites. Most of the port business consists of imported raw materials, particularly petroleum (from both overseas and coastwise traffic) and iron ore (largely from Venezuela and Canada). In comparison with New York, the Delaware River port complex is minor as an exporter.

Metropolitan Philadelphia is a major manufacturing center. Its industrial structure is extremely well balanced compared to other urban centers of Megalopolis; indeed, it is considered the most diversified in the United States. Philadelphia has large apparel and publishing industries, but unlike New York City, they do not dominate the manufactural scene. Also important are the electrical and nonelectrical machinery industries, which together provide jobs for almost one-fourth of the area's factory workers. Total manufacturing employment has declined for several years, but a steady increase in nonmanufacturing jobs has somewhat compensated for the decline.

The urban problems and opportunities of Anglo-America in general and Megalopolis in particular are clearly demonstrated in Philadelphia. Here is a classic example of an older industrialized city with a dwindling manufacturing job base, a city torn by racial antipathy, a city of run-down neighborhoods left to decay in the wake of white flight to the suburbs, a city of substandard housing for many minority residents, and a city with a decaying central business district and a ballooning municipal budget deficit.

Yet a combination of public works and private-sector investment has started to revitalize the city. The downtown area experienced some of the most efficient urban renewal on the continent. Modern bridges replaced ferries across the Delaware; acres of decaying buildings, just north of Independence Hall, were removed in favor of a graceful mall; and the modern complex of Penn Center buildings now stands where the "Chinese Wall" muddle of elevated railway tracks once congested downtown traffic flow. Several inner-city residential neighborhoods broke the grip of decay through private gentrification. Portions of the waterfront are being refurbished. There is even a perceptible increase in high-technology employment. Philadelphia seems to be on the way back.

Metropolitan Baltimore

Baltimore, a seaport on Chesapeake Bay, is located in what is almost the remotest corner of the bay (nearly 200 miles from its mouth) where the estuary of the Patapsco River provides a useful deep-water harbor (Fig. 8–17). It was founded relatively late (1729) at a Fall Line power site and its exports of agricultural produce (largely flour and tobacco) soon ensured its preeminence over Annapolis, an administrative center already eight decades old. By the end of the eighteenth century it was the fourth-ranking port in the nation; ever since it has maintained a high position among American ports, based in part on a productive hinterland, in part on excellent railway connections to the interior, and in part on prosperous local manufacturing. Even its distance from the open sea has been reduced by the construction of the Chesapeake and Delaware Canal across the narrow neck of the Delmarva Peninsula, and its relatively inland location, in comparison with other North Atlantic ports, gives it a comparative freight-rate advantage to the Midwest.

Although its exports (particularly grain and coal) are not inconsiderable, Baltimore is primarily an importer of bulk materials from both foreign and Gulf Coast sources; many of these materials are processed locally. Copper, sugar, and petroleum refining, as well as fertilizer manufacture, are significant port-related industries, but most prominent in this regard is the Sparrows Point steel complex. Baltimore is the nation's leader in the import of iron ore. In an effort to expand general cargo traffic, Baltimore's container-handling facilities have been greatly expanded, making Baltimore the second-ranking East Coast container port.

The city, which has been called "the most southerly northern city and the most northerly southern city," sprawls widely in all directions but is not restricted by adjacent urban centers and has considerable room for expansion. Although relatively near the larger Washington conurbation to the southwest, Baltimore's quite different economic orientation has minimized the disadvantages of the shadow effect except in the matter of air-transport development.

In common with many other northeastern cities, Baltimore experienced a period of demoralizing decay, epitomized by the fact that no new hotel or office building was constructed for a quarter of a century. But within the past two decades Baltimore made one of the most striking recoveries of any Anglo-American city, fueled by a $1 billion facelift initiated by the business community rather than the municipal government, and proceeded with a focus on neighborhood revival rather than downtown rehabilitation. Showpieces are Harborplace (a $450 million waterfront complex of shops and restaurants), Charles Center (a 13-block downtown development), a new subway system, the new national aquarium, and an enormous convention center, but the most innovative approach is "homesteading" (buy an old house from the city for $1, provided that you renovate it and live in it) and "shopsteading" (buy a defunct store for $100, provided that you restore it and live in it).

The dramatic revitalization of Baltimore has been very uneven, both in terms of area and people. The highly visible Inner Harbor development contrasts starkly with many decaying inner-city neighborhoods. Baltimore remains a highly segregated city; two-thirds of the census tracts are either more than 90 percent white or more than 90 percent nonwhite.

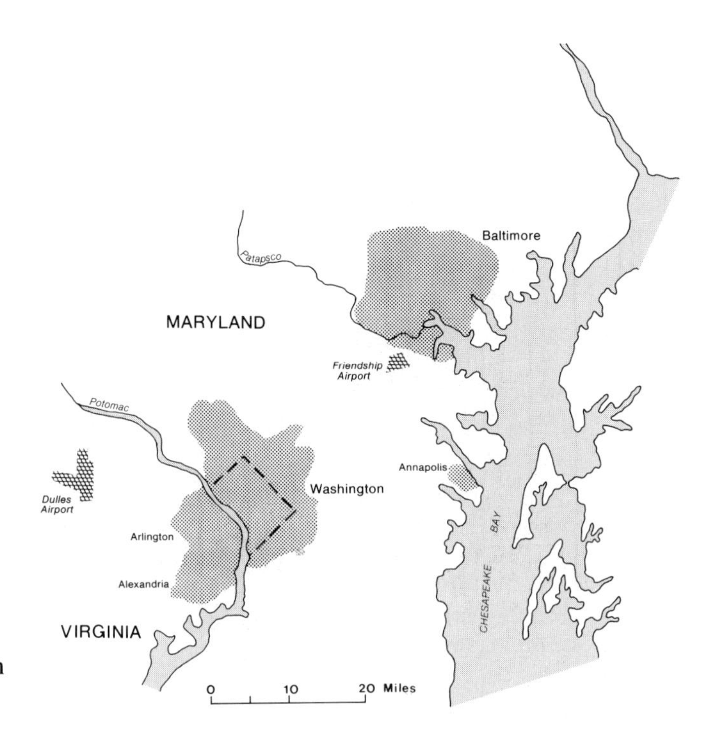

FIGURE 8-17 Metropolitan Baltimore and Washington.

Metropolitan Washington

It is only 35 miles (56 km) from Baltimore to Washington; yet the rural area in between is still far from being completely suburbanized and the two cities are remarkably unlike. Originally laid out in the 1790s, Washington was designed as a governmental center that was relatively centrally located to the population distribution of that time. The original District of Columbia was a square of land ceded approximately equally from Maryland and Virginia. The latter state was allowed to reannex its portion in 1846 so that the present 69 square miles (180 km^2) of the District is all on the Maryland side of the Potomac River.

Much of the original site was low lying and swampy, but extensive drainage and an elaborate municipal plan produced a city of orderly pattern and impressive appearance. The original street pattern combined the regular form of a rectangular grid with a diagonal network of avenues and traffic circles. Massive complexes of government buildings are mixed with attractive parks between the Potomac River and the central business district, with the landscaped mall that connects Capitol–White House–Washington Monument as the hub of activity. Both government offices and commercial districts are scattered over the metropolis, although square mile after square mile of residences, ranging from stately old mansions to modern high-rise apartments to monotonous row houses, dominate the surroundings (Fig. 8–18). In the last four decades runaway growth overwhelmed orderliness and the metropolitan area sprawls well beyond the District's boundaries into the two adjoining states.

Overcrowding and congestion are a trademark of the Washington scene. The Capitol Beltway around the city is helpful, but routes in and out become traffic quagmires during rush hours and inadequate bridging of the Potomac aggravates the problem. Washington is largely a white-collar city and most employment opportunities are downtown, not in the suburbs. Consequently, the daily commutation flow, on a per capita basis, is higher than anywhere else in Megalopolis. Attempts were made to alleviate this situation by locating some government complexes in the suburbs and by developing an elaborate new rapid transit system, the Washington Metro. The Metro is attractive, comfortable, and efficient, but extraordinarily costly.

An increasing proportion of the population resides in suburbs outside the District. In all suburban counties the population is growing rapidly, and the metropolitan area now houses some 4 million people. Within the city proper the population is 70 percent black; public school enrollment is more than 90 percent black. Unlike most large cities, however, there is an increasing flow of blacks to the suburbs, particularly southeastward into Prince Georges County.

The economic function of Washington is unlike that of any other major North American metropolis. The employment structure is unusual in four important respects:

1. Most obvious is the remarkably high percentage of federal government workers, amounting to about 45 percent of the District's work force. It is much higher than for any other major city on the continent.
2. The proportion of females in the work force is abnormally high, reflecting the large number of clerical jobs in government offices.
3. The proportion of blacks in the work force is also higher than in other large cities; the lack of hiring discrimination in federal employment accounts for much of the black population influx.
4. Only 4 percent of the jobs in metropolitan Washington are in manufacturing, which is by far the lowest proportion in the nation, and half of that 4 percent is employed in printing and publishing, indicating the great flood of government publications.

FIGURE 8-18 Moderate-height office buildings and multiunit residences are commonplace in and around Washington's central business district (TLM photo).

Two other facets of the city's economy should be mentioned. Its financial community has been expanding, largely as a result of increasing federal controls and regulations; in no other part of Megalopolis outside New York has the *relative* growth of financial services been as marked. Also, Washington continues to be a major tourist attraction; its many government offices, monuments, and museums are high on the list of sights to see for American and foreign visitors alike.

Metropolitan Richmond

The rapidly growing capital of Virginia is a Fall Line city on the James River, some 75 miles (120 km) south of Washington (Fig. 8–19). Its inclusion in the Megalopolis Region cannot be justified by contiguous urban land use, although the strip between

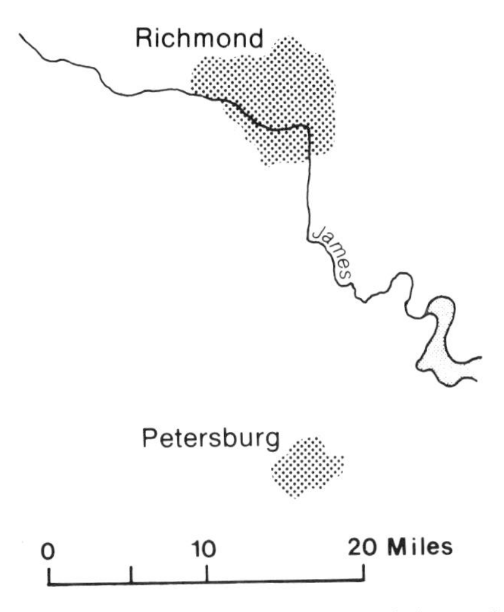

FIGURE 8-19 Metropolitan Richmond.

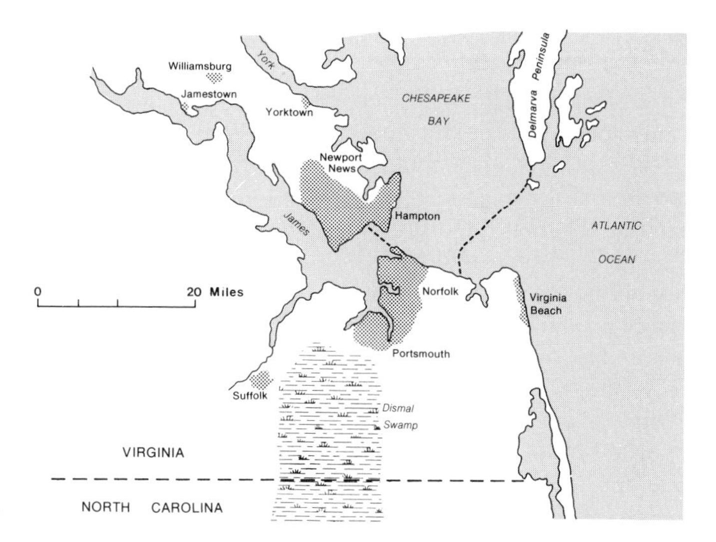

FIGURE 8-20 Hampton Roads.

Richmond and Washington is rapidly being suburbanized from both ends. Rather, it is the city's business orientation to the north and east and its efficient transport connections with Washington and Norfolk that suggest a logical extension of Megalopolis as far southwest as Richmond.

The city was founded in early colonial days and became a bastion of southern culture and Confederate politics. It experienced rapid growth as a commercial and, to a lesser extent, financial center in recent decades. Its moderate level of industrial development is highly specialized toward production of cigarettes and chemicals, which together provide nearly half of total manufactural jobs.

Hampton Roads

Just inland from the mouth of Chesapeake Bay is the commodious harbor of Hampton Roads, situated where the north-flowing Elizabeth River enters the mouth of the James River estuary and slightly upstream of the latter's confluence with Chesapeake Bay (Fig. 8–20). The resultant natural harbor is one of the finest in the Western Hemisphere and around it developed one of the largest maritime, military, and ship-building complexes in the world. Clustered about the mouth of the Elizabeth River are the cities of Norfolk (Virginia's largest), Chesapeake, and Portsmouth, and on the tip of the peninsula across the mouth of the James River are Newport News and Hampton.

Military and port-related activities dominate the economy of this group of cities, sometimes nicknamed "Tidewater." More than a dozen military bases are in or adjacent to the urbanized areas and their payrolls constitute a large share of the total metropolitan income. The strategic location in the middle of the nation's Atlantic Coast and the splendid harbor make Hampton Roads a logical center for protective naval and air facilities.

The export of coal is the most prominent shipping activity of the Hampton Roads ports. The coal-handling facilities at Norfork and Newport News are among the largest and most automated in the world, and Hampton Roads handles about half of all U.S. coal exports.

Manufacturing, with one exception, is not of major importance in the Hampton Roads cities. The exception is shipbuilding, which employs more than half the industrial workers in the area and is dominated by the massive shipyards at Newport News.

Nearby recreational attractions on three sides of Hampton Roads generate much tourist as well as local business. A few miles northwest of Newport News are three outstanding historic restorations: Williamsburg, Yorktown, and Jamestown. East of Norfolk are the tidewater beach areas, focusing on Virginia Beach. To the northeast are the waters of Chesapeake Bay, with their fishing and boating interest, and the bridge-tunnel route that crosses the bay mouth to the relatively unspoiled southern reaches of the Delmarva Peninsula.

THE RURAL SCENE

Despite the predominance of an urban lifestyle, population, and economy, most of the land in Megalopolis is actually rural. It is estimated that some 80 percent of the total land area of the region supports nonurban land uses. This rural land is important for many things. It supplies a great volume of foodstuffs to the cities of the region, although most of the food for Megalopolis is actually produced elsewhere; its rivers, streams, springs, and lakes provide much of the region's water supply, although many urban watersheds extend far beyond the confines of Megalopolis. Perhaps most important of all, the rural lands provide breathing space and recreational areas for the 45 million megalopolitans. The significance of green spaces becomes greater with the construction of each new high-rise apartment block, with each passing day of expanding urban sprawl.

The Coast and the Coastal Plain

The most pervasive aspect of the regional environment is the sea and its interface with the land. The long axis of Megalopolis parallels the Atlantic and there are many hundreds of miles of coastline along the irregular shore. People turn to the ocean and its edge for much of their commerce and recreation and for some of their food. No part of Megalopolis is more than 100 miles (160 km) from the coast.

The most striking characteristic of the coast is its irregularity. Few parts of the Anglo-American coastline, and certainly no section with such a sizable population, have such an uneven, embayed, island-studded outline. Different parts of the present shoreline have varied origins and their diversity of form is striking.

The principal embayments each have different shapes and different patterns of river flow into them. From north to south they include Cape Cod Bay, protected from the stormy North Atlantic by the hooked peninsula of Cape Cod; Narragansett Bay, Rhode Island's island-dotted waterway; Long Island Sound, sandwiched between the Connecticut coast and the North Shore of Long Island; Lower New York Bay, the broad entryway to the Hudson lowland, between New Jersey and Long Island; Delaware Bay, the extensive estuary of the Delaware River; and Chesapeake Bay, the continent's most complex and second largest estuary (Fig. 8–21).

There are three prominent peninsulas along this stretch of coast. Cape Cod is a long, low-lying sandy hook that is world famous as a summer recreational area. Cape May is a peninsula at the southern tip of New Jersey that shelters Delaware Bay from the open ocean. The Delmarva Peninsula, largest on the East Coast north of Florida, encompasses parts of the three states for which it is named.

The numerous offshore islands, with one outstanding exception, are crowded summer vacationlands that are largely depopulated in winter. Most famous are the islands off the southeastern coast of Massachusetts: Nantucket, Martha's Vineyard, the Elizabeth Islands, and Block Island. Many of the sandbar islands off the Long Island, New Jersey, and Delmarva coasts are also popular holiday spots. Long Island itself, of course, is much larger and more complex in its function.

The coastal plain is underlain by relatively unconsolidated sediments, most of which are of geologically recent vintage (Tertiary and Quaternary), although the inner margin of the plain has some older (Cretaceous) deposits. Beneath the sediments is a base complex of ancient igneous and metamorphic rocks, which reaches the surface in a very complicated pattern in southern New England, the New York City area, and northern New Jersey.

After a long period of gentle uplift of the continental shelf in Tertiary time, in which a series of broad open valleys developed, a significant drowning of the coastal plain occurred, presumably as a combined result of the weight of the Pleistocene ice load and the postglacial rise in sea level attributable to glacial meltwater.[5] Glaciation modified the northern part of the coastal plain, primarily by the laying down of extensive glacial and glaciofluvial deposits; indeed, the size of Long Island, Cape Cod, Martha's Vineyard, and Nantucket Island was significantly increased by the deposit of terminal moraines. The most recently developed terrain features of the region are ephemeral beaches, coastal dunes, and sandbars, with their associated shallow lagoons.

[5] William D. Thornbury, *Regional Geomorphology of the United States* (New York: John Wiley & Sons, 1965), p. 36.

FIGURE 8-21 Aerial parallel spans of the Lane Memorial Bridge cross Chesapeake Bay to connect the Eastern and Western shores of Maryland near Annapolis (courtesy Photo Researchers, Inc.).

The resulting topographic pattern throughout the Megalopolis Region is one of exceedingly flat land sloping gently toward the sea. Along the inner (western) margin of the region there is a rise toward rolling land or occasional steep-sided hills, but in the plain itself only minor prominences appear above the uniform level; they are mostly in the north and are chiefly related to accumulations of glacial deposition. Hard-rock bluffs (the Palisades) lining the lower course of the Hudson River are exceptions to this generalization.

The megalopolitan coast is a well-watered region: its coastal plain is crossed by a large number of important rivers flowing southward or southeastward to the Atlantic. These rivers originate in the interior uplands of the Appalachian system and move swiftly off the crystalline rocks of the Piedmont onto the softer sedimentaries of the coastal plain. The lithologic change from hard rock to softer rock is usually marked by a steeper gradient, producing a rapid or small waterfall, and a line drawn on a map to connect these sites is often referred to as the *Fall Line*.[6]

Climate

The climate is classified as humid midlatitude and semimarine. The coast has a leeward location on the continent, relative to the general west-to-east movement of weather systems, which mutes the maritime influence. Still, the effect of the adjacent ocean is to

[6] Several cities of varying sizes have riverside sites along the Fall Line between New Jersey and Georgia. Some scholars believe that a causative relationship is involved, for the rapids often served as both an early source of power generation and as head of navigation on the coastal plain rivers, thus providing economic incentive for town sites. This contention has been challenged, particularly by Roy Merrens, who has marshaled an impressive array of evidence to refute the notion that any functional relationships between river rapids and town sites along the Fall Line do, in fact, exist. See H. Roy Merrens, "Historical Geography and Early American History," *William and Mary Quarterly,* 3rd series, 22 (October 1965), 529–548.

ameliorate both summer and winter temperatures, to lengthen the growing season, to ensure that the region's harbors remain ice free (with rare exceptions), and to increase the moisture content of the atmosphere.

Precipitation is the most consistently prominent feature of the climate. Most of the region receives between 40 and 50 inches (1016 and 1270 mm) of moisture annually. It is generally well distributed throughout the year, with a slight maximum in summer.

Storms of various kinds are not uncommon and most bring considerable precipitation. Thundershowers are prevalent during the warmer months and late summer occasionally brings a "northeaster," a rainstorm with high wind that may last for several days. Every few years in late summer or fall the region may be visited by an errant tropical hurricane that has worked its way north from the Caribbean; heavy rain and roaring winds may whip up the sea sufficiently to cause considerable coastal damage and occasional loss of life. Winter snowstorms in the northern part of the region are often abrupt and the deep snow is occasionally paralyzing.

"Wild" Vegetation

Most of what is now the Megalopolis Region was originally covered with a mixture of woodland and forest in which deciduous species predominated but coniferous trees sometimes occurred in significant concentrations, especially on areas of sandy soil. Much of the forest and woodland was cleared for farming and other purposes by the middle of the nineteenth century. Since then, there has been a resurgence of trees, largely because of farm abandonment. The proportion of land in trees and brush has increased steadily for several decades—Connecticut, for example, was one-third wooded in 1850 but is two-thirds wooded today—and the trend seems likely to continue for some time.

The present vegetation associations of woods and brush are difficult to classify as natural vegetation; there has been too much human interference, both deliberate and accidental, in most areas. Many exotic species, especially shade trees and weeds, were added to the floristic inventory. Still, there are vast sectors where woods and forest predominate and in a few localities the natural vegetation has not been changed or altered by humans.

Of particular interest is the so-called pine barrens area of New Jersey. Located in the most densely populated state and just southeast of the most densely traveled traffic corridor in the world, these 650,000 acres (260,000 ha) of pitch pine and oak have a population density of less than 25 people per square mile. The area has remained virtually intact for several hundred years, apparently because of lack of agricultural potential. It also has an extensive aquifer of pure, soft water. Many real estate schemes have threatened the barrens, but so far only the edges have been nibbled away.

Soils

The soils of Megalopolis vary widely in characteristics. Still, three principal kinds can be noted: sand or sandy loams in the better-drained parts of the coastal plain, hydromorphic soils in the numerous areas of marsh and swamp, and varied residual soils in scattered localities.

Spodosols, with their subsurface accumulations of iron, aluminum, and organic matter, dominate in southern New England. Thin, light-colored Inceptisols are typical of the New York–northern New Jersey area. Clayey or sandy Ultisols predominate in the southern part of the region.

Agriculture is well entrenched but not primarily because of the soils; indeed, few soils are inherently very fertile for crop growing, and high agricultural productivity is usually associated with heavy use of fertilizer. In many farming areas the natural soil is simply a medium through which to feed the crops. Even the relatively productive soils of the Connecticut Valley and parts of New Jersey are heavily fertilized wherever cash crops are grown.

Specialized Agriculture

The past few decades have seen a continuing attrition of farmland in the Megalopolis Region. Urban sprawl pushed many farms out of production and woodland replaced cultivated land where marginal farms were abandoned for agriculture. The long-run trend has been for general farming to decline and specialty farming to increase and for total farm acreage to decrease while total value of output continues to rise. In other words, with the pressures from expanding urban land use on one side and increasing costs of farming on the other, the megalopolitan farmer has tended to abandon the poorer land and the less valuable products and concentrate on high-value output. The remaining farms tend to be efficient and specialized, emphasizing the production of perishables either for local consumption or immediate canning or freezing.

Although the region has a generally favorable climate for agriculture and soils that are only moderately fertile but responsive to fertilizer, it is not the physical advantages that underlie farming in Megalopolis; indeed, there is practically nothing in the region that could not be grown as well or better elsewhere in the country. The great advantage is market, and most agricultural production is market oriented.

The typical farm of contemporary Megalopolis is engaged either in raising vegetables and fruits in a market-gardening type of operation or in dairying. Important specialty production involves meat-type chickens (broilers, fryers) on the Delmarva Peninsula (Fig. 8–22), and horticulture (nursery and greenhouse plants) in a multiplicity of near-urban locations.

Commercial Fishing

Although the commercial fishing industry of Megalopolis has a long history of successful operation, its significance, both absolutely and relatively, continues to decline through the years. Generally, fishing

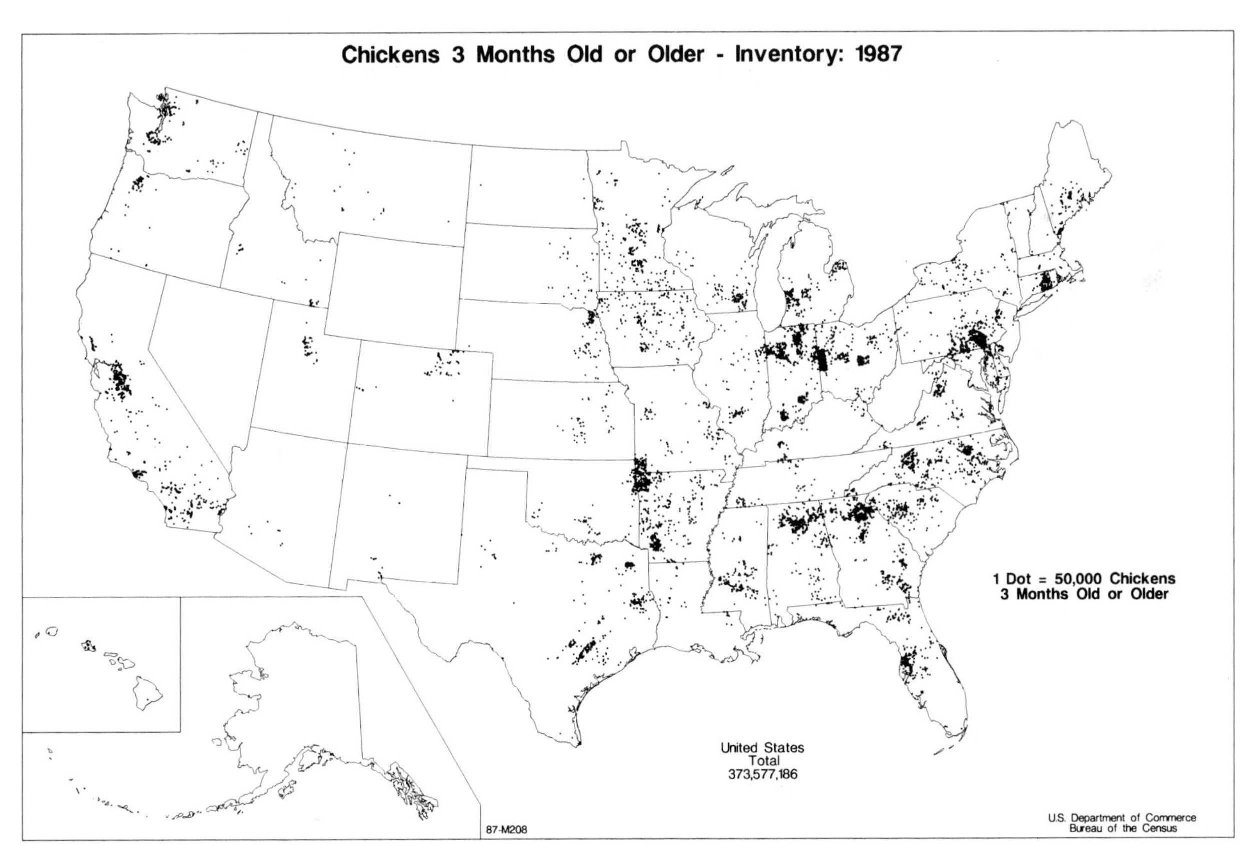

FIGURE 8-22 Distribution of commercial chicken-raising in the United States. Broilers account for most of the sales. The Delmarva Peninsula is one of the several major production areas.

fleets operate relatively close to home, in near North Atlantic waters, although some boats—particularly those from New England ports—regularly visit the "banks" area. Shellfish (primarily oysters, clams, and crabs), menhaden, and flounder are the principal catch (Fig. 8–23).

Menhaden, a fish little known to the general public, is taken in far greater volume by U.S. commercial fishermen than any other product of the sea. In most years it amounts to nearly 30 percent by weight of the total U.S. catch, which is about five times as much as pollack, the second variety; but it is a low-value fish, rarely providing more than 5 percent of the total value of the U.S. catch. About one-third of the menhaden caught in the United States comes from the waters off the Atlantic coastal plain. Partly because of its extreme oiliness, menhaden is not generally considered a food fish. It is used mainly as a source of oil and meal in making stock and pet feed and commercial fertilizers.

The mid-Atlantic coast has a long history as a leading fishery for oysters and crabs, particularly from Chesapeake Bay. However, in the last few years the oyster catch has declined precipitously, owing to pollution, overfishing, and a mysterious disease. Chesapeake Bay still produces more crab meat than the rest of the nation combined, but this, too, is on a downward trend.

FIGURE 8-23 A successful fishing trip for menhaden. The hold of the boat is full and fish are piled more than a foot deep on the deck. The port is Gloucester, Massachusetts (TLM photo).

The total commercial fishery of the Megalopolis Region yields about one-fourth of the U.S. catch, whether measured by weight or value. Virginia is the third-ranking state in volume of catch, largely because of menhaden landings. Massachusetts is the third-ranking state in value of catch. The leading fishing ports of Megalopolis are Gloucester and New Bedford in Massachusetts.

RECREATION AND TOURISM

Tourist and recreational attractions of Megalopolis are numerous and varied, as might be expected in such a populous region. Local people spend much of their holiday or vacation time within the region; in addition, a great many visitors come from other regions and overseas. Tourist interest focuses mainly on three general categories of attractions: cities and their points of interest, coastal areas, and historical sites.

Without doubt it is the urban attractions of Megalopolis that are most beguiling to resident and visitor alike. New York City is the number-one tourist goal on the continent and Washington is not far behind. Here, as in other large cities of Megalopolis, there are a host of things to see and do, although famous landmarks, such as the Empire State Building, United Nations, National Capitol, and the White House, seem to rank first in popularity.

The long seacoast and numerous islands of Megalopolis are important summer playgrounds, but their recreational significance in winter is limited. There are scores of beach resorts, the more famous being Cape Cod (Fig. 8–24), Nantucket Island, Martha's Vineyard, Asbury Park, Atlantic City, Ocean City, and Virginia Beach.

The almost-continuous beaches of southern New Jersey attract the greatest flood of patronage; some 40 million people (including repeats) are estimated to visit this area each summer. The resort towns, from Point Pleasant in the north to Cape May in the south, are built right along the beach, usually separated from the sand by only a boardwalk. Real estate prices continue to soar and construction of motels, condominiums, town houses, and other types of housing units flourish, catering to the immense nearby markets of Philadelphia, New Jersey, and New York.

FIGURE 8-24 Cape Cod has some of the busiest beaches of Megalopolis (courtesy Massachusetts Department of Commerce and Development).

A special attraction of coastal New Jersey is Atlantic City, once the nation's most storied beach resort but now the leading gambling town and the single most visited travel resort in the country (Fig. 8–25). The first casinos were opened in 1978, and by 1984 Atlantic City had surpassed Las Vegas in both visitors and money wagered.

The historical attractions of Megalopolis may occupy the visitor for a shorter period of time than the city or shore, but the great number of historic sites and the frequently intriguing nature of their presented interpretation make many of them hard to pass up. Most states of the region have capitalized on "selling" history to tourists; but Virginia (with hundreds of permanent historical markers along its highways), Pennsylvania, and Massachusetts have done the most thorough jobs. Battlefield sites of the Revolutionary and Civil wars are particularly notable. An interesting and popular trend is the integrated restoration of early settlements on a large-scale basis—for example, the Pilgrim village of Plymouth, Massachusetts; the restored seaport of Mystic, Connecticut; and the colonial town of Williamsburg, Virginia.

THE ROLE OF THE REGION

The significance of Megalopolis is much greater than its area or even its population would indicate. It is not enough to say that within the region are one-sixth of the people, one-fifth of the factory output,

FIGURE 8-25 A crowded boardwalk scene in Atlantic City on a sunny winter day. High-rise casinos appear in the background (courtesy Photo Researchers, Inc.).

and one-fourth of all wholesale sales of the nation. It must be reiterated that Megalopolis is also the financial heart of the country—six of the eight largest banks in the nation are in New York City. Of even greater importance, Megalopolis is the brain and nerve center of the nation. A large share of the decisions that shape the economy and government of the United States, thus significantly influencing economic and political decisions and events over most of the world, is made here.

It is impossible to quantify the magnitude of the decision-making role of Megalopolis, but some indication of this factor can be seen by tabulating the locations of the headquarters of the major business corporations of the nation. According to the annual *Fortune* magazine survey for 1990, one-third of the largest industrial corporations in the nation were headquartered in Megalopolis. The region also housed headquarters for 28 percent of the largest retailing companies, 44 percent of the largest life insurance companies, and 60 percent of the largest diversified financial companies. Add to this the Megalopolitan location of both the seat of government of the United States and the United Nations, as well as a host of state, city, and local government centers, and clearly much of the economy and politics of the nation is guided from Megalopolis—for better or worse.

Another way in which the importance of Megalopolis transcends its size is in its manner of coping with urban problems. This is an urban region with an unmatched intensity of living. Its urban problems, too, are unmatched in their complexity and magnitude. Equally abundant within the region are workers, brainpower, skills, and technology with which to face the problems. Jean Gottmann called Megalopolis the "cradle of the future," implying its role, both actual and potential, in meeting the challenge of an urbanized world.

There is yet another way in which Megalopolis has a role to play. What can it show the world in the matter of despoliation or protection of the environment? As cities grow and green spaces shrink, how does our relationship to nature change and how does this change affect the quality of life for people and all other living things? Must air pollution continue to worsen, for there are increasing numbers of machines and people to expel pollutants into the atmosphere? Will New York Harbor eventually be so choked with garbage scows that there will be no room for ocean liners? Must Chesapeake Bay become as sterile as Newark Bay, where in some reaches even coliform (intestinal) bacteria cannot survive? The inhabitants, abruptly and increasingly conscious of ecological relationships, watch with interest and apprehension.

SELECTED BIBLIOGRAPHY

ALEXANDER, LEWIS M. *The Northeastern United States* (2nd ed.). New York: D. Van Nostrand Company, 1976.

ALFORD, JOHN J., "The Chesapeake Oyster Fishery," *Annals,* Association of American Geographers, 65 (1975), 229–239.

BERGMAN, EDWARD F., AND THOMAS W. POHL, *A Geography of the New York Metropolitan Region*. Dubuque, IA: Kendall/Hunt Publishing Company, 1975.

CAREY, GEORGE W., *A Vignette of the New York–New Jersey Metropolitan Region*. Cambridge, MA: Ballinger Publishing Company, 1976.

CONZEN, MICHAEL P., AND GEORGE K. LEWIS, *Boston: A Geographical Portrait*. Cambridge, MA: Ballinger Publishing Company, 1976.

DANSEREAU, PIERRE, ED., *Challenge for Survival: Land, Air, and Water for Man in Megalopolis*. New York: Columbia University Press, 1970.

DILISIO, J. E., *Maryland*. Boulder, CO: Westview Press, 1983.

GOTTMANN, JEAN, *Megalopolis: The Urbanized Northeastern Seaboard of the United States*. New York: The Twentieth Century Fund, 1961.

———, *Megalopolis Revisited: 25 Years Later*. College Park: University of Maryland, Institute for Urban Studies Monograph Series No. 6, 1987.

———, *Virginia in Our Century*. Charlottesville: University of Virginia Press, 1969.

HEATWOLE, CHARLES A., AND NIELS C. WEST, "Mass

Transit and Beach Access in New York City," *Geographical Review,* 70 (April 1980), 210–217.

HEKMAN, JOHN S., "What Attracts Industry to New England?" *New England Economic Indicators,* (December 1978), A3–A5.

KANTROWITZ, N., *Ethnic and Racial Segregation in the New York Metropolis: Residential Patterns Among White Ethnic Groups, Blacks and Puerto Ricans.* New York: Praeger Publications, 1973.

KIERAN, JOHN, *A Natural History of New York City.* New York: Fordham University Press, 1982.

LIPPSON, ALICE JANE, *The Chesapeake Bay in Maryland: An Atlas of Natural Resources.* Baltimore: Johns Hopkins University Press, 1973.

MCMANIS, DOUGLAS R., *Colonial New England: A Historical Geography.* New York: Oxford University Press, 1975.

MCPHEE, JOHN A., *The Pine Barrens.* New York: Farrar, Straus & Giroux, Inc., 1968.

MEYER, DAVID R., *From Farm to Factory to Urban Pastoralism: Urban Change in Central Connecticut.* Cambridge, MA: Ballinger Publishing Company, 1976.

MITCHELL, ROBERT D., "American Origins and Regional Institutions: The Seventeenth-Century Chesapeake," *Annals of the Association of American Geographers,* 73 (September 1983) 404–420.

MULLER, PETER O., KENNETH C. MEYER, AND ROMAN A. CYBRIWSKY, *Metropolitan Philadelphia: A Study of Conflict and Social Cleavages.* Cambridge, MA: Ballinger Publishing Company, 1976.

OLSON, SHERRY, *Baltimore.* Cambridge, MA: Ballinger Publishing Company, 1976.

———, *Baltimore: The Building of an American City.* Baltimore: Johns Hopkins University Press, 1980.

PROCTER, MARY, AND BILL MATUSZESKI, *Gritty Cities: A Second Look at Allentown, Bethlehem, Bridgeport, Hoboken, Lancaster, Norwich, Paterson, Reading, Trenton, Troy, Waterbury, Wilmington.* Philadelphia: American University Press Services, 1978.

ROBICHAUD, BERYL, AND MURRAY F. BUELL, *Vegetation of New Jersey: A Study of Landscape Diversity.* New Brunswick, NJ: Rutgers University Press, 1973.

SHANKLAND, GRAEME, "Boston—The Unlikely City," *Geographical Magazine,* 53 (February 1981), 323–327.

STANSFIELD, CHARLES A., *New Jersey.* Boulder, CO: Westview Press, 1983.

THOMPSON, DEREK, ed., *Atlas of Maryland.* College Park: University of Maryland, Department of Geography, 1977.

THOMAS, JEAN-CLAUDE MARCEAU, "Washington," in *Contemporary Metropolitan America, Vol. 4, Twentieth Century Cities,* John S. Adams, ed., pp. 297–344. Cambridge, MA: Ballinger Publishing Company, 1976.

WACKER, PETER O., *Land and People: A Cultural Geography of Pre-industrial New Jersey: Origins and Settlement Patterns.* New Brunswick, NJ: Rutgers University Press, 1975.

WARF, BARNEY, "The Port Authority of New York–New Jersey," *Professional Geographer,* 40 (August 1988), 288–296.

9

THE APPALACHIANS AND THE OZARKS

The hill-and-mountain country of eastern North America is noncontiguous. The highlands of the Appalachians are separated from the uplands of the Ozark–Ouachita section by more than 300 miles (480 km) of exceedingly flat terrain. Yet the physical and cultural characteristics of these disconnected slopeland areas are generally similar; so it is logical to include them within the same broad region despite their separated positions.

The Appalachian highlands extend from central New York State, south of the Mohawk Valley, to central Alabama, where they terminate at the edge of the Gulf coastal plain. The eastern boundary of this subregion follows the topographic distinction between the southern Piedmont and the Blue Ridge–Great Smoky Mountains, from Alabama to central Virginia (Fig. 9–1). Northward of this area other factors seem to be more significant in locating the boundary; the northeastern margin of the subregion is deemed to lie along the western edge of Megalopolis, where rural land use replaces urban land use. Thus the northern part of the Appalachian Piedmont is included within the Appalachian subregion. The western boundary of the subregion is more indistinct physically but is conceptually situated along the western edge of the Appalachian "plateaus."

The uplands of the Ozark–Ouachita subregion extend in a general southwestward direction from southeastern Missouri through northwestern Arkansas into southeastern Oklahoma. The boundary of this section on all sides is determined primarily by the topographic pattern: the prevalence of slopelands within and the prevalence of relatively flat land without.

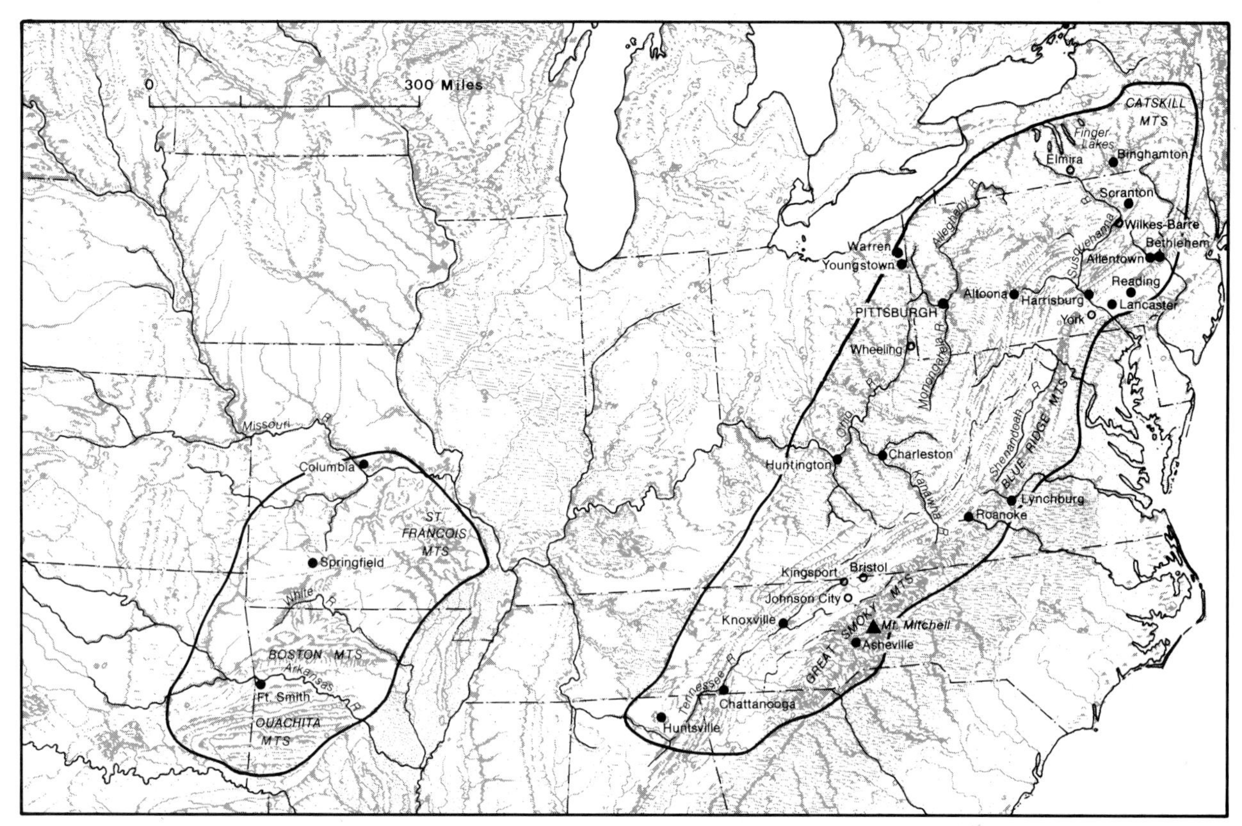

FIGURE 9-1 The Appalachians and Ozarks Region (base map copyright A.K. Lobeck; reprinted by permission of Hammond, Inc.).

THE REGIONAL CHARACTER

As with any large region, generalizations about predominant characteristics abound with exceptions. There are many differences in the various parts of the region. Some authorities would say that the northern Appalachians (roughly from the Kanawha River northward) are quite distinct from the southern Appalachians; others would note that the main differences are between coal-mining areas and noncoal-mining areas; and still others would point up differences between the Appalachians and the Ozarks. Throughout the region there are also pronounced variations in lifestyle among the citizens: between the black factory worker in Chattanooga and the white subsistence farmer on Hickory Ridge, between the family-centered Pennsylvania Dutchman of Lancaster County and the externally oriented resort operator on Lake of the Ozarks.

It has been tempting to paint the region as a hill people haven, replete with colorful speech, a charming folk culture, and hidden moonshine stills. This exaggerated image belongs to another era if, indeed, it ever pertained. Hill people and mountain folk certainly exist today in various parts of the region, but with few exceptions their isolation and distinctiveness are gone forever.

Certain elements contribute to geographical generalizations about the region. Most of the land is in slope and the slopes are often steep. As a result, life is focused in the valleys; settlements, transportation routes, and industrial developments compete with river or stream for the limited amount of flat land on the valley floor. Another pervasive aspect of the region is forest; almost all the area was originally forest covered and most of it is still clothed with virgin or second-growth trees.

Another significant facet of regional character

is recent population trends. During the middle decades of the twentieth century—the 1940s and 1950s—there was an actual population decline in most parts of the region, a situation that pertained to no other populous portion of Anglo-America. Appalachian birthrates, long among the highest on the continent, declined rapidly. More significant, however, was the impact of out-migration, predominantly teenagers and young adults, that characterized most of the region until the mid-1960s. Beginning in the late 1960s, however, more people moved into the region than left. This pattern accelerated in the 1970s and the traditionally declining areas of southern Appalachia and the Ozarks–Ouachitas are now experiencing a population resurgence that is not restricted only to urban areas but also includes a great many rural counties.

The population growth is indicative of greater economic opportunities in the region, but it does not necessarily indicate a reversal of the long-standing economic plight of the people of Appalachia and Ozarkia. The 1930 Census of Agriculture showed that the Appalachians encompassed the highest proportion of low-income farms in the country. Later the significant decline in coal-mining employment added another dimension to economic difficulty, for farming and coal-mining were long the principal occupations of the people of Appalachia; thus this region came to be recognized as the number-one long-run problem area (in a geographical sense) in the nation's economy. Any discussion of the region's geography must consider the low incomes and restricted economic opportunities of a large proportion of the population.

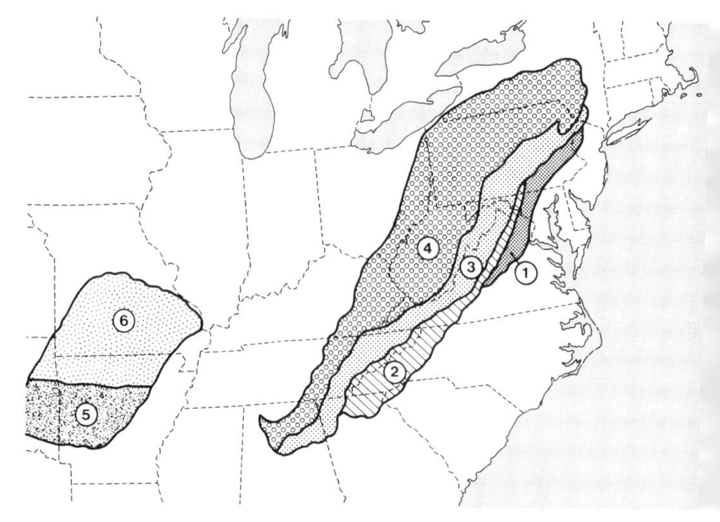

FIGURE 9-2 Topographic subdivisions of the Appalachians and Ozarks Region: (1) the Northern Piedmont, (2) the Blue Ridge–Great Smoky Mountains, (3) the Ridge and Valley section, (4) the Appalachian Plateaus, (5) the Ouachita Mountains and Valleys, and (6) the Ozark Plateaus.

THE ENVIRONMENT

Although the Appalachian and Ozark portions of the region are separated from one another by several hundred miles, their geologic and geomorphic affinities are so marked that there is general agreement that Appalachian lithology and structure continue westward beneath coastal-plain sediments to reappear at the surface in the Ozark–Ouachita subre-

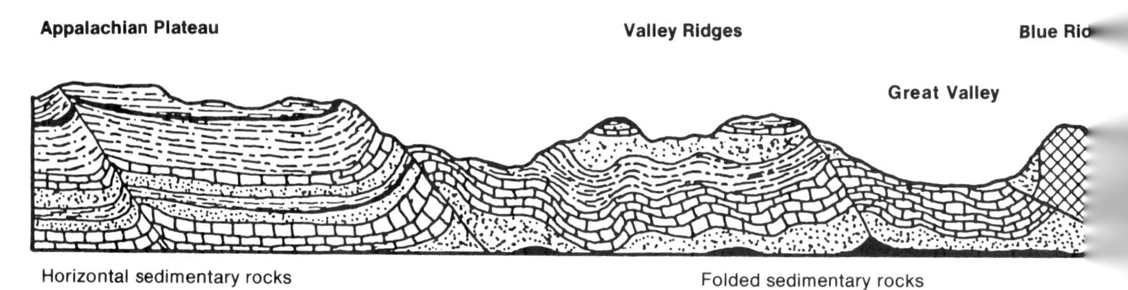

FIGURE 9-3 Generalized geologic cross section of the Appalachian subregion.

gions.[1] The result is notable similarity in rock types, structure, and landform patterns, which give the Appalachian and Ozark Region topographic distinctiveness that provides a convenient rationale for delineating physical subregions. The prominent subregions are

1. The Northern Piedmont
2. The Blue Ridge–Great Smoky Mountains
3. The Ridge and Valley section
4. The Appalachian Plateaus, often further subdivided into the Allegheny Plateau in the north and the Cumberland Plateau in the south
5. The Ouachita Mountains and Valleys
6. The Ozark Plateaus (Fig. 9–2)

The Northern Piedmont

The Northern Piedmont occupies an area extending from central Virginia across central Maryland and southeastern Pennsylvania into northern New Jersey. Its gently undulating surface reaches to only a few hundred feet above sea level in its highest parts. The underlying metamorphic and plutonic rocks are crystalline and ancient (Fig. 9–3).

Along the inner (western) margin of the Northern Piedmont is a lengthy lowland tract that developed on sedimentary rocks of mostly Triassic Age. This so-called Triassic Lowland encompasses some of the finest agricultural land on the continent, in part because of the soils that developed there.

[1] William D. Thornbury, *Regional Geomorphology of the United States* (New York: John Wiley & Sons, 1965), p. 262.

The Blue Ridge–Great Smoky Mountains

The Blue Ridge consists largely of crystalline rocks of igneous and metamorphic origin: granites, gneisses, schists, diorites, and slates. Extending from Pennsylvania to Georgia, the Blue Ridge exceeds all other mountains in the East in altitude. North of Roanoke it consists of a narrow ridge cut by numerous gaps; south of Roanoke it spreads out to form a tangled mass of mountains and valleys more than 100 miles (160 km) wide. The mountains are steep, rocky, and forest covered. The highest peak is Mount Mitchell in North Carolina (6684 feet or 2005 m), in the range known as the Great Smoky Mountains.

Because of heavy rainfall, the Blue Ridge was covered with magnificent forests, originally consisting of hardwood trees, especially oak, chestnut, and hickory. (Fig. 9–4) The greater part of the original forest was logged years ago, part of it being converted into charcoal for iron furnaces in the nineteenth century. Much of the area has been cut repeatedly and a great deal has been burned; however, most of the subregion is still cloaked with extensive forest.

The Ridge and Valley

This subregion consists of a complex folded area of parallel ridges and valleys.

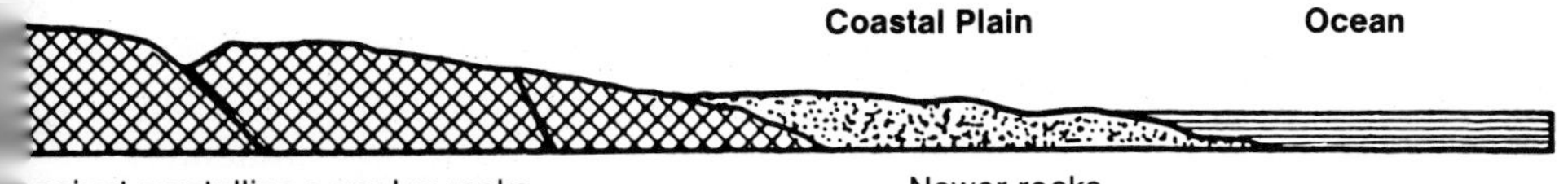

FIGURE 9-4 A view of the Blue Ridge in Virginia's Shenandoah National Park (TLM photo).

The Great Valley This nature-chiseled groove that trends northeast-southwest from the Hudson Valley of New York to central Alabama is one of the world's longest mountain valleys. Its flattish floor is divided into separate sections by minor cross ridges that serve as watershed separations for the several streams that drain the valley. Different names are applied to various segments of the Great Valley. Near the Delaware it is called the Lehigh Valley; north of the Susquehanna, the Lebanon Valley; south of the Susquehanna, the Cumberland Valley; in northern Virginia, the Shenandoah; in Virginia as a whole, the Valley of Virginia; and in Tennessee, the Valley of East Tennessee.

The Great Valley has long been the north-south highway in the Appalachian region as well as one of the most productive agricultural areas in the East. It has never been of outstanding industrial importance, although several of its cities (particularly in Pennsylvania and Tennessee) have important manufacturing establishments.

The eastern and western confines of the Great Valley are definite: the knobby wooded crest of the Blue Ridge towers above the valley floor on the east and the wild, rugged, though less imposing, Appalachian ridges bound it on the west.

The Ridge and Valley Section West of the Great Valley is a broad series of roughly parallel ridges and valleys that developed along parallel folds in Paleozoic sediments. Topographical development results primarily from differential erosion, with the configuration based on differences in the resistance of the bedrock. The ridgemakers are mostly hard sandstones, and the valleys usually indicate outcrops of softer limestones and shales. The ridges vary in size and shape, but most valleys are narrow, flattish, and cleared for agriculture. From the air the forested ridges show up as dark parallel bands and the cleared valleys as light ones.

This area gets less rainfall than other parts of the Appalachians—about 40 inches (1016 mm). This difference is ascribed to the sheltering effect of the Blue Ridge on the one side and the Allegheny–Cumberland Escarpment on the other. The seasonal distribution is quite even, although autumn is somewhat more dry than spring. About half the precipitation falls during the growing season, whose average length varies from 176 days in the north to more than 200 days in the south.

The natural vegetation of the valley lands consisted primarily of oak and hickory, with sycamore, elm, and willow near the streams. On the ridges it was similar to that found on the Blue Ridge Mountains to the east. Some of this magnificent forest was destroyed by the pioneers who settled in the area, but the ridges are still tree covered even though much of the cover is second and third growth. There were also areas of grassland in the broader river valleys—the result of burning by the Indians.

The Appalachian Plateaus

The western division of the Appalachian Highlands is a broad belt of land known as the Appalachian Plateaus. Along its eastern edge it has a high, bold escarpment, the Allegheny Front, that is so steep that roads and railroads ascend it with difficulty. Most of the layers of rocks that form the plateaus lie flat, one on the other. This subregion extends from the Catskill Mountains to north-central Alabama and from the Allegheny Front to the Interior Lowland. The term *plateau* is applicable in a structural sense, but the present topography is chiefly hill country, with only accordant summits as reminders of any previous plateau surface. The area can be subdivided into the Allegheny Plateau (glaciated and nonglaciated sections) and the Cumberland Plateau.

The Glaciated Allegheny Plateau Rounded topography typifies most of this area, which on the north is bounded by the Mohawk Valley and the Ontario Plain. The nearly flat-lying sandstones and shales are much dissected by streams that have cut down and back into the plateau. In the northern part—the Finger Lakes country—six slender lakes, trending north-south, occupy the valleys of preglacial streams that were modified by ice erosion and blocked by ice deposition. Here is an area of rolling terrain. In northeastern Ohio and adjacent northwestern Pennsylvania the plateau was modified by the ice and the relief is gentle, with broad divides. Northeastern Pennsylvania consists of a hilly upland with numerous streams, lakes, and swamps.

The Unglaciated Allegheny Plateau Topographically this plateau is more rugged than its glaciated neighbor to the north. Most of the area might properly be regarded as hill country, for the plateau has been maturely dissected. In the Kanawha Valley, the streams lie 1000 to 1500 feet (300 to 500 m) below the plateau surface. Some valleys are so narrow and canyonlike as to be uninhabited; others have inadequate room even for a railroad or highway.

The Cumberland Plateau This "plateau" is mostly rugged hill country that is so maturely dissected that practically none of its former plateau characteristics remains except locally, as in parts of Tennessee. No sharp boundary separates it from the Allegheny Plateau to the north; thus the Cumberland Plateau is regarded here as beginning in southern Kentucky (the upper reaches of the Kentucky River) and extending to the Gulf Coastal Plain. It includes parts of southeastern Kentucky, eastern Tennessee, and northern Alabama.

The Ozark–Ouachita Uplands

The Ozark–Ouachita subregion consists of three major divisions:

1. The Ozark section, consisting mainly of eroded plateaus, such as the Salem and Springfield plateaus and two hilly areas, the St. Francois Mountains in Missouri and the Boston Mountains in Arkansas (Fig. 9–5).

FIGURE 9-5 The heavily forested Ozark Mountains in northern Arkansas. (courtesy Arkansas Department of Parks and Tourism).

2. The broad structural trough of the Arkansas River Valley.
3. The Ouachita Mountains, whose strongly folded and faulted structures result in ridge and valley parallelism similar to the Ridge and Valley section of the Appalachians. The Ouachita Mountains reach their highest elevation—about 2800 feet (840 m) above sea level—near the Arkansas–Oklahoma border.

SETTLEMENT OF THE APPALACHIAN HIGHLANDS

It was a century and a half after the founding of Atlantic seaboard colonies before settlement began to push into the hill country of the Appalachians, although the Northern Piedmont was settled 50 years earlier. Welsh Quakers and Scotch-Irish Presbyterians were in the vanguard that funneled through William Penn's Philadelphia and spread out in southeastern Pennsylvania. It was German Protestants, however, who were the bulk of the settlers, attracted by both economic opportunity and the promise of religious freedom. By 1750 half of the population of the Pennsylvania colony consisted of immigrant Germans.[2]

From the Pennsylvania Piedmont through the Triassic Lowland was the only easy route to the Great Valley of the Appalachians. Into this valley in the early part of the eighteenth century came pioneer settlers from Pennsylvania; later they were joined by a trickle and then a flood of people moving more directly west from tidewater Virginia and Maryland. A road was finally cut through the western mountains via the Cumberland Gap, providing access to the interior. This "Wilderness Road" furnished the only connection to the infant settlements in Kentucky for several decades, later being supplemented by the old national road west from Baltimore.

In the latter part of the eighteen century and the early part of the nineteenth the coves and valleys of the Appalachians began to be occupied. "Many of the early settlers were hardy Ulster Scots, descendants of Scots who had been settled in the ancient Irish province of Ulster more than a century before they had moved to the New World."[3] People of English and German stock were also numerous and there was a minority of French Huguenot and Highland Scots as well. The Great Valley, with its prime agricultural land, was fairly densely occupied before the settlers began to push into the mountains in large numbers. By the 1830s following the eviction of Cherokees from their homes in northern Georgia, that area was occupied by whites and the settlement of the Appalachians was more or less complete.

For many years most of the Appalachians south of Pennsylvania were a sort of landlocked island, isolated from the rest of the country. Most rivers were too swift to be used for transportation and the rugged terrain inhibited railroad and highway construction. Even today most of the railways are branch lines built solely to exploit the coal and timber resources.

When the attractions of urban living became known to the hill people, particularly during World War II, an out-migration began. The availability of industrial employment and urban amenities drew tens of thousands of inhabitants from the Appalachians eastward to Washington, southward to Atlanta and Birmingham, and especially westward and northward to Nashville, Louisville, Cincinnati, Pittsburgh, Cleveland, Detroit, and Chicago. There they often settle in relatively close-knit neighborhood communities, maintaining a strong flavor of hill country living in the midst of the city until time eventually wears away the traditions.

The Great Valley has not significantly shared in the egress; rather, its attractions of cities, industry, and better farming areas produced a population increase. Indeed, most parts of the subregion experienced population expansion during the past decade, a striking turnaround from the declining trend that prevailed for more than half a century.

The present population of the Appalachians is preponderantly native-born and white. Blacks constitute a significant minority in the Pittsburgh area,

[2] Ezra Bowen and Editors of Time-Life Books, *The Middle Atlantic States* (New York: Time-Life Books, 1968), p. 35.

[3] John Fraser Hart, *The South* (Princeton, NJ: D. Van Nostrand Company, 1976), p. 52.

in some of the other Pennsylvania and Ohio industrial towns, and in parts of the Piedmont; elsewhere they live mostly in Chattanooga and Charleston.

SETTLEMENT OF THE OZARK–OUACHITA UPLANDS

The earliest white settlements in the Ozark–Ouachita Uplands were those of the French along the northeastern border of Missouri. Such attempts were feeble and were based on the presence of minerals, especially lead and salt. Because silver, the one metal they wanted, was lacking, the French did not explore systematically or try to develop the subregion. The first recorded land grant was made in 1723. The French, who never penetrated far into the Upland, were reduced to a minority group by English-speaking colonists toward the close of the eighteenth century. After the purchase of this territory in 1803 by the United States, settlement proceeded rapidly.

After the initial occupance of the Missouri, Mississippi, and Arkansas river valleys, succeeding pioneers entered the rougher and more remote sections of the Upland and remained there. The hilly, forested habitat, which provided so amply for the needs of the pioneers, later retarded their development by isolating them from the progress of the prairies.

World War I was an important factor in the breakdown of isolation in this subregion. The draft and the appeal for volunteers caused many ridge dwellers to leave the hills to join the armed forces. High wages in the cities during the war also enticed many from their mountain fastnesses. After the war, those who returned brought back new ideas. But the Great Depression of the 1930s had a regressive effect, for many who went to the cities lost their jobs and returned to the Upland and a large part of the plateau population went on relief when various social service agencies were developed under the New Deal.

World War II repeated the effect of the earlier war. Selective Service and high wages, particularly in munitions and aircraft factories in St. Louis, Kansas City, Tulsa, Dallas, and Fort Worth, attracted many younger persons away from the hills.

Modern highways opened most of the subregion to much greater interaction with the surrounding lowlands. In the last few years the long-term trend of out-migration was reversed. Increasing numbers of people are settling in the Ozarks and Ouachitas, not only in the urban areas but also throughout the rural counties. The attraction is partly the availability of jobs in a more diversified industrial scene, but, more importantly, it is based on tourism and retirement, reflecting the recreational, climatic, and scenic amenities of the subregion.

AGRICULTURE

Most early settlers of the Appalachian and Ozark Region depended on crop growing and livestock raising for their livelihood, and farming continues as a prominent factor throughout most of the region today. In general, however, agriculture has not been a very prosperous occupation. Niggardly environmental provisions of flat land and fertile soils combined with a lack of accessible markets to circumscribe the agricultural opportunities.

Historically farms were small and cropped acreage per farm was limited. Even in recent decades of increasing average farm size elsewhere, many parts of Appalachia experienced an opposite trend. Farm mechanization also lags behind the rest of the country. Generally farm income and living standards continue to be the lowest in the nation.

A common generalization is that the characteristic Appalachian land-use pattern is one of forested slopes and cleared valley bottoms. Such a pattern clearly prevails in many of the more productive areas, such as the ridge-and-valley portion of Pennsylvania. For most of the region, however, this generalization is invalid; the countryside appears disorganized and patternless, with a hodge-podge of land use on both slopeland and bottomland. One astute student of Appalachia described this situation as follows:

> As you travel through the core of Appalachia . . . you quite literally do not know what to expect when you turn the next corner or go over the crest of the next hill. . . . Field, for-

A CLOSER LOOK *The Pennsylvania Dutch—A World Apart*

In southeastern Pennsylvania lies a portion of the Northern Appalachian Piedmont that is remarkably different from all other parts of the Appalachians and Ozarks Region. The area is generally referred to as Pennsylvania Dutch country. It centers on Lancaster County but also includes most of Berks County to the east and York County to the west (Fig. 9–a). The original settlers were mostly *Deutsch*, or German, in nationality, and the term was corrupted to *Dutch*. Other settlers hailed from Switzerland, Austria, France, and Holland, but whatever their origin the majority shared a common bond of piety and came to America largely to escape religious persecution. Their descendants have clung remarkably to tradition as well as to land. Unlike farmers in many areas, generations pass through the same homesteads; sons follow fathers on the same soil.

The industrious, devout, old-fashioned farm population of the area can be divided into two groups:

1. The *plain Dutch* are mostly pious Old Order Amish folk, but include some Mennonites, Quakers, and Brethren (Dunkards). They are dedicated to a traditional way of life that preserves the peaceful, family-centered, home-oriented style of living for which the Pennsylvania Dutch are justly famous. For the most part, both adults and children dress uniformly in plain, solid-colored clothing that is without ornaments, buttons (they use hooks-and-eyes), collars, or embroidery. Women always wear aprons and plain bonnets and often capes. Married men wear full beards but not mustaches (because of their historical association with the military), whereas the women never cut their hair and wear it in a braided bun at the nape of the neck. They are opposed to such modern conveniences as electric-

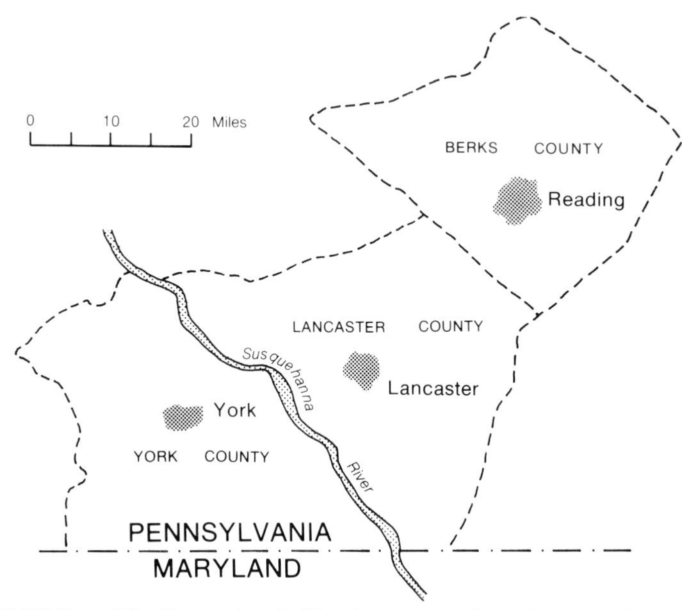

FIGURE 9-a The Pennsylvania Dutch country of southeastern Pennsylvania.

FIGURE 9-b An Amish farmer on the way to town (TLM photo).

FIGURE 9-c A representative Amish farm near Lancaster, Pennsylvania. Tobacco and corn are prominent crops (TLM photo).

ity, telephones, farm machinery, television, indoor plumbing, bicycles, and automobiles. They use horse-and-buggies for transport (Fig. 9–b, and are ingenious in their use of windmills and waterwheels for power. They seek a self-sufficient isolation from the temptations of the modern world and are totally opposed to such civilized delights as Social Security, insurance, lawyers, courts, and war. Their insistence on a close-knit education has been upheld, and their children generally attend a nearby, parochial, one-room school through the eighth grade (an average of five new schools was built for Old Order children every year during the 1980s), after which education is handled in a family setting. Most plain Dutch people are trilingual, speaking Pennsylvania Dutch, English, and High German. Despite having large families, the proportion of plain Dutch continues to diminish because of rapid growth of other components of the population. In the early 1990s less than 10 percent of the three-county population is Old Order.

2. A significant segment of the farm population has drifted away from the strict lifestyle of the plain Dutch; they are referred to as the *gay Dutch*. Although they retain many aspects of the traditional values, their pattern of living is more outgoing and they have embraced many aspects of modernity, such as automobiles, electricity, modern dress, and farm machinery.

Few farming areas in Anglo-America have been as richly endowed by nature and as well handled by people as the Pennsylvania Dutch country. The gently rolling terrain, abundant summer rainfall, relatively long growing season, and fertile, limestone-derived soils provide a splendid environment for agriculture. The Pennsylvania Dutch farmers are legendary for their skill and industriousness, although at least one authority has challenged the widespread belief that the eighteenth-century *Deutsch* were better farmers than those from the British Isles.[a] In any event, the farm landscape provides ample evidence that this is an area of rich and productive agriculture. The fields are well tended, the abundant livestock is well fed, scientific techniques are widely used (with some ingenious devices to compensate for the lack of machinery on plain Dutch farms), and most farmsteads are virtual showplaces. Lancaster County is not only Pennsylvania's leading farm county but is also the nation's leading nonirrigated farm county.

Livestock is the principal agricultural output of the Pennsylvania Dutch country, with dairy products, cattle, eggs, and broilers providing (in order)

[a] James T. Lemon, "The Agricultural Practices of National Groups in Eighteenth-Century Southeastern Pennsylvania," *Geographical Review*, 56 (October 1966), 467–496.

(*continued*)

the principal farm income. For many years tobacco was the principal cash crop of the area, but it has been in a severe decline recently. The principal crops today are corn, oats, potatoes, and hay.

The Pennsylvania Dutch country is an enclave of remarkable agricultural productivity in a region not noted for prosperous farming (Fig. 9–c). It is also an area of unique lifestyle in a region that contains a variety of unusual ones. The striking dichotomy of the area is that the plain Dutch people have become a major tourist attraction, which is a great annoyance to them[b] but an economic boon to many of their less conservative neighbors. Their self-imposed isolation has persisted for two-and-a-half centuries, but is increasingly difficult to maintain.

TLM

[b] The plain Dutch believe that photographs in which individuals can be recognized violate the biblical commandment against graven images; hence, they are unalterably opposed to being photographed. Tens of thousands of camera-carrying tourists, on the other hand, find the plain Dutch to be irresistibly photogenic. These incompatible viewpoints bring about frequent confrontations—always nonviolent.

est, and pasture are scattered across flat land and hillside alike, with no apparent logic, and the slope of the land seems to have scant power to predict how man will use it. Steep slopes are cultivated but level land is wooded. The tiny tobacco patches stick to the more or less level land, but rows of scraggly cornstalks march up some treacherously steep hillsides, often just across the fence from stands of equally scraggly trees. On the far side of the woods, as like as not, and on the selfsame slope, a herd of scrawny cattle mopes through a gulley-scored "pasture" choked with unpalatable grasses, unclipped weeds, blackberry briers, sumac bushes, and sprouts of sassafras, persimmon, thornapple, and locust.[4]

The seemingly random and nondescript character of land use can be explained in part by the small size of many holdings and the relatively high rate of farm abandonment that prevailed for sometime. More fundamental, however, is the more or less inadvertent cycle of land rotation that is traditional in much of the region: land is cleared and cultivated as cropland for some years and then allowed to lapse into a state of less intensive use (pasture) or disuse (woodland) for a period. This cycle is usually lengthy, the farmer clearing, or abandoning, a particular piece of land only once in a working lifetime (perhaps without realizing that he is reclearing a parcel that had previously been cleared by his father or grandfather).[5] The long-range nature of such a cycle makes it difficult to delineate and comprehend.

Crops and livestock are varied over this extensive region. It is primarily a general farming area and most farms yield a variety of products, generally in small quantities. Corn is the most common row crop, although it is usually not grown for commercial purposes. Hay growing is widespread and occupies the greatest acreage of cropland. Tobacco is the typical cash crop (Kentucky, with one-fourth of the national crop, ranks second only to North Carolina in tobacco production, although much of the output is east of the Appalachians), but average acreages are quite small. Livestock are significant and usually provide the greater part of farm income (Fig. 9–6).

Product specialization is notable in some areas, the most widespread specialties being dairying and apple growing. Dairy farming is particularly notable in the northern part of the region (New York and northern Pennsylvania) and in the Springfield Plateau of the Ozark section. Apple orchards are most prominent in and near the Great Valley, throughout most of its length. Five states of the region (New York, Pennsylvania, Virginia, West Virginia, and North Carolina) are among the nine leaders in apple production, and most of their output is associated with the Great Valley or with other ridge-and-valley locations.

The most productive agricultural portion of the region by far is the Northern Piedmont, especially in southeastern Pennsylvania (see vignette).

[4] John Fraser Hart, "Land Rotation in Appalachia," *Geographical Review*, 67 (April 1977), 148.

[5] Ibid., p. 154.

FIGURE 9-6 A Shenandoah Valley farm scene in Rockingham County, Virginia. Hereford cattle are in the foreground, with large chicken houses behind (TLM photo).

FOREST INDUSTRIES

Almost all of this upland region was originally forested and supported logging or lumbering activities at one time or another. Well over half the area is now tree covered, although much is second growth. Relatively valuable hardwoods constitute most of the total stand, sometimes in relatively pure situations but often intermingled with softwoods.

A large proportion of total forested acreage is within the boundaries of national forests. In some districts, however, extensive tracts are owned by major forest products companies and there are a great many small private holdings, particularly in Pennsylvania and Virginia.

In recent years there has been a modest resurgence of forest industries in the Appalachians. With increasing farmland abandonment, it is obvious that woodland acreage is expanding. The replacement process has been described as follows:

> A hard-scrabble, hillside farm is finally abandoned. The first summer, weeds quickly take over. The next summer some grass gets a foothold under the weeds, and blackberry seedlings make their appearance. After several more years, clusters of trees push up above the brambles. The trees may be gray birch from wind-borne seeds, and Eastern red cedar trees from seeds dropped by birds that had dined on red cedar berries. Ultimately maples and oaks crowd the birches and cedar for sunlight. By and by the oaks and maples predominate.[6]

Forestry is generally minor in the total economy of the region, but almost every county has at least a little of it and its local significance may be great even though it is usually characterized by low wages and temporary employment. Lumber is the principal product in most areas, but there has been a continuing increase in pulping operations. No part of the region is a major lumber product, but, taken as a whole, about one-third of the nation's hardwood lumber comes from the Appalachians and Ozarks.

MINERAL INDUSTRIES OF THE APPALACHIANS

Various ores and other economic minerals are produced in the Appalachians, but coal is much more important than all the rest combined. No other activity has made such an imprint on the subregion and is so intimately associated with the "problem" of Appalachia.

[6] Evan B. Alderfer, "a Jogtrot Through Penn's Woods," *Business Review*, Federal Reserve Bank of Philadelphia (February 1969), p. 11.

Bituminous Coal

A large proportion of the Appalachian subregion is underlain by seams of bituminous coal that occur in remarkable abundance. This area has been the world's most prolific source of good-quality coal and still yields about two-thirds of the nation's total output. Every state in the region except New York and North Carolina is a producer of bituminous coal.

The history of the industry is one of ups and downs, but the energy crisis of the 1970s demonstrated conclusively that the United States must rely heavily on its domestic resources, and its reserves of coal are far larger than its supplies of all other practical energy sources combined.

Coal mining had been declining from the end of World War II until the early 1960s, losing its two biggest markets, railway steam engines and domestic heating, and experiencing a national employment decrease from 400,000 to 150,000. Increasing use of coal in electric power generation halted the decline in the early 1960s, but there was only a slow growth in output and almost no increase in employment until the accelerated demand of the early 1970s. By 1975 domestic bituminous coal production had surpassed the previous record-year output (1947) and a new peak was reached every year until 1982, after which another declining trend set in (Fig. 9–7).

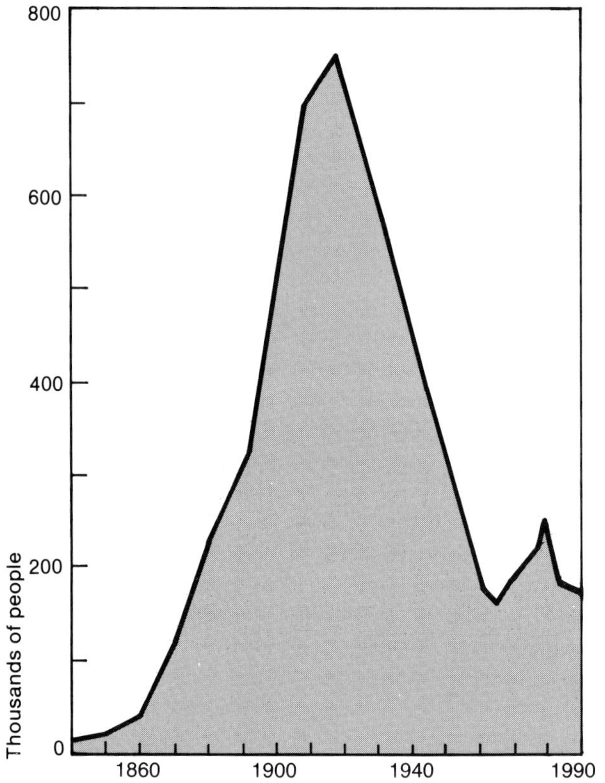

FIGURE 9-7 Historical fluctuations in coal mining employment in the United States. The half-century downtrend has now ceased, but the resurgence has been erratic.

There are more than 5000 coal mines in the Appalachian and Ozark Region, employing about 150,000 miners. Many mines are small operations ("dogholes") with only a few workers, but most of the output comes from large mines owned by only a dozen or so companies, most of which are subsidiaries of giant corporations.

About 60 percent of the coal is extracted from strip mines, in which gigantic power shovels, the largest machines ever to move on land, exploit seams that are close to the surface. Mechanization is virtually complete in the larger underground shaft mines as well. Mechanization has meant a dramatic increase in productivity per miner but a significant decrease in the number of miners needed.

The strongest production record has been in the southern Appalachians, partly because of the higher proportion of more desirable coals with low sulfur content. West Virginia continues as the leading Appalachian coal-producing state, and five of the six leading coal mining states are in the region (Kentucky, Pennsylvania, Virginia, and Ohio are the others).[7]

Coal has been a mixed blessing for Appalachia in the past; the same holds true today. Relatively minor market fluctuations can have a significant impact in the coal counties. The long-range prognosis, however, is more favorable than in decades. With current technology and at contemporary levels of production, more than three centuries worth of coal is still in the ground.

[7] Kentucky is actually the single leading coal-producing state in the nation, but about one-third of its output is from the western part of the state, which is not in the Appalachian and Ozark Region.

Anthracite Coal

In the northern end of the ridge-and-valley country in northeastern Pennsylvania lies the once important anthracite coal field. Anthracite, a very high-quality coal, is quite expensive to mine because of its narrow and highly folded seams. For over a century this was a prominent mining activity, employing nearly 180,000 miners in the peak year of 1914. Competition with lower-cost bituminous coal and other fuels overwhelmed the anthracite industry. Despite massive reserves, fewer than 3000 miners are currently employed, and production is negligible.

Petroleum

Petroleum has long been important in Pennsylvania, Kentucky, and New York. The nation's first oil well was put down near Titusville in northwestern Pennsylvania in 1859 and this state led all others in the production of crude oil until 1895. Although still productive, the area is outstanding more for the high quality of the lubricants derived from the oil than for the quantity of production. Today less than 2 percent of the country's petroleum is produced in the Appalachians and Ozarks, and only one of the 60 leading oil fields (the Bradford–Allegheny field in Pennsylvania and New York) lies in the region.

Copper and Zinc

Numerous deposits of metallic ores yielding copper, zinc, and small quantities of other metals are worked along the western flanks of the Great Smoky Mountains in Tennessee. This state is the leading zinc producer of the nation, with nine mines in the area just east of Knoxville. In the extreme southeastern corner of Tennessee are five underground copper mines, the only significant source of copper in the eastern United States.

MINERAL INDUSTRIES OF THE OZARK–OUACHITA UPLANDS

For an upland area, the Ozark–Ouachita subregion is poor in minerals—with two exceptions. Southeastern Missouri is one of the oldest mining areas of the United States, having produced almost continuously since 1725. It is the principal source of lead ore in the nation, with much of the output from recently developed or expanded mines. The Tri-State District where the Missouri, Kansas, and Oklahoma borders meet is another underground mining district that has been prominent during the past half century. A leading zinc producer in the past, the Tri-State District now yields chiefly lead. Combined output from mines in these two districts makes Missouri the leading lead-producing state, normally with more than 90 percent of national output.

CITIES AND INDUSTRIES

The Appalachian and Ozark Region is not highly urbanized. Urbanization is a much more recent phenomenon here than in most of the rest of the nation, and in many extensive areas rural dwellers are in the majority (see Table 9–1 for the region's largest urban places).

Pittsburgh is the only major metropolis in the entire region (Fig. 9–8). Including associated industrial satellites in the valleys of the lower Monongahela and Allegheny rivers and the upper valley of the Ohio and the nearby complexes of *Youngstown* and *Wheeling,* the Pittsburgh district was one of the great metal-manufacturing areas of the world. Its longtime economic focus was on the production of primary iron and steel and on further processing in fabricated metals and machinery industries. However, in the 1980s the American steel industry virtually died, and the Pittsburgh area probably suffered more than any other, losing some 150,000 high-paying manufacturing jobs. It is no longer among the leading two dozen manufacturing cities of the nation. The city responded with an imaginative program called Renaissance II, which has revived the urban economy with downtown renovation, the attraction of high-tech businesses, and neighborhood revitalization.

Besides this notable urban agglomeration, there are four general areas in which loose clusters of medium-sized cities occur.

1. In northeastern Pennsylvania and adjacent parts of southern New York State are sev-

A CLOSER LOOK River Basin "Development"

The "taming" or "harnessing" of a river has become one of the favorite tools for local or regional economic development in the twentieth century. All across the United States, from the Penobscot to the Sacramento, rushing rivers were turned into quiet reservoirs by straightening channels, constructing levees, and, particularly, building dams. In most cases, such development meant multiple use because of the advantages of flood control, decreased erosion, improved navigation, expanded hydroelectricity potential, increased water availability, and broadened recreational opportunities. Opponents of river development schemes point out that the economic logic of such projects is nearly always suspect at best and that often ecological problems arise.

The controversy between proponents and opponents of river development is an old one. Meanwhile, most rivers of the nation experienced some sort of development and plans are in process for the rest. In the Appalachian and Ozark Region lie major portions of two of the most integrated and best known of all river-basin development schemes: the Tennessee Valley Authority and the Arkansas River Waterway.

THE TENNESSEE VALLEY AUTHORITY

To date, the TVA is perhaps the greatest experiment in regional socioeconomic planning and development carried out by the federal government. It was created in the 1930s to aid in controlling, conserving, and using water resources of the area. It

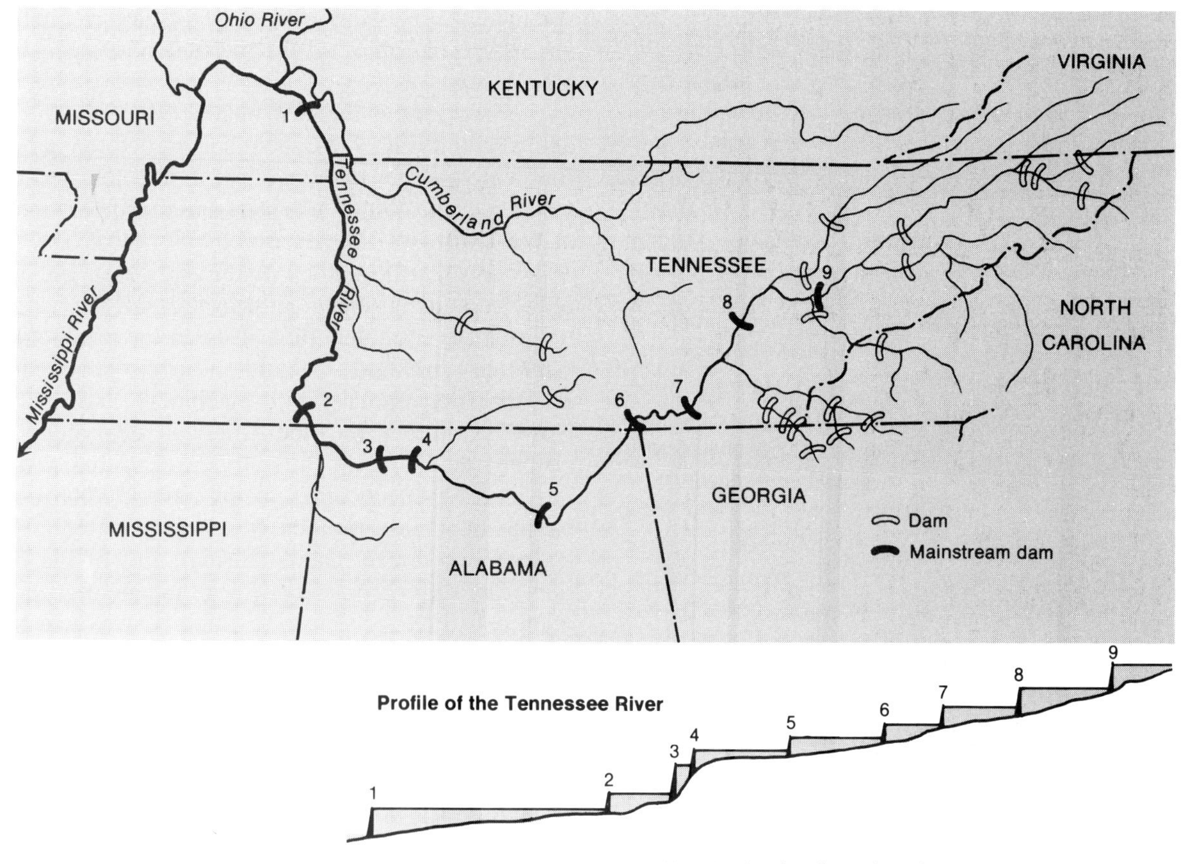

FIGURE 9-d The dams of the Tennessee Valley Authority (based on data supplied by the Tennessee Valley Authority).

deals with such diverse matters as flood control, power development and distribution, navigation, fertilizer manufacturing, agriculture, afforestation, soil erosion, land planning, housing, and manufacturing.

The area of the Tennessee Valley Authority encompasses the watershed of the Tennessee River and its tributaries—more than 40,000 square miles (104,000 km^2) in parts of seven states (Fig. 9-d). This area was selected because it was the most poverty-stricken major river basin in the country. Except that it is a drainage basin, it is not a unified region because land utilization, agriculture, manufacturing, transportation, and the distribution of power all cut across the drainage boundary.

There are no longer any free-flowing stretches of the river except in its upper headwaters; it has been dammed into a series of quiet lakes for its 650-mile (1040 km) length from Knoxville to its confluence with the Ohio River near Paducah. Besides 9 mainstream dams, there are more than 24 others on tributaries. Hydroelectricity is generated at each dam, but most of the TVA's power output now emanates from a dozen thermal electric plants that it controls. This comprehensive electric system was a great boon to the valley, but it is also the most controversial feature of the TVA's operation. Private power companies contend that their very existence is threatened by the cheaper TVA power, part of the cost of which is federally subsidized. Furthermore, many complaints about destructive strip mining in the region are aimed at the TVA, which is the single largest purchaser of coal.

A major initial purpose of the TVA was to control flooding in the valley, a sporadic, major hazard. This goal was largely achieved, but some critics point out that flood damage was prevented by permanently flooding much of the best valley land.

Other benefits attributable to the TVA include attraction of industry to the region, alleviation of much accelerated soil erosion, provision of new water recreational areas, and maintenance of a permanent 9-foot (3-m) navigation channel. The benefits that resulted are unquestionable, but there is considerable difference of opinion about costs in relation to benefits. Critics claim that the money could

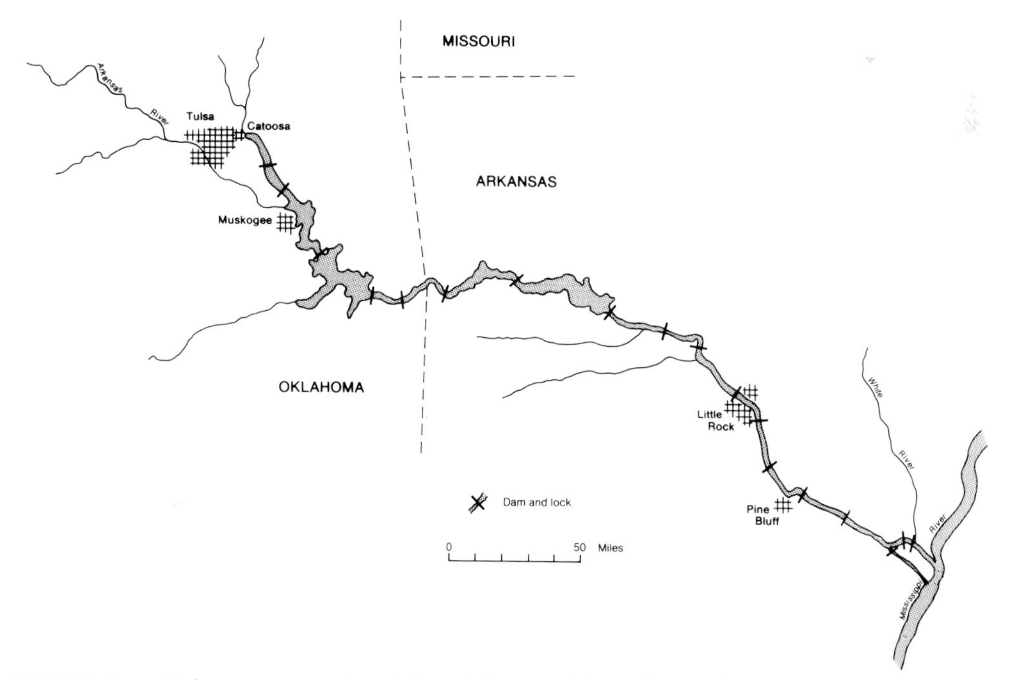

FIGURE 9-e Major components of the Arkansas River Navigation System.

(continued)

have been used more effectively in other ways.

The controversy will doubtless continue; nevertheless, many lessons can be learned from the experiment. The TVA functions at a level between centralized federal government and fragmented local authorities and yet on an interstate basis. It is blessed by some and cursed by others but is often referred to, particularly by foreigners, as a comprehensive and relatively successful example of functional regional planning. On the other hand, many residents of the region consider the TVA to be little more than a large electric power utility.

THE ARKANSAS RIVER NAVIGATION SYSTEM

On a somewhat smaller scale, but not at a significantly lower cost, is the Arkansas River Navigation System, dedicated in 1971 after nearly two decades of construction. Its purpose was to produce a 440-mile (700-km) navigable channel (9 feet or 3 m in depth) up the Arkansas River from its confluence with the Mississippi, via Pine Bluff, Little Rock, Fort Smith, and Muskogee, to the head of navigation near Tulsa. Eighteen dams have "stabilized" the river, thus adding flood control and hydroelectricity production to the project's benefits (Fig. 9–e).

There is no doubt that the local areas benefited from cheaper transportation, power generation, and industrial attraction; however, the expense of the undertaking, which was approximately equal to the amount spent on construction of the Great Lakes–St. Lawrence Seaway, makes it one of the most costly public works projects in the history of the nation. Critics who contend that "it would be cheaper to pave it" are speaking only partly in jest.

TLM

TABLE 9-1

Largest urban places of the Appalachians and Ozarks Region

Name	*Population of Principal City*	*Population of Metropolitan Area*
Allentown, Pa.	105,200	677,100
Altoona, Pa.	52,800	132,500
Asheville, N.C.	61,220	173,100
Bethlehem, Pa.	72,490	
Binghamton, N.Y.	51,100	260,200
Charleston, W. Va.	55,730	260,800
Charlottesville, Va.	41,600	123,800
Chattanooga, Tenn.	162,670	438,100
Columbia, Mo.	64,330	105,800
Cumberland, Md.	23,120	102,400
Elmira, N.Y.	32,600	91,700
Fayetteville, Ark.	40,730	110,600
Fort Smith, Ark.	74,600	180,700
Hagerstown, Md.	34,680	117,800
Harrisburg, Pa.	51,720	591,100
Hickory, N.C.		222,100
Huntington, W. Va.	56,300	322,300
Huntsville, Ala.	159,450	236,700
Jamestown, N.Y.	33,930	141,300
Johnson City, Tenn.	43,350	442,300
Johnstown, Pa.	30,420	250,600
Joplin, Mo.	41,630	136,000
Knoxville, Tenn.	172,080	599,600
Lancaster, Pa.	58,980	414,100
Lynchburg, Va.	69,800	145,500
Parkersburg, W. Va.	37,550	154,400
Pittsburgh, Pa.	375,230	2,094,300
Reading, Pa.	76,550	329,100
Roanoke, Va.	97,700	221,600
Scranton, Pa.	81,250	736,600
Sharon, Pa.		122,400
Springfield, Mo.	142,690	234,300
State College, Pa.	41,300	115,700
Steubenville, Ohio	22,350	147,700
Warren, Ohio	51,640	
Wheeling, W. Va.	38,940	171,500
Wilkes-Barre, Pa.	47,020	
Williamsport, Pa.	31,820	118,300
York, Pa.	45,370	410,400
Youngstown, Ohio	101,150	501,700

eral cities whose economy significantly depended on coal mining or specialized manufacturing for several generations. The decline of both anthracite coal mining and the once prominent textile and apparel manufacturing industries has been spectacular. Employment in the tertiary sector, particularly service industries, has more than doubled in the last quarter century, easing the economic hardship, but long-run population decline continues.

FIGURE 9-8 Looking northward across the Monongahela River into downtown Pittsburgh (courtesy Paolo Koch, Rapho/Photo Researchers, Inc.).

The principal Pennsylvania urban complexes are *Allentown–Bethlehem–Easton* on the lower Lehigh River and *Scranton* and *Wilkes-Barre* in the upper valley of the Susquehanna River. *Binghamton* and *Elmira* are separate urban nodes in southern New York that have fairly well-balanced economies and have absorbed considerable in-migration from the Pennsylvania anthracite area.

2. Several other old, medium-sized cities are scattered over the Piedmont of southeastern Pennsylvania. *Reading, Lancaster,* and *York* are commercial centers for the Piedmont in general and the Pennsylvania Dutch country in particular; however, they also have an important industrial function, especially in metalworking and machinery production. *Harrisburg* is larger than the other three cities and much less dependent on manufacturing. It is an important railway center, and as the state capital, its governmental employment is sizable.
3. The middle valley of the Ohio River and the adjacent valley of the lower Kanawha have long been centers of heavy industry, particularly metallurgical and chemical factories. The lure of firm power has been particularly important as an industrial attraction; large nearby coal supplies are much more reliable than hydroelectricity with its seasonal vagaries. Inexpensive river transportation and a nearby surplus of suitable labor are other recognizable assets for industry.

   The principal urban nodes are the tricities of *Huntington* (West Virginia)–*Ashland* (Kentucky)–*Ironton* (Ohio) on the Ohio River, and *Charleston* on the Kanawha (Fig. 9–9). But many of the most spectacular industrial developments in this area took place away from the older centers, often in the splendid isolation of a completely rural riverside setting some distance from any urban area.
4. The valleys of eastern Tennessee have had considerable urban and industrial growth, much of it associated with TVA power development. The principal nodes are *Chattanooga, Knoxville, Johnson City, Kingsport,* and *Bristol.*

FIGURE 9-9 Large factories, especially those producing chemical products, crowd the valley of the Kanawha River around Charleston, West Virginia (courtesy Charleston Chamber of Commerce).

RESORTS AND RECREATION

The highlands of the northeastern Appalachians are not particularly spectacular or unusually scenic; they consist of pleasant, forested hills, with many rushing streams and a number of lakes. They also contain, particularly in the Catskill and Pocono mountains, one of the densest concentrations of hotels, inns, summer camps, and resorts to be found on the continent. The great advantage is that they are literally next door to Megalopolis and provide a relatively cool summer green space for urban millions to visit.

The principal tourist attractions of the Pennsylvania Piedmont and the Shenandoah Valley are historical. This was an area of almost continual conflict during the Civil War; in few other areas are historical episodes presented so clearly and accurately to the visitor, most notably at Gettysburg. The Pennsylvania Dutch culture is another major attraction of the area. Tourists are disliked by the plain Dutch, but they and their dollars are welcomed by most other residents of the area.

Farther south the national parks attract large numbers of tourists. Shenandoah National Park, with its beautifully timbered slopes and valleys, is well known for picturesque Skyline Drive. Great Smoky Mountains National Park is a broad area of lofty (by eastern standards) mountains clothed with dense forests of pine, spruce, fir, and hardwoods. In most years it receives more visitors than any other national park on the continent.

The recreational possibilities of the Ozark–Ouachita Uplands were recognized by Congress as early as 1832,[8] but the resort industry as it exists today is a recent development. Although Hot Springs and other centers became important locally in the 1890s, the present development had to wait until better railroads and highways were built into the mountains of the area and until the urban centers in surrounding regions attained sufficient size to support a large nearby resort industry. Both goals have been achieved and today the Ozark–Ouachita Uplands occupy the unique position of being the only hilly or mountainous area within a few hours' drive of such populous urban centers as Kansas City, St. Louis, Memphis, Little Rock, Dallas, Fort Worth, Oklahoma City, and Tulsa.

As in other parts of the Southeast and Gulf Southwest, some of the most successful recreational areas developed around the large, branching reservoirs that were constructed in various river valleys. Water sports, in the form of boating, fishing, swimming, and skiing, are now very much a part of holiday living for hundreds of thousands of families in an area where natural lakes are almost nonexistent. Most important as a recreational center is Lake of the Ozarks in Missouri but also notable are Lake O' the Cherokees in Oklahoma, Lake Ouachita in Arkansas, and Bull Shoals Reservoir on the Arkansas–Missouri border.

[8] An area around Hot Springs was set aside as a federal preserve some four decades before Yellowstone, normally considered to be the first of the national parks, was established.

THE OUTLOOK

The plight of the Appalachian and Ozark Region is celebrated in song and story. It has been recognized as a major negative economic anomaly, an extensive region of poverty in the heart of the richest nation on earth. Its way of life has been called a "culture of despair." The reasons underlying such a situation are complex and imperfectly understood but certainly include a litany of environmental difficulties and a variety of questionable economic approaches and negative human attitudes.

Into a region of small, hill-country farmers came three waves of economic development, each largely financed by "outside" entrepreneurs, each largely sending the profits outside the region, and each despoiling the environment to a notable (sometimes disastrous) degree. The story of logging and mining, the first two waves, is well known. Some local people made money from sale of land or resources and many jobs were provided; however, many sales were at relatively low prices and most jobs were low paying, part time, or both. Recreation and tourism, the third wave, is more recent but is nearly as massive and sudden as the other two. Developers, usually corporate and often from outside the region, purchased large acreages of high-amenity (scenic or waterside) land on which to build massive recreational and housing projects. Lack of integrated planning and zoning allows development that often does not conform to the landscape.

Federal and state governments attempted to alleviate the situation with massive infusions of capital and ambitious development programs. Such efforts, most notably represented by the TVA, the Arkansas River Waterway, and the Appalachian Regional Commission, provided many advantages to the region but usually at a cost-benefit ratio of depressing proportions.

Until fairly recently the popular solution was graphically shown by migration statistics; most parts of the region experienced a massive and continuous out-migration and population decline for several decades. Beginning in the late 1960s, however, and continuing to the present, new demographic trends have appeared. Many sections, particularly in southwestern Appalachia (Kentucky, Tennessee, and Alabama) and the Ozarks of Arkansas and Missouri, experienced an upsurge of population growth. This growth is based partly on expanded employment opportunities in manufacturing (new factories) and recreation (mostly reservoir-related services) that stemmed the prevailing out-migration of working-age people and partly on an influx of older people who opted for retirement in the pleasant rural surroundings offered by the hill country.

Some parts of the region have functioned for a considerable period as pockets of prosperity—for example, the Poconos, the Pennsylvania Dutch country, and much of the Great Valley. Economic stimulation has been provided by long-term government installations at such places as Oak Ridge, Tennessee, and Huntsville, Alabama. And many larger cities, such as Pittsburgh, Chattanooga, Knoxville, and Springfield, continued to experience "normal" urban growth patterns based on their diversified economies.

The renewed importance of coal in the national energy scene augurs well for the economy of most of the coal-mining areas despite production and employment fluctuations of the 1980s.

It seems likely that the Appalachians and Ozarks are partly free of the traditional syndrome of regional poverty and depression. Agricultural specialization, especially in beef cattle and poultry, will result in more efficient and profitable output in many farming areas. Industrial and recreational developments will continue to make contributions to economic diversification. Population statistics will be watched with interest to see if the recent short-term growth in rural areas develops into a long-term trend.

Many areas, however, did not escape the patterns of the past. There are still too many areas of poor farms, eroded soil, and marginal mines with underpaid workers interspersed among the districts of improving conditions.

SELECTED BIBLIOGRAPHY

BATTEAU, ALLEN, ed., *Appalachia and America: Autonomy and Regional Dependence*. Lexington: University Press of Kentucky, 1983.

BENHART, JOHN E., and MARJORIE E. DUNLAP, "The Iron and Steel Industry of Pennsylvania: Spatial Change and Economic Evolution," *Journal of Geography*, 88 (September–October 1989), 173–184.

BINGHAM, E., "Appalachia: Underdeveloped, Overdeveloped, or Wrongly Developed?" *Virginia Geographer*, 7 (Fall–Winter 1972), 9–12.

CUFF, DAVID J., et. al., *The Atlas of Pennsylvania*. Philadelphia: Temple University Press, 1989.

FEGLEY, RANDALL, "Plain Pennsylvanians Who Keep Their Faith," *Geographical Magazine*, 52 (December 1981), 968–975.

FORD, THOMAS R., ed., *The Southern Appalachian Region: A Survey*. Lexington: University of Kentucky Press, 1962.

HART, JOHN FRASER, "Land Rotation in Appalachia," *Geographical Review*, 67 (1977), 148–166.

KARAN, P. P., and COTTON MATHER, eds., *Atlas of Kentucky*. Lexington: University of Kentucky Press, 1977.

LEMON, JAMES T., *The Best Poor Man's Country; A Geographical Study of Southeastern Pennsylvania*. Baltimore: Johns Hopkins Press, 1972.

MARSH, BEN, "Continuity and Decline in the Anthracite Towns of Pennsylvania," *Annals*, Association of American Geographers, 77 (September 1987), 337–352.

———, "Environment and Change in the Ridge and Valley Region of Pennsylvania," *Journal of Geography*, 88 (September–October 1989), 162–166.

MILLER, E. JOAN WILSON, "Ozark Superstitions as Geographic Documentation," *Professional Geographer*, 24 (1972), 223–226.

———, "The Ozark Culture Region as Revealed by Traditional Materials," *Annals*, Association of American Geographers, 58 (1968), 51–77.

MILLER, E. WILLARD, *Socioeconomic Patterns of Pennsylvania: An Atlas*. Harrisburg: Pennsylvania Department of Commerce, 1975.

MILLER, E. WILLARD, "The Anthracite Region of Northeastern Pennsylvania: An Economy in Transition," *Journal of Geography*, 88 (September–October 1989), 167–172.

MITCHELL, ROBERT D., *Commercialization and Frontier: Perspectives on the Early Shenandoah Valley*. Charlottesville: University Press of Virginia, 1977.

———, "The Shenandoah Valley Frontier," *Annals,* Association of American Geographers, 62 (1972), 461–486.

RAFFERTY, MILTON D., *Missouri: A Geography*. Boulder, CO: Westview Press, 1983.

———, *The Ozarks: Lands and Life*. Norman: University of Oklahoma Press, 1980.

RAITZ, KARL B., RICHARD ULACK, and THOMAS LEINBACH, *Appalachia: A Regional Geography*. Boulder, CO: Westview Press, 1983.

STRAHLER, A. H., "Forests of the Fairfax Line," *Annals,* Association of American Geographers, 62 (1972), 664–684.

VERNON, PHILIP H., and OSWALD SCHMIDT, "Metropolitan Pittsburgh: Old Trends and New Directions," in *Contemporary Metropolitan America. Vol. 3: Nineteenth Century Inland Centers and Ports,* ed. John S. Adams, pp. 1–59. Cambridge, MA: Ballinger Publishing Company, 1976.

WALLACH, BRET, "The Slighted Mountains of Upper East Tennessee," *Annals of the Association of American Geographers,* 71 (September 1981), 359–373.

ZELINSKY, WILBUR, "The Pennsylvania Town: An Overdue Geographical Account," *Geographical Review,* 67 (1977), 127–147.

10
THE INLAND SOUTH

For many generations "the South" has been an evocative term that applies to a distinctive section of the continent, roughly the southeastern quarter of the United States. Although its boundaries were never easily delimited, it had certain cultural (that is, economic, demographic, political, and social) characteristics that tended to set it apart. Referred to in the past as "Cotton Belt" or "Old South," this broad section of the nation experienced notable changes in many aspects of its geography in recent decades and is now more properly conceptualized as the "New South" or "Eastern Sun Belt."

In this text, the South, like New England, is depicted more accurately as two regions than as one. There are important differences, both physical and cultural, between the interior and coastal portions. The actual boundary between the two regions—Inland South and Southeastern Coast—is determined primarily on the basis of topography, hydrography, and land use. As delineated in Figure 10–1, the Inland South Region occupies most of the broad coastal plain southeast, south, and southwest of the Appalachian–Ozark Region. The immediate littoral zone, including much of Louisiana and all of Florida, is excluded from the Inland South Region.

The northern boundary of the Inland South is demarcated by the southern margin of the two upland areas (Appalachians and Ozark–Ouachitas). Between and on either side of the uplands—in the Mississippi River lowland, in Virginia, and in Oklahoma–Texas—the extent of the Inland South is determined by variation in crop patterns. In each of these three areas there is a transition from a southern pattern featuring cotton to some other pattern: a wheat belt pattern in Oklahoma–Texas, a corn belt pattern in the Mississippi Valley, and a less distinctive general farming pattern in Virginia.

The southern and eastern boundaries of the Inland South are in a transition zone that roughly parallels the Gulf and South Atlantic coasts, some 75

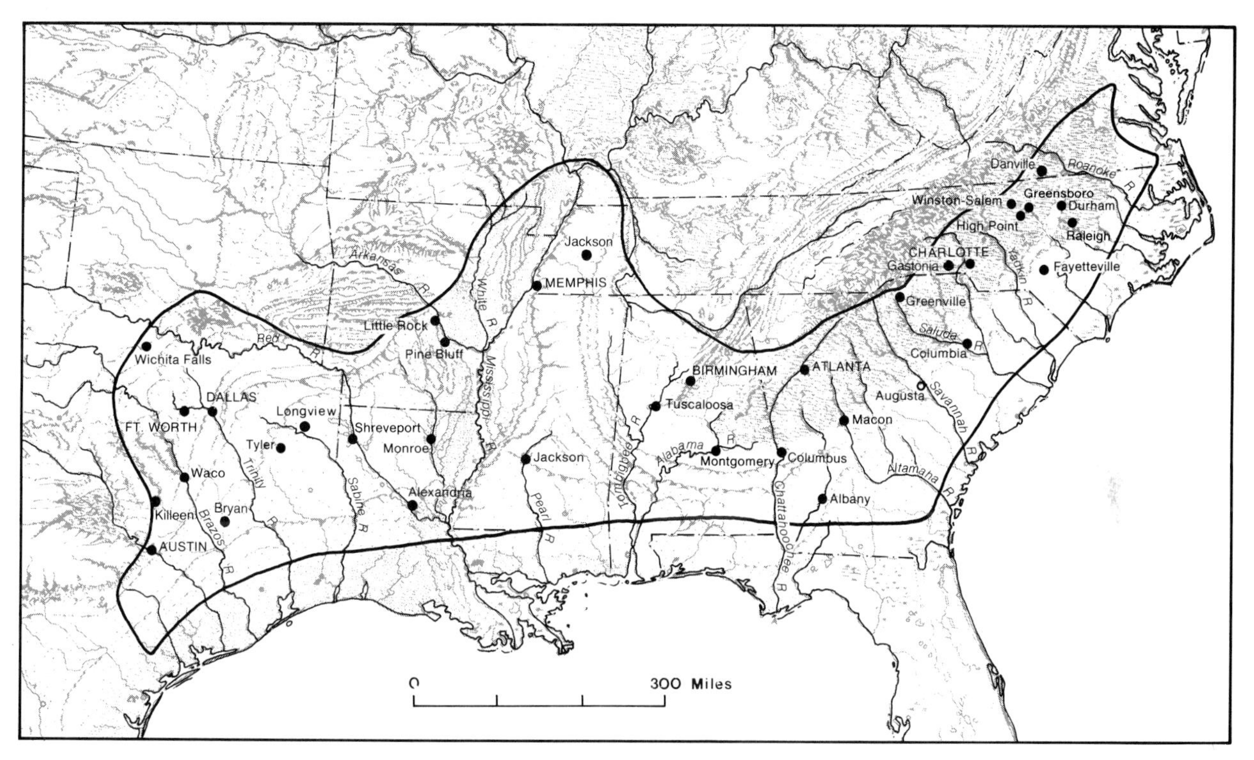

FIGURE 10-1 The Inland South Region (base map copyright A. K. Lobeck; reprinted by permission of Hammond, Inc.).

to 100 miles (120 to 160 km) inland. Dense forest, poor drainage, and spotty agriculture predominate coastward; more open woodland, better drainage, and less discontinuous agriculture are found interiorward. The littoral zone of the southeastern states is prominently oriented toward the sea; its port activities, industrial development, and seaside recreation set it apart from the Inland South.

The western boundary of the region is less clear-cut than the others. Differences in land use seem to be the most important criteria. In south central Texas the change from flat land to the short but steep slopes of the "hill country" is accompanied by a change from crop farming to pastoralism. North of the "hill country," however, there is only an indefinite transition from relatively small general farms eastward to somewhat larger specialty farms (grain sorghums, wheat, irrigated cotton, and cattle) westward.

THE PHYSICAL ENVIRONMENT

There is widespread uniformity of physical attributes in the Inland South, several of which provide conspicuous elements of unity for the region. The land surface is quite flat in most places, the drainage pattern is broadly centrifugal and functionally simple, summer and winter climatic characteristics are grossly uniform, and red and yellow soils of only moderate fertility predominate. It is, however, the prevalence of forest and woodland over almost all of the region that is the most noticeable feature of the regional landscape.

The Face of the Land

The Inland South is primarily a coastal plain region; the flat or gently undulating land surface is only occasionally interrupted by small hills or long, low ridges. The underlying rocks consist of relatively thick beds of unconsolidated sediments of Tertiary or Cretaceous age. These beds have a monoclinal dip that is gently seaward so that progressively older rocks are exposed with increasing distance from the coast.

Along the interior margin of the region in the east is an extensive section of older crystalline rocks that form the Southern Appalachian Piedmont. The topography here is somewhat more irregular than in the coastal plain, although slopes are not steep and the landscape is best described as gently rolling. Between the rocks of the Piedmont and those of the coastal plain is a continuation of the Fall Line. The Southern Piedmont extends in an open arc from central Virginia to northeastern Alabama.

That portion of the Inland South Region that is north of the Gulf of Mexico exemplifies a pattern of landform development known as a belted coastal plain. The sedimentary beds outcrop in successive belts that are arranged roughly parallel to the coast. Some of these strata are less susceptible to erosion than others and thus stand slightly above the general level of the land as long, narrow, resistant ridges called cuestas (Fig. 10–2). The overall pattern is a series of broad lowlands developed on weaker limestones and shales, which are bounded on the seaward side by the low but abrupt scarps that mark the inward edge of the resistant sandstone cuestas (Fig. 10–3). In some cases, the cuestas extend for hundreds of miles. They are more numerous in the west Gulf coastal plain of Texas and Louisiana but are slightly bolder and more conspicuous in the east Gulf coastal plain of Mississippi and Alabama.

Between the two belted zones of the coastal plain is the broad north-south alluvial plain of the Mississippi Valley. This alluvial lowland varies in width from 25 to 125 miles (40 to 200 km) and is

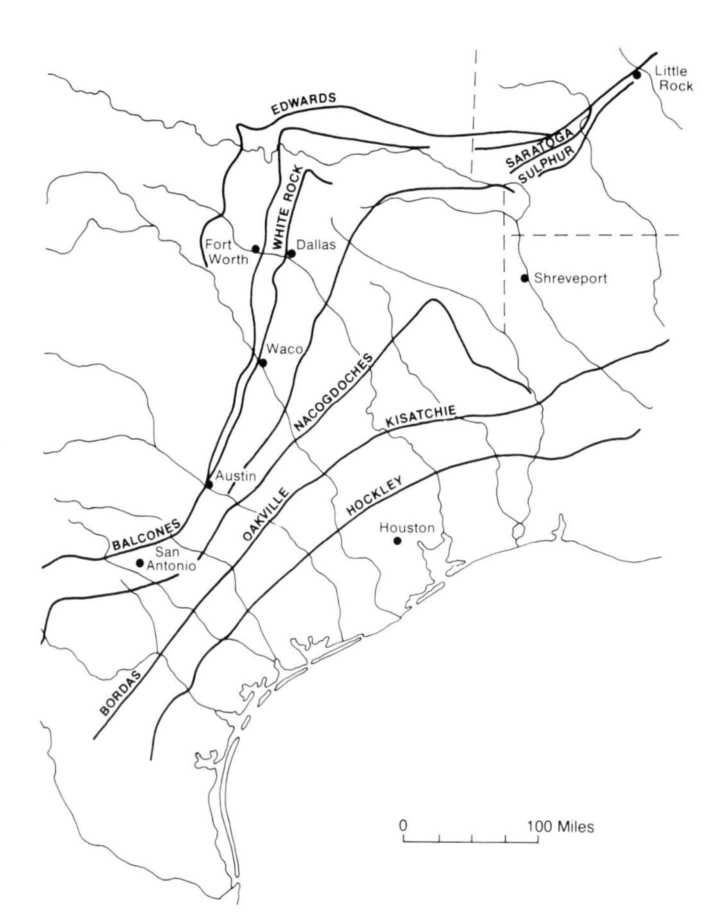

FIGURE 10-2 Principal cuestas making up the belted coastal plain of the western Gulf area. (From N. M. Fenneman, *Physiography of Eastern United States*. [New York: McGraw-Hill Book Company, 1938]. Reprinted by permission.)

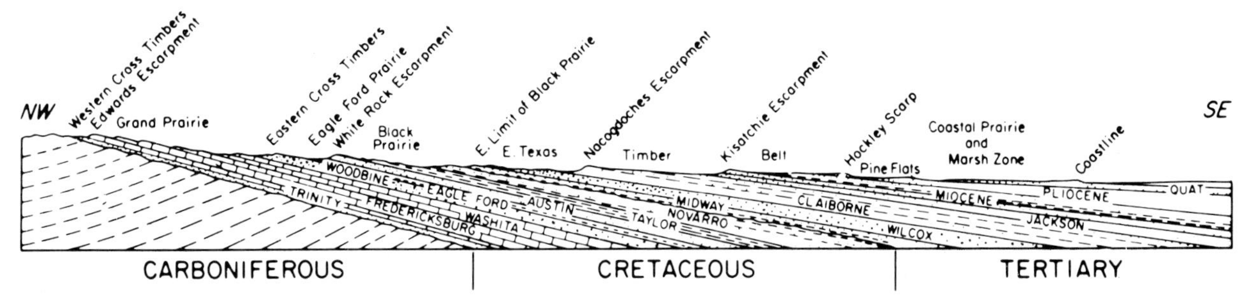

FIGURE 10-3 Generalized geologic cross section of the western Gulf Coastal Plain, extending northwest-southeast across Texas. (From Fenneman, *Physiography of Eastern United States*. Reprinted by permission.)

bounded on the eastern and western sides by prominent bluffs that rise as much as 200 feet (60 m) above the lowland. Although a few residual upland ridges interrupt the valley floor, most of the lowland is extraordinarily flat and has a southward slope averaging only about 8 inches (20 cm) per mile. With such a gentle gradient, the Mississippi River and its low tributaries meander broadly and produce many oxbow lakes and winding scars.

The general drainage pattern of the region consists of a number of long, relatively straight rivers that flow sluggishly toward the ocean from interior upland areas (Appalachians, Ozarks, Ouachitas, and high plains of West Texas). The normally dendritic pattern of their tributaries is interrupted in several places by the cuestas, resulting in right-angle bends and trellising. Very few natural lakes, other than small oxbows, are found along the rivers, but swamps, and bayous, indicative of poor drainage, are widespread.

Climate

The climate of the Inland South is sometimes described as humid subtropical, but such terminology is belied by occasional severe winter cold spells. Nevertheless, summer is clearly the dominant season in this region. It is a long period of generally high humidity that is hot by day and warm by night. Summer is also the time of maximum precipitation, with most rainfall coming in brief convective downpours.

Winter is a relatively short season, but it is punctuated by sporadic sweeps of continental polar air across the region that push the normally mild temperatures well below the freezing point. In no part of the region is snow unknown and most sections can anticipate one or more snowfalls each winter, although the length of time of snow cover is usually measured in hours rather than days. Winter is only slightly less moist than summer and most winter precipitation falls in protracted drizzles rather than brief showers. Total annual rainfall in the region varies from 55 inches (1400 mm) in the east to 20 inches (500 mm) on the western margin.

Spring and fall are relatively long transition seasons, marked by pleasant temperatures. The former is a notably windy time of the year and, in the western part of the region, is often characterized by major dust storms carried on westerly winds.

Soils

The most widespread soils of the Inland South are Ultisols. They are primarily red or yellowish-gray in color, indicating a considerable degree of leaching and the subsequent concentration of insoluble iron and aluminum as well as accumulation of a clay horizon. With careful management, they can be agriculturally productive soils; unfortunately, careful management was often lacking and some of the nation's worst examples of accelerated soil erosion can be found in the region.

In the Black Belt of Alabama–Mississippi and the Black Waxy Prairie of central Texas are extensive areas of soils derived from underlying limestone and marl. Classed as Mollisols, they are among the most naturally fertile soils anywhere; but they, too,

FIGURE 10-4 The relentless spread of the kudzu vine has overwhelmed this landscape in central Mississippi. It is kept from encroaching onto the highway by the use of chemical pesticides (TLM photo).

have been subjected to severe erosion and in many localities the black topsoil has been stripped away, revealing the lighter-colored subsoil.

The rich alluvial soils of the Mississippi Valley are also dark in color, rich in organic matter, and highly productive. They are chiefly classed as Inceptisols and Alfisols.

Natural Vegetation

The Inland South was originally a timbered region characterized by southern yellow pines on most of the interfluves and southern hardwoods (gums, oaks, cypress) in the stream valleys, with a proportion roughly half pine and half hardwood. There are 11 species in the group called southern yellow pines, of which seven are prominent in the Inland south. The most widespread pine species in the region are loblolly, shortleaf, and longleaf.[1] The principal concentrations of hardwoods are in the Mississippi River lowland and scattered widely over the northern portion of the state of Mississippi. In the natural state, three parts of the region apparently were relatively treeless. The Black Belt and Black Waxy Prairie were probably covered with prairie grasses[2] and the extreme western part of the region had a mixed cover of grassland, low open woodland, and scrubby brush.

Although not part of the natural vegetation, an introduced species has become so prominent in the vegetational landscape of the Inland South that it deserves special mention. The kudzu plant (*Peuraria lobata*) is a leguminous, climbing vine that was imported from the Orient for the dual purpose of preventing soil erosion and providing livestock forage. It flourished remarkably in its adopted environment in the southeastern states and is now considered a pest species in many localities because of its propensity to climb on and inundate anything (trees, telephone poles, barns, houses, even "slow-moving Southerners") not protected from it (Fig. 10–4). Throughout the Inland South, from Virginia to Texas, the kudzu covers an inordinate amount of surface.

PEOPLING AND PEOPLE OF THE INLAND SOUTH

The aboriginal inhabitants of the Inland South consisted of several strong and well-organized Indian tribes and a number of minor ones. Most important were the "Five Civilized Tribes"—Cherokee, Choctaw, Chickasaw, Creek, and Seminole—that originally occupied most of the area between the Mississippi River and the Atlantic Coast. West of the Mississippi, the Caddo, Osage, and Apache were important; later the Comanche moved down from the northern plains and dominated the western frontier of the region for several decades.

During the early years of European contact conflict with the Indians was relatively limited, primarily because profitable trading relationships were established between tribesmen and coastal merchants and because pressure for European settlement in the region was slow in building up. Once the Europeans began to move inland significantly, however, the days of the Indian were numbered regardless of the treaties that were frequently pro-

[1] Elbert L. Little, Jr., and William B. Critchfield, *Subdivisions of the Genus Pinus (Pines)* U.S. Department of Agriculture Miscellaneous Publication no. 1144 (Washington, D.C.: Government Printing Office, 1969).

[2] As in the case with many seemingly "natural" grasslands, there is considerable debate as to whether the prairie association of the Black Belt was indeed natural or had been induced by repeated burning by Indians. See, for example, Erhard Rostlund, "The Myth of a Natural Prairie Belt in Alabama: An Interpretation of Historical Records," *Annals*, Association of American Geographers, 47 (December 1957), 392–411.

A CLOSER LOOK *The Fire Ant—An Exotic Scourge*

The conspicuous presence of an exotic plant, the kudzu, is more than matched by the insidious invasion of an exotic animal, the South American fire ant, which has become a scourge throughout the southeastern United States over the last half century. A native of the floodplain of the Paraguay River, the "invincible" fire ant (*Solenopsis invicta*) was introduced inadvertently via cargo ship to the port of Mobile, Alabama, in the 1930s. From that foothold it has spread to the farthest reaches of the Southeast, presumably being halted only by cold on the north and dryness on the west.

These tiny (one-eighth inch or 3.5 mm long) reddish-brown or black creatures congregate in colonies that have up to 200 queens and half a million worker ants. Each colony lives in a largely subterranean mound, and an acre of infested land may hold up to 400 mounds.

The virtues of fire ants have yet to be discovered, but the problems they cause are legion. Their durable mounds damage plows and other farm machinery. They nibble on the insulation of utility cables, causing power outages and other breakdowns in services. But their most daunting impact is the result of their voraciously carnivorous appetites. They possess formidable mandibles and an acutely venomous sting. They prey particularly on other invertebrates, but will attack larger creatures if they can get to them. Reptiles, rodents, and birds are particularly susceptible, but such larger animals as raccoons, deer, goats, and cattle are recurrent victims. Humans, too, are attacked with frequency. Only a handful of human deaths has been ascribed to fire-ant attacks, but thousands of people have suffered from their stings, and in areas of major infestation humans modify their behavior to avoid fire-ant locales.

Chemical warfare has been waged against the invading fire ants for more than three decades, at a cost of hundreds of millions of dollars. Such deadly chemicals as dieldrin, heptachlor, chlordane, and mirex are effective pesticides, but they wipe out many components of the ecosystem in addition to the fire ants. When these ecologically unsound chemicals are prohibited, the fire ants rebound quickly.

The outlook for comprehensive fire-ant control is dismal. Decades of research have produced no feasible answers. It is likely that we will just have to learn to live with them as we have done with such unpleasant but more tolerable pests as cockroaches and mosquitos. The discouraging prognosis is summed up in the words of a Texas entomologist: "Basic research may eventually give us some answers. We may come pretty close to at least understanding why we can't control them."

TLM

mulgated and just as frequently dishonored by the colonial and federal governments.

Indian wars became commonplace in the early 1800s and before long most of the recalcitrant tribes had either been wiped out or shifted to new homes west of the Mississippi. An acerbic contemporary critic described the process in which "the most grasping nation on the globe" would take the Indians "by the hand in brotherly fashion and lead them away to die far from the land of their fathers . . . with wonderful ease, quietly, legally, and philanthropically. . . . It is impossible to destroy men with more respect for the laws of humanity."[3]

European settlement in the eastern part of the Inland South was little more than a tiny trickle until the eighteenth century and the occupance of most of the region dates from the early nineteenth century.

[3] Alexis de Tocqueville, *Democracy in America*, trans. George Lawrence (New York: Harper & Row, 1966), p. 312.

Five generalized tides of settlement can be discerned:

1. From the early colonial coastal settlements of Virginia and Maryland freemen moved west and southwest, joined by settlers coming directly from Europe but funneled through the Chesapeake Bay ports; this stream was augmented by a flow up the Shenandoah Valley into eastern Tennessee and beyond.
2. From the South Atlantic coastal ports, particularly Charleston and Savannah, more European settlers were channeled into the interior.
3. A third route of settler flow was southward from the Midwest and the upper South (Kentucky and Tennessee) and, in part, down the Mississippi Valley.

4. A fourth stream moved northward through the Gulf Coast ports of New Orleans and Mobile.
5. There was also an early movement of settlers of Spanish ancestry northeastward from Mexico into central and eastern Texas; this flow was circumscribed first by contact with the French in Louisiana and later by the persistent movement of Anglos into Texas from the East and Northeast.

Except for those coming from Mexico, nearly all settlers of the region were Northwest European in origin. It is probable that most initial settlers were American-born, but many came directly from Europe and certainly the great majority were of recent European ancestry. British people (English, Welsh, Scots, and Scotch-Irish) were the most numerous, but Germans, French, Swiss, and Irish were also significant.

Cotton as a Settlement Catalyst

By the 1790s most of the Carolina portion of the Inland South and part of eastern Georgia had been settled and there were other small settled districts in the lower Mississippi Valley and central Texas. In 1793 young Eli Whitney, visiting on a Georgia plantation near Savannah, developed a vastly improved cotton gin. It was the first practical machine for separating "green seed" cotton lint from seed. With remarkable suddenness, cotton was adopted as the commercial crop of settlement, for it could be produced on small farm or large plantation, provided that cheap labor was available. Thus cotton production boomed, settlement spread, and slave importation expanded.

By 1820 eastern and central Georgia were fully occupied, the good lands of central and western Alabama were settled, much of the Mississippi Valley was being farmed, and settlements in eastern Texas were expanding. Within another three decades nearly all the Inland South was under settlement except for much of the alluvial Mississippi valley. Plantations, with large slaveholdings, were widespread. In addition, many small farmers were engaged in limited commercial enterprises with only a few slaves.

Yeoman farmers, who operated without labor assistance, were commonplace, especially in areas of less productive soil or steeper slope. A few freed slaves were farming in Virginia and North Carolina. The total slave population amounted to more than 3 million in 1850 compared to a white population of twice that number.

The Contemporary Population

The present population of the Inland South is more homogeneous than that of most regions. It has two principal components: some 27 million whites, who are mainly of Anglo-Saxon ancestry and Protestant religious affiliation, and 11 million blacks, who are also mostly Protestant (Fig. 10–5). The largest denominational affiliations in the region are Baptist and Methodist, although various evangelical Protestant groups are strongly entrenched and are experiencing a rapid growth in membership.

Compared to the total population of the region, adherents of Roman Catholicism are relatively few in number except among the large Hispano minorities of several Texas cities. Jews constitute an ex-

FIGURE 10-5 The Inland South is sometimes referred to as the "Bible Belt" because of numerous and enthusiastic adherents of various forms of Protestantism (TLM photo).

tremely small proportion of the population but have a disproportionately large impact because of their prominence in economic and civic affairs.

In terms of population mobility in the region, there are three recognizable trends in partial opposition to one another:

1. A large share of the people who live in the Inland South were born there. For example, more than 80 percent of the population of Mississippi, Alabama, Georgia, and South Carolina are still living in the state of their birth in contrast to the national average of less than 70 percent.
2. The urban areas, particularly the larger ones, are significant foci of in-migration, particularly of people from outside the region. Newly arrived whites tend to settle in the suburbs whereas black newcomers mainly go to the central cities.
3. The long-run high rate of net out-migration of blacks from the Inland South has now been reversed. Today there is a significant net inflow of blacks to the region from the Northeast and north-central states (but not from the West). Even so, nearly 90 percent of all migrants into or out of the Inland South are nonblacks and the net impact of migration on the racial composition of the region's population is to make it whiter rather than blacker.

THE CHANGING IMAGE OF THE INLAND SOUTH

Perhaps in no other part of Anglo-America is the sense of regional identity so pervasive as in the Inland South. The southern way of life is recognized by Southerner and non-Southerner alike as being regionally distinctive, partially in tangible ways and partially as a state of mind. For many people, this regional identity arouses strong emotions, ranging from reverence to abhorrence. To some, the Southern way of life is "genteel"; to others, "decadent." The strongly entrenched and nondiversified economic base of the southern past is part of the image, but more a part of contemporary consciousness and feelings about social conditions, particularly in regard to the black minority of the population.

The origin of the southern way of life is complex and beyond the scope of this presentation. Briefly, a generation or two ago there was a regional character to the Inland South that was simple of generalization even if imperfect of image: the region was economically depressed and socially divided. Agriculture, the traditional basis of the southern economy, was straitjacketed by corn for subsistence and by cotton and tobacco for cash. The unholy duo of tenancy and soil erosion had a stranglehold on rural life. Industry was present, but it was undiversified and paid low wages. Per capita income was well below the national average and poverty was relatively widespread. Average educational attainment was low and illiteracy was a problem. Practically speaking, only one political party was extant. Class distinctions were strong and the black was universally accorded—by white and black alike—the bottom rung of the social and economic ladder. There was more to the image—veneration of womanhood, religious piety, economic pluck, and patriotic valor; only inadequacies have been emphasized. The point is that the Inland South displayed certain significant disadvantages and faced certain significant handicaps.

What of its regional character today? Certainly many of the "old" elements of regional distinctiveness still prevail. Regional speech characteristics, for example, show no significant change; the southern drawl is as distinctive today as it ever was despite population diversification and the influence of mass media. Dietary preferences have been modified, particularly in the large cities, but there is still a pronounced regional "flavor" to both choice of food and method of cooking; corn bread, hominy grits, hush puppies, pot likker, black-eyed peas, turnip greens, and other delicacies are distinctly southern. Protestant religious fundamentalism continues to be strong despite weakening trends in the cities, and the term "Bible Belt" is still quite apt. And "southern hospitality" is more than a cliché; there is a regional claim to personal warmth and friendliness that is readily discernible.

A CLOSER LOOK Redistribution of the Black Population

The redistribution of the black population within the United States has been paralleled by a local redistribution within the regions of the plantation South where blacks traditionally have lived. In 1940, 6,191,000 of the 9,905,000 blacks in the South were rural, and 71 percent of the rural black population resided on farms. Despite out-migration, several million blacks still reside in the old plantation regions, and some 3 million live in rural areas. Less than 5 percent of the rural black population, however, now resides on farms.

The Yazoo Delta—the floodplain of the Mississippi River and its tributaries in northwestern Mississippi—illustrates local redistribution of the black population. In 1940 the Yazoo Delta was one of the South's most important cotton plantation areas. The region's 313,200 blacks comprised 73 percent of the population. Eighty-one percent of the blacks were rural farm (see Box Table 1). Only 8 percent were rural nonfarm, and only 11 percent were urban. Almost all of the blacks were members of tenant families. Plantations were subdivided into 10- to 40-acre tenant farms with a house located on each. The rural settlement pattern was one in which blacks were dispersed across the landscape.

The mechanization and reorganization of tenant plantations as "neoplantations" initiated a redistribution of the Yazoo Delta's black population. Today, the majority of the 178,300 blacks in the region do not live on plantations or work in agriculture. Only 3 percent of them now reside on farms (Box Table 1). The urban population, namely persons living in incorporated places of 2,500 or more, is 46 percent. The largest population category, however, is rural nonfarm.

Nucleation is the dominant local trend in black population redistribution. In addition to urban concen-

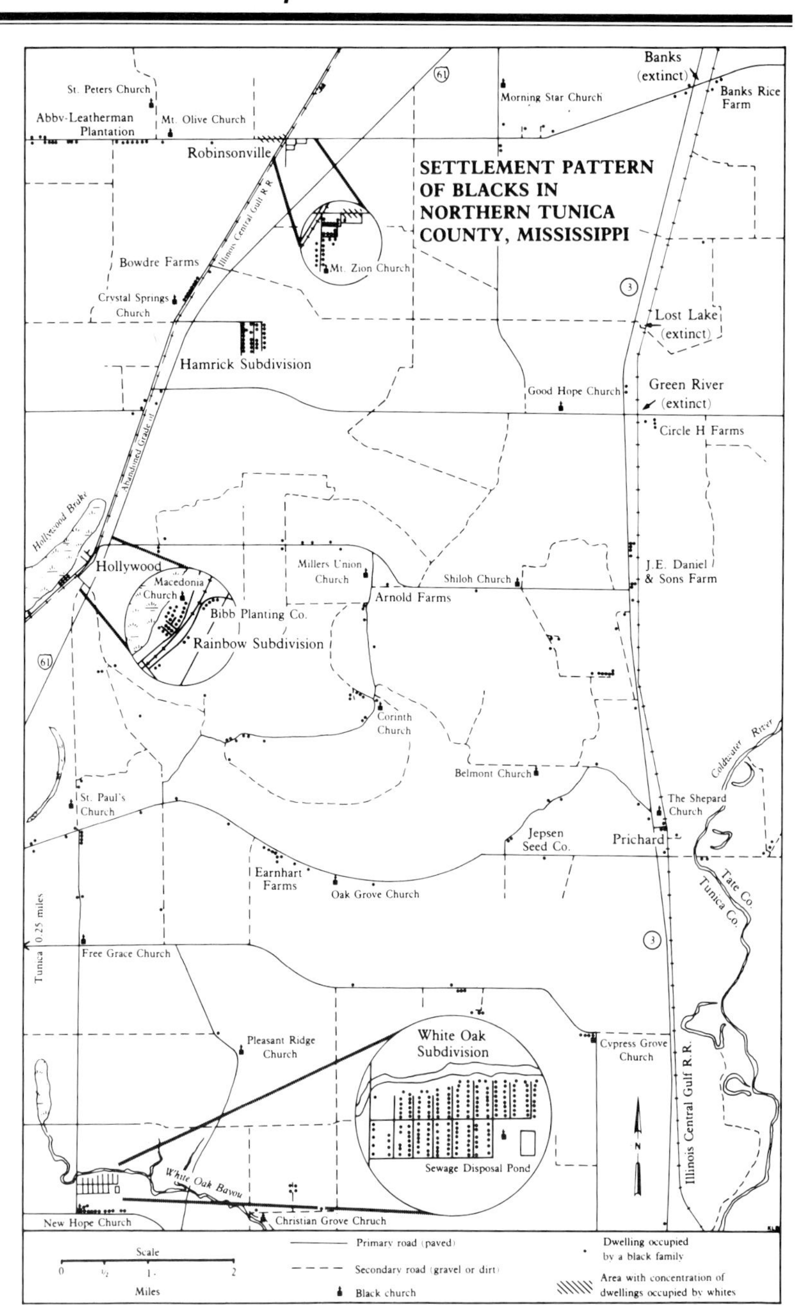

FIGURE 10-a

BOX TABLE 1

Location of the Yazoo Delta's black population, 1940 and 1980

	1940		*1980*	
	No.	%	*No.*	%
Urban	34,958	11.2	82,094	46.0
Rural Farm	253,014	80.8	5,473	3.1
Rural Nonfarm	25,269	8.0	90,729	50.9
Total	313,241	100.00	178,296	100.0

Source: Charles S. Aiken, "Race as a Factor in Municipal Underbounding," *Annals of the Association of American Geographers,* 77 (December 1987), 566.

tration, the rural farm and nonfarm populations also have become nucleated. An area of Tunica County, Mississippi, which is located in the northern portion of the Yazoo Delta, illustrates the contemporary settlement pattern of blacks (Figure 10–a). With the reorganization of tenant plantations as neoplantations, the dispersed settlement pattern of the tenant farmer era was replaced by a nucleated one. The remaining black farm population is concentrated in clusters of houses on plantations. As plantations were mechanized, the houses of tenant farmers scattered across fields were razed. New dwellings for salaried machinery operators were built in rows near plantation headquarters. The neoplantation settlement pattern in Tunica County is illustrated by Abby-Leatherman Plantation, Bowdre Farms, and the Bibb Planting Company.

The Yazoo Delta's large rural nonfarm population is concentrated in new hamlets in the countryside, old unincorporated plantation communities, and the fringes of incorporated municipalities. Two types of hamlets have emerged in the countryside—unplanned assemblages of dwellings and planned subdivisions. Unplanned hamlets often are comprised of extended families. They evolve on farms owned by blacks through the creation of house sites for relatives. Large unplanned hamlets usually have developed to a second stage in which lots are sold to nonrelatives. Some unplanned hamlets consist of new well-kept houses, but most are a hodgepodge of shacks, inexpensive shell houses, mobile homes, and brick-veneered dwellings. In northern Tunica County, a small unplanned hamlet is located at Corinth Church.

Planned hamlets are ones in which lots are plotted and the development is officially registered as a subdivision. Usually developers are local entrepreneurs who perceive market potential produced by the growth of the rural nonfarm black population. Unless building standards are enforced, a subdivision evolves into a mishmash of dwellings that often is almost indistinguishable from an unplanned hamlet (Figure 10–b).

A substantial proportion of the new federally sponsored housing in the Yazoo Delta has been built in planned subdivisions, including ones located in the countryside. Beginning in the 1960s with the War on Poverty, an increase occurred in federal funding to construct new housing for nonmetropolitan low-income persons.

FIGURE 10-b A portion of a rural subdivision occupied by blacks in Tunica County, Mississippi (photo by Charles Aiken).

(continued)

The largest rural hamlet of blacks in Tunica County is White Oak, a planned subdivision sponsored by the Farmers Home Administration, an agency of the U.S. Department of Agriculture. White Oak, which was built during the 1970s, is located 3 miles from the town of Tunica. More than 900 individuals, almost one-seventh of Tunica County's black population, live in this 180-house development.

Old unincorporated plantation communities are another type of place in which the Yazoo Delta's black rural nonfarm population is concentrated. Robinsonville, Hollywood, and Prichard are examples of such communities, which lost their traditional retail and administrative functions. Prichard is now virtually extinct, and Robinsonville and Hollywood would be extinct had not the black population surged with the demise of tenant farming on surrounding plantations. The creation of Rainbow subdivision for blacks by a local white entrepreneur reestablished and redefined Hollywood.

Almost 20 percent of the blacks in the Yazoo Delta live within a mile of the political limits of municipalities. This large black fringe population can be explained primarily on the basis of the white-controlled municipal governments' refusal to annex black residential areas. During the last decade of the nineteenth century and first decade of the twentieth century, southern states disenfranchised blacks. In 1960 only 31 percent of the adult blacks in the seven states that comprise the plantation South were registered to vote. In Mississippi only 7 percent were registered. Ever since Congress passed the 1965 federal Voting Rights Act, southern blacks have come to play a greater role in national, regional, and local politics.

Locally, annexation of black residential areas dilutes white voting strength in municipalities. Not all municipalities in the Yazoo Delta have refused to annex black residential areas, but enough of them have that a large proportion of the region's black population that moved from farms to municipalities have not been incorporated into them. In 1980 the town of Tunica had a population of 1361, 30 percent of whom were blacks. The fringe population within a mile of the political limits was 2474, 82 percent of whom were blacks.

Professor Charles S. Aiken
University of Tennessee
Knoxville

Race relations in the Inland South, as in most of the nation, are far from tranquil. The servile Negro of yesterday is the proud black of today who demands equality and seeks redress. Racist feelings of the white majority have considerably softened and overt discrimination is either eliminated or notably reduced. Public as well as most private facilities are desegregated, and blacks are accorded the rights and responsibilities of full citizenship in most places and activities. Black legislators have been elected in Georgia, black mayors in Mississippi, and black sheriffs in Alabama. Perhaps more important, state and local administrations in such key states as North Carolina and such major cities as Atlanta, Birmingham, and Dallas have become more cognizant of and more responsive to the needs of their black citizens.

The economy of the Inland South has undergone expansion, diversification, and substantial strengthening. Industrial growth is notable and widespread. In every state of the region, value added by manufacturing far exceeds the value of agricultural production. As recently as 1947 Atlanta was the only city in the region that ranked among the 50 leading industrial centers of the nation, whereas in the early 1990s the region had four urban areas (Dallas, Atlanta, Charlotte, and Greensboro–Winston-Salem) among the top 20 industrial cities. The tertiary and quaternary components of the economy also expanded and diversified, with greatly increased employment opportunities in trade, services, finance, construction, transportation, and government.

Perhaps, however, the changing character of the South is best exemplified by the agricultural scene. The cotton-and-corn, plantation-oriented duoculture (never merely a cotton monoculture) that characterized primary production in the region for several generations gave way to a diversification of significant proportions. This diversification is most conspicuously demonstrated by the dethroning of "King Cotton."

THE RISE AND DECLINE OF COTTON IN THE INLAND SOUTH

When the American Southeast was first settled by Europeans, the colonists felt that whites could do little sustained physical labor in its tropical summers. Consequently, Charleston and Savannah became great slave-importing cities. The first crops raised with the help of slave labor were rice, indigo, sugarcane, tobacco, and some cotton. In 1786 long-fibered Sea Island cotton was introduced and successfully grown along the coastal lowlands.

In 1793 Whitney's invention of the cotton gin revolutionized the cotton industry. European demand expanded, settlement spread westward, cotton acreage enlarged, and many more slaves were brought in to work the fields. By the middle of the nineteenth century the American South was supplying 80 percent of the world's export cotton.

The Civil War and emancipation of the slaves were big setbacks for the cotton industry, as was the spread of the boll weevil across the South in the early 1900s. In both cases, however, the industry rebounded and grew to new heights. By the late 1920s the United States was growing well over half of all the world's cotton and exporting an all-time high of more than 11 million bales annually.

With the onset of the economic depression of the 1930s, however, cotton production in the Inland South went into a steady decline. Cotton acreage diminished from a peak of about 17 million acres to a nadir of about 3 million. By the mid-1980s the old Cotton Belt was growing only about 40 percent of all U.S. cotton; most production had shifted to irrigated lands farther west, primarily in west Texas, Arizona, and California (Fig. 10–6). Texas, for decades the undisputed leader in cotton output, was surpassed by California in 1982. Moreover, four-fifths of the Texas production is west of the borders of the Inland South Region.

In the late 1980s, however, there was a cotton renaissance in the region, occasioned by a variety of factors, but particularly by a dramatic rise in demand for cotton products and a federal farm program that bolstered market prices. By the early

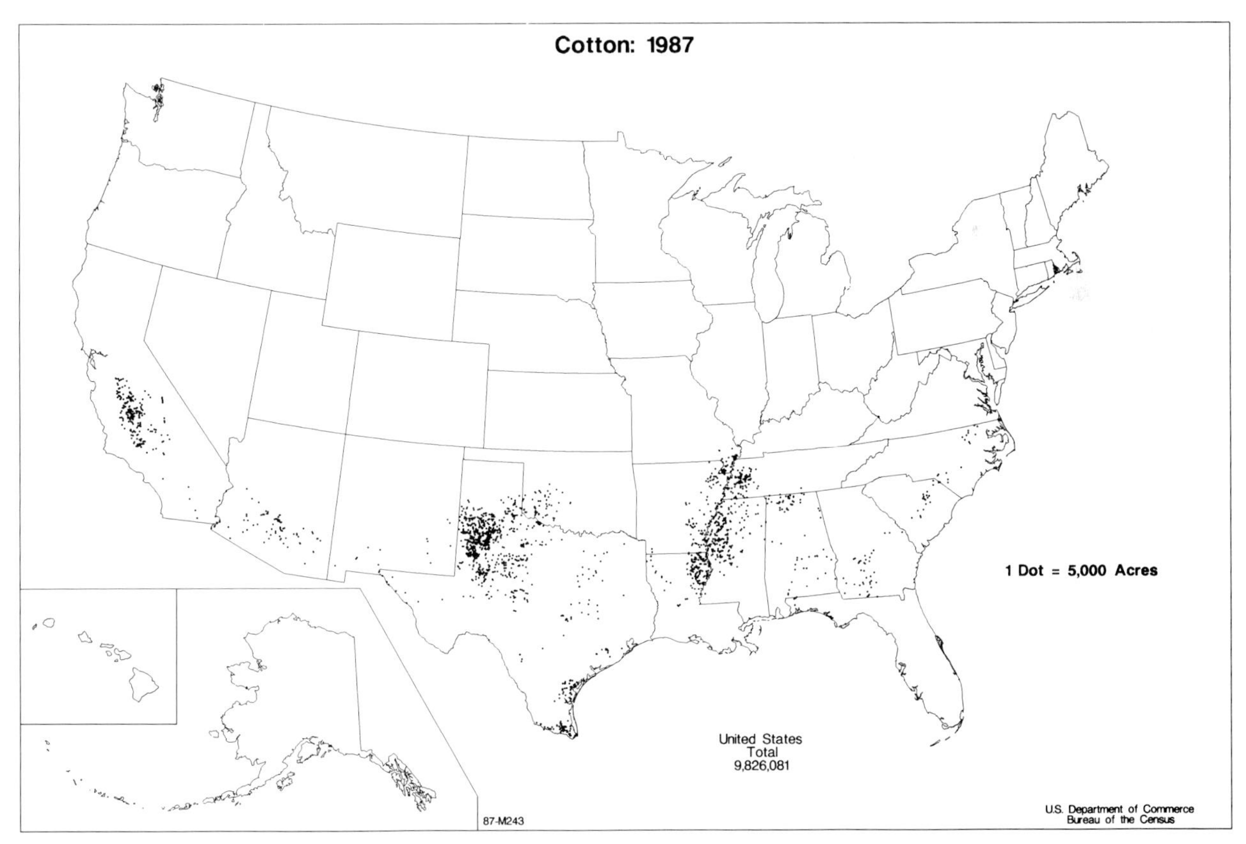

FIGURE 10-6 Distribution of cotton production in the United States. The only part of the old Cotton Belt that is still a major producing area is the middle Mississippi Valley.

1990s Texas was again the leading producer, with one-third of the national crop (though mostly in west Texas). Mississippi ranks third (after California), followed by Arizona and then the southeastern states of Louisiana, Arkansas, Tennessee, and Alabama.

FARMING IN THE INLAND SOUTH: PRODUCTIVE DIVERSITY

For several decades social scientists often referred to the "changing South," listing tendencies and prognosticating trends in economic and social matters that will make a "New South" out of the "Old South." Many predicted changes have long since materialized and in no field have they been more striking than in agriculture.

Crop rotation, improved soil management and fertilization, supplemental irrigation, variation in field crops, and increased mechanization are all part of the scheme, but the most pronounced aspect is the shift toward mixed farming, with pastures for cattle as the dominant feature of the landscape. Cropped acreage, especially of cotton, has decreased and many eroded fields have been restored to useful production by the planting of forage and pasture.

With the decline of cotton as undisputed king, many attendant evils have also diminished. Soil erosion and fertility loss decreased. The established and practical but often corrosive pattern of absentee ownership and farm tenancy has been broken—presumably forever.

If southern agriculture has diversified and cotton acreage has declined, what replaced cotton and what are the elements of diversification? In terms of gross acreage, the answer is pasture and woodland. A great deal of former cropland, especially cotton land, has been returned to a grass or tree cover and is now used mainly for grazing cattle and, to a lesser extent, as a source of pulpwood. The major component of agricultural diversification has been a shifting emphasis from crops to livestock, particularly beef and dairy cattle and, locally very important, poultry. The most notable cropping diversification has been the expanding cultivation of soybeans.

Soybeans

The soybean became significant in the Mississippi Valley during the early 1930s when its introduction was encouraged by the Soil Conservation Service as a legume soil-building crop. Yields per acre were low and the soybean was not in any way considered a rival to cotton. Cotton farmers, attracted by the adaptability of soybeans to mechanized farming and by their value as a cash crop, began to increase acreage when faced with a shortage of labor during World War II. Continued research to improve the quality and productivity of the bean and to extend its use resulted in such an increase in cultivation that the middle Mississippi Valley is now the second largest soybean-producing area in the United States (after the Corn Belt), although production declined in the 1980s because of increased competition from lower-cost midwestern producers (Fig. 10–7).

Corn

The principal system of farming in the region was long based on two crops, cotton and corn, and virtually every farmer grew both. Since 1938, however, total corn acreage has decreased. This decrease has been greatest in the western areas where corn was largely replaced by grain sorghums and small grains. During the same period the yield per acre materially increased and further increase on an even more reduced acreage seems to be the trend for the future.

Beef Cattle

Since the early 1950s, livestock yielded more income than cotton, or any other crop, to farmers of the Inland South, with beef cattle in the forefront (Fig. 10–8). In association with cattle raising, the acreage of grass and legume meadows for pasturage was greatly increased; however, a considerable amount of grazing is carried on in the fairly open forest that characterizes more than half the region. It is estimated that five-sixths of the forested area is grazed. The most widespread beef breed is the Hereford, but Angus, Brahma, and Santa Gertrudis are also popular.

Many cattle enterprises of the Inland South are operated more or less as an avocation by urban business executives who derive satisfaction and a tax shelter from a ranch of their own. The quality of

FIGURE 10-7 A field of soybeans near Columbia, South Carolina (TLM photo).

their livestock is often high and their willingness to experiment and innovate makes them a powerful force for upgrading the entire industry even though their ranch may be more for outdoor recreation than for livestock operation.

FIGURE 10-8 The raising of beef cattle is a major enterprise throughout the Inland South. This pure-bred Brahma herd is near San Augustine, Texas (TLM photo).

Poultry

In the past, poultry raising, particularly chickens, was of importance for home consumption and local market only; today there are certain areas in the South where the raising of poultry for commercial consumption dominates farm activities. There is some emphasis on producing eggs, frying chickens, and turkeys, but the principal product is broilers, which constitute more than two-thirds of all poultry consumed in the United States.

Much of this industry is handled under a contractual agreement whereby a feed merchant or meatpacker provides chicks, feed, vitamins, medicines, scientific counsel, and a market, and the farmer supplies only housing and labor. The contractee is benefited by needing little capital and having an assured market; the contractor, by having an assured supply of dependable quality; and the consuming public, by lower poultry prices. Skyrocketing poultry consumption resulted, but an insecure national market, labor disputes at packing plants, and cutthroat competition combined to inhibit stability in the industry. Nevertheless, major centers of commercial broiler raising are found in the Southern Piedmont of Georgia and the Carolinas, northern Alabama, central Mississippi, southwestern Arkansas, and easternmost Texas.

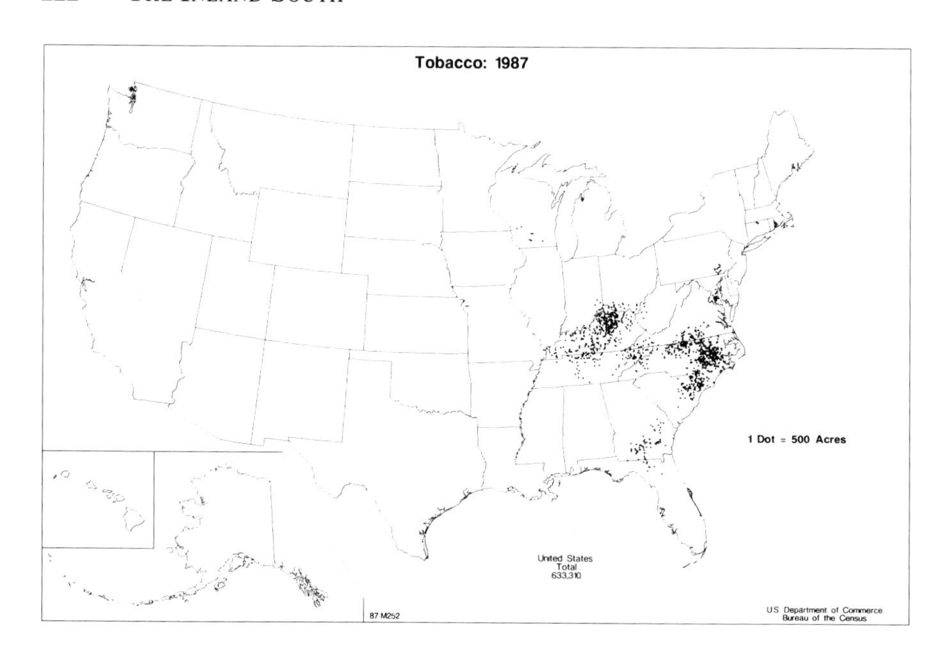

FIGURE 10-9 The leading area of tobacco growing in the United States is on the Piedmont and Inner Coastal Plain of the Carolinas and Virginia.

Tobacco

An outstanding specialty crop in the northeastern portion of the region is tobacco. More than half the nation's output comes from eastern and central North Carolina and adjacent parts of southern Virginia and northern South Carolina (Fig. 10–9). Tobacco was introduced here in the early days of settlement by in-migrants from the tobacco section of eastern Maryland, but it did not become a popular crop until the time of the Civil War. Since then it has been the most important cash crop in the area and has attracted to the North Carolina Piedmont the greatest concentration of cigarette factories to be found anywhere in the world.

Of the seven classes of tobacco cultivated in the United States, this area concentrates almost totally on one that is synonymously called "flue-

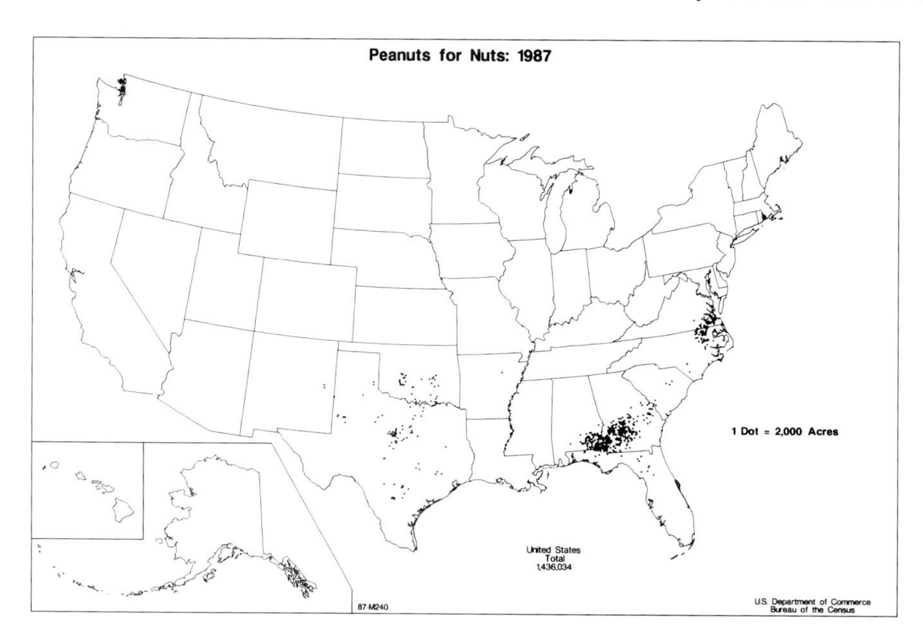

FIGURE 10-10 Peanuts are grown widely in the Inland South Region, with the major concentration in southwestern Georgia and southeastern Alabama.

cured," "Virginia," and "bright." Most tobacco is grown on small- to medium-sized farms, which are distinguished by the presence of an array of shiny new metal bulk-curing barns that have replaced thousands of antiquated traditional flue-curing barns in the last few years. Tobacco harvesting extends over a period of four to eight weeks during which the pickers comb the field about once a week, selecting ripe leaves from the bottom of the stalks. Mechanization has become prominent in both harvesting and curing, with the result that the previous high density farm population has diminished significantly since about 1970.

Other Crops

Various other crops are grown in quantity in the region. Nearly one-half of the national production of *rice* comes from the bottomlands of eastern Arkansas and adjacent parts of Mississippi. More than half the national output of *peanuts* is in southwestern Georgia and southeastern Alabama and much of the rest is from northeastern North Carolina and southeastern Virginia and from the scattered localities in central Texas and Oklahoma (Fig. 10–10). Three-fourths of the national output of *pecans* comes from this region, especially Georgia. South Carolina and Georgia rank second and third among the states producing *peaches*. Other vegetable and fruit crops are widespread.

FOREST PRODUCTS

The Inland South Region lies in the southern part of the eastern forest, which once was one of the greatest stretches of timber in the world. Originally almost every portion of the region was forested. Demand for cotton land coupled with reckless burning reduced the supply of standing timber long before lumbering began on a commercial basis. The early days of forest exploitation were characterized, as in most parts of the nation, by a thoughtless and wasteful "cut out and get out" approach. But the long growing season and heavy rainfall of the region produce rapid tree growth, and forest industries have become well established in every state. A forest planted with pine seedlings can be thinned for pulpwood after only 12 to 15 years, and a mature crop of sawtimber can be harvested within four decades.

Over the last several years there has been continually increasing logging in the region, with the result that timber, especially softwood timber, is now being harvested faster than it is being grown. Moreover, a growing deficiency in pine regeneration is resulting in substantial replacement of pines by hardwoods. These trends produce concerns about long-range productivity, which have partly been allayed by the development in the late 1980s of the federal Conservation Reserve Program, in which landowners receive an annual payment for idling farmland on a long-term basis. Much of this idled land is being planted to trees.

Softwood Lumbering

Cutting southern yellow pines for lumber was the first forest industry to be widespread in the region. Small operations, feeding small sawmills, have been characteristic. Prior to the 1960s there had been a long-term decline in regional lumber output. The trend, however, was reversed in that decade and production has increased, partly because of tree farming and partly because of increased scale of operations and the establishment of larger sawmills. The region yields about one-third of the nation's softwood lumber and the same proportion of its plywood.

Hardwood Industries

Hardwood logging is much more localized. The greatest concentration is in the Mississippi River lowland, where cutting is a seasonal operation because of the boggy conditions caused by winter rains. Hardwood output is also notable in the Carolina Piedmont, where it is carried on year-round. Southern hardwoods—oak, hickory, cypress, gums—are in great demand for furniture, veneers, and shingles.

Pulp and Paper Industry

This branch of forest industries is the leading source of income in the region, having grown rapidly in the last few decades. The wood from southern pine has

FIGURE 10-11 A typical pulp log woodyard in southwestern Mississippi (TLM photo).

A CLOSER LOOK Sunbelt vs. Snowbelt

The term "Sunbelt" was coined by journalists in the late 1970s to refer to the Florida–Texas–California axis of mild-winter states within which there had been, and presumably would continue to be, population and economic growth considerably above the norms for the United States as a whole, thus implying a hitherto unprecedented concentration of political power as well. The appropriateness of this concept and the catchiness of its appellation were affirmed by the rapidity with which both idea and name were adopted by scholars and laypersons alike.

No precise boundaries have been attached to the Sunbelt, although most references seem to accept the parallel of 37° north latitude (roughly the northern border of several states from North Carolina to Arizona) as approximately its northern limit. Nor does the Sunbelt have clear-cut attributes. Despite the name, its principal environmental earmark is not the presence of sunshine but the relative absence of cold weather. Concomitant characteristics include significant population growth and economic well-being.

The Sunbelt concept goes hand-in-hand with its antithesis—a Snowbelt or Frostbelt—which vaguely refers to the northeastern quarter of the United States. This was the "core" of the nation and its economic heartland throughout history. Although there have been precursors of significant economic/population/political reorientation for years, it was not until the 1970s that the trends became sufficiently persistent and vigorous to induce the widespread acceptance of the Sunbelt idea.

There is no question that the established patterns have experienced significant change. The Sunbelt has been gaining, both absolutely and relatively, over the Snowbelt in terms of both population growth and economic prosperity. The causes of the changes are complex and variable, but they include at least four factors: (1) lower production costs in the Sunbelt; (2) favorable developments in transportation, communication, and industrial technology in the Sunbelt; (3) a decrease in the significance of many of the cost efficiencies once enjoyed in the Snowbelt; and (4) the increasing importance of amenity factors (including mild winters) as an attraction for choosing a place of residence.

Thus the rise of the Sunbelt is very real. It should be cautioned, however, that the Sunbelt has never in actuality been a monolithic area of growth and prosperity; it had (and continues to have) significant areas of economic stress and population decline. Moreoever, economic advantages are relative things, and in the late 1980s there emerged prominent areas of economic progress in the Snowbelt (witness the resurgence of the Boston area) whereas some of the prodigy areas of the Sunbelt have suffered severely (as the "oil patch" of Texas/Louisiana/Oklahoma). Still, the Sunbelt concept is a notion whose time has come, and it is now well established in public perception.

TLM

long fibers, which are preferred for making heavy duty paper and paperboard. The region yields nearly two-thirds of the total pulpwood cut; Georgia is the national leader in production of pulpwood, pulp, and paper, with Alabama a close second in pulpwood output.

Throughout the Inland South, a conspicuous feature of the landscape is a clearing along a railway or highway that serves as a depot (called a *woodyard*) for stacking pulpwood cut in nearby areas (Fig. 10–11). From the woodyard the pulpwood is shifted, usually by rail, to a pulp mill. Many mills, especially the larger ones, are located outside the Inland South in the Southeastern Coast Region, often at a seaport.

MINERALS AND MINING

Apart from hydrocarbons and bauxite, the region is poor in economic mineral resources. Moreover, the abrupt decline of the national petroleum industry in the 1980s impacted the western part of the Inland South cataclysmically.

Petroleum

The great oil boom of the region dates from 1930 when the East Texas Field, the most prolific in North American history, was brought in. At maximum extent, the field was 42 miles (67 km) wide and 9 miles (14 km) long. Within an area of about 300 square miles (777 km^2) were more than 27,000 producing wells. This field yielded more than 5 percent of all oil ever produced in the United States, nearly three times as much as any other field.

The western states of the Inland South area all among the national leaders in petroleum production: Texas, first; Louisiana, third; Oklahoma, fifth. Other than the East Texas Field, there are three principal areas of production: north-central Texas and south-central Oklahoma, centering on Wichita Falls and Ardmore; the northwestern corner of Louisiana and the southwestern corner of Arkansas, with adjacent portions of northeastern Texas, centering on Shreveport and El Dorado; and northeastern Louisiana and adjacent portions of Mississippi.

The drastic decline in demand for domestic petroleum during the 1980s was a severe blow to the economy of the oil-producing states. Unemployment rates soared, and Louisiana even experienced a population decrease, unprecedented in the Sunbelt states. As of this writing, recovery is very slow and erratic.

Natural Gas

Production and consumption of natural gas increased dramatically in Anglo-America over the last three decades. Proved reserves are now at critically low levels, however, and production trends continue downward despite accelerated exploration. Texas and Louisiana are by far the leading producers of natural gas, although most of their production is outside the Inland South Region. In recent years natural gas production from these two states amounted to about three-fourths of total national output.

Bauxite

Although seven-eighths of the bauxite ores used in the United States are imported (largely from the Caribbean area), domestic production is also important. All domestic supplies now being worked lie within the boundaries of the Inland South Region. About 85 percent of the output comes from several mines, all but one of which are open pit, in Saline County of central Arkansas. Three counties in Alabama and Georgia supply the remainder.

URBAN-INDUSTRIAL DYNAMISM

An important part of the new look of the Old South is furnished by the dynamic urban and industrial growth of the region (see Table 10–1 for the region's largest urban places). From an area of sparse and specialized manufacturing in small cities it has developed into a region of notable industrial diversity and strength in booming metropolises.

While the Northeast experienced a significant reduction in manufacturing employment in recent years, the Southeast had a large increase in its industrial base. The traditional "southern" industries in the Inland South Region grew only slowly; it is the "northern" industries of the region that expanded rapidly. Thus textiles, apparel, and tobacco prod-

TABLE 10-1
Largest urban places of the Inland South Region

Name	Population of Principal City	Population of Metropolitan Area
Albany, Ga.	83,540	116,300
Alexandria, La.	50,180	137,800
Anderson, S.C.	29,460	143,100
Anniston, Ala.	29,130	123,300
Arlington, Tex.*	257,460	
Athens, Ga.	41,420	144,700
Atlanta, Ga.	420,220	2,736,600
Augusta, Ga.	42,830	396,400
Austin, Tex.	464,690	748,500
Birmingham, Ala.	277,280	923,400
Bossier City, La.*	55,810	
Bryan, Tex.	60,410	116,600
Burlington, N.C.	37,610	105,800
Carrollton, Tex.*	69,190	
Charlotte, N.C.	367,860	1,112,000
Columbia, S.C.	94,810	456,500
Columbus, Ga.	177,680	246,900
Dallas, Tex.	987,360	2,475,200
Danville, Va.	53,400	108,100
Decatur, Ala.	46,090	132,700
Durham, N.C.	115,430	
Fayetteville, N.C.	75,470	255,700
Florence, Ala.	34,850	135,500
Fort Worth, Tex.	426,610	1,290,900
Gadsden, Ala.	44,890	102,900
Garland, Tex.*	180,450	
Gastonia, N.C.	53,260	
Grand Prairie, Tex.*	98,900	
Greensboro, N.C.	181,970	924,700
Greenville, S.C.	59,190	621,400
High Point, N.C.	67,240	
Irving, Tex.*	133,000	
Jackson, Miss.	201,250	396,200
Jackson, Tenn.	53,320	78,200
Killeen, Tex.	64,930	239,600
Little Rock, Ark.	180,090	513,100
Longview, Tex.	71,970	166,600
Macon, Ga.	117,940	286,700
Memphis, Tenn.	645,190	979,300
Mesquite, Tex.*	93,120	
Monroe, La.	54,520	144,000
Montgomery, Ala.	193,510	300,800
N. Little Rock, Ark.*	62,410	
Pine Bluff, Ark.	61,230	90,800
Plano, Tex.*	118,790	
Raleigh, N.C.	86,720	683,500
Richardson, Tex.*	77,080	
Sherman, Tex.	31,020	97,900
Shreveport, La.	218,010	359,100
Texarkana, Tex.-Ark.	33,110	119,400
Tuscaloosa, Ala.	74,100	145,400
Tyler, Tex.	74,740	152,600
Waco, Tex.	103,420	188,000
Wichita Falls, Tex.	97,870	124,600
Winston-Salem, N.C.	148,690	

* A suburb of a larger city.

ucts had an inconspicuous growth; major expansion was focused in such durable goods categories as metal fabrication, transportation equipment, and electronics-electrical equipment. As a result, there is a convergence of the industrial structure of region and nation; that is, the Inland South is becoming less distinctive economically and is developing a balanced industrial structure that is reflective of the national pattern. Despite continuing absolute growth, manufacturing in the Inland South, as in the nation as a whole, is becoming relatively less important in the economy because of the rapid expansion of the tertiary sector.

Primacy of the Gateway Cities

At the top of the urban hierarchy of the Inland South are two dominant metropolises that epitomize the concept of the gateway city. *Atlanta* in the east and *Dallas* in the west serve as regional capitals in terms of commerce, finance, transportation, and other economic aspects. These two cities are the principal funnels and nerve centers through which extraregional goods, services, ideas, and people are channeled into the Inland South, and to a lesser extent, they accommodate the reverse flow of regional output to the nation. This gateway function is best shown by the magnitude of wholesale sales, the concentration of financial institutions, especially banks and insurance companies, the large number of national and regional corporate headquarters, and the daily passenger flow through the respective airports.

FIGURE 10-12 The major cities of the Inland South have experienced booming industrialism in recent years. This extensive electronics plant is a few miles north of Dallas (courtesy Texas Instruments, Incorporated).

Although approximately the same in size and function, the two cities have quite different origins and histories. Industrial growth in recent decades was spectacular in both urban nodes, particularly Dallas, which now ranks as the tenth largest industrial center in the nation (Fig. 10–12). In both metropolises most industrial expansion occurred in the related fields of aerospace and electronics, although their industrial structures are generally well diversified. The economic function of these gateway cities, however, is much more significantly commercial than industrial, and the impressive skyline of Dallas and the progressive atmosphere of Atlanta suggest the improving economic and cultural image of the region (Fig. 10–13).

Secondary Regional Centers

One step lower in the regional hierarchy are the medium-sized cities of Memphis, Fort Worth, and Birmingham, each of which is also growing rapidly. *Memphis* is the traditional river city of the middle Mississippi basin, dominating the area between the spheres of New Orleans and St. Louis.

Fort Worth, although near enough to have a twin-city relationship with Dallas, is remarkably different from its neighbor.[4] Whereas Dallas's eastward orientation emphasized commerce, finance, oil, and electronics, Fort Worth is clearly oriented to the west, focusing more on cattle, railways, and agricultural processing.

[4] For statistical purposes, the Bureau of the Census has declared Dallas and Fort Worth to be part of a single metropolitan area that encompasses 11 small north Texas counties. The local media coined the inelegant term "Metroplex" to refer to this conurbation.

FIGURE 10-13 The skyline of Atlanta on a typical humid, hazy summer day (TLM photo).

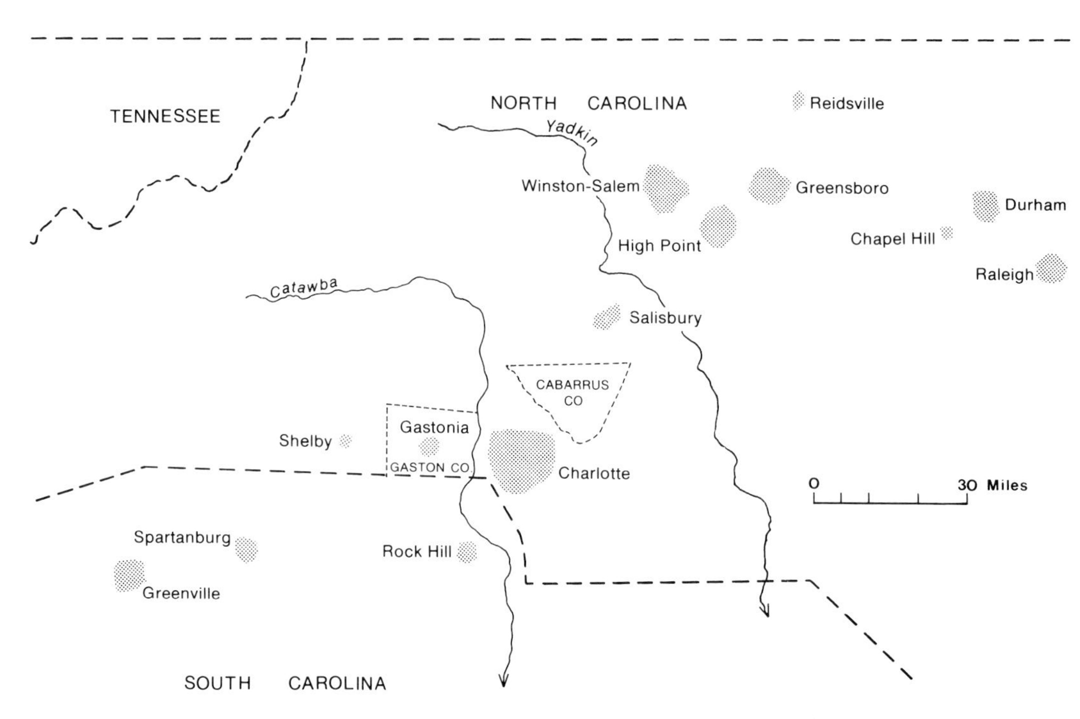

FIGURE 10-14 The urban-industrial complex of the Carolinas.

The story of *Birmingham* is unusual; it has long been the only heavy-industry center in the South. Local deposits of iron ore and coal made it the least expensive place in the nation to manufacture steel. Distance from major markets inhibited its growth and the nearby ore deposits were depleted, but Birmingham has developed a variety of manufactures and is also an important commercial center.

The Carolina Urban-Industrial Complex

Although lacking a major metropolis, there is a zone of notable urban and industrial development centered on the Piedmont of North Carolina (Fig. 10–14). This zone extends northward into southern Virginia, southwestward across South Carolina into northern Georgia, and eastward onto the North Carolina coastal plain. The heart of the district is the North Carolina Piedmont, with its medium-sized cities of *Charlotte, Winston-Salem,* and *Greensboro.* It is a highly industrialized district; many factories are located in smaller towns, such as Reidsville and Shelby, and in nonurban areas, especially in Gaston and Cabarrus counties. New industries tended to locate in rural areas or in small towns rather than in cities, partly because of a tight labor market. Manufacturing is not diversified in the district, being mainly tobacco, textiles, chemicals, and furniture. The area contains the greatest concentrations of cigarette, cotton-textile, and furniture factories to be found in the nation.

The importance of manufacturing to the local economy is shown by the fact that North and South Carolina have greater proportions of their labor forces employed in factories (more than 30 percent) than do any other states.

This Piedmont industrial district contains about half as many industrial facilities and employees as the six New England states combined. Moreover, three of the urban nodes, centered on Greensboro, Greenville, and Charlotte, are among the three dozen leading industrial centers in the nation despite their relatively small populations.

Other facets of urbanism are also notable in the district. For example, in the area encompassing *Raleigh, Durham,* and *Chapel Hill* is the "Research Triangle," probably the most successful corporate research center in the nation. It is related to the three major universities of the area but oriented toward industrial and environmental research.

THE OUTLOOK

The Inland South is a region with a distinctive cultural heritage, but the only really unifying physical phenomena are climate and forest. It is a region that has traditionally produced a few staple commercial crops but where an agricultural revolution has taken place. It is a region that has been traditionally rural and agrarian, but it has become significantly urbanized. It is a region of slow population growth and out-migration that has become the nation's principal focus of interregional in-migration.

Livestock raising, particularly beef cattle and chickens, probably will persist as dominating features of the agricultural scene, despite the recent resurgence of cotton growing and continuing crop diversification. Supplemental irrigation will become commonplace to carry crops vigorously through short dry spells and to increase overall yields.

The continued decrease of rural population, both white and black, will probably be a blessing rather than a curse to the region. The displaced marginal agriculturalists will probably find more satisfactory urban employment, both within and without the Inland South; and either machinery or migrant workers will take up the labor slack on the fewer but larger remaining farms.

Production from timberlands will become increasingly important for both lumber and pulp. The trend toward clearcutting will probably spread throughout most of the larger pine holdings, although its efficiency as a management technique will be partly counterbalanced by environmental and aesthetic objections. The inadequate replacement of pines should become a serious problem in the 1990s.

The economic outlook in nonagricultural fields is generally promising, particularly in the western part of the region. The attraction of the Inland South to manufacturing industries is now solidly based and industrial investment will probably continue to be strong in the near future, even though the relative role of manufacturing in the economy is now diminished. Low-skilled and low-paid industries will continue to decline, thus shrinking the region's competitive advantage, but worker retraining and entrepreneurial leadership should overcome this problem.

Urban growth and associated commercial expansion seem ensured. Some of the nation's fastest-growing medium-sized cities are in this region and the two gateway metropolises continue as leaders of expansion. At the other end of the urban spectrum, however, many smaller towns seem destined to wither.

The population mix will probably become more diverse with continued influx of people from other regions. The rate of in-migration may slow down, but it is unlikely to diminish significantly. The concentration of blacks in the larger cities, particularly the inner-city districts of Atlanta, Birmingham, and Memphis, will provide increased political power to black voters.

All things considered, the passage of time will continue to blur the image of the Inland South as a distinctive region even as the region's economic and social role in the nation becomes more significant.

SELECTED BIBLIOGRAPHY

ADKINS, HOWARD G., "The Imported Fire Ant in the Southern United States," *Annals*, Association of American Geographers, 60 (1970), 578–592.

BOUNDS, JOHN H., "The Alabama-Coushatta Indians of Texas," *Journal of Geography*, 70 (1971), 175–182.

CLAY, JAMES W., and DOUGLAS M. ORR, JR., eds., *Metrolina Atlas*. Chapel Hill: University of North Carolina Press, 1972.

CLAY, JAMES, W., DOUGLAS M. ORR, JR., and ALFRED W. STUART, eds., *North Carolina Atlas: Portrait of a Changing Southern State*. Chapel Hill: University of North Carolina Press, 1976.

CONWAY, DENNIS, et al., "The Dallas-Fort Worth Region," in *Contemporary Metropolitan America, Vol. 4, Twentieth Century Cities*, ed. John S. Adams, pp. 1–37. Cambridge, MA: Ballinger Publishing Company, 1976.

CROMARTIE, JOHN, and CAROL B. STACK, "Reinterpretation of Black Return and Nonreturn Migration to the South, 1975–1980," *Geographical Review*, 79 (July 1989), 297–310.

CROSS, R. D., et al., *Atlas of Mississippi*. Jackson: University of Mississippi Press, 1974.

DOSTER, JAMES F., and DAVID C. WEAVER, *Tenn-Tom Country*. Tuscaloosa: University of Alabama Press, 1987.

FALK, WILLIAM W., and THOMAS A. LYSON, *High Tech, Low Tech, No Tech: Recent Industrial and Occupational Change in the South*. Albany: State University of New York Press, 1988.

FOSCUE, EDWIN J., "East Texas: A Timbered Empire," *Journal of the Graduate Research Center*, Southern Methodist University, 28 (1960), 1–60.

FUSSELL, MICHARD, *A Demographic Atlas of Birmingham*. Tuscaloosa: University of Alabama Press, 1975.

HAMLEY, W., "Research Triangle Park: North Carolina," *Geography*, 67 (January 1982), 59–62.

HART, JOHN FRASER, "Land Use Change in a Piedmont County," *Annals of the Association of American Geographers*, 70 (December 1980), 492–527.

———, "Migration to the Blacktop: Population Redistribution in the South," *Landscape*, 25 (1981), 15–19.

———, "The Demise of King Cotton," *Annals*, Association of American Geographers, 67 (1977), 307–322.

HARTSHORN, TRUMAN A., et al., "Metropolis in Georgia: Atlanta's Rise as a Major Transaction Center," in *Contemporary Metropolitan America. Vol. 4: Twentieth Century Cities*, ed. John S. Adams, pp. 151–225. Cambridge, MA: Ballinger Publishing Company, 1976.

HAYES, CHARLES R., *The Dispersed City: The Case of Piedmont North Carolina*. Chicago: University of Chicago, Department of Geography, 1976.

HEATWOLE, CHARLES A., "The Bible Belt: A Problem in Regional Definition," *Journal of Geography*, 77 (February 1978), 50–55.

HODLER, THOMAS W., and HOWARD A. SCHRETTER, eds., *The Atlas of Georgia*. Athens: Institute of Community and Area Development of the University of Georgia, 1986.

JORDAN, TERRY G., "Antecedents of the Long-Lot in Texas," *Annals*, Association of American Geographers, 64 (1974), 70–86.

———, "Early Northeast Texas and the Evolution of Western Ranching," *Annals*, Association of American Geographers, 67 (1977), 66–87.

———, *German Seed in Texas Soil: Immigrant Farmers in Nineteenth Century Texas*. Austin: University of Texas Press, 1967.

———, *Trails to Texas, Southern Roots of Western Cattle Ranching*. Lincoln: University of Nebraska Press, 1981.

JORDAN, TERRY G., JOHN L. BEAN, and WILLIAM M. HOLMES, *Texas*. Boulder, CO: Westview Press, 1984.

KOVACIK, CHARLES F., and JOHN J. WINBERRY, *South Carolina: The Making of a Landscape*. Columbia: University of South Carolina Press, 1989.

LAMB, ROBERT BYRON, *The Mule in Southern Agriculture*. Berkeley: University of California Press, 1963.

LINEBACK, NEAL G., ed., *Atlas of Alabama*. Tuscaloosa: University of Alabama Press, 1973.

———, ed., "Manufacturing the South" *Southeastern Geographer*, 14 (November 1974), entire issue.

MCGREGOR, JOHN R., "A Delimitation of Manufacturing Regions within the Southeastern United States," *Professional Paper No. 9*, Indiana State University, Terre Haute, Department of Geography and Geology (1977), pp. 13–31.

MEINIG, DONALD W., *Imperial Texas: An Interpretive Essay in Cultural Geography*. Austin: University of Texas Press, 1969.

POSTON, DUDLEY, and ROBERT WELLER, eds., *The Population of the South*. Austin: University of Texas Press, 1981.

PRUNTY, MERLE C., and CHARLES S. AIKEN, "The Demise of the Piedmont Cotton Region," *Annals*, Association of American Geographers, 62 (1972), 283–306.

REED, JOHN SHELTON, *One South: An Ethnic Approach to Regional Culture*. Baton Rouge: Louisiana State University Press, 1982.

SINIARD, L. ARNOLD, "Dominance of Soybean Cropping in the Lower Mississippi River Valley," *Southeastern Geographer*, 15 (May 1975), 17–32.

SWANTO, JOHN R., *The Indians of the Southeastern United States*. Washington, D.C.: Smithsonian Institution Press, 1979.

THOMAS, BILL, "Congaree: Last of the Bottomland Forests," *The Living Wilderness*, 40 (July–September 1976), 16–21.

TRIMBLE, STANLEY W., *Man-Induced Soil Erosion on the Southern Piedmont, 1700–1970*. Ankeny, IA: Soil Conservation Society of America, 1974.

WHEAT, LEONARD F., *Urban Growth in the Nonmetropolitan South*. Lexington, MA: Lexington Books, 1976.

WHEELER, JAMES O., "Studies of Manufacturing Location in the Southern United States," *Geographical Review*, 65 (1975), 270–272.

WINBERRY, JOHN J., and DAVID M. JONES, "Rise and Decline of the 'Miracle Vine': Kudzu in the Southern Landscape," *Southeastern Geographer*, 13 (November 1973), 61–70.

11
THE SOUTHEASTERN COAST

Attenuated along the "subtropical" shores of eastern United States is the Southeastern Coast Region. It is long and narrow in shape, dynamic in recent population and economic development, and represents a curious mixture of the primitive and the modern. It is a region that contains some of the most pristine natural areas as well as some of the most completely unnatural areas to be found anywhere on the continent. Burgeoning cities and busy factories nestle side by side with quiet backwaters where fin and feather and fur dominate life, a juxtaposition restricted to places where a low-lying coastal environment permits an infinite variety of land-and-water relationships. Sophisticated urbanites live close to backwoods trappers and some of the nation's most scientific and mechanized agriculture is carried on only a stone's throw from areas of hard-scrabble subsistence farms.

The region extends from the Dismal Swamp of southeasternmost Virginia along the littoral zone of the Atlantic Ocean and Gulf of Mexico to the international border at the Rio Grande; thus is abuts the Inland South Region along most of its interior margin, brushing the Great Plains Region in southern Texas and touching the Megalopolis Region in southern Virginia (Fig. 11–1).

The inland margin of the Southeastern Coast Region is not clearly defined. Physically it is marked by a transition zone that separates the poorly drained coastal lands from the better-drained country interiorward. Culturally it can be considered to be inland of the seacoast cities and their immediate

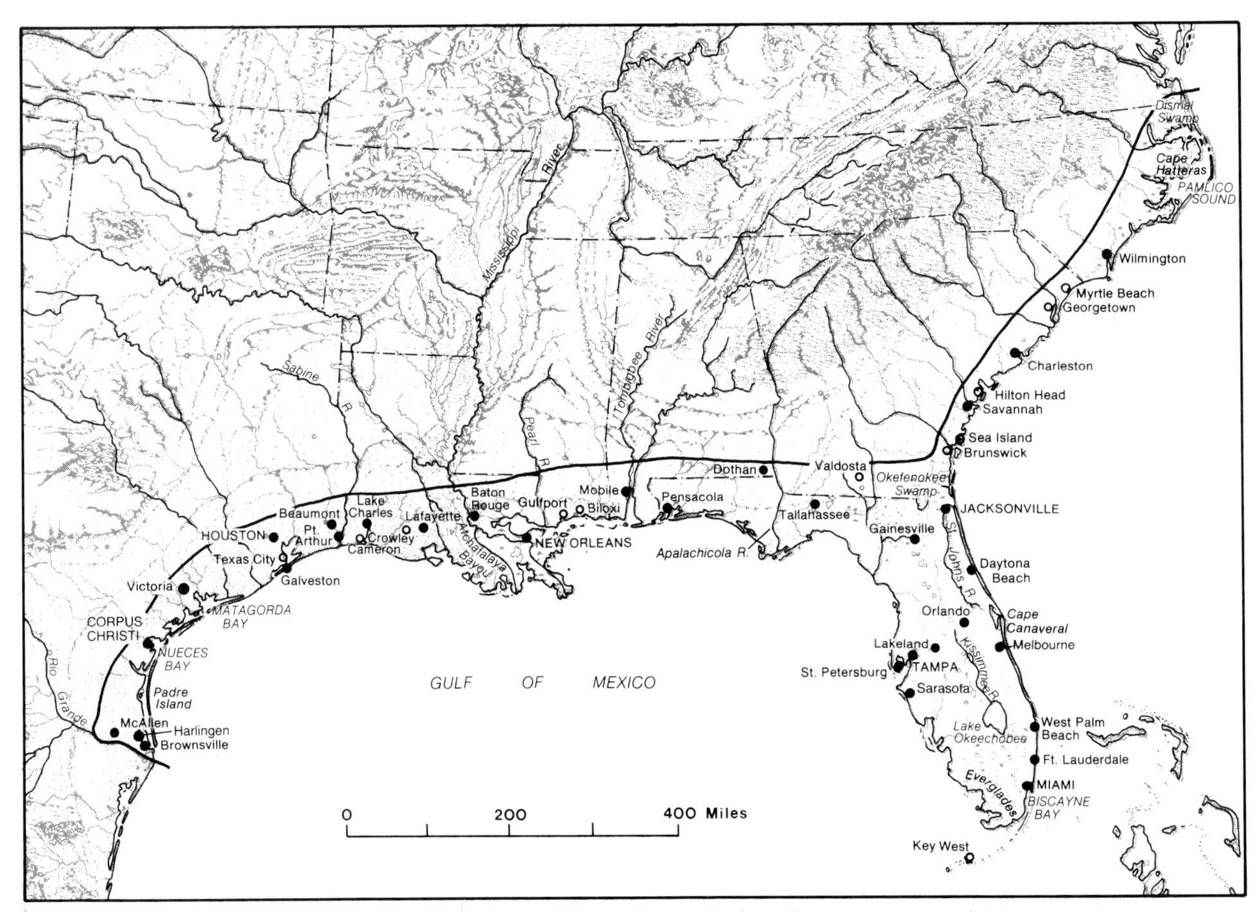

FIGURE 11-1 The Southeastern Coast Region (base map copyright A. K. Lobeck; reprinted by permission of Hammond, Inc.).

city-serving farmlands. As thus defined, this region encompasses a relatively narrow coastal strip of Virginia, the Carolinas, Georgia, Alabama, Mississippi, and Texas; it also includes about half of Louisiana and all of Florida.

THE PHYSICAL SETTING

Relief of the Land

Although almost uniformly characterized by flatness and impeded drainage, the regional topography is varied in detail and primarily reflects variations in geomorphic history. Except in much of Florida, where Tertiary formations predominate, almost the entire region is underlain by relatively unconsolidated Quaternary sediments of marine origin. These beds extend seaward to constitute also the continental shelf. The shelf offshore of this region is quite broad—generally more than 75 miles (120 km) in width—everywhere except off the southeastern corner of Florida and near the delta of the Mississippi River. Along almost all the coast the ocean is very shallow for a considerable distance offshore; in many localities a swimmer must wade out tens or even hundreds of yards to find water deep enough to swim in.

Low-lying, sandy islands occur off the coast of most of the region except parts of Louisiana and the Gulf of Mexico side of peninsular Florida. The almost continuous chains of islands off the North Carolina and Texas coasts represent barrier sandbars at an extreme stage of lengthwise development; they

UNIVERSITY OF CALIFORNIA - LOS ANGELES

TO

DATE	TIME	RECEIVED BY

WHILE YOU WERE OUT

CALLER'S NAME

FROM

TELEPHONE NUMBER EXT.

○ TELEPHONED	○ PLEASE PHONE	○ CAME TO SEE YOU
○ RETURNED YOUR CALL	○ WILL PHONE AGAIN	○ URGENT

MESSAGE

CALL SLIP 71465-134 (8/82)

— The schedule

— The exam

— The [illegible] days

are long, narrow, and low and encompass lagoons and sounds that are in the process of being filled in as the coastline progrades outward. Other offshore islands (along the Georgia and Mississippi coasts, for example) are more varied in origin, ranging from erosional remnants to old beach ridges, with gradations in between. The Florida Keys are totally different in formation. The eastern keys are mostly uplifted coral reefs whereas the western keys represent limestone shoals upraised from an earlier sea bottom.

With the important exceptions of peninsular Florida and southeastern Louisiana, the topography of the mainland is generally uniform throughout the region. The land is exceedingly flat, sloping gently toward the sea. A great many rivers meander across this lowland. The lower reaches of most streams were drowned by coastal subsidence; thus the coastline is studded with estuaries and bays in addition to the lagoons and sounds previously mentioned. Inadequate drainage is, therefore, common and there are extensive stretches of swampland as well as some large marshes (Fig. 11–2). The two largest and most famous of these poorly drained lands are Okefenokee Swamp in Georgia–Florida (Fig. 11–3) and Dismal Swamp on the Virginia–North Carolina border.

The topography of peninsular Florida is distinctive from that of the rest of the region. The peninsula is a recently emerged mass of carbonate rocks, largely limestone, characterized by karst features in the north and immense areas of inadequate drainage in the south. The caves and sinks of typical karst regions are modified in Florida by the uniformly high water table; most caverns are water-filled and most sinks have become small lakes. There are so many sinkhole lakes in central Florida that this area is referred to as the "lakes district."

FIGURE 11-2 Tall-grass marshes are widespread in this region. The scene is near Hilton Head, South Carolina (TLM photo).

Associated with the karst is the most extensive artesian system in Anglo-America. Some components of the system are deep and discharge on the sea bottom many miles offshore. Shallower components often discharge as artesian springs. Florida has the greatest concentration in the country and many serve as the source of short rivers. Silver Springs is the largest known of the artesian outflows, with an average daily discharge of nearly half a billion gallons (1.9 gigaliters) of water.[1]

The exceedingly flat terrain of southern Florida has a natural drainage system that is both complex and delicate. Lake Okeechobee, with a surface area of about 750 square miles (1900 km^2) and a maximum depth of less than 20 feet (6 m), is fed in part by overflow from the Kissimmee Lakes to the north. Lake Okeechobee, in turn, overflows broadly southward with a 50-mile (80 km)-wide channel that maintains the natural water supply of the Everglades and Big Cypress Swamp. Human interference with this drainage system caused major problems.

The area around the mouth of the Mississippi, the continent's mightiest river, is generally flat and ill-drained and its detailed pattern of landforms is notably complex. Essentially all southeastern Louisiana within the Southeastern Coast Region is a part of the Mississippi River deltaic plain. During the past 20 centuries the lower course of the river shifted several times, producing at least seven different subdeltas that are the principal elements of the present complicated "bird's foot" delta of the river. The main flow of the river during the past 600 years or so has been along its present course extending southeast from New Orleans. This portion of the delta was built out into the Gulf at a rate of more than 6 miles (9.6 km) per century in that period.

[1] William D. Thornbury, *Regional Geomorphology of the United States* (New York: John Wiley & Sons, 1965), p. 47.

FIGURE 11-3 Moss-draped cypress and pines tower over Okefenokee Swamp on the Georgia–Florida boundary (courtesy Georgia Department of Industry and Trade).

The Atchafalaya distributary is the principal secondary channel of the river. It carries about 25 percent of the Mississippi's water; and under the normal pattern of deltaic fluctuation the main flow of the river would now be shifting to the shorter and slightly steeper channel of the Atchafalya. An enormous flood-control structure was erected above Baton Rouge, however, in an effort (thus far successful) to prevent the Mississippi from abandoning its present channel and delta.

The terrain of the Mississippi deltaic plain consists of many bayous and swamps, with marshes occupying all the immediate coastal zone. In addition to the natural bayous, the coastal marshlands are criss-crossed by a maze of artificial waterways, mostly shallow, narrow ditches (called *trainasse*) that were crudely excavated by local people to provide small-boat access. Many of these simple canals are now established elements of the drainage systems, as are more recent canals dredged to provide boat access to oil and gas wells. The only slightly elevated land is along natural levees, which parallel the natural drainage channels, and on low sandy ridges (*cheniers*) that roughly parallel the coast of southwestern Louisiana.

Climate

This region is typical of the warmer, more humid phase of the humid subtropical climate. It is characterized by a heavy rainfall and a long growing season (from 240 days in the north to almost frostless areas of southern Florida). The total annual precipitation decreases from more than 60 inches (1520 mm) in southern Florida and along the eastern Gulf Coast to less than 30 inches (760 mm) in the southwest. Over most of the region the rainfall is evenly distributed throughout the year, but the maximum comes during

FIGURE 11-4 A view of the Everglades in a wet year (TLM photo).

the summer and early autumn, which are thunderstorm and hurricane seasons. The torrential rains (most stations in this region have experienced more than 10 inches (250 mm) of rain in a 24-hour period) coupled with high-velocity winds have caused considerable damage to crops and wrought great destruction to many coastal communities at some time or other.

From a psychological standpoint mild, sunny winters largely offset the hurricane menace and thus enable this region to capitalize on climate in its agriculture and resort business. But severe, killing frosts affect part of the region almost every year.

Soils

The region has a considerable variety of soil types, but most are characterized by an excess of moisture, quartz particles, or clay. Most widespread are hydromorphic (poorly drained) varieties, but sandy soils are also common. From a taxonomic standpoint the soils of the Southeastern Coast Region are the most complex of the continent. All soil orders except Aridisols are extensively represented, often in a complicated patchwork pattern.

Natural Vegetation

Because of heavier rainfall, the eastern part of the region is largely covered with forests, mainly yellow pines and oaks on the better-drained, sandy lands, with cypress and other hardwoods dominating the swamps and other poorly drained areas. Hardwoods were probably more widespread long ago, before the Indians began burning the woodlands in winter to aid their hunting. Periodic burning favored the pines and oaks, for they are more resistant to fire than other trees and shrubs. Until fairly recently burning was commonplace in the region, generally done on a casual and indiscriminate basis in order to "improve" grazing for scrub cattle. A common feature of the region is the so-called Spanish moss, which festoons live oaks and cypress but is much less abundant on pines except in swampy areas.

Some extensive grasslands, usually marshy, are found along the coast. Dotted irregularly through these marshlands are bits of elevated ground, usually called *islands* or, in Louisiana, *cheniers,* which are generally covered with big trees laden with moss. From southwestern Louisiana westward most of the natural vegetative cover is coastal prairie grassland; it is partially replaced by scrubby brush country south of Corpus Christi.

Most of Florida south of Lake Okeechobee is part of the Everglades (literally "sea of grass") where tall sawgrass dominates the landscape (Fig. 11–4). Patches of woodland, slightly elevated above the surrounding Everglades marsh, are called *hammocks;* they are the favorite haunts of Florida's most vicious wildlife, mosquitos. The littoral zone of the coastal Everglades has Anglo-America's only extensive growth of tangled mangroves.

Wildlife

The Southeastern Coast Region contains the only part of the conterminous states that is a major wintering ground for migratory birds, particularly wa-

A CLOSER LOOK Major Wetland Problem Areas

Coastal wetlands are among the most critical and most fragile environments. Where they are subjected to rapid human population growth and pell-mell development, the ecological ramifications may shortly become overwhelming. Much of the Southeastern Coast Region is susceptible to such problems, but two major areas are of particular concern.

THE EVERGLADES

The semitropical flatlands of southern Florida contain a unique ecosystem that depends on a delicate balance of natural factors, the cornerstone of which is an unusual natural drainage system that includes nearly one-fifth of the state's total area. Human disruption of this drainage system is the root of the problem.

Under normal circumstances there is a steady and reliable flow of water from the Kissimmee chain of lakes into Lake Okeechobee, from which the natural overflow sends a very broad and very shallow sheet of water southward to sustain Big Cypress Swamp and the Everglades before eventually draining into the Gulf of Mexico (Fig. 11–a). As more and more settlers were attracted to southern Florida, farms and cities were established and grew apace. The relatively frequent minor, and occasional major, natural floods brought hardships to farmers and urbanites alike and a number of artificial drainage canals were constructed for flood control purposes and to "improve" the land for agricultural and urban expansion.

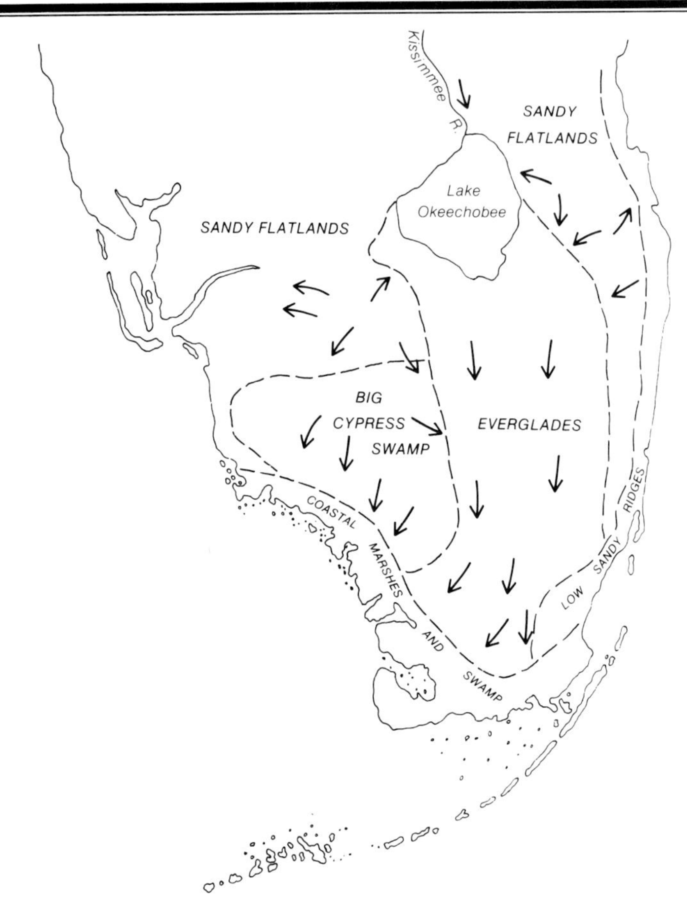

FIGURE 11-a The natural drainage pattern of southern Florida is generally southwestward from Lake Okeechobee through the Everglades (after G. G. Parker. U.S. Geological Survey Water Supply Paper 1255).

As metropolitan Miami burgeoned, specialty farmers of southern Florida were displaced and the only direction they could move was westward, into the 'Glades. Two highways, the Tamiami Trail and Alligator Alley, were built east-west across the area. Further land development, primarily for residential and second-home purposes, was undertaken in Big Cypress Swamp, as a sort of subsidiary of the west coast urban clusters of Fort Myers and Naples. Even the Kissimmee River was channeled into a straight-line, stagnant canal.

These varied developments were increasingly opposed by individuals and groups from many parts of the country and the "conservation coalition," gaining in strength and political muscle, succeeded at some points. Everglades National Park was established in 1947 and in 1974 the federal and state governments authorized the purchase of more than half a million acres (200,000 ha) of Big Cypress Swamp to stop piecemeal development and set up a watershed reserve.

The ecological threat, however, persists. Much of the land north and east of Everglades National Park and the Big Cypress Reserve is in private ownership and pressures for development have not abated. They are mostly "upstream" areas in the watershed and any disruptions in the normal drainage pattern may have serious repercussions in the downstream sections. Moreover, most original drainage canals are still functioning, thus cutting off much of the natural southward flow, which is particularly serious in minimum rainfall years.

The problem is to maintain a flow of water throughout the Everglades—Big Cypress area from the Kissimmee drainage southward. Dechannelization of the Kissimmee River has been "authorized," but no funds are available to implement the project. Meanwhile, most of the normal southward flow from Lake Okeechobee can be (and usually is) legally diverted.

In 1990, for the first time, the Army Corps of Engineers (which built most of the flood-control and drainage structures) and the South Florida Water Management District (which operates the drainage system) began planning a comprehensive project to resolve the major problems by restoring water to the Everglades in a systematic fashion. The plan is very controversial in the state, and its ultimate price tag would be hundreds of millions of dollars, so the salvation of the 'Glades is far from assured.

THE MISSISSIPPI DELTA

The problem is more complex in southeasternmost Louisiana, but the basic cause—human disruption of the natural drainage system—is the same. In order to provide a dry surface for human settlement and activities, the land was increasingly drained and the rivers channeled for two centuries, a process that was greatly accelerated in recent decades. Artificial levees and flood-control structures keep the Mississippi and its distributaries in relatively narrow channels, thus denying both silt and freshwater to the marshes. In addition, a maze of canals (extending for an estimated 10,000 miles [16,000 km]), dredged mostly to gain access to marshland oil wells, accelerates erosion and allows saltwater to encroach.

The result is a continuing diminishment of land that is unprecedented on the North American coastline (Fig. 11–b). The delta is both washing away and sinking. The amount of marsh continues to decrease and the amount of open water in the delta continues to expand. Over the last three decades the amount of land loss averaged about 40 square miles (104 km^2) annually, or an acre every 15 minutes. The freshwater and brackish ecosystems suffer and some of the higher land containing human settlements is sinking.

The relevant authorities agree that large-scale introduction of sediment-laden river water is urgently needed to assuage the problem and that canal dredging should be strictly curtailed. But the economic imperative of more shipping and more oil development

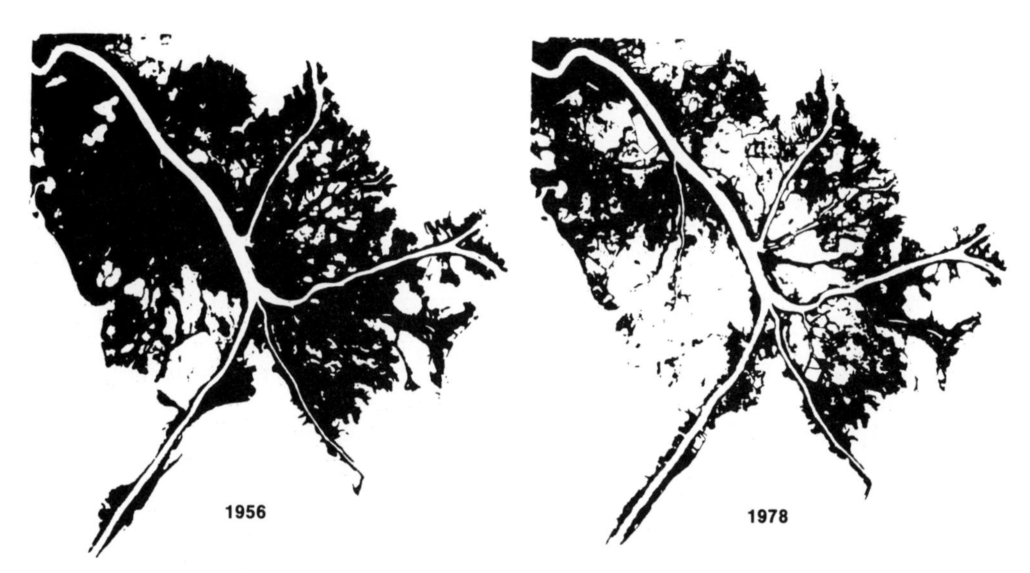

FIGURE 11-b Land loss in the lower Mississippi delta. Much of the marsh land shown in the left diagram (1956) has sunk or been washed away by the intruding ocean in the right diagram (1978).

(continued)

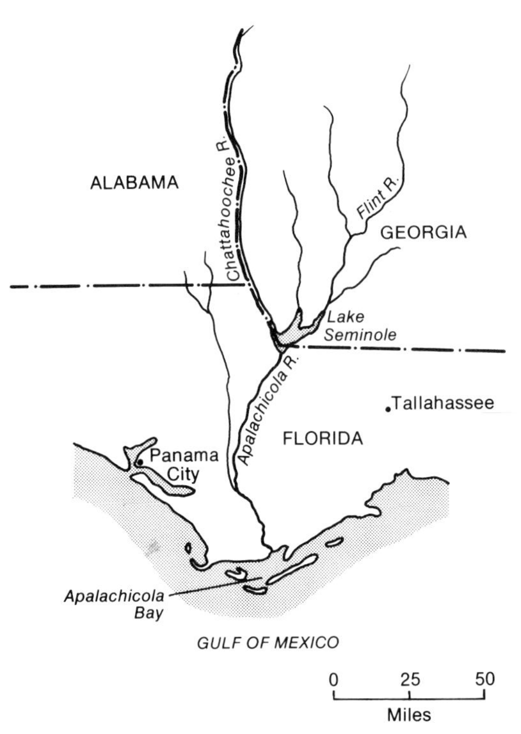

FIGURE 11-c The Apalachicola River area of Florida, Georgia, and Alabama.

continue to cloud the issue. Indeed, there are plans to dredge the river 15 feet (4.5 m) deeper to accommodate larger bulk carriers.

THE APALACHICOLA RIVER

The "threat" to wetlands is widespread in the region, but a hopeful trend in the opposite direction is shown by the situation in the Apalachicola drainage of Florida (Fig. 11–c). Formed by the confluence of the Chattahoochee and Flint rivers, the Apalachicola itself is only 100 miles (160 km) long, but it is among the ten largest U.S. rivers in volume of flow.

It was planned to dam the river at four locations and to make it into a barge waterway. Citizen opposition, however, persuaded both state and federal governments to act in the late 1970s to acquire much of the land fronting the lower river course and Apalachicola Bay, and the affected counties have prepared master zoning plans that will severely limit construction in the wetlands. Perhaps this step will set a precedent for coordinated wetlands management.

TLM

terfowl. From Florida to Texas coastal marshes and swamps provide a last stronghold for the wintering (for example, whooping cranes) or breeding (various egrets and spoonbills) of numerous rare and endangered birds as well as seasonal or permanent homes for many other avian species.

The region's poorly drained wild areas also provide an important habitat for a number of native quadruped species. Mustelids, such as mink, otter, and skunk, are prominent, but muskrats and raccoons are the most numerous of the region's native furbearers. Here, too, is the last stronghold for the cougar in eastern United States.

The American alligator, Anglo-America's only large reptile, is found exclusively in the Southeastern Coast Region. Heavily persecuted in the past, it responded to stringent protection in the 1960s and is once again abundant in Louisiana and Florida and increasing in suitable areas from the Big Thicket of Texas to the Dismal Swamp of Virginia.

The vast expanses of marshes and other types of wilderness and the relatively mild winters in this region make the Southeastern Coast an attractive habitat for a host of exotic animals that were introduced either deliberately or inadvertently. It is in Florida, where natural ecosystems have been most disturbed and where winter weather is most permissive, that the greatest variety of alien species has become established. Siamese walking catfish (Fig. 11–5), Mexican armadillos, African cattle egrets, Indian rhesus monkeys, South American giant toads, and Australian parakeets represent only a sampling of the international Noah's ark that Florida is becoming.

The exotic species that became most conspicuous in the regional biota, however, is the South American nutria (*Myocastor coypu*). It was deliberately brought to Louisiana in the 1930s as a dual-purpose introduction: to provide another source of furs for trappers of the area and to help in controlling the rapid expansion of water hyacinth (another exotic from South America) in the state's bayous and lakes. The nutria is larger than the muskrat and its fur is richer, more akin to beaver fur; so nutria pelts are considerably more valuable than muskrat. The first nutria brought to the United States were kept on fur farms, and today there are several hundred operating nutria farms in Louisiana. In the 1940s, however, nutria were released or escaped from farms and are now very numerous in Louisiana and adjacent parts of Texas and Mississippi. The animal has now become well established, is a major quarry for trappers, and has found many items that it prefers to eat other than water hyacinths.

PRIMARY INDUSTRIES

In a region that is close to nature numerous primary economic activities are likely to be carried on, provided that basic resources are present. The Southeastern Coast is such a region and primary production significantly contributes to both the regional economy and the regional image.

Trapping

Although swamps and marshes are widespread in the region, Louisiana is the only major source of wild fur; it is the leading producer among the 50 states, with an annual take twice as great as any other state. Over 1 million muskrats are trapped every year in Louisiana. Mink, otter, skunk, and other mustelids are also trapped there, but the most important species is nutria, which accounts for half the state's fur harvest.

Some 3500 trappers are licensed in Louisiana, but it is a full-time occupation for relatively few of

FIGURE 11-5 A walking catfish making progress and obeying signs (courtesy Charles Rapho/Photo Researchers, Inc.).

them. Generally it is a sideline for fishermen, shrimpers, and farmers. Trapping is of less importance elsewhere in the region, although it is carried on from Virginia to Texas.

Commercial Fishing

Both the volume and the value of the commercial catch from South Atlantic–Gulf of Mexico waters are greater than those from any of the other three regions (North Atlantic, Pacific, Alaska) recognized by the U.S. National Maritime Fisheries Service. Menhaden are by far the bulk of the catch, although their value is comparatively small.

The shrimp fishery is easily the most important of the region. Shrimp are the second most valuable variety of ocean product in the United States and more than half the total national shrimp landings are made at Southeastern Coast Region ports, particularly in Texas and Louisiana. Shrimping is so widespread in this region that almost every estuary and bay has at least one little fishing hamlet and a small complement of shrimp boats (Fig. 11–6). Larger port towns and cities, such as Corpus Christi, Pascagoula, and Savannah, are likely to have a more numerous shrimping fleet.

Louisiana is second only to Alaska in volume of seafood landings; Texas ranks third and Mississippi is fifth. Menhaden accounts for a large share of the tonnage in all three states. In terms of value of catch, Louisiana is also second and Texas ranks fifth. Half the leading fishing ports of the nation are located along the Gulf of Mexico or South Atlantic coasts (Fig. 11–7), particularly in Louisiana, where such tiny towns as Cameron, Empire, Venice, Dulac, and Chauvin are notable.

Lumbering and Forest Products Industries

Except for the prairie sections of Texas and Louisiana and the Florida Everglades, most of this region was originally forested. Consequently, such forest products as lumber and pulpwood were important. The picture in this region is similar to that in the Inland South, with the additional fact that some larger pulp mills have tidewater sites, as at Mobile and Houston.

Another forest product that is of only minor significance in the Inland South but of major importance in the Southeastern Coast Region is naval stores. These are the tar and pitch that were used before the advent of metal ships to caulk seams and preserve ropes of wooden vessels. When it was discovered that turpentine and resin could be distilled from the gum of southern yellow pine, many new uses for these substances were found in such products as paints, soaps, shoe polish, and medicines.

FIGURE 11-6 There are dozens of small fishing ports in the region, and almost every one has its shrimping fleet. This harbor scene is at Pass Christian, Mississippi (TLM photo).

FIGURE 11-7 A crab packing plant at Brunswick, Georgia (TLM photo).

The normal method of obtaining resin is to slash a tree and let the gum ooze into a detachable cup that can be emptied from time to time. A more recently developed and popular method is to shred and grind stumps and branches (formerly considered as waste) and put them through a steam-distillation process. A considerable amount of naval stores is also obtained as byproduct from the sulfate process of paper making.

Extensive forests and high-yielding species of trees permit this region to produce more than half the world's resin and turpentine. The principal area of production is around Valdosta in southern Georgia and in adjacent portions of northern Florida.

Agriculture

Farming is by no means continuous throughout the region. Large expanses of land have little arable land, particularly such poorly drained areas as the Dismal and Okefenokee swamps, the Everglades, the coastal flatwoods zone of western Florida and adjacent Alabama and Mississippi, the coastal marshlands of Louisiana, and the drier coastal country between Corpus Christi and the Lower Rio Grande Valley. Although terrain hindrances are nil, agriculture is frequently handicapped by drainage problems. Also, soils tend to be infertile; considerable fertilizer, strongly laced with trace elements, must be applied, especially in Florida.

Regional agriculture faces many problems: insect pests thrive in the subtropical environment, frost damage is sometimes heavy, marketing is often complex, and overproduction sometimes occurs. A long growing season, adequate moisture, and relatively mild winters compensate, with the result that a considerable quantity of high-quality, and often high-cost, crops is grown. The grains that are the staple crops of most Anglo-American farming areas are virtually absent in this region. Instead, farmers concentrate on growing specialty crops of various kinds. The region's major agricultural role is the output of subtropical and off-season specialties.

Citrus Fruits If any crop typifies both the image and the actuality of specialty production in the Southeastern Coast Region, it is probably citrus (Fig. 11–8). Central Florida and the Lower Rio Grande Valley of Texas, two of the four principal citrus areas of the country, are in this region (Fig. 11–9).

Florida is the outstanding citrus producer, supplying about 70 percent of the national orange and tangerine crops, about 80 percent of the grapefruit, and even larger shares of limes and tangelos. Commercial groves have always been concentrated in the gently rolling, sandy-soiled lakes district of central Florida (Fig. 11–10), but in the last few years there has been a distinct shift southward in order to avoid the devastating effects of freezes that were so prominent during the 1980s. Citrus acreage has declined by about one-third over the past two decades, due to frost damage and vigorous competition from Brazil.

The exacting requirements of modern citrus production make it increasingly practical for farmers, particularly if their holdings are not large, to contract all operations to a production company. Such farmers may never set foot on the land; indeed, many live tens or hundreds of miles away from their orchards. Many growers are members of cooperatives, which usually also have their own production crews and equipment and operate their own processing plants for citrus concentrate.

Marketing arrangements are often complicated. More than half of all citrus harvested in Florida is sold as processed products, particularly frozen concentrated juice. Citrus waste (pulp and peeling) is generally fed to cattle.

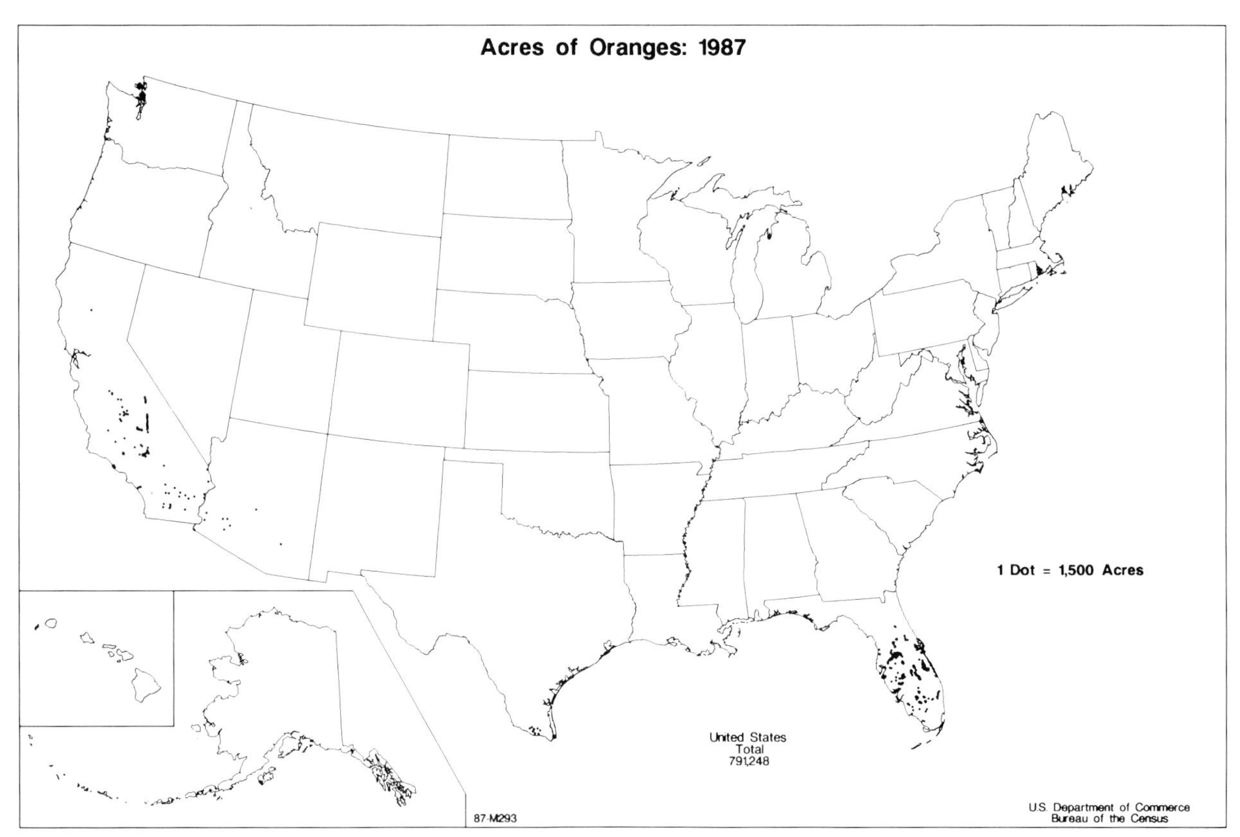

FIGURE 11-8 Florida is the principal source of oranges in the United States.

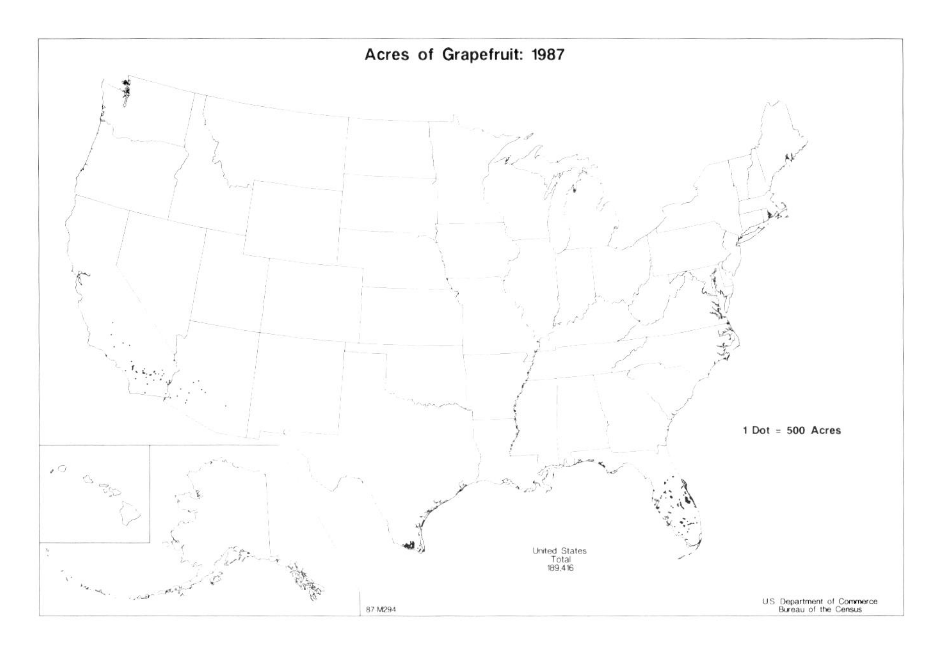

FIGURE 11-9 Central Florida is the principal producer of grapefruit, and the Lower Rio Grande Valley of Texas is an important secondary producer.

FIGURE 11-10 The lake-and-orchard district of central Florida. The orderly geometry of citrus groves is interrupted by the patchy pattern of shallow lakes (courtesy Florida Citrus Commission).

The Texas citrus area is located on the terraces of the Lower Rio Grande delta. Almost all production is based on irrigation water diverted from the river. Severe frost occasionally damages the groves, sometimes so badly that new trees must be planted. On such occasions, the horticulturists usually raise cotton or vegetables for a few years until the groves become reestablished. Seedless pink grapefruit is the specialty of the valley.

Truck Farming With increasing demands for fresh vegetables, the Southeastern Coast has become one of the major regions for early truck products of the continent. Aside from parts of southern California and the gardens under glass in the North, nearly all the nation's winter vegetables grown for sale come from this region, particularly from Florida and the Lower Rio Grande Valley.

Climate is the dominant factor affecting the growth of early vegetables, and soils help to determine the specific locations. In Florida the best truck crops are produced on muck or other lands having a higher organic content than the sandy soils that characterize much of the state. Vegetable growing in the Lower Rio Grande Valley is largely confined to the alluvial lands of the delta.

The climate in most of the region is suited to the production of early vegetables. In the southern parts of Florida and Texas some winters have no killing frosts; however, even here an occasional cold spell, or "norther," may sweep down from the interior and kill the more sensitive vegetable crops.

Sugarcane Sugarcane growing in the delta country of Louisiana began in 1751 when Jesuits introduced the crop from Santo Domingo. Most Louisiana cane production today is on alluvial soil or drained land west of New Orleans, where there is a shorter growing season than in the other domestic sugarcane states; consequently, both yields per acre and average sugar content are relatively low. Production and harvesting are almost entirely mechanized.

Florida has been a sugarcane producer since 1931, but the output was relatively limited until the 1960s when the United States ceased buying Cuban sugar. Production is concentrated on the organic soils just south of Lake Okeechobee where very high yields are attained. These soils, however, oxidize and "evaporate" when exposed to the atmosphere, which causes subsidence at a rate of about 12 inches (30 cm) per decade. This factor creates major water-control problems and the expanding industry must move to sandy soils farther from the lake. Even so, Florida cane planting has been on a steady upward trend for several years and now provides nearly half the national cane crop (as compared with less than 25 percent from Louisiana).

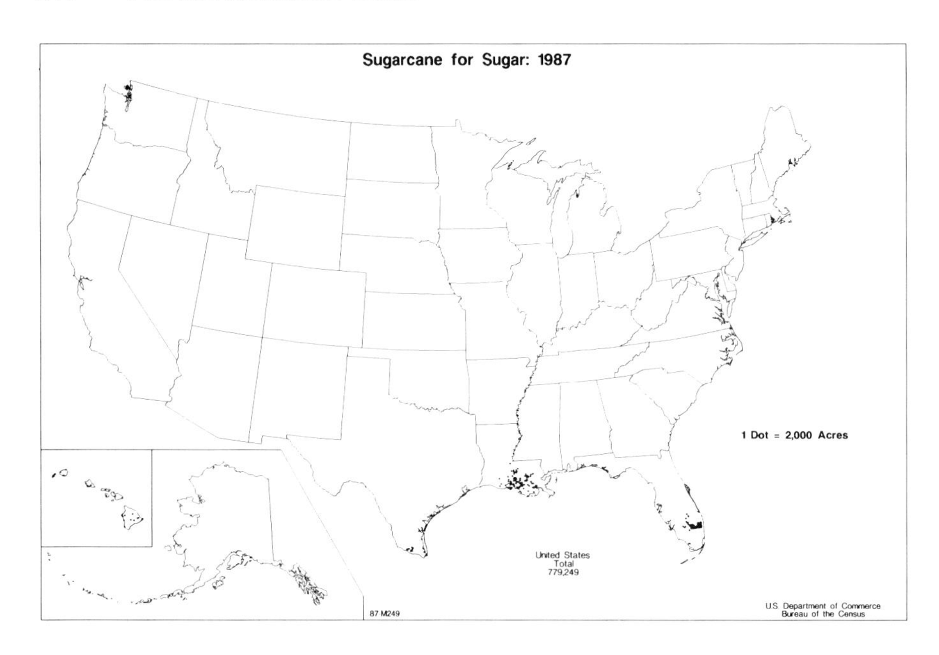

FIGURE 11-11 All sugar cane in the conterminous states is grown in the Southeastern Coast Region—in southern Florida, southern Louisiana, and the Lower Rio Grande Valley of Texas.

Most of the harvesting is accomplished by Jamaicans who are flown in to work during the November–April cutting season.

The newest sugarcane area is the Lower Rio Grande Valley of Texas, which had almost a century of commercial cane production until its total demise in the 1920s. The rise in sugar prices, however, tempted valley farmers to begin planting cane again in 1973. Results were encouraging and now about 5 percent of the national output comes from "the Valley" (Fig. 11–11).

Rice In colonial times the coastal areas of South Carolina and Georgia produced large quantities of rice, but it did not become an important commercial crop in Louisiana until about a century ago when the introduction of harvesting machinery permitted large-scale farming on the prairies in the southwestern part of that state and adjacent areas in Texas.

Mills, concentrated mainly in Crowley, Lake Charles, and Beaumont, are equipped with complicated machinery for drying, cleaning, and polishing the rice and for utilizing its byproducts. Favorable geographical conditions and complete mechanization enable this region to grow rice at a low per-acre cost.

Rice production in the United States has been on an upward trend for more than two decades. This expansion is predicated mostly on overseas markets; normally the country grows less than 2 percent of the world rice crop but supplies more than 20 percent of total world rice exports. Output increased more in Arkansas and California than in this region, however. About 25 percent of the U.S. rice crop is grown here, divided about equally between Texas and Louisiana.

Other Crops Other prominent crops in the Southeastern Coast Region include cotton and grain sorghums in the Lower Rio Grande Valley and the Coastal Bend area (around Corpus Christi) of Texas, tobacco in southern Georgia and northern Florida, and peanuts in southwestern Georgia–southeastern Alabama.

The Livestock Industry

The Southeastern Coast Region has been an important producer of beef cattle since French and Spanish colonial times. But after the Great Plains were opened to grazing in the 1870s, the poorer pastures of the Gulf Coast fell into disfavor. For several decades cattle raising was all but abandoned except in the Acadian French country of southwestern Louisiana, the coastal grasslands of southwestern Texas, and the open-range cattle country of northern and central Florida. Through the introduction of new breeds, particularly the Brahman, the cattle industry expanded in recent years to other parts of the coastal region (Fig. 11–12).

Brahman, or Zebu, cattle (*Bos indicus*) were brought from India to the Carolina coastal country more than a century ago because cattlemen thought they would be better adapted to the hot, humid, insect-ridden area than the English breeds common in the rest of the country. The oversized, hump-shouldered, lop-eared, slant-eyed, flappy-brisketed newcomer has been a thorough success and is often crossbred to produce special-purpose hybrids.

One of these hybrids, the Santa Gertrudis, meticulously developed on Texas's gigantic King Ranch, is considered to be the first "true" cattle breed ever developed in Anglo-America. It is five-eighths Shorthorn and three-eighths Brahman. Although there are many purebred Brahmans, as well as Herefords, Santa Gertrudis, and other breeds in the region, most of the beef cattle are of mixed ancestry and their physiognomy usually reveals the presence of some Brahman inheritance.

Mineral Industries

The economically useful minerals of the region are few in number but occur in enormous quantities.

Phosphate Rock The United States produces more than one-third of the world's phosphate rock and about 75 percent of the output is from central Florida. A little hard-rock phosphate is dug, but most production comes from unconsolidated land-pebble phosphate deposits east of Tampa Bay. There are mines, all open-pit, in three counties (Fig. 11–13). Large holes (often ponded because of the high water table), spoils banks, and considerable smoke from the processing plants (a council on air-pollution control has been established) are characteristic landscape features in the mining area. Phosphate mining began in 1888 in Florida and abundant reserves still exist. Most of the output is used in fertilizer manufacture, but there is considerable export, especially to Japan.

Salt The entire region from Alabama westward is dotted with deposits of rock salt. The deposits occur mostly in salt domes of almost pure sodium chloride. Mining is limited to Louisiana and Texas, the two leading salt-mining states in the nation. The reserves are enormous. Although only a few domes are mined in the region at present, they yield great quantities of salt and none is approaching exhaustion.

The mine usually lies at the top of the salt dome, 600 feet (180 m) or more below the surface. A shaft is driven down and large chambers are exca-

FIGURE 11-12 A typical mixed-breed herd of Florida cattle, near Winter Haven (TLM photo).

FIGURE 11-13 Phosphate quarries near Lakeland in central Florida (TLM photo).

vated. Mine props need not be used because large supporting columns of salt are left in place. The chambers are sometimes more than 100 feet (30 m) high. Some salt is extracted in brine solution by a modified Frasch process. Although the Gulf Coast has no monopoly in salt production, it has sufficient reserves to make it an important producer for a long time. There are many known salt domes (some of them are now producing petroleum or sulfur) where salt can be obtained when needed. Moreover, half a dozen of the domes have been hollowed out (by solution mining) to serve as enormous storage reservoirs for crude oil. Some 750 million barrels of oil were stockpiled here during the 1980s as a cushion against a future world oil shortage.

Sulfur Coastal Louisiana and Texas produce about two-thirds of the continent's sulfur. Deposits are found in the cap rock overlying certain salt domes (Fig. 11–14), where extraction is accom-

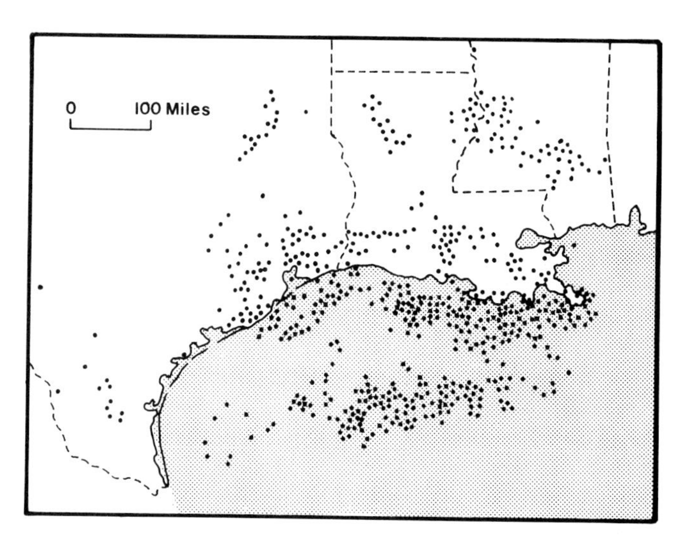

FIGURE 11-14 Distribution of salt domes on and near the Gulf Coastal Plain. (After William D. Thornbury, *Regional Geomorphology of the United States* [New York: John Wiley & Sons, Inc., 1965], p. 67. Reprinted by permission.)

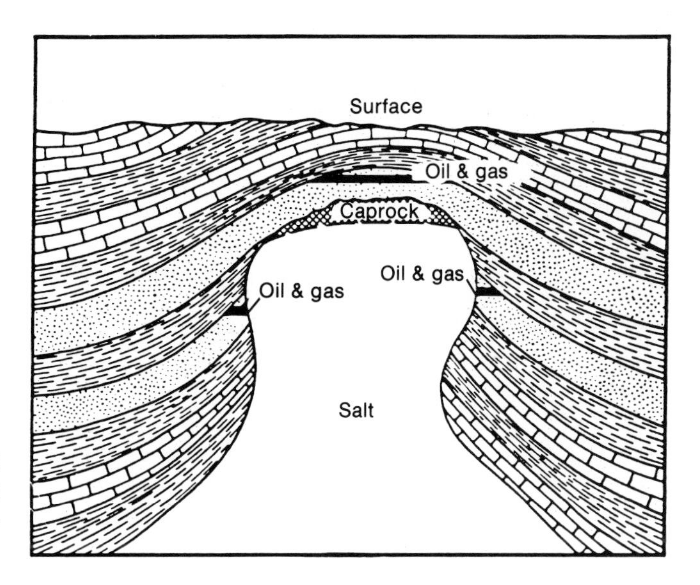

FIGURE 11-15 Generalized geologic cross section of a salt dome, showing typical reservoirs of oil and gas.

plished by the Frasch process. Wells are drilled into the deposits on top of the salt plugs Fig. 11–15). Superheated water is pumped into the deposit and the sulfur is converted into a liquid. Compressed air is used to force the molten material to the surface. The pipes are heated to prevent the sulfur from solidifying until it reaches huge temporary vats on the surface. From the vats the molten sulfur is transported by pipeline or allowed to harden and then broken into fragments for shipping.

Petroleum and Natural Gas The Southeastern Coast Region encompasses most of the area known as the Gulf Coast Petroleum Province, one of the continent's major oil producers. It normally yields about one-fifth of the U.S. total flow. The oil fields, almost exclusively in association with salt domes, are scattered along the west Gulf Coast Plain from the mouth of the Mississippi to the mouth of the Rio Grande.

Because much crude oil is transported by pipelines today, the Gulf Coast port cities have become the termini of most pipelines from the Gulf Coast and midcontinent oil provinces. These pipelines have permitted the rapid development of the refining industry along the west Gulf Coast between Baton Rouge and Corpus Christi. Enormous quantities of both crude and refined petroleum are also shipped by tankers to the North Atlantic seaboard.

The continental shelf off the Gulf Coast is broad and shallow and underlain by geologic structures of the same types as those of the coastal plain. Consequently, salt domes, with their associated petroleum, natural gas, and sulfur, are widespread in the "tidelands," as the shelf has come to be called. Many favorable structures were discovered geophysically and numerous wells were sunk.

Offshore drilling has been extended farther and farther out into the Gulf. More than 25,000 offshore wells have been drilled, the vast majority directly south of Louisiana. These operations are extremely expensive, primarily because of the elaborate, self-contained drilling platforms used. As many as 60 wells can be drilled from a single platform, using whipstocking techniques to angle the wells in various directions.

The Southeastern Coast Region is also a major source of natural gas, which is distributed by pipeline to consumers in many parts of the country. Reserves of gas in the tidelands area are thought to be gigantic, but as yet there is little production of natural gas from the continental shelf because of the high cost of drilling.

Shells Oyster and clam shells have been used for many years as a road-building material, but the big development in this industry did not begin until the heavy chemical industry started its phenomenal

FIGURE 11-16 Mountains of oyster shells that have been dredged from Matagorda Bay along the Texas coast (TLM photo).

growth in the 1940s. Today one sees large mounds of gray, gravelly material piled up at many industrial establishments; they are oyster shells that have been dredged from the shallow bays along the Gulf Coast (Fig. 11–16). With the nearest source of limestone in Texas more than 200 miles (320 km) from the coast, oyster shells from dead reefs provide industry with a cheap and abundant supply of lime.

The extraction operation is totally mechanized. Dredges of various sizes suck up the dead oyster shell, mostly from the Virginia oyster (*Crassostrea virginica*), clean it, and expel it onto large barges floating alongside, which, in turn, transport it to the numerous markets in the coastal area. The devastating nature of a dredging operation causes significant controversy in an area where a number of competing activities are relatively incompatible. Commercial fishermen, sports fishermen, tourists, and nature lovers frequent the same bays as the shell dredgers; petroleum extraction, commercial shipping, and pleasure boating add to the activities in these shallow waters.

MANUFACTURING

For several decades the Southeastern Coast Region has experienced booming industrial growth. Dozens of new factories, many of them large and highly mechanized, were opened, particularly along the western coast. The region offers many attractions to industrial firms, but the principal locational advantages are

1. Large supplies of basic industrial materials, such as petroleum, natural gas, sulfur, salt, and lime (oyster shells).
2. An abundance of natural gas for fuel.
3. A tidewater location providing cheap water transport rates for bulky manufactured products.

Other factors, such as available labor, a growing local market, and local venture capital, are also important but of less overall significance in attracting major industries.

In most parts of the region the industrial sector is specialized rather than diversified, on the basis of specific resources or other locational attractions. The principal types of manufacturing are limited in variety, although often enormous in magnitude.

The basic specialized manufacturing industry of the region is petroleum refining. From Pascagoula to Corpus Christi there are numerous nodes of refineries, inevitably connected by a network of pipelines to other industrial facilities as well as to storage tanks and shipping terminals (Fig. 11–17). Much of the crude oil is obtained within the region, but a considerable proportion of the refinery input comes from oil fields to the north of the coastal zone and from imported oil. The entire region contains approximately 40 percent of the nation's refinery capacity. The principal refining centers are Houston–Texas City and Beaumont–Port Arthur in southeastern Texas.

In conjunction with refining, a vast complex of petrochemical industries developed. These are manufacturing operations in which various "fractions" (components) are separated from petroleum or natural gas and used to make other marketable products, ranging from butane (for heating gas), butadiene (for synthetic rubber), and toluene (for explosives) to more sophisticated industrial hydrocarbons.

Petrochemical plants are usually located near refineries. Often a complex pattern of pipelines conducts liquid or gaseous products from one plant to another, the finished product of one factory serving as the raw material ("feedstock") for another. Pet-

FIGURE 11-17 The Humble refinery complex at Baton Rouge, located on the east bank of the Mississippi River. Much of the refinery's output moves by water transport on the river, both by inland waterway barges and by ocean-going tankers (courtesy Humble Oil and Refining Company).

rochemical plants are scattered along the coast from Mobile to Brownsville, but the main concentrations are in the major refining areas: along the Houston Ship Canal, around Texas City, in the vicinity of Sabine Lake, around Lake Charles, and along the Mississippi River between Baton Rouge and New Orleans.

Other prominent specialized manufacturing industries of the region include plastics (particularly polyethylene and styrene), inorganic chemicals (most notably sulfuric acid and soda ash), alumina (more than 90 percent of domestic alumina output comes from coastal plants in Texas and Louisiana), pulp and paper, rice mills, sugar refineries, and seafood processing.

In consonance with the situation all across the country, manufacturing has declined in importance in this region during the past decade. However, it continues to be a significant and distinctive component of the regional economy.

URBAN BOOM IN THE SPACE AGE

No other part of Anglo-America except Arizona and California experienced such notable urban growth in recent years as the Southeastern Coast Region (see Table 11–1 for the region's largest urban places). A relative abundance of employment opportunities combines with the attraction of mild winters and a coastal environment to make this a region of rapid, continuing in-migration from other parts of the country. The location of several major space-related facilities in Texas, Florida, Louisiana, and Mississippi contributed significantly to the boom in the region.

TABLE 11-1
Largest urban places of the Southeastern Coast Region

Name	Population of Principal City	Population of Metropolitan Area
Baton Rouge, La.	235,270	536,500
Baytown, Tex.	62,530	
Beaumont, Tex.	114,210	363,900
Biloxi, Miss.	45,260	205,000
Boca Raton, Fla.	61,620	
Bradenton, Fla.	40,090	186,900
Brazoria, Tex.		184,600
Brownsville, Tex.	104,510	264,000
Charleston, S.C.	81,030	510,800
Clearwater, Fla.	97,640	
Coral Springs, Fla.*	66,690	231,800
Corpus Christi, Tex.	260,930	358,000
Daytona Beach, Fla.	60,560	348,400
Dothan, Ala.	53,820	131,100
Florence, S.C.	32,380	118,000
Fort Lauderdale, Fla.	145,610	1,187,000
Fort Myers, Fla.	44,150	309,100
Fort Pierce, Fla.		183,100
Fort Walton Beach, Fla.	22,800	150,600
Gainesville, Fla.	86,230	207,600
Galveston, Tex.	56,720	210,000
Harlingen, Tex.	56,420	
Hialeah, Fla.*	162,080	
Hollywood, Fla.*	120,140	
Houma, La.	33,850	
Houston, Tex.	1,698,090	3,247,000
Jacksonville, Fla.	635,430	898,100
Kenner, La.*	75,120	
Lafayette, La.	86,230	209,600
Lake Charles, La.	72,480	172,400
Lakeland, Fla.	64,350	395,800
Largo, Fla.*	63,260	
McAllen, Tex.	87,270	387,900
Melbourne, Fla.	59,690	388,300
Miami, Fla.	371,100	1,813,500
Miami Beach, Fla.	92,590	
Mobile, Ala.	208,820	485,600
Naples, Fla.	20,430	138,500
New Orleans, La.	531,710	1,306,900
North Charleston, S.C.*	74,730	
Ocala, Fla.	47,850	189,800
Orlando, Fla.	155,950	971,200
Panama City, Fla.	35,690	125,500
Pasadena, Tex.*	116,880	
Pascagoula, Miss.	30,930	128,100
Pensacola, Fla.	62,780	349,900
Plantation, Fla.*	61,130	
Pompano Beach, Fla.	68,640	
Port Arthur, Tex.	59,610	
St. Petersburg, Fla.	235,450	
Sarasota, Fla.	53,280	260,600
Savannah, Ga.	145,980	244,400
Tallahassee, Fla.	125,640	228,600
Tampa, Fla.	281,790	1,995,100
Victoria, Tex.	55,330	74,300
W. Palm Beach, Fla.	73,050	785,800
Wilmington, N.C.	55,810	117,300

* A suburb of a larger city.

Gulf Coast Metropolises: Houston and New Orleans

Houston and New Orleans are the well-established metropolitan centers of the region, although the former long since surpassed the latter in terms of both absolute and relative growth. Houston, perhaps more than any city east of the Rockies, epitomizes the story of Anglo-American urban expansion in recent decades whereas New Orleans is an older city that has grown with unspectacular vigor.

Houston is the largest city in the South and no city in its size-category has grown at a faster rate in the last 25 years (Fig. 11–18). Its image is one of a brash, sprawling, oil-oriented boom town. It is noted for low taxes, no zoning, a high crime rate, a very unsatisfactory municipal transportation system, and unfettered economic boosterism. It is the classic American example of the application of the private enterprise ethic to a metropolis.

Houston is one of the busiest ports in the nation, although its traffic is largely coastal shipment of bulk materials. Its artificial harbor is located at the end of a narrow, dredged, 52-mile (83-km) channel, which is one of the most dynamic and most dangerous (owing to the great quantity of volatile materials being shipped on a cramped waterway) canals in the world. As the site of the Manned Spacecraft Center of the National Aeronautics and Space Administration (NASA), Houston's preeminent place in the space age is ensured. Nearby elements in the Houston conurbation include residential Pasadena and

FIGURE 11-18 The glittering skyline of Houston (courtesy Greater Houston Convention and Visitors Council).

Baytown, industrial Texas City, and the island port and beach resort of Galveston.

New Orleans is one of those uncommon Anglo-American cities that is distinguished by uniqueness of character. Its wretched and crowded site between the Mississippi River and Lake Pontchartrain,[2] its splendid location near the mouth of the continent's mightiest river, its remarkable past that encompasses the full span of southern history, and its unusual French cultural flavor that is unparalleled south of Montreal combine to set New Orleans apart. In most years its foreign trade is exceeded only by that of New York; indeed, New Orleans depends much more heavily on income from oceanborne commerce than any other major port in the country. Conversely, the industrial base is limited; proportional employment in manufacturing is only half the national average for large cities.

The Boom Cities of Florida

Florida experienced several periods of rapid population and economic expansion in its history, but the growth trends of recent years are unparalleled in continuity and stability. This growth was predominantly in urban areas. During the 1980–1990 decade 10 of the 20 fastest-growing metropolitan areas in the nation were in Florida.

The state's permier metropolis, *Miami,* now sprawls widely from its focus on Biscayne Bay and the resort suburb of Miami Beach merges imperceptibly northward with Fort Lauderdale. The urbanized zone extends almost uninterruptedly to Palm Beach, which is 50 miles (80 km) north of the Miami central business district. Miami's rapid growth makes it the second largest city in the region and fourth largest (after Houston, Dallas, and Atlanta) in the South. Its economic orientation is essentially commercial and recreational, but its attraction as a tourist center is waning, presumably because Miami's large urban area and unusual ethnic mix engendered a host of social problems (for example, the highest rate of major crimes in the nation) and the city has lost some of its charm as a tourist goal. Construction of new hotels was virtually nonexistent for more than a decade and attempts to restore the image of a swinging resort by obtaining legalized casino gambling failed.

Manufacturing is of relatively minor importance in the economy and maritime commerce is relatively limited, although increasing. Nevertheless, the city's role as gateway to Latin America is stronger than ever, its airport is one of the nation's busiest for both domestic and international flights, and Miami has displaced New York as the leading

[2] See Peirce Lewis's list of the 10 major environmental handicaps that give New Orleans what is probably the most miserable site of any Anglo-American metropolis in *New Orleans: The Making of an Urban Landscape* (Cambridge, MA: Ballinger Publishing Co., 1976), pp. 31–32.

FIGURE 11-19 There are now many ethnic neighborhoods in the cities of Florida. Cuban districts are particularly extensive in greater Miami (courtesy James R. Curtis).

U.S. port of departure for international sea passengers (with nearby Port Everglades ranking third). A significant "hidden" component of the local economy is drug smuggling; it is estimated that the total value of drugs brought into the United States annually through Dade County (the county in which Miami is located) is only slightly less than the county income generated by tourism.[3] This factor presumably explains why the Dade County branch of the Federal Reserve Bank has a greater cash surplus than any other branch in the country.

The great influx of refugees from the Caribbean (chiefly Cubans and Haitians) in the 1970s–1980s placed an enormous strain on the economic and political infrastructure of the metropolitan area and required major social adjustments that significantly heightened ethnic tensions. So many Cubans settled in Miami that the city enacted an ordinance officially establishing bilingualism in 1973, but in 1980 an emotional backlash prompted a referendum that repealed the bilingual ordinance. Still, more than 18,000 businesses in the county are owned by Cubans and Miami continues to be unofficially bilingual (Fig. 11–19).

Florida's other principal urban areas are also growing rapidly, although the rate of increase is greater in the cities of central Florida than in those of the northern part of the state. The *Tampa–St. Petersburg* metropolitan area now has a population of 2 million; its diversified economy and busy phosphate port continue to attract many newcomers. *Jacksonville* is one of Florida's oldest cities, but its recent growth pattern has been more erratic than that of cities farther south. The burgeoning *Orlando–Winter Park* area of central Florida is the state's major inland metropolis. It shared significantly in the space boom because of its relative proximity to the Kennedy Space Center on Cape Canaveral, and it is the nearest city to North America's leading theme amusement park, the Walt Disney World-Epcot Center, which records more than 23 million visitors annually.[4] The *Lakeland–Winter Haven* area, between Tampa and Orlando, is also growing rapidly; a midstate megalopolis is in the making here, stretching from St. Petersburg to Orlando.

Smaller Industrial Centers

Several small- to medium-sized cities are highly industrialized and represent, in microcosm, the recent economic history of the region. All are ports and all depend heavily on one or more aspects of the petroleum industry.

[3] James Stolz, "Dade County: The Melting Pot Boils," *Geo*, 3 (August 1981), 110.

[4] This figure represents 10 million more people than the second most heavily visited theme park (Disneyland in California).

FIGURE 11-20 Ancient architecture (both colonial and antebellum) and cobblestone streets are part of the charm of Charleston, South Carolina (TLM photo).

Baton Rouge, situated on the Mississippi River more than 150 miles (240 km) from the Gulf of Mexico is a beehive of petrochemical activity and oil refining. *Corpus Christi* is the busy port of the central Texas coast. As such, its commercial function is more notable than its manufacturing, although it is the site of a significant concentration of factories, many of which are petroleum oriented. In the extreme southeastern corner of Texas are the Sabine Lake cities of *Beaumont*, noted for oil refining, rice milling, and wood products manufacturing; *Port Arthur*, petrochemicals and synthetic rubber; and *Orange*, shipbuilding.

The Older Port Cities

Several of the region's older ports followed the trend of economic and population expansion at a somewhat slower rate. *Charleston* and *Savannah*[5] had glory days during colonial and Confederate times but experienced limited and erratic growth during most of the past century (Fig. 11–20). Today their river-mouth harbors are busy again, their fishing fleets are active, and their rate of industrial expansion (pulp and paper, shipbuilding, and food processing) is impressive.

As the only major Gulf Coast port serving much of the Southeast, *Mobile* enjoys a privileged position. It is the ore-import port for the Birmingham steel industry and serves as the funnel for goods moving in and out of Alabama's inland waterway system. A few major industrial plants—shipbuilding, paper making, alumina reduction—are notable. Several small cities serve as central places in the Lower Rio Grande Valley, but only *Brownsville* has a deep-water harbor and effectively functions as a port. Urban growth in the valley is related primarily to agriculture, fishing, tourism, and services rather than to manufacturing or port activities.

INLAND WATERWAYS

Inland waterway freight transportation is notable in the Southeastern Coast Region. Most of the traffic is by means of barges, flat-bottomed craft that usually are connected in tandem and pushed or pulled by tugs. Barges provide the least expensive form of transport, but they can only be used in quiet waters because of their susceptibility to swamping or capsizing.

[5] As of the late 1980s, Savannah and Charleston were the third- and fourth-ranking East Coast container ports, exceeded only by New York, Baltimore, and Montreal.

The region is particularly well suited to barge traffic because of the large number of available canals and other dredged interior channels and because of its direct connection with the Mississippi River waterway system. Flat land and a high water table make the construction of inland waterways in the region relatively inexpensive; thus an extensive system of interconnected channels exists. Most of the region's important ports are set inland some distance from the open sea and can function as ports only through channels that were dredged along stream courses in the flat coastal plain. Houston, with its dredged Buffalo Bayou, is the prime example, but the same process has been carried out elsewhere. The port of New Orleans, for example, had a second channel cut directly east from the city, thus shortening the distance to the Gulf considerably.

The most remarkable of the region's water routes is the Gulf Intracoastal Waterway, which extends from Brownsville to the Florida panhandle just inland of the coast, using lagoons behind barrier islands where possible and cutting through the coastal plain where necessary. For most of its 1100-mile (1760-km) length it has only a 12-foot (3.6-km) depth; thus it is designed primarily for barge traffic (Fig. 11–21). It was only partially completed by World War II but was particularly useful during that conflict because of the activities of German submarines in the Gulf of Mexico.

The waterway was finally completed in 1949. It is most heavily used in its central portion; two-thirds of total traffic is in the section between Houston and the Mississippi River. Sulfur, cotton, grain, and other bulk goods are important commodities

A CLOSER LOOK *The Dynamic Demographics of Florida*

Although most parts of the Southeastern Coast Region experienced significant population changes of various kinds in recent years, the state of Florida is particularly notable for its dynamic demographics. Some characteristics of its population are unusual, some are unique, and some probably represent portents of the future for larger areas of the United States.

For the state as a whole, the population increase of the last few decades, particularly the last few years, was remarkable.[a] At midyear of the twentieth century Florida ranked 20th among the states, with a population of 2.7 million. By 1960 it ranked 10th, with 4.9 million people. Its 1980 population of 9.7 million gave it seventh place among the states. With 13 million people in 1990, Florida had become the fourth most populous state.

During the most recent census decade, the 1980s, Florida was the only state that ranked among the leaders in both total and relative growth. Its total decennial increase was exceeded only by California and Texas, and its proportional increase also gave it third ranking, behind only Nevada and Arizona.

The most striking aspect of Florida's population expansion is the fact that almost all of it is due to net in-migration rather than to natural increase. In terms of natural increase (excess of births over deaths), Florida has almost reached the zero threshold; its annual natural increase rate is 0.2 percent, the lowest among the 50 states. The components of this equation are simple—Florida has the highest death rate of all states and it ranks 39th in birthrate.

Thus Florida has been nurtured by a sustained high rate of in-migration for many years, particularly during the 1970s, when 92 percent of the state's total population increase was the result of the excess of immigration over emigration. This represents a long-continuing 2 percent annual population growth due to net migration.

This population expansion extended over most of the state, although it is most prominent in the central and southern parts and least notable in the northern panhandle. During the 1970s three-fourths of Florida's counties experienced "fast growth" (22 percent or more) and the remainder were in the "moderate-" or "slow-" growth category; none had a population decline. In no other southern state did as many as one-third of the counties experience fast growth.

People who move to Florida come from many different places and represent a great diversity of backgrounds and characteristics. Florida has a positive migration balance with almost every state, but the great bulk of its domestic immigrants come from Megalopolis and from the Heartland. Only a relatively small proportion of these immigrants are nonwhites; of the 2.7 million net immigrants of the 1970s, it is estimated that only 100,000 were black (which nevertheless represents the first decade in a century that Florida experienced net black immigration).

Probably the most conspicuous feature of the immigrant cohort is its relatively advanced age. Fully 20 percent of the immigrants are retirees compared with a national average of shipped, but the most important are petroleum and petroleum products. Heavy use of the waterway is shown by the fact that 9 of the 15 largest U.S. seaports (in tonnage) are located along its course.

Of major significance to the Gulf Intracoastal Waterway is its connection with the Mississippi River system. Barge traffic from the waterway can move into the Mississippi via two major channels in southeastern Louisiana, which means that barges from the Gulf Coast can be towed to the far-flung extremities of the Mississippi navigation system—up the Tennessee, Ohio, or Missouri rivers and even into the Great Lakes via the Mississippi and Illinois rivers.

The Alabama, Tombigbee, and Black Warrior rivers of the state of Alabama were also dredged and stabilized to provide navigation systems for barges.

FIGURE 11-21 Barge tows moving on a busy stretch of the Gulf Intracoastal Waterway near Morgan City, Louisiana (courtesy U.S. Army Corps of Engineers, New Orleans District).

about 5 percent. Thus the continuing flow of immigrants contributes particularly to the "graying" of the state's total population. More than 18 percent of Florida's population is now 65 or over, which is half again greater than the national average. The Florida population is easily the oldest of any state, which primarily explains the state's low birthrate and high death rate.

From an ethnic standpoint Florida's population becomes proportionally whiter year by year. In 1940 some 27 percent of the population was black; by 1990 the proportion had declined to 14 percent despite the fact that a record number of blacks (1.8 million) were residents of the state. It should be noted that the black component of the population is a major contributor to the natural increase rate; nearly three-fourths of the state's natural increase during the past two decades was accounted for by blacks.

The "internationalization" of Florida's population since the 1960s has received much attention, as well it should. It is epitomized by the flood of Cuban immigration to southern Florida; nearly 60 percent of the people of Cuban ancestry in the United States live in Florida and the majority are concentrated in metropolitan Miami. There have also been notable influxes from Haiti, Jamaica, and other parts of Latin America. Even so, the magnitude of the domestic migration to Florida is such that the proportion of Hispanics in the state population is only slightly higher than the national average.

Prospects for the future look like more of the same. The ambience and reputation of Florida as a Sunbelt refuge for snow-weary Northerners are clearly well established and Florida seems to be both functionally and psychologically more accessible to the emigrant states of the Northeast than are the states of the Texas-to-California axis. Moreover, Florida attracts people of working age because it has an expanding job market; it attracts business people and retirees because it has no state income tax (only five other states have no income tax) and its sales tax is still relatively moderate.

Planners in other states are watching with interest as Florida copes with its dynamic population growth and unusual population mix. The age structure of Florida's population in the 1980s is likely to be replicated in the nation as a whole in the 1990s as our total population ages. Will an elderly population support needed capital expenditures for roads, sewers, schools, and other infrastructural elements that will be needed in the future? And in greater Miami there is the conspicuous example of another language finding a semiequal footing with English in a situation where the second language is not simply that of an ethnic enclave excluded from the power structure but is actually a language of big business and international trade.

TLM

[a] Much of the data presented here is based on research by John W. Stafford of the University of South Florida and was kindly provided by him.

FIGURE 11-22 Hundreds of hotels and apartment buildings line the ocean front almost continuously from Miami Beach to Fort Lauderdale (courtesy Miami Beach Tourist Development Authority).

A modest amount of traffic traverses these rivers, mainly between Birmingham and Mobile. Much more ambitious is the Tennessee–Tombigbee Waterway (called Tenn-Tom), which connects the previously dredged Tombigbee River with the previously dammed Tennessee River, thus providing a barge route directly north from Mobile to the Tennessee River system. This, the nation's largest and costliest ($1.8 billion) navigation project was opened for business in 1985, to disappointingly low traffic usage.

Along the Atlantic Coast is the Atlantic Intracoastal Waterway, which is theoretically analogous to the Gulf Intracoastal Waterway. The former, however, does not have the advantage of tapping a major region of bulk production or of connecting to the Mississippi navigation system. Consequently, commercial traffic on the Atlantic Intracoastal Waterway is exceedingly sparse except in a few short reaches. It is used much more extensively for pleasure boating.

A Cross-Florida Canal has long been proposed to join the two intracoastal waterways. Construction actually began in the late 1960s but halted in 1970 because of anticipated ecological problems and the questionable economics of the project.

RECREATION

The recreation and tourist industry of the Southeastern Coast Region is a major segment of the regional economy. The mild winter climate is part of the attraction, as are an abundance of shows, amusement parks, and historical sites. It is the coastline and its beaches, however, that draw visitors to the region. There is a greater total mileage of usable beaches in this region than in the rest of Anglo-America combined. And except in the vicinity of the largest cities or principal resorts, the beaches are likely to be both clean and uncrowded—even on a hot Sunday afternoon in August.

The coast of Florida is most heavily used, of course (Fig. 11–22). The Atlantic margin of that state consist of an almost-continuous beach from the St. Johns estuary in the north to Biscayne Bay in the south. Beaches are spaced more irregularly on Florida's Gulf Coast but are splendid in quality.

The Texas coast also contains hundreds of miles of beaches, although most are on offshore islands and not all are readily accessible. The principal beach resort is Galveston Island, near the Houston conurbation. The Corpus Christi area also has fine beaches, and Padre Island, extending for 100 miles (160 km) south from Corpus Christi, has the longest stretch of undeveloped beach in the nation.

The Mississippi coastline, from Pascagoula to Bay St. Louis, is one continuous beach and has an interesting pattern of development. In the area from Pass Christian to Ocean Springs, a stretch of some 35 miles (56 km), there is an almost unbroken line of lovely old homes set on the beach ridge, a few feet higher than the magnificent stretch of white sand, with the busy coastal highway intervening. Wherever this pattern is interrupted, the string of big homes is replaced by a newer resort or commercial development, as at Biloxi.

Beaches are intermittent and usually located on offshore islands along the South Atlantic Coast north of Florida. The most famous resort area is near Brunswick, where each of the three nearby offshore islands has a different use pattern. Sea Island is the classic resort island; its relatively small acreage is devoted to pretentious summer homes, many of which are mansions. Just to the south is St. Simons Island. This is a much larger landmass that shows evidence of sporadic development over a long period of time, ranging from summer homes to modest motels. Of more recent vintage is the development of Jekyll Island; most of this island is maintained as uncrowded stretches of white sand, but in the center is a prominent concentration of modern motels and convention facilities. There are similar developments on Hilton Head Island in South Carolina and the 60-mile (96-km) beach centering at Myrtle Beach (South Carolina) is one of the most heavily used "camping" beaches in the country, with over 11,000 developed campsites. Much of the North Carolina mainland shore is without beaches, but projecting as a curved crescent into the Atlantic are the remarkable sand bar islands of Cape Lookout and Cape Hatteras, with a continuous oceanside beach.

The region has many places of historical interest, but the most notable are the four urban centers that have preserved the architectural flavor of yesteryear. The Vieux Carré (French Quarter) of New Orleans is the preeminent attraction of this type and sets the theme—with its Royal Street shops and Bourbon Street nightspots—for one of North America's most distinctive tourist centers (Fig. 11–23). Both Savannah and Charleston are cities of similar and unusual historic interest. Their extensive areas of eighteenth- and nineteenth-century architecture are unmatched elsewhere and are nicely counterpointed by Charleston's waterfront Battery area and Savannah's delightful system of city squares. In northeastern Florida the Jacksonville–St. Augustine area has maintained a smaller but more varied sampling of historic architecture.

In a region where recreation and tourism are big business it is not surprising that there has been heavy capital investment in constructed or "modified" attractions to which the public is invited for an admission charge. Florida, again, is the leader in such development. Most early endeavors were associated with some sort of water show featuring swimmers, skiers, or fish at one of the numerous artesian springs or lakes. Then the scope of the attractions broadened and now runs the gamut from specialized commercial museums (circus, vintage car, and so on) to the variety of Walt Disney World and its associated Epcot (Experimental Prototype City of Tomorrow) Center.

Florida is the most popular tourist destination

FIGURE 11-23 New Orleans' Vieux Carré (French Quarter) is an extremely popular tourist destination, although frequent rain may keep the visitors off the streets from time to time (TLM photo).

in North America and its tourist industry is a major source of income. In the southern half of the state it flourishes primarily in the winter, with a secondary smaller peak in summer; in the northern half business is greatest in the summer. Visitors come from almost every state and many foreign countries, but the majority are urban dwellers from east of the Mississippi and north of the Ohio and Potomac rivers. Avoidance of winter is one of the major stimuli for coming to Florida, in a pattern that began nearly a century ago.

The tourist industry in Florida experienced more than one major setback and is still overextended at times. But today it is solidly based and the most important revenue-producing activity in the state. The principal concentration of resorts is along the southeastern coast, the best known being the extravagant and yet decaying hostelries of Miami Beach. Peripheral attractions include Everglades National Park, the various islands of the Keys, and tourist-oriented Seminole Indian settlements.

THE OUTLOOK

The Southeastern Coast Region remained one of the most backward parts of Anglo-America until almost the beginning of the present century. Prior to 1900 its chief activities were forest exploitation, agriculture, and ranching. The few rundown ports were in need of modernization.

With the discovery of oil at Spindletop and the development of salt, phosphate, lime (oyster shells), and sulfur, the coastal region began to attract industry even though no major development took place until the 1940s. In this region industry has found a favorable habitat; indeed, few parts of Anglo-America present a brighter outlook for manufacturing than this area that is the world's most extensively industrialized subtropical region.

Petroleum overshadows all other factors in the economy of the western part of the region, and its serious decline in the 1980s was a staggering blow. Texas experienced net out-migration for a couple of years in the mid-1980s, and Louisiana actually lost population for the first time in several decades. In the early 1990s the oil industry is making a slow recovery, although it is still far from robust, and Louisiana, in particular, continues to exhibit a fragile economy.

Conventional port business is expanding throughout the region, notably led by the export of bulk foodstuffs (particularly soybeans) through New Orleans and manufactured goods destined for Latin America through Miami. Port facilities are being enlarged significantly at Houston, New Orleans, and Mobile.

Specialized farming will continue to occupy an important place in the economy, led by citrus, sugarcane, and vegetables. Intensification of beef cattle raising is probable.

The space boom of the 1960s has significantly ebbed, but the period of hardship resulting from the sharp decline seems to have passed and a steadier but less spectacular era of space-related activities will probably become established at the prominent centers of Cape Canaveral–Titusville in Florida, Houston, and the Bogalusa–Picayune border area of Louisiana–Mississippi.

The attraction of mild winters and outdoor living remains compelling for Northeasterners; thus people are likely to continue to pour into the region at a rapid rate to visit, work, or retire, especially in Florida and the Lower Rio Grande Valley. Rapid population growth and a continually expanding tourist trade will be major elements in the economic and social geography of the region for some time to come. Increasingly too, Florida serves as a way station for tourists on the way to the Bahamas or the Caribbean.

The striking juxtaposition of natural areas and constructed landscapes in this region makes it a prominent place for ecological confrontations. Several major battles have already occurred: the Everglades jetport has been delayed if not canceled; the Cross-Florida Canal has at least been postponed: "development" plans for Padre Island and Cape Hatteras are being contested; the ramifications of pesticides are being heatedly debated in southern Louisiana; the maintenance of a water supply for the Everglades remains an unresolved issue. In this region the "developers" and the "preservationists" are in serious conflict and with the passage of time the arena of combat is certain to widen as the need for rational land-use plans becomes increasingly pressing.

SELECTED BIBLIOGRAPHY

Buczacki, Stefan T., "Florida for Fruit and Vegetables," *Geographical Magazine,* 51 (November 1978), 100–102.

Carter, L. J., *The Florida Experience: Land and Water Use Policy in a Growth State.* Baltimore: Johns Hopkins University Press, 1975.

Catling, Patrick Skene, "Down South in New Orleans," *Geographical Magazine,* 50 (April 1978), 472–475.

Davis, Donald W., "*Trainasse,*" *Annals,* Association of American Geographers, 66 (1976), 349–359.

Hart, John Fraser, and Ennis L. Chestang, "Rural Revolution in East Carolina," *Geographical Review,* 68 (October 1978), 435–458.

Hilliard, Sam B., ed., "Man and Environment in the Lower Mississippi Valley," *Geoscience and Man,* 19 (June 1978), 1–165.

Jenna, William, *Metropolitan Miami: A Demographic Overview.* Coral Gables: University of Miami Press, 1972.

Jordan, Terry G., John L. Bean, and William M. Holmes, *Texas.* Boulder, CO: Westview Press, 1984.

Kniffen, Fred B., *Louisiana, Its Land and People.* Baton Rouge: Louisiana State University Press, 1968.

Lewis, Peirce F., *New Orleans: The Making of an Urban Landscape.* Cambridge, MA: Ballinger Publishing Company, 1976.

Longbrake, David B., and Woodrow W. Nichols, Jr., *Sunshine and Shadows in Metropolitan Miami.* Cambridge, MA: Ballinger Publishing Company, 1976.

Marcus, Robert B., and Edward A. Fernald, *Florida: A Geographical Approach.* Dubuque, IA: Kendall/Hunt Publishing Company, 1974.

Muller, Robert A., and James E. Willis, *New Orleans Weather 1961–1980: A Climatology by Means of Synoptic Weather Types.* Baton Rouge: Louisiana State University Press, 1983.

Newton, Milton B., Jr., *Louisiana: A Geographical Portrait.* (2nd ed.). Baton Rouge, LA: Geoforensics, 1987.

Padgett, Herbert R., "Physical and Cultural Associations on the Louisiana Coast," *Annals,* Association of American Geographers, 59 (1969), 481–493.

Palmer, Martha E., and Marjorie N. Rush, "Houston," in *Contemporary Metropolitan America. Vol. 4: Twentieth Century Cities,* ed. John S. Adams, pp. 107–149. Cambridge, MA: Ballinger Publishing Company, 1976.

Randall, Duncan, P., "Wilmington, North Carolina: The Historical Development of a Port City," *Annals,* Association of American Geographers, 58 (1968), 441–451.

Rudzitis, Gundars, "Resolution of an Oil-Shrimp Environmental Conflict," *Geographical Review,* 72 (April 1982), 190–199.

Stafford, John W., "The Impact of Offshore Oil Exploration of Florida," *The Mississippi Geographer,* 6 (Spring 1978), 38–43.

Stansfield, Charles A., Jr., "Changes in the Geography of Passenger Liner Ports: The Rise of the Southeastern Florida Ports," *Southeastern Geographer,* 17 (May 1977), 25–32.

Toner, M. F., "Farming the Everglades," *National Parks and Conservation Magazine,* 50 (August 1976), 4–9.

Turner, R. Eugene, and Edward Maltby, "Louisiana Is the Wetland State," *Geographical Magazine,* 55 (February 1983), 92–97.

Wood, Roland, and Edward A. Fernald, *The New Florida Atlas: Patterns of the Sunshine State.* Tampa: Trend Publications, 1974.

Ziegler, John M., "Origin of the Sea Islands of the Southeastern United States," *Geographical Review,* 49 (1959), 222–237.

12
THE HEARTLAND

The Heartland is a broad region of moderately high population density and enormous economic productivity. It includes all of three states (Indiana, Illinois, Iowa) and parts of twelve others as well as a significant portion of the province of Ontario. For comparative purposes, it can be noted that the regional population is only two-thirds that of Britain and France combined, but its gross output (industrial, agricultural, and mineral) is more than twice as great.

It is a region with a remarkably favorable combination of physical factors for the development of agriculture. Along with intelligent farm management and the considerable application of inanimate energy, they make it the largest area of highly productive farmland in North America, if not the world. The Heartland Region is also the industrial core of the continent. Although containing less than one-third of Anglo-America's population, it has over half the manufactural output of both countries.

This widespread and well-balanced regional economy provides much of the stability and diversity that permitted the people of the United States and Canada to enjoy one of the world's highest standards of living. And yet it is not economic muscle alone that gives this region its "heartland" appellation. This is in many ways the "core" region of Anglo-American society. Here are exemplified the ideas, attitudes, and institutions that are most representative of the way of life in Anglo-America. Here the population amalgam is most thoroughly distilled and the "average" American or Canadian (not, however, the average French Canadian) is most likely to be found. The region possesses an inordinate amount of political power, partly because the American political system has given rural people a greater proportional representation than urbanites in legislative bodies and the rural population is larger here than in other parts of the continent.

The Heartland is the transportation and communication hub of the continent. Its productive lands and relatively level terrain stimulate and make feasible a dense network of highways and railroads. Waterway traffic on the lower Great Lakes and the major midwestern rivers—Ohio, Mississippi, Missouri, and Illinois—continues to increase. Air transportation among the cities of the region is remarkably dense. Indeed, Chicago is the leading railway center of the world and has the busiest commercial airport in Anglo-America.

Several important parts of the two countries are not located in the Heartland: the great decision-making centers of New York, Washington, and Montreal; the rapidly growing cities of the West and South; the totality of French Canada; and many others. Nevertheless, the role of the Heartland Region in the life of Anglo-America is critically important. In this regard Robert McLaughlin noted, "Had any of America's other regions been detached over the course of history, the United States would not be the nation as we know it—but it would still exist. Without the Heartland, however, the United States would be inconceivable."[1] Wilbur Zelinsky declared the region to be "justly regarded as the most modal, the section most nearly representative of the national average."[2]

The Heartland encompasses a smaller portion of Canada, but that portion (southernmost Ontario) contains such a concentration of wealth, power, and population that it is often regarded as the "norm" of Canadian life and has had a more important policy-making role than any other part of the country.[3] The "Americanism" of southern Ontario has attracted much attention from scholars, as has its cultural affinity with the U.S. Midwest.[4]

The Heartland, then, is a broad region that includes what is generally referred to as the Midwest. It covers, however, somewhat more than the cultural Midwest: western Kentucky and the Nashville Basin are more properly "southern" in culture, western New York is only marginally midwestern, and the term is rarely used in southern Ontario.

[1] Robert McLaughlin and the Editors of Time-Life Books, *The Heartland* (New York: Time Incorporated, 1967), p. 16.

[2] Wilbur Zelinsky, *The Cultural Geography of the United States* (Englewood Cliffs, NJ: Prentice-Hall, 1973), p. 128.

[3] John Warkentin, "Southern Ontario: A View from the West," *Canadian Geographer,* 10 (1966), 157.

[4] See, for example, Andrew H. Clark, "Geographical Diversity and the Personality of Canada," in *Readings in Canadian Geography,* ed. Robert M. Irving (Toronto: Holt, Rinehart & Winston of Canada, 1972), pp. 9–10; and Zelinsky, *Cultural Geography of the United States,* p. 128.

EXTENT OF THE REGION

The Heartland occupies only part of the vast interior plain of North America; it is set off from adjoining regions, particularly to the west and north, by imprecise and transitional boundaries (Fig. 12–1).

The eastern and southern margins of the region are relatively definite because of land-use contrasts that are associated with physiographic differences. The interior lowlands support more intensive, diversified, and prosperous agricultural activities, as well as more and bigger cities, than do the slopelands of the Appalachians and Ozarks; thus the regional boundary generally follows the topographic trend from northernmost New York State south and west to the Missouri–Kansas border, with a southerly salient to include parts of western Kentucky and Tennessee.

The northern boundary of the Heartland Region approximately follows the southern edge of the Canadian Shield, where differences in bedrock geology are accompanied by related vegetation and agricultural variations. On the west the Heartland merges with the Great Plains; scanty rainfall separates the prairie margin from the short-grass country of the plains at about the 98th meridian. Here precipitation is inadequate for the profitable production of unirrigated corn and this crop tends to be replaced by those that are more drought resistant. The western boundary of the Heartland Region is generalized as the transition from corn-dominated farming on the east to wheat-grain sorghum-pasture-dominated land use on the west.

The Heartland is by no means the largest region in Anglo-America; several exceed it in areal extent. It has, however, the greatest population of any major region as well as the largest economic output. Despite the magnitude of these factors, it is difficult to subdivide the region in a satisfactory manner, for there is an essential homogeneity of

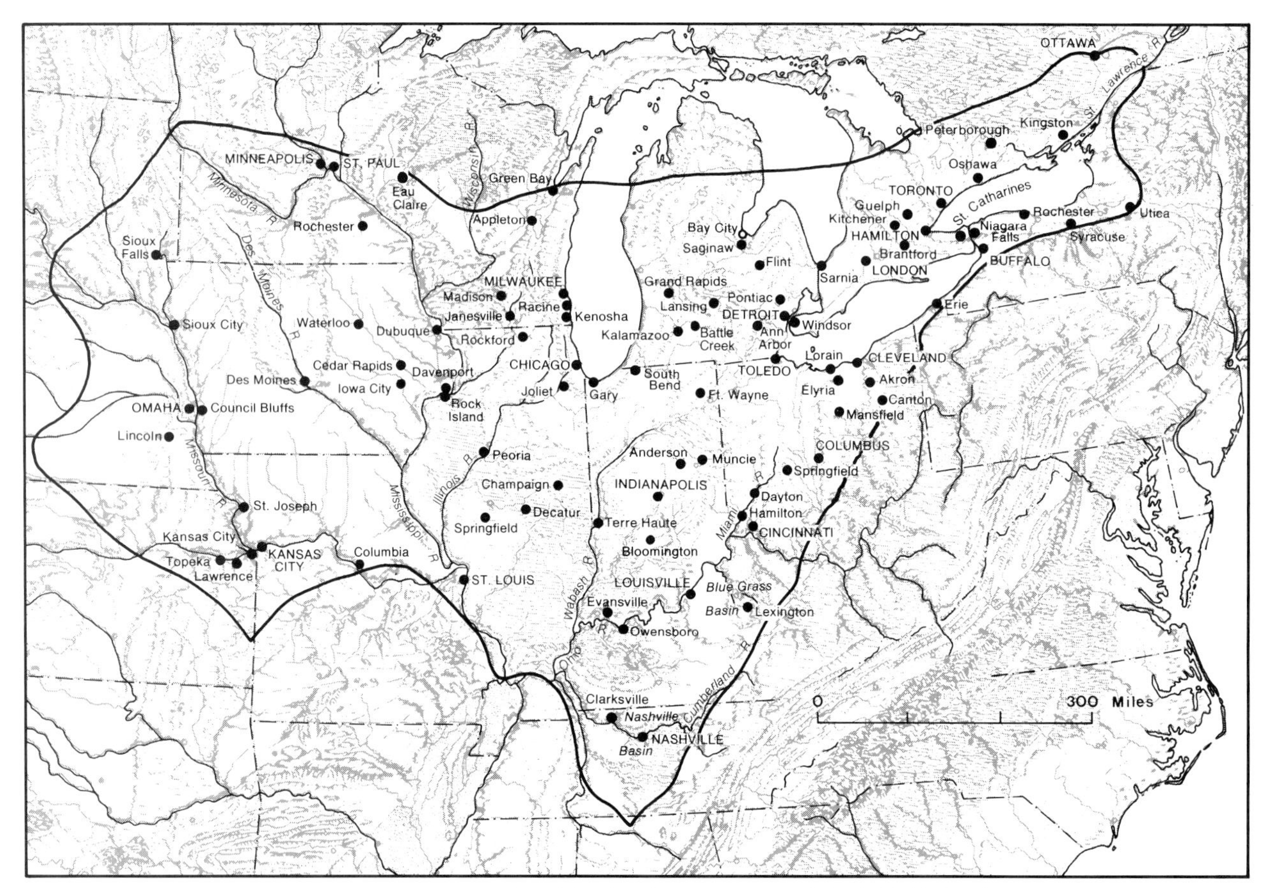

FIGURE 12-1 The Heartland Region (base map copyright A. K. Lobeck; reprinted by permission of Hammond, Inc.).

pattern and interdependence of relationship of most elements of geographical significance. It is, however, possible to delineate meaningful subregions on the basis of limited factors—for example, agriculture.

The Corn Belt is a widely accepted term applied to the core of the Heartland; it is an east-west band extending from central Ohio to eastern Nebraska in which the predominance of corn in the agricultural scene has long been recognized. North of the Corn Belt, and most prominent in Minnesota–Wisconsin–Michigan, is the Dairy Belt. Southeast of the Corn Belt is a tobacco and general farming area. But if subdivision by agricultural pattern is meaningful, on a more general geographical basis it is not unless a relatively large number of subregions is recognized. Accordingly, we will not be concerned with delimiting subregions.

THE LOOK OF THE LANDSCAPE

The landscape of the Heartland has a certain similarity of appearance throughout its length and breadth, which is another element of unity in the region. The conspicuous environmental features are absence of slopeland, prevalence of trees, and abundance of small bodies of water. The terrain is flat to undulating throughout and extends to a relatively featureless horizon in all directions. Despite the extensive acreage of cropland and pasture, trees are present in many portions of the region: wide-branching oaks, maples, and other hardwoods in the forests and woodlots, and tall cottonwoods and poplars along the stream courses. Large lakes, with the exception of the Great Lakes, are uncommon, but thousands of small marshes and ponds dot the landscape everywhere north of the Ohio and Missouri rivers.

FIGURE 12-2 Large, well-kept farmsteads are typical of the Heartland, as in this example from southern Minnesota (TLM photo).

No other human activity is nearly so conspicuous as agriculture; crops and pastures cover more of the land than does anything else. Appearance, of course, varies greatly with the season. In spring, for example, deep green fields of winter wheat, pale green pastures, and still paler green blocks of oats are just breaking through the earth and black, brown, and tan, depending on the type of soil, squares of land are plowed and ready to be planted to corn and soybeans.

Of great prominence in the landscape, particularly if viewed from the air, is the rectangularity of areal patterns. Primarily as a result of the systematic rectangular land survey of the eighteenth century, landholdings in most of the region have right-angled boundaries. Thus fields and farms appear as a gigantic checkerboard intersected by a gridiron pattern of roads, which chiefly cross at 1-mile (1.6-km) intervals. Modern interstate highways may appear as diagonal scars superimposed on the grid, but the functional roadway network of the rural Heartland is almost as straight and angular as ever.

Conspicuous farmsteads dot the patterned rectangles of the farmlands. Although large and pleasant, the farmhouses (Fig. 12–2) may suffer in comparison with the architectural gems of New England, New York, eastern Pennsylvania, and Maryland, but the total farmstead is usually large and impressive. The two-story farmhouse, normally ringed with trees, is dwarfed by imposing and generally well-kept outbuildings, such as huge barns, corncribs, silos, machine sheds, dairy buildings, and chicken houses.

Villages and small towns occur at regular intervals over the land, delicately sprawling around a road junction or alongside a railway line. They lie flat against the earth, with only a water tower or grain elevator rising above the leafy green trees. Businesses are clustered along one or two main streets, but commercial bustle is usually lacking, for most small towns are stagnating or withering anachronisms in an era of metropolitan expansion and transportation ease. The residential sections are characterized by big old homes, sometimes frame and sometimes brick, whose unused front porches look across broad lawns to tree-lined streets (Fig. 12–3).

Large and small cities are spaced more irregularly across the region. In contrast to the serenity of small towns, they are places of constant movement. These are representative North American cities,

FIGURE 12-3 The residential tranquility of a small Heartland town, as represented by Fairfield in eastern Iowa (TLM photo).

sprawling outward around the periphery and rising upward in the center. Above all, the urban centers are foci of activity: hubs of transport routes that converge from all directions with a steady stream of incoming and outgoing traffic.

THE PHYSICAL SETTING

Few, if any, large regions of the world have a more favorable combination of climate, terrain, and soils for agriculture. J. Russell Smith's classic accolade, "The Corn Belt is a gift of the gods,"[5] also almost equally applies to other parts of the Heartland Region.

Terrain

Almost all the Heartland Region is in the vast central lowland of North America. Throughout the lowland the land is mostly level to gently undulating, with occasional steeper slopes marking low hills, ridges, or escarpments. The entire region is underlain by relatively horizontal sedimentary strata, one of the most extensive expanses of such bedrock to be found on earth.[6] These various sedimentaries—mainly limestone, sandstone, shale, and dolomite—are relatively old, originating primarily in Paleozoic time. Their structural arrangement is generally subdued, consisting of broad shallow basins and low domes.

The limited relief and gentle slopes of the region are partly ascribed to the underlying structure, but are predominantly the result of glacial action during Pleistocene time. The several ice advances of the last million years had lasting effects, for they leveled the topography from its preglacial profile. There was some planing of hilltops and gouging of valleys, but generally ice action in this region was depositional rather than erosional in nature;[7] thus it was the filling of valleys rather than the wearing down of hills that produced the present land surface.

[5] J. Russell Smith and M. O. Phillips, *North America* (New York: Harcourt, Brace and Co., 1942), p. 360.

[6] Wallace E. Akin, *The North Central United States* (Princeton, NJ: D. Van Nostrand Company, 1968), p. 8.

[7] William D. Thornbury, *Regional Geomorphology of the United States* (New York: John Wiley & Sons, 1965), p. 218.

Over most of the region the bedrock has been buried many tens or hundreds of feet by glacial debris; for the Corn Belt section glacial drift is thought to average more than 100 feet in depth. The greater part of the Heartland is, therefore, indebted in no small measure to the Ice Age for its flat lands and productive soils.

Impaired drainage and loess are two other significant legacies of the Pleistocene in the region. The areas of more recent glaciation, the so-called Wisconsin stage, contain many marshes, bogs, ponds, and lakes because "normal" drainage patterns have not as yet had time to develop since the most recent ice recession (about 8000 years ago). Beyond (south of) the margin of Wisconsin glaciation are extensive areas that are mantled with deep deposits of loess, which is fine-textured windblown silt, believed to have originated from the grinding action of ice on rock, which often produces fertile soils.

There are several relatively discrete topographic sections in the Heartland Region, with characteristics sufficiently distinctive to warrent separate mention (Fig. 12–4).

The northeastern part of the region, from the Upper St. Lawrence Valley in Ontario–New York to south-central Wisconsin, is the *Great Lakes* (1) section where large and small lakes dominate the landscape. Two of the Great Lakes are entirely and two are partially within this section, as are thousands of smaller bodies of water. There are many prominent examples of glacial or glaciofluvial deposition: drumlins, eskers, outwash plains, and, particularly, the long irregular ridges of terminal moraines. Several significant scarps occur in this section, the most conspicuous of which is the Niagara Escarpment. Its abrupt cliffs of gray dolomite overlook the gentler glaciated plain between lakes Erie and Ontario, then arc around the north side of lakes Huron and Michigan, and reappear in Wisconsin.

In southwestern Wisconsin and adjacent Minnesota, Iowa, and Illinois lies the *Driftless Area* (2), an unglaciated island in a sea of glacial drift. Missed by the continental ice sheets, it differs topographically and economically from the surrounding territory. Its landscape, consisting mostly of low but steep-sided hills, is similar to the one that existed before the glacier came.

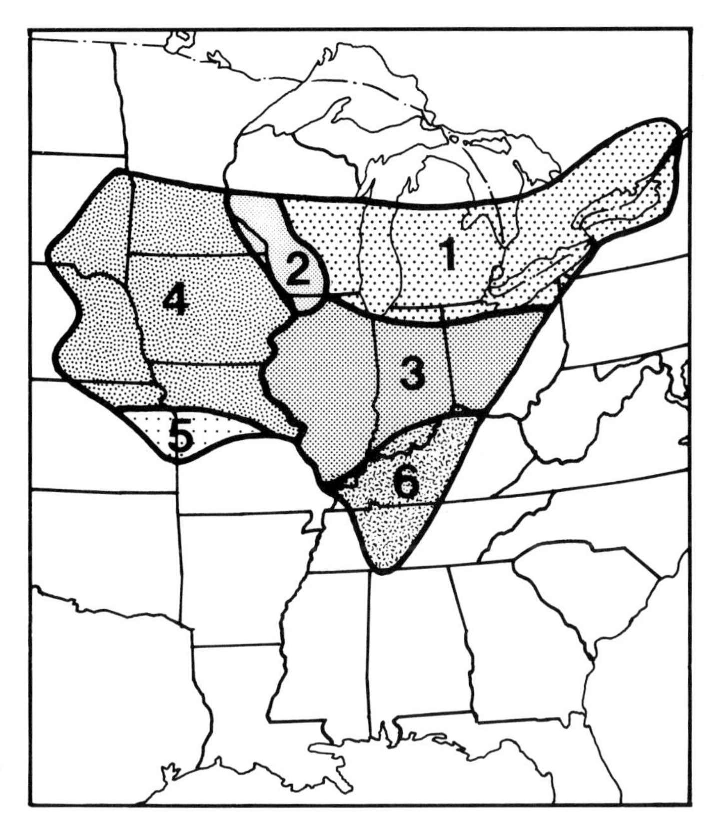

FIGURE 12-4 Topographic subdivisions of the Heartland: (1) Great Lakes, (2) Driftless Area, (3) Till Plain, (4) Dissected Till Plain, (5) Osage Plains, and (6) Interior Low Plateaus.

The central portion of the Heartland, from central Ohio to eastern Iowa, is the *Till Plain* (3) section. The terrain varies from extremely flat to gently rolling, the remarkable lack of relief presumably resulting from cumulative deposition of at least three ice advances. Drainage is much better integrated here than in the Great Lakes section and there are fewer lakes and marshes.

The *Dissected Till Plain* (4) occupies the northwestern portion of the region, extending from Minnesota to Missouri. It is covered with till from earlier glacial advances and there has been more time for stream erosion to modify the surface; hence there is a greater degree of dissection and a general absence of lakes as well as a lack of terminal moraines.

In the southwestern corner of the Heartland is the *Osage Plains* (5) section that extends from Missouri southwestward into Texas. This is an area that was essentially unaffected by glaciation and is represented by a nearly featureless plain developed on horizontal sedimentary beds.

In the southeastern portion of the Heartland Region is a section known as the *Interior Low Plateaus* (6), although most of the terrain has the appearance of low hill country. This area is structurally and topographically complex, with a series of cuestas, escarpments, basins, major fault lines, and some intrusive vulcanism. Much of the area is referred to as the Highland Rim; it consists mostly of low hills and scarp ridges and has a great many karst features, particularly sinkholes and caverns. There are two famous and fertile basins in this section, the Nashville Basin and the Bluegrass Basin, that are flat lands surrounded mostly by infacing escarpments.

Climate

The region's interior location in the eastern part of the continent results in a humid continental climate that is noted for significant seasonal and abrupt day-to-day changes in weather conditions. Climatic patterns are broad and transitional across the region, for there is relatively uniform relief and no significant topographic barriers. Thus temperatures and length of growing season increase more or less uniformly from north to south and precipitation generally decreases from east to west.

Summer is a time of hot days and relatively warm nights. Humidity is generally high and well over half the total annual precipitation falls during this season. Thunderstorms are common in summer and tornadoes are frequent.

Although summer conditions are quite benign for crop growing, winter is generally severe. The cold weather is not continuous, for there are periodic spells of relative warmth with the passing of low pressure centers. But cold fronts are frequent, often accompanied by blizzards and followed by anticyclonic spells of clear and very cold weather. Snow is commonplace throughout the region, although only in the far north does it remain unmelted on the ground for long periods (Fig. 12–5).

The transitional seasons, spring and fall, are usually brief but dramatic, with rapid shifts of temperature and abrupt periods of storminess. Flooding is a perennial natural hazard in spring and early summer, sometimes resulting from heavy spring rains falling on frozen ground but often simply the result of prolonged rain with rapid runoff. In any event, the densely populated floodplains of the region's major rivers are frequently threatened with inundation.

Natural Vegetation

The region's original vegetation consisted of forest and grass. The eastern part was forested. Ohio, Indiana, southern Michigan, southern Wisconsin, and southern Illinois all were a part of the oak–hickory southern hardwood forest and the Kentucky Bluegrass and the Nashville Basin were a part of the chesnut–oak–yellow poplar southern hardwood forest. The forest near the northern boundary of the region was characterized by conifers on the sandy soils and by magnificant stands of hardwoods on the clay lands. Elsewhere the forest consisted wholly of hardwoods.

Southern Minnesota, all of Iowa, central Illinois, northern Missouri, and eastern Nebraska and Kansas formed the prairie, a vast billowy sea of virgin grass without timber except along the streams. It was tall grass with long blades and stiff stems, growing to a height of 1 to 3 feet (0.3 to 0.9 m) and sometimes 6 to 8 feet (1.8 to 2.4 m). Trees growing along streams were chiefly cottonwoods, oaks,

FIGURE 12-5 Winter is a prominent season in most of the Heartland Region. This farm scene is in southeastern Wisconsin, near Racine (TLM photo).

and elms in the western portion, with occasional sycamores and walnuts farther east.

The true prairie extended from Illinois (small patches existed in western Ohio and northern Indiana) to about the 98th meridian, where it was gradually replaced by the short grass of the steppe. The boundaries of the prairie were never sharply defined. They were not the meeting place of two contrasted vegetation belts; rather they were broad mobile zones that moved with pronounced changes in precipitation. Many interesting theories for the origin of the prairie were advanced, but none as yet is wholly acceptable to botanists, plant ecologists, and plant geographers.

Soils

Nowhere else on the continent is there such a large area that combines generally fertile soils with a humid climate. This combination is at its best in the region's core, the Corn Belt. The Bluegrass area and the Nashville Basin are also highly productive. Some soils, however, as in central Michigan and central Wisconsin, in the Driftless Area, and in the Highland Rim of Kentucky and Tennessee, are far from rich.

Most of the region is characterized by Alfisols and Mollisols. Although the former develop under deciduous forest in the milder of the humid continental climates, the fact that they are forest soils means that, in general, they are less productive than the dark-brown to black Mollisols of the prairie. Still, some forest soils, such as those in western Ohio and north-central Indiana, yield about as well per acre as prairie soils. It may be said that forest soils that develop on calcareous till or on limestone, granite, gneiss, and schist are superior to those that evolve from shale and sandstone. All forest soils, however, are permanently leached, acid in reaction, and poor in humus.

The true prairie soils are generally fertile. They develop in cool, moderately humid climates under the influence of grass vegetation and are characterized by a dark-brown to black topsoil underlain by well-oxidized subsoils. Being relatively well supplied with moisture, they are moderately leached and acid in reaction and they lack a zone of lime accumulation. They are primarily silt loams and clay loams in texture and are derived largely from glacial till.

In summary, much of the eastern portion of the Heartland is dominated by Alfisols, which are gray-brown in color and have subsurface clay accumulation. In the west there is a preponderance of Mollisols, which tend to be black in color and rich in organic matter.

HUMAN OCCUPANCE OF THE HEARTLAND

The Heartland has always been a productive region for goods that were valuable in each period of its history: game, furs, crops, minerals, and factory output. Thus it has been a region that was coveted and struggled over by a diversity of peoples who learned of its riches. Three great nations, as well as a dozen major Indian tribes, fought for supremacy here, and the early history of the region is punctuated by battles, massacres, wars, and alliances.

Aboriginal Occupance

Relatively little is known about the earlier aboriginal inhabitants of the region. Prehistoric Indians occupied many areas for perhaps a century, beginning about 300 B.C. They are generally referred to as "Mound Builders" and just about the only landscape evidence of their presence is the large number of burial mounds and other scattered earthen structures.

At the time of European contact many well-organized tribes existed in what is now the Heartland Region. They were chiefly forest dwellers of the Algonkian linguistic group. There were, however, Iroquoian tribes on the northeastern fringe, including the important Hurons in southern Ontario and Siouan tribes (especially the Sioux and the Osage) on the western prairie margin. The forest Indians were semisedentary in pattern, their economy combining hunting with farming (corn, beans, squash, and tobacco). In some areas, as with the Hurons north of Lake Erie, a considerable section of forest was cleared for agriculture, although in most cases such cleared land had again reverted to woodland between the times the Indians were expelled and the white settlers arrived in any numbers.

French Exploration and Settlement

Most early explorers and pioneering fur traders in the region were French. During the seventeenth and the early part of the eighteenth centuries various French individuals and expeditions explored most of the major waterways of the Heartland. They were primarily interested in fur and wanted to monopolize the fur trade; consequently, they helped various Indian tribes to keep colonial settlers east of the Appalachians.

The French made little attempt at colonization, but eventually they founded a number of settlements that were the first towns of the region. Most started out as trading posts, forts, or missions and were located along the Mississippi River, the Wabash River, or near the Great Lakes. Cahokia and Kaskaskia, in what is now Illinois, date from 1699 and became thriving wilderness towns in a short while.[8] Vincennes, on the Wabash River, was founded soon afterward. In 1701 the French established a fort where Detroit now stands, but it was not incorporated as a village for another century. The only original French settlement in southern Ontario that still exists was also on the Detroit River.[9] St. Louis, founded in 1764, was another French trading post; it soon became the principal settlement in the Upper Mississippi Basin.

The Opening of the Midwest to Settlement

Britain gained title to most of the present Heartland Region in 1763 by overwhelming the French in Canada. An Indian alliance, led by the Ottawa tribe, was immediately formed to keep the British out of the region. The tribes destroyed eight British forts and laid siege to Detroit. Eventually the siege was raised, but the British government agreed to reserve that part of the continent between the Appalachians and the Mississippi solely for Indian occupance—a highly impractical resolution, considering the sentiment and politics of the time. Within two decades the American colonies successfully revolted against the Crown and the new nation inherited control of the trans-Appalachian Midwest by virtue of the surrender of land claims by the states of the eastern seaboard.

The first Congress of the United States drew up ordinances in 1785 and 1787 to provide for the systematic survey and disposition of lands in the "Northwest Territories"—the territory northwest of the Ohio River—that proved to be some of the most enduring legislation ever promulgated. A grid system, based on principal meridians and baseline parallels, was staked out to divide the entire area into a township and range pattern, a township to consist of 36 sections of 640 acres (one square mile) each (Fig. 12–6). Thus the land could be accurately surveyed and realistically sold on a sight-unseen basis. The result of this surveying system can be seen in the field, land ownership, settlement, and road patterns of most of the Heartland today.

Provision was soon made for a territory to be admitted to the Union as a state, in all respects equal to the original state, as soon as its population reached 60,000. Ohio was admitted under this provision in 1803, although both Kentucky and Tennessee had already become states in the 1790s.

Prior to the American Revolution, white settlement had begun on a small scale in the Bluegrass portion of Kentucky and parts of Tennessee; these areas attracted a small flood of settlers immediately after the Revolution. There was also a considerable influx into Upper Canada (southern Ontario) at this time, mostly people whose property had been confiscated in New York and other revolutionary colonies; some 10,000 such Tories settled along the upper St. Lawrence and at either end of Lake Erie by 1783.

The Ohio River was a major artery of movement during this period. Fort Duquesne had already evolved into Pittsburgh. Louisville was founded at the falls of the Ohio in 1779. Cincinnati began a decade later and soon became the principal river town.

Ohio did not attract many settlers until the recalcitrant Indians had been dealt with. During the early 1790s several thousand soldiers fought a series of battles with the "lords of the forest," finally achieving a decisive defeat of the Miamis and their allies in 1794. This action opened a floodgate of in-migration and within less than a decade Ohio had

[8] Akin, *The North Central United States*, p. 43.

[9] Jacob Spelt, "Southern Ontario," in *Canada: A Geographical Interpretation*, ed. John Warkentin (Toronto: Methuen Publications, 1968), p. 340.

enough inhabitants to become a state. Cleveland was founded in 1796. By the War of 1812 there were more than a quarter of a million people in Ohio, although the only places of significant settlement farther west were in southern Indiana and the Mississippi Valley below St. Louis.

Westward Expansion

During the War of 1812 Tecumseh, the Shawnee chief, tried to organize a Great Lakes-to-Gulf Indian confederation to fight the United States. He was only partly successful in his mission but did persuade many tribes, from the Creeks in Alabama to the Chippewas in the Lake Superior country, to join the alliance. The venture ended in 1813 in a battle in Ontario; Tecumseh was killed and the confederation collapsed. After the war most Heartland Indians were deported west of the Mississippi, where they were promised that the land would be theirs forever.

Settlement in the Heartland basically flowed from three fountainheads: (1) New England, whose Puritans came by way of the Mohawk Valley; (2) the South, whose frontiersmen broke through the mountains of Kentucky and Tennessee via Cumberland Gap; and (3) Pennsylvania, whose Scotch-Irish and Germans came via Pittsburgh and the Ohio River country as well as by way of Cumberland Gap. Only 1 million Americans were living west of the Appalachians in 1800 (less than 50,000 of this total in Ontario), but their numbers had increased to 2.5 million by 1820 and to 3.5 million by 1830.

After the War of 1812, settlement rapidly expanded in the forested portions of Indiana and Illi-

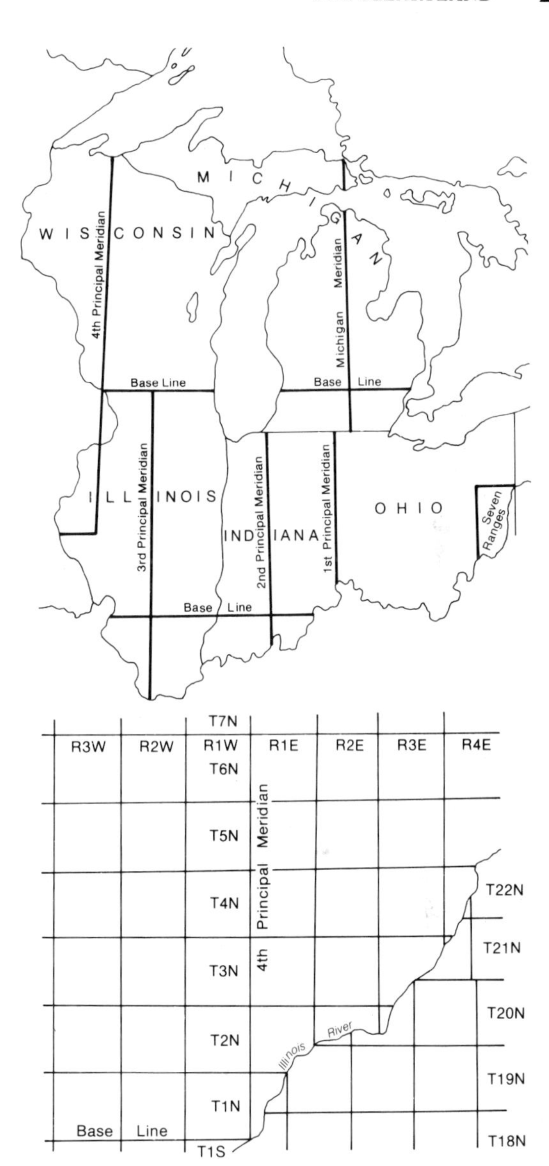

Township 4N, Range 2E

6	5	4	3	2	1
7	8	9	10	11	12
18	17	16	15	14	13
19	20	21	22	23	24
30	29	28	27	26	25
31	32	33	34	35	36

FIGURE 12-6 A range-and-township pattern of land survey was begun in the lands west of the Appalachians in 1785. A rectangular grid system, surveyed from principal meridians and base line parallels, was started in seven rows of townships ("the seven ranges") in eastern Ohio and gradually extended westward. All of the basic survey of the area shown here was completed within half a century. Detail of the range and township pattern in a portion of western Illinois is shown in the second drawing as an example of the whole. The third drawing shows the 6-mile-by-6-mile grid of a typical township, with the sequential numbering of the sections.

nois; these two states were admitted to the Union in 1816 and 1818, respectively. The Driftless Area of Wisconsin attracted settlers because of discoveries of lead ore. The Missouri Valley attracted enough settlement so that Missouri became a stae in 1821.

The tide of settlement soon shifted to the southern Great Lakes area. Fort Dearborn (Chicago) was founded in 1816; Toledo in 1817. The opening of the Erie Canal in 1825 was the harbinger of a reorientation of the regional transportation pattern from north-south along the rivers to an east-west axis, which was furthered by the building of more canals and the beginning of railway transportation in the 1840s. The major flow of settlers was then via the Mohawk corridor (Erie Canal route) from New York and New England.

On reaching the forest–prairie margin, in such areas as northern Illinois and eastern Iowa, migrant settlers were puzzled by the fact that the land was clothed with grass rather than forest. They were even suspicious, reasoning that soils bearing no timber must be inferior, and they usually avoided the prairie.[10]

Pioneers who did settle on the prairie often chose tracts that were contiguous to forest land, which they made the real base of the farm establishment. The taming of the prairie was not easy, for the heavy soil stuck to the iron plows then in use until the plow could not move in the furrow and many plows broke. The prairie was not really conquered until 1837, when John Deere, a blacksmith living in the tiny village of Grand Detour, Illinois, invented the steel plow. It soon became apparent that the good crop yields of the deep prairie soil would readily pay for the cost of breaking the sod. And so Illinois and Missouri was almost totally settled by 1850, as were southern Wisconsin and southern Iowa.

The more westerly cities in the region were founded at this time: Milwaukee and St. Paul in the 1840s, and Minneapolis and Kansas City in the 1850s. Intensification of settlement proceeded rapidly; southern Ontario, for example, had more than 40 percent of Canada's population by 1861. But as the frontier moved westward beyond the Heartland, a continual out-migration followed it; if the Heartland was easy to move into, it was also easy to move out of. This was particularly true of the Ontario portion of the region.

[10] In 1836 Alby Smith, an Ilinois pioneer, lost an election because the voters decided that anyone so stupid as to settle in the prairie should not be entrusted with the responsibilities of public office. See Harlan H. Barrows, "Geography of the Middle Illinois Valley," *Illinois Geological Survey, Bulletin 15* (Urbana: Illinois Geological Survey, 1910), p. 78.

Immigrants to the Cities

After about 1880 there was a greatly increased flow of European immigrants to the Heartland, which continued through the early decades of the twentieth century. By 1920 one-sixth of the region's population was foreign-born. Generally, these immigrants settled in urban areas, often giving them a distinctive ethnic flavor, such as Germans in Milwaukee or Scandinavians in Minneapolis.

In many ways, the Heartland has become the true melting pot of Anglo-America. The midwestern blend of culture, speech, and lifestyle is the one most readily identified as the standard for the United States; similarly, southern Ontario represents Anglo-Canada. The large-scale and relatively recent in-migration of blacks, mostly from southern states, to Heartland cities gives a discordant note to the melting-pot concept. Prejudice and ghettoization are commonplace in the region, although numerous encouraging examples of racial harmony exist.

THE INCREDIBLE OPULENCE OF HEARTLAND AGRICULTURE

Although the economy of the region is not primarily agricultural and the regional population is not predominantly rural, the Heartland is easily the preeminent producer of crops and livestock on the continent. Its large expanse of productive agricultural land is unparalleled in North America; and despite the continuing downtrend in farm numbers and farm population, the level of farm output continues to rise. By far the greatest concentration of agricultural counties in the United States is found in this region; this is less so for Canada (Fig. 12–7).

Throughout the greater part of this region the country appears to be under almost complete cultivation, with four or five farmsteads to each square

FIGURE 12-7 Environmental conditions are favorable for agriculture in most parts of the Heartland. Productive floodplain soils, as represented here along the Illinois River, are particularly significant (TLM photo).

mile in the eastern portion and two or three in the western portion. The farms are based on the subdivision of sections into half-sections, quarter-sections, and 40-acre (16-ha) plots. Corn, winter wheat, soybeans, oats, and hay are almost universal crops. Tobacco and fruits are locally important and much land is in pasture. The region not only grows tilled crops but also supports the densest population of cattle and swine in North America.

FARM OPERATIONS

The typical Heartland farm has been a family operated enterprise of modest size (a few hundred acres) that is highly mechanized, highly productive, and yields a variety of staple commodities in a system referred to as "mixed" (i.e., crops and livestock) farming. Grains, particularly corn, have been the dominant crops, with much of the output being fed to livestock in a relatively small feedlot that was the cornerstone of the operation.

Many of these characteristics still pertain; however, some notable changes have occurred in the last few years. Mixed farming, for example, is much less prominent. Cash grain farming is now the principal farming system, and many farms no longer are fenced because there are no livestock to exclude from the cropped fields.

The total area in farms has been on a declining trend for decades in both Canadian and U.S. portions of the region. The number of farms is also decreasing. Average farm size, on the other hand, has doubled in the past third of a century, as the more successful (or more dedicated) farmers buy and especially rent more land. Family farms continue to dominate the region's agrarian scene, although the formation of family held corporations is increasingly common.

Changing technology made crop diversification and rotation less necessary in the region. The trend to specialize more and diversify less is especially pronounced on the better lands, involving, in particular, a concentration on corn and soybeans at the expense of hay crops and small grains.

In common with agriculturalists elsewhere on the continent, Heartland farmers fell on hard times during the 1980s, owing to a variety of factors, many of them international in scope. Low incomes, declining asset values, and bankruptcy became widespread. Like it or not, North American farmers are relatively high-cost producers by world standards, and the income derived from sale of their commodities partly reflects this fact. Conditions were improving for Heartland farmers in the early 1990s, but only slowly.

Crops

Corn Corn (maize) thrives under the favorable conditions of hot, humid summer weather; fertile, well-drained, loamy soils; and level to rolling terrain. No other country has this favorable combination of growing conditions over such a wide territory; thus the United States produces more than one-third of the world's corn, most of which is grown in this region.

Genetic and agronomic technological improvements have made crop rotation and diversification less necessary, and today corn is seldom grown in a rotational cycle in this region. The seeds, furthermore, are planted in much greater numbers in more closely spaced rows and fertilizer and herbicide chemicals are heavily used. The resulting yields in many areas are more than 150 bushels per acre; for the Corn Belt as a whole, the average yield in most years is nearly 100 bushels per acre.

The principal corn-growing areas are still in the heart of the Corn Belt, from central Indiana to eastern Nebraska; this section also achieves the highest average yields. Corn continues as the dominant crop in most of the Heartland; it occupies nearly twice as much acreage as any other crop in the Corn Belt (Fig. 12–8). About 75 percent of all grain corn grown

A CLOSER LOOK A Corn Belt Farm

In 1982, Charlie Beiser, 51, and his sons, Steve, 25, and Andy, 24, cultivated a thousand acres of land in the Miami Valley of southwestern Ohio, where the Corn Belt farming system originated. They had a legal partnership, but they planned to form a family held corporation. They had invested in buildings, machinery, and equipment rather than in land. Charlie had inherited a 58-acre farm 7 miles northeast of Hamilton that they used as a base, and they rented 14 other farms, some on shares, some for cash. The farthest rented farms were 17 miles north and 15 miles south.

"It's real hard to find good land to rent," Charlie began. "Each year we lose about a hundred acres and have to find another hundred to replace it. What land you rent depends on who you know. We got one real good farm because Andy played guard on the high school football team, and the mother of the guy that played tackle next to him told me about it.

"This farm I own is soon going to be too valuable to farm or even own," Charlie continued, "because it's too close to the city. That's why last year we paid $1900 an acre for a new 160-acre farm 10 miles farther out. I would like to put buildings on it some time and move out there, but right now I can't even afford to think about it.

"My dad had a regular rotation of corn, wheat, and hay, but soybeans have pretty much taken the place of hay and even wheat. We haven't needed any hay since we stopped using horses and mules and switched over to tractors. We still raise 125 acres of wheat because we need the straw for bedding. Some years wheat is a good cash crop, but last year we just gave it away. We double-crop wheat with soybeans. We sow wheat in October, after we have combined the beans, and we harvest it the last of June. Then we immediately plant beans in the wheat stubble, and combine them in November. I have done better than 30 bushels of beans double-cropping, but on my 175 acres of full-season beans I shoot for 50.

"We raise 700 acres of corn," Charlie explained. "We shoot for 200 bushels an acre, and I can remember not long ago when I wouldn't even have believed hundred-bushel corn. We have gone as high as 225 bushels, but one real wet year we only made 70. We shell all the corn in the field with a combine and dry it here at the farm. We use the same combine with a different head to harvest wheat and beans. We sell all of our wheat and beans and about a third of our corn at the local co-op. We need the rest of the corn to feed our hogs and cattle.

"We have a hundred sows, and expect each one to have two litters a year. We breed bunches to farrow every six or seven weeks. We market a couple thousand hogs a year at two and a quarter [hundred pounds]. A change of only one cent in the price of hogs will change my gross annual income $4500. That's why I listen to the farm prices on the radio every day. Some of the farms we rent have steep land that is in pasture, and we have to use it. We buy 125 head of lean feeder cattle in the spring, run them on the pasture all summer, and finish [fatten] them on corn after harvest in fall."

Charlie's farmstead is an impressive complex of large white buildings, including a comfortable two-story brick farmhouse more than a hundred years old; there are also two metal machine sheds with a well-equipped in the United States is in the Heartland; the comparable figure for Canada is about 80 percent.

Soybeans This shallow-rooted legume is popular because it yields a heavy crop of beans; is valuable for meal and oil; makes good hay, silage, and pasturage; has few diseases; and is not attacked by pests.

No crop in twentieth-century Anglo-America experienced such expanded production as the soybean. The United States is now the world's leading producer (75 percent) and leading exporter (90 percent). Soybeans are second only to corn as a source of cash farm income to American farmers and are usually the leading agricultural export of this country.

Three-quarters of the national output of soybeans emanate from this region, and the seven leading states are all Heartland states (Fig. 12–9). Nearly all of Canada's soybeans are grown in the extreme southern part of Ontario, just north of Lake Erie.

Alfalfa The legume alfalfa is well adapted to the region, especially to the prairie portion where winter rainfall is less abundant and the soils less leached and hence higher in calcium. It thrives best on soils rich in lime. The crop has greatly increased in importance in the Corn Belt and the Dairy Belt since 1920, even in the eastern part of the Heartland.

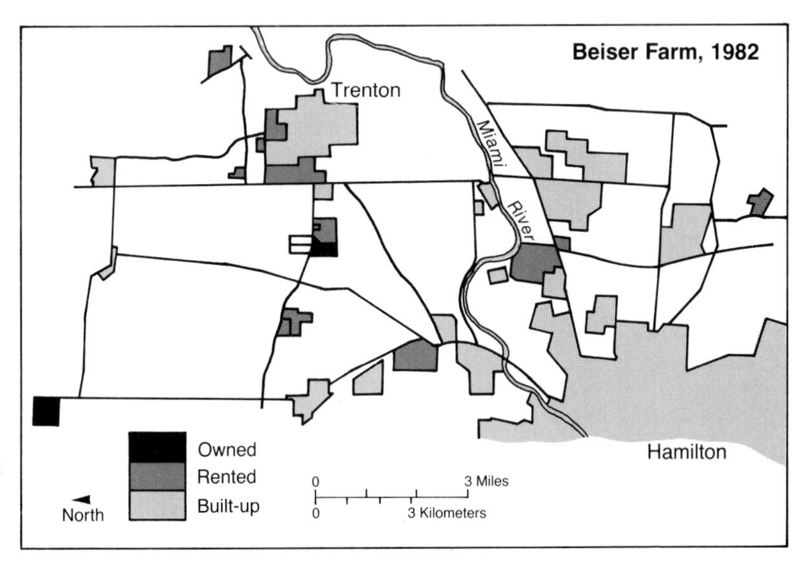

FIGURE 12-a The Beiser farm. (Drawn by Philip Schwartzberg and Tanya Meyer. Reproduced by permission of *Annals*, Association of American Geographers.)

repair shop; a wooden forebay barn that has been converted into eight farrowing sheds where sows can give birth to their litters; a one-story metal building with 10 more farrowing sheds; a well-ventilated metal hog house with 19 pens that hold 40 fattening hogs each; and a battery of five cylindrical metal grain storage bins that can hold 10,000 bushels each, capped by a spiderlike network of slim metal tubes, with a central "leg" enclosing the elevator that carries grain to the top, where a distributor head can direct it through the tubes to any one of the bins or to the drier.

Charlie was skeptical about the map of the farm, because it changes every year, but even so the map (Fig. 12–a) hints at some of the problems of trying to put together and operate a farm of reasonable size in the shadow of a major metropolitan area: finding enough land to rent in an area where most farmers own only 80 to 120 acres; farming many small scattered parcels; moving farm machinery along country roads and city streets; wasting time traveling up and down the highways; and having to listen patiently to complaints from city neighbors about such normal and necessary farm operations as spreading manure and spraying pesticides.

Professor John Fraser Hart
University of Minnesota
Minneapolis

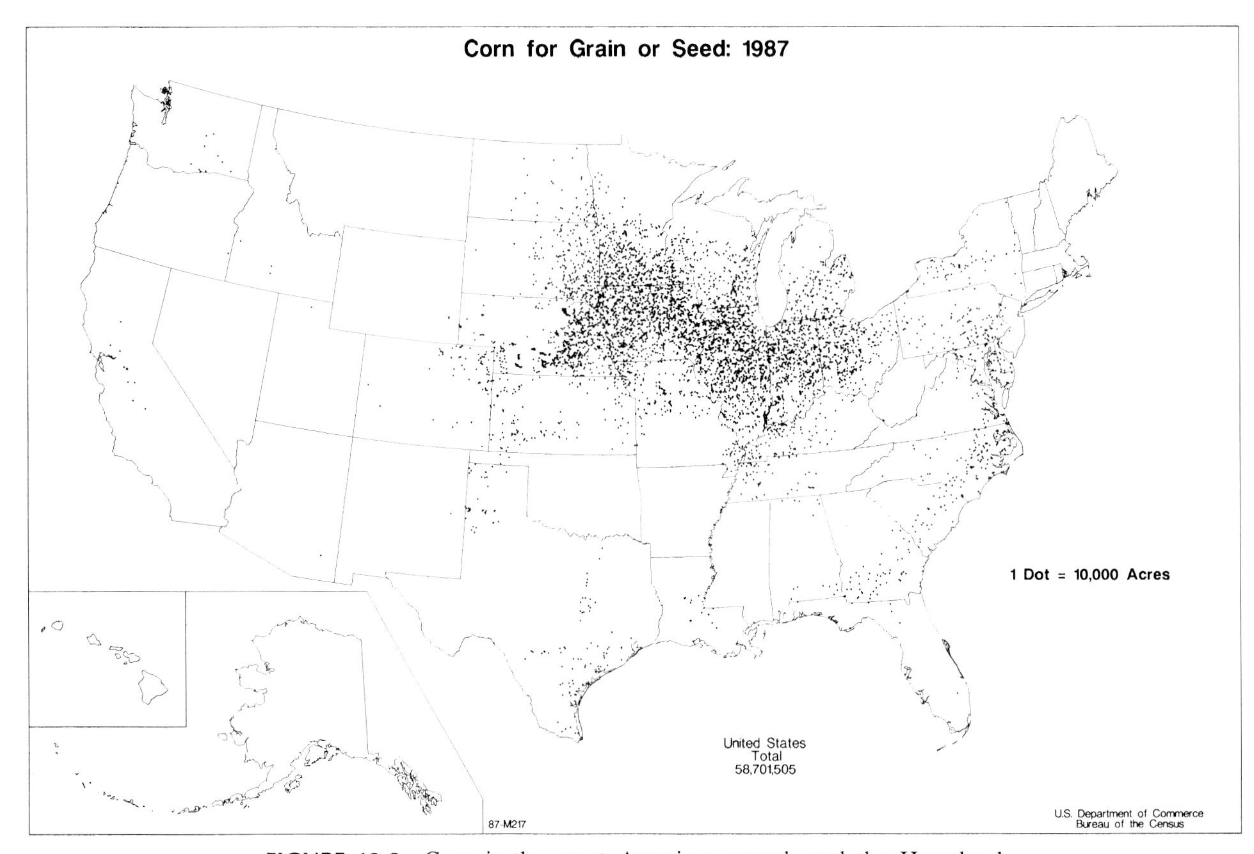

FIGURE 12-8 Corn is the great American cereal, and the Heartland grows the bulk of it. The Corn Belt subregion alone produces nearly one-half of the national total.

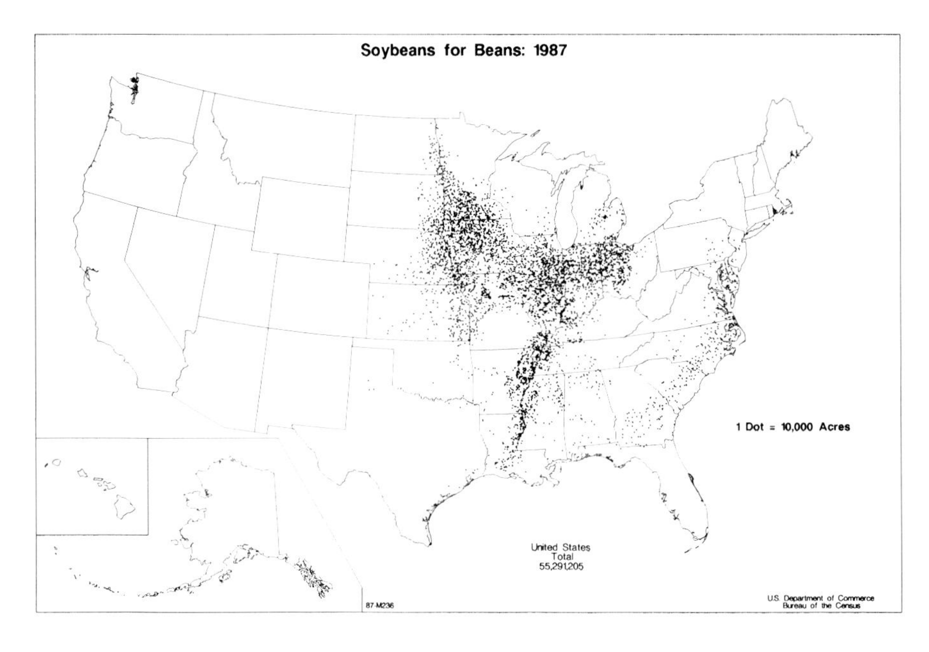

FIGURE 12-9 The United States is the world's leading producer of soybeans, and the Heartland is its outstanding growing region.

Because alfalfa is harvested several times each season and recovers quickly after cutting, the per-acre yield exceeds that of any other hay crop. In the short-summer dairy portion of the region alfalfa becomes a very significant crop. Wisconsin, for example, grows almost as much alfalfa as corn. In most years Wisconsin and Minnesota vie with California as the leading grower of alfalfa.

Other Hay Crops Numerous other crops are also planted for hay production—the so-called tame hays. None approaches alfalfa in acreage or output, but their combined total in the region is approximately double that of alfalfa. *Clover, timothy,* and *clotim mixtures* are the most widespread types of tame hay after alfalfa. Emphasis on nonalfalfa hay production is greatest in the southern part of the region, especially Missouri and Kentucky.

Tobacco Kentucky is second only to North Carolina in tobacco growing, and output is also significant in and around the Nashville Basin in Tennessee. Both burley and dark-fired tobacco are produced. Although generally tobacco is not considered an important Canadian crop, it is the leading one in several counties bordering on Lake Erie, particularly around Norfolk. Tobacco is the most valuable cash crop in Ontario.

Fruit Commercial growing of fruit is not a widely distributed enterprise but is concentrated in definite localities. Tree fruits, more exacting in climate than in soil requirements, frequently suffer from extremes of temperature; so the best suited areas are those with a minimum of danger from late spring and early fall frosts, notably peninsulas, hillsides, and leeward sides of lakes. This tempering effect of a large body of water gave rise to a fruit belt just east of Lake Michigan (particularly apples and cherries), just south of Lake Erie (primarily grapes), and just south of Lake Ontario (especially apples and cherries).

The Niagara Peninsula in Canada, benefiting from the same climatic principle, is famous for fruits, particularly grapes and peaches. The narrow lake plain between Hamilton and the Niagara River is one of only two major areas in Canada where tender fruit crops can be produced.

The Livestock Industry

Livestock raising and feedlot operations are widespread throughout the region (Fig. 12–10). As a generalization, however, it can be stated that livestock feeding dominates in the west, particularly in Iowa where both beef cattle and hog feedlots are ubiquitous. Feedlots are also notable in the east, especially in Indiana where hogs dominate the scene. In the central part of the region (Illinois and vicinity) livestock feeding is less important than cash-grain farming.

Beef Cattle Beef cattle are most numerous in the western part of the region—the old prairie portion—for, unlike swine, they are essentially grass eaters. Contrary to the common notion that the range states provide only the grazing and breeding lands and the Corn Belt the fattening areas, the fact is that about two-thirds of the animals slaughtered within the Corn Belt are bred in it.

Feeding is a major agricultural enterprise in the western part of the region, the animals being carried through the winter on hay and other home-grown feeds and fattened on corn. There are more cattle on feed in Iowa than in any other state; Nebraska ranks next. Most cattle feedlots in this region are relatively small, fattening less than 100 animals annually. This is in marked contrast to the situation in other regions, particularly the West, where there is a strong trend toward ever-larger feedlots.

Swine Of all large domesticated animals, swine most efficiently and rapidly convert corn into meat. Spring shoats, for example, are ready for market in approximately eight months. During the feeding period they gain in weight from 1 to 1¼ pounds (0.45 to 0.56 kg) per day. Two-thirds of the swine in the United States are raised in this region.

The generally smaller farms in the eastern part of the region often emphasize pork rather than beef production, for swine require less space than cattle (Fig. 12–11). Even so, the largest numbers of swine are found where there is the greatest production of

A CLOSER LOOK *Preserving Canada's Fruitlands—The Niagara Fruit Belt and the Okanagan Valley*

SOFT FRUIT-GROWING AREAS IN CANADA ARE LIMITED

The areas in Canada that can grow grapes and soft fruits such as apricots, peaches, and sweet cherries are severely limited by climatic conditions.

The Niagara Fruit Belt, a narrow strip of land along Lake Ontario, stretching from Hamilton to Niagara Falls, lays claim to fame as the best grape and soft-fruit-growing region in all of Canada (Fig. 12–b). The water of Lake Ontario retains its warmth well into the middle of the winter and moderates the cold air masses that sweep in from the north. In spring, the lake is slow to warm up and thus keeps the weather cool enough to retard the opening of fruit blossoms until after most risk of frost is over. The slope of the land provides air drainage (cold air runs downhill just like water), which further protects against frost damage on clear, cold nights. It may surprise you to learn that Niagara has less risk of spring frost damage to peaches than has the "peach" state of Georgia.

The Okanagan Valley, tucked in the mountain and plateau country of south-central British Columbia, rates second only to the Niagara Fruit Belt for the growing of grapes and soft fruits. It is also internationally renowned for its production of apples, particularly the red Delicious variety. The winters of the Okanagan are moderated by a great deal of cloud cover, and the sloping sites of the broad terraces above the valley floor provide excellent air drainage to protect against spring frosts.

URBAN SPRAWL IN CANADA'S FRUIT BELTS

In Canada, cities are located on the country's best agricultural land. In fact, 50 percent of Canada's population lives on the best 5 percent of its farmland. To make matters worse, urban growth has occurred in a low-density urban sprawl pattern that consumes, and sterilizes for agricultural production, much more land than would be required with more orderly and compact development. Nowhere is this low-density urban sprawl more evident than in the two major soft-fruit-growing regions.

In the Niagara Peninsula, several major cities and numerous towns are located on the prime fruit-growing strip along Lake Ontario (Figure 12–c). The Queen Elizabeth Way (QEW) not only takes up valuable land but also has facilitated the sprawling of urban land uses right across the fruitland. The greatest density of urban uses is found on the well-drained sandy soils that are required for crops such as peaches and sweet cherries and vinifera and hybrid grapes. Several studies have shown that by directing urban growth into non-fruit-growing areas and requiring contiguous compact urban development, there is room in the Niagara Peninsula for a great amount of urban development without completely destroying the land resource for fruit-growing.

The degree and pattern of urban sprawl in the Okanagan Valley is strikingly similar to that of the Niagara region. Considering the fact that the Okanagan has a population of only about one-fifth that of Niagara, the degree of low-density sprawl is relatively greater in the Okanagan than in the Niagara case. As in Niagara, most of the urban sprawl is occurring on the best fruitlands. Also, as in Niagara, a major highway (Route 97) goes right through the best fruitlands, thus facilitating urban sprawl on those lands. However, unlike Niagara, in the Okanagan the terrain does not provide for feasible alternative locations for a highway and urban development.

Visible urban expansion can be mapped; the indirect effects on the fruit-growing industry are more difficult to assess. Although we cannot measure quantitatively the indirect impact of urbanization, interviews with hundreds of Niagara and Okanagan fruit growers yield the following list: vandalism, crop pilfering, trespass, complaints from nonfarm residents about spraying and noisemakers (to scare off birds), high land prices that preclude farm expansion, and high land taxes to pay for urban-type services.

Besides the problems associated with urbanization, the fruit-growing industry faces many other hazards and problems: vicissitudes of weather (frost, hail, poor weather for bee pollination at blossom time, drought, windstorms, rain at harvest time), disease, insects, and lack of skilled workers. All of the above are serious, but they are usually taken in stride by the growers; they are part of the business of farming.

All fruit growers agree that their most serious problem is the ever-tightening cost-price squeeze. Costs of things such as energy, fertilizer, chemicals, and equipment have skyrocketed in the last several decades. On the other hand, prices received for fruit have been kept low by competition from foreign countries where costs are lower and the agricultural industry is more heavily subsidized. The General Agreement on Tariffs and Trade (GATT) prevents Canada from protecting the fruit-growing industry with trade tariffs. The Canada–U.S. Free Trade Agreement (1989) is providing even stiffer competition for Canadian fruit growers.

When growers are not receiving a reasonable return on their investment of capital, management, and labor, it is not surprising that many of them welcome the opportunity of selling their land at high prices for urban pur-

FIGURE 12-b Vineyards and orchards in Ontario's Niagara Fruit Belt. The Niagara Escarpment looms in the background (photo by Ralph Krueger).

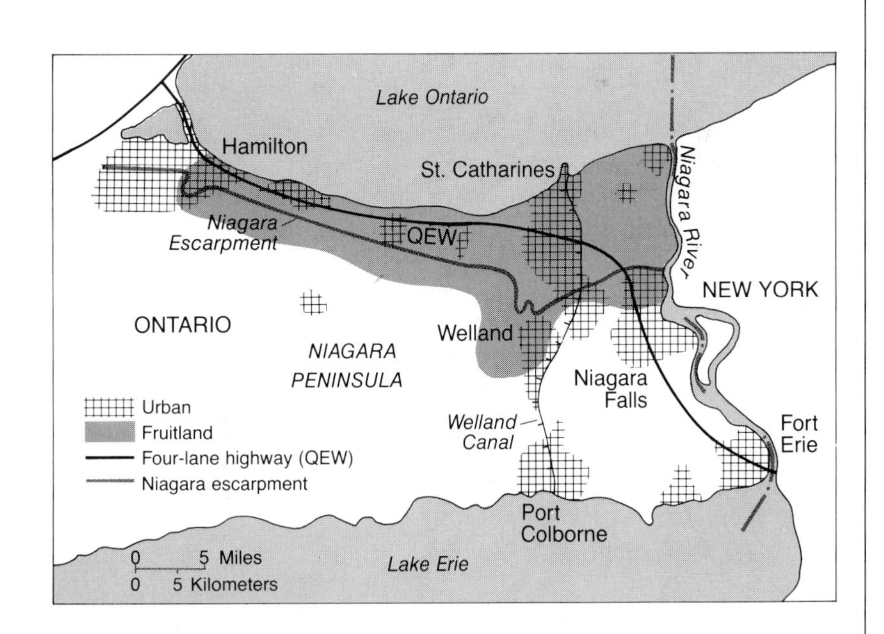

FIGURE 12-c The urbanization of fruitlands in the Niagara Fruit Belt.

poses. By 1991, about one-half of the Niagara and Okanagan fruitlands had been ruined for agricultural purposes by urban development.

If the fruit-growing and related processing and wine industries were to collapse, there would be serious economic dislocation in the Niagara Peninsula and the Okanagan Valley. Conservationists are even more concerned that the collapse of the fruit-growing industry would result in the

(*continued*)

destruction of scarce and irreplaceable land resources that can produce the widest range of crops in Canada.

FRUITLAND PRESERVATION MEASURES

In Canada, the management of resources and the planning of land uses are responsibilities of the provincial governments. Ontario has a policy statement related to preserving agricultural land, the *Food Land Guidelines* (1978). In the late 1970s, when the Regional Municipality of Niagara proposed a land-use plan that violated the *Food Land Guidelines*, a group of local citizens challenged the plan. After two years of hearings, the Ontario Municipal Board (a tribunal appointed by the Ontario government) ruled that the area designated for urban development in the Niagara plan be rolled back to save some 4000 acres of fruitland, and that future urban development be directed to less valuable agricultural land. The result has been a significant deescalation of the urban conversion of fruitland since 1981.

In British Columbia, the provincial government passed strong legislation to preserve agricultural land with its *British Columbia Land Commission Act* (1973). The act resulted in the establishment of Agricultural Land Reserves in which all good agricultural land is preserved for farming and related uses. The *Land Commission Act* was accompanied by the *Farm Income Assurance Act*, which provides for the payment of indemnities to farmers when the price of produce falls below the cost of production. Despite a number of successful appeals to have land excluded from the Agricultural Land Reserves, the provincial legislation has succeeded in greatly reducing the urbanization of fruitland in the Okanagan Valley.

Although conservationists have won some major battles in the struggle to preserve the Niagara and Okanagan fruitlands, it is not safe to assume that these resources have been preserved in perpetuity. Governments, policies, and legislation can change. The long-term preservation of these and other renewable resources depends upon continuously informed and alert citizens who must insist that governments implement the necessary conservation measures.

Professor Ralph R. Krueger
University of Waterloo
Waterloo, Ontario

corn; thus Iowa has more than twice as many hogs as the second-ranking state, Illinois.

Dairy Cattle Dairying is widespread in the Heartland but is dominant only in the northern part of the region and around the major cities (Fig. 12–12). In Wisconsin and Minnesota the principal products are manufactured items—butter, cheese, evaporated milk, dried milk—to combat the high cost of shipping fluid milk. Dairy farms tend to be relatively small and often involve marginal land that is not well suited for crop farming but is satisfactory for pasture. Many dairy farmers now also raise and fatten dairy animals for sale as beef; this activity has to some extent replaced the secondary raising of hogs on dairy farms, which was a traditional activity.

The National Government and the Farmer

Today in both Canada and the United States agriculture, that "last stronghold of free enterprise," is ever more dependent on the government. Within the last six decades a bureaucracy of astounding magnitude evolved, doing many worthwhile things for the farmer but enmeshing agriculture in an endless series of quotas, regulations, and artificial conditions.

Direct government influence on farming occurs in many ways and differs somewhat in the two countries. The geography of agriculture, however, is most likely to be affected by the following practices.

Cost Sharing of Conservation Practices. Interested farmers can obtain advice, plans, and technical assistance for soil conservation from the government. Also, the government pays approximately half the cost of establishing certain conservation practices on farmers' lands. In addition, farmers can take some acres out of cultivation and receive direct payments for not growing crops. All these measures were beneficial for overused and eroded land. But at the same time the high productivity of intensified farming on the better-cultivated land offsets the tendency toward decline in total production that would be an anticipated effect of conservation practices.

Farm Credit. U.S. and Canadian governments make loans to their qualifying farmers who cannot obtain credit from conventional sources.

Crop Insurance. Both governments set up crop

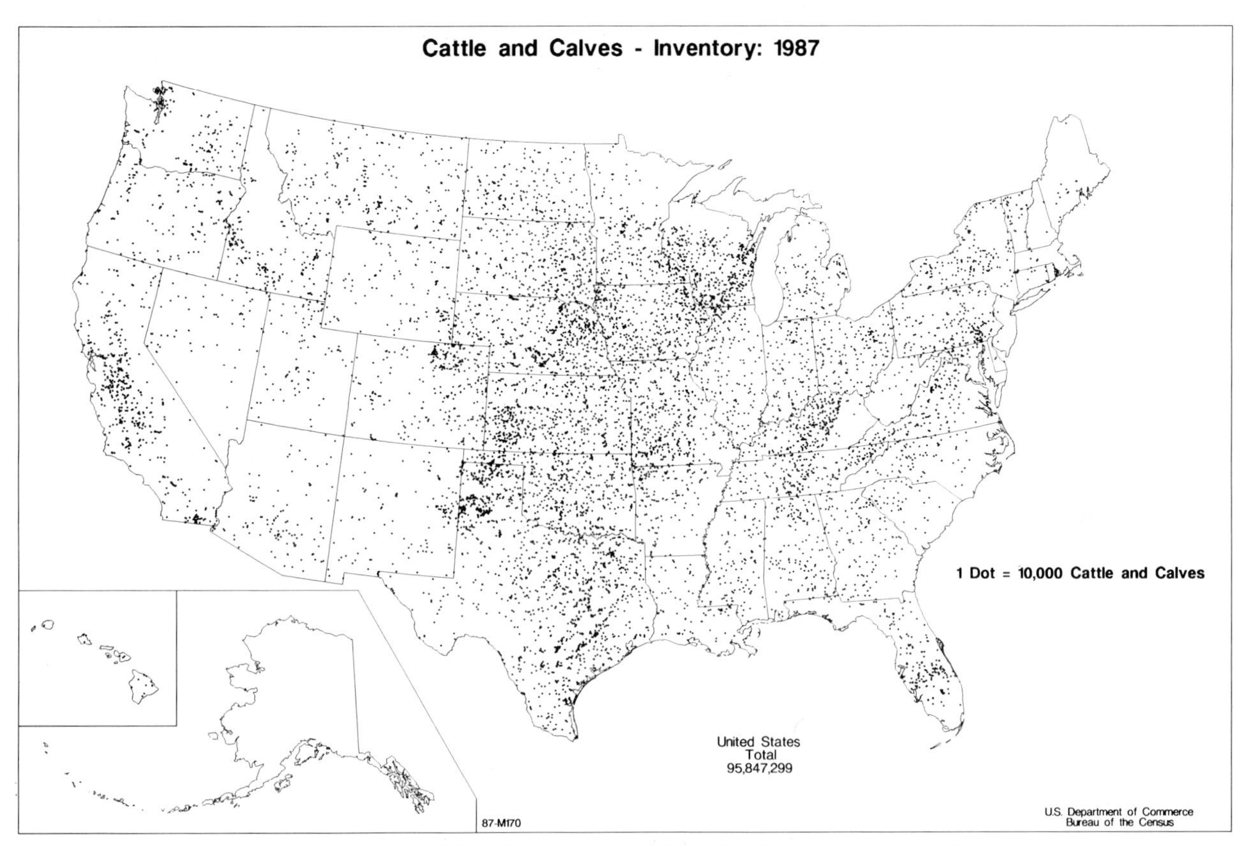

FIGURE 12-10 Although cattle are widely distributed over the nation, the greatest concentrations occur when there is an overlap of Corn Belt beef cattle with dairy cattle of the Hay and Dairy Belt.

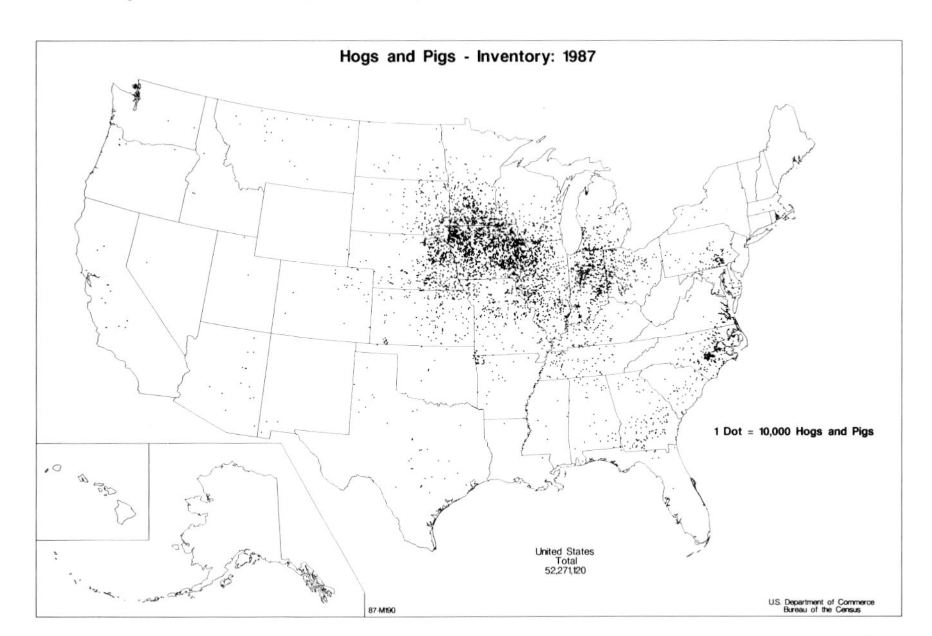

FIGURE 12-11 Corn is a major feedstuff for swine, so the distribution pattern of hogs is similar to that of corn.

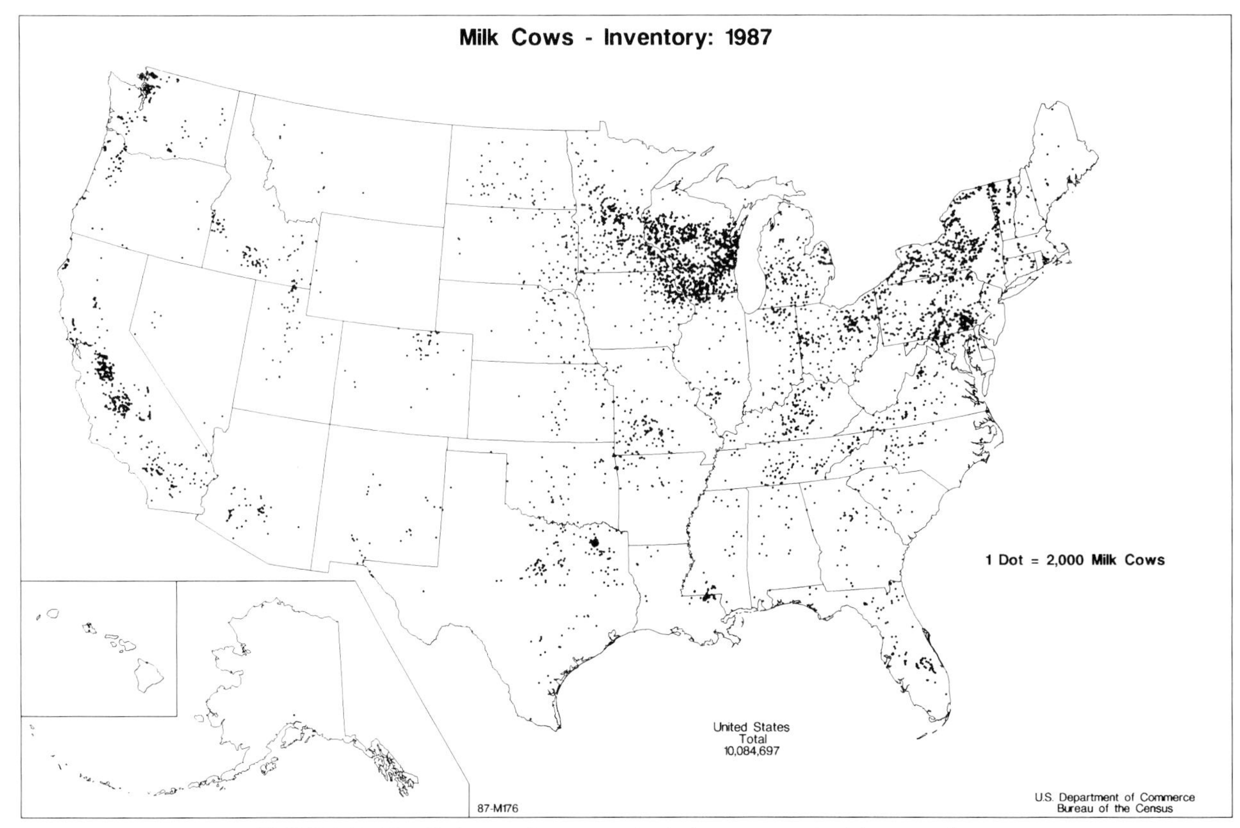

FIGURE 12-12 Dairy cattle are particularly concentrated in a northern belt from Minnesota to New England, with the largest numbers in the western lakes states of Wisconsin and Minnesota.

insurance agencies to indemnify farmers for crop losses due to "acts of God."

Acreage Allotments and Marketing Quotas. Most major crops are now grown under some sort of allotment and quota system whereby maximal acreages are designated for each year's harvest.

Price Supports. Several basic agricultural commodities in both countries achieve what amounts to a guaranteed sale on the basis of federal price supports, regardless of national or international market conditions. In the early 1980s the U.S. government embarked on an innovative but costly program designed to make agriculture less reliant on government intervention and more market oriented. Instead of direct payments to farmers, the government provides grain (corn, wheat, sorghum, or rice) or cotton from government surplus stocks to farmers, who reduce their own plantings of these crops. This process is referred to as "payment in kind" or PIK. The farmer is then able to sell these commodities on the free market. More than one-third of all eligible farmers opted to participate in this program in its first year of operation (1983). Government-owned crop surpluses diminished rapidly and cash outlays were reduced, but foreign competitors complained vociferously about commodity "dumping," and overall costs to the U.S. government actually increased.

Export Controls. In a highly fluctuating pattern the two governments sporadically impose export controls on certain basic commodities. An aggressive program to restimulate export demand for farm products was started in the United States, primarily through "blended credit," which has the effect of reducing the interest cost component of credit sales.

What to do about the complex federal farm programs (which are grossly oversimplified in this brief discussion) is enigmatic. Friend and foe alike agree that they are too big, too all-pervading, and too expensive. And yet neither Republicans nor Democrats nor Conservatives nor Liberals nor Social Crediters nor Independents have been able to devise a scheme for getting government out of agriculture without wrecking the farm economy.

As the leading producer, consumer, and exporter of grain in the world, the United States exerts an inordinate influence on international grain trade. Its long-established philosophy of management by exception (that is, intervening in response to perceived needs) never worked too satisfactorily and the likelihood of a satisfactory program is increasingly remote as the agricultural sector of the economy becomes ever more complex and heterogenous.

MINERALS

Although relatively inconspicuous in this region of remarkable agricultural and industrial output, mining is also an important enterprise in many localities.

Coal

Coal deposits are widely distributed and one of the nation's leading provinces, the Eastern Interior, is in this region. The Eastern Interior Province is second (although distantly) only to the Appalachian Province as a coal producer. Illinois is the fourth-ranking coal state, with about 10 percent of total national output. Mines in western Kentucky and southern Indiana together yield slightly more coal than Illinois and their output is increasing. In most years Muhlenberg County in Kentucky produces nearly twice as much bituminous coal as any other county in the nation. Most of the coal in the Eastern Interior Province is obtained from strip mines.

Petroleum and Natural Gas

Important oil and gas reserves have been tapped in Michigan, Ohio, Indiana, Kentucky, and Illinois. Production is scattered, however, and the region as a whole yields only 2 percent of total U.S. output; about half comes from Illinois.

The only oil and gas of any consequence in eastern Canada are found in this region, somewhat scattered in extreme southern Ontario. Oil output amounts to less than 1 percent of the national total, but further modest discoveries are anticipated. The proportion of the national output of natural gas is somewhat higher.

Limestone

In south-central Indiana, in the Bedford–Bloomington area, are the famous limestone quarries that supply a superior limestone used in buildings throughout the East and Midwest. About three-fifths of the dimension (block) limestone of the country comes from these quarries. The building-stone business is declining, however, for it suffers from competition with cheaper concrete, brick, and lumber.

Salt

Salt occurs widely in Anglo-America, but major deposits are in the vicinity of the southern end of Lake Huron. Michigan is one of the leading states and Ontario is the leading province in salt output of their respective countries. The salt is obtained in solid form by underground digging and in the dissolved state by the modified Frasch process. There are several mines in Michigan, but the principal output is actually beneath metropolitan Detroit. The Ontario mines are at Goderich, Watford, Sarnia, and Windsor.

HEARTLAND MANUFACTURING

Much of the character, prestige, and reputation of the Heartland Region was derived from two outstanding components of its economy—agriculture and manufacturing—that have been in decline for some years. Of the two, manufacturing has suffered the most severe and long-lasting deterioration. The Heartland was the preeminent industrial region of the continent for many decades, but both the relative significance and absolute significance of manufacturing have plummeted. Industrial output is still

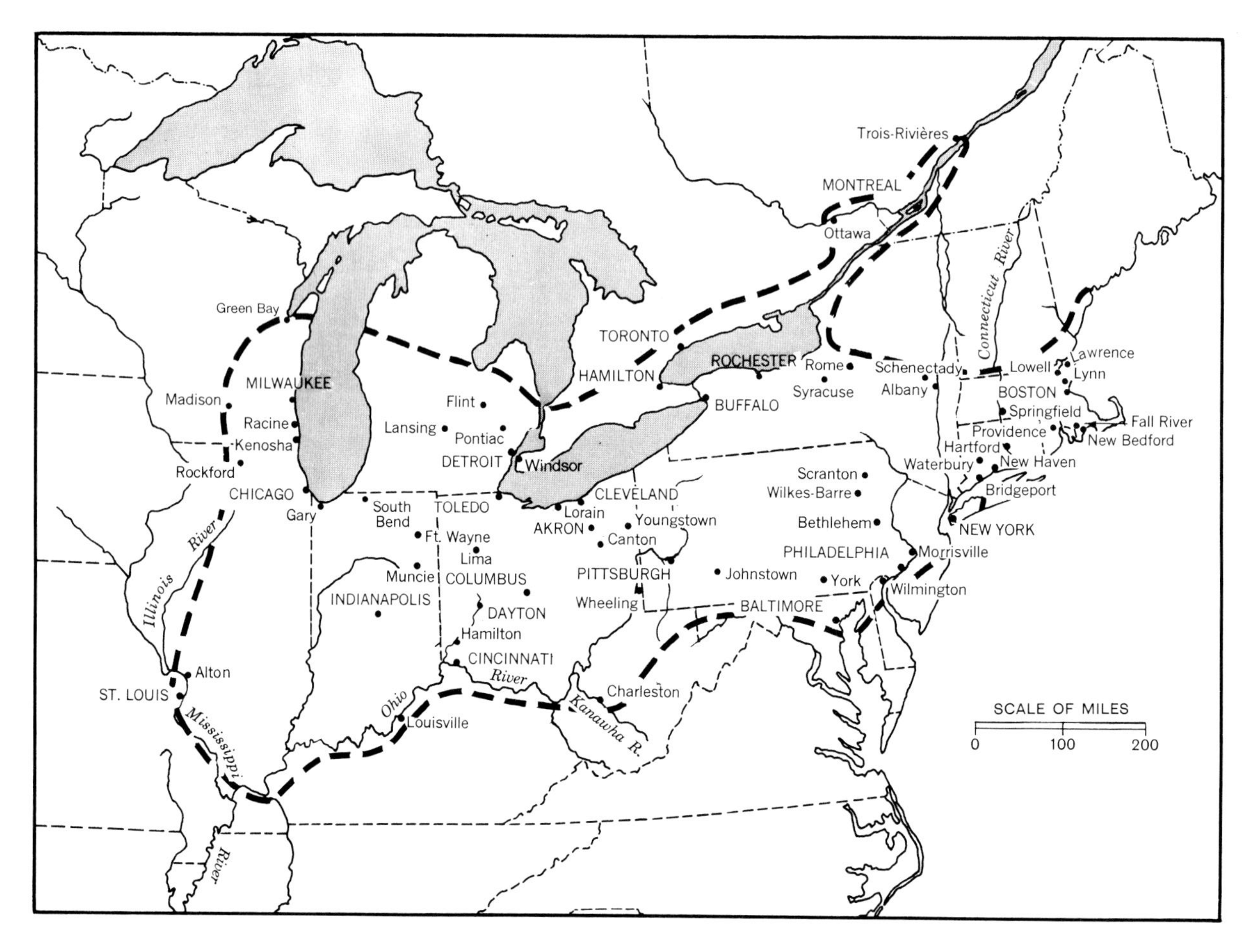

FIGURE 12-13 The American Manufacturing Belt.

of great importance to the regional economy, but, as is true in most regions, its dominant role has been superseded by the tertiary sector (services and trade).

The American Manufacturing Belt

The American Manufacturing Belt is a broad zone of industrial-urban concentration in parts of northeastern United States and adjacent southern Ontario and Quebec (Fig. 12–13). As in most of Anglo-America, manufacturing here is clustered in and near large cities.

Although referred to as a belt, the actual factory distribution is spotty and occupies only a small fraction of the land. Here the concept of a belt involves the relative continuity of significant industrialization from one urban node to the next in the broad area extending from Portland and Baltimore on the East Coast to Green Bay and St. Louis in the Midwest; most of the land, however, is in nonindustrial usage. In short, there are large expanses of meager manufactural activity among numerous centers of industry.

The pattern of secondary industry varies significantly in different portions of the American Manufacturing Belt. In order to gain more than a very generalized understanding of the belt, it is necessary to subdivide it into districts (Fig. 12–14). Four of these districts and a portion of four others lie outside the Heartland Region and were discussed in preceding chapters.

Generally, the more westerly districts of the American Manufacturing Belt experienced more economic vigor in recent years than the easterly ones. Although the belt as a whole suffered a fairly severe decline (see box, "A Closer Look"), western districts declined more slowly and a few areas actually grew slightly, although the trends are anything but uniform. These western districts are in the Heartland Region and provide the principal industrial components to its economy.

Central New York The Mohawk Valley and the Ontario Plain occupy the great water-level route from Troy to Buffalo and are traversed by major land (highway, freeway, railway) and water (barge canal) thoroughfares.

The district contains many cities. The first arose in response to the stimulus of the Erie Canal, built in 1825; the later ones developed near a series of short railroads that paralleled the river and canal. Later these lines were consolidated and the area became one of the major traffic arteries of Anglo-America. Factories sprang up all along the route.

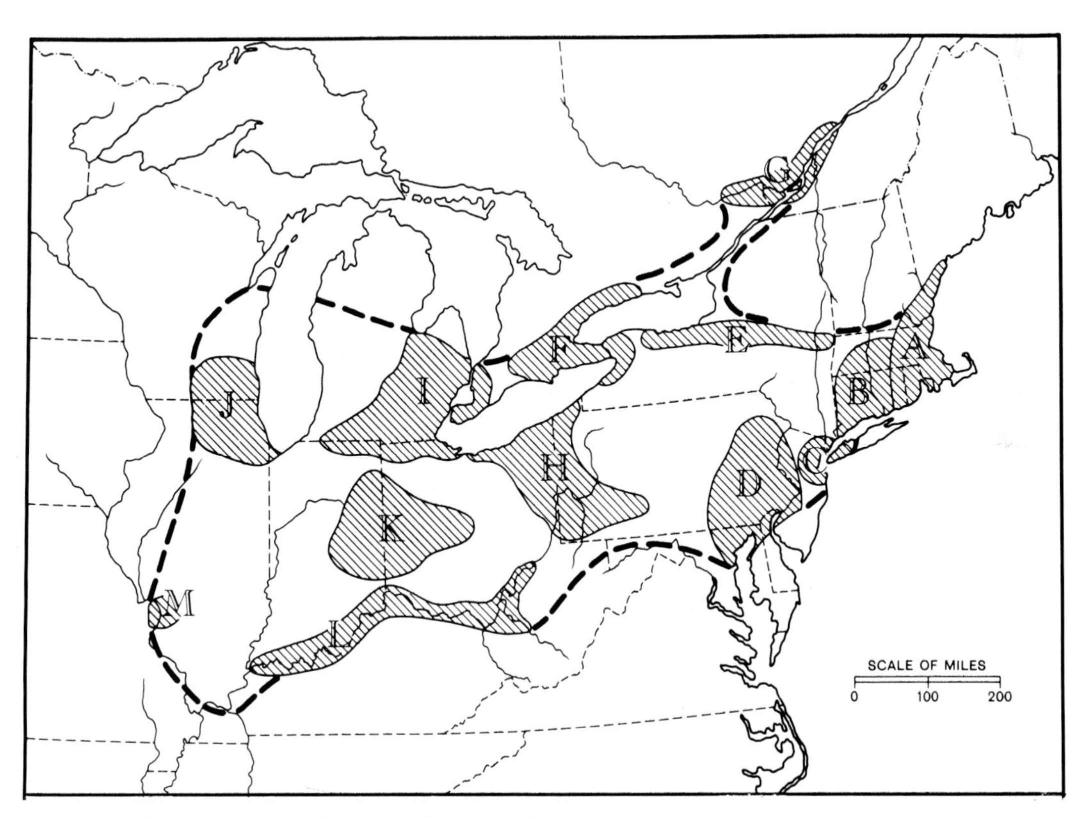

FIGURE 12-14 Principal manufacturing districts:
A Southeastern New England,
B Southwestern New England,
C Metropolitan New York,
D Philadelphia-Baltimore district,
E Central New York,
F Niagara Frontier,
G Middle St. Lawrence district,
H Pittsburgh-Cleveland district,
I Southern Michigan automotive district,
J Chicago-Milwaukee district,
K Inland Ohio-Indiana district,
L Middle Ohio Valley,
M St. Louis district.

A CLOSER LOOK The Diminishing Significance of the American Manufacturing Belt

For more than a century the American Manufacturing Belt was recognized as one of the great industrial domains of the world. Within the roughly shaped parallelogram that describes its boundaries were found some two-thirds of the total factory production of Anglo-America. Its metropolises—Chicago, Toronto, Detroit, Philadelphia, Cleveland, St. Louis, Pittsburgh, Milwaukee—were the classic industrial cities of the continent, blue-collar urban nodes with a multiplicity of workshops large and small.

By the 1980s, however, the situation had changed and is continuing to change. The industrial dominance of the belt is waning rapidly, challenged by vigorous industrial growth in the South and West and by the continent-wide decline of manufacturing in favor of service industries. The bellwether industries of the American Manufacturing Belt—steel and automobiles—fell on hard times, experiencing reduced demand that resulted in shutdowns and layoffs throughout the Northeast. Most other types of manufacturing did not suffer such sharp declines, but almost all were affected negatively.

Precise statistics are hard to find, but the magnitude of the industrial decline is easy to see.

In the last three decades the American Manufacturing Belt's share of continental factory production declined from two-thirds to two-fifths of the total.

Total manufacturing employment in the belt diminished by more than 2 million in the last two decades, a drop of nearly 20 percent.

Fifteen years ago 8 of the 10 leading industrial centers of the nation were in the American Manufacturing Belt; today there are only 2.

FIGURE 12-d A huge machine shop directly controlled by computers operates in the McDonnell Douglas plant in St. Louis, producing aircraft parts (courtesy McDonnell Douglas).

A recent Federal Reserve Bank study showed that 15 states experienced "deindustrialization" (a diminution of manufacturing output) between 1967 and 1986. All of the Heartland states except those of the western tier (Minnesota and Wisconsin to Missouri) were among the 15.[a]

Although there had been a decentralizing flow of industry from the American Manufacturing Belt for several decades, it was not until the mid-1960s that the massive concentrations of manufacturing in the belt began to weaken. Until then innovation and new industry growth allowed the belt to maintain its dominance. Then began a phase in which the belt experienced continuous comparative (although not absolute) industrial decline. Absolute losses presumably were avoided because of the belt's ability to continue functioning as a seedbed in which new high-technology and capital-intensive industries offset closures. Moreover, conventional location factors continued to be significant for some types of manufacturing, thereby fostering some growth (Fig. 12–d).

Later, however, innovative capacity diminished and absolute losses be-

[a] Gerald A. Carlino, "What Can Output Measures Tell Us About Deindustrialization in the Nation and Its Regions?," *Business Review*, Federal Reserve Bank of Philadelphia (January–February 1989), 19.

gan to occur. The development of high-technology manufacturing in other localities was particularly significant because these industries (electronics, computers, aerospace, instrumentation, weapons, drugs, etc.) had been flourishing at twice the rate of that of manufacturing as a whole. To be sure, high-technology industries expanded somewhat in the belt, particularly in Massachustts, but the major high-technology growth areas are in California, Texas, Florida, and other places far removed from the industrial heartland.

Meanwhile, the "core" industries of the American Manufacturing Belt are severely depleted (Fig. 12–e). Most notable is the decline of steel production. Because of foreign competition, high-cost output, diminished domestic demand, and other factors, steel production in the United States and Canada nosedived. At the beginning of the 1990s, employment in U.S. and Canadian steel mills was less than one-third what it was 30 years earlier and one-half the total of just 10 years previously. Most of this drastic decline occurred in the American Manufacturing Belt—from Buffalo and Hamilton through Pennsylvania and Ohio to Michigan and Illinois. The automotive industry also suffered notably. Massive layoffs took place throughout the prime manufacturing areas, particularly in Michigan, Ontario, and Ohio.

FIGURE 12-e A steel mill in Detroit, with ore boats at the unloading docks (courtesy Detroit Department of Public Information).

Studies of this shifting industrial phenomenon show that there is no wholesale exodus of firms from the belt to nonbelt locations. Instead, branch plants were established or expanded at nonbelt locales and new firms tended to locate elsewhere than in the old industrial core region. The attractiveness of peripheral regions for new industry is sometimes included in the concept of "favorable business climate." One study that ranked states on the basis of their business climate had the following conclusion: of the 14 states entirely or largely within the American Manufacturing Belt, only 1 was in the upper half of the ranking and it was not in the top 20.[b]

These trends show little sign of abating. Most contemporary growth industries are not significantly constrained by the availability of production materials and energy supplies and they are relatively unresponsive to market accessibility. Skilled labor is an important locational factor, but it does not negate the significance of a favorable business climate and low production costs. It seems likely that both the relative and the absolute significance of the American Manufacturing Belt will continue to diminish.

TLM

[b] James S. Fisher and Dean M. Hanink, "Business Climate: Behind the Geographic Shift of American Manufacturing," *Economic Review,* Federal Reserve Bank of Atlanta, 67 (June 1982), 23.

Each city tends to specialize in several products: *Rochester* makes cameras, electrical machinery, optical goods, and men's clothing; *Rome*, copper and brass; and *Syracuse*, machinery and alkalies.

The Niagara Frontier This subregion clusters around the western end of Lake Ontario and the eastern end of Lake Erie, on both sides of the international boundary that separates New York and Ontario. Power generated at Niagara Falls is widely used on both sides of the border and has played an important part in the development of chemical, metallurgical, and other industries. Because of the international border, there are two industrial capitals: Buffalo on the United States side, and Toronto on the Canadian.

Flour milling was long the symbolic industry of Buffalo and the city still ranks as one of the world's leading milling centers. The total industrial structure of the district, however, is quite diversified, especially on the Canadian side. The automotive industry is well established in the Niagara Frontier, with many parts factories and a few assembly plants in Toronto, Hamilton, and Buffalo. Electrochemical and electrometallurgical facilities have been established on both sides of the border to use Niagara Falls power. Fabricated metals (especially in Buffalo), agricultural machinery (particularly in Toronto), and steel-making at Hamilton are also products of note.

Canada's leading industrial center is *Toronto*, which ranks about twelfth among the major manufacturing cities of Anglo-America. *Hamilton* has only one-fourth as much industrial output as Toronto and yet is the fourth largest Canadian manufactural center. *Buffalo* declined significantly as an industrial center in the past few years and no longer ranks among the top 30.

The Pittsburgh-Cleveland District The Pittsburgh–Cleveland district is the continent's outstanding producer of iron and steel. It is also a leading producer of metal products in their secondary and tertiary stages—fabricated metals and machinery. Some other industries, such as rubber and glass products, are important, but this is predominantly a district of metalworking.

The area is strategically located for heavy industry, for it lies between Lake Erie on the north, over whose waters move millions of tons of iron ore and limestone, and the productive northern Appalachian coal field on the south.

Pittsburgh, *Youngstown*, and *Cleveland* were traditionally among the leading steelmakers of the continent, but all experienced significant declines in the 1970s and 1980s, particularly Youngstown. The steel mills of the district are typically old and medium-sized and occupy cramped valley-bottom sites. As the national steel industry suffers, these relatively antiquated facilities are particularly hard hit.

Motor vehicle manufacturing and assembly is also a major industry in this district; it is particularly notable in Cleveland, Lorain, and Youngstown. Akron was the historic center of rubber manufacturing (particularly automobile tires) in the nation; its output virtually vanished over the past two decades, but some slack was taken up by expansion in plastics manufacturing.

Other major manufacturing industries in the district are mostly producers of durable goods that use steel as an important component of their product: fabricated metals, machinery of various kinds, machine tools, and electrical machinery.

The Southern Michigan Automotive District The automobile-manufacturing district contains the heart of the world's automotive industry. It includes, besides the metropolitan areas of *Detroit* and *Windsor*, the inner and outer rings of cities. Included in the semisuburban inner ring are *Pontiac*, *Ann Arbor*, *Ypsilanti*, and *Monroe*. Included in the outer ring are nine cities in the orbit of Detroit's great industry: *Port Huron*, *Bay City*, *Saginaw*, *Flint*, *Oshawa*, *Lansing*, *Jackson*, *Adrian*, and *Toledo*.

Detroit became the first great automobile center by a geographical accident—it is the hub of a circle within which were located the pioneers of the industry, such as Henry Ford and Ransom Olds. Ford was particularly instrumental in the rise of the industry in Detroit: he developed an automobile cheap enough for the average family; he adapted the assembly line to the industry; and he introduced standardization and interchangeable parts, thereby making mass production possible.

The Canadian motor vehicle industry, which is mainly centered at Windsor and Oshawa in Ontario, is the most important industry in this province. As in the United States, however, factories in many widely scattered cities and towns provide the innumerable parts.

The chemical industry is also a distinctive activity of this district. Most of the chemical plants, drawing on huge deposits of salt that underlie Detroit, are located down the Detroit River from the Rouge or just beyond the northern edge of the city.

Sarnia, situated where Lake Huron empties into the St. Clair River, is an outstanding Canadian chemical manufacturing center. Oil refining, based on local petroleum, got its start late in the nineteenth century and expanded considerably since then, first with pipelined oil from Ohio and later with the completion of a pipeline to bring in vast supplies from Alberta. The local refineries, as well as nearby salt deposits, stimulated the attraction of petrochemical, synthetic rubber, plastics, fertilizer, and other types of chemical factories.

Detroit also has a steel industry of note, based on the extensive local market (auto manufacturing) and economical assembly of raw materials. Following the national trend, this industry has declined since the 1970s.

The Chicago–Milwaukee District The Chicago–Milwaukee district occupies the western and southwestern shores of Lake Michigan from Gary to Wauwatosa and includes satellite towns and cities extending a short distance inland. Heavy industry predominates. The fountainhead of all manufacturing was the great primary iron and steel industry at the southern tip of Lake Michigan. South Chicago–Gary is strategically located for making iron and steel: iron ore and limestone can be brought directly to the blast furnaces by lake carrier, and coal is not far distant in central and southern Illinois, although most coking fuel is brought from West Virginia and Kentucky by rail or by the combination of rail and lake carrier. This district has the best balance between production and consumption of any iron and steel area in the United States.

As the domestic steel industry faltered in recent years, Chicago was less severely affected than most other centers. Several of its mills are relatively new and large and some steel companies tended to concentrate their production here at the expense of older, smaller mills in other locations.

Although this district is justly famous for its primary steel industry now and its meatpacking industry in the past, its prominent types of manufacturing today—whether measured by employment or by value of production—are machinery output and metal fabrication. Chicago is the continent's leading center in the fabrication of metals, that relatively prosaic heavy industry in which primary metal (mostly sheets, bars, and rods of steel) is shaped and fashioned into pipes, screws, wire, beams, cans, and other products of specific use.

For the district as a whole and Chicago in particular, however, machinery production is the leading type of manufacturing. Electrical machinery, such as communication equipment, radio and television sets, electronic equipment, electric lighting and wiring equipment, is particularly notable. This is the largest single industry in the district. Nonelectrical machinery is also produced in great quantity, particularly metalworking, construction, and industrial machinery.

Most manufacturing in Wisconsin and Illinois is confined to this district. *Chicago* is the third largest industrial center on the continent and *Milwaukee* ranks in the top 20. Smaller centers of note include *Rockford, Joliet, Racine,* and *Kenosha*.

Inland Ohio–Indiana District This district is situated between the Great Lakes and the Ohio River; thus it can draw on both for transportation but suffers in cost by being adjacent to neither. It lies between coal fields to the east and productive farmlands to the west. Its industries are diversified, inluding machine tools, cash registers, refrigerators, soaps, meat, tobacco, iron and steel, beer, shoes, radios, and clothing. The most intensely industrialized part is the Miami Valley from Springfield to Hamilton. *Indianapolis* is the major center in the western part; *Dayton* and *Columbus* in the eastern.

Middle Ohio Valley One of the more dynamic heavy manufacturing districts in the world extends for 500 miles (800 km) along the valley of the Ohio River. The district benefits by its river location, but cheap water transportation is only one factor that attracted many large industrial enterprises in the last few years.

The availability of large, reliable supplies of coal from nearby mines is an important feature. The seasonal vagaries of hydroelectricity caused many industrialists to consider coal as a power source again and the Ohio Valley has a plentiful supply. In addition, earlier the valley had an abundance of land at reasonable prices, although flat land is now more expensive.

Within the Heartland portion of the district are two older, established industrial centers that have shared in the new boom. *Cincinnati,* largest city in the valley, has added electrical equipment to the machine tools, auto parts, and aircraft engines that were its previous stock in trade. *Louisville* is famous for its distilleries, cigarette factories, and chemical plants, but its principal recent growth has been in machinery production, particularly household appliances.

St. Louis District The St. Louis district lies mostly in Missouri but partly in Illinois. The largest urban center between Chicago and the Pacific Coast, St. Louis is a commercial and transportation hub and the fifth largest industrial center in the region (after Chicago, Detroit, Cleveland, and Toronto). It has a well-diversified industrial structure, with an emphasis on aircraft and automotive production.

Heartland Manufacturing Outside the American Manufacturing Belt

West of the American Manufacturing Belt, several large and medium-sized cities function as isolated nodes of industrial concentration. As might be expected, the bulk of the manufacturing in these urban areas is oriented toward agriculture, either by using farm products as raw materials or by producing items for sale to farmers.

Meatpacking is probably the principal type of food processing, following the well-established trend of decentralization toward the source of raw materials—in this case toward the areas of cattle and hog raising. Omaha long ago displaced Chicago as the premier slaughtering center of the continent and it, in turn, has, been surpassed by Sioux Falls (South Dakota) and Sioux City (Iowa). Other major meatpacking localities in the Heartland include South St. Paul, Omaha, Waterloo, St. Joseph, and Kansas City.

Flour milling is the other outstanding food-processing activity in the region, part of the wheat coming from the Heartland but most of it from the Great Plains to the west. Minneapolis and Kansas City have maintained major milling enterprises for decades.

The farm populace also serves as a market. Major manufacturing centers of farm machinery are Minneapolis, the Tri-Cities (Davenport-Rock Island-Moline), Des Moines, Waterloo, and Kansas City.

Diversified manufacturing, primarily for the wholesale markets, is also characteristic of Minneapolis–St. Paul, Kansas City, and Nashville.

TRANSPORTATION

The extensive and busy Heartland Region is well served by transportation facilities. Gentle relief and the lack of topographic barriers made the construction of surface transport lines relatively easy, partly accounting for the dense networks of roads, railways, and pipelines in the region. In addition, the two outstanding natural inland waterway systems of North America, the Great Lakes and the Mississippi River drainage, are largely within the region and numerous canals have at one time or another helped to augment the waterway transport system.

Railways

The first railroads in the Midwest were not built as competitors to navigable waterways but as links connecting them. The rapid extension and improvement of railway facilities after 1850, however, profoundly changed agriculture and revolutionized the whole course of internal trade. By 1860 railroads had triumphed over inland waterways and since then port rivalries have been expressed in the competition of the railroads serving them. Railroads, by

FIGURE 12-15 A major railway bridge crosses the Mississippi River just below a dam near St. Louis (TLM photo).

spanning the great interior with a network of steel rails, also stimulated the growth of cities and the development of manufacturing.

The present railway network is dense, but its major flow is east-west (Fig. 12–15). The main lines principally connect the eastern metropolises of New York, Philadelphia, and Toronto–Montreal with the major midwestern hubs of Chicago and St. Louis. The principal north-south traffic is associated with the Mississippi Valley axis of Chicago–St. Louis–New Orleans.

Since World War II there has been a long-term downtrend in railway usage in the Heartland, as over most of North America. Passenger traffic is limited. Many short lines were abandoned and trackage was considerably consolidated. The role of the railroad today, more than ever before, is to take bulk commodities, particularly mineral and agricultural products, on long hauls. Most of the short-haul freight business, as well as a considerable amount of long-haul traffic, has been lost to truckers.

Roads

The truck and the automobile revolutionized transportation in the twentieth century almost as much as the railway did in the nineteenth. Motor trucking operations benefit from low capacities, high speed, and flexibility of route; therefore they can provide frequent shipments at relatively low cost. Their principal advantage over railroads is on short hauls, although they can often compete on intermediate and even long hauls.

Most parts of the Heartland are well served by highways and roads, for both long-distance and local travel. Several major toll roads were built in this region in the 1940s and 1950s in an effort to speed cross-country traffic and alleviate congestion around cities. Later, however, the idea of toll roads was virtually abandoned because of the construction of the national interstate highway system, which was essentially completed in 1978 (42,000 miles—67,000 km—at a cost of $70 billion). All large and most medium-sized cities in the nation are connected by this system, normally with a four-lane, divided, controlled-access highway. A large portion of the new roadway system is located in the Heartland Region.

Inland Waterways

River, lakes, and canals have been and, in many cases, continue to be of considerable importance for transportation in the Heartland Region.

Rivers Rivers, the principal routeways of pioneer days, were used whenever possible in preference to hard and slow overland routes. Their chief advantages as highways were low cost and convenience. Many a stream that now seems too small or shallow for transportation was extensively used and many a settlement would have died out had there been no stream over which to float products to market. Almost all large western communities in the period from 1800 to 1850 were located on the Ohio or the Mississippi.

The Ohio River The channel of the Ohio River was not navigable during the droughts of late summer until the federal government established a permanent 9-foot (2.7 m) stage with a system of four dozen dams that back the water into a succession of lakes deep enough for navigation. Locks permit boats to get around the dams.

The Ohio thus has become one of the continent's leading carriers of water-borne freight. Thousands of commodious barges, shackled in tows, are propelled over the river at all seasons except for several days in spring when the water is too high or in winter when ice is hazardous. The tonnage is several times greater now than it was at the height of the steamboat period. Bulk products constitute 95 percent of the total freight: coal, coke, ore, sand and gravel, stone, grain, pig iron, and steel.

The dams are several decades old and their locking systems are slow and cumbersome. Replacement construction has begun, however, according to a master plan that calls for a 60 percent reduction in the number of dams and locks on the river, which will result in longer pools and considerable time saving for the carriers.

The Mississippi and Missouri Rivers Barge traffic is also significant on these two rivers, located in the western Heartland. Minneapolis is the head of navigation on the Mississippi, and Sioux City is the effective head of navigation on the Missouri. There are more than half a hundred flood-control and power-generation dams on the two rivers, with locks to let the barges pass through. Most of these dams or locks are also antiquated; during the busy shipping season bottlenecks create significant traffic jams.

Canals During the nineteenth century literally dozens of canals were in use in the Heartland. Most were small and short, serving as connectors between rivers or lakes. A few, however, were of considerable length. Most were long since abandoned, but a handful still play a vital role in freight transportation.

The longest of the Heartland canals is still in use. The Erie Canal crossed New York state from east to west to connect Lake Erie with the Hudson River, following the only practical low-level route through the Appalachian barrier. It was opened in 1825, became the busiest inland waterway in the world, and was a major factor in making New York City the dominant Atlantic port. Railway competition eventually caused it to decline in importance. The entire waterway was reconstructed as the New York State Barge Canal, which was completed in 1918. Traffic today is only moderate.

Great Lakes–St. Lawrence Seaway System The Great Lakes constitute the most valuable system of inland waterways in the world. Since 1829, when the Welland Canal was completed as a bypass route around Niagara Falls, the lakes have been open to oceangoing traffic via their natural outlet, the St. Lawrence River (Fig. 12–16). Only small freighters could be accommodated by the canal, however, even after enlargement during the 1913–1931 period.

In the late 1950s the governments of the United States and Canada finally agreed on, and actually constructed, the St. Lawrence Seaway project (Fig. 12–17). It consisted primarily of deepening and augmenting the between-lake connections, particularly the Welland and Sault St. Marie canals, and the canals that bypass several rapids in the St. Lawrence River between Montreal and Lake Ontario. A 27-foot (8-m) channel is now maintained for 2350 miles (3760 km) from the Atlantic Ocean to the western end of Lake Superior.

Despite high costs, commerce has been stimulated in three dozen lake ports and the economy of many parts of the Heartland has prospered as a result. The Seaway is not deep enough, however, for containerships and other large oceangoing vessels; so there has been a disappointing increase in general

FIGURE 12-16 A lift bridge on the Welland Canal, near Niagara Falls, Ontario (courtesy Ontario Ministry of Industry and Tourism).

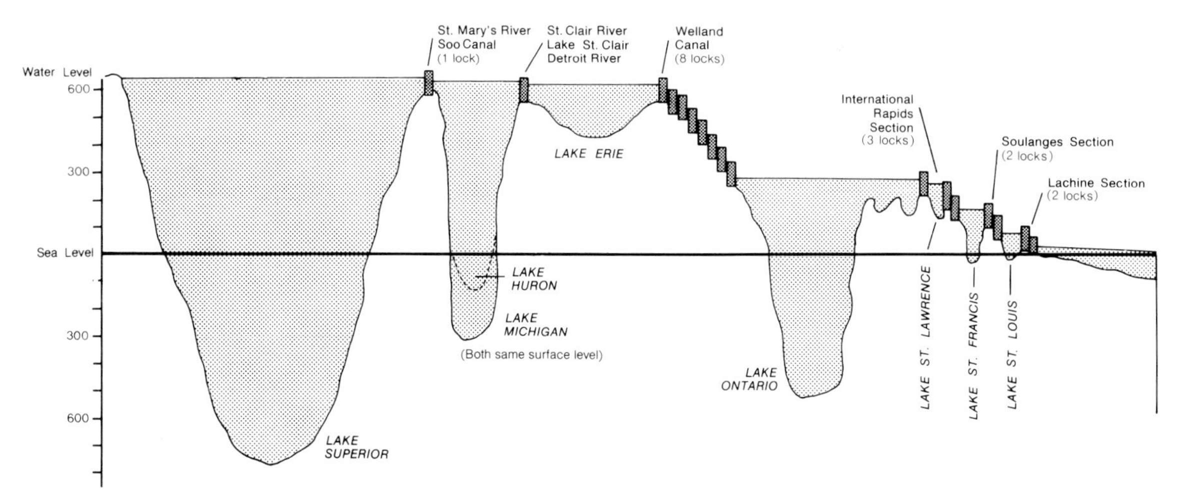

FIGURE 12-17 Compressed diagrams of the Great Lakes–St. Lawrence Seaway System, showing dams and locks.

cargo. The Seaway is primarily a limited bulk-cargo waterway; only about 10 percent of the traffic is general cargo. Iron ore, grain, and coal are the principal items carried, by far, but even these are on a gently declining trend.

THE URBAN SYSTEM OF THE HEARTLAND

In no other major region of Anglo-America is there such regular development of an urban hierarchy as in the Heartland (see Table 12–1 for a listing of the region's largest urban places). The concept of city and hinterland is prominently displayed in most parts of the region and there is a "nested hierarchy" of at least half a dozen levels of magnitude from metropolis to village. The almost classical regularity of the pattern in the Corn Belt and the western margins of the region is interrupted by three major factors that counterpoint the scheme:

1. The Great Lakes, which constrict the areal pattern, provide an important functional connection with overseas regions and stimulate urban development on lakeshores to an unusual degree.
2. Major deposits of economic mineral resources (coal and petroleum) distort the pattern with regard to the distribution of smaller cities and towns.
3. The presence of an important international boundary essentially divides the pattern into two separate systems, one of which has a significant influence on the other.

It would require a detailed analysis of the economic base and function of each individual city to determine its exact position in the hierarchy; even then the position would depend on the stated limits of the scheme. Here we can note only some major components and guess about the position of individual cities.

1. Although the economic influence of New York City is felt to some extent throughout the region, particularly in the eastern portion, *Chicago* is clearly the dominant metropolis. Its position as the wholesaling

TABLE 12-1

Largest urban places of the Heartland Region

Name	Population of Principal City	Population of Metropolitan Area
Akron, Ohio	221,510	653,500
Anderson, Ind.	60,720	131,820
Ann Arbor, Mich.	108,440	267,800
Appleton, Wis.	66,310	312,900
Arlington Heights, Ill.*	73,320	
Aurora, Ill.*	91,150	355,400
Battle Creek, Mich.	55,060	139,200
Barrie, Ont.	48,900	
Benton Harbor, Mich.	14,550	166,600
Bloomington, Ill.	48,860	124,700
Bloomington, Ind.	54,850	103,100
Bloomington, Minn.*	87,090	
Brampton, Ont.*	190,200	
Brantford, Ont.*	78,400	
Buffalo, N.Y.	313,570	958,700
Burlington, Ont.*	117,800	
Cambrdige, Ont.*	80,300	
Canton, Ohio	86,030	401,400
Cedar Rapids, Iowa	110,300	171,500
Champaign, Ill.	59,190	172,100
Chicago, Ill.	2,977,520	6,216,300
Cicero, Ill.*	61,670	
Cincinnati, Ohio	370,480	1,448,800
Clarksville, Tenn.	72,620	158,900
Cleveland, Ohio	521,370	1,845,000
Columbus, Ohio	569,570	1,344,300
Council Bluffs, Iowa	56,700	
Davenport, Iowa	97,140	364,200
Dayton, Ohio	178,000	948,000
Dearborn, Mich.*	86,180	
Dearborn Heights, Mich.*	60,840	
Decatur, Ill.	88,220	123,700
Des Moines, Iowa	192,910	391,800
Des Plaines, Ill.*	55,490	
Detroit, Mich.	1,035,920	4,352,400
Dubuque, Iowa	59,360	90,900
Eau Claire, Wis.	55,030	138,400
Elgin, Ill.*	69,810	
Elyria, Ohio	56,850	
Elkhart, Ind.	45,250	151,100
Erie, Pa.	112,800	277,000
Etobicoke, Ont.*	305,400	
Euclid, Ohio*	55,320	
Evanston, Ill.*	69,910	
Evansville, Ind.	128,210	281,200
Farmington Hills, Mich.*	68,270	
Flint, Mich.	141,620	430,700
Florissant, Mo.*	60,560	
Fort Wayne, Ind.	179,810	367,400

(continued)

TABLE 12-1 Continued

Name	Population of Principal City	Population of Metropolitan Area
Gary, Ind.	132,460	612,200
Grand Rapids, Mich	185,370	665,200
Green Bay, Wis.	95,640	191,200
Guelph, Ont.	79,700	
Hamilton, Ohio	65,500	279,700
Hamilton, Ont.	307,100	557,000
Hammond, Ind.*	84,630	
Independence, Mo.*	115,090	
Indianapolis, Ind.	727,130	1,236,600
Iowa City, Iowa	50,770	86,700
Jackson, Mich.	37,640	149,500
Joliet, Ill.	74,540	379,200
Janesville, Wis.	51,250	136,300
Kalamazoo, Mich.	76,310	217,900
Kankakee, Ill.	26,840	97,900
Kansas City, Kans.	160,630	
Kansas City, Mo.	438,950	1,575,400
Kenosha, Wis.	76,170	122,600
Kettering, Ohio*	60,080	
Kingston, Ont.	56,100	
Kitchener, Ont.	152,300	311,200
Kokomo, Ind.	43,950	98,900
La Crosse, Wis.	48,360	95,500
Lafayette, Ind.	44,290	125,400
Lakewood, Ohio*	58,340	
Lansing, Mich.	124,960	428,400
Lawrence, Kans.	59,460	76,500
Lexington, Ky.	225,660	347,900
Lima, Ohio	46,240	156,700
Lincoln, Nebr.	187,890	211,600
Livonia, Mich.*	101,100	
London, Ont.	272,200	342,300
Lorain, Ohio	71,710	270,500
Louisville, Ky.	281,880	967,000
Madison, Wis.	178,180	352,800
Mansfield, Ohio	51,640	129,000
Milwaukee, Wis.	599,380	1,398,000
Minneapolis, Minn.	344,670	2,387,500
Mississauga, Ont.*	379,100	
Muncie, Ind.	73,320	120,100
Muskegon, Mich.	40,360	161,300
Naperville, Ill.*	79,620	
Nashville, Tenn.	481,380	971,800
Nepean, Ont.*	97,200	
Niagara Falls, N.Y.	62,640	216,900
Niagara Falls, Ont.	73,100	
North York, Ont.*	558,000	
Oak Lawn, Ill.*	57,480	
Omaha, Nebr.	353,170	621,600

(continued)

TABLE 12-1 Continued

Name	Population of Principal City	Population of Metropolitan Area
Oshawa, Ont.	126,100	203,500
Ottawa, Ont.	302,500	819,200
Overland Park, Kans.*	106,860	
Owensboro, Ky.	57,340	87,800
Parma, Ohio*	89,440	
Peoria, Ill.	109,560	340,400
Peterborough, Ont.	61,700	
Pontiac, Mich.	71,800	
Racine, Wis.	82,510	173,800
Rochester, Minn.	60,300	101,000
Rochester, N.Y.	229,780	980,100
Rockford, Ill.	134,500	282,200
Royal Oak, Mich.*	64,120	
Saginaw, Mich.	71,650	406,200
St. Catharines, Ont.	123,900	343,300
St. Clair Shores, Mich.*	70,210	
St. Cloud, Minn.	45,000	181,200
St. Joseph, Mo.	73,490	85,400
St. Louis, Mo.	403,700	2,466,700
St. Paul, Minn.	259,110	
Sarnia, Ont.	49,900	
Scarborough, Ont.*	492,500	
Sheboygan, Wis.	47,410	103,000
Sioux City, Iowa	79,240	115,700
Sioux Falls, S. Dak.	99,610	125,500
Skokie, Ill.*	58,580	
South Bend, Ind.	106,190	244,200
Southfield, Mich.*	71,870	
Springfield, Ill.	99,860	191,700
Springfield, Ohio	69,550	183,885
Sterling Heights, Mich.*	114,720	
Syracuse, N.Y.	153,610	650,300
Taylor, Mich.*	71,640	
Terre Haute, Ind.	56,330	132,600
Toledo, Ohio	340,760	616,500
Topeka, Kans.	122,360	164,800
Toronto, Ont.	619,900	3,426,900
Troy, Mich.*	68,700	
Utica, N.Y.	66,180	312,600
Warren, Mich.	145,410	
Waterloo, Iowa	68,050	147,800
Waterloo, Ont.*	59,700	
Waukegan, Ill.*	72,610	
Westland, Mich.*	181,490	
Windsor, Ont.	193,900	254,000
Waukesha, Wis.*	55,250	
West Allis, Wis.*	64,020	
Wyoming, Mich.*	62,410	
York, Ont.*	141,300	

* A suburb of a larger city.

A CLOSER LOOK The Twin Cities

The Twin Cities of Minneapolis and St. Paul are at the confluence where the Minnesota River, which drains the prairies to the southwest, joins the Mississippi, which flows south from the boreal forest. The river was the lifeline that first brought people and goods into the area, but its deep trenchlike valley, a glacial spillway, was and still remains a major barrier to movement overland. Sheer bluffs 100 to 200 feet (30 to 60 m) high frame a milewide marshy bottomland that floods when heavy spring rains coincide with rising temperatures that melt the winter's accumulation of snow.

At the end of the glacial epoch the Mississippi River tumbled over the brink of the bluff in a magnificent waterfall that had eroded its way 7 miles (11 km) upstream to the present site of downtown Minneapolis by the time the first white explorers entered the area. The Falls of St. Anthony represented the largest waterpower site west of Niagara. Downstream from the falls the river flows through a deep gorge, but just upstream it is divided into two shallow channels by Nicollet Island, where the first bridge ever built across the Mississippi anywhere was opened in 1855.

Downtown St. Paul is perched on the bluffs overlooking the floodplain of the Mississippi. For all practical purposes it is the head of navigation on the river. Just to the east a broad tributary valley, now an industrial area, gives easy access to the level uplands back of the bluffs. St. Paul was the focus of the lucrative fur trade over the ox cart trail that ran northwestward along the edge of the boreal forest to the Red River Valley and Winnipeg, and in 1858 St. Paul was made the state capital of Minnesota.

The railroads that began pushing into the upper Midwest after the Civil War could ignore neither the power site and bridging point nor the capital and head of navigation, but the two places were 9 miles (4 km) apart, too far to be served from a single central station in the horse-and-buggy era. The railroads built complete sets of facilities at each location, and they were largely responsible for the development of two separate, almost equal, and fiercely competing central cities, which the automobile still has not succeeded in blending into one. Minneapolis claims to be more progressive, Scandinavian, and Lutheran, whereas St. Paul is more traditional, German and Irish, and Catholic.

In 1848 the Falls of St. Anthony were harnessed to saw logs floated downstream from the northern pineries, and sawmilling flourished for 60 years, but the last sawmill was closed in 1919 because the forest had all been depleted. The wheat boom on the prairies after the Civil War sparked the growth of flour milling, which dominated the economy for half a century, but after World War I it too began a gradual decline when wheat production moved westward. Fortunes from the milling business were invested in the growth industry of the times, namely the manufacture of electrical controls, which laid a foundation for the development of the contemporary electronics and computer industries.

Today no single firm or type of activity dominates the economy of the Twin Cities, which have become a diversified regional capital comparable to Kansas City, Dallas, or Atlanta, the "front office" for a large but sparsely populated hinterland that stretches westward to the Montana Rockies. The population of the Twin Cities is predominantly white, middle-class, and northwest European. The only real cluster of people of southern or eastern European ancestry is in the industrial area of northeast Minneapolis.

The streets of the two central business districts often seem deserted, because in cold or hot weather most pedestrians use the glass-enclosed "skyways" that lace together office buildings, stores, and hotels at the second-story level. Perhaps also for climatic reasons, the Twin Cities were a national leader in developing large enclosed shopping malls in suburban areas. Narrow commercial strips follow former streetcar lines in the central cities and arterial highways in the suburbs. The commercial strip along the interstate highway west of the airport has become a serious rival to the two downtown areas.

The first industrial areas were near the Mississippi River for power and cheap transportation. Hulking flour mills and great batteries of grain elevators still line the river downstream from the Falls, but most are now derelict or have been converted to other uses. Later industrial areas spread along railroad lines, and the newest planned industrial districts, with pleasant structures on landscaped lots, are near major highways; a nice example is the 3-M Company complex north of the interstate highway east of St. Paul. The Twin Cities have no heavy "smokestack" industries, and the principal "nuisance" industries are the steel mill, former stockyards, chemical plants, and oil refineries well downriver from St. Paul.

The area around downtown Minneapolis is the only part of the Twin

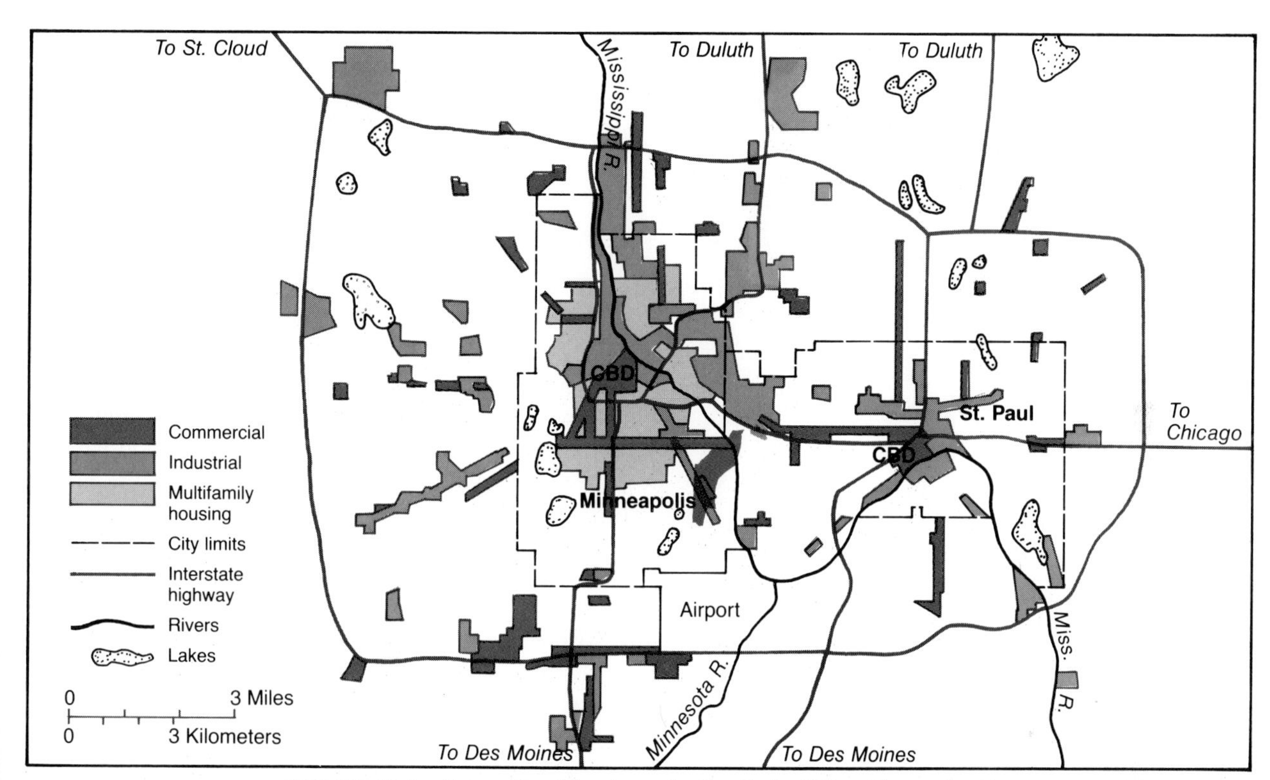

FIGURE 12-f Generalized land use in the Twin Cities (drawn by Don Pirius).

Cities that is densely built-up, with significant numbers of multifamily dwellings, but apartment buildings also buffer some shopping strips (Fig. 12–f). The rest of the built-up area has single-family homes on individual lots that are interspersed with schools, churches, parks, lakes, and streams. The flat, sandy outwash plains have large tracts of inexpensive houses, and the more expensive houses are in the rolling glacial moraines, especially near parks and water bodies. The southern part of the metropolitan area tends toward white-collar workers and office jobs, whereas most blue-collar workers live closer to the industrial areas in the north.

Before 1940 the built-up area was fairly well contained within the two central cities, but since World War II it has spread well beyond the encircling interstate highway bypass. The Twin Cities have enjoyed most of their growth during the automobile era, and it is impossible to design any sensible public transit system to serve them. Their built-up area is more than half as large as the built-up area of Los Angeles, but it has less than one-fifth as many people; it sprawls three times as much as the classic city of sprawl.

Professor John Fraser Hart
University of Minnesota
Minneapolis

FIGURE 12-18 Toronto, its skyline punctuated by the tallest structure on the continent, is one of three subregional metropolises in the Heartland Region (courtesy Ontario Ministry of Industry and Tourism).

center and transportation hub is an obvious indicator of its primacy in any hierarchical urban system for the Midwest.

2. Subregional metropolises, at the second level of magnitude, would probably be represented by Toronto, Detroit, and St. Louis. *Toronto's* sphere is essentially limited to the Ontario portion of the Heartland, but as one of the two primate cities of Canada, it serves many of the same functions as Chicago (Fig. 12–18). *Detroit* would probably be a major subregional metropolis even without the automobile industry because of its outstanding situation on a strategic isthmus alongside the principal waterway of the continent; the addition of the automotive industry adds another major dimension to its significance. *St. Louis* is a more straightforward example of the hierarchical pattern, serving as the principal gateway city for much of the midcontinent.
3. The third level of the hierarchy should probably include such major industrial cities and Great Lakes ports as *Cleveland*, *Buffalo*, and *Milwaukee*, as well as such sectional gateway cities as *Cincinnati*, *Minneapolis–St. Paul*, and *Kansas City*.
4. The fourth level should probably include cities of major intrastate influence, such as *Ottawa*, *Hamilton*, *Rochester*, *Columbus*, *Dayton*, *Indianapolis*, *Louisville*, *Nashville*, and *Omaha*.
5. Succeeding levels in the hierarchy would successively enumerate smaller urban centers with successively less extensive fields of influence.

THE OUTLOOK

The importance of the Heartland to Anglo-America cannot be overstated. It includes a large share of the population and economic power of the United States and it includes that relatively small part of Ontario that provides much of the economic and political leadership of Canada. Thanks to benign nature and enterprising people, the Heartland generally has had a prosperous past; its near future is less sanguine, but the long-run advantages remain.

Although the farmers of the region occasionally suffer from the caprices of nature—drought, flood, tornado, hail, freezes, insect pests—agricultural problems are much more associated with marketing than with production. Free-market prices tend to be soft and erratic and only the continuance of considerable government support, as distasteful as it is to all concerned, is likely to keep the farm economy viable. Current trends of fewer farms, larger farms, decreasing acreage, and increasing yields will probably continue—at least in the short run. The levels of accumulated crop surpluses may fluctuate from time to time, partly as a result of the international market and partly with changes in federal agricultural policies. Without strict government controls, however, increasing yields and falling prices would be a predictable scenario.

Despite the prominence of the agricultural sector and the drift of population away from the metropolises, the tempo of the Heartland is geared to the city, not to rural areas. Metropolitan expansion, although slowed, is the norm in the United States and Canada and this region reflects the pattern. Economic indicators in recent years have been distressing. In the 1980s, for the first time on record, the Heartland fell below the national income average.

Still, the economic significance of the Heartland cannot be denied.

Not that stagnation, poverty, and other problems will not occur from place to place and from time to time. Any contemporary discussion of "distressed cities" would certainly feature Detroit, Cleveland, St. Louis, Akron, Dayton, Hamilton, and other Heartland examples. A depressed steel industry and a struggling automotive industry undermine important pillars of the regional economy. Yet the region's automotive industry already appears to be making a comeback; for example, 60 percent of the nation's active auto assembly plants are now in the Heartland. Adjustment to the various economic problems and situations may not come easily, but it is difficult to imagine that the inherent environmental and societal advantages of the region will not prevail in the long run.

SELECTED BIBLIOGRAPHY

ABLER, RONALD, JOHN S. ADAMS, and JOHN R. BORCHERT, *The Twin Cities of St. Paul and Minneapolis*. Cambridge, MA: Ballinger Publishing Co., 1976.

AKIN, WALLACE E., *The North Central United States*. Princeton, NJ: D. Van Nostrand Company, 1968.

ALEXANDER, CHARLES S., and NELSON R. NUNNALLY, "Channel Stability on the Lower Ohio River," *Annals*, Association of American Geographers, 62 (1972, 411–417.

BAERWALD, THOMAS J., "The Twin Cities: A Metropolis of Multiple Identities," *Focus*, 36 (Spring 1986), 10–15.

BAINE, RICHARD P., and A. LYNN MCMURRAY, *Toronto: An Urban Study*. Toronto: Clarke Irwin & Co., Ltd., 1970.

BERRY, BRIAN J. L., ET AL., *Chicago: Transformation of an Urban System*. Cambridge, MA: Ballinger Publishing Company, 1976.

BORCHERT, JOHN R., and NEIL C. GUSTAFSON, *Atlas of Minnesota: Resources and Settlement* (3rd ed.). Minneapolis: University of Minnesota and the Minnesota State Planning Agency, Center for Urban and Regional Affairs, 1980.

BURNS, NOEL M., *Erie: The Lake That Survived*. Totowa, NJ: Rowman and Allanheld, 1985.

CARLSON, ALVAR W., "Specialty Agriculture and Migrant Laborers in Northwestern Ohio," *Journal of Geography*, 75 (1976), 292–310.

CHAPMAN L. J., and D. F. PUTNAM, *The Physiography of Southern Ontario*. Toronto: University of Toronto Press, 1966.

CUTLER, IRVING, *Chicago: Metropolis of the Mid-Continent* (2nd ed.). Dubuque, IA: Kendall/Hunt Publishing Company, 1976.

———, *The Chicago Metropolitan Area: Selected Geographic Readings*. New York: Simon & Schuster, 1970.

———, *The Chicago-Milwaukee Corridor*. Evanston, IL: Northwestern University, Department of Geography, 1965.

DARDEN, JOE T., RICHARD CHILD HILL, JUNE THOMAS, and RICHARD THOMAS, *Detroit: Race and Uneven Development*. Philadelphia: Temple University Press, 1987.

DAVIS, ANTHONY M., "The Prairie-Deciduous Forest Ecotone in the Upper Middle West," *Annals, Association of American Geographers*, 67 (1977), 204–213.

DEAR, M. J., J. J. DRAKE, and L. G. REEDS, eds., *Steel City: Hamilton and Region*. Toronto: University of Toronto Press, 1987.

EHRHARDT, DENNIS K., "The St. Louis Daily Urban System," in *Contemporary Metropolitan American. Vol. 3: Nineteenth Century Inland Centers and Ports*, ed. John S. Adams, pp. 61–107. Cambridge, MA: Ballinger Publishing Company, 1976.

EICHENLAUB, VAL, *Weather and Climate of the Great Lakes Region*. Notre Dame, IN: University of Notre Dame Press, 1979.

ELFORD, JEAN, "The St. Clair River: Centre Span of the Seaway," *Canadian Geographical Journal*, 86 (January 1973), 18–23.

FRAMPTON, ALYSE, "Toronto's Harbourfront: An Exciting Blueprint for Urban Renewal," *Canadian Geographic*, 104 (December 1984–January 1985), 62–69.

FULLERTON, DOUGLAS, "Whither the Capital?," *Canadian Geographic*, 107 (December 1987–January 1988), 8–19.

GAYLER, H. J., "The Problems of Adjusting to Slow Growth in the Niagara Region of Ontario," *The Canadian Geographer*, 26 (1982), 165–172.

GENTILCORE, LOUIS, ed., *Ontario*. Toronto: University of Toronto Press, 1972.

GORRIE, PETER, "Tobacco Alternatives: New Crops Offer Hope for Hard-Pressed Ontario Growers," *Canadian Geographic*, 108 (June–July 1988), 58–65.

HART, JOHN FRASER, "The Middle West," *Annals*, Association of American Geographers, 62 (1972), 258–282.

———, "Change in the Cornbelt," *Geographical Review*, 76 (January 1986), 51–72.

———, "Small Towns and Manufacturing," *Geographical Review*, 78 (July 1988), 272–287.

JOHNSON, HILDEGARD BINGER, *Order upon the Land: The U.S. Rectangular Land Survey and the Upper Mississippi Country*. New York: Oxford University Press, 1976.

KARAN, PRADYUMNA K., *Kentucky: A Regional Geography* (2nd ed.). Dubuque, IA: Kendall/Hunt Publishing Company, 1976.

KEATING, MICHAEL, "Fruitlands in Peril: We're Covering Them with Houses, Factories, and Asphalt," *Canadian Geographic*, 106 (October–November 1986), 26–35.

KILBOURN, WILLIAM, "The New Toronto: A Great Modern City," *Canadian Geographical Journal*, 96 (April–May 1978), 10–19.

KINGSBURY, ROBERT C., *An Atlas of Indiana*. Bloomington: Indiana University, Department of Geography, 1970.

KRUEGER, RALPH R., "The Struggle to Preserve Specialty Cropland in the Rural-Urban Fringe of the Niagara Peninsula of Ontario, *Environments*, 14 (1982), 1–10.

———, "Urbanization of the Niagara Fruit Belt," *Canadian Geographer*, 22 (Fall 1978), 179–194.

LANGMAN, R. C., *Poverty Pockets: The Limestone Plains of Southern Ontario*. Toronto: McClelland and Stewart, Ltd., 1975.

LIGHTBOURN, ARTHUR, "The Perils of Preserving the Niagara Scarp," *Canadian Geographic*, 102 (August–September 1982), 10–17.

MACLACHLAN, IAN, and DEBORAH YOUNG, "Toronto's Danforth Speaks with Many Tongues," *Canadian Geographic*, 102 (December 1982–January 1983), 20–25.

MALCOMSON, ROGER, "The Niagara River in Crisis," *Canadian Geographic*, 107 (October–November 1987), 10–19.

MARTIN, VIRGIL, *Changing Landscapes of Southern Ontario*. Toronto: Boston Mills Press, 1989.

MATHER, COTTON, et al., *Upper Coulee Country*. Prescott, WI: Trimbelle Press, 1975.

MAYER, HAROLD M., and RICHARD C. WADE, *Chicago: Growth of a Metropolis*. Chicago: University of Chicago Press, 1969.

MAYER, HAROLD M., and THOMAS CORSI, "The Northeastern Ohio Urban Complex," in *Contemporary Metropolitan American. Vol. 3: Nineteenth Century Inland Centers and Ports*, ed. John S. Adams, pp. 109–179. Cambridge, MA: Ballinger Publishing Company, 1976.

NELSON, RONALD E., *Illinois: Land and Life in the Prairie State*. Dubuque, IA: Kendall/Hunt Publishing Company, 1977.

NOBLE, ALLEN G., and GAYLE A. SEYMOUR, "Distribution of Barn Types in Northeastern United States," *Geographical Review*, 72 (April 1982), 155–170.

NORRIS, DARRELL A., "Ontario Fences and the American Scene," *The American Review of Canadian Studies*, 12 (Summer 1982), 37–50.

PLATT, RUTHERFORD H., "Farmland Conversion: National Lessons for Iowa," *Professional Geographer*, 33 (February 1981), 113–121.

POWLEDGE, FRED, "Profiles: Louisville, City in Transition," *The New Yorker* (September 9, 1974), pp. 42–83.

RAITZ, KARL B., *The Kentucky Bluegrass: A Regional Profile and Guide*. Chapel Hill: University of North Carolina, Department of Geography, 1980.

———, "Kentucky Bluegrass," *Focus*, 37 (Fall 1987), 6–11.

REES, JOHN, "Technological Change and Regional Shifts in American Manufacturing," *The Professional Geographer*, 31 (February 1979), 45–54.

ROBINSON, ARTHUR H., and JERRY B. CULVER, *The Atlas of Wisconsin*. Madison: University of Wisconsin Press, 1974.

ROSEMAN, CURTIS C., ANDREW J. SOFRANKO, and JAMES D. WILLIAMS, eds., *Population Redistribution in the Midwest*. Ames: Iowa State University, North Central Regional Center for Rural Development, 1981.

RUBENSTEIN, JAMES M., "Changing Distribution of the American Automobile Industry," *Geographical Review*, 76 (July 1986), 288–300.

RUSSWURM, LORNE H., and BALESHWAR THAKUR, "Hierarchical and Functional Stability and Change in a Strongly Urbanized Area of Southwestern Ontario," *Canadian Geographic*, 25 (1981), 149–166.

SANTER, RICHARD, A., *Michigan: Heart of the Great Lakes*. Dubuque, IA: Kendall/Hunt Publishing Company, 1977.

SHORTRIDGE, JAMES R., *The Middle West: Its Meaning in American Culture*. Lawrence: University Press of Kansas, 1989.

SIMMONS, JAMES, "How much Growth Can Toronto Afford?" *Canadian Geographical Journal,* 92 (March–April 1976), 4–11.

SINCLAIR, ROBERT, and BRYAN THOMPSON, "Detroit," in *Contemporary Metropolitan America. Vol. 3: Nineteenth Century Inland Centers and Ports,* ed. John S. Adams, pp. 285–354. Cambridge, MA: Ballinger Publishing Company, 1976.

SOMMERS, LAWRENCE M., ed., *Atlas of Michigan.* Grand Rapids, MI: William B. Eerdmans Publishing Company, 1977.

SOMMERS, LAWRENCE M., JOE T. DARDEN, JAY R. HARMAN, and LAURIE K. SOMMERS, *Michigan: A Geography.* Boulder, CO: Westview Press, 1984.

SPELT, JACOB, *Toronto.* Don Mills, Ont.: Collier-MacMillan Canada, 1974.

———, *Urban Development in South-Central Ontario.* Toronto: McClelland & Stewart, 1972.

SQUIRES, GREGORY, LARRY BENNETT, KATHLEEN MCCOURT, and PHILIP NYDEN, *Chicago: Race, Class and the Response to Urban Decline.* Philadelphia: Temple University Press, 1987.

STANLEY, JAMES, "Salmon Revival in Lake Ontario," *Canadian Geographer,* 101 August–September 1981), 46–51.

VOGELER, INGOLF, ed., *Wisconsin: A Geography.* Boulder, CO: Westview Press, 1986.

WALKER, D. F., and J. H. BATER, eds., *Industrial Development in Southern Ontario.* Waterloo, Ont.: University of Waterloo Press, 1974.

WARREN, K., *The American Steel Industry 1850–1970: A Geographic Interpretation.* New York: Oxford University Press, 1972.

YEATES, MAURICE, *Main Street: Windsor to Quebec City.* Toronto: MacMillan Company of Canada, 1975.

———, "The Extent of Urban Development in the Windsor–Quebec City Axis," *Canadian Geographer,* 31 (Spring 1987), 64–69.

13

THE GREAT PLAINS AND PRAIRIES

Cutting through the center of North America in a south-north orientation is one of the most distinctive and widely recognized regions of the continent—the Great Plains and Prairies. As identified in this volume, the region corresponds roughly with the Great Plains physiographic province that has been described by many geomorphologists and geographers.[1] In the United States the region is universally referred to as "the Great Plains." In Canada, however, that term is seldom used; instead the Canadian portion of the region is called "the prairies" or occasionally "the interior plains."[2] To accommodate both local usages, "The Great Plains and Prairies" is the name chosen for use here.

The extent of the region as recognized here is broader than the extent of the physiographic province. On the basis of land-use and cropping patterns, the delimitation of the region has been pushed somewhat to the south to include ranching and irrigated farming country in south-central Texas and to the northeast to encompass the agricultural area of the Red River Valley (Fig. 13-1). The entire eastern boundary of the region varies somewhat from the physiographic boundary, again on the basis of land-use patterns.

The problem of delimitation of the regional boundaries of the Great Plains was discussed in some detail in Chapter 5. In summary, the western boundary is marked fairly abruptly by the rise of the frontal ranges of the Rocky Mountains; the northern

[1] For example, Wallace W. Atwood, *The Physiographic Provinces of North America* (Boston: Ginn & Company, 1940); Nevin N. Fenneman, "Physiographic Divisions of the United States," *Annals*, Association of American Geographers, 18 (1928), 261–353; Charles B. Hunt, *Natural Regions of the United States and Canada* (San Francisco: W. H. Freeman & Company, 1973); and William D. Thornbury, *Regional Geomorphology of the United States* (New York: John Wiley & Sons, 1965).

[2] Note J. Brian Bird, *The Natural Landscapes of Canada* (Toronto: Wiley Publishers of Canada, 1972); P. J. Smith, ed., *The Prairie Provinces* (Toronto: University of Toronto Press, 1972).

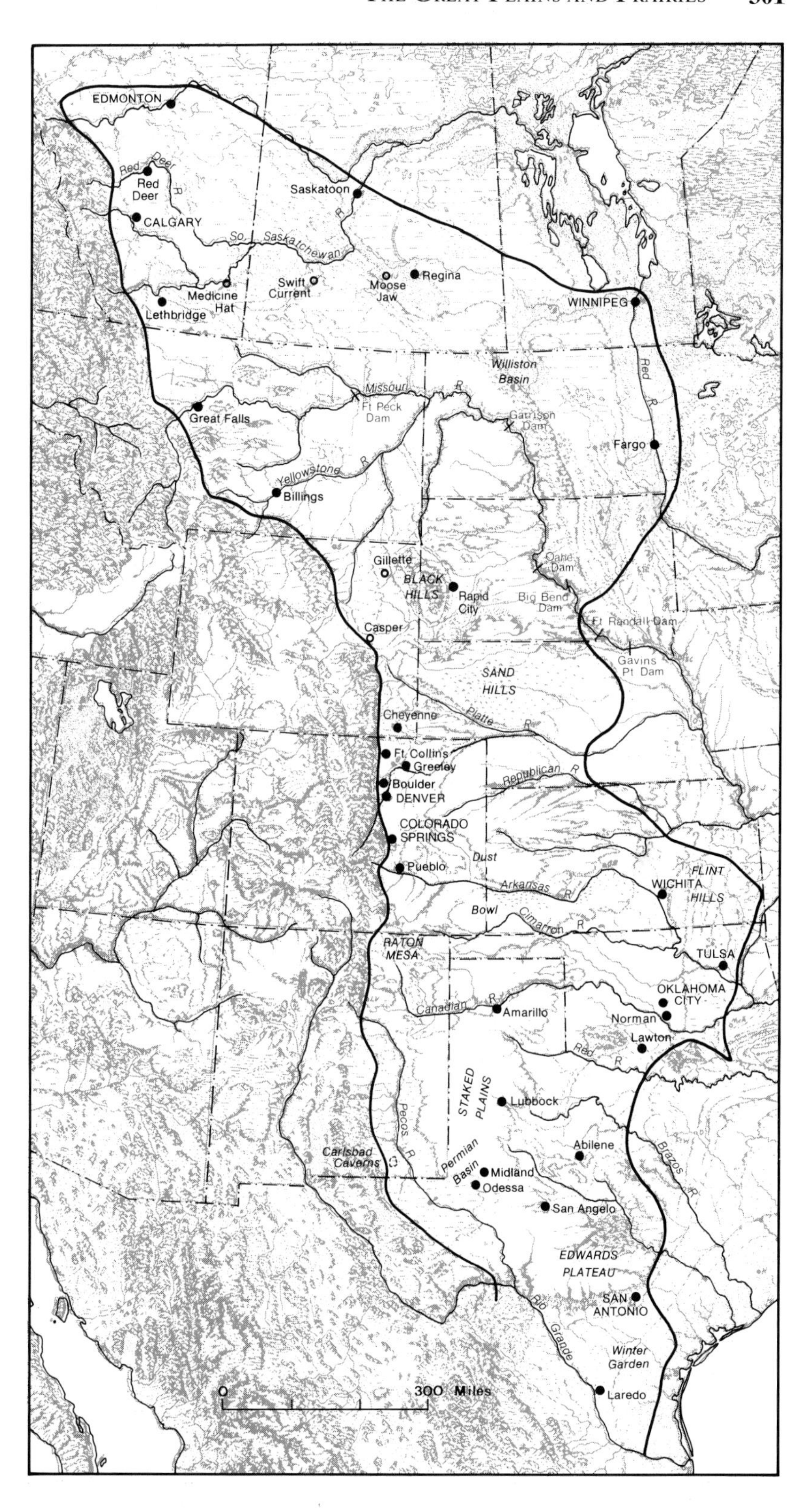

FIGURE 13-1 The Great Plains and Prairies Region (base map copyright A. K. Lobeck; reprinted by permission of Hammond, Inc.).

boundary represents the southern margin of the boreal forest; and the eastern boundary is transitional between the extensive wheat-farming systems to the west and corn-and-general-farming systems to the east.

THE CHANGING REGIONAL IMAGE

The dramatic weather and unpredictable climate of the vast interior plains region have defied accurate assessment by its occupants. As a result, fluctuating patterns of land use and economic well-being occurred but never changed the region's basic role as a quantity producers of selected agricultural and mineral resources for the continent.

The Great Plains and Prairies Region contains some of the best soils and potentially most productive farmlands of Anglo-America; yet crop failures have alternated with crop surpluses and accelerated soil erosion is commonplace. Relatively deep, dark-colored soils, which contain a considerable amount of organic matter and lime, are widespread. These soils, low relief, and much summer sunshine combine to provide several of the necessary ingredients for productive agriculture, but erratic precipitation, sometimes ill-advised farming practices, and the vagaries of the marketplace are inhibitory factors. Average annual precipitation is on the minimal side for crop production and its usefulness is further limited by considerable fluctuations from the average in any given years, by the cloudburst nature of much of the rain, and by spring floods caused by rapid snowmelt and heavy showers.

FIGURE 13-2 The characteristic Great Plains landscape consists of extensive cropland (strip-cropped wheat here) and pasture, dotted with small towns that are marked by tall grain elevators. This is Cowley, Alberta (Alberta Government photo).

An expanding international market for wheat, and occasionally for other crops, in the past persuaded farmers to attempt cultivation of land that should not have been plowed. Crop failures and soil abuse resulted, most notably in dust bowl conditions. In such a subhumid-to-semiarid region the dangers of accelerated soil erosion are great and wind erosion, in particular, has left its mark on extensive areas. The history of the region is thus marked by occasional monumental crop failures, although in many years bumper crops resulted in stupendous surpluses.

Relatively few people within the region are engaged in the primary production of relatively few products, but their output is tremendous. The total farm and ranch population of the region is only a small percentage of the Anglo-American total; yet well over half the continent's annual output of wheat, grain sorghums, barley, rye, flax, canola, mohair, and potash originates here. In addition, there is notable production of coal, cotton, cattle, wool, petroleum, and natural gas. Gathering, storing, transporting, and sometimes processing these products are major activities that employ a large number of people, although there are marked seasonal fluctuations. Ultimately the great majority of the output leaves the region for the most of its processing, as well as for final disposition to consumers.

Although cities are growing and many villages are stagnating, there is a strong rural orientation to life. As in the rest of the continent, increasing urbanization is characteristic of the Great Plains and Prairies. Many small towns and villages, however, do not show marked growth tendencies (Fig. 13-2); the larger cities are the growth centers. Overall, even though the majority of the populace is urban, primary production has always been such a significant backbone of the economy that rurally oriented points of view, values, opinions, and judgments are often prevalent.

The character of the region encompasses a curious mixture of the drab and the grandiose. Flat land, endless horizons, blowing dust, colorless vistas, withering towns, and workaday tasks emphasize the former. But many facets of the regional scene are on an heroic scale: sweeping views, dramatic weather, natural calamities, stupendous production, staggering problems, and immense distances.

In the past the Great Plains stereotype was a vast land of dry, treeless plains, sparsely settled and given over primarily to raising cattle. But technological change came to this region as it has to most of North America. Improved dry-farming techniques, development of large-scale irrigation enterprises, establishment of state-of-the-art livestock feedlots, expanded output of energy minerals, and rapid growth of significant urban-industrial nodes (Fig. 13-3) combined to reshape the distinctive image of the region.

THE PHYSICAL SETTING

It is convenient to think of the physical geography of the plains–prairies as being uniform, the flat land engendered basic homogeneity in other physical aspects. There is some validity to this concept of broad regional unity in terms of gross patterns; in any sort of detailed consideration of the region, however, there is notable variety of contrasts—in physical as well as cultural geography.

Terrain

The region has the basic topographic unity of an extensive plains area (Fig. 13-4), but in detail the plains character is only true in the broadest sense owing to significant variations from area to area. The underlying structure is a broad geosyncline composed of several basins separated by gentle arches, the surface bedrock consisting mostly of gently dipping sedimentary strata of Cretaceous and Tertiary age. The surface expression is an extensive plain that is highest in the west and gradually descends to the east at an average regional slope of about 10 feet

FIGURE 13-3 A modern boomtown—Colstrip, Montana—in the Great Plains (courtesy Montana Power Company).

FIGURE 13-4 The image of the Great Plains is one of flatness, and reality is much like the image in many parts of the region. This scene is in southeastern Alberta, just north of the international border (TLM photo).

per mile (1.9 m per km). Near the western margin the plain is more than 6000 feet (1800 m) above sea level in some places; at the eastern edge the average altitude is less than 1500 feet (450 m). There are great aprons of alluvial deposits near the Rocky Mountain front and along the major river valleys, and glacial deposits thinly cover the surface north and east of the Missouri River.

Several prominent physiographic subdivisions can be recognized on the basis of their landform associations. From south to north, they are as follows.

1. The Rio Grande Plain is flattish throughout, with some incised river valleys.
2. The Central Texas Hill Country consists of a broad crescent of low but steepsided hills that form the dissected margin of the Edwards Plateau on its eastern and southern sides and somewhat to the north. Associated features include an eroded dome of Precambrian rocks and a number of large fault-line springs that discharge around the edge of the hills.
3. The High Plains section of the Great Plains occupies most of the area from the Edwards Plateau northward to Nebraska. Much of it is extraordinarily flat except where crossed by one of the major eastward-flowing rivers. The surface rock is chiefly a thick mantle of Tertiary sediments. The extreme flatness is partly a result of a concentration of carbonates (caliche) in a "cap rock" layer that resists erosion and is partly due to surface formations that are sandy and thus highly porous. In both cases, water erosion is at a minimum except along the escarpment-like edges of the cap rock and where certain rivers, particularly the Canadian and Red, have cut down through the resistant surface. In west Texas and eastern New Mexico is the Llano Estacado (Staked Plains) where there are some 30,000 square miles (78,000 km^2) of almost perfect flatness, essentially unmarked by stream erosion. The Edwards Plateau, extending southeastward from the Llano Estacado, is geologically different but topographically identical.
4. The longitudinal valley of the Pecos River is a gentle trough lying below the level of the Llano Estacado; it is characterized by karst features and gravel-capped terraces.
5. The Raton Mesa section along the New Mexico–Colorado border consists of a series of mesas and buttes supported by basalt flows, along with a few cinder cone volcanoes.
6. The Colorado Piedmont is an irregularly shaped zone extending along the Rocky Mountain front, from the Arkansas Valley to the Platte Valley, where much of the Tertiary alluvium has been eroded, causing the surface to be lower than the High Plains to the east. Topography here is strongly controlled by stream dissection.
7. The Nebraska Sand Hills cover much of the western and central portions of that state. The area is a maze of sand dunes

and ridges that rise to several hundred feet in height and are separated by numerous small basins.

8. The Unglaciated Missouri Plateau occupies most of the northern Great Plains north of Nebraska and south of the Missouri River. There is considerable variety to the topography, but most is gently undulating. There are conspicuous badlands in South Dakota, North Dakota, and Montana as well as notable outliers of the Rocky Mountains.

9. The Glaciated Missouri Plateau section, north and east of the Missouri River, demonstrates many features of glacial origin on its surface, particularly moraines and ponds.

10. The Lake Agassiz Basin encompasses the valley of the Red River of the North as well as much of southern Manitoba and eastern Saskatchewan. Lake Agassiz was the largest of the late Pleistocene ice marginal lakes, and its ancient lake bed is extremely flat and deeply floored with silty clay. Several dozen beach lines are identifiable.

In the northwestern portion of the region several isolated ranges are offset from the Rocky Mountains. Although most are topograpically and geologically related to the Rockies, their outlying position makes them a part of the Great Plains and Prairies Region. The largest and most conspicuous of the outliers is the Black Hills; others are shown in Fig. 13-5.

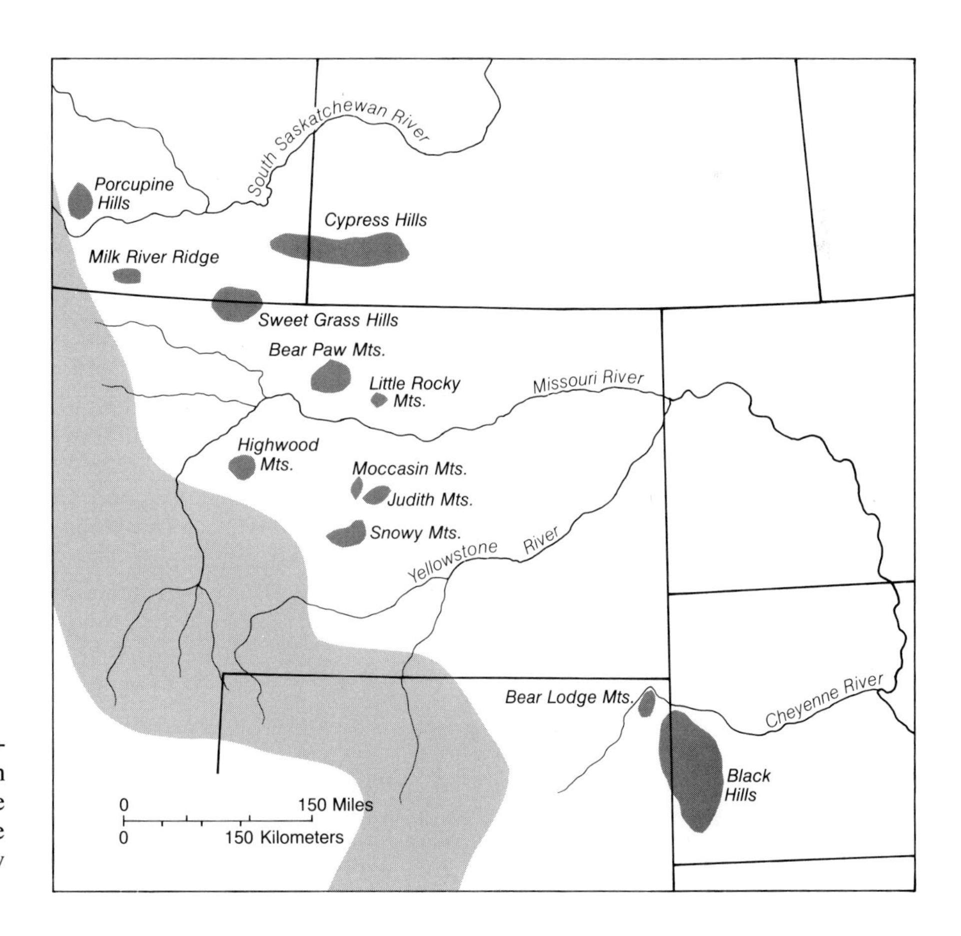

FIGURE 13-5 Principal mountain outliers in the northern Great Plains and Prairies. The extensive shaded area on the left represents the Rocky Mountains.

FIGURE 13-6 Late spring and early summer constitute a time of flooding for many of the rivers in the region. This is a June scene on the Arikaree River near Haigler in central Nebraska (TLM photo).

Drainage and Hydrography

There is generally good drainage throughout the region, with some significant exceptions. Much of the area north and east of the Missouri River is dotted with small lakes and marshes, which are chiefly the result of Pleistocene glacial deposition. Drainage in the Nebraska Sand Hill area is also irregular, with many small basins and pockets that do not have exterior drainage outlets.

The basic stream-flow pattern of the region is from west to east; the rivers rise in the Rocky Mountains and flow down the regional slope to join the Mackenzie, Hudson Bay, Mississippi, or Gulf of Mexico systems (Fig. 13-6). The only two significant variants from this pattern are the Red River of the North, which flows northward into Lake Winnipeg, and the Pecos River, which flows generally southward to become a tributary of the Rio Grande. Most of the principal river valleys are conspicuous as narrow strings of irrigated agriculture, denser rural settlement, and urban clusters.

Large or medium-sized lakes are virtually unknown in the region. A number of large reservoirs were constructed, however, and more are planned. The most prominent are those along the Missouri River, where there is now little free-flowing water from eastern Montana to northeastern Nebraska.

Climate

Such a latitudinally extended region has considerable variation in climate, particularly in temperature. The essential characteristics, however, are clear-cut: moisture conditions are subhumid to semiarid, with evaporation usually exceeding precipitation; there are pronounced seasonal extremes; and much drama and violence occur in day-to-day weather conditions (Fig. 13-7).

The climate of the plains and prairies is continental; precipitation ranges from 15 inches (380 mm) in the northwest to 35 (890 mm) in the southeast and varies greatly from year to year. There are periods of dry years when the westerly margins become almost desertic. The growing season varies from 120 days in the north to 300 days in the south. Summers are normally very hot, although the duration of high temperatures is much shorter in the northern part of the region than the southern.

Winters are bitterly cold[3] and dry and there-

[3] The city of Winnipeg serves as the locus of many folk tales about the extremes of winter temperatures. It is said to have "ten months of winter and two months of bad ice." Steffanson, the last of the great Arctic explorers, has been quoted as saying that if you can live in Winnipeg you can live anywhere in the Arctic, as regards winter discomfort. See George Jacobsen, "The Northern

FIGURE 13-7 Dramatic weather is almost commonplace in the Great Plains and Prairies Region. These dark cumulonimbus clouds presage a hail storm near High River in southern Alberta (TLM photo).

fore very hard on such perennials as cultivated hay and fruit trees. The differences in temperature between winter and summer are so great as to give this region the distinction of having one of the greatest seasonal ranges of any region.

Summer winds are so hot and dry and those of winter so biting and cold that most farms have trees planted as a windbreak to reduce surface wind velocities. In the western part of the region, however, from Colorado northward, the winter weather is sporadically ameliorated by *chinook* winds. These warming, drying, downslope winds from the Rockies bring periods of relative mildness that are a welcome relief to both people and lifestock.

The Great Plains and Prairies experience the highest incidence of hail of any region in North America. The crop-destroying nature of the ice pellets is so intense that hail insurance is important to most farmers, particularly in the northern portion.

Above all, this is a region of violent weather conditions and abrupt day-to-day or even hour-to-hour weather changes. The horizon may be flat and dull in the Great Plains and Prairies, but the skies are often turbulent and exciting. Cold fronts, warm fronts, tornadoes, thunderstorms, blizzards, heat waves, hail storms, and dust storms are all part of the annual pageant of weather in this region.

Urban Scene," in *Canada's Changing North,* ed. William C. Wonders (Toronto: McClelland and Steward Limited, 1971), p. 292.

Soils

The soils of the wheat belts, among the most fertile in North America, are mostly Mollisols. They have a lime zone, a layer of calcium carbonate a few inches or a few feet beneath the surface within reach of plant roots. Because of the scanty rainfall, these soils have not had the lime leached from them. Their fertility—when combined with greater rainfall—makes them the most productive, broadly distributed soils in the world, although there is less humus than in the grassland soils to the east. They are characterized by being dark colored, rich in organic matter, well supplied with chemical bases, and usually containing a subsurface accumulation of carbonates, salts, and clay.

In drier localities Entisols and Aridisols are dominant, particularly in eastern Colorado, Wyoming, and Montana and in western Nebraska. These soils contain little organic matter and are either dry and clayey or dry and sandy, although their level of natural fertility is generally high.

Natural Vegetation

Between the forests on the east and the mountains on the west lie the prairies and the steppe. The prairie, whose grasses usually attain a height of 1 to 3 feet (0.3 to 1 m), characterizes areas with 20 to 25 inches (500 to 635 mm) of precipitation in the north and 35 to 40 inches (890 to 1016 mm) in the south. Merging with the prairie on the semiarid fringe to the

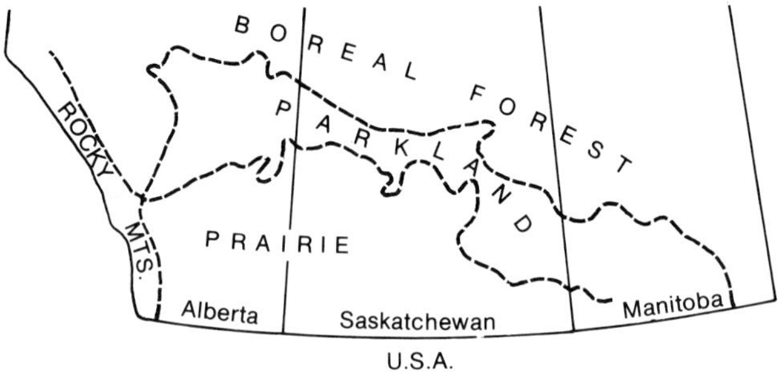

FIGURE 13-8 An area of "parkland" vegetation functions as a transition zone between the boreal forest to the north and the prairie to the south.

west is the steppe, whose grasses are of low stature and where rainfall is less than 20 inches (500 mm).

The native vegetation of the semiarid grazing portion of the Great Plains is primarily short grass, with grama and buffalo grasses most conspicuous. Before the introduction of livestock in the latter half of the nineteenth century, luxuriant native grasses (mainly western wheat grass) covered extensive areas. Overgrazing and extension of wheat farming into unsuitable areas reduced thousands of square miles to a semidesert.

The entire region is not grass covered, of course. The isolated upland enclaves are mostly forested, primarily with Rocky Mountain conifers, plus aspen and willow. The largest forest area is in the Black Hills, but tree cover is also dominant in the Raton Mesa area, the so-called Black Forest between Colorado Springs and Denver, portions of the Nebraska Sand Hills, most of the Montana mountain outliers, and almost every hill in the southern Prairie Provinces. A scrubby juniper woodland also covers much of the central Texas hill country and some cap rock escarpment faces in that same state.

Along the northern fringe of the region is a heterogeneous mixture of grasses and trees that serves as a transition zone between the prairies to the south and the boreal forest to the north. It is known as a "parkland" area (Fig. 13-8) with an erratic variation of dominant species—willows, aspen, conifers, and various grasses.

The major stream valleys of the region are usually marked by a narrow band of riparian timber, nearly all of which consists of cottonwood, willows, poplars, and similar deciduous species.

During the past century much of central and southern Texas had a massive invasion by a deep-rooted, scrubby tree called mesquite (*Prosopis juliflora*); it is native to the area but has greatly expanded its range, presumably as the result of overgrazing, short-term climatic fluctuations, and cessation of recurrent grassland fires.[4] This invasion significantly reduced the grazing forage and encouraged ranchers to undertake stringent control campaigns, involving poisons, burning, and especially uprooting with heavy equipment (Fig. 13-9).

Also within the last century junipers (*Juniperus* spp.) similarly expanded their range over more than 25 million acres of what had been mostly grassland in central and western Texas. There are nine species of these hardy, scrubby, fragrant conifers, which are often inaccurately referred to as "cedars," in the southern plains. Although useful for fence posts and as a source of oil to add an aroma to household detergents, junipers, like mesquite, are generally considered pastoral pests.

Wildlife

The Great Plains and Prairies Region was the principal habitat of the American bison, with an estimated 50 million of these magnificent beasts occupying the region at the coming of the white man. Once white penetration of the region got underway in earnest, almost all the vast herds were exterminated in less than a decade.

Other hoofed animals—pronghorn antelope, deer, elk, and mountain sheep—were also common. They suffered a lesser fate than the bison, mostly being pushed into the mountains to the west as settlement advanced.

Although this is a subhumid region, furbearers

[4] David R. Harris, "Recent Plant Invasions in the Arid and Semi-Arid Southwest of the United States," *Annals*, Association of American Geographers, 56 (September 1966), 408–422.

FIGURE 13-9 This scene, near Abilene in west Texas, shows two floral "invaders" in the southern part of the Great Plains. Both mesquite and broomweed are natives to the region, but in recent decades they have spread widely, presumably as a response to overgrazing by livestock (TLM photo).

were numerous along the streams. Beaver, muskrat, mink, and otter attracted the trappers and fur traders, who were, with the exception of a few explorers, the first whites to penetrate the region.

A tremendous number of small, shallow marshes and ponds dot the glaciated terrain of the Dakotas and Prairie Provinces. These poorly drained areas provide an excellent muskrat habitat and are used as summer breeding ground for myriad waterfowl. It is estimated that about half of all the ducks in North America breed in these ponds.

Several exotic species have been introduced to this region, generally as additional prey for hunters. Most important by far is the ring-necked pheasant (*Phasianus colchicus*), which has become well established in every state and province from Colorado and Kansas northward. Because of the money spent by nonresident hunters, pheasant hunting has become so important to the economy of South Dakota that it is one of the prime economic and political factors in this state.

SEQUENT OCCUPANCE OF THE GREAT PLAINS AND PRAIRIES

The human saga of the region, with its varied stages of occupance and settlement and its diversified attempts at satisfactory and profitable land use, is a dramatic and interesting one. Only a few of the highlights are recounted here, with emphasis on sequential occupance.

The Plains Indians occupy a special place in North American history because of their relationship to the Wild West era and their midcontinent position athwart the axis of the westward flow of empire. At the time of European contact Indians of the Great Plains consisted of about two dozen major tribes, most of which were scattered in small semisedentary settlements over a particular territory. Their livelihood was based partly on hunting, especially buffalo, and partly on farming, particularly corn; their chief avocation was combat with other tribes; and one of their major problems was lack of transportation over the vastness of the plains.

By the middle of the eighteenth century essentially all the Plains Indians had obtained horses and most had become expert in their use. They became much more mobile, much more effective as hunters, and much more deadly as warriors. Some tribes, such as the Dakota (Sioux) and Blackfeet in the north and the Comanche and Apache in the south, became very powerful and for many years exerted a strong influence over parts of the region. Their dominance was eventually challenged and overthrown, however, in part by eastern tribes that were displaced to the plains by whites, in part by the virtual extermination of bison, the principal food supply, but mainly by the overwhelming superiority of white soldiers and settlers.

The last stronghold of Indians in the Great Plains was the Indian Territory established between Texas and Kansas (Fig. 13-10). Displaced tribes from the Southeast were settled there in the early 1800s, and midwestern and Plains tribes were relegated there after the Civil War. Eventually Indian Territory became Oklahoma and all its reservations were dissolved.

The first significant movement of white settlers into the region came from the south, from Mexico. Very early in the eighteenth century Spanish settlers moved north of the Rio Grande, following missionaries who had come to the Tejas Indians in 1690. Their major bastion in this region was San Antonio, founded in 1718 (the same year that New Orleans was founded by the French). For several decades Spanish and later Mexican settlers trickled into what is now southern Texas. They were soon joined by Anglos, who were attracted by Mexico's initially generous land-grant policies. Most of the early Anglo settlement of Texas, however, was to the east of the Great Plains Region.

The first white settlers in the central and northern plains were not slow to follow the explorers and fur traders of the early nineteenth century, but most of the major early parties were moving across the region to Oregon, Utah, and California. Meanwhile, prairies of the eastern part of the region were being occupied by farmers who moved out of the forested Midwest, the major thrust being into Kansas. The westward flow of settlement into the region was soon in full swing, to be interrupted only partially by the Civil War.

After the war the cowboy era came to the Great Plains. The extensive diamond-shaped area of

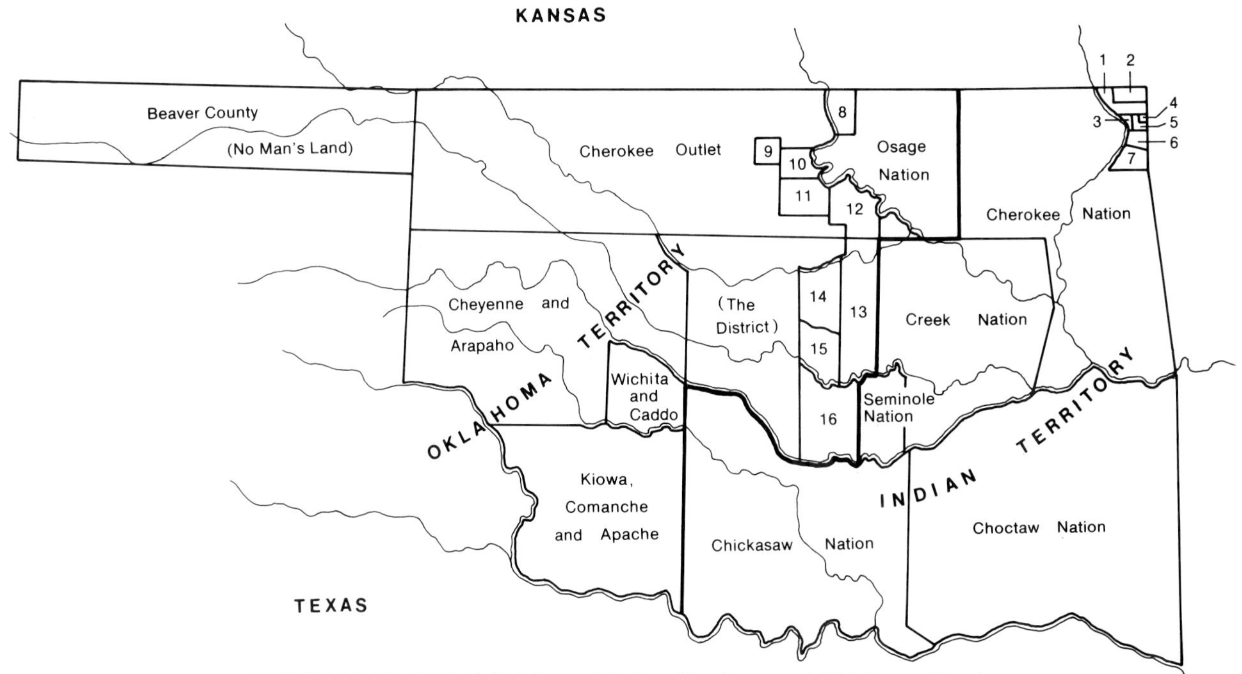

FIGURE 13-10 Tribal divisions of Indian Territory and Oklahoma Territory, generalized for the late 1800s: (1) Peoria, (2) Quapaw, (3) Ottawa, (4) Modoc, (5) Shawnee, (6) Wyandot, (7) Seneca, (8) Kansa, (9) Tonkawa, (10) Ponca, (11) Oto and Missouri, (12) Pawnee, (13) Sauk and Fox, (14) Iowa, (15) Kickapoo, and (16) Potawatomi and Shawnee. (From *A Guide to the Indian Tribes of Oklahoma* by Muriel H. Wright. Copyright 1951 by the University of Oklahoma Press.)

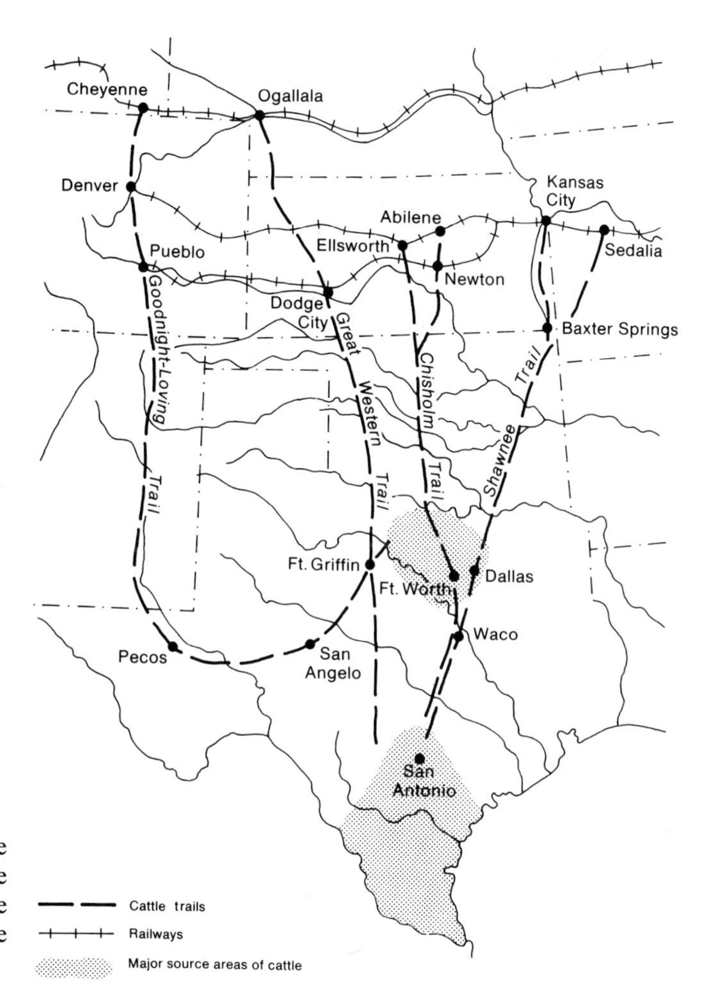

FIGURE 13-11 Major cattle trails and railways of the southern plains during the trail-driving era, from the 1860s to 1880s.

south Texas, between San Antonio and Brownsville, was the home of literally millions of wild Longhorn cattle and thousands of wild Mustang horses (Fig. 13-11). To the penniless returning Confederate soldiers who were hard working enough, here was the raw material of a livestock enterprise that lacked only one significant factor, a market. This problem was soon solved by trail driving the cattle northward to railheads in Missouri and Kansas. Trail driving started in 1866 and lasted for barely two decades, long enough to establish an enduring legend. During this period much of the northern plains area was also stocked by Longhorns that were overlanded to Wyoming and Montana from Texas.

The open-range trail drives were soon replaced by established ranches, which were made possible by the use of barbed wire for fencing and windmills to augment water supply. A vast cattle kingdom of large ranches soon spread the length and breadth of the region, with sheep introduced into some areas.

Railways provided the next catalyst to settlement. It was hoped that they would accelerate all phases of development of the region, but as it turned out, their main function was to provide access to markets for the products of the plains. The railways received huge land grants and sold much of the land to settlers.[5] But only in the Canadian portion of the region did the railway, the Canadian Pacific, actually colonize; the others merely sold land.

[5] The Union Pacific, for example, received 20 square miles (52 km^2) of land for every mile of track laid.

The flood tide of settlement in the region was most prominent in the last three decades of the nineteenth century. During the 1870s Nebraska's population more than tripled to nearly half a million; Kansas reached the million mark early in the 1880s. The Great Plains were "tamed," the age of the farmer began, and the frontier moved westward to other regions.

There was only a slight increase in cattle on the plains after 1890. The encroachment of the wheat farmer curtailed the amount of land available for cattle ranches and overgrazing on the drier western parts still further reduced the area.

After 1910 several new influences arose in the region. The development of the tractor, combine, and other power machinery made feasible the planting and harvesting of a much larger acreage of land. Numerous drought-resistant crops, especially wheat and the grain sorghums, were planted on the plains.

The skyrocketing demand for wheat during World War I ushered in a two-decade period of planting on land that should never have been cultivated. In the spring of 1934 winds began to blow the soil. Great clouds of dust swept eastward from this land largely devoid of anchoring vegetation. Thus the nation paid a high price for having grown wheat on grazing land. And yet it is what might be expected from a people who had inherited the idea that in America land is practically unlimited and soil is inexhaustible.

The Dust Bowl at its greatest extent covered 16 million acres (6,400,000 ha). During December to May, the blow season, fine fertile soil particles were whisked hundreds of miles away, forming "black blizzards." The heavier particles remained as drifts and hummocks. Sand dunes attained heights of 20 feet (6 m). The atmostphere was choked with dust; in some areas people had to put cloth over their faces when going out of doors. The vegetation in the fields was coated and rendered inedible for cattle and whole groups of counties became almost unlivable.

Although modified, Dust Bowl-like conditions have returned occasionally since the 1930s, the soils of the region were generally treated much more carefully in ensuing decades. Contour plowing, strip cropping, stubble mulching, and a host of other conservation farming techniques were introduced, particularly by the Soil Conservation Service. Most importantly, however, much marginal land that never should have been plowed was returned to grass. Problems of soil depletion still occur in the plains and prairies, but the greatest uncertainties today focus on the use and overuse of water.

CONTEMPORARY POPULATION OF THE GREAT PLAINS AND PRAIRIES

A current map of population distribution would show a fairly open and regular pattern, generally decreasing from east to west. In detail, the irrigated valleys stand out as distinct strings of denser occupance. The topographically unfavorable areas, such as the Sand Hills and various badlands and mountain outliers, are mostly barren of population.

There is considerable ethnic homogeneity to the population of the region. Most people are of European origin and are the vast majority are Anglo-Saxon. Asians are almost nonexistent in the populace and blacks are a smaller minority than in any other major region of the United States. In the Dakotas, for example, less than 0.5 percent of the population is black.

The principal ethnic "minority" consists of Hispanics, who are prominent in portions of Texas, New Mexico, and Colorado. The major concentration is in southern Texas, where more than half the citizenry of San Antonio (second largest city in the region) is of Hispanic extraction.

There is a varied European ethnic mix in the Canadian portion of the region. The only concentration of French Canadians in western Canada is found in the St. Boniface suburb of Winnipeg, but Ukrainians and other Eastern Europeans are prominent in many parts of the Prairie Provinces.

Native Americans constitute a significant proportion of the population in Oklahoma and South Dakota. The 200,000 citizens of Indian extraction in the former represent some three dozen tribes and make up 6 percent of the state's population. South Dakota has about 50,000 Native Americans, constituting 7 percent of its population total. There are large Indian reservations in South Dakota and Montana and the Blood Reserve in Alberta is the largest in Canada.

The most significant characteristic of the population of the Great Plains is probably its urbanity. Despite the prominence of a rural way of life in most of the region, the population is nearly 70 percent urban.

CROP FARMING

Although crop growing does not occupy as much acreage as livestock raising, it is the most conspicuous use of the land in the region. The close-knit precision of the rows of irrigated cotton in the Texas High Plains, the gargantuan linearity of strip-cropped wheat in central Montana, and the immense cultivated circles serviced by center pivot sprinklers in western Kansas are representative of the grandiose geometry of Great Plains farming that is obvious even to the traveler jetting 40,000 feet (12,000 m) above the region.

Despite marginal precipitation, the natural advantages of the region for growing certain kinds of crops are not to be denied. Flat land and fertile soil combined with some sort of water source make this a region of prodigious production for grains, oilseeds, and some irrigated crops. Wheat is the keystone on which fortunes are made and governments are elected, but active efforts toward diversification have also made other crops significant.

The region as a whole is characterized by a mixture of intensive and extensive farming. Irrigated vegetables, sugar beets, and cotton typify the former; wheat, other small grains, and oilseeds, the latter. Storage, transportation, and processing facilities are conspicuous in every farming area—cotton gins here, sugar mills there, and grain elevators looming on the flat horizon in every town.

Wheat Farming

Wheat is widely grown throughout the United States and Canada, but the two most important areas are the winter wheat belt in Kansas, Nebraska, Colorado, Oklahoma, and Texas and the spring wheat belt in the Dakotas, Montana, western Minnesota, and the Prairie Provinces (Fig. 13-12).

The two wheat-growing areas are not contiguous. Between them (in southern South Dakota and northern Nebraska) is a belt where little wheat is grown. Much of this is Sand Hill country, a disordered arrangement of grass-covered slopeland that is unsuited to cultivation.

The Winter Wheat Area The seasonal rhythm of winter wheat cultivation begins with planting in late summer or fall. The seeds germinate and growth begins. By the time winter sets in, the green wheat seedlings have raised their shoots several inches above the ground. Winter is a period of dormancy, but the shoots can begin growth with the first warm days of spring. The crop is ready for harvesting by late May or early June in the southern part of the region and even in Montana it can normally be harvested before August.

Winter wheat is increasingly cultivated in the northern part of the region because, where it can survive winter, it generally gives larger and more valuable yields than spring wheat. Improved varieties of seed are more tolerant of cold weather and allow a northward shift. Today hardly any spring wheat is grown in eastern Wyoming. In South Dakota spring wheat is still the leader, but winter varieties are rapidly gaining in flavor. In the plains area of Montana more winter than spring wheat is actually grown, and even in southwestern Alberta there is increasing cultivation of winter wheat.

The Spring Wheat Area Only in North Dakota, western Minnesota, and most of the Prairie Provinces does spring wheat still hold undisputed sway. Spring wheat is planted as soon as the ground thaws and dries in late spring or early summer. It grows during the long days of summer and the harvest takes place in August, September, or occasionally October. Unlike winter wheat, which is cut and threshed in a single combined operation, spring wheat normally is harvested in two steps. The stalks are first cut and raked into long windrows, where they are left for a few days to dry out. Spring wheat usually contains so much moisture that it would mildew in storage if not dried first. After drying, it is safe to collect and thresh the wheat.

Natural Rainfall vs. Dry Farming Most wheat in the region is grown under natural rainfall conditions. One of the great advantages of wheat growing

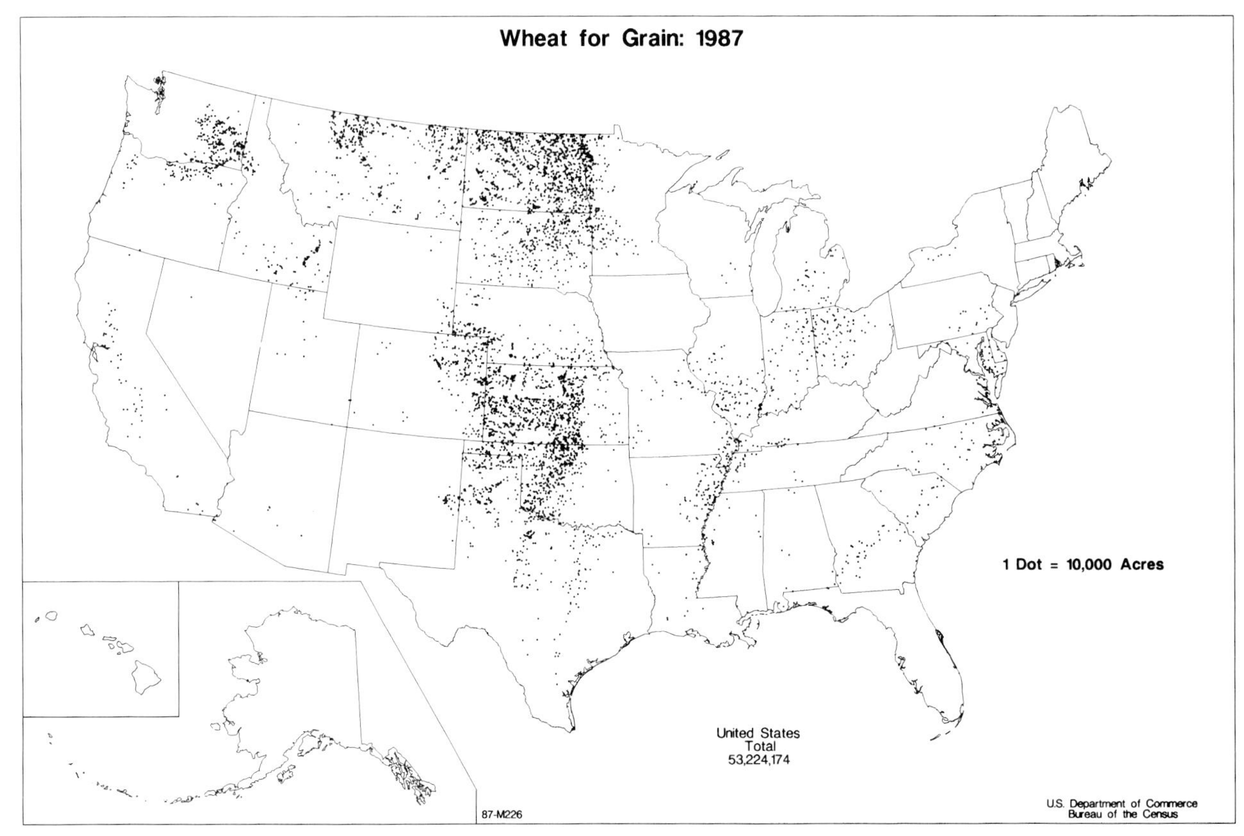

FIGURE 13-12 The Winter Wheat Belt is centered on Kansas, and the Spring Wheat Belt focuses on North Dakota.

is that it is capable of producing a pleintiful harvest on a minimum of rainfall. In some parts of the region satisfactory yields are obtained where the annual rainfall is only 11 inches (280 mm).

To maximize the effectiveness of the scanty precipitation, it is characteristic of wheat farmers, particularly those in the western part of the region, to use special "dry-farming" techniques. The simplest and most widespread of these techniques is strip-cropping, in which strips of wheat are alternated with strips of fallow land (Fig. 13-13). Care is taken to see that nothing grows on the fallow strips; the year's rainfall is thus "stored" in the soil of those strips so that it can be used the next year when the fallow strips are cropped and the cropped strips left fallow. One of the most distinctive landscape patterns in the Great Plains is the expansive crop-and-fallow stripping of dry-farmed wheat areas. The strips are usually oriented north-south, which is at right angles to the prevailing westerly winds, so as to minimize damage from wind erosion.

Harvesting Migrant workers have been involved in wheat harvesting for many decades. They operate as custom-combining crews, working northward throughout the long summer and returning to their homes and families, usually in Texas, for the winter. After World War II there was a decided decline in custom combining, for wheat farmers had accumulated sufficient wealth and land to make it feasible for most of them to own their own combines. Since the mid-1950s, however, federal wheat acreage controls have caused a swing back to custom combining. Some 16,000 people work as combine crews, following the northward harvest trail. More than one-third of the total wheat crop is combined by migratory crews; the proportion is twice as great in much of the Winter Wheat Belt.

FIGURE 13-13 The geometrical precision of strip-cropped wheat near Great Falls, Montana (TLM photo).

Other Extensive Farming Crops

In addition to wheat, several other grains and some oilseeds are important crops in the extensive farming system of the region. All are grown under irrigation in some instances, but the great bulk of their production is under natural rainfall conditions.

Grain Sorghums The principal area of grain sorghum production of the continent is the winter wheat area on the High Plains of Texas, Kansas, Nebraska, and Oklahoma (Fig. 13-14). These drought-resistant crops, introduced from semiarid parts of Africa, have grown in importance in the last quarter century. There is now a greater acreage in sorghums than in wheat in both Nebraska and west Texas and it is a major crop in Kansas.

The principal use of grain sorghums (Fig. 13-15) is for stock feed. Some are considered to be 90 percent as good as corn for feeding and fattening. The more important types of sorghums grown on the High Plains include the milos, the kafirs, feterita, darso, and hegari. They provide a good substitute for wheat, corn, and cotton in areas of questionable water.

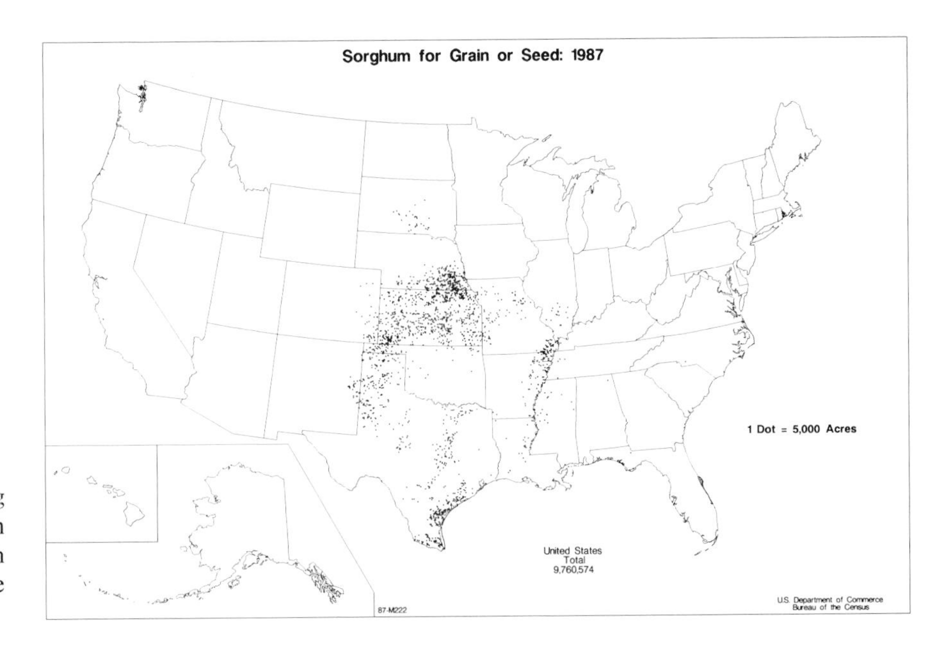

FIGURE 13-14 The growing of grain sorghums is mostly in the Winter Wheat Belt, with special concentration in the Texas High Plains.

FIGURE 13-15 Grain sorghums on the High Plains of west Texas, near Brownfield (TLM photo).

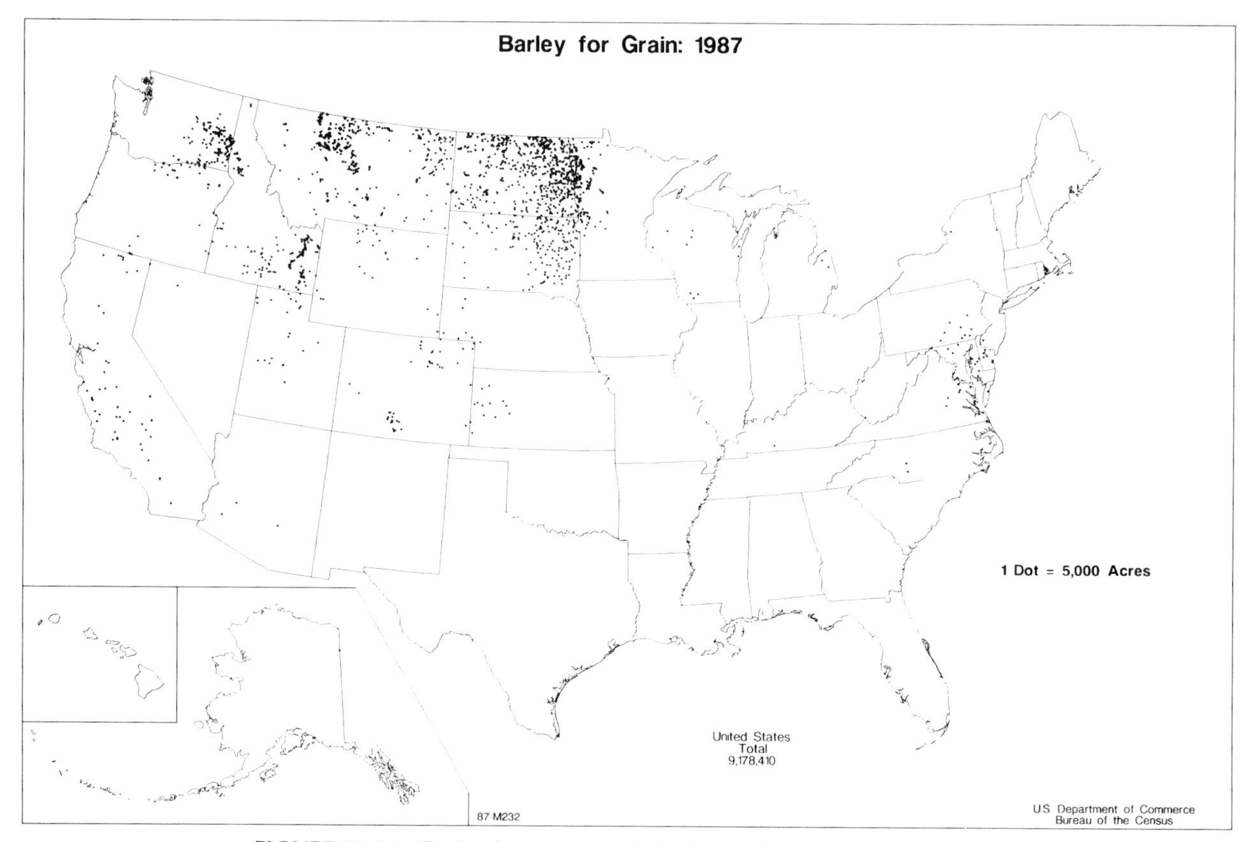

FIGURE 13-16 Barley is grown mostly in the northern plains, especially in the valley of the Red River of the North.

FIGURE 13-17 Wheat (on the left) and canola growing side by side north of Edmonton, Alberta. Canola is bright yellow in color (TLM photo).

Small Grains The northern plains constitute North America's major producing areas of barley (Fig. 13-16) and rye and rank a close second to the Heartland as a producer of oats. The greatest concentration of production of all three of these feed grains is on the Lake Agassiz Plain of North Dakota, Minnesota, and Manitoba.

Oilseeds The northern plains have long been the stronghold of flax production in North America; 95 percent of the U.S. crop and 99 percent of the Canadian crop are grown here. Flax is grown almost exclusively for its seed, from which linseed oil is extracted. Sunflower seed is also cultivated extensively in the Lake Agassiz Plain.

The most dynamic new development in oilseed production was the development of canola in the mid-1970s by Canadian agronomists, and its rapid adoption as a major crop in the Prairie Provinces. Rapeseed had been an important oilseed crop in the prairies since World War II, but concerns about its high level of erucic acid led to the development of canola (Fig. 13-17) as a derivative that compares favorably with other oilseeds in protein quality. It is used as a major ingredient in the production of cooking oil and margarine and is a staple of protein meal for livestock feed. Already it is the leading vegetable oil used in Canada and export markets are developing rapidly. Thus far Canada is the world's only producer, with output particularly concentrated in the prairies and parkland zone of Saskatchewan and Alberta.

Irrigation Agriculture

Irrigation farming is widespread in the region and has been developed in a variety of ways. Most spectacular have been the large-scale government projects developed by the federal Bureau of Reclamation in the United States and by various national and provincial authorities in Canada. The largest single scheme in the Great Plains is the Colorado–Big Thompson Project in northeastern Colorado, which depends primarily on transmountain diversion of water from the western slope of the Rocky Mountains to the valley of the South Platte River where more than 600,000 (240,000 ha) acres are irrigated. Also particularly notable are the projects associated with the six huge dams of the Pick-Sloan Plan on the upper Missouri River.

A great deal of irrigation in the region, however, is not related to such giant schemes but involves smaller local projects or individual farms that have their own water sources, often from wells. Many different types of irrigation are practiced. In sprinkler irrigation, by far the most popular in recent years, the water is sprayed on the land from lightweight aluminum pipes that can be shifted from place to place manually or that move automatically in response to motors or piston drive. Of the various sprinkler irrigation techniques, the center pivot is the most prominent recent development.

Center pivot irrigation uses large self-propelled sprinkling machines that water circular patches with great precision and efficiency. Their rapid proliferation in the last three decades produced a spectacular change in the otherwise rectangular geometry of field patterns in the region, as well as fundamental changes in cropping patterns, water use, energy use, and land ownership. Center pivot systems have been most widely adopted in Nebraska, Kansas, eastern Colorado, and western Texas, but they are used considerably throughout the region. The most conspicuous cropping change associated with center pivots is the expansion of irrigated corn production in the region, especially in Nebraska. There are serious implications, however, to the heavy water requirements of center pivot systems, no matter how efficiently water is used.

The Major Irrigated Crops The length of the growing season within this region varies from more than 280 frost-free days in the Winter Garden and Laredo districts of south Texas to less than 120 in

A CLOSER LOOK *The "mining" of the Ogallala*

Ogallala is a Sioux Indian word that means "to scatter one's own." It was notable historically as a designation for one of the major branches of the Teton Sioux Indian nation and a town (in western Nebraska) has since been graced with the name. Today, however, the word is most widely recognized as the proper name of a geological formation that underlies a large part of the central Great Plains. The Ogallala is a calcareous and arenaceous sedimentary formation of late Tertiary age that is relatively near the surface under some 225,000 square miles (585,000 km^2) of the High Plains stretching from southern South Dakota to west Texas.

The distinctiveness of the Ogallala lies in the fact that it is an enormous *aquifer*—that is, a porous bed that absorbs water and holds it because it is underlain by impervious strata. The Ogallala Aquifer functions as a giant underground reservoir that ranges in thickness from a few inches in parts of Texas to more than 1000 feet (300 m) under the Nebraska Sand Hills; "it is like a 500-mile-(800-km)-long swimming pool with the wading end in west Texas and the deep end in northern Nebraska."[a]

Water has been accumulating in the aquifer for some 30,000 years. By the middle of the twentieth century, it was estimated that it held a total amount roughly equivalent to that of one of the larger Great Lakes. The rate of accumulation is very gradual; it is recharged only by rainfall and snowmelt that trickles down from above, and the climate of the High Plains provides relatively little of both (Fig. 13-a).

If input is tediously slow, outgo is distressingly rapid. Irrigation from this underground source began about half a century ago in the early 1930s. Before the end of that decade it was noticed that the level of the water table was already dropping. After World War II the development of high-capacity pumps, sophisticated sprinklers, and other technological advances encouraged the rapid expansion of irrigation based on Ogallala water. Irrigated acreage expanded more than fourfold in a quarter century.

The results of this accelerated usage were spectacular. Above ground occurred a rapid spread of high-yield farming into areas never before cultivated (especially in Nebraska) and a phenomenal increase in irrigated crops (particularly corn, cotton, sorghums, grain, hay, and vegetables) in all nine Ogallala states. Beneath the surface, however, the water table sank ever deeper and the rate of extraction could be likened to a mining enterprise because a finite resource was being removed with no hope of replenishment.

Farmers who had obtained water from 50-foot (15-m) wells now must bore to 150 or 250 feet (45 or 75 m); and as the price of energy skyrockets, the sheer cost of pumping increases operating expenses enormously. Perhaps 100,000 wells tap the Ogallala; many are already played out and almost all the rest must be deepened annually.

Anguish over this continually deteriorating situation is widespread. Some farmers are shifting to crops that require less water (for example, less corn and more sorghums). Others are adopting water-conserving and energy-conserving measures that range from a simple decision to irrigate less frequently to the installation of sophisticated technology that uses water in the most efficient fashion. Many agriculturalists have faced or will soon face the prospect of abandoning irrigation entirely. Millions of dollars have already been spent in a plethora of studies to assess the details and ramifications of the problem and to recommend mitigating measures.

Many farmers, however, have taken another approach. As the cost of water and energy rises, they reason that it is time to concentrate on high-value crops before it is too late. In Lubbock County (Texas), for example, grain sorghum acreage decreased by 80 percent during the 1970s whereas cotton acreage expanded by 50 percent. Moreover, federal tax laws lessen the need for water-saving tech-

[a] James Aucoin and Anne Pierce, "A Curious Piece of Real Estate," *Audubon*, 85 (September 1983), 88.

northern Montana and southern Canada. The most widely grown crop is *alfalfa*, which is the basic hay crop throughout the West. Alfalfa occupies the largest acreage of any irrigated crop from southern Colorado northward.

The Winter Garden and Laredo areas of southern Texas produce early *Bermuda onions, spinich*, and other *winter truck crops*. On the several irrigated areas of the Pecos Valley in southern New Mexico, cotton and alfalfa are the dominant crops; *peanuts* and *grain sorghums* are also important.

Probably the outstanding irrigated area in the region is the Texas High Plains section in the vicinity of Lubbock, where *cotton* is the principal source of farm income. Irrigation in this district depends on groundwater wells whose water-yielding aquifers are rapidly being depleted. Nevertheless, prodigious yields of cotton, sorghums, wheat, and corn encour-

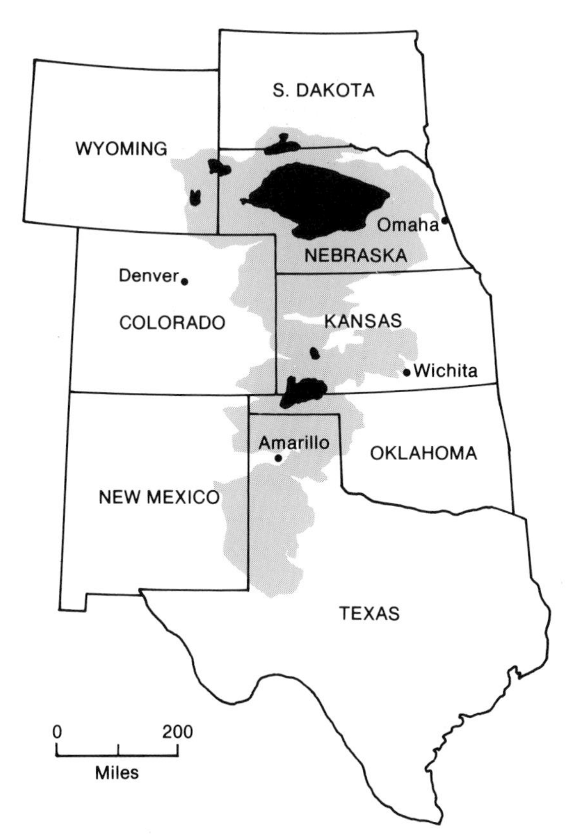

FIGURE 13-a The Ogallala aquifer. Darker areas indicate greater thickness of the water-bearing strata.

niques by allowing irrigators to deduct the yearly cost of exhausted Ogallala water from their gross income, in the form of a "depletion allowance" similar to that given mining companies. Water conservation is further complicated by the obvious fact that groundwater is no respecter of property boundaries. A farmer who is very conservative in his water use must face the reality that his less careful neighbors are pumping from the same aquifer and that their profligacy may seriously diminish the water available to him.

In the final analysis, there is no way to escape the inevitable. Ogallala water is a finite resource and sooner or later it will disappear throughout the subregion no matter which conservation techniques are used. Doomsday is a decade away in some areas, perhaps half a century in others Irrigation is still on an upward trend in areas where the aquifer is deepest and less used, as in much of Nebraska. But for most of the High Plains subregion Ogallala water is decreasingly available and increasingly costly. Ambitious schemes have been proposed to replace the diminishing groundwater supply by importing surface water from somewhere—the Missouri and Arkansas rivers, for example. Such proposals are multibillion dollar ones, however, and clearly their development would be a future project, if ever.

TLM

FIGURE 13-18 The sophisticated machinery of center-pivot irrigation is widespread in the Great Plains and Prairies Region. This scene is near Loveland in northeastern Colorado (TLM photo).

age farmers to continue, and even expand, irrigation despite the cost and the specter of depletion.

The Arkansas Valley in eastern Colorado and western Kansas is noted for sugar beets, feed grains, alfalfa, and *cantaloupes*.

In northeastern Colorado, associated with the Colorado–Big Thompson Project and the South Platte River, is the second outstanding irrigated area in the Great Plains. The chief specialty crop is *sugar beets,* but a great variety of other crops is also grown, and Weld County (the Greeley area north of Denver) is a national leader in total value of agricultural output. Vegetables, dry beans, corn, alfalfa, and feed grains also occupy large acreages.

The rapid expansion of cattle feedlots and dairying in northeastern Colorado in the last few years has stimulated production of feed grains. Much *corn* is grown under irrigation and most of it is cut green for silage. Corn is also a major crop all across southern Nebraska, particularly in association with center pivot systems (Fig. 13-18).

Irrigation is also widespread in the northern plains and prairies, but in lesser concentration than northeastern Colorado or the Texas High Plains. Most river valleys have irrigated sections along their flood plains; more than one million acres are irrigated in Alberta, for example, almost all within the drainage area of the South Saskatchewan River. Throughout this northern portion hay crops occupy the bulk of the irrigated acreage, with a considerable share also devoted to grains (particularly wheat), sugar beets, potatoes, and oilseeds.

LIVESTOCK RAISING

The Great Plains and Prairies Region has long been famous for its range livestock industry. Some of the world's largest, best-run, and most productive ranches are located here. Both beef cattle and sheep are widespread; in some cases, both species are raised on the same ranch. By and large, however, the better lands are used by cattle; sheep tend to be restricted mostly to rougher, drier, or otherwise less suitable country. Cattle are distributed relatively uniformly over the region, although densities decrease northward and westward (Fig. 13-19). Sheep are much more irregularly scattered, with the principal concentration being in the central Texas hill country and Edwards Plateau; lesser concentrations occur in northeastern Colorado (primarily on irrigated pastures), eastern Wyoming, and adjacent

FIGURE 13-19 Hungry Herefords awaiting a delivery of hay on a cold February morning near Denver (TLM photo).

parts of South Dakota and Montana, and the western margin of the prairie in Alberta.

Summer grazing is chiefly on natural grasses, although there is increasing replacement with more nutritive artificial pastures. The most noted natural grasslands of the region are probably in the so-called bluestem belt: the Flint Hills of eastern Kansas and Osage Hills of northeastern Oklahoma. In winter, however, artificial feeding—mostly with hay—is necessary over much of the region because of the long period of snow cover.

Cattle Feedlots

The most dynamic development in the livestock industry of the plains and prairies has been the rapid proliferation of cattle feedlots. In the past nearly all range cattle were shipped out of the region, usually to the Corn Belt, for fattening. More recently the feedlot capacity of the Great Plains has expanded phenomenally and thus beef production has been vertically integrated in the region, from raising on the range through fattening in feedlots to slaughtering in local packing houses.

The Chicago–St. Louis axis of the "beef belt" has now shifted considerably westward and can be considered as being oriented along a Sioux Falls–Amarillo line. Feedlot operations are widespread in Nebraska and Kansas but are particularly prominent in west Texas and northeastern Colorado; furthermore, many new feedlots are highly mechanized and have very large capacities. In the Texas Panhandle, for example, more than a million cattle can be accommodated at one time on large feedlots; feedlot capacity in the three principal counties of northeastern Colorado is more than half a million.

FIGURE 13-20 Angora goats in the central Texas hill country near Fredericksburg (TLM photo).

Angora Goats

The hill country of central Texas yields more than 85 percent of the nation's mohair. Goat ranching is similar to sheep ranching except that goats, being browsing animals, can subsist on pastures not good enough for sheep. Most of the pasture land used by Angora goats is in the brush country where scrub oak and other small trees and shrubs supply browse (Fig. 13-20).

MINERAL INDUSTRIES

Some of the most flamboyant history in the region revolved around the discovery (1874) and production of gold in the Black Hills. The deep underground Homestake Mine was the only functioning gold mine in the United States until the upsurge of world gold prices in the 1970s made it economical to open (or reopen) several other mines in the West. Even so, the Homestake Mine yielded about one-tenth of all the gold ever produced in the nation and still has the largest output of any single mine.

It is, however, the nonmetallic minerals that are the mainstay of the region's mining industry.

Petroleum

Because of its extensive size and because most of it is underlain with sedimentary rocks that may contain oil, the Great Plains and Prairies Region has many producing oil fields. Among the 14 states and provinces with portions included in this region, only Minnesota lacks significant production.

Nearly half of all Texas petroleum production —amounting to one-sixth of the national output—is from west Texas fields, particularly the Permian Basin. Major yields are also obtained in parts of Oklahoma and Kansas; in the increasingly prolific fields of Wyoming, particularly the Elk Basin; in scattered

Montana localities; in the Williston Basin of North Dakota; and in bountiful production areas of Alberta and Saskatchewan.

Natural Gas

The Prairie Provinces are also prominent producers of natural gas, particularly Alberta, whose output is nearly nine-tenths of the Canadian total. Natural gas is also a major product from Wyoming, Kansas, Oklahoma, and west Texas; the Panhandle field of the last three states is the largest producer of natural gas in the world. In the late 1970s an outstanding new field was developed in the Laredo area of southern Texas, with indications that it might eventually equal the Panhandle field in output.

Helium

The major helium-producing area in the world is located within the Panhandle gas fields. Originally helium was used mostly as a lifting gas for dirigibles, but now is has a multitude of atomic, spacecraft, medical, and industrial uses (particularly for helium-shielded arc welding). Of the 12 plants in the United States producing helium, 11 are in the Great Plains Region. The long-range helium outlook is for increasing demand and decreasing supply.

Coal

It was long known that there were enormous deposits of relatively low-grade bituminous and lower-grade lignite coal in the northern Great Plains, but they were largely unexploited until the energy crisis of the 1970s. Although some mining has been carried on for decades in such places as Montana's Judith Basin and lignite has been extracted in southeastern Saskatchewan for almost a century, it is only in the last few years that coal mining has abruptly begun to change the landscape, economy, and lifestyle in parts of the Plains (Fig. 13-21).

Most of the coal deposits are extensive and near the surface and so are strip-mined. The deleterious environmental consequences of strip mining are well-known, but the relevant states have enacted strict regulations concerning restoration and revegetation of plundered lands, and in some locations cattle are already grazing on rolling grassy swales

FIGURE 13-21 An open-pit coal mine in southern Montana (courtesy Montana Power Company).

where there was a yawning open-pit coal mine only three years previously.

The abrupt and ambitious plans for mining in Wyoming, Montana, and the Dakotas have occasioned unprecedented economic and population growth and enormous growing pains. Land values skyrocketed. Sleepy villages became bustling boom towns (Gillette, Wyoming, for example, experienced a fivefold population increase in eight years) and "instant" towns sprouted, with the help of acres of mobile homes, on the barren steppe (for example, the population in Colstrip, Montana, grew from 0 to 3000 in two years).

Regional coal reserves are stupendous. Montana and Wyoming have the largest known reserves of any states and Wyoming is already the third-ranking (after Kentucky and West Virginia) coal producer in the nation. Generally Great Plains coal is of low heating value, but it also has a low sulfur content, which makes it highly desirable, especially for electricity generation.

The major area of action is the Powder River Basin of north-central Wyoming and south-central Montana. The only incorporated town in Wyoming's Campbell County, Gillette, has become the classic example of the modern boom town. A dozen enormous mines operate in the vicinity. Every day 40 huge unit coal trains roll in or out of town. Residential subdivisions appear miraculously on the windswept sagebrush flats, but a large proportion of the populace still resides in mobile homes. Debate is endless about the maintenance of a "traditional" ranching way of life *vis-à-vis* the easy wealth of selling out to the coal companies. Perhaps nowhere else North America is there a more clear-cut collision between yesterday and tomorrow.

Potash

The world's two largest suppliers of mineral potash are located at opposite ends of the Great Plains and Prairies Region. Near Carlsbad, New Mexico, on both sides of the New Mexico–Texas boundary, are extensive deposits of polyhalite and sylvite, which have been a major source of potash for many years.

Considerable technological assistance from the New Mexico producers has helped establish a prominent potash mining industry in southern Saskatchewan in the last quarter century. With ten producing mines and perhaps two-thirds of the world's known potash reserves, Saskatchewan could easily supply total world needs for the next 1000 years. It is a high-cost producing area, however, because of waterlogging problems in the overlying and surrounding rock. The government of Saskatchewan is heavily involved in the enterprise, although more than half the operation is in private hands.

THE EBB AND FLOW OF URBANIZATION IN THE GREAT PLAINS AND PRAIRIES

The Great Plains and Prairies is a region that exemplifies many contemporary trends of North American geography, not the least of which is the pattern of urbanization. Here, better than in any other region, is demonstrated the decay of the small town in juxtaposition with the rapid growth of larger urban centers.

Withering Towns

Proportionally, no other region has as many small towns that have registered population declines during the last four census periods (Fig. 13-22). This decline is partly the result of the changing economy of the Great Plains and Prairies, but in large measure it reflects the specialized origin of many towns and the trend toward larger farm size and smaller farm population.

Originally numerous towns were established along the advancing tentacles of westering railway lines. The situation is most prominent in the Prairie Provinces, where settlement did not significantly precede the railways and the urban pattern was preconceived and superimposed on the land, but it is shown to some extent throughout the region because a large share of the towns grew up along the east-west railway routes. With regard to the prairies, the scheme has been described as follows:

> The building of the railways accompanied or preceded settlement. Settlement in advance of the railway merely anticipated the railway already projected or under construction, and such towns as were built were usually estab-

A CLOSER LOOK Arena, North Dakota

"Arena, North Dakota, Population 0" is how the welcoming sign at the city limits should read. But there is no sign. A thirsty traveler heading down the two-lane blacktop highway might see Arena's school or its church steeple from the hill north of town and detour onto the gravel road leading to Main Street, expecting to find a cafe or a filling station. The buildings are all there: stores, houses, grain elevators, churches, and a brick school; but Arena is totally deserted (Fig. 13-b). Its last few inhabitants have gone off to live somewhere else and left the town to fade back into the little piece of North Dakota prairie that was first proclaimed "Arena" in a flurry of excitement little more than 75 years ago. The place has become a ghost town, although not of the type usually associated with the American West. Arena had no gaudy dance halls or shoot-'em-up saloons and it never played host to gambling cowboys or roughneck gold miners. Its history was far less exciting than that.

Arena was a trade-center town, one of thousands created in the Middle West and Great Plains during the era of railroad building in the late nineteenth and early twentieth centuries. The student of geography might well ask, "Why were so many towns created?" and "Why were they so small?" The two questions are actually one since the answers to both come from understanding the role of railroads in shaping the settlement pattern of the interior plains region of Anglo-America.

The railroad's influence was dominant because it was the new technology of the period when the settlement frontier reached the nation's midsection. Investors in cities like Chicago and Minneapolis saw the possibility of creating a weblike arrangement of railroad lines that would funnel the millions of bushels of wheat and corn produced on the prairies and plains to their cities' mills or export docks. It was to be an agricultural factory of unprecedented scale. Hundreds of thousands of farmers were eager for the income that cash-grain sales would bring them; and the rising industrial cities of the Middle West saw the profits that would come from their role as millers, bankers, and wholesalers to the system. The railroad offered cheap transportation, linking producers to the market.

What was lacking and needed to be supplied immediately was a series of collection points along the railroad where farmers would bring their grain. The horse-team-and-wagon technology of the times dictated that these sites could not be too far apart because there was a limit to the amount of time farmers were willing to spend hauling crops they had grown. Farmers participating in the cash-grain economy also needed stores where they could purchase goods they did not produce, banks where they could arrange the necessary loans that carried them from one harvest to the next, and many other services that even the rudimentary

FIGURE 13-b The central business district of Arena, North Dakota (photo by John C. Hudson).

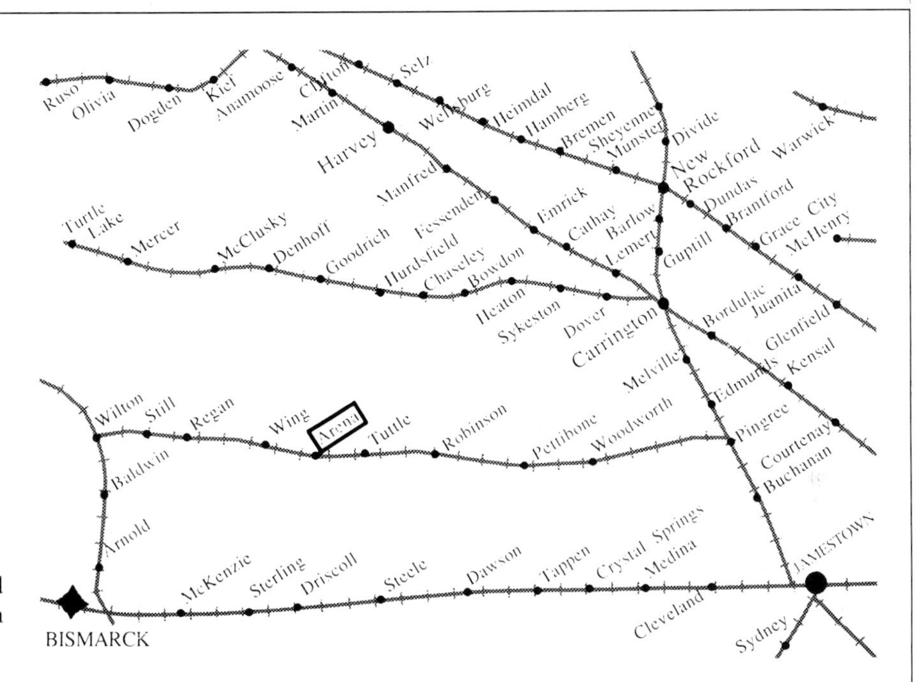

FIGURE 13-c Towns and railway lines in central North Dakota.

way of life on a prairie grain farm required.

The result was the creation of hundreds of small trade-center towns, spaced at 7- to 10-mile (11- to 16-km) intervals along the railroad track, that were designed to attract and keep the trading allegiance of farmers in their vicinity (Fig. 13-c). Sometimes these towns were platted on land the railroad already owned, but most such towns were created by independent businessmen who had a working relationship with the railroad company. They purchased a tract of land where the railroad designated a town would be built and then lured to the new town as many would-be merchants as their advertising wiles could manage. Many new towns were staked out, the lots were sold at auction, and the first business buildings were erected in a matter of weeks. The towns were thus born almost full grown; for many, the first few months were the best times there would ever be.

Arena, North Dakota, like every other town in its vicinity, was created in this manner. And, like nearly all of its neighbors, Arena's population peak came early in its life, before economic and social changes began to render its existence pointless. Automobiles and trucks made it possible for farmers to bypass their local town's small array of businesses in favor of larger, more distant trade centers that had more to offer. The farm population declined steadily, even as the output of local agriculture increased, because fewer farmers were able to use technology to work more acres and produce more per acre than ever before. Arena became a place where the old folks lived—retired farmers and a few older merchants clinging to what remained of businesses that young folks once might have purchased and operated. By 1989, the last of Arena's inhabitants had passed away or moved away.

Many of the movers went only a short distance, to the state capital of Bismarck, a city that has boomed because of energy developments in the Missouri River basin that now make this region an important source of fossil fuels (sub-bituminous coal, lignite, oil) and a center for electricity generation utilizing new technology to transmit power long distances. The fate of the many small, trade-center towns like Arena is thus not the fate of North Dakota or the Great Plains region. But there will be more ghost towns in the years to come. Fewer farmers means fewer retired farmers, and thus fewer retirees to seek a place to live in town. Even the grain elevators in towns like Arena are closing because they are too small to service the 100-ton (0.9-t) capacity grain cars the railroads now use. All of these reasons, plus too few children to require a school, too few parishioners to keep a church alive, add up to no reason for anyone to live there anymore. The system that once required Arena and its many look-alike neighbors no longer needs such places. They have become museum pieces of a way of life that probably never will return.

Professor John C. Hudson
Northwestern University
Evanston, Illinois

FIGURE 13-22 Many small towns in the region are stagnating or dying. For example, more than half the commercial buildings fronting the town square of Paducah, Texas, are now empty (TLM photo).

lished in anticipation of the line passing through them. When thwarted, these settlements were frequently moved across the prairie to sites adjoining the line. . . . The distance between the towns was determined by the economic distance for hauling grain at a time when local transportation was horse-drawn. Whenever possible, the railway companies brought production areas to within 10 miles of their lines and, by placing elevators and sidings 7 to 10 miles apart, created a maximum hauling distance of 12 to 15 miles, the upper limit depending on the location and direction of the roads. These transshipment centres became the distributing points for supplies, e.g., agricultural implements, coal, lumber, and general merchandise. . . . The result of this method of settlement is that the towns are arranged along the railway like beads on a string. They appear, heralded by elevators, as regularly as clockwork. . . . Frequently, the names as well as the sites of the towns were chosen by the railway companies, since there were few existing place names to recognize. This task, too, was executed with characteristic dispatch. One solution was simply to arrange the names in alphabetical order down the line as, for example, on the Grand Trunk railway, which runs from Atwater, Bangor, Cana, to Xena, Young, and Zelma.[6]

With the decreasing importance of railway transportation, most such towns no longer have any functional significance and are left to decay. Other towns that grew up along roads and highways were subsequently bypassed by new interstate or interprovincial highway construction and they, too, have become anachronistic and stagnant.

Burgeoning Cities

More auspiciously located market towns, located farther apart and with a more diversified economic function, prospered at the expense of the smaller places; thus railway division points, separated from one another by 100 miles (160 km) or so, were able to maintain their vigor by expanding their functional hinterland. Even division points that were sited more or less arbitrarily by the railway companies have been growth nodes—for example, Moose Jaw,

[6] Ronald Rees, "The Small Towns of Saskatchewan," *Landscape*, 18 (Fall, 1969), 30.

TABLE 13-1
Largest urban places of the Great Plains and Prairies Region

Name	Population of Principal City	Population of Metropolitan Area
Abilene, Tex.	109,110	121,800
Amarillo, Tex.	166,010	196,300
Arvada, Colo.*	90,980	
Aurora, Colo.*	218,720	
Billings, Mont.	78,020	116,400
Bismarck, N. Dak.	47,740	85,800
Boulder, Colo.	75,990	217,900
Brandon, Man.	39,100	
Calgary, Alta.	672,400	672,400
Casper, Wyo.	42,680	64,700
Cheyenne, Wyo.	54,010	75,200
Colorado Springs, Colo.	283,110	393,900
Denver, Colo.	492,190	1,640,000
Edmonton, Alta.	589,500	785,800
Enid, Okla.	46,620	58,300
Fargo, N. Dak.	69,780	148,400
Fort Collins, Colo.	77,120	182,000
Grand Forks, N. Dak.	48,430	70,500
Great Falls, Mont.	58,280	78,200
Greeley, Colo.	57,430	136,200
Hutchinson, Kans.	40,850	
Lakewood, Colo.*	119,340	
Laredo, Tex.	124,730	128,900
Lawton, Okla.	83,650	119,300
Lethbridge, Alta.	59,500	
Lubbock, Tex.	188,090	226,800
Medicine Hat, Alta.	42,300	
Midland, Tex.	95,880	107,300
Midwest City, Okla.*	52,130	
Norman, Okla.	78,280	
Odessa, Tex.	95,010	124,700
Oklahoma City, Okla.	434,380	963,800
Pueblo, Colo.	101,070	127,600
Rapid City, S. Dak.	55,780	82,000
Red Deer, Alta.	56,100	
Regina, Sask.	178,200	186,500
Roswell, N. Mex.	43,230	
Salina, Kans.	43,240	
San Angelo, Tex.	87,340	99,300
San Antonia, Tex.	941,150	1,323,200
Saskatoon, Sask.	180,800	200,600
Tulsa, Okla.	368,330	727,600
Westminster, Colo.*	73,890	
Wichita, Kans.	295,320	483,100
Winnipeg, Man.	598,900	625,300

* A suburb of a city.

Swift Current, Medicine Hat, and Calgary along the Canadian Pacific.

The larger urban centers of the region are few and far between (see Table 13-1 for a listing of the region's largest urban places). Their prosperity in each case represents an extensive trading territory or a local area of high productivity. In almost every instance, their recent rate of population growth has been higher than the national average.

Denver is by far the largest city in the region; it is the commercial and distributing center for much of the plains as well as an extensive portion of the mountain West (Fig. 13-23). In addition to being the state capital of Colorado, it is a major regional center for federal offices and has a rapidly increasing industrial component. Although situated on the plains, it is the tourist gateway to the mountain recreation areas of the Southern Rockies.

FIGURE 13-23 Despite having seemingly endless land over which to sprawl, the cities of the region are marked by a notable vertical dimension. This is Denver (TLM photo).

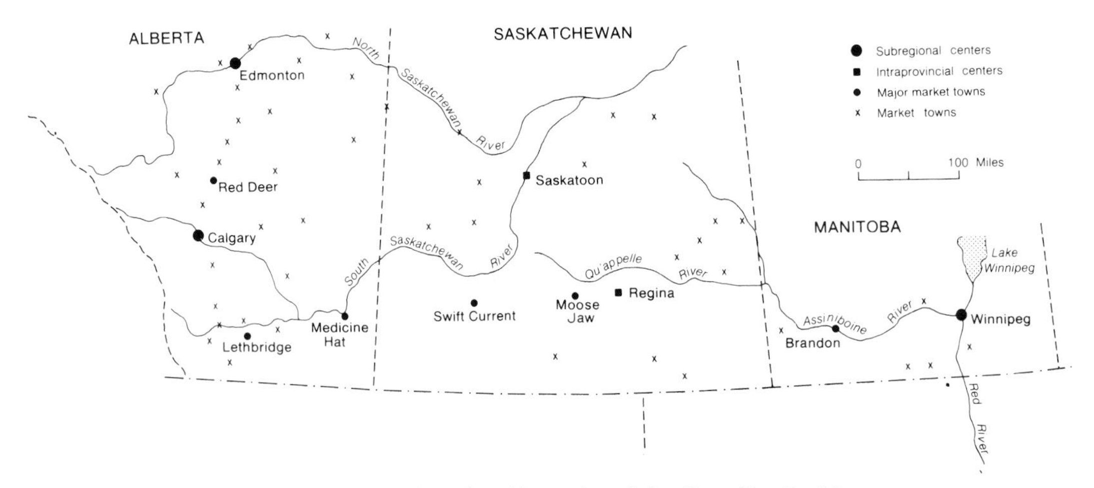

FIGURE 13-24 The urban hierarchy of the Canadian Prairies.

The other large urban centers of the southern plains have significant functional differences. *San Antonio* is the oldest city in the region; it is the principal commercial city of southern Texas and is surrounded by an unusually large number of military bases. *Oklahoma City* and *Tulsa* are both prominent in the petroleum industry; the former's larger size is at least partly related to its role as a governmental and financial center. *Wichita* has a prosperous agricultural hinterland, but its special claim to fame is an aircraft manufacturing center.

The northern plains in the United States have no large cities. Apparently the overlapping hinterlands of Denver, Minneapolis, and Seattle have militated against metropolitan growth in this subregion.

The metropolitan pattern of the Prairie Provinces is still evolving (Fig. 13-24). *Winnipeg* achieved its early dominance on the basis of its splendid gateway location at a river crossing at the apex of an extensive, fan-shaped potential hinterland in the prairies. But Winnipeg never became the primate city of the Prairie Provinces, indicating that a peripheral location can perhaps be a handicap despite its gateway opportunity. After World War II, *Edmonton* and *Calgary* rapidly became the foci of the western prairies and both are now larger than Winnipeg. Curiously, the centrally placed cities of *Regina* and *Saskatoon* have always been the smallest of the prairie metropolitan areas.

A TRANSIT LAND

The Great Plains and Prairies Region has served primarily as a transit land. Most of the freight and passenger traffic passes through the region en route to and from the Pacific Coast; thus the transportation lines of the region show a predominance of east-west railways, highways, and airways. The only important north-south traffic flows along the western edge, at the foot of the Rocky Mountains, where there are some significant population clusters. The cities of this western margin, such as Roswell, Pueblo, Colorado Springs, Denver, Cheyenne, Billings, Great Falls, Calgary, and Edmonton, serve both the Great Plains and the Rocky Mountains, although all are located in the former region.

Despite the rapid growth of the larger urban areas of the region, the historic pattern of transportation—through rather than to—has been maintained. The major interstate and interprovincial highways continue to have an east-west orientation, and flow patterns are predominantly latitudinal.

LIMITED TOURISM

The Great Plains and Prairies Region, with its extensive area of level to gently rolling lands and its continental climate of hot summers and very cold winters, offers few attractions for the tourist. Thus it is not important as a resort region despite the fact that each summer it is crossed by throngs of tourists seeking a vacation in the mountains to the west. Three areas within the region, however, are of significance: the Black Hills, the Carlsbad Caverns, and the hill country of southwestern Texas.

Of these areas, the most scenic and most important is the Black Hills. With forest-clad mountains, the attractive Sylvan Lake area, Wind Cave National Park, Custer State Park and its abundant wildlife, Mt. Rushmore Memorial, and the only legal gambling casino (opened in 1989 at Deadwood, South Dakota) between Nevada and Atlantic City, the Black Hills area annually attracts hundreds of thousands of tourists (Fig. 13-25).

In southeastern New Mexico, in an area where the surface of the land is harsh, lies the world-famous Carlsbad Caverns National Park with its extensive subterranean caves. As one of the most popular tourist attractions of North America, it is visited by great numbers of people each year. The area, however, does not encourage the tourist to stay long; in most cases, it is visited for a short time by persons on their way to other places.

Several cities on the western margin of the Great Plains have become important summer tourist centers because they are gateways to the mountains beyond. Colorado Springs, Denver, Cheyenne, and Calgary are particularly notable in this respect.

FIGURE 13-25 One of the very few major tourist destinations in the region is Mt. Rushmore, with its presidential heads carved in granite, in the Black Hills of South Dakota (TLM photo).

Hunting and fishing activities are limited in this region, with three important exceptions: the central Texas hill country contains one of the largest and most accessible deer herds in North America, the introduced pheasant population of the central and northern plains is a leading hunter's quarry from Colorado northward, and Wyoming's enormous antelope population is a classic example of how a combination of good management and good luck can restore a wildlife species to primeval levels of abundance.

THE OUTLOOK

Despite varying degrees of economic diversification, the prosperity of the Great Plains and Prairies Region has always been significantly tied to agricultural conditions, and particularly to the balance between grain output and the world market. Output will always fluctuate on the basis of growing-season weather conditions, but it depends even more on the vagaries of the international marketplace and the agricultural policies of the two national governments.

The 1980s were times of bin-busting records for most grain and oilseed crops in the region. This did not make for prosperity, however, as similar conditions prevailed over much of the world, and export markets shriveled. Wheat, the region's leading crop, was hit hardest, as the United States normally exports half of its wheat harvest and Canada usually sells 80 percent of its crop overseas. The region has enormous capability for crop growing, but future agricultural prosperity is anything but assured.

The area under irrigation will diminish, slowly at first, in the southern part of the region where water from the Ogallala is so important. From Nebraska northward, however, an expansion of irrigation farming can be expected, for water resources are not yet being used to full capacity. The spread of irrigation will be particularly notable (at least on a relative basis) in the Canadian prairies, where much surface water is available. The established irrigated districts in southern Alberta and Saskatchewan will be the scene of most of the expansion. Manitoba has relatively little irrigation and much of it depends on groundwater, which is already being withdrawn at a worrisome rate.

Farms will continue to become larger as small operators sell their land and large farmers consolidate their holdings. The complexity of farm ownership will probably increase, with more part-owners, absentee owners, and nonagricultural investors—except in the Prairie Provinces, where foreign investment is severely curtailed and there are even some restrictions on corporate farmland purchases.

The frantic pace of energy-related developments in the region has subsided abruptly, and even such previously booming locales as Wyoming and Alberta have experienced out-migration. Coal continues to be in demand, but the outlook for oil and gas is less sanguine.

Alberta can be expected to enjoy many positive residual effects of the recent oil boom due to the farsighted establishment of a provincial "Heritage Fund." A large share of the petroleum income has been channeled into this fund (which now amounts to more than $5 billion), whose purpose is to maintain economic growth and improve the quality of life of the citizenry. Saskatchewan has established a similar Heritage Fund on an obviously more modest basis.

Prosperity can be expected to wax and wane, but the region will maintain its essential character. The landscape will be dominated by a mixture of extensive and intensive agriculture, with notable centers of mineral activity; the population will never be very dense except in a few urban areas; and it will always be an important transit land, lying as it does across all transcontinental routeways.

SELECTED BIBLIOGRAPHY

ARREOLA, DANIEL D., "The Mexican American Cultural Capital," *Geographical Review*, 77 (January 1987), 17–34.

BALTENSPERGER, BRADLEY H., *Nebraska*. Boulder, CO: Westview Press, 1984.

BARR, BRENTON M., ed., *Calgary: Metropolitan Structure and Influence*, Western Geographical Series, Vol. 11. Victoria: University of Victoria Department of Geography, 1975.

BEATY, CHESTER B., *The Landscapes of Southern Alberta: A Regional Geomorphology*. Lethbridge, Alta.: University of Lethbridge, Department of Geography, 1975.

BLOUET, BRIAN W., and FREDRICK C. LUEBKE, eds., *The Great Plains: Environment and Culture*. Lincoln: University of Nebraska Press, 1979.

BORCHERT, JOHN R., "Climate of the Central North American Grassland," *Annals*, Association of American Geographers, 40 (1950), 1–39.

———, "The Dust Bowl in the 1970s," *Annals*, Association of American Geographers, 61 (1971), 1–22.

BOWDEN, MARTYN J., "Fashioning the American Landscape: Creating Cowboy Country," *Geographical Magazine*, 52 (July 1980), 693–701.

BRADO, EDWARD, B. Mennonites Enrich the Life of Manitoba," *Canadian Geographic*, 99 (August–September 1979), 48–51.

———, "Red Deer River, Another Alberta Treasure," *Canadian Geographic*, 98 (June–July 1979), 22–27.

BROWN, ROBERT HAROLD, *Wyoming: A Geography*. Boulder, CO: Westview Press, 1980.

CARLYLE, WILLIAM J., "Farm Population in the Canadian Parkland," *Geographical Review*, 79 (January 1989), 13–35.

———, "The Management of Environmental Problems of the Manitoba Escarpment," *Canadian Geographer*, 24 (Fall 1980), 238–247.

CHAKRAVARTI, A. K., "The June–July Precipitation Pattern in the Prairie Provinces of Canada," *Journal of Geography*, 71 (1972), 155–160.

COURTENAY, ROGER, "New Patterns in Alberta's Parkland," *Landscape*, 26 (1982), 41–47.

DORT, W., and J. K., JONES, eds., *Pleistocene and Recent Environments of the Central Great Plains*. Lawrence: University of Kansas Press, 1970.

EVERITT, JOHN, "Social Space and Group Life-Styles in Rural Manitoba," *The Canadian Geographer*, 24 (Fall 1980), 237–254.

Fairbairn, Kenneth J., "Alberta and Oil," *Yearbook of the Association of Pacific Coast Geographers,* 42 (1980), 89–99.

Farney, Dennis, "The Last of the Tallgrass Prairie," *Defenders,* 50 (1975), 308–316.

Fidler, V., "Cypress Hills: Plateau of the Prairie," *Canadian Geographical Journal,* 87 (September 1973), 28–35.

Georgianna, Thomas D., and Kingsley Haynes, "Competition for Water Resources: Coal and Agriculture in the Yellowstone Basin," *Economic Geography,* 57 (July 1981), 225–237.

Griffiths, Meland and Lynnell Rubright, *Colorado: A Geography*. Boulder, CO: Westview Press, 1983.

Harris, David R., "Recent Plant Invasions in the Arid and Semi-Arid Southwest of the United States," *Annals,* Association of American Geographers, 56 (1966), 408–422.

Haryett, Clifford R., "Potash—Our World Class Reserve," *Geos,* 12 (Winter 1983), 19–21.

Hewes, Leslie, *The Suitcase Farming Frontier: A Study in the Historical Geography of the Central Great Plains*. Lincoln: University of Nebraska Press, 1973.

Hudson, John D., "Towns of the Western Railroads," *Great Plains Quarterly,* 2 (Winter 1982), 41–54.

Jankunis, Frank J., "Perception, Innovation, and Adaptation: The Palliser Triangle of Western Canada," *Yearbook, Association of Pacific Coast Geographers,* 39 (1977), 63–76.

Karpan, Robin, and Arlene Karpan, Should the Souris Be Dammed?," *Canadian Geographic,* 108 (February–March 1988), 12–20.

Laatsch, William G., "Hutterite Colonization in Alberta," *Journal of Geography,* 70 (1971), 347–359.

Laut, Peter, *The Geographical Analysis and Classification of Canadian Prairie Agriculture*. Winnipeg: University of Manitoba Department of Geography, 1974.

Lawson, Merlin P., Kenneth F. Dewey, and Ralph E. Neild, *Climatic Atlas of Nebraska*. Lincoln: University of Nebraska Press, 1977.

Lehr, J. C., "The Sequence of Morman Settlement in Southern Alberta," *Albertan Geographer,* 10 (1974), 20–29.

———, "Ukranian Presence on the Prairies," *Canadian Geographic,* 97 (October–November 1978), 28–33.

Lonsdale, Richard E., ed., *Economic Atlas of Nebraska*. Lincoln: University of Nebraska Press, 1977.

Luebke, Frederick C., eds., *Ethnicity on the Great Plains*. Lincoln: University of Nebraska Press, 1980.

Lynch, Wayne, "Prairie Grasslands Preserved in Latest Park," *Canadian Geographic,* (February–March 1982), 10–19.

MacKenzie, Robert C., "Prairie Elevators 'Go to Town'," *Canadian Geographic,* 98 (February–March 1979), 36–39.

McKnight, Tom L., "Centre Pivot Irrigation: The Canadian Experience," *Canadian Geographer,* 23 (Winter 1979), 360–367.

Mather, E. Cotton, "The American Great Plains," *Annals,* Association of American Geographers, 62 (1972), 237–257.

Munro, David A., "The Prairies and the Ducks," *Canadian Geographical Journal,* 75 (July 1967), 2–13.

Reed, Andrew N., "Economic Considerations as an Explanation of Summer Fallowing on the Canadian Prairies," *The Albertan Geographer,* 18 (1982), 43–60.

Rees, Ronald, *New and Naked Land: Making the Prairies Home*. Saskatoon: Western Producer Prairie Books, 1988.

Richards, J. H., and K. I. Fung, eds., *Atlas of Saskatechewan*. Saskatoon: University of Saskatchewan, 1969.

Rogge, J., eds., *The Prairies and Plains: Prospects for the '80s*. Winnipeg: University of Manitoba, Department of Geography, 1981.

Rosenvall, L. A., "The Transfer of Mormon Culture to Alberta," *The American Review of Canadian Studies,* 12 (Summer 1982), 51–63.

Ryan, John, "The Agricultural Economy of Manitoba Hutterite Colonies," *Annals,* Association of American Geographers, 69 (December 1979), 651–652.

Sherman, W. C., *Prairie Mosaic: An Ethnic Atlas of Rural North Dakota*. Fargo: Institute for Regional Studies, 1983.

Shortridge, James R., *Kaw Valley Landscape*. Lawrence: University Press of Kansas, 1988.

Sims, John, and Thomas Frederick Saarinen, "Coping with Environmental Threat: Great Plains Farmers and the Sudden Storm," *Annals,* Association of American Geographers, 59 (1969), 677–686.

Smith, P. J., ed., Edmonton: *The Emerging Metropolitan Pattern*. Victoria: University of Victoria, Department of Geography, 1978

———,"Edmonton and Calgary: Growing Together," *Canadian Geographical Journal,* 92 (May–June 1976), 25–33.

———, ed., *The Prairie Provinces*. Toronto: University of Toronto Press, 1972.

———, and DENIS B. JOHNSON, *The Edmonton–Calgary Corridor*. Edmonton: University of Alberta, 1978.

———, and L. D. MCCANN, "Residential Land Use Change in Inner Edmonton," *Annals of the Association of American Geographers,* 71 (December 1981), 536–551.

STARK, MALCOLM, "Soil Erosion Out of Control in Southern Alberta," *Canadian Geographic,* 107 (June–July 1987), 16–25.

TIESSEN, H., "Mining Prairie Coal and Healing the Land," *Canadian Geographical Journal,* 90 (January 1975), 29–37.

VOGELER, INGOLF, and TERRY SIMMONS, "Settlement Morphography of South Dakota Indian Reservations," *Yearbook,* Association of Pacific Coast Geographers, 37 (1975), 91–108.

WALLACH, BRET, "Sheep Ranching in the Dry Corner of Wyoming," *Geographical Review,* 171 (January 1981), 51–63.

WATTS, F. B., "The Natural Vegetation of the Southern Great Plains of Canada," *Geographical Bulletin,* 14 (1960), 24–43.

WEIR, THOMAS R., ed., *Atlas of Winnipeg*. Toronto: University of Toronto Press, 1978.

———, ed., *Manitoba Atlas*. Winnipeg: Department of Natural Resources, 1984.

———, *Maps of the Prairie Provinces*. Toronto: Oxford University Press, 1971.

WILLIAMS, JAMES H., and DOUG MURFIELD, eds., *Agricultural Atlas of Nebraska*. Lincoln: University of Nebraska Press, 1977.

WILKINS, CHARLES, "Amazing Flax: Such a Versatile Crop on the Prairies," *Canadian Geographic,* 108 (October–November 1988), 38–45.

———, "Winnipeg: Tough, Self-Reliant, a Truly Canadian City," *Canadian Geographic,* 104 (December 1984–January 1985), 8–19.

WOODCOCK, DON, "How Big Farms Are Changing the Prairies," *Canadian Geographic,* 103 (April–May 1983), 8–17.

WOTJIW, L., "The Climatology of Hailstorms in Central Alberta," *Albertan Geographer,* 13 (1977), 15–30.

14
THE ROCKY MOUNTAINS

In the western interior of North America is a great cordillera that constitutes the Rocky Mountain Region. It encompasses one of the more conspicuous highlands of the world, rising between the flatness of the interior plains and the irregular topography of the intermontane West. It is a region of steep slopes, rugged terrain, and spectacular scenery; of extensive forests, abundant wildlife, and deep snows; and of sparse settlements, decaying ghost towns, and busy ski trails. Mine output has long been notable, forest products are significant in some areas, and there are scattered pockets of productive agriculture. The chief functions of the region, however, are as a place for outdoor recreation and as a source of water for most rivers of the West.

EXTENT OF THE REGION

The Rocky Mountain Region, as recognized in this book, includes the lengthy extent of the Rockies from northern New Mexico to the Yukon Territory as well as the various mountains, valleys, and plateaus of interior British Columbia. This latter section, although much of it is not mountainous, is included within the Rocky Mountain Region because its patterns of vegetation, occupance, land use, and economic activities are more akin to those of the Rockies than to those of either the Intermontane Region to the south or the North Pacific Coast Region to the west.

As thus delimited, the eastern boundary of the

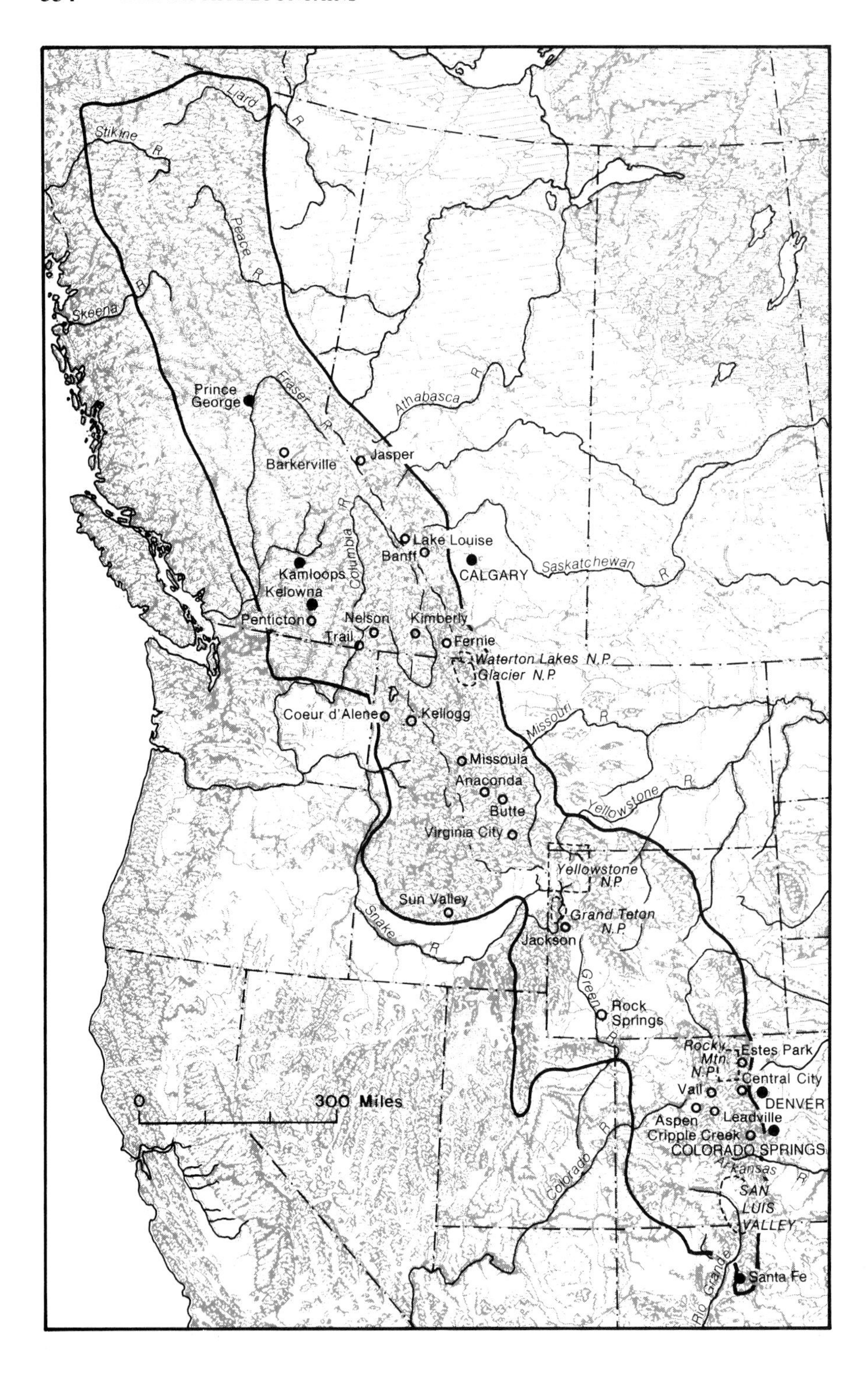

FIGURE 14-1 The Rocky Mountain Region (base map copyright A. K. Lobeck; reprinted by permission of Hammond, Inc.).

Rocky Mountain Region is marked by the break between the Great Plains and the Rocky Mountains (Fig. 14-1). From central New Mexico to northern British Columbia the mountains rise abruptly from the flatlands, except in central Wyoming where the Wyoming Basin merges almost imperceptibly with the Great Plains. The western boundary, which marks the transition from the Rocky Mountain Region to the Intermontane Region, is fairly disinct in the southern and middle sections. In much of British Columbia, however, the demarcation between the interior mountains and plateaus to the east and the Coast Mountains to the west consists of an indefinite transition zone, particularly in north-central British Columbia where the Omineca and Skeena mountains provide a "bridge" across the interior plateaus to connect the Coast Mountains with the Rockies. The northern boundary of the region is also transitional except where the broad lowland of the middle Liard River Valley makes an abrupt interruption in the topographic pattern.

It should be noted that the name "Rocky Mountains" has a different usage in Canada, normally being applied only to a specific range along the Alberta–British Columbia border.

ORIGIN OF THE ROCKY MOUNTAINS

During the Cretaceous Period most of the area of the Rocky Mountain Region and the Great Plains was covered by a shallow sea that extended from the Gulf of Mexico to the Arctic Ocean. At the close of that period, the Rocky Mountain area was uplifted and the waters drained off. Sediments with a thickness of perhaps 20,000 feet (6000 m) were involved in this first great uplift.

A long period of erosion accompanied and followed this early uplift, during which time much material was removed from the summits and deposited in the basins. In the later Tertiary Period the Rocky Mountains were subjected to another period of uplift, accompanied by considerable volcanic activity and followed by still another period of leveling. The region's master streams, flowing over sediments that had buried the mountain roots, established courses that they continued to hold after they cut into older rocks, forming major gorges and canyons through many ranges. In the more recent uplifts many ranges have been raised so high that erosion has stripped away much of the sedimentary cover, exposing the ancient Precambrian rocks (usually granitic) of the mountain core. The sedimentaries that originally extended across the axis of the uplifts are now mostly found as uptilted edges on the flanks of the ranges or as downfolded or downfaulted basins within the mountain masses.

The geologic history varies from range to range, but overall it is a region of crustal weakness and young mountains; consequently, the relief is great and slopes are steep. In many areas, however, there are fairly extensive tracts of land at high elevation—10,000 to 11,000 feet (3000 to 3300 m) in much of the Colorado Rockies, for example—which apparently represent old erosion surfaces that have been uplifted without significant deformation and appear as gently rolling upland summits or accordant ridge crests. These smooth upland surfaces are often referred to as peneplains or pediplains, but there is considerable dispute among geomorphologists as to their origin.[1]

Essentially all the high country was severely reshaped by glaciation during the Pleistocene Epoch. In more southerly latitudes abrupt **U**-shaped valleys, broad cirques, and horn peaks are the most prominent results of mountain glaciation; farther north the glacial features are more complex, for the mountain glaciers were larger and some continental glaciation also impinged on the ranges.

It goes without saying that the entire region is not mountainous. There are many valleys and some extensive areas of relative flatness. But mountains are everywhere on the horizon and alpine country almost universally dominates the landscape. The most extensive nonmountainous area is the so-called Wyoming Basin, which represents an extension of Great Plains topography into the Rocky Mountains in southern and central Wyoming. This "basin" is topographically quite heterogeneous,

[1] For a discussion of peneplanation versus pediplanation, see W. W. Atwood and W. W. Atwood, Jr., "Opening of the Pleistocene in the Rocky Mountains of the United States," *Journal of Geology* 46 (1938), 239–247; William D. Thornbury, *Regional Geomorphology of the United States* (New York: John Wiley & Sons, 1965), 328–329.

varying from alluviated plains to badlands to steep hills.

Scattered about in the Southern Rockies are a large number of relatively flat-floored basins, generally called *parks,* that are mostly not timbered and present a distinct change in landscape from the surrounding mountainous terrain. The largest is the San Luis Valley. Other notable basins include South Park, Middle Park, and North Park, and there are numerous smaller parks.

North of Wyoming the circular basin is much less common; more distinctive landscape features in this subregion are long linear valleys of structural origin. The most conspicuous is the Rocky Mountain Trench, which extends for 1000 miles from the vicinity of Montana's Flathead Lake to the Liard Plain in northeastern British Columbia.

MAJOR GEOMORPHIC SUBDIVISIONS

In terms of geomorphic association and geographical proximity, the Rocky Mountain cordillera can be subdivided into five principal sections (Fig. 14-2).

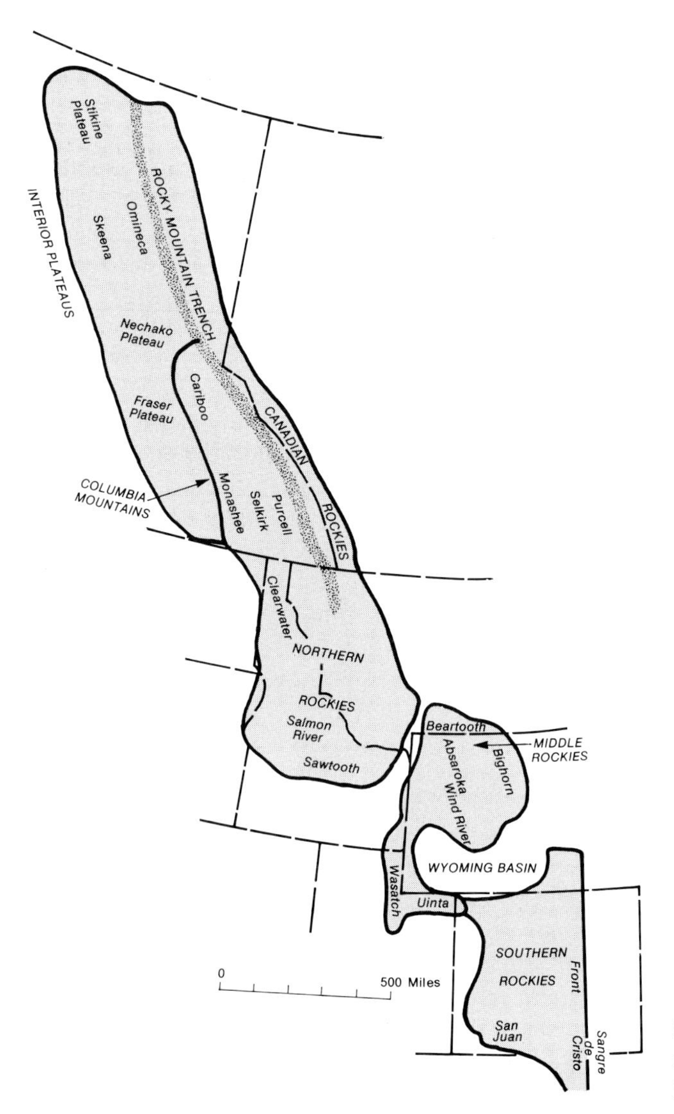

FIGURE 14-2 Principal geomorphic units of the Rocky Mountain Region.

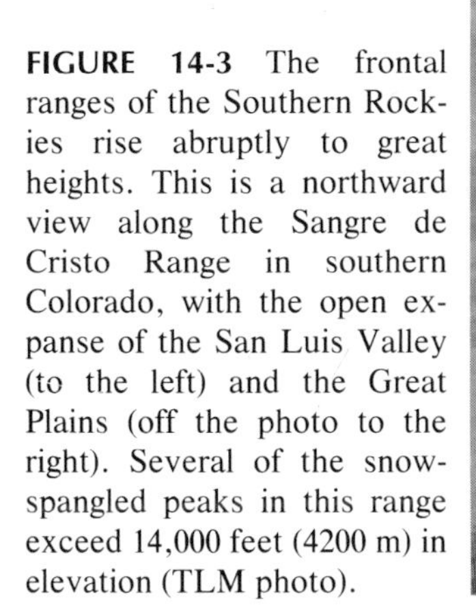

FIGURE 14-3 The frontal ranges of the Southern Rockies rise abruptly to great heights. This is a northward view along the Sangre de Cristo Range in southern Colorado, with the open expanse of the San Luis Valley (to the left) and the Great Plains (off the photo to the right). Several of the snow-spangled peaks in this range exceed 14,000 feet (4200 m) in elevation (TLM photo).

1. Southern Rockies, mostly in the state of Colorado
2. Middle Rockies, primarily in Utah and Wyoming
3. Northern Rockies, in Montana and Idaho
4. Columbia Mountains, in southeastern British Columbia
5. Canadian Rockies along the Alberta-British Columbia boundary

The Southern Rockies

The Southern Rockies include a series of linear ranges that extend from north-central New Mexico northward into southern Wyoming as well as a number of mountain masses in central and western Colorado that are less orderly in arrangement. The frontal ranges rise abruptly from the western edge of the Great Plains with just a narrow foothill zone that consists chiefly of uptilted sedimentary strata, often in the form of hogback ridges. In southern Wyoming these mountains are called the Laramie or Snowy ranges, in most of Colorado they are simply named the Front Range, and in southern Colorado and northern New Mexico they are known as the Sangre de Cristo Mountains (Fig. 14-3).

The origin of these frontal ranges is complex, involving huge batholithic intrusions, erosion of the sedimentary cover, further uplift, perhaps peneplanation, and extensive alpine glaciation. The general pattern of the present topography is a high-altitude subdued upland, with peaks up to 14,000 feet (4200 m) standing above the rolling upland surface and deep canyons trenching it.

Just west of the frontal ranges is an interrupted line of the four large-sized basins mentioned, extending from Wyoming into New Mexico. The origin of these extensive "parks" is varied, but all four are flat-floored and broad.

The ranges of central and western Colorado are generally similar in appearance and height to the frontal ranges. Their patterns are much more amorphous, however, and there are significant variations in both structure and origin (Fig. 14-4). The most complicated highland mass is the San Juan Mountains in southwestern Colorado.

The Southern Rockies reach the greatest altitudes of the entire Rocky Mountain cordillera, although greater local relief may be found in some parts of the Middle Rockies. There are no low passes through the Southern Rockies. These ranges functioned much more significantly as a barrier to trans-

FIGURE 14-4 High altitude, steep slopes, and rugged rockiness characterize the Maroon Bells in central Colorado (TLM photo).

portation than did most of the ranges farther north in the region.

The Middle Rockies

The Middle Rockies include the mountains of western Wyoming, northern Utah, and adjacent parts of Idaho and Montana. Four major separate ranges are involved:

1. The Uinta Range in Utah is the only significant range in the entire cordillera that extends east-west rather than north-south. It is a broad and massive range and is less rugged than the others.
2. Utah's Wasatch Range is a massive linear feature with considerable variation in the height of its crest line. It rises fairly gently on the eastern side and several streams have cut relatively low passes through it. Its precipitous western face, the Wasatch Front, however, is remarkable for its combination of height, steepness, and continuity.
3. The Wind River Mountains in western Wyoming encompass some of the most rugged granitic wilderness that can be found. It is a little-known area characterized by steep slopes and high relief. It was heavily glaciated in the past and still contains more than 60 living glaciers, constituting a greater total ice area than all other United States Rocky Mountain glaciers combined.[2]
4. The Bighorn Mountains of north-central Wyoming are a massive range rising between the Great Plains on the east and the Bighorn Basin on the west. There are a steep eastern slope, a broad subdued upland surface, ramparts of higher glaciated country, another broad subdued upland, and a less rugged western slope.

In addition to these major units, a series of smaller ranges extends from northern Utah into northern Wyoming, arranged with a conspicuous north-south linearity. These relatively minor highlands culminate northward in the majestic Grand Tetons of west-central Wyoming, whose block-faulted and glaciated eastern scarp presents an even

[2] Stephen F. Arno, "Glaciers in the American West," *Natural History*, 78 (February 1969), 88.

more spectacular face than the Wasatch Front (Fig. 14-5).

Also a part of the Middle Rockies is the broad, jumbled mountain mass that makes up the Absaroka Range in Wyoming and the Beartooth Mountains in Montana. They are fairly rugged and steep-sided but with no discernible pattern and no notable crest line.

The lava plateau of Yellowstone Park and the north-trending valley of the upper Yellowstone River provide a relatively complete break between the Middle and Northern Rockies, although this breach had no transportational significance in the history of the West.

FIGURE 14-5 The spectacular eastern fault scarp of the Grand Tetons rises abruptly to glaciated peaks above the flat floor of Jackson Hole (TLM photo).

The Northern Rockies

The Northern Rockies occupy essentially all of western Montana and central and northern Idaho. The eastern half consists of a number of discrete linear ranges separated by broad, flat-bottomed, structural valleys. It is almost ridge-and-valley topography on a grand scale. These are often referred to as the Broad Valley Rockies because the spacious intermontane valleys occupy so much (sometimes half) of the total area. This "broad valley" pattern also extends into parts of central Idaho. Some of the mountains are exceedingly rugged, but altitudes are lower than in the Middle and Southern Rockies (Fig. 14-6).

The western portion of the Northern Rockies is a massive jumble of mountains that is almost patternless. Nearly all the land is in slope with many deep, narrow river valleys. Generalized nomenclature refers to the southern part as the Salmon River Mountains and the Sawtooth Mountains; most of the northern section is considered part of the Clearwater Mountains. The Sawtooths are the highest and most rugged, but no part is easily crossed by transportation routes. The largest of Idaho's numerous lakes—Coeur d'Alene, Pend d'Oreille, and Priest—are found in the extreme north, where continental glaciation occurred.

The Columbia Mountains

The Columbia Mountains occupy an area in southeastern British Columbia that is quite broad in the south but tapers northward until it pinches out at about the latitude of Edmonton and Prince George (54° north latitude). There are four major ranges: the north-south trending Purcell, Selkirk, and Monashee ranges in the south and the knot of the Cariboo Mountains in the north. Unlike the true Canadian Rockies, these mountains consist largely of crystalline rocks. The narrow north-south trenches among the ranges are chiefly occupied by long, beautiful bodies of water, such as Kootenay Lake and Lake Okanagan.

FIGURE 14-6 The Northern Rockies are lower in elevation than those farther south, but their spectacular scenery is unparalleled. This is Mt. Jackson and St. Mary Lake in Montana's Glacier National Park (TLM photo).

FIGURE 14-7 Looking north up the Rocky Mountain Trench near Windermere, British Columbia (TLM photo).

The Rocky Mountain Trench

In many ways, the most remarkable topographic feature of the entire cordillera is the Rocky Mountain Trench (Fig. 14-7). It is a depression that extends in a direct line from Montana almost to the Yukon Territory with regular boxlike sides. Its exact origin is uncertain and variable, although parts are clearly related to faulting.[3] The trench bottom is flat to rolling and even hilly in spots, with four low drainage divides between major north- and south-flowing streams. Eleven rivers, including the upper reaches of both the Columbia and the Fraser, occupy some portion of the trench. New dams now impound two immense reservoirs in the trench: Williston Lake behind W. A. C. Bennett Dam on the Peace River and McNaughton Lake behind Mica Dam on the Columbia River.

The Canadian Rockies

Along the Alberta–British Columbia border is a lengthy northern continuation of the cordillera known as the Canadian Rockies. Although not as high as the Middle and Southern Rockies, there is great local relief and the narrow ranges are more rugged and steep-sided. There was alpine glaciation during the Pleistocene, with extensive present-day mountain glaciers and even some upland icefields—the largest is the Columbia Icefield—from which lengthy glaciers extend down-valley.

The mountains consist primarily of conspicuously layered sedimentary rocks that were uplifted and, in some cases, extensively folded and faulted. Ridges and valleys are more or less uniform in altitude and tend to parallel one another for long stretches in a general northwest-southeast trend. These ranges provide some of the most spectacular scenery of the continent (Fig. 14-8).

Interior Plateaus and Mountains of British Columbia

The interior section of British Columbia consists of an extensive area of diversified but generally subdued relief. The Fraser Plateau in the south and the Nechako Plateau in the center are characterized by moderately dissected hills, occasional mountain protuberances, and a few large entrenched river valleys. Most notable of these entrenchments are in portions of the Fraser and Thompson valleys and in the Okanagan Trench with its series of beautiful lakes.[4]

North of the Nechako Plateau is a relatively complicated section of mountains and valleys—the Skeena Mountains to the west and the Omineca Mountains to the east—that flattens out into a series of dissected tablelands, the Stikine Plateau, in the far north. The Stikine country is underlain by lava and surmounted by several volcanic peaks.

[3] J. Lewis Robinson, "The Rocky Mountain Trench in Eastern British Columbia," *Canadian Geographical Journal*, 77 (October 1968), 132.

[4] This lengthy linear valley extends southward across the international border into Washington State. It has the same name in both countries but different spellings. In Canada the word is spelled "Okanagan"; in the United States "Okanogan."

FIGURE 14-8 The jagged peaks of the Canadian Rockies are world famous. This is Mount Edith Cavell in Jasper National Park (Alberta Government photo).

VERTICAL ZONATION: THE TOPOGRAPHIC IMPERATIVE

In a landscape dominated by slopes and considerable relief the nature of other environmental elements is significantly dictated by the topography; thus climate and vegetation are markedly affected by altitude and less so by exposure and latitude. Within the Rocky Mountain Region are found some of the classic examples of vertical zonation in vegetation patterns, with major variations occurring in short horizontal distances because of significant vertical differences.

Climate

The paradox of the Rocky Mountain climate is that summers tend to be semiarid and winters relatively humid. Summer is a time of much sunshine even though there may be a characteristically brief thundershower each afternoon. Windiness, generally low humidity, and sunshine quickly evaporate the rainfall, partially negating the effectiveness of the precipitation. Summer is therefore a time of relative dryness; after the main runoff of the melting snowpack has subsided, dust is much more characteristic than mud on the mountain slopes except in the Canadian portion of the region where the summers are rainier.

The inland location of this region means that the prevailing westerly air masses have had to pass over other mountains before reaching the Rockies and that part of their available moisture was dropped there. The western slopes of the high country may receive a considerable amount of summer rain, for the forced ascent squeezes out more moisture from the westerlies. But the lowland valleys, parks, and trenches of the region may experience almost desertlike conditions. So the center of Colorado's San Luis Valley receives only 6 inches (152 mm) of moisture annually and Challis, on Idaho's Salmon River, records 7 inches (178 mm); Ashcroft, in the Thompson River Canyon, with 7 inches (178 mm), is probably the driest nonarctic locality in Canada.

The low temperatures of winter make dry weather seem less dry because the precipitation that does fall is normally in the form of powdery snow; once it is on the ground, it may not melt until spring, even in the basins. Storm tracks are also shifted southward in winter, with the result that stormy conditions are more numerous and frontal precipitation is more commonplace. Usually snow lies deep and long over most of the region, particularly at higher altitudes (Fig. 14-9).

The general pattern, then, is that the lower elevations are quite dry, but with increasing altitude there is increasing precipitation up to some critical level (generally between 9000 and 11,000 feet [2700 and 3300m]) above which there is once again a decreasing trend. West-facing slopes normally receive more rain and snow than comparable levels on east-facing slopes because the prevailing winds are from

FIGURE 14-9 Winter is long and bitterly cold in the high country. This is Apache Peak and Long Lake in Colorado's Front Range (TLM photo).

the west and make their forced ascent on that side. Also, south-facing slopes receive more direct sunlight than north-facing ones, which makes for more rapid evaporation on the former and reduces the effectiveness of the rainfall or snowmelt received.

Natural Vegetation

The region is basically a forested one with coniferous species predominant, but there are many areas in which trees are absent. All the valleys and basins in the Southern and Middle Rockies and some in more northerly localities are virtually treeless except for riparian hardwoods along stream courses. A sagebrush association is widespread throughout the Wyoming Basin and in many lower parks and valleys as far north as southern British Columbia. The higher valleys and basins are more likely to be grass covered.

At the other altitudinal extreme, on the mountaintops, trees are also absent as a rule. These higher elevations have the low-growing but complex plant associations of the alpine tundra. Thus many mountain ranges of the region have a double tree line: one at lower elevation that marks the zone below which trees will not grow because of aridity and one at higher elevation above which trees cannot survive because of low temperatures, high wind, and short growing season (Fig. 14-10).

The relationship of altitude to latitude is shown quite clearly by the variation in elevation of the upper tree line in the Rocky Mountains. At the southern margin of the region, at 36° north latitude in New Mexico, the upper timberline occurs at 11,500 to 12,000 feet (3450 to 3600 m) above sea level. This elevation progressively decreases northward: about 9500 feet (2850 m) at 45° in Yellowstone Park; about 6000 feet (1800 m) at 49° at the international border; and about 2500 feet (750 m) at 60° at the northern extremity of the region near the Yukon border. Exposure, drainage, and soil characteristics influence the details of this pattern, but the general principle is clear-cut.

The varying elevation of the upper tree line represents only one facet of the broader design of vertical zonation in vegetation patterns. As a result of abrupt altitudinal changes in short horizontal distances, various plant associations tend to occur in relatively narrow bands or zones on the slopes of the ranges (Fig. 14-11). The principle of vertical zonation is ever present in the Rockies; only the details vary.

FIGURE 14-10 There is no question about the direction of the prevailing wind in this area. The irresistible effect of wind shear on tree growth near the tree line is clearly shown in this Colorado scene. Branches are able to sprout and survive on this subalpine fir only on the direct leeward side of the tree trunk (TLM photo).

THE OPENING OF THE REGION TO SETTLEMENT

The Rocky Mountains Region has always been characterized by sparse human population. It was no more attractive as a permanent habitation for aboriginal Indians than it is for contemporary Americans and Canadians; the climate was too rigorous and the resources too few.

There was a scattering of semipermanent Indian settlements in the Rockies—particularly in what is now Idaho and British Columbia—before the arrival of the first Europeans, but such settlements always were small and marginal. During summers there were frequent incursions into the Rockies by Plains Indians from the east and by Great Basin and Plateau Indians from the west, but almost always on a temporary basis. Thus, prior to the coming of Europeans various Indian tribes had considerable knowledge of the region, primarily in terms of summer hunting grounds and routes of passage. As a whole, however, the region was marginal to their way of life.

Except in the extreme southern part of the region that was penetrated by the Spaniards at the end of the sixteenth century, the first white men to see the Rocky Mountains were French fur traders who sporadically worked their way up the Great Plains rivers during the eighteenth century. Beginning in the late 1700s, there were several significant exploring expeditions in various parts of the region; some were government sponsored and others were backed as commercial ventures.

Alexander Mackenzie was the first explorer of note to cross the Rockies; he did so in 1793 by penetrating the Peace River Valley in the far north of the region. The Lewis and Clark Expedition of 1803–1804 made its way up the Missouri River and crossed the Northern Rockies on the way to the mouth of the Columbia. While Lewis and Clark were still encamped on the Pacific shore, Simon Fraser, who was affiliated with the North West Company as was Mackenzie, established the first trading post west of the Rockies, at Fort McLeod on a tributary of the Peace River.

Fur trapping and trading became a way of life throughout the region (Fig. 14-12), with large periodic trading rendezvous at such places as Jackson's Hole at the eastern base of the Grand Tetons in Wyoming and Pierre's Hole in eastern Idaho. This colorful period, however, did not last long because the value of beaver fur declined in the 1840s, following changes in the style of men's hats. From that time until the discovery of gold in the late 1850s the Rocky Mountains served only as a barrier to the westward movement, and pioneers pushed through the lowest mountain passes as rapidly as possible on their way to the Oregon Territory or California.

The 1849 gold rush to California encouraged many people to cross the Rockies. While doing so, some prospected for gold in the stream gravels and found traces of the precious metal. Although most seekers of gold went on to California, many returned within a few years to prospect further in the numerous mountain gulches.

A CLOSER LOOK A Forest Epidemic—The Example of the Mountain Pine Beetle

The forests of the Rocky Mountain Region are presently in the throes of an insect epidemic. The infestation is not restricted to this region; it extends west into Oregon and Washington and as far east as the Black Hills of South Dakota and the Cypress Hills of Alberta and Saskatchewan. The culprit is a tiny black bug called the mountain pine beetle. It is an endemic inhabitant of pine forests, especially those dominated by lodgepole pine. In an ordinary year pine beetles kill millions of trees in western United States and Canada. In the last two decades, however, a full-fledged epidemic developed that vastly accelerated the mortality and damage to pines in every state and province of the Rocky Mountain Region.

The epidemic may have started in several places, but it is particularly pinpointed to northwestern Wyoming in the late 1960s. Since then it has become conspicuous as far south as central Colorado and as far north as central British Columbia. It has reached "crisis" proportions in British Columbia, Alberta, Idaho, and Wyoming and is approaching that level in other locations. Lodgepole pines are particularly affected, but other pine species, especially ponderosa and white pines, are also attacked severely.

Under "normal" conditions the beetles infest pines in moderate numbers, in relatively peaceful coexistence. Adult beetles bore through the brittle outer bark and lay eggs in the moist, tender inner bark (phloem). The eggs hatch in a week or so and the larvae live on the phloem, eating and maturing. On reaching adulthood, they bore out, seek mates, and start the cycle again. A healthy tree secretes pitch to repel the invader, but a damaged or aged tree is often unable to maintain adequate defenses.

Epidemics generally begin when extensive pine stands reach relatively advanced ages (100 to 130 years). The beetle carries a fungus that disrupts the water transport system of the tree, strangling it slowly so that the needles become dessicated, turn red, and eventually drop off, leaving the tree a dead hulk.

Treatment of the infestation is difficult. Aerial spraying of insecticides has limited effect because so little of the spray reaches the tree trunk. Control programs generally involve concentrating efforts at critical locations—spraying individual trees from the ground, felling infested trees, or even sanitation felling of large blocks of trees. All such actions are costly in money and effort.

In the United States the Forest Service is the principal control agency; its efforts consist primarily of holding actions and planning to minimize future outbreaks. An interagency Committee on the Mountain Pine Beetle was established in Canada in 1981, with members representing various federal and provincial (British Columbia and Alberta) agencies; its mandate is to stop the spread of the infestation. Several major national parks—Yellowstone, Grand Teton, Waterton Lakes, Banff, Kootenay, Yoho—have been severely affected, but management philosophy there is not to interfere with natural events except in critical areas where the beetles may spread beyond park boundaries.

In the broad view, it must be realized that insects constitute but one of many natural mortality factors to which trees are susceptible. And there are many tree-killing insects. A severe infestation of Engelmann spruce bark beetles threatened most of the conifers in the Southern Rockies in the 1950s, for example. That epidemic, like most, eventually dissipated but left millions of acres of dead trees in north-central Colorado as its legacy.

At present, two other insects—the western spruce budworm and the Douglas fir beetle—also are in epidemic infestation in parts of the Rocky Mountain Region.

When someone realizes that a healthy green forest is turning brown and dying, it is difficult to accept the long-term ecological view that such events have occurred for millions of years and that the forest ecosystem is adapted to them. Such a perspective is particularly improbable if the observer happens to own the trees or the land or the "scenery" that is affected.

Silvicultural research with regard to insect infestations is now concentrating on preepidemic mitigation. How can potential infestation sites best be managed to minimize future epidemics? The answers seem to lie in judicious thinning of stands before the pines begin to reach overmature ages and in clearcutting under some circumstances.

TLM

It was not until 1859, however, that gold in paying quantities was found in the Rocky Mountains. At almost the same time a gold rush began to Central City, Colorado, and another to the western flanks of the Cariboo Mountains in British Columbia. By the early 1860s the Cariboo gold fields had the largest concentration of people in western Canada; and Barkerville, center of the find, was the largest western town north of San Francisco. By the early 1870s Central City had grown to become the largest urban center in the Rocky Mountain Region.

Within a short time more people had settled in

FIGURE 14-11 The vertical zonation of vegetation is clearly recognizable on Sierra Blanca in the Sangre de Cristo Mountains of Colorado (TLM photo).

the mountain country than in all its previous history. Practically every part of the region was prospected and many valuable mineral deposits were found, especially in the Southern Rockies. Boom towns sprang up in remote valleys and gulches of the high country, which, in turn, led to the development of a series of narrow-gauge railroads, built at great expense per mile, for hauling out gold ore. As the higher-grade ores became exhausted, production declined in these camps; in time most became ghost towns (Fig. 14-13).

Lumbering and logging, grazing activities, irrigation agriculture, and the tourist trade later brought additional population to the mountains, but none was so significant in the early peopling of the region and in bringing its advantages to the attention of the rest of the country as gold mining.

THE MINING INDUSTRY

Wherever rich mineralized zones were found, mining camps developed. Colorado was especially important, with its Central City, Ouray, Cripple Creek, Victor, Leadville, Aspen, Georgetown, and Silver Plume. Wyoming and New Mexico were relatively insignificant, but farther north there were major discoveries around Virginia City in southwestern Montana, in the vicinity of Butte and Anaconda, in the Coeur d'Alene area of northern Idaho, and in the Kootenay and Cariboo districts of British Columbia. Although gold was the mineral chiefly sought, valuable deposits of silver, lead, zinc, copper, tungsten, and molybdenum were also found.

Mining Today

The region is highly mineralized. A few of the major mining districts are discussed here.

The Leadville District One of the oldest and most important mining areas of the Rocky Mountain Region is located at Leadville, near the headwaters of the Arkansas River at an elevation about 10,000 feet (3000 m). Placer gold was discovered in this remote valley in 1860, and within four months more than 10,000 people were in the camp. Silver–lead mining began a dozen years later, and zinc was discovered in 1885. Soon after the turn of the century the value of zinc output exceeded that of any of the other three metals.

FIGURE 14-12 The presence of beaver attracted the first whites to the Rocky Mountains Region. These industrious rodents are still widespread, as evidenced by their lodges and dams (TLM photo).

FIGURE 14-13 Remnants of hundreds of ghost towns still exist in the region. This is Zincton in the Columbia Mountains of British Columbia (TLM photo).

The Leadville area continues as an important mining district, producing all its former metals as well as molybdenum and tungsten (Fig. 14-14). The molybdenum comes from Anglo-America's highest major mining complex (11,000 feet [3300 m]) at Climax, which is 13 miles (21 km) northeast of Leadville and is the world's largest producer of this valuable ferroalloy ore.

The Butte–Anaconda District In western Montana is located one of the most famous of all Anglo-America's mining districts but one that has now ceased operations in a "temporary suspension" that may well become permanent. With mines at Butte and a gigantic smelter 23 miles (37 km) away at Anaconda, this area was one of the great copper producers of the world and also yielded significant amounts of silver, gold, lead, and zinc. Butte's "richest hill on earth" is honeycombed with tunnels that once had as many as 20,000 workers digging ore from 150 mines, although in recent years most production was concentrated in a single large open-pit operation.

For more than a century the Anaconda Minerals Company dominated the economy, and often the politics, of the state, besides owning most daily newspapers in Montana. In 1983, however, the Atlantic Richfield Company (which now owns Anaconda) decided that the $1 million per week losses were too great to bear and shut down the operation. Many mining enterprises in Anglo-America have suffered severe financial stress in recent years, but the Butte–Anaconda district represents the most spectacular example of the basic economic equation of rising mining costs and "soft" market prices for the market.

The Coeur d'Alene District This mining area, one of the richest of the Rocky Mountains, lies in northern Idaho. The two dozen or so underground mines of the district have yielded more than $2 billion worth of various ores in their century of operation. The history of mining in northern Idaho is a story of technological triumphs. The structural geology of the district is complicated and the mineralogy of the ores is complex. Techniques were devised to meet every challenge, however, in terms of mining, concentrating, and smelting.

Output from this district makes Idaho easily the leading silver-producing state and gives it second ranking in lead and third in zinc. Antimony, copper, gold, and other byproducts are also produced. The principal urban place of the district is Kellogg, which is also the site of the major smelter for the local ores.

FIGURE 14-14 Leadville, Colorado, was a mining town that experienced several periods of notable prosperity and now continues as a shadow of its former self. The mine dumps in the foreground are clear evidence of past economic history. Mount Massive in the background is the second highest peak in the entire Rocky Mountain cordillera, but its 14,418-foot (4325 m) peak is not particularly prominent because even the valley bottoms in this area are more than 10,000 feet (3000 m) above sea level (courtesy Colorado Department of Public Relations).

The Kootenay District In the southeastern corner of British Columbia lies one of Canada's foremost mining districts. The great Sullivan mine at Kimberly has been dominant, but a dozen other mines are active. The most important minerals have always been lead and zinc, with silver as the usual associated mineral. Other by-products are tin, gold, bismuth, antimony, sulphur, and cadmium.

Concentrates from the Kootenay district, as well as from other parts of western Canada and some foreign sources, are treated in an enormous refinery at Trail, which is one of the world's greatest nonferrous metallurgical works (Fig. 14-15).

Recent Developments Mining is inherently an unstable economic activity. As technologies change and demands fluctuate, there are almost immediate responses from the mineral industries. What is a valueless sulphide may become a valuable ore. Today's boom can become tomorrow's bust. The rise

FIGURE 14-15 One of the largest metal refining complexes on the continent is found adjacent to the Columbia River at Trail, British Columbia (TLM photo).

in value of one mineral may generate a frenzy of exploration and development whereas a declining market for another produces a ghost town.

These characteristics are prominent in recent developments in the Rocky Mountain Region. Many established mining centers have retrenched their operations significantly and some, like Butte, actually closed down. On the other hand, many new mineral extraction activities have begun and the chance of more valuable deposits being discovered is considerable. Major emphasis has been on energy minerals (coal, petroleum, natural gas), but not exclusively.

Southwestern Wyoming is a major focus of development. Open-pit coal and iron mines are now in production and the continent's largest supply of trona (soda ash) is being mined on a large scale. Oil and gas extraction was begun in the 1970s, but collapsed in the 1980s.

The Canadian portion of the region has seen significant new development. Coal mining expanded remarkably at several places in the Alberta foothills and at scattered localities in British Columbia, with the old coal-mining area of Crow's Nest Pass–Fernie (situated on the Alberta–British Columbia border) leading the way. Copper, however, is British Columbia's most valuable mineral product, with several new copper mines opening in the central part of the province. Significantly increased output of molybdenum has also occurred in British Columbia.

FORESTRY

Although there has long been small-scale local cutting of timber for mine props, firewood, cabin construction, and other miscellaneous purposes, large-scale commercial forestry was restricted to certain sections of the Rocky Mountain Region, particularly the northern half. For example, less than 2 percent of total U.S. timber cut is from the Southern and Middle Rockies whereas nearly 10 percent comes from the Northern Rockies. Large sawmills and pulp mills operate at several localities in western Montana and northern Idaho.

The traditional logging area of western Canada has been coastal British Columbia, but lately increasing timber supplies have come from the interior of the province. Sawmilling is a major activity at such places as Fernie, Cranbrook, Nelson, Golden, and McBride. Pulping, although growing, is much more limited, but there are large mills at Prince George, Kamloops, and Castlegar. New pulping operations have also been started at several locations in the Alberta (eastern) foothills of the Rockies, in some cases processing aspen rather than conifers.

AGRICULTURE AND STOCK RAISING

In any consideration of Rocky Mountain agriculture and ranching, it must be remembered that the region consists chiefly of forested or bare rocky slopes. The occasional level areas are used for irrigation farming, dry farming, or ranching and thus attain an importance out of proportion to their size.

The earliest agricultural use of Rocky Mountain valleys involved livestock grazing and subsistence farm plots. Generally agricultural and pastoral settlement was late to occur because of inaccessibility, harsh winters, a short growing season, and Indian opposition. The southern San Luis Valley became two large Spanish land grants before the mid-nineteenth century, but no significant settlement arose there until after the Mexican War. Elsewhere in the region, farming and ranching in the valleys and basins were even later developments.

Today there are livestock enterprises in most valleys of the region. Both beef cattle and sheep ranches (Fig. 14-16) are scattered throughout the Rockies, but the former are much more widespread. The typical ranching operation involves *transhumance:* during the summer the animals are taken (either by walking or by truck) to high-country pastures while valley ranch lands are used for growing hay; in early autumn the stock is returned to the ranch where the hay is used as basic or supplemental feed. Most summer grazing is done under purchased permit on national forest lands; a long-established policy by the two principal land management agencies of the West—the Forest Service and the Bureau of Land Management—affirms the principle of grazing by permit on federal land, thereby maintaining continuity and coherence in these ranching operations.

Although a certain amount of crop growing is done under natural rainfall conditions in the valleys

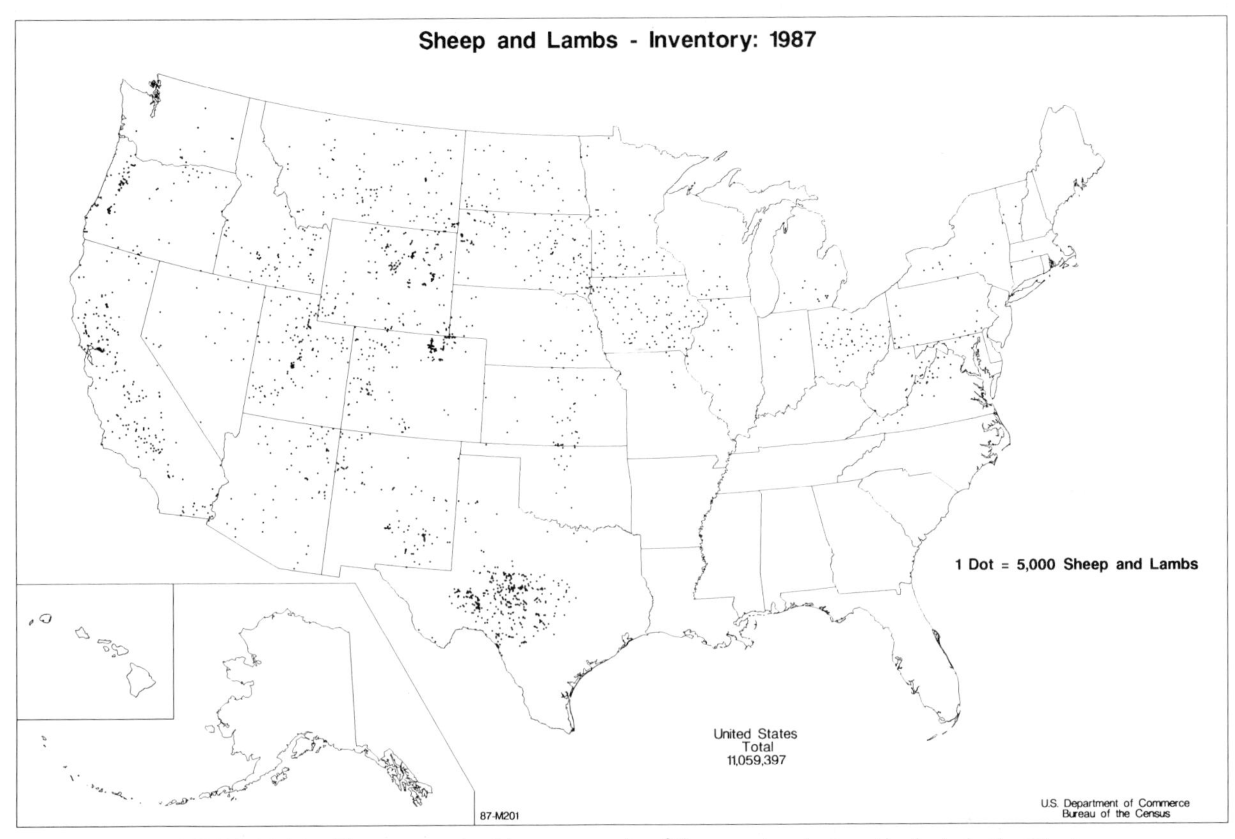

FIGURE 14-16 Sheep are raised in many parts of the country, but particularly in the West.

of the region, most farming involves irrigation. Most of this irrigation is on a small scale and simply involves diversion of water from the valley-bottom stream into adjacent fields. The basic crop throughout the region is hay. Irrigated valleys yield valuable supplies of alfalfa and other hay crops for winter feeding of livestock (Fig. 14-17). Some specialty crops are also grown; pinto beans and chili peppers in New Mexico, miscellaneous vegetables in Colorado and Utah, sugar beets and grains in

FIGURE 14-17 Dairying is an important activity in the "broad valley" Rockies of western Montana. This is the Mission Valley near Ronan (TLM photo).

FIGURE 14-18 The Okanagan Valley, western Canada's premier fruit-growing area. Shown here are Lake Osoyoos, the town of Osoyoos, and orchards (darker areas) covering most of the valley bottom lands (courtesy British Columbia Department of Travel Industry).

Idaho and Montana, fruits around Flathead Lakc and in some Kootenay valleys, and wheat in the Kootenay River Valley and around Cranbrook. But hay occupies the principal cultivated acreage almost everywhere.

The most notable irrigated valley in the Rocky Mountain Region, and one of Canada's most distinctive speciality crop areas, occupies the long narrow trench of the Okanagan Valley (Fig. 14-18). Extending north for 125 miles (200 km) from the international border, the valley is only 3 to 6 miles (5 to 10 km) wide except where it broadens a bit into tributary valleys in the north. Large lakes occupy most of the valley floor and farming is limited to adjacent terraces. Irrigation is necessary in the desertlike conditions (annual rainfall, 9 inches [230 mm] and temperatures up to 110°F [43°C]) of the southern part of the valley, but general farming can be carried on under natural rainfall in the north (precipitation, 17 inches [430 mm]).

Various field crops, feed crops, and vegetables are grown, but fruit growing is the distinctive activity in the valley. The Osoyoos section in the south produces the earliest fruits in Canada. Apples are the major crop (occupying about two-thirds of the valley's orchard area and yielding about one-third of Canada's total output), but the valley is particularly noted as one of only two areas in the nation that has a sizeable production of soft fruits—peaches, plums, pears, cherries, and apricots. Lately there has been a significant increase in the acreage devoted to vineyards.

Because of its relatively mild climate and abundant sunshine, the Okanagan is an attractive place to live. Consequently, increasing population pressure exacerbates problems of congestion and pollution and raises land prices in an ever-higher spiral that makes it difficult for small farmers to continue farming.

The other conspicuous farming area in the region is the San Luis Valley. Although it has a short growing season because of its high elevation (above 7000 feet [2100 m]) and its rainfall totals are low, much of the earlier ranch land has been converted to irrigated farming. An underground water resource has been easily tapped by artesian wells. Most of the valley's farmers have been attracted to center pivot technology, with the result that the San Luis Valley now has one of the greatest concentrations of center pivot irrigation to be found anywhere in the world. Crop options are limited because of the short growing season and harsh winters. Although hay is the most widely grown crop, potatoes constitute the principal source of farm income. Vegetables and barley are also grown in quantity.

WATER "DEVELOPMENT"

The Rocky Mountain Region is sometimes spoken of as the "mother of rivers" because so many of the major streams of western North America have their headwaters on the snowy slopes of the Rockies. This water-collecting function is generally considered one of the two leading economic assets (tourism is the other) of the region. From the Rocky Mountains the Rio Grande and Pecos flow to the south; the Arkansas, Platte, Yellowstone, Missouri, South Saskatchewan, North Saskatchewan, Athabasca, Peace, and Liard flow to the east; and the Stikine, Skeena, Fraser, Columbia, Snake, Green, and Colorado flow to the west.

The capriciousness of river flow has long been recognized and the alternation between flood and low-water stages has been deplored. Dam building is the chief tool to smooth these imbalances of flow and make the waters more "usable" for various purposes. In the past, most "development" of Rocky Mountain rivers has been deferred to downstream locations, particularly in the Great Plains and Intermontane regions. But later dam building came to the high country, modified, in some cases, by wholesale water-diversion schemes.

The principle of transmountain diversion is that "unused" water, generally from western slope streams, is taken from an area of surplus by means of an undermountain tunnel to an area of water deficit, normally on the eastern slope of the Rockies or in the western edge of the Great Plains. Potential western-slope users are compensated for this loss by building storage dams there to catch and hold flood-stage flow for later release when the river is low.

Until the 1960s there had been almost no dam building in the Canadian portion of the region except for five small hydroelectric dams along the lower Kootenay River to supply power to the huge smelter at Trail. In 1964, however, the United States and Canada ratified the Columbia River Treaty, which provided for the construction of major dams on the upper Columbia River and its tributaries to control floods and permit increased hydroelectric power generation in both countries. The dams were paid for by the United States, as an advance against power generated in the future in Canada but sold in the United States.

An even more grandiose scheme has been constructed on the upper reaches of the Peace River near Finlay Forks, British Columbia. The W. A. C. Bennett Dam backs up a reservoir for 70 miles (110 km) on the Peace and another 170 miles (270 km) on two major tributaries, the Finlay and the Parsnip rivers. Much of the power is transmitted 600 miles (960 km) to metropolitan Vancouver, the largest population concentration in western Canada.

The long, narrow, structural valleys of the Columbia Mountains and the Rocky Mountain Trench are thus increasingly filled with reservoirs, resulting in undeniable economic benefits and equally undeniable environmental degradation.

THE TOURIST INDUSTRY

The other major role of the Rocky Mountain Region is in providing an attractive setting for outdoor recreation. The tourist industry is undoubtedly the most dynamic segment of the regional economy.

Summer Tourism

With its high rugged mountains, spectacular scenery, extensive forests, varied wildlife, and cool summer temperatures, the Rocky Mountain Region is a very popular summer vacationland. The location of the region between the Great Plains on the east and the intermontane and Pacific coastal areas on the west places it directly across lines of travel. Nearby flatlanders flock to the mountains for surcease from the summer heat of Dallas, Kansas City, or even Denver. And people come from greater distances to sample the scenic delights of Banff (more than 3 million tourists per year) or Yellowstone (200,000 visitors in a midsummer week, a number equal to more than one-third of Wyoming's resident population).

Throughout the Rockies there are places of interest for visitors whose interests and activities are varied, but spectacular scenery is the principal attraction. Many of the outstanding scenic areas have been reserved as national parks, which generally function as the key attractions of the region. Seven of the most popular tourist areas in the region, in terms of numbers of visitors, are described next.

New Mexico Mountains The southernmost portion of the Rocky Mountains, located in northern New Mexico, is not particularly rugged or spectacular and consists primarily of pleasant forested slopes. Its summers, however, are considerably cooler than those of the parched plains of the

Southwest. It is an area with a rich historical heritage that is manifested in the pervasive Indian and Hispanic character of the cultural landscape. As a result, the narrow twisting streets of Santa Fe (the principal focal point of the area) are jammed with visitors' vehicles during the summer. The Taos area is a center for dude ranch and youth-camp activities and Red River has developed into a year-round tourist resort.

Pikes Peak Area Colorado Springs is a city of the Great Plains, but its site is at the eastern base of one of the most famous mountains in America. Pikes Peak, with a summit elevation of 14,110 feet (4233 m), is far from being the highest mountain in Colorado, but its spectacular rise from the plains makes it an outstanding feature. The surrounding area contains some of the most striking scenery (Garden of the Gods, Seven Falls, Cheyenne Mountain, Cave of the Winds, Rampart Range) and some of the most blatantly commercial (Manitou Springs) tourist attractions in the region. Few tourists visit the Southern Rockies without at least a brief stop in the Pikes Peak area, as overcrowding of even the unusually wide streets of Colorado Springs gives eloquent evidence.

Denver's Front Range Hinterland The Front Range of the Southern Rockies rises a dozen miles (16 km) west of Denver and the immediate vicinity provides a recreational area for the residents of the city as well as visitors from more distance places. Denver maintains an elaborate and extensive group of "mountain parks," which are actually part of the municipal park system. There are thousands of summer cabins for rent; many streams to fish; deer, elk, mountain sheep, and bear to hunt; dozens of old mining towns to explore; and countless souvenir shops in which to spend money.

Rocky Mountain National Park Following many years of agitation by the people of Colorado for the establishment of a national park in the northern part of the state to preserve the scenic beauty of that section of the Continental Divide, a rugged area of 400 square miles (1036 km^2) was reserved by Congress in 1915 as Rocky Mountain National Park (Fig. 14-19). It includes some of the highest and most picturesque peaks, glacial valleys, and canyons of the region, as well as extensive forested tracts.

Spectacular Trail Ridge Road traverses the park, connecting the tourist towns of Grand Lake and Estes Park, and reaches an elevation of 12,185 feet (3656 m). Automobile touring is the principal activity in the park, but hiking, climbing, and trail riding are also popular.

Yellowstone–Grand Teton–Jackson Hole In the northwestern corner of Wyoming is an extensive forested plateau of which a 3500-square-mile (9065 km^2) area has been designated as Yellowstone National Park. It was established in 1872 as the first national nature preserve in the United States. It is lacking in spectacular mountains but contains a huge high-altitude (elevation 7700 feet [2300m]) lake, magnificent canyons and waterfalls, and the most impressive hydrothermal displays—geysers, hot springs, fumaroles, hot-water terraces—in the world (Fig. 14-20). A few miles to the south is Grand Teton National park, a smaller and more recently reserved area that encompasses the rugged grandeur of the Grand Teton Mountains, a heavily glaciated fault block that rises abruptly from the flat floor of Jackson Hole. Jackson Hole was an early fur-trappers' rendezvous that is now the winter home of the largest elk herd on the continent.

The Grand Tetons are particularly attractive to hikers and climbers; Yellowstone is a motorists' park. The ubiquitous moose and a great variety of other species of wildlife add to the interest of the area. In spite of its relatively remote location and the brevity of the tourist season, the Yellowstone–Grand Teton country annually attracts more than 3 million visitors.

Waterton–Glacier International Peace Park Glacier National Park in Montana and Waterton Lakes National Park in Alberta are contiguous and have similar scenery. The mountains are typical of the Canadian Rockies, with essentially horizontal sediments uplifted and massively carved by glacial action. The area is a paradise for hikers, climbers, horseback riders, and wildlife enthusiasts, and contains one of the most spectacular automobile roads on the continent, the Going-to-the-Sun Highway that traverses Glacier Park from east to west.

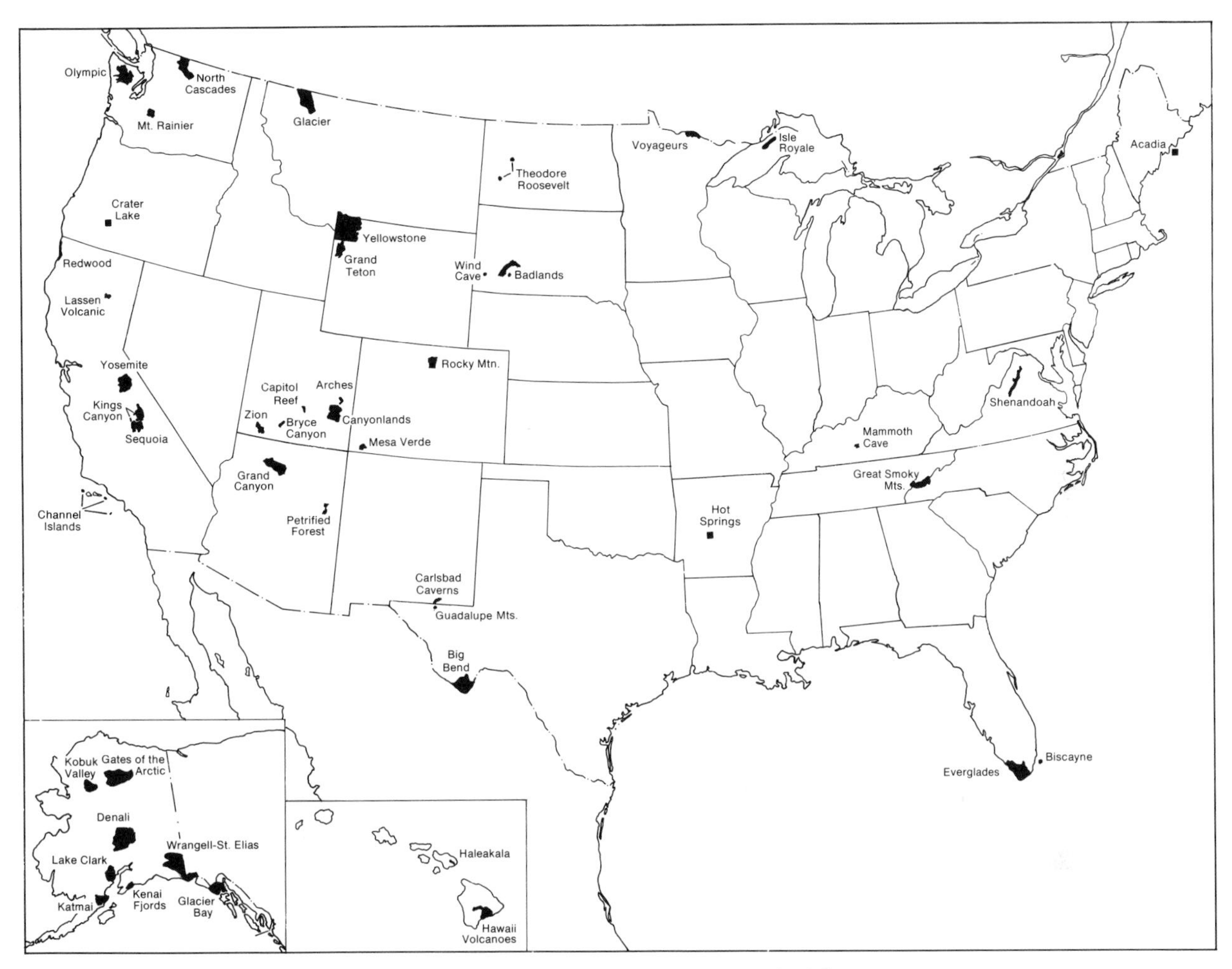

FIGURE 14-19 National parks of the United States.

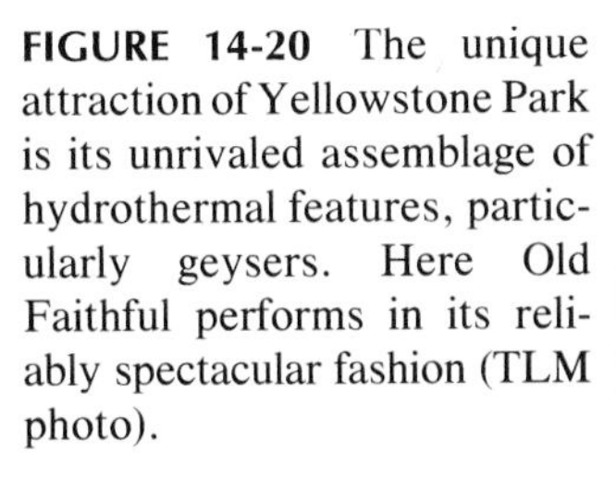

FIGURE 14-20 The unique attraction of Yellowstone Park is its unrivaled assemblage of hydrothermal features, particularly geysers. Here Old Faithful performs in its reliably spectacular fashion (TLM photo).

FIGURE 14-21 National parks of Canada.

The Canadian Rockies The most magnificent mountains in the region are found west of Calgary and Edmonton on the Alberta–British Columbia boundary. The national park system of Canada was started in 1885, when a small area in the vicinity of the mineral hot springs at Banff was reserved as public property (Fig. 14-21). The famous resorts of Banff, Lake Louise, and Jasper were developed by the two transcontinental railways, and the Canadian Pacific and Canadian National are still major operators in the area. Today there are four national parks and six provincial parks with a contiguous area totaling nearly 11,000 square miles (28,500 km^2), probably the largest expanse of frequently visited nature recreational areas in the world (Fig. 14-22). Heavily glaciated mountains, abundant and varied wildlife, spectacular waterfalls, colorful lakes, deep canyons, the largest ice field in the Rockies, and luxurious resort hotels characterize the area (Fig. 14-23).

Winter Sports

The region possesses superb natural attributes for skiing, and ever-increasing numbers of skiers are being attracted to the Rocky Mountains during the period of heavy snows. High elevation ensures a long period of snow cover; many areas can provide skiing from Thanksgiving to mid-May. Because win-

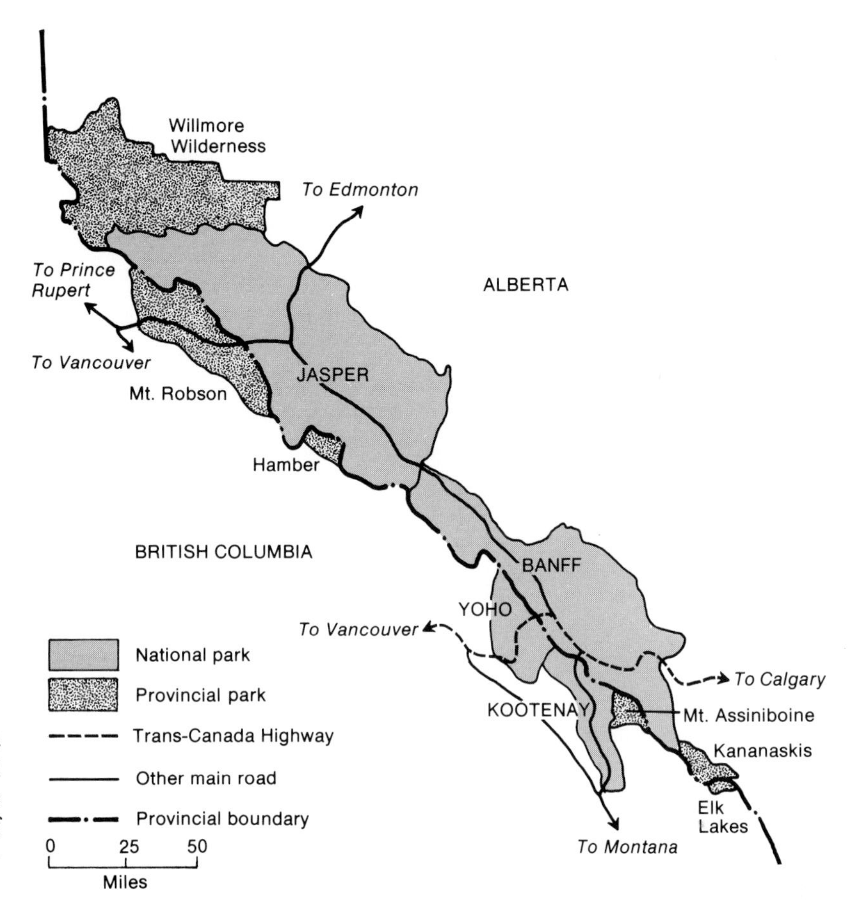

FIGURE 14-22 There is a splendid collection of national and provincial parks in close juxtaposition along the trend of the Canadian Rockies in Alberta and British Columbia.

ter storms have lost much of their moisture by the time they reach the Rockies, the snow is often of the fine, powdery variety preferred by skiers. And there is an abundance of different degrees of slopeland to accommodate all classes of skiers.

In the past, the region was relatively remote from large population centers and few skiers came to the Rockies. To some extent this is still true, but many regional ski areas, especially in the Southern Rockies, have experienced a rapid increase in the

FIGURE 14-23 Spectacular mountain scenery abounds in the Canadian Rockies. This is Mt. Sarbach above the North Saskatchewan River in Banff National Park (TLM photo).

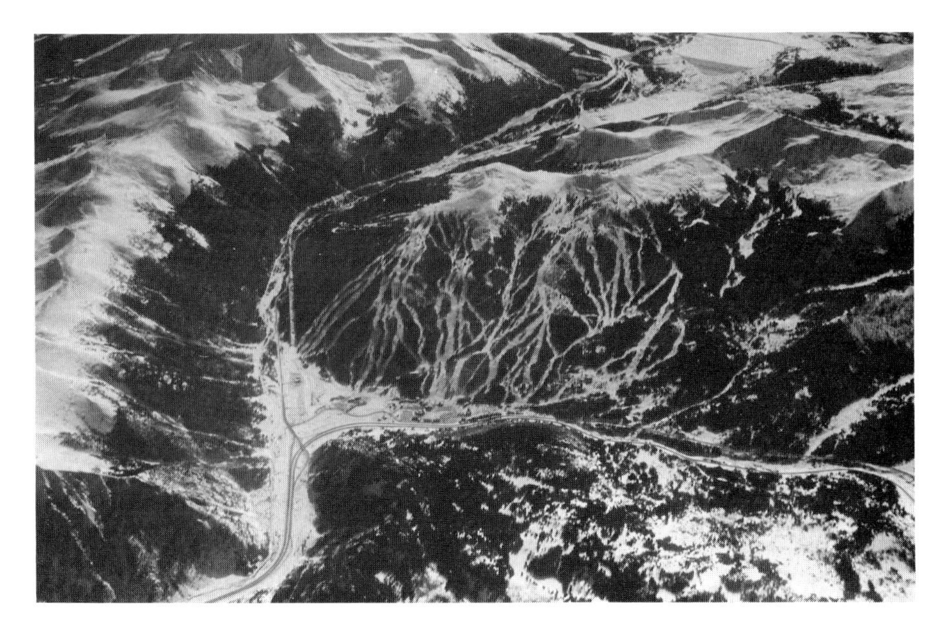

FIGURE 14-24 There are many winter sports areas in the Rocky Mountains Region, with the greatest concentration in north-central Colorado. One of the many in that district is Copper Mountain (TLM photo).

number of users. Nearby populations have grown and skiers travel much greater distances to ski than they did in the past. About 20 percent of the citizens of Colorado are skiers and some two-thirds of the users of ski areas in the state are local people (Fig. 14-24). In the other large winter sports areas of the region about half the users are nonresidents.

Fishing and Hunting

No other generally accessible region in North America provides such a variety of faunal resources to tempt the hunter and fisherman. The fishing season normally lasts from May to September, with various species of trout as the principal quarry. Only diligent artificial stocking can maintain the resource in the more accessible lakes and streams, where overfishing is rampant. The hunting season lasts from August to December in various parts of the region. The list of legal game is extensive, running from cottontail rabbit to grizzly bear and from pronghorn antelope to mountain goat. In an average year more than a third of a million big-game animals and two dozen hunters are shot in this region.

Problems

The flocking of visitors to these high-country scenic areas is a mixed blessing. By its very nature, a pleasurable outdoor experience can be ruined by overcrowding of people and overdevelopment of facilities to cater to the crowds. The national parks and other prime scenic attractions of the Rockies have become centers of controversy between the advocates of wilderness preservation on one hand and developers on the other. Important, precedent-setting decisions are now being made. Should a complete summer–winter resort town site be constructed at Lake Louise? Should the roadway system of Yellowstone Park be converted to one-way traffic? Do we want national parks or national parking lots?

TRANSPORTATION

The Rocky Mountains have always functioned as a conspicuous barrier to east-west travel. The principal early trails and later routeways across the Rockies either passed around the southern end in New Mexico or crossed through the Wyoming Basin between the Southern and Middle Rockies. The first "transcontinental" railroad, the Union Pacific, used the Wyoming Basin route. It continues as the busiest east-west rail line, with as many as 75 freight trains per day in operation. The paralleling highway, Interstate 80, vies with Interstate 40 (which passes through New Mexico around the southern end of the Rockies) as the most heavily used east-west roadway. Only the Denver and Rio Grande Railway built a line across the high-altitude Southern Rockies and it did not become an all-weather route until the 6-mile (10-km)-long Moffat Tunnel was constructed west of Denver in 1927.

North of the Wyoming Basin there are six railway routes across the Rockies. Three cross the Northern Rockies in Montana and Idaho and three are Canadian lines. The Canadian Pacific built the pioneer route westward from Calgary through Banff; its descent on the western side of the Canadian Rockies is through the famous spiral tunnels down into the Rocky Mountain Trench. This route continues westerly through the precipitous Selkirk Mountains by a long tunnel and then continues toward Vancouver along river valleys and through canyons. A second Canadian Pacific line was constructed subsequently over the more southerly Crow's Nest Pass route. The government-owned Canadian National Railway crosses the Rockies by means of the northerly but low-level Yellowhead Pass in Jasper National Park.

Canada's most ambitious road-building program was the Trans-Canada Highway, which extends from the Atlantic Ocean to the Pacific. The most expensive and difficult section of the highway to construct was that crossing the Rockies, especially in the Selkirk Mountains where the Rogers Pass segment is blocked several times a year by avalanches despite the protection offered by lengthy concrete snowsheds.

SETTLEMENT NODES

This is virtually a cityless region. No urban agglomeration in the Rocky Mountains has as many as 80,000 people and no single "city" in the United States portion of the region contains as many as 50,000. The relatively modest existing population nodes are chiefly associated with major lumbering, pulping, and mining-smelting activities or with agricultural valleys. The greatest concentration of population in the entire region is in the Okanagan Valley, where a dense farming population is clustered around the three urban centers of Kelowna, Penticton, and Vernon in an area that is also popular with summer vacationers (Table 14-1).

TABLE 14-1
Largest urban places of the Rocky Mountain Region

Name	*Population of Principal City*	*Population of Metropolitan Area*
Bozeman, Mont.	23,530	
Butte, Mont.	32,490	
Coeur d'Alene, Idaho	24,040	
Durango, Colo.	13,050	
Kalispell, Mont.	11,960	
Kamloops, B. C.	61,800	
Kelowna, B. C.	61,900	
Missoula, Mont.	35,640	
Prince George, B. C.	67,700	
Rock Springs, Wyo.	20,870	
Santa Fe, N. Mex.	59,300	112,500

During the busy tourist season the population of some resort towns swells to many times the normal size. Estes Park in Colorado, Jackson in Wyoming, and Banff in Alberta are prime examples.

THE OUTLOOK

Permanent settlements in the region are based mostly on mining, forestry, limited agriculture, and tourism. Farming and ranching are developed almost to capacity and cannot be expected to change to any great extent. Logging activities have also probably reached their limit except in British Columbia, which has considerable capability for expansion, provided that demand is sufficient. The British Columbia lumber industry depends considerably on the home-building market in the United States whereas its pulp and paper industry is particularly affected by varying competition from Scandinavia.

Mining will undoubtedly fluctuate in different areas, in a boom-and-bust syndrome that has great historical precedent in the region. The abrupt hydrocarbon expansion of the 1970s was followed by an equally sudden collapse in the 1980s.

Contrasting prosperity in mining highlights the instability of the regional economy, and is counterpointed by gyrating population fluctuations. Regional population growth was three times the national average (in the United States) in the 1970s, and well below the national average in the 1980s. Wyoming, for example, was the fastest-growing state (proportionally) in the 1970s (expanding by 40 percent), owing to in-migration, and virtually stagnated in the 1980s, owing to out-migration.

The dynamic future of the Rocky Mountains appears to be intimately associated with that seasonal vagabond, the tourist. Natural attractions are almost limitless; recreational developments on government lands have been accelerated by the National Park Service, the Forest Service, and other federal, state, and provincial agencies; and improvements in transportation facilities and accommodations are being made haphazardly but continually. Despite the high cost of gasoline, tourism continues to be a growth industry.

An offshoot of tourism is the growth of "second home" developments, which are occurring with increasing frequency in recreational or scenic areas throughout the continent and can be expected to increase in the future. There are many foci of such developments in the Rockies; notable examples include the area around Taos (New Mexico), many prominent ski resorts (especially Vail, Aspen, and Steamboat Springs), Estes Park (Colorado), Jackson (Wyoming), the Flathead Lake area and the Bitterroot Valley of Montana, the Banff area of Alberta, and Golden (British Columbia).

A major philosophical controversy with significant social, political, and economic overtones has arisen in the region: how is it possible to reconcile the pressures of rapidly expanding development while maintaining an environment that visitor and resident alike can enjoy? The problem surfaces most conspicuously in connection with open-pit mining, mineral boom towns, large-scale recreational developments, and national parks in general. The Rocky Mountains constitute a region of remarkable aesthetic appeal, but the balance of effective use without destructive abuse seems increasingly difficult to attain.

SELECTED BIBLIOGRAPHY

ALWIN, JOHN A., *Western Montana: A Portrait of the Land and Its People*. Helena: Montana Magazine, 1983.

ARNO, STEPHEN F., "Glaciers in the American West," *Natural History*, 78 (1969), 84–89.

CHENG, JACQUELINE R., "Tourism: How Much Is Too Much? Lessons for Canmore from Banff," *The Canadian Geographer*, 24 (Spring 1980), 72–80.

CROWLEY, JOHN M., "Ranching in the Mountain Parks of Colorado," *Geographical Review*, 65 (1975), 445–460.

———, "The Rocky Mountain Region: Problems of Delimitation and Nomenclature," *Yearbook*, Association of Pacific Coast Geographers, 50 (1988), 59–68.

FARLEY, A. L., *Atlas of British Columbia: People, Environment, and Resource Use*. Vancouver: University of British Columbia Press, 1979.

GARDNER, JAMES S., "What Glaciers do for Western Canada," *Canadian Geographical Journal*, 96 (February–March 1978), 28–33.

GEORGE, RUSSELL, "More Energy for British Columbia in Peace River Coal," *Canadian Geographic*, 100 (February–March 1980), 26–33.

GRIFFITHS, MEL, and LYNNELL RUBRIGHT, *Colorado*. Boulder, CO: Westview Press, 1983.

HARRINGTON, LYN, "The Columbia Icefield," *Canadian Geographical Journal*, 80 (June 1970), 202–205.

IVES, JACK D., ed., *Geoecology of the Colorado Front Range*. Boulder, CO: Westview Press, 1980.

———, et al., "Natural Hazards in Mountain Colorado," *Annals*, Association of American Geographers, 66 (1976), 129–144.

KRUEGER, RALPH R., and MAGUIRE, N. G., "Protecting Specialty Cropland from Urban Development: The Case of the Okanagan Valley, B.C.," *Geoforum*, 16 (1985), 287–300.

NELSON, J. G., *Man's Impact on the Western Canadian Landscape*. Toronto: McClelland and Stewart, 1976.

Province of British Columbia Lands Service, *The Atlin Bulletin Area*. Victoria: Queen's Printer, 1967

———, *The Fort Fraser-Fort George Bulletin Area*. Victoria: Queen's Printer, 1969

———, *The Kamloops Bulletin Area*. Victoria: Queen's Printer, 1970.

———, *The Okanagan Bulletin Area*. Victoria: Queen's Printer, 1968.

———, *The Quesnel-Lillooet Bulletin Area*. Victoria: Queen's Printer, 1968.

ROBINSON, J. LEWIS, "The Rocky Mountain Trench in Eastern British Columbia," *Canadian Geographical Journal,* 77 (1968), 132–141.

ROE, JOANN, "Our Beautiful Okanagan," *Canadian Geographic,* 102 (December 1982–January 1983), 25–26.

SHEWCHUK, MURPHY, "Will B.C. Develop its Unused Coal Wealth?" *Canadian Geographic,* 97 (December 1978–January 1979), 32–37.

SLAYMAKER, H. O., and H. J. MCPHERSON, eds., *Mountain Geomorphology: Geomorphological Processes in the Canadian Cordillera,* British Columbia Geographical Series 14. Vancouver: Tantalus Research Ltd., 1972.

TODHUNTER, RODGER, "Banff and the Canadian National Park Idea," *Landscape,* 25 (1981), 33–39.

TRENHAILE, A. S., "Cirque Elevation in the Canadian Cordillera," *Annals,* Association of American Geographers, 65 (1975), 517–529.

———, "Cirque Morphometry in the Canadian Cordillera," *Annals,* Association of American Geographers, 66 (1976), 451–462.

WARDEN, GEOFF, "Black Pits and Vanishing Hills: Coal Development in British Columbia," *B.C. Outdoors,* 32 (August 1976), 34–38.

15
THE INTERMONTANE WEST

The western interior of the United States makes up the Intermontane West Region. As the term *intermontane* implies, it primarily encompasses the vast expanses of arid and semiarid country between the Rocky Mountain cordillera on the east and the major Pacific ranges (Sierra Nevada and Cascade) on the west (Fig. 15-1).

The boundary of the region is relatively clear-cut in most sections, for there are prominent geomorphic units that are usually associated with obvious land-use changes. In only two areas is the regional boundary indistinct.

1. In northwestern Colorado there is an east-west transition from Rocky Mountains to Intermontane Region that is broad and indistinct.
2. In southern California there is a very clear environmental boundary between the desert portion of the Intermontane Region and the various Transverse and Peninsular mountain ranges of the California Region. But the spillover of urban population from the Los Angeles basin into the Palm Springs area of the Colorado Desert and the Antelope Valley section of the Mojave Desert is so pronounced and the resultant functional connection of these two areas with the Los Angeles metropolis is so strong that it seems clear to this observer that these two areas should be considered outliers of the southern California conurbation and thus part of the California Region. The boundary of the Intermontane Region is therefore drawn east of these two areas.

The vast extent of the Intermontane Region and its topographic diversity have led some regionalists to consider it as several regions rather than

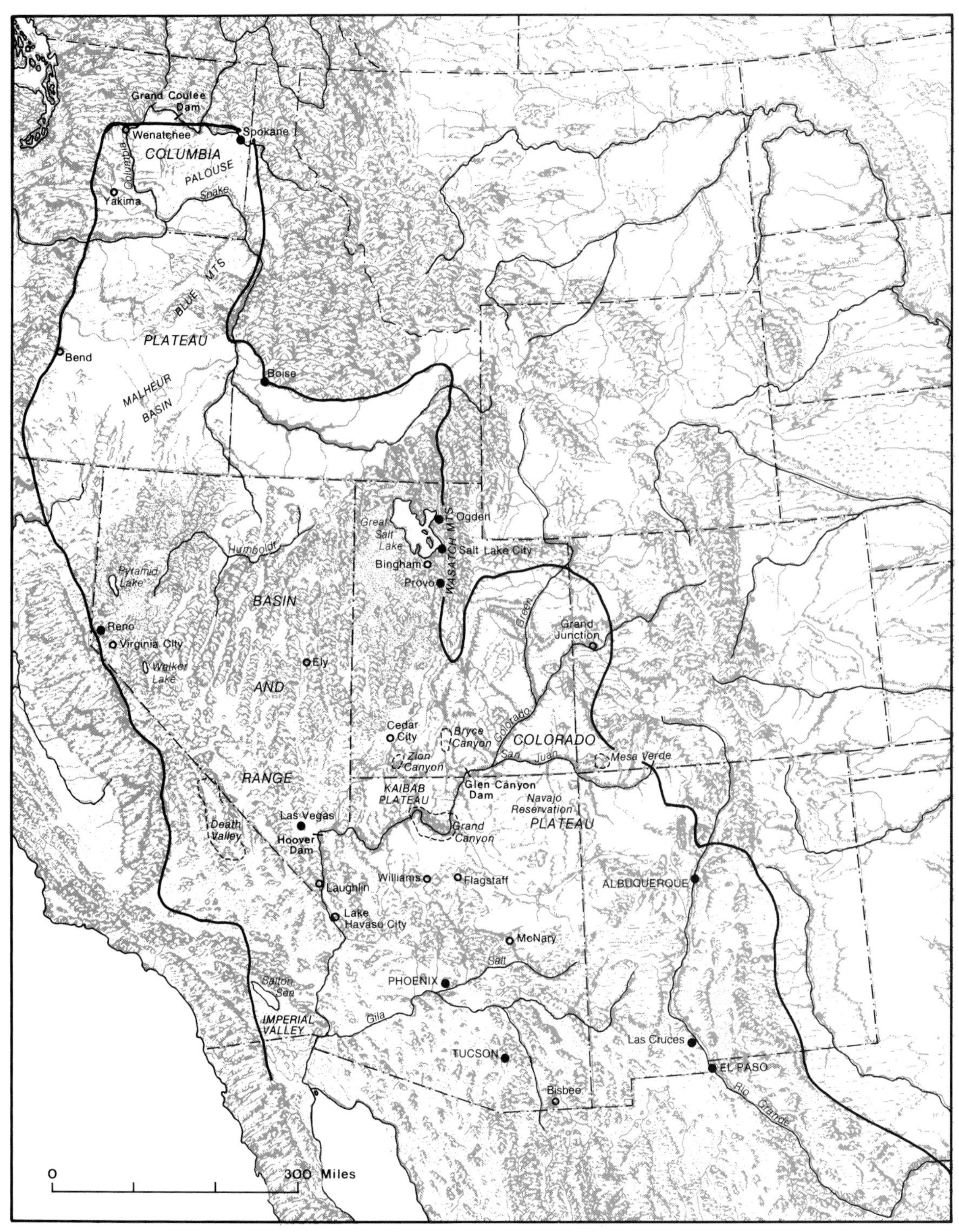

FIGURE 15-1 The Intermontane West (base map copyright A. K. Lobeck; reprinted by permission of Hammond, Inc.).

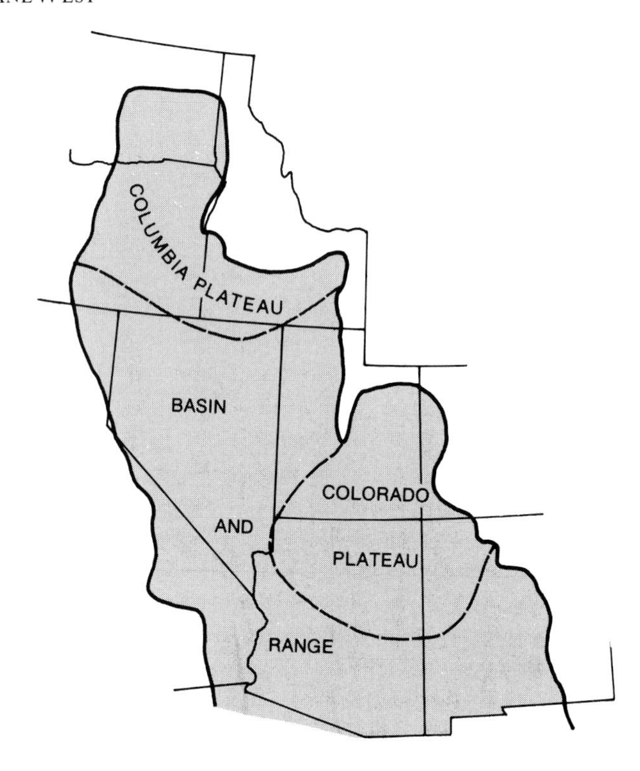

FIGURE 15-2 Major topographic subdivisions of the Intermontane Region.

one. For this book, however, it is felt that a single regional designation is warranted. The three prominent subregions identified in Figure 15-2 include the Columbia Plateau in the north; the Colorado Plateau, occupying parts of four states, in the eastern portion of the region; and the Basin-and-Range section, the largest subregion, extending in a crescent from southern Oregon to western Texas.

ASSESSMENT OF THE REGION

In broad generalization, the Intermontane West can be thought of as a sparsely populated region whose vast extent, relatively isolated inland location, rugged but varied terrain, and paucity of freshwater make it best suited to serve the nation as a limited source of primary resources and as a recreational ground.

Where people have attempted economic endeavor in this region, the bulldozer has been their instrument, the cloud of dust their symbol, and a drastically altered and "tamed" landscape the result. Where humans are found in any numbers, there are the ordered fields of an irrigated farming development, the giant amphitheater of an open-pit copper mine, the massive wall of a major dam, the extensive runways of a military airfield, or the bright lights of a gambling town.

The region's economy is based partly on primary production, especially from irrigated agriculture, pastoralism, and mining, and partly on such tertiary activities as tourism and government expenditures. Where water is available for irrigation and land is sufficiently level, intensive farming prevails as an oasis type of development. Sheep and beef cattle are grazed widely throughout the region, usually on extensive ranches, and the region, in general, is well endowed with economic mineral resources. The greater part of the feeding and slaughtering of livestock, refining of ores, and marketing of both occurs outside the region. It is significant that most of the land in the region is still in the public domain and government expenditures for management, development, construction, and exploitation are a major contribution to the local economy.

Inaccessibility was long a hallmark of the region. The terrain is characterized by deep gorges, abrupt cliffs, and steep mountainsides. The climate is noted for temperature extremes. And almost everywhere there is a scarcity of water. Civilization was slow to reach many areas. The last Indian fights occurred just over a century ago. Some post offices were still being served by packhorses as recently as 60 years ago. The last sizable portion of the country to be provided with all-weather roads was northeastern Arizona–southeastern Utah in the early 1960s.

The historical movement of population in the region has been from both east and west, generally *across* rather than *into*. The region served as a barrier to westward expansion and only incidentally as a goal for settlement except in the Great Salt Lake basin. In recent years, however, this pattern has been modified. The interior West still serves in part as a transit region, but more and more it is the chosen destination of automobile nomads. Tourists, hunters, skiers, and other recreationists visit the region for a few days or weeks of vacation, and increasingly people are settling on a long-term or permanent basis, particularly in the sunny southern section.

The explosive rate of population growth, combined with rapid economic "development" of various kinds, ushered in an era of unprecedented conflict in the region. The equation is simple: Attractive landscape plus fragile environment plus large-scale development schemes equal unending controversy.

If there is an eternal verity for the Intermontane Region, it is scarcity of water. The fact that evaporation exceeds precipitation throughout the region is critical to all forms of life. John Wesley Powell, the most notable explorer of the inland West, probably said it best: "All the great values of this territory have ultimately to be measured to you in acre-feet."

TOPOGRAPHIC VARIETY

There is a great deal of topographic variety in the Intermontane Region. Each of the three subregions has its distinctive geomorphic personality, which can be easily recognized and described.

The Columbia Plateau

The Columbia Plateau lies between the Cascade Mountains on the west and the Rocky Mountains on the east and north and grades almost imperceptibly into the Basin-and-Range section to the south. Although called a plateau, which popularly suggests a rather uniform surface, the area has quite varied relief features of mountains, plateaus, tilted fault blocks, hills, plains, and ridges. In general, this intermontane area is covered with basalt lava flows that originally were poured out over a nearly horizontal landscape and interbedded with a considerable quantity of silts that were deposited in extensive lakes. After the outpouring of the sheets of lava and the deposition of the lake beds, the surface of much of the region was strongly warped and faulted so that the present surface of the lava varies from a few hundred feet above sea level to nearly 10,000 feet (3,000 m) in elevation (Fig. 15-3).

FIGURE 15-3 A lava landscape in southern Idaho. This is a portion of the lengthy gorge of the Snake River (TLM photo).

In central Washington, steep-sided, flat-floored, streamless canyons, eroded by glacial melt-water floods following Pleistocene glaciation, cut the plateau into a maze known as the *channeled scablands*. In eastern Washington is the rolling Palouse hill country, deeply mantled with loess.

Northern Oregon has an irregular pattern of faulted and folded mountains, generally referred to as the Blue and Wallowa mountains. Southeastern Oregon and southern Idaho have variable terrain, ranging from the lava-covered flatness of the Snake River Plain in southeastern Idaho to the irregular basins and hills of the Malheur Basin in south-central Oregon to the spectacularly deep canyons of the lower Snake drainage.

The Colorado Plateau

This large area consists of several strongly differentiated parts but has sufficient unity to justify separation from adjacent subregions. It stretches outward from the Colorado River and its tributaries in Colorado, Arizona, New Mexico, and Utah. The greater part consists of a series of flattish summit areas slightly warped or undulating as a result of earlier crustal movements and interrupted by erosion scarps in the eastern portions and fault scarps in the western parts. Physiographically the area is distinguished by the following features.

1. All the subregion except the bottoms of canyons and the highest peaks has an elevation of 4000 to 8000 feet (1200 to 2400 m). Some high plateau surfaces reach 11,000 feet (3300 m), and a few mountain ranges have still higher peaks.
2. Hundreds of remarkable canyons (Fig. 15-4) thread southeastern Utah, northern Arizona, and the Four Corners country in general. They make this subregion the most dissected and difficult to traverse part of the country.
3. Numerous arroyos, which cut some parts of the subregion into mazes of steep-sided chasms, are dry during most of the year but filled from wall to wall during the rare rains.
4. Mesas, flat-topped islands of resistant rock, rise abruptly from the surrounding land.

The basic topographic pattern might be described as mesa and scarp—that is, flat summits bordered by near-vertical cliffs. Some summit areas, such as the Kaibab Plateau in northern Arizona and Mesa Verde in southwestern Colorado, are remarkably extensive. The scarps, too, sometimes extend

FIGURE 15-4 Canyon de Chelly (near Chinle in northeastern Arizona) is a classic example of mesa-and-scarp terrain (TLM photo).

FIGURE 15-5 The basin-and-range subregion consists mostly of alternating mountains and valleys in parallel arrangement. This is the Santa Rosa Range in northern Nevada (TLM photo).

to great lengths; the Book Cliffs of Colorado–Utah, for example, are more than 100 miles (160 km) long.

Throughout the Colorado Plateau subregion the land is brilliantly colored, particularly in the exposed sedimentary surfaces of the scarp cliffs. The Painted Desert of northern Arizona, which is badlands terrain, is especially noted for its rainbow hues, but throughout the mesa-and-scarp country the landscape is marked by colorful rocks and sand.

The Basin-and-Range Section

To the northwest, west, southwest, south, and southeast of the Colorado Plateau, from southern Oregon to western Texas, is a vast expanse of desert and semidesert country that has notable physiographic similarity. Throughout this extensive area the terrain is dominated by isolated mountain ranges that descend abruptly into gentle alluvial piedmont slopes and flat-floored basins (Fig. 15-5).

The mountain ranges are characteristically rough, broken, rocky, steep sided, and deep canyoned. They tend to be narrow in comparison with their length, distinctly separated from one another, and often arranged in parallel patterns. Although their origins are somewhat diverse, most consist of tilted and block-faulted masses of previously folded and peneplained rocks. The canyons and gullies that drain them are waterless most of the time, harboring intermittent streams only after a rain.

Near the base of the mountains there is normally an abrupt flattening out of the slopes. As the streams reach the foot of the mountains, their gradient is sharply decreased so that they can no longer carry the heavy load of silt, sand, pebbles, and boulders that they have brought down from the highlands, and considerable deposition takes place. (Although the streams flow only intermittently, they are subject to violent floods, and the amount of erosion that they can accomplish is tremendous.) This piedmont deposition generally occurs in fan- or cone-shaped patterns (called alluvial fans) that become increasingly complex and overlapping (piedmont alluvial plains) as the cycle of erosion progresses.

The fans become increasingly flatter at lower elevations and eventually merge with the silt-choked basin floors. The basins themselves frequently are without exterior drainage. Shallow lakes, mostly intermittent, may fill the lowest portion of the basins. They are saline because they have no outlet and because the streams that feed them, like all streams, carry minute amounts of various salts. As the lake waters evaporate, the salts become more concentrated; the complete disappearance of the water leaves an alkali flat or salt pan.

There are several large and relatively permanent salt lakes in the subregion, particularly lakes Walker and Pyramid in Nevada and Great Salt Lake in Utah (Fig. 15-6). The last is a shrunken remnant of prehistoric Lake Bonneville, a great body of fresh water that was as large as present Lake Huron. Although Lake Bonneville and other Pleistocene lakes in the region began to shrink and disappear thousands of years ago, old beach lines still remain strikingly clear on the sides of the surrounding mountains. The highest Bonneville shoreline lies about 1000 feet (300 m) above Great Salt Lake. The present lake expands and contracts according to the

FIGURE 15-6 Look, readers; no hands! The renowned bouyancy of Great Salt Lake is its principal attraction for swimmers. Less attractive are the salt flies that abound, and the salt itches that result. In the background the slopes of the Oquirrh Mountains are partly blotted out by the fumes from the nonferrous metal smelter at Garfield (TLM photo).

variation in precipitation in the mountains, its water source, and according to the rate at which irrigation water is drawn off. Because the lake is shallow (average depth, 14 feet [4 m]), its area fluctuates remarkably; the known areal extremes are 2400 square miles (6200 km^2) in 1873 and 1000 square miles (2600 km^2) in 1963. In the 1980s the lake was in an expanding phase, creating havoc for adjacent settlements and transportation routes.

There are only a few permanent streams in the region and generally they can be classified as "exotic" because the bulk of their water supply comes from adjacent regions. Most conspicuous are the Colorado River and the Rio Grande. The former and its left-bank tributary, the Gila, provide a significant amount of water for irrigation and domestic use. The Salton Basin in southeastern California was partially flooded in 1906 when attempted irrigation permitted the Colorado River to get out of control. The river was reestablished in its original channel the following year, but the Salton Sea still exists as a permanent reminder of the incident.

AN ARID, XEROPHYTIC ENVIRONMENT

The greater part of the region is climatically a desert or semidesert and the vegetation shows a variety of xerophytic (drought-resisting) characteristics.

Climate

Moisture is the most critical element of the climate. On the basis of precipitation-evaporation ratios, there are four moisture realms: (1) the subhumid, (2) the semiarid, (3) the moderately arid, and (4) the extremely arid.

The *subhumid* portion of the region occurs only in limited highland areas, primarily Washington and Oregon. More precipitation on the upland slopes and less evaporation because of lower summer temperatures result in a climate that shows little evidence of precipitation deficiency. Winters are long and cold; summers are short and cool. Precipitation is concentrated in summer or is evenly distributed.

The *semiarid* climate is typical of most of the Columbia Plateau. Precipitation ranges from 10 to 20 inches (250 to 500 mm) per year and falls chiefly in late autumn, winter, and spring.

The *moderately arid* climate, characteristic of most of the Great Basin, has periodic rainfalls that are fairly regular, although limited, and during which vegetation bursts into life and the water table is replenished. The precipitation at Elko, Nevada, a typical station, is 9 inches (230 mm). The frostless season varies from 100 to 180 days.

In the *extremely arid* climate the rainfall is episodic, coming largely in summer at irregular intervals and usually as cloudbursts. The Mojave–Gila Desert exemplifies this type. Its annual precipitation is less than 5 inches (125 mm), too little even for grazing. Almost the entire annual rainfall may come in a single downpour lasting but a few moments. So much water falls so quickly that little can penetrate the soil deeply.

The diurnal range of temperature throughout the region is high. The days are generally hot to very hot in summer, but radiational cooling in the dry atmosphere decreases the temperature rapidly at night except at low elevations in the southern part of the region, which has the highest summer nighttime temperatures to be found in Anglo-America. In winter the nights are usually quite chilly, following daytime temperatures that may be relatively mild or even warm.

Natural Vegetation

In such a large area and in one varying so greatly in landforms marked differences in the natural vegetation occur. On the whole, however, low-growing shrubs and grasses predominate.

Forests Ponderosa pine and Douglas fir forests are mostly confined to higher elevations where rainfall is relatively heavy (Fig. 15-7). Where precipitation is somewhat less, forest is replaced by woodland, a more open growth of lower trees, particularly piñon and juniper.

Grasslands More extensive than might be supposed, grasslands characterize the uplands of southeastern Arizona, New Mexico, and the Columbia Basin. Mesquite grass grows where temperatures are high, evaporation excessive, and annual precipitation low. Short grass characterizes large areas in the high plateaus of New Mexico and Arizona, as does bunchgrass in the Columbia Plateau. The noxious cheatgrass is almost everywhere.

Desert Shrub Xerophytic plants dominate the deserts. *Sagebrush,* the principal element in the vegetation complex of the northern part of the region, grows in pure stands where soils are relatively free from alkaline salts. It is especially abundant on the bench lands that skirt mountains and on the alluvial fans at the mouths of canyons. *Shadscale,* a low, gray spiny plant with a shallow root system, grows on the most alkaline soils but never in dense stands. Much bare ground lies between the plants. It is especially prominent in Utah and Nevada. *Greasewood,* bright green in color and occupying the same general region as sagebrush and shadscale, grows from 1 to 5 feet (0.3 to 1.5 mm) in height and is tolerant of alkali. *Creosote bush,* dominating the southern Great Basin as sagebrush does the northern, draws moisture from deep down under the surface. Creosote bush is a large plant, attaining a height of 10 to 15 feet (3 to 4.5 m).

Within the last century, a number of woody plant species greatly expanded their range in the arid and semiarid Southwest, mostly at the expense of grassland communities. As in the southern Great Plains, *mesquite* has occupied the greatest area of new territory, particularly in the Rio Grande and Tularosa valleys of New Mexico and the Colorado, Gila, Santa Cruz, and San Pedro valleys of Arizo-

FIGURE 15-7 On the higher mountains in the Intermontaine Region are found forests and woodlands. This ponderosa pine scene is in the Sheep Mountains north of Las Vegas, Nevada (TLM photo).

na.[1] There has also been considerable expansion of the acreage of native *juniper* and introduced *tamarisk*, the latter having extensively colonized islands and sandbars along most southwestern rivers, especially the Colorado and its tributaries.

Various types of *cacti* are widespread in the more arid portions of the regions, especially Arizona. The giant saguaro, symbol of the desert, is most prominent, although many smaller species of cactus are more numerous (Fig. 15-8).

Exotics The sparser plant associations (grasses and shrubs) of the region are highly susceptible to invasion and replacement by alien annual species, usually grasses, that have been introduced, chiefly from Asia. Severe livestock-grazing pressure in much of the region produced a "biological near-vacuum" that in many areas has become dominated by such grasses as downy brome (*Bromus tectorum*) and such weeds as Russian thistle and tumble mustard.[2] Restoring these degraded range lands has proved difficult, even if the grazing livestock were to be removed.

[1] David R. Harris, "Recent Plant Invasions in the Arid and Semi-Arid Southwest of the United States," *Annals*, Association of American Geographers, 56 (1966), 409.
[2] J. A. Young, R. A. Evans, and J. Major, "Alien Plants in the Great Basin," *Journal of Range Management*, 25 (1972), 195.

FIGURE 15-8 Even the most stressful environments often contain distinctive and conspicuous plants. The largest of the cacti—organ pipe on the left and saguaro on the right—prosper in the parched Sonoran Desert of southern Arizona (TLM photo).

FIGURE 15-9 Larger animals are generally scarce in the desert, but no part of the region is devoid of wildlife. This desert bighorn ram is in the Kofa Mountains of southwestern Arizona (TLM photo).

Fauna

In spite of considerable barrenness and scarcity of water, the Intermontane Region has a surprisingly varied fauna. It was never an important habitat for bison, but the plains-dwelling American antelope, or pronghorn, is still found in considerable numbers in every state. In the mountains and rough hills other ungulates are notable, including deer, elk, desert mountain sheep, feral burros (particularly in California and Arizona), feral horses (especially in Nevada, Utah, and Oregon), and javelinas (in Arizona). (Fig. 15-9).

Furbearers are common only in forested portions of the northern half of the region. Most large predators (cougar, coyote, bobcat, and fox) are scarce as a a result of a systematic poisoning and trapping. The relatively few rivers and lakes provide important nesting and resting areas for migratory waterfowl.

SETTLEMENT OF THE REGION

The pre-European inhabitants of the Intermontane Region were extraordinarily varied. Most Indian tribes in the northern part of the region eked out a precarious existence as seminomadic hunters. Yet in the arid Southwest some of the highest stages of Indian civiliation to be found in Anglo-America developed, mostly in the form of sedentary villages based on self-contained irrigated farming. These settled tribes, the Pueblos, Hopis, Zuñis, and Acomas, were islands of stability in an extensive sea of nomadic hunting and raiding tribes, most notably Apaches and Utes.

Arrival of the Spanish

The Spanish, the first European arrivals, were brought to the area by tales of great wealth. They pushed up the Rio Grande almost four centuries ago and established several settlements throughout what is now New Mexico. The Spanish explored widely and ruled most of the Southwest for more than 200 years. The major early Spanish settlements were in the Socorro–Albuquerque–Santa Fe–Taos area in the Upper Rio Grande Valley, with another important concentration in the El Paso oasis. Many decades later and at a much lower level of intensity they occupied that part of southern Arizona called *Pimería Alta,* mostly in the Santa Cruz Valley as far north as Tucson.

The Spaniards left an indelible influence on both the history of the Southwest and on American civilization. Their livestock formed the basis of the later American cattle and sheep industry and their horses gave mobility to the Indians, the importance of which can hardly be overestimated. Small Spanish settlements and trading posts, such as Albuquerque, housed most of the Caucasian population of the Southwest until the middle of the nineteenth century.

Explorers and Trappers

British and American explorers began to filter into the region in the early nineteenth century. Lewis and Clark entered the Pacific Northwest in 1804–1805. The Astorians were active in 1811–1813. Smith penetrated the Great Basin in 1826; and Wyeth, the Pacific Northwest in 1832–1833. Bonneville, in 1832 and 1836, traded furs and casually explored the area drained by the Bear River. Fremont, in 1845–1846, entered the Salt Lake Basin by way of the Bear River, becoming the first white man to examine it systematically. They are but a few of the many who explored the region.

Trapping, a powerful incentive to exploration, was the main object of many of the men who explored the West in the early nineteenth century. The trappers were a special breed—self-reliant, solitary, largely freebooters—who strove to outwit their rivals, to supplant them in the goodwill of the Indians, and to mislead them in regard to routes. They lasted until fashion suddenly switched from beaver to silk for men's hats. The trappers were then through; nevertheless, they played a major role in the region's history.

The Farmer Invasion

The outstanding example of farmer invasion was the Mormon migration to the Salt Lake Basin in 1847. The Mormons had trekked from New York into Ohio and Illinois, and then Missouri to escape persecution and find a sanctuary where they might maintain their religious integrity. To do so, they felt impelled to establish themselves on the border of the real American Desert. The agricultural fame of the Deseret colony was soon known far and wide.[3] Utah is the only state in the Union that was systematically colonized. The leader, Brigham Young, sent scouts into every part of the surrounding area to seek lands suitable for farming. Throughout the latter half of the nineteenth century Mormon pioneers participated in a major colonizing effort that established settlements, usually based on irrigated agriculture, in valleys and oases throughout the Intermontane West.

The California Gold Rush

Following the explorers, trappers, and farmers came the gold seekers of 1849. So large was the movement that it led to the establishment of trading posts and stations where the migrants rested and refreshed themselves. The Salt Lake Oasis especially became a stop for the weary and exhausted.

[3] *Deseret* is a word from the Book of Mormon, meaning honeybee and symbolizing the hard work necessary for the success of their desert settlements. The Mormons organized the State of Deseret, but it was not accepted by Congress, which later formulated the Territory of Utah.

A CLOSER LOOK America's Wild Horses

There are some 50,000 wild or feral horses in the desert portions of eight western states (Fig. 15-a). Since 1971 the horses have been under the custody of the Department of Interior, Bureau of Land Management (BLM). In that year Congress declared that "free-roaming horses and burros are living symbols of the historic and pioneer spirit of the West, and they contribute to the diversity of life forms within the Nation and enrich the lives of the American people."

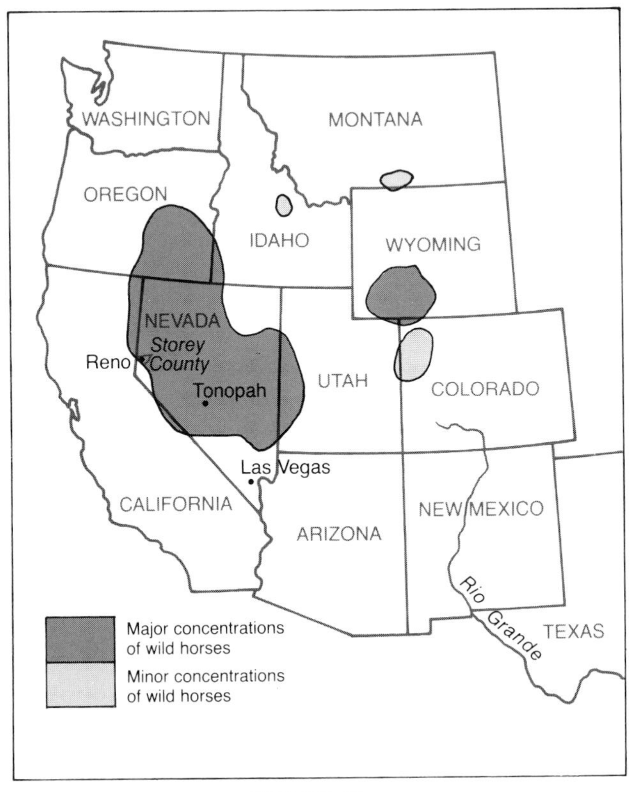

FIGURE 15-a

At the turn of the century it was alleged that there were more than a million feral horses in the American West. Then they lived pretty much at the mercy of ranchers and professional mustangers (those who caught wild horses for a living). Many ranchers liked to harvest the horses from time to time; they were an extra source of income. Some animals were deliberately crossed with Belgians, Percherons, or good ranch studs. A few ranchers saw genuine range value in the animals; they dispersed seed and kept water holes open in winter for cattle. Some ranchers, like people everywhere, took aesthetic pleasure in the horses. But when the cattle market was good, or when ranchers thought there were too many horses eating too much range forage and drinking too much water, livestock took precedence. Then it was time to shoot the horses or round them up. The horses were killed for government bounties, or sent to war fronts or rodeos, or broken for ranch use. Many were shipped to rendering plants for their hides and hair or to be used as chicken feed.

Strong demand for feral horses began in the early 1920s when chicken-feed processors in California began buying them by the railroad carload. Before long entrepreneurs came up with the idea that pet dogs and cats needed a more balanced diet—meaning more protein. By 1935 scores of companies in the United States were producing dog and cat food made from horsemeat. Within a decade the slaughter of horses—not all of them wild—had risen from 150,000 pounds a year to a high of 30 million. Europeans, especially Germans, French, and Belgians, were buying millions of pounds of horsemeat. They liked it cured, chopped, smoked, as steaks, in sausages. Americans were not then, and never have been, similarly inclined.

By the 1950s there were cries that something had to be done lest wild horses disappear completely from the American West. Soon a crusade was afoot to save the horses, led by a diminutive, polio-stricken woman by the name of Velma Johnston (to western ranchers and others she was known as "Wild Horse Annie"). Mrs. Johnston claimed that there were fewer than 20,000 wild horses left in the West. (The real number was double, perhaps triple, that figure.) Wild Horse Annie also complained that the BLM, the presumed enforcer of the 1934 Taylor Grazing Act (legislation designed to ensure more prudent use of public rangelands) was at best indifferent, at worst in cahoots with ranchers who saw feral horses as pests—voracious consumers of scarce water and grasses that "belonged" to cattle and sheep.

Largely through Wild Horse Annie's efforts, her home state of Nevada passed a law that made it illegal to chase feral horses with an airplane or helicopter on public lands. Thereafter she worked to get similar legislation passed for all public lands in the United States. By the late 1960s there was so much national support for the wild-horse cause that only the Vietnam War was generating more mail to members of Congress.

After the BLM took control of feral horses in 1971, it soon became evident that there were more horses on public lands than Wild Horse Annie and her supporters had guessed. It also became apparent that, left to the procreative dictates of natural selection and without interference by humans, the animals multiply at amazingly high rates. With the horses now a potential threat to ranching and the brittle desert, enlightened ranchers and conservationists began to highlight environmentally destructive horse habits. Sensible conservationists were also quick to point out that most western rangelands have far too many sheep and cattle on them.

Because ranchers could no longer profit from the horses or control their numbers as whim dictated, they began demanding that the BLM either remove all horses or reduce herd numbers to what they had been prior to enactment of the 1971 federal law. The BLM rounded up thousands (Fig. 15-b), and in the mid-1970s it set up a national adoption program to find homes for the animals. Virtually anyone with an acre or two and a pile of timber for a fence and a romantic hankering for a piece of the American West could adopt a feral horse for a minimal fee. By 1990 the BLM had found homes for more than 80,000 animals.

The Sierra Club, the largest and most important conservation group in the United States, has never been a willing participant in the feral-horse issue. Afraid to polarize its diverse membership, and unwilling to recognize the wild horse as true wildlife (horses were introduced to the desert Southwest by Spaniards in the sixteenth century), the Sierra Club has left it up to a few of its Nevada chapter members to formulate local policies on the horses. Sierra Club members in Nevada, the state with more than half of the nation's wild horses, have come out in favor of reducing horse herds, and simultaneously cattle herds, to numbers compatible with a diverse and healthy ecosystem.

Desertification of western rangelands is widespread, and the BLM, in many ways a typical government bureaucracy, has not been nearly as wise as it might have been in managing the horses. It has done a poor job of completing environmental impact statements on the condition and carrying capacity of western rangelands. Some of those it has completed have been challenged and shown to be inadequate, which has left the BLM largely defenseless against rancher claims that their ranges are not overstocked.

Wild horses continue to be a volatile rangeland issue. At one extreme are those who love horses, are opposed to population control, and, all evidence to the contrary, maintain that those which are feral will soon be extinct. At the other extreme are some ranchers and others who view the range in terms of livestock productive capacity and seek to shoot horses when the opportunity presents itself. These radical voices tend to overshadow those who would like to see the delicate desert ecosystem preserved at some cost to both horse and livestock numbers.

Dr. Richard Symanski
Champaign, Illinois

FIGURE 15-b Bureau of Land Management wild horse round-up, central Nevada (Bureau of Land Management photo).

Other precious-metal discoveries had a significant influence on early settlement. Outstanding were the silver lodes of western Nevada, dating from 1859.

The Graziers

Much of this region was marginally favorable for the grazier. For some years after the Spaniards came, cattle raising was almost the only range industry, although Navajo Indians and Mexican colonists herded some sheep. Northward in Utah and Idaho, as well as in the Oregon country, cattle raising held sway in nonfarming areas. In fact, the Columbia grasslands were major cattle-surplus areas for many years and shared the stocking of the Northern Great Plains ranges with Texas.

In Utah the self-sufficing Mormons raised sheep for homespun and as early as the 1850s nearly every farmer possessed a few head.

In the 1870s and early 1880s bands of Spanish and French Merino sheep were driven into the Southwest from California, furnishing a fine short-staple wool in sharp contrast to the coarse long wool of Navajo sheep. Transhumance (seasonal movement of livestock) was practiced: In Arizona the cool northern mountains were used from May until August; then the flocks were moved to the lower desert ranges. Late spring found them once more in the mountain pastures.

Spread of Settlement in the Southwest

The movement of people into the Southwest was erratic and variable and extended over a long period of time. Least noticed but fundamentally very important was the gradual influx of Hispanics.

> The gradual contiguous spread of Hispano colonists during the nineteenth century is a little-known event of major importance. Overshadowed in the public mind and regional history by Indian wars, cattle kingdoms, and mining rushes, this spontaneous unspectacular folk movement impressed an indelible cultural stamp upon the life and landscape of a broad portion of the Southwest. It began in a small way in Spanish times, gathered general momentum during the Mexican period, and continued for another generation, interrupted but never really stemmed until it ran head on into other settler movements seeking the same grass, water, and soil.[4]

In the latter part of the nineteenth century the influx of Anglos from Kansas, Colorado, California, and especially Texas was the major force in the region, quickly dominating both the economy and the political pattern. Mining camps and pastoral enterprises were particularly prominent, but the coming of the two major east-west railroad corridors—a northerly one through Albuquerque and Flagstaff, and a southerly route through El Paso and Tucson—signaled the beginning of a more diversified economy and the growth of urban nodes.

LAND OWNERSHIP IN THE INTERMONTANE REGION

A striking feature of the geography of the Intermontane West is the large amount of land that is in the public domain. In the 11 conterminous western states more than half the land is owned by the federal government.[5] And in the Intermontane Region the proportion is much higher. To illustrate, in Utah, Arizona, and Nevada, the three states that are almost wholly within the region, more than 75 percent of the land is owned by either federal or state government.

The basic reason that so little land is in private ownership is that nonirrigated agriculture is impractical over most of the region; thus there was little opportunity for dense rural settlement and little practical demand for freehold ownership. Although successful homesteading occurred in many localities, the homestead laws (which made land available either without cost or inexpensively to legitimate rural settlers) were designed to apply to more humid regions and did not function well in the arid West.

The two principal categories of public land are national forests, which include the great majority of

[4] D. W. Meinig, *Southwest: Three Peoples in Geographical Change, 1600–1970* (New York: Oxford University Press, 1971), p. 30.

[5] East of the Mississippi River, only Florida and New Hampshire have as much as 10 percent of their land in federal ownership.

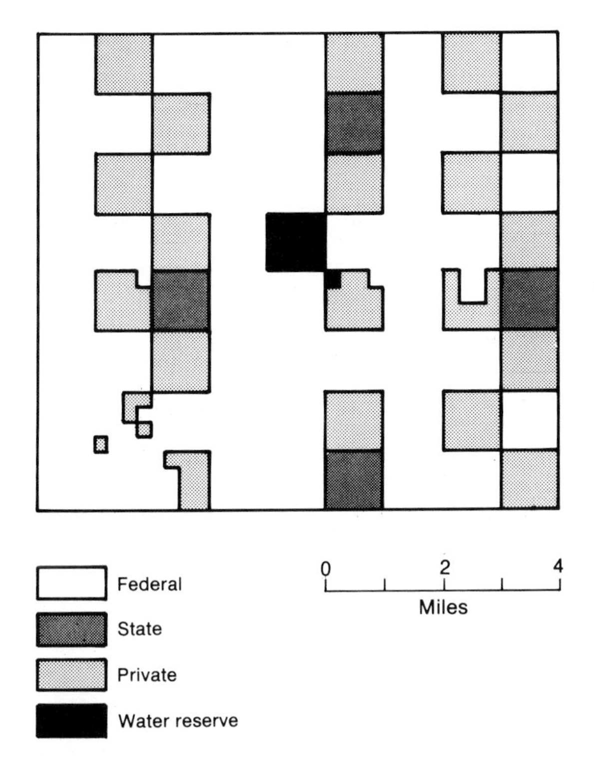

FIGURE 15-10 In much of the Intermontane West, the land ownership pattern resembles a gigantic, imperfect checkerboard. This sample is from San Bernardino County, California, near the Nevada border. The federal land mostly is administered by the Bureau of Land Management. Private land includes that owned by railroad companies, other corporations, and individuals. (Data source: Amboy Quadrangle of Bureau of Land Management map series, 1978).

all forest land in the region, and Taylor grazing lands, which were withdrawn from homesteading and are reserved for seasonal grazing use. Also notable in the region are Indian and military reservations.

A major problem associated with the large amount of government-owned land is the great complexity of the ownership pattern. Much of the land that is not federally owned occurs in a bewildering checkerboard arrangement (derived from granting scattered but designated sections of land to states, railroad companies, and other institutions either as reward or to generate revenue or as a stimulus to rural settlement) within a broad matrix of public domain (that is, federal) land. This fragmentation often precludes any rational development of use (Fig. 15-10).

The largest category of scattered, nonfederal land is owned by state governments. When western states were originally formed, the federal government made extensive land grants to them (generally four disconnected sections of land out of each township—4 square miles out of each 36) as trust land to support public education. Some 42 million acres (16.8 million ha) of this school trust land (in all states except Nevada, which sold all its acreage) remain scattered across the West, particularly in the Intermontane Region.

A concerted legal effort is now underway in Utah to rationalize this hodgepodge ownership pattern by a complicated series of land exchanges between federal and state governments that would consolidate the holdings into large, coherent blocks. Other western states are watching these negotiations with great interest, for the results might set a precedent that would significantly alter both land ownership and land-use patterns across the West.

THE CONTEMPORARY POPULATION: VARIED AND RAPIDLY INCREASING

What population there is in the Intermontane West congregates mostly in "islands" where (1) precipitation is adequate, (2) water is available for irrigation farming, (3) ore deposits permit commercial mining, (4) transportation routes converge, (5) some special recreational attraction exists, or (6) there is Sunbelt retirement. Cities are few and scattered, but rapidly growing. The principal nonmetropolitan population concentrations are in the middle Salt River Valley and parts of the lower Colorado River Valley in Arizona, the Imperial Valley of southernmost California, along the western Wasatch Piedmont in Utah, at various places along the Snake River in Idaho, and along the middle Columbia Valley in central Washington.

This is a region of considerable population movement—migration into, out of, within, and across. Many of the inhabitants are both restless and mobile. Numerous population clusters are characterized by an extraordinary number of mobile homes, travel vans, campers, and recreational vehicles—which emphasizes their impermanence. Even in the metropolitan areas there is frequent movement from one home to another. For the region as a whole, some 60 percent of the population changes domicile at least once every 5 years.

The net rate of population growth has been very rapid in the southern part of the region but somewhat slower in the north. More than 2 million inhabitants were added to the regional total during the 1980s. Nevada and Arizona, the two states wholly within the region, grew by 37 percent and 36 percent, respectively, during that decade.

There are three significant and readily identifiable "minority" elements in the contemporary population of the Intermontane Region. All three represent subcultures that are more prominent here than in any other region in the United States. There is a Mormon culture realm centered in Utah, an Hispanic-American borderland along the southern margin of the region, and Indian lands covering vast areas, particularly in Arizona and New Mexico.

Mormon Culture Realm

As the earliest white settlers in the central part of the Intermontane Region, Mormons (members of the Church of Jesus Christ of Latter-day Saints) have dominated the human geography of the Deseret area for nearly a century and a half. Their cohesive and readily distinguishable culture is manifested in various social patterns, in economic organization and development, and in certain aspects of settlement.[6] Today most Mormons, like most other Anglo-Americans, are urbanites; nevertheless, distinctive cultural characteristics set them apart and set their realm apart as a cultural subregion.

[6] The traditional Mormon town was a small, nucleated settlement with large lots, extraordinarily wide streets, a network of irrigation canals alongside the streets, relic agricultural features, unpainted barns, and houses of Greek Revival style constructed of bricks. Such settlements were totally unique in western Anglo-America but actually represented a re-creation of the "New England nucleated village and the persistence of nineteenth century structures and . . . patterns in the twentieth century" (Richard H. Jackson, "Religion and Landscape in the Mormon Cultural Region" [unpublished manuscript, 1977]). Small towns in which these characteristics persist today are relatively rare and occur in remoter parts of the Mormon culture realm.

The Mormon culture realm (Fig. 15-11) was defined in Meinig's classic 1965 study and has since been refined and updated in the *Atlas of Utah*.[7] The *core* is that intensively occupied and organized section of the Wasatch oasis that focuses on Salt Lake City and Ogden; it is the nodal center of Mormonism. The *domain*, which encompasses most of Utah and southeastern Idaho, includes the area where Mormonism is dominant but with less intensity and complexity of development. The *sphere* is defined as an area in which Mormons live as important nucleated groups enclaved within Gentile (non-Mormon) country. Despite the continuing expansion of Mormonism, its Utah focus is shown by the fact that more than three-fourths of the total county population in 20 of the state's 29 counties are members of the Mormon church whereas in only 5 other counties in the nation (all in southern Idaho) does such a proportional membership exist.

Hispanic-American Borderland

Along the southern margin of the Intermontane Region, from the Imperial Valley in California to the Pecos River in Texas, there are concentrations of varying intensity of Hispanic people. Their presence is numbered in the millions. Their proportional size is so great that they are in the majority in some towns and counties, including El Paso, the third largest urban center of the Intermontane Region.

The legacy of Hispanic settlement in the Southwest is long and notable. Architecture, settlement patterns, language, and cuisine are but a few of the more prominent elements of this heritage.[8] The continuing high rate of immigration from Mexico and the rapidity of increase among Hispanic-Americans ensure that this portion of the Intermontane Region should maintain its Hispanic subculture indefinitely. Thus from a demographic and cultural

[7] D. W. Meinig, "The Mormon Culture Region: Strategies and Patterns in the Geography of the American West, 1847–1964," *Annals*, Association of American Geographers, 55 (June 1965), 191–220; Deon C. Greer et al., *Atlas of Utah* (Provo, UT: Brigham Young University Press, 1981), pp. 140–143.

[8] For additional details on the concept of an Hispanic-American borderland, see Richard L. Nostrand, "The Hispanic-American Borderland: Delimitation of an American Culture Region," *Annals*, Association of American Geographers, 60 (1970), 638–661.

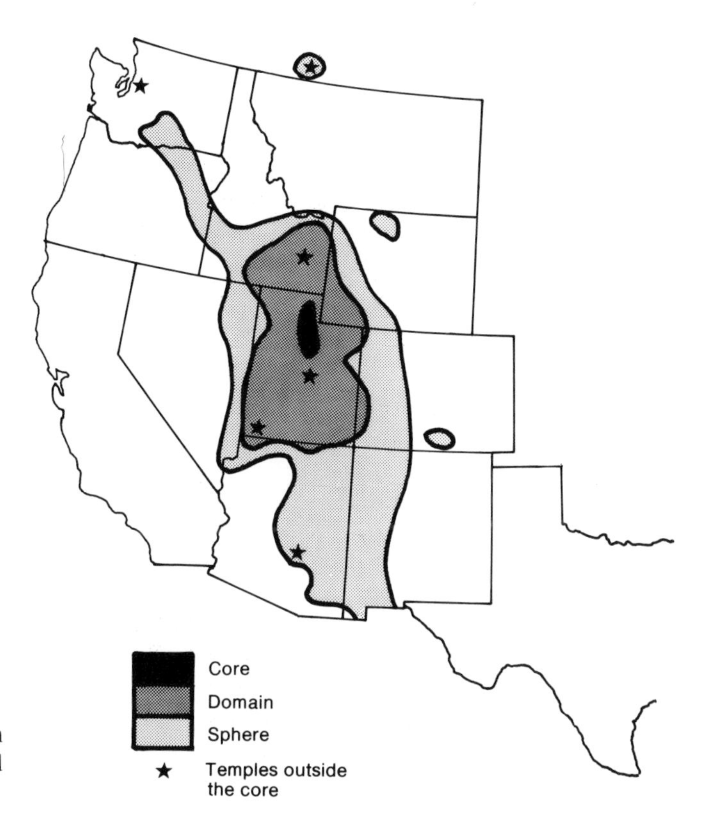

FIGURE 15-11 The Mormon culture realm (after Meinig and *Atlas of Utah*).

standpoint Latin America merges northward into the United States. In contrast, however, the virtual absence of "Americans" south of the international boundary suggests that Anglo-America ends abruptly at that line.[9] In a real sense the United States–Mexico border is a frontier where First World meets Third World. As J. W. House has noted, "Along its entire length, the US–Mexican boundary is one of the most remarkable and abrupt culture contact-faces in the world. . . . Nowhere else are there such steep economic and social gradients across an international boundary."[10] The accompanying box ("The U.S.–Mexico Border") highlights the situation.

[9] Richard L. Nostrand, "The Hispanicization of the United States," unpublished manuscript, 1983.

[10] J. W. House, *Frontier on the Rio Grande: A Political Geography of Development and Social Deprivation*. Oxford: Oxford University Press, 1982, p. 37.

Land of the Indian

More than in any other large portions of Anglo-America outside the Subarctic, the Intermontane West is the land of the Indian. Some 300,000 Native Americans of various tribal affiliations are scattered over the region, although predominantly in Arizona and New Mexico, where some 7 percent of the bistate population is of Indian origin. There are several large reservations in the northern portion of the region: Yakima in Washington, Umatilla in Oregon, Pyramid Lake and Walker River in Nevada, and Uintah and Ouray in Utah (Fig. 15-12). It is around and south of the Four Corners country that Indian lands are most prominent, however. The Navajo Reservation is by far the largest, but there also are extensive reservations for the various Apache tribes, the Papagos, the Hualapais, the Hopis, and the Utes. In addition, there are many smaller reservations in the Intermontane Region; some are densely populated, particularly the Pueblo reservations in north-central and northwestern New Mexico.

In general, the Indians of the Intermontane Region have been economically poor, socially deprived, and politically inactive. On reservations they have usually maintained cohesive tribal identities, although their livelihood is often near or below the poverty level. Those who have left the reservation—as all Indians are free to do—often find that adjusting to life in a harsh Anglo world is difficult.

There are, however, many pleasant exceptions to this generally depressing picture. Many off-reservation Indians have adjusted to living in southwestern cities, as the rapidly growing Indian populations of Los Angeles, Phoenix, and Albuquerque attest; furthermore, economic and social conditions

A CLOSER LOOK The U.S.—Mexico Border—A Line, or a Zone?

The Mexico–U.S. border, or *la frontera* (the frontier) as it is called in Mexico, is popularly if somewhat derisively referred to as the "Tortilla Curtain." This moniker not only suggests the role the border plays as a convenient symbolic divide between Anglo-America and Latin America but also between the First and Third worlds. Few international boundaries separate, at least politically, such fundamentally different countries and prevailing cultures.

Born of conflict and increasing U.S. domination of the region during the first half of the nineteenth century, the border was created by the Treaty of Guadalupe Hidalgo, which concluded the war between the two countries in 1848. From the Pacific Ocean to the Gulf of Mexico, the border runs for 2,076 miles (3,322 km) over granite-studded mountains, deserts, high plateaus, and coastal plain; it passes through land as inhospitable and desolate as any in either country, but also bisects some of the fastest-growing, most dynamic urban centers in the Western Hemisphere. All along its path it generates political complexities and controversies as it cuts through 25 counties in four U.S. states (California, Arizona, New Mexico, and Texas) as well as 35 *municipios*—municipalities that administratively are like counties in the United States—in six Mexican states (Baja California, Sonora, Chihuahua, Coahuila, Nuevo Leon, and Tamaulipas). For approximately 1325 miles (2120 km) it follows the course of the Rio Grande, or Rio Bravo del Norte as the river is known in Mexico. Westward from El Paso and Ciudad Juárez it is marked by barbed wire and wire mesh fence, and 258 white marble obelisks. Yet while the location of this invisible boundary is precise, the perception of it is not. In general, it is viewed from one of two perspectives—as being either a line, or a zone.

Those who support the former concept argue that the border is an abrupt demarcation, a sort of cultural fault, separating two countries with vastly contrasting material and nonmaterial elements of culture. They point to the differences in history, tradition, religious affiliation, values and symbols, language, ethnic composition, patterns of social organization, lifestyles, political and legal systems, architecture and design, cuisine, music, and levels of economic development. To support this position, they suggest one need only cross the border, in either direction, to immediately perceive the distinctions: The look, the smell, the character, the sense of the two places are quite different. These differences in the cultural landscape—the composite of all manufactured features—are not merely cosmetic but rather, it is argued, reflect the deeper cultural-historical currents of the respective societies. That the landscapes of Matamoros, say, are for the most part unlike those of neighboring Brownsville, or that the landscapes of Tijuana would not likely be confused with those of San Diego, are both readily apparent, and culturally meaningful.

Conversely, there are others who see the border as a zone that straddles the international boundary. It functions, so the argument goes, as a kind of linear third country or special domain with its own identity and character. This zone of "overlapping territorality," as it has been labeled, has produced a hybrid culture, one that is part Mexican and part American, similar to yet different from the cultural mainstreams found in the interiors of the two nations. Here, for instance, Spanish and English words are liberally and spontaneously mixed together in everyday conversations, creating a dialect and a language usage unique to this land between, it might be called. It is suggested that the so-called *fronterizos,* the border people, share a common experience and are tied not only by geographical proximity but also by interdependence, mutual interests, and transborder concerns, including a host of social, economic, and environmental issues.

Perhaps the most persuasive evidence in support of the border-as-zone position is the large and increasing volume of goods, services, people, and ideas that flow to and fro over the boundary from adjacent communities. Agricultural products on many reservations have been improving rapidly. The Apaches of the Fort Apache, San Carlos, and Mescalero reservations have developed prosperous logging industries and have shrewdly organized outdoor recreational advantages to attract tourists. The 30,000 Pueblo Indians have generally been able to adjust to the pressures of modern civilization because their ancestral lands were legally restored to them and each Pueblo village is thus surrounded by a protective girdle of farmland that reinforces its insularity.

Of all the large Indian tribes of the subcontinent, it is the Navajos who have led the way in adapting to a capitalistic society while still maintaining their cultural and tribal integrity. To do so, they have had wise leaders, but they also have the im-

FIGURE 15-c At many places along the border, as here in Tecate near San Diego, the fence separating the United States and Mexico serves better as a clothesline than as a barrier (photo by James Curtis).

and manufactured goods move in both directions. The populations on both sides, often linked by family ties and long-standing friendships, intermingle as they cross the border to work, shop, and play. Yearly, there are over 200 million *legal* crossings from Mexico to the United States, and a majority of these, it is estimated, are made by border residents. For them, if not for others, this highly permeable border is indeed, for all practical purposes, a zone of binational interaction and bicultural attractions (Fig. 15-c).

So whether the border is a line or a zone depends finally on what perspective is taken. Seen from the ground (a micro-level view), the empirical evidence, especially the cultural landscape characteristics, tend to support the former; whereas seen from above (a macro-level view), the complex functional patterns of transborder movement and interdependence become dominant, thus supporting the latter. Places, it should be remembered, are not one-dimensional entities that can be easily classified.

Professor James R. Curtis
Oklahoma State University
Stillwater

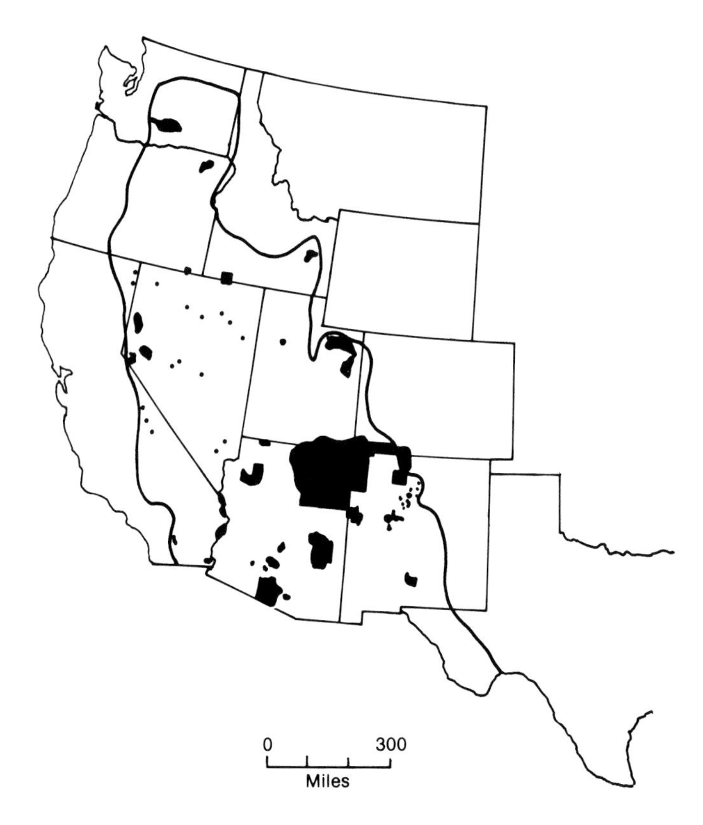

FIGURE 15-12 There is a greater extent of Indian reservations in the Intermontane Region than in the rest of the country combined.

measurable advantage of valuable natural resources to exploit. The massive and abrupt effort to establish Navajo lands as a functional part of modern America, without loss of the traditional Navajo culture, has been rife with problems; the effort continues and the complexities increase.

The Navajo Reservation of Arizona, New Mexico, and Utah is the largest (15 million acres) on the continent and the Navajo tribe is also the largest (more than 150,000 people) and fastest growing. Until recently, pastoralism was the dominant occupation of the Navajos. Their mixed flocks of sheep and goats range widely over the reservation, usually tended by women or children. In the last few years, however, exploitation of minerals in payable quantities has changed the basis of Navajo life. Coal, petroleum, natural gas, uranium, and helium are being extracted, many jobs have been created, and the Navajo Tribal Council now has an annual budget of more than $100 million with which to work. A forest products company has been organized, factories have been attracted to or near the reservation, and tourist facilities have been expanded.

A major and longstanding problem is the acrimonious relationship between the Navajos and their closest Indian neighbors, the Hopis. Some 6500 Hopis occupy a reservation that is virtually in the center of, and completely surrounded by, the Navajo Reservation (Fig. 15-13). Most Hopis live in agricultural villages that are situated on three high mesas. They are much more farming oriented than the Navajos, but the Hopis, too, engage in extensive pastoralism with sheep and goats on and around their mesas. There are many facets to the Navajo–Hopi dispute, but it centers around the use of grazing lands. Each tribe claims land that is used by the other and much stock that strayed beyond disputed boundaries has been confiscated by both Navajo and Hopi police who patrol the disputed areas.

There are various other internal controversies in Navajo-land, but much more visible to outsiders are the problems associated with the exploitation of

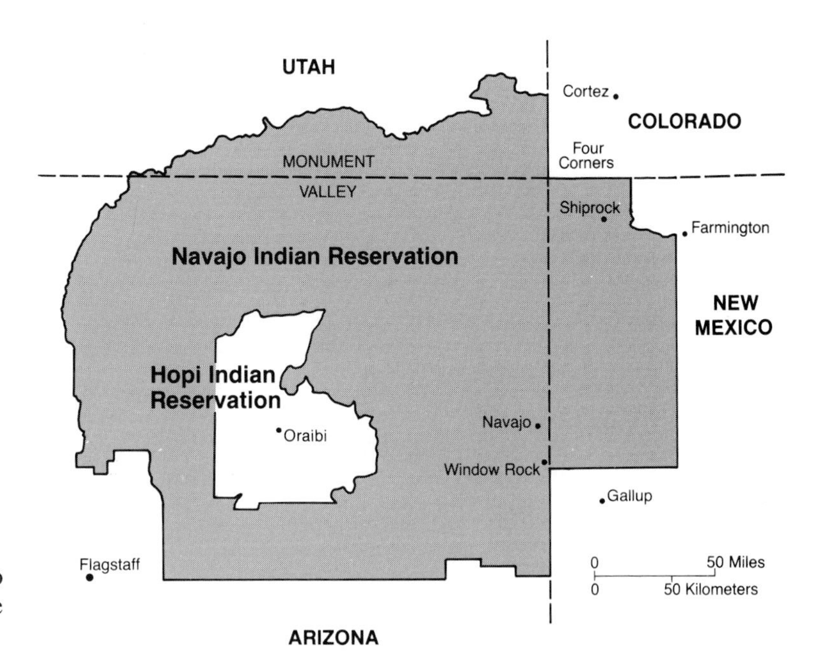

FIGURE 15-13 The Navajo and Hopi reservations of the Four Corners country.

energy resources, particularly coal. During the 1960s the Tribal Council signed several long-term contracts with mineral and utility companies for the mining and transport of coal and the construction of power-generating facilities to use the coal. Although the Navajos receive large sums of money for the leases and royalties involved, many believe that the value received is inadequate for the resources given up and the environmental deterioration that results. The coal is strip-mined, and thousands of acres of admittedly poor grazing land are lost as a result. The coal is used to generate electricity and produce coal gas in several large plants that spew enormous amounts of conspicuous pollutants into the heretofore pristine air of the Four Corners country. The electricity and gas produced are transported outside the region, mostly to southern California, for sale. Probably the most serious problem, however, is the huge amount of water required for these operations. Ultimately water is the most critical of all resources in the arid reaches of the Navajo Reservation and the long-term commitment of this precious commodity, as specified in the contracts, may cause problems in the future.

THE WATER PROBLEM

The limitations imposed by paucity of water are felt throughout the Intermontane Region. Limited rainfall makes stream water desirable, but rivers are scarce and often located in deep gorges, making their water relatively inaccessible. Well water has been obtained in some areas, but the principal hope for increasing the natural water supply of the region has always been river catchment and diversion.

In this region, as in other parts of the West, the federal government has not hesitated to become "enlisted on the side of The People vs. The Desert" by building dams and blocking the rivers to create tiny islands of moisture availability in the sea of aridity.[11] Flood control and irrigation have been the twin purposes of most river development projects in the region, although some dams are more specifically designed to provide an urban water supply or generate hydroelectricity.

[11] Walter Prescott Webb, "The American West: Perpetual Mirage," *Harper's Magazine*, 214 (May 1957), 28.

The Dammed Columbia

Of all rivers in North America, the annual flow of the Columbia is exceeded only by the Mississippi and the St. Lawrence. The Columbia, however, has a steeper gradient and hence the greatest hydroelectric potential on the continent. The Columbia's average runoff is about ten times that of the Colorado, but only one-third that of the Mississippi. Along its 740-mile (1184-km) course in the United States, the Columbia descends 1290 feet (387 m); it has now been so completely dammed that only 157 feet (47 m) of this "head" is still free flowing.

Beginning with Rock Island Dam in 1929–1931, 11 dams have been built along the Columbia in Washington and Oregon. Except for Grand Coulee, the dams are primarily for hydroelectricity generation and navigation on the lower Columbia. Grand Coulee is a dam of superlatives (Fig. 15-14). It houses the largest hydroelectric power plant in the world; impounds the sixth largest reservoir, F. D. Roosevelt Lake, in the United States; is the fifth highest dam in the nation; and is designed to irrigate more than 1 million acres (400,000 ha).

Several other dams were built on tributaries of the Columbia, particularly the lower Snake (for hydroelectricity and navigation) and several short streams issuing from the Cascade Mountains (for irrigation).

The Continuing Controversy of the Colorado

Although many rivers in Anglo-America carry more water, the Colorado is particularly important because it is the only major river in the driest part of the subcontinent. The river flows for 1400 miles (2240 km), its drainage basin encompassing about one-twelfth of the area of the conterminous states. Seven states and Mexico clamor to use the Colorado's waters.

The first major dam in the Colorado watershed was the Roosevelt Dam on the Salt River near Phoenix, which was begun soon after Congress passed the Reclamation Act of 1902 authorizing the Department of the Interior to establish large-scale irrigation projects (Fig. 15-15). Before long it was realized that basin-wide planning was needed for efficient use of the waters of the basin. The Colorado River Compact was finally hammered out, taking effect in 1929; its main provision apportioned the use of Colorado River water between the upper basin states

FIGURE 15-14 Rocky Reach, near Wenatchee, is one of the smaller of the Columbia River dams (TLM photo).

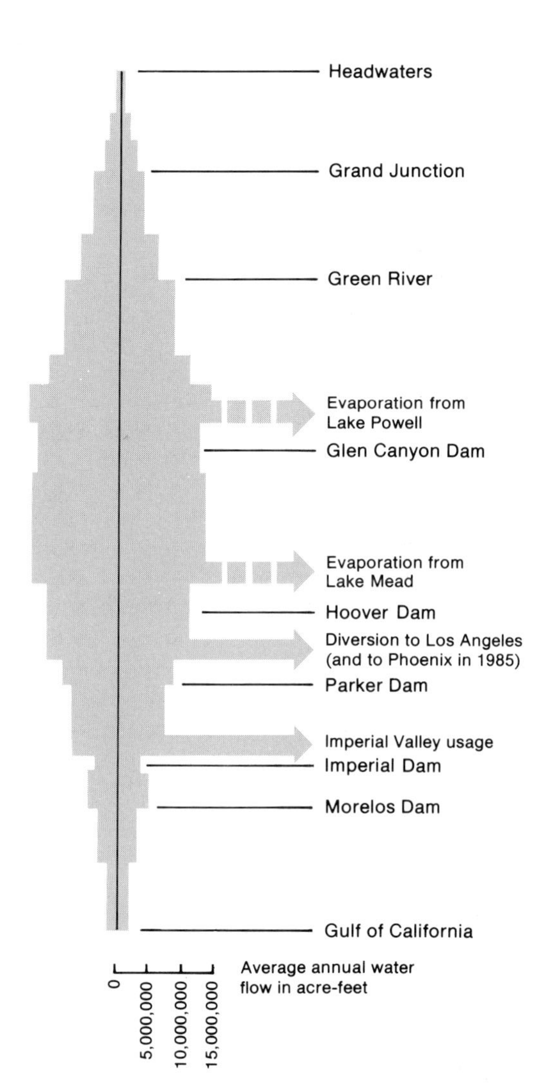

FIGURE 15-15 Most rivers increase in volume of flow from headwaters to mouth. In desert regions, however, evaporation, seepage, and diversion usually cause a diminished flow downstream. In the case of the Colorado River, the flow diminishes almost to nothingness.

(Colorado, Wyoming, Utah, and New Mexico) and the lower basin states (Arizona, Nevada, and California) on a 50–50 basis.

Later emendations provided a share for Mexico and subdivided the upper basin total among the four states involved. The lower basin states, however, could not agree on division of their share and complex litigation finally ended with the Supreme Court subdividing the Lower Basin allotment.

Several major problems persisted, not least the fact that the Colorado River was bankrupt; the various agreements called for an annual use of 3 million more acre-feet than the river normally carried. Four dams, starting with the mammoth Hoover Dam, were built along the lower course of the Colorado for various purposes but particularly to stabilize the river's flow and provide maximum usage in California and Arizona (Fig. 15-16).

After years of planning, construction began on the Central Arizona Project in 1973 and the first water was delivered to the Phoenix area in 1985. Plans call for completion of the $2.5 billion project (the largest ever undertaken by the U.S. Bureau of Reclamation) by 1992, but no one seriously believes that it will be accomplished by then. Meanwhile, the reality of the situation has been recognized in southern California. Prior to 1985 the Central Arizona Project's share of Colorado River water was being diverted to Los Angeles and vicinity. Now that water will increasingly go to Arizona.

AGRICULTURE

Farming is sparse and scattered in this vast, dry region, but it is nevertheless the most prominent activity in most nonurban settled areas. Only a fraction (3 percent in Utah, for example) of the total land area is in farms and little of this is actually in crops (Fig. 15-17).

Precipitation is so sparse that growing crops under natural rainfall conditions is restricted mostly to the Columbia Plateau subregion. Most important by far is the Palouse country of eastern and central Washington and adjacent parts of northern Oregon. This area has rich prairie soils enhanced by the deep accumulation of loess (windblown silts), which makes it the highest-yielding wheat-growing locale on the continent. Whitman Country, just north of the Snake River in southeastern Washington, is the leading wheat-producing county in the United States, primarily because of the very high yields per acre. The bulk of the output is soft white winter wheat, which is commonly used for pastry,

FIGURE 15-16 Many segments of the Colorado River are dammed, thereby creating lengthy reservoirs. The largest of the reservoirs is Lake Mead, near Las Vegas, Nevada (TLM photo).

crackers, and cookies rather than bread. Normally more than three-quarters of the output is exported, particularly to Japan and India.

The Palouse area also leads the nation in production of dry peas and lentils and has considerable acreage in barley, clover, and alfalfa. The Blue Mountains district of northeastern Oregon and southeastern Washington is the leading source of green peas in the nation.

There are no other major areas of dry-land farming in the region. Scattered patches of dry-land wheat are found in Idaho, Oregon, and Washington. Dry beans are raised without irrigation in central New Mexico, southwestern Colorado, and southeastern Utah where there is sufficient summer rain.

Most crop-growing in the region, however, is dependent on irrigation, which has been expanding ever since it was introduced by the Mormons in Deseret in 1847. Some 10 million acres (4 million ha) of cropland are now under irrigation, of which about one-third is in Idaho and one-sixth in Washington (Fig. 15-18). The principal areas of irrigated farming are summarized below.

The Salton Trough This hot, flat, below-sea-level valley is occupied in part by the saline waters of the Salton Sea and in part by two highly intensive irrigated farming areas—Imperial Valley to the south and Coachella Valley to the north (Fig. 15-19).

FIGURE 15-17 Dry farming is practiced on a small scale in many parts of the region. This is Indian corn in southeastern Utah's Monument Valley (TLM photo).

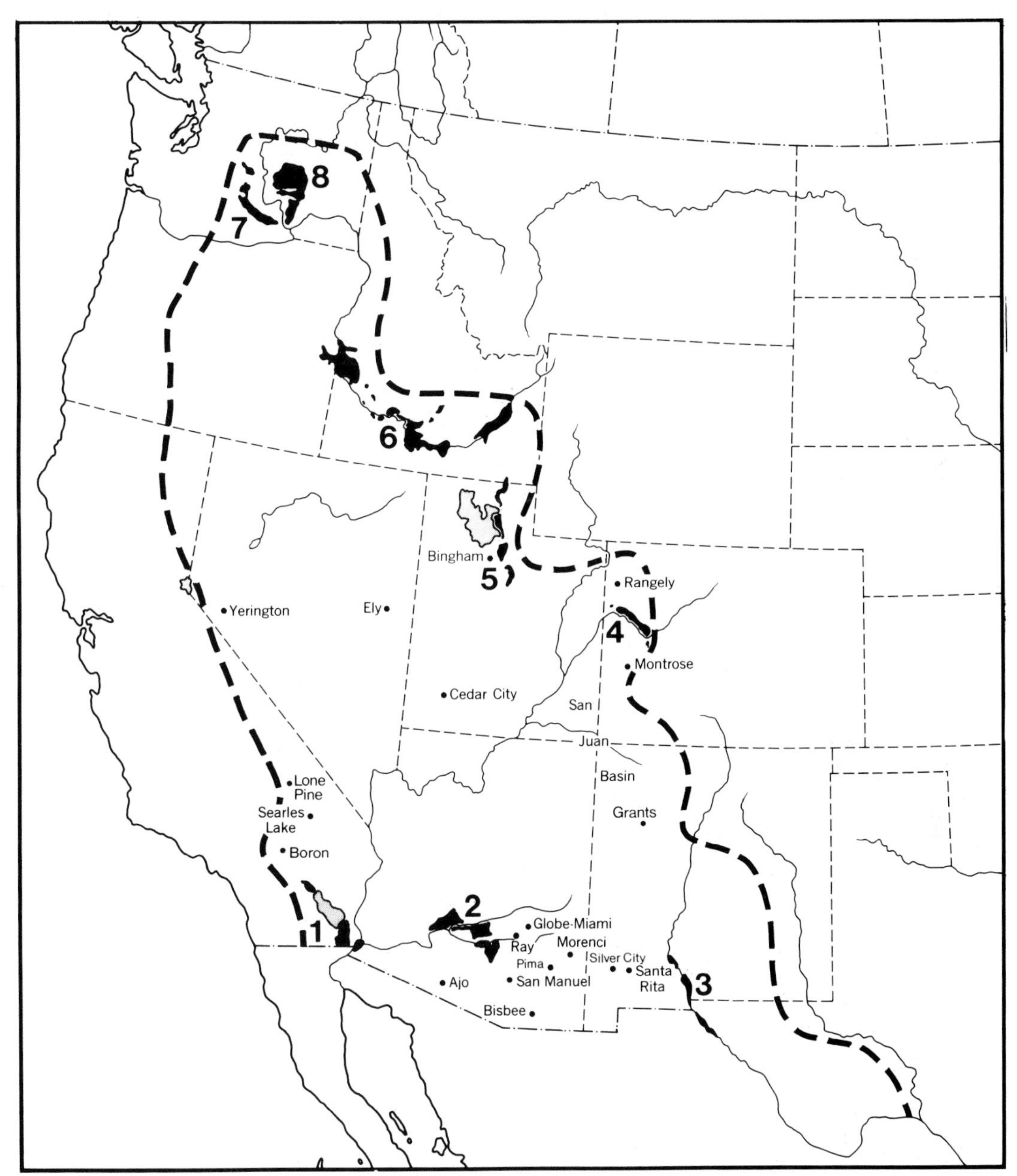

FIGURE 15-18 Major irrigated areas and mining towns in the Intermontane Region: (1) Salton Trough, (2) Salt River Valley, (3) Rio Grande Project, (4) Colorado's Grand Valley, (5) Salt Lake Oasis, (6) Snake River Plain, (7) Columbia Plateau Fruit Valleys, and (8) Columbia Basin Project.

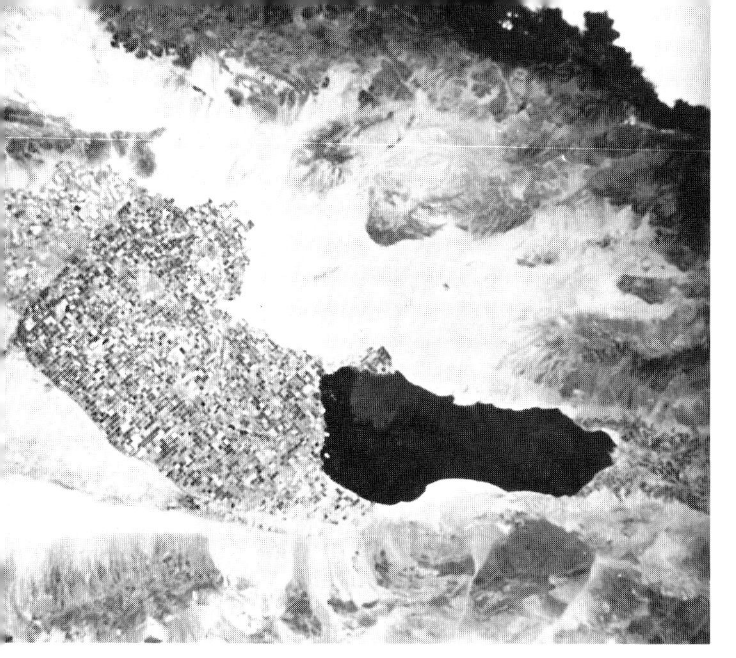

FIGURE 15-19 Probably the most famous space image of Anglo-America is this scene in Southern California and adjacent Mexico. The checkerboard pattern of intensive, irrigated agriculture south (Imperial Valley) and north (Coachella Valley) of the Salton Sea contrasts strongly with the surrounding barrenness of desert basins and mountains. The international boundary shows as an east-west line at the south end of the Imperial Valley; its abruptness is caused by a different intensity of land use in Mexico, as well as seasonal differences in planting and harvesting practices (Landsat image).

The 470,000 irrigated acres (190,000 ha) of Imperial Valley yield about 750,000 acres (300,000 ha) of crops each year as a result of the widespread adoption of double-cropping. A great deal of labor is required on most farms, and the area depends heavily on migratory workers. The valley is watered from the Colorado River via the All-American Canal and produces a remarkable variety of crops, ranging from high-value iceberg lettuce (which dominates the winter market in the United States) to mundane alfalfa (up to seven cuttings a year). It is a major producer of sugar beets and cotton, but its most valuable output is beef from cattle that are fattened in the area before marketing.

Coachella Valley also obtains its water from the Colorado River and grows a variety of crops, but the bulk of farm income comes from the four-level agricultural pattern of vegetables (especially carrots), vineyards (mostly table grapes), grapefruit (California's principal area), and dates (the only significant commercial source in the nation).

The Salt River Valley This was the first major federal irrigation project and one of the most successful economically. The chief cash crop is short-staple cotton, but there is considerable acreage in a great variety of other crops, especially hay, wheat, barley, sorghums, citrus, and safflower (Fig. 15-20).

Much irrigation water is also obtained from wells, which have been so depleted by heavy pumping that the water table has dropped alarmingly. No other state has such deep irrigation wells, on the average, as Arizona. Indeed, Arizona has the highest-cost irrigation of any state, but it has outstanding yields to compensate.

The Rio Grande Project The Middle Rio Grande Valley, which is above and below El Paso, constitutes one of the oldest irrigated areas on the continent, having been initiated by pre-Columbian Indians. Sporadic private irrigation diversions in the late nineteenth and early twentieth centuries severely diminished the water available below El Paso, which was an important Mexican irrigated district. After considerable international negotiation, the Rio Grande Project was developed under federal auspices. The key structure is Elephant Butte Dam on the Rio Grande in southern New Mexico. Some 180,000 irrigated acres (72,000 ha) are included within the project. The greatest acreage is devoted to hay and feed grains for cattle fattening. Other prominent farm enterprises involve cotton, poultry, pecans, and grapes.

Colorado's Grand Valley West central Colorado has several major irrigated areas that use water flowing westward from the Rocky Mountains in such rivers as the Colorado and the Gunnison. Most notable is the Grand Valley project near Grand Junction. Many different field crops, such as corn, small grains, alfalfa, sugar beets, potatoes, and vegetables, are grown. The area's reputation, however, is based on its fruit crop, primarily peaches but also other orchard fruits.

Salt Lake Oasis The valley of the Great Salt Lake was occupied by Mormon pioneers in the 1850s. Within the first decade of occupance, the

FIGURE 15-20 High-quality cotton is a major product of the irrigated valleys in central Arizona. This cotton gin, with acres of bales and mountains of seeds, is near Phoenix (TLM photo).

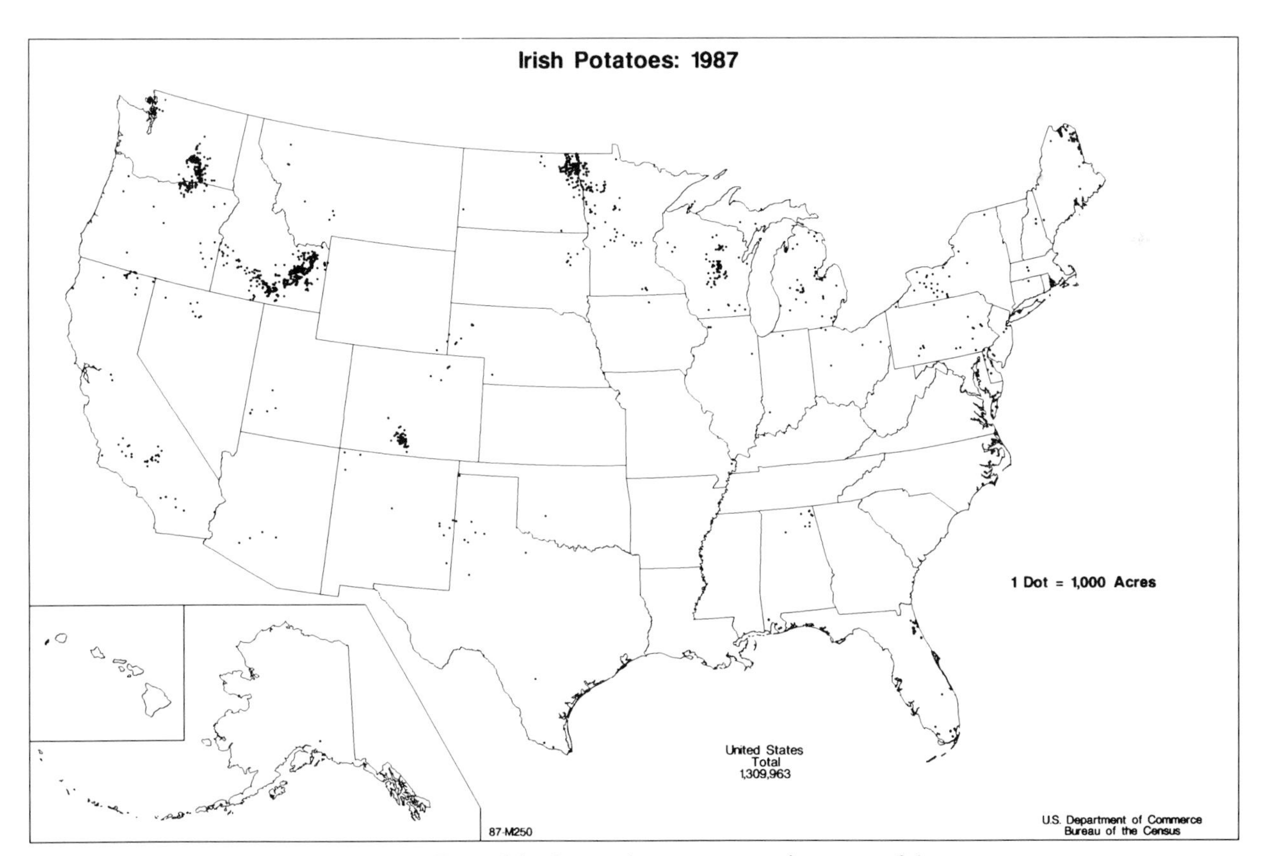

FIGURE 15-21 Two of the four major potato-growing areas of the country are in the northern part of the Intermontane Region.

land-use pattern was established and relatively little expansion of irrigated farming occurred since then.

The lofty Wasatch Mountains tower above the oasis on the east, their snow-clad slopes providing life-giving water for the dry lands at their base. At the mouth of almost every stream canyon, as it emerges from the Wasatch, is located a city or village girdled by green fields and adorned by orchards and shade trees. The cropping pattern is extremely diverse, although the greatest acreage is devoted to hay and grains (especially wheat). Other notable crops are sugar beets and fruits (particularly apples, peaches, and cherries). Livestock are also abundant and varied, most notably cattle and chickens.

The Snake River Plain Southern Idaho contains a lengthy sequence of irrigated areas scattered across the sagebrush flats above the abrupt gorge of the Snake River. The basic Idaho crops are hay (by far the most acreage), potatoes (Idaho produces more than one-fourth of the national crop), and sugar beets (Idaho is the third-ranking producer).

The three principal irrigation projects are the Minidoka in southeastern Idaho and the Boise and Owyhee in southwestern Idaho and adjacent southeastern Oregon. The most dynamic irrigation developments in the West in recent years, however, have been extensive, mostly privately financed operations relying on electric or gas-operated pumps in various locations in the Snake River Plain. The water is pumped from 400 to 600 feet (120 to 180 m) from wells, or a comparable vertical rise from the Snake River. These new farms tend to be enormous in size, corporate in structure, and quite capital intensive, making heavy use of sophisticated equipment and other farm machinery (Fig. 15-21).

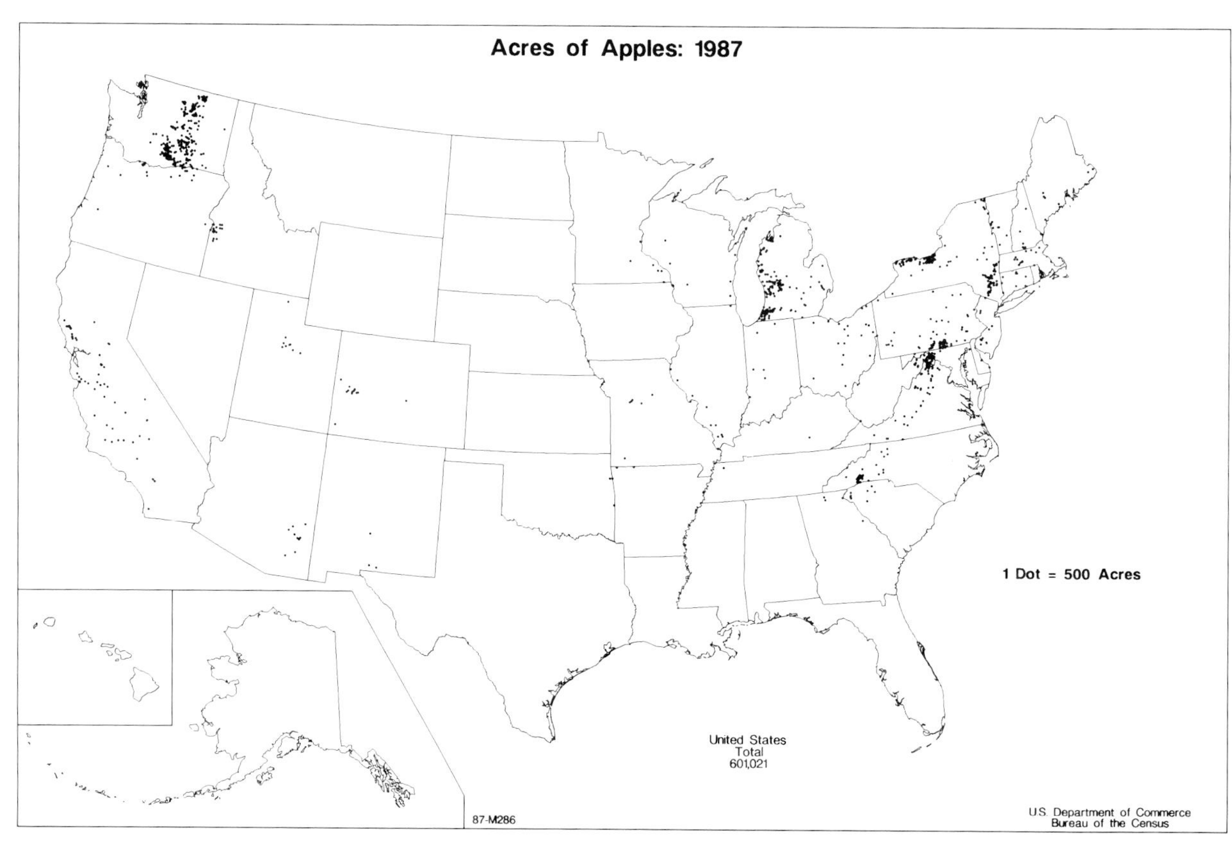

FIGURE 15-22 Central Washington is by far the leader in United States apple production.

FIGURE 15-23 Circular irrigation on a grand scale, a multiplicity of center-pivot systems near Moses Lake in central Washington (courtesy Valmont Industries, Inc.).

Columbia Plateau Fruit Valleys In the rain shadow of the Cascade Mountains lies a series of Columbia River tributaries whose valleys—especially the Yakima and Wenatchee—produce two-fifths of the national apple crop (Fig. 15-22). There is also notable output of hay and field corn for cattle feeding, hops, potatoes, and enormous recent expansion of vineyards.

The Columbia Basin Project Grand Coulee Dam was built in the 1930s as the first high dam on the Columbia River. Water from its impounded reservoir is diverted to central Washington where it is planned that 1.2 million acres (480,000 ha) of semiarid sagebrush country will be irrigated. Only two-thirds of the project was completed by the early 1990s, and the agricultural pattern is still developing. The most recent major development is the introduction of center pivot systems to disperse the water (Fig. 15-23). This has become a leading center pivot area, with a notable expansion of output of potatoes and corn.

PASTORALISM

In this region of rough terrain, light rainfall, sparse vegetation, and poor soils most of the land (if it is to be used at all) must serve as range for livestock. Pronounced differences in elevation cause differences in precipitation and vegetation, which, in turn, are reflected in the seasonal utilization of the range. Mountain pastures are strictly summer pastures; deserts are used mostly in winter, when snowfall provides water for sheep and occasionally cattle. Oasis pastures and feedlots are handling more and more animals throughout the year.

The establishment of federal grazing districts by the Taylor Grazing Act of 1934 had a significant effect on the pastoral pattern of the region. This

A CLOSER LOOK Sheep, Basques, and the American Southwest

A great part of the American Southwest consists of desert basins, semidesert plateaus, and mountains rising from desert basins through brush-covered slopes to forested summits. In their original state, these areas are not usable agriculturally, were poorly endowed for the grazing of cattle, but do fairly well as grazing areas for sheep and goats.

The original inhabitants, various tribes of American Indians, had no domesticated animals and lived largely by hunting and gathering, with limited rudimentary agriculture. But soon after the first whites occupied the area, herds of cattle and sheep were introduced and soon became a common sight throughout the region. The cattle were limited to the plains and lower plateaus, but the sheep traveled everywhere. Goats never became important except among some Indian groups.

Sheep raising quickly became a commercial enterprise—at first for the wool, and later, when markets became accessible, for meat (always called "lamb" in the United States, despite the age of the animal—never "mutton" as it is in much of the British world).

From the outset, the grazing of sheep took on a form of transhumance—the movement of flocks from one ecological area to another as the change in seasons causes a change in the available forage. In the winter, the flocks would feed in the lowlands: in the San Joaquin Valley of California on natural feed standing in the pastures; in the desert basins, largely on hay made in summer and stored for winter use. When new grass appeared on the lower mountainsides in spring, the sheep were moved to it, and in the weeks that ensued, they followed the retreating snowline and advancing grass higher and higher into the ranges. In autumn, there was a rapid descent of flocks to the valleys to avoid entrapment by early snowfalls. All of this movement was done afoot; "sheep lanes" were carefully demarcated and are still visible in some parts of California, Nevada, and Utah. In the early days, all of the lands grazed were in the public domain, and available to all comers.

Today, in essence, the same situation prevails, but the details are different. Movement of animals is largely by double-decked trucks, to reduce the weight loss that goes with long-distance walking of the sheep. Lowland grazing in winter is usually on privately owned irrigated pastures, or on temporarily empty vegetable fields, where the animals are fed on cull vegetables (those not fit to sell, dry, or can), trimmings, and the like. Other grazing is on land owned by the federal government (Forest Service, Bureau of Land Management, etc.) for which the sheep rancher pays a fee of so many cents per day per animal. Truck hauls of hundreds of miles are not uncommon—as from the summer range in the Sierra Nevada to the warm winter desert lowlands near the Mexican border—the Imperial and Coachella valleys.

Grazing has always been in large mobile bands, with from a few score to a couple of thousand animals in each band. Continuous human supervision is required: in early days to guard against predators such as wolves; today against theft, to keep the animals on the lands to which they have been allocated, and to prevent agitation leading to hysterical flight, to which sheep are quite susceptible. A few small operators handle their own sheep, but all big ranchers hire sheepherders (the term *shepherd* is not commonly used in the American West), one individual for each band of sheep. Each herder is accompanied by two or three dogs, who herd, head, and drive the sheep, barking at their heads, nipping (but not biting) at their heels, in response to shouted or whistled commands from the herder.

The job requires a very special type of person—one who enjoys being alone, seldom seeing other people, working seven days a week for months on end in mountains, deserts, or empty agricultural lands, and moving repeatedly from one location to another, with only the dogs for company.

To fulfill these requirements, a curious tradition has developed, involving a little-known people from far away. For well over a century, most of the sheepherders in the American Southwest have been Basques. They come from small mountain villages in the Pyrenees between France and Spain. Opportunities are few at home, and this employment (with no opportunity for spending money) offers a chance to accumulate a substantial cash base. They "sign on" for a long period—14 years is common—obtaining a special visa the U.S. State Department has just for them. And some, after returning home and getting married, have returned to the West as regular immigrants, and from them have stemmed some of the fine ranch families of the West.

But who are the Basques? A good question. Their ethnologic origins are unknown. Their language is unlike any other in the world, with no affiliation with any other language of Western Europe. They are a very independent people—yet because of events beyond their control they split between France and Spain. Currently, they are much in the news as they actively, sometimes violently, demand autonomy from Spain.

Professor Richard F. Logan
University of California, Los Angeles

FIGURE 15-24 Cowboys and cattle on the sagebrush plains of eastern Oregon (TLM photo).

legislation put an end to unrestricted grazing on public lands and helped stabilize the balance between forage resources and numbers of stock. Ranchers may lease portions of a grazing district for seasonal use. It is up to the Bureau of Land Management, the administering agency, to harmonize the carrying capacity of the range with the economic realities of the ranchers.

Ranches for both sheep and cattle are widespread in the region (Fig. 15-24). In recent years there has been a rapid proliferation of feedlots, mostly for cattle. Feedlots have become a big business in the Phoenix area, the Imperial Valley, around Yuma, and in several parts of Utah, southern Idaho, and central Washington.

MINING

From the Wasatch to the Sierra Nevada and from Canada to the Mexican border the region is dotted with communities located solely to tap the mineral resources. These communities enjoy viability as long as the mines produce but decline precipitously and become ghost towns once the ores are worked out or relative price changes make mining unprofitable.

Copper

The most notable mineral resource of the Intermontane Region throughout most of history has been copper. Long the world production leader, the United States now ranks second (to Chile) owing to a drastic decline in demand for the high-priced domestic ore in the 1980s. About half the copper mines in the Intermontane Region were closed, and the others are mining at a reduced capacity.

Arizona has been the leading copper mining state for more than a century. It has large open-pit mines at *Morenci, Globe-Miami, Ray, Ajo, Pima, Bisbee,* and near *Tucson,* as well as underground mines at *San Manuel.* All these communities were hard-hit by closures and diminished production. Utah is the second-ranking state, with most production from the largest individual copper mine in the country, at *Bingham,* until it was "temporarily" closed in the 1980s (Fig. 15-25). New Mexico is the third leading state; it experienced rapid production increases in the 1970s from several mines in the *Santa Rita–Silver City* district, but these mines were among the most severely affected by recent closures. Nevada's principal copper mines are at *Ely* and *Yerington.*

Other Metals

There has been a spectacular increase in precious-metal mining in the Intermontane Region ever since prices began to rise significantly in the 1970s, with most of the action in Nevada. Nevada gold output rose from virtually nothing to an annual output of nearly 400,000 troy ounces (120,000 kg) by the early 1990s, which is more than half of the national total. Nevada also produces more than one-third of the national output of silver.

Other prominent mining of metal ores in the region include iron in southwestern Utah and tungsten from several locales.

Salts

Several places in the Intermontane deserts yield salts of one kind or another, with major output from southeastern California. Borax minerals are obtained from a voluminous pit at *Boron,* borate and potash compounds are scraped from the surface and extracted from brine wells at *Searles Lake,* various sodium and calcium salts are mined in *Death Valley,* and soda ash has been taken from the dry bed of Owens Lake near *Lone Pine*. Various potash salts are obtained from the *Bonneville Salt Flats* west of Great Salt Lake.

FIGURE 15-25 The open-pit copper mine in Bingham Canyon, west of Salt Lake City, has been producing ore for more than three-quarters of a century, and is the largest of its kind in the world (courtesy Salt Lake Valley Convention and Visitors Bureau).

Coal

This region is fairly well endowed with coal. Among the states included, Utah ranks first in reserves, production, and the importance of coal in a state's total economy. There has been production for several decades in Carbon and Emery counties, an area with extensive reserves.

Coal output has also been expanding in the Four Corners country, with mines in Arizona, New Mexico, and southern Utah. Most of the production is used by thermal electric generating plants or coal gasification plants. These facilities are heavy users of precious water resources and create an abundance of obvious air pollution.

Petroleum

Although the map of oil lands in this region is expanding and the amount of drilling is increasing, the Intermontane Region contributes less than 2 percent of the national output. Principal production comes from the *Rangely* field in northwestern Colorado, several fields in the *Uinta Basin* of northeastern Utah, and northwestern New Mexico.

Oil Shale Some day oil shale may serve as a great source of petroleum. It is widely scattered over that part of the region in Utah south of the Uinta Mountains, in adjacent west-central Colorado, and in southern Wyoming (partly outside the Intermontane Region). However, high costs and potential production problems have shelved more than a dozen heralded development projects, and at present only a small amount of oil shale is actually being converted to crude oil. Oil shale certainly will be an important energy source in the future, but the future may not yet be here.

Uranium

The history of uranium mining in the United States is one of remarkable ebbs and flows. The frantic boom of the early 1950s and the great decline of the late 1950s were followed in the late 1970s by another major boom. About half the nation's uranium is found in a 100-mile strip of northwestern New Mexico, centering on the town of *Grants*. There is scattered production elsewhere in New Mexico as well as in western Colorado and eastern Utah.

FORESTRY

Forests are generally absent from this region except on the higher mountains and in the north; so logging is not a major activity. Only two areas are notable: central Arizona and the intermontane fringe areas in Oregon, Washington, and Idaho.

The high plateaus and mountains of the Mogollon Rim country and the Coconino Plateau of Arizona are clothed with forests of ponderosa pine, Douglas fir, and other coniferous species. Exploitation is limited mostly to a few large timber-cutting operations and their associated sawmills, although a pulp mill has been brought into operation.

A considerable amount of relatively open forest is found around the margins of the Intermontane Region in the three northern states. The principal species involved is ponderosa pine. Logging here is usually on a small scale, except in a few instances, such as at Bend, Klamath Falls, and Burns, Oregon, which are major pine sawmilling centers.

TOURISM

Tourism has become a major activity in the Intermontane Region in recent years. Its scenic splendors are unmatched on the continent, and there are many other types of attractions for visitor interest. The region also has become a major battleground of environmental concerns. Its scenery, particularly in the Colorado Plateau section, is unique, spectacular, and fragile. The construction of open-pit coal mines and their accompanying thermal-fired plants creates major visual pollution as well as other detrimental environmental impacts. The conflicts are clear-cut and the controversies will undoubtedly proliferate.

Scenery

The Intermontane Region abounds in scenic grandeur. Within it are the Grand Canyon (Fig. 15-26), Monument Valley, Zion Canyon, Cedar Breaks, Bryce Canyon, Death Valley, the Painted Desert, Utah's Canyonlands, the Petrified Forest, Great Salt Lake, gorges of the Snake and Columbia rivers, and the Big Bend canyons of the Rio Grande. Most of these and many other beauty spots have been set aside as protected reserves by federal or state governments. The Colorado Plateau section is particularly magnificent and contains the greatest concentration of national parks (eight) in the nation.

Colorado River Dams and Reservoirs

Although environmental purists may shudder at the thought, the reservoirs created by damming the Colorado River at various locations have become

FIGURE 15-26 The mighty Grand Canyon has been deeply and abruptly incised into the level-surfaced plateaus of northern Arizona. This view is from Mather Point on the South Rim (TLM photo).

FIGURE 15-27 The most extensive and most famous cliff dwellings in the nation are in Mesa Verde National Park in southwestern Colorado. This is Cliff Palace (TLM photo).

major tourist attractions. Lake Powell behind Glen Canyon Dam, Lake Mead behind Hoover Dam, and Lake Havasu behind Parker Dam are prominent foci for water-oriented recreation—boating, fishing, and skiing—and the dams themselves serve as educational attractions of some note.

Other Tourist Attractions

This region of colorful and spectacular scenery has an almost unlimited number of natural attractions. Many, however, are difficult to reach and most tourists invest only a limited amount of time in sightseeing off the beaten track. Consequently, various constructed attractions, which almost invariably are reached by excellent roads, draw considerable attention from visitors to the region.

Many ancient *cliff dwellings* and cliff-dwelling ruins are to be found in the Southwest. The most elaborate and best interpreted are in Mesa Verde National park in southwestern Colorado (Fig. 15-27). Many others are preserved (and sometimes restored) in national monuments in Arizona and New Mexico. The present-day *Indians,* with their traditional culture and handicrafts, also draw visitors from other regions. The Navajo Reservation and several of the Pueblo villages attract the largest crowds.

There are many *historic towns* in the region, remnants of that romanticized and immortalized period in American history, "the Old West." Some, such as Bisbee, Tombstone, and Jerome in Arizona or Virginia City in Nevada, have capitalized on their heritage and built up a steady trade in historically minded tourists.

The remarkable history of *Mormonism* in Utah and the continued importance of Salt Lake City as headquarters of the Mormon Church are compelling tourist attractions. Various edifices and monuments in and around Salt Lake City are visited by hundreds of thousands of visitors annually (Fig. 15-28).

Although Nevada has a colorful past, its present is in many ways even more flamboyant. As the only state that has systematically used *gambling* as a major source of revenue, it has developed games of chance in amazing proportions. Almost every town in the state has its cluster of "one-armed bandits" and mini-casinos, but the chief centers are Las Vegas (52 casinos) and Reno, although the greatest recent development has been in the tiny town of Laughlin, at the southern tip of the state (and therefore the closest casino location to the major population clusters of Los Angeles, San Diego, and Phoenix). As added attractions these gambling cities also specialize in glamorous entertainment, quick marriages, and simple (although not speedy) divorces (Fig. 15-29).

SPECIALIZED SOUTHWESTERN LIVING

The rapid population growth of the southern part of the Intermontane Region in recent years, matched only by Florida and southern California, is an obvious tribute to sunshine and health. Many present-day Americans feel that sunny mild winters and informal outdoor living provide sufficient satisfaction to counteract the problems of moving to a distant locality, even if that locality is characterized by scorching summers. Sufferers from respiratory afflictions also derive some real and some imagined health benefits from the dry air of the Southwest.

In Arizona, southern Nevada, southern New Mexico, and southeastern California particularly, the ordinary summer tourist is a relatively minor element in comparison with the frequent winter visitor, the new resident eagerly anticipating opportunity in a growing community, and the retired couple content to spend their last years in sunny relaxation.

FIGURE 15-28 The international headquarters of the Church of Jesus Christ of Latter Day Saints towers high above the more famous Mormon Temple in Salt Lake City (TLM photo).

It is on these three groups that the social and, to a considerable extent, the economic structure of the Southwest is turning. The significance of these groups is demonstrated nowhere quite as pointedly as in the growth of suburban Phoenix. Scottsdale, on the northeast, is a semiexclusive residential and resort suburb whose luxury hotels and elaborately picturesque shops and restaurants are geared specifically to the winter visitor. Mesa, on the east, is a sprawling desert community scattered with large factories and ambitious subdivisions for the migrant from eastern states. Sun City, on the northwest, is a specifically planned retirement community without facilities for children but with abundant amenities for senior citizens. Litchfield Park, on the west, is a grand design for a totally planned community in which last year's cotton fields will give way to next decade's city of 100,000 people.

SUBURBIA IN THE SUN: THE SOUTHWEST'S RUSH TO URBANISM

The extremely rapid population expansion of the southern part of the Intermontane Region is primarily manifested in the burgeoning of cities and

FIGURE 15-29 Las Vegas at night. The flamboyance of gambling casinos and the relatively inexpensive power of nearby Hoover Dam combine to give Fremont Street the brightest lights in the Intermontane Region. In a three-block frontage there are some 43 miles (69 km) of neon and more than 2 million light bulbs (TLM photo).

TABLE 15-1

Largest urban places of the Intermontane Region

Name	*Population of Principal City*	*Population of Metropolitan Area*
Albuquerque, N. Mex.	378,480	493,100
Boise City, Idaho	111,030	200,700
Chandler, Ariz.*	81,080	
El Paso, Tex.	510,970	585,900
Glendale, Ariz.*	140,170	
Henderson, Nev.*	59,310	
Idaho Falls, Idaho	44,250	
Las Cruces, N. Mex.	56,000	132,000
Las Vegas, Nev.	210,620	631,300
Mesa, Ariz.*	280,360	
North Las Vegas, Nev.*	51,450	
Ogden, Utah	66,320	
Orem, Utah*	64,420	
Phoenix, Ariz.	790,183	2,029,500
Pocatello, Idaho	43,520	
Provo, Utah	73,250	242,700
Reno, Nev.	115,130	239,700
Richland, Wash.	32,490	146,400
Salt Lake City, Utah	152,740	1,065,000
Sandy, Utah*	73,120	
Scottsdale, Ariz.*	121,740	
Sparks, Nev.*	54,550	
Spokane, Wash.	170,900	356,400
Tempe, Ariz.*	140,440	
Tucson, Ariz.	385,720	636,000
West Valley, Utah*	93,030	
Yakima, Wash.	49,280	185,500

* A suburb of a larger city.

extensive urban sprawl (see Table 15-1 for a listing of the region's largest urban places). The spectacular growth of southwestern cities in the last three decades is readily apparent (Fig. 15-30).

Metropolitan Area Population Increase, 1960–1990 (%)	
Albuquerque	110
El Paso	133
Las Vegas	605
Phoenix	311
Tucson	191

If the magnitude of recent southwestern urban growth has been remarkable, it is the character and form of this growth that have been even more eye-catching.

> Their physical structure is looser than in older cities; their average density is low, they consist mostly of detached single-family houses or garden apartments, they expand rapidly at their edges, and they often enclose a crazy-quilt pattern of unbuilt-upon land. Not even the slum areas, backward as they might be, approach traditional urban densities. Mass transportation is inadequate or nonexistent. . . .
>
> These cities have not only grown to maturity in the time of the automobile, they live by the automobile—and it is for the most part a pleasant and convenient way of life. Traffic jams are rare; parking, if not always well designed, is at least plentiful and inexpensive. Because of the automobile, the strip has often long ago replaced downtown as a center of business. Not just a competitor of the central core, it has become the vital economic area—if not in terms of quantity of money handled, then certainly in terms of daily shopping activity.[12]

Many "new" cities have been condemned by urban planners; they are different and do not resemble the "old" and great cities of the world. They lack the important attributes of high population and building density and a centric orientation.

> But downtown in these new cities is not the same downtown that we remember from other places and times. To revitalize or preserve a downtown that contains excellent stores and restaurants, museums and schools as well as banks and offices, a downtown that is served by an adequate or expandable rapid-transit system, and that has an emotional meaning to the people of a city is one thing. *Creating* a downtown in an area having no good stores

[12] Robert B. Riley, "Urban Myths and the New Cities of the Southwest," *Landscape*, 17 (Autumn 1967), 21.

and few good restaurants, no cultural or education facilities, an area in which even the movie theaters are second-rate with the cinerama-size screens located in the suburbs, where the only unique facilities are more old and cheap office space, bank headquarters, and the bus and railroad stations, a downtown located in a city of a density too low to support mass transit, in a city whose inhabitants' nostalgic memories are of a downtown in a far-off place that they have left—this is a very different matter.

. . . This new form of urban structure . . . is a result of increasing affluence and mobility, vastly improved communication, greater flexibility of transportation, and the increased importance of amenity in residential, commercial, and site location. . . .

What is happening [in Southwestern cities] . . . is precisely what is happening in the megalopolises of the eastern and western seaboard and the urban regions of the Midwest—with one important difference. In the latter areas the new developments take place over, around, or between strong and still vital industrial urban forms, forms which both dampen and distort the growth of radically new patterns. In the Southwest, where no such strong earlier forms exist, the new forms, as yet neither fully developed or understood, can at least be seen more clearly and studied for what they are or want to become.[13]

One of the most conspicuous features of the "new" southwestern cities is their expansive sprawl. Their booming growth has a spatial expression of erratic, leapfrogging development (for instance, nearly 40 percent of the land area within the Phoenix city limits is open space), their population density is quite low (Phoenix has a population density that is only about one-third as much as Los Angeles, long the epitome of low density), and their areal extent is likely to be enormous (Tucson, for example, encompasses twice as many square miles as Boston).

[13] *Ibid.*, p. 23.

Apart from this "new" form of urban development in the region, two types of specialized communities have become prominent:

1. Paired towns have grown up at a number of locations on opposite sides of the international border with Mexico. There are 17 pairs of these international twins, 11 of which are on the southern margin of the Intermontane Region.[14] Although the individual towns of each pair have a different cultural origin, they comprise a single spatial unit with a symbiotic economic and social relationship (Fig. 15-31). In almost all cases, the Mexican town is more popu-

FIGURE 15-30 The central areas of most of the larger Intermontane cities have been revitalized and metamorphosed by ambitious and imaginative building projects in recent years. This splashy scene is in Albuquerque's Civic Center (TLM photo).

[14] From west to east, they include Tecate, CA/Tecate, Baja California; Calexico, CA/Mexicali, Baja California; San Luis, AZ/San Luis Rio Colorado, Sonora; Lukeville, AZ/Sonoita, Sonora; Sasabe, AZ/Sasabe, Sonora; Nogales, AZ/Nogales, Sonora; Naco, AZ/Naco, Sonora; Douglas, AZ/Agua Prieta, Sonora; Columbus, NM/Las Palomas, Chihuahua; El Paso, TX/Ciudad Juarez, Chihuahua; and Presidio, TX/Ojinaga, Chihuahua.

FIGURE 15-31 Looking south across downtown El Paso into the smog-covered city of Juarez, Mexico (TLM photo).

lous than its American counterpart (Ciudad Juarez, for instance, has nearly twice the population of El Paso), but the central shopping district for both twins is in the American community. The symbiotic relationship has been particularly enhanced since 1965 when the Mexican government initiated its *maquiladora* (twin-plant) border industry program, which was designed to encourage the establishment of U.S. assembly plants just south of the border, using inexpensive Mexican labor but with essentially all the finished products being shipped to the United States for marketing. More than 500 such plants have been established in Mexican border towns, about half of them in communities adjacent to the Intermontane Region. The largest *maquiladora* development has been in Cuidad Juarez.

2. Specialized recreation-retirement towns are blossoming widely in the Intermontane Region. In every case, their site has some physical attraction—characteristically, the shoreline of some water body, otherwise a mountain location or a mild-winter desert spot. These towns experience relatively heavy tourist traffic, but stability is provided by a growing number of "permanent" residents. There are many examples in the region, such as Ruidoso (New Mexico) and Show Low (Arizona), but the prime examples of a self-sustaining, nongovernmentally developed, recreational-retirement new town are Lake Havasu City, Arizona, which grew from nothing to a population of nearly 20,000 in two decades, making it the largest urban place in its county, and upriver Bullhead City (Arizona's fastest-growing community in the late 1980s), opposite Laughlin, Nevada's newest casino center.

THE OUTLOOK

People have accomplished much in this restrictive environment. No one can stand on the steps of the State Capitol Building at Salt Lake City and gaze at the green island that is the Oasis without being impressed. Nevertheless, there is a limit to what human beings can accomplish against a stubborn and relentless nature. Because water, which means life, is scarce and much of the terrain is rugged, the greater part of the region is destined to remain one of the emptiest and least used on the continent.

Agriculture should become more important, but the development of additional large reclamation projects is unlikely, simply because most feasible dam sites in the region have already been developed (except at very controversial locations in the Grand Canyon). Irrigated crop acreages will expand most in central Washington and southern Idaho, but modest expansions will be widespread, particularly in Oregon and Arizona, often associated with center pivot technology.

Livestock raising will probably increase. More feeding will be carried on, both at local ranches and at centralized feedlots. Hay and sorghum feeding will continue to dominate, but grains, often brought into the region, will increase in importance. More attention will be paid to breeding, too, with improved Hereford and Angus strains in the north and more emphasis on Santa Gertrudis and Charolais in the south.

Mining is an industry of fluctuating prosperity in the region and will continue to be so. Copper, historically the most important intermontane mineral, often experienced instability due to an erratic market and antagonistic labor relations, a situation that is unlikely to change. Prospecting for, and mining of, energy minerals was the booming activity of the 1970s, but in the 1980s this was dramatically superseded by precious-metal mining.

Manufacturing may continue to grow in a few urban places. Although such localities are limited, they are the population centers and continued industrial growth will be impressive in the overall economy. High-technology industries have begun to flock to Arizona, which augurs well for its industrial future. For the region as a whole, it is a measure of the stability of the current population expansion that industrial growth is keeping pace.

Tourism in this region, as in most, is bound to expand. Summer is tourist time in the northern three-fourths of the Intermontane West; winter visitors are more important in the southern portion. An abundance of natural allurements, a variety of constructed attractions, and improving transportation routes combine to ensure a steady flow of tourists.

Population increase in the northern half of the region was slow in the 1980s, expressed principally by net outmigration. The southern, or Sunbelt, portion of the region continued to grow apace. These trends probably will persist. Mild winters, few clouds, dry air, informal living patterns, and the mysterious attraction of the desert will continue to exert their magnetic effects on dissatisfied, snow-shovel-weary citizens of the northern states.

Such a migration-fostered population growth will probably become overextended at times, outstripping a sound economic base. Generally, however, it is likely to grow with soundness, for capital will accompany people in the migration. Water may be a long-run limiting factor, but in the short run it is no barrier; urban growth is often at the expense of irrigated agriculture, and the former uses less water than the latter.

The southern Intermontane Region, then, is in functional transition from desert to metropolis. Today one can find smog in Phoenix that would make a Los Angeleno proud, traffic jams in Albuquerque that would do credit to Chicago, and tension-induced psychiatric treatments in Salt Lake City that would be suitable for New York. Indeed, the "new" cities of the interior West are already embarking on imaginative urban renewal projects to revitalize their downtowns. The developments are most striking in Tucson and Albuquerque, but almost every city of note from Spokane to El Paso has made at least a start on a mall–park–fountain complex in association with sparkling modern high-rise buildings in the heart of the central business district.

Rapid growth, however, is accompanied by growing pains. Cities are not immune to urban problems merely because they are new and different. Moreover, it is in rural areas that the major conflicts and controversies are likely to occur. Development versus preservation marks a battle line throughout North America, but perhaps in no other region are the opposing interests and values so clear-cut and the opportunities for compromise so limited.

SELECTED BIBLIOGRAPHY

BEHEIRY, SALAH A., "Sand Forms in the Coachella Valley, Southern California," *Annals*, Association of American Geographers, 57 (1967), 25–48.

CARLSON, A. W., "Long-Lots in the Rio Arriba," *Annals*, Association of American Geographers, 65 (1975), 48–57.

COMEAUX, MALCOLM, *Arizona, A Geography*. Boulder, Co: Westview Press, 1981.

COOKE, RONALD U., and RICHARD W. REEVES, *Arroyos and Environmental Change in the American South-West*. London: Oxford University Press, 1976.

DUNBIER, ROGER, *The Sonoran Desert: Its Geography, Economy, and People*. Tucson: University of Arizona Press, 1968.

DURRENBERGER, ROBERT W., "The Colorado Plateau,"

Annals, Association of American Geographers, 62 (1972), 211–236.

Francaviglia, Richard V., *The Mormon Landscape.* New York: AMS Press, 1979.

Gilbert, Bil, "Is This a Holy Place?" *Sports Illustrated,* 58 (May 1983), 76–90.

Goodman, James, M., *The Navajo Atlas.* Norman: University of Oklahoma Press, 1982.

Green, Christine, and William Sellers, *Arizona Climate.* Tucson: University of Arizona Press, 1964.

Greer, Deon C., et al., *Atlas of Utah.* Provo, UT: Brigham Young University Press, 1981.

Hecht, Melvin E., "Climate and Culture, Landscape and Lifestyle in the Sun Belt of Southern Arizona," *Journal of Popular Culture,* 11 (Spring 1978) 928–947.

———, and Richard W. Reeves, *The Arizona Atlas.* Tucson: University of Arizona, Office of Arid Lands Studies, 1981.

———, Richard Reeves, and Nat de Gennaro, "Agriculture: Its Historic and Contemporary Roles in Arizona's Economy," *Arizona Review,* 27 (November 1978), 7–12.

Hundley, Norris, Jr., *Dividing the Waters.* Berkeley: University of California Press, 1966.

———, *Water and the West: The Colorado River Compact and the Politics of Water in the American West.* Berkeley: University of California Press, 1975.

Jaeger, Edmund C., *The California Deserts.* Stanford, CA: Stanford University Press, 1965.

Jett, Stephen C., "The Navajo-Homestead: Situation and Site," *Yearbook of the Association of Pacific Coast Geographers,* 42 (1980), 101–117.

Kersten, Earl W., "Nevada Then and Now: Forging an Economy," *Yearbook,* Association of Pacific Coast Geographers, 47 (1986), 7–26.

McGregor, A. C., *Counting Sheep: From Open Range to Agribusiness on the Columbia Plateau.* Seattle: University of Washington Press, 1983.

McKnight, Tom L., "The Feral Horse in Anglo-America," *Geographical Review,* 49 (1959), 506–525.

Meinig, D. W., *The Great Columbia Plain—An Historical Geography,* 1805–1910. Seattle: University of Washington Press, 1968.

———, "The Mormon Culture Region: Strategies and Patterns in the Geography of the American West, 1847–1964," *Annals,* Association of American Geographers, 55 (1965), 191–220.

———, *Southwest: Three Peoples in Geographical Change, 1600–1970.* New York: Oxford University Press, 1971.

Morris, John W., *The Southwestern United States.* New York: Van Nostrand Reinhold Company, 1970.

Nostrand, Richard L., "The Hispanic-American Borderland: Delimitation of an American Culture Region," *Annals,* Association of American Geographers, 60 (1970), 638–661.

———, "The Hispano Homeland in 1900," *Annals of the Association of American Geographers,* 70 (September 1980), 382–396.

———, "Spanish Roots in the Borderlands," *The Geographical Magazine,* 52 (December 1979), 202–210.

Pease, Robert W., "Modoc County, A Geographic Time Continuum on the California Volcanic Tableland," *University of California Publications in Geography,* 17 (1965), 1–304.

Rogers, Garry F., *Then and Now: A Photographic History of Vegetation Change in the Central Great Basin Desert.* Salt Lake City: University of Utah Press, 1982.

Ruhe, Robert V., "Landscape Morphology and Alluvial Deposits in Southern New Mexico," *Annals,* Association of American Geographers, 54 (1964), 147–159.

Sellers, William D., and Richard H. Hill, *Arizona Climate, 1931–1972.* Tucson: University of Arizona Press, 1974.

Smith, David Lawrence, "Superfarms vs. Sagebrush: New Irrigation Developments on the Snake River Plain," *Proceedings,* Association of American Geographers, 2 (1970), 127–131.

Symanski, Richard, *Wild Horses and Sacred Cows.* Flagstaff, AZ: Northland Press, 1985.

Vale, Thomas R., "Forest Changes in the Warner Mountains, California," *Annals,* Association of American Geographers, 67 (1977), 28–47.

———, and Geraldine R. Vale, *Western Images, Western Landscapes: Travels Along U.S. 89.* Tucson: University of Arizona Press, 1989.

Wheeler, Sessons S. *The Nevada Desert.* Caldwell, ID: Caxton Printers, 1971.

16

THE CALIFORNIA REGION

The California Region, one of the smallest major regions of the continent is probably the most diverse. This diversity is shown in almost every aspect of the region's physical and human geography, from landforms to land use and from soil patterns to social patterns.

The region's location in the southwestern corner of the conterminous states has been a major long-range determinant of its pattern and degree of development (Fig. 16-1). The location is remote from the area of primary European penetration and settlement of Anglo-America and from the heartland of the nation that emerged after colonial times. Adjacency to Mexico has been significant from early days, even though the region is relatively distant from the heartland of that country, too. The region is well positioned for contact across the Pacific, but the Pacific has been, until recently, the wrong ocean for significant commercial intercourse. California has thus been denied ready access to the core regions of the United States and Mexico, and even to the Pacific Northwest, by pronounced environmental barriers (mountains and deserts). This factor has had an important effect on the population and economic development of the region; it is remarkable that it has not had an even greater effect.

The California Region encompasses most settled parts of the state, excluding only the northern mountains (north Coast Ranges, Klamath Mountains, southern Cascades), the northeastern plateaus and ranges, and the southeastern deserts except the Antelope Valley and Palm Springs areas, which are functionally integrated with the southern California conurbation and are therefore considered part of this region. The region, then, is essentially a California region and includes the intrinsic California.[1] More than 97 percent of the inhabitants of the most popu-

[1] Richard Logan, a leading California geographer, emphasizes the "atypical characteristics of the northeastern, northwestern, and southeastern portions of the state by referring to them as "unCalifornia." David Lantis, another prominent California specialist, calls these areas "the other Californias."

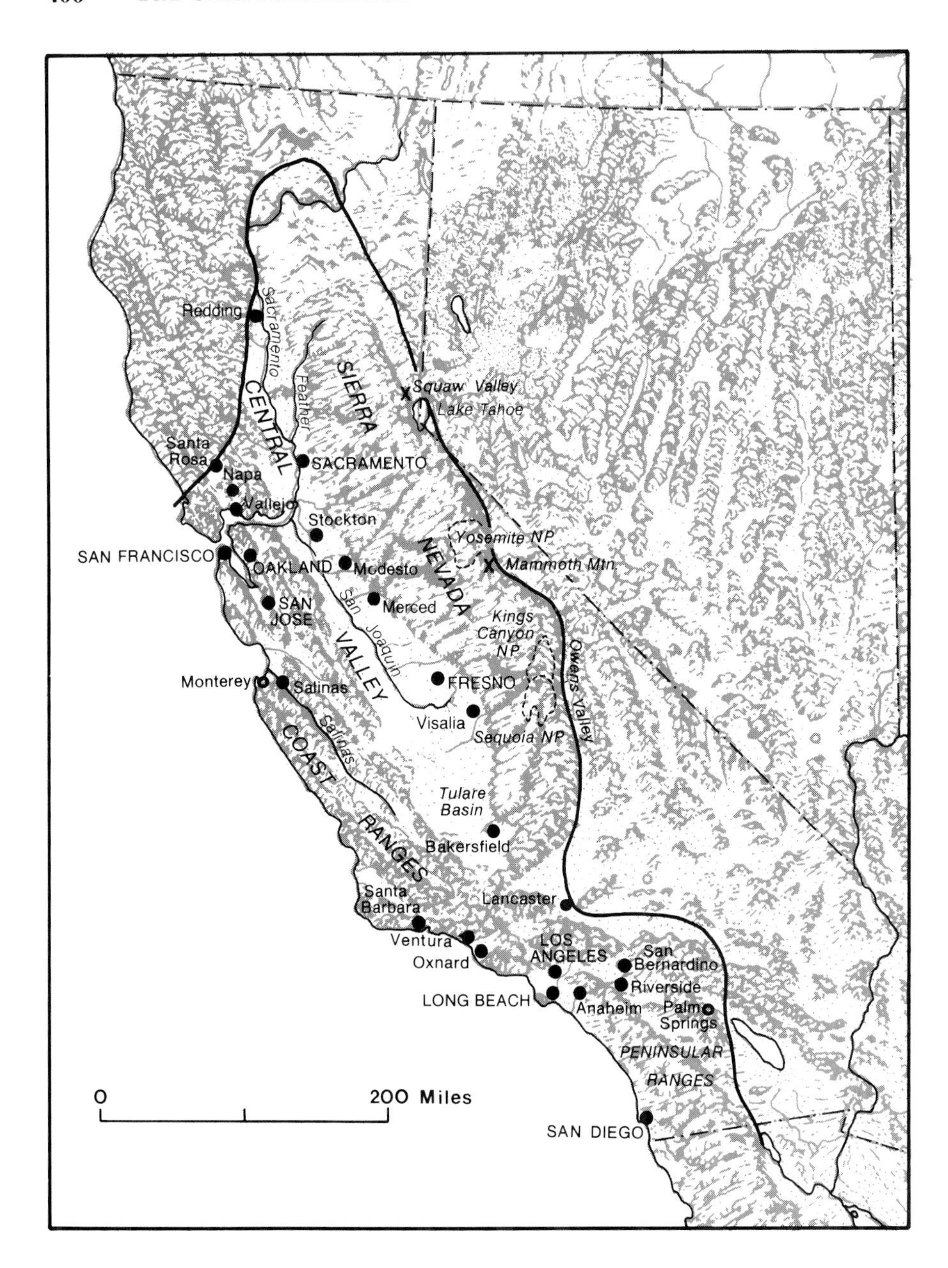

FIGURE 16-1 The California Region (base map copyright A. K. Lobeck; reprinted by permission of Hammond, Inc.).

lous state, numbering about 30 million persons in the early 1990s, reside within this region.

THE CALIFORNIA IMAGE: BENIGN CLIMATE AND LANDSCAPE DIVERSITY

The image of the region is synonymous with the image of the California lifestyle. It is the never-never land of contemporary American mythology. It is focused on Hollywood, Disneyland, and the Golden Gate; flavored with equal parts of glamour and smog; and populated by a mixture of sun-bronzed beach lovers and eccentric night people.

That the actuality is less exciting than the image is not too important. The fact that most Californians live in the same sort of suburban tract homes as other Anglo-Americans, watch the same television shows, vote for the same political candidates, and complain about the same taxes is convenient to overlook, for the California lifestyle has been sugarcoated, packaged, and marketed to the world as a thing apart, a destiny with a difference.

The image does, of course, have some substance. It is due in part to the relatively late development of the region's urban economy; in part to the boom-and-bust psychology and flamboyant nature of some of the staple industries—for example, gold mining, oil drilling, real estate promotion, motion picture industry, aircraft and spacecraft production; in part to the unusual natural endowments of this southwestern corner of the country; and in part to its residents, people who are drawn from every corner of the globe and who come seeking an elusive opportunity that they failed to find in their homeland and that they expect to discover in California.

Of utmost importance, both physically and psychologically, is the regional climate. This is the only portion of the continent with a dry-summer subtropical climate, generally called "mediterranean." Its basic characteristics are simple and appealing: abundant sunshine, mild winter temperatures, absolutely dry summers, and relatively dry winters. No other type of climate is so conducive to outdoor living.

And the diverse characteristics of the California outdoors multiply the opportunities and enhance the appeal of the region. High mountains are adjacent to sandy beaches and dramatic sea cliffs; dense forests rise above precipitous canyons that open into fertile valleys. Nearby to the east is the compelling vastness of the desert, and to the south is the charm of a different culture in a foreign land. And yet there is much more to the regional character than a flamboyant image, a benign climate, and a diverse landscape.

The California Region is outstanding in agriculture, significant in petroleum, unexcelled in aerospace and electronics, important in design, trend-setting in education, innovative in urban development, and increasingly significant in decision making. It is also the world champion in air pollution, the national leader in earthquakes and landslides, preeminent in both traffic movement and traffic jams, and the destination of a population inflow that is unparalleled in the history of the continent.

THE ENVIRONMENTAL SETTING

The diverse terrain of the California Region engenders much variety in most other environmental components.

Structure and Topography

The region occupies an area of great crustal instability, due primarily to its location at the interface of two major tectonic plates (North American and Pacific). Thus there has been, and continues to be, much diastrophic movement in the region, with a widespread occurrence of active faults. These faults are many and varied, but have two prominent characteristics: (1) The principal ones mostly have a NNW-SSE trend; and (2) many of them are extraordinarily lengthy. As a result, fault lines and fault scarps are prominent in the landscape (Fig. 16-2), and the general trend of most topographic lineaments (including the coastline) is NNW-SSE. Moreover, much of the topography has a "youthful" appearance, with a prevalence of steep slopes and high relief.

In gross pattern there are three broad topographic complexes within the region—coastal mountains and valleys, Central Valley, and Sierra Nevada.

The western portion of the region contains a series of mountain ranges with interspersed valleys. The Coast Ranges (roughly from Santa Maria northward) are strikingly linear in arrangement and are separated by longitudinal valleys of similar trend. The topography is structurally controlled, with prominent fault lines—of which the San Andreas is the most conspicuous and famous—tilted fault blocks, and some folding of the predominantly sedimentary strata. The nearly even crests of the ranges average 2000 to 4000 feet (600 to 1200 m) above sea level and are notable obstacles to the westerly sea breezes (Fig. 16-3).

South of the Coast Ranges are higher and more rugged mountains (the Transverse and Peninsular ranges) that are less linear and parallel in pattern.

The Central Valley consists of a broad downwarped trough averaging about 50 miles (80 km) in width and more than 400 miles (640 km) in length. The trough is filled with great quantities of sand, silt,

FIGURE 16-2 This originally straight sidewalk and wall down a residential street in Hollister provide dramatic evidence of recent movement along the San Andreas Fault (TLM photo).

and gravel (to a depth of more than 2000 feet [600 m] in places) washed down from the surrounding mountains. The Sacramento River flows south through the northern half of the Central Valley and the San Joaquin flows north through the southern half. They converge near San Francisco Bay and empty into the Pacific Ocean through the only large break in the Coast Ranges (Fig. 16-4). Their delta originally consisted of a maze of distributaries, sloughs, and low islands. It has now been diked and canalized and is one of the state's leading truck-farming and horticultural areas.

The Sierra Nevada is an immense mountain block 60 to 90 miles (100 to 150 km) wide and 400 miles (640 km) long, situated just east of the Central Valley. It was formed by a gigantic uplift that tilted the block westward. The eastern front is marked by a bold escarpment that rises 5000 to 10,000 feet (1500 to 3000 m) above the alluvial-filled basins of the Intermontane Region to the east; this escarpment marks one of the most definite geographical boundaries on the continent (Fig. 16-5). The western slope, although more gentle, is deeply incised with river canyons and was greatly eroded by glaciers, forming such magnificent canyons as those of Yosemite (Fig. 16-6) and Tuolumne.

The summits of the Sierra Nevada suffered severe glacial erosion and consist of a series of interlocking cirques. Complex faulting, mountain glaciation, and stream erosion account for most details of the mountain terrain. Some block-faulted valleys contain lakes, the most noted of which is Lake Tahoe; smaller but often spectacularly sited lakes occupy some glaciated valleys.

The long western slope of the range contains many magnificent glaciated valleys in which the glacial debris was completely washed away by the subsequent meltwater and deposited in the Central Valley. On the abrupt eastern slope, however, drier conditions resulted in less ice and less melting, with the result that glacial till accumulated in great moraines and now often encloses lovely alpine lakes. The high country of the Sierra Nevada still contains some 70 glaciers, all small.

FIGURE 16-3 Along most of the coast of central California the Coast Ranges drop precipitously to the sea. This is the Big Sur section (TLM photo).

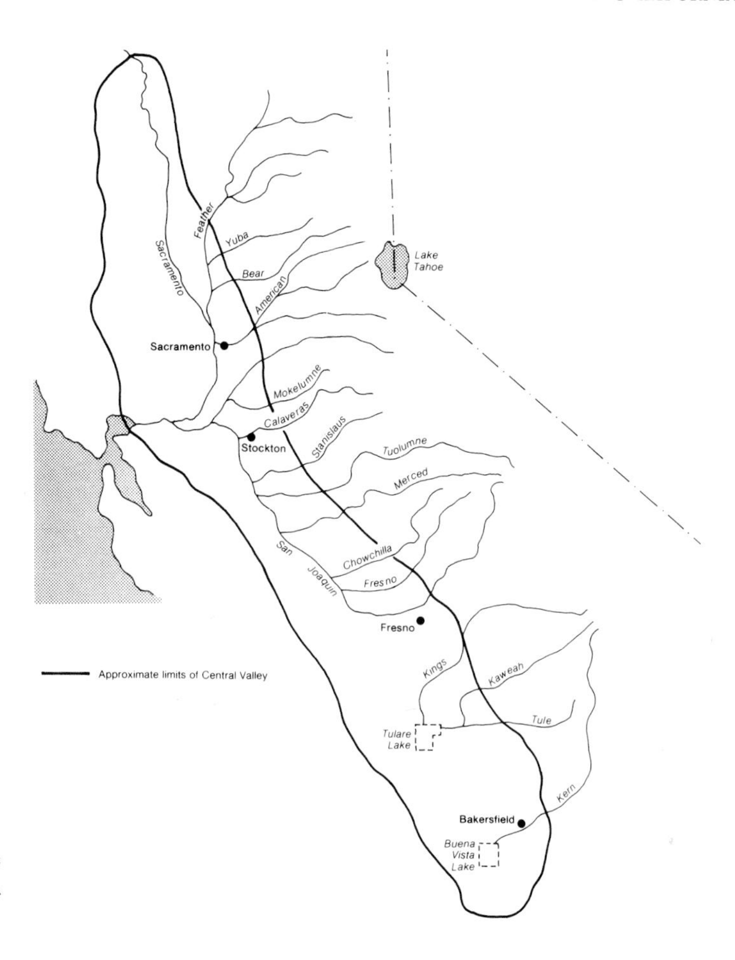

FIGURE 16-4 The drainage pattern of the Sierra Nevada and Central Valley.

Climate

The distinctive characteristics of the climate of the California Region are mild temperatures and a mediterranean precipitation regime.

Summer is dominated by a subsiding, diverging air flow that results in calm, sunny, rainless days and mild rainless nights. Temperature inversions are frequent, and inland locations can experience much hot weather.

Winter is the rainy season; this is due to recurrent cyclonic disturbances, but lowland areas receive only limited amounts of precipitation. The annual total increases northward from 10 inches (250 mm) in San Diego to 20 inches (500 mm) in San Francisco. West-facing mountain slopes receive much more rain and, particularly in the Sierra Nevada, considerable snow (Fig. 16-7).

Natural Vegetation

There is much localized variation in the vegetation pattern, but basic generalizations can be made. The original plant association of most of the immediate coastal zone was dominated by low-growing shrubs of coastal sage. Chaparral (a close growth of various tall, broadleaved, evergreen, resinous shrubs) was the characteristic vegetation of most of the valleys and lower mountain slopes of southern and central California, although there were also extensive areas of grassland dotted with oak trees. The Central Valley itself was mostly a natural grassland. The middle and upper slopes of the mountain ranges were (and still are) mostly forested, largely with a variety of conifers. Ponderosa pine, lodgepole pine, and various firs are the principal species. The world-famous giant sequoias occur in about 75 scattered groves at

FIGURE 16-5 The abrupt eastern escarpment of the Sierra Nevada, as seen from the Owens Valley. Mount Whitney is the peak in the center of the photograph (TLM photo).

middle elevations (4500 to 7500 feet [1350 to 2250 m]) along the western side of the Sierra Nevada (Fig. 16-8).

Natural Hazards: Shake and Bake in California

Unstable Earth The region is in a very unstable crustal zone, seamed in profusion by faults. Thus, earthquakes are experienced, sporadically and unpredictably, but frequently. Some tremors have had spectacular and tragic results (for example, the Loma Prieta/San Francisco quake of 1989 killed 65 people and caused $8 billion worth of property damage), but despite a high public consciousness of the potential danger, only a minuscule proportion of homeowners carry earthquake insurance.

Other types of earth slippages, on a much smaller scale, occur with greater frequency. The steep hills and unstable slopes of the subregion often afford spectacular views but precarious building sites. Every year there are dozens of small slumps and slides, and a few unfortunate residents "move down to a new neighborhood" whether they want to or not.

FIGURE 16-6 The long western slope of the Sierra Nevada contains spectacular canyons that were shaped by Pleistocene glaciers, the most striking of which is Yosemite Valley (TLM photo).

Flood and Fire The relatively small amount of precipitation falls almost entirely in the winter. As water flows down the steep hill slopes into urban areas, which are so extensive that they are inadequately supplied with storm drains, destructive floods and mud slides often result. During the rainless summers, on the other hand, forest and brush fires are ever imminent. The tangled chaparral, chamisal, and woodland vegetation of the abrupt hills and mountains that abut and intermingle with the urbanized zones is readily susceptible to burning. Only carefully enforced fire precautions, a network of fire breaks and fire roads, and efficient suppression crews are able to hold down the damage. And where burning strips away vegetation in the summer, flooding becomes even more likely in the following winter.

FIGURE 16-7 The Sierra Nevada normally is deeply inundated with snow every winter. Here a snow-blower clears the road in Giant Forest of Sequoia National Park (TLM photo).

SETTLEMENT OF THE REGION

The region's physical and cultural diversity is further reflected in its sequential occupance pattern. The prominent waves of Indian, Spanish, Mexican, and Anglo settlers have been more recently augmented by an influx of blacks and Asians.

California Indians

In contrast to the prosperous and attractive way of life in the region today, the aboriginal inhabitants found California to possess limited resources of value to them and thus they had restricted livelihood opportunities. Essentially they were hunters and gatherers and the basic foods for most tribes were either seafoods or plants that gave seeds and nuts that could be ground into flour and then boiled as gruel or baked into bread.

Tribal organization was loosely knit and generally involved small units. Often just a dozen or so families, perhaps related in some sort of patriarchal lineage, would form a wandering band. There were few large tribes or strong chiefs and internecine warfare was the exception rather than the rule. This made it possible for California to support a relatively high density of Indian population compared to most other parts of the country.

There was considerable linguistic, cultural, and economic diversity among the Indians within the constraints of their rather limited resources. Most, however, could be classed as relatively primitive and were easily degraded on contact with a more advanced culture. Thus most California In-

FIGURE 16-8 Giant sequoias are found in several dozen discrete groves at middle elevations on the western slope of the Sierra Nevada. This is Crescent Meadow in Sequoia National Park (TLM photo).

dians tamely gave way to white occupance of the region and left few marks on the landscape to signal their passing. And yet all the earliest recorded history of California revolved about them and the entire pre-Anglo period was shaped by Indian relations.

The Spanish Mission Period

The European "discoverer" of the region was Rodriguez Cabrillo, on a maritime mission from Mexico in 1542. Following this expedition came a long era of

A CLOSER LOOK The Santa Ana River—Sleeping Giant of Orange County

The Santa Ana River, about 100 miles (160 km) long, is a mere stripling compared with the great rivers of the North American continent. In 1769, Father Crespi of the Portola expedition named it Rio del Dulcissimo Nombre de Jesus (River of the Sweetest Name of Jesus). Now called the Santa Ana, it is viewed neither as "dulcissimo" nor "santa": Indeed, according to the U.S. Army Corps of Engineers, it poses one of the greatest flooding threats west of the Mississippi. Serious flooding occurred in 1938, when 45 lives were lost, and in 1969, 1978, and 1980, each time with major property damage. As population growth and urban sprawl engulf the Santa Ana Valley, there is little recognition of the local geography.

The Santa Ana and its headwaters rise in the San Bernardino and San Jacinto mountains, whose highest peaks top 10,000 feet (3000 m). The river drains a huge watershed of 3200 square miles (8300 km^2). The upper basin in San Bernardino and Riverside counties, with its major tributaries, Lytle Creek and the San Jacinto River, narrows into the Santa Ana Canyon between the Santa Ana Mountains and the Chino Hills. Thereafter the channel winds southwest across Orange County through a large alluvial floodplain, across which in the past the river and its major tributary, Santiago Creek, wandered excessively during flood stages. Now confined to a concrete and rip rap-sided single channel, it emerges to the Pacific Ocean at the city boundary between Huntington Beach and Newport Beach (Fig. 16-a).

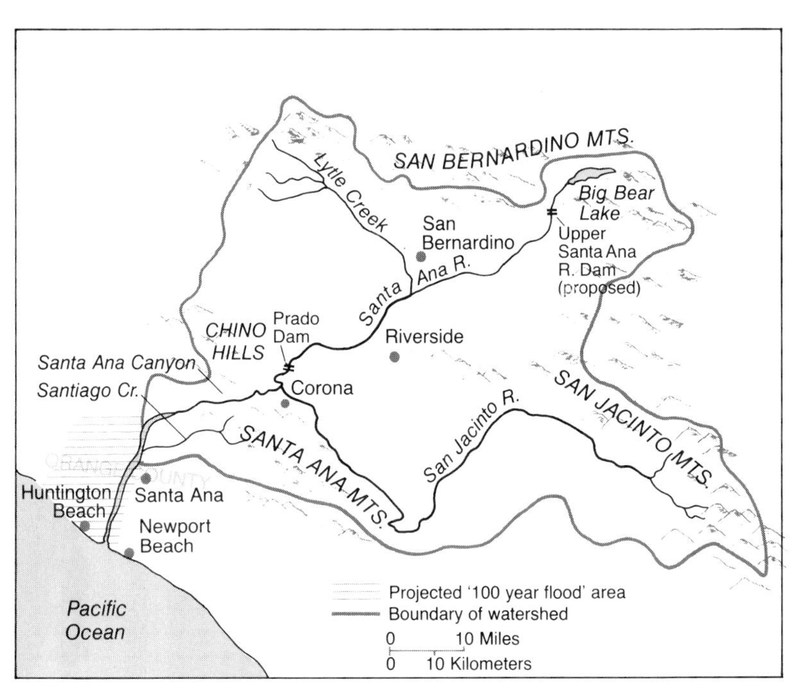

FIGURE 16-a

Geologically, the core of the mountain source region consists of metamorphic granodiorites and granites, which in the upper middle basin are buried beneath young rocks, poorly consolidated. These boulders and gravels, clays and shales are fan glomerates, the remains of earlier alluvial fans and extremely liable to slippage. Debris flows are common after rain, and a myriad of earthquake faults, especially associated with the Elsinore and San Jacinto faults, further contribute to this slope instability. The alluvial plain of Orange County is a "layer-cake" of riverine and marine silts, sands and gravel deposits, clay beds and widespread peat deposits within which the water table is high, often 2 to 4 feet (0.6 to 1.2 m) below the surface.

Across this geological mosaic, the flow of the river is extremely variable, deriving its water from the notoriously seasonal rainfall of this semidesert Mediterranean climate. Winter fronts from the Pacific bring rainfall with marked orographic effects: Newport Beach, at sea level, has a mean annual rainfall of 12.06 inches, Santa occasional exploration, but more than two centuries passed before there was any attempt at colonization of Alta California by the Spaniards. California seemed to offer little attraction to the adventurous, gold-seeking conquistadors. Renewed interest was stimulated not by the possibilities of the region itself but by the rapid advance of Russian domination of the Pacific Coast south from Alaska. Although Spain considered the California area economically worthless, it wanted a buffer state to prevent possible

Ana, at 133 feet (40 m) above sea level, 13.85 inches, while the inland mountains average 33 to 39 inches. What is more significant, however, is the annual variability of rainfall totals: Santa Ana over a period of 53 years of records shows a maximum annual total of 25.35 inches and a minimum of 4.85 inches: Big Bear in the San Bernardino Mountains over 28 years varied between 51.41 inches maximum and 13.3 inches minimum. In addition, frontal storms may often closely follow each other within a brief period. In January 1983, for example, the four-storm sequence was as follows: January 22/23, 1.53 inches; January 23/24, 1.20 inches; January 26/27, 1.95 inches; and January 28/29, 1.39 inches. Runoff therefore peaks rapidly.

Rapid runoff is further complicated by the characteristics of the local natural vegetation. Apart from the highest points of the watershed, the slopes are covered with chaparral scrub with very high volatility. Santa Ana winds spread accident- or arson-caused fires (only 10 percent are caused by lightning) relentlessly: Hundreds of acres can be burned off in a brief time, leaving little but a blanket of ash. Add to this the waxy residue left just under the surface by the burning of these oily xerophytic species, and the potential for rapid runoff of intense storms is maximized.

Thus the disaster scenario for floods: fall and early winter Santa Ana winds and fires; heavy winter/spring rains of Pacific storms; rapid snow melt off the high watersheds; onshore winds and high spring tides along the coast to counter the outflow and impact the runoff. But there is more! First, the sudden reduction in gradient as the river leaves the Santa Ana Canyon to flow across the Orange County floodplain. Velocity is checked, river load is deposited, channels are clogged, and levees are overtopped by the rising water. Second, there is a significant gradient fall from the canyon to the marine outlet: The drop here in a distance of 30 miles (48 km) is equivalent to a similar gradient drop in the lower Mississippi over 600 miles (960 km)! The flood wave would move through here in 8 hours, compared with the two weeks it would take to cover the same gradient distance on the Mississippi. Such alarming flood predictions are further complicated by the "concretization" of urban growth, earlier in coastal Orange County and now very rapidly in the "Inland Empire" from Corona to Riverside. No longer absorbed by natural soils and vegetation, the runoff sheds extremely rapidly over urban surfaces and concentrated peak flows result.

The worst predicted scenario is one called the Standard Project Flood. Such a flood has a statistical chance of occurring once every 200 years and would cover 100,000 acres (40,000 ha) of central Orange County to an average depth of 3 feet (7 m). It is estimated that such a major flood could cost 3000 lives and cause over $15 billion damage. A 100-year flood would cover much of the lower basin from the city of Garden Grove to the coast.

So what is to be done? The Orange County Flood Control District was formed in 1927. Local property taxes are the main source of funding for flood-control maintenance and improvements. After the 1938 flood, Prado Dam near Corona was completed by the Army Corps of Engineers in 1941 to hold back flood water from the upper basin and feed it in a controlled fashion downstream. In spite of its partial efficacy in subsequent floods, damage and levee breaks occurred downstream, and it is recognized that Prado Dam and reservoir do not have the capacity for major flood control. Finally, in 1986, the Santa Ana River Project was signed into law by President Ronald Reagan. Agreement on details among the counties of Orange, Riverside, and San Bernardino was finally reached in December 1989. The $1.1 billion All-River Plan has three major components (which will take 7 to 20 years to complete). First, an Upper Santa Ana River Dam in the San Bernardino Mountains must be built to withstand a major earthquake. Second, the Prado Dam will be raised by 30 feet (9 m) to 136 feet (41 m) to increase its holding capacity to withstand a 170-year flood. Finally, flood channel improvements along the Santa Ana River and Santiago Creek will include relining, grading, and widening. Does this plan mean that the giant is tamed, or will it awaken before the plan is completed? Only time—and geography—will tell.

Dean Sheila Brazier
Golden West College
Huntington Beach, California

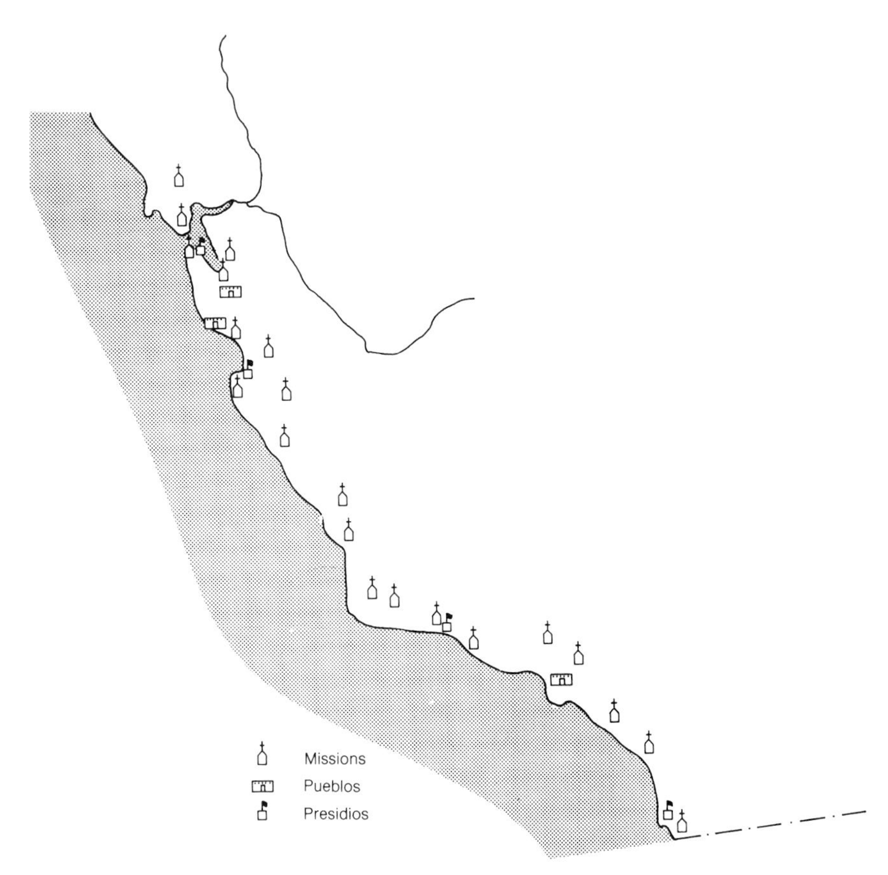

FIGURE 16-9 Early Spanish settlements in California were along or near the coast.

Russian encroachment on the more valuable colony of Mexico.

The first of Spain's famous missions in California was founded in 1769 at San Diego and within three years an irregular string of missions, both in protected coastal valleys and directly on the coast, had been established as far north as Monterey. Eventually 21 continuing missions were established, the last in 1823 (Fig. 16-9). The purpose of the mission system was to hold the land for Spain and to Christianize the Indians. These steps were accomplished by settling the nearby Indians at each mission to carry on a completely self-sufficient, sedentary crop-growing and livestock-herding existence. Each of the missions garnered an attached population of from 500 to 1500 Indians.

The basic results of the system were that some 5,000 Indians were "civilized," at least that many and probably more were exterminated by the inadvertent introduction of exotic diseases, various European plants and animals were introduced to the region, and a number of settlement nuclei and transportation routes was established.

Two other types of settlements, *presidios* and *pueblos,* were established during the mission period. Four presidios were set up as army posts for protection against Indians and pirates; each developed as a major settlement nucleus (San Diego, Monterey, San Francisco, and Santa Barbara). Pueblos, three in number, were planned farming villages; Los Angeles and San Jose survived as future major cities.

Generally the missions reached the height of their prosperity in the 1820s in spite of problems with fire, flood, earthquake, and unbelievers. The mission–presidio–pueblo system, however, did not succeed in attracting many settlers to California. Land grants were few, small, and allotted chiefly to retired soldiers.

The Mexican Period

Mexico, including California, achieved independence from Spain in 1822 and the missions were gradually secularized. The result was a rapid alienation of the land into private ownership, neglect of property and buildings, and lapsing of the mission Indians into degraded poverty or demoralized frontier life. The Mexican government embarked on a large-scale scheme of land grants to individuals; within only a few years most of the land from Marin and Sonoma counties (just north of San Francisco Bay) southward along the coast was held in *ranchos*.

Ranchos quickly replaced missions as the focus of life in California. They were primarily self-sufficient cattle empires, with beef as the chief food and hides and tallow as the principal exports. Towns were few, small, and far apart. Los Angeles was the largest; Monterey was the capital and the chief seaport. Most of the inland portion of the region remained unsettled except for a few ranchos in the Central Valley.

The Early Anglo-American Period

California Territory was annexed to the United States at the end of the Mexican War in 1848. By a remarkable coincidence, gold was discovered in the Sierra Nevada foothills only two months before the peace treaty was signed. The impact of gold discovery was sensational. The total population of California in 1845 was about 5000, of whom less than 8 percent were Anglo-Americans. Half a decade later the population had reached nearly 100,000, of whom 90 percent were Anglos.

It was San Francisco that prospered most as a result of gold. Its population of 800 almost vanished in the first rush to the gold fields. Its outstanding location soon gave it ascendancy, however, and in a short time it was the largest city in western Anglo-America. The population increased to 35,000 in two years and was more than 50,000 in 1860 despite having been leveled by fires five times during that period. The population of California was 380,000 by then.

Los Angeles at that time had less than 5000 people; the influence of the gold strike had been much less significant in the "cow counties" of southern California. Most of the land was devoted to cattle raising and, except for the pueblo of Los Angeles and settlements around a few presidios and missions, there were no towns at all.

The Beginnings of the Commercial-Industrial Era

The first California land boom started slowly but grew abruptly in the 1880s. The boom spread from Los Angeles to San Diego and Santa Barbara and was felt over much of the region. The population of Los Angeles County tripled to 100,000 in that decade. Separate colonies, based on irrigation farming of specialty crops, were founded in many places. There were orange colonies at Anaheim, Pasadena, Riverside, and Redlands; grape colonies in the Fresno area; and other varied colonies in different parts of the San Joaquin Valley.

The availability of "transcontinental" railway connections provided access to distant markets and proved a major solution to marketing problems. The first transcontinental line, from Omaha to San Francisco, was completed in 1869 and a railway was extended from San Francisco to Los Angeles in 1876.

Irrigated agriculture spread and intensified rapidly in the Los Angeles lowland, in the numerous coastal valleys, and in the Central Valley. Agricultural progress was accelerated by the development of refrigerated railway cars, canning and freezing of produce, improved farm machinery, pumps and pipes for irrigation and drainage, and the beginnings of agribusiness enterprises.

Industrial growth began with petroleum discoveries toward the end of the nineteenth century. The emergence of the motion picture business was a major catalyst, both physical and promotional, to economic and population growth and urban expansion. Manufacturing was stimulated during World War I and experienced a phenomenal growth during World War II.

The present occupance and land-use pattern of the region, however, was basically in existence by the early years of the twentieth century. Changes since that time have mostly involved the extension of irrigated acreage in lowland areas and the never-ending sprawl of the cities.

POPULATION: SENSATIONAL GROWTH

The keynote of the geography of the California Region during most of this century was the rapid growth of its population. The rate of natural increase (excess of births over deaths) was only average, but the trend of net in-migration (excess of in-migrants over out-migrants) has been nothing short of sensational.

Although there have been fluctuations, for most of the past half century California's population has been increasing at an annual average rate approximately twice as great as that for the nation as a whole. As of the early 1990s, this has amounted to about 600,000 people annually, or 2 percent each year. More than half the population increment now is the result of net in-migration and most of it comes from foreign countries rather than other states. The key fact is the long-term continuity of rapid growth. For many decades both the amount and the rate of increase have been exceptional, and there is no indication of impending change.

The ethnicity of the regional population mix is also extraordinarily varied. Of the 50 major ancestry groups tabulated in the 1980 census, California contained the largest population of 21 of them.[2] People of nonEuropean origin are represented in unusually large proportions; more than a third of the national population of each of the following ancestry groups resides in California: Armenian, Chinese, Filipino, Iranian, Japanese, Mexican, and Vietnamese. As a corollary, it might be noted that in California only about three-quarters of the residents speak only English in their homes whereas, for the nation as a whole, more than 90 percent of all people speak only English at home.

One other prominent characteristic of the regional population is its urbanness (Table 16-1). More than 92 percent of all Californians live in urban areas. This figure is exceeded proportionately only

[2] These included American Indian, Armenian, Chinese, Danish, Dutch, English, Filipino, French, German, Iranian, Irish, Japanese, Korean, Lebanese, Mexican, Portuguese, Scottish, Spanish-Hispanic, Swedish, Swiss, and Vietnamese.

TABLE 16-1

Largest urban places of the California Region

Name	*Population of Principal City*	*Population of Metropolitan Area*
Alameda*	75,080	
Alhambra*	72,230	
Anaheim	244,670	2,257,000
Bakersfield	157,650	520,000
Baldwin Park*	65,280	
Bellflower*	60,230	
Berkeley*	103,660	
Buena Park*	66,650	
Burbank*	91,960	
Carson*	89,380	
Cerritos*	58,520	
Chico	35,000	174,500
Chula Vista*	126,240	
Compton*	95,060	
Concord*	108,040	
Costa Mesa*	90,710	
Daly City*	85,810	
Downey*	86,520	
El Cajon*	88,240	
El Monte*	97,640	
Escondido*	95,390	
Fairfield*	73,250	
Fontana*	72,430	
Fountain Valley*	56,310	
Fremont*	166,590	
Fresno	307,090	614,800
Fullerton*	109,740	
Garden Grove*	135,310	
Glendale*	161,210	
Hawthorne*	64,730	
Hayward*	103,600	
Huntington Beach*	186,880	
Inglewood*	103,920	
Irvine*	100,130	
Lakewood*	76,200	
Lancaster	76,380	
Long Beach	415,040	
Los Angeles	3,352,710	8,587,800
Merced	50,270	170,000
Modesto	148,670	341,000
Montebello*	56,790	
Monterey Park*	61,920	
Moreno Valley*	88,840	
Mountain View*	61,850	
Napa	57,320	
Newport Beach*	69,060	
Norwalk*	91,410	

(continued)

TABLE 16-1 *(Continued)*

Name	*Population of Principal City*	*Population of Metropolitan Area*
Oakland	356,860	2,006,300
Oceanside*	111,030	
Ontario*	123,380	
Orange*	105,710	
Oxnard	130,080	647,300
Palo Alto*	55,970	
Pasadena*	132,010	
Pico Rivera*	56,210	
Pomana*	120,470	
Rancho Cucamonga*	98,340	
Redding	55,400	139,700
Redondo Beach*	63,890	
Redwood City*	60,030	
Rialto*	62,750	
Richmond*	81,220	
Riverside	210,630	2,277,600
Sacramento	338,220	1,385,200
Salinas	101,090	348,800
San Bernardino	148,420	
San Diego	1,070,310	2,370,400
San Francisco	731,650	1,590,100
San Jose	738,420	1,432,000
San Leandro*	66,790	
San Mateo*	82,980	
Santa Ana*	239,540	
Santa Barbara	78,170	343,100
Santa Clara*	89,830	
Santa Clarita*	95,590	
Santa Cruz	47,770	226,700
Santa Monica*	94,060	
Santa Rosa	108,220	366,000
Simi Valley*	98,890	
South Gate*	81,670	
Stockton	190,680	455,700
Sunnyvale*	116,180	
Thousand Oaks*	101,530	
Torrance*	137,940	
Upland*	62,410	
Vacaville	63,880	
Vallejo	100,730	420,700
Ventura	88,900	
Visalia	66,070	297,900
Walnut Creek*	60,780	
West Covina*	97,120	
Westminster*	73,320	
Whittier*	73,630	
Yuba City	22,910	118,400

* A suburb of a larger city.

by two small states in the Megalopolis Region and compares with a national average of 74 percent urban. For the first time in modern California history, however, rural areas are growing at a faster rate than urban areas. In the 1980s the rate of rural population increase was three times that of the cities. Even so, the vast majority of the region's population continues to be urban in location, character, and orientation.

THE RURAL SCENE

Although most of the population of California is urban and most employment is in urban-oriented activities, three prominent nonurban activities are significant contributors to the regional economy. Each is discussed in turn.

Agriculture

Although many other states harvest greater acreages of crops than California, throughout the past half century the total value of farm products sold has been higher in California than in any other state (Fig. 16-10). Fertile soil, abundant sunshine, supplies of irrigation water, a long growing season, and careful farm management result in high yields of high-value products. California leads the nation in the production of 58 different crops and livestock products, including the total national production of 10 crops, more than 90 percent of the output of another 10 crops, and more than half the yield of another 16 farm products.

By almost any measure, California is the leading agricultural state. It produces more than 40 percent of the national total of fresh fruits and vegetables, for instance, which is more than the combined total of the next three ranking states. With 2 percent of the nation's farms, it earns nearly 10 percent of gross national farm receipts. Eight of the 10 most productive agricultural counties in the nation are in the California Region.

Irrigation is the backbone of the region's agriculture, and California is by far the leading irrigated state. Irrigation projects and techniques are many and varied. Complete integration and totally effective utilization of water supply are difficult to achieve but have been more closely approached in

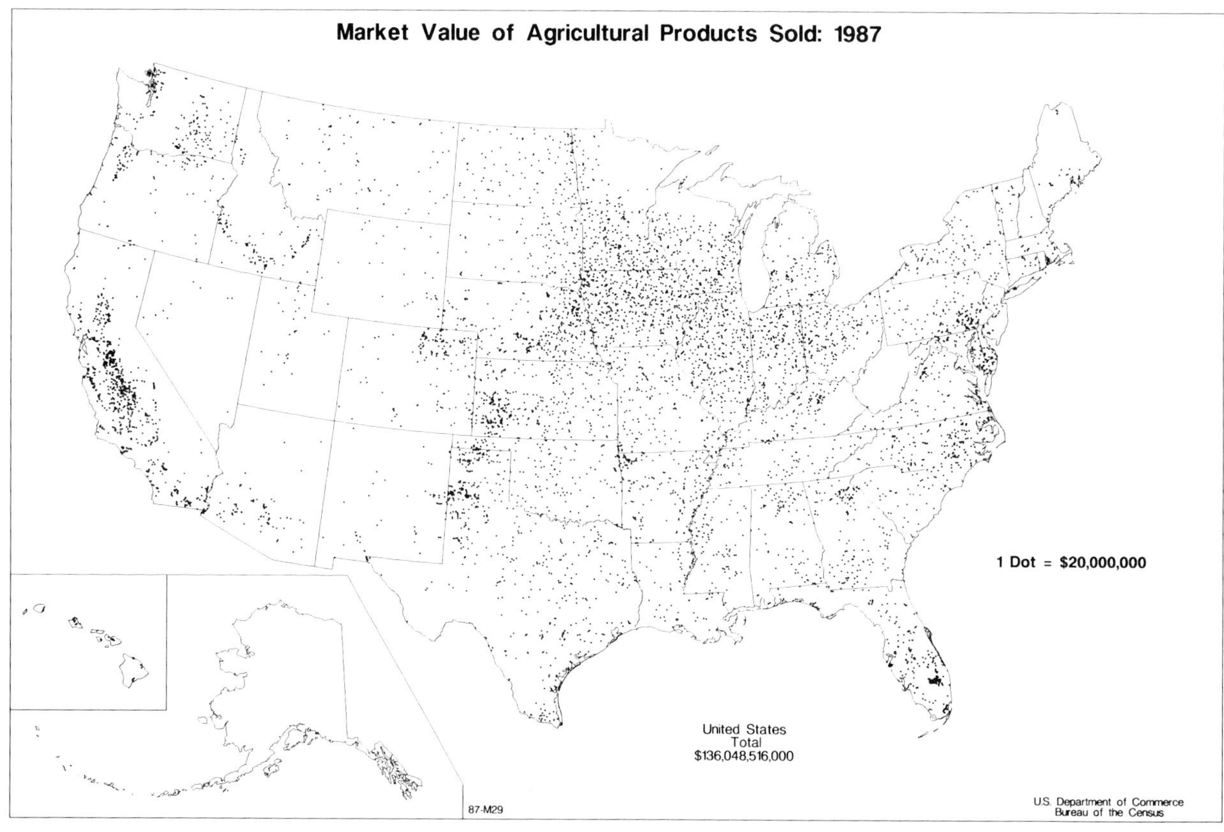

FIGURE 16-10 Gross farm income is higher in the California Region than in any other. The Central Valley concentration is particularly marked.

this region than just about any irrigated area in the world. The most ambitious projects—Central Valley Project and California Water Plan—both transfer water from the surplus north to the deficient south. The principle of operation is that surplus Sacramento River water is stored and then diverted by canal to the San Joaquin Valley to the south. Many smaller valleys have integrated water projects, and there is widespread pumping from wells.

Agricultural areas Three-fourths of the farm output of the California Region come from the *Central Valley*. The *Sacramento Valley* (which comprises the northern third of the Central Valley) receives more rainfall and therefore requires less intensive irrigation; its major products are rice, almonds, tomatoes, sugar beets, wheat, hay, cattle, milk, and prunes. The heavily irrigated *San Joaquin Valley* comprises an agricultural cornucopia that has major output of dozens of farm products, particularly cattle, milk, grapes, cotton, hay, oranges, and almonds. (Fig. 16-11). The *Salinas Valley,* south of Monterey, specializes in such sensitive vegetables as artichokes, lettuce, and brussels sprouts. The *Oxnard Plain,* between Los Angeles and Santa Barbara, concentrates on vegetables and citrus fruits. A dozen other coastal valleys and basins are noted for high-quantity output of high-value crops—ranging from nursery products on the San Diego terraces to the famed vineyards of the Napa Valley just north of San Francisco—largely grown under irrigation.

Agricultural products *Milk* is the most valuable agricultural commodity produced in the California Region, and California ranks second only to Wisconsin as a dairy state. Production is widespread, but comes principally from the San Bernardino–Riverside area and the San Joaquin Valley. *Cattle* comprise the second most valuable farm product, and come most notably from the San Joaquin Val-

FIGURE 16-11 The variety of crops grown in the Central Valley is remarkable. This is a field of cantaloupes near Bakersfield (TLM photo).

ley. The region's third-ranking farm product and leading crop is *grapes,* of which California produces nearly 90 percent of the national total, primarily from the San Joaquin Valley (Fig. 16-12). California is also the nation's leader in *nursery products,* the region's fourth-ranking farm product, with most production from Los Angeles, Orange, and San Diego counties. *Cotton,* primarily from the San Joaquin Valley, ranks fifth, but is on a declining trend because of a cutback in government price supports. *Hay* is the sixth-ranking farm product, and California is the leading producing state. Other major

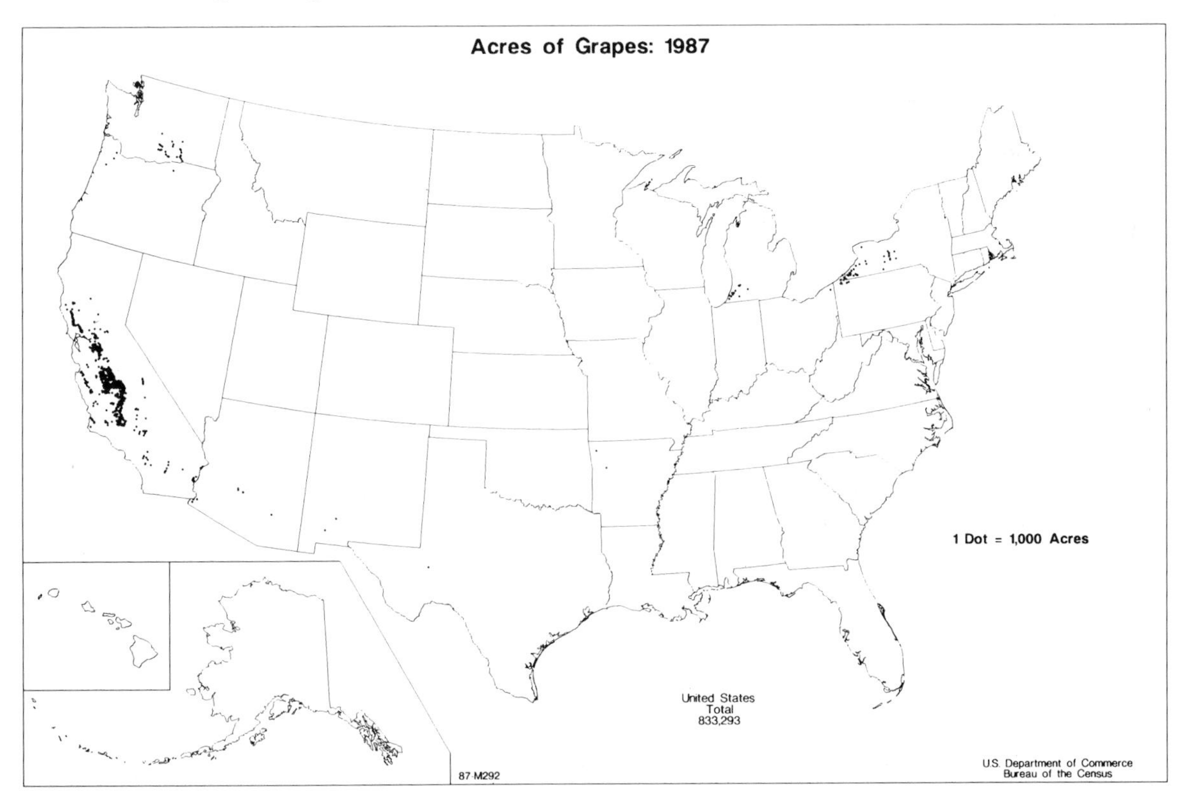

FIGURE 16-12 California, especially the Central Valley, produces most of the nation's grapes.

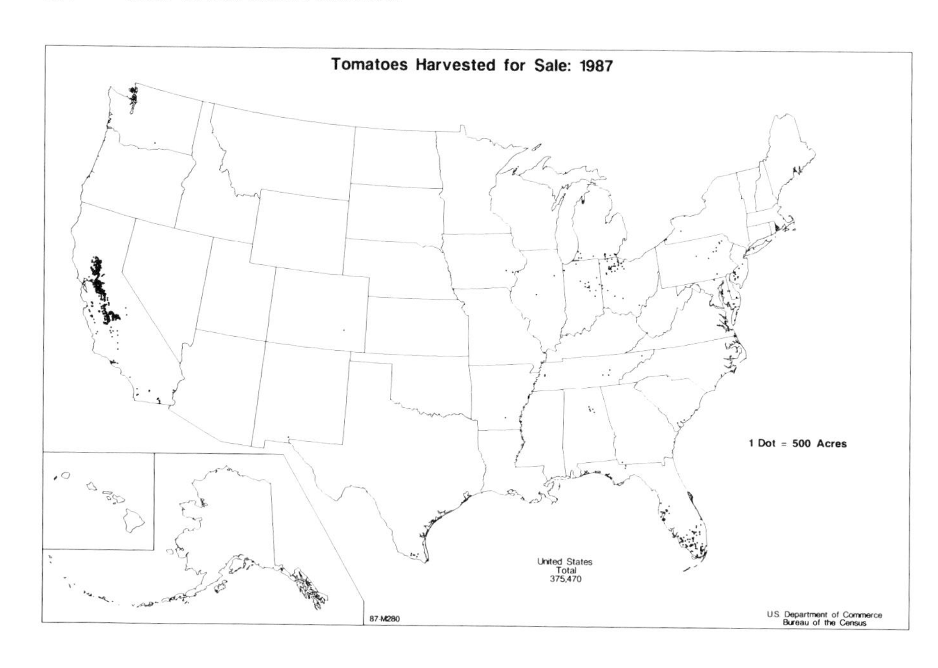

FIGURE 16-13 The delta area and adjacent parts of the Sacramento and San Joaquin valleys grow the majority of the nation's tomatoes.

farm products from the California Region are *flowers, lettuce, almonds, processing tomatoes, strawberries, oranges,* and *chickens* (Fig. 16-13).

Petroleum and Natural Gas

Only three other states exceed California in output of oil and natural gas. Most production is in the southern California coastal area, but there are also two dozen small oil fields in the southern part of the San Joaquin Valley. Four of the 10 leading oil fields in the United States, in terms of cumulative production, are in southern California. The Wilmington Field is second only to the fabulous East Texas Field in output, the Huntington Beach Field ranks sixth, Long Beach is seventh, and Ventura is tenth.

There are large known and suspected reserves of petroleum in the southern California offshore waters, but exploitation was limited until 1965 when the city of Long Beach sold the first major leases to oil companies. Several hundred new wells have been drilled since then, most of them whipstocked from four islands that were built in Long Beach harbor. The wells have been totally camouflaged by a covering of pastel shielding and the artificial islands have been made as attractive as possible by planting palm trees, creating waterfalls, and night lighting (Fig. 16-14).

Offshore drilling is more expensive in southern California than in the Gulf of Mexico because the producing horizons are deeper. Furthermore, several costly and controversial leaks—most notably in the Santa Barbara Channel—occurred and resulted in protracted litigation and stringent regulation. Nevertheless, the reserves of this offshore area are among the largest in North America. By the early 1990s some 2000 wells were producing in the Santa Barbara Channel, whipstocked from 24 drilling platforms and 7 artificial drilling islands. This area is second only to Alaska's Prudhoe Bay in recent production expansion.

Commercial Fishing

Major fishing grounds extend from southern California waters southward to Peru and westward to Hawaii. Los Angeles is the only prominent fishing port in the region, and its significance is on the decline, as lower-cost Asian fishing and fish-processing activities provide harsh competition for West Coast operators.

Tuna is easily the most important catch of southern California fishermen, although anchovies (for fish meal, oil, and bait), mackerel (especially for pet food), and bonito are also significant. In overall landings, California ranks fourth among the states both in quantity and value of its fishing industry.

FIGURE 16-14 A cluster of offshore oil wells near Long Beach, California. For aesthetic purposes, the wells are disguised to look like colorful buildings (TLM photo).

URBANISM

The California Region is second only to Megalopolis in its degree of urbanism. Some 94 percent of its inhabitants are urban dwellers, and most of them live in large metropolitan areas.

The Southern California Metropolis

The development of a southern California "megalopolis" is now an accomplished fact (Table 16-1). The urban assemblage is an extensive one: The east-west axis from Palm Springs to Santa Barbara is 170 miles (270 km); the north-south extent from Lancaster, in the Antelope Valley, to the Mexican border is approximately the same distance; and the southeast-northwest coastal frontage from San Diego to Santa Barbara covers more than 200 miles (320 km). Urbanization is by no means complete within these dimensions. Massive hills and even a few mountain ranges are encompassed; urban areas are restricted mostly to the lowlands and valleys except where favored hill slopes are being carved and terraced for residential construction.

Characteristics of the Urban System Urban southern California consists of a loosely knit complex of people, commerce, and industry—all fused in a single system by a highly developed freeway network, a common technology, a common economic interest, and by numerous other shared values. Los Angeles is the major center and its name serves as a toponymic umbrella for most of the urban system; however, Los Angeles cannot properly be classified as the focus of the metropolis, for this sprawling urban complex really has no focus. Its development has been polynuclear, and with each passing year the other nuclei become proportionately stronger and more self-contained.

The Los Angeles–Long Beach node is the largest and most prominent; the other major nuclei, in descending rank-order of population, are San Diego, Anaheim–Santa Ana, Riverside–San Bernardino, Oxnard–Ventura, and Santa Barbara–Santa Maria–Lompoc. All these nodes are interrelated within the subregional economy and within the general sphere of Los Angeles influence, but they are also essentially separate entities that dominate their own cluster of lesser cities. Separating the major nodes is a mixture of nonurban land, attenuated and irregular string-street commercial development, and low-density residential sprawl.

The complexities of the polynuclear pattern are varied. Important commercial centers, for example, are not limited to central business district locations but are widely dispersed over the metropolis. A visual indication of this dispersion is provided by the relatively large number of localities—more than a dozen—where there are concentrations of high-rise commercial buildings. Although southern California skyscrapers do not rise as high as those in many eastern cities, their clustering is much more thoroughly disseminated, which has a markedly different effect on traffic patterns and subregional economic relationships.

The Economic Base The southern California economic structure is highly diversified, as with most metropolises, but it has certain distinctive elements. Most notable is perhaps the major role played by certain types of manufacturing in this area, the second largest manufacturing district of the continent. Ever since the early years of World War

II, the local economy has been significantly geared to defense production. Shipbuilding was once an important facet of this, but the aircraft industry, which has now developed into the *aerospace* industry, has always been dominant (Fig. 16-15).

Southern California has been the world's leading aircraft manufacturing center since the 1930s and maintains that distinction today in aerospace despite the recent development of spacecraft production facilities in other parts of the country. Roughly one-third of the subregion's total manufactural employment is in aerospace. The huge windowless plants of the prime contractors are invariably located adjacent to an airport—Los Angeles International, Santa Monica, Long Beach, Burbank, Palmdale, San Diego's Lindbergh Field—but the numerous subcontracting firms are widespread over the subregion.

Other notable or distinctive types of manufacturing in southern California include electronics; sports clothes; furniture, particularly outdoor furniture; automobile assembly; rubber products, especially tires; and petroleum refining. The industrial structure has also been strengthened in recent years

A CLOSER LOOK With a Landbridge, Who Needs the Panama Canal?

Each year the Panama Canal becomes increasingly less useful to the United States, and American harbors such as the port of Los Angeles become progressively more important. One factor involved in the growing disinterest in the Panama Canal is that modern American aircraft carriers are simply too big to fit through this waterway. The same is true for the giant supertankers known as "ultra large crude carriers" (ULCC), which move great quantities of oil so effectively. Much of the oil delivered by ULCCs to North American ports on the Pacific Coast is shipped eastward by pipeline. (One of those pipelines even crosses the Isthmus of Panama.) The greatest volume of shipping business being drawn away from the canal, however, is that which is carried by "container" vessels. These days, for example, much of the Far East cargo bound for the East Coast of the United States is off-loaded in harbors such as Los Angeles and then shipped across the country by rail on the "landbridge."

At the heart of this "intermodal" system is containerization, which has revolutionized cargo handling for long-distance shipping. Formerly, processed or manufactured goods were boxed, crated, or baled and then loaded onto trucks, rail cars, or ships by forklift trucks or small cranes. A great deal of time and labor was invested every time a carrier had to be loaded or unloaded or whenever a shipment had to be transferred from one transportation mode to another. Today, a shipper can load the cargo into large aluminum containers at the production site and not handle the individual boxes, crates, or bales again until the final destination is reached. It does not matter how many times those containers change modes of transportation along the route; the cargo remains sealed inside until the end of the trip.

The containers are designed to fit quickly and securely onto trucks, rail cars, and ships. The average container

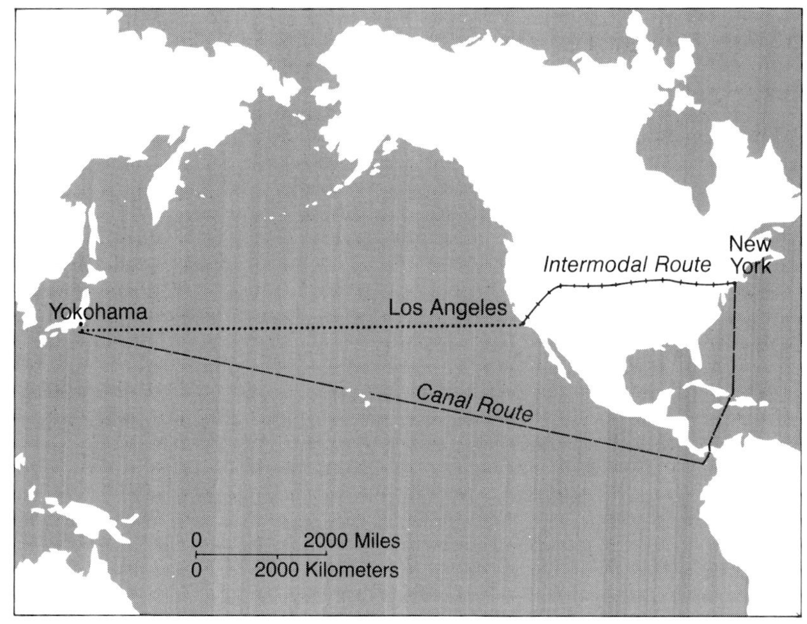

FIGURE 16-b The function of the continental landbridge.

by the growth of several traditional "eastern" industries. During the 1970s, for example, New York City (the leader) lost 50,000 jobs in the garment industry whereas Los Angeles (now ranking second) added 30,000. During the same decade employment in printing and publishing declined by one-fourth in New York City (the leader) while it was expanding by one-third in Los Angeles (which ranks third).

The most distinctive major economic activity in the subregion, however, is the *motion picture and media entertainment industry*. Hollywood was the birthplace and long-time capital of movie making and is still a focal point, although there has been decentralization of the major studios and network production facilities, both within the metropolitan area and far afield. Still, more than one-third of America's total employment in film production and distribution is in southern California and more than 90 percent of the world's recorded entertainment is produced within 5 miles (8 km) of the corner of Hollywood and Vine.

The list could go on and on. There are now more people employed in health care in southern California than in any other single industry, includ-

is the size of a trailer-truck van. All that is needed is a large crane to mount each container on the chassis of a trailer and that truck is ready for the road. Special rail cars capitalize on the strength of these containers by carrying one on top of another. These "double-stack" trains haul up to 250 containers at a time. The hold of an ocean-going container ship consists of dozens of cells, arranged to accommodate stacks of containers. This type of ship is also configured to carry hundreds more containers outside, on the deck. The largest ship can carry almost 2000 (40-foot [12 m]) containers.

For many shipping firms and their clients, time has become the cost factor of overriding importance. Certain types of cargo simply must be delivered to market faster than others. Additionally, since newer ships are so much more expensive to build than their predecessors, the amount of time a vessel devotes to each shipment is now a greater cost consideration to a shipping line than fuel or labor.

The intermodal alternative to the Panama Canal saves shipping time in several ways. One is that, for cargo traveling from the Far East to the East Coast of the United States, the intermodal route is simply several thousand miles shorter than the canal route (Fig. 16-b). Since large cranes can off-load a container vessel two to three times faster than a conventional freighter of equivalent capacity, the time lost in transferring the cargo from one transportation mode to another is negligible (Fig. 16-c). Once loaded onto rail cars, the containers can be hauled rapidly along any one of several different cross-country railroad routes that comprise the landbridge. The rail component of the intermodal system operates most efficiently when entire trainloads of containers can be earmarked for express delivery to the East Coast on a regularly scheduled basis. Such double-stacked "liner trains" take six days or less to travel the landbridge from Los Angeles to New York. All told, the intermodal system can get cargo delivered from the Far East to the East Coast a week and a half faster than ships taking the Panama Canal route. One shipping company serving Los Angeles is so committed to the intermodal alternative that it has commissioned several new container ships that are too wide to fit through the Panama Canal.

FIGURE 16-c A modern container ship (courtesy American President Lines).

Professor Joseph P. Beaton
California State Polytechnic University
Pomona

FIGURE 16-15 Large aerospace factories are numerous in Southern California. This is the Rockwell International plant at Anaheim (courtesy Rockwell International).

ing aerospace. Los Angeles has even become the dominant financial center in the West; its 200,000 workers in financial institutions are half again as many as in San Francisco, the traditional banking center.

The Population Mix The spectacular long-run population growth of southern California has slowed a bit in the most recent years but is still very impressive. The 1990 population of southern California (15,500,000) was larger than that of 47 of the 50 states.

Within the last three decades there has been a remarkable upsurge of inmigration of ethnic minority peoples (Fig. 16-16). The proportion of blacks and Hispanic-Americans in the subregional population mix increased dramatically and an even greater percentage gain was registered by Asians and Native Americans, although their total numbers are much smaller than the blacks or Hispanics. Los Angeles itself is a "minority" city, with no major ethnic group amounting to as much as half of the total (46% Anglo, 29% Hispanic, 17% black, 8% other). Indeed, Los Angeles has displaced New York as the principal embarkation point for foreign migrants; more than 2 million foreign immigrants settled in greater Los Angeles during the 1980s.

The Changing Urban Pattern For many years the form and pattern of urban areas of southern California, in general, and Los Angeles, in particular, have been acclaimed as unique, their develop-

FIGURE 16-16 Los Angeles has become a melange of diverse ethnicity. At the time this photo was taken the Kosher Burrito shop was owned by a Japanese (TLM photo).

ment diverging from that of other North American metropolises. The most conspicuous abnormalities have included a lack of focus on the central business district, emphasis on detached single-family housing, low population density, and overwhelming dependence on the automobile for local transportation (Fig. 16-17). More recently, however, the distinctiveness of southern California urbanism has been significantly muted. This reflects, in part, the fact that the older cities of North America have been exhibiting some of these same characteristics—declining CBD focus, increasing urban sprawl, greater dependence on automobiles—in their recent development.

More pertinent to southern California, however, is the fact that opposite tendencies have become apparent within the subregion. Population density is increasing, multifamily housing is expanding rapidly, and CBDs are developing "new" central functions. In short, a centripetal trend is setting in. In and around the principal urban cores of southern California multifamily housing units, often high-rise, are now being constructed in greater numbers than are single-family detached units. Rising land costs and an expanding population are (perhaps inevitably) producing a housing and business profile, and possibly an overall urban profile, that is increasingly similar to cities in the East and Midwest.

These centripetal tendencies are particularly prominent in Los Angeles but can also be seen in San Diego, Santa Monica, Long Beach, Santa Barbara, and even Pasadena. The functional complexity of the CBD increases. There is no intensification of the commercial function, but there are an increasing focalization and centralization of corporate headquarters, financial services, government offices, and cultural facilities. Furthermore, population and building densities are increasing in residential areas, owing to high-rise apartments, smaller subdivisions, and other factors. By the mid-1970s, for example, suburban Los Angeles had a population density comparable to that of suburban Chicago or Philadelphia.

The southern California metropolis, long a harbinger of things to come in urbanism, has thus lost much of its distinctiveness. In varying ways the urban area has evolved into a reasonably "average" metropolis. This is not to say that uniformity has set in but merely that some tendencies have been changed and even reversed.

FIGURE 16-17 The world's first four-level freeway interchange was "The Stack" in Los Angeles. It is still the focal point of the city's freeway system (TLM photo).

There are still many unique and perhaps futuristic characteristics to the urban pattern and lifestyle of southern California. Where else does one find such a conspicuous example of cellular urban development on a polynuclear framework, spatially integrated by a highly developed freeway system, and seemingly capable of indefinite expansion? Where else can one find a meticulously detailed plastic Matterhorn rising 146 feet (44 m) above the floor of Disneyland, only to be topped by the $21 million, 230-foot-high (69-m) scoreboard of a nearby baseball stadium, which is, in turn, overshadowed by the 250-foot (75 m) flourescent cross of a neighboring drive-in church?

Smog Air pollution is a persistent fact of life in most cities, but its most infamous occurrence is in the southern California metropolis. Los Angeles is the undisputed champion, with more than twice as many "smog alert" days as any other city in Anglo-America, but Riverside, San Bernardino, and San Diego are also among the leaders.

The frequent high level of smogginess in Los Angeles is due primarily to the unique physical characteristics of the metropolitan site. The Los Angeles lowland is sandwiched between desert-backed mountains and a cool ocean. At this latitude there is persistent subsidence of air from upper levels, which

FIGURE 16-18 A typical late summer inversion lid over the Los Angeles lowland. Air pollutants are trapped below the lid (marked by the shallow cloud layer), with clear skies above (TLM photo).

acts as a stability lid over the lowlands, inhibiting updrafts and keeping vertical air motions at a minimum (Fig. 16-18). Significant horizontal air motion is also limited by the combination of hot desert, high mountains, and cool waters, with the result that Los Angeles has a much lower average wind speed than any other major metropolitan area in Anglo-America. Local winds—land and sea breezes, mountain and valley breezes—help to break the stability from time to time, but the most prominent feature of the air over the metropolis is its relative lack of movement. This stagnant condition enables air pollution to build up with annoying frequency.

Many varieties of pollutants are scattered into the stagnant air from factory smokestacks, electricity generation, human lungs, and especially automobile exhausts. It is estimated that some 90 percent of all pollutants emanate from moving sources, primarily automobiles and trucks.

The Bay Area Metropolis

The remarkable urban metropolis that grew up around San Francisco Bay had its antecedents in the Gold Rush period but reached its present form and significance in association with the rapid population expansion of the last few decades. The bayside location has been an outstanding economic advantage from earliest days. As the best large natural harbor on the West Coast, it functioned as the nation's funnel to the Pacific, serving an extensive western hinterland but especially oriented toward the Mother Lode Country and then the Central Valley.

San Francisco was the largest city in the western United States from the time of the Gold Rush until World War I and served as the dominant focus of the Bay Area metropolis until well after the Golden Gate and Oakland Bay bridges were completed in the 1930s; indeed, San Francisco is still "The City" to most people in northern California (Fig. 16-19).

Since World War II, however, the metropolitan focus has become much more diffused. The metropolis has become noncentric; the three generalized foci of development are San Francisco, the East Bay district, and San Jose, around each of which further diffused growth has taken place. The movement of workers and commuters is not predominantly from the periphery to a "core" but rather from one outlying area to another.

San Francisco itself is a place of charm and variety and usually ranks high on lists of favorite cities for people all across the country (Fig. 16-20). The beauty of its setting (sloping streets readily providing extensive bay or ocean views) combines with the unusual nature of its weather patterns (brisk breezes and the alternation of brilliant sunshine and moving fog), the rich diversity of its culture, and the cosmopolitan lifestyle to make it one of the few cities in Anglo-America with a valid claim to urban uniqueness.

Its economic base is also unusual, primarily in the relatively limited role of manufacturing (San Francisco is one of the least industrialized cities—proportional to its size—in the nation) and the heavy dependence on government employment (San Francisco is second only to Washington in number of federal employees). Further distinction is provided by a style of residential development that is different from other western cities. Most homes are packed

FIGURE 16-19 The imposing skyline of San Francisco (TLM photo).

next to one another; the tall, narrow, white stucco row houses are set close to the street, with a tiny backyard, minuscule frontyard, and no sideyard. Population density is significantly higher in San Francisco than in any city west of Chicago.

The East Bay area consists of a number of separate but adjacent communities, with Oakland as the largest political entity. This is a very heterogenous urban complex with several important commercial cores, many affluent residential hillsides, prominent black ghettos, major port facilities (the port of Richmond greatly outranks the port of San Francisco in total tonnage because of the volume of petroleum shipments, and Oakland, one of the largest container ports in the world, surpassed San Francisco in general cargo handling more than two decades ago), notable counterculture complexes, and an endless horizon of middle-class subdivisions.

San Jose is the focal point of South Bay urban sprawl. This Santa Clara County area has experienced a phenomenally rapid transition from an agricultural and food-processing economy to one based on durable goods manufacture and related urban services. The so-called Silicon Valley of San Jose and vicinity is the principal U.S. center for the production of computers and other high-technology equipment. San Jose has a greater share of its work force employed in factories than any other major city in the region and is one of the 10 leading industrial centers in the nation. The rapid outward spread of urban sprawl has triggered much controversy with regard to both the philosophy and practicality of maintaining large green spaces around mushrooming metropolises.

Concern about urban sprawl is only one of many catalysts that led to innovative changes in social attitude and lifestyle. As southern California

FIGURE 16-20 The Golden Gate Bridge is a famous landmark in San Francisco. It is a critical (and spectacular) element in the transportation network of the Bay Area (TLM photo).

FIGURE 16-21 Fresno is a typical medium-sized, fast-growing, sprawling California city. In this residential scene there are no less than 105 backyard swimming pools (TLM photo).

has been a trend setter in recent years, so the Bay Area may be destined to make a major impact on American society in the near future. In the words of one observer,

> Here seems to be the most concentrated awareness of many national problems and concern for solutions or alternatives. From here has come the main impetus of the new environmental consciousness. . . . Here, certainly, is the major hearth of the "counter-culture" which has mounted a comprehensive critique of American society and markedly influenced national patterns of fashion, behavior, and attitudes. . . . Although national in scope, the impact of such movements is regionally varied, and their prominence and power in [the Bay Area] serves to set that . . . diverse metropolitan area apart from other Western regions, reinforcing earlier cultural distinctions.[3]

Central Valley Urbanism

Urban places are dotted with relative uniformity over the established agricultural sections of the Central Valley. The eastern portion and the central "trough" have, therefore, a hierarchical scattering of small and large market towns. The drier western portion, where agriculture was insignificant until recently, has a much more open urban network and the towns are mostly associated with a crossroad location or petroleum production.

Nine urbanized areas, from Redding in the north to Bakersfield in the south, exceed the statewide population growth rates. Although their functions are broadening beyond "farm town" image, they mostly are immature as metropolitan areas.

The largest city of the Central Valley, *Sacramento,* is the commercial center for the northern third of the subregion. Its commercial function is almost overshadowed by the administrative importance of being the capital of the most populous state. There is also a significant amount of manufacturing in Sacramento, particularly aerospace production.

In the San Joaquin Valley are three prominent urban centers, spaced equally apart. *Fresno,* in the center, is the largest (Fig. 16-21), but *Stockton* and *Bakersfield* are of the same magnitude and rapidly growing.

Tourism

The California Region, with its mild and sunny climate, is an American "Riviera." It offers tourists mountains, beaches, citrus groves, the Golden Gate,

[3] D. W. Meinig, "American Wests: Preface to a Geographical Interpretation," *Annals,* Association of American Geographers, 62 (1972), 182.

A CLOSER LOOK A State of Deterioration

A definite deterioration of California as a place to live and work seems to have occurred in the past half century. In days long gone the Native Americans and then the Hispanic colonials altered the state only slightly.

Since 1930 the population of California has expanded fivefold; it now exceeds 28 million. This growth has produced changes that have detracted from the attractiveness of the Golden State. Accelerated change came with expansion of shipyards and aircraft plants at the beginning of World War II. Soon some attractive market towns were converted into less-attractive industrial residence places.

Considerable alteration of the natural landscape actually began much earlier, especially in the Central Valley, with widespread planting of wheat in the second half of the nineteenth century. This was followed, in the late nineteenth century, by irrigation agriculture and diversified cropping, which affected the Central Coast and Southern California as well as the Central Valley.

Decimation of fauna also began before the twentieth century. Expansion of agriculture led to decline of such grazing animals as elk in the San Joaquin Valley. Canning of salmon began near Sacramento in 1864; excessive catches in the Sacramento River eventually led to the cessation of commercial fishing.

After World War II population growth accelerated and construction of a multilane road network facilitated sprawl. The Santa Clara Valley, in the San Francisco Bay Area, represents a good example of the result. As increasing numbers of people endeavored to live in favored areas, land costs rose. Hence, many people spend an hour to 2 hours commuting each way to and from work daily.

From an occupance standpoint much of the growth has occurred in two small areas, Southern California and the San Francisco Bay Area. Yet, to a varying degree, much of the state has felt the impact of metropolitan increase. For retirees and others not bound to the 9-to-5 work day, trips to popular destinations are less critical. But for many California residents, visits to places like Yosemite or Lake Tahoe or Palm Springs on three-day weekends have become frustrating. Not only is vehicular traffic heavy, but facilities at destinations are oftentimes crowded. In 1950 several hundred visitors climbed Mt. Whitney during the entire summer. Now that many may ascend the summit in a single weekend. Not too distant, Mammoth Mountain is now the most frequented ski area in the United States.

While cultural changes have been more momentous than physical changes, the latter have also occurred. Smog, principally a handicap to Southern California, the Bay Area, and the Central Valley, has resulted from added emission into the atmosphere, mostly from hydrocarbons from automotive exhausts. Expansion in urban areas of blacktopped surfaces and unshaded buildings has aggravated the effects of "heat islands," sometimes as much as ten degrees Fahrenheit.

The effect of population increase upon the topography has been limited. Yet in some instances human interference with stream runoff, as from damming of streams, has accelerated shoreline erosion.

While humankind does not directly bring about changes due to earthquakes, damage to people and property has been worsened by construction of facilities unable to withstand violent tremors. This has been particularly evident in the San Francisco Bay Area, where construction of residences and other structures has taken place upon unconsolidated landfill. The results were particularly evident with the 1906 and 1989 earthquakes. Too much construction has taken place unwisely on unconsolidated surfaces.

While cultivation had led to soil improvement, people have sometimes worsened soil conditions. For example, in the southwestern San Joaquin Valley concentrations of salts have resulted from irrigation practices.

"Mining of the forest" has depleted extensive stands of trees. A good example is the Coast redwood. Except in preserves, chiefly state parks, over 90 percent of the stands have been depleted. Extensive areas of yellow pine on private holdings have also been cut without replenishment.

Street and highway congestion is one of California's most frustrating problems. No other state approaches California in total number of private automobiles. In Los Angeles County alone there are more than 6 million cars. And in San Francisco the number of automobiles exceeds the available parking spaces.

For a few years in the 1960s, as immigration into California from the other 49 states lessened, it seemed that population stabilization might occur. Then a surge of newcomers from Latin America and eastern Asia especially led to renewed growth. California is on the verge of becoming the nation's first Third World state. The size and rapidity of this new migration has led to considerable social disharmony and high unemployment.

The decreasing attractiveness of metropolitan California has prompted a number of residents of the Golden State, especially older Anglos, to move elsewhere. More popular destinations include Oregon, Washington, and Colorado.

Professor David Lantis
California State University, Chico

FIGURE 16-22 Disneyland in southern California is the first and most famous of the world's theme amusement parks (TLM photo).

Disneyland (Fig. 16-22), and Hollywood. Generally speaking, the natural attractions of the region (mountains, beaches, islands) are the haunts of local residents on vacation; the constructed attractions draw the out-of-state visitors. Disneyland, for example, has become a tourist goal that ranks with the Grand Canyon, Yellowstone Park, and Niagara Falls in popularity. Movie and television studios are always swarming with visitors, and such specialized commercial spots as Knott's Berry Farm, the San Diego Zoo, and Sea World are nationally famous.[4]

The great extent of the Sierra Nevada, its height, deep canyons, waterfalls, fish-laden streams and lakes, forests, historical interests, good roads, and heavy winter snows at well-developed ski resorts, and the tremendous nearby urban population make the mountains a great attraction for tourists and vacationists. In the Sierra Nevada are located three of the nation's famous national parks: Sequoia, Kings Canyon, and Yosemite. All are noted for their magnificent groves of giant sequoia trees, spectacular glaciated valleys, waterfalls, and wildlife. Another particularly scenic area is Lake Tahoe, which nestles in a large pocket between the double crest of the Sierra Nevada.

Another major tourist attraction is San Francisco, a city of great beauty and charm. Its hill-and-water site, its cosmopolitan air, and its many points of scenic and historic interest have given it an international reputation as a place to visit.

The central coast is stirringly beautiful when not fogbound and is a favorite holiday area, particularly for Californians. Principal interest focuses on the Monterey Peninsula, where spectacular scenery, marine fauna (sea lions, sea otters, water birds), and a reputation for glamour (Monterey, Carmel, Big Sur, Pebble Beach) exist in close conjunction (Fig. 16-23).

THE OUTLOOK

The future of this region, which includes all but the wettest and driest parts of California, seems bright. Blessed in climate—possibly the most highly publicized in the world—as well as the bases for a thriving agriculture, and with a rich heritage in forests, minerals, and scenic attractions, California has long experienced an absolute population growth greater than any other state. In every population census since 1920 California has outstripped all other states.

Year after year California ranks as the leading agricultural state and many counties are in the vanguard in national standing. More and more, the agriculture of this region will be devoted to the production of specialized fruit and truck crops and to dairy products.

Farmers are innovative and imaginative in their choice of crops, and one can anticipate that the list of current favorites (almonds, pistachios, kiwis, pecans, table grapes) will vary considerably from year to year. Displacement of farm land by urban sprawl seems to be a permanent feature of the land-use pattern of the region. Pressure to subdivide farm acreage undoubtedly will intensify and expand, particularly in southern California and the Bay Area. The Central Valley will continue to be the principal recipient of displaced agriculture, although increased irrigation and intensified farming will also be experienced in various coastal valleys. A spread of

[4] Disneyland, Knott's Berry Farm, and Universal Studios normally receive more visitors than other other commercial attractions in Anglo-America except Florida's Walt Disney World.

FIGURE 16-23 The spectacular scenery of California's central coast is epitomized by the Pinnacle at Point Lobos (TLM photo).

sophisticated, water-conserving irrigation techniques can be anticipated. California already contains nearly half of North America's drip irrigation acreage, primarily in tree crops, and more will undoubtedly be added. Center pivot technology, so popular in most other irrigation states, has been little adopted in California, but its even more sophisticated offshoot (linear move) literally was invented for San Joaquin Valley conditions and is rapidly spreading in the region.

The expansive sprawl of urbanization will continue its steady march in the region. Its most conspicuous impact should be in the coastal zone between Los Angeles and San Diego, where the ambitious planned development of the huge Irvine Ranch and many lesser individual schemes will eventually result in an unbroken conurbation except for the Camp Pendleton Marine base in San Diego County. Other major growth areas will be in the hills and valleys between the San Fernando Valley and Ventura, in the desert margin of the Antelope Valley, in the Santa Clara lowland, and in the Concord–Walnut Creek area east of the Berkeley Hills.

The decline in the region's leading manufacturing activity, aerospace, has ended, but the industry still depends heavily on government expenditures and thus has an uncertain future. Aerospace manufacturers are increasingly diversifying into consumer and commercial electronics production as a hedge against this problem. Still, California's dominance in high-technology and aerospace industries is not as great as it once was and it is likely to continue to diminish from a relative standpoint. The region still has the basic advantages of skilled workers and excellent academic institutions, but it has the increasing disadvantages of high costs for land, labor, housing, and taxes. Moreover, numerous environmental regulations and anti-growth restrictions have been imposed in most parts of California, with the result that an increasing stream of manufacturers is moving out of the state.

From the broad economic point of view the region has been an important trend setter for the nation and its shift to a sequence of slower growth thus has implications for the broader national economy as well as for its own future. California has declined in economic importance relative to the rest of the nation since the mid-1960s, but it still generates about one-ninth of total national income and would be among the world's 10 largest countries in personal income if it were a nation in itself.

Continued population growth and accompanying urban expansion are predictable, at least in the near future. Interstate migration to California has slowed a bit, but the international flow has accelerated. A recent government study discovered that fully one-fourth of all immigrants to this country intended to settle in California! Clearly the region's population will become increasingly ethnic, particularly Hispanic and Asian.

Eventually there must be a limit to frantic urban expansion. Perhaps the potential water problems can be solved by desalinization of seawater; but smog, transportation difficulties, urban crowding, energy problems, and the sheer mass of humanity may combine to destroy the "California way of life" and, with it, the principal reasons for continued long-term growth.

SELECTED BIBLIOGRAPHY

ANTHROP, DONALD F., "The Peripheral Canal and the Future for Water in California," *Yearbook,* Association of Pacific Coast Geographers, 44 (1982), 109–128.

ARON, ROBERT H., "The Changing Location of California Almond Production," *California Geographer,* 28 (1988), 69–94.

ARREOLA, DANIEL D., "The Chinese Role in Creating the Early Cultural Landscape of the Sacramento-San Joaquin Delta," *California Geographer,* 15 (1975), 1–15.

BAILEY, HARRY P., *The Climate of Southern California.* Berkeley: University of California Press, 1966.

BANHAM, REYNER, *Los Angeles: The Architecture of the Four Ecologies.* New York: Harper & Row, Publishers, 1971.

BARBOUR, MICHAEL G., and JACK MAJOR, *Terrestrial Vegetation of California.* New York: John Wiley & Sons, 1977.

BLAND, WARREN, R., and RONALD YACHNIN, "Inadequate Airport Capacity: A Developing Transportation Crisis in the Los Angeles Region," *California Geographer,* 27 (1987), 97–106.

BUCZACKI, STEFAN T., "The Land that is the North American Salad Bowl," *Geographical Magazine,* 53 (February 1981), 297–299.

CANTOR, LEONARD M., "The California State Water Project: A Reassessment," *Journal of Geography,* 79 (April–May 1980), 133–140.

COOK, DOUGLAS D., "The Fight to Conserve California's Coast," *Geographical Magazine,* 54 (November 1982), 623–629.

DILSAVER, LARY M., "Taking Care of the Big Trees," *Focus,* 37 (Winter 1987), 1–8.

FORD, LARRY, and ERNST GRIFFIN, "The Ghettoization of Paradise," *Geographical Review,* 60 (April 1979), 140–158.

GODFREY, BRIAN J., *Neighborhoods in Transition: The Making of San Francisco's Ethnic and Nonconformist Communities.* Berkeley: University of California Press, 1988.

GRANGER, ORMAN E., "Climatic Variations and the Raisin Industry," *Geographical Review,* 70 (July 1980), 300–313.

———, "Increasing Variability in California Precipitation," *Annals* of the Association of American Geographers, 69 (December 1979), 533–543.

HALLINAN, TIM S., "River City—Right Here in California," *Yearbook,* Association of Pacific Coast Geographers, 51 (1989), 49–64.

HANNES, GERALD, "Summertime Coastal Air Flow in Northern California," *The California Geographer,* 21 (1981), 18–29.

HEIZER, ROBERT F., ed., *Handbook of North American Indians, California.* Vol. 8: Washington, D.C.: Smithsonian Institution, 1978.

HORNBECK, DAVID, *California Patterns: A Geographical and Historical Atlas.* Palo Alto, CA: Mayfield Publishing Company, 1983.

———, "Mexican-American Land Tenure Conflict in California," *Journal of Geography,* 75 (1976), 209–221.

LANTIS, DAVID W., RODNEY STEINER, and ARTHUR E. KARINEN, *California: The Pacific Connection.* Chico, CA: Creekside Press, 1989.

MCKNIGHT, TOM L, "Center Pivot Irrigation in California," *Geographical Review,* 73 (January 1983), 1–14.

MEDDERS, STANLEY, "California's Channel Islands," *National Parks and Conservation Magazine,* 49 (October 1975), 11–15.

MILLER, CRANE, and RICHARD HYSLOP, *California: The Geography of Diversity.* Palo Alto, CA: Mayfield Publishing Co., 1983.

NELSON, HOWARD J., *The Los Angeles Metropolis.* Dubuque, IA: Kendall/Hunt Publishing Co., 1983.

———, "The Spread of an Artificial Landscape over Southern California," *Annals,* Association of American Geographers, 49 (1959), 80–100.

NELSON, HOWARD J., and W. A. V. CLARK, *The Los Angeles Metropolitan Experience: Uniqueness, Generality, and the Goal of the Good Life.* Cambridge, MA: Ballinger Publishing Company, 1976.

PARSONS, JAMES J., "A Geographer Looks at the San Joaquin Valley," *Geographical Review,* 76 (October 1986), 371–389.

PRYDE, PHILIP R., ed., *San Diego: An Introduction to the Region.* Dubuque, IA: Kendall/Hunt Publishing Company, 1976.

STEINER, RODNEY, "Large Private Landholdings in California," *Geographical Review,* 72 (July 1982), 315–326.

———, *Los Angeles: The Centrifugal City.* Dubuque, IA: Kendall/Hunt Publishing Co., 1980.

VALE, THOMAS R., "Vegetation Change and Park Purposes in the High Elevations of Yosemite National Park, California," *Annals,* Association of American Geographers, 77 (March 1987), 1–18.

VANCE, JAMES E., JR., "California and the Search for the Ideal," *Annals,* Association of American Geographers, 62 (1972), 185–210.

VANCE, JEAN, "The Cities by San Francisco Bay," in *Contemporary Metropolitan America. Vol: 2: Nineteenth Century Ports,* ed. John S. Adams, pp. 217–307. Cambridge, MA: Ballinger Publishing Company, 1976.

VAN KAMPEN, CAROL, "From Dairy Valley to Chino: An Example of Urbanization in Southern California's Dairy Land," *California Geographer* 17 (1977), 39–48.

VANKAT, JOHN L., "Fire and Man in Sequoia National Park," *Annals,* Association of American Geographers, 67 (1977), 17–27.

WALKER, RICHARD A., and MATTHEW J. WILLIAMS, "Water from Power: Water Supply and Regional Growth in the Santa Clara Valley," *Economic Geography,* 58 (April 1982), 95–119.

WALLACH, BRET, "The West Side Oil Fields of California," *Geographical Review,* 70 (January 1980), 50–59.

WILVERT, CALVIN, "San Diego/San Francisco–Sansan," *Geographical Magazine,* 53 (January 1981), 269–272.

17
THE HAWAIIAN ISLANDS

The Hawaiian Islands form the smallest major region of Anglo-America. Situated some 2100 miles (3360 km) southwest of California in the Pacific Ocean, Hawaii is tied to the mainland by political affiliation and commercial dependence, which are strong enough to prevail over the volcanic base, tropical climate, Polynesian history, and Oriental population that in the past have tended to set the islands apart from Anglo-America.

The state of Hawaii encompasses a 1600-mile (2560-km) string of islands, islets, and reefs (except for the Midway Islands, which are administered by the U.S. Navy), that extends westward across the Pacific from the island of Hawaii (at 155° west longitude) to Kure (at 178° west longitude). For practical purposes, the term *Hawaiian Islands* is normally restricted to a group of two dozen islands that stretch 400 miles (640 km) from Hawaii to Niihau (Fig. 17-1). The total land area of the Hawaiian Islands is approximately 6500 square miles (16,835 km^2)—about the same area as New Jersey—nearly all of which is made up of eight major islands, listed here in order of size: Hawaii (4021 square miles or 10,415 km^2), Maui, Oahu, Kauai, Molokai, Lanai, Niihau, and Kahoolawe (45 square miles, or 115 km^2). The area of Hawaii is nearly twice the combined area of the other seven islands.

THE PHYSICAL SETTING

Origin and Structure of the Islands

The Hawaiian islands represent isolated tops of a submarine mountain range that were built up so much by volcanic action that they protrude above sea level, where their surface was modified by further vulcanism and subaerial erosion. The volcanoes are of the quiescent shield type that develop by emission of lava rather than by explosive ejection of

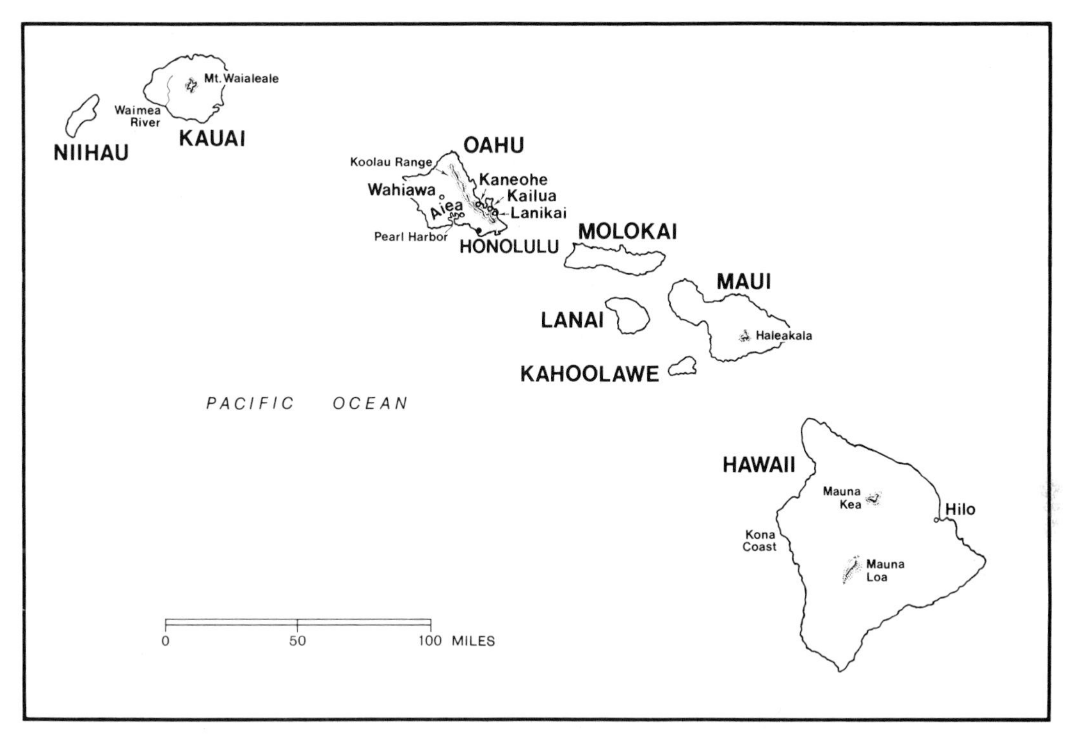

FIGURE 17-1 The Hawaiian Islands Region.

rock fragments. Essentially the larger islands are basaltic domes in various stages of dissection. The two highest peaks of the islands, Mauna Kea (elevation 13,784 feet, or 4135 m) and Mauna Loa (elevation 13,679 feet, or 4104 m) are sometimes described as the highest in the world from base to summit, for their bases are set some 18,000 feet (5400 m) below sea level on the floor of the Pacific. Mauna Loa is the world's largest active volcano. It is an extraordinarily massive mountain; the volume of Mauna Loa above its seafloor base is estimated to be 125 times greater than that of California's Mount Shasta.

Vulcanism continues to the present; there are two major active volcanoes on "the Big Island" (Hawaii) and several smaller areas of geothermal activity. Periodic lava flows are actually expanding the area of the Big Island, forming new peninsulas on the southeast coast. A single eruption of Kilauea in 1960 added 500 acres (200 ha) of new land to the island. Lava ejections from the northeast rift zone on Mauna Loa's flank pose a continuing potential danger for Hilo, the island's largest city. Some 24 flows have entered the present area of Hilo in the last 2000 years, and it is predicted that an average of one flow per century can be expected to penetrate Hilo in the future. Indeed, Kilauea's continuing eruptions (it is called the world's "most active volcano") expelled lava that destroyed more than 100 homes in 1990.

Coralline limestone has been uplifted in a few places to form flattish coastal plains of modest size. In addition, submerged fringing coral reefs partially but not completely surround most of the islands. Coral sand has frequently been washed up to form beaches in bays that are sheltered between rocky lava headlands.

Surface Features

The islands are all dominated by slopeland and most are distinctly mountainous. Sheer cliffs, called *pali,* and rugged, steep-sided canyons provide the most abrupt changes in elevation. Flat land is scarce, even around the coastal fringes.

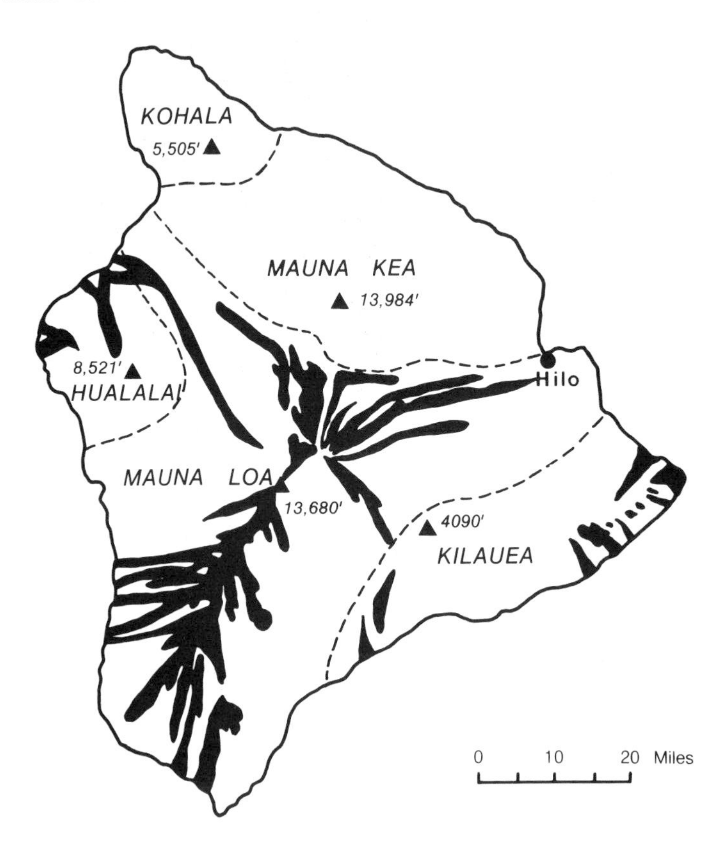

FIGURE 17-2 Shield volcanos of the Big Island, showing the extent of major lava flows within historic time.

The Big Island was formed by the overlapping union of five gently sloping volcanic cones (Fig. 17-2), of which Mauna Kea and Mauna Loa are the highest and most massive (Fig. 17-3). Only Kilauea and Mauna Loa are currently in an active phase. There are a few short rivers on the northern and eastern sides of the island, some of which drop into the sea as waterfalls over sheer cliffs.

Maui is composed of two volcanic complexes separated by a narrow lowland isthmus. The extinct volcano Haleakala (elevation 10,032 feet, or 3000 m) towers over the eastern part of the island. Its gigantic eroded summit depression, which resembles a giant caldera but is not, contains several dormant cinder cones and other volcanic forms. The West Maui Mountains are much lower but are rugged.

Oahu consists of two mountain ranges separated by a rolling plain. The Waianae Mountains parallel the west coast and the more extensive Koolau Range parallels the east coast. The Nuuanu Pali of the latter range is one of the most spectacular terrain features in Anglo-America, its sheer cliffs descending from cloud-shrouded peaks to a fertile coastal fringe (Fig. 17-4). The lowland between the mountains is abruptly dissected by deeply incised, steep-sided gorges in several places and at its southern end is the embayment of Pearl Harbor, one of the largest and finest harbors in the North Pacific. Honolulu's two conspicuous natural landmarks, Diamond Head and Punchbowl, are the stumps of extinct volcanoes.

Kauai is completely dominated by Mount Waialeale, which rises to 5170 feet (1550 m) in the center of the island. Heavy rainfall results in numerous short rivers plunging coastward, often cutting deep canyons, of which Waimea Canyon is the most spectacular. The coastal fringes have little flat land,

FIGURE 17-3 Mauna Loa (on the left) and Mauna Kea rise to great heights on the Big Island, but their slopes are long and relatively gentle (TLM photo).

with the steep Na Pali of the northwest coast prohibiting the building of a complete circumferential roadway.

Kahoolawe, a barren hilly island with a maximum elevation of nearly 1500 feet (450 m), is uninhabited and used only as a military firing-range target. Lanai is also hilly. The long, narrow island of Molokai consists of a rugged mountain mass on the east and a broad sandy plateau on the west. Niihau consists of a moderately high tableland in the center, with low plains at either end of the island.

Climate

In its basic characteristics, the climate of the Hawaiian Islands is controlled by three factors:

1. Its tropical fringe location in a vast ocean accounts for generally mild, equable temperatures and an abundance of available moisture.
2. The northeast trade winds blow almost continually across the islands during the summer but are less pronounced, although persistent, during the winter.
3. The terrain's height and orientation are the major determinants of temperature and rainfall variations.

The dominance of the trade winds throughout the region means that usually there are some clouds in the sky, the humidity is relatively high on the average, temperatures are mild to warm, there is some wind movement that hastens evaporation, and showers of brief duration are to be expected with some frequency.

The most striking aspect of the climate is the variation in rainfall from place to place. In general, a windward (essentially northeast in this region) location receives considerably more precipitation than a leeward one. As moisture-laden air rises over a topographic barrier, it expands and cools and becomes incapable of retaining all the moisture it contains; rainfall results. As the same air descends the lee side of a mountain, it contracts and becomes warmer and can hold more moisture; thus rainfall is unlikely. Specifically, it can be seen that the higher mountains receive most rain on the northeast flanks at moderate elevations of 2000 to 4000 feet (600 to 1200 m), whereas the lower mountains receive most rain along or near the crest line.[1] This difference is the result of the movement of the onshore trades that blow *over* the lower mountains but *around* the higher ones.

[1] David I. Blumenstock, "Climate of Hawaii," *Climates of the States,* U.S. Weather Bureau Publication (Washington, D.C.: Government Printing Office, 1961), p. 8.

FIGURE 17-4 The stark and spectacular cliffs of Nuuanu Pali on the windward side of Oahu (TLM photo).

This windward-leeward relationship results in extraordinary rainfall variations within short horizontal distances. A weather station on the northeast slope of Mount Waialeale on Kauai (said to be the rainiest spot on earth) records an average of 476 inches (12,090 mm) of rain annually, but 15 miles (24 km) away the average is only 20 inches (518 mm). Variations of the same order of magnitude also occur on Oahu, Molokai, Maui, and Hawaii. Only the three small and relatively low-lying islands do not have similar situations. Within the urbanized area of Honolulu it is possible to choose a building site with a 93-inch (2410-mm) average rainfall or one with a 25-inch (650-mm) average only 5 miles (8 km) away.[2]

Temperatures in the lowlands are uniformly mild. Honolulu's January average of 72°F (23°C) is close to its July average of 78°F (25°C), and the highest temperature ever recorded in the city is only 88°F (31°C), in comparison with an absolute minimum of 57°F (13°C). Only in locales of great altitude do temperatures drop markedly below the mild range; for example, the Mauna peaks of the Big Island are sometimes snowcapped.

Soils

Heavy rain and steep slopes have been the principal determinants of soil development. In general, the slopelands have only a thin cover of soil and the flattish lands have deep soil development. Most of the mature soils are lateritic in nature, having been leached by percolating water. The average soil in agricultural areas is red in color, moderately fertile, and relatively permeable so that irrigation is often necessary even in places of considerable rainfall.

Biota

The Hawaiian chain is the most remote group of high volcanic islands in the world—remote from any continent or any other islands of appreciable size. This isolation engendered a native flora and fauna that were not only limited in variety but also exhibited many unusual characteristics. To cite but two prominent examples among many:

1. More so than in any other area in the world, many Hawaiian plant groups have developed arborescence, an evolution of growth form from small nonwoody plants into large shrubs, even trees.
2. There is an exceptionally high proportion of flightless insects.

The fragility of island ecosystems is well known and Hawaii is a classic example of this principle. More than 95 percent of the biotic species originally endemic to Hawaii were found nowhere else in the world; yet within a relatively short time the vast majority were exterminated, significantly endangered, or thoroughly displaced. The statistics are overwhelming: "24 of the 69 species of birds known from Hawaii are extinct, and another 26 are either rare or endangered. Both native species of mammals—the hoary bat and the monk seal—are endangered, and half the native land mollusks are extinct. A partial list of endangered insects contains over 250 entries; the list of endangered plants includes 227 species."[3] This situation is chiefly the result of human carelessness or indifference and the introduction of exotic plants and animals that often proliferated at the expense of native biota.

For example, when Captain James Cook arrived in 1778, there were perhaps 1500 species of seed-bearing plants on the islands. Since then, more than 4000 other varieties have been introduced, and many became naturalized.

As a result, Hawaii accounts for more than 70 percent of all plant and animal extinctions that have taken place in the United States, and it harbors more than one-fourth of the nation's endangered birds and plants.

Mild temperatures and abundant precipitation provide conditions for lush vegetation. The better-watered areas are noted for thick growth of tropical trees and shrubs. Most areas of thick forest, however, have been denuded by commercial logging or overgrazing. In areas of intermediate rainfall the flora often reflects more arid conditions because the highly permeable volcanic soil permits water to percolate rapidly to great depths—frequently beyond the reach of plant roots. Xerophytic plants characterize such areas. The introduction of exotic plants has been particularly characteristic of this region so

[2] Loyal Durand, Jr., "Hawaii," *Focus*, 9 (May 1959), p. 2.

[3] Warren King, "Hawaii: Haven for Endangered Species?" *National Parks and Conservation Magazine*, 45 (October, 1971), p. 9.

that now a large proportion of the total vegetative cover represents earlier imports.

Animal life has always been limited on the islands. Native fauna was mostly restricted to insects, lizards, and birds. The most conspicuous wildlife today consists of feral livestock (livestock that reverted to the wild). Tens of thousands of feral sheep, goats, and pigs roam the islands and there are considerable numbers of feral cattle, dogs, and cats. These animals destroy many plants and other animals and are a major nuisance in some areas, particularly on the Big Island. But they provide an important recreational resource for sport hunting. Other exotic species, such as mouflon, axis deer, and mongoose, are also present in some profusion.

POPULATION

Early Inhabitants

Little is known of the pre-Polynesian inhabitants of the Hawaiian Islands except that they are reported to have been short in stature and peaceful by nature. Presumably they were either destroyed or assimilated by waves of Polynesian settlers, the first of which is thought to have arrived from the western Pacific between A.D. 750 and 1000. After several hundred years of isolation, another great Polynesian immigration occurred in the fourteenth and fifteenth centuries, followed by a second lengthy period of insular seclusion.

These people have been known through recent history as Hawaiians and are characterized by bronze skin, large dark eyes, heavy features, and dark-brown or black hair. Although their social and community life was intricately complicated by restrictions and regulations, making a living was relatively simple. They had domesticated pigs and chickens and a variety of cultivated food plants. Fruits and vegetables were common, but dietary staples were fish and poi (the cooked and pounded root of the taro plant).

European Penetration

The islands were officially discovered by Captain James Cook of England in 1778; however, it is thought that a Spanish captain named Gaetano had been there more than two centuries previously and still other seafarers may have touched the islands before Cook. It is clear, nevertheless, that Cook's visit opened up the "Sandwich Islands," as he called them, to the world. Before long the islands became important bartering, trading, and refreshment stops for merchant vessels of England, France, Spain, Russia, and the United States and for whalers and pearlers of many nationalities.

British influence was strong for many years, but few colonists were attracted. Missionaries from New England arrived during the 1820s and these dedicated people became very influential by the 1840s. Moderate but increasing numbers of United States settlers migrated to the islands during the nineteenth century.

Other Immigrants

Asiatics came and were brought to Hawaii in considerable numbers during the last half of the nineteenth century, usually in response to a need for cheap labor. The first Chinese were brought to work on the sugar plantations in 1852. The Japanese began to arrive in 1868, first as fishermen blown off course and later as plantation workers. In spite of restrictions, Japanese immigration greatly exceeded that from any other country. The first Filipino sugar workers came in 1906 and many more were brought in during succeeding years. Other immigrants came from Korea, Samoa, other Pacific islands, and Portugal's Atlantic islands (Azores and Cape Verde).

The Contemporary Melting Pot

The present population of the islands is more complex and varied than that of any other region in Anglo-America except perhaps California. All ethnic groups have intermarried, particularly in recent decades.

Hawaiian society is increasingly open and has an unusual degree of social and economic mobility. Although there are visible cracks in the region's widely acclaimed racial harmony, Hawaii still stands as Anglo-America's most successful melting pot. Caucasians (called *haole*), mostly from mainland North America, were significantly outnumbered by Japanese in the past, but in the last few years a large *haole* influx has made Caucasians the most numerous element in the region's population, more than one-fourth of the total. Japanese are almost as numerous; Hawaiians and part-Hawaiians constitute about 20 percent; Filipinos, sparked by rapid immigration in recent years, about 15 percent; and Chinese, about 6 percent. Nearly half the population is of Asian ancestry; in no other state is that figure as much as 6 percent.

CENTURIES OF POLITICAL CHANGE

Throughout this region's early history the islands were politically fragmented. Various kings and chiefs ruled different islands and parts of islands with a sporadic pattern of warfare and change. The uniting of the region under one ruler was accomplished by Kamehameha I, but it required 28 years of war, diplomacy, and treachery. In 1782, he began a bloody civil war on the Big Island that lasted for 9 years. Later he conquered Maui and Molokai and overwhelmed Kalanikupule's army to seize control of Oahu. The other islands came under his rule by 1810.

Six kings and a queen successively carried on the monarchy after Kamehameha's death in 1819. The first constitution was promulgated during the reign of Kamehameha III and the kingdom became a constitutional monarchy. Most influential Hawaiians became increasingly interested in some sort of liaison with Britain, but a location near the United States and increased trade with California (sugar, rice, and coffee) in the 1850s and 1860s, as well as the establishment of regular mail service with San Francisco, foreshadowed the manifest destiny of annexation by the United States. The monarchy declined after 1875 and Queen Liliuokalani was deposed in 1891.

An American immigrant served as president of the interim republic until annexation was completed in 1900. For Hawaii, the basic motive for annexation was to increase trade, especially in sugar; and for the United States, to secure a major Pacific naval base. In spite of much agitation for statehood, the islands remained a territory for nearly six decades. In 1959 Hawaii was admitted as the fiftieth state.

THE HAWAIIAN ECONOMY: SPECIALIZED, LIVELY, ERRATIC

Economic opportunities were always limited in the region because of its insular position and lack of natural resources. Although bauxite and titanium have been discovered, no mineral deposits have ever provided a significant income. Commercial fishing has been only partially successful. Logging of sandalwood was once very important and formed the basis of a thriving trade with the Orient, but the stands of sandalwood have long since been depleted.

Although only one-tenth of the land is arable, the soil is generally productive, there is no danger of frost, and natural rainfall and abundant groundwater provide sufficient moisture. The first foreign cash crop was tobacco, which flourished during the first half of the nineteenth century and then faded out. During the Gold Rush period in California, foodstuffs of various kinds were exported to the West Coast. This trade was significant to the Hawaiian economy for only about a decade and then virtually ceased. The provisioning of whalers and other ships bolstered the production of local crops from the 1820s until the 1870s, but it also declined. Eventually it became evident that the best hope for agriculture was to specialize in growing subtropical crops for the mainland market.

The growing of specialized plantation crops carried the brunt of the regional economy for a long time. Nevertheless, the role played by federal government expenditures, primarily for the construction and maintenance of military facilities, was the single largest component of the economy for many decades. Today the relative significance of agriculture has declined and the importance of tourism has soared. The latter, however, is an erratic producer of income, for it reacts quickly to the ebbs and flows of the national economy. In recent years, however, fluctuations have been less pronounced because of the dramatic impact of increased tourism from Japan.

Cane and "Pines"

Sugarcane has been the leading crop of Hawaii for more than a century and pineapples have ranked as a strong second crop for almost as long. Indeed, the Hawaiian image has long been almost as much "sugar and pineapple" as it has been "grass skirts and Waikiki." At present, however, the steady decline of both staple crops clearly signals their relegation to a position of only supplementary importance, fading perhaps to even less than that.

Other Crops

For many years there was talk of the need to diversify the region's agriculture, for almost all foodstuffs must be imported from a distant source.[4] Nevertheless, diversification was slow to occur and most efforts were aimed at growing other tropical specialties for export rather than producing food for local consumption (Fig. 17-5).

[4] The only staples in which the region is self-sufficient are sugar and milk.

The third-ranking crop (and some sources place it second only to sugar) on the islands in terms of revenue is probably marijuana, or *pakalolo,* as it is called locally. Officially, however, macadamia nuts are considered to be most valuable after sugar

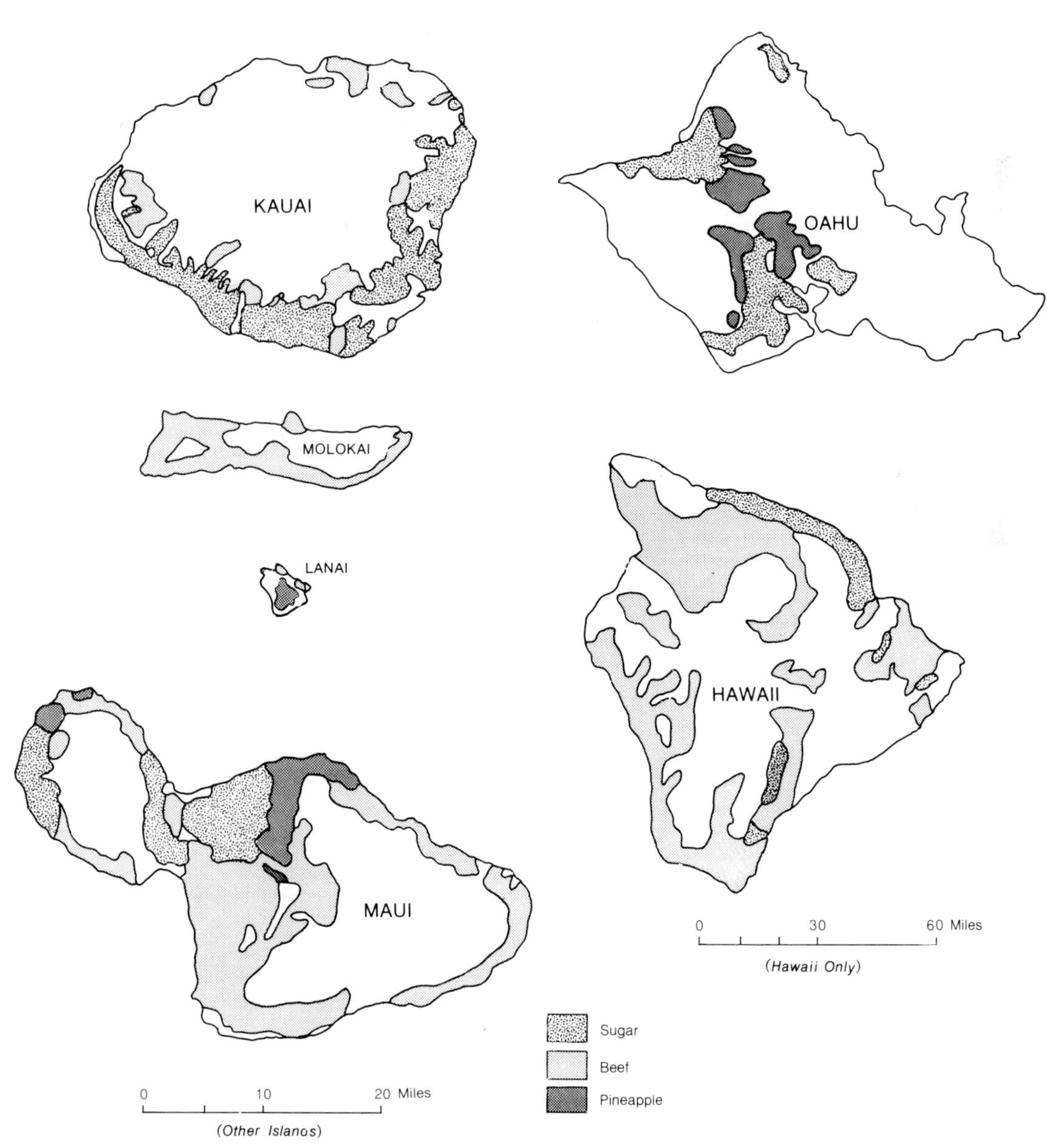

FIGURE 17-5 The land-use pattern of the principal islands. Cattle ranching and the intensive growing of sugarcane and pineapple are striking.

A CLOSER LOOK The De-Sweetening of Hawaii

Few crops have been as intimately associated with both the image and the economic well-being of a region as sugarcane and pineapples with the Hawaiian Islands. Brought to Hawaii by Polynesian migrants, cane was being produced commercially in the 1830s. Cultivation was primarily by smallholders until the 1850s when a swelling influx of contract workers (initially from China; later from Japan, Portugal, the Philippines, and other places) provided the labor force needed for large-scale plantation farming. By the 1860s sugarcane had become the economic mainstay of Hawaii, a position it retained for three-quarters of a century.

Pineapples were also introduced from elsewhere in Polynesia, but there was no significant commercial production until 1890. The industry became notable after Dole's development of improved varieties in 1903 and Ginaca's invention of a mechanical peeling, coring, and slicing machine 10 years later.

The islands had a few critical physical advantages for these crops, most important of which was the semitropical climate. Low temperatures are never a problem, nor is searing heat for any length of time. Both crops are heavy users of water (it takes about 1 ton (0.9 t) of water to grow enough cane to produce 1 pound (0.45 kg) of raw sugar), but the Hawaiian growing areas either receive adequate rainfall or can be supplied by irrigation from adjacent mountains (Fig. 17-a). The deep lateritic soils, although not particularly fertile, are certainly not impoverished, and take well to fertilizers.

An insufficient labor supply was an early disadvantage, but a continuing flow of contract workers from various countries soon eliminated this problem. There was no local market to speak of, but the relative closeness of the United States provided a rich market potential and aggressive advertising did the rest. Thus, during the first half of the twentieth century sugarcane and pineapple were dominating elements in the Hawaiian economy.

During the 1930s there were more than 50,000 workers in the cane fields of the islands, another 5000 in the pineapple plantations, and perhaps 10,000 employees in the sugar mills and pineapple canneries. This total represented nearly 40 percent of all jobs in the Hawaiian Islands. After World War II the number of workers declined steadily, but output continued to rise, as did acreage. The area devoted to pineapples peaked at about 70,000 acres (28,000 ha) in 1948 (Fig. 17-b); the greatest sugarcane acreage—250,000 (100,000 ha)—was cultivated in 1969.

Then came the decline. Pineapple experienced the most disastrous reversal (Fig. 17-c). Hawaii had been easily the world's dominant producer for years; in the early 1940s some 80 percent of the world's pineapples were grown on the islands. By the

FIGURE 17-a Harvested cane is trucked to this sugar mill near Waimea on the island of Kauai (TLM photo).

FIGURE 17-b A pineapple plantation on the island of Oahu (TLM photo).

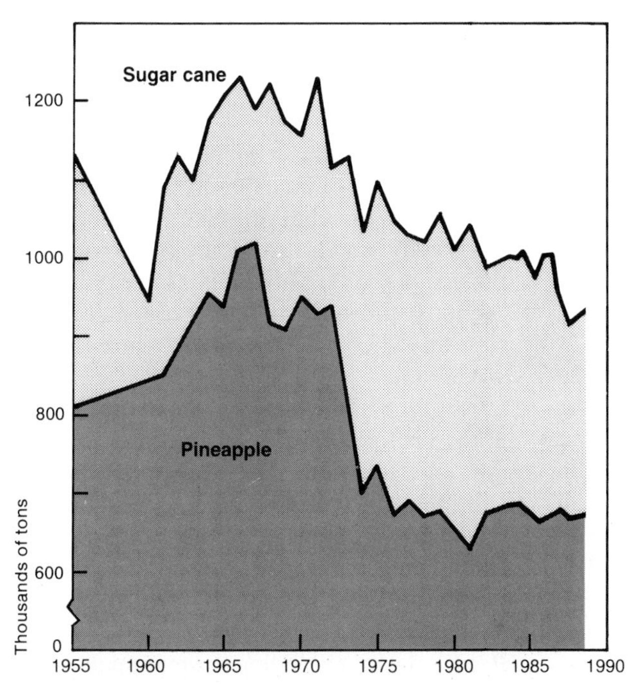

FIGURE 17-c Both sugarcane and pineapple are on a long-term downward production trend in Hawaii.

early 1990s this share had dropped to less than 25 percent. Various factors were involved in the decline, but the most important was the cost of labor. New plantations in the Philippines, Taiwan, Malaysia, and other tropical countries pay wages that are only 5 to 10 percent as much as those paid to pineapple field workers in Hawaii; the disparity of cannery workers' wages is even greater. The principal producing companies, chiefly large conglomerates, such as Del Monte and Dole, operate in foreign countries as well as Hawaii. So it is in their interest to shift production to lower-cost areas.

By the 1980s both pineapple acreage and employment were less than half as much as in peak years. Two (Kauai and Molokai) of the five islands where pineapple was once prominent are no longer producers and output has diminished on all three of the others (Lanai, Maui, and Oahu). Only two canneries are still in operation compared with nine a half

(*continued*)

century ago. In fact, if not for vigorous marketing of pineapple as a fresh fruit item, the decline would have been even more precipitous. Until the last decade almost all Hawaiian pineapple was sold in cans; now, however, more than three-fourths of the fruit is marketed fresh.

The downturn in sugar is more recent and less abrupt but no less real. In 1974 the U.S. Congress failed to renew the Sugar Act, which for four decades had provided a strong measure of protection for domestic sugar producers against foreign imports. Higher production (primarily labor) costs in Hawaii mean that many foreign producers have a competitive advantage, and thus sugar acreage, production, and employment have all been on a downward trend in this region. In the early 1980s Congress enacted some price supports for domestic sugar that mitigated, but did not stop, the decline.

The region now produces only one-fourth of the total U.S. sugar cane output. The principal cane-growing areas are on the windward coast of the Big Island, in the lowland of Maui, in several places on Oahu, and around the coastal margin of Kauai. Most production comes from a dozen large sugar ranches. The refining, most of which is done in California, and the marketing of the raw sugar are handled by a producers' cooperative, the California and Hawaiian Sugar Corporation.

Although growing and processing sugar and pineapple are still significant in Hawaii, their relative importance has diminished notably. Receipts from sales of these two products combined are far below the income generated from any major nonagricultural activities (government expenditure, tourism, commerce, services, manufacturing, transportation, construction) of the region. And total employment in all sugar- and pineapple-related activities now amounts to only 1 percent of the labor force.

TLM

and pineapple; an enormous expansion of this unique Hawaiian crop has occurred in the last few years. Other crops that increased in importance are horticultural specialties (potted plants, cut flowers, and lei flowers) and fruits. The bulk of all these specialties is cultivated on the Big Island. The well-known Kona coffee of the leeward coast of Big Island has declined since the 1960s owing to a variety of problems, most notably the cost and availability of labor.

Livestock

Cattle are the principal domestic livestock in the region and ranches occupy more than three-quarters of the agricultural land (Fig. 17-6). Most ranches are large—one is reputed to be among the five largest in Anglo-America—and concentrate on the raising of beef cattle. Grain feeding is uncommon and, although some hay is produced, generally the animals subsist on pasturage. Nearly all the meat is consumed on the islands, but this amount satisfies less than one-half the local demand for beef. Hides are exported. Cattle ranches are most notable on the Big Island but also occupy much of Maui and Molokai, and the entire island of Niihau is owned and operated as a single ranch.

Both dairying and poultry raising are expanding on the four larger islands.

Tourism

The most rapidly expanding sector of, and the largest source of income for, the Hawaiian economy is tourism. Beaches, climate, scenery, ceremonies, hospitality, and trans-Pacific crossroads location are the major assets. These items are exploited by one of the most thorough and best-organized publicity efforts anywhere; the renown of a holiday in Hawaii is worldwide.

Much of the business life of the region is geared to the visitor. Companies that cater to sleeping, eating, entertainment, and transportation services are continually expanding their operations. Hawaii is the major stopping point for trans-Pacific passengers. The great majority of all passenger ships and planes crossing the Pacific call at Honolulu. It is unrivaled as the major terminal city within the entire Pacific Basin, excluding the marginal centers of California, Japan, and Australia. Summer is the busiest season, with June decidedly the peak month. A smaller secondary peak occurs in December and January.

FIGURE 17-6 Cattle ranching is widespread and a cowboy culture is well established. Here some of the locals are having a Sunday afternoon competition in calf-roping on the Big Island (TLM photo).

The Waikiki area of Honolulu is the unquestioned center of island tourism (Fig. 17-7). It contains more than half the region's hotel rooms and is a seething hive of restaurants, elegant shops, sparkling beaches, and fashionable sunburns. Most visitors, however, also manage to see some other parts of the region. Interisland air carriers offer frequent and convenient service among the six larger islands and it is estimated that more than three-fourths of the visitors go to at least one other island in addition to Oahu (Fig. 17-8). The volcanic features of the Big Island, the exceptional scenic beauty of Kauai, and the outstanding beaches of Maui are the principal attractions among the "outer islands."

The visitor industry, as tourism is referred to officially, has experienced booming growth in recent years. The annual number of visitors to the region is approaching 6.5 million, which is more than five times as large as the permanent population. On any given day, more than 125,000 tourists are in the region. Three-fourths of the visitors are "westbound"—that is, arrivals from the North American mainland, with the majority originating in California. "Eastbound" visitors have increased significantly in recent years, three-quarters of them from Japan. On the average, westbound visitors stay 2½ times as long, but eastbound visitors spend 2½ times as much money per day.

FIGURE 17-7 The premiere attraction of Hawaii is Waikiki Beach (in the left foreground). Downtown Honolulu is farther left, the traditional landmark of Diamond Head is in the right foreground, and the cloud-capped Koolau Range towers in the distance (TLM photo).

FIGURE 17-8 A portion of the Napili Bay Resort on the island of Maui (TLM photo).

Federal Government Expenditures

Second only to the visitor industry as a generator of wealth in the islands is the federal government (Fig. 17-9). Because of the strategic value of its mid-Pacific location, Hawaii contains some of the nation's largest military bases. It is the headquarters for the U.S. Pacific Command and the administrative center for the Pacific operations of each of the three individual services.

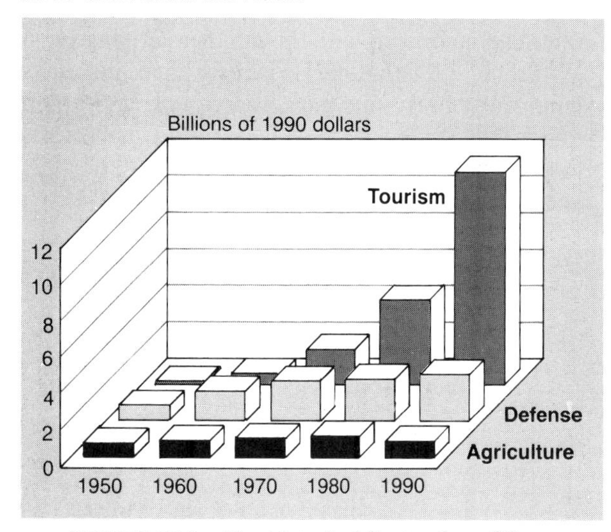

FIGURE 17-9 The historical fluctuation of three basic sectors of the Hawaiian economy, 1950–1990. (After Bank of Hawaii Annual Economic Report, 1990, p. 8.).

Approximately one of every five members of the region's labor force, including military personnel, is employed by the federal government, mostly in military establishments. The U.S. Armed Forces annually spend more than $2 billion in the islands; however, in every year since 1975 the volume of federal nonmilitary spending has been even greater than that figure, primarily because of the vast increase in direct health, education, and welfare benefit payments.

URBAN PRIMACY: A ONE-CITY REGION

No other region in Anglo-America is as thoroughly dominated by a single city as Honolulu dominates the Hawaiian Islands (Table 17-1). Not only does it contain most of the region's population—some 80 percent of the total—but it also has the great preponderance of all economic, political, and military activities.

There has been a long-continued drift of population from the outer islands in general and rural areas in particular to Honolulu. Since 1930 outer-island population has been on a declining trend, which only recently was slightly reversed. Oahu's share of the Hawaiian population total has grown from just over half in 1930 to more than four-fifths in the early 1990s. Honolulu's population continues to

TABLE 17-1

Largest urban places of the Hawaiian Islands

Name	Population of Principal City	Population of Metropolitan Area
Aiea*	34,200	
Hilo	37,100	
Honolulu	376,110	838,500
Kailua*	38,900	
Kaneohe*	31,100	
Pearl City*	47,700	
Waipahu*	30,900	

* A suburb of a larger city.

boom whereas other parts of the region show only irregular and sporadic growth.

Despite its exotic location, Honolulu is much like other rapidly growing Anglo-American urban areas and exhibits both the best and the worst of urban patterns and trends. There are many delightful residential areas and various parks and beaches provide almost unparalleled recreational opportunities. Transportation routes, however, are congested and some Honolulu traffic jams would be impressive in cities twice in size. The cost of living is high, most people cannot afford to own their own homes, unregulated high-rise construction has blighted the most cherished views of Waikiki and Diamond Head, and the city's raw sewage still pours undiminished into the blue Pacific. Urban sprawl is extensive, although constricted by the steep slopes of the Koolau Range. The extent of the urbanized area has rapidly spread eastward toward Koko Head and westward partially to surround Pearl Harbor and has jumped the Koolau Range to encompass the shores of Kaneohe Bay on the east coast (Fig. 17-10).

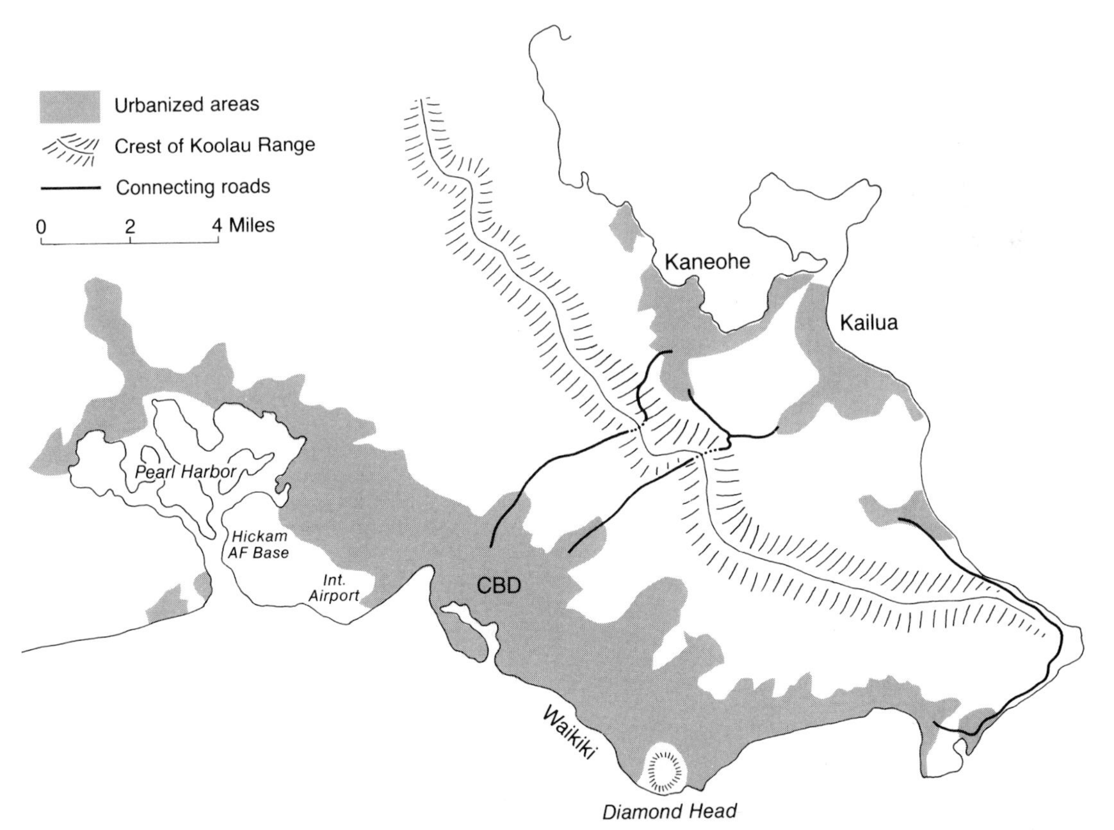

FIGURE 17-10 Honolulu's urban sprawl has encompassed much of the leeward coast of the island and jumps across the Koolau Range to occupy much of the Kaneohe lowland.

PROBLEMS AND PROSPECTS

Its unique environment and location compared to the mainland of the United States helped to engender in Hawaii an unusual pattern of economic-social-political relationships that have significant effects on the geography of the region.

Population Explosion

Overshadowing all else today is the high rate of population growth. Although not overwhelming in relation to other states (Hawaii's population increased by 23 percent during the 1980s, a rate exceeded by 11 other states), associated facts make such growth of great concern.

The burgeoning population is largely focused in Honolulu and it is there that the unpleasantness of overcrowding has an almost smothering effect. In portions of the Ala Moana–Waikiki census tract the population density exceeds 60,000 people per square mile. Although such numbers would not be unusual on Manhattan Island or in Chicago's Loop, they represent an abrupt shattering of an image of tranquil Hawaii. And the concomitant deterioration of the quality of life shows clearly.

Vehicular congestion is a particular nuisance. The region's population has doubled in the last quarter century whereas the number of motor vehicles has tripled. Hawaii has fewer miles of streets than the District of Columbia, for example, but twice as many cars.

Another disturbing aspect of the population increase is the escalating welfare load. Nearly 10

A CLOSER LOOK Urban Honolulu and Waikiki—A Love-Hate Relationship

From Hawaii-kai and Diamond Head to Honolulu International Airport and from *mauka* (mountain) of Waikiki to *makai* (seaward) of Punchbowl, urban Honolulu is a clear affirmation for those who believe that cities contain the best and worst of humankind. Poised mid-Pacific and populated by Euro-Americans, Japanese Americans, Chinese Americans, Filipino Americans, Hawaiian Americans, Korean Americans, African Americans, Samoan Americans, and a mix of all these and others, urban Honolulu has a love-hate relationship with itself.

Leeward of the Pali, early Honolulu offered sunny skies, constant winds, safe harbor, and marshy lowlands inhabited with a wide selection of plants and animals. It was the preserve of Hawaiian royalty. Through time, the marshes were transformed into canals and living space, houses were planted up the slopes and within the steep valleys, and Pearl Harbor and Honolulu International Airport formed the infrastructure that would link Honolulu to the circum-Pacific. Even as these changes occurred, the people of Hawaii, of the "Aloha spirit," gave flower leis for special occasions, offered ample food and hospitality at luaus, celebrated Aloha Friday, and ate home-grown pineapples, sugarcane, and macadamia nuts.

However, late in the twentieth century the now dominant tourist industry had considerably restructured urban Honolulu. Specifically, it was the tourists and investors from Japan that caused the changes. As of 1990, 22 percent of the tourists came from Japan, accounting for 43 percent of the tourist dollar spent in Hawaii—most of it within the confines of urban Honolulu. The Waikiki district has become a mecca of services and retail stores, hotels, restaurants, and elegantly appointed shops and gourmet-food stores—no longer just the provider of plastic flowers, Don Ho concerts, and polyester muumuus! Gucci's and Tiffany are among a few of the newcomers. Entire Waikiki shops have directed their retail business to the Japanese. And locals now have access to goods once found only on mainland (U.S.) shopping trips.

Restoration of landmark hotels and creation of architectural wonders, open to the sky and sea, have altered the landscape of Waikiki. *Kamaainas* (locals) will refuse to admit ever setting foot in Waikiki, yet they can be seen spending long evenings at the newly restored Moana Hotel or at Sunday brunches at the Royal Hawaiian Hotel! Senior proms at the Ilikai or Sheraton hotels are events that tie up traffic in Waikiki for longer periods than any invasion of tourists could ever create. Any morning of the week, locals can be found at Ala Moana Beach or Kapiolani Park where they swim and enjoy family luaus and feasts. Alongside the locals are the tourists engaging in snorkling and diving, sailing, or surfboarding. The beaches of the southern side of Oahu suffer most from the influx of tourists.

The development of the Ala Moana Shopping Center, Wards Center, and Kahala Mall—to name a few—are the effects of a burgeoning urban population and eager tourists percent of the total population is on welfare and welfare costs accelerated ninefold in the most recent decade. Sample studies show that only 30 percent of the welfare cases are Hawaii-born; most represent relatively recent in-migrants. The high cost of living in Hawaii is generally unanticipated by potential immigrants, whether foreigners or mainlanders.

Although crowding, congestion, pollution, staggering welfare costs, and related problems are not restricted to Hawaii, they pose a particularly menacing situation in a region that is both insular and isolated. Hawaii has limited resources and a finite land base, but the crux of this whole complex of problems is that this region, unlike any other in Anglo-America, must face the problems alone, without feasible interaction with neighboring regions or states. A mainland region can supplement its own resources by bringing products in from elsewhere; water or gas or electricity can be obtained by pipeline or transmission line. This option, of course, is unavailable for Hawaii. Ships and planes can move goods into and people out of the region to alleviate the situation, but for Hawaii even these normal flows are greatly complicated by cost and distance.

Palliatives, such as land-use restrictions and zoning, have been instituted, but they do not get to the root of the problem. Overpopulation can only be solved by some type of limitation on growth. Yet freedom of interstate (and international) migration is a cherished U.S. principle that is clearly upheld by the Constitution.

Hawaii has long been noted for its activist, liberal stance on socioeconomic issues. Today, however, authorities face the need to consider some

who covet chic services and goods. The small food stalls once scattered throughout the Ala Moana Center have been relocated to the Makai Center where modern neon-lighted shops offer an assortment of circum-Pacific foods and drinks. Tourists and locals both gravitate to the Ala Moana to eat, shop, talk, and, coupons in hand, to capture the latest bargain at Long's Drugstore.

Away from the Ala Moana, islanders' food consumption is eclectic: Spam is an island favorite—it is consumed in 3½ times the amounts of any other state. Equally important are *saimin* (noodles with anything added), *manapua* (steamed buns with pork), vienna sausages, *pupus* (hors d'oeuvres of any kind), *haole saimin* (noodles with hot dogs), and pork in any form; however, the staple is still rice with additions such as pork and beans; or spam and eggs; or fish sauce; or macaroni; or tuna; or *kim chee* (Korean cabbage). As the influx of mainlanders increases, fern bars and brie and salad bars have been added to the list of foods consumed.

Hawaii is plagued by the high cost of imported fresh foods and skyrocketing housing prices in an environment that has limited room to grow and where quality of life and resource utilization are thorny issues. The islanders suffer from underemployment resulting from the primacy of the tourist industry and the need to expand the technological fields—all of which makes the disparity even greater between the islanders and the foreign and mainland visitors.

Urban Honolulu is beginning to gentrify its older downtown and to seek limited or no-growth restrictions on hotels in favor of affordable housing for its underemployed population. But room is still the limiting factor, and investors from outside can and will pay handsomely for land and buildings. Out *Ewa* way (or west beyond the airport), Honolulu is planning to build its new central business district—a planned CBD with accompanying affordable housing, parking, and open spaces.

Yet Hawaii is the only state that sings its state song in a language other than English; where the state seal includes a crown; where the Kamehameha Interstate Highway begins and ends on the island of Oahu; where almost everyone came from somewhere else; where the major holidays are a potpourri of the circum-Pacific's cultures; and where geckos, myna birds, and "poi" dogs are the local animal life. The dichotomous geography of urban Honolulu surpasses any other region except perhaps Alaska.

Does anyone leave? Few—instead the rate of immigration grows and the attraction of the primate city continues as urban Honolulu stretches out into the suburbs, across the Pali, inland through the isthmus to Waipahu, windward to Kailua and Kaneohe, and westward to Ewa. The Aloha spirit continues to dominate both the *kaamainas* and new arrivals as does the need for services, the romance of the Trade Winds, and the glow of the sun and traffic on urban Honolulu.

Professor Joan Clemons
Los Angeles Valley College
Van Nuys

very repressive and conservative approaches to population limitation. What is the best way to react to the immutable problem of overpopulation in a fragile island environment, exacerbated by the region's unique position with relation to mainland United States and, increasingly, to Asia?

Development Versus Preservation

A prominent and continually escalating controversy in the region revolves around the desirability and propriety of development. Development in the Hawaiian context chiefly involves building resort complexes. Many projects are in the already super-crowded environs of Waikiki, but increasingly the emphasis is on the construction of condominium-hotel resorts, with their associated infrastructure of roads and golf courses, in remoter locations on Oahu and in the outer islands.

The visitor accommodation inventory (hotel rooms and condominiums) in the region increased from 5000 in 1960 to 70,000 in 1990, but the occupancy rate continues to be high even in recession years. Thus the demand for more rooms and facilities appears to be unabated. Can the fragile environment of the island state withstand such pressures (Fig. 17-11)?

FIGURE 17-11 For all its problems, Hawaii continues to be a major destination for visitors from all over the world, and particularly from continental United States. Warm water, fine beaches, and a tropical climate provide the basis for its attractiveness. This is Poipu Beach on the island of Kauai (TLM photo).

Hawaii was the first state to enact a statewide land-use plan (1961),[5] but its principal thrust was to preserve agricultural land. New laws, including a constitutional amendment, in 1978 addressed a different set of land-related concerns—those associated with rapid development in general and urbanization in particular. Moreover, many areas have local regulations that strictly control zoning, building heights, parking, landscaping, and building design. Still, it would appear that, without an agonizing reappraisal of land-use plans and more regulatory safeguards, the charm of uncluttered beaches and valleys, not to mention the perpetuation of local lifestyles, may be dissipated or destroyed by the ambitious schemes of land developers, highway lobbies, and construction unions.

Land Ownership

One of the most unusual aspects of Hawaiian geography involves the system and pattern of land ownership. Approximately 42 percent of the total land area is under government control. It is a much smaller proportion than in many states west of the Mississippi, but, in addition, another 47 percent is held by a mere 70 estates, trusts, and other large owners. Less than 11 percent of the land is therefore subject to general private ownership. Many plans were advanced, particularly by the state legislature, to enable individuals to obtain small parcels of land. As a result, the number of farms is increasing and the average farm size is decreasing, both in direct opposition to the trends on the mainland. More than half the farms in Hawaii are now less than 10 acres (4 ha) in size.

Homesteads proliferated on Molokai, Hawaii, and Oahu, but generally the land is in large estates. Traditionally these estates do not pass freely to heirs; instead, trusts of various types are set up to administer them. This step resulted in "freezing" the ownership and the land is leased in large blocks rather than being sold. One-half the private land in the islands is owned by nine major estates. Such a situation is not inherently unsavory, but the long-

[5] The 1961 Land-Use Law put the state government in charge of all land use administration, regardless of ownership of the land.

range effect may prove deleterious to economic growth as well as social and political conditions.

Transportation

Here, as for other islands far removed from the mainland, the problem of transportation is always notable. Because the region consists of a group of islands, the physical matter of moving people and goods from one place to another within the region can be intricately complicated. Factories in Honolulu must have materials that originate on the other islands and citizens of the outer islands need goods produced in or shipped through Honolulu.

Most tourists arrive by air, but the shipment of food and other commodities is handled by surface transport, which is slow, expensive, and subject to disruption by labor disputes, weather, and other factors. The problem of transportation is one of the immutable facts of life in the region.

THE OUTLOOK

Perhaps in no other region are the hazards and the potentials so clear-cut. From an economic standpoint there is cause for concern, but there is also reason for optimism. The sugar industry is well-established, but it is caught in a cost–price squeeze that portends an increasingly precarious existence. The future for pineapples has also been dark, and despite the expansion of the fresh pineapple market, the outlook is not favorable.

Much discussion has concerned agricultural diversification, but little was done to diminish the vast need for importing basic foodstuffs. Expanded production of subtropical specialties, such as macadamia nuts and papayas, is likely to continue.

Hawaii's cost of living is significantly higher than that of any other state except Alaska and the state budget is strained in many directions. The unemployment rate has been at a high level for some time even though such occupations as garment workers and coffee harvesters are in short supply. Manufacturing, in general, shows some prospects for growth but primarily in fields that are not basic to the total economy—that is, those that supply goods for the Hawaiian market.

As agriculture declined in recent years, the tourist boom took up much of the economic slack in this resource-poor region. The "visitor industry" dramatically expanded during the 1980s so that its revenues doubled from 20 percent to more than 40 percent of the gross state product. It is anticipated that visitor numbers will continue to climb, but that the length of stay (and thus expenditure) will be reduced. More exotic destinations are increasingly attractive. Nevertheless, a prominent lesson of recent years is that Hawaiian tourist visitation is more dependent on inexpensive air fares than on general economic prosperity. If the public perceives that it is relatively cheap to go to Hawaii, then the public will go to Hawaii in considerable numbers.

Although tourism passed military expenditures as a source of regional revenue two decades ago, Hawaii is still to a considerable extent a garrison state. Federal defense spending probably will continue a slow absolute increase, but its relative importance to the economy is likely to diminish.

From a social standpoint Hawaii has been an American showcase for racial assimilation. Continued intermarriage will probably blur individual ethnic strains into a more widespread Hawaiian blend.

The region's population should grow faster than the national average because of a relatively high birthrate and continued immigration of people attracted by the prospect of island living. Greatly expanded urbanization is likely, particularly on Oahu, where Greater Honolulu, along with its extended suburbs of Lanikai–Kailua–Kaneohe, will spread north and south on both sides of the Koolau Range.

Hawaii's ancient motto is "The life of the land is perpetuated in righteousness." But how is it possible to preserve righteousness toward the land in a tropical island milieu that is being overwhelmed with the incessant pressures of civilization? The delicate balance between an economy dependent on boom-growth tourism and the maintenance of a pleasant environment and attractive lifestyle to lure tourists is an almost imponderable dilemma.

SELECTED BIBLIOGRAPHY

ARMSTRONG, R. WARWICK, ed., *Atlas of Hawaii*. Honolulu: University Press of Hawaii, 1973.

BRYAN, E. H., JR., *The Hawaiian Chain*. Honolulu: Bishop Museum Press, 1954.

CARLQUIST, SHERWIN JOHN, *Hawaii: A Natural History*. Garden City, NY: Natural History Press, 1970.

CUDDIHY, LINDA W., and CHARLES P. STONE, *Alteration of Native Hawaiian Vegetation: Effects of Humans, Their Activities and Introductions*. Honolulu: University of Hawaii Press, 1989.

FARRELL, BRYAN H., *Hawaii, the Legend that Sells*. Honolulu: The University Press of Hawaii, 1982.

FOSBERG, F. R., "The Deflowering of Hawaii," *National Parks and Conservation Magazine*, 49 (October 1975), 4–10.

FROST, MARVIN, D., "Savannas on the Island of Hawaii," *The California Geographer*, 19 (1979), 29–48.

GAGNE, W. C., "Hawaii's Tragic Dismemberment," *Defenders*, 50 (1975), 461–470.

KAY, E. ALISON, ed., *A Natural History of the Hawaiian Islands: Selected Readings*. Honolulu: University of Hawaii Press, 1972.

KING, WARREN, "Hawaii—Haven for Endangered Species?" *National Parks and Conservation Magazine*, 45 (1971), 9–13.

LIN, GONG-YUH, "Spectacular Trends of Hawaiian Rainfall," *Proceedings*, Association of American Geographers, 8 (1976), 12–14.

MACDONALD, GORDON A., and AGATIN T. ABBOTT, *Volcanoes in the Sea: The Geology of Hawaii*. Honolulu: University of Hawaii Press, 1970.

MCDOUGALL, HARRY, "Volcanoes of Hawaii," *Canadian Geographical Journal*, 80 (June 1970), 208–217.

MORGAN, JOSEPH R., *Hawaii: A Geography*. Boulder, CO: Westview Press, 1983.

NORDYCKE, ELEANOR C., *The Peopling of Hawai'i*. Honolulu: University of Hawaii Press, 1989.

WALTHAM, TONY, and CHRIS WOOD, "Fiery Tunnels of Kilauea," *Geographical Magazine*, 53 (September 1981), 766–771.

18

THE NORTH PACIFIC COAST

Extending latitudinally for more than 2000 miles (3200 km) along the northwestern fringe of the continent is the North Pacific Coast Region. The region encompasses the continental margin from northern California to southwestern Alaska, nowhere penetrating inland more than 200 miles (320 km) from the sea.

The attenuated, coast-hugging shape of the region is due largely to the topographic pattern (Fig. 18-1). The major mountain ranges of far western Anglo-America are oriented parallel to the coastal trend, lying directly athwart the prevailing currents of midlatitude atmospheric movements and severely restricting the longitudinal penetration of oceanic influences. The interior (eastern and northern) boundary of the region is thus approximately coincidental with the crest of the principal mountains: the Cascade Range in the conterminous states, the Coast Mountains in British Columbia, the St. Elias Mountains in the Yukon Territory, and the Wrangell and Alaska ranges in Alaska. The southern margin of the region is just north of the San Francisco Bay Area conurbation in California and the western extremity is in the Alaska Peninsula where forest is replaced by tundra.

Such coastal proximity ensures that the influence of the sea is pervasive throughout the region, although it is somewhat subdued in such sheltered lowlands as Oregon's Willamette Valley and Vancouver Island's eastern coastal plain. Human activities and the physical environment are significantly affected by the maritime influence, which is most conspicuously reflected in climatic characteristics. Winters are unusually mild for the latitude and summers are anomalous in their coolness.

The most memorable climatic characteristics are associated with moisture relationships. Abundant precipitation, remarkably heavy snowfalls in the mountains, high frequency of precipitation, considerable fogginess, and the widespread and rela-

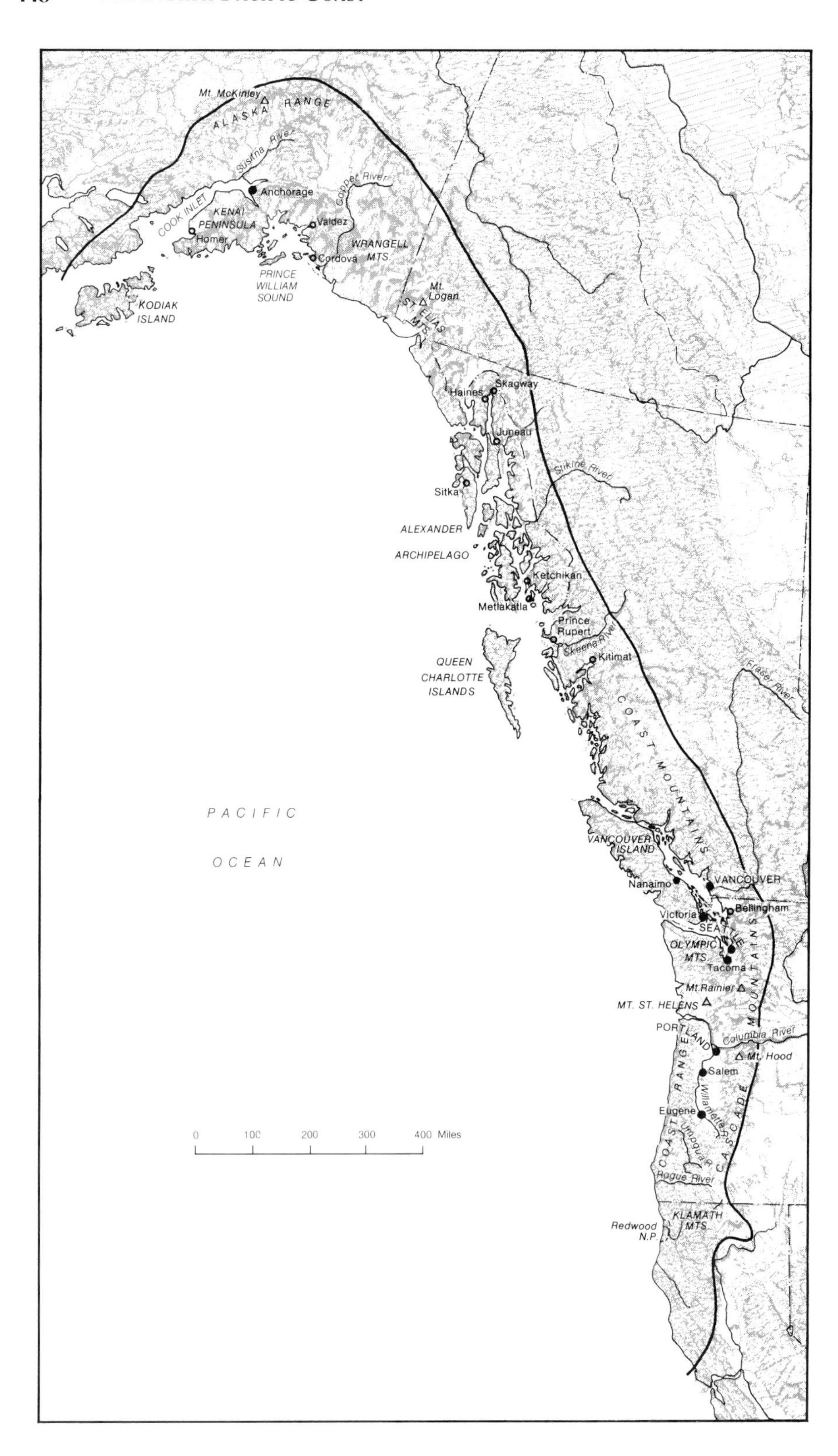

FIGURE 18-1 The North Pacific Coast Region (base map copyright A. K. Lobeck; reprinted by permission of Hammond, Inc.).

tively continuous occurrence of overcast cloudiness produce a climatic regime that, although not extreme, is exceedingly drab.

Another prominent characteristic of the North Pacific Coast Region is that its natural resources occur in limited variety but often in great quantity. Partly as a result, the economy of the region is not diversified but is dependent on specialties of production; furthermore, the limited resource base is a continual arena of controversy. The exploitation and development of resources frequently involve major conflicts of interest. How to dam the rivers for hydroelectricity generation without ruining the salmon fishery? How to exploit the timber resources without despoiling the unparalleled scenery? How to develop the national parks for visitor convenience without destroying the wilderness?

This is a region in which the people have had to live with remoteness—which is a joy to some but despair for others. The North Pacific Coast is remote from the heartland of both nations and is separated by significant topographic barriers from all external population centers. Access to the interior is limited to a relatively few routes and north-south connections are difficult. No railroad runs along any part of this coast and no highway is found along most of the British Columbia and Alaska coastline. This difficulty of access and connectivity has been a significant deterrent to economic growth, resulting in high transportation costs for goods brought into the region and in one of the continent's highest-cost-of-living indexes.

Another effect of remoteness has been psychological, a provincialism of attitude that rivals the more celebrated parochialism of Texans, Californians, New Englanders, French Canadians, or Southerners.[1] But this aspect of the regional character is changing as improved communications tend to smooth out regional differences throughout North America.

[1] A classic example of this attitude was the headline in the *Vancouver Sun* a few years ago when a severe blizzard had disabled major highways, both railway lines, and all wired communications extending east from Vancouver. The headline read, "Canada Cut Off."

THE TERRAIN: STEEP AND SPECTACULAR

The entire region is dominated by mountains. They vary in height from the comparatively low coastal ranges of northern California, Oregon, and Washington, to the higher ranges of the Cascade Mountains with their superb volcanic peaks, and to the great alpine ranges of western Canada and Alaska surmounted by Mount McKinley[2] (elevation 20,300 feet, or 6090 m), the highest peak on the North American continent.

The general topographical pattern consists of three very long landform complexes, northwest-southeast in trend and generally parallel to one another throughout the region. The westernmost zone consists of low mountains that become higher, more rugged, and more severely glaciated toward the north. Just to the east is a longitudinal trough that is prominent from Oregon to the Yukon. The easternmost zone consists of complex mountain masses surmounted by spectacular volcanic peaks in the south and extensive ice fields in the north.

Coastal Ranges

The coastal ranges within the United States portion of this region are a series of somewhat distinct mountain areas. They include the Coast Ranges of northern California, the Klamath Mountains that tie these ranges to the southern Cascades, the Coast Range of Oregon and Washington, the Olympic Mountains of Washington, the Vancouver Island Mountains, the Queen Charlotte Ranges, and the mountains of the Alexander Archipelago in southeastern Alaska.

California's northern Coast Ranges have the same parallel ridge-and-valley structure as those of the central coast of the state. The topography is strongly controlled by structure, with folding and faulting dominant. Ridge crests are even, if discontinuous; slopes are generally steep.

The Klamath Mountains mass appears as a complicated and disordered complex of slopeland. It

[2] Although the former Mt. McKinley National Park now is officially designated Denali National Park, the name of the mountain itself has not been changed. At the time of this writing, the official name is still Mount McKinley.

FIGURE 18-2 The rugged crest of the Olympic Mountains (TLM photo).

is wild and rugged country extending inland to connect with the southern part of the Cascade Range. Peaks reach to almost 9000 feet (2700 m) and much of the high country was heavily glaciated in Pleistocene time.

The Coast Range of Oregon and Washington has relatively low relief; crest lines average about 1000 feet (300 m) in elevation, with some peaks another 1000 feet (300 m) above that. North of the Columbia River the range is much more subdued. Unlike the California Coast Ranges, it is crossed by a number of prominent transverse river valleys—notably those of the Columbia, Rogue, and Umpqua rivers.

The Olympic Mountains are a massive rugged area of high relief and steep slopes. Although their highest peak (Mount Olympus) is less than 8000 feet (2400 m), heavy snowfall has given rise to more than 60 active glaciers in the range (Fig. 18-2). The margins of the Olympic Mountains are abrupt and precipitous, particularly on the east side.

The Strait of Juan de Fuca, which separates the Olympic Peninsula from Vancouver Island, provides a broad passage across the coastal mountain trend. Most of Vancouver Island is composed of a complex mountain range that descends steeply to the sea on the southwestern side, where the coastline is deeply fiorded and embayed. This range, too, has been heavily glaciated and contains a number of active glaciers, although its maximum elevation is less than 7500 feet (2250 m).

The mountains of the Queen Charlotte Islands and the Alexander Archipelago are relatively low and less rugged. Farther north, the coastal mountains unite with the inland ranges in the massive and spectacular knot of the St. Elias Range.

Interior Trough

Between the Cascades to the east and the Olympic and Coast ranges to the west lies the structural trough of the Willamette Valley and Puget Sound. The trough was formed by the sinking of this landmass at the time the Cascades were elevated. In glacial times a large lobe of ice advanced down Puget Sound and was instrumental in shaping the basin that holds that body of water. Today the Willamette Valley is a broad alluvial plain, 15 to 30 miles (24 to 48 km) wide and 125 miles (200 km) long; the Puget Sound lowland is somewhat smaller in area, for a large part of it has been submerged. North of Puget Sound the structural trough persists as the Strait of Georgia and Queen Charlotte Strait, which becomes a continuous waterway (the "Inside Passage") almost to the Yukon Territory.

Inland Ranges

The Cascade Range, extending from Lassen Peak in northern California to southern British Columbia, is divided into a southern and a northern section by the deep gorge of the Columbia River. The relief features of the southern part are predominantly volcanic in origin, with a subdued and almost plateaulike crest on which several conspicuous volcanic cones were superimposed. From south to north, prominent peaks are Mount Lassen, one of only two volcanos in the conterminous states to have erupted in this century; the 14,162-foot

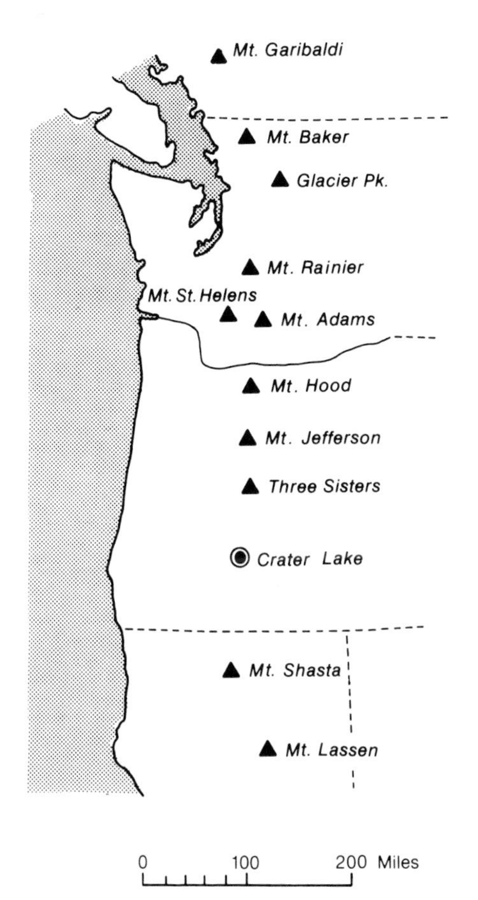

FIGURE 18-3 The major volcanic peaks of the Cascade Range.

(4250-m) Mount Shasta; collapsed Mount Mazama, whose caldera is now occupied by Crater Lake, one of the world's deepest lakes; The Three Sisters; Mount Jefferson; and the 11,225 (3368-m) Mount Hood.

The northern Cascades are granitic rather than volcanic and are much more rocky and rugged. Surmounting the mass of the range are five prominent, old, icecapped volcanoes: Mount Adams, Mount St. Helens, Mount Rainier, Glacier Peak, and Mount Baker (Fig. 18-3). Heavy snowfall and cool summers combine to produce extensive glaciation (Fig. 18-4 and 18-5). Some 750 glaciers exist in the north Cascades, which is two-thirds of all the active glaciers in the conterminous states (Fig. 18-6). Indeed, more than 40 glaciers occur on Mount Rainier alone.[3]

North of the Fraser Valley, the ranges that correspond to the Cascades are known as the Coast Mountains. They average 100 miles (160 km) in width, are nearly 900 miles (1440 km) in length, and have been severely eroded by mountain icecaps and glaciers. Deep canyons of the Fraser, Skeena, Stikine, and Taku rivers have cut across the range, forming features similar to the Columbia Gorge.

[3] Glaciers have developed wholly within the forest zone in some parts of the North Cascades, a phenomenon that apparently does not occur elsewhere in the Northern Hemisphere. See Stephen F. Arno, "Glaciers in the American West," *Natural History*, 78 (1969), 86.

FIGURE 18-4 Mt. St. Helens (foreground) and Mt. Adams in the tranquil days prior to the former's 1980 eruption (TLM photo).

FIGURE 18-5 Mt. St. Helens (foreground) and Mt. Adams after the eruption. The changed shape and size of Mt. St. Helens are notable (TLM photo).

These mountains plunge directly down to the coast without a fringing lowland, where they are incised by a remarkable series of long fiords and inlets. The high country contains innumerable glaciers and some ice fields that are hundreds of square miles in extent (Fig. 18-7).

Inland from the Gulf of Alaska, where Alaska, the Yukon Territory, and British Columbia come together, is the ice-and-rock wilderness of the St. Elias Mountains. This mountain fastness contains Canada's highest peak, Mount Logan (19,850 feet, or 5955 m), and has more than a dozen peaks that are higher than any in the conterminous states or elsewhere in Canada (Fig. 18-8). These are the highest coastal mountains in the world, and they encompass the world's most extensive glacial environment outside the polar regions (Fig. 18-9).

West and northwest from the St. Elias massif the mountain trend again bifurcates. Along the coast is the long, remarkable rugged, and heavily glaciated Chugach Range, which eventually gives way to the less extensive but equally rugged Kenai Mountains. The inland ranges include the massive Wrangell Mountains and the Alaska Range, which is the continent's highest, culminating in Mount McKinley. The Alaska Range is crescent shaped, with its western extremity terminating at the base of the Alaska Peninsula. Between the Alaska Range and the Kenai Peninsula is the only extensive lowland in the northern part of the region; about half of it is occupied by the broad bay of Cook Inlet and the remainder consists of the valleys of the Susitna and Matanuska rivers.

FIGURE 18-6 The high country of the North Cascades includes dozens of jagged peaks. This scene is near Rainy Pass (TLM photo).

FIGURE 18-7 There are dozens of high mountain icefields in this region. The Juneau Ice Field, pictured here, straddles the Alaska–British Columbia border for nearly 100 miles (160 km) (TLM photo).

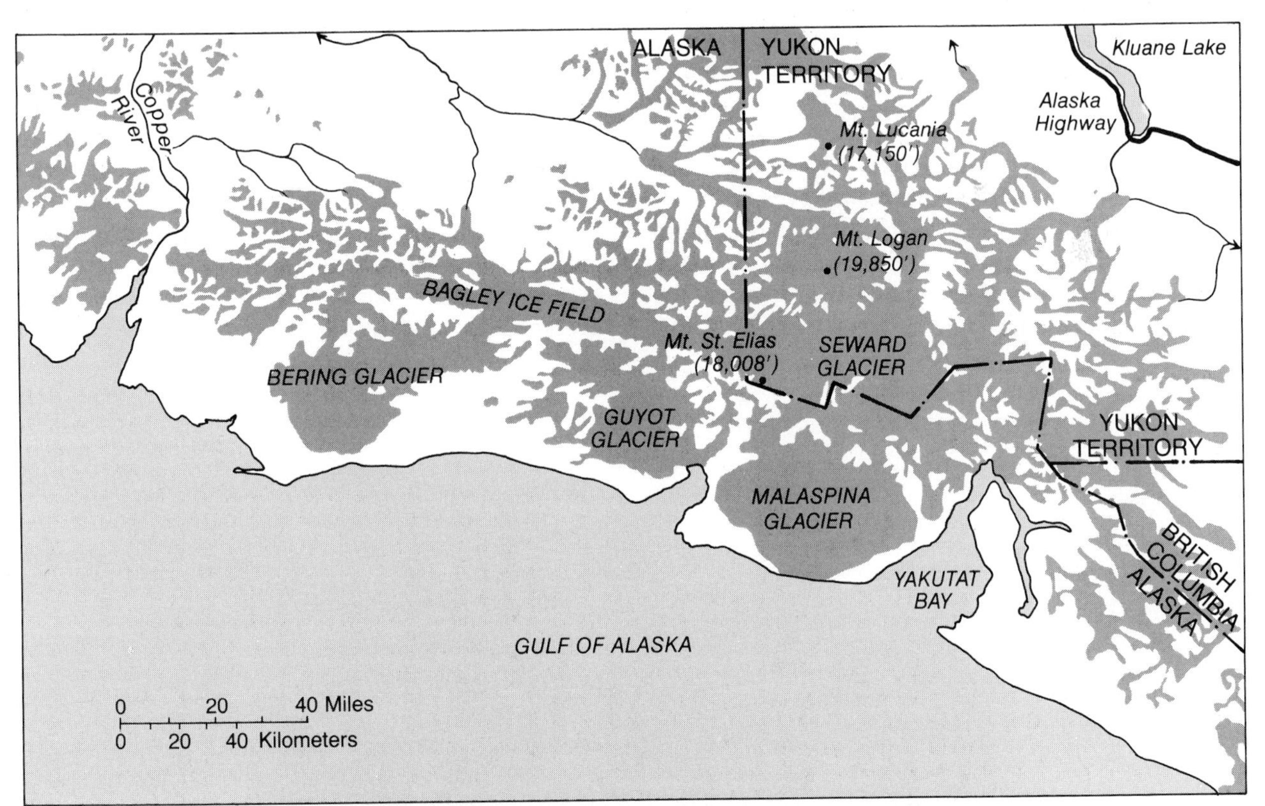

FIGURE 18-8 Anglo-America's most spectacular corner, the St. Elias–Wrangell–Malaspina area where Alaska, British Columbia, and the Yukon Territory come together. The shaded areas represent glaciers and icefields.

FIGURE 18-9 Hundreds of glaciers flow down valleys in southern and southeastern Alaska. Here the surface of Casement Glacier (near Haines) is candy-striped by the longitudinal debris of several medial moraines (TLM photo).

CLIMATE: MOIST AND MONOTONOUS

The North Pacific Coast Region has a temperate marine climate in which the downwind relationship with the ocean markedly ameliorates temperatures. High summer temperatures are almost unkonwn, and low winter temperatures occur with frequency only in the lowlands of south-central Alaska and in highlands throughout the region.

Precipitation is characteristically abundant, but this factor is sharply modified by altitude and exposure. In general, there is a fairly even seasonal regime in the north but a decided winter maximum in the south. During the winter gigantic cyclonic storm systems, which migrate eastward across the Pacific Basin, bring simultaneous rains for 1000 miles (1600 km) north and south along this coastal region. Mountainous terrain influences the areal distribution of precipitation, southwest slopes receiving copious rainfalls and northeast sides receiving scant ones. The southwest flank of the Olympic Mountains saturated in winter, has an average annual precipitation of 150 inches (3800 mm), the maximum for the conterminous United States. In contrast, the northeast side of these mountains, only 75 miles (120 km) away, has an annual rainfall of 16 inches (400 mm), which is too little to support even good pastures without the aid of irrigation. The northwestern corner of Vancouver Island records 250 inches (6350 mm) annually and is considered the wettest spot in continental Anglo-America. Snow accumulates to great depth, especially on exposed

FIGURE 18-10 Low clouds and fog are commonplace in the North Pacific Coast Region. This foggy harbor scene is at Ketchikan, Alaska (TLM photo).

mountain slopes; some localities experience annual snowfalls in excess of 80 feet (24 m).

Modified by the terrain, the east-west precipitation pattern falls into four easily recognizable belts:

1. The coastal strip, with abundant rainfall and little snow.
2. The windward side of the coastal ranges, with excessive precipitation.
3. The leeward side of the coastal mountains and the interior trough, with only a moderate rainfall and sporadic snows.
4. The western slope of the Cascades in the United States and Coast Mountains in Canada, with nearly 100 inches (2540 mm) of precipitation—mainly winter snows.

The winter season is cloudy, monotonously damp, and protected from chilling continental winds by a double barrier of mountains to the east. Summer—the dry season—is characterized by mild temperatures, light surface winds, coastal fogs (Fig. 18-10)), and low clouds. Throughout most of the region, the average number of clear days each year is less than 100; Juneau, for example, records only 45.

THE WORLD'S MOST MAGNIFICENT FORESTS

Except in some of the interior valleys, the heavy precipitation throughout this region makes it a land of forests. In the northern California Coast Ranges the dominant tree is the redwood and within this area may be found some of the most magnificent forests of the world. Along the coasts of Oregon and Washington and in the Cascade Mountains the Douglas fir is dominant, constituting one of the major lumber trees of the continent (Fig. 18-11). Other notable trees within this area are western hemlock, western red cedar, and Sitka spruce. Douglas fir is the most valuable species, followed by hemlock, in the Canadian section. The Sitka spruce is the leading tree in southeastern Alaska. The forests of this region are almost exclusively coniferous.

A marked contrast in natural vegetation exists between the Willamette Valley of Oregon and Puget

FIGURE 18-11 Massive trees dominate the forests of the region. This Douglas fir scene is near Chemainus on Vancouver Island (TLM photo).

Sound area of Washington. In the former, the barrier nature of the Coast Range tends to produce a light summer rainfall; in the latter, the rainfall is slightly heavier in summer because the gaps in the Coast Range allow more rain-bearing winds to enter from the Pacific. The cooler temperature of the Puget Sound area also causes it to be somewhat more humid. As a result, the native vegetation of the Puget Sound area was a dense stand of giant Douglas fir trees with limited expanses of prairie; the Willamette Valley's original vegetative cover apparently was largely prairie grass, although some scholars believe that the original prairies were artifically generated and maintained by deliberate annual fires set by the Indians.[4] This contrast of vegetation types at the time of European contact profoundly influenced the settlement of the two areas.

Much of the region, particularly in the north, is above the tree line; in such areas, alpine tundra rock, and ice constitute the ground cover.

OCCUPANCE OF THE REGION

Aboriginal inhabitants of the North Pacific Coast Region consisted of a great variety of Indian tribes: small tribes of Algonkian speakers in northern California; various Chinook groups in the Columbia River area; Salishan, Nootka, Kwakiutl, Bellacoola, and Tsimshian in British Columbia; and the redoubtable Haida and Tlingit in southeastern Alaska. From the Indian standpoint this was a region rich in resources: salmon and other seafoods in great quantity, berries in profusion in summer, and great forests of useful trees. The Indians built impressive buildings of evergreen planks, fashioned large dugout canoes from the native cedars, and, particularly in the north, became skilled woodcarvers—with giant totem poles as their most lasting achievements.

The Chinooks and Bellacoolas did considerable trading with the encroaching whites, but for the most part Native Americans of the region had minimal effect on European penetration and settlement, with the important exception of the Tlingits in the north. The Tlingits were proud, well organized, and resourceful; for several decades they exerted both political and economic hegemony in the Alaska panhandle area, controlling trade between the more primitive Athabaskan Indians of the interior and the European mercantile interests. But even the Tlingits were soon swept aside in the flurry of prospecting and settlement.

Today Native Americans are insignificant in numbers in the southern part of the region. In north coastal British Columbia and southeastern Alaska, however, they constitute much of the work force in fishing, canning, and forestry.

The voyages of Vitus Bering between 1728 and 1742 led to the advance of Russian trappers and fur traders southward along the Alaskan coast. In 1774 the Spaniard Juan Perez sailed as far north as latitude 55°. Another important voyage was that of the English Captain James Cook in 1778, who explored the coast between latitudes 43° and 60° north, and further complicated the claims to this strip of coast. In 1792 a New England trading vessel reached the mouth of the Columbia River and established the fourth claim to the region. Thus by the end of the eighteenth century Spain, Russia, Great Britain, and the United States had explored and claimed in whole or in part the Pacific Coast of North America from San Francisco Bay to western Alaska.

Permanent Spanish settlements were never established north of San Francisco Bay, but by 1800 Russia was entrenched on Baranof Island in southeastern Alaska and had its seat of colonial government at Sitka. Further settlements were made to the south, but agreements were signed with the United States and Great Britain in 1824 and 1825 limiting the Russians to territory north of the 54°40′ parallel. Spain abandoned its claim to all land north of the 42nd parallel. This left the Oregon country—between the Spanish settlements in California and the Russian settlements in Alaska—to the United States and Great Britain.

Following the Lewis and Clark expedition, the American Fur Company in 1810 established a trading post at Astoria, at the mouth of the Columbia River, but this settlement was seized by agents of the Hudson's Bay Company during the War of 1812. British forts on the lower Columbia dominated the

[4] For example, see Carl L. Johannessen et al., "The Vegetation of the Willamette Valley," *Annals*, Association of American Geographers, 61 (June 1971), 302.

area until 1818, when an agreement was reached for joint occupance by English and American traders.

At first the only Americans who reached this far-off land were a few trappers and traders, but in the early 1830s New England colonists came overland via the Oregon Trail, and soon a mass migration began. The great trek along the Oregon Trail took place in the early 1840s. These pioneers, determined to establish a Pacific outlet for the United States, had the slogan "Fifty-four forty or fight." Most of them located in the prairie land of the Willamette Valley of Oregon and by 1845 there were 8000 Americans in the Oregon country. In the settlement of the "Oregon Question" in 1846 the United States got the lands south of the 49th parallel (except for Vancouver Island) and Great Britain got the land between there and Russian America. The final status of the San Juan Islands, now a part of the state of Washington, was not decided until 1872.

Victoria was chosen as a settlement site by the Hudson's Bay Company in 1843, but there was little activity in the British part of this region until the great Cariboo gold rush that began in 1858. Victoria prospered as the transshipment point and funnel for the British Columbia gold fields in the same way that San Francisco did for the California mining areas. In the 1850s two other settlements were founded in southwestern British Columbia: Nanaimo on Vancouver Island had the only tidewater coal field on the North American Pacific coast, and New Westminster in the lower Fraser Valley became the mainland commercial center. British Columbia joined the Canadian confederation as a province in 1871, but there was little population or economic growth until the arrival of the transcontinental railway in 1886 and the founding of Vancouver as its western terminus. Within a decade of its origin, Vancouver had surpassed Victoria as the largest city in western Canada (population 25,000).

After furs became depleted in the 1840s, the Russians began to lose interest in their far-off American possession. Although they had leased or sold some of their posts to the Hudson's Bay Company, they were loath to sell Alaska to Great Britain because of the Crimean War and therefore offered it to the United States. The purchase was made in 1867 for the sum of $7.2 million, or less than 2 cents per acre.

The Puget Sound country still was remote from populous centers of the continent and until the completion of the Northern Pacific Railroad in 1883 its only outlet for bulky commodities of grain and lumber was by ship around Cape Horn. Between 1840 and 1850 a number of small sawmills were erected in the area to export lumber to the Hawaiian Islands and later to supply the mining camps of California. The Canadian Pacific Railway reached its Vancouver terminus in 1886; in 1893 the Great Northern completed its line across the mountains to Puget Sound; and sometime later the Chicago, Milwaukee, St. Paul, and Pacific Railroad built into the region. Meanwhile, the Union Pacific established direct connection with Portland and the Southern Pacific linked Portland with San Francisco. These rail connections made possible the exploitation of the great forest resources, which became important about the beginning of the present century and also contributed to the industrial development and urban growth of that part of the region. Not until 1914 was the other Canadian transcontinental railway, originally called the Grand Trunk Pacific and now the Canadian National, completed to its terminus at Prince Rupert, although a subsequent terminal at Vancouver was much more important to the Canadian National.

When gold was discovered in the Klondike in 1897 and at Nome in 1898, a stampede began that closely rivaled the California rush of 1849. It was a long, hard, dangerous trip; the most direct route to the Klondike field was by ship through the Inside Passage from Seattle to Dyea and later Skagway, thence over Chilkoot Pass or White Pass to the headwaters of the Yukon River, and finally by riverboat or raft about 500 miles (800 km) downstream to Dawson. When gold was found in the beach sands at Nome, the trip was made entirely by ship, but in each case Seattle and Vancouver profited by being the nearest ports with railway connections to the rest of the continent.

Settlement significantly expanded and intensified in older centers of the region—that is, the Willamette Valley, Puget Sound lowland, lower Fraser Valley, and the Victoria area. The most spectacular relative growth in recent decades, however, has been in the Cook Inlet area of south coastal Alaska. Anchorage and its immediate hinterland, the Matanuska Valley and the Kenai Peninsula, attracted a great many settlers from various parts of the United States. Nearly half the people of Alaska live within 25 miles (40 km) of Anchorage.

WOOD PRODUCTS INDUSTRIES: BIG TREES, BIG CUT, BIG PROBLEMS

The North Pacific Coast Region, with its temperate marine climate, contains the most magnificent stand of timber in the world. The trees decrease in size from the giant redwoods of northwestern California and the large Douglas firs and western red cedars of Oregon, Washington, and British Columbia to the smaller varieties of spruce, hemlock, and fir along the coast of Alaska. At one time probably 90 percent of the region was covered by these great forests.

The Douglas fir has a greater sawtimber volume than any other tree species on the continent, the size of the individual trees and the density of the stand being exceeded only by sequoias and redwoods. Douglas fir attains its best development in western Oregon, Washington, and British Columbia and constitutes about half of both the sawtimber volume and cut of this region.

The North Pacific Coast Region has long been dominant in lumber production on the continent, although its share has been diminishing over the past few years. Oregon, California, and Washington continue to be the three leading states in lumber production. British Columbia normally produces more than two-thirds of Canada's lumber output, although much of the production is from the interior of the province (Fig. 18-12).

FIGURE 18-12 A logger felling a Douglas fir on Vancouver Island (TLM photo).

This region is also a major producer of plywood, particleboard, and pulp and paper, normally ranking just behind the Inland South in each category. In recent years there has been a great increase in "production" of logs and wood chips from the North Pacific Coast region; the major destination for both products is Japan.

Logging Operations

Most timber harvesting in this region has long been done by the clearcutting technique, in which every tree in a designated section of an area is removed but none is removed from surrounding sections until later years. A typical acreage may be divided into 70 sections, with one section being clearcut each year and immediately afforested. At the end of the seventieth year all sections have been cut and the first section, now supporting a mature, even-aged stand, is ready for harvest again (Fig. 18-13).

Today the logging industry is highly mechanized. Giant diesel-powered tractors or bulldozers "snake" the logs out of the forest. These logs are then hauled by large trucks or by rail to the sawmills or to streams where they can be floated to the mills. Another common logging system uses two strong spar trees, one at the top of a slope and the other at the bottom, with a cable fastened between them high above the ground. A traveling carriage runs along the cable, or skyline and from it a steel cable is fastened to the logs. The logs, with one end elevated, are dragged to the lower spar tree. By the use of skylines, logs can be taken from mountainous forests more cheaply than by any other method.

More imaginative techniques are sometimes used for log removal in remoter locations. Helicopter transport is increasingly popular. A long cable attached to a hovering copter is fastened to three or four big logs and the helicopter quickly whisks them over intervening hills to a collection depot where the logs are uncoupled while the copter hovers (Fig.

18-14). In some cases, tethered, helium-filled balloons give logs an airborne ride out of difficult terrain.

Most of the timberland in the region is government owned. In the United States the Forest Service is the principal administering agency; it conducts regular timber sales and tries to cooperate with the forest products companies in coordinating the areas of operation for both efficient utilization and sustained yield production. Similar management is carried out on both provincially and federally owned timberlands in British Columbia.

Most privately owned timberland in the region is either owned outright by major forest products corporations or is leased by them. Many large tracts have been organized into tree farms for perpetual forestry production.

Conservation Controversies

Despite a generally good record of forest conservation policies, the logging interests in the North Pacific Coast Region have long been the subject of attack for despoiling the environment. Much ill will is engendered by the technique of clearcutting because it is perceived differently by different people.

> To the forest products manufacturer, clearcutting is the cheapest and most efficient method to cut down timber and open the way for intensive new growth of certain species. To the professional forester, it is an effective tool of scientific management of a renewable resource. To the wildlifer, it creates new kinds of habitat which attract greater numbers and varieties of wildlife. To the outdoors enthusiast, it jars the serene landscape with stripped, ravaged surfaces that in some areas cause soil erosion and water polution. To the ecologist, the single-species of tree-growth . . . that follows

FIGURE 18-13 Forest regeneration in the Pacific Northwest. The three photos were taken from virtually the same location in 1940, 1955, and 1970. The logging camp building (lower right in the top photo) can be seen in the other two photos as well (courtesy Weyerhaeuser Company).

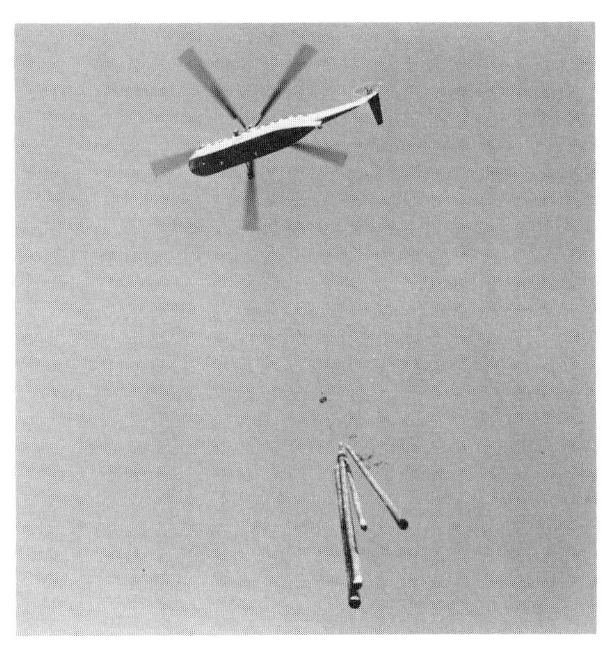

FIGURE 18-14 Logs can be airlifted by helicopters from difficult sites without the necessity of building haul roads (TLM photo).

clearcutting can be more vulnerable to pests or fire than a diversified, multiaged forest.[5]

The principle of clearcutting is considered by most foresters as a sound silvicultural practice in stands of big, shallow-rooted trees, such as the Douglas fir (Fig. 18-15); nevertheless, it is a practice that lends itself to abuse in many instances (Fig. 18-16 and 18-17).

Clearcutting is a long-established practice in the North Pacific Coast Region, and recently it has become widespread in other parts of the continent. It is an "even-aged management system" that has several attractive attributes, but it is coming under increasing attack, particularly by environmentalists, because of various concerns, most noticeably reflected by its scarring effect on the landscape. Accordingly, forest managers in the region, for the first time, are now planning the use of other types of management systems on a large scale. Some alternatives are indicated in Figure 18-18.

1. *Clearcut.* Entire stands of trees are removed at one time.
2. *Seedcut.* Five to 10 trees per acre are left to provide source of seeds.
3. *Shelterwood cut.* Up to 30 trees per acre are left to shelter seedlings.
4. *Selective cut.* Mature trees are harvested individually, leaving behind trees of varying ages. This has been the most common management system outside the North Pacific Coast Region, until recently.

In many parts of the region clearcutting has been the only system in use. Responding to public outcry, however, other approaches are now being taken. In northern California, for example, the Forest Service announced in 1990 that clearcutting would be used on only about half of all timber operations, and that more attention would be given to preservation and to the maintenance of "biodiversity" (perpetuation of all plant and animal species naturally found in forests).

The U.S. Forest Service finds itself under increasing criticism in the region for three other management practices: allowing "too much" timber to be cut, selling timber at prices below market value, and building "too many" roads in the forests for use of timber harvesters. These objections apply particularly to the situation in Tongass National Forest

[5] "Clearcutting Moves into Congress," *Conservation News*, National Wildlife Federation, 41 (1 May 1976), 10.

FIGURE 18-15 Clearcutting normally is done in relatively small patches. This scene is in the Umpqua River drainage of Oregon (TLM photo).

FIGURE 18-16 This clearcutting of immense proportions, far from prying eyes in the back country of central Oregon, is a highly questionable approach (TLM photo).

(the nation's largest national forest) in southeastern Alaska.

Throughout the region timber-cutting controversies abound. The local citizenry generally favors accelerated harvesting because the economy is so dependent upon this resource, but the wisdom of prudent long-range planning for sustained yield and scenic preservation is hard to deny.

Another focal point of controversy is in the Alaskan panhandle, where the Forest Service is auctioning off timber stands in Tongass National Forest. Despite the guidelines under which this government agency operates, there is considerable criticism that timber harvesting is being permitted in reckless fashion, without adequate survey and planning.

FIGURE 18-17 Clearcutting leaves the landscape with a devastated look, particularly when done over a large area, as in this scene from the central part of Vancouver Island (TLM photo).

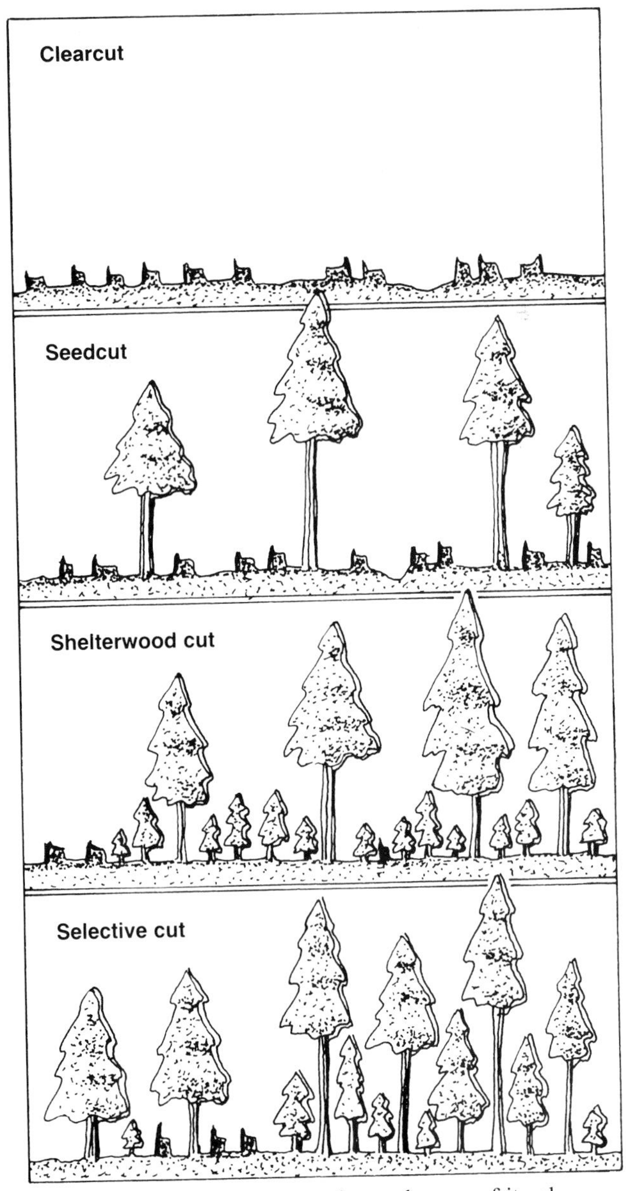

FIGURE 18-18 Clearcutting and some of its alternatives.

Forest Products Industries

Although most large sawmills of the region have always been at coastal locations, in the early days of the industry many small sawmills were located on remoter, landlocked sites. As time passed, the trend was to phase out the small mills and concentrate activities at ever-larger sawmills situated on tidewater or at strategic inland locations (Fig. 18-19).

A prominent pulp and paper industry has also developed in the region, particularly in recent years. Although a considerable quantity of pulp is shipped out of the region, local paper mills are consuming an increasing quantity in the manufacture of kraft paper and newsprint.

Big corporations and huge mills are characteristic of the wood products industries in this region. Vertical and horizontal integration of production facilities is commonplace. It enables many economies of scale and more efficient use of each log.

Integration of logging, transportation, processing, and manufacturing operations is widespread in western Oregon and Washington but is most notable in the Strait of Georgia area of British Columbia, where functional interconnection in terms of both areal space and corporate organizations is tightly knit.

Economic Trends

As a major income generator of the region, the forest products industry is critically important to economic well-being. Yet it has experienced hard times for much of the last two decades. Home construction, particularly in the United States, is the largest market for the lumber producers of both the United States and Canada, and the housing demand has been weak for some years. Expansion of pulping and plywood production has helped to ease the problem, but it has failed to take up all the slack.

This industrywide decline would have been even more serious if not for the increasing role of Japanese purchases of logs, lumber, and other wood products. Japanese capital has been infused in several Canadian and Alaskan forestry enterprises, and Alaskan lumber mills now ship more than 90 percent of their output to Japan. Whether the Japanese market will be so significant in the future is conjectural, but its short-run impact has been favorable.

Overall the forest products industry of the North Pacific Coast Region continues in a severe recession. The situation verges on the disastrous in some areas, such as northwestern California. The only broad solution may be an increase in building construction, which seems unlikely. Clearly this

FIGURE 18-19 Large sawmills are scattered through the region. This huge mill is near Enumclaw in western Washington (TLM photo).

critical regional industry has a precarious economic balance.

AGRICULTURE: SPARSE AND SPECIALIZED

The North Pacific Coast is still largely in timber, relatively little of the land being suited to agriculture. Dairying, the dominant farming activity of the region, accounts for a large part of the agricultural land being in pasture. Hay and oats occupy more than half of the land in crops.

Some of the locally important agricultural areas that contain most of the crop land are

1. The Umpqua and Rogue River valleys of southwestern Oregon
2. The Willamette Valley
3. The Cowlitz and Chehalis valleys and the lowlands around Puget Sound in Washington
4. The Bellingham lowland
5. The lower Fraser Valley of British Columbia
6. Southeastern Vancouver Island.

Of these areas, the Willamette Valley, with more than 2 million acres (800,000 ha) in crops is by far the largest and best developed.

Agriculture in Oregon and Washington

The older settled parts of the Willamette Valley and Puget Sound lowlands present an agricultural picture of a mature cultural landscape that can be found in few places in the West. The Willamette Valley has miles of fruit farms that grow prunes, cherries, berries; hop fields; fields of wheat and oats; excellent pastures; and specialized farms that produce commercial grass seeds or mint (for the oil). The Willamette Valley, occupied by farmers of the third or fourth generation on the same farms, is the old, long-settled prosperous heart of Oregon that grows most of the fruit, berry, vegetable, and grain crops of the North Pacific Coast Region. Dairying is the principal occupation, but diversified horticultural and general farms are also common (Fig. 18-20).

In the Puget Sound lowlands, where considerable land is diked or drained, are the region's best dairy and pasture lands. Market gardening is an important agricultural activity that has increased in ratio with the growth of the large urban centers. Vegetables are grown in this area for the local urban markets, but the surplus is shipped to other parts of the United States.

The lower floodplains and deltas of four medium-sized rivers, often collectively referred to as the Bellingham Plain, occupy most of the eastern fringe of Puget Sound north of Seattle. Dairy farm-

FIGURE 18-20 Many fields in the Willamette Valley are burned after the late-summer harvest (TLM photo).

ing is the principal rural activity in this area, but there is a major concentration of green pea cultivation; many kinds of vegetables and various berries are also grown here.

Agriculture in British Columbia

Farming in the British Columbia portion of this region is concentrated on the floodplain and delta of the Fraser River or on southeastern Vancouver Island. Agriculture here serves the urban markets of metropolitan Vancouver and Victoria, but the small area of suitable farmland is inadequate to produce sufficient food for the urban population. General mixed farming and dairying are carried on in these areas; there are also many specialty crops, such as fruits, berries, vegetables, and flowering bulbs. The leading agricultural industry of the Lower Fraser Valley is the production of whole milk for the Vancouver market. Pasture occupies the largest proportion of farm land.

Other important agricultural activities include horticulture, especially berries; poultry and beef cattle raising; and hops. Major farm problems in the valley include small and uneconomic farm size, too little summer rain (the annual average is 70 inches,

FIGURE 18-21 A Matanuska Valley farm scene near Palmer, Alaska (TLM photo).

or 1780 mm, but dry summer periods often cause crops to wilt) occasional floods, and inadequate seasonal labor supply.

The southeastern lowlands of Vancouver Island contain about 50,000 acres (20,000 ha) of cultivated land. Temperatures are similar to those of the Fraser delta, but rainfall is considerably lower. In addition to dairying, poultry raising, and the cultivation of fruits and vegetables, this area has specialized in growing spring flowers for eastern Canadian markets. Many farms are marginal and Vancouver Island is not even self-sufficient in dairy products.

The Matanuska Valley of Alaska

The Matanuska Valley, a fairly extensive and well-drained area of reasonably fertile silt-loam soils, lies at the head of Cook Inlet inland from Anchorage. Nearly two-thirds of Alaska's cropland acreage is in the valley, but it comprises only a few dozen farms, many of them marginal (Fig. 18-21). Production costs are high and crop options are limited, with the result that Matanuska farming, never very important, continues to decline in significance.

THE UPS AND DOWNS OF COMMERCIAL FISHING

The North Pacific Coast is one of the major fishing regions of the world and ranks, in total catch, with the banks fisheries of the North Atlantic. The bulk of the catch consists of only a few varieties of fish, but they are taken in tremendous quantities, and the contribution of the industry to the total economy of the region is a major one. Although salmon are the commercially dominant fish in North Pacific waters (amounting to almost half of the total value of all landings in the region), large quantities of herring, halibut, pollack, cod, and a few other species are also caught. Moreover, a gigantic shellfish industry has developed in the last two decades.

Ranking second to salmon in both quantity and value among North Pacific finfish are herring. They are chiefly netted from large schools during winter, especially in the Puget Sound–Strait of Georgia waters. Historically most of the herring catch was processed into fishmeal and oil rather than being marketed for human food. Overfishing, however, so depleted stocks that catching herring for reduction purposes was totally banned. The fishery now has recovered and new foreign markets, especially in Japan, have been developed so that there is increasing demand.

Halibut is the largest and most valuable of the bottom fish of the North Pacific. They are long lived (up to 35 years) and can reach enormous size (up to 500 pounds [225 kg]). They are caught by longline hooks or by power trawling. After many years of declining catches, an International Pacific Halibut Commission, consisting of representatives of the United States and Canada, was established and the fishery has been stabilized at a productive level with strict quotas enforced.

Although oysters, crabs, and other shellfish had been taken commercially for a long time in regional waters, only in recent years has an outstanding shellfishery developed. This situation was stimulated primarily by greatly increased catches of king crabs in the Gulf of Alaska. These demersal giants weigh up to 25 pounds (11 kg) and may have a "claw spread" of 6 feet (1.8 m). They are caught in baited crab pots or taken by trawling. The king crab fishery declined markedly in the 1980s, but other shellfish (notably shrimp, clams, and other species of crabs) took up most of the slack.

POWER GENERATION

Because of mountainous terrain and heavy rainfall, the region has one of the greatest hydroelectric potentials of any part of North America. Some of the potential was developed early; Victoria, for example, had electric street lights in 1882, just one year after Edison's developments in New York, and a commercial hydroelectric power plant in 1895, two years after the first such plant at Niagara Falls. Some large facilities were established to provide power to the Trail smelter shortly after the turn of the century, but Grand Coulee and Bonneville dams were the first major projects on the U.S. side of the border.

Most major power-generating facilities are associated with dams situated to the east of the North Pacific Coast Region, particularly on the Columbia River and some of its tributaries. In addition, the continent's largest atomic-generating plant is adjacent to the Columbia at Hanford, Washington. The availability of "firm" power was greatly enhanced by the Columbia River Treaty whereby water stored behind Canadian dams can be released at low water periods and thus diminish seasonal fluctuations in stream flow and power generation.

The principal markets for Columbia Basin power have been the large cities of Washington and Oregon and major aluminum factories, most of

A CLOSER LOOK *The Saga of the Salmon*

The Pacific salmon is the leading commercial fish in North America in terms of value of catch. It is easily the most important fish caught in Alaska, the leading fishing state, and British Columbia, the leading fishing province.

Because of its remarkable life cycle, vast numbers, and susceptibility to entrapment and depletion, it is both a major resource of the region and a focal point of controversy and dispute. There are five species of Pacific salmon: chinook (*Oncorhynchus tshawytscha*), coho (*O. kisutch*), sockeye (*O. nerka*), pink (*O. gorbuscha*), and chum (*O. keta*). All five species are anadromous—that is, they spend most of their life in the ocean but migrate up freshwater streams to spawn. Only a few weeks or months after being hatched, salmon fingerlings swim downstream to the sea. They spend from one to five years in saltwater before returning to their place of birth (up the same river, the same tributary, and the same creek) to spawn. After the female lays eggs in the stream gravel and the male fertilizes them, the adults die.

When in the ocean, salmon must face the normal hazards of the sea and increasing pressure from oceangoing fishermen with their incredibly long and effective driftnets. However, it is after the spawning instinct begins to govern their behavior that they cluster in immense numbers and become liable to almost total entrapment. As salmon congregate in estuaries and bays at the mouths of rivers and particularly as they begin to move upstream in singleminded response to the urge to propagate, it is relatively easy to capture entire runs by using nets, seines, and fishtraps.

In addition to the potential for overfishing, the concentration in rivers of migrating fish (both adults moving upstream and fingerlings going downstream) produces the risk of severe depletion owing to damming. Although mature salmon are incredibly persistent and tenacious in swimming up rapids and jumping over small obstructions in rushing streams, only the smallest of dams is needed to prohibit their upstream progress totally. In most cases where dams were built on salmon rivers, fish ladders have been constructed to ease upstream migration, and fish boats have been designed to transport fingerlings downstream, thus bypassing the dams. These are costly schemes that are sometimes very successful but not always.

Damming on the Columbia River, the principal salmon river in the conterminous states, has a continuing adverse effect. Fish ladders permit salmon to pass around low dams like the Bonneville, but high dams, such as Grand Coulee, are apparently impassable. Even more serious is the staggering mortality rate among fingerlings, which must either cascade over the spillways or be sucked through the turbines as they move downstream to the sea. There are now 11 major dams on the Columbia River; it has only a single 50-mile (80-km) stretch of free-flowing water in its entire 750-mile (1200-km) length between the Canadian border the Pacific Ocean. More than half the natural spawning area of the Columbia Basin has been denied to anadromous fish by dams on the Columbia and its tributaries (Fig. 18-a).

The conflict of interests between advocates of hydroelectricity generation and salmon fishermen is a major long-term cause of disharmony in the region. It is increasingly clear that the two resources are virtually incompatible and that one can be developed only at the expense of the other. In the United States, the dam builders have the upper hand. In Canada, however, provincial regulations prohibit dam-building on major salmon rivers, such as the Fraser (which has the largest natural salmon run in the world), the Stikine, and the Skeena.

Regardless of varied controversies, however, the saga of the salmon represents one of the most blatant examples of fishery mismanagement imaginable. Here is one of the most valuable fishery resources in the world, a resource that can be exploited at relatively low cost, that is capable of sustained yield management, and that has a strong market demand that can be translated into high prices for the product. In rational economic terms, this situation should lead to a stable and prosperous fishing industry. What has happened, however, is the opposite: the history of the salmon industry consists of alternations of sporadic bursts of expansive prosperity and long, dragging periods of economic hardship.

The resource is not constrained by property rights, and almost anyone can become a commercial salmon fisherman with a moderate investment in equipment and licenses. As a result, the fishery has consisted mostly of a frantic scramble to take fish which are in the North Pacific Coast Region. The rapid recent increase in generation capacity has led to a search for new markets and the establishment of intertie facilities that link the electric systems of 11 western states in the largest electrical transmission program ever undertaken in the United States. Thus the circular cycle continues: build more dams to supply more power and then seek new markets requiring more power that call for more dams, and so on.

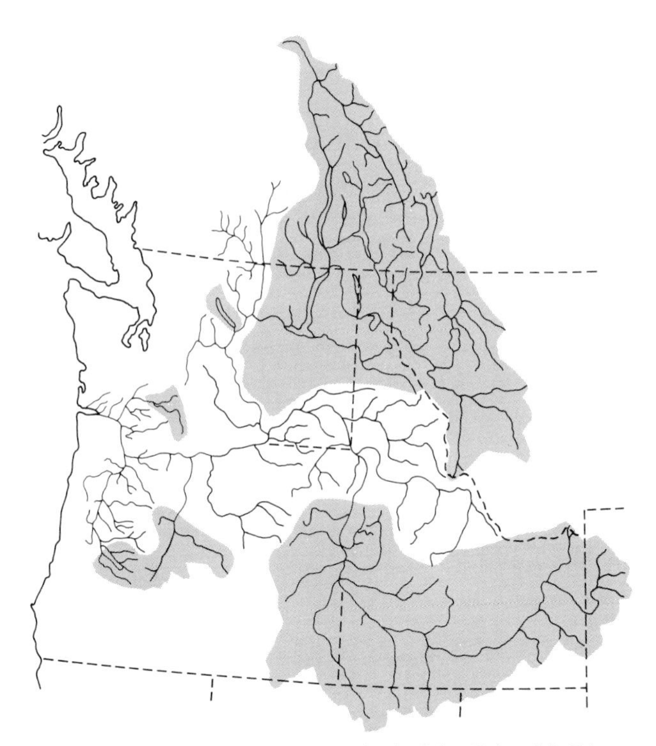

FIGURE 18-a Many parts of the watershed of the Columbia River and its tributaries are now inaccessible to salmon because of dams. The shaded areas represent those portions of the drainage basin from which salmon have been excluded. (After Ed Chaney, "Too Much for the Columbia River Salmon," *National Wildlife*, Vol. 8, April/May 1970, p. 20.)

quickly before someone else takes them. Although increasingly complex fishing regulations were introduced in both Canada and the United States, they were planned primarily to ensure sufficient escapement so that the salmon stocks could be maintained. To accomplish this objective, most regulations were specifically designed to reduce the efficiency of vessels and gear, which had some biological merit but was economically disastrous. Consequently, wasteful duplication of capital and labor occurred in an industry that was already overdeveloped.

There has been a long-term downward trend in salmon numbers particularly in the southern part of their range (California, Oregon, Washington). For example, the Columbia watershed, by far the most important salmon area in the conterminous states, has experienced a decline of more than 75 percent from its peak several decades ago. In the early 1990s the annual salmon run in the Columbia consisted of only about 2.5 million fish. Moreover, less than one-fourth of these are "wild" salmon; the remainder are hatchery stock, which consistently are weaker and more vulnerable.

More enlightened approaches to management have been followed in recent years and indications are that it may be possible to have both biological and economic stability in the salmon fishery. Starting in 1968, Canada led the way with an innovative change in management strategy that phased in a program of both limiting and reducing the number of licensed fishermen and boats in the industry. Alaska, with a more complex situation, followed suit in 1973. Later Washington adopted a similar, but less comprehensive, program. There are still many burdensome regulations that restrict the efficiency of the fishermen, for maintenance of a sustained yield resource must still be the keystone to any management program. Nevertheless, at least a start has been made at balancing harvesting capabilities with resource productivity. Finally, after nearly three decades of negotiations, the United States and Canada signed a bilateral agreement about joint and separate harvesting of salmon coming into the Strait of Georgia area. This had been a matter of serious dispute for some time.

TLM

MINERAL INDUSTRIES

Various discoveries of gold, silver, lead, iron ore, coal, and copper have led to moderate flurries of mining activity in southern Alaska and on Vancouver Island in the past, but almost all the operations are of historical interest only. Even the huge gold mine at Juneau, with its nearly 100 miles (160 km) of tunnels and shafts (said to be the world's largest low-grade gold mine), has been shut down for half a century.

Petroleum is another matter. Alaska's dynamic petroleum industry has been a focal point of excitement and controversy for nearly two decades. Until recently, attention was focused primarily on the North Slope-producing area and the pipeline corridor, which are largely outside the North Pacific Coast Region. However, in 1989 an oil tanker collided with an underwater reef in Prince William Sound, near the southern terminus (Valdez) of the pipeline, producing the most massive oil spill in American history. This accident resulted in immense damage to the fragile environment, an unprecedented clean-up effort, hundreds of lawsuits, and notable political and economic repercussions that will continue for years.

Almost lost in the excitement of North Slope petroleum is the realization that major oil and gas extraction have been occurring beneath the waters of Cook Inlet and on the nearby Kenai Peninsula (both areas just southwest of Anchorage) since 1957. Although production has declined in recent years, it continues at a relatively steady pace.

Offshore areas of Alaska are thought to be the most promising prospects for future petroleum development in the United States. Various offshore districts will eventually be prospected, but the first tangible development is taking place in the nearshore waters of the Gulf of Alaska, southwest of the St. Elias mountain knot. The first exploratory wells were completed in the early 1980s, but exploitation has proved to be both slow and difficult. The waters of the Gulf of Alaska are tumultuous at best and supremely hazardous during frequent winter storms. Moreover, these coastal waters are some of the most biologically prolific in the world, and the potential environmental damage from oil spills or leaks is a major point of contention, particularly since the 1989 catastrophe in Prince William Sound.

URBANISM: MAJOR NODES AND SCATTERED POCKETS

Most residents of the North Pacific Coast Region—even the Alaskan portion—are urban dwellers (Table 18-1). There are five prominent metropolitan nodes and a number of smaller cities and towns, many of which exist in relative isolation (Fig. 18-22). Four nodes and several smaller centers serve conspicuously as coastal gateways to river or pass openings through the mountain barrier(s) that blockades access to the interior.

Burgeoning Metropolitan Centers

Portland (gateway: Columbia River barge route with paralleling railways and highways) is the dominant commercial center of the lower Columbia and Willamette valleys. The city's dual harbor facilities, part on the Willamette River and part on the Columbia, are among the most modern in the nation. Bolstered by the transshipment of grain and ores that are barged down the Columbia, Portland's bulk

TABLE 18-1

Largest urban places of the North Pacific Coast region

Name	*Population of Principal City*	*Population of Metropolitan Area*
Anchorage, Alaska	218,500	218,500
Bellevue, Wash.*	84,710	118,700
Bellingham, Wash.*	46,310	180,900
Bremerton, Wash.	37,320	
Corvallis, Oreg.	39,770	
Eugene, Oreg.	108,030	270,100
Everett, Wash.*	62,740	
Gresham, Oreg.*	58,130	
Medford, Oreg.	45,610	145,900
Nanaimo, B. C.	50,100	
New Westminster, B. C.*	40,100	
Olympia, Wash.	31,660	156,600
Portland, Oreg.	418,470	1,188,000
Salem, Oreg.	97,210	269,800
Seattle, Wash.	502,200	1,861,700
Springfield, Oreg.*	39,200	
Tacoma, Wash.	163,960	559,100
Vancouver, B. C.	434,200	1,380,600
Vancouver, Wash.*	45,680	226,200
Victoria, B. C.	67,300	256,300

* A suburb of a larger city.

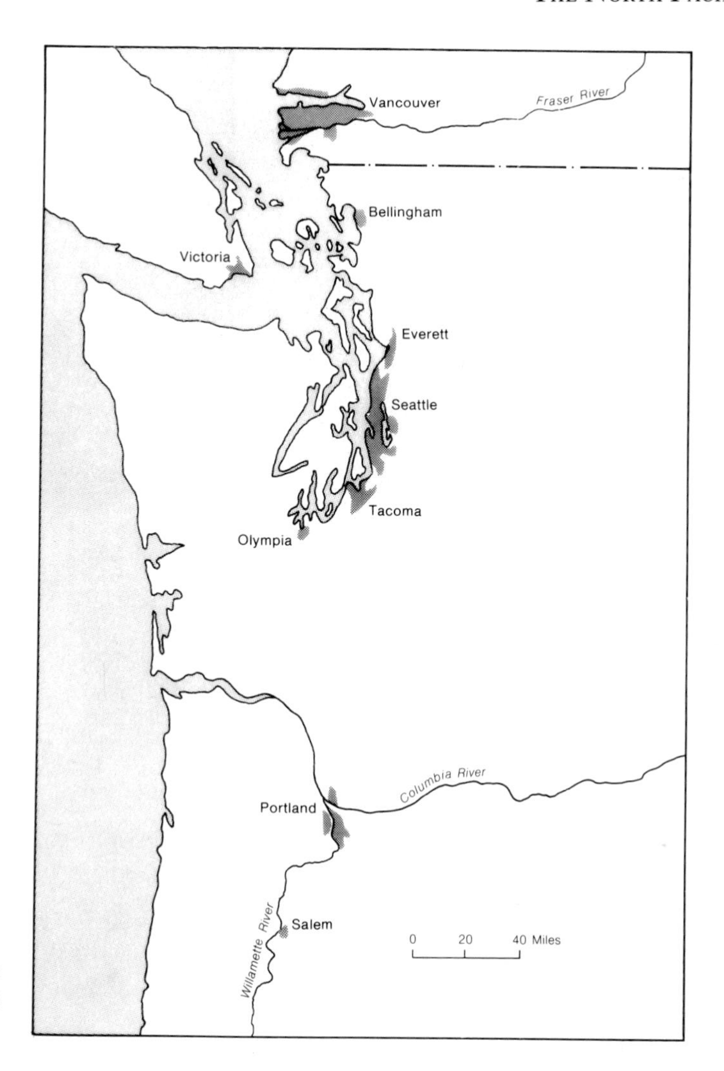

FIGURE 18-22 Metropolitan complexes of the Pacific Northwest.

ocean freight business is approximately equal to that of Seattle. Wood products and food processing dominate the industrial structure.

Seattle (gateway: Cascade Mountains passes with rail and highway routes to the interior) is the focus of a highly urbanized zone that fronts the eastern shore of Puget Sound from Bellingham and Everett on the north to Tacoma and Olympia on the south (Fig. 18-23). Seattle itself is located on a hilly isthmus between the sound and Lake Washington, which gives it a constricted central area in contrast to the sprawling suburbs to the north and south. The deep and well-protected harbor requires no dredging and Seattle has long been a major Pacific-oriented port as well as the principal gateway to Alaska. Seattle has outstanding container-handling facilities and is one of the 10 largest container ports in the world; two-thirds of its port traffic is containers.

The Boeing Company has been the mainstay of the city's economy since the 1920s, although its employment has fluctuated wildly with defense contracts, ranging between 50,000 and 100,000 over the past decade. Wood products and shipbuilding are other major industries in the metropolis (Fig. 18-24). The population of the extended metropolitan area approximates 2 million.

Vancouver (gateway: Fraser and Thompson River canyons, Trans-Canada Highway, and both transcontinental railway lines) overcame its early rivalry with Victoria for economic dominance of British Columbia and has grown to become Canada's second-ranking port. As the principal western

FIGURE 18-23 A high-altitude view of the Seattle isthmus. Puget Sound is on the left, Lake Washington is on the right, and the waterway connecting the two is near the top of the picture (TLM photo).

terminus of Canada's transcontinental transport routes, Vancouver is clearly the primate city of western Canada as well as the nation's fourth largest industrial center (Fig. 18-25). The development at Roberts Bank, 25 miles (40 km) to the south, of a bulk products superport that is primarily for exporting coal and ores to Japan, has added to the city's urban primacy.

Victoria, British Columbia's capital, is the second largest city in the province. It is perhaps the continent's most attractive city and experiences Canada's mildest climate, which makes it a major goal for tourists and retired people.

Anchorage (gateway: Alaska Railroad and two highways to the interior) is the largest city in Alaska, with nearly half the state's population in its metropolitan area. It has been the dominant growth center for both population and economic activity in the state. Despite oil to the south, farming to the north, and its role as an international air-transport hub, the city's economy is primarily dependent on government, especially military, activities.

Urbanization in Isolation: The Case of the Alaska Panhandle

In the long stretch of coastland between Vancouver and Anchorage most urban places are isolated and remote. Only the dual towns of *Prince Rupert* and

FIGURE 18-24 Seattle is one of the leading shipbuilding centers in the nation (TLM photo).

FIGURE 18-25 Vancouver's skyline, as seen from the south. The Coast Mountains rise abruptly in the distance (TLM photo).

Kitimat, on the north coast of British Columbia, have useful surface transport connections with the rest of the continent; the former is a railway terminus and fishing port and the latter is one of the world's largest aluminum-refining localities.

In the Alaska panhandle are seven small urban centers that exist in remarkable isolation in an area of magnificent scenery and persistent rain (Fig. 18-26). Only the two northernmost, *Haines* and *Skagway,* have land transport connections with the rest of the world; the other five, *Ketchikan, Wrangell, Petersburg, Juneau* (Fig. 18-27), and *Sitka,* are all situated either on islands or on mountain-girt peninsulas; yet the narrow streets of these hilly

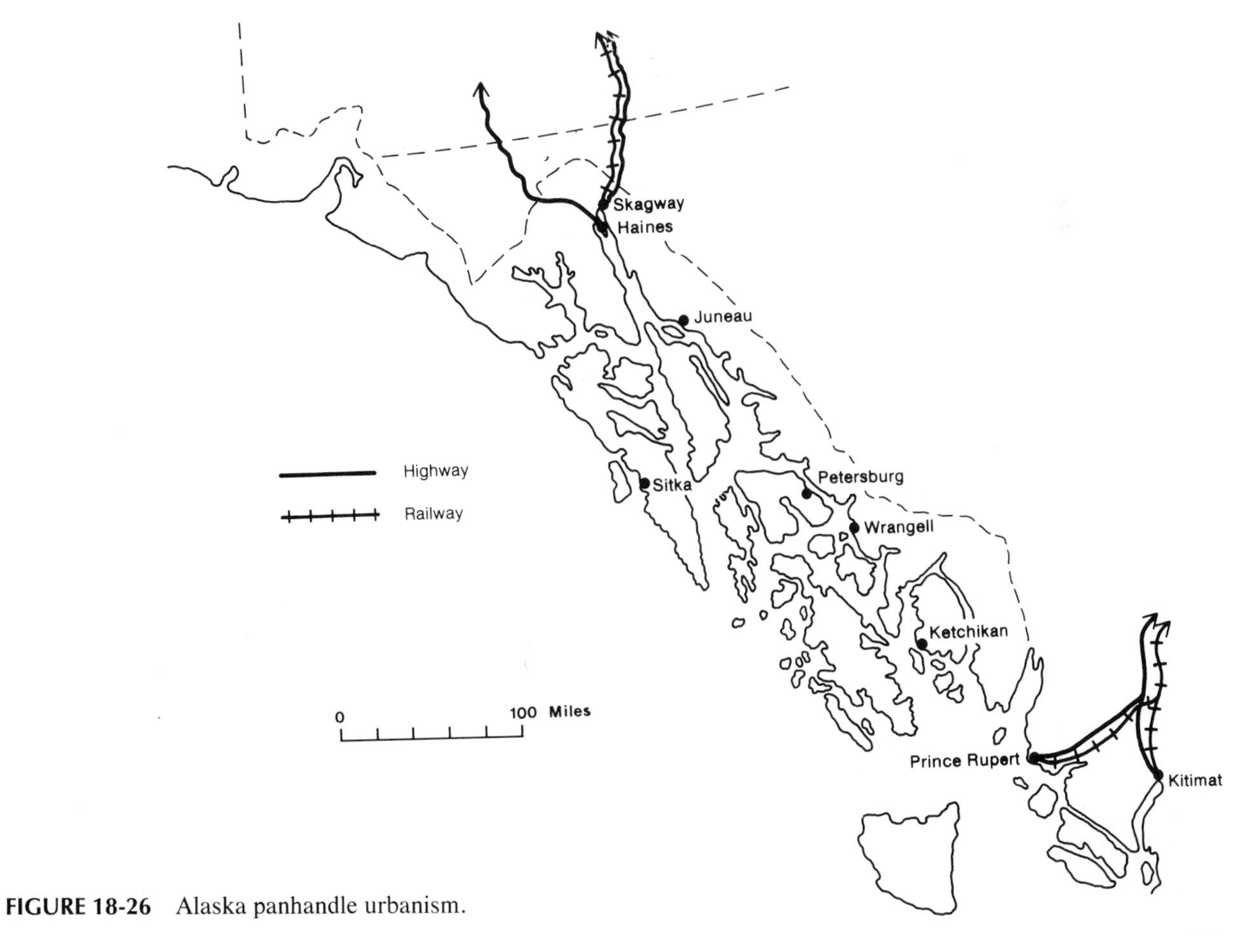

FIGURE 18-26 Alaska panhandle urbanism.

A CLOSER LOOK Vancouver B.C.—Changing Land Use and Functions

The site of Vancouver was occupied by Europeans in the 1870s as one of many single-industry resource-based towns that were to become characteristic of the hinterland of Canada later in the twentieth century. Sawmills were built on both sides of the excellent, sheltered harbor of Burrard Inlet. The tall, straight trees of a luxuriant and untouched West Coast coniferous forest were cut on the gentle mountain slopes around this harbor and northward along the island-fringed, indented and unoccupied West Coast. Huge logs were felled from temporary logging camps along the coastal hinterland and towed through sheltered coastal waterways to large "central place" sawmills near the delta mouth of the Fraser River. Sailing vessels and later steamships, came to the deep harbor for long, high-grade lumber and took it to growing urban settlements in western United States and East Asia.

This was one of the beginnings of a resource-based economy that was to dominate the development of British Columbia for the next century. Other tiny villages of British settlers dotted the southwestern corner of British Columbia, clustered around Georgia Strait, which linked Vancouver Island to the mainland of Canada. These "outposts of Empire" were dependent on fish canneries, mines, fur-trading posts, and sawmills—all using the untouched resources of the natural environment and sending their partially processed products to urban markets elsewhere. These resource-based settlements were typical of much of the Canadian, including British Columbia, economy into the early twentieth century.

These settlements preceded the arrival of the Canadian Pacific transcontinental railway to Vancouver in 1886. That little sawmill town became a major Canadian, and world, port. But these changes did not happen immediately. Large foreign markets were far away, in both time and distance, before the opening of the Panama Canal during World War I. Vancouver developed into an industrial city in the first half of the twentieth century, primarily supplying consumer goods to its own rapidly increasing population and to the million people concentrating into southwestern British Columbia. In one sense, local industries were "protected" by long distances and high transport costs from the industrial producers of eastern Canada and northeastern United States.

Industrial land use spread along three belts in Vancouver (Fig. 18-b). Land around the Burrard Inlet harbor was occupied by industries served by both the Canadian Pacific railway and ocean or coastal shipping. These included lumber mills, a sugar refinery, ship-building, fish canneries, and later grain elevators and oil refineries. This industrial land use spread eastward into adjoining Burnaby, Port Moody, and to North Vancouver on the north side of the harbor. Their primary markets were Pacific centers; much of the raw materials came from the Prairie Provinces to the eastward or from northward along the coast.

A second industrial belt, consisting of wood-based industries such as sawmills, shingle and furniture plants, and ship-building, clustered around the shallow water of False Creek, south of the city's original commercial core (see the accompanying map). After the Canadian National Railway (a second transcontinental railway) and the Great Northern Railway (from the United States) came during and after World War I to the filled-in, reclaimed land at the east end of False Creek, more secondary industries spread eastward along rail and highway routes into Burnaby. Much of this production was for local Vancouver and British Columbia markets.

A third industrial zone spread eastward into New Westminster, along the distributary arms of the Fraser River, on the south side of Vancouver. These industries included sawmills and fish canneries, and like the Burrard Inlet harbor, were served by railways and shallow-draft ocean transport. Modest small wooden houses on narrow lots were built nearby by industrial workers, creating a working-class "village" known as Marpole, along the Fraser River, far from the Vancouver city center.

Small wooden homes of industrial workers were also built in the northeastern part of Vancouver and North Burnaby, near the areas of industrial land use. These eastern parts of the city were "working man's" Vancouver and included many immigrants from central and southern Europe and from China and Japan. The latter were low-income commercial and service workers. In social and economic contrast, business, management, and commercial workers built larger and nicer homes on the west side of the city, near the growing commercial core south of the harbor and southwest of the original railway station. Most of these residents were of British origin who flooded into the west side of Vancouver between 1920 and 1940, mostly from eastern Canada and the Prairie Provinces. Vancouver was becoming a city of ethnic, economic, and occupation contrasts between its east and west sides.

The industrial characteristics of Vancouver began to change after about 1950. When the population of the metropolitan area exceeded 1 million in 1971, it became a large internal market for a wide range of secondary manufacturing and small consumer-product plants. Primary industries such as big sawmills, and major employers such as ship-building, moved from the Burrard Inlet harbor and around False Creek. The old homes of industrial workers around

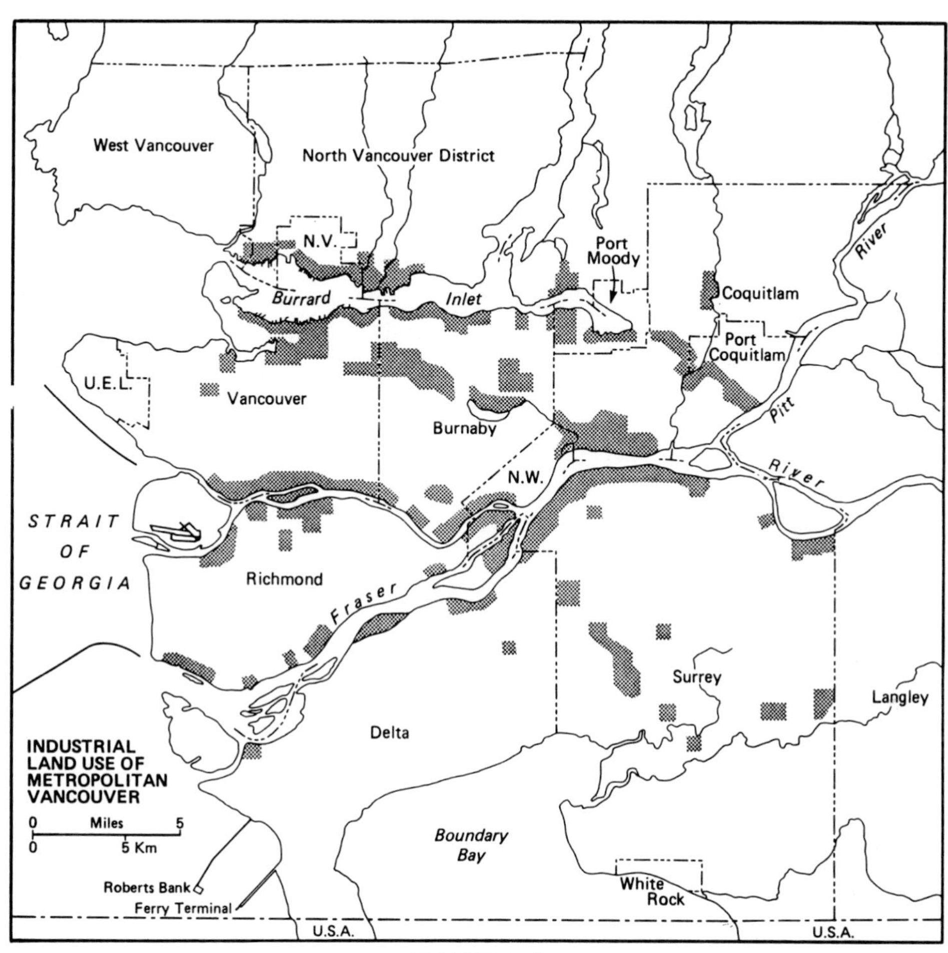

FIGURE 18-b

False Creek were torn down and replaced by attractive apartments and condos. This dirty, polluted area of the 1930s had become an attractive residential area for commercial, professional and service workers by the 1980s.

Industrial land use around the Burrard Inlet harbor changed functions from manufacturing to transfer and transport uses. Raw materials such as coal, sulphur, chemicals, potash and wheat, mainly from the western Prairie Provinces, arrived by rail to storage and container facilities around the harbor, in addition to lumber by water transport from coastal British Columbia. These products are nearly all for export to Pacific and world markets and need no additional processing in Vancouver. The dock workers operating modern machinery there are not low-income employees.

Manufacturing and assembly for the growing internal consumer market of Vancouver and British Columbia spread to cheaper land in the suburbs around Vancouver (see the accompanying map). These industrial areas are dispersed, related mainly to truck transport, and therefore so are these industrial workers. Vancouver itself became a commercial, business, management, service, and entertainment city. The industrial areas within the city decreased and changed functions.

Vancouver is an example of changing land use and functions,

(continued)

characteristic of many North American cities. It started as a single-industry resource-based town and then developed dirty wood-processing industries for local and Pacific markets. It became Canada's main West Coast port, through which funnelled many raw materials of Western Canada for Pacific and world markets. New, secondary industries for local markets occupied land in the suburbs; people in Vancouver became employed more in commercial, management, and service occupations, many working in high-rise office towers in the downtown core—and facing the same congestion, transport, and parking problems of other large commercial cities of North America.

Professor J. Lewis Robinson
University of British Columbia
Vancouver

FIGURE 18-27 Juneau is nestled at the foot of the mountain front on the shore of Gastineau Channel. A bridge connects Juneau with the town of Douglas on Douglas Island (left foreground). The mass of light-colored material in the channel just south (this side) of Juneau is waste rock from the famous Alaska-Juneau gold mine (courtesy Alaska Travel Division).

towns are crowded with autos. All have highly specialized economies, mostly oriented toward commercial fishing or forestry.

Juneau's principal claim to fame, the fact that it is the state capital, is presumably a temporary distinction. The voters of Alaska, in statewide referendums in 1974 and 1976, decreed that the capital would be shifted to a much more accessible site 70 miles (112 km) northeast of Anchorage, where a new planned city would be built. The cost of such an endeavor, however, will be astronomical and the voters have twice (in 1978 and 1982) rejected approval of a bond issue to finance the operation.

FIGURE 18-29 Mt. Rainier on a rare cloudless day (TLM photo).

SPECTACULAR SCENERY

The North Pacific Coast is perhaps the continent's most scenic region and is hailed as one of its most desirable outdoor recreation areas.[6] Remoteness from large population centers and the inaccessibility of many scenic spots, however, have retarded the development of tourism. Improved transport by road, air, and ferry have significantly stimulated the tourist business.

The section south of the international border is best developed, with six magnificent national parks; one of the finest combined scenic-and-sandy coastlines in the world, the Oregon coast (Fig. 18-28); splendid forests; and a plentitude of accessible winter sports areas. The Canadian section and the Inside Passage of southeastern Alaska rank among the world's most scenic areas, with their spectacular mountains, glaciers, and fiords (Fig. 18-29). In south coastal Alaska, however, is found the continent's most magnificent landscape, thousands of square miles of ice and rock and forest, culminating in the grandeur of Denali National Park (Fig. 18-30).

It is to be expected that land-use controversies are frequent and ecological confrontations numerous in a region of such scenic splendor. The conflicts

[6] The scenery, however, is frequently shrouded in fog, mist, or rain. The famous Seattle weather forecast still pertains: "If you can see Mount Rainier, it's going to rain; if you can't, it is raining."

FIGURE 18-28 The Oregon coast is very scenic from one end to the other. This view is near Port Orford in the south (TLM photo).

FIGURE 18-30 Mount McKinley is the highest peak on the continent and is the focal attraction of Denali National Park (TLM photo).

vary but focus particularly on commercial timber cutting and potential oil spills.

One of the most notable developments in the tourist industry of the North Pacific Coast Region has been the rapid proliferation of cruise ship visitation. This activity is focused on the "Inside Passage" to Alaska, normally involving a 7- or 10-day voyage that begins and ends at either Seattle or Vancouver and visits several ports or scenic spots in route (always Juneau and Glacier Bay; sometimes Prince Rupert, Ketchikan, Misty Fiords, Sitka, Skagway). Each ship carries several hundred passengers, and more than 200 cruises ply the Inside Passage each summer. Indeed, the visitation has become so heavy that advance reservations are needed for a cruise ship to enter the waters of Glacier Bay so as to avoid too much disturbance to the marine fauna.

THE VITAL ROLE OF FERRIES IN THE REGION

Highways and railways, the more prosaic forms of transportation, are well established from the Fraser Valley southward despite topographic hindrances. In the northern two-thirds of the region, however, both road and rail routes are chiefly limited to lines extending inland from the ports of Prince Rupert, Skagway, Haines, Cordova, Valdez, Seward, and Homer. Air transport thus becomes very important throughout the region.

As in no other part of North America, ferries play a specialized and vital role in the transportation of this region. There are three large government ferry networks and a few small private systems (Table 18-2). The emphasis in each is on passenger traffic, with roll-on/roll-off facilities for automobiles and trucks. (Most freight is carried by regular ocean-going vessels or tug-propelled barges.)

The principal systems are as follows:

1. The Washington State Ferry system operates throughout the Puget Sound area. Most of its service radiates from Seattle, but there is also interisland service in the San Juan Islands, plus several individual

TABLE 18-2

Selected Statistics of Major Ferry Systems, 1989

	Washington State Ferry System	*British Columbia Ferry Corporation*	*Alaska Marine Highway System*
Vessels in service	23	38	8
Ports served	20	43	31
Passengers carried	19,642,600	19,228,600	388,600
Vehicles carried	8,409,000	7,499,200	105,500

FIGURE 18-31 One of the ferries of the Alaska Marine Highway system passing through Discovery Passage between Vancouver Island and the continental mainland (TLM photo).

connections between island and peninsula ports.

2. The British Columbia provincial government maintains an excellent ferry network in the Strait of Georgia, interconnecting mainland and Vancouver Island ports, with other service extending northward along the mainland coast and into the Queen Charlotte Islands.
3. The Alaska Marine Highway System has two discrete route networks. The largest serves the 15 principal settlements of the panhandle, with external connections to Prince Rupert and Seattle (Fig. 18-31). The other connects about a dozen ports in south-central Alaska (Prince William Sound, Kenai Peninsula, Anchorage, and Kodiak Island). In the near future there are plans to connect these two networks with service across the turbulent waters of the Gulf of Alaska.

THE OUTLOOK

The characteristics and relative significance of the North Pacific Coast Region are likely to change little in the near future. It will continue to be a region of specialized economy, Pacific Ocean orientation, magnificent scenery, and conservation controversy.

The forest industries, long outstanding in the economy, are caught up in both international economic price squeezes and conservation controversies. These factors are likely to cause a diminution of cut, even from the present depressed levels. It is clear that the prosperity of the wood products industries will continue to fluctuate, and the long-term outlook is not too favorable. An increased housing demand would be a great boon, but this scenario is not likely. Meanwhile, the higher-quality and more accessible timber supplies grow scarcer, which means that the costs of logging continue to rise. Communities heavily dependent on the forest industries are, with a few exceptions, unlikely to flourish.

Agriculture should become even more specialized than it is today, although dairying should be unrivaled as the principal farm activity throughout the region. There is little likelihood of expanding farm output.

The commercial fishing industry will probably be characterized by considerable fluctuation in the annual catch of the various species. Year-to-year variations in the availability of fish reflect both natural factors and the erratic results of overexploitation. Fishermen have demonstrated remarkable versatility and resiliency in the past, shifting with great rapidity from an overfished species to an underutilized one. As fish processing and marketing facilities develop a similar measure of versatility, the entire industry will become more stabilized. Problems of overexploitation and conflicts of interest, however, will continue to cloud the scene, particularly in regard to salmon.

The impact of the oil spill in Prince William Sound is likely to have continuing and far-reaching reverberations. Oil-happy developers in Alaska will find diminishing enthusiasm among the citizenry, and environmental safeguards probably will be expanded and extended in a multiplicity of arenas.

Commercial ties with Asia in general and with Japan in particular will undoubtedly grow rapidly. The bustling Japanese economy should provide a significant market for the quantity products of the North Pacific Coast's primary industries and the products from the continental interior that are shipped from the region's ports. A continuation of the high level of Japanese capital investments in the region is to be anticipated.

The established urban areas of the southern portion of the region will probably remain the major centers of population and economic growth. Most of the region, however, will remain largely a wilderness, dominated by a few extractive industries and continually beckoning Anglo-Americans for all forms of outdoor recreation.

SELECTED BIBLIOGRAPHY

"Admirality—Island in Contention," *Alaska Geographic*, 1 (1973), entire issue.

"Alaska's Salmon Fisheries," *Alaska Geographic*, 10 (1983), entire issue.

"Anchorage and the Cook Inlet Basin," *Alaska Geographic*, 10 (1983), entire issue.

ANDRUS, A. PHILLIP, et al., "Seattle," in *Contemporary Metropolitan America. Vol. 3: Nineteenth Century Inland Centers and Ports,* pp. 425–500. Cambridge, MA: Ballinger Publishing Co., 1976.

BARKER, MARY L., "Heart of B.C.: The Strait of Georgia Region," *Canadian Geographical Journal,* 92 (January–February 1976), 28–35.

BEATTY, ROBERT A., "Pacific Rim National Park," *Canadian Geographical Journal,* 92 (January–February 1976), 14–21.

BOWEN, WILLIAM A., *The Willamette Valley: Migration and Settlement on the Oregon Frontier*. Seattle: University of Washington Press, 1978.

BROWNING, R. J., "Fisheries of the North Pacific," *Alaska Geographic,* 1, no. 4 (1974), entire issue.

CAREY, N. G., "Queen Charlottes: Recovery, Rediscovery," *Canadian Geographical Journal,* 89 (October 1974), 4–15.

CULLEN, BRADLEY, T., "Changes in the Size and Location of Northwestern California's Wood Products Industry," *California Geographer,* 25 (1985), 45–64.

DICKEN, SAMUEL N., and EMILY F. DICKEN, *Oregon Divided: A Regional Geography*. Portland: Oregon Historical Society, 1982.

DOWNIE, BRUCE K., "Kluane, One of our Most Exciting National Parks," *Canadian Geographic*, 100 (April–May 1980), 32–38.

ELLIS, DEREK V., ed., *Pacific Salmon: Management for People,* Western Geographical Series, Vol. 13. Victoria: University of Victoria Department of Geography, 1977.

ERICKSON, KENNETH A., "The Tillamook Burn," *Yearbook,* Association of Pacific Coast Geographers, 49 (1987), 117–138.

EVENDEN, L., ed., *Vancouver: Western Metropolis*. Victoria: University of Victoria, Department of Geography, 1979.

FARLEY, A. L., *Atlas of British Columbia: People, Environment, and Resource Use*. Vancouver: University of British Columbia Press, 1979.

FORWARD, CHARLES N., ed., *British Columbia: Its Resources and People*. Victoria: Department of Geography, University of Victoria, 1987.

———, *The Geography of Vancouver Island.* Victoria: University of Victoria Department of Geography, 1979.

FOSTER, HAROLD D., ed., *Victoria: Physical Environment and Development,* Western Geographical Series, Vol. 12. Victoria: University of Victoria, Department of Geography, 1976.

FRANKLIN, JERRY F., and C. T. DYRNESS, *Natural Vegetation of Oregon and Washington*. Corvallis: Oregon State University Press, 1988.

GASTIL, RAYMOND D., "The Pacific Northwest as a Cultural Region," *Pacific Northwest Quarterly,* 64 (1973), 147–162.

GIBSON, E. M., *The Urbanization of the Strait of Georgia Region*. Ottawa: Lands Directorate, 1976.

HALEY, DELPHINE, "Puget Sound at the Crossroads," *Defenders,* 50 (1975), 317–325.

HARDWICK, WALTER G., *Vancouver*. Toronto: Collier-MacMillan of Canada, 1974.

HAYES, DEREK W., "Fog and Cloud in British Columbia," *Canadian Geographical Journal,* 83 (December 1971), 200–203.

JOHANNESSEN, CARL L., et al., "The Vegetation of the Willamette Valley," *Annals,* Association of American Geographers, 61 (1971), 286–302.

KENNEDY, DES, "Fraser Delta in Jeopardy," *Canadian Geographic,* 106 (August–September 1986), 34–43.

———, "Whither the Gulf Islands? Nature's Largesse Spawns a B.C. Dilemma," *Canadian Geographic,* 105 (October–November 1985), 40–49.

LEY, DAVID, "Inner-City Revitalization in Canada: A Vancouver Case Study," *Canadian Geographer,* 25 (1981), 124–148.

LINS, HARRY F., "Energy Developments at Kenai, Alaska," *Annals,* Association of American Geographers, 69 (June 1979), 289–303.

LOY, WILLIAM G., et al., *Atlas of Oregon.* Eugene: University of Oregon Press, 1976.

LYNCH, DONALD F., DAVID W. LANTIS, and ROGER W. PEARSON, "Alaska, Land and Resource Issues," *Focus,* (January–February 1981), 1–8.

O'NEIL, ERNEST, "Science Doubles Number of Spawning Salmon," *Canadian Geographic,* 99 (December 1979–January 1980), 62–65.

POWERS, RICHARD L., et al., "Yakutat: The Turbulent Crescent," *Alaska Geographic,* 2 (1975), entire issue.

"Prince William Sound," *Alaska Geographic,* 2, no. 3 (1975), entire issue.

PROVINCE OF BRITISH COLUMBIA LANDS SERVICE, *Lower Coast Bulletin Area.* Victoria: Queen's Printer, 1970.

———, *Vancouver Island Bulletin Area.* Victoria: Queen's Printer, 1970.

ROBINSON, J. LEWIS, ed., *British Columbia.* Toronto: University of Toronto Press, 1972.

———, "How Vancouver Has Grown and Changed," *Canadian Geographical Journal,* 89 (October 1974), 40–48.

———, "Sorting Out All the Mountains in British Columbia," *Canadian Geographic,* 107 (February–March 1987), 42–53.

———, and WALTER G. HARDWICK, *British Columbia: 100 Years of Geographical Change.* Vancouver: Talonbooks, 1973.

ROSENFELD, CHARLES, and ROBERT COOKE, *Earthfire: The Eruption of Mount St. Helens.* Cambridge: MIT Press, 1982.

SCHREINER, JOHN, "The Port of Vancouver: Its Health Is Vital to Western Canada's Economy," *Canadian Geographic,* 107 (August–September 1987), 10–21.

THEBERGE, J. B., "Kluane: A National Park Two-thirds Under Ice," *Canadian Geographical Journal,* 91 (September 1975), 32–37.

———, ed., *Kluane: Pinnacle of the Yukon.* Toronto: Doubleday Canada Ltd., 1980.

———, and J. A. KRANLIS, "The Saint Elias: Our Highest, Youngest, and Iciest Mountains," *Canadian Geographic,* 105 (December 1985–January 1986), 36–45.

TOWLE, JERRY C., "Man and the Pacific Salmon: A New Era?" *California Geographer,* 22 (1982), 13–32.

VAN OTTEN, GEORGE A., "Changing Spatial Characteristics of Willamette Valley Farms," *Professional Geographer,* 32 (February 1980), 21–31.

WAHRHAFTIG, CLYDE, *Physiographic Divisions of Alaska,* Geographical Survey Professional Paper 482, Washington, D.C.: Government Printing Office, 1965.

"Wrangell-Saint Elias: International Mountain Wilderness," *Alaska Geographic,* 8 (1981), entire issue.

YOUNG, CAMERON, "The Last Stand: Timber Running Low as Confrontations Continue in B.C.," *Canadian Geographic,* 106 (February–March 1986), 8–19.

19
THE BOREAL FOREST

The largest region of the continent sprawls almost from ocean to ocean across the breadth of the continent at its widest point, in subarctic latitudes. The Boreal Forest Region is primarily a Canadian region, occupying almost half of that nation's areal extent, although it also encompasses the Upper Lakes States and central Alaska in the United States.

It is a region of rock, water, and ice, but particularly of forest—interminable, inescapable forest (Fig. 19-1). The forest extends for hundreds of miles over flattish terrain with relatively little variation in either appearance or composition. Its very endlessness arouses a feeling of monotony in some people. Edward McCourt, in describing a traverse in western Ontario, emphasized this point:

> In Canada there is too much of everything. Too much rock, too much prairie, too much tundra, too much mountain, too much forest. Above all, too much forest. Even the man who passionately believes that he shall never see a poem as lovely as a tree will be disposed to give poetry another try after he was driven the Trans-Canada Highway.[1]

This is a region in which nature's dominance has been but lightly challenged by humankind. The landscape is largely a natural one and people are only sporadic intruders. Much of the region consists of the Canadian Shield, a land of Precambrian crystalline rock, rounded hills almost devoid of soil, fast-flowing rivers, and innumerable lakes, swamps, and muskegs. Long, cold winters characterize most of the region; its short growing season makes agriculture only locally important. From boundary to

[1] Edward McCourt, *The Road Across Canada* (Toronto: Macmillan of Canada, 1965), p. 110.

FIGURE 19-1 The Boreal Forest is a region of trees and water. This scene shows a small portion of Lake of the Woods in southwestern Ontario (TLM photo).

boundary most of the inhabitants are engaged overwhelmingly in *extractive* pursuits.

Most North Americans have never seen this region and the vast majority will never set foot in it. Yet it is a region that has enriched the autochthonous literature and folk music of both countries; the sagas of Paul Bunyan, the stories of Jack London, the poetry of Robert Service, and many other literary contributions have brought the "North Woods" into public consciousness. This is particularly true in Canada, where the forest syndrome has played a major role in the concept of the Canadian character. Subarctic expert William Wonders pointed out this emotional attachment.

> Forest, rock, water—these are the elements which exert a near-irresistible call for most Canadians. Even if most of the year is spent in a very different setting, the traditional northern ingredients are sought out for rest and relaxation. Though they may live to the south of it and rail against its harshness, many Canadians probably find in the Subarctic the "emotional heartland" of their nation.[2]

The boundaries of the Boreal Forest Region are nowhere clear-cut but can easily be conceptualized (Fig. 19-2). The northern margin is the tree line, which separates the Subarctic (Boreal Forest Region) from the tundra of the Arctic Region. As with most such vegetation boundaries, the idea of a line is misleading; the interfingering of forest and tundra is complex and the northern margin of the region is a transition zone rather than a line. This northern floristic boundary is further reinforced by its approximate coincidence with the 50°F (10°C) July isotherm and its almost absolute separation of Dene (to the south) and Inuit (to the north) settlements.

The region's southern boundary is more complicated and should be visualized as follows. In the east it is approximately coincidental with the limit of close agricultural settlement north of the St. Lawrence and Ottawa rivers. Across southeastern Ontario it follows the southern margin of the Canadian Shield. It crosses northern Michigan and central Wisconsin and Minnesota as a transition zone separating essentially agricultural land on the south from essentially forest land on the north.

From western Minnesota to western Alberta the boundary coincides with the change from prairie landscape to forest landscape. In western Canada and Alaska the boundary is largely determined by topography; the mountainous terrain of the Rockies and the North Pacific Coast Region ranges mark the southern limits of the Boreal Forest Region in this part of the continent.

The eastern (Labrador) and western (Alaska) margins of the region are generalized as separating the zones of coastal settlement from the almost unpopulated interior, with the added distinction of a forest-tundra ecotone in western Alaska.

[2] W. C. Wonders, "The Forest Frontier and Subarctic," in *Canada: A Geographical Interpretation,* ed. John Warkentin (Toronto: Methuen Publications, 1968), p. 477.

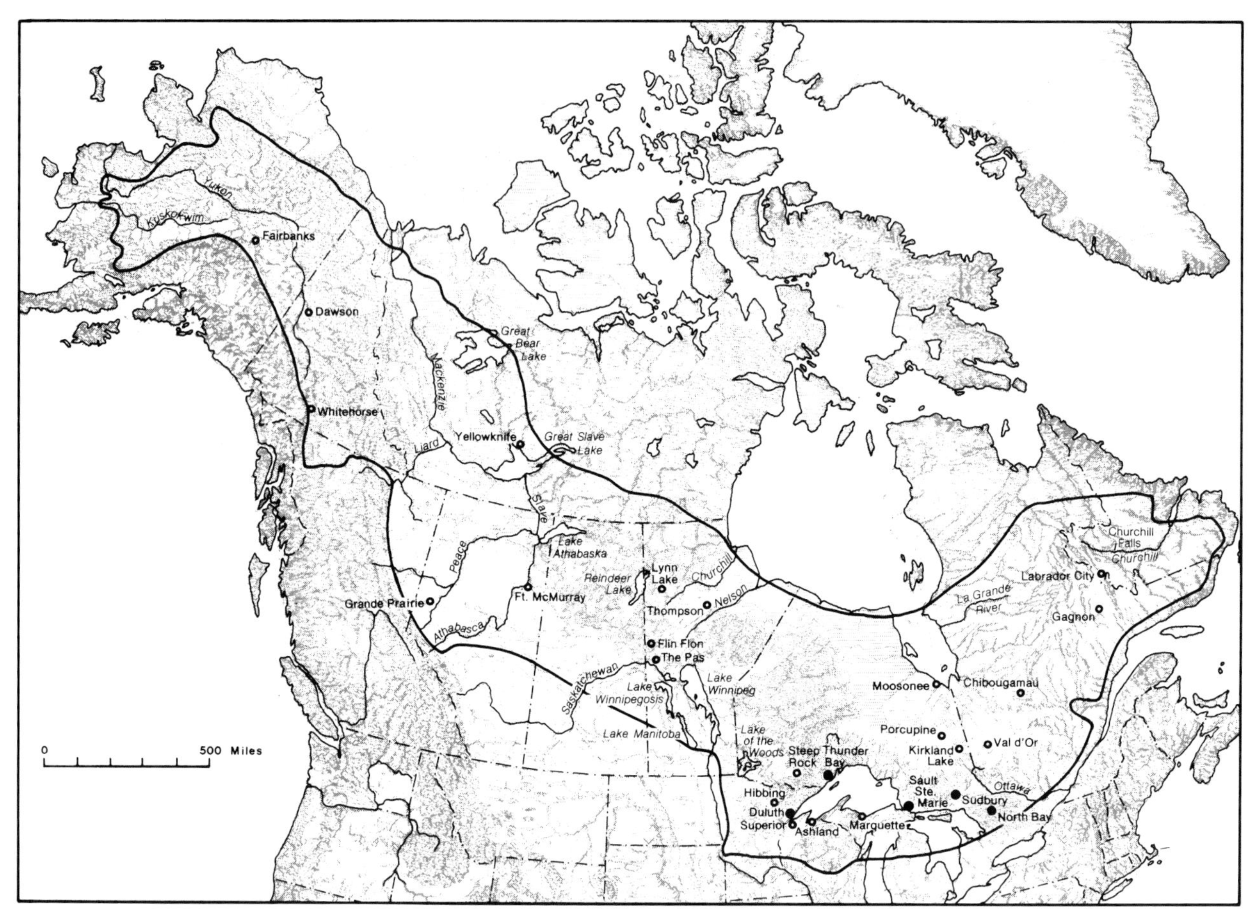

FIGURE 19-2 The Boreal Forest Region (base map copyright A.K. Lobeck; reprinted by permission of Hammond, Inc.).

A HARSH SUBARCTIC ENVIRONMENT

From a human point of view the natural environment of the Boreal Forest Region is a harsh one. It is a land dominated by winter, and winter temperatures are the most severe to be found on the continent. The surface is frozen for many months; during the summer poor drainage inhibits land transportation. The rocky forested landscape, with its filigree of lakes and rivers, was so inhospitable that Canadian settlement expansion was delayed and deflected. The Canadian Shield stood as a barrier between eastern and western Canada, denying the nation the strength and vigor that come with geographical and political unity. Thin soils, poor drainage, hostile climate, and pestiferous insects made penetration and settlement anywhere in the region a difficult and expensive undertaking.

Terrain

The eastern two-thirds of the region is underlain largely by the Canadian Shield, a vast, gently rolling surface of ancient crystalline rocks that has been scraped and shaped by the multiple glaciations of the Pleistocene Epoch. In Labrador and Quebec are several sprawling ranges of mountains and hills that reach elevations exceeding 4000 feet (1200 m). The southern edge of the Shield in Quebec is marked by the spectacular and complex Laurentide Escarpment; other less obtrusive scarps occur in other parts of that province. Elsewhere on the Shield the topography is more gently undulating and elevations are generally well under 2000 feet (600 m). The remarkable sameness of terrain over vast expanses of the Shield can be explained by the constancy of rock types; more than 80 percent of the surface consists of predominantly gneissic granitic rocks.[3]

Scouring by Pleistocene ice sheets remolded the surface and removed most of the preexisting soil. Drainage was totally disarranged, bare rock was left exposed on most of the upland surfaces, and glacial and glaciofluvial debris was deposited in countless scoured valleys. In some places the accumulation of glacial debris and deposits on the floors of old glacial lakes mantled the underlying Precambrian rocks to considerable depth, most notably in the large lowland surrounding James Bay and the western side of Hudson Bay; such areas have exceedingly flat surfaces.

To the west of the hardrock Shield is a broad lowland of vast extent that is underlain by softer sedimentary materials. It has been build up by deposition of sands and silts washed off the Rockies and the Shield. Consisting for the most part of a plain with scattered hilly districts, it is almost as poorly drained as the Shield, with the result that water features (rivers, lakes, muskegs) are commonplace. Some of the largest rivers (Mackenzie, Slave, Athabaska, Saskatchewan) and lakes (Great Bear, Great Slave) in North America are found here. Most of the sedimentary plain, sometimes referred to as the forested northern Great Plains, drains northward to the Arctic Ocean via a complicated hydrographic system dominated by the Mackenzie River.

In the Yukon Territory the region encompasses an area of more complicated geology, greater relief, and greater variety of landscape. There are several rugged mountain ranges and deeply incised plateau surfaces as well as two lengthy structural trenches.

In central Alaska the drainage basin of the Yukon River occupies the broad expanse of land between the Alaska Range to the south and the Brooks Range to the north, widening ever more broadly westward. The lower basin of the Yukon and the equally extensive basin of the Kuskokwim River to the south consist of broad flat lowlands that are quite marshy in early summer.

[3] J. Brian Bird, *The Natural Landscapes of Canada* (Toronto: Wiley Publishers of Canada Ltd., 1972), p. 136.

Hydrography

Water is abundant in most of the region during the summer, chiefly as a result of glacial derangement of drainage patterns and the fact that the subsoil is at least partly frozen throughout much of the region, thus preventing downward percolation of surface moisture (Fig. 19-3). There are

> water bodies of every conceivable shape and size, from small ponds to some of the largest lakes in the world. [Water] spills from one to the next in swift, dashing streams and sweeps along in major rivers which reach such proportions that in the case of the Mackenzie, the width of its lower reaches is measured in miles. Rapids and falls are common features on almost all rivers.[4]

Lakes are so numerous that parts of the Shield might almost be described as water with occasional land (Fig. 19-4).

Most rivers originate within the region and flow outward from it: some to the Atlantic via the Great Lakes–St. Lawrence system; some in a centripetal pattern into Hudson Bay; some directly from Quebec or Labrador into the Atlantic; and some to join the Mackenzie, Yukon, or Kuskokwim systems in their path to the Arctic Ocean or the Bering Sea. Only in the so-called Nelson Trough area of northern Manitoba are there rivers that flow *across* the region; the Nelson and Churchill systems originate in the Rocky Mountains and Prairies to the west before flowing eastward into Hudson Bay.

Swamps and marshes are also widespread, as is that peculiar northern feature, muskeg (poorly drained flat land covered with a thick growth of mosses and sedges). These areas severely inhibit overland transportation and provide an extensive habitat for the myriad insects that swarm over the region during the brief summer.

[4] Wonders, "The Forest Frontier and Subarctic," p. 474.

FIGURE 19-3 The dominant landscape elements in the region are endless forests and active water. This is the Kenebec River near Elliot Lake, Ontario (TLM photo).

Climate

In so enormous an area important differences of climate from north to south or east to west might be expected, but actually the differences are relatively slight. Everywhere the climate is continental; this is mainly the result of interior location, great distance from oceans, and the barrier effect of the western cordilleras.

This continentality is the dominating feature of the region's natural environment. Winters are long, dark, and bitterly cold. Temperatures occasionally drop to 60°F below zero (−51°C) over most of the region, and in the northwest readings of less than −70°F (−57°C) have been recorded.

The region is saved from recurrent glaciation only by the warmth of summer; summer temperatures are generally mild, but the occasional incursion of warm air masses sometimes produces decidedly warm weather. The transition seasons are short but stimulating. Spring is characterized by a high degree of muddiness, which makes it the most difficult season for overland transportation.

Precipitation is relatively light over most of the region but is quite effective because evaporation is scanty. Summer is the period of precipitation maximum; the entire landscape is covered with snow throughout the winter (Fig. 19-5). In the east, Quebec and Labrador, precipitation totals are much higher.

FIGURE 19-4 The region contains hundreds of thousands of lakes. This is Lesser Slave Lake in northern Alberta (TLM photo).

Soils

Soils of the region are characteristically acidic, severely leached, and poorly drained. Spondosols, Inceptisols, and Histosols predominate. Where farming is carried on, it is invariably in an area of more productive soil, generally a clay-based soil that was derived from glacial lake sediments.

Permafrost

Roughly half the surface area of Canada and about three-fourths of that of Alaska are underlain with permafrost, or permanently frozen subsoil (Fig. 19-6). In the zone of continuous permafrost the *ac-*

FIGURE 19-5 Snow does not accumulate to great depths over most of the region, but it stays on the ground for many months. This scene is in central Alaska near Nenana (TLM photo).

tive layer (that part that freezes in winter and thaws in summer) of the soil is only 1.5 to 3 feet (0.45 to 0.9 m) in depth.[5] Beneath it is the permafrost layer, which is usually several hundred feet thick and has been measured to depths of 1300 feet (390 m) in both Canada and Alaska. In the zone of discontinuous permafrost various factors contribute to the presence of sporadic unfrozen conditions; where permafrost occurs, the layer is much thinner and the

[5] R. J. E. Brown, "Permafrost Map of Canada," *Canadian Geographical Journal*, 76 (February 1968), 57.

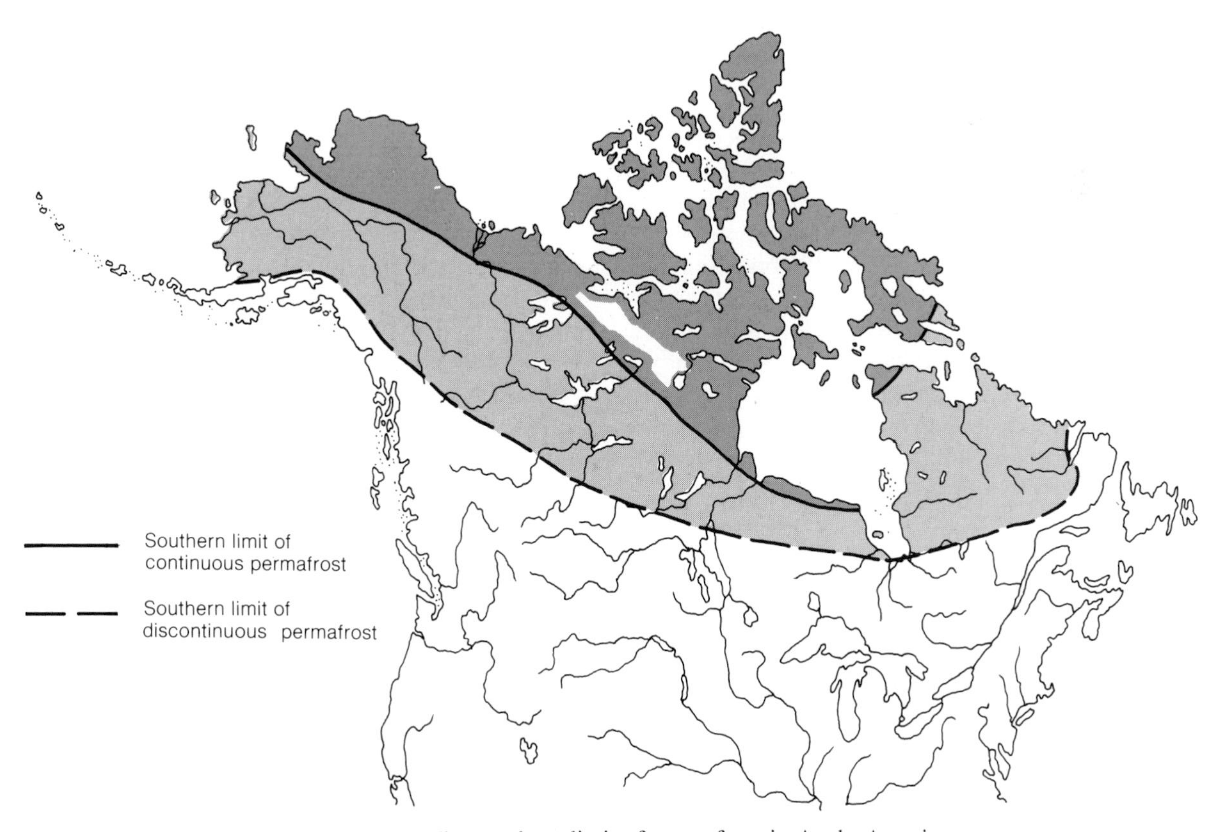

FIGURE 19-6 The southern limit of permafrost in Anglo-America.

FIGURE 19-7 In many parts of the region the taiga seems limitless. This scene is in central Alaska near Fairbanks (TLM photo).

overlying active layer may be more than 10 feet (3 m) in depth.

The presence of permafrost creates many engineering difficulties. Although providing a solid base for supporting structures (buildings, roadbeds, railways) when frozen, permafrost readily thaws when its insulating cover is removed and then loses its strength to such a degree that it will not support even a light weight; this condition results in buckling and displacement of any structure built on it. Various engineering techniques were developed to overcome these difficulties; most important is to remove the insulating cover (vegetation, surface soil) several months before actual construction begins, thus allowing permafrost conditions to stabilize to a new equilibrium.

Natural Vegetation

The distinctive natural vegetation of the Shield is coniferous forest with a deciduous admixture. From south to north the trees become smaller, those in the extreme north being small and scraggly. In this region of slow growth the trees never reach great heights, although they are normally close growing except where interrupted by bedrock or poor drainage.

The boreal forest is often referred to as *taiga;* a similar expanse of northern continental forest extends across Eurasia in similar latitudes (Fig. 19-7). Conifers, growing in relatively pure stands, dominate the plant associations of the taiga. The principal species are white spruce, black spruce, tamarack, balsam fir, and jack pine. Associated with the dominant coniferous species is a small group of hardwoods that are quite widespread; most notable are white birch, balsam poplar, aspen, black ash, and alder.

Although relatively pure stands of conifers are very extensive, hardwoods are very common in the southern part of the region. A mixed forest (coniferous and deciduous), with deciduous species predominating, is prominent in the southern part of the region in eastern Canada and the Lake States, whereas across the central part of the Prairie Provinces the forest is almost completely deciduous.

In the northern part of the boreal forest, there is much natural hybridization of tree species (cf., lodgepole pine and jack pine; black cottonwood and balsam poplar; Alaska birch and paper birch), so that identification of actual species often is difficult.

Fire is a prominent ecological force in the Boreal Forest Region, often burning thousands of square miles before its is finally quenched;

moreover, the incidence of fire in the taiga has greatly increased in recent years. Burned areas regenerate slowly in these subarctic latitudes. Hardwood species, especially birch and aspen, attain their greatest areal extent as an initial replacement for the conifers after a forest fire; they are eventually superseded by conifers as the climax association becomes reestablished.

Native Animal Life

Native animal life was originally both varied and abundant; the variety is still great, although abundance has considerably diminished. Furbearing animals have had the greatest geographical and economic influence because of the trapping carried on by natives and outsiders alike. The beaver, distributed throughout the region, has been the most important species, but also notable are the muskrat, various mustelids (ermine, mink, marten, otter, fisher, and wolverine), canids (wolf and fox), the black bear, and the lynx. Ungulates are another significant group. Woodland caribou occupy most of the region and vast herds of barrenground caribou spend the winter in the taiga. Moose and deer are also found, as well as forest-dwelling bison.

Bird life is abundant and varied in summer when vast hordes of migratory birds, especially waterfowl, come to the region to nest. Only a handful of avian species, however, winters in the Boreal Forest Region.

Insects are superabundant in summer. For two or three months of the year mosquitoes, black flies, no-see-ums (biting midges), and other tiny tormentors rise out of the muskeg and the sphagnum moss in veritable clouds and make life almost unbearable for people and animals alike.

THE OCCUPANCE

Native Peoples

Before the arrival of Caucasians, the Boreal Forest Region was occupied by widely scattered bands of seminomadic Indians. Algonkian speakers, particularly Crees and Ojibwas, were in the eastern portion, and various Athabaskan-speaking tribes lived in the west. The material culture and economy of these widely dispersed tribes were quite uniform, presumably reflecting the homogeneity of the taiga environment. They were hunters and fishermen, depending primarily on the wandering caribou for their principal food supply. The relative scarcity of inhabitants was due to a warlike social tradition and to the fact that hunting and trapping always preclude a dense population, for they entail the great disadvantage of uncertainty. Life was precarious, poverty extreme, and starvation not uncommon.

With the coming of whites, fur trapping offered significant trading opportunities. Many Indian women, particularly Ojibwas and Crees whose hunting territory was where most of the fur trade was carried on, intermarried with whites, especially in the earlier years of contact. The offspring of such unions are called *Metis*. They normally acquired a cultural background similar to that of the Indians and lived like Indians. The Metis component has grown rapidly through the years so that it is probable that there are more Metis than full-blooded Indians in the region today, although no accurate statistics on the Metis population are available.

White Penetration and Settlement

Over most of the Boreal Forest Region the evolution of white settlement has been associated with exploitive activities: trapping, mining, and forestry, or with governmental functions. Only in relatively limited areas and under often marginal conditions has settlement on a broader scale or greater diversity been attempted.

In the eighteenth and early nineteenth centuries furs were almost the only stimulus to white penetration of the region. By the middle of the nineteenth century, however, lumbermen were actively at work on the southern margin of the taiga, especially in the Upper Lakes States. At about the beginning of the present century exploitive activities were accelerated, with the initiation of pulping, hydroelectricity generation, and large-scale mining. Although Lake Superior iron ores were being mined as early as the 1850s and there were other isolated mineral discoveries in the region before 1900, mining in the region is largely a twentieth-century activity.

The agricultural frontier advanced into the Boreal Forest Region only slowly and hesitantly.

Initial farming settlement consisted mostly of a northward overflow from the lower St. Lawrence Valley. Pioneer farmers pushed into various sections of northern Michigan, Wisconsin, and Minnesota, more or less concurrently with logging in those areas. More intensive agricultural settlement came later in the Clay Belt. Still later, the farmlands of Alaska's Tanana Valley were settled. Finally, about the time of World War I, the Peace River block began to be occupied with some intensity.

White settlement of the Boreal Forest can thus be seen as a push from the east and the south, one primarily motivated by a search for furs, timber, and ores. Agrarian colonies came later and often persited on the economic margin. Urban settlement was sporadic and, for the most part, urban prosperity depended on either the stability of mineral output or some specialized transportation function.

Present Population

Most of the population of the Boreal Forest Region is located on its southern fringe and nowhere is there a significant density. The few nodes of moderate density are associated with some sort of economic opportunity, such as farming in the Peace River district or mining in northern Minnesota.

The population of the region is a little over 4 million, amounting to less than 2 percent of the Anglo-American total. Although occupying half of Canada's areal extent, the Canadian portion of the region contains only about 10 percent of that nation's populace.

Indians and Metis constitute a significant proportion of the total; they are less numerous on the southern margin of the region, although even Wisconsin and Minnesota have a considerable number of Indians. Their rate of natural increase is very high; in Canada the Indians and Metis are the fastest-growing ethnic group in the nation, increasing at a rate of 3 percent annually.

The Boreal Forest Region has an economy that is primarily resource oriented, which means that it is sensitive to national and international economic trends. Until the last few years there was a rapid rate of population increase, based on a high birthrate among Indians and Metis and a prominent influx of outsiders who were lured north by the availability of high-paying jobs for which local people lacked requisite skills. Since the mid-1970s, however, the regional economy has stagnated, resulting in a net out-migration (mostly whites who had moved North temporarily). This factor, coupled with a downturn in the birthrate, produced a zero population growth rate for the region in recent years.

THE ECONOMY

Most of the Boreal Forest Region remains economically underdeveloped, but that portion lying immediately north of the St. Lawrence River and around Lake Superior has been settled for many decades, and small but increasing numbers of people have found their way into other parts of the region, especially the Mackenzie Valley. In most areas the economy is based on a single exploitive activity, but in three widely separated portions of this vast region a broader development pattern has unfolded, resulting in a more diversified economic base, a higher density of settlement, and a more stabilized, although not necessarily prosperous, economy.

Areas of Broader Development

The Upper Lakes Area Surrounding lakes Superior and Huron are parts of northern Michigan and Wisconsin, northeastern Minnesota, and southern Ontario that are functionally focused on the Great Lakes waterway, along which are transported vast quantities of the area's ore output (Fig. 19-8). The long-established mineral and transportation industries of the Upper Lakes provide a base for the most diversified subregional economy in the Boreal Forest Region.

Mining The Upper Lakes area has a limited variety of mineral resources, but they occur in tremendous quantities. Southwest and northwest of Lake Superior lie the continent's largest and most favorably located iron ores. On Michigan's Keweenaw Peninsula are long-mined but diminishing copper deposits. On the west shore of Lake Huron are valuable beds of metallurgical limestone.

Iron mining has long been the outstanding economic enterprise in the subregion, hundreds of millions of tons of red hematite ore having been ex-

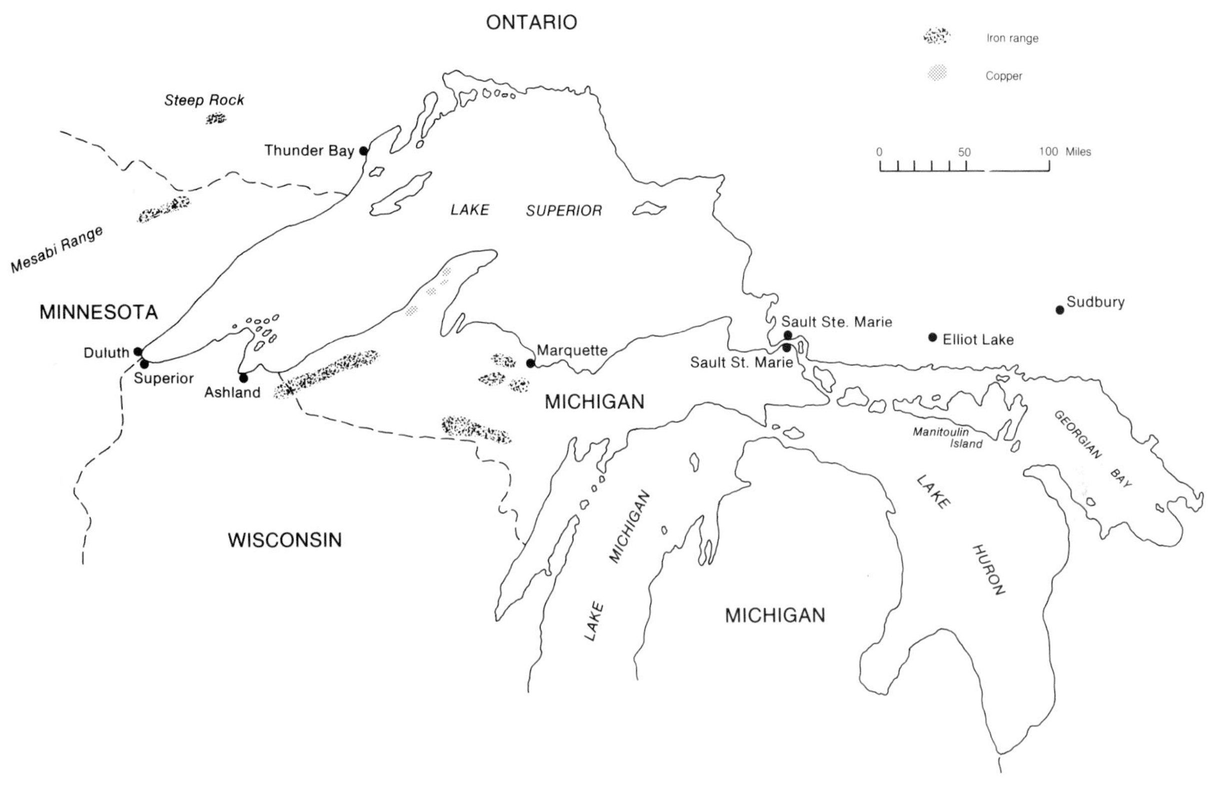

FIGURE 19-8 The Upper Lakes area.

tracted and shipped down the Lakes. The area, which accounts for more than nine-tenths of United States production, is spread over 11 counties (3 in Minnesota, 5 in Michigan, and 3 in Wisconsin). Minnesota alone contributes two-thirds of the total, chiefly from its world-famous Mesabi Range (Fig. 19-9). Moreover, significant iron-ore deposits were long exploited at Canada's Steep Rock mine, 140 miles (225 km) northwest of the head of Lake Superior.

FIGURE 19-9 An open-pit iron mine on the Mesabi Range (TLM photo).

FIGURE 19-10 A typical ore dock (at Marquette, Michigan), with an ore train moving slowly into position to dump (TLM photo).

Present (and recent) production is generally of low-grade taconite and jasper ores, which must undergo an expensive pelletizing process to yield a higher-grade concentrate before shipping. The pellets are loaded into rail cars, which are assembled into trains and taken to the various Lake Superior ore ports (particularly Duluth, Thunder Bay, and Marquette). The trains move onto the loading docks that jut out into the lake like huge peninsulas and dump their ore into "pockets," from which it can be dropped through hatches into the holds of lake vessels (Fig. 19-10). The lake carriers operate a busy one-way traffic, carrying a vast tonnage of ore across Lake Superior, through the Soo Canals, and down to the Lower Lakes ports. The return trip is often without cargo despite lower rates offered for upbound freight (Fig. 19-11). In most years the ore-shipping season on the lakes is about 10 months long, beginning in April and being closed by ice in January.

Although mining and concentrating facilities of the iron ranges are modern and efficient, the beneficiation process adds a notable cost factor. Higher-grade direct shipping ores from some foreign sources are available at much lower costs. Moreover, the North American steel industry has faltered conspicuously because of foreign competition; so it requires only about half as much ore as it did a decade ago. Consequently, the Lake Superior ore producers have closed their more expensive operations and increased the operating levels of their more efficient facilities. The net result in this area is a stifling unemployment rate, with production at less than one-fourth its previous level.

Logging The Upper Lakes subregion was a major source of lumber during the latter half of the

FIGURE 19-11 An ore boat returning to the harbor at Duluth to pick up another load of iron ore. The return journey usually is made without cargo, but in this case the deck of the boat is loaded with new automobiles picked up in Detroit (TLM photo).

nineteenth century. Heavy exploitation so denuded the forests that they diminished to relative insignificance for several decades. Second- and even third-growth timber is now being exploited. There is a considerable, although scattered, forestry industry today in the three states and one province that constitute the subregion. Pulpwood production is more important than logging for lumber.

Fishing Fishing was a flourishing enterprise in lakes Huron, Michigan, and Superior, antedating lumbering and mining. Whitefish and lake trout in 1880 made up 70 percent of the catch. Since then they have steadily declined as a result of overfishing, depredation by the destructive sea lampreys, destruction of immature fish, fouling of waters by city sewage and industrial waste, and changing physical conditions in the lakes.

An unusual addition to the recreational and commercial fishery of the Great Lakes is the coho salmon. In the late 1960s young salmon were deliberately released in Michigan streams and soon established their anadromous life cycle in the Upper Lakes. A large niche in the lakes ecosystem was unoccupied and the coho filled this vacuum and occupied a place in the food chain without much competition to native species. Such exotic introductions must be approached carefully, but early results of the coho experiment seem positive.

Farming Although a great deal of land in the northern part of the Upper Lakes States was cleared for farming, most of it was only of marginal value. Short growing season, lack of sunshine, poor drainage, and mediocre soils combine to produce very limited areas for prosperous crop growing or animal husbandry. Dairying is the principal farm activity in the area and grains and root crops are cultivated.

Farming is somewhat more prosperous on the Canadian side, where pockets of good soil are used and the poor areas have not been settled. In the Rainy River district north of Minnesota, for example, farms average 50 to 100 acres (20 to 40 ha) cleared and produce good crops of oats, hay, and flax. The chief source of agricultural income is from the sale of livestock and livestock products.

Urban Activities This portion of the Boreal Forest Region has several important urban nodes, each of which is a lake port of significance. Duluth–Superior is the funnel at the western end of the Great Lakes waterway on the U.S. side of the border; Thunder Bay performs an analogous function on the Canadian side. The twin cities of Sault Ste. Marie, Ontario, and Sault St. Marie, Michigan, are at the crossroads of the east-west land and water route and the north-south land route of the Upper Lakes. The former city is also one of Canada's leading iron and steel manufacturing centers.

The Clay Belt The so-called Clay Belt of the Quebec–Ontario borderland actually consists of two areas: the Great Clay Belt, which extends about 100 miles (160 km) both east and west from Lake Abitibi; and the Little Clay Belt, which occupies a more restricted area northeast and northwest of Lake Timiskaming (Fig. 19-12). Although some of the more productive soils are clay derivatives from glacial lake sediments, much of the soil development is from glacial and lacustrine deposits that contain only minor clay elements.[6]

Agricultural settlement in the area is a product of this century. Many immigrants from overseas came into the Ontario portions whereas most settlers on the Quebec side of the border were French Canadians who moved from the more closely settled portions of Quebec.

Agriculture in the area was fostered from early days by mineral discoveries. Silver and cobalt ores were discovered in 1903 and the town of Cobalt became the mining center. A major gold discovery in 1909 soon evolved into the Timmins-Porcupine complex. A second locale of high-grade gold deposits, around Kirkland Lake, began to be developed two years later. Several dozen mines were opened in the next two decades, mostly producing gold. The mines provided a local market for farm products, which enabled farmers to achieve some stability of production despite the high cost of their operations and the limited cropping possibilities.

Most mines closed down during World War II and farming became an increasingly profitless occupation. However, a mining rejuvenation occurred in the 1960s, with the development of zinc, copper, and silver ores near Porcupine and the opening of a large

[6] For more details on this topic and the resulting confusion, see J. Lewis Robinson, *Resources of the Canadian Shield* (Toronto: Methuen Publications, 1969), pp. 99–106.

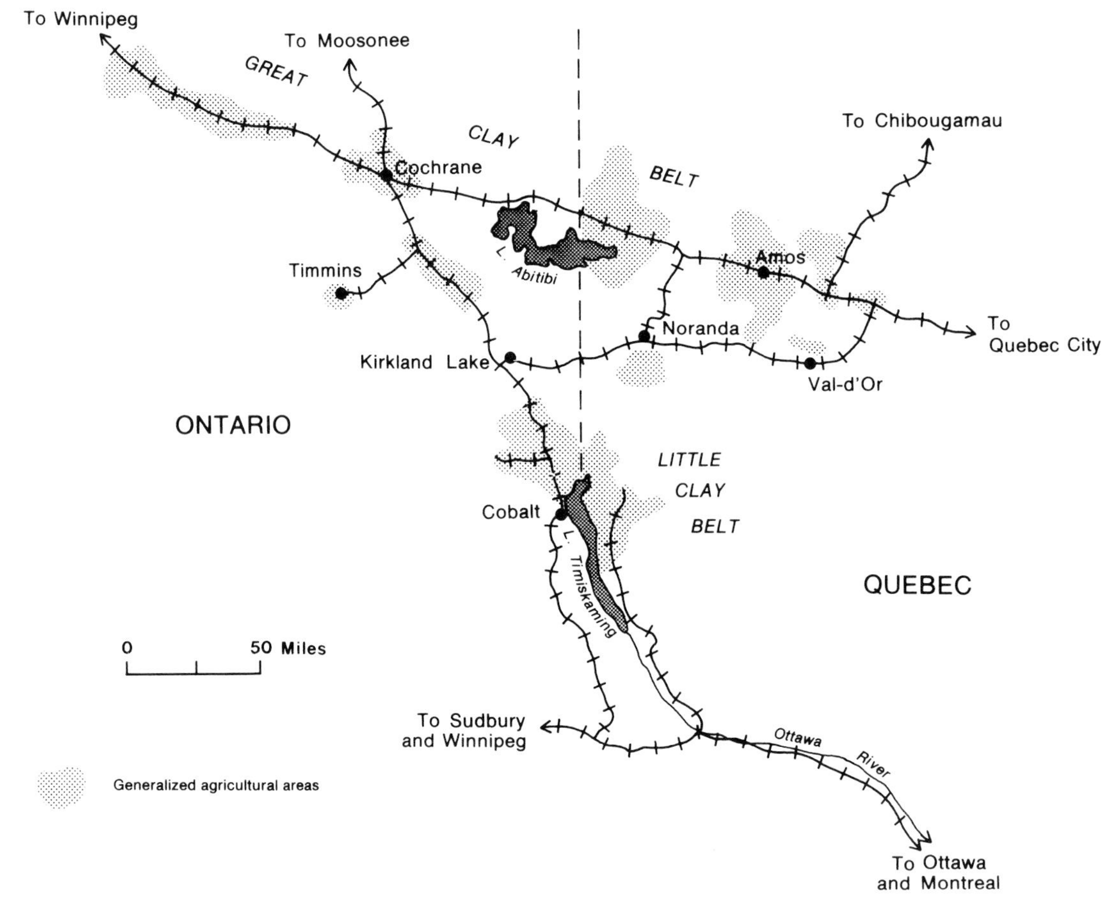

FIGURE 19-12 The Clay Belt.

iron ore mine at Kirkland Lake. In the late 1970s higher prices for gold stimulated the reopening of several old gold mines and the development of several new ones. Although farming continues to be marginal in the area, a semblance of subregional prosperity has returned.

The Peace River District In the northwestern part of Alberta and adjacent British Columbia is an enclave of black soil and grassy parkland that has some 25,000 square miles (65,000 km^2) of potentially cultivable land (Fig. 19-13). This is Canada's northernmost area of satisfactory commercial agriculture. The Peace River district was first entered by settlers as long ago as 1879, but it was only sparsely occupied until well into the twentieth century, particularly after a rail line was laid from Edmonton to Grande Prairie in 1916.

Mixed farming is characteristic, with an emphasis on cash grains (especially barley) in the older areas, and livestock (particularly beef cattle and hogs) in the newer ones. Lately there has been a phenomenal increase in canola production, a continuing increase in barley, and an accompanying decrease in both wheat and oats.

Completion of a Vancouver railway connection in 1958 gave the Peace River farmers access to the markets of southern British Columbia for the first time. This improved transportation has accounted for a considerable increase in production of feed grain for sale to ranchers west of the Rocky Mountains and for an acceleration of forestry exploitation.

The district is no longer an area of frontier farming. Even though the pioneer fringe of agriculture is being pushed slowly northward toward Great Slave Lake, commercial agriculture is now an accomplished fact. A lot of land clearing is associated with the improvements of already existing farms.

Lumber and pulp are being produced in increasing volumes, primarily from white spruce and lodgepole pine. Petroleum exploration is continuing and several fields are now in production. Improved transportation by road and rail have stimulated growth and the larger towns of the district are rapidly expanding.

Faunal Exploitation: Hunting and Trapping

Both Indians and Metis led seminomadic lives in the past, depending on wildlife and fish for subsistence and sometimes trapping as a sideline for trading. The taiga today still has a moderate number of these activities, but nomadism is a thing of the past. The introduction of modern arms and ammunition had the predictable effect of making it easier to kill game

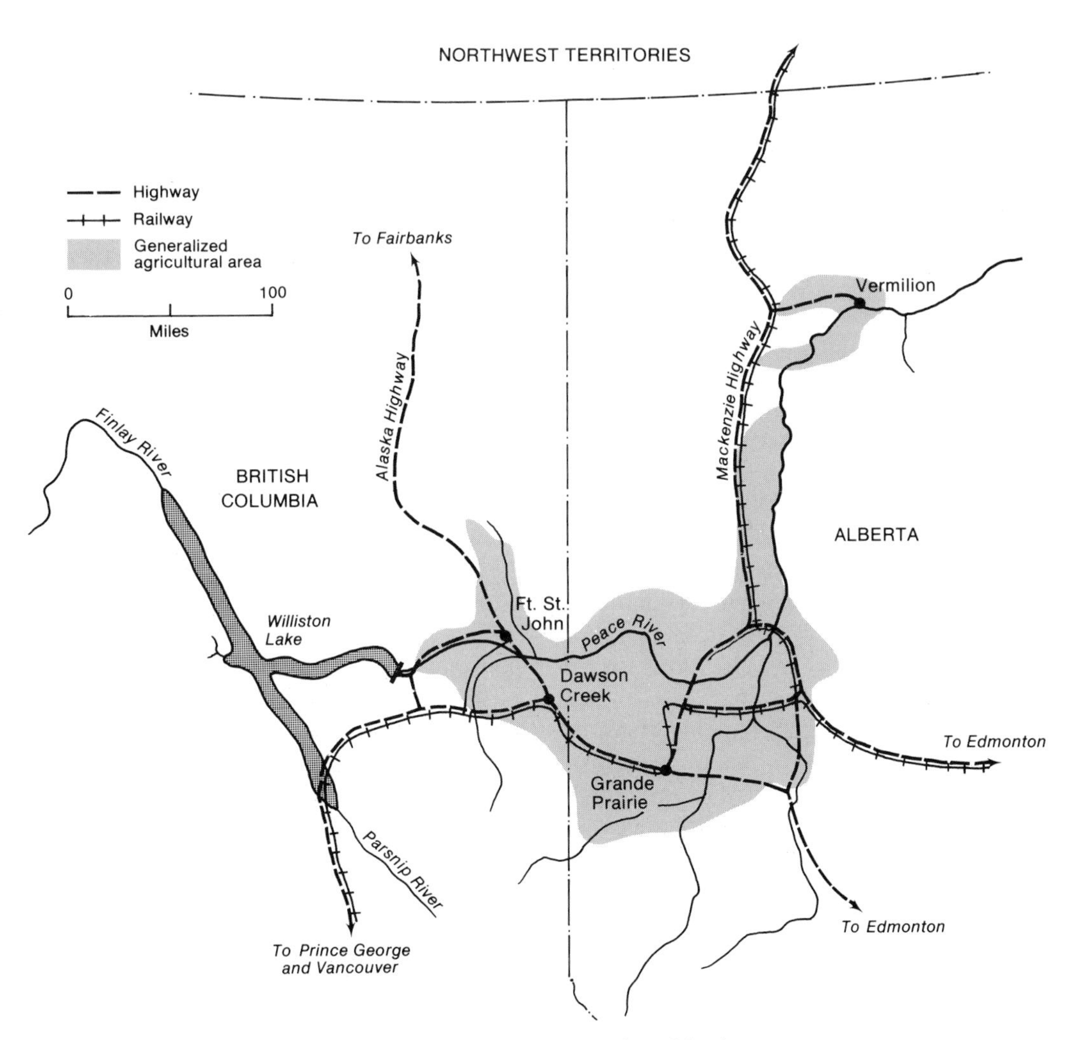

FIGURE 19-13 The Peace River District.

FIGURE 19-14 An Indian sets a trap in northern Manitoba (courtesy Hudson's Bay Company).

at first, but the deer, moose, and caribou that once supplied most of the meat are greatly depleted.

Until recent years trapping was a major source of income for many native inhabitants of the region (Fig. 19-14). Within the last three decades, however, its importance has declined and in many areas trapping has almost completely disappeared. The availability of wage labor and government social service payments has removed much of the incentive for the trapper to spend long, hard weeks in the wilderness, and antitrapping campaigns organized by various humane and animal-rights organizations have greatly diminished the market for furs.

Federal and provincial government programs to bolster the wild-fur industry have been promulgated, especially in Manitoba. The establishment of a Registered Trapline System brought order to a previously chaotic endeavor. Controlled harvesting of muskrats on "fur rehabilitation blocks" has brought both stability and increased productivity in various floodplain and delta areas. Other programs are aimed at trapper training, trapline development grants, and fur marketing improvements.

Fur Farming

In recent years the fur-farming trend in Canada has been markedly upward. The proportion of pelt value from fur farms of the total pelt value (wild and farmed) has steadily increased from about one-third a half century ago to two-thirds today.

Foxes, mainstay of the early fur farmers, are rarely raised today. Mink farms make up more than 90 percent of the total farms and chinchillas most of the remainder. Fur farms are found in every region and every province of Canada, but they are characteristic of the Boreal Forest Region, with Ontario as the leading producer.

Forestry

Logging and lumbering started early in the accessible edges of the Boreal Forest Region, although only in the Upper Lakes area was there intensified exploitation. Once lumbering got underway in any locale it was usually only two or three decades before the entire stand had been cut-over, loggers moved on, and settlements were abandoned or stagnated.

The pulp and paper industry came later and has continued as a prominent activity in many parts of the region, although it, too, is concentrated on the southern margin. This is Canada's leading industry; the nation is second only to Sweden as an exporter of pulp and produces nearly half the world's newsprint. Much of the industry is controlled by U.S. capital.

Cheap power is an important factor in the location of pulp plants, as is shown by their distribution (Fig. 19-15). A string of mills (all water driven) lines the southern edge of the forest from the mouth of the St. Lawrence to Lake Winnipeg. Farther west in the region, sawmills and pulp mills are usually large, but they are fewer in number and much more scattered in distribution.

Logging depends on snow in this region of long, cold winters. The cutters begin their work in autumn because the logs must be moved out while the ground is frozen. Tractors pull the sleds laden with logs to rivers where the cut is piled to await the spring thaw and the freshets, which transport the logs by the hundreds of thousands to downriver

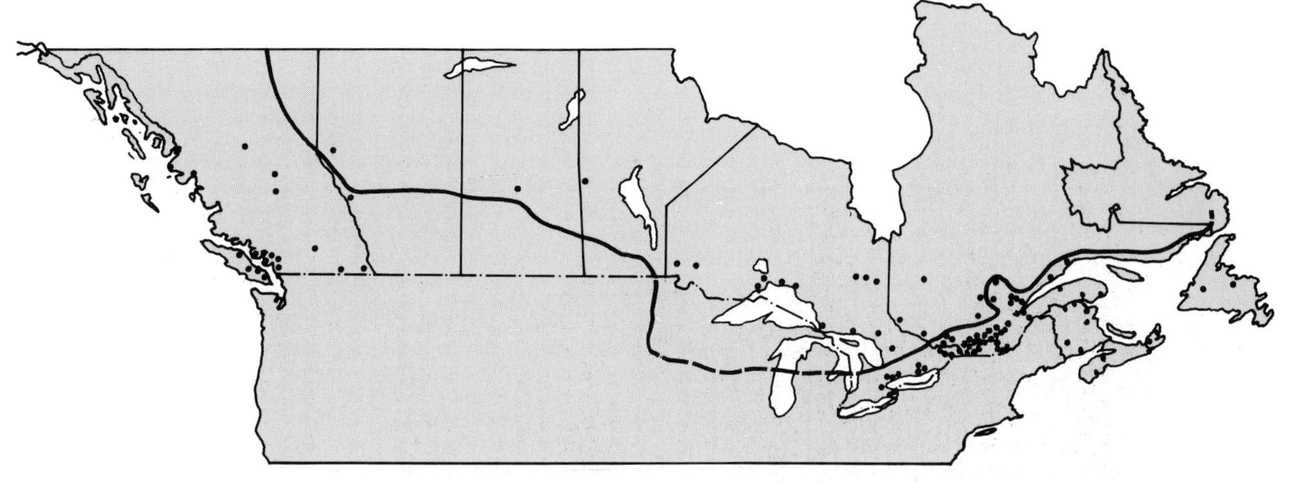

FIGURE 19-15 The location of pulp mills in Canada. Although most of the pulp logs are cut in the Boreal Forest Region, most of the mills are located south of the region. (The southern boundary of the Boreal Forest Region is shown by a black line.)

mills. River driving is an economical method of transporting logs (Fig. 19-16).

Commercial Fishing

In the last three decades commercial fisheries have developed in several large lakes of the western taiga. The industry is on a small scale compred with oceanic fishing but adds measurably to the local economy. It is a year-round activity, but emphasis is on winter ice-fishing and gill-netting. Nets are set in winter with the aid of an ice jigger, a simple machine that walks along the undersurface of the ice. Transportation is significantly simplified during the cold

FIGURE 19-16 Loggers use long-handled peaveys to guide floating logs to an outlet of Lake of the Woods near Kenora, Ontario (TLM photo).

months when tractor trains can reach the remoter lakes over frozen ground that would be soddenly impassable in summer.

Great Bear Lake is too deep, cold, and barren to provide a commercial fishery, but Great Slave Lake has an expanding industry centered at Hay River. Whitefish and lake trout are caught, primarily for export to the eastern United States. Lake Winnipeg has been a major fishing ground, but commercial operations are sometimes suspended there because of mercury contamination in the fish.

Several other large lakes in the region support commercial fishing on a regular basis and a recent summer innovation, pulse fishing, has made it economically feasible to conduct concentrated fishing operations on small lakes without overusing the resource. Pulse fishing involves rotational use of small lakes—that is, heavy fishing for one season followed by no fishing at all for five to seven years; this practice also allows for economical use of portable ice-making and refrigerated fish-holding facilities.

Mining

Mineral industries have been the mainstay of the regional economy throughout most of its history. Mining has formed the basis for most of the urban settlements. Today the outstanding significance of mining continues despite the fluctuating fortunes that accompany variations in local supply and worldwide demand. The hardrock nature of the Shield provides an unfavorable basis for agriculture but abundant opportunities for mineralization in economic quality and quantity. Also, the sedimentary formations of the Northwest are favorable for hydrocarbon accumulation.

Ore bodies are erratically distributed over the region, but several mining districts can be recognized on the basis of relative location.

Labrador Trough Ore bodies had been known in the Labrador Trough—which extends from near the estuary of the St. Lawrence northward to Ungave Bay—for decades before exploitation was finally initiated in the mid-1950s. Transportation was the big problem until completion of the initial railway. So difficult is the terrain that during construction bulldozers and other heavy equipment, food supplies, and people had to be flown into the area. The 360-mile (575-km)-long railway follows the winding Moisie River for part of the way from the mines at Schefferville to Sept Isles on the shore of the Gulf of St. Lawrence.

Four principal mining centers were developed. *Schefferville,* at the end of the long railway from the major ore port of Sept Îsles, was the first and largest. *Labrador City–Wabush* is connected by spur to the same railway line. *Gagnon,* farther southwest in Quebec, is connected by a 200-mile (320-km) rail line to Port Cartier on the St. Lawrence. Farther east are the massive ilmenite (titanium) deposits of *Lac Allard,* which are sent by rail 25 miles to Havre-St.-Pierre for shipment.

As with other iron ore producers in America, the Labrador Trough mines have been plagued by falling demand and high costs in the last few years. As a result, the mines at Schefferville were closed, presumably forever, and output decreased at the other three locales.

Sudbury–Clay Belt District Considering the magnitude of the Shield, this is a district of rather closely spaced mining activity. Many metals are recovered, but mostly nickel, copper, and gold.

Modern mining began in Canada after the discovery of the Sudbury nickel–copper ores in 1883. It is the outstanding nonferrous metal mining center in the nation, producing about one-fifth of the world's supply of nickel and considerable amounts of copper, silver, cobalt, and platinum. Both open-pit and shaft mining are used and most of the smelting is done locally. Fumes from the smelter smokestacks have denuded the nearby countryside of trees, giving a moonscape appearance that is only slowly being alleviated by intensive regreening efforts.

In the Clay Belt country straddling the Ontario–Quebec boundary is a major gold–copper zone that extends from Timmins on the west 200 miles (320 km) to Val d'Or on the east. After more than a decade of sporadic inactivity, more than a dozen gold mines are now operating again and base metal production has been expanding.

Two hundred miles (320 km) northeast of Val d'Or is the still-developing mining complex of Chibougamau. Lead, zinc, copper, silver, and gold are produced from nearly a dozen mines, the area being served by rail connections with both the Clay Belt and the Lake St. John lowland.

At Elliott Lake, on the north shore of Lake Huron's North Channel, is situated one of the world's largest deposits of uranium ore. After a booming few years of production in the 1950s and 1960s, the mines have been only erratically active because of fluctuating international demand for uranium.

Northern Manitoba District Although copper–zinc ore was discovered in 1915 at Flin Flon, exploitation did not begin until a rail line was built from The Pas in 1928. Flin Flon, a well-established center, is a major copper producer and also yields considerable zinc, gold, and silver.

Other producing centers are located northeast of Flin Flon and north of the railway line to Churchill. Lynn Lake yields nickel and copper. Snow Lake produces gold and copper. But the outstanding development is at the planned town of Thompson, a "Second Sudbury" 200 miles (320 km) northeast of Flin Flon, which began producing nickel in 1961. It is the world's second-largest nickel producer.

Isolated Mining Centers At several other localities in the region are more isolated mines. However, they open and close with such frequency that any list or map is out of date within a few weeks.

Hydrocarbons In the northwestern part of the region there are vast possibilities for production of mineral fuels. Coal of inferior quality is widespread in central Alaska, but the small population, lack of manufacturing, and limited railroad mileage keep production low. It is, however, important locally because transportation costs are high and the winters bitterly cold. Most coal mined in Alaska is from the Healy River Field not far from Fairbanks. Ladd and Eielson Air Force bases are the principal consumers of the coal.

Significant production of oil and gas has thus far come only from the Peace River district, where the Rainbow and Zama Lake oil fields are being intensively developed.

For many years it was known that within the boreal forest of Alberta a fabulous petroleum potential is locked up in tar sands (Fig. 19-17). Recoverable reserves are estimated to be half as much as the total reserves of conventional crude oil in the entire world. There are four major tar sands accumulations, generally 200 to 250 miles (320 to 400 km) northeast and northwest of Edmonton.

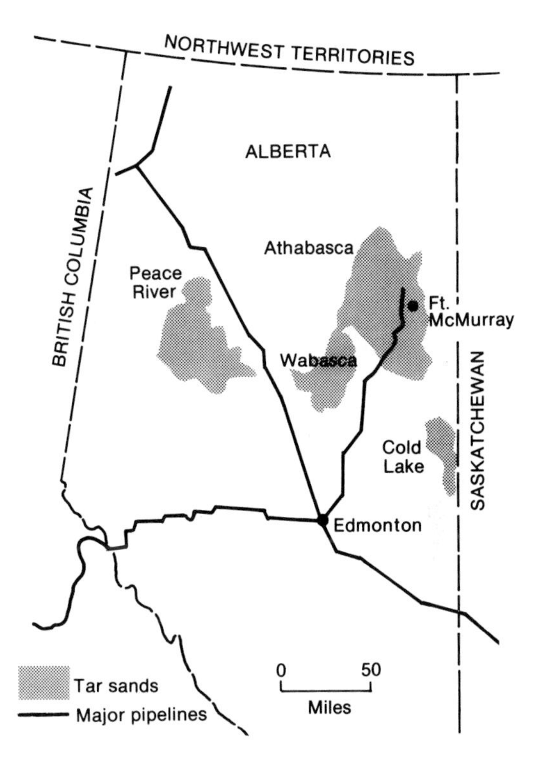

FIGURE 19-17 The major tar sands deposits of Alberta.

Complex technological requirements and extraordinary capital investment costs have deterred development. It is both difficult and expensive to separate the oil from its matrix of imprisoning sand. Two small plants have thus far come into operation, both near Fort McMurray in the heart of the Athabaska deposit, which is the largest, at a cost of about $3 billion. Three other plants are under construction—two at Fort McMurray and one at Cold Lake—and each will cost at least $10 billion. Fort McMurray has burgeoned from a village of 1000 to a city of more than 50,000, making it the largest urban place in the western part of the region. Although political and environmental objections to the development of the tar sands continue to be heard, the ultimate economic success of these ventures seems ensured. Production is increasing and will become more significant as Alberta's conventional oil fields are depleted.

Productive Water Usage

Hydroelectricity is the principal source of electric power in Canada, accounting for about 70 percent of the total. More than half the installed hydroelectric-generating capacity of the nation is within the Boreal Forest Region. Hydroelectric power facilities have been critical to the success of most of the pulp milling, mining, smelting, and refining that occur within the region. In addition, a great deal of power has been transmitted southward for use in southern Quebec, southern Ontario, the prairies, southwestern British Columbia, and even the United States.

The continually increasing demand for power has led to the near maximal development of hydroelectric power sources on the tributaries to the St. Lawrence in Quebec and Ontario, with the greatest development on the Ottawa, St. Maurice, and Saguenay rivers. Now that long-distance, high-voltage power transmission technology has been perfected, it can be anticipated that there will be still further extension of hydroelectric development to the more remote sections of the region.

Presently there are four major development schemes in operation, or nearly so, each of which is a great distance from the ultimate market for the power:

1. The Churchill Falls project in Labrador, which delivered its first power in 1972, was at that time the largest construction undertaking in Canadian history. Most of its power is used in southern Quebec.
2. In 1973 the government of Quebec initiated construction of a massive scheme in the area east of James Bay. The total plan involves damming and/or diversion of six rivers, building two new towns, constructing several hundred miles of highways and railways, and installing a dozen power-generating facilities (Fig. 19-18). The first phase, on the La Grande River, went into full operation in 1985; it alone is the largest hydroelectric installation in North America.
3. In northern Manitoba a complex scheme is under construction on the Nelson River, supplemented by a large diversion from the Churchill River, that will presumably supply all that province's power needs until the end of the century and provide a surplus for export.
4. The Peace River project in British Columbia, now in operation, sends most of its power to the Vancouver area, nearly 600 miles (960 km) away.

Agriculture

Despite the unattractiveness of the Boreal Forest Region for agriculture, farmers were settling along the edge of the Shield in Quebec and Ontario by the middle of the nineteenth century. Farming settlements were attempted in many places, generally reaching a peak of expansion in the 1930s. Frontier farming on the edge of the Shield declined for the next four decades, although there was some agricultural expansion in the western portion of the Boreal Forest Region, notably in the Peace River district.

The Clay Belt and Peace River district, the only real agricultural areas, were discussed earlier. Otherwise farming is scattered and marginal, with a few minor concentrations in such places as the Delta Junction area southeast of Fairbanks (where a large, state-supported farm scheme is being developed), the Tanana Valley near Fairbanks, and the delta of the Saskatchewan River near The Pas, Manitoba. Most frontier farms emphasize beef production and most of their cultivated acreage is in hay; other typical products are barley, oats, hogs, potatoes, and cool-season vegetables, such as cabbages and cauliflowers.

In the last few years there has been some expansion of fringe agriculture along the southern margin of the Boreal Forest Region from Manitoba to Alaska. Most of this activity does not involve the establishment of "new" farms; rather, it is the enlargement of already existing farming enterprises.

FIGURE 19-18 The first of the James Bay project dams (courtesy Quebec Government House).

As such, it does not imply pioneering in the traditional sense. A long-time researcher on the agricultural fringe has noted:

> The . . . fringe has lost much of its frontierlike appearance; increasingly it resembles established farming districts. . . . Many log cabins and crude barns [have been] replaced by more elaborate and substantial structures or sometimes with modern mobile homes. . . . The conversion of raw land to cultivation . . . still involves a degree of experimentation. . . . Abandoned farms here and there attest that success is not guaranteed.[7]

[7] Burke G. Vanderhill, "The Passing of the Pioneer Fringe in Western Canada," *Geographical Review*, 72 (April 1982), 217.

SUBARCTIC URBANISM: ADMINISTRATIVE CENTERS AND UNIFUNCTIONAL TOWNS

Although cities are scarce in the Boreal Forest Region, intimations of increased urbanism can be recognized (see Table 19–1 for a listing of the region's largest urban places). The population continues to cluster as bush dwellers settle in small settlements and residents of small settlements move to larger centers. The availability of wage-labor opportunities and the advantage of having a stable mailing address for government social service checks provide the major attractions, but the well-recognized amenities of town life are inducements in the Subarctic just as in Megalopolis.

Most urban places in the region are essentially unifunctional, depending largely on a single type of economic activity for their livelihood. In some

TABLE 19-1
Largest urban places of the Boreal Forest region

Name	Population of Principal City	Population of Metropolitan Area
Bemidji, Minn.	10,660	
Chibougamau, Que.	10,800	
Dawson Creek, B.C.	12,500	
Duluth, Minn.	81,850	241,400
Escanaba, Mich.	13,800	14,355
Fairbanks, Alaska	28,410	22,645
Fort McMurray, Alta.	40,600	
Grande Prairie, Alta.	25,200	
Hibbing, Minn.	18,400	
Marquette, Mich.	21,000	
North Bay, Ont.	50,900	
Owen Sound, Ont.	21,100	
Prince Albert, Sask.	32,300	
Rouyn, Que.	17,100	
Sault St. Marie, Mich.	13,910	
Sault Ste. Marie, Ont.	81,100	
Sudbury, Ont.	89,200	148,900
Superior, Wis.	26,980	
Thompson, Man.	15,100	
Thunder Bay, Ont.	114,200	122,200
Timmins, Ont.	46,800	
Traverse City, Mich.	16,670	
Val-d'Or, Que.	20,900	
Wausau, Wis.	34,530	113,400
Whitehorse, Y. T.	16,200	

cases, they are heterogeneous bush towns that have grown up haphazardly around a mine, a mill, or a transportation crossroads. The modern mining towns, however, are planned communities, designed for a specific size.

Two of the three largest urban centers in the region, Duluth-Superior and Thunder Bay, are Lake Superior ports whose well-being is intimately associated with the shipping of bulk products on the Great Lakes (Fig. 19-19). The twin cities of Sault Ste. Marie, Ontario, and Sault Ste. Marie, Michigan, have a crossroads function, although the big steel mill in the former center provides another dimension to its economy. Sudbury is an example of a mining-smelting town that grew into a subregional commercial center. Other places that serve significantly as somewhat diversified subregional centers include Timmins and Val d'Or in the Clay Belt, North Bay (Ont.), Fort McMurray (Alta.), and Fairbanks. Smaller mining or mining–smelting centers, such as Thompson and Hibbing, are more clearly unifunctional.

It should be noted that income from government sources, in the form of both wages and social service payments, also significantly contributes to the local economy for many urban places in the region. Whitehorse and Yellowknife have become territorial capitals, whereby their economy has profited. Many other towns in the region also serve as modified administrative centers, with district or subdistrict offices of various government agencies headquartered there. Often the economic base for small, subarctic settlements can be summarized as "government-supported economy supplemented by trapping." The wisdom of diffusing the administrative infrastructure into communities that no longer have a viable economic base, thereby perpetuating stagnant communities that often have serious economic and social problems, continues to be questioned.

TRANSPORTATION: DECREASING REMOTENESS AND INCREASING ACCESSIBILITY

Transportation difficulties have often been cited as the principal deterrent to economic prosperity in the region. With increasing technological sophistication, however, it is now clear that mere remoteness is no longer a major handicap. If the economic prize is sufficiently promising, transportation can be provided.

Early transport in the region was by canoe in summer and dog team and sledge in winter. The maze of waterways with only short portages enabled the canoe to go unbelievable distances and the accumulation of snow, because of the absence of thaws in winter, made sledging relatively easy.

Water transport became notable from the earliest days of European penetration of the region, utilizing the long rivers, large lakes, and intricate network of streams. The Quetico area (that part of Ontario and Manitoba between Lake Superior and the edge of the prairie) was traversed by a varied flotilla of watercraft, for instance, and "during cer-

FIGURE 19-19 Canada's lakehead port, Thunder Bay, is marked by an array of huge shoreline grain elevators (courtesy Ontario Ministry of Industry and Tourism).

tain periods, the volume of traffic made this one of the busiest regions of interior North America."[8]

The Yukon and Mackenzie rivers were heavily used waterways, and the latter is still used by barges. Many other rivers and lakes had, and have, considerable usage, but the Great Lakes are the major waterway of commercial importance. Their traffic consists almost entirely of bulk products—iron ore, grain, and limestone downbound, and coal upbound—although considerable St. Lawrence Seaway shipping traverses the Huron–Michigan route destined for Chicago.

Canada's two transcontinental railway lines cross the Shield through Ontario, providing critical transport links for the nation and encouraging mineral and forestry exploitation. Two rail lines were built across the region to Hudson Bay, reaching Churchill in 1929 and Moosonee in 1932. Most other railway construction in the region was designed as feeder lines to connect mining localities with the outside world: the Labrador Trough lines, Chibougamau, the Clay Belt, the various Manitoba mining centers, and the Great Slave Lake Railway to Pine Point.

In the far northwest are two railway routes of different origin. The *White Pass and Yukon Railway* was built in 1898 during the feverish boom days of the Klondike. It is 111 miles (178 km) long and extends from Skagway over White Pass to Whitehorse. The *Alaska Railroad,* extending 470 miles (750 km) from Seward to Fairbanks, was built by the United States government to help develop and settle interior Alaska. It was completed in 1923.

Although vital to the economy of their local areas, neither route is profitable. Indeed, the former ceased operating, at least temporarily, in the early 1980s. The Alaska Railroad is still owned by the federal government, but negotiations are underway to transfer its ownership to the state of Alaska.

The relatively few miles of roads and highways in the region are being rapidly expanded (Fig. 19-20). The Upper Lakes area and Clay Belt have been fairly well served for years, but only recently has roadway construction been accelerated elsewhere. The *Alaska Highway,* extending 1500 miles (2400 km) from the Peace River district to Fairbanks, has been an important connection ever since its con-

[8] Bruce M. Littlejohn, "Quetico Country: Part I," *Canadian Geographical Journal,* 71 (August 1965), 41.

FIGURE 19-20 A scene along the Dempster Highway in the Yukon Territory near Dawson City (TLM photo).

struction in 1941–1942; it is an all-weather road. The *Mackenzie Highway* is a 650-mile (1040 km) link from the Peace River district to Great Slave Lake, extending as far north as Yellowknife. The *Dempster Highway* has been built from the southern Yukon northward to the Mackenzie delta, but a planned roadway down the full length of the Mackenzie River in the Northwest Territories has been suspended, presumably awaiting settlement of native land claims in the area. In Alaska an all-weather road, the *Dalton Highway,* has been built along the route of the Trans-Alaska Pipeline from Fairbanks to the Arctic coast, but it is not open to the general public north of the Brooks Range. In addition, roads usable only in winter have been bulldozed to provide seasonal access in many remoter localities.

Perhaps the most important development in the history of the region was the introduction of the airplane in about 1920. It revolutionized communications and accelerated the pattern of economic and social progress. There are increasing networks of scheduled airline service, particularly in the Mackenzie Valley and central Alaska, and non-scheduled bush-pilot flying is significant. Light aircraft can be equipped with pontoons, skis, nose-warmers, and other devices to permit operation into almost any area in both summer and winter. Short-Take Off-and-Landing (STOL) modifications make it possible to use mere puddles or meadows as an airstrip and helicopters do not even require that much room.

"Personal" transportation in the bush was greatly enhanced by the proliferation of motorized ATVs (all-terrain-vehicles) and snow machines. Although expensive to purchase and operate, these machines provide speed and flexibility never before available.

TOURISM

The recreational activities of the region are of two principal types: brief winter and summer visits from nearby population centers and more extensive expeditions involving considerable travel.

A number of major population centers lie just beyond the southern margin of the region and short-term visitors from these areas constitute the major component of tourism in the Boreal Forest Region. The Upper Lakes area is within driving distance from many midwestern cities whose summer weather is sufficiently uncomfortable to urge "North Woods" vacations on its populace. Thus many Detroiters visit northern Michigan; people from Chicago and Milwaukee go to northen Wisconsin; and Twin Cities residents travel to the Minnesota north country.

A similar situation prevails for urbanites in southern Canada. Principal areas of attraction include the Laurentides Park area, particularly for the people of Quebec City; the Central Laurentians, favored for skiing as well as summer activities, especially by Montrealers; Algonquin Park near Toronto, notable for fishing and canoeing; the lake country of southern Manitoba and Riding Mountain National Park, for Winnipeg; Prince Albert National Park near Saskatoon; and Elk Island National Park, a wildlife preserve within picnicking distance of Edmonton.

The abundant fish and wildlife resources of the taiga are major attractions for hunters on both sides of the international boundary. They make long trips, frequently by air, to fish in remote lakes for trout, whitefish, pike, perch, and muskie, and to hunt moose, caribou, bear, and deer.

NATIVE LAND CLAIMS IN THE NORTH

The native inhabitants—Indian, Inuit, Aleut, and Metis—of the North have emerged from decades of relatively unobtrusive docility to assert, with increasing stridency, claims for much greater control of their own social, economic, and political destiny. These claims focus particularly on land rights and have been unusually effective in attracting attention in both Ottawa and Washington because they have arisen at least partially in direct opposition to previously discussed energy developments.

Judge Berger stated with reference to the Mackenzie Valley:

> [Our] society has refused to take native culture seriously. European institutions, values and use of land were seen as the basis of culture. Native institutions, values and language were rejected, ignored or misunderstood and . . . [we] had no difficulty in supposing that native people possessed no real culture at all. Education was perceived as the most effective instrument of cultural change: so, educational systems were introduced that were intended to provide the native people with a useful and meaningful cultural inheritance, since their own ancestors had left them none.[9]

Native people constitute the bulk of the permanent population in northern Anglo-America and a large part of the white population is temporary. Thus it seems logical that the future of the North should be determined to some degree by its permanent residents. Such is the underlying thought behind native claims.

The first "settlements" of claims were made in the early 1970s. Since then, however, the nature of the claims has become much more complex and clearly native people are now seeking a fundamental reordering of the relationship between native and nonnative components of contemporary society.

After seemingly endless negotiations, the seven "comprehensive" (as opposed to "specific" claims that deal with relatively small areas claimed by individual tribes) claims in the North have either been settled or are very close to settlement at this writing. (It should be noted that several of the "specific" claims, especially in British Columbia, may become of major significance.)

Alaska Native Claims Settlement Act (ANCSA), 1971

Representatives of the Alaskan native populace (Indian, Inuit, and Aleut) formed the Alaskan Federation of Natives in 1966, which instituted lawsuits claiming native ownership of three-fourths of the state's 375 million acres (150 million ha) of land (Fig. 19-21). Congress, goaded by the fact that the litiga-

[9] *Northern Frontier, Northern Homeland: The Report of the Mackenzie Valley Pipeline Inquiry,* Vol. 1 (Ottawa: Minister of Supply and Services Canada, 1977), p. xviii.

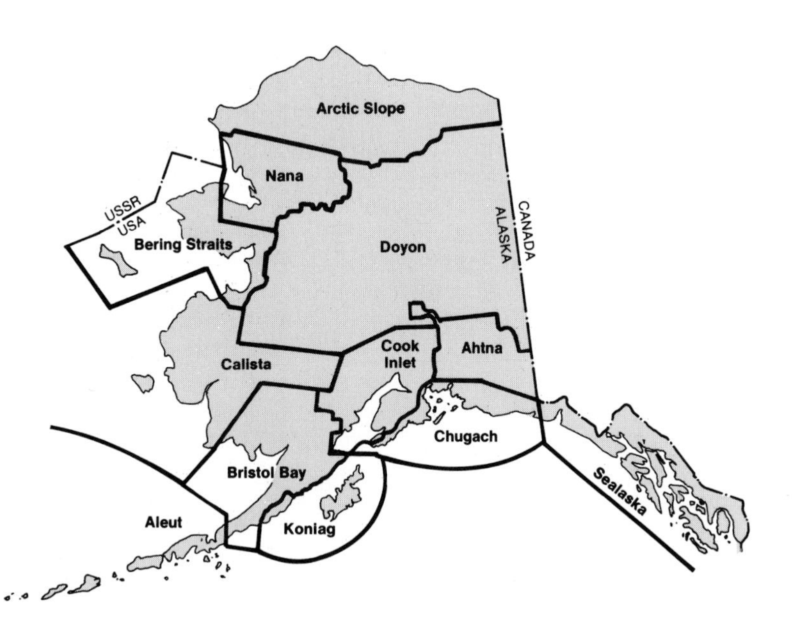

FIGURE 19-21 The areas served by the Alaska Native Regional Corporations.

A CLOSER LOOK *Tourism as an Economic Alternative—The Case of Michigan*

With a growth rate through the 1980s of between 4.5 and 6 percent annually, and with similar prospects for the future, tourism, along with recreation, has become the world's most rapidly expanding industry. Many less-developed countries rely on tourism revenues to a remarkable degree, whereas in richer, more mature economies (such as those of North America) tourism is often the only component seen likely to expand steadily well into the next century.

As an economic activity, there are two particularly attractive characteristics of tourism. First, tourism revenues are generally seen as "clean" money. That is, visitors to a given region or attraction do not (usually) cut down trees, remove mineral deposits, or otherwise diminish the economic base. Additionally, although there are substantial exceptions, tourism is normally a less environmentally threatening activity than is mining, manufacturing, or power generation. Finally, tourists customarily make few demands on such expensive areas of the social infrastructure as health care and Social Security.

The second great advantage of tourism is its unique ability to provide some economic base for regions that the geographic laws of locational advantage have rendered unattractive for other activities. In other words, what constitute disadvantages for agriculture and industry are often positive features for tourism: Rugged topography, climatic patterns, and sheer isolation from the stresses of urban life may provide exactly the sort of setting most appealing to many tourists. Moreover, in such regions, tourism revenues may finance many infrastructural improvements and support commercial and recreational activities that enhance the quality of life of local residents.

In few parts of North America have these economic advantages of tourism been more clearly evident than in Michigan. The dramatic decline in manufacturing employment as a result of downturns in the automotive and other traditional industries dealt a severe blow to the state's economy by 1980; but fiscal austerity and the promotion of alternative activities—prominent among them tourism—have resulted in a dramatic turnaround. In 1988, tourism generated nearly $16 billion for the Michigan economy, supporting 328,000 jobs in the process. Although these figures are well below the totals posted by such tourist meccas as California or Florida, they are sufficient to rank Michigan among the top six or eight tourism states.

Tourism in the state naturally includes substantial business and convention travel focused on Detroit and other cities in the urbanized southeast; but its greatest proportional impact has been in the northern two-thirds of the state. Here, the often thoughtless exploitation of the natural resource base by humans over a century and a half has left many traditional economic supports—lumbering, mining, commercial fishing—shadows of their former importance. Moreover, manufacturing has found little favor in these remote counties, far from market centers and often with an inadequate transportation network.

Enter tourism, and the disadvantages become advantages. The cool summer climate of the north has been an attraction to residents of sweltering cities farther south ever since the pre-air-conditioning days of the late nineteenth century. In much the same way that wealthy New Yorkers built extravagant "summer cottages" at Newport, Rhode Island, or Bar Harbor, Maine, so rich folk from cities such as Chicago and Cincinnati, as well as Detroit, built summer homes in the Traverse Bay region, along the Lake Huron shore, or in Michigan's Upper Peninsula.

By the 1930s, and especially since World War II, the democratization of tourism (stimulated by the paid vacation negotiated into union contracts) led to ever-larger waves of less-affluent visitors who were interested mainly in escaping the heat and oppression of the cities for a week or two in the cooler, relatively unspoiled north. At about the same time, a dramatic nationwide increase in winter sports created additional opportunity for northern Michigan, where the rough topography of glacial moraines and abundant snowfall (generated by the "lake effect" as air masses move across the open waters of Lakes Superior and Michigan) have stimulated development of the most important winter sports industry in the Midwest. While clearly not in the class of the famous New England or Colorado resorts, Michigan's winter sports facilities are within inexpensive weekend range of a large and increasing clientele.

In any event, the major attraction of the north is the outdoor life. Hunting and sport fishing are important seasonal activities (Michigan is reputed to have a larger deer population than any state except Pennsylvania and Texas), and the generate substantial revenues. Water-related activities

FIGURE 19-a Most of the summer recreation in northern Michigan is water-oriented. This is the pleasure boat docking area at Harbor Springs (photo by James R. McDonald).

of all types are a big draw. Michigan's Great Lakes coast added up to a longer shoreline than any state except Alaska: It exceeds in length the entire Atlantic seaboard. When this is combined with the thousands of inland lakes left as a legacy by the glaciers, it is easy to understand why Michigan leads the states in boat registrations: more than 830,000 (Fig. 19–a). In 1987, some 5.2 million campers formed part of the 24.2 million visitors to Michigan's 85 state parks.

Tourism income in Michigan provides another advantage in the form of enhanced interest in environmental quality. "Unspoiled" landscapes usually must be protected if their beauty preserved, and this is naturally easier to accomplish if local interests can appreciate the financial windfall that increased tourism can represent. This synthesis of tourism and environmental concern has done much to promote creation of the Pictured Rocks National Lakeshore in the Upper Peninsula, and the Sleeping Bear Dunes National Lakeshore in the northwestern part of the Lower Peninsula.

Tourism is of course not without its problems. The seasonal character of much travel and recreation means that—even where a winter sports industry has developed—there is a boom-and-bust cycle in many regions and their appeal to tourists are to be dependent on tourism. Additionally, because tourism is both a seasonally and spatially concentrated activity, heavy environmental pressure is often placed on waste-disposal and water-supply facilities, while demands for provision of services (fire, police, road maintenance) may exceed the capabilities of local authorities. Nonetheless, the image of tourism remains positive, and its financial contribution to the state's economy is vital. Without its support, the quality of life in most of northern Michigan would be seriously diminished.

Professor James R. McDonald
Eastern Michigan University
Ypsilanti

tion prevented construction of the Trans-Alaska Pipeline, passed ANCSA in 1971, remanding to Alaska's estimated 50,000 native people the title to 44 million acres (18 million ha) of land (equivalent in area to the state of Washington) and a cash settlement of nearly $1 billion.[10]

This legacy is managed by an elaborate system of native corporations that hold title to the lands and receive the financial benefits. The state was divided into 12 regions on the basis of common native cultural heritage and interests and a regional corporation was established in each (a 13th corporation was added later for natives who had moved away from Alaska). In addition, some 250 village corporations were formed. Each native residing in a village and a region in 1971 automatically became a stockholder in that village corporation and that regional corporation.

Village corporations were entitled to select a stipulated acreage (prorated on the basis of population) of land relatively near the village; regional corporations were allowed to choose vast acreages within their regional boundaries and they also own the subsurface mineral rights to all village land within their regions. An important provision of ANCSA is that no corporation land can be either sold or taxed for 20 years (until 1991), in order to give even the youngest stockholder a voice in management or disposition of the land resource.

The ultimate goal of ANCSA is assimilation of native people into the general society and capitalistic economy of Alaska. After 1991 there will be no special rights or guarantees for them or their land. Meanwhile, regional corporations are using their patrimony in various ways in attempts to better the economic and social conditions of their stockholders. Some corporations have already lost great sums of money whereas others have made remarkable profits. Throughout Alaska, however, the native people now have greatly expanded opportunities—as well as temptations.

[10] The ANCSA settlement amounted to more land than is held in trust for all other American Indians and four times the amount of money that all other Indian tribes have received from the U.S. Indian Claims Commission since it was founded.

James Bay and Northern Quebec Agreement, 1975

Some 10,000 natives (mostly Cree Indians and some Inuit) who inhabit northern Quebec agreed to relinquish all claim to lands affected by the James Bay hydroelectric project in exchange for $225 million, exclusive right to 5000 square miles (12,950 km^2) of land, and hunting-fishing-trapping rights to another 60,000 square miles (155,000 km^2). In 1990 the Crees entered into litigation to invalidate the agreement, claiming that the governments of Quebec and Canada had not fulfilled all legal obligations (Fig. 19-22).

Northeastern Quebec Agreement, 1978

This is supplementary to the James Bay agreement and deals with the Naskapi tribe of the Schefferville area.

Western Arctic (Inuvialuit) Claims Settlement Act, 1984

This agreement was between the federal government and the Committee for Original Peoples' Entitlement (COPE), representing the Inuvialuit (native people) of the Western Arctic. The Inuvialuit received land, cash ($45 million), a $10 million Economic Enhancement Fund, wildlife harvesting and management rights, and Inuvialuit participation on advisory boards dealing with land-use planning and environmental management. About 4500 individuals are involved.

Dene Nation/Metis Agreement, 1990

This agreement provides the 8000 status Indians and 5000 non-status Indians and Metis of the Mackenzie Valley with surface rights and exclusive control over development on 70,000 square miles (180,000 km^2) and shared management of an area five times as large.

Council for Yukon Indians, 1990

The 7000 Yukon Indians will receive entitlements similar to those of the Dene/Metis agreement.

Tungavik Federation of Nunavut, 1990

This federation represents the 15,000 Inuit of the central and eastern Arctic. An agreement-in-

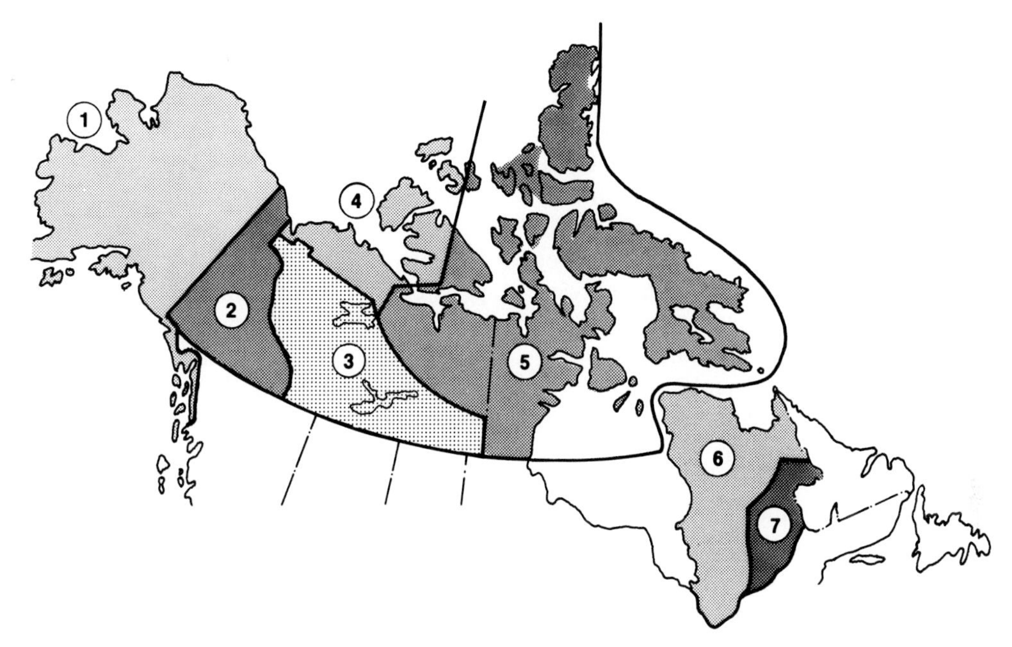

FIGURE 19-22 The areas encompassed by major northern native land claims in Canada and Alaska:
1 Alaska Native Claims Settlement Act (ANCSA)
2 Council for Yukon Indians
3 Dene Nation/Metis
4 Committee for Original People's Entitlement (COPE)
5 Tungavik Federation of Nunavut (TFN)
6 James Bay/Northern Quebec
7 Northeastern Quebec

principle was signed in 1990, and final negotiations continue at the time of this writing.

It seems clear that extensive portions of subarctic and arctic North America are, or soon will be, governed at the regional and local level by public bodies controlled by native peoples. The implications of these developments are tremendous.

THE OUTLOOK

The Boreal Forest Region is primarily an empty area as far as human occupancy is concerned and logically will remain so. It represents, however, the northern frontier and as long as there are frontiers there will be people trying to push them back. Yet today many parts of the region are emptier than they were 50 or 100 years ago, as bushdwellers increasingly congregate in larger, permanent settlements.

The region's population is youthful. The Northwest Territories and Yukon Territory have a smaller proportion of old people than any Canadian province, and Alaska has a smaller proportion than any of the other 49 states. This preponderance of young adults results in a high birthrate; the two territories have the highest in Canada and Alaska's is the highest in the United States. But the population is not a stable one. There is considerable seasonal fluctuation, with a summer peak and a winter ebb.

Many people enter the region for temporary summer jobs and then leave at the end of the season. Even many year-round residents, attracted by high wages in "hardship" posts, have come to the region on a temporary basis; they are sojourners, not settlers. Although improved transportation makes it easier for them to "go south" for a holiday, it also makes it easier to go south permanently.

For the region as a whole, mineral industries have been and will continue to be the keystone to the economy. More major ore deposits will undoubtedly be found and developed. It is unlikely that there will ever be a second Sudbury, but more Scheffervilles and Thompsons can be anticipated, which only emphasizes the boom-and-bust syndrome that is characteristic of such activities and communities.

Large-scale energy developments—oil and gas exploration and production, oil sand development, pipeline projects, and dam building—will dominate the scene in many areas. The ramifications of these major projects are far-reaching and long-lasting. Now that native land claims in the Mackenzie Valley appear to be settled it seems likely that major transportation ventures will be initiated, perhaps including a highway, a gas pipeline, and an oil pipeline.

The role of native people should become increasingly significant in most parts of the region. Their political power is increasing and financial settlements of native claims are providing them with some economic leverage for the first time. Still, a high proportion of the native population will continue to be socially and economically disadvantaged. Elements of "white" material culture and values have been introduced throughout the region. In many cases, the native villages have neither the physical nor the organizational foundations to support these introductions, and the results are likely to be severe social-psychological maladaptations. Many young-adult natives of the region suffer from an identity crisis that is often manifested as a generational conflict within the village or tribe.

As with so many "wilderness" regions, there will be an increasing emphasis on tourism and recreation. Genuine tourists will begin to appear where now only fishermen and hunters go. There is only a short timelag in North America between the completion of any fairly negotiable road and the appearance of motels, trailer camps, picnic tables, roadside litter, and new money.

SELECTED BIBLIOGRAPHY

Adams, George F., "Commercial Fishing in Northern Ontario," *Canadian Geographic,* 97 (August-September 1978), 62–69.

"Alaska's Great Interior," *Alaska Geographic,* 8 (1980), entire issue.

Bradbury, John H., and Isabelle St. Martin, "Winding Down in a Quebec Mining Town: A Case Study of Schefferville," *The Canadian Geographer,* 27 (Summer 1983), 128–144.

Brook, G. A., and D. C. Ford, "Nahanni Karst: Unique Northern Landscape," *Canadian Geographical Journal,* 88 (June 1974), 36–43.

Brown, R. J. E., *Permafrost in Canada.* Toronto: University of Toronto Press, 1970.

———, "Permafrost Map of Canada," *Canadian Geographical Journal,* 76 (1968), 56–63.

Cornwall, Peter, C., and Gerald McBeath, eds., *Alaska's Rural Development.* Boulder, CO: Westview Press, 1982.

Darragh, Ian, "Gatineau Park," *Canadian Geographic,* 107 (December 1987–January 1988), 20–29.

Dixon, Mim, *What Happened to Fairbanks?* Boulder, CO: Westview Press, 1978.

Elliott-Fisk, Deborah L., "The Stability of the Northern Canadian Tree Limit," *Annals of the Association of American Geographers,* 73 (December 1983), 560–576.

Fahlgren, J. E. J., and Geoffrey Mathews, *North of 50°: An Atlas of Far Northern Ontario.* Toronto: University of Toronto Press, 1985.

Fuller, W. A. "Canada's Largest National Park," *Canadian Geographical Journal,* 91 (December 1975), 14–21.

Gibbens, R. G., "How Hydro Power Brought Aluminum to Canada," *Canadian Geographical Journal,* 93 (August-September 1976), 18–27.

Gill, Don, "Modification of Northern Alluvial Habitats by River Development," *Canadian Geographer,* 17 (1973), 138–153.

Gorrie, Peter, "The Bruce: Our Newest National Park," *Canadian Geographic,* 107 (October–November 1987), 62–71.

Hare, F. Kenneth, and J. C. Ritchie, "The Boreal Bioclimates," *Geographical Review,* 62 (1972), 333–365.

Harrington, Lyn, "Thompson, Manitoba: Suburbia in the Bush," *Canadian Geographical Journal,* 81 (November 1970), 154–163.

______, "Thunder Bay: The Lakehead City," *Canadian Geographical Journal,* 80 (January 1970), 2–9.

______, "The Yukon River in Canada," *Canadian Geographical Journal,* 85 (1972), 200–209.

KAKELA, PETER J., "The Superiority of Low-Grade Iron Ore," *Professional Geographer,* 33 (February 1981), 95–102.

LANTIS, DAVID W., DONALD MEARES, and VALENE SMITH, "The Alaskan Bush in Transition: An Overview," *Yearbook,* Association of Pacific Coast Geographers, 51 (1989), 125–134.

MCKAY, DONALD, *Heritage Lost: The Crisis in Canada's Forests.* Toronto: Cross Canada Books, 1985.

MEREDITH, THOMAS C., "The George River Caribou Herd," *Canadian Geographer,* 29 (Winter 1985), 364–366.

NICHOLSON, N. L., "The U.S. Northwest Angle: East of Manitoba," *Canadian Geographical Journal,* 96 (February–March 1978), 54–59.

OLSON, ROD, FRANK GEDDES, and ROSS HASTINGS, eds., *Northern Ecology and Resource Management.* Edmonton: University of Alberta Press, 1984.

Province of British Columbia Lands Service, *The Peace River Bulletin Area.* Victoria: Queen's Printer, 1968.

PUGH, DONALD E., "Ontario's Great Clay Belt Hoax," *Canadian Geographical Journal,* 90 (January 1976), 19–24.

"Richard Harrington's Yukon," *Alaska Geographic,* 2, no. 2 (1974), entire issue.

ROBERGE, ROGER A., "Resource Towns: The Pulp and Paper Communities," *Canadian Geographical Journal,* 94 (February–March 1977), 28–35.

ROBINSON, J. LEWIS, *Resources of the Canadian Shield.* Toronto: Methuen Publications, 1969.

SEABORNE, ADRIAN A., and PATRICIO N. LARRAIN, "Changing Patterns of Trade Through the Port of Thunder Bay," *Canadian Geographer,* 27 (Fall 1983), 285–290.

SHORTRIDGE, JAMES R., "The Collapse of Frontier Farming in Alaska." *Annals,* Association of American Geographers, 66 (1976), 583–604.

STRUZIK, ED, "Yellowknife and Whitehorse: Sister Cities North of Sixty," *Canadian Geographic,* 106 (June–July 1986), 24–33.

SYMINGTON, FRASER, *Tuktu: A Question of Survival.* Ottawa: Canadian Wildlife Service, 1965.

TYNER, GERALD E., and JUDITH A. TYNER, "Tourism in Canada's Northwest Territories: Aspects and Trends," *The California Geographer,* 18 (1978), 137–149.

USHER, PETER J., "Unfinished Business on the Frontier," *The Canadian Geographer,* 26 (Fall–Autumn 1982), 187–190.

VANDERHILL, BURKE G., "The Passing of the Pioneer Fringe in Western Canada," *Geographical Review,* 72 (April 1982), 200–217.

WILLIAMS, PETER J., *Pipelines and Permafrost: Physical Geography and Development in the Circumpolar North.* Toronto: Longmans–New Academic Press Canada, 1980.

WINTERHALDER, KEITH, "The Re-greening of Sudbury," *Canadian Geographic,* 103 (June–July 1983), 23–29.

WONDERS, WILLIAM C., ed., *Canada's Changing North.* Toronto: McClelland and Steward Limited, 1971.

______, "The Canadian Northwest: Some Geographical Perspectives," *Canadian Geographical Journal,* 80 (May 1970), 146–165.

______, "Japan's Role in Land Resource Development in the Canadian Northwest," *Journal of Geography,* 75 (1976), 200–208.

______, ed., *The North.* Toronto: University of Toronto Press, 1972.

WRIGHT, ALLEN A., "Yukon Hails Opening of the Dempster Highway," *Canadian Geographic,* 98 (June–July 1979), 16–21.

20
THE ARCTIC

Enormous in size but sparse in population, the Arctic Region spraws across the vastness of the northern edge of North America. In few parts of the earth is nature more niggardly, more unyielding, or more unforgiving, and nowhere else are people's ways of living more closely atuned to the physical environment.

This is primarily a region of the Inuit and the Aleut; yet they occur only in small numbers and in scattered settlements. Although parts of this region have been known to nonnatives for six centuries, only a few "outsiders" have come to the Tundra and only a tiny fraction of that few have been more than visitors or sojourners for a relatively short period of time. Conversely, only a small number of Inuit or Aleuts have departed from the region on anything other than a temporary basis.

Such urban places as those in the midlatitudes are virtually nonexistent. In the more remote areas a settlement nucleus may include only a handful of families. Wherever they are and whatever their size, the settlements of the region are always dominated by the immensity of the environment. The inhabited places are separated by great distances of trackless and treeless land or by equally barren water or ice. Within a settlement, the buildings and artifacts of people take on a peculiarly aggressive significance. There are no trees or shrubs to cover mistakes, provide transitions, or ease the exposed rawness. The sparse, slow-growing, unobtrusive vegetation survives with difficulty, and where it has been ripped away by human endeavor, the nakedness persists for a long time. Settlements are inevitably scars on the fragile landscape.

This, then, is a region in which nature thoroughly dominates humankind. Such items as ice thickness, windchill factor, permafrost depth, caribou migration route, hours of daylight, blizzard frequency, abundance of harp seal, and formation of fast ice are critical to human existence.

Conversely, what people do in this region has remarkably long-lasting effects on the environment. Although the surface of the land is rock-hard through the long winter months, it is extremely susceptible to the impress of human activities during the brief summer period. The structural fragility of the ground-hugging tundra plants and the spongy soil beneath them is such that any type of compression leaves a mark that is not soon erased. The scrape of a bulldozer blade will leave a scar for generations, the track of a wheeled vehicle will be visible for years, and even a single footprint may be obvious for months.

The Arctic Region includes the part of North America that extends from the Bering Sea on the west to the Atlantic Ocean on the east and from the Boreal Forest on the south to the Arctic Sea on the north; it also includes the vast Arctic Archipelago north of the Canadian mainland and the far-flung chain of Aleutian islands that arcs westward from the southwestern corner of mainland Alaska (Fig. 20-1).

Although some basic characteristics of the region are broadly uniform, there are many aspects of heterogeneity. It is possible to generalize validly about the region, but, as in any extensive region, many exceptions and variations must be considered. Particularly notable in this respect is the fact that the eastern and western extremities—coastal Labrador in the east and the Bering seacoast–Aleutian Island area in the west—have a pronounced orientation toward commercial fishing that is quite unlike the situation over most of the region. This is a function of the availability of exploitable marine resources, which is governed particularly by climatic differences.

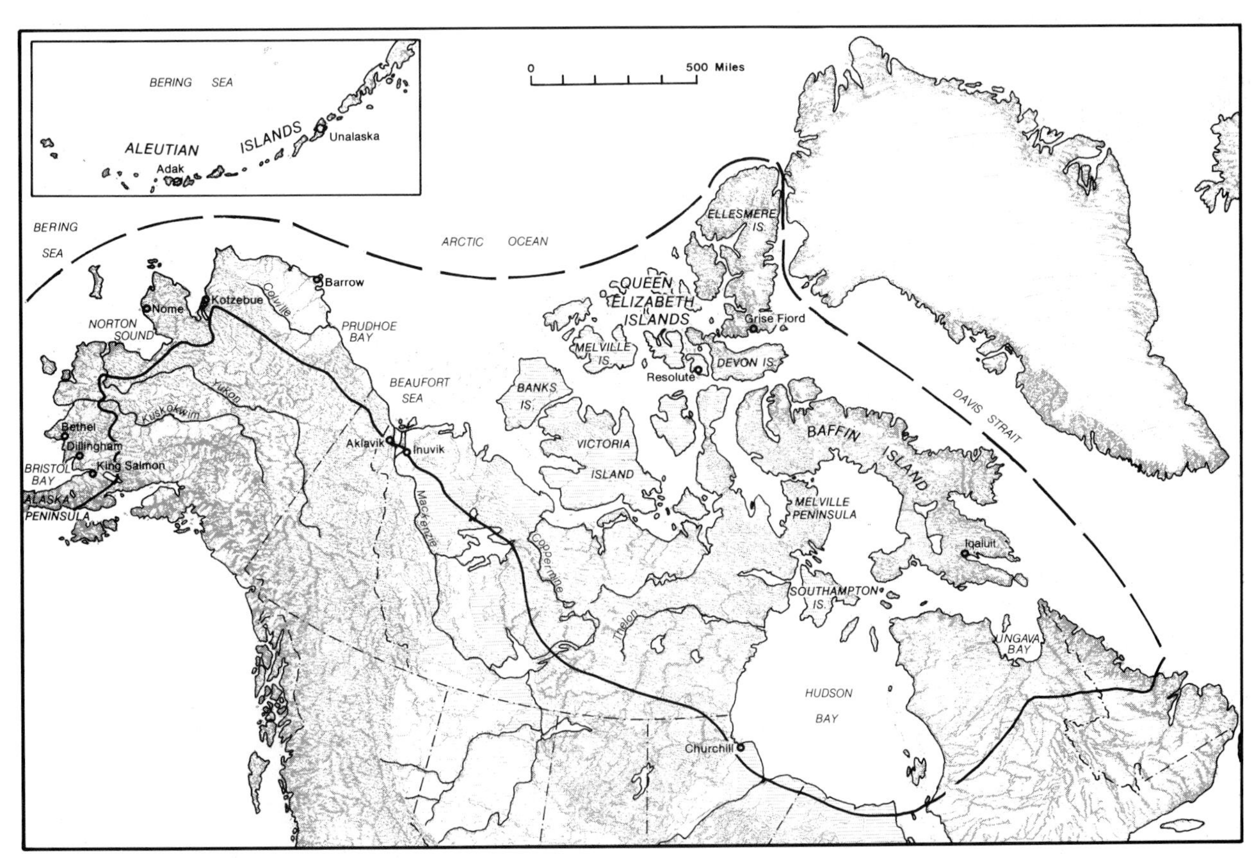

FIGURE 20-1 The Arctic Region (base map copyright A. K. Lobeck; reprinted by permission of Hammond, Inc.).

As a remote and unpopulous region, the Arctic has generally received little attention from most citizens of Canada and the United States. Lately, however, it intrudes into our consciousness more often. Vast energy resources are being discovered; environmental and ecological concerns are being raised; political, social, and economic demands of the native people are being voiced more stridently. Public interest in the Arctic Region is rising, and both Washington and Ottawa have taken note.

THE PHYSICAL SETTING

Climate

The Arctic is not, as novelists would have it, a land of perpetual ice and snow. Winter temperatures are low, but they are higher than in the taiga to the south. Point Barrow has yet to record winter temperatures as low as those characterizing certain stations in North Dakota and Montana. Extremes become greater south of the coast, for the country increases in altitude and is more remote from the ameliorating effects of the ocean. Thus the temperature range at Allakaket, 350 miles (560 km) south of Point Barrow, is much greater; whereas the lowest and highest temperatures at Point Barrow are −56°F (−49°C) and 78°F (26°C), respectively, those at Allakaket are −79°F (−67°C) and 90°F (32°C). But everywhere in the Arctic Region winters are long and summers short.

Although the temperature range at Point Barrow is the more limited, the growing season is only 17 days; at Allakaket, however, it is 54. Snow may be absent for two to four months. The growing season along much of the Arctic coast is less than 40 days.

Air at low temperatures cannot absorb or retain much water vapor; so the precipitation is light and varies over much of the region from 5 to 15 inches (125 to 380 mm). In the far east and far west there are higher totals, but part of the High Arctic is the most arid area of North America; most of Ellesmere Island, for example, is a frigid desert that receives less than 2 inches (50 mm) of moisture annually. The precipitation that does fall in the region is mostly fine dry snow or sleet.

Winds, especially in winter, are very strong and frequently howl day after day. They greatly affect the sensible temperature; thus on a quiet day a temperature as low as −30°F (−34°C) is not at all unpleasant if one is suitably clothed, but on a windy day a temperature of zero (−18°C) may be quite unbearable. The wind sweeps unobstructed across the frozen land and sea and packs the snow into drifts so hard that they often take no footprints, and no snowshoes are required for human locomotion. Winter windchill not only discourages people and animals from moving about but also significantly contributes to the slow growth of plants. But it is not always cold. The long daylight hours of summer combine with continual reflection off water surfaces to produce heat that can occasionally become intolerable.

The coastal areas of Labrador, the Bering Sea, and the Aleutian Islands experience widespread overcast conditions, considerable fogginess, and more storminess than other parts of the region. The relative warmth of the Aleutians contributes to much heavier precipitation there; persistent mist and rain are characteristic. On the average, the Aleutians experience only two clear days per month.[1]

Perhaps the climatic phenomenon of greatest significance to humans is the seasonal fluctuation in length of days and nights. Continual daylight in summer and continual darkness in winter persist for lingering weeks, and even months, in the northerly latitudes.

Terrain

The gross topographic features of the Arctic Region are similar to those of other parts of the continent; only in relatively superficial details does the unique stamp of the arctic environment appear. Most typical are flat and featureless coastal plains, which occupy much of the Canadian central Arctic, as well as

[1] The region as a whole is noted for its inhospitable climate, but the Aleutian Islands area is particularly famous for its wild weather. Survivors of a visit to the Aleutians have been known to report that "the islands have two seasons: fog and storm." Native inhabitants of the Aleutians, however, claim that a more valid statement is "the Aleutians have two seasons: this winter and next winter."

FIGURE 20-2 The snowy labyrinth of the Brooks Range in winter (courtesy Alyeska Pipeline Company).

most of the northern and western coasts of Alaska. Prominent mountains occur in several localities. The massive Brooks Range separates the Yukon and Arctic watersheds in northern Alaska (Fig. 20-2). The eastern fringe of arctic Canada, from Labrador to northern Ellesmere Island, is mountain-girt with numerous peaks over 6000 feet (1800 m), rising to the 8544-foot (2563-m) level of Barbeau Peak in the far north.

Of special interest is the 1000-mile (1600-km) long chain of the Aleutian Islands, extending westward in a broad arc from the tip of the Alaska Peninsula. These treeless, desolate, fog-shrouded islands are essentially a series of volcanoes built on a prominent platform of older rocks. The chain contains about 280 islands, mostly small, but is more notable in that it has more than twice as many active volcanoes (46) as the rest of North America combined.

The coastline adjacent to Baffin Bay and the Labrador Sea is notably embayed and fiorded as a result of glacial modification of the numerous short, deep, preglacial valleys that crossed the highland rim. Along the east coast of the three islands (Ellesmere, Devon, and Baffin) and Labrador are innumerable fiords, some of which penetrate inland for more than 50 miles (80 km). Offshore is a fringe of rounded, rocky islets called skerries.

A number of large rivers flow into the Arctic Ocean and Bering Sea from the continental mainland, but the Arctic islands have no streams of importance. Most notable of the rivers are the Mackenzie and the Yukon, both of which form extensive deltas; that of the former river contains literally thousands of miles of distributary channels and as many as 20,000 small lakes.[2] Other major rivers of the region are the Kuskokwin and Colville in Alaska and the Coppermine and Thelon in the Northwest Territories.

Lakes, large and small, abound in the region, including the Arctic islands. For example, Lake Hazen (included in the new Ellesmere Island National Park Reserve), at latitude 82°, is 45 miles (72 km) long and 900 feet (270 m) deep.

Distinctive Topographic Features of the Arctic

Three types of distinctive landform features in the Arctic Region are limited to this harsh environment.

Icecaps Icecaps and glaciers constitute less than 5 percent of the ground cover of arctic Canada, but many are quite large in size.[3] Baffin Island contains two icecaps that are larger in area than the province of Prince Edward Island, and the islands north of Baffin (Bylot, Devon, Axel Heiberg, and Ellsemere) each contain icecaps that are still more extensive in size (Fig. 20-3). Almost all these ice features have been diminishing in area and thinning in depth during the twentieth century, and the evidence of glacial recession in conspicuous.

[2] J. Ross Mackay, "The Mackenzie Delta," *Canadian Geographical Journal*, 78 (May 1969), 148.

[3] J. Brian Bird, *The Physiography of Arctic Canada* (Baltimore: Johns Hopkins Press, 1967), p. 23.

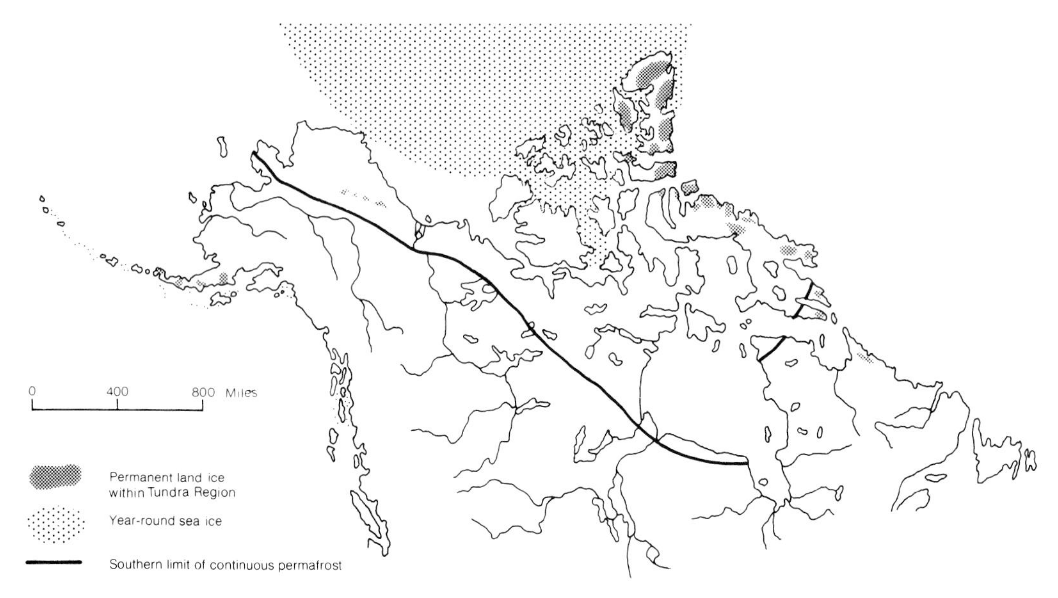

FIGURE 20-3 Ice conditions in the Arctic Region.

Ground newly uncovered by the retreating ice front is raw, light-coloured, unvegetated, in contrast to the ground beyond the trimline which was the position reached by the glacier at its greatest recent extent. Many small glaciers have completely disappeared since the turn of the century, larger ones have become 300 to 500 feet thinner and in some instances their snouts have retreated several miles.[4]

Raised Gravel Beaches Adjacent to many portions of the present coastline are relics of previous sea levels. Characteristically, they appear as gravelly beaches that may extend inland for several hundred feet above the contemporary coastline. These subaerial beaches indicate emergence of the land after the great weight of Pleistocene ice sheets was removed by melting. The postglacial recovery of the land from ice depression varies from 100 (30 m) to as much as 900 vertical feet (270 m) in some places.

Patterned Ground The most unique and eye-catching of tundra terrain is patterned ground, the generic name applied to various geometric patterns that repeatedly appear over large areas in the Arctic. The patterns, consisting of circles, ovals, polygons, and stripes, are of apparently varied but still unknown origin, although it generally accepted that frost action is instrumental in their formation.

> A very distinctive type of patterned ground is the tundra polygon . . . [which resemble] enormous mud cracks, such as those of a dried-up muddy pool, but with diameters of from 50 to 100 feet. The tundra polygons may be nearly as regularly shaped as the squares on the checkerboard, but most are irregular, somewhat like the markings on turtle shells. The boundary between two adjacent polygons is a ditch. Beneath the ditch there is an ice wedge of whitish bubbly ice which tapers downwards, like the blade of an axe driven into the ground. Some ice wedges are more than ten feet wide at the top and are tens of feet deep. . . . On a smaller scale, the ground observer may see stones arranged in circles or garlands a few feet across, like stone necklaces; or the ground may have stripes trending downhill . . . . Of particular interest to people in the western Arctic are the conical ice-cored hills called pingos, an Eskimo word for hill [Fig. 20-4]. The pingos are most numerous

[4] J. D. Ives, "Glaciers," *Canadian Geographical Journal*, 74 (April 1967), 115.

FIGURE 20-4 A pair of pingos near the northern edge of the Mackenzie delta. These ice-cored hills are conspicuous in an otherwise featureless landscape (courtesy Canadian Government Travel Bureau).

> near the Mackenzie Delta, where there are nearly 1,500 of them. The pingos may reach a height of 150 feet and so are prominent features in the landscape. They are found typically in shallow or drained lakes and are believed to have grown as the result of the penetration of permafrost into a thawed lake basin. Each pingo has an ice core of clear ice. If the ice core should melt, a depression with a doughnut-shaped ring enclosing a lake is left behind.[5]

The principal significance of patterned ground is that it demonstrates the mobility of tundra terrain, emphasizing the role of soil ice in producing geomorphic processes that are largely unknown farther south.[6]

[5] J. Ross Mackay, "Arctic Landforms," in *The Unbelievable Land,* ed. I. Norman Smith (Ottawa: Department of Indian Affairs and Northern Development, n.d.), p. 62.

[6] J. Brian Bird, *The Natural Landscapes of Canada: A Study in Regional Earth Science* (Toronto: Wiley Publishers of Canada, Ltd., 1972), p. 160.

Permafrost

Most, but not all, of the Arctic Region is underlain with permafrost. In the Aleutian Islands and part of the Alaska Peninsula it is unknown. In most of the region, however, permafrost is both continuous and thick. It has been measured to a depth of 1600 feet (480 m) in some places.

Natural Vegetation

Most of the Arctic as considered here refers to that part of North America lying north of the tree line, the great coniferous forest belt. The line on the map separating the tundra[7] from the forest symbolizes a zone within which the trees gradually become smaller and more scattered until they disappear altogether. It coincides rather closely with the 50° (30°C) isotherm for the warmest month. This zone, in most instances, lies south of the Arctic Circle, even

[7] *Tundra* is a Finnish word meaning "barren land."

reaching as far south as the 55th parallel on the west side of James Bay.

From area to area, however, the boundaries between taiga and tundra differ, and the extent to which the taiga penetrates the tundra seems to depend on a combination of low temperatures, wind velocity, and availability of soil moisture. The forest boundary extends farthest north in the valley of the Mackenzie River, where a forest of white spruce reaches into the southern part of the delta, at about the 68th parallel. There are also significant, although not large, forest outliers in the valleys of the Thelon and lower Coppermine rivers.

The characteristic vegetation association of this region is tundra. The tundra consists of a great variety of low-growing and inconspicuous plants that belong to seven principal groups:

1. Lichens, which grow either on rocks or in mats on the ground where they form the principal food supply of the migrating caribou herds
2. Mosses
3. Grasses and grasslike herbs
4. Cushion plants
5. Low shrubs
6. Dwarf trees
7. Flowering annuals

This is one of the world's harshest floristic environments, and the growing season is so short that there is simply not time during the brief summer for the life processes of annual plants to be completed. Instead, the plant cover consists of hardy perennials, which can remain dormant for as long as 10 months and then spring to life for an accelerated annual cycle during the abbreviated summer period. This means that arctic plants have a very slow weight increase. Studies of arctic willow (*Salix artica*) on Cornwallis Island, for example, show an annual increment of only one-third of total plant weight; in the midlatitudes such an increase can take place in a week.[8]

[8] Patrick O. Baird, *The Polar World* (New York: John Wiley & Sons, 1964), p. 112.

In the drier parts of the region (and some areas have an annual precipitation of less than 2 inches [50 mm]) conditions for plant growth are ever more trying. Vegetation "is restricted largely to poorly drained areas, such as small depressions, or to areas where remnant snow patches provide local seepage during much of the melt season."[9]

However limited the tundra vegetation may be, it is of critical importance to animal life. It nourishes the entire terrestrial food chain, from mosquitoes and muskoxen on to the carnivores.

Native Animal Life

The regional ecosystems are characterized by simplicity and sensitivity. Only a few animal species inhabit the lands and waters of the region. Although they must be hardy to survive the rigorous environment, at the same time they are vulnerable, largely because of a limited food supply, a slow growth rate that delays maturing and restricts reproductive potential, and wide oscillations in population abundance that frequently results in local extinction of species.

In no other region are the fauna so important to people. There has long been an intimate association between the abundance or scarcity of animal life and the welfare of the natives of the region. This close relationship is diminishing but is still pronounced.

Aquatic mammals have long been the mainstay of Inuit livelihood. Several varieties of *seals* range throughout the region and they are the most common quarry of Inuit hunters. The ringed seal is the most successful and widespread of the marine mammals because it can use breathing holes to live under fast ice (solid surface ice) all winter.

The *walrus* is a ponderous, slow-breeding creature (the female does not reproduce until the sixth year and then has only one pup every other year) that is verging on extermination over much of its range. It must live in an area of strong currents that keep the sea ice moving all winter, for unlike the seal it does not gnaw a breathing hole through the ice. The Atlantic walrus inhabits most of the eastern Arctic Ocean, with particular concentration around

[9] John England, "Ellesmere Island Needs Special Attention," *Canadian Geographic*, 103 (June–July 1983), 14.

the Melville Peninsula and Southampton Island. The Pacific walrus inhabits the Bering Sea; there is a 1000-mile (1600-km) gap between the ranges of the two.

Whales of various species are much sought by the Inuit, but by far the most common is the white whale, or *beluga*. They occur in considerable numbers throughout the Arctic, chiefly in saltwater but often going up the larger rivers (they have been seen as far up the St. Lawrence as Quebec City). They gather in remarkable abundance upon occasion; for example, up to 5000 cluster off the Mackenzie delta in summer. The *narwhal* is much less common but is highly prized because of the ivorylike horn of the male. Varieties of the large whales, such as the *bowhead,* are limited and sporadic in distribution but are much prized by Inuit hunters. With limited exceptions, the larger whales are now fully protected.

The *polar bear* ranges widely in the Arctic, roaming primarily on sea ice except for land denning to give birth. They are sometimes found several tens of miles from the nearest land. There is an unusual denning concentration along the Hudson Bay lowland between James Bay and Churchill, where the Ontario government has established a large provincial park for their protection. Unlike other bears, which are omnivorous, polar bears are almost wholly carnivorous.

The outstanding land animal is the *barren-ground caribou,* which numbered in the millions a few decades ago (Fig. 20–5). There are several large herds, each of which makes an annual migration from taiga in winter to tundra in summer. Thousands of Indians and Inuit have depended—some still do—significantly on caribou meat as a dietary staple. There has been a general and, in some cases, precipitate downtrend in caribou numbers over the last half century, particularly in the Western Arctic herd of northern Alaska and the Keewatin herd of the Barren Grounds west of Hudson Bay. In the 1980s, however, the Keewatin herd experienced a rapid increase in numbers, as did the George River herd in northeastern Quebec. In the early 1990s the George River herd was the largest in the world. The reasons for these significant population fluctuations are not well understood.

The Arctic islands have a scattering of barren-ground caribou, particularly on Baffin Island. Several northern islands also contain a limited population of the smaller *Peary caribou*.

FIGURE 20-5 Caribou grazing on Alaska's North Slope with an oil drilling rig operating in the background (courtesy Alyeska Pipeline Company).

The only other native ungulate of the Arctic Region is the *muskox*.[10] In general, they are protected from hunters, and their numbers have been rebuilding from low points reached in the 1930s and 1940s. Prominent concentrations have long been found in the Thelon Valley of the Keewatin district and in the Lake Hazen area of northern Ellesmere Island. In recent years there has been considerable increase in numbers on several Arctic islands, particularly Banks Island. Muskox were exterminated in Alaska long ago but were reintroduced in the 1930s to Nunivak Island in the Bering Sea, where they have flourished enough to provide seed stock for reintroduction to the Alaskan mainland.

Furbearers of note include the *arctic fox*, a prolific breeder whose population seems to run in cycles; the *lemming*, a queer nocturnal burrowing rodent noted for its seemingly pointless migrations; the *arctic wolf*, principal large predator of the region; and the *arctic hare*, whose population pattern follows wildly fluctuating cycles.

Fish life is not abundant, nor is it particularly varied.

Of the birds, the snow owl, ptarmigan, gyrfalcon, raven, and snow bunting are year-round inhabitants. Others, summer residents, arrive by the millions to breed in the seclusion and security of the tundra. They are also attracted by the prolific insect life.

Because of the abundance of poorly drained land, insects find this region a paradise during the short summer season. More than 1000 insect species occur north of the tree line, and about half of them belong to the order *Diptera* (two-winged flies), which includes mosquitoes, blackflies, and midges. According to one observer, "On a warm, cloudy, windless day, the insect life on the Barrens defies description."[11] Happily such days are rare; most days are windy and cause the insect hordes to lie low.

[10] Moose and Dall sheep sometimes venture into the southern edge of this region, particularly on the north side of the Brooks Range.

[11] Eric W. Morse, "Summer Travel in the Canadian Barren Lands," *Canadian Geographical Journal*, 74 (1967), 162.

THE PEOPLE

This vast region has fewer than 80,000 inhabitants, two-thirds of whom are Inuit. Perhaps 6 percent is Aleut and most of the rest are Caucasians. These people reside chiefly in widely scattered small settlements, of which there are about 50 in Canada and 130 in the Alaskan portion of the region. No settlement has a population of more than 4000, and less than half a dozen exceed 2500.

The Inuit

The Inuit are one of those rare races readily identifiable by all three basic anthropological criteria: physical characteristics, culture, and language. They have well-marked physical homogeneity, a distinctive culture, and a language spoken by themselves and no one else.

The term "Eskimo" (or a translation thereof) has traditionally been used in all major languages of the world to designate the tundra dwellers of the North American Arctic and Greenland. The name, originally applied by Algonkian Indians to their northern neighbors, means "eaters of raw meat." In recent years the "Eskimo" people have increasingly sought a new generic term for self-designation. The choice was not easily reached, for there were different ethnic designations in common use in different areas. Thus in Greenland the "Eskimos" refer to themselves as Kalaallit; in the eastern Canadian Arctic, as Inuit; in the western Canadian Arctic, as Inuvialuit; in northern Alaska, as Inupiat; and in southwestern Alaska, as Yupik. These names all mean approximately the same thing—"the people." In the late 1970s a pan-Eskimo conference was held in Point Barrow for the purpose of choosing one official name for the "Eskimo" race and its culture. The choice was "Inuit," which has now been generally accepted by all native tundra dwellers of Alaska, Canada, and Greenland as a replacement for "Eskimo." It should be noted that "Inuit" is the plural form; an individual is referred to as "Inuk" whereas the "Eskimo" language is known as "Inuktitut."

In their aboriginal condition the Inuit have shown remarkable ingenuity in adapting themselves to an almost impossible environment. They live in one of the coldest and darkest parts of the world and in one that is among the poorest in available fuel; yet they have not only survived but have also enjoyed life in self-sufficient family groups. Originally their entire livelihood depended on fishing and hunting whereas practically all Eurasian tundra people were herders.

Inuit culture history revolves around successive waves of people and cultural innovations spreading eastward from the Bering Sea area. The earliest Inuit, or "proto-Inuit," of North America presumably derived their culture in Siberia and originally migrated from there. This culture stage—called the Arctic Small Tool Tradition or simple Pre-Dorset—had an indefinite tenure in Alaska but lasted in Canada until about 800 B.C. when it was replaced by the eastward-spreading Dorset Culture. After about 20 centuries the Dorset Culture was, in turn, replaced by the Thule Culture, which also spread eastward from Alaska.

Thule Culture had two principal attributes lacking in the Dorset era: domesticated dogs to aid in hunting and in pulling sleds and a full range of gear for hunting the great baleen whales, a major food source not available to Dorset people.[12] Thule Culture evolved into the contemporary Inuit culture in about the eighteenth century, in a transition stage marked by the decline of whale hunting and the almost complete depopulation of the northern Canadian islands.

The coming of whites to Inuit country marked the beginning of the end for their way of life. European diseases, midlatitude foods, liquor, rifles, motorboats, and a new set of mores were introduced; the overall result was more often bad than good.

The Inuit today has drifted toward "civilization." Relatively few exist by subsistence hunting and fishing. More and more take temporary jobs on construction projects, in salmon canneries, and at other white outposts in the Arctic. The trading post, the DEW (distant early warning) line station, the tuberculosis sanitarium, and the welfare check are now well established in their way of life.

[12] William E. Taylor, Jr., "The Fragments of Eskimo Prehistory," *The Beaver*, Outfit 295 (Spring 1965), 14.

The Aleuts

The origin of the Aleuts is unclear, although generally they are considered to derive from Inuit, or proto-Inuit, who settled in a maritime environment in southwestern Alaska and developed a livelihood based almost entirely on fishing and sea hunting. Aleuts occupied the Aleutian Islands and the Alaska Peninsula in considerable numbers (perhaps 25,000) at the time of Russian contact in the eighteenth century. They were killed and enslaved by the Russian fur seekers and in relatively few years their number was reduced by 90 percent.

Today there are some 6000 Aleuts in southwestern Alaska, most of mixed blood. They are primarily commercial fishermen, sealers, and workers in salmon fishing and canning in the Bristol Bay area. In addition, they provide most of the labor for the fur-sealing industry of the Pribilof Islands, where the village of Saint Paul is largest single Aleut settlement in existence.

The Indians

There are fewer than 1000 Indians in the entire Arctic Region and almost all are in Canada, where they are called *Dene*. Generally speaking, the tree line has served as the northern boundary of Dene occupance, just as it has served as the southern border of Inuit settlement. Dene and Inuit live adjacently in any numbers in only four localities: Aklavik and Inuvik in the Mackenzie delta, Churchill, and Poste-de-la-Baleine on the eastern shore of Hudson Bay. The Dene livelihood in this region is based on hunting, trapping, fishing, and temporary construction work.

The Whites

Whites living in the region now number some 30,000, a total that is growing. Well over half are in Alaska, where Nome is the largest white settlement. Most whites who live in the region today are in control of the defense and weather installations or are government officials, oil company employees, prospectors, fur traders, fishermen, or missionaries.

THE DIMINISHING SUBSISTENCE ECONOMY OF THE REGION

Throughout history most of the population of the Arctic Region was involved in essentially subsistence activities: fishing, hunting, and trapping. Since World War II there has been a decline in these pursuits and their replacement by a money economy. Later this trend was markedly accelerated all across the region, from the Aleutian Islands to the High Arctic. Despite the downturn in subsistence activities, however, they still contribute significantly to the well-being of the people in providing "country food" (the name given to subsistence food obtained from the land or water), material for clothing and implements, and furs for sale or trade.

A CLOSER LOOK Linguistic Development Among the Canadian Inuit

When a racial or ethnic group begins to emerge from the obscurity of a colonial existence into the complexities of modern civilization, many cultural-social changes must be handled in accelerated fashion. Initially the people are likely to be relatively passive recipients of changes foisted on them by external influences. Before long, however, elements of the affected populace marshall their resources and begin to act rather than react.

The Canadian Inuit are now in this stage of development. They are coming to grips with a variety of cultural and economic factors that will be influential on their future. One of the most prominent involves language standardization and retention.

Although there is a single common language (Inuktitut) for the Inuit of Canada, it has not been standardized in either spoken or written form. The development of a written language was a slow and tedious process, carried out by government agents, missionaries, and traders over a long period of time in various ways, with differing results in different parts of Arctic Canada.

Since the 1950s the Canadian government has pursued a small but continuing effort to clarify and reform the various orthographies of Inuktitut. An Inuit Language Commission was formed in 1974 and soon a commitment was made to establish a standardized dual orthography, one using the Roman alphabet and one using syllabics, that would be compatible and interchangeable. Representatives of all major Inuit groups accepted the principle and ratified the initial dual orthography that was developed (Fig. 20-a).

In practice, however, there were many objections. There are six broad areas of dialect variation and the reaction differed in each (Fig. 20-b). In Labrador, for example, syllabics are rarely used and the people generally prefer the long-established Moravian

ENGLISH	INUKTITUT (Syllabics)	INUKTITUT (Roman Orthography)
Ever since the building of the great tower of Babel, humanity has been trying to bridge communication gaps among the many languages of the world. It is therefore satisfying to learn about the inventiveness of certain individuals, such as those early Europeans who laboured among our people to come up with a workable written language. Syllabics in particular has become a unique writing system for many Canadian Inuit.	ᑕᐃᒪᖓᓂᑦ ᐸᐃᐸᓕᒥ ᐃᓄᒃᓱᒡᔪᐊᓕᐅᓚᐅᕐᒪᑕᒥᑦ ᐱᒋᐊᕐᓂᖃᖅᓄᓂ. ᓯᓚᕐᔫᑉ ᐃᓄᖏᑦ ᐃᑳᕈᓯᐅᕋᓱᒃᓯᒪᓕᖅᐳᑦ ᑐᑭᓯᐊᑐᕋᐅᑎᔪᒪᓂᕐᒧᑦ ᐅᖃᐅᓯᖃᖃᑎᒌᖏᑦᑐᓂ ᐊᒥᓱᓂ ᓯᓚᕐᔪᐊᕐᒥ. ᑭᓯᐊᓂᓕ ᖁᐱᐊᓇᖅᑰᕗᖅ ᖃᐅᔨᒪᑉᓗᓂ ᐃᓄᐃᑦ ᐃᓚᖏᑦ ᐃᖅᑲᖅᓴᒃᑲᐅᓐᓂᕐᒪᑕ ᑕᐃᑉᑯᐊ ᑕᕆᐅᕐᔫᑉ ᐊᑭᐊᓂᑦ ᐃᐅᕈᑉᒥᐅᑦ ᑎᑭᓚᐅᖅᓯᒪᔪᑦ ᐊᒃᓱᕈᖅᓄᑎᒃ ᐱᓕᕆᕙᓚᐅᕐᓂᕐᒪᑕ ᐃᓄᖁᑎᑉᑎᖕᓂ ᐊᑐᖅᑕᐅᔪᓐᓇᖅᑐᓂᒃ ᑎᑎᕋᐅᓯᓕᐅᓚᐅᕐᓂᕐᒪᑕ. ᓯᓛᐱᒃᔅᑎᑐᑦ ᑎᑎᕋᐅᓯᖅ, ᐱᓗᐊᖅᓄᒍ, ᑎᑎᕋᐅᓯᑦᓯᐊᒻᒪᕆᐅᕗᖅ, ᐊᑐᖅᑕᐅᓕᖅᑐᖅ ᐃᓄᖕᓂᑦ ᐊᒥᓱᓂᑦ ᑲᓇᑕᒥᐅᑕᐅᔪᓂᑦ.	Taimanganit Babyloniami inuksugjualiulaurmatamit pigiarniqaq&uni, silarjuup inungit ikaarusiurasuksimaliqput tukisiaturautijumanirmut uqausiqaqatigiingittuni amisuni silarjuarmi. Kisianili quvianaqtuuvuq qaujimapluni inuit ilangit iqqaqsakkaunnirmata taipkua tariurjuup akianit iurupmiut tikilauqsimajut aksuruq&utik pilirivalaurnirmata inuqutiptingni atuqtaujungnaqtunik titirausiliulaurnirmata. Silaapikstitut titirausiq, piluaq&ugu, titirausitsiammariuvuq, atuqtauliqtuq inungnit amisunit kanatamiutaujunit.

FIGURE 20-a An illustration of the three forms of writing used in Canadian Inuit country (from special supplement to *Inuktitut* magazine, September 1983, p. 2).

Hunting

Hunting for sea and land animals is widespread (Fig. 20-6). In some areas where winter conditions persist almost year-round, such as in the Queen Elizabeth Islands, ice hunting for sea mammals is a specialty. In other areas, such as the Colville River country of Alaska, inland groups specialize in caribou hunting. But most native settlements are on the coast and there are well-marked seasonal changes in hunting patterns. A typical "quarry rotation" might be seals in early fall, walrus in late fall, bear and fox in winter, seals in spring, fish and beluga in early summer, and caribou in later summer. In hunting areas, late winter and spring are the most important seasons for resource use. Midwinter is too dark and cold; ice

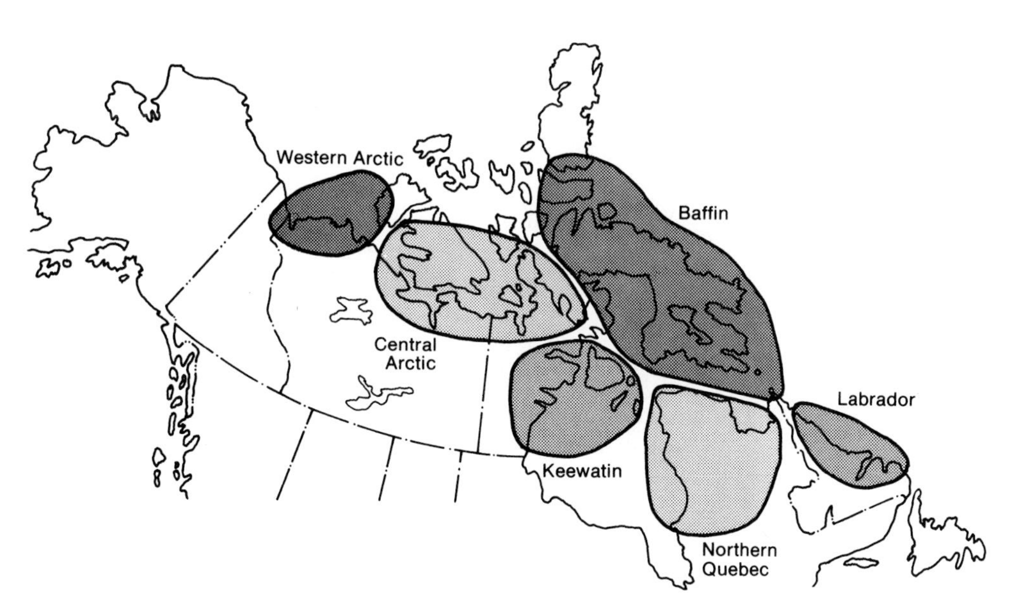

FIGURE 20-b Major Inuktitut dialect areas in Canada.

orthography (introduced by missionaries) to the standardized Roman orthography. The only system used in northern Quebec is syllabics, and the people generally are opposed to any change. In the Western Arctic an unenthusiastic response and considerable questioning of the entire project resulted.[a] The most positive response came from the Keewatin and Baffin areas.

Despite the objections, most Inuit leaders seem in favor of standardization. Clearly one important aspect of Inuit culture is a language that may be dialectally varied and orthographically inconsistent but is common to all. Efforts toward acceptable standardization continue; orthographic reform is a dynamic endeavor.

Meanwhile, there are strong efforts to retain commitment to native language retention. In the Northwest Territories, kindergarten and junior grades are now taught in the students' mother tongue throughout. There are native-language radio and television broadcasts, simultaneous translation (English/Inuktitut) during all Northwest Territories Legislative Assembly sessions, and many native-language articles in northern newspapers and periodicals.

TLM

[a] A representative comment by an inhabitant of the Mackenzie delta noted, "Here in the delta the only two languages you need are English and broken English." Quoted in Anonymous, "Inuktitut Writing Systems: The Current Situation," *Inuktitut,* 53 (September 1983), 64.

and fog handicap long trips by small boat in summer; and autumn is stormy and newly forming ice is hazardous.

Some animals, such as caribou, birds, and hares, are hunted on land. Others, such as walrus and narwhal, are taken from boats. In many cases, the most important hunting is done on the sea ice, largely for seals and polar bear.

Modern technology has made it much easier for the Inuit to hunt. Repeating rifles, improved ammunition, snow machines, all-terrain-vehicles, and motorboats have provided greater mobility and killing power. The long-term result has been a decline in the number of potential quarry, particularly caribou. Concepts of conservation and sustained yield are only slowly being understood.

The Inland Inuit of the Barren Grounds are now but a footnote to history. Apart from coastal settlements, the vast expanse of tundra in the Keewatin district is now totally unpopulated except for a small settlement at Baker Lake, largely because of the caribou decline.

Fishing

Subsistence fishing is carried on wherever possible. Stone fish traps are constructed along some streams, but most fishing is done in the ocean. Summer fishing is easier, but ice fishing also takes place throughout the winter. Subsistence fishing is less important now than in the past because sled dogs, which are major consumers of fish, are much more scarce.

Trapping

Besides supplying meat and clothing, trapping has long been the major source of money income for natives of the Arctic Region. Various animals are trapped. Most often caught is the muskrat (about 100,000 are trapped annually in the Mackenzie delta alone), but the most important is the white arctic fox. The most reliable fox-trapping areas in the Arctic are in the southwesternmost portion of the Arctic Archipelago, particularly on Banks Island but also on Victoria Island. Other major trapping areas include northern Labrador, the Grise Fiord area of southern Ellesmere Island, and the deltas of the Yukon and Kuskokwim rivers.

Trappers are licensed and carefully regulated in both countries. Only natives are allowed to trap in the Canadian Arctic and only a very few white trappers are permitted in Alaska.

Trapping in the Arctic Region has been on a general decline for several decades, but in the 1980s it decreased dramatically. Humane, animal rights, and antitrapping advocates mounted an international campaign that destroyed virtually the entire seal-fur market and had a lesser but notable effect on trapping of other furbearers.

FIGURE 20-6 A just-killed seal is about to be skinned (courtesy Information Canada Photothèque).

The ability of Inuit and Dene to rely on trapping as a source of income has been severely compromised, but trapping, hunting, and fishing continue as important subsidiary activities for a considerable proportion of the native population.

THE RISE OF A MONEY ECONOMY AND AGGLOMERATED SETTLEMENTS

The most important trend in the contemporary geography of the Arctic Region is the increasing concentration of the population in fewer and larger settlement centers where life is based on a money economy. This pattern is being followed throughout the region. Wage labor is now the preferred means of livelihood, and more and more families are abandoning their seminomadic hunting camps and settling down to a sedentary existence in a settlement node where the younger generation grows up relatively ignorant of the techinques requisite to a subsistence economy.

As a result, large areas of the tundra, such as the Keewatin Barrens, the Colville delta, and much of the northern Labrador coast, are now almost unpopulated. Dozens, and sometimes hundreds, of Inuit have moved into settlements that previously had very small populations. The possibility of a steady job and the lure of stores, medical facilities, and perhaps housing are proving irresistible (Fig. 20-7).

Unfortunately, the rate of natural increase in the population is very high—generally the highest on the continent—and the supply of jobs is inadequate. Construction work has been the principal provider of wage labor in the past. During and after World War II there was a great deal of construction of defense and meteorological installations: radar stations, communications bases, and airfields. Such construction activity inevitably decelerated, but it gave the people of the region a taste of sedentary regularity that they cannot forget.

Population clusters have clearly been a mixed blessing. They made it possible for the Inuit to enjoy generally improving housing, a well-developed school system, access to medical facilities and social services, rudimentary community government, opportunities for the growth of local cooperatives, introduction to "outside" amenities (ranging from A&W Root Beer stands to color television brought by satellite transmission to communities that had never known newspapers or magazines), and an awakening cultural awareness.

Negative effects, however, are also notable. Agglomerated living with limited employment opportunities and the incursion of outside temptations has caused all indexes of social pathology to skyrocket. Intrafamily violence, divorce, mental illness, suicide, and drug abuse are increasing rapidly, and alcoholism has become a ubiquitous menace. Several dozen communities across the region have enacted prohibition regulations, but even most of these settlements have major problems with alcoholism. And despite access to improved medical facilities, health problems have increased dramatically. Many settlements are hygienically as well as socially grim. Providing water and removing waste are problems almost everywhere. Dependence on packaged foods has dramatically increased the incidence of tooth decay and malnutrition.

The agglomeration of population and the consequent dependence on modern technology result in a much higher cost of living. Purchased food is expensive and much of it must be air-freighted in, which adds markedly to the cost. But the greatest expense is for power and fuel. Most Arctic settlements are overwhelmingly dependent on oil—a necessity for fueling home stoves, heating dwellings, powering village electric generators, and propelling the ubiquitous snowmobiles and motorized boats.[13] In many areas more than half the family income goes for electricity and fuel costs.

Associated with agglomeration has been the growth of cooperatives. Prior to 1960 all Inuit art in Canada was marketed through Hudson's Bay Company or the Department of Indian Affairs and Northern Development. Since then cooperatives have proliferated so that they are now the largest single employer of Inuit labor in the Canadian Arctic. In addition to the production and marketing of arts, crafts, and sculpture, they are often involved in pro-

[13] One of the unsavory ironies of life in Alaskan villages is that the oil they use makes a 5000-mile (8000 km) round trip before it is received. Pumped from the North Slope fields on the shore of the Arctic Ocean, the crude oil is taken to California for refining and then the heating oil is returned to Alaska for sale.

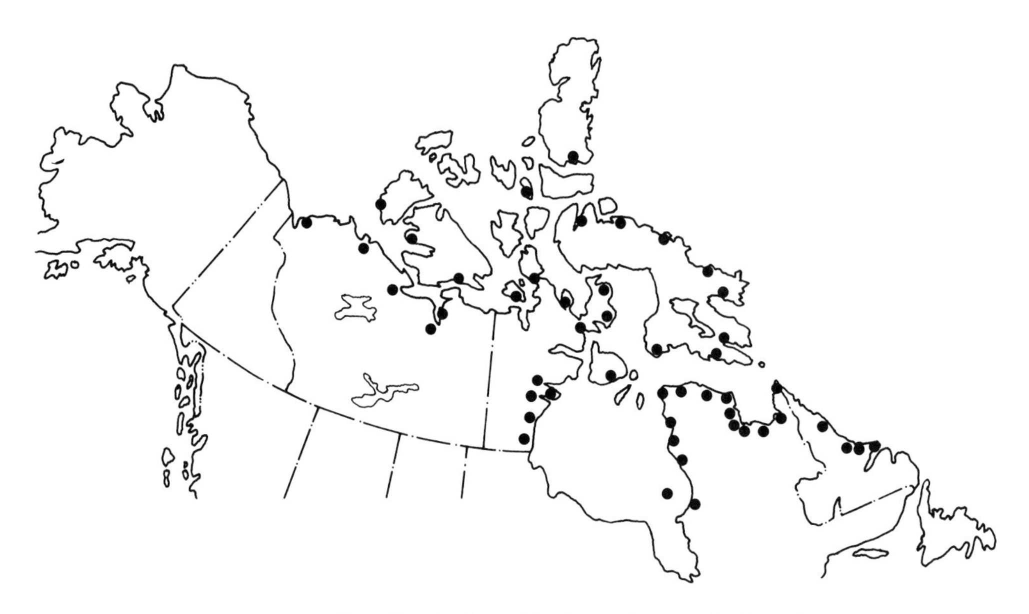

FIGURE 20-7 The distribution of Inuit settlements in Canada.

viding municipal services, construction, commercial fishing, retailing, and even mining.

Cooperatives continue to be vital economic and social components for almost all Canadian Inuit communities, but during the 1980s their business, and thus their vitality, went into a serious slump that is cause for great concern in the region. Cooperatives are much less common in the Alaskan tundra, primarily because each village has a corporation of its own under the Alaska Native Claims Settlement Act.

One of the most important factors in the change from a subsistence to a commercial economy in the Arctic has been the adoption of personal motorized travel vehicles (Fig. 20-8). Generally called "snowmobiles" in Canada and "snow machines" in

FIGURE 20-8 Inuit riding snowmachines on Little Diomede Island in the Bering Strait (courtesy State of Alaska, Division of Tourism).

Alaska, these small, versatile, noisy vehicles have displaced dog sleds and sled dogs throughout most of the region. A snowmobile can cover 100 to 150 miles (160 to 240 km) in a day in contrast to 20 or 30 miles (32 to 48 km) by a dog sled. As the population increases, nearby game becomes scarcer and so the distance needed to travel for even the basic necessities of hunting or trapping is much greater. Moreover, the need to find food for hungry dogs all year is obviated, although that of finding cash to buy fuel becomes acute. In some areas snow machines are being replaced by all-terrain-vehicles, which provide the flexibility of year-round transport. And in a few localities dog sleds are making a comeback, based on the high cost of fuel and the fact that if one becomes stranded, "one can't eat a snowmobile."

NODES OF SETTLEMENT

Because the people of the Arctic Region increasingly tend to cluster, a nodal pattern of population distribution is becoming apparent. Most settlements lack many attributes of towns in other parts of the continent: Their form is often sprawling, amorphous, and unregimented to a street pattern; their urban functions are extremely limited; their buildings are often raised above ground level to keep the permafrost from thawing and buckling the foundations; and provision of utilities is generally primitive or totally absent (sometimes the piped conduits for water, sewerage, and gas are in "utilidor" or "servidor" systems raised above the ground in heavily insulated tunnels).

Despite the agglomerating tendencies, there are only a few settlements of any size (Table 20-1). The larger settlements are listed below.

TABLE 20-1
Largest urban places of the Arctic Region

Name	*Population of Principal City*
Adak, Alaska	3575
Barrow, Alaska	4225
Churchill, Man.	1205
Iqaluit, N.W.T.	3150
Inuvik, N.W.T.	3470
Kotzebue, Alaska	2595
Nome, Alaska	2750

1. *Adak and Unalaska* are the largest communities in the Aleutian Islands. The former is primarily a naval base, but dependents are permitted to live there; so it has a relatively normal urban form and function. Unalaska, which has incorporated the across-bay settlement of Dutch Harbor, is now a major commercial fishing port.
2. *Bethel and Kotzebue* are Bering Sea towns of primarily Inuit population but some diversity of function. Bethel is a commercial fishing center and Kotzebue has become an attraction to fly-in tourists who want to see a "real" Inuit village.
3. Despite its relative isolation, *Nome* has managed to function as a subregional commercial center (Fig. 20-9). Only about one-third of its population is Inuit.
4. *Barrow* is the administrative center of the North Slope Borough, an oil-rich local government unit that has built new schools, sewers, and water lines, but that has run up an incredible $1 billion debt. It is an oil operations center and probably has the largest Inuit population of any settlement in the region.
5. Twin towns of the Mackenzie delta are *Aklavik* (the old settlement) and *Inuvik* (the newer planned town and administrative center for the Western Arctic). Both are situated on the edge of the boreal forest but have strong functional relationships with the Arctic Region.
6. *Iqaluit* (previously known as *Frobisher Bay*), on Baffin Island, is the administrative center for Canada's Eastern Arctic and is distinguished by an impressive town center, a large federal building, a high-rise apartment block, a relatively busy commercial airfield, and a weekly newspaper. It contains the largest Inuit community in Canada.

FIGURE 20-9 The historic town of Nome, on the Bering Sea (courtesy Alaska Travel Division).

7. *Churchill* is a busy grain-shipping port during its three-month navigation season, for its ocean route for transporting Prairie wheat to Europe is 1000 miles (1600 km) shorter than the route via eastern Canada. Churchill experiences a flurry of tourist visitation in the fall owing to the attraction of polar bears migrating right through the town.

ECONOMIC SPECIALIZATION

The commercial economy of the Arctic Region involves only a few specialized activities on a generally limited scale.

Commercial Fishing

There are two areas where commercial fishing is of considerable significance. In both places it is primarily a seasonal activity, and many participants are outsiders who come into the region for only a few weeks during the fishing season. The Bristol Bay area of southwestern Alaska is one of the state's principal salmon fisheries. Canneries operate at a feverish pitch during late summer and fall at King Salmon, Dillingham, and other localities. There is also a prominent Bering Sea fishery for halibut and crabs.

Along the north coast of Labrador there has long been an important commercial fishing venture, primarily for cod but also supplying char and salmon for sale. Many Newfoundlanders come to this area as "stationers" or "floaters" in the summer to join the Inuit and "liveyeres" (white settlers of Labrador) who fish these coastal waters.

Small and sporadic commercial fisheries are also in operation at other places in the region, most notably for arctic char, a freshwater species.

Commercial Reindeer Herding

Reindeer were introduced into Alaska from Siberia in the 1890s, with the idea of establishing a viable pastoral enterprise among the Inuit. The Canadian government subsequently brought some Alaskan reindeer to the Mackenzie delta in the late 1920s to serve as a nucleus for eventual dispersal.

In neither case, however, has the reindeer experiment been very successful. Although the Alaskan herds increased to a total of more than 600,000 reindeer in the 1930s, problems developed and the numbers diminished rapidly. There are now about 30,000 reindeer in Alaska in 17 separate herds, mostly on the Seward Peninsula. Five of the herds are noncommercial, with slaughtering for home use only, and not one of the others has yet become profitable.

The Mackenzie delta herd reached a peak of about 9000 animals in 1942, but interest waned and the herd has declined to under 3000 today (Fig. 20-10).

Mineral Industries

Considerable mineral wealth is known to exist in this region, but transportation is so expensive that mining is virtually nonexistent at present (Fig. 20-11). Exploration for oil and gas has been extensive and is continuing. The only commercially feasible discovery thus far has been the tremendous reserves of Alaska's North Slope, centering on Prudhoe Bay. The North Slope now provides about 20 percent of all U.S. oil production, and Alaska ranks second only to Texas among producing states. Only a relatively few wells are involved (about 1000), but the investment is enormous. Anticipated expenses to develop the North Slope, including construction of the pipeline, amount to $32 billion, said to be larger

FIGURE 20-10 A herd of domesticated reindeer in a temporary corral in the Mackenzie River delta country (courtesy Information Canada Photothèque).

A CLOSER LOOK *Good-Bye Northwest Territories?*

The map of Canada soon may lose one of its most historic names, with the possible disappearance of the "Northwest Territories" (NWT). Originally, the name was applied to the vast area acquired in 1870 from the Hudson's Bay Company and Great Britain (Rupert's Land and the North-Western Territory), which lay northwest of central Canada, to which the Arctic Islands were added by Great Britain in 1880. It thus made up most of Canada's national area. Since then large portions of the NWT have been carved off from time to time, to be added to individual provinces and to create new provinces and the Yukon Territory (Fig. 20-c). Nevertheless, the Northwest Territories still are the largest political subdivision within Canada, 34 percent of the national area.

The possibility of the disappearance of this historic name reflects a unique situation amongst Canadian primary political subdivisions: Only in the NWT do native peoples form the majority of the population (some 55 percent of the total 52,241 in 1986) and also the majority of elected members in the legislature. With devolution of government from federal to territorial control, native voters and native legislators in the Territories now are able to ensure that political and economic decisions reflect their values and their views rather than those of the dominantly white Canadian population at large. Their legal position has been strengthened further by the recent inclusion of "aboriginal rights" in the Canadian Constitution, even though definition of those rights remains to be established. Finally, it now appears that after almost two decades of negotiations, comprehensive land claims of the native peoples soon will be settled with the federal government, resulting in major acquisitions of financial and natural resources.

FIGURE 20-c The changing boundaries of the Northwest Territories through the years.

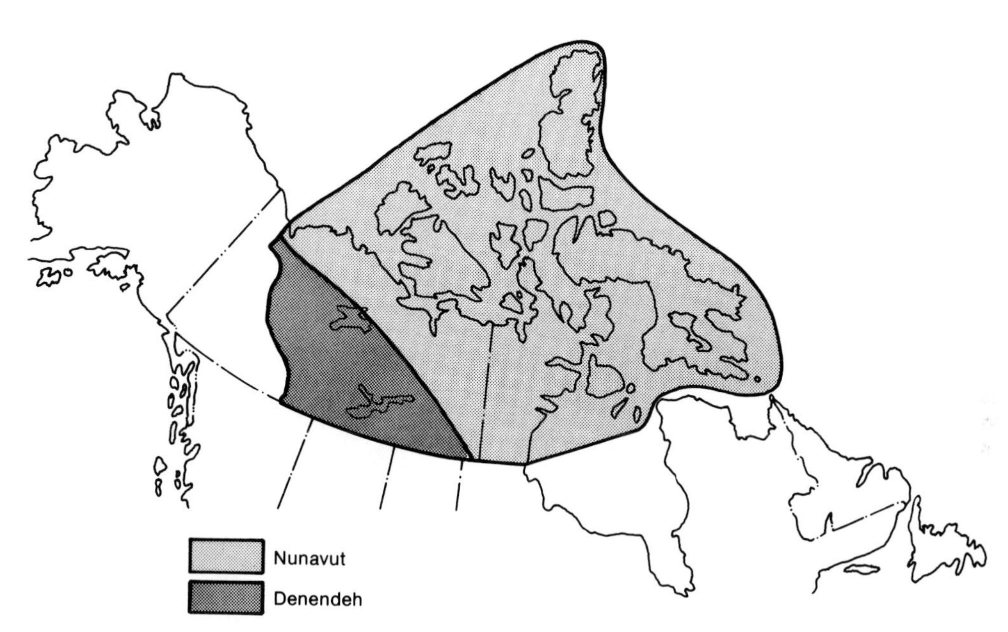

FIGURE 20-d Proposed new territories in the Canadian North.

In this situation of new-found political strength the natives are confronted with difficult decisions. They seek to preserve and strengthen their traditional culture, yet in many ways they have become integrated into the wider Canadian society at least as far as social services are concerned—educational and health services, transportation and communication links, modern housing, etc. The only realistic way to pay for these services would seem to be exploitation of natural resources, essentially mineral, of the NWT, a process that is environmentally destructive to some degree at least, thereby threatening the traditional way of life of the local people. The fur trade, which for centuries was the major economic support of the natives, has been severely reduced by the animal rights movement of the outside world, further compounding the problem.

The population of the NWT is the most youthful in Canada, with a median age of 24 in 1986, as compared with 32 for the national average. There is great urgency to provide employment for the young and increasingly better educated people. Although administration and service jobs exist, only resource development can increase employment opportunities sufficiently.

Somehow, the native peoples of the NWT seek to strike a balance between the realities of the modern world and their traditional cultural values. The Inuit of the Arctic have argued that only through creation of their own territory (Nunavut—"Our Land") is this possible (Fig. 20-d). The remaining subarctic western mainland (tentatively identified by the Dene as Denendeh—"Our Land") inhabited by Dene, Metis, and most of the NWT whites, thus far has been much less supportive of the division, although the NWT legislature has approved it in principle. Discussions to this end continue, but it must be approved by a majority of territorial residents in a plebiscite before it would be put into effect, thereby removing the "Northwest Territories" from Canada's political map. In the meantime, signs of change already can be noted. Some English place names in Canada's Arctic have been officially replaced by Inuit names—Frobisher Bay on Baffin Island is now Iqaluit; Eskimo Point on the west coast of Hudson Bay is now Arviat. With five Dene languages presently recognized officially by the NWT, place names in the subarctic sector potentially become a much more complex problem in this new and possible final era of the Northwest Territories.

Professor William C. Wonders
University of Alberta
Edmonton

FIGURE 20-11 Mining is a particularly challenging venture in the far north. This is a base metal mine on Little Cornwallis Island (courtesy Canadian Pacific).

than the investment in the entire U.S. automobile industry. Some 6000 people work on the North Slope in summer; about half that many in winter.

About 125 miles (200 km) east of Prudhoe Bay is another potentially commercial petroleum deposit that is currently the site of the bitterest environmental controversy in Alaska. The oil underlies the coastal plain of the Arctic National Wildlife Refuge. Enormous dispute about the economic and ecological desirability of "development" of this deposit appeared about to be resolved by governmental permission for drilling when the 1989 *Exxon Valdez* oil spill in Prince William Sound provoked agonizing reappraisal of the enterprise. At the time of this writing, no final decision has been reached, but it seems likely that "development" will win out over "preservation" in this case.

Considerable oil, especially in the Mackenzie delta–Beaufort Sea area, and natural gas, particularly on Melville and King Christian islands, have been discovered in the Canadian Arctic, but the indicated reserves thus far are considerably smaller than those of the North Slope. Moreover, production costs are very high (drilling expense in the Beaufort Sea is about $30 million per hole) and the massive problem of transportation awaits future solution.

Prospecting and proving of metallic ore deposits have been continuing in various locations, but most seem economically questionable. Nevertheless, several mines have been opened on the northern islands since 1977.

TRANSPORTATION

Inadequate transportation is the outstanding deterrent to economic progress in the Arctic. All-weather roads have been almost nonexistent and the only railway line is that connecting Churchill with the Prairie Provinces. In other words, conventional forms of land transportation are almost totally lacking. Tractor trains, temporary winter roads, and snowmobiles provide minimal transport during the cold season and some useful riverboat service exists during summer. By and large, however, regional transport depends on oceangoing vessels, which have a short navigation season, or on aircraft, which are expensive.

There is now a great deal of scheduled air service in the region. Resolute on Cornwallis Island, for example, receives regular air traffic from both Montreal and Edmonton. A surprising number of small settlements are served at least occasionally by scheduled flights. Nonscheduled flying fills in many gaps. Construction of an airfield has therefore become an essential for most population nodes and frequently the airfield is the location of the most modern and desirable facilities to be found locally. The light aircraft has become an almost ubiquitous link between the inhabitants of the region and the products and services of the postindustrial civilization on which they are dependent.

Many settlements, however, still receive their bulk supplies from government patrol boats that often cover thousands of miles on a single summer trip, bringing foodstuffs, fuel, and other materials that must last until the ship returns the following summer. This service is particularly characteristic of the eastern Canadian Arctic and the northern coast of Labrador.

THE OUTLOOK

The region doubtless will continue as a land of great distances and few people where nothing more than a scanty livelihood is obtainable by trapping, hunting, fishing, or grazing. Cities in the true sense will, as now, be nonexistent. Fur trapping will continue to be important to the natives, but it is too dependent on the vagaries of fashion, pressures from the animal rights movement, fluctuation of prices, and biological cycles to provide a steady means of livelihood for large numbers of people.

Expansion of commercial fishing offers considerable promise in a few localities, but in most of the region its growth possibilities seem quite limited. Expansion of reindeer herding seems logical in theory, but the lesson of seven decades of history is thoroughly negative.

Family and community life continue to deteriorate over most of the Arctic Region, as the population agglomerates in fewer settlement nodes and acquires a taste for an alien way of life but has scant opportunity to make a living. The traditional and the modern are in a state of continual collision; the children apathetically watch an American soap opera on television while their grandmother chews a caribou skin to soften it.

Efforts are being made to retain traditional skills, virtues, and values. Some villages restrict both TV and alcohol use. The government of the Northwest Territories has introduced an "outpost camp programme" to encourage hunters to live at least part-time in outposts in order to pursue a traditional way of life; more than 100 such camps have been established.

Despite such efforts, it is likely that most natives of the region will make the transition to a "civilized" way of life, and the transiton probably will be a difficult and painful one. They love the North and are adjusted to northern living, and so could become the backbone of northern development if they could acquire new skills without losing their identities. But with the perishing of the "old" way of life, employment opportunities must be made available or the native is likely to sink into a slough of apathy and degradation, as the monumental problem of alcoholism clearly demonstrates. Is it possible to develop a sound economic base for the native people of the Arctic without endangering their cultural survival?

Mineral exploitation will almost surely provide the major economic development stimulus for the region. The actualities of Prudhoe Bay and the potentialities of the Beaufort Sea area and the Mackenzie Valley corridor will be the principal foci of activity. What will this do for local inhabitants? Will they receive "early, visible, and lasting benefits" from such developments, as is the Canadian government's stated ambition? Oil drilling and pipeline operation can be either a boon or a burden (or both) to the region.

The regional and village corporations in the Alaskan portion of the region now have considerable capital to work with and it is quite likely that native claim payments will significantly augment the economy in the Canadian tundra as well (as is already the case with the James Bay Cree). The wise use of these windfall monies could do much to alleviate the bleakness of the long-term prospects for the native people.

Even in this thinly populated and rarely visited region questions of environmental preservation are being raised and debated at length. The major focus of concern involves oil and gas activities, which have generally been subjected to very restrictive environmental protection regulations. Significant areas were set aside as conservation reserves of one sort or another in the Alaskan tundra (see Chapter 19). Such actions have thus far been much more limited in the vaster expanses of the Canadian tundra, but an extensive inventory of the land resource has been accomplished, with the result that six major wilderness parks have been proposed and some 136 "significant conservation" areas have been recognized.[14]

Decision-makers of both nations are faced with the difficult dilemma of dealing with a region that has a fragile resource base, a rapidly changing geopolitical scenario, a marginal economy, and heavy pressure for exploitation of energy minerals. It is in many ways a classic conflict of colonialism.

SELECTED BIBLIOGRAPHY

Area Economic Surveys (Ottawa: Industrial Division, Department of Indian Affairs and Northern Development)
1958. Ungava Bay
1962. Southampton Island
1962. Tuktoyaktuk–Cape Parry
1962. Western Ungava
1963. The Copper Eskimos
1963. Keewatin Mainland
1963. Yukon Territory Littoral
1965. Banks Island
1965. Northern Foxe Basin
1966. East Coast–Baffin Island
1966. Frobisher Bay
1966. The Mackenzie Delta
1966. Rae-Lac La Martre
1967. Central Mackenzie
1967. Lancaster Sound
1967. South Coast–Baffin Island
1967. South Shore–Great Slave Lake
1968. Central Arctic
1968. Keewatin Mainland Reappraisal

BIRD, J. BRIAN, *The Physiography of Arctic Canada*. Baltimore: Johns Hopkins Press, 1967.

BRITTON, M. E., *Alaskan Arctic Tundra*. Washington, D.C.: Arctic Institute of North America, 1973.

"The Brooks Range: Environmental Watershed," *Alaska Geographic,* 4, no. 2 (1977), entire issue.

[14] L. C. Munn et al, *Canada's Special Places in the North: An Environment Canada Prespective for the '80's* (Ottawa: Environment Canada, 1983).

BRUEMMER, FRED, *The Arctic*. Englewood Cliffs, NJ: Prentice-Hall, 1974.

______, "Churchill: Polar Bear Capital of the World," *Canadian Geographic,* 103 (December 1983–January 1984), 20–27.

COURTNEY, J. L., "Airports, Route Service, and Remote Area Access: The Case of Northern Canada," *Canadian Geographer,* 25 (1981), 111–123.

COWAN, IAN MCTAGGART, "Ecology and Northern Development," *Arctic,* 22 (1969), 3–12.

CROWE, KEITH J., *A History of the Original Peoples of Northern Canada*. Montreal: Arctic Institute of North America, McGill-Queen's University Press, 1974.

DANIELSON, ERIC W., JR., "Hudson Bay Ice Conditions," *Arctic,* 24 (1971), 90–107.

DAVIS, JAMES H., "Barrow, Alaska: Technology Invades an Eskimo Community," *Landscape,* 21 (1977), 21–25.

DEAR, MICHAEL, and SHIRLEY CLARK, "Planning a New Arctic Town at Resolute," *Canadian Geographic,* 97 (December 1978–January 1979), 46–51.

DEY, B., "Seasonal and Annual Variations in Ice Cover in Baffin Bay and Northern Davis Strait," *Canadian Geographer,* 24 (1980), 368–384.

ENGLAND, JOHN, "Ellesmere Island Needs Special Attention," *Canadian Geographic,* 103 (June–July 1983), 8–17.

EVERETT, K. R., "Summer Wetlands in the Frozen North," *Geographical Magazine,* 55 (October 1983), 510–515.

FLETCHER, ROY J., "Settlement Sites Along the Northwest Passage," *The Geographical Review,* 68 (January 1978), 80–93.

FRENCH, HUGH, "Why Arctic Oil Is Harder to Get than Alaska's," *Canadian Geographical Journal,* 94 (June–July 1978), 46–51.

HARDING, L., "Mackenzie Delta: Home of Abounding Life," *Canadian Geographical Journal,* 88 (May 1974), 4–13.

HARRY, DAVID G., "Banks Island: Gem of the Western Arctic," *Canadian Geographic,* 102 (October–November 1982), 40–49.

HEMSTOCK, R. A., "Transporting Petroleum in the North," *Canadian Geographical Journal,* 90 (April 1975), 42–49.

HERRINGTON, CLYDE, *Atlas of the Canadian Arctic Islands.* Vancouver: Shultoncraft Publishing Co., 1969.

"Islands of the Seals: The Pribilofs," *Alaska Geographic,* 9 (1982), entire issue.

KELLEY, J. J., JR., and D. F. WEAVER, "Physical Processes at the Surface of the Arctic Tundra," *Arctic,* 22 (1969), 424–437.

LAMONT, JAMES, "Pangnirtung: Gateway to Our Only Arctic Park," *Canadian Geographic,* 100 (August–September 1980), 34–37.

MACKAY, J. ROSS, "The Mackenzie Delta," *Canadian Geographical Journal,* 78 (1969), 146–155.

______, "The World of Underground Ice," *Annals,* Association of American Geographers, 62 (1972), 1–22.

MAXWELL, J. B., *The Climate of the Canadian Arctic Islands and Adjacent Waters,* Vols. I and II. Hull, Quebec: English Publishing, 1980.

MCCANN, S. B., P. J. HOWARTH, and J. G. COGLEY, "Fluvial Processes in a Periglacial Environment: Queen Elizabeth Islands, N.W.T., Canada," *Transactions,* Institute of British Geographers, 55 (1972), 69–82.

MULLER, FRITZ, *The Living Arctic.* New York: Methuen, 1981.

"Nome, City of the Golden Beaches," *Alaska Geographic,* 11 (1984), entire issue.

PYNN, LARRY, "The Dempster," *Canadian Geographic,* 109 (June–July 1989), 32–39.

REEVES, RANDALL R., "Narwals: Another Endangered Species," *Canadian Geographical Journal,* 92 (May–June 1976), 12–19.

RITCHIE, J. C., *Past and Present Vegetation in the Far Northwest of Canada.* Toronto: University of Toronto Press, 1984.

SMITH, I. NORMAN, ed., *The Unbelievable Land.* Ottawa: Queen's Printer, 1964.

STIX, JOHN, "National Parks and Inuit Rights in Northern Labrador," *The Canadian Geographer,* 26 (Winter 1982–1983), 349–354.

USHER, P. J., "The Growth and Decay of the Trading and Trapping Frontiers in the Western Canadian Arctic," *Canadian Geographer,* 19 (1975), 308–320.

WALKER, H. J., and M. C. BREWER, "Patterns on the Alaskan Tundra," *The Geographical Magazine,* 50 (October 1977), 42–45.

WEBB, KENNETH, "What Made the Hudson Bay Arc?" *Canadian Geographical Journal,* 92 (May–June 1976), 20–25.

WOODING, FREDERICK H., "Muskoxen Comeback: They Even Return to Siberia," *Canadian Geographic,* 98 (February–March 1979), 18–23.

"Yukon–Kuskokwin Delta," *Alaskan Geographic,* 6 (1978), entire issue.

INDEX

A

B

C

D

G

H

I

J

M

N

O

P

S

T

U

V

W

Y

Z

MAJOR POLITICAL UNITS OF THE UNITED STATES AND CANADA

	ABBREV.	AREA MI²	AREA KM²	AREA RANK-ORDER	POPULATION TOTAL	POPULATION RANK-ORDER	CAPITAL
Alabama	AL	51.6	133.6	29	4,063	22	Montgomery
Alaska	AK	586.4	1518.8	1	552	49	Juneau
Arizona	AZ	113.9	295.0	6	3,678	24	Phoenix
Arkansas	AR	53.1	137.5	27	2,362	33	Little Rock
California	CA	158.7	411.0	3	29,839	1	Sacramento
Colorado	CO	104.2	269.9	8	3,308	26	Denver
Connecticut	CT	5.0	13.0	48	3,296	27	Hartford
Delaware	DE	2.1	5.4	49	669	46	Dover
Florida	FL	58.6	151.8	22	13,003	4	Tallahassee
Georgia	GA	58.9	152.6	21	6,508	11	Atlanta
Hawaii	HI	6.5	16.8	47	1,115	40	Honolulu
Idaho	ID	83.6	216.5	13	1,012	42	Boise
Illinois	IL	56.4	146.1	24	11,467	6	Springfield
Indiana	IN	36.3	94.0	38	5,564	14	Indianapolis
Iowa	IA	56.3	145.8	25	2,787	30	Des Moines
Kansas	KS	82.3	213.2	14	2,486	32	Topeka
Kentucky	KY	40.4	104.6	37	3,699	23	Frankfort
Louisiana	LA	48.5	125.6	31	4,238	21	Baton Rouge
Maine	ME	33.2	86.0	39	1,233	38	Augusta
Maryland	MD	10.6	27.5	42	4,799	19	Annapolis
Massachusetts	MA	8.3	21.5	45	6,029	13	Boston
Michigan	MI	58.2	150.7	23	9,329	8	Lansing
Minnesota	MN	84.1	217.8	12	4,387	20	St. Paul
Mississippi	MS	47.7	123.5	32	2,586	31	Jackson
Missouri	MO	69.7	180.5	19	5,138	15	Jefferson City
Montana	MT	147.1	381.0	4	804	44	Helena
Nebraska	NE	77.2	200.0	15	1,585	36	Lincoln
Nevada	NV	110.5	286.2	7	1,206	39	Carson City
New Hampshire	NH	9.3	24.1	44	1,114	41	Concord
New Jersey	NJ	7.8	20.2	46	7,749	9	Trenton
New Mexico	NM	121.7	315.2	5	1,522	37	Santa Fe
New York	NY	49.6	128.5	30	18,045	2	Albany
North Carolina	NC	52.6	136.2	28	6,658	10	Raleigh
North Dakota	ND	70.7	183.1	17	641	47	Bismarck
Ohio	OH	41.2	106.7	35	10,887	7	Columbus
Oklahoma	OK	69.9	181.0	18	3,158	28	Oklahoma City